CORROSION

Volume 1

Metal/Environment Reactions

CORROSION

Volume 1

Metal/Environment Reactions

Edited by

L. L. Shreir, PhD, FRIC, FIM, FICorrT, FIMF

Head of Department of Metallurgy and Materials,
City of London Polytechnic

NEWNES-BUTTERWORTHS

LONDON BOSTON

Sydney–Wellington–Durban–Toronto

THE BUTTERWORTH GROUP

ENGLAND
Butterworth & Co (Publishers) Ltd.
London: 88 Kingsway, WC2B 6AB

AUSTRALIA
Butterworths Pty Ltd.
Sydney: 586 Pacific Highway, NSW 2067
Also at Melbourne, Brisbane, Adelaide and Perth

SOUTH AFRICA
Butterworth & Co (South Africa) (Pty) Ltd.
Durban: 152–154 Gale Street

NEW ZEALAND
Butterworths of New Zealand Ltd.
Wellington: 26–28 Waring Taylor Street, 1

CANADA
Butterworth & Co (Canada) Ltd.
Toronto: 2265 Midland Avenue, Scarborough, Ontario M1P 4S1

USA
Butterworth (Publishers) Inc.
Boston: 161 Ash Street, Reading, Mass. 01867

First published by George Newnes Ltd., 1963

Second impression 1965

Second edition published 1976 by Newnes-Butterworths

ISBN 0 408 00109 7

Filmset by Oliver Burridge Filmsetting Ltd., Crawley, Sussex
Printed and bound in England by Hazell Watson & Viney Ltd., Aylesbury, Bucks

Foreword by J. B. Cotton, Esq.,

Chairman, Department of Industry Corrosion Committee (1972–75)

The fact that the first edition of this book met with such success demonstrates that its initial publication filled a long standing need. Since its first appearance in 1963, knowledge of corrosion science and technology has increased enormously and several significant developments have occurred that have made it essential to produce a revised, up-dated edition.

An event of the utmost significance was the attempt to survey, comprehensively and quantitatively, the cost of corrosion to the United Kingdom. This task, sponsored by the then Department of Trade and Industry in 1969, was undertaken by a Committee, chaired by Dr T. P. Hoar. The results, contained in the Report of the Committee on Corrosion and Protection, published by HMSO in 1971, put the annual cost of corrosion in the UK at £1365M. More importantly, it was estimated that of this sum, £310M could be saved if the available knowledge on corrosion prevention was fully applied in practice. It is fair to say that the new edition of *Corrosion* sets down the basic knowledge that needs to be applied in recouping this loss.

The selection of the most appropriate system for corrosion prevention and its translation into industrial practice is, however, often a complicated procedure, in which the true economic benefit has to be assessed and appreciated before investment in a protective scheme can be justified. Although the book is sufficiently comprehensive to provide the answer to many corrosion problems, an individual system may often require a level of judgement and expertise that it is not possible to include in any work of reference, however detailed, and for the non-expert this book also indicates areas where it is advisable to seek further advice.

The major finding of the Hoar Report was the need for (a) better dissemination of information on corrosion and protection, (b) more education in corrosion and protection, and (c) an increased awareness of the hazards of corrosion, and it was to pursue these objectives in the economic interests of the nation that the Department of Industry Committee on Corrosion was established in 1973. A comprehensive work of reference, such as this book, is an essential aid to achieving the Committee's objectives and had a book of sufficient scope not existed it would have been necessary to produce one. Fortunately, the up-dating and revision of a book that has already established itself as an authoritative source of information on the subject has been completed so that publication can take place almost coincidentally with the opening of the National Corrosion Service at Teddington. These developments, together with the expansion of advisory facilities that is also taking place at existing centres, should ensure that industry in the UK is not lacking in the provision of advice upon protection against corrosion.

Corrosion knows no national boundaries, and this book should be of great value, therefore, to countries throughout the world and in particular to those where advisory services are non-existent.

PREFACE TO 2ND EDITION

Corrosion was first published in 1963 and with the appearance of the 2nd edition it is of interest to look back over the intervening years and to consider the progress that has been made in the science and technology of the corrosion and protection of metals. The word 'corrosion' signifies 'destroy gradually, decay, disease' and over the years attempts have been made to find an alternative that would have a more positive connotation and that would be more attractive to the young student with a leaning towards the applied sciences. Unfortunately, an appropriate word has proved to be elusive, although in this respect other technologies have succeeded where 'the corrosionist' has failed; *tribology* and *terotechnology*, to name but two, have without doubt the imposing sound that *corrosion* lacks. The difficulty is overcome to some extent by using expressions such as 'corrosion control', 'corrosion protection', etc. but the unfortunate word 'corrosion' cannot be dispensed with.

Strangely, and in spite of its name, 'corrosion' continues to flourish and has now become an established academic discipline, and if the quality and quantity of scientific publications and books are criteria of its importance, then the story of corrosion over the past decade or so is one of outstanding success. In the United Kingdom there are now Professors of Corrosion Science at the University of Manchester and Imperial College, London, and there are also a number of others, who although not designated as such have made their reputation by their work in this field. There are now a number of universities and polytechnics with departments that specialise in corrosion and provide courses in the subject for undergraduate and postgraduate students; there are three MSc courses devoted solely to corrosion and protection and a number of others in which it is given some prominence.

The importance of the subject is also reflected in the establishment of a Corrosion and Protection Centre at the University of Manchester Institute of Science and Technology, under the direction of Professor G. C. Wood, the first Professor of Corrosion Science in this country, and more recently a special advisory centre has been established at the National Physical Laboratory.

Progress has also been made in establishing corrosion protection as a reputable professional activity, and the recent merger of the Institute of Corrosion Technology and the Corrosion and Protection Association to form the Institute of Corrosion Science and Technology represents an important step in that direction. It is hoped that the new Institute will follow in the footsteps of the well established National Association of Corrosion Engineers in stimulating courses, publications, standards and research in the field of metals protection.

This progress, which will undoubtedly continue, stems from a number of factors and due acknowledgement must be made to the early workers in this field who did so much to give corrosion its present status. Of more recent origins is the awareness by government of the cost of corrosion to the economy of the country as a whole, to industry, and to the individual consumer, and it is evident that the technology for the preservation of metals from the ravages of corrosion will become of increasing importance.

The present edition of *Corrosion* represents a complete revision of the 1st edition which was published in 1963, and although the philosophy and structure of the book have remained unchanged (see Preface to 1st edition) several new sections have been added, which should make the work more comprehensive. Over the past 13–14 years the scope of the subject has increased enormously, and this means that a detailed treatment of each aspect of corrosion is not possible even in a work that extends to two volumes, and contains approximately 2300 pages. Nevertheless, it is hoped that *Corrosion* in its revised form will make a valuable contribution to 'better dissemination of existing knowledge' in this important field of materials conservation.

L.L.S.

PREFACE TO 1ST EDITION

The enormous scope of the subject of corrosion follows from the definition which has been adopted in the present work. Corrosion will include all reactions at a metal/environment interface irrespective of whether the reaction is beneficial or detrimental to the metal concerned—no distinction is made between chemical or electropolishing of a metal in an acid and the adventitious deterioration of metal plant by acid attack. It follows, therefore, that a comprehensive work on the subject of corrosion should include an account of batteries, electrorefining, chemical machining, chemical and electrochemical polishing, etc.

The fact remains, nonetheless, that the environmental reaction of a metal used as a construction material is the most important type of corrosion reaction, and the one of most concern to the engineer. The technological and economic consequences of the wastage of metals by corrosion are now fully appreciated, and figures have been published which show the enormous financial losses, both to the individual organisation and to the economy of the country as a whole, resulting from the deterioration of metals. The need for conserving metals has been publicised by Dr U. R. Evans, Dr J. C. Hudson, Mr T. H. Turner, Professor H. H. Uhlig, Dr W. H. J. Vernon, and others, and the 'corrosion consciousness' which prevails today is largely due to their efforts.

In the light of what has been said above, little further explanation of the implications of the title of the present work is required. Its treatment of the subject of corrosion will centre round the *control* of the environmental interactions of metals and alloys used as materials of construction.

The effective control of corrosion reactions must be based on an understanding of the mechanism of such reactions and on the application of this knowledge to practical problems. The work, regarded as a whole, represents an attempt, therefore, to present the subject of corrosion as a synthesis of corrosion science and corrosion engineering. Thus in the planning of the content an attempt has been made to strike a suitable balance between the primarily scientific and the primarily practical aspects, and so the nature of individual sections ranges from the fundamental and theoretical to the essentially practical.

It is hoped that this approach has resulted in a work that will be of some value to the student, the corrosion worker, and the engineer in the field of corrosion.

Corrosion represents the joint effort of over 100 authors, all of whom have been free, within the necessary limitations of length, to express their own views. Grateful acknowledgements are made to the individual authors from Great Britain, the United States, and Canada for their valuable and enthusiastic co-operation.

The task of the editor in finding suitable authors for various topics was considerably lightened by the fact that the majority of corrosion specialists in this country belong either to the Corrosion Group of the Society of Chemical Industry or to the Institute of Metal Finishing, and acknowledgements are made to Mr S. C. Britton (then Secretary, Corrosion Group) and to Dr S. Wernick and Mr I. S. Hallows (Hon. Secretary and Assistant Editor respectively of the I.M.F.). The editor wishes to express his appreciation of the considerable assistance received from Dr E. C. Rhodes and Dr G. L. J. Bailey of INCO (Mond) in providing authors from this organisation.

The editor also acknowledges with pleasure the encouragement and assistance he has received from Mr L. W. Derry (Head of Department of Metallurgy) and Dr D. M. A. Leggett (Principal) of the Battersea College of Technology, and from Dr A. M. Ward (Principal) of Sir John Cass College.

Throughout the course of this work the content and subject matter have been discussed with various workers in the field, and the editor would like to take this opportunity to thank Mr S. C. Britton, Professor C. W. Davies, Dr T. P. Hoar, Dr E. C. Potter, and others for their advice and constructive criticism.

He would also like to take this opportunity to express his appreciation to Dr U. R. Evans and Dr W. H. J. Vernon for assistance given when he first contemplated entering the field of corrosion, and for their encouragement and advice in connection with the present work.

Finally, grateful acknowledgements are made by the editor to Mr T. F. Saunders and Mrs N. E. Orna, M.A., of George Newnes (Technical Books) for their kind co-operation at all stages of the work.

L.L.S.

1963

ACKNOWLEDGEMENTS

This work is the joint effort of over 100 authors, who have approached the task of writing a short section covering a very wide field with diligence and enthusiasm, and I would like to express my sincere appreciation to them for their very kind co-operation.

My colleagues in the Department of Metallurgy have been very helpful in commenting on various sections in the book and in making contributions to it; and in particular I would like to thank Dr C. J. L. Booker, Dr M. Clarke, Mr E. J. Easterbrook, Dr E. Gillam, Mr R. A. Jarman, Dr E. F. I. Roberts and Dr S. Wach. I would also like to acknowledge the assistance I have received from Mrs J. Sadowska-Mazur and Mr P. Woods, who were research students in the Department during the time the book was being compiled, and from the Secretary of the Department, Mrs M. Evans. I would also like to express my appreciation to Dr P. Boden, Mr H. S. Campbell and Dr V. F. Lucey for their assistance and helpful comments.

L.L.S.

CONTRIBUTORS

K G Adamson, AMCST, LIM
Development Officer,
Magnesium Elektron Ltd., Manchester

D F Aitken, MA
British Rubber Manufacturers' Association (Retired)

J C B Alcock, ARCS, DSc, PhD, CChem, FRIC
Professor and Chairman,
Dept. of Metallurgy and Materials Science,
University of Toronto, Canada

D Ames, MICorrT
Associate, Spencer and Partners,
Consulting Engineers

K F Anderson
Morganite Carbon Ltd. (Retired)

J E Antill, PhD, BSc
Head, Chemical Metallurgy Group,
Materials Development Division, UKAEA,
Harwell

D J Arrowsmith, MA, PhD
Reader in Industrial Metallurgy,
University of Aston, Birmingham

J C Bailey, BSc, FIM
Formerly Deputy Director (Technical),
Aluminium Federation (Retired)

W E Ballard, CChem, FRIC, FIM
Consultant,
Formerly Managing Director, Metallisation Ltd.

T A Banfield, PhD, DIC, ARCS, CChem, FRIC, FICorrT, FTSC
Deputy Manager,
Group Research Laboratory,
Berger Jenson and Nicholson Ltd.

G E Barrett, BSc, ARIC, PhD
Technical Director, Plastic Coatings Ltd.

E W Beale, ARIC
Senior Scientific Officer,
Materials Quality Assurance Directorate,
Ministry of Defence

J Bentley, BSc, DipChemEng, CEng, MIChemE
Principal Chemist, Wastes Division
Directorate General, Water Engineering,
Dept. of the Environment

W Betteridge, DSc, FInstP, FIM
Consultant,
Formerly of International Nickel Ltd.

P J Boden, PhD, CChem, FRIC, MIM, FICorrT
Senior Lecturer, Dept. of Metallurgy and Materials Science,
Nottingham University

C J L Booker, BSc, PhD, ARIC, FICorrT
Senior Lecturer in Corrosion Science,
Dept. of Metallurgy and Materials,
City of London Polytechnic

J W L F Brand, MITE, TEng(CEI), MICorrT
Divisional Manager,
Corrosion Control Division,
Corrosion and Welding Engineering Ltd.

C F Britton, LRIC, MICorrT
Product Manager, Instruments,
Magnachem (UK) Ltd.

S C Britton, MA, CChem, FRIC, FIM, FIMF, FICorrT
Tin Research Institute (Retired)

J A Brydson, FPRI, ANCRT
Deputy Head,
National College of Rubber Technology

W Bullough, BSc, ARIC
Principal Research Officer,
BSC Research Centre,
Strip Mills Division

V E Carter, FICorrT, FIMF
Corrosion and metal finishing consultant,
Formerly of BNFRA

J E Castle, BSc, PhD, CChem, FRIC, FICorrT
Reader in Chemical Metallurgy,
University of Surrey, Guildford

K A Chandler, BSc, ARSM, FICorrT
Head, Corrosion Advice Bureau, BSC

A R L Chivers, MA
Senior Technical Officer,
Zinc Development Association,
London

B Chatterjee, PhD, AIM, ARIC
Research Scientist,
British Aluminium Co.

M Clarke, BSc, PhD, DSc, CChem, FRIC, FIM, FICorrT, FIMF
Principal Lecturer, Dept. of Metallurgy and Materials,
City of London Polytechnic, and Consultant Technologist

R J Clarke, MA, CEng, FIChemE, FIFST
Hon. Visiting Lecturer in Food Engineering,
Queen Elizabeth College, London

H G Cole, BSc, FIMF, FICorrT
Principal Scientific Officer, Ministry of Defence (Procurement Executive)

H H Collins, BSc, CChem, FRIC
Superintendent, Chemistry Research,
Stanton & Staveley/BSC

J B Cotton, AMCT, ARIC, FICorrT
IMI, Kynoch, Birmingham (Retired)

D P Dautovich, MSc, PhD
Corrosion Engineer, Research Division,
Ontario Hydro, Canada

K Julyan Day, FICorrT, FTSC, MBIM
Anti-Corrosion Consultant

J Dodd, BSc, FIM
Manager, Cast Iron Development,
Climax Molybdenum Company, USA

P D Donovan, MSc, ARIC, FIM
Principal Scientific Officer,
Ministry of Defence

C W Drane, BSc, CChem, FRIC
Technical Manager, Water Specialities and Services,
Industrial Chemicals Division, Albright and Wilson Ltd.

F G Dunkley, FICorrT
Consultant,
Formerly of British Rail, Derby

E J Easterbrook, BSc(Eng), ARSM, AMIMM, MIM
Principal Lecturer,
Dept. of Metallurgy and Materials,
City of London Polytechnic

T E Evans, BSc, ARIC, FICorrT
Principal Technologist,
International Nickel Ltd., Birmingham

G N Flint, BSc, CChem, FRIC, FIM
Co-ordinator, Technical Development,
International Nickel Ltd., Birmingham

D Fyfe, MA, PhD
Senior Marketing Engineer,
Chemetics Ltd.,
Montreal, Canada

P J Gay, BSc, FTSC, FICorrT
Consultant

J S Gerrard, AMIEE
Formerly Joint General Manager,
Metal and Pipeline Endurance Ltd. (d. 1974)

P T Gilbert, BSc, PhD, CChem, FRIC, FIM, FIMarE, FICorrT, CEng
Technical and Metallurgical Director,
Yorkshire Imperial Metals Ltd.

E W F Gillham, MSc, PhD, CChem, FRIC, CEng, FInstF
Consultant in Land Restoration,
Formerly of CEGB, Midland Region

D J Godfrey, BSc, PhD
Principal Scientific Officer,
Admiralty Materials Laboratory,
Poole, Dorset

V R Gray, BA, MA, PhD, FIWSc, MNZIC
Senior Scientist, Chemistry Division,
DSIR, New Zealand

T B Grimley, BSc, PhD
Reader in Theoretical Chemistry,
Dept. of Inorganic, Physical and Industrial Chemistry,
University of Liverpool

J O Harris, PhD
Professor of Bacteriology,
Kansas State University

A C Hart, BSc, ARIC, AIMF
Senior Technologist,
International Nickel Ltd.

A A B Harvey, MSc, CChem, FRIC, FIMF
Senior Technical Officer,
British Standards Institution

Manfred Hess, CChem, FRIC, FIMF, FTSC
Member of the Association of Consulting Scientists
Principal of Manfred Hess,
Consulting Chemist and Paint Technologist

D K Hill, DSc, PhD, FSGT
Formerly Technical Manager, British Indestructo Glass Ltd. (Retired)

J Hines, MA, PhD
North West Region Materials Group Manager, ICI Ltd.

T P Hoar, MA, PhD, ScD, BSc, CChem, FRIC, FIM, FIMF, FICorrT
Consultant, Emeritus Reader in Chemical Metallurgy, University of Cambridge

H Howarth, AMICorrT, AMet
Investigator, Production Metallurgy Section, Special Steels Division, BSC

J C Hudson, DSc, DIC, ARCS, FIM
Consultant, Formerly of BISRA

D E Hughes, MA, DSc, PhD, FIBiol
Professor of Microbiology (Head of Department), University College, Cardiff

M L Hughes,
Consultant, Formerly of BISRA (d. 1963)

D Inman, BSc, ARCS, FRIC, PhD, DSc, DIC, MIMM
Reader in Chemical Metallurgy, Nuffield Fellow in Extraction Metallurgy, Dept. of Metallurgy and Materials Science, Imperial College, London

R A Jarman, MSc, PhD, CEng, MIProdE, MWeldE
Senior Lecturer in Welding, Dept. of Metallurgy and Materials, City of London Polytechnic

D A Jones, BS, MS, PhD(Rensselaer Polytechnic Inst.)
Senior Research Engineer, Research Laboratory, U.S. Steel Corporation, Penn., USA

L Kenworthy, MSc, ARCS, CChem, FRIC, FIM, FICorrT
Consultant, Formerly Navy Dept. (Ministry of Defence)

F LaQue, BSc, LLD, Past President, Nat. Assoc. Corrosion Eng., Am. Soc. Test and Mat., Electrochemical Society
Senior Lecturer, Scripps Institution of Oceanography, University of California

D N Layton, PhD, MSc, ARCS, DIC, MInstP, FIMF
Managing Director, Fredk. Mountford (Birmingham) Ltd.

D A Lewis, BSc(Eng), FICorrT
Partner of Spencer and Partners, Consulting Engineers

E L Littauer, BSc, PhD, MIM, AMIMM
Manager, Electrochemistry and Environmental Sciences, Lockheed Missiles and Space Co., California, USA

G O Lloyd, BSc
Principal Scientific Officer, Division of Materials Applications, National Physical Laboratory

N A Lockington, MA, PhD, ARIC, FIM
Metallurgist, Director, The Chrome-Alloying Co. Ltd.

G L Long, PhD, CChem, FRIC
Principal Scientific Officer, Energy Technology Support Unit, UKAEA, Harwell

W B Mackay, BSc, AIMF
Director, B.K.L. Alloys Ltd.

C A May, MSc, PhD
Lecturer, Dept. of Metallurgy and Materials, City of London Polytechnic

J E O Mayne, DSc, ARCS, DIC, CChem, FRIC, FICorrT
Dept. of Metallurgy and Materials Science, University of Cambridge

A D Mercer, BSc
Principal Scientific Officer National Physical Laboratory

N S C Millar, CChem, FRIC, FICeram, MICorrT, MBIM
General Manager, Thermovitrine Ltd.

D S Oliver, BSc, PhD, FIM, FInstP
Group Director of Research and Development, Pilkington Bros. Ltd.

M W O'Reilly, Dip Tech, LRIC
Decorative Paints Market Team Leader, ICI Paints Division

S Orman, BSc, PhD, FICorrT, CChem, FRIC
Senior Principal Scientific Officer, AWRE, Aldermaston

R N Parkins, BSc, PhD, DSc, FIM
Professor and Head of Department, Metallurgy and Engineering Materials, University of Newcastle upon Tyne

A W Pearson, MIM
Research Division, British Aluminium Co. Ltd.

R Pinner, BSc, FICorrT, FIMF
Consultant

P Poole, BSc, PhD
Senior Scientific Officer, Royal Aircraft Establishment

J K Prall, PhD, BSc, CChem, FRIC
Section Manager,
Unilever Research Laboratory

F C Porter, MA, FIM, FICorrT, FIMF
Zinc Development Association, London

R P M Procter, MA, PhD, MICorrT
Lecturer in Corrosion Science,
Corrosion and Protection Centre, UMIST

E F Redknap, BSc
Retired

F H Reid, BSc, CChem, FRIC, FIMF
Consultant,
Formerly of International Nickel Ltd.

R G Robson, BSc(Eng), MIEE
Chartered Engineer

D van Rooyen, BSc, PhD
Advisory Scientist, Westinghouse Bettis Atomic Power Labs., USA

J C Rowlands, FICorrT
Admiralty Materials Laboratory,
Poole, Dorset

J Sadowska-Mazur, MSc(Gdansk)
Research Assistant, City of London Polytechnic

F W Salt, BSc, PhD, FIMF
Head, Basic Surface Studies Section
BSC Research Centre,
Strip Mills Division

B A Scott, ARCS, BSc, PhD, CChem, FRIC
Deputy Information Officer,
Group Technical Information Service,
British Aluminium Co. Ltd.

J C Scully, MA, PhD, FIM, FICorrT
Reader in Corrosion Science,
University of Leeds

H J Sharp, PhD, MSc, CChem, FIM, FPRI
Director of International Associates

R E Shaw, BSc, FIM, FIMF
ICI Paints Division (Retired)

G S Shipley, FICeram
Technical Advisor, Hathernware Ltd.

L L Shreir, PhD, CChem, FRIC, FIM, FICorrT, FIMF
Head, Dept. of Metallurgy and Materials,
City of London Polytechnic

H Silman, BSc, CEng, CChem, FRIC, FIChemE, FIM, FIMF
Consultant

E W Skerrey, BSc, CChem, FRIC, FICorrT, AIM
Assistant Manager, Application Technology Department, Research Division,
British Aluminium Co. Ltd.

R A Smith, BSc, PhD, FIM
Manager,
Research Laboratory,
International Nickel Ltd.

J F Stanners, BSc, FICorrT
Head of Corrosion Research,
Inter-Services Laboratory, BSC

W H Tatton, ARIC, FIMF, FTSC
Technical Officer,
British Standards Institution

J G N Thomas, BSc, PhD, ARCS, DIC
Corrosion Section, Division of Materials Applications,
National Physical Laboratory

J E Truman, AMet
Technical Liaison Representative,
Firth Brown Ltd., Sheffield

R Walker, BSc, DipEd, MSc, MSc(Eng), PhD
Lecturer, Metallurgy and Materials Technology Dept, University of Surrey,
Guildford

G W Walkiden, BSc, CChem, FRIC, MIM
Consultant, Ever Ready Central Laboratories

J R Walters,
Consultant to British Post Office

R B Waterhouse, MA, PhD, FIM, FICorrT
Reader in Metallurgy,
Dept. of Metallurgy and Materials Science,
University of Nottingham

K O Watkins, FIM, FICorrT
Corrosion Advice Bureau, BSC

H C Wesson, MA, BSc, CChem, FRIC
Formerly Technical Manager,
Lead Development Association (Retired)

S A Watson, BSc, PhD, CChem, FRIC, FIMF
Senior Development Officer,
International Nickel Ltd.

E E White, CChem, FRIC, FIM, CEng, MIMM, FCS, MIInfSc, FICorrT, FIMF
Consultant,
Inter-Services Laboratory, BSC

R W Wilson, MA, PhD, FIM
Head, Metallurgical Research and Services Division,
Shell Research Ltd.

P A Woods, BSc
HM Inspector of Factories

K H R Wright, BSc, PhD, MInstP
Senior Principal Scientific Officer,
Materials Group, National Engineering Laboratory

CONTENTS

Volume 1. Metal/Environment Reactions

Volume 2. Corrosion Control

1 PRINCIPLES OF CORROSION AND OXIDATION

1.1 Basic Concepts of Corrosion

Modern technology has at its disposal a wide range of constructional materials—metals and alloys, plastics, rubber, ceramics, composites, wood, etc. and the selection of an appropriate material for a given application is the important responsibility of the design engineer. No general rules govern the choice of a particular material for a specific purpose, and a logical decision involves a consideration of the relevant properties, ease of fabrication, availability, relative costs, etc. of a variety of materials; frequently the ultimate decision is determined by economics rather than by properties, and ideally the material selected should be the cheapest possible that has adequate properties to fulfil the specific function.

Where metals are involved, mechanical, physical and *chemical* properties must be considered, and in this connection it should be observed that whereas mechanical and physical properties can be expressed in terms of constants, the chemical properties of a given metal are dependent entirely on the precise environmental conditions prevailing during service. The relative importance of mechanical, physical and chemical properties will depend in any given case on the application of the metal. For example, for railway lines elasticity, tensile strength, hardness and abrasion resistance will be of major importance, whereas electrical conductivity will be of primary significance in electrical transmission. In the case of heat-exchanger tubes, good thermal conductivity is necessary, but this may be outweighed in certain environments by chemical properties in relation to the aggressiveness of the two fluids involved—thus although the thermal conductivity of copper is superior to that of aluminium brass or the cupronickels, the alloys are preferred when high velocity sea water is used as the coolant, since copper has very poor chemical properties under these conditions.

While a metal or alloy may be selected largely on the basis of its mechanical or physical properties, the fact remains that there are very few applications where the effect of the interaction of a metal with its environment can be completely ignored, although the importance of this interaction will be of varying significance according to circumstances; for example, the slow uniform wastage of steel of massive cross section (such as railway lines or sleepers) is of far less importance than the rapid perforation of a buried steel pipe or the sudden failure of a vital stressed steel component in sodium hydroxide solution.

The effect of the metal/environment interaction on the environment itself is frequently more important than the actual deterioration of the metal (*see* Section 2.7). For instance, lead pipes cannot be used for conveying plumbosolvent waters, since a level of lead $>0{\cdot}1$ p.p.m. is toxic; similarly, galvanised steel may not be used for certain foodstuffs owing to the toxicity of zinc salts (*see* Section 2.8). In many chemical processes selection of a particular metal may be determined by the need to avoid contamination of the environment by traces of metallic impurities that would affect colour or taste of products or catalyse undesirable reactions; thus copper and copper alloys cannot be used in soap manufacture, since traces of copper ions result in colouration and rancidification of the soap. In these circumstances it will be essential to use unreactive and relatively expensive metals, even though the environment would not result in the rapid deterioration of cheaper metals such as mild steel. A further possibility is that contamination of the environment by metals' ions due to the corrosion of one metal can result in the enhanced corrosion of another when the two are in contact with the same environment. Thus the slow uniform corrosion of copper by a cuprosolvent domestic water may not be particularly deleterious to copper plumbing, but it can result in the rapid pitting and consequent perforation of galvanised steel and aluminium that subsequently comes into contact with the copper-containing water (Sections 4.1, 4.2 and 4.7).

Finally, it is necessary to point out that for a number of applications metals are selected in preference to other materials because of their visual appearance, and for this reason it is essential that brightness and reflectivity are retained during exposure to the atmosphere; stainless steel is now widely used for architectural purposes, and for outdoor exposure the surface must remain bright and rust-free without periodic cleaning (Section 3.3). On the other hand, the slow-weathering steels, which react with the constituents of the atmosphere to form an adherent uniform coating of rust, are now being used for cladding buildings (Section 3.2), in spite of the fact that a rusty surface is usually regarded as aesthetically unpleasant.

The interaction of a metal or alloy (or a non-metallic material) with its environment is clearly of vital importance in the performance of materials of construction, and the fact that the present work is largely confined to a detailed consideration of such interactions could create the impression that this was the sole factor of importance in materials selection. This, of course, is not the case although it is probably true to say that this factor is the one that is the most neglected by the design engineer.

Definitions of Corrosion

In the case of non-metallic materials, the term *corrosion* invariably refers to their deterioration from chemical causes, but a similar concept is not necessarily applicable to metals. Many authorities[1] consider that the term *metallic corrosion* embraces all interactions of a metal or alloy (solid or liquid) with its environment, irrespective of whether this is deliberate and beneficial or adventitious and deleterious. Thus this definition of corrosion, which for convenience will be referred to as the *transformation* definition, will include, for example, the deliberate anodic dissolution of zinc in cathodic protection and electroplating as well as the spontaneous gradual wastage of zinc roofing sheet resulting from atmospheric exposure.

On the other hand, *corrosion* has been defined[2] as 'the undesirable deterioration' of a metal or alloy, i.e. an interaction of the metal with its environment that adversely affects those properties of the metal that are to be preserved. This definition—which will be referred to as the *deterioration* definition—is also applicable to non-metallic materials such as glass, concrete, etc. and embodies the concept that corrosion is always deleterious. However, the restriction of the definition to undesirable chemical reactions of a metal results in anomalies which will become apparent from a consideration of the following examples.

Steel, when exposed to an industrial atmosphere, reacts to form the reaction product rust, of approximate composition $Fe_2O_3.H_2O$, which being loosely adherent does not form a protective barrier that isolates the metal from the environment; the reaction thus proceeds at an approximately linear rate until the metal is completely consumed. Copper, on the other hand forms an adherent green patina, corresponding approximately with bronchantite, $CuSO_4.3Cu(OH)_2$, which is protective and isolates the metal from the atmosphere. Copper roofs installed 200 years ago are still performing satisfactorily, and it is apparent that the formation of bronchantite is not deleterious to the function of copper as roofing material—indeed, in this particular application it is considered to enhance the appearance of the roof, although a similar patina formed on copper water pipes would be aesthetically objectionable.

The rapid dissolution of a vessel constructed of titanium in hot 40% H_2SO_4 with the formation of Ti^{4+} aquo cations conforms with both definitions of corrosion, but if the potential of the metal is raised (anodic protection) a thin adherent protective film of anatase, TiO_2, is formed, which isolates the metal from the acid so that the rate of corrosion is enormously decreased. The formation of this very thin oxide film on titanium, like that of the relatively thick bronchantite film on copper, clearly conforms with the *transformation* definition of corrosion, but not with the *deterioration* definition, since in these examples the rate and extent of the reaction is not significantly detrimental to the metal concerned. Again, magnesium, zinc or aluminium is deliberately sacrificed when these metals are used for the cathodic protection of steel structures, but as these metals are clearly not required to be maintained as such, their consumption in this particular application cannot, according to the *deterioration*, be regarded as corrosion. Furthermore, corrosion reactions are used to advantage in technological processes such as pickling, etching, chemical and electrochemical polishing and machining, etc.

The examples already discussed lead to the conclusion that any reaction of a metal with its environment must be regarded as a corrosion process irrespective of the extent of the reaction or of the rates of the initial and subsequent stages of the reaction. It is not illogical, therefore, to regard *passivity*, in which the reaction product forms a very thin protective film that controls rate of the reaction at an acceptable level, as a limiting case of a corrosion reaction. Thus both the rapid dissolution of *active* titanium in 40% H_2SO_4 and the slow dissolution of *passive* titanium in that acid must be regarded as corrosion processes, even though the latter will not be detrimental to the metal during the anticipated life of the vessel.

It follows that in deciding whether the corrosion reaction is detrimental to a metal in a given application, the precise form of attack on the metal (general, intergranular, etc.), the nature of the reaction products (protective or non-protective), the velocity and extent of the reaction and the location of the corrosion reaction must all be taken into account. In addition, due consideration must be given to the effect of the corrosion reaction on the environment itself. Thus corrosion reactions are not always detrimental, and our ability to use highly reactive metals such as aluminium, titanium, etc. in aggressive environments is due to a limited initial corrosion reaction, which results in the formation of a rate-controlling corrosion product. Expressions such as 'preventing corrosion', 'combating corrosion' or even 'fighting corrosion' are misleading; with the majority of metals corrosion cannot be avoided and 'corrosion control' rather than 'prevention' is the desired goal. The implication of 'control' in this context is that (*a*) neither the form, nor the extent, nor the rate of the corrosion reaction must be detrimental to the metal used as a constructional material for a specific purpose, and (*b*) for certain applications the corrosion reaction must not result in contamination of the environment. The scope of corrosion control is considered in more detail in the Introduction to Volume II, but it is relevant to mention here that it must involve a consideration of materials, availability, fabrication, protective methods and economics in relation to the specific function of the metal and its anticipated life. At one extreme corrosion control in certain environments may be effected by the use of thick sections of mild steel without any protective system, at the other the environmental conditions prevailing may necessitate the use of platinum.

The scope of the term 'corrosion' is continually being extended, and Fontana and Staehle have stated[3] that '*corrosion* will include the reaction of metals, glasses, ionic solids, polymeric solids and composites with environments that embrace liquid metals, gases, non-aqueous electrolytes and other non-aqueous solutions'.

Vermilyea, who has defined corrosion as a process in which atoms or molecules are removed one at a time, considers that evaporation of a metal into vacuum should come within the scope of the term, since atomically it is similar to other corrosion processes[4].

Evans[5] considers that corrosion may be regarded as a branch of chemical thermodynamics or kinetics, as the outcome of electron affinities of metals and non-metals, as short-circuited electrochemical cells, or as the demolition of the crystal structure of a metal.

These considerations lead to the conclusion that there is probably a need for two definitions of corrosion, which depend upon the approach adopted:

1. Definition of corrosion in the context of Corrosion Science: the reaction of a solid with its environment.
2. Definition of corrosion in the context of Corrosion Engineering: the reaction of an engineering constructional metal (material) with its environment with a consequent deterioration in properties of the metal (material).

Methods of Approach to Corrosion Phenomena

The effective use of metals as materials of construction must be based on an understanding of their physical, mechanical and chemical properties. These last, as pointed out earlier, cannot be divorced from the environmental conditions prevailing. Any fundamental approach to the phenomena of corrosion must therefore involve consideration of the structural features of the metal, the nature of the environment and the reactions that occur at the metal/environment interface. The more important factors involved may be summarised as follows:

1. *Metal*—composition, detailed atomic structure, microscopic and macroscopic heterogeneities, stress (tensile, compressive, cyclic), etc.
2. *Environment*—chemical nature, concentrations of reactive species and deleterious impurities, pressure, temperature, velocity, impingement, etc.
3. *Metal/environment interface*—kinetics of metal oxidation and dissolution, kinetics of reduction of species in solution; nature and location of corrosion products; film growth and film dissolution, etc.

From these considerations it is evident that the detailed mechanism of metallic corrosion is highly complex and that an understanding of the various phenomena will involve many branches of the pure and applied sciences, e.g. metal physics, physical metallurgy, the various branches of chemistry, bacteriology, etc. although the emphasis may vary with the particular system under consideration. Thus in stress-corrosion cracking (*see* Section 8.1) emphasis may be placed on the detailed metallurgical structure in relation to crack propagation resulting from the conjoint action of corrosion at localised areas and mechanical tearing, while in underground corrosion the emphasis may be on the mechanism of bacterial action in relation to the kinetics of the overall corrosion reaction (*see* Section 2.6).

Although the mechanism of corrosion is highly complex the actual control of the majority of corrosion reactions can be effected by the application of relatively simple concepts. Indeed, the Committee on Corrosion and Protection[6] concluded that 'better dissemination of existing knowledge' was the most important single factor that would be instrumental in decreasing the enormous cost of corrosion in the U.K.

Corrosion as a Chemical Reaction at a Metal/Environment Interface

As a first approach to the principles which govern the behaviour of metals in specific environments it is preferable for simplicity to disregard the detailed structure of the metal and to consider corrosion as a heterogeneous chemical reaction which occurs at a metal/non-metal interface and which involves the metal itself as one of the reactants (cf. catalysis). Corrosion can be expressed, therefore, by the simple chemical reaction:

$$aA + bB = cC + dD \qquad \ldots(1.1)$$

where A is the metal and B the non-metal reactant (or reactants) and C and D the products of the reaction. The non-metallic reactants are frequently referred to as *the environment* although it should be observed that in a complex environment the major constituents may play a very subsidiary role in the reaction. Thus in the 'atmospheric' corrosion of steel, although nitrogen constitutes approximately 75% of the atmosphere, its effect, compared with that of moisture, oxygen, sulphur dioxide, solid particles, etc. can be disregarded (in the high-temperature reaction of titanium with the atmosphere, on the other hand, nitrogen is a significant factor).

One of the reaction products (say, *C*) will be an oxidised form of the metal, and *D* will be a reduced form of the non-metal—*C* is usually referred to as the *corrosion product*, although the term could apply equally to *D*. In its simplest form, reaction 1.1 becomes

$$aA + bB = cC \qquad \text{...(1.2)}$$

e.g.

$$4Fe + 3O_2 = 2Fe_2O_3$$

where the reaction product can be regarded either as an oxidised form of the metal or as the reduced form of the non-metal. Reactions of this type which do not involve water or aqueous solutions are referred to as 'dry' corrosion reactions.

The corresponding reaction in aqueous solution is referred to as a 'wet' corrosion reaction, and the overall reaction (which actually occurs by a series of intermediate steps) can be expressed as

$$4Fe + 2H_2O + 3O_2 = 2Fe_2O_3.H_2O \qquad \text{...(1.3)}$$

Thus in all corrosion reactions one (or more) of the reaction products will be an oxidised form of the metal, aquo cations (e.g. Fe^{2+}(aq.), Fe^{3+}(aq.)), aquo anions (e.g. $HFeO_2^-$(aq.), FeO_4^{2-}(aq.)), or solid compounds (e.g. $Fe(OH)_2$, Fe_3O_4, $Fe_3O_4.H_2O$, $Fe_2O_3.H_2O$), while the other reaction product (or products) will be the reduced form of the non-metal. Corrosion may be regarded, therefore, as a *heterogeneous redox reaction at a metal/non-metal interface in which the metal is oxidised and the non-metal is reduced*. In the interaction of a metal with a specific non-metal (or non-metals) under specific environmental conditions, the chemical nature of the non-metal, the chemical and physical properties of the reaction products, and the environmental conditions (temperature, pressure, velocity, viscosity, etc.) will clearly be important in determining the form, extent and rate of the reaction.

Environment

Environments are considered in detail in Chapter 2, but some examples of the behaviour of normally reactive and non-reactive metals in simple chemical solutions will be considered here to illustrate the fact that corrosion is dependent on the nature of the environment; the thermodynamics of the systems and the kinetic factors involved are considered in Sections 1.4 and 1.9.

Gold is stable in most strong reducing acids, whereas iron corrodes rapidly, yet finely divided gold can be quickly dissolved in oxygenated cyanide solutions which may be contained in steel tanks. A mixture of caustic soda and sodium nitrate can be fused in an iron or nickel crucible, whereas this melt would have a disastrous effect on a platinum crucible.

Copper is relatively resistant to dilute sulphuric acid but will corrode if oxygen or oxidising agents are present in the acid, whereas austenitic stainless steels are stable in this acid only if oxygen or other oxidising agents are present. Iron will corrode rapidly in oxygenated water but extremely slowly if all oxygen is removed; if, however, oxygen is brought rapidly and simultaneously to all parts of the metal surface the rate will become very slow, owing to the formation of a protective oxide film. Lead will dissolve rapidly in nitric acid, more slowly in hydrochloric acid, and very slowly in sulphuric acid. These examples show that the corrosion behaviour of a metal cannot be divorced from the specific environmental conditions prevailing, which determine the rate, extent (after a given period of time) and form of the corrosion process.

Metal

Heterogeneities associated with a metal have been classified in Table 1.1 as atomic (*see* Fig. 1.1), microscopic (visible under an optical microscope), and macroscopic, and their effects are considered in various sections of the present work. It is relevant to observe, however, that the detailed mechanism of all aspects of corrosion, e.g. the passage of a metallic cation from the lattice to the solution, specific effects of ions and species in solution in accelerating or inhibiting corrosion or causing stress-corrosion cracking, etc. must involve a consideration of the detailed atomic structure of the metal or alloy.

The corrosion behaviour of different constituents of an alloy is well known, since the etching techniques used in metallography are essentially corrosion processes which take advantage of the different corrosion rates of phases as a means of identification, e.g. the grain boundaries are usually etched more rapidly than the rest of the grain owing to the greater reactivity of the disarrayed metal (*see* Sections 1.3 and 9.2).

Table 1.1 Heterogeneities in metals

1. *Atomic* (as classified by Ehrlich and Turnbull[7], see Fig. 1.1).
   - (*a*) Sites within a given surface layer ('normal' sites); these vary according to the particular crystal plane (Fig. 1.2).
   - (*b*) Sites at edges of partially complete layers.
   - (*c*) Point defects in the surface layer: vacancies (molecules missing in surface layer), kink sites (molecules missing at edge of layer), molecules adsorbed on top of complete layer.
   - (*d*) Disordered molecules at point of emergence of dislocations (screw or edge) in metal surface.

2. *Microscopic*
   - (*a*) Grain boundaries—usually, but not invariably, more reactive than grain interior.
   - (*b*) Phases—metallic (single metals, solid solutions, intermetallic compounds), non-metallic, metal compounds, impurities, etc.—heterogeneities due to thermal or mechanical causes.

3. *Macroscopic*
   - (*a*) Grain boundaries.
   - (*b*) Discontinuities on metal surface—cut edges, scratches, discontinuities in oxide films (or other chemical films) or in applied metallic or non-metallic coatings.
   - (*c*) Bimetallic couples of dissimilar metals.
   - (*d*) Geometrical factors—general design, crevices, contact with non-metallic materials, etc.

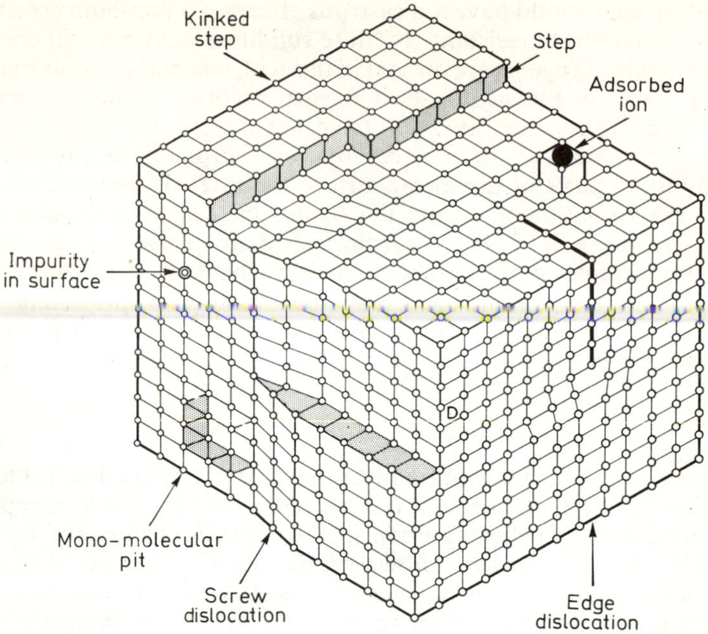

Fig. 1.1 Surface imperfections in a crystal (after Erlich and Turnbull[7])

Fig. 1.2 Hard-sphere model of face-centred cubic (f.c.c.) lattice showing various types of sites. Numbers denote Miller indices of atom places and the different shadings correspond to differences in the number of nearest neighbours (courtesy Erlich and Turnbull[7])

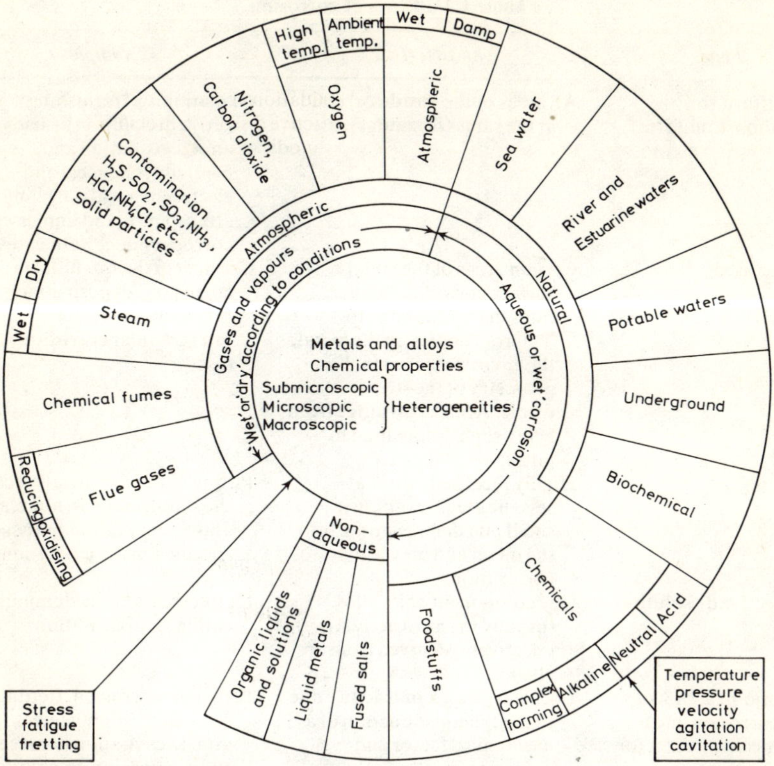

Fig. 1.3 Environments in corrosion

Macroscopic heterogeneities, e.g. crevices, discontinuities in surface films, bimetallic contacts etc. will have a pronounced effect on the location and the kinetics of the corrosion reaction and are considered in various sections throughout this work. Practical environments are shown schematically in Fig. 1.3, which also serves to emphasise the relationship between the detailed structure of the metal, the environment, and external factors such as stress, fatigue, velocity, impingement, etc.

Types of Corrosion

Corrosion can affect the metal in a variety of ways which depend on its nature and the precise environmental conditions prevailing, and a broad classification of the various forms of corrosion in which five major types have been identified, is presented in Table 1.2. Thus an 18Cr–8Ni stainless steel will corrode uniformly during polishing, active dissolution or passivation, but will corrode locally during intergranular attack, crevice corrosion or pitting; in certain circumstances selective attack along an 'active path' in conjunction with a tensile stress may lead to a transgranular fracture. Types of corrosion are dealt with in more detail in Appendix 1.1A.

Table 1.2 Types of corrosion

Type	*Characteristic*	*Examples*
1. Uniform (or almost uniform)	All areas of the metal corrode at the same (or similar) rate	Oxidation and tarnishing; active dissolution in acids; anodic oxidation and passivity; chemical and electrochemical polishing; atmospheric and immersed corrosion in certain cases
2. Localised	Certain areas of the metal surface corrode at higher rates than others due to 'heterogeneities' in the metal, the environment or in the geometry of the structure as a whole. Attack can range from being slightly localised to pitting	Crevice corrosion; filiform corrosion; deposit attack; bimetallic corrosion; intergranular corrosion; weld decay
3. Pitting	Highly localised attack at specific areas resulting in small pits that penetrate into the metal and may lead to perforation	Pitting of passive metals such as the stainless steels, aluminium alloys, etc., in the presence of specific ions, e.g. Cl^- ions
4. Selective dissolution	One component of an alloy (usually the most active) is selectively removed from an alloy	Dezincification; dealuminification; graphitisation
5. Conjoint action of corrosion and a mechanical factor	Localised attack or fracture due to the synergistic action of a mechanical factor and corrosion	Errosion–corrosion, fretting corrosion, impingement attack, cavitation damage; stress corrosion cracking, hydrogen cracking, corrosion fatigue

Ideally, the metal selected, or the protective system applied to the metal, should be such that no corrosion occurs at all, but this is seldom technologically or economically feasible. It is necessary, therefore, to tolerate a rate and a form of corrosion that will not be significantly detrimental to the properties of the metal during its anticipated life. Thus, providing the corrosion rate is known, the slow uniform corrosion of a metal can frequently be allowed for in the design of the structure; for example, in the case of a metal that shows an active/passive transition the rate of corrosion in the passive region is usually acceptable whereas the rate in the active region is not. It follows that certain forms of corrosion can be tolerated and that corrosion control is possible, providing that the rate and form of the corrosion reaction are predictable and can be allowed for in the design of the structure.

Pitting is regarded as one of the most insidious forms of corrosion, since it frequently leads to perforation and to a consequent corrosion failure. In other cases pitting may result in loss of appearance, which is of major importance when the metal concerned is used for decorative architectural purposes. However, aluminium saucepans that have been in service for some time are invariably pitted, although the pits seldom penetrate the metal, i.e. the saucepan remains functional and the pitted appearance is of no significance in that particular application.

These considerations lead to the conclusion that the relationship between corrosion and deterioration of properties of a metal is highly complex, and involves a consideration of a variety of factors such as the rate and form of corrosion and the specific function of the metal concerned; certain forms of corrosion such as uniform attack can be tolerated, whereas others such as pitting and stress corrosion cracking that ultimately lead to complete loss of function, cannot.

The implications of the terms *predictable* and *unpredictable* used in the context of corrosion require further consideration, since they are clearly dependent on the knowledge and expertise of the engineer, designer or corrosion designer who takes the decision on the metal or alloy to be used, or the procedure to be adopted, to control corrosion in a specific environmental situation. On this basis a corrosion failure (i.e. failure of the function of the metal due to corrosion within a period that is significantly less than the anticipated life of the structure) may be the result of one or more of the following possibilities:

1. *Predictable*. (*a*) The knowledge and technology are available but have not been utilised by the designer; this category includes a wide variety of design features such as the wrong choice of materials, introduction of crevices and bimetallic contacts etc., and is the most frequent cause of corrosion failures. (*b*) The knowledge and technology are available, but have not been applied for economic reasons; e.g. inadequate pretreatment of steel prior to painting and the use of unprotected mild steel for silencers and exhaust systems of cars.

2. *Unpredictable*. (*a*) The design has been based on specific environmental conditions, which have subsequently changed during the operation of the process; in this connection it should be noted that small changes in the chemical nature of the environment, temperature, pressure and velocity may lead to significant changes in the corrosion rate and form; the catastrophic oxidation and failure of steel bolts in nuclear reactors in the U.K. resulting from an increase in the temperature of the carbon dioxide is an example of an unpredictable failure due to a change in environmental conditions. (*b*) There is insufficient knowledge and experience of the metal, alloy or the environment to predict with certainty that failure will not occur; examples could be quoted of new alloys that have been subjected to an extensive series of carefully planned corrosion tests, but have failed in service.

Professor M. Fontana[8] has made the statement that "Virtually all premature corrosion failures these days occur for reasons which were already well known and these failures can be prevented". It is apparent from this statement, and from the conclusions reached by the Committee on Corrosion and Protection, that category 1 is responsible for the majority of incidents of corrosion failure that could have been avoided if those responsible were better informed on the hazards of corrosion and on the methods that should have been used to control it.

Principles of Corrosion

It has been stated that metallic corrosion is an art rather than a science and that, at present, insufficient knowledge is available to predict with any certainty how a particular metal or alloy will behave in a specific environment[4]. It should be appreciated that the decision to use a particular metal or alloy in preference to others in a given environment or to employ a particular protective system is based usually on previous experience and empirical testing (*see* Chapter 20) rather than on the application of scientific knowledge —the technology of corrosion is without doubt in advance of corrosion science and many of the phenomena of corrosion are not fully understood. Thus the phenomena of passivity which was first observed by Faraday in 1836 is still a subject of controversy, the specific effect of certain anions in causing stress-corrosion cracking of certain alloy systems is not fully understood, and dezincification of brasses can be prevented by additions of arsenic (or other elements such as antimony or phosphorus) but no adequate theory has been submitted to explain the action of these elements (*see* Section 4.2).

An understanding of the basic principles of the science of metallic corrosion is clearly vital for corrosion control, and as knowledge of the subject advances the application of scientific principle rather than an empirical approach may be used for such purposes as the selection of corrosion inhibitors, formulation of corrosion-resisting alloys, etc.

Terminology

The classification given in Table 1.2 is based on the various forms that corrosion may take, but the terminology used in describing corrosion phenomena frequently places emphasis on the environment or cause of attack rather than the form of attack. Thus the broad classification of corrosion reactions into 'wet' or 'dry' is now generally accepted, and the nature of the process is frequently made more specific by the use of an adjective that indicates type or environment, e.g. concentration–cell corrosion, crevice corrosion, bimetallic corrosion and atmospheric corrosion,

Table 1.3 Terminology in corrosion

Type of attack	*Environmental*	*Cause of attack*	*Mechanical factors*	*Corrosion product*
general (uniform)	wet*	concentration cell	stress	rusting
localised	dry	bimetallic cell	fretting	tarnishing
pitting (or intense)	atmospheric	active–passive cell	fatigue	scaling
intergranular	immersed	stray current (electrolysis)	cavitation	green rot
transgranular	underground	hydrogen evolution	erosion	tin pest
selective	sea water	oxygen absorption	impingement	
parting	chemical	impingement		
catastrophic	fused-salt	hydrogen embrittlement		
layer	flue-gas	caustic embrittlement		
filiform	biochemical			
	bacterial			
	high-temperature			
	liquid-metal			

*See Appendix to this section.

high-temperature corrosion, sea-water corrosion, etc. Alternatively, the phenomenon is described in terms of the corrosion product itself—tarnishing, rusting, green rot. The terminology used in corrosion is given in Table 1.3 and is considered in more detail in Appendix 1.1A.

L. L. SHREIR

REFERENCES

1. Hoar, T. P., *J. Appl. Chem.*, **11**, 121 (1961); Vernon, W. H. J., *The Conservation of Natural Resources*, Instn. of Civil Engrs., London, 105 (1957); Potter, E. C., *Electrochemistry*, Cleaver-Hume, London, 231 (1956)
2. Uhlig, H. H. (Ed.), *The Corrosion Handbook*, Wiley, New York and Chapman and Hall, London (1948); Uhlig, H. H., *Corrosion and Corrosion Control*, Wiley, New York (1971); Fontana, M. G. and Greene, N. D., *Corrosion Engineering*, McGraw-Hill (1967)
3. Fontana, M. G. and Staehle, R. W., *Advances in Corrosion Science and Technology*, Plenum Press, New York (1970)
4. Vermilyea, D. A., *Proc. 1st International Congress on Metallic Corrosion*, London, 1961, Butterworths, London, 62 (1962)
5. Evans, U. R., *The Corrosion and Oxidation of Metals*, Arnold, London, 12 (1960)
6. *Report of the Committee on Corrosion and Protection*, Department of Trade and Industry, H.M.S.O. (1971)
7. Ehrlich, G. and Turnbull, D., *Physical Metallurgy of Stress Corrosion Fracture*, Interscience, New York and London, 47 (1959)
8. Fontana, M. G., *Corrosion*, **27**, 129 (1971)

1.1A Appendix—Classification of Corrosion Processes

Existing Classifications

A logical and scientific classification of corrosion processes, although desirable, is by no means simple, owing to the enormous variety of corrosive environments and the diversity of corrosion reactions, but the broad classification of corrosion reactions into 'wet' or 'dry' is now generally accepted, and the terms are in common use. The term 'wet' includes all reactions in which an aqueous solution is involved in the reaction mechanism; implicit in the term 'dry' is the absence of water or an aqueous solution.

These terms are evidently ambiguous; for example, it is not always clear whether 'wet' is confined to aqueous solutions—the 'wetting' of solids by mercury indicates that liquid-metal corrosion should be classified as 'wet'. Even if the term is restricted to aqueous solutions, the difficulty arises that the mechanism of growth of magnetite scale during the reaction of the interior of a boiler drum with dilute caustic soda at high temperatures and pressures is best interpreted in terms of a 'dry' corrosion process. Similar considerations apply to the reactions of aluminium and zirconium with high-temperature water.

Considering oxidation as a typical 'dry' reaction it follows from Fig. 1.A1*a* that at the *interfaces*:

$$M \rightleftharpoons (M^{z+}\bigcirc/O) + z(e\bigcirc/O)$$

where $M^{z+}\bigcirc$ is an interstitial metal ion, $e\bigcirc$ an interstitial electron and /O indicates the metal/oxide interface (Section 1.8).

If the metal dissolves to enter a vacant site, then

$$M \rightleftharpoons (M^{z+}\square/O + ze\square/O)$$

where $M^{z+}\square$ represents a cation vacancy and $e\square$ a positive hole.

At the gas/oxide interface the O_2 gas ionises

$$(\tfrac{1}{2}O_2/\text{ads.}) + 2(e/X) \rightleftharpoons (O^{2-}/\text{ads.})$$

where /X indicates the gas/oxide interface.

By definition, these *interfaces* can be considered as anodes and cathodes respectively.

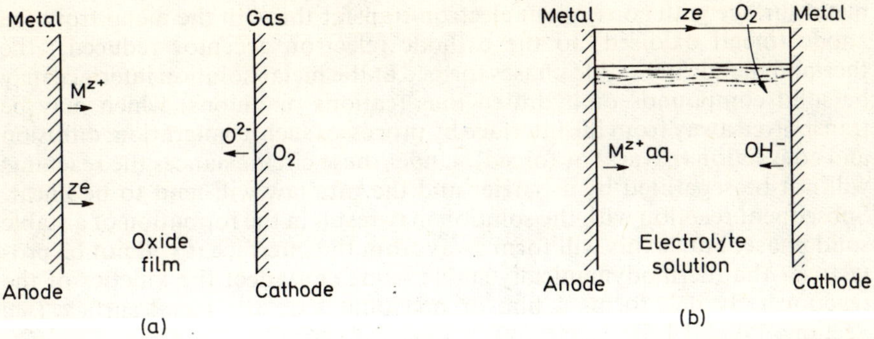

Fig. 1.A1 Anodes and cathodes in corrosion processes. (*a*) 'Dry' corrosion and (*b*) 'wet' corrosion

The corresponding 'wet' corrosion half-reactions (Fig. 1.A1*b*) are:

$$[M^{z+}+ze]_{\text{lattice}}+H_2O \rightarrow M^{z+}\text{aq.}$$

and

$$O_2+2H_2O+4e = 4OH^-$$

or

$$O_2+4H^++4e = 2H_2O$$

'Dry' Corrosion

These are generally metal/gas or metal/vapour reactions involving non-metals such as oxygen, halogens, hydrogen sulphide, sulphur vapour, etc. and oxidation, scaling and tarnishing are the more important forms. A characteristic of these reactions is that the initial oxidation of the metal, reduction of the non-metal, and formation of compound must occur at one and the same place at the metal/non-metal interface. Should the compound be volatile or discontinuous, further interaction at the interface (or through a thin film of constant thickness) is possible and in most cases the reaction rate will tend to remain constant with time (linear law). If the film is continuous it will present a barrier to the reactants and further interaction will necessitate passage of the reactants through the film by (*a*) diffusion of the non-metal or (*b*) diffusion and migration of ions of the reactants. The detailed mechanisms of these reactions are considered in Sections 1.8–1.10, but it is appropriate to observe that the formation of a continuous film of reactant product at a metal/non-metal interface will result in a growth rate which, when the film becomes sufficiently thick to be rate determining, decreases as the film thickens, i.e. parabolic, logarithmic, asymptotic, cubic, etc.

'Wet' Corrosion

In 'wet' corrosion the oxidation of the metal and reduction of a species in solution (electron acceptor or oxidising agent) occur at different areas on the metal surface with consequent electron transfer through the metal from the anode (metal oxidised) to the cathode (electron acceptor reduced); the thermodynamically stable phases formed at the metal/solution interface may be solid compounds or hydrated ions (cations or anions) which may be transported away from the interface by processes such as migration, diffusion and convection (natural or forced). Under these circumstances the reactants will not be separated by a barrier and the rate law will tend to be linear. Subsequent reaction with the solution may result in the formation of a stable solid phase, but as this will form away from the interface it will not be protective—the thermodynamically stable oxide can affect the kinetics of the reaction only if it forms a film or precipitates on the metal surface (*see* Sections 1.4 and 1.5).

Further points which distinguish 'wet' from 'dry' corrosion are:

1. In 'wet' corrosion the metal ions are hydrated—the hydration energy of most metal ions is very large and thus facilitates ionisation (*see* Section 1.9).
2. In 'wet' corrosion ionisation of oxygen to hydroxyl must involve the hydronium ion or water.
3. In 'dry' corrosion the direct ionisation of oxygen occurs.

Corrosion in Organic Solvents

Corrosion reactions in aggressive organic solvents are becoming a more frequent occurrence owing to developments in the chemical and petrochemical industries, and these reactions can lead to the deterioration of the metal and to undesirable changes in the solvent. This aspect of corrosion has recently been the subject of an extensive review by Heitz[1] who has considered the mechanisms of the reactions, the similarities between corrosion in organic solvents and in aqueous solutions, the methods of study and the occurrence of the phenomenon in industrial processes.

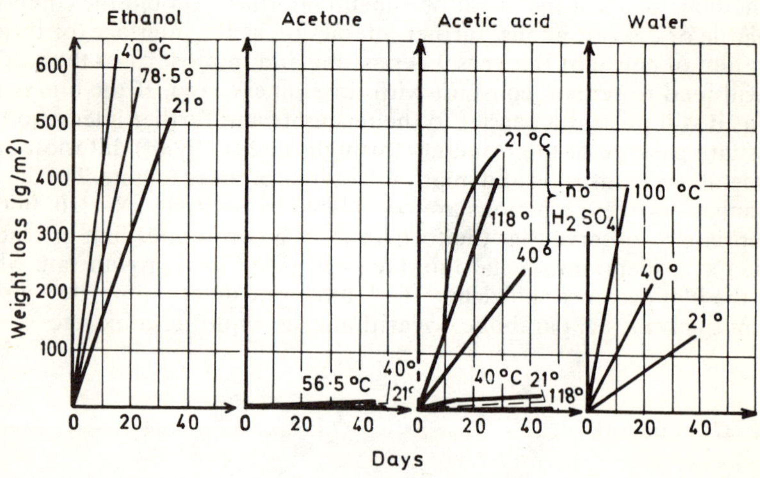

Fig. 1.A2 Corrosion of nickel in different solvents containing 0·05 wt.% H_2SO_4 at various temperatures (after Heitz[1])

Figure 1.A2 shows the weight loss against time curve for nickel in various solvents containing 0·05 wt.% H_2SO_4 at various temperatures, and illustrates the unpredictable nature of corrosion in organic solvents. Thus the corrosion rates in ethanol are far greater than those in the aqueous acid whereas in acetone the rate is practically zero; even more surprising is the fact that in acetic acid the addition of 0·05% H_2SO_4 actually decreases the corrosion rate.

Heitz classifies corrosion reactions in organic solvents into

1. Electrochemical reactions, which follow a similar mechanism to those in aqueous solution.
2. Chemical reactions, which involve direct charge transfer between the metal atom in the lattice of the metal and the oxidising species.

In the case of electrochemical reactions the partial anodic reaction results in the formation of a solvated metal cation $M^{z+}_{\text{solv.}}$, a charged or uncharged metal complex MX^- or a solid compound MX_z, where X is a halogen ion, organic acid anion, etc.

The cathodic partial reactions are as follows:

(*a*) Reduction of a solvated proton to H_2 gas

$$\text{H}^+_{\text{solv.}} + e \rightarrow \tfrac{1}{2}\text{H}_2$$

(*b*) Reduction of acidic hydrogen of a proton donor

$$\text{H}A + e \rightarrow \tfrac{1}{2}\text{H}_2 + A^-$$

where A^- is a carboxylic acid anion, alcoholate ion, etc.

(*c*) Reduction of an oxidising gas Y

$$Y_z + zme \rightarrow zY^{m-}$$

where Y can be O_2, Cl_2, F_2, Br_2, O_3, N_2O_4, etc.

(*d*) Reduction of oxidising ions such as Fe^{3+}, Cu^{2+}, MnO_4^-, ClO_3^-, etc.

It is evident from the above that in many systems the reaction of a metal with an organic solvent follows a mechanism that is similar to the electrochemical mechanism of corrosion in aqueous solution.

Non-electrochemical processes may be represented by the general equation

$$M + 2\text{C}_x\text{H}_yX_z \rightarrow MX_2 + \text{C}_{2x}\text{H}_{2y}X_{2z-2}$$

where X is a halogen and M is a divalent metal, e.g. the Grignard reaction

$$\text{Mg} + \text{CH}_3\text{Cl} \rightarrow \text{CH}_3\text{MgCl}$$

A further type of chemical process, which is analogous to high-temperature corrosion, is the reaction of metals with organic sulphur compounds, which follow the equation

$$2M + 2R\text{SH} \rightarrow 2M\text{S} + \text{H}_2 + R_2$$

Heitz quotes a number of case studies of corrosion of metals in organic solvents and concludes that the phenomenology indicates no specific differences from that experienced in aqueous corrosion. Thus general corrosion, pitting, crevice corrosion, intergranular corrosion, erosion-corrosion cracking, hydrogen embrittlement, etc. can all occur in organic solvents. The methods of control also follow that used for corrosion in aqueous solutions, although there are certain differences. Thus cathodic and anodic protection are seriously limited by the resistivity of the solvent, and paint coatings deteriorate rapidly in contact with the solvent.

Suggested Classification and Nomenclature

On a basis of the preceding discussion, the classification and nomenclature outlined in Table 1.A1 is suggested as a possible alternative to the accepted classification of corrosion reactions into 'wet' and 'dry'.

Table 1.A1 Classification of spontaneous corrosion reactions

1. *Film-free Chemical Interaction*
   - (*a*) Metal/gas—oxide or compound volatile (e.g. reaction of molybdenum with oxygen, reaction of iron or aluminium with chlorine).
   - (*b*) Metal/liquid.
     Reactions of solid metals with liquid metals (e.g. dissolution of aluminium in mercury).
     Dissolution of metal in their fused halides (e.g. lead in lead chloride).
     Dissolution of metals in non-aqueous solutions (e.g. reaction of aluminium with carbon tetrachloride).
2. *Electrochemical*
   - (*a*) *Inseparable anode/cathode type* (*insep. A/C*)
     Reactions with aqueous solutions. Uniform dissolution or corrosion of metals in acid, alkaline or neutral solutions (e.g. dissolution of zinc in hydrochloric acid or in caustic soda solution; general corrosion of zinc in water or during atmospheric exposure).
     Reactions with non-aqueous solution (e.g. dissolution of copper in a solution of ammonium acetate and bromine in alcohol).
     Reactions with fused salts.
   - (*b*) *Separable anode/cathode type* (*sep. A/C*)
     All reactions of metals in aqueous or non-aqueous solutions or in fused salts where one area of the metal surface is predominantly anodic and the other is predominantly cathodic so that the sites are physically identifiable.
   - (*c*) *Interfacial anode/cathode type in which the metal surface is filmed*
     - (i) Metal/gas and metal/vapour reactions
       All reactions in which charge is transported through a film of reaction product on the metal surface—the film may or may not be rate determining (e.g. parabolic, logarithmic, asymptotic, etc. or linear growth laws, respectively).
     - (ii) Metal/solution reactions
       All reactions involving the uniform formation and growth of a film of reaction product (e.g. reaction of metals with high-temperature water, reaction of copper with sulphur dissolved in carbon disulphide).

It is considered that the main types of corrosion reactions can be classified as follows:

1. Film-free chemical interaction in which there is direct chemical reaction of a metal with its environment. The metal remains film-free and there is no transport of charge.

2. Electrochemical reactions which involve transfer of charge across an interface. These electrochemical reactions can be further subdivided into:

(*a*) Inseparable anode/cathode type (insep. A/C). The anodes and cathodes cannot be distinguished by experimental methods although their presence is postulated by theory, i.e. the *uniform* dissolution of metals in acid*, alkaline or neutral aqueous solutions, in non-aqueous solution, or in fused salts.

(*b*) Separable anode/cathode type (sep. A/C). Certain areas of the metal can be distinguished experimentally as predominantly anodic or cathodic, although the distances of separation of these areas may be as small as fractions of a millimetre. In these reactions there will be a macroscopic flow of charge through the metal.

(*c*) Interfacial anode/cathode type (interfacial A/C). One entire interface will be the anode and the other will be the cathode. Thus in Fig. 1.A1*a* the metal/metal oxide interface might be regarded as the anode and the metal/oxygen interface as the cathode.

It is apparent that, in general, 2(*a*) and 2(*b*) include corrosion reactions which are normally classified as 'wet', while 2(*c*) includes those which are normally classified as 'dry'.

The terminology suggested can be illustrated by reference to the corrosion behaviour of iron:

1. Reaction of iron with oxygen at room temperature or with oxygen or water at high temperatures—interfacial A/C type.
2. Reaction of iron with oxygenated water or with reducing acids—inseparable A/C type.
3. Reaction of iron containing a discontinuous magnetite scale with oxygenated water, crevice corrosion, water-line attack, 'long-line' corrosion of buried iron pipes, etc.—separable A/C type.

Although it is realised that this classification and terminology has certain limitations, it represents a preliminary attempt to provide a more rational classification of corrosion processes than that based on 'wet' and 'dry'.

Acknowledgement

Grateful thanks are due to Dr. W. B. Jepson, Dr. M. Pryor and Mr. J. N. Wanklyn for helpful discussions during the preparation of this Appendix.

L. L. SHREIR

REFERENCES

1. Heitz, E., 'Corrosion of Metals in Organic Solvents', *Advances in Corrosion Science and Technology* (ed. M. G. Fontana and R. W. Staehle), Vol. 4, Plenum Press, 149 (1974)

*Dr. Pryor considers that in certain cases of uniform dissolution of metals in acids (e.g. Al in hydrochloric or sulphuric acid) or alkalis a thin film of oxide is present on the metal surface—the film is not rate-determining but its presence would indicate that reactions of this type should be classified under 2 (*c*).

1.2 Nature of Films, Scales and Corrosion Products on Metals

The study of corrosion is essentially the study of the nature of the metal reaction products (corrosion products) and of their influence on the reaction rate. It is evident that the behaviour of metals and alloys in most practical environments is highly dependent on the solubility, structure, thickness, adhesion, etc. of the solid metal compounds that form during a corrosion reaction. These may be formed naturally by reaction with their environment (during processing of the metal and/or during subsequent exposure) or as a result of some deliberate pretreatment process that is used to produce thicker films or to modify the nature of existing films. The importance of these solid reaction products is due to the fact that they frequently form a kinetic barrier that isolates the metal from its environment and thus controls the rate of the reaction; the protection afforded to the metal will, of course, depend on the physical and chemical properties outlined above.

In general, reaction products (films*, scales and corrosion products) may be formed under the following environmental conditions.

(*a*) Direct reaction with a gas (O_2, CO_2, CO, H_2+O_2, H_2S, etc.) at temperatures that range from ambient to very high (1 000–2 000°C).

(*b*) Direct reaction with an aqueous solution with the formation of a thin invisible film (passivation) or of a thick visible corrosion product (protective or non-protective).

(*c*) By the deliberate formation of thick oxide films (e.g. anodising) at elevated potentials or by changing the nature of existing films by chemical treatments (e.g. chromating or phosphating).

For example, in a dry atmosphere a reactive metal such as aluminium may carry a natural protective oxide film of only some 3 nm (30 Å) thickness, while for increased corrosion resistance aluminium may be anodised to give a coating 10^4 times thicker (*see* Section 16.1). However, thickness alone does not provide a criterion of protection; and although a thick protective layer of millscale is formed on iron and steel during processing it is not continuous owing to spalling, and the attack on the exposed substrate at the discontinuities is far greater than if the surface was bare. Thus the kinetics of attack

*The distinction between a film and scale is not well defined, but it is usual to use the former when referring to a *thin* continuous layer of reaction product (visible or invisible) whilst the latter is normally used for *thick* high-temperatures layers (always visible).

will be related to a variety of other factors such as composition, structure, continuity, adhesion to the substrate, cohesion, mechanical properties, etc. of the film or scale of reaction products.

This section represents an attempt to survey in general terms the variation in the nature of very thin films originating in the initial reaction of a metal with its environment and their progression to the thicker overgrowths that control the kinetics. Recent developments in instrumental techniques have led to significant advances in the characterisation of these film- and scale-forming systems, and a summary of the experimental approaches available is provided at the end of the section. It is appropriate to consider first the products of reaction formed by a gaseous oxidising atmosphere and then to proceed to a consideration of the effect of water and aqueous systems.

Initial Surface Reaction States

The application of ultra-high vacuum techniques to low-energy electron diffraction (L.E.E.D.) studies of very clean metal surfaces in low-pressure oxidising and sulphidising atmospheres over a range of temperatures above ambient has provided detailed information on the initial states of interaction[1,2]. The following sequence of events is generally observed in the case of exposure to oxygen:

1. Rapid physical adsorption of molecular oxygen.
2. Chemisorption of atomic oxygen to form a partial or complete monolayer.
3. Further chemisorption of atomic oxygen into a second layer and/or further physical adsorption of O_2.

In Stage 2 a distinct structural modification to an expanded lattice at sub-monolayer coverages has been observed on nickel, indicating that the oxygen ions become progressively incorporated into the metal lattice. These two-dimensional crystals then gradually transform into a three-dimensional nickel oxide lattice as more oxygen becomes incorporated. Subsequent exposure to high-temperature conditions ($>1\,000°C$) has confirmed the extreme stability of the Stage 2 state.

Similarly, under low-temperature conditions ($<25°C$) three stages have been recognised and defined[3] as follows:

1. Physical adsorption of oxygen resulting in the formation of one or more monolayers of oxide and requiring no activation energy.
2. Electron tunnelling through the stable oxide film to the adsorbed oxygen which sets up a potential and causes ion drift, thus resulting in logarithmic oxide growth.
3. Film rearrangement resulting in the formation of oxide subgrain and grain boundaries; these paths of easy ion migration promote the formation of oxide 'islands' and result in an increase in the growth rate of the oxide.

Oxide films formed at low temperatures are initially continuous and amorphous, but may undergo local crystallisation with the incorporation of the oxide 'islands', a process that is facilitated by water, heat, high electric fields and mechanical stress[4].

Thin-Film Region

Studies of thermally grown oxides in the thin-film region (<100 nm) have revealed[5] on single crystal substrates interesting details of epitaxy, stress generation, mosaic structure and film topography, and oxidation rate anisotropic behaviour. Mismatch between the oxide lattice and the metal substrate gives rise to stresses which may find relief in the generation of mosaic structures consisting of small crystallites (5–100 nm diameter) whose lattices are slightly twisted or tilted with respect to one another. Their boundaries represent potential paths of easy diffusion through the oxide.

The uniformity of film thickness is dependent upon temperature and pressure. The nucleation rate rises with pressure, such that at pressures above atmospheric the high rate of nucleation can lead to comparatively uniform oxide films, while increase in temperature reduces the density of oxide nuclei, and results in non-uniformity. Subsequently, lateral growth of nuclei over the surface is faster than the rate of thickening until uniform coverage is attained, when the consolidated film grows as a continuous layer[2].

Growth of oxide nuclei may also be accompanied by the appearance of whiskers and platelets under certain conditions[6]. It has been demonstrated that oxidation of iron in air at about 200°C initially leads to nuclei of Fe_3O_4 developing to form a porous layer. Over this homogeneous oxide layer, nuclei of α-Fe_2O_3 appear and spread over the Fe_3O_4, but no γ-Fe_2O_3 is observed. After 30 days whiskers of α-Fe_2O_3 appear, ultimately reaching a length of 1 μm. At higher temperatures, too, whiskers of α-Fe_2O_3 appear and subsequently develop into crystallographic platelets. In general, products of this nature occur as fine features developing from otherwise protective films.

Scale-Forming Situations

In considering film growth at higher temperatures, a changeover to diffusion control, which is dependent on concentration gradients, tends to give rise to parabolic and paralinear kinetics as substantial scales form at thicknesses of 1–100 μm or more. This is the area of vital concern in the development and application of engineering alloys for high-temperature resistance, and is in distinct contrast to the thin-film régime. Nevertheless, the initial state of the metal surface can still influence subsequent oxidation behaviour. Thus, different oxidation patterns may be observed depending upon whether the surface is electropolished, hydrogen reduced, mechanically abraded or cathodically pretreated. When metals of variable valency become subjected to the oxidising potential gradient across the scale, a duplex or multiple series of layers forms. The classical case of iron oxidised above 600°C has been well established[6], and it has been shown that the system consists of $Fe/FeO/Fe_3O_4/Fe_2O_3/O_2$. In these situations film thickening occurs by transport of cations, anions, vacancies and electrons across the various phase boundaries, which is possible owing to the non-stoichiometric composition of the various coexistent oxides (*see* Sections 1.8, 1.9 and 7.2).

A rather different situation arises when mild steel is exposed to liquid water or dilute sodium hydroxide at 300–360°C. Here a duplex Fe_3O_4 scale is formed, consisting of an inner adherent protective film in contact with an outer poorly adherent crystalline layer of magnetite (*see* Section 1.10).

In alloy systems the course of events is complicated by such factors[7,8] as:

1. The affinity of the component metals for each other and for the non-metal.
2. The diffusion rates of atoms in the alloy and of ions in the compounds.
3. The mutual solubilities of the products present in the oxidation layers.
4. The formation of ternary compounds, e.g. spinels (*see* Table 1.4).
5. The relative volumes of the various phases.

In practice, thermal cycling rather than isothermal conditions more frequently occurs, leading to a deviation from steady state thermodynamic conditions and introducing kinetic modifications. Lattice expansion and contraction, the development of stresses and the production of voids at the alloy–oxide interface, as well as temperature-induced compositional changes, can all give rise to further complications. The resulting loss of scale adhesion and spalling may lead to *breakaway oxidation*[9,11] in which linear oxidation replaces parabolic oxidation (*see* Section 1.10).

Examination of the structural consequences of these complex interacting factors is now being elucidated in considerable detail by systematic application of electron optical and X-ray analysis techniques[9], as well as by a range of other methods[10].

In certain systems the oxidation reactions may lead to a particularly protective single phase being formed at the surface, e.g. magnetite (Fe_3O_4) in the case of iron and steel and γ-Al_2O_3 in the case of aluminium. The 'spinel' ($MgAl_2O_4$) lattice is important in relation to the protection it affords to alloys used at high temperatures, and such structures often occur with a continuously varying stoichiometry as a 'double oxide' phase, which may provide an effective kinetic barrier to the oxidation process (e.g. $NiO.Cr_2O_3$ spinel in Cr–Ni–Fe alloys). Some examples are given in Table 1.4.

The spinel structure is of especial significance in the corrosion behaviour of iron and alloy steels both at high temperatures and in aqueous environments. Its crystallographic unit cell can be represented as $8XY_2O_4$ (or $X_8Y_{16}O_{32}$) in which the valencies of the metal ions X and Y may be (*a*) X^{II}, Y^{III}; (*b*) X^{IV}, Y^{II} (giving rise to the so-called '2–3 spinels' or the '4–2 spinels'; and (*c*) X^{VI}, Y^{I}. The structure is based on a cell containing 32 oxygen atoms in a close-packed cubic arrangement. This provides for the incorporation of the X atoms in eight equivalent tetrahedral sites and the Y atoms in 16 equivalent octahedral sites. 'Inverse' spinels follow a different arrangement, represented by $Y(XY)O_4$, in which half of the Y atoms are located tetrahedrally, while the remaining Y atoms together with the X atoms are randomly arranged among the 16 octahedral positions. More generally, some spinels exist with a fraction λ of Y cations in tetrahedral sites where $0 > \lambda < \frac{1}{2}$.

Table 1.4 Spinel phases encountered in alloy oxidation

n-type	$MgFe_2O_4$	$ZnCo_2O_4$
	$NiFe_2O_4$	$MgAl_2O_4$
	$ZnFe_2O_4$	$ZnAl_2O_4$
p-type	$MgCr_2O_4$	$ZnCr_2O_4$
	$FeCr_2O_4$	$CoAl_2O_4$
	$CoCr_2O_4$	$NiAl_2O_4$

It should be noted that single metal oxides such as Fe_3O_4 and Co_3O_4 are inverse spinels, while Mn_3O_4 is a normal spinel. The spinel structure is prominent in the oxides on iron and aluminium[18,19]. The oxides M_2O_3 (and also the hydroxides and oxy-hydroxides $M(OH)_3$ and $MO.OH$) exist in the α and γ forms. Corundum and haematite represent the isostructural α forms, while the γ forms have cubic spinel-like structures deficient in metal ions. For example, in γ-Fe_2O_3 there are only $21\frac{1}{3}$ Fe^{3+} ions per unit cell of 32 O^{2-} ions, and these are randomly distributed among the eight tetrahedral and 16 octahedral 'available' sites. In magnetite, represented as $Fe^{3+}(Fe^{2+}Fe^{3+})O_4$, one third of the cations are Fe^{2+} and continuous interchange of electrons between Fe^{2+} and Fe^{3+} ions in the 16-fold positions accounts for its extremely high electronic conductivity. Careful oxidation of Fe_3O_4 yields γ-Fe_2O_3, which may be converted back into Fe_3O_4 by heating *in vacuo* at 250°C. Because wüstite (FeO) ideally has the NaCl-type structure (f.c.c. anion lattice), with four Fe^{2+} and four O^{2-} ions per unit cell, deviations from stoichiometry lead to not every octahedral site being filled in the metal deficient lattice (e.g. at 570°C $Fe_{0\cdot93}O$ contains cation vacancies and compensating Fe^{3+} ions). At lower temperatures disproportionation occurs:

$$4FeO \rightleftharpoons \alpha\text{-}Fe + Fe_3O_4$$

Therefore the relationship between these interconvertible structures originates from a cubic anion lattice of 32 O^{2-} ions in the cell. With 32 Fe^{2+} ions in the octahedral holes stoichiometric FeO is formed. Replacement of a number of Fe^{2+} ions with two-thirds of their number of Fe^{3+} ions maintains electrical neutrality but provides non-stoichiometric $Fe_{1-x}O$. Continual replacement in this way to leave 24 Fe atoms in the cubic cell produces Fe_3O_4, and further exchange to an average of $21\frac{1}{3}$ Fe^{3+} ions leads to γ-Fe_2O_3

$$Fe_{1-x}O \rightarrow Fe_3O_4 \rightarrow \gamma\text{-}Fe_2O_3$$

In actual oxidation, the cubic anion lattice becomes extended by the addition of new layers of close-packed O^{2-} ions into which Fe atoms migrate to give rise to the appropriate stable structures.

The defect γ-structures may be stabilised by the presence of Li^+ or H^+ ions (e.g. $LiFe_5O_8$). Cation diffusion rates in these and other lattices developed on metal surfaces play an important rôle in governing corrosion behaviour.

Surface Reaction Products Formed in Aqueous Environments

Whereas a film formed in dry air consists essentially of an anhydrous oxide and may reach a thickness of 3 nm, in the presence of water (ranging from condensed films deposited from humid atmospheres to bulk aqueous phases) further thickening occurs as partial hydration increases the electron tunnelling conductivity[3]. Other components in contaminated atmospheres may become incorporated (e.g. H_2S, SO_2, CO_2, Cl^-), as described in Sections 2.2 and 3.1.

Films may thus range from thin transparent oxides (passive films on Al, Cr, Ti and Fe–Cr alloys), or thin visible sulphides (on Cu and Ag) to thicker visible films, which may be compact, adherent and protective (anodic oxide

Table 1.5 Variations in the nature and thickness of the product formed on aluminium under different conditions

Formation conditions	*Nature of oxide film*	*Thickness* (nm)
Dry air or O_2	Amorphous Al_2O_3	1–2
Humid atmosphere	$AlOOH + Al_2O_3.3H_2O$	50–100
Boiling water	$AlOOH$ (or $Al_2O_3.H_2O$)	500–2000
Chemical conversion	$AlOOH$ + anions of solution	1000–5000
Anodic oxidation (barrier films)	Amorphous + crystalline Al_2O_3 + anions of solution	1000–3000

films on Al and Ti, $PbSO_4$ films on Pb, etc.) or bulky, poorly adherent and non-protective (rust on steel, 'white rust' on Zn). In some cases, fairly precise limits can be placed on the nature and thickness of the products formed under different conditions, as with aluminium illustrated in Table 1.5. In other cases, the undesirable wastage of the basis metal (e.g. the rusting of steel) is of more significance than the thickness of the corrosion product, although the nature of the latter may provide information useful in interpreting the mechanism of its formation.

Thus in industrial atmospheres the presence of $FeSO_4.4H_2O$ has been identified in combination with α- and γ-FeO.OH, and the two latter incorporate free water in excess of the composition $Fe_2O_3.H_2O$. Furthermore, although some of the corrosion product may be adherent, most of it is not[12] (Sections 3.1 and 3.2).

In the fully immersed situation where the corrosion product is produced by a secondary reaction such as $M^{2+} + 2H_2O \rightarrow M(OH)_2 + 2H^+$, as in the case of iron or zinc in dilute aqueous aerated chloride solutions, the sites of the anodic and cathodic processes are separated, and widely so in the partially immersed condition. Thus OH^- ions are formed at the cathode and $M^{2+}_{aq.}$ ions at the anode, giving rise to dispersed $M(OH)_2$ where they meet and react; under these circumstances the corrosion product cannot influence the kinetics. If chloride or sulphate is present, a basic compound $M_x(OH)_y(X)_z$ may form whose range of stability will depend upon the concentration of the anion pX and the pH of the solution; diagrams with axes pX and pH have been constructed that show the range of stability of these basic compounds. In the case of iron, the $Fe(OH)_2$ formed initially is subsequently oxidised to yellow FeO(OH) or $Fe_2O_3.H_2O$, or in low oxygen conditions black Fe_3O_4 is formed containing green reduced corrosion products. Vertical surfaces allow ready detachment of the products formed, while they may settle on a horizontally corroding surface and provide some blanketing action, restraining access of oxygen to the surface. Precise identification of the products and a knowledge of the pH at their location on the surface may provide information on the conditions of formation[13].

Thin Passive Films

In considering passivity and passivation (Sections 1.4 and 1.5), the nature of the surface product (the passivating film) entering into the process between the curve for active dissolution and that for the onset of film breakdown or oxygen evolution, assumes considerable significance.

As the system passes from the active to the passive state the initial interaction depends on the composition of the aqueous phase[14]. An initial chemisorbed state on Fe, Cr and Ni has been postulated in which the adsorbed oxygen is abstracted from the water molecules[2]. This has features in common with the metal/gaseous oxygen interaction mentioned previously. With increase in anodic potential a distinct 'phase' oxide or other film substance emerges at thicknesses of 1–4 nm. Increase in the anodic potential may lead to the sequence

$$M \rightarrow \underset{\text{monolayer}}{M\text{—OH}} \rightarrow \underset{\text{multilayer}}{M(\text{OH})_2} \rightarrow \underset{\text{phase oxide}}{MO}$$

which has been suggested for Ni in acid solutions, and Cd and Zn in alkaline solutions. On the other hand, Fe in strong H_2SO_4 first forms a layer of $FeSO_4$ crystals, which at higher potentials is replaced by an Fe_2O_3 film, the normal product formed during anodic polarisation in dilute acid[15]. In near-neutral solutions the passive film on Fe (2–6 nm thick) has been characterised as the so-called cubic oxide γ-Fe_2O_3 overlying a thin film of Fe_3O_4 on the metal surface[16].

The nature of γ-Fe_2O_3 in passive films is very significant and has been reviewed in detail[17]. Here again a spinel structure is prominent (derived from magnetite). Its structure is considered to be cation defective with protons (H^+) progressively replacing Fe^{2+} ions in the Fe_3O_4 spinel, and leading to a continuous series of solid solutions of which Fe_3O_4 and Fe_2O_3 are the end products. In some cases an HFe_5O_8 composition is indicated in which some Fe^{2+} ions have been replaced by protons. The implication of this mechanism of replacement of Fe^{2+} ions is that water is incorporated into the passive film by a process of oxidative hydrolysis of the initial Fe_3O_4 substrate as the potential of the metal is progressively raised.

An important feature of such films is their low ionic conductivity that restricts cation transport through the film substance. Electronic semi-conduction, however, permits other electrode processes (oxidation of H_2O to O_2) to take place at the surface without further significant film growth. At elevated anodic potentials adsorption and entry of anions, particularly chloride ions, may lead to instability and breakdown of these protective films (Sections 1.5 and 1.6).

Thick Anodic Films

Where the electronic conductivity of the film substance is low, as in the case of the 'valve' metals (Al, Nb, Ta, Zr, Ti), an increase in anode potential gives rise to a high electric field across the passive layer. Under these circumstances ion transport occurs and film growth continues to several hundred volts with thicknesses rising to hundreds of nanometres. At low voltages an amorphous or microcrystalline 'barrier' oxide is formed, which may recrystallise thermally or by the action of a high field to γ-Al_2O_3, β-Ta_2O_5 or TiO_2, etc. A 'mosaic' structure has been attributed to these amorphous films[17] to account for their high field conduction properties. In the case of a valve metal with variable valency a number of anodic oxides may form over a range of anodic potential, e.g. Ti in strongly oxidising conditions gives TiO_2, while anodic passivation at lower potentials leads to Ti_2O_3. 3–4 TiO_2, or even Ti_3O_5. Furthermore, different structural modifications can be produced depending on the precise conditions of formation. For example, with Al[20] and Ti[21] high temperatures and high formation voltage tend to favour crystalline modifications as compared with the more commonly observed amorphous oxides. While, in general, anodic films produced represent those expected from thermodynamic data, significant free-energy gradients may exist across the film substance. Such situations may lead to complex geometrical arrays of different compounds as shown by Burbank[22] for anodic

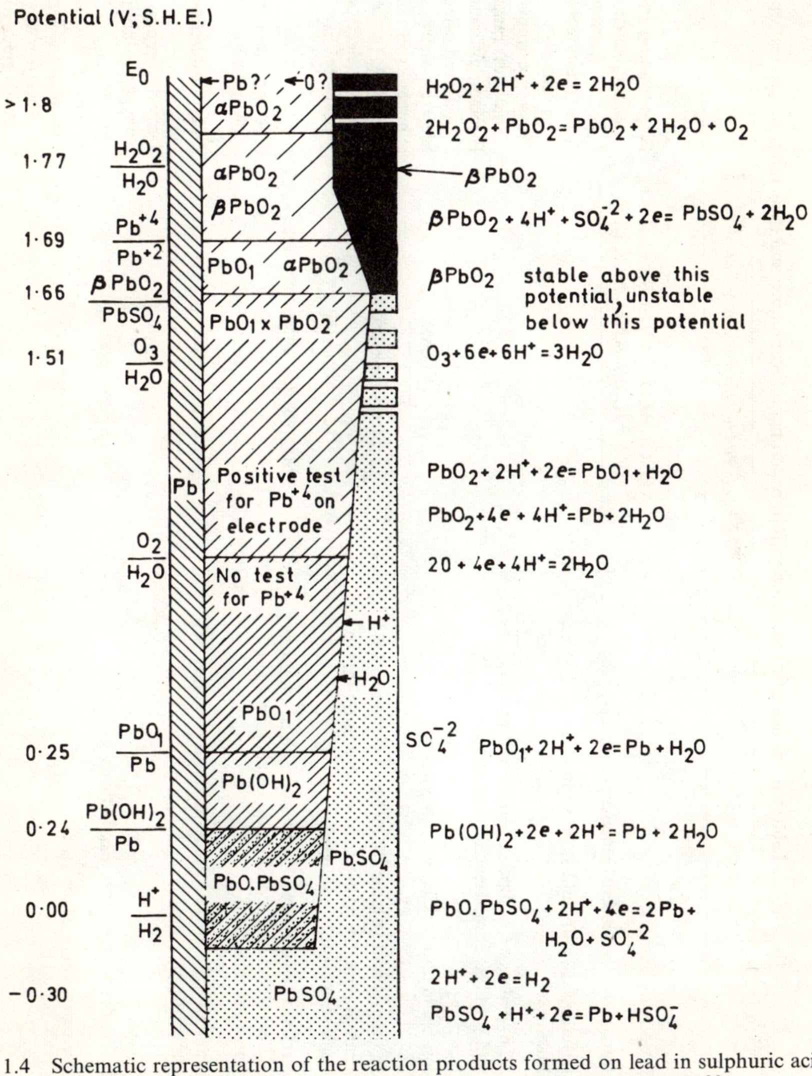

Fig. 1.4 Schematic representation of the reaction products formed on lead in sulphuric acid and their distribution over a range of anodic potentials (after Burbank[22])

Table 1.6 Schematic representation of experimental techniques and their range of application (after Wood[10])

Technique		*Thickness range (approximate) of technique*	*Technique*
Ellipsometry (kinetics)	1 nm ↓	ADSORBED LAYERS, VERY THIN FILMS AND NUCLEI	Field ion and emission microscopy, and allied techniques (morphology; analysis, etc.)
Electrometric reduction (kinetics; thickness)			L.E.E.D. (structure)
Interference colours and spectrophotometry (kinetics; thickness)			Adsorption techniques (e.g. flash desorption for bonding)
A.C. impedance (thickness; conduction mechanisms and profiles; compactness; crystallinity)			Mass spectrometry (analysis—applications in various thickness ranges)
			Electron diffraction (structure; epitaxy; grain size; preferred orientation)
Electrical methods (kinetics; thickness)		FILMS	Electron microscopy (topography; local thickness; nuclei; substructure—all thickness ranges)
Manometric and volumetric methods (kinetics)		Stress measurements	E.M.M.A. (structure; topography; analysis)
Thermogravimetry (kinetics from very thin films to thick scales; stoichiometry)		Adhesion	Scanning electron microscopy (thickness; topography; porosity; barrier layers; fracture sections)
Electrical conductivity of oxides and allied methods (defect structures; conduction mechanisms; transport numbers)		Stress/strain characteristics	
	200 nm ↓	Creep	Electron probe microanalysis (composition; diffusion profiles; local thickness)
Radioactive tracers and allied methods (kinetics; self diffusion; markers)		Hardness (oxide mechanical properties; oxygen solution in metal)	X-ray fluorescence analysis (composition; thickness)
			X-ray diffraction (thickness; structure; grain size; oxygen solubility in metal; preferred orientation; stress)
Inert markers (transport mechanisms)		Thermal cycling tests	
Gas adsorption (surface area)	1 μm ↓	SCALES	Optical microscopy (local thickness; topography; nucleation; general morphology; internal oxidation)
			I.R. spectroscopy (specialised analysis and applications)
			Spectrographic analysis (trace element analysis)
			Chemical analysis (analysis; stoichiometry)
	100 μm		Vacuum fusion analysis (oxygen solubility in metal)

films of lead in sulphuric acid (Fig. 1.4) in which it can be seen that the nature and thickness of the oxidation products are highly dependent on the anodic potential.

In the particular case of aluminium in acid electrolytes, an initially formed thin barrier film breaks down to give a porous coating which can be grown to a considerable thickness (Table 1.5). The voltage remains low as the porous anodic coating continues to thicken. Significant amounts of the acid anion (SO_4^{2-}, PO_4^{3-}, CrO_4^{2-}) may be incorporated into the oxides so produced, together with protons to provide a degree of hydration (*see* Section 16.1). These features can significantly influence the structure and properties of the coatings obtained.

Techniques of Examination

This limited survey has indicated the wide range of chemical compounds, particularly oxides, which may be formed on a metal surface as a result of a corrosion process. The nature of such films and scales needs to be carefully characterised. Fortunately, a wide spectrum of experimental techniques is now available to provide such valuable information, and others are under development. A convenient summary[10] is provided in Table 1.6.

In this scheme the nature of the surface product is arbitrarily divided into (*a*) adsorbed layers, very thin films and nuclei (1–200 nm thickness); (*b*) thin films (200 nm–1 μm), and (*c*) scales (above 1 μm). The principal techniques are located as appropriately as possible to indicate their areas of useful application. The spectrum thus ranges from the régime of very clean metal surfaces to grossly thick scales which may result from exposure to industrial oxidising atmospheres. Initial interaction may be studied by field-ion or field-emission spectroscopy and low energy electron diffraction, after which time the kinetics of the growth process may be followed by such techniques as ellipsometry, thermogravimetry, or electrometric reduction, while the structure may be examined by electron microscopy, electron diffraction or X-ray microanalysis. Stoichiometric and defect characteristics may be examined by a number of electrical methods. As the thickness approaches scale dimensions less sensitive techniques become applicable. Information on stress distribution, hardness, porosity, adhesion as well as thermal cycling characteristics also become accessible. Chemical analysis, scanning electron microscopy, X-ray diffraction techniques and gas adsorption data may provide further information on the composition, structure and porosity of thick scales, while electron probe microanalysis permits detailed examination of the concentration profiles across specimen sections. Many of these techniques are equally applicable to films formed under aqueous electrochemical conditions.

C. J. L. BOOKER

REFERENCES

1. Benard, J., 'Adsorption of Oxidant and Oxide Nucleation', in *Oxidation of Metals and Alloys*, Seminar, 1970; American Society for Metals, Ohio, 1 (1971)
2. Uhlig, H. H., Proceedings of the Third International Congress on Metallic Corrosion, Moscow, 1966, Vol. 1, 25 (1969); also *Corros. Sci.*, **7**, 325 (1967)
3. Fehlner, F. P. and Mott, N. F., 'Oxidation in the Thin-Film Range', as Reference 1, 37 (1971)
4. Fehlner, F. P. and Mott, N. F., *Oxid. Metals*, **2**, 59 (1970)
5. Cathcart, J. V., 'The Structure and Properties of Thin Oxide Films', as Reference 1, 17 (1971)
6. Grauer, R. and Feitknecht, W., *Corrosion Sci.*, **6**, 301 (1966)
7. Hauffe, K., *Metalloberflache*, **8**, 97 (1954)
8. Kubaschewski, O. and Hopkins, B. E., *Oxidation of Metals and Alloys*, Butterworths, London, 114 (1967)
9. Wood, G. C., 'The Structures of Thick Scales on Alloys', as Reference 1, 201 (1971)
10. Wood, G. C., in *Techniques in Metals Research*, Rapp, R. A. (Assoc. Ed.), Vol. 4, Interscience, New York, 494 (1970)
11. Douglas, D. L., 'Exfoliation and the Mechanical Properties of Scales', as Reference 1, 137 (1971)
12. Evans, U. R., *Corrosion and Oxidation of Metals*, First Supplementary Volume, Arnold, London, 194 (1968)
13. Feitknecht, W. and Keller, G., *Z. Anorg. Chem.*, **262**, 61 (1950); Feitknecht, W., Weidman, H. and Haberli, E., *Helv. Chim. Acta.*, **26**, 1911 (1943), **32**, 2294 (1949) and **33**, 922 (1950)
14. Brusic, V., 'Passivation and Passivity', in *Oxides and Oxide Films* (Ed. Diggle, J. W.), Marcell Dekker, New York (1972)
15. Evans, U. R., as Reference 12, 98 (1968)
16. Bloom, M. C. and Goldberg, L., *Corros. Sci.*, **5**, 623 (1965)
17. Dignam, M. J., as Reference 14, 91 (1972)
18. Wells, A. F., *Structural Inorganic Chemistry*, 3rd edn, Clarendon Press, Oxford (1962)
19. Greenwood, N. N., *Ionic Crystals, Lattice Defects and Nonstoichiometry*, Butterworths, London, 92, 101 (1968)
20. Diggle, J. W., Downie, T. C. and Goulding, C. W., *Chem. Rev.*, **69**, 365 (1969)
21. Aladjem, A., *J. Mat. Sci.*, **8**, 688 (1973)
22. Burbank, J., *J. Electrochem. Soc.*, **106**, 369 (1959)

1.3 Effects of Metallurgical Structure on Corrosion*

The objective of this section is to show by means of specific examples how the various crystalline defects and structural features described in Section 9.2 can affect the form, location and kinetics of the corrosion of metals and alloys.

Effect of Crystal Defects on Corrosion—General Considerations

Before considering specific examples it is appropriate to note that there are, in principle, two quite distinct ways in which crystal defects can affect corrosion behaviour.

Firstly, they might be expected to have an effect when corrosion occurs under conditions of active (film-free) anodic dissolution and is not limited by the diffusion of oxygen or some other species in the environment. However, if the rate of active dissolution is controlled by the rate of oxygen diffusion, or if, in general terms, the rate-controlling process does not take place at the metal surface, the effect of crystal defects might be expected to be minimal.

Secondly, crystal defects might be expected to affect the corrosion behaviour of metals which owe their corrosion resistance to the presence of thin passive or thick protective films on their surface. The crystal defects and structural features discussed in Section 9.2 might, in principle, be expected to affect the thickness, strength, adhesion, porosity, composition, solubility, etc. of these surface films, and hence, in turn, the corrosion behaviour of the filmed metal surfaces. Clearly, this is the more common situation in practice.

Finally, it should be noted that in both cases the effect of crystal defects and microstructural features must, in general, be to tend to make the corrosion less uniform and more localised.

*The basic concepts of physical metallurgy are considered in Section 9.2, which should be regarded by those who are not conversant with the subject as an introduction to this section. Some of the diagrams referred to here will be found in Section 9.2.

Active Dissolution and Crystal Defects—Energy Considerations

The crystal defects described in Section 9.2 are all regions of higher energy than the adjacent perfect crystal lattice; they are therefore all inherently more chemically active and hence are potential sites for preferential attack under conditions of active dissolution. This preferential attack is, however, masked in highly aggressive environments, when there is very rapid dissolution and severe general corrosion, and it is not observed when the corrosion rate is controlled either by oxygen diffusion or some other process not occurring at the metal/environment interface. Furthermore, although the energy associated with the various defects may be quite large in metallurgical terms, when converted to a potential difference it is quite small in electrochemical terms, being not more than a few millivolts, at the most.

Etching of Single Crystals and Polycrystals

There is no evidence that any particular crystal structure is more readily corroded than any other. For example, the difference in the corrosion behaviour of austenitic and ferritic stainless steels is, of course, due to compositional rather than structural differences.

Using single crystals it has been shown that different low-index crystal faces (*see* Section 9.2) exhibit different corrosion rates. However, the relative corrosion rate of the different faces varies with the environment and these structural effects are of little practical significance. On the other hand, the fact that polycrystal grains of different crystallographic orientation may corrode at different rates, is of some importance.

A freshly polished metal surface appears quite featureless even when viewed at high magnification, while on etching different grains are attacked to differing degrees, as shown in Fig. 9.25 (bottom). The surface of grain *B* in Fig. 9.33*a* corresponds to a low-index low-energy plane while the surfaces of grains *A* and *C* correspond to high-index high-energy planes. In fact, the surfaces of grains *A* and *C* actually consist of low-index terraces separated by ledges with kinks in them. Since dissolution occurs most readily from kinks and ledges, owing to the lower co-ordination number of atoms at such sites, grains *A* and *C* will be attacked more rapidly than grain *B*, as illustrated schematically in Fig. 1.5*a*. It must be emphasised, however, that this is primarily a laboratory effect, albeit an important one. In practice, preferential corrosion of grains of a particular crystallographic orientation is not generally a problem. One possible exception to this is the etching of coarse-grained brass door-handles by sweaty hands!

Dislocations, Etch Pits and the Effect of Cold Work on Corrosion

Preferential corrosion or attack at many other types of crystal defect may also be best illustrated during the etching of metallographically polished

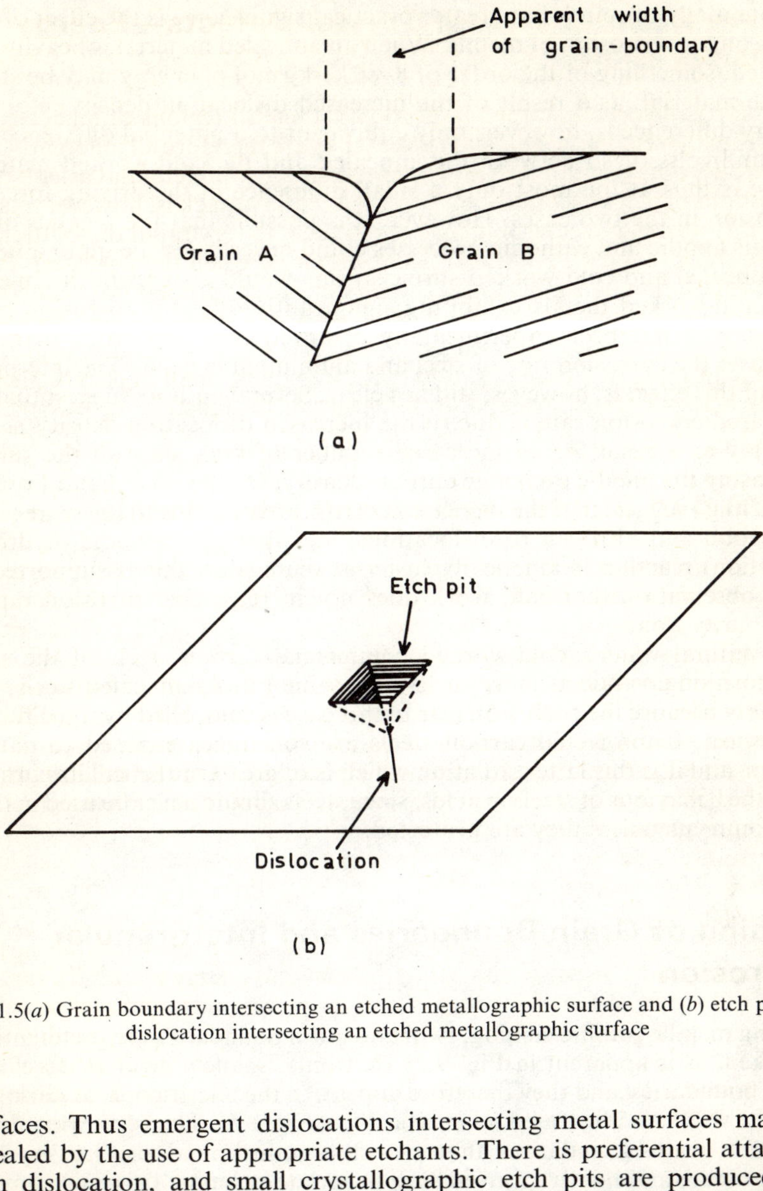

Fig. 1.5(*a*) Grain boundary intersecting an etched metallographic surface and (*b*) etch pit at a dislocation intersecting an etched metallographic surface

surfaces. Thus emergent dislocations intersecting metal surfaces may be revealed by the use of appropriate etchants. There is preferential attack at each dislocation, and small crystallographic etch pits are produced[1] as shown schematically in Fig. 1.5*b*. However, this effect is again of little practical significance, and the development of etch pits is used primarily as a research technique in the study of dislocations. Moreover, the technique must be used with some caution with metals, since there is often doubt as to whether there is a one-to-one correlation between etch pits and emergent dislocations, and because etch pits can also develop at other defects in surface films. Finally, there are many instances where it is thought that etch pits are produced only as a result of segregation of impurities to the dislocations. 'Clean' dislocations may not result in etch pits.

Potentially of somewhat greater practical significance is the effect of cold work on the corrosion of metals. When an annealed material is heavily cold worked, something of the order of 8–80 kJ/kg mol of energy may be stored in the material, as a result of the increased dislocation density, etc. This energy difference is, however, only equivalent to a potential difference of a few millivolts or so between the annealed and the cold-worked material. There is thus at the most only a small difference in the driving force for corrosion in the two cases. However, it is possible that the kinetics of the various anodic and cathodic processes could nevertheless be quite different on annealed and cold-worked surfaces; this would also result in annealed and cold-worked metals exhibiting significantly different corrosion rates. Certainly it has been experimentally observed that cold work markedly increases the corrosion rate of steel and aluminium in acids. The interpretation of this effect is, however, still not clear. Several authors suggest that the increased corrosion rate is due to the increased dislocation density *per se*, possibly as a result of an increased number of kink sites on the surface increasing the anodic exchange current density. On the other hand Foroulis and Uhlig[2] suggest that the increased corrosion rate is due to the segregation of carbon and nitrogen to dislocations, and that the cathodic (hydrogen evolution) reaction is kinetically easier at these sites; this is supported by their observation that cold work does not increase the corrosion rate of high-purity iron.

In natural waters, cold-worked commercial carbon steels of the same composition corrode at more or less the same rate as annealed steels, presumably because the corrosion rate in this case is controlled by the diffusion of oxygen. Unprotected carbon steels are sometimes exposed to natural waters, and it is this latter situation which is of greater practical importance than the behaviour of steels in acids, since steels should never be used in these environments unless they are protected.

Etching of Grain Boundaries and Intergranular Corrosion

During metallographic etching, twin and grain boundaries are preferentially attacked, as is apparent in Fig. 9.25 (bottom). Shallow grooves develop at these boundaries, and they therefore appear, in the microscope, as dark lines of finite width, as illustrated schematically in Fig. 1.5*a*. The best experimental evidence available indicates that even the grain boundaries in very high purity metals are slightly grooved by appropriate etchants. This is presumably due to the grain boundaries being more active than the adjacent crystal lattice, as implied by the energy associated with grain boundaries in metals. However, the grain boundaries in impure metals and alloys are generally much more readily etched, presumably as a result of segregation to them of the impurities and alloying additions. In this context it is important to note that grain-boundary regions may be preferentially attacked either because segregation makes them more base or because segregation makes them more noble; in the latter case the grain boundary itself acts as a local cathode, and the region immediately adjacent to the grain boundary is preferentially attacked. The subject of segregation and preferential attack at grain boundaries has recently been reviewed by Aust and Iwao[3].

Again it must be emphasised that preferential etching of twin and grain boundaries is predominantly a laboratory effect. There are no practical instances of significant corrosion problems resulting from the preferential attack of twin boundaries. In practice, grain-boundary effects in metals and alloys are *usually* of little or no consequence in the corrosion of metals. Severe intergranular corrosion (in the absence of tensile stress) is generally observed as a practical problem only when there is very gross segregation or solute depletion at grain boundaries, or, in certain instances, when there is marked intergranular precipitation, as discussed below.

Intergranular Corrosion of Austenitic Stainless Steels

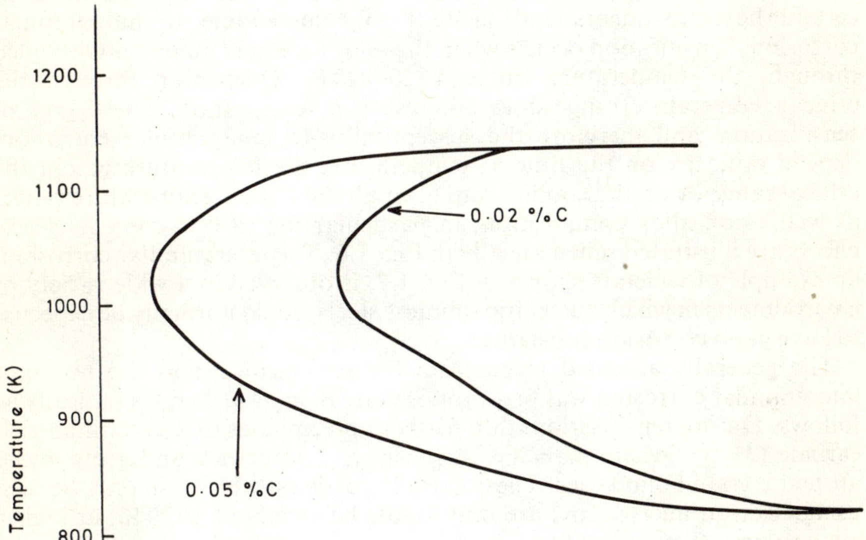

Fig. 1.6 Curves of the effect of temperature on the time required to sensitise two austenitic stainless steels of different carbon content

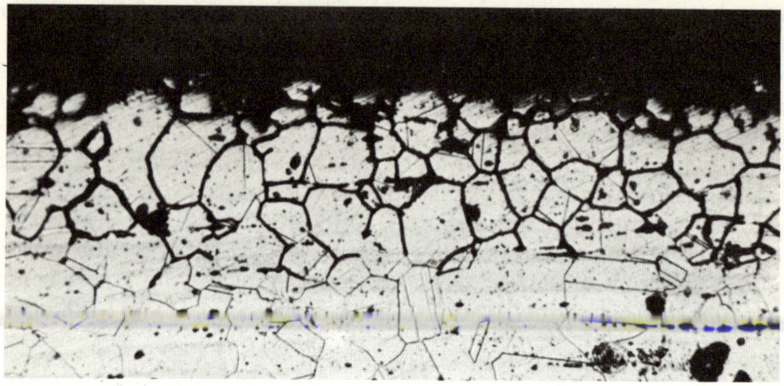

Fig. 1.7 Light micrograph showing intergranular corrosion of a sensitised austenitic stainless steel; ×200

As is well known, certain austenitic stainless steels may be 'sensitised' by certain heat treatments and made highly susceptible to intergranular corrosion. Sensitisation occurs when the alloys are held in, or slowly cooled through, the temperature range 1 120–820 K. Quenching through the critical temperature range does not result in sensitisation. The degree of sensitisation, and therefore the susceptibility to intergranular corrosion, depend critically on the time at temperature, the temperature within the critical range or on the cooling rate through the critical temperature range, as well as on alloy composition, in particular the carbon content. These effects are illustrated schematically in Fig. 1.6. The intergranular corrosion, an example of which is shown in Fig. 1.7, is observed in a wide variety of environments in which austenitic stainless steels would normally be expected to have good corrosion resistance.

The generally accepted mechanism for sensitisation and the resultant intergranular corrosion was first proposed by Bain, *et al.*[4] and is basically as follows. During sensitisation, thin feathery precipitates of a chromium-rich carbide ($M_{23}C_6$ where $M = Fe_{0\cdot2-0\cdot3}Cr_{0\cdot8-0\cdot7}$) nucleate and grow in the austenite grain boundaries. These carbide particles, which can only be seen using electron microscopy, are only stable below about 1 120 K; at higher temperatures they do not form, or if already present, tend to dissolve. On the other hand, below about 820 K, the diffusion rate of chromium in steels is too low for precipitation of the chromium-rich carbide to occur within a practical time scale. During precipitation of the carbide, which contains 70–80 wt.% Cr, the austenite matrix adjacent to the grain boundaries becomes depleted of chromium. In particular, the chromium level in these regions falls below the approximately 12% Cr required in solid solution to confer corrosion resistance, i.e. to permit the formation of a complete and protective passive film on the steel surface. The regions adjacent to the grain boundaries are therefore no longer passive and hence corrode preferentially. This mechanism is illustrated schematically in Fig. 1.8. The preferential attack of the non-passive chromium-depleted regions adjacent to the grain boundaries will be accelerated by the fact that these regions will be less noble than both the carbide precipitates in the grain boundary and the passive grain interiors.

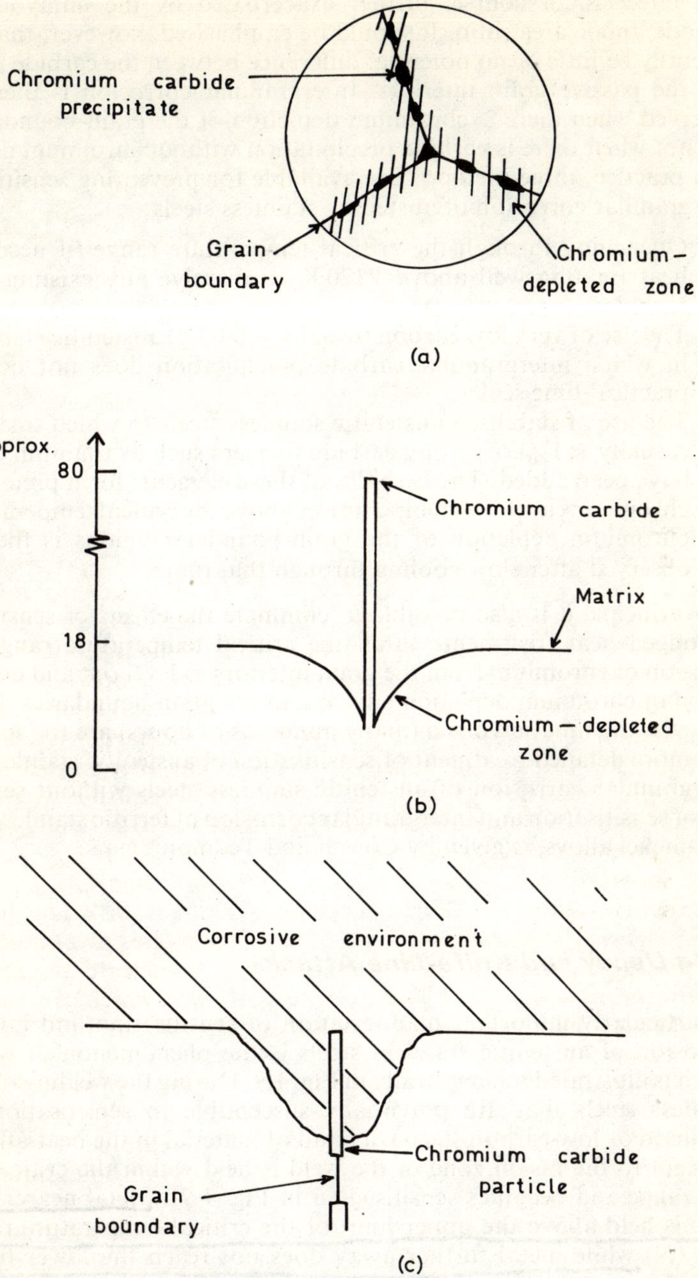

Fig. 1.8(*a*) Intergranular precipitation of chromium carbide particles in a sensitised austenitic stainless steel and the consequent chromium-depleted zones adjacent to the grain boundaries, (*b*) variation of the chromium content across a grain boundary in a sensitised austenitic stainless steel (18Cr) and (*c*) intergranular corrosion of a sensitised austenitic stainless steel

This effect is, of course, further exacerbated by the unfavourable high cathode/anode area ratio. It should be emphasised, however, that there will generally be little or no potential difference between the carbide precipitates and the passive grain interiors. Intergranular corrosion is therefore only observed when there is chromium depletion of the grain-boundary regions and not when there is carbide precipitation without chromium depletion.

In practice, three methods are available for preventing sensitisation and intergranular corrosion of austenitic stainless steels:

1. Quenching through the critical temperature range (if necessary after heat treating well above 1 120 K to dissolve any existing chromium carbides).
2. The use of very low carbon (usually <0·03%) austenitic stainless steels, in which intergranular carbide precipitation does not occur within practical time scales.
3. The use of stabilised austenitic stainless steels to which small amounts (usually <1%) of strong carbide formers such as titanium or niobium have been added. The carbides of these elements form preferentially to chromium carbide at temperatures above the critical temperature range; chromium depletion of the grain-boundary regions is therefore not observed after slow cooling through that range.

In principle it is also possible to eliminate the effects of sensitisation by prolonged heat treatment within the critical temperature range to allow diffusion of chromium from the grain interiors to level out and eliminate the region of chromium depletion adjacent to the grain boundaries. In practice, however, the times involved (many hundreds of hours) are too long.

A more detailed treatment of sensitisation of austenitic stainless steels, of intergranular corrosion of austenitic stainless steels without sensitisation, and of sensitisation and intergranular corrosion of ferritic stainless steels and high-nickel alloys, is given by Cowan and Tedmon[5].

Weld Decay and Knife-Line Attack

A particularly important manifestation of sensitisation and intergranular corrosion of austenitic stainless steels is the phenomenon of weld decay, which is illustrated schematically in Fig. 1.9. During the welding of austenitic stainless steels that are potentially susceptible to sensitisation (i.e. not stabilised or low-carbon steels), a band of material in the heat-affected zone adjacent to the fusion zone of the weld is held within the critical temperature range and becomes sensitised (*B* in Fig. 1.9). Metal nearer the fusion zone is held above the upper limit of the critical temperature range (*A* in Fig. 1.9) while metal further away does not reach the lower limit of the critical temperature range (*C* in Fig. 1.9). An intermediate band of material parallel to, but away from, the fusion zone is therefore susceptible to severe intergranular corrosion, as shown in Fig. 1.9*b*. Weld decay is more severe with gas welding than with electric arc welding, and with thick plates rather than thin sheets, owing basically to different temperature–time profiles (*see also* Section 10.5).

A somewhat similar phenomenon is knife-line attack which may be observed after welding titanium or niobium stabilised austenitic stainless steels. In this case there is a very narrow band of severe intergranular attack along the interface between the parent metal and the fusion zone. During welding, the parent metal immediately adjacent to the fusion zone is heated to just below the melting point and both chromium carbides and niobium or titanium carbides dissolve completely. On cooling rapidly, a situation that prevails when relatively thin sections are welded, neither chromium carbide nor niobium or titanium carbide have time to precipitate. If the weld is now

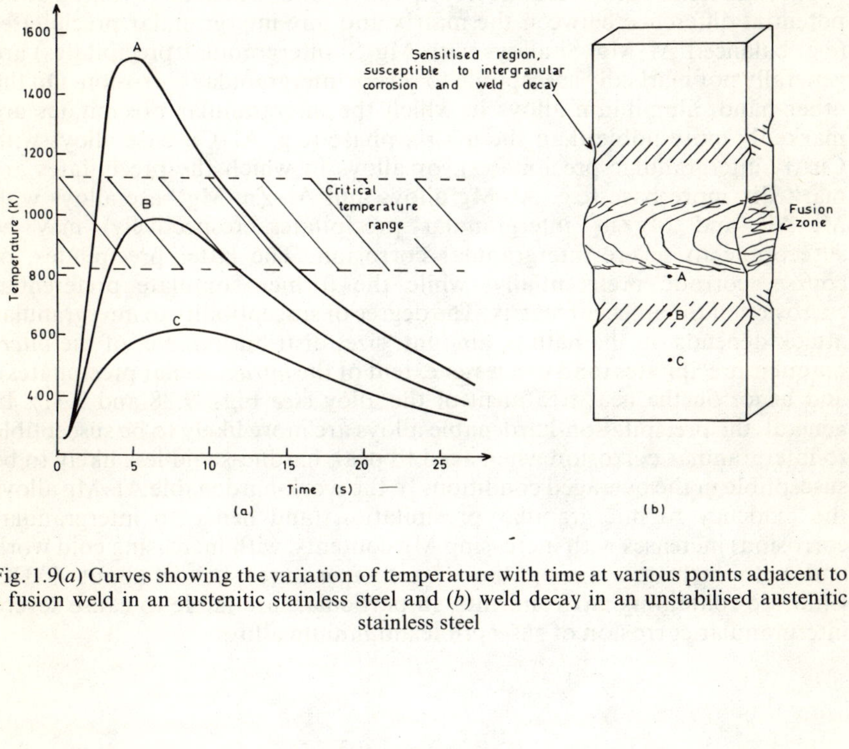

Fig. 1.9(*a*) Curves showing the variation of temperature with time at various points adjacent to a fusion weld in an austenitic stainless steel and (*b*) weld decay in an unstabilised austenitic stainless steel

heated to within the critical temperature range (e.g. to relieve residual stresses or during multi-pass welding) chromium carbide will precipitate rapidly but the temperature will be too low to precipitate niobium or titanium carbide (other than very slowly). The narrow band of material which during welding reached a temperature high enough to dissolve the stabilising carbides (>1 500 K) thus becomes sensitised and susceptible to intergranular corrosion. The remedy is to heat the fabrication to about 1 340 K after welding so that the chromium carbide dissolves and titanium (or niobium) carbide forms; following this solution treatment the rate of cooling is not important.

Intergranular Corrosion in Aluminium Alloys

A number of aluminium alloys may also, depending on their metallurgical structure, be susceptible to severe intergranular corrosion. The alloys concerned are primarily the precipitation-hardenable Al–Cu and Al–Zn–Mg based alloys and the work-hardenable Al–Mg alloys containing more than 3% Mg. Since there is much less inherent difference in the corrosion resistance of the grain-boundary regions and the grain centres than there is in the case of sensitised austenitic stainless steels, the mechanism of the intergranular corrosion in these alloys is primarily electrochemical, involving local cell action between grain-boundary precipitates and the adjacent matrix.

Aluminium alloys in which intergranular precipitation is not observed (e.g. commercial purity Al and Al–Mn alloys) or in which there is little or no potential difference between the matrix and any intergranular precipitates (e.g. balanced Al–Mg–Si alloys with Mg_2Si intergranular precipitates) are generally not markedly susceptible to severe intergranular corrosion. On the other hand, aluminium alloys in which the intergranular precipitates are markedly more noble than the matrix phase (e.g. Al–Cu base alloys with $CuAl_2$ intergranular precipitates), or alloys in which the precipitates are markedly more base (e.g. Al–Mg alloys and Al–Zn–Mg base alloys with Mg_2Al_3 and $MgZn_2$ intergranular precipitates, respectively) may be susceptible to severe intergranular corrosion. The latter precipitates, of course, corrode preferentially, while the former stimulate preferential corrosion of the adjacent matrix. The degree of susceptibility to intergranular attack depends on the nature, amount, size, distribution, etc. of the *inter*granular precipitates (and to a lesser extent of the *intra*granular precipitates), and hence on the heat treatment of the alloy (*see* Figs. 9.28 and 9.31). In general, the precipitation-hardenable alloys are more likely to be susceptible to intergranular corrosion when aged to peak hardness and less likely to be susceptible in the overaged condition. In the work-hardenable Al–Mg alloys the tendency to intergranular precipitation (and hence to intergranular corrosion) increases with increasing Mg contents, with increasing cold work and with increased ageing times at temperatures below about 400 K. Chloride-containing environments, in particular, are liable to cause severe intergranular corrosion of susceptible aluminium alloys.

Intergranular Corrosion in Other Alloy Systems

A number of other alloy systems may also be susceptible to intergranular corrosion. For example, zinc die-casting alloys containing aluminium may be susceptible to intergranular attack in steam- and chloride-containing environments. Stray currents often result in intergranular corrosion of lead cable sheaths. These instances are, however, relatively unimportant compared to the intergranular corrosion of sensitised stainless steels and, to a lesser extent, to intergranular corrosion of intermediate- and high-strength aluminium alloys.

Effect of Grain Structure on Corrosion

The grain structure of alloys, as well as intergranular precipitation, can also markedly affect their corrosion behaviour. For example, the corrosion resistance of certain wrought metals may be less on surfaces perpendicular to the hot- or cold-working direction than on surfaces parallel to this direction. Typically there may be severe localised corrosion starting on the faces perpendicular to the working direction and proceeding into the metal in the working direction, while the surfaces parallel to the working direction remain relatively unattacked. Such *end-grain* attack, which is basically the result of the grain structure being elongated in the working direction, has been observed in austenitic stainless steels, titanium alloys and mild steel.

Layer Corrosion

However, the most marked effect of grain structure on corrosion is observed in wrought aluminium alloys. These alloys generally do not recrystallise during heat treatment after rolling, extrusion, etc. mainly because their grain boundaries are pinned by inclusions; they therefore exhibit the elongated pancake-shaped grain structure shown in Fig. 9.31. As a result of this structure, these alloys may be susceptible to *exfoliation* (also known as *layer* or *lamellar*) corrosion. The attack proceeds along a number of narrow planar paths (usually but not necessarily intergranular) parallel to the working direction. The corrosion products formed force the layers apart and cause the metal to swell and, in severe instances, to disintegrate into separate sheets of metal (i.e. to exfoliate). Exfoliation is most common and severe in Al–Cu, Al–Zn–Mg and Al–Mg based alloys, but mild exfoliation also occurs in Al–Mg–Si alloys. Since the exfoliation is normally intergranular it is clear that exfoliation and intergranular corrosion are associated, and exfoliation is usually affected by intergranular precipitation and hence by heat treatment. However, aluminium alloys that are susceptible to intergranular attack will not be susceptible to exfoliation corrosion if they have an equiaxed grain structure. Transgranular exfoliation is thought to be the result of segregation in the original ingot persisting in the wrought alloy.

Stress-corrosion Cracking

Grain structure also affects the stress-corrosion behaviour of high-strength age-hardenable aluminium alloys. Cracking in these alloys is always exclusively intergranular. When they are stressed in the short transverse direction (*a* in Fig. 1.10) their highly elongated pancake-shaped grain structure ensures that an easy path for crack propagation is readily available. On the other hand, when stressed in the long-transverse or the longitudinal direction (*b* and *c*, respectively in Fig. 1.10) the possible intergranular crack paths are clearly complex and difficult. Many high-strength aluminium alloys are therefore quite susceptible to stress-corrosion cracking when stressed in the short-transverse direction but quite resistant or immune when stressed in the long-transverse or longitudinal directions. This result is of

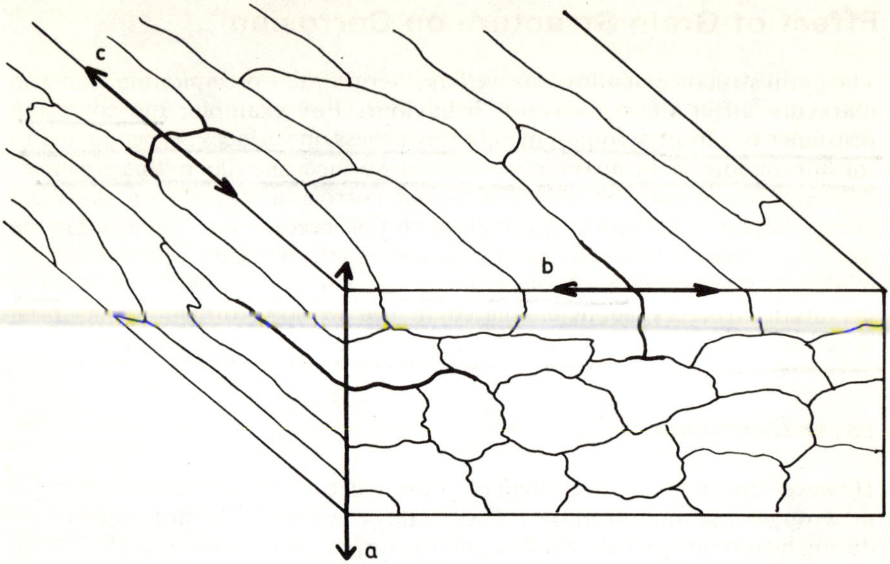

Fig. 1.10 Grain structure of a wrought high-strength precipitation-hardening aluminium alloy showing potential intergranular stress-corrosion crack paths

considerable practical importance: high-strength aluminium alloys can often be used in sheet form (when the short-transverse tensile stresses are generally negligible) in tempers in which they cannot normally be used in thick sections of forgings (when the tensile stresses in the short-transverse direction may be high). The distinction is, for example, between an aircraft's wing-skin and its landing gear; with a susceptible alloy stress-corrosion cracking is likely to be a problem in the latter instance, but not in the former.

Corrosion of Impure Metals and Single-phase Alloys

Many of the forms of corrosion already discussed have been caused or affected not only by metallurgical structure but also by the segregation of impurities or alloying additions to dislocations, grain boundaries, precipitates, etc. However, the presence of impurities or alloying elements in homogeneous solid solution can also markedly affect corrosion behaviour, without any segregation effects. In the context of this section it is the deleterious effects of soluble impurities and alloying additions that are relevant, rather than the beneficial effects, such as the addition of chromium and nickel to iron to give stainless steels, or the addition of nickel and small concentrations of iron to copper to give cupronickels.

As is well known, high-purity zinc corrodes much less rapidly in dilute acids than commercial purity material; in the latter instance, impurities (particularly copper and iron) are exposed on the surface of the zinc to give local cathodes with low hydrogen overpotentials; this result is of practical significance only in the use of zinc for sacrificial anodes in cathodic protection or for anodes in dry cells. In neutral environments, where the cathodic reaction is oxygen reduction, there is very little difference in the corrosion rates of pure and impure zinc.

In contrast, the selective dissolution or leaching-out by corrosion of one component of a single-phase alloy is of considerable practical importance. The most common example of this phenomenon, which is also referred to as 'parting', is dezincification, i.e. the selective removal of zinc from brass (*see* Section 1.6). Similar phenomena are observed in other binary copper-base alloys, notably Cu–Al, as well as in other alloy systems.

Corrosion and Selective Dissolution in Two-phase Alloys

In principle the selective dissolution of the less noble component of a single-phase alloy would perhaps be expected and is in fact observed (dezincification of an α-brass, etc.) even though the details of the mechanism by which it occurs is not yet fully understood. In contrast, the preferential attack of the less noble phase of a two-phase alloy is not only expected and observed—the mechanism by which it occurs in practice is also quite clear. Selective dissolution of the more active phase of a two-phase alloy is best exemplified by the graphitic corrosion or *graphitisation* of grey cast iron.

Graphitisation

Cast irons, although common, are in fact quite complex alloys. The iron–carbon phase diagram exhibits a eutectic reaction at 1 420 K and 4·3 wt.% C (*see* Fig. 9.41). One product of this eutectic reaction is always austenite; however, depending on the cooling rate and the composition of the alloy, the other product may be cementite or graphite. The graphite may be in the form of flakes which are all interconnected (although they appear separate on a

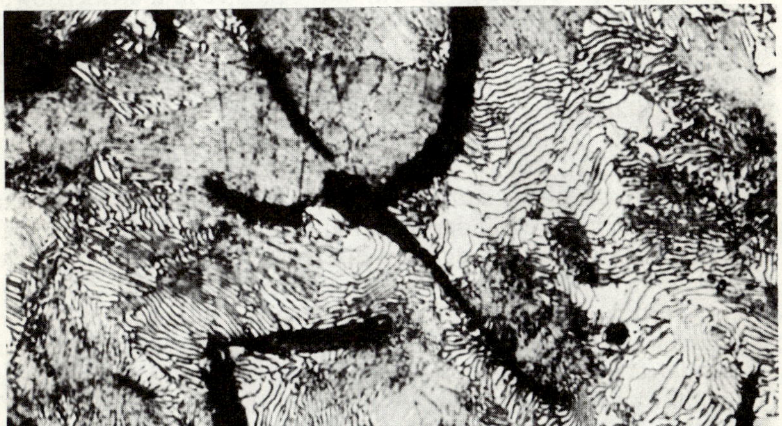

Fig. 1.11 Light micrograph of the microstructure of a pearlitic grey cast iron; ×720

metallographic section), or spheroids, which are all separate. The resultant alloys are known as white, grey and spheroidal graphite cast irons, respectively. During subsequent cooling the austenite may transform (as already mentioned) to ferrite, pearlite or martensite, or it may, in the case of high-nickel irons, be retained to room temperature. The microstructure of a grey cast iron with a predominantly pearlitic matrix is shown in Fig. 1.11.

Graphite is, of course, more noble than any of the other phases in cast iron and is a very good cathode material; highly effective galvanic cells therefore exist between the graphite and the surrounding, less noble matrix. In grey cast irons, the matrix therefore corrodes preferentially, leaving behind the network of interconnected graphite flakes; this is, of course, very porous and weak. The attack is often not readily apparent on superficial inspection. White cast irons are not susceptible to graphitisation since they contain no graphite; spheroidal graphite cast irons are also not susceptible to graphitisation, since although they do contain graphite it is in the form of discrete spheroids which have a limited effect, instead of the interconnected graphite flakes in grey cast iron. Thus not only the existence but also the distribution of a cathodic phase is important (*see also* Section 3.6).

Influence of Structure on Surface Films—Pitting Corrosion

Metals which owe their good corrosion resistance to the presence of thin, passive or protective surface films may be susceptible to pitting attack when the surface film breaks down locally and does not reform. Thus stainless steels, mild steels, aluminium alloys, and nickel and copper-base alloys (as well as many other less common alloys) may all be susceptible to pitting attack under certain environmental conditions, and pitting corrosion provides an excellent example of the way in which crystal defects of various kinds can affect the integrity of surface films and hence corrosion behaviour.

In general, pitting corrosion may be divided into two stages, pit initiation and pit propagation. During pit initiation the passive film breaks down and does not reform. During pit propagation, the small active sites formed during the initiation stage propagate, often very rapidly, to form pits. The most recent ideas on the mechanism of pit initiation and propagation are dealt with in some detail in Reference 6. The propagation of pits is relatively well understood and is comparatively insensitive to the structure of the metal (*see* Sections 1.5 and 1.6).

On the other hand, pit initiation which is of course the necessary precursor to propagation, is less well understood but is probably far more dependent on metallurgical structure. A detailed discussion of pit initiation is clearly beyond the scope of this section. The two most widely accepted models are, however, as follows. Heine, *et al.*[7] suggest that pit initiation on aluminium alloys occurs when chloride ions penetrate the passive oxide film by diffusion via lattice defects. McBee and Kruger[8] indicate that this mechanism may also be applicable to pit initiation on iron. On the other hand, Evans[9] has suggested that a pit initiates at a point on the surface where the rate of metal dissolution is momentarily high, with the result that more aggressive anions are attracted to the point and produce a local environment favourable to further dissolution, i.e. an autocatalytic process similar to that operative in pit propagation. This view has recently found increasing support, since there is now evidence that pits initiate at flaws or discontinuities in the passive film which result from mechanical, geometrical or compositional inhomogeneities in the metal surface[10,11]. The latter model, in particular, predicts a strong influence of metallurgical structure on the integrity of the passive film and hence on susceptibility to pitting corrosion.

In practice many metallurgical factors do appear to affect pitting corrosion. For example, severe cold work increases the pitting susceptibility of austenitic stainless steels, while molybdenum alloying additions, in particular, reduce it. Pitting is less likely to occur on smooth, polished surfaces than on rough, etched, ground or machined surfaces. Austenitic stainless steels are more susceptible to pitting if they have been held briefly in the sensitising temperature range. Pure aluminium is much more resistant to pitting than impure metal and alloys, particularly those containing copper. In general, the more homogeneous a metal surface the better is the resistance of passive films on that surface to pitting. In austenitic stainless steels, pits have been observed to initiate at grain boundaries and also at certain sulphide inclusions. These effects are all evidence of the fact that crystal defects and metallurgical structure and composition affect the thickness, strength, solubility, porosity, etc. of passive films, and hence the susceptibility of those films to localised breakdown and pitting.

Effect of Mechanical Stresses on Corrosion

The presence of stresses does not usually affect the general corrosion behaviour of metals and alloys to any very significant extent. However, two extremely important forms of localised corrosion may occur when metals are simultaneously exposed to stress and a corrosive environment. Metals subjected simultaneously to alternating stresses and any corrosive environment may be subject to corrosion fatigue, while certain alloys exposed simultaneously to tensile stresses and fairly specific corrosive environments may fail by stress-corrosion cracking. Other sections (*see* Chapter 8) deal specifically with the mechanism and phenomenology of corrosion fatigue and stress-corrosion cracking of various alloy systems, and it is not the intention to duplicate that material in this section. However, the susceptibility of many alloys to stress-corrosion cracking is determined not only by the presence of tensile stresses and specific environmental conditions, but also by the metallurgical structure of the metals. These instances will be discussed briefly, by way of further examples of the effect of structure on the corrosion of metals.

Stress-corrosion Cracking of Copper-base Alloys

Single-phase α-brasses are susceptible to stress corrosion in the presence of moist ammonia vapour or certain ammonium compounds[12]. Here the predominant metallurgical variable is alloy composition, and in practice brasses containing less than 10–15% Zn seldom fail by stress corrosion; above about 15% Zn, the stress-corrosion susceptibility increases with zinc content. Other structural factors are secondary: cold-worked brass, in practice, is more likely than annealed material to fail by stress-corrosion cracking, but this is probably only a reflection of the fact that the residual stresses are likely to be higher in cold-worked than in annealed alloys. The stress-corrosion susceptibility of α-brasses increases with increasing grain size. There is also evidence that decreasing stacking-fault energy results in a transition from inter- to transgranular cracking in a number of binary copper-base alloys[13] and that the presence of order (*see* Section 9.2) increases the susceptibility of certain complex copper-base alloys[14]. It must be emphasised, however, that these are only secondary factors which merely tend to increase or decrease fairly marginally the stress-corrosion susceptibility.

Stress-corrosion Cracking of Aluminium-base Alloys

In contrast to brasses, metallurgical structure plays a predominant rôle in determining the susceptibility of high-strength aluminium alloys in the presence of tensile stresses and moist chloride-containing environments. Under these conditions these alloys may vary from highly susceptible, to practically immune, to intergranular stress-corrosion cracking, depending on their microstructure, as determined by heat treatment[15] (*see* Section 8.5).

The effect of grain shape on the stress-corrosion behaviour of aluminium alloys has already been discussed. The effect of heat treatment on the stress-corrosion susceptibility of high-strength precipitation-hardenable Al–Zn–Mg alloys is illustrated schematically in Fig. 1.12. In the solution-heat-treated and quenched condition, these alloys are very resistant to stress-corrosion cracking but, of course, they are also too weak to be of much use in this condition. On ageing, the alloys become progressively stronger (*see* Section 9.2), but also increasingly susceptible to stress corrosion, as shown in Fig. 1.12. Maximum stress-corrosion susceptibility is observed in the intermediate-strength, under-aged condition; thereafter the alloys become increasingly more resistant to stress-corrosion cracking. Thus, as shown in Fig. 1.12 the highest-strength, peak-aged condition is moderately susceptible to stress corrosion, while the intermediate-strength, over-aged condition is relatively resistant. In practice, therefore, there is a choice between maximum strength alloys with moderate stress-corrosion susceptibility and somewhat lower strength alloys with little stress-corrosion susceptibility. As suggested above, the former condition or temper might be selected for thin sheet applications while the latter heat treatment would be specified for thick sections. There is some doubt as to whether the effect of heat treatment on the stress-corrosion susceptibility of precipitation-hardenable aluminium alloys results from variations in the precipitate-free zone width and the intergranular precipitate morphology, or from variations in the interaction of intragranular precipitates with dislocations. The effect of microstructure on the stress-corrosion susceptibility of Al–Cu, although again very substantial, is somewhat less straightforward than in the case of Al–Zn–Mg alloys[15].

Microstructure also plays a predominant rôle in determining the stress-corrosion susceptibility of the work-hardenable Al–Mg alloys. The Al–Mg

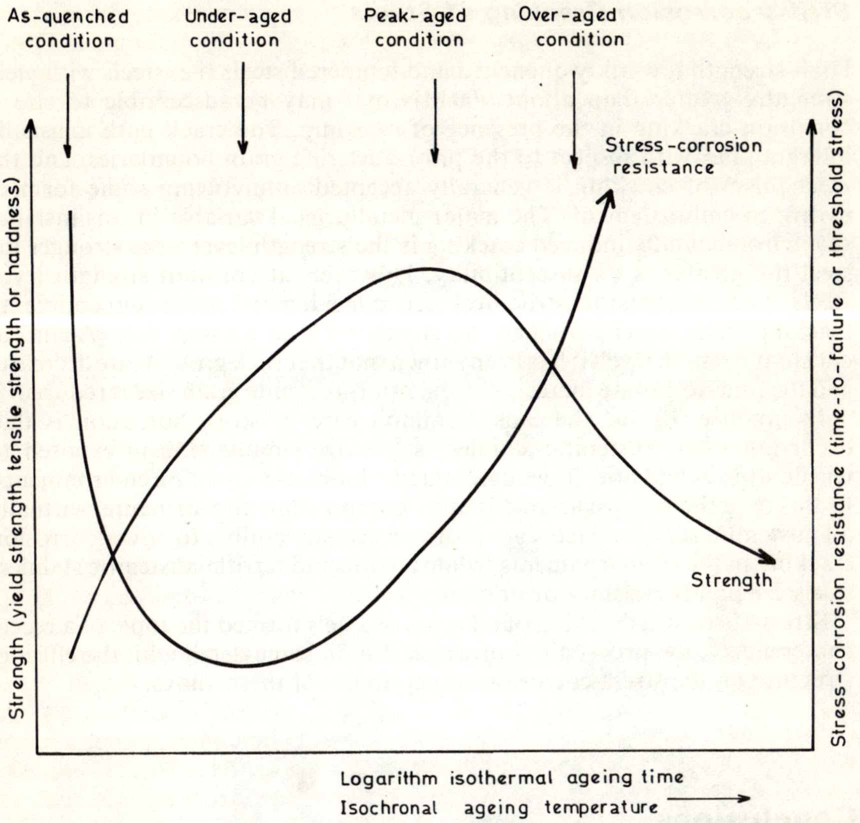

Fig. 1.12 Curves showing the relationship between strength, stress-corrosion susceptibility and heat treatment for a high-strength precipitation-hardening aluminium alloy

system, like the Al–Cu system, exhibits decreasing solubility with decreasing temperature, and on ageing a solution-heat-treated and quenched alloy, precipitation of Mg_2Al_3 is observed. However, the alloy is not strengthened by this precipitation, as it occurs either as very coarse, widely dispersed intragranular precipitates (which do not interact with dislocations), or as a more or less continuous intergranular film. These alloys can therefore only be strengthened by cold working. Nevertheless, the precipitate morphology controls the stress-corrosion susceptibility; alloys which exhibit continuous films of intergranular precipitate are highly susceptible to stress-corrosion cracking (and to exfoliation and intergranular corrosion), while those which exhibit coarse intragranular precipitates (or no precipitation at all) are generally much less susceptible or resistant. The higher the magnesium level, the greater the degree of cold work, and the lower the ageing temperature, the more likely is the formation of a continuous intergranular precipitate and therefore the greater is the potential stress-corrosion susceptibility.

Stress-corrosion Cracking of Steels

High-strength low-alloy quenched and tempered steels (i.e. steels with yield strengths greater than about 900 MN/m^2) may be susceptible to stress-corrosion cracking in the presence of moisture. The crack path is usually intergranular with respect to the prior-austenite grain boundaries, and the mechanism of cracking is generally accepted as involving some form of hydrogen embrittlement. The major metallurgical variable in this instance of environmentally induced cracking is the strength level—the stronger the steel the greater is its susceptibility. However, at constant strength level, steels with martensitic structures are considerably more susceptible to cracking than steels with bainitic structures (*see* Section 9.2). Again, at constant strength level, it has been shown that the crack growth rate decreases and the time-to-failure increases as the prior-austenite grain size is reduced[16].

In practice, by far the most common case of stress corrosion is that occurring when austenitic stainless steels are simultaneously exposed to tensile stresses and hot, aqueous, aerated, chloride-containing environments. In this case the major variable is alloy composition and structure; virtually all austenitic stainless steels are more or less susceptible to stress-corrosion cracking in these environments, while ferritic and ferritic/austenitic stainless steels are highly resistant or immune.

Stress-corrosion cracking of all types of steels formed the topic of a recent conference[17], the proceedings of which deal in some detail with the effect of structure on the stress-corrosion susceptibility of these alloys.

Conclusions

This discussion on the relationships between structure and corrosion should not be taken as exhaustive. For example, the stress-corrosion cracking behaviour of titanium-base alloys in a variety of environments is affected to differing degrees by the microstructure of the alloys[18]. Again, the effect of the changes in structure produced by welding on the corrosion behaviour of metals and alloys is of great practical importance[19]. This topic has been considered above in relation to stainless steels, but it is also of considerable importance in the welding, brazing and soldering of other alloy systems. In certain instances, the corrosion resistance of a weld is markedly affected by the structure of the weld metal and the adjacent heat-affected zone. Further examples of the effect of metallurgical structure on corrosion phenomena are provided by (*a*) the possible rôle of emergent dislocations and slip-steps in the mechanism of stress-corrosion cracking of austenitic stainless steels, (*b*) by the rôle of δ-ferrite in the corrosion of certain austenitic stainless steels, (*c*) by the rôle of local spheroidisation in 'ringworm' corrosion of mild steel and (*d*) by the rôle of manganese sulphide and other inclusions in the pitting of mild steel.

Since corrosion is essentially a reaction between a metal and its environment, the very significant effect of crystal defects and metallurgical structure on certain corrosion phenomena is to be expected. It is no more possible to neglect the metallurgical aspects of a corrosion problem than it is to overlook the environmental and electrochemical factors.

R. P. M. PROCTER

REFERENCES

1. Ives, M. B., in *Proc. U. R. Evans Internat. Conf. on Localised Corrosion*, to be published by N.A.C.E. (Houston)
2. Foroulis, Z. A. and Uhlig, H. H., *J. Electrochem. Soc.*, **111**, 522 (1964)
3. Aust, K. and Iwao, O., in *Proc. U. R. Evans Internat. Conf. on Localised Corrosion*, to be published by N.A.C.E. (Houston)
4. Bain, E. C., Aborn, R. H. and Rutherford, J. B., *Trans. Amer. Soc. Steel Treating*, **21**, 481 (1933)
5. Cowan, R. L. and Tedmon, C. S., in *Advances in Corrosion Science and Technology*, Vol. III, Plenum Press, New York (1973)
6. *Proc. U. R. Evans Internat. Conf. on Localised Corrosion*, to be published by N.A.C.E. (Houston)
7. Heine, M. A., Keir, D. S. and Pryor, M. J., *J. Electrochem. Soc.*, **112**, 29 (1965)
8. McBee, C. L. and Kruger, J., in *Proc. U. R. Evans Internat. Conf. on Localised Corrosion*, to be published by N.A.C.E. (Houston)
9. Evans, U. R., *Corrosion*, **7**, 238 (1951)
10. Richardson, J. A. and Wood, G. C., *Corr. Sci.*, **10**, 313 (1970)
11. Ashworth, V., Boden, P. J., Leach, J. S. Ll. and Nehru, A. J., *Corr. Sci.*, **10**, 481 (1970)
12. Pugh, E. N., Craig, J. V. and Sedriks, A. J., in *Proc. Internat. Conf. on Fundamental Aspects of Stress-corrosion Cracking*, N.A.C.E., Houston (1969)
13. Ohtani, N. and Dodd, R. A., *Corrosion*, **21**, 161 (1965)
14. Popplewell, J. M., Procter, R. P. M. and Ford, J. A., *Corr. Sci.*, **12**, 193 (1972)
15. Speidel, M. O. and Hyatt, M. V., in *Advances in Corrosion Science and Technology*, Vol. II, Plenum Press, New York (1972)
16. Procter, R. P. M. and Paxton, H. W., *Trans. A.S.M.*, **62**, 989 (1969)
17. *Proc. Internat. Conf. on Stress-corrosion Cracking and Hydrogen Embrittlement of Iron Base Alloys* (Firminy, France, 1973) to be published by N.A.C.A. (Houston)
18. Blackburn, M. J., Feeney, J. A. and Beck, T. R., in *Advances in Corrosion Science and Technology*, Vol. III, Plenum Press, New York (1973)
19. Lancaster, J. F., *Metallurgy of Welding, Brazing and Soldering*, George Allen and Unwin, London (1970)

1.4 Corrosion in Aqueous Solutions

In this section the interaction of a metal with its aqueous environment will be considered from the viewpoint of thermodynamics and electrode kinetics, and in order to simplify the discussion it will be assumed that the metal is a homogeneous continuum, and no account will be taken of submicroscopic, microscopic and macroscopic heterogeneities, which are dealt with elsewhere (*see* Sections 1.3 and 9.2). Furthermore, emphasis will be placed on uniform corrosion since localised attack is considered in Section 1.6.

Aqueous environments will range from very thin condensed films of moisture to bulk solutions, and will include natural environments such as the atmosphere, natural waters, soils, body fluids, etc. as well as chemicals and food products. However, since environments are dealt with fully in Chapter 2, this discussion will be confined to simple chemical solutions, whose behaviour can be more readily interpreted in terms of fundamental physico-chemical principles, and additional factors will have to be considered in interpreting the behaviour of metals in more complex environments. For example, iron will corrode rapidly in oxygenated water, but only very slowly when oxygen is absent; however, in an anaerobic water containing sulphate-reducing bacteria, rapid corrosion occurs, and the mechanism of the process clearly involves the specific action of the bacteria (*see* Section 2.6).

All corrosion reactions in aqueous solutions are characterised by the following features:

1. The electrified interface between the metal and the electrolyte solution (the metal surface may be film-free or partially or completely covered with films or corrosion products).
2. Transfer of positive charge from the metal to the solution with consequent oxidation of the metal to a higher valency state.
3. Transfer of positive charge from the solution to the metal with consequent reduction of a species in solution (an electron acceptor) to a lower valency state.
4. Transfer of charge through the solution and corroding metal.

It follows that corrosion is an electrochemical reaction in which the metal itself is a reactant and is oxidised (loss of electrons) to a higher valency state, whilst another reactant, an electron acceptor, in solution is reduced (gain of electrons) to a lower valency state. This may be regarded as a concise expression of the 'electrochemical mechanism of corrosion'.

Thermodynamics and Kinetics of Corrosion Reactions

Thermodynamics provides a means of predicting the *equilibrium state* of a system of specified components, but provides no information on the detailed course of the reaction nor of the rate at which the system proceeds to equilibrium. With the aid of suitable catalysts the reaction may be made to proceed by different mechanisms and at different rates, but the final position of equilibrium, which can be predicted unequivocally by thermodynamics, remains unchanged. At elevated temperatures reactions proceed to equilibrium far more quickly than at ambient temperatures, and for this reason metal/gas equilibria at elevated temperatures can be defined entirely in terms of the thermodynamics of the systems under consideration (*see* Section 7.6). The position is frequently quite different with systems at ambient temperatures, since although a reaction may have a very pronounced tendency to proceed in a given direction this may be entirely nullified by the very slow rate at which it proceeds to its equilibrium state. Thus although the interaction of hydrogen and oxygen to form water at 25°C is accompanied by a large decrease in the standard free energy (*see* Appendix 9.1A)

$$H_2(g.) + \tfrac{1}{2}O_2(g.) \rightarrow H_2O(g.) \qquad \Delta G^{\ominus}_{298} = -237\,\text{kJ}$$

the rate of reaction is so slow that the reaction may be regarded as not occurring at all; however, in the presence of a catalyst, e.g. platinum black, the reaction proceeds instantaneously with explosive violence. It follows that although the change in free energy ΔG provides a quantitative measure of the *tendency* of a reaction to proceed in a given direction, it gives no information on the rate of the reaction.

It is indeed fortunate that the *rates* of the majority of corrosion reactions are very slow, since it is this factor that permits metals to be used as constructional materials. Considering metals in contact with oxygen at room temperatures, it follows from the standard free energies of formation of metal oxides that, with the exception of gold, the oxide of the metal is the thermodynamically stable state; the free energy decreases with temperature but the common metals are thermodynamically stable in an oxygen atmosphere only at very elevated temperatures (Fig. 7.43, Section 7.6). However, in practice, highly reactive metals like aluminium, magnesium, tantalum, niobium, etc. are relatively stable in oxygen whereas other reactive metals such as the alkali and alkaline-earth metals oxidise rapidly and completely. In these examples, the kinetics of the reaction is determined by the nature of the oxide film formed on the metal surface, which forms a protective barrier in the former group of metals and a non-protective one in the latter. Nevertheless, it should be noted that when the surface area is very large compared to the volume, the thermodynamic tendency for oxidation will outweigh kinetics, and metals such as aluminium, magnesium, titanium and iron in the form of very fine powders, can ignite spontaneously at ambient temperatures.

Most metals (copper, silver, gold, mercury and the platinum metals are the exceptions) are found in nature in the combined form as minerals, which are then reduced to the metal by the expenditure of sufficient energy (chemical, electrical or thermal) to reverse the natural spontaneous reaction. It follows that most metals are thermodynamically unstable and will tend to revert back to the thermodynamically stable combined state when exposed to the atmosphere, but the rate at which the reaction proceeds may be so slow that for all practical purposes the metal is stable. It is of some significance that very reactive metals such as tantalum, niobium and titanium, which have high affinities for oxygen and are reduced from their oxides with the expenditure of considerable chemical energy, are highly stable in the majority of environments; in fact, it is unlikely that these metals will ever revert back to oxides during atmospheric exposure owing to the highly protective nature of the thin oxide film. However, similar considerations do not apply to the less reactive metals iron and mild steel, since the oxide formed on these metals (rust) is not so protective as that formed on certain of the more thermodynamically reactive metals.

These examples show that the thermodynamic tendency of metals to corrode is frequently outweighed by kinetic factors that control the rate of the process so that the final position of equilibrium, i.e. complete conversion to corrosion product, is attained only slowly or not at all. Although the formation of oxide films, or films of other corrosion products, are of vital importance in controlling the rate of a corrosion reaction in aqueous solution, it must be emphasised that in certain circumstances the activation energy of the process or the rate of diffusion of species to and from the metal surface may be more significant than film formation. Thus the slow rate of corrosion of pure zinc in sulphuric acid, in which the oxide film is thermodynamically unstable, is dependent on the high activation energy needed for the reduction of hydrogen ions to hydrogen gas. In neutral solutions, on the other hand, diffusion of oxygen to the metal surface may be the rate-determining factor.

The magnitude of ΔG, the free energy change, of a specific corrosion reaction provides a measure of the spontaneity of the reaction and of the extent to which it will proceed before equilibrium is attained; if $\Delta G \ll 0$

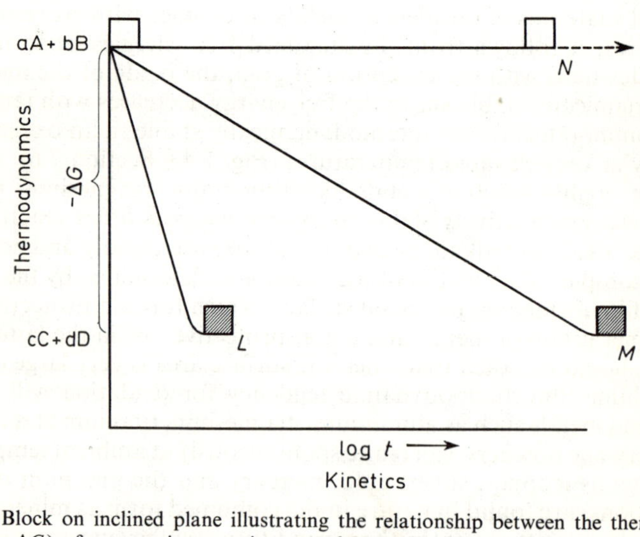

Fig. 1.13 Block on inclined plane illustrating the relationship between the thermodynamic tendency ($-\Delta G$) of a corrosion reaction $aA+bB \rightarrow cC+dD$ and the rate of conversion of metal A to corrosion products C and D

(negative) the tendency for a metal to react with species in solution will be high, but whether or not the reaction proceeds to any extent will depend on kinetic factors; if $\Delta G > 0$ (positive) the metal is stable and no further consideration need be given to kinetics; if $\Delta G = 0$ the system is at equilibrium and will not proceed in either direction. The relationship between the free energy decrease $-\Delta G$ and the rate of the reaction

$$aA + bB = cC + dD$$

where A is the metal, B is a reactant in solution, and C and D are the reaction (or corrosion) products, is illustrated in Fig. 1.13, which shows that although the thermodynamic tendency is the same for L, M and N the rates are different. This illustration may be exemplified by regarding L as an unprotected steel pipe buried in a corrosive soil, M the pipe protected by a bituminous coating and N the pipe coated and also cathodically protected; in each case the free energy change for the corrosion reaction

$$2Fe + H_2O + \tfrac{3}{2}O_2 \rightarrow Fe_2O_3.H_2O$$

is the same, but the rate at which the reaction proceeds to equilibrium (complete conversion of iron to $Fe_2O_2.H_2O$) varies significantly.

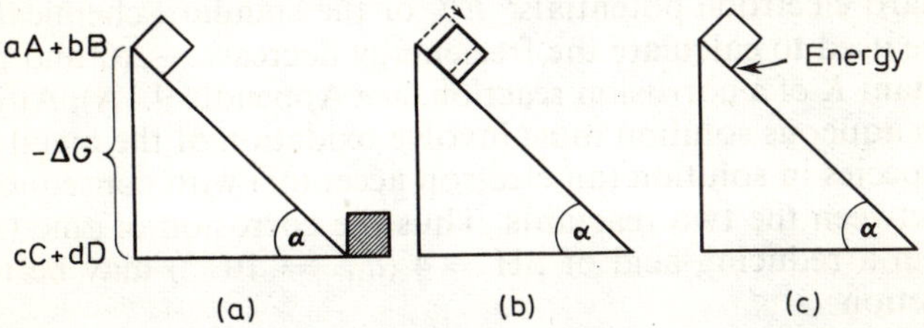

Fig. 1.14 Block on inclined plane illustrating the significance of (*a*) corrosion, (*b*) passivity and (*c*) immunity

Figure 1.14 illustrates the concepts implicit in the terms *corrosion*, *passivity* and *immunity*, which are of particular significance in both the thermodynamic and kinetic approach to corrosion reactions. In this analogy the metal A is in the form of a cube that rests on the top of an inclined plane and is in contact with the environment B. The slope of the plane α may be regarded as a measure of the tendency of the metal to slide down the plane with progressive formation of corrosion product $(-\Delta G)$ until equilibrium is attained at the bottom of the plane, when all the metal is converted to corrosion product. The rate at which the block slides down the plane, which in this analogy corresponds to the rate of the reaction, will depend both on α and on the frictional resistance between the metal, or the metal filmed with corrosion product, and the plane. In Fig. 1.14*a* the frictional resistance of both metal and corrosion product is assumed to be low, and the metal proceeds rapidly down the plane with a correspondingly high corrosion rate. In Fig. 1.14*b* the corrosion product is assumed to form a continuous adherent film on the metal surface that has a high frictional resistance; it follows that progress of the block down the plane will be decreased to a very low value as soon as a thin film of oxide is formed, and although some corrosion will occur it will not affect the bulk properties of the metal block significantly.

In this analogy Fig. 1.14*a* represents a corrosion rate that could not be tolerated in practice, whereas Fig. 1.14*b* represents the phenomenon of *passivity* in which the rate of corrosion is controlled at an acceptable level by virtue of the kinetic barrier of corrosion product. It should be noted, however, that the metal in the passive state is still thermodynamically unstable, and any environmental factor that results in the removal of the oxide film, either locally or generally, will result in rapid corrosion; a characteristic of passive metals is that in certain environmental situations localised breakdown of the oxide film occurs with consequent pitting of the exposed metal, the filmed surface surrounding the pit remaining unattacked. Finally, no corrosion will occur if the block can be maintained at the top of the plane by a continuous supply of energy of a magnitude that is equal to or greater than the energy liberated when the reaction proceeds spontaneously. The analogy in Fig. 1.14*c* represents the phenomenon of *immunity* that forms the basis of the important practical method of corrosion control known as *cathodic protection* in which the potential of the metal is depressed sufficiently to prevent it oxidising (*see* Chapter 11).

Thermodynamic Approach to Corrosion Reactions*

The standard electrode potentials[1] $E^{\ominus}$, or the standard chemical potentials $\mu^{\ominus}$, may be used to calculate the free energy decrease $-\Delta G$ and the equilibrium constant K of a corrosion reaction (*see* Appendix 9.1A). Any corrosion reaction in aqueous solution must involve oxidation of the metal and reduction of a species in solution (an electron acceptor) with consequent electron transfer between the two reactants. Thus the corrosion of zinc ($E^{\ominus}_{Zn^{2+}/Zn} = -0{\cdot}76\,V$) in a reducing acid of pH = 4 ($a_{H^+} = 10^{-4}$) may be represented by the reaction:

$$Zn + 2H^+(a_{H^+} = 10^{-4}) = Zn^{2+} + H_2 \ (p_{H_2} = 1) \qquad \text{...(1.4)}$$

which consists of the two half reactions

$$Zn = Zn^{2+}(aq.) + 2e \ (\text{oxidation of } Zn \rightarrow Zn^{2+}) \qquad \text{...(1.5)}$$

and

$$2H^+ + 2e = H_2 \qquad (\text{reduction of } H^+ \rightarrow H_2) \qquad \text{...(1.6)}$$

Since $\Delta G = -zFE$ and since 1 F (F is a Faraday) ≈ 96 500 C

$$\Delta G_{Zn^{2+}/Zn} = -2 \times 96\,500 \times -(-0{\cdot}76)\,J\,mol^{-1}\,Zn \qquad \text{...(1.7)}$$

and

$$\Delta G_{H^+/H_2} = -2 \times 96\,500 \times (-4 \times 0{\cdot}59)\,J\,mol^{-1}\,H_2 \qquad \text{...(1.8)}$$

and for the overall reaction

$$\Delta G_{reaction} = -2 \times 96\,500 \times (0{\cdot}76 - 0{\cdot}24) = -100{\cdot}3\,kJ\,mol^{-1}\,Zn$$

which shows that the reaction of zinc with hydrogen ions at $a_{H^+} = 10^{-4}$ g ion/l is spontaneous and proceeds in the direction as written in equation 1.4. However, in spite of the large free-energy decrease for equation 1.4 the rate of corrosion, which in this particular system is controlled by the rate of hydrogen evolution (equation 1.6), is extremely slow.

(At this point it should be noted that the free energies of half reactions can be added algebraically to evaluate $\Delta G_{reaction}$ and similar considerations apply

*Thermodynamic data are given in Chapter 21, Tables 21.5 and 21.6.

to the evaluation of E_{reaction}, providing the number of electrons z in the two half reactions and in the complete reaction are the same (*see* example just given). However, should z differ, then it is essential to sum the *free energies* in order to evaluate E_{reaction}, and this can be illustrated by calculating the value of $E^{\ominus}_{Cu^{2+}/Cu^{+}}$ from $E^{\ominus}_{Cu^{2+}/Cu} = 0{\cdot}346\text{ V}$, and $E^{\ominus}_{Cu^{+}/Cu} = 0{\cdot}522\text{ V}$:

$$Cu^{2+} + 2e = Cu, \quad -\Delta G^{\ominus} = 2\,F \times 0{\cdot}346 = 0{\cdot}792\,F \qquad \text{...(1.9)}$$

$$Cu^{+} + e = Cu, \quad -\Delta G^{\ominus} = 1\,F \times 0{\cdot}522 = 0{\cdot}522\,F \qquad \text{...(1.10)}$$

Subtracting equation 1.10 from equation 1.9

$$Cu^{2+} + e = Cu^{+}, \quad -\Delta G^{\ominus} = (0{\cdot}792\,F - 0{\cdot}522\,F) = 0{\cdot}170\,F$$

$$\therefore\ 1\,F\,E^{\ominus}_{Cu^{2+}/Cu^{+}} = 1\,F \times 0{\cdot}170, \text{ and } E^{\ominus}_{Cu^{2+}/Cu^{+}} = 0{\cdot}17\text{ V}$$

Direct subtraction of the standard electrode potential would have given the incorrect value $E^{\ominus}_{Cu^{2+}/Cu} = 0{\cdot}346 - 0{\cdot}522 = -0{\cdot}176\text{ V}$.)

In this example of the corrosion of zinc in a reducing acid of pH = 4, the corrosion product is Zn^{2+}(aq.), but at higher pHs the thermodynamically stable phase will be $Zn(OH)_2$ and the equilibrium activity of Zn^{2+} will be governed by the solubility product of $Zn(OH)_2$ and the pH of the solution; at still higher pHs ZnO_2^- anions will become the stable phase and both Zn^{2+} and $Zn(OH)_2$ will become unstable. However, a similar thermodynamic approach may be adopted to that shown in this example.

Table 1.7 Half reactions involving the oxidation of a metal in aqueous solutions

Equation number	*Half reaction*	
1	$M \rightarrow M^{z+}(\text{aq.}) + ze$	Oxidation to aquo cations
2(a)	$M + zH_2O \rightarrow M(OH)_z + zH^+ + ze$	Oxidation to metal hydroxide
or 2(b)	$M + zOH^- \rightarrow M(OH)_z + ze$	
3(a)	$M + zH_2O \rightarrow MO_z^{z-}(\text{aq.}) + 2zH^+ + ze$	Oxidation to aquo anions
or 3(b)	$M + zOH^- \rightarrow MO_z^{z-}(\text{aq.}) + zH^+ + ze$	

Table 1.7 shows typical half reactions for the oxidation of a metal M in aqueous solutions with the formation of aquo cations, solid hydroxides or aquo anions[2]. The equilibrium potential for each half reaction can be evaluated from the chemical potentials of the species involved (*see* Appendix 9.1A) and it should be noted that there is no difference thermodynamically between equations 2(*a*) and 2(*b*) nor between 3(*a*) and 3(*b*) when account is taken of the chemical potentials of the different species involved.

In view of the importance of the hydronium ion, H_3O^+, and dissolved oxygen as electron acceptors in corrosion reactions, some values of the redox potentials E and chemical potentials μ for the equilibria

$$2H^+ + 2e = H_2, \ E = 0{\cdot}00 - 0{\cdot}059\text{pH} - 0{\cdot}030 \log p_{H_2} \qquad \text{...(1.11)}$$

and $\frac{1}{2}O_2 + H_2O + 2e = 2OH^-, \ E = 1{\cdot}23 - 0{\cdot}059\text{pH} + 0{\cdot}15 \log p_{O_2}$ ...(1.12)

at 25°C are given in Table 1.8. The following should be noted:

1. Dissolved oxygen has a higher redox potential than the hydrogen ion at all values of pH, i.e. it is a more powerful oxidant.

Table 1.8 Potentials of the $H^+/\frac{1}{2}H_2$ and O_2/OH^- electrodes ($p_{H_2} = p_{O_2} = 1$ atm*)

Equilibrium	*Activity*	pH	*E* (V)	$\mu = 2 \times 96\,500\,E$ kJ
$2H^+ + 2e = H_2$ (see equation 1.11)	$a_{H^+} = 1$	0	0·00	0·00
	$a_{H^+} = 10^{-7}$	7	−0·414	−79·89
	$a_{H^+} = 10^{-14}$	14	−0·828	−159·8
$\frac{1}{2}O_2 + H_2O + 2e = 2OH^-$ (see equation 1.12)	$a_{OH^-} = 1$	14	0·401	77·4
	$a_{OH^-} = 10^{-7}$	7	0·815	157·3
	$a_{OH^-} = 10^{-14}$	0	1·229	237·5

*1 atm = $1{\cdot}013 \times 10^5$ Pa (Pascal).

2. In both equilibria *E* decreases with increase in pH
3. A plot of *E vs.* pH for each equilibrium gives a linear curve of slope 0·059 (*see* curves *ℓ* and *m* in Fig. 1.15 (bottom)).
4. A decrease in p_{O_2} will displace curve *m* in Fig. 1.15 (bottom) in the negative direction, whereas a decrease in p_{H_2} will displace curve *ℓ* in the positive direction (equations 1.11 and 1.12).

Table 1.9 shows the values of $\Delta G^{\ominus}$, *K* and the equilibrium activities of metal cations and pressures (fugacities) of hydrogen gas for the reaction of typical metals with a reducing acid ($a_{H^+} = 1$) to form metal cations M^{z+}(aq.). Equilibria such as Au^{3+}/Au, Ag^+/Ag and Cu^{2+}/Cu have redox potentials $>0{\cdot}0$ V, and the metal (the reduced species of the equilibrium) may be regarded as being thermodynamically stable in a reducing acid, since the equilibrium activities of metal cations are negligible; thus in these systems no further consideration of the rates of the reaction are necessary. On the other hand, the metals *M* of M^{z+}/M equilibria that are negative to the H^+/H_2 equilibrium are thermodynamically unstable, and the high values of the equilibrium activities of M^{z+} indicate that the reaction will proceed to completion, although at a rate that cannot be predicted from thermodynamic consideration alone. Thus pure zinc corrodes slowly in hydrochloric acid, more quickly if the zinc contains impurities of lead and more quickly still when the impurities are copper; in each of these three cases the thermodynamic tendency and the final position of equilibrium (*see* Table 1.9) are the same, although the rates at which equilibrium is achieved are markedly different. (Note that each equilibrium must be written so that $-\Delta G^{\ominus} > 0$ (positive), i.e. in the direction in which it proceeds spontaneously so that $E^{\ominus}_{\text{reaction}}$ is always positive although $E^{\ominus}_{M^{z+}/M}$ may be either positive or negative.)

Table 1.9 Thermodynamics of the reaction of metals with acid solutions ($a_{H^+} = 1$, $E^{\ominus}_{H^+/\frac{1}{2}H_2} = 0{\cdot}00$ V)

Reaction	$E^{\ominus}_{M^{z+}/M}$ (V)	$E^{\ominus}_{\text{reaction}}$ (V)	$-\Delta G^{\ominus}$ (kJ)	*K*	$a_{M^{z+}}$ (when $p_{H_2} = 1$)	p_{H_2} (when $a_{M^{z+}} = 1$)
$\frac{2}{3}Au^{3+} + H_2 = 2H^+ + \frac{2}{3}Au$	1·50	1·50	289·4	$7{\cdot}2 \times 10^{50}$	$8{\cdot}5 \times 10^{-77}$	$1{\cdot}4 \times 10^{-51}$
$2Ag^+ + H_2 = 2H^+ + 2Ag$	0·79	0·79	152·4	$6{\cdot}0 \times 10^{26}$	$4{\cdot}1 \times 10^{-14}$	$1{\cdot}7 \times 10^{-27}$
$Cu^{2+} + H_2 = 2H^+ + Cu$	0·337	0·337	65·0	$2{\cdot}6 \times 10^{11}$	$3{\cdot}8 \times 10^{-12}$	$3{\cdot}8 \times 10^{-12}$
$Sn + 2H^+ = H_2 + Sn^{2+}$	−0·136	0·136	26·2	$4{\cdot}0 \times 10^{4}$	$4{\cdot}0 \times 10^{4}$	$4{\cdot}0 \times 10^{4}$
$Fe + 2H^+ = H_2 + Fe^{2+}$	−0·440	0·440	84·9	$8{\cdot}1 \times 10^{14}$	$8{\cdot}1 \times 10^{14}$	$8{\cdot}1 \times 10^{14}$
$Zn + 2H^+ = H_2 + Zn^{2+}$	−0·763	0·763	147·2	$7{\cdot}2 \times 10^{25}$	$7{\cdot}2 \times 10^{25}$	$7{\cdot}2 \times 10^{25}$
$\frac{2}{3}Al + 2H^+ = H_2 + \frac{2}{3}Al^{3+}$	−1·66	1·66	320·3	$1{\cdot}9 \times 10^{56}$	$2{\cdot}6 \times 10^{84}$	$1{\cdot}9 \times 10^{56}$
$Mg + 2H^+ = H_2 + Mg^{2+}$	−2·37	2·37	457·3	$2{\cdot}2 \times 10^{80}$	$2{\cdot}2 \times 10^{80}$	$2{\cdot}2 \times 10^{80}$

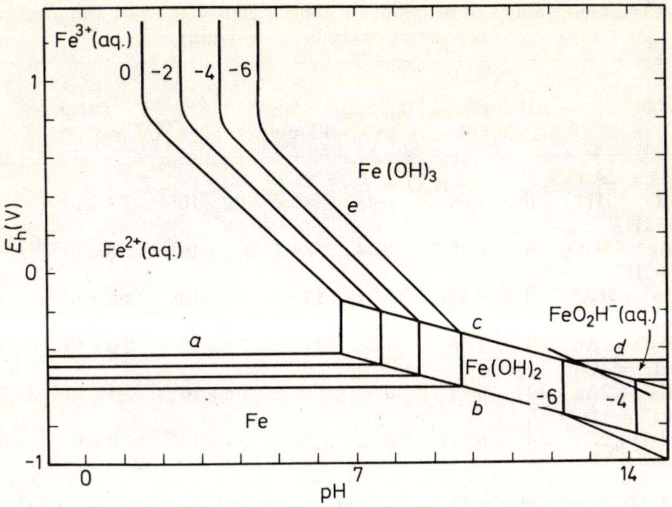

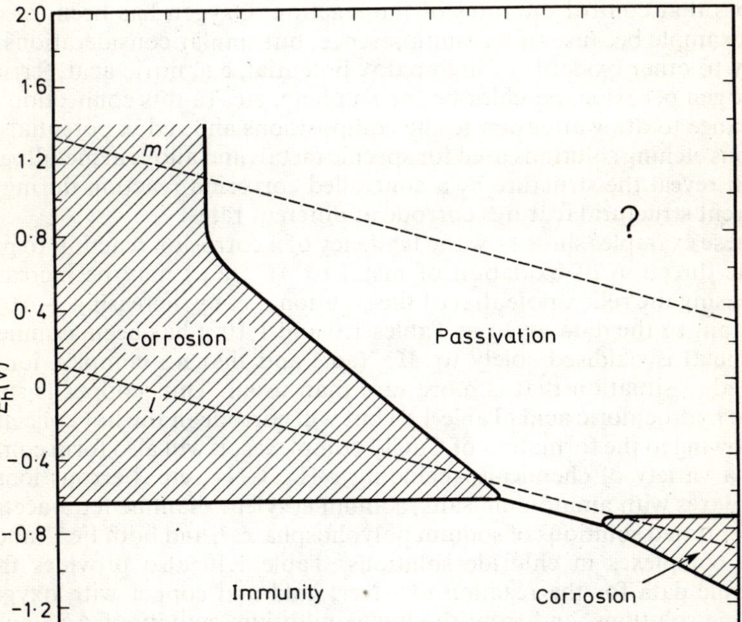

Fig. 1.15(top) Equilibrium potential–pH diagram for the Fe–H_2O system showing the zones of stability of cations, anions and solid hydroxides (after Deltombe and Pourbaix[7]) and (bottom) simplified version showing zones of corrosion, immunity and passivity (curve ℓ is the H_2O/H_2 equilibrium at $p_{H_2} = 1$ and curve m is the O_2/H_2O equilibrium at $p_{O_2} = 1$)

Table 1.10 Thermodynamics of reactions of noble metals with acids and alkaline solutions containing oxygen (p_{O_2} = 1 atm)

Reaction	pH	$E^{\ominus}_{M^{z+}/M}$ (V)	$E^{\ominus}_{reaction}$ (V)	$-\Delta G^{\ominus}$ (kJ mol^{-1})	K	$a_{M^{z+}}$ (when p_{O_2} = 1)	p_{O_2} (when $a_{M^{z+}}$ = 1)
$2Au^+ + H_2O = 2Au + \frac{1}{2}O_2 + 2H^+$	0	1·68	0·45	86·8	$1{\cdot}7\times10^{15}$	$2{\cdot}4\times10^{-8}$	$2{\cdot}9\times10^{30}$
$2Ag + \frac{1}{2}O_2 + 2H^+ = 2Ag^+ + H_2O$	0	0·79	0·44	84·9	$8{\cdot}1\times10^{14}$	$2{\cdot}8\times10^{7}$	$1{\cdot}5\times10^{-30}$
$2Cu + \frac{1}{2}O_2 + 2H^+ = 2Cu^+ + H_2O$	0	0·521	0·71	137·0	$1{\cdot}1\times10^{20}$	$1{\cdot}0\times10^{10}$	$8{\cdot}2\times10^{-41}$
$2Au^+ + 2OH^- = \frac{1}{2}O_2 + H_2O + 2Au$	14	1·68	1·28	247·0	$2{\cdot}4\times10^{43}$	$2{\cdot}0\times10^{-22}$	$5{\cdot}8\times10^{86}$
$2Ag^+ + 2OH^- = \frac{1}{2}O_2 + H_2O + 2Ag$	14	0·79	0·39	75·2	$1{\cdot}6\times10^{13}$	$2{\cdot}5\times10^{-7}$	$2{\cdot}6\times10^{26}$
$2Cu^+ + 2OH^- = \frac{1}{2}O_2 + H_2O + 2Cu$	14	0·521	0·12	23·1	$1{\cdot}2\times10^{4}$	$1{\cdot}0\times10^{-2}$	$1{\cdot}4\times10^{8}$

Table 1.10 shows the effect of raising the redox potential of the acid by means of dissolved oxygen (p_{O_2} = 1 atm), and it can be seen that the noble metals (metals more positive than 0·0 V) silver and copper are now thermodynamically unstable, and due consideration must be given therefore to the factors that control the rates of the reaction. Oxygen has been selected in this example because of its omnipresence, but similar considerations would apply to other oxidants of high redox potential, e.g. nitric acid, ferric salts, hydrogen peroxide, perchlorate, persulphate, etc. In this connection it is of relevance to draw attention to the compositions and redox potentials of the various etching solutions used for specific metals and alloys in metallography, which reveal the structure by a controlled corrosion reaction during which different structural features corrode at different rates.

These examples show how the tendency of a corrosion reaction to proceed in the direction of oxidation of metal to M^{z+}(aq.) may be increased by increasing the redox potential of the solution, i.e. by increasing $-\Delta G$ of the reaction. In the data given in Tables 1.9 and 1.10 it has been assumed that the metal is oxidised solely to M^{z+}(aq.) and that no complex ions have formed, a situation that is more rare than usual. Thus copper is stable in dilute hydrochloric acid (Table 1.9) but will corrode in the hot concentrated acid owing to the formation of $CuCl_2^-$ complexes; it will also form complexes with a variety of chemicals, including NH_3, CN^-, etc. Ferrous ions form complexes with ammonium salts, sodium ethylene diamine tetra-acetates or concentrated solutions of sodium polyphosphates[3], and both Fe^{2+} and Fe^{3+} form complexes in chloride solutions. Table 1.10 also provides thermodynamic data for the reaction of silver, gold and copper with oxygenated alkaline solutions, and from the low equilibrium activity of Ag^+ and Au^+ it can be assumed that these metals are stable in that environment. However, in the presence of CN^- anions, cyano-complexes are formed and there is an equilibrium between M^+ and $M(CN)_2^-$ that is defined by the instability constant (Table 1.11)[4], and although the a_{M^+} (which is dependent upon the redox potential and pH of the solution and is unaffected by CN^-) is very low, the $a_{M(CN)_2^-}$ is significantly high. Thus addition of CN^- to an oxygenated sodium hydroxide solution will result in the corrosion of gold, and this reaction is utilised in extractive metallurgy for the leaching of gold from its ores by means of an oxygenated solution of sodium cyanide.

The data given in Tables 1.9 and 1.10 have been based on the assumption that metal cations are the sole species formed, but at higher pH values oxides, hydrated oxides or hydroxides may be formed, and the relevant half reactions will be of the form shown in equations 2(a) and 2(b) (Table 1.7). In these circumstances the $a_{M^{z+}}$ will be governed by the solubility product of the solid compound and the pH of the solution. At higher pH values the solid compound may become unstable with respect to metal anions (equations 3(a) and 3(b), Table 1.7), and metals like aluminium, zinc, tin and lead, which form amphoteric oxides, corrode in alkaline solutions. It is evident, therefore, that the equilibrium between a metal and an aqueous solution is far more complex than that illustrated in Tables 1.9 and 1.10. Nevertheless, as will be discussed subsequently, a similar thermodynamic approach is possible.

Finally, it is necessary to observe that the values of activities and fugacities calculated are thermodynamic quantities that cannot always be realised in practice, e.g. very high activities of metal ions cannot be attained because of solubility consideration and very low activities have no physical significance.

Table 1.11 Equilibrium between M^+ and $M(CN)_2^-$ in cyanide solutions ($a_{OH^-} = 1$ (pH = 14); $a_{CN^-} = 1$; $p_{O_2} = 1$)

Equilibrium	*Instability constant*	a_{M^+}	$a_{M(CN)_2^-}$
$Au(CN)_2^- = Au^+ + 2CN^-$	10^{-39}	2×10^{-22}	10^{17}
$Ag(CN)_2^- = Ag^+ + 2CN^-$	10^{-20}	$2{\cdot}5 \times 10^{-7}$	10^{13}
$Cu(CN)_2^- = Cu^+ + 2CN^-$	10^{-16}	$9{\cdot}1 \times 10^{-8}$	10^{9}

Potential–pH Equilibrium Diagrams

Pourbaix and his co-workers[5,6] have calculated the phases at equilibrium for M/H_2O systems at 25°C from the chemical potentials of the species involved in the equilibria, and have expressed the data in the form of *equilibrium diagrams* having pH *vs.* E, the *equilibrium potential* (*vs.* S.H.E.) as ordinates. These diagrams, which are analogous to the composition–temperature diagrams of alloy systems, provide a thermodynamic basis for the study of corrosion reactions, although, as emphasised by Pourbaix, their limitations in relation to practical problems must be appreciated. Since these diagrams are referred to throughout this work, some emphasis is given to their significance in this section; reference should be made to the numerous publications of Pourbaix and his co-workers for a fuller account of the subject.

It should be emphasised that potential–pH diagrams can also be constructed from experimental E_p–I curves, where E_p is the *polarised potential* and I the current[6]. These diagrams, which are of more direct practical significance than the equilibrium potential–pH equilibrium diagrams constructed from thermodynamic data, show how a metal in a natural environment (e.g. iron in water of given chloride ion concentration) may give rise to general corrosion, pitting, perfect or imperfect passivity, or to immunity, depending on the pH and potential (*see* Section 1.6, Fig. 1.56).

Construction of Potential–pH Diagrams

Pourbaix has classified the various equilibria that occur in aqueous solution into homogeneous and heterogeneous, and has subdivided them according to whether the equilibria involve electrons and/or hydrogen ions. The general equation for a half reaction is

$$aA + mH^+ + ze = bB + cH_2O \qquad \ldots(1.13)$$

which shows that the equilibrium between the reactant A and product B depends on a_{H^+} (pH) and the electrode or redox potential E; if neither hydrogen ions nor electrons are involved then the equilibrium is independent of pH and E. From the Nernst equation (*see* page **9**.69), and substituting for $-\log a_{H^+} = \text{pH}$

$$E = E^{\ominus} - 0{\cdot}059\frac{m}{z}\text{pH} - \frac{0{\cdot}059}{z}\log\frac{a_B^b}{a_A^a} \qquad \ldots(1.14)$$

It follows from equation 1.14 that for any constant ratio of a_B^b/a_A^a the E *vs.* pH relationship will be linear with a slope $-0{\cdot}059m/z$, and that when $a_B^b = a_A^a = 1$ the intercept of the curve on the E axis (i.e. pH = 0) will be $E^{\ominus}$, the standard equilibrium potential, which by definition is the potential when the species involved in the equilibrium are at unit activity.

Pourbaix has evaluated all possible equilibria between a metal M and H_2O (*see* Table 1.7) and has consolidated the data into a single potential–pH diagram, which provides a pictorial summary of the anions and cations (nature and activity) and solid oxides (hydroxides, hydrated oxides and oxides) that are at equilibrium at any given pH and potential; a similar approach has been adopted for certain M–H_2O–X systems where X is a non-metal, e.g. Cl^-, CN^-, CO_2, SO_4^{2-}, PO_4^{3-}, etc. at a defined concentration. These diagrams give the activities of the metal cations and anions at any specified E and pH, and in order to define corrosion in terms of an equilibrium activity, Pourbaix has selected the arbitrary value of 10^{-6} g ion/l, i.e. corrosion of a metal is defined in terms of the pH and potential that give an equilibrium activity of metal cations or anions $>10^{-6}$ g ion/l; conversely, passivity and immunity are defined in terms of an equilibrium activity of $<10^{-6}$ g ion/l. (Note that g ion/l is used here because this is the unit used by Pourbaix; in the S.I. the relative activity is dimensionless.)

Fe/H_2O system Figure 1.15 (top) is a simplified version[7] of the Fe–H_2O potential–pH equilibrium diagram [the region of stability of magnetite (Fe_3O_4) is not included] and it is instructive to consider some of the more important equilibria involved:

Curve *a*, $Fe^{2+} + 2e = Fe$; $E = -0{\cdot}440 + 0{\cdot}030 \log a_{Fe^{2+}}$ ...(1.15)

Curve *b*, $Fe(OH)_2 + 2H^+ + 2e = Fe + 2H_2O$;

$$E = -0{\cdot}047 - 0{\cdot}059\text{pH} \qquad \ldots(1.16)$$

Curve *c*, $Fe(OH)_3+H^++e = Fe(OH)_2+H_2O$;

$$E = 0{\cdot}271-0{\cdot}059\text{pH} \qquad \ldots(1.17)$$

Curve *d*, $Fe(OH)_3+e = FeO_2H^-+H_2O$;

$$E = -0{\cdot}81-0{\cdot}059 \log a_{FeO_2H^-} \qquad \ldots(1.18)$$

Curve *e*, $Fe(OH)_3+3H^++e = Fe^{2+}+3H_2O$;

$$E = 1{\cdot}060-0{\cdot}177\text{pH}-0{\cdot}059 \log a_{Fe^{2+}} \qquad \ldots(1.19)$$

Since equation 1.15 does not involve H^+ it is pH independent, and variation of $a_{Fe^{2+}}$ will result in a series of curves parallel to the pH axis that extend across the diagram until the pH is sufficiently high to reduce the $a_{Fe^{2+}}$ to $<10^{-6}$ g ion/l by formation of $Fe(OH)_2$. The relevant equilibrium is:

$$Fe^{2+}+2H_2O = Fe(OH)_2+2H^+$$

for which values of $\mu^\ominus$ (J) for the species involved are Fe^{2+}(aq.), $-84\,760$; H_2O, $-236\,700$; $Fe(OH)_2$, $-482\,000$, and since

$$\log K = \frac{-\Sigma v_i \mu_i^\ominus}{2{\cdot}303RT} = \frac{-(\Sigma v_i \mu_{i\,(\text{products})}^\ominus - \Sigma v_i \mu_{i\,(\text{reactants})}^\ominus)}{2{\cdot}303RT} \qquad \ldots(1.20a)$$

and at 25°C

$$\log K = \frac{-[-482\,000-2(-236\,700)-(-84\,760)]}{2{\cdot}303\times 8{\cdot}315\times 298{\cdot}16} = -13{\cdot}37 \qquad \ldots(1.20b)$$

Since $$K = \frac{a_{H^+}^2}{a_{Fe^{2+}}}, \text{ then } \log K = \log a_{H^+}^2 - \log a_{Fe^{2+}} \qquad \ldots(1.21)$$

and $$\log a_{Fe^{2+}} = 13{\cdot}27-2\text{pH} \qquad \ldots(1.22)$$

Thus at $a_{Fe^{2+}} = 10^{-6}$ the equilibrium pH will be 9·7 and the $a_{Fe^{2+}}$ will be $<10^{-6}$ at any higher pH value; it follows that the formation of a new solid phase $Fe(OH)_2$ at a sufficiently high pH must limit the zone of corrosion as defined by Pourbaix.

Whereas lowering the potential results in a decrease in $a_{Fe^{2+}}$, the converse applies when the potential is raised. However, this increase in activity is again limited by the formation of a solid phase. Thus curve *e* of Fig. 1.15 (top) gives the equilibrium between $Fe(OH)_3$ and Fe^{2+} at any predetermined activity of the latter in the range 10^0–10^{-6}. At $a_{Fe^{2+}} = 10^{-6}$ g/ion/l, $E = [1{\cdot}06+(-6\times 0{\cdot}059)]-0{\cdot}177\text{pH}$ which defines the boundary between corrosion and passivity at high potentials (equation 1.19).

At high pH values and low potentials, Fe, Fe^{2+}, Fe^{3+}, $Fe(OH)_2$ and $Fe(OH)_3$, etc. will be thermodynamically unstable with respect to FeO_2H^- and a further limited zone of corrosion will appear on the right-hand side of the diagram.

Significance of Zones in Potential–pH Diagrams

The above outline of the method adopted in the construction of the potential–pH diagram of the Fe–H_2O system serves to illustrate the essentially thermodynamic nature of diagrams of this type, which therefore cannot provide any information on the rates of corrosion processes. However, on the basis of certain assumptions that have no thermodynamic significance, it is possible to separate the Fe–H_2O diagram into the following zones [Fig. 1.15 (bottom)]

Corrosion: activity of Fe^{2+}, Fe^{3+}(aq.) or FeO_2^-(aq.)$>10^{-6}$ g ion/l (0·06 p.p.m. of Fe in the case of the cations).

Passivity: $Fe(OH)_2$, Fe_3O_4 or $Fe(OH)_3$ in equilibrium with metal ions at an activity $<10^{-6}$ g ion/l.

Immunity: Fe metal in equilibrium with Fe^{2+}(aq.) or FeO_2H^-(aq.) at an activity $<10^{-6}$ g ion/l.

It should be noted that Fig. 1.15 (top) is based entirely on thermodynamic data and is therefore correctly described as an equilibrium diagram, since it shows the phases (nature and activity) that exist at equilibrium. However, the concepts implicit in the terms *corrosion*, *immunity* and *passivity* lie outside the realm of thermodynamics, and, for example, passivity involves both thermodynamic and kinetic concepts; it follows that Fig. 1.15 (bottom) cannot be regarded as a true equilibrium diagram, although it is based on one that has been constructed entirely from thermodynamic data.

In Fig. 1.15 (bottom) curves ℓ and m show the potential–pH relationships for the reversible hydrogen and oxygen electrodes at $p_{H_2} = p_{O_2} = 1$ atm respectively. Within the area confined by the curves ℓ and m, H_2O is thermodynamically stable and $p_{H_2}<1$ and $p_{O_2}<1$; whereas below ℓ and above m, $p_{H_2}>1$ atm, and $p_{O_2}>1$ atm, respectively (*see* equations 1.11 and 1.12). Thus the diagram shows the solid phases of iron, the activities of metal ions and the pressures of hydrogen and oxygen gas that are at equilibrium at any given potential and pH when pure iron reacts with pure water.

This can be illustrated by considering the changes that will *tend* to occur when iron with a coating of rust ($Fe_2O_3.H_2O$) is immersed in oxygenated water at pHs and potentials that correspond with the various zones in the Fe–H_2O diagram[6].

Immunity Any Fe_2O_3 on the surface (or any Fe^{2+} in solution) will be reduced to metal, and $a_{Fe^{2+}}<10^{-6}$ g ion/l; water will be reduced to hydrogen and $p_{H_2}>1$ atm; any dissolved oxygen present will be reduced to OH^-, and $p_{O_2}\ll 1$ atm.

Corrosion (in acid solutions) At low potentials iron will be oxidised to Fe^{2+} and Fe_2O_3 reduced to Fe^{2+}, and the $a_{Fe^{2+}}$ will be $>10^{-6}$ g ion/l; water will be reduced to hydrogen or remain stable, depending upon whether E is below or above curve ℓ. At high potentials iron will be oxidised to Fe^{3+} and Fe_2O_3 will dissolve to form Fe^{3+} ($E^{\ominus}_{Fe^{3+}/Fe^{2+}} = 0{\cdot}76$ V); water will be stable or will be oxidised to oxygen, depending upon whether E is below or above curve m, respectively.

Passivation According to Fig. 1.15 (top) *all* the Fe will be converted to Fe_2O_3, whilst the rust originally present will be unaffected. According to Fig. 1.15 (bottom) the rust will be unaffected, whilst the iron surface exposed to the solution through pores in the rust will be passivated by a protective film of Fe_2O_3. Water will be stable except at high potentials where it will be oxidised to O_2.

Thus the *tendency* for an electrochemical reaction at a metal/solution interface to proceed in a given direction may be defined in terms of the relative values of the actual electrode potential E (experimentally determined and expressed with reference to the S.H.E.) and the reversible or equilibrium potential E_r (calculated from $E^\ominus$ and the activities of the species involved in the equilibrium).

When $E > E_r$ the reaction can only proceed in the direction of oxidation.
When $E < E_r$ the reaction can only proceed in the direction of reduction.
When $E = E_r$ the reaction is at equilibrium.

This can be summarised by the relationship

$$(E - E_r)I \geqslant 0 \qquad \ldots(1.23)$$

in which I, the reaction current, is regarded as positive in the case of oxidation and negative in the case of reduction. The parameter $(E - E_r)$, which may be positive, negative or zero, is termed the *overpotential* or *affinity*, and gives the tendency for the reaction to proceed in the direction of oxidation or reduction or to be at equilibrium, respectively. However, the precise magnitude of I will depend upon kinetic factors, which will be considered subsequently.

Advantages and Limitations of Diagrams

Although the zones of corrosion, immunity and passivity are clearly of fundamental importance in corrosion science it must be emphasised again that they have serious limitations in the solution of practical problems, and can lead to unfortunate misconceptions unless they are interpreted with caution. Nevertheless, Pourbaix and his co-workers, and others, have shown that these diagrams used in conjunction with E–i curves for the systems under consideration can provide diagrams that are of direct practical use to the corrosion engineer. It is therefore relevant to consider the advantages and limitations of the equilibrium potential–pH diagrams.

The M–H_2O diagrams present the equilibria at various pHs and potentials between the metal, metal ions and solid oxides and hydroxides for systems in which the only reactants are metal, water, and hydrogen and hydroxyl ions; a situation that is extremely unlikely to prevail in real solutions that usually contain a variety of electrolytes and non-electrolytes. Thus a solution of pH 1 may be prepared from either hydrochloric, sulphuric, nitric or perchloric acids, and in each case a different anion will be introduced into the solution with the consequent possibility of the formation of species other than those predicted in the M–H_2O system. In general, anions that form soluble complexes will tend to extend the zones of corrosion, whereas anions that form insoluble compounds will tend to extend the zone of passivity. However, provided the relevant thermodynamic data are available, the effect of these anions can be incorporated into the diagram, and diagrams of the type M–H_2O–X are available in Cebelcor reports and in the published literature.

The effect of anions on the zones of corrosion and passivation can be exemplified by a comparison of the Pb–H_2O and Pb–H_2O–SO_4^{2-} equilibrium diagrams (*see* Section 4.3, Figs. 4.13 and 4.14) and it can be seen that in the presence of SO_4^{2-} the corrosion zone corresponding with stability of Pb^{2+}

is completely replaced by $PbSO_4$ so that passivation is possible in the acid region owing to the thermodynamic stability of $PbSO_4$.

Similar considerations apply to the potential E which can be varied by means of an auxiliary electrode and an external source of e.m.f., or by varying the redox potential of the solution, and in the case of the latter a given redox potential may be achieved by using different oxidants. Thus at pH 7 and $E = -0{\cdot}44$ V, iron will be in the zone of corrosion ($a_{Fe^{2+}} = 1$), and the potential could be raised into the passive region by either dissolved oxygen, potassium chromate or potassium perchlorate. However, the effects produced will depend upon a variety of factors, and whereas passivation can be achieved if chromate is present in sufficient concentration, it may cause pitting at lower concentrations. Perchlorate will tend to cause pitting, and dissolved oxygen can result in localised attack and will passivate iron only if it is brought rapidly and simultaneously to all parts of the metal surface.

A further serious limitation is that diagrams evaluated from thermodynamic data at 25°C have little relevance in high-temperature aqueous corrosion, but it is now possible to construct diagrams[8,9] that are applicable at elevated temperatures from data obtained at 25° (*see* Section 2.1).

Pourbaix[6] has studied the behaviour of iron in city water (Brussels) at

Table 1.12 State of environments and iron in Brussels water; see also Fig. 1.16 (after Pourbaix[6])

Experiment	*Sample number*	*Solution*		pH	E_H (V)	*State of metal**	*Gas*
a	1	H_2O distilled		8·1	−0·486	●	—
	2	NaCl	1 g/l	6·9	−0·445	●	—
	3	H_2SO_4	1 g/l	2·3	−0·351	●	H_2
	4	$NaHSO_4$	1 g/l	6·4	−0·372	●	—
	5	NaOH	1 g/l	11·2	+0·026	○	—
	6	K_2CrO_4	1 g/l	8·5	+0·235	○	—
	7	K_2CrO_4 + NaCl	1 g/l	8·6	−0·200	◑	—
	8	$KMnO_4$	0·2 g/l	6·7	−0·460	●	—
	9	$KMnO_4$	1 g/l	7·1	+0·900	○	—
	10	H_2O_2	0·3 g/l	5·7	−0·200	●	—
	11	H_2O_2	3·0 g/l	3·4	+0·720	○	O_2
	12	Brussels city water		7·0	−0·450	●	—
b	13	NaOH 40 g/l degassed		13·7	−0·810	●	H_2
c	14	city water–iron–copper		7·5	−0·445	●	—
	15	city water–iron–zinc		7·5	−0·690	○	H_2
	16	city water–iron–magnesium		7·5	−0·910	○	H_2
	17	city water–iron–platinum		7·5	−0·444	●	—
c′	14′	city water–iron–copper		7·8	−0·385	●	—
	15′	city water–iron–zinc		7·7	−0·690	○	H_2
	16′	city water–iron–magnesium		8·7	−0·495	○	H_2
	17′	city water–iron–platinum		—	—	●	—
d	18	$NaHCO_3$ 0·1 M Pole −		8·4	−0·860	○	H_2
	19	$NaHCO_3$ 0·1 M Pole +		8·4	−0·350	●	—
	20	$NaHCO_3$ 0·1 M Pole −		8·4	−0·885	○	H_2
	21	$NaHCO_3$ 0·1 M Pole −		8·4	+1·380	○	O_2
e	22	$NaHCO_3$ 0·1 M Pole −		8·4	−0·500	●	—
	23	$NaHCO_3$ 0·1 M Pole +		8·4	+1·550	○	O_2
	24	$NaHCO_3$ 0·1 M Pole −		8·4	−1·000	○	H_2
	25	$NaHCO_3$ 0·1 M Pole +		8·4	+1·550	○	O_2

*● General corrosion; ◑ local corrosion; ○ absence of corrosion.

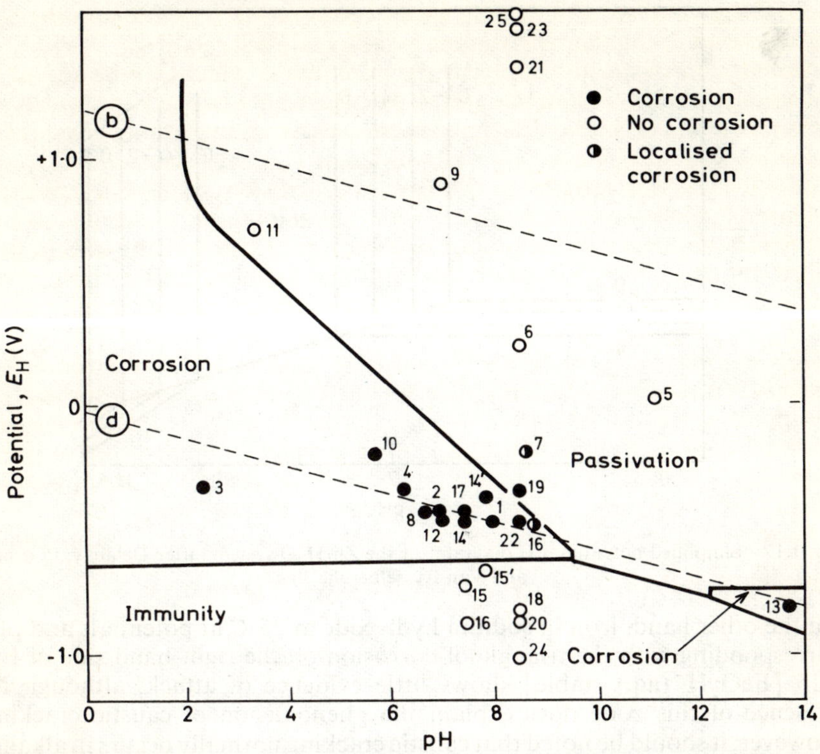

Fig. 1.16 Potential–pH diagram for the Fe–H_2O system in which results obtained for the behaviour of iron in Brussels water have been inserted (*see* Table 1.12) (after Pourbaix[6])

various pHs and potentials (Table 1.12); the study of the former was effected by adding acids and alkalis and the latter by applying an external e.m.f., by coupling the iron to either more positive (Cu, Pt) or more negative metals (e.g. Zn, Mg), or by adding oxidants (e.g. K_2CrO_4, $KMnO_4$, H_2O_2). The corrosion rate of the iron has been determined and results have been inserted in the potential–pH diagram for Fe–H_2O (Fig. 1.16), and it can be seen that in this particular water there is good agreement between the predictions of the diagram and the corrosion behaviour of the iron. However, it does not follow that this correlation would necessarily apply to all fresh waters or to sea-water.

Zones of Corrosion

The Zn–H_2O (Fig. 1.17) diagram[10] shows that extensive corrosion zones exist at both low and high pH values (compare the very restricted corrosion zone in the Fe–H_2O diagram at high pHs); similar zones in the region of low and high pH are obtained with other amphoteric metals such as aluminium, lead and tin. The diagram for Zn–H_2O predicts with some accuracy the behaviour of the metal in practice, where it has been established that zinc corrodes rapidly outside the range pH 6–12·5 but is passive within it.

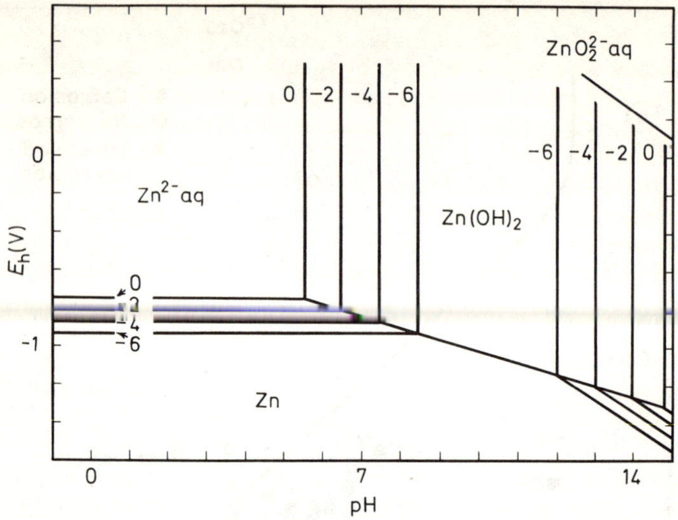

Fig. 1.17 Simplified potential–pH diagram for the Zn/H_2O system (after Delahay, Pourbaix and Van Rysselberghe[10])

On the other hand, iron in sodium hydroxide at 25°C at potentials and pHs corresponding with the triangle of corrosion on the right-hand side of Fig. 1.15 [FeO_2H^-(aq.) stable] shows little evidence of attack, although the presence of this zone does explain the phenomenon of caustic cracking. However, it should be noted that caustic cracking normally occurs in alkaline waters at elevated temperatures and that temperature will have a marked effect on both the thermodynamics and kinetics of the reaction (*see* Section 2.1).

Finally, it should be noted that although the arbitrary activity of 10^{-6} g ion/l represents a very low concentration of metal ions it could be significant in certain circumstances, e.g. lead at that concentration would render potable water toxic. It should also be noted that if the equilibrium is continuously disturbed, e.g. by a flowing solution, significant amounts of metal will corrode even at an equilibrium activity of 10^{-6} g ion/l.

Zone of Immunity

The region of immunity [Fig. 1.15 (bottom)] illustrates how corrosion may be controlled by lowering the potential of the metal, and this zone provides the thermodynamic explanation of the important practical method of cathodic protection (Section 11.1). In the case of iron in near-neutral solutions the potential $E = -0{\cdot}62$ V for immunity corresponds approximately with the practical criterion adopted for cathodically protecting the metal in most environments, i.e. $-0{\cdot}52$ to $-0{\cdot}62$ V (*vs.* S.H.E.). It should be observed, however, that the diagram provides no information on the rate of charge transfer (the current) required to depress the potential into the region of immunity, which is the same ($< -0{\cdot}62$ V) at all values of pH below 9·8. Consideration of curve ℓ for the H_2/H_2O equilibrium shows that as the pH decreases the thermodynamic tendency of water to become reduced to hydrogen increases, and although theoretically cathodic protection is feasible in the acid region it would be economically impracticable, since hydrogen evolution at a very high rate would occur at the potentials required to achieve immunity. Thus cathodic protection is normally confined to near-neutral solutions.

In the case of the Al–H_2O diagram system (Section 4.1, Fig. 4.41) immunity in the near-neutral region can be achieved only at potentials $< -1{\cdot}82$ V, which cannot be attained in practice owing to the hydrogen evolution reaction, which is the thermodynamically preferred process. However, owing to the presence of a surface oxide film the potential of aluminium in practical environments is far more positive than the reversible potential, e.g. the corrosion potential in sea-water is $-0{\cdot}55$ V compared to $E^{\ominus}_{Al^{3+}/Al} = -1{\cdot}7$ V, and cathodic protection may be achieved in practice by making the potential 100 mV more negative than the corrosion potential. This is because aluminium in neutral chloride-containing environments corrodes by pitting, and the criterion of cathodic protection is thus the critical pitting potential (*see* Sections 1.5 and 1.6) and not the zone of immunity of the potential–pH diagram, and similar considerations apply to the cathodic protection of stainless steels.

The Al–H_2O diagram does show, however, the danger that may arise due to an increase in pH when the metal is cathodically protected in near-neutral solutions; indeed, the possibility of alkaline corrosion has seriously limited the use of cathodic protection for aluminium structures.

Zone of Passivity

Although thermodynamics can predict the region of pH and potential in which solid oxides, hydroxides and other compounds are stable, it can provide no other information; thus on the basis of these considerations alone a metal in the passive region should be completely converted to a solid compound by reacting with water with a consequent loss of properties.

Implicit in the concept of passivity is the assumption that the solid compound forms a kinetic barrier between the reactants so that further interaction becomes very slow. Whether this occurs in practice will depend on the position of formation of the oxide (an oxide produced by the sequence $M \rightarrow M^{2+}(aq.) \rightarrow M_2O_3.H_2O$ is likely to precipitate away from the metal surface owing to the mobility of the M^{2+}(aq.) ion and to be non-protective, and an oxide produced directly, e.g. $M \rightarrow M_xO_y$ is likely to form on the metal surface and to be protective), the adhesion of the oxide to the metal, the solubility of the oxide, its cohesion, crystal form, etc. Thus iron in a neutral chloride solution maintained in the region of passivation by dissolved oxygen will corrode owing to the fact that the hydrated oxide $Fe_2O_3.H_2O$ precipitates away from the metal surface and is therefore non-protective. Similarly, metals such as magnesium, aluminium and zinc, which according to the relevant potential–pH diagram are all passive in near-neutral solutions at elevated potentials, can be used as sacrificial anodes in sea-water, since the presence of the chloride ion precludes passivation; in fact in this particular application it is essential to ensure that the metal does not passivate, and in the case of aluminium and zinc, additions of mercury may be used to prevent the formation of a protective film thus facilitating uniform corrosion (*see* Section 11.2).

A number of metals and alloys can be passivated in the acid region at elevated potentials, although this phenomenon is not evident from the pH–potential diagrams, which give the impression that the metal will corrode. Thus iron is passivated in fuming nitric acid, and aluminium in nitric acid at concentrations $> \approx 70\%$; iron, nickel and cobalt can be passivated in sulphuric acid by raising the potential by applying an external source of e.m.f. The reason for this behaviour is that although the passive zone is based on the thermodynamic stability of solid compounds it is possible for these compounds to exist as metastable phases outside the regions defined by thermodynamic data. Under these circumstances the rate of corrosion of the metal will be controlled by the rate of transport of metal cations through the film and by the dissolution of the oxide in the solution (*see* Section 1.4).

The fact that oxides can exist as metastable phases is illustrated by the Ni–H_2O diagram (Fig. 1.18) in which the curves for the various oxides of nickel have been extrapolated into the acid region of Ni^{2+} stability, and this diagram emphasises the fact that nickel can be passivated outside the region of thermodynamic stability of the oxides[11].

The converse situation occurs when complexants are present, and this can be exemplified by the Cu–H_2O–NH_3 system (Fig. 1.19)[12] in which the zones

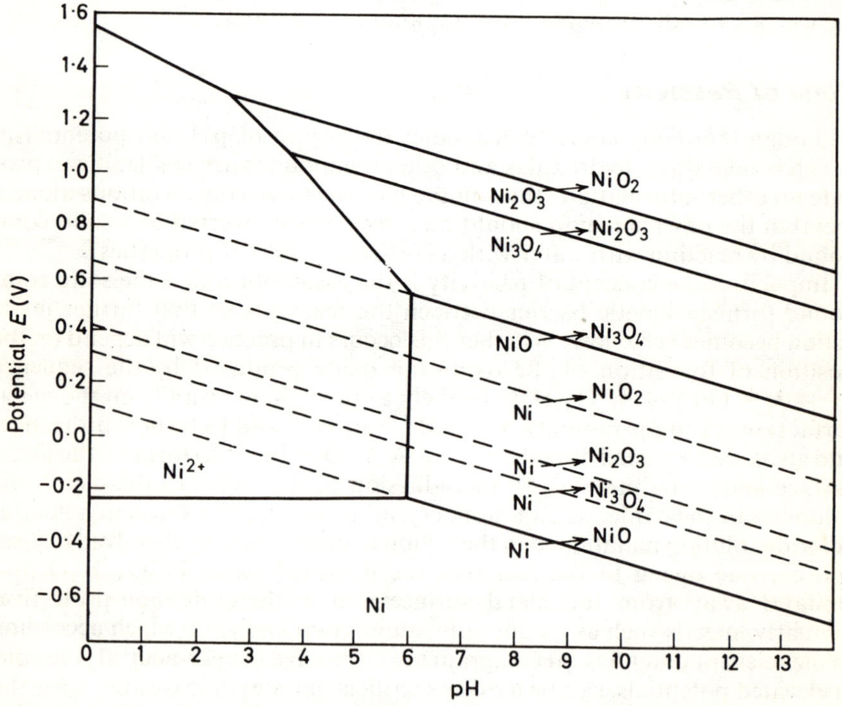

Fig. 1.18 Modified potential–pH diagram for the Ni–H_2O system; the curves showing the stability of the nickel oxides have been extrapolated into the acid region to indicate the formation of metastable oxides (after De Gromoboy and Shreir[11])

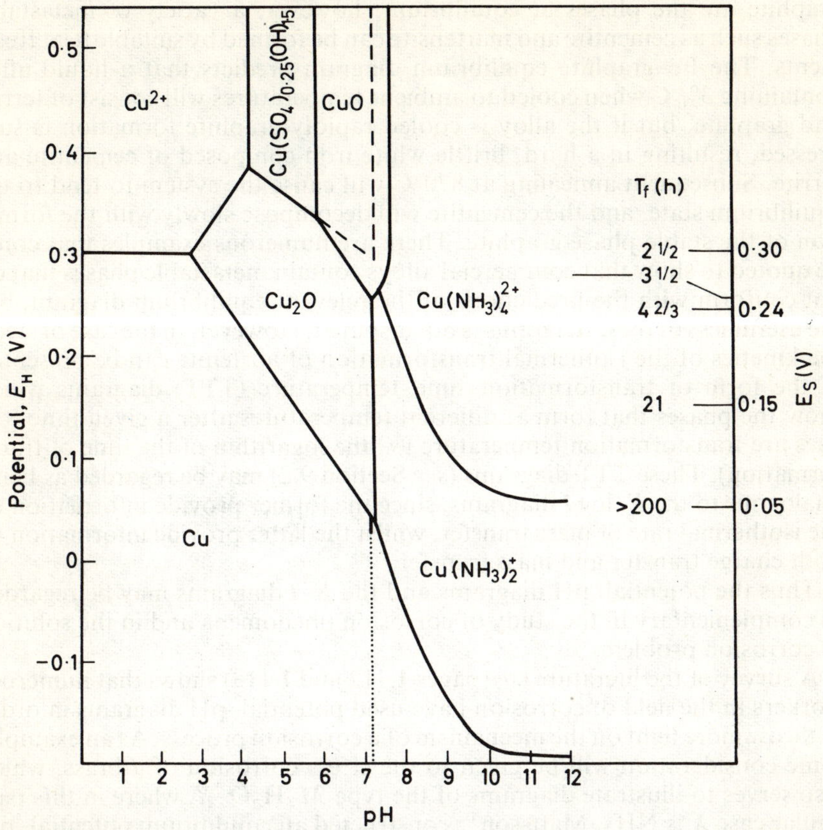

Fig. 1.19 Potential–pH diagram for copper in solutions containing Cu^{2+} and $(NH_4)_2SO_4$ (after Mattsson[12]) with superimposed times to fracture T_f of direct-loaded α-brass wires held at various potentials in the solution of pH 7·2; the specimen without external polarisation had $T_f = 3\frac{1}{2}$ h (after Hoar and Booker[13])

of passivation due to the stability of Cu_2O and CuO are confined to a narrow region of pH, whereas in the Cu–H_2O water system passivation can be achieved at all pHs $> \approx 6{\cdot}8$ (Section 4.2, Fig. 4.10).

Importance of Potential–pH Diagrams in Corrosion Science

Since the remainder of this section will be confined to a consideration of rates of corrosion reactions, it is appropriate to conclude this review of potential–pH diagrams with an assessment of their significance in corrosion science. In this connection, it is relevant to consider the composition–temperature equilibrium phase diagrams of alloy systems, which provide the foundation for the study of the structure of metals and alloys, although it is recognised that these diagrams have serious limitations owing to the fact that many commercial alloys are not in a state of equilibrium. This can be exemplified by the Fe–C equilibrium diagram, in which ferrite (α-Fe) and graphite are the phases at equilibrium; however, a variety of metastable phases such as cementite and martensite can be formed by suitable heat treatments. The Fe–graphite equilibrium diagram predicts that a liquid alloy containing 3% C when cooled to ambient temperatures will consist of ferrite and graphite, but if the alloy is cooled rapidly graphite formation is suppressed, resulting in a hard, brittle white iron composed of cementite and ferrite. Subsequent annealing at 870°C will cause the system to tend to the equilibrium state, and the cementite will decompose slowly with the formation of the stable phase graphite. There are numerous examples that could be quoted to show that commercial alloys contain metastable phases that do not conform with the predictions of the relevant equilibrium diagram, but the usefulness of these diagrams is not disputed. However, in the case of steels the kinetics of the isothermal transformation of austenite can be presented in the form of transformation–time–temperature (TTT) diagrams which show the phases that form at different temperatures after a given time (the axes are transformation temperature *vs.* the logarithm of the time of transformation). These TTT diagrams (*see* Section 9.2) may be regarded as being analogous to the E–$\log I$ diagrams, since the former provide information on the isothermal rate of mass transfer, whilst the latter provide information on both charge transfer and mass transfer.

Thus the potential–pH diagrams and the E–I diagrams may be regarded as complementary in the study of corrosion phenomena and in the solution of corrosion problems.

A survey of the literature (*see* pages **1**.112 and **1**.113) shows that numerous workers in the field of corrosion have used potential–pH diagrams in order to throw more light on the mechanism of a corrosion process. As an example, some consideration will be given to the stress corrosion of α-brass, which also serves to illustrate diagrams of the type M–H_2O–X, where in this particular case X is NH_3. Mattsson[12] constructed an equilibrium potential–pH diagram for copper in solutions of $(NH_4)_2SO_4$ at various pHs (Fig. 1.19), and also studied experimentally the rate of cracking of brass in dilute solutions containing NH_3, Cu^{2+} ions and SO_4^{2-} ions at various pH values of from 2 to 11. It was found that intergranular cracking occurred mainly in the range pH 6·3–7·7, and was most rapid in the range pH 7·1–7·3. The diagram shows that, thermodynamically, Cu_2O formation becomes increasingly easy up to pH 7·3, and although previous workers in this field had observed that a black film of corrosion product (subsequently identified as Cu_2O) accompanied cracking, it was regarded as of little significance. Mattsson suggested that the cuprous oxide stimulated cracking, and this was confirmed by the subsequent work of Hoar and Booker[13] who studied the time to failure (T_f) of stressed brass wires at various pHs and potentials (E_s) corresponding with significant zones in the potential–pH diagram. They found that rapid fracture occurred in the pH range 7·1–7·3, and that at fracture the potential of the brass had risen from 0·15 to 0·25 V, corresponding with the anodic formation of Cu_2O from Cu. At potentials below $-0·05$ V, at which Cu_2O cannot form, cracking was arrested. It is not appropriate here to consider the precise rôle of Cu_2O in the cracking process, but this example does serve to illustrate the usefulness of potential–pH diagrams in providing information that can assist in establishing the mechanism of corrosion phenomena[13–15].

Electrochemical Mechanism of Corrosion

The rate (or kinetics) and form of a corrosion reaction will be affected by a variety of factors associated with the metal and the metal surface (which can range from a planar outer surface to the surface within pits or fine cracks), and the environment. Thus heterogeneities in a metal (*see* Section 1.3) may have a marked effect on the kinetics of a reaction without affecting the thermodynamics of the system; there is no reason to believe that a perfect single crystal of pure zinc completely free from lattic defects (a hypothetical concept) would not corrode when immersed in hydrochloric acid, but it would probably corrode at a significantly slower rate than polycrystalline pure zinc, although there is no thermodynamic difference between these two forms of zinc. Furthermore, although heavy metal impurities in zinc will affect the rate of reaction they cannot alter the final position of equilibrium.

The essential features of the electrochemical mechanism of corrosion were outlined at the beginning of the section, and it is now necessary to consider the factors that control the rate of corrosion of a single metal in more detail. However, before doing so it is helpful to examine the charge transfer processes that occur at the two separable electrodes of a well-defined electrochemical cell in order to show that since the two half reactions constituting the overall reaction are *interdependent*, their rates and extents will be equal.

Electrochemical and Electrolytic Cells

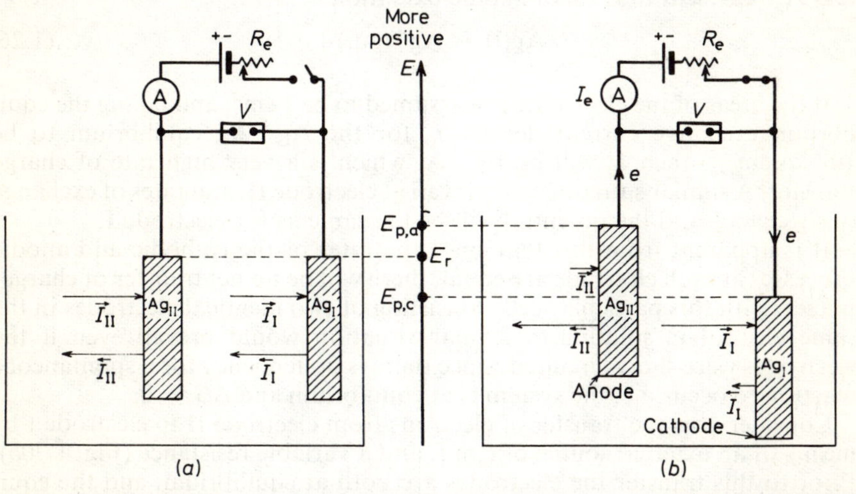

Fig. 1.20 Cell consisting of two reversible Ag^+/Ag electrodes (Ag in $AgNO_3$ solution). The rate and direction of charge transfer is indicated by the length and arrow-head as follows: gain of electrons by $Ag^+ + e \rightarrow Ag \longrightarrow$; loss of electrons by $Ag \rightarrow Ag^+ + e \longleftarrow$. (*a*) Both electrodes at equilibrium and (*b*) electrodes polarised by an external source of e.m.f.; the position of the electrodes in the vertical direction indicates the potential change. (*V*, high-impedance voltmeter; *A*, ammeter; *R*, variable resistance)

An electrochemical cell is a device by means of which the enthalpy (or heat content) of a spontaneous chemical reaction is converted into electrical energy; conversely, an electrolytic cell is a device in which electrical energy is used to bring about a chemical change with a consequent increase in the enthalpy of the system. Both types of cells are characterised by the fact that during their operation charge transfer takes place at one electrode in a direction that leads to the oxidation of either the electrode or of a species in solution, whilst the converse process of reduction occurs at the other electrode.

For simplicity a cell consisting of two identical electrodes of silver immersed in silver nitrate solution will be considered first (Fig. 1.20*a*), i.e. $Ag_I/AgNO_3/Ag_{II}$. On open circuit each electrode will be at equilibrium, and the rate of transfer of silver ions from the metal lattice to the solution and from the solution to the metal lattice will be equal, i.e. the electrodes will be in a state of dynamic equilibrium. The rate of charge transfer, which may be regarded as either the rate of transfer of silver cations (positive charge) in one direction, or the transfer of electrons (negative charge) in the opposite direction, in an electrochemical reaction is the current I, so that for the equilibrium at electrode I

$$\vec{I}_{Ag,I} = \overleftarrow{I}_{Ag,I} = I_{0,I} \qquad \ldots(1.24)$$

In this equation $I_{0,I}$ is the equilibrium exchange current, and the arrow convention adopted is that $\vec{I}_{Ag}$ represents the rate of cathodic reduction

$$Ag^+(aq.)+e \rightarrow Ag(l) \qquad \ldots(1.25)$$

and $\overleftarrow{I}_{Ag}$ represent the rate of anodic oxidation

$$Ag(l) \rightarrow Ag^+(aq.)+e \qquad \ldots(1.26)$$

If the areas of the electrodes are assumed to be 1 cm^2, and taking the equilibrium exchange current density i_0 for the Ag^+/Ag equilibrium to be $10^{-2}\,A\,cm^{-2}$, then I_0 will be 10^{-2} A, which is a very high rate of charge transfer. A similar situation will prevail at electrode II, and rates of exchange of silver ions and the potential will be the same as for electrode I.

It is apparent from this that since the rates of the cathodic and anodic processes at each electrode are equal, there will be no net transfer of charge; in fact, with this particular cell, consisting of two identical electrodes in the same electrolyte solution, a similar situation would prevail even if the electrodes were short-circuited, since there is no tendency for a spontaneous reaction to occur, i.e. the system is at equilibrium and $\Delta G = 0$.

Consider now the transfer of electrons from electrode II to electrode I by means of an external source of e.m.f. and a variable resistance (Fig. 1.20*b*). Prior to this transfer the electrodes are both at equilibrium, and the equilibrium potentials of the metal/solution interfaces will therefore be the same, i.e. $E_I = E_{II} = E_r$, where E_r is the reversible or equilibrium potential. When transfer of electrons at a slow rate is made to take place by means of the external e.m.f., the equilibrium is disturbed and the rates of the charge transfer processes become unequal. At electrode I, $\vec{I}_{Ag,I} > \overleftarrow{I}_{Ag,I}$, and there is now a net *cathodic* reaction (equation 1.25). At electrode II, $\overleftarrow{I}_{Ag,II} > \overrightarrow{I}_{Ag,II}$, and there is now a net *anodic* reaction (equation 1.26).

The rate of transfer of electrons in the external circuit I_e, which is the rate actually measured by the ammeter, is the difference between rates of the dominant or forward reaction and the subsidiary or reverse reaction at each electrode, and it follows that

$$I_e = \overrightarrow{I}_{Ag,I} - \overleftarrow{I}_{Ag,I} = \overleftarrow{I}_{Ag,II} - \overrightarrow{I}_{Ag,II} \qquad ...(1.27)$$

By definition, electrode II at which oxidation is the predominant reaction is the *anode*, whereas electrode I at which reduction is the predominant reaction is the *cathode*. It is apparent that the removal of electrons from Ag_{II} will result in the potential of its interface becoming more positive, whilst the concomitant supply of electrons to the interface of Ag_I will make its potential become more negative than the equilibrium potential:

$$E_{p,c} < E_r < E_{p,a} \qquad ...(1.28)$$

where $E_{p,c}$ and $E_{p,a}$ are the polarised potentials of the cathode and anode, respectively.

If the resistance in the external circuit is decreased sufficiently so that

$$E_{p,c} \ll E_r \ll E_{p,a} \qquad ...(1.29)$$

the rates of the reverse subsidiary reactions may be disregarded and

$$\overrightarrow{I}_{Ag,c} = \overleftarrow{I}_{Ag,a} = I_e \qquad ...(1.30)$$

Concentration Cells

In the previous example of an electrolytic cell the two electrodes were immersed in the same solution of silver nitrate, and the system was therefore thermodynamically at equilibrium. However, if the activities of Ag^+ at the electrodes differ, the system is unstable, and charge transfer will occur in a direction that tends to equalise the activities, and equilibrium is achieved only when they are equal.

Figure 1.21 shows a reversible concentration cell in which $a_{Ag,I} > a_{Ag,II}$, and for the equilibrium $Ag^+ + e \rightleftharpoons Ag$ the e.m.f. of the cell will be

$$E_{r,cell} = \frac{RT}{zF} \ell n \frac{a_{Ag,I}}{a_{Ag,II}} = 0{\cdot}059 \log \frac{a_{Ag,I}}{a_{Ag,II}} \text{ at } 25°C \qquad ...(1.31)$$

in which $E_{r,cell}$ must be positive since $a_{Ag,I} > a_{Ag,II}$. If $a_{Ag,I} = 1{\cdot}0$ and $a_{Ag,II} = 0{\cdot}1$, then $E_{r,cell} = 0{\cdot}059$ V and electrode I will be the cathode [$E_r = (0{\cdot}79 + 0{\cdot}006)$V] and electrode II the anode [$E_r = (0{\cdot}079 - 0{\cdot}060)$V]. If the cell operates spontaneously, charge transfer takes place until $a_{Ag,I} = a_{Ag,II}$ when ΔG becomes zero, i.e. the system is at equilibrium.

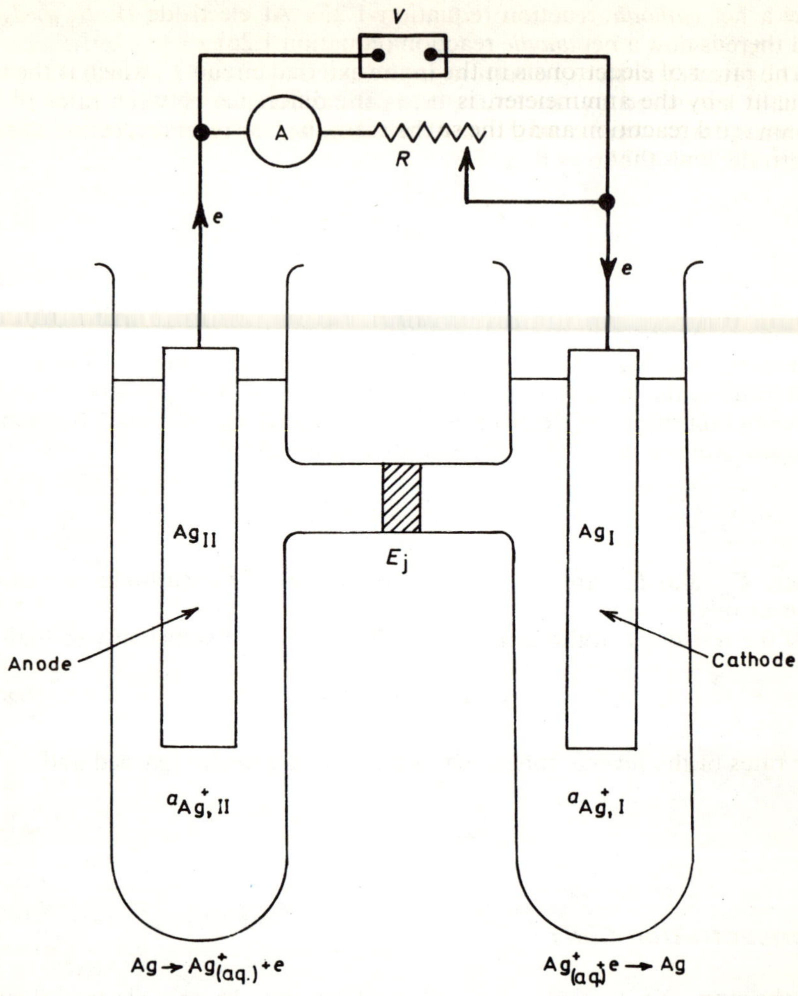

Fig. 1.21 Concentration cell in which $a_{Ag^+,II} < a_{Ag^+,I}$ so that charge transfer occurs spontaneously and proceeds until the activities are equal (E_j is the liquid junction potential at the sintered glass plug that is used to minimise mixing of the two solutions)

Rates and Extents

The current I is the rate of charge transfer, and a rate of $1\ A \equiv 1\ Cs^{-1}$. The charge on the electron is $1{\cdot}602\,0 \times 10^{-19}$ C, and since 1 mol of an element contains $6{\cdot}023\,5 \times 10^{23}$ atoms (Avogadro's number) the cathodic reduction of a univalent metal ion M^+ will require $6{\cdot}023\,5 \times 10^{23} \times 1{\cdot}602\,0 \times 10^{-19} = 96\,494$ C. This statement is essentially Faraday's law and 1 faraday $= 96\,494\ C \approx 96\,500$ C.

In the more general case of the anodic oxidation or cathodic reduction of a species by change transfer, zF coulombs will be required for 1 mol where

z is the number of electrons required to carry out one act of the electron transfer process. Thus the *rate* of charge transfer of an electrochemical reaction is given by

$$k_e = \frac{I}{zF} \quad (\text{mol s}^{-1}) \qquad \ldots(1.32)$$

or

$$k_e = \frac{IM}{zF} \quad (\text{kg s}^{-1}) \qquad \ldots(1.33)$$

where M is the molar mass (kg mol^{-1}) of the species involved. In these examples of cells the areas of the electrodes have been disregarded, but in electrode kinetics it is the rate per unit area of the electrode that is of significance. The rate per unit area is the current density i, and

$$i = IS^{-1} \qquad \ldots(1.34)$$

where S is the area, and i therefore has units of A cm^{-2}, A m^{-2}, mA cm^{-2}, etc. Thus equation 1.33 can be expressed in terms of a rate per unit area

$$k_e = \frac{iM}{zF} \quad (\text{kg cm}^{-2}\,\text{s}^{-1}) \qquad \ldots(1.35)$$

where i is in A cm^{-2}. (Note that although the metre is the recommended S.I. unit, the centimetre is so widely used that it will be retained in certain units in this book.)

The *extent* of an electrochemical reaction is the quantity of charge Q (coulombs) transferred in a given time t, and

$$Q = It \qquad \ldots(1.36)$$

where t is the time (s). Since 1 mol of charge transfer requires zF coulombs

$$Q = \frac{It}{zF} \quad (\text{mol}), \quad \text{or } Q = \frac{it}{zF} \quad (\text{mol cm}^{-2}) \qquad \ldots(1.37)$$

alternatively,

$$Q = \frac{ItM}{zF} \quad (\text{kg}), \quad \text{or } Q = \frac{itM}{zF} \quad (\text{kg cm}^{-2}) \qquad \ldots(1.38)$$

It is apparent (Fig. 1.21) that at potentials removed from the equilibrium potential (*see* equation 1.30) the rate of charge transfer of (*a*) silver cations from the metal to the solution (anodic reaction), (*b*) silver aquo cations from the solution to the metal (cathodic reaction) and (*c*) electrons through the metallic circuit from anode to cathode, are equal, so that any one may be used to evaluate the rates of the others. The rate is most conveniently determined from the rate of transfer of electrons in the metallic circuit (the current I) by means of an ammeter, and if I is maintained constant it can also be used to evaluate the extent. A more precise method of determining the quantity of charge transferred is the coulometer, in which the extent of a single well-defined reaction is determined accurately, e.g. by the quantity of metal electrodeposited, by the volume of gas evolved, etc. The reaction $Ag^+(\text{aq.}) + e = Ag$ is utilised in the silver coulometer, and provides one of the most accurate methods of determining the extent of charge transfer.

Partial Cathodic and Anodic Reactions

The reversible cells described are characterised by the fact that the same charge transfer process $Ag^+(aq.) \rightleftharpoons Ag(l)$ takes place at each electrode. However, this type of cell is not typical, and the fact that the exchange processes usually differ can be exemplified by considering an electrolytic cell consisting of two platinum electrodes immersed in a solution of deoxygenated Na_2SO_4, in which water is decomposed into hydrogen and oxygen

$$H_2O \rightarrow H_2 + \tfrac{1}{2}O_2$$

by the interdependent reactions

$$2H_2O + 2e = H_2 + 2OH^- \quad \text{(cathode)} \qquad \ldots(1.39)$$

and

$$H_2O = \tfrac{1}{2}O_2 + 2H^+ + 2e \quad \text{(anode)} \qquad \ldots(1.40)$$

In contrast to the cell consisting of two reversible Ag^+/Ag electrodes, charge transfer cannot occur until the e.m.f. of the equivalent reversible cell is exceeded ($>1{\cdot}23$ V)*, and in view of the high overpotential for oxygen evolution the rate will not be significant until the e.m.f. is more than about 1·6 V. If the deoxygenated Na_2SO_4 solution is replaced by an oxygenated sodium chloride solution, the following *additional* reactions are also possible

$$O_2 + 2H_2O + 2e = 2OH^- \quad \text{(cathode)} \qquad \ldots(1.41)$$

$$2Cl^- \rightarrow Cl_2 + 2e \quad \text{(anode)} \qquad \ldots(1.42)$$

It is not appropriate here to consider the kinetics of the various electrode reactions, which in the case of the oxygenated NaCl solution will depend upon the potentials of the electrodes, the pH of the solution, activity of chloride ions, etc. The significant points to note are that (*a*) an anode or cathode can support more than one electrode process and (*b*) the sum of the rates of the *partial cathodic reactions* must equal the sum of the rates of the *partial anodic reactions*. Since there are four exchange processes (equations 1.39–1.42) there will be eight partial reactions, but if the reverse reactions are regarded as occurring at an insignificant rate then

$$\vec{I}_{O_2} + \vec{I}_{H_2} = \overleftarrow{I}_{Cl_2} + \overleftarrow{I}_{O_2} = I_e \qquad \ldots(1.43)$$

This leads to the fundamental concept that irrespective of the number of electrode processes or whether they occur on one or more than one electrode surface

$$\Sigma I_c = \Sigma I_a \qquad (C\,s^{-1}) \qquad \ldots(1.44)$$

i.e. during charge transfer the sum of the partial cathodic currents must equal the sum of the partial anodic current.

From Faraday's law

$$\sum \frac{I_c}{zF} = \sum \frac{I_a}{zF} \qquad (\text{mol}\,s^{-1}) \qquad \ldots(1.45)$$

which means that the sum of the cathodic processes must equal the sum of the anodic processes with respect to both the *rate* at any instant of time and the *extent* after any period of time. This law applies equally to these cells, in which

*The value of 1·23 V follows from Table 1.8 and Fig. 1.15 (bottom).

the electrodes are separable, and to those corrosion cells in which the electrodes cannot be distinguished physically (*see* Table 1.A1).

Electrode Area and Current Density

It is now necessary to consider equation 1.45 in terms of the rates per unit area of the electrode surfaces, which may be equal or unequal depending on circumstances.

When the area of the cathode equals the areas of the anode, equation 1.44 is applicable, and I can be replaced by i, the current density:

$$\Sigma i_c = \Sigma i_a \qquad (\mathrm{C\,cm^{-2}\,s^{-1}}) \qquad \ldots(1.46)$$

and

$$\sum \frac{i_c}{zF} = \sum \frac{i_c}{zF} \qquad (\mathrm{mol\,cm^{-2}\,s^{-1}}) \qquad \ldots(1.47)$$

which may be regarded as the criterion for the *uniform corrosion* of a single metal. However, if the area of the anode is smaller than that of the cathode, then $i_a > i_c$ and

$$\Sigma i_a > \Sigma i_c \qquad \ldots(1.48)$$

whereas if the cathode area is smaller, $i_c > i_a$ and

$$\Sigma i_c > \Sigma i_a \qquad \ldots(1.49)$$

and it follows that the criterion for *localised attack* (or pitting) will be

$$\frac{\Sigma i_a}{\Sigma i_c} > 1 \qquad \ldots(1.50)$$

and that the greater the magnitude of this ratio the more intense will be the attack. Equation 1.50 is referred to as the pitting ratio. The current density i requires a knowledge of both the current I and the area of the electrode S, and the latter is seldom equal to the geometrical or superficial area. In fact, it is possible to distinguish the following types of areas in relation to electrode processes that take place on metal surfaces: *geometrical*, *true*, *active* and *effective*.

In uniform corrosion the *superficial* or *geometrical* area of the metal is used to evaluate both the anodic and cathodic current density, although it might appear to be more logical to take half of that area. However, surfaces are seldom smooth and the true surface area may be twice to three times that of the geometrical area (a cleaved crystal face or an electropolished single crystal would have a true surface area that approximates to its superficial area). It follows, therefore, that the true current density is smaller than the superficial current density, but whether the area used for calculating i_a and i_c is taken as either equal to or half of the superficial area, is unimportant compared to the fact that they are equal. Furthermore, the number of metal atoms at the metal surface that are active in the processes of metal dissolution or cathodic reduction are usually far less than the total number available. Thus Hoar and Notman[16] have shown during the anodic dissolution of nickel that the current density at any given potential is increased by a factor of 10 by cold working the annealed nickel. Similar considerations apply to the hydrogen evolution reaction, in which the coverage of the atoms on a metal surface with adsorbed hydrogen atoms can range from very small to about 100%, depending on the mechanism of the reaction (*see* Section 9.1). Thus the active surface area (the number of atomic sites that participate in the electrode process) may be appreciably less than the total number of sites available.

Finally, it is important to point out that although in localised corrosion the anodic and cathodic areas are physically distinguishable, it does not follow that the total geometrical areas available are actually involved in the charge transfer process. Thus in the corrosion of two dissimilar metals in contact (bimetallic corrosion) the metal of more positive potential (the predominantly cathodic area of the bimetallic couple) may have a very much larger area than that of the predominantly anodic metal, but only the area adjacent to the anode may be effective as a cathode. In fact in a solution of high resistivity the effective areas of both metals will not extend appreciably from the interface of contact. Thus the effective areas of the anodic and cathodic sites may be much smaller than their geometrical areas.

It follows from equation 1.45 that the corrosion rate of a metal can be evaluated from the rate of the cathodic process, since the two are faradaically equivalent; thus either the rate of hydrogen evolution or of oxygen reduction may be used to determine the corrosion rate, providing no other cathodic process occurs. If the anodic and cathodic sites are physically separable the rate of transfer of charge (the current) from one to the other can also be used, as, for example, in evaluating the effects produced by coupling two dissimilar metals. There are a number of examples quoted in the literature where this has been achieved, and reference should be made to the early work of Evans[17] who determined the current and the rate of anodic dissolution in a number of systems in which the anodes and cathodes were physically separable.

More recently, Fontana and Greene[18] measured the current between a pit in stainless steel and the surrounding metal; the pit was allowed to form, and cut out from the surrounding metal (the cathode), its edge was insulated and it was then replaced in the hole with a suitable connection for measuring the current flow between the pit and the surrounding metal. These workers showed that under certain conditions i_a was about a thousand times i_c.

When the anodic and cathodic sites are inseparable the corrosion current cannot be determined directly by an ammeter, but it can be evaluated electrochemically by the *linear polarisation* technique (*see* Sections 20.1–20.3).

Electrochemical Cells and Corrosion Cells

One of the most well-known electrochemical cells that is used for the conversion of chemical energy into electrical energy is the Daniell cell

$$Zn|ZnSO_4(aq.)|CuSO_4(aq.)|Cu \qquad ...(1.51)$$

in which the spontaneous reaction

$$Cu^{2+}(aq.) + Zn \rightarrow Zn^{2+}(aq.) + Cu \qquad ...(1.52)$$

takes place. The half reactions that constitute the overall reaction are

$$Zn \rightarrow Zn^{2+}(aq.) + 2e \quad \text{(anodic reaction)}$$

$$Cu^{2+}(aq.) + 2e \rightarrow Cu \quad \text{(cathodic reaction)}$$

and electrons are transferred from the zinc to the copper through the metallic circuit.

During the operation of the cell (or during the direct interaction of zinc metal and cupric ions in a beaker) the zinc is oxidised to Zn^{2+} and corrodes, and the Daniell cell has been widely used to illustrate the electrochemical mechanism of corrosion. This analogy between the Daniell cell and a corrosion cell is perhaps unfortunate, since it tends to create the impression that corrosion occurs only when two dissimilar metals are placed in contact and that the electrodes are always physically separable. Furthermore, although reduction of $Cu^{2+}(aq.)$ does occur in certain corrosion reactions it is of less importance than reduction of H_3O^+ ions or dissolved oxygen.

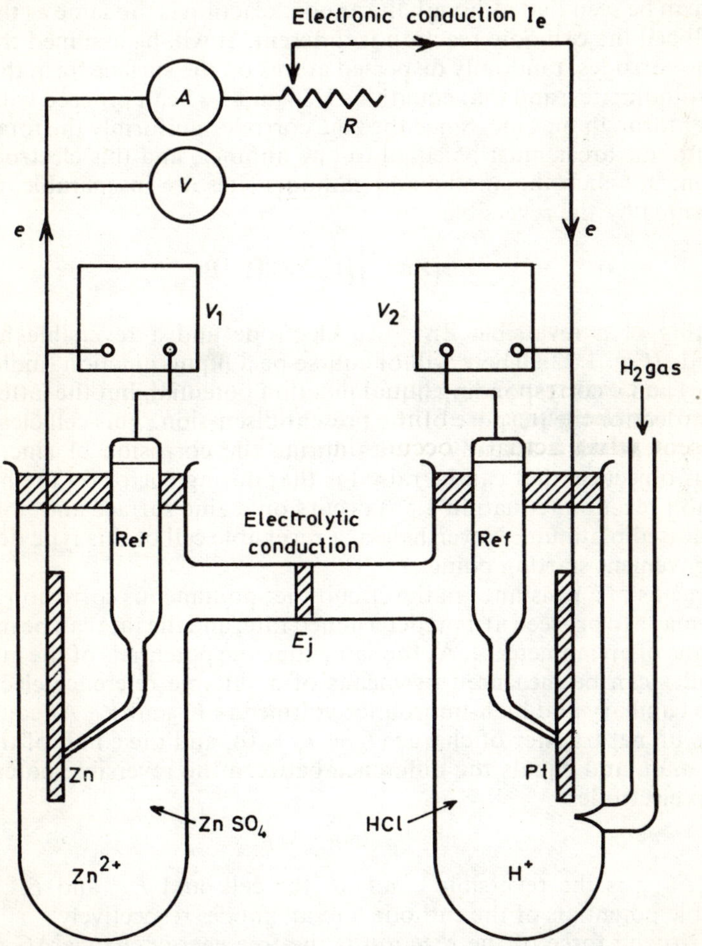

Fig. 1.22 Spontaneous corrosion of zinc in acid illustrated by the reversible cell $Zn|Zn^{2+}|H_3O^+, H_2|Pt$. The individual potentials of the electrodes are determined by a reference electrode (Ref) and a Luggin capillary to minimise the *IR* drop in the solution

For these reasons a somewhat different approach will be adopted here, and an attempt will be made to show how a corrosion reaction may be represented by a well-defined reversible electrochemical cell, although again there are a number of difficulties. Consider the corrosion of metallic zinc in a reducing acid

$$Zn + 2H_3O^+ \rightarrow Zn^{2+}(aq.) + H_2 + H_2O \quad ...(1.53)$$

which occurs spontaneously when zinc is immersed in hydrochloric acid; the dissolution of the zinc is usually quite uniform so that there is no means by which anodic and cathodic sites can be identified physically. The half reactions involved are

$$Zn \rightarrow Zn^{2+} + 2e \quad \text{(anodic reaction)} \quad ...(1.54)$$

$$2H^+ + 2e \rightarrow H_2 \quad \text{(cathodic reaction)} \quad ...(1.55)$$

and it can be seen that although the anodic reaction is the same as that in the Daniell cell the cathodic reaction is different. It will be assumed that when the zinc corrodes, randomly dispersed atoms on the surface form the anodic and cathodic sites, and that equations 1.54 and 1.55 can proceed with charge transfer through the zinc. Since the zinc corrodes uniformly the total anodic and cathodic areas must be equal to one another, and this electrochemical reaction, in which the anodic and cathodic sites are inseparable, could be represented by the reversible cell

$$Zn|Zn^{2+}\Big|H_3O^+, H_2|Pt \quad ...(1.56)$$

consisting of a reversible Zn^{2+}/Zn electrode and a reversible hydrogen electrode (Fig. 1.22). There will of course be a liquid junction (indicated by the line) and a corresponding liquid junction potential, but the latter will be disregarded for the purpose of the present discussion. This cell clearly does not represent what actually occurs during the corrosion of zinc, and an obvious objection that can be raised is that during corrosion the hydrogen evolution reaction (equation 1.55) occurs on a zinc surface and not on one of platinised platinum. Nevertheless, a reversible cell of this type does serve as a convenient starting point.

By means of a resistance in the circuit the spontaneous corrosion reaction can be made to proceed at a predetermined rate, and the rate can be measured by means of an ammeter A. At the same time the potentials of the individual electrodes can be measured by means of a suitable reference electrode, a Luggin capillary and high-impedance voltmeters V_1 and V_2. At equilibrium there is no net transfer of charge ($I_c = I_a = 0$), and the e.m.f. of the cell is a maximum and equals the difference between the reversible potentials of the two electrodes

$$E_{r,cell} = E_{r,c} - E_{r,a} \quad ...(1.57)$$

where $E_{r,cell}$ is the reversible e.m.f. of the cell, and $E_{r,c}$ and $E_{r,a}$ are the reversible potentials of the cathode and an anode, respectively.

The driving force of the reaction is the free energy change ΔG which is related to the reversible or equilibrium e.m.f. of the cell by the relationship

$$\Delta G = -zFE_{r,cell} \quad ...(1.58)$$

and, as emphasised in Appendix 9.1B, both ΔG and $E_{r,cell}$ are thermodynamic quantities that provide a means of evaluating the equilibrium constant K, and hence the activities of the reactants and products when the reaction comes to equilibrium (Table 1.9).

If now the resistance in the external circuit is decreased slightly the reaction will proceed at a finite rate, and the electrodes constituting the cell will become mutually polarised and displaced from their equilibrium values, i.e. the polarised potential of the anode (Zn^{2+}/Zn) will become more positive, whilst that of the cathode ($2H^+/H_2$) will become more negative (Fig. 1.23).

The displacement of the potential of an electrode from its reversible value is the *overpotential* η, and

$$\eta = E_p - E_r \qquad \text{...(1.59)}$$

where E_p is the polarised potential and E_r the reversible or equilibrium potential. Since $E_{p,c} < E_{r,c}$ (more negative)

$$\eta_c = E_{p,c} - E_{r,c} < 0 \qquad \text{...(1.60)}$$

and the cathode overpotential η_c is always *negative*, although $E_{p,c}$ may be positive or negative depending on the sign of E_r and the magnitude of η_c. (If the potential of a Cu^{2+}/Cu electrode, where $E_r = 0{\cdot}34$ V, is polarised

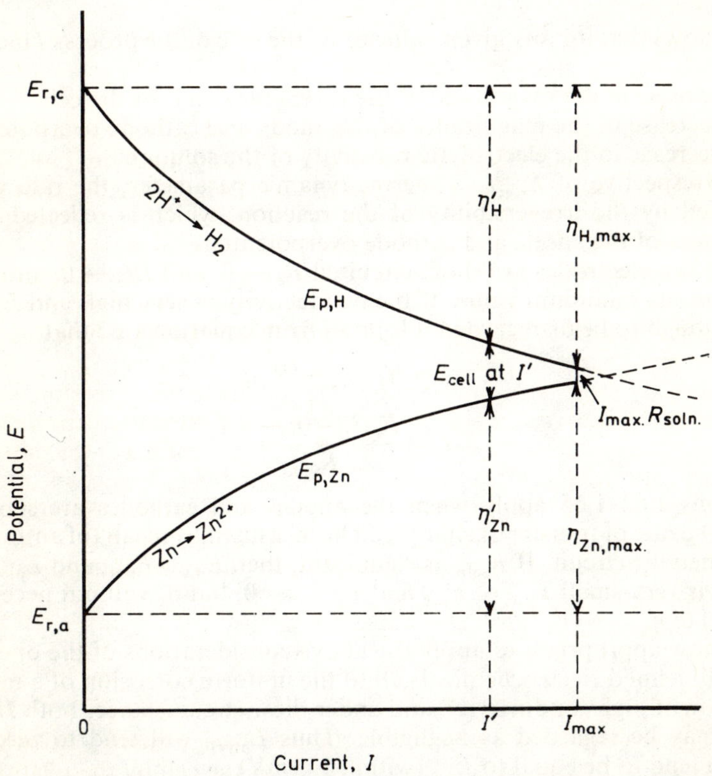

Fig. 1.23 E–I curves for the corrosion of zinc (*see* Fig. 1.22) showing the relationship between E_r, E_p and η for the cathodic and anodic half reactions

cathodically to 0·32 V, then $\eta_c = -0{\cdot}020$ V; if the same procedure is adopted with a Zn^{2+}/Zn electrode where $E_r = -0{\cdot}76$ V, E_p will be $-0{\cdot}78$ V, but η_c will still be $-0{\cdot}020$ V.) Conversely, $E_{p,a} > E_{r,a}$ (more positive), and since

$$\eta_a = E_{p,a} - E_{r,c} > 0 \qquad \text{...(1.61)}$$

the anode overpotential is always *positive*.

It should be noted that whereas E is always relative to a specified reference electrode this will not apply to the overpotential (*see* equation 1.59).

As the rate of charge transfer is increased by decreasing the resistance R_e in the circuit, the magnitudes of η_c and η_a increase thus decreasing the magnitude of the polarised e.m.f. of the cell $E_{p,cell}$. It follows from Fig. 1.23 that for any given rate of charge transfer I

$$E_{p,cell} = E_{r,cell} - (\eta_a + \eta_c + IR_{soln.}) \qquad \text{...(1.62)}$$

where η_a and η_c are the *magnitudes* of the overpotentials (the negative sign for η_c must be omitted) corresponding to the rate I, and $R_{soln.}$ is the electrolytic resistance of the solution.

Since $$E_{p,cell} = IR_e$$

$$I = \frac{E_{r,cell} - (\eta_a + \eta_c + IR_{soln.})}{R_e} \qquad \text{...(1.63)}$$

which shows that for any given value of R_e the rate of the process I increases with

(*a*) Increase in the magnitude of the reversible e.m.f. of the cell.
(*b*) Decrease in the magnitudes of the anode and cathode overpotentials.
(*c*) Decrease in the electrolytic resistivity of the solution.

Thus, irrespective of $E_{r,cell}$, a thermodynamic parameter, the rate will be controlled by the irreversibility of the reaction, which is reflected in the magnitudes of the anode and cathode overpotentials.

If the two electrodes are short-circuited $R_e \to 0$, and $IR_e \to 0$, and $E_{p,cell}$ will attain its minimum value. If the conductivity is very high and $E_{p,cell}$ is small enough to be disregarded it follows from equation 1.62 that

$$E_{r,cell} = \eta_a + \eta_c + IR_{soln.} \qquad \text{...(1.64)}$$

and $$I = \frac{E_{r,cell} - (\eta_a + \eta_c)}{R_{soln.}} \qquad \text{...(1.65)}$$

Equations 1.62–1.65 apply when the anodes and cathodes are separable so that the rate of transfer of charge can be measured by means of an ammeter in the metallic circuit. If $R_{soln.}$ is significant, then $E_{p,c} > E_{p,a}$, and $E_{p,cell} > 0$; if $R_{soln.}$ is very small $E_{p,c} \sim E_{p,a}$ and $E_{p,cell} \to 0$, but η_c will not necessarily be equal to η_a.

It is now appropriate to apply the above considerations of the operation of a well-defined electrochemical cell to the uniform corrosion of a metal in a solution of high conductivity, and under these circumstances both IR_e and $IR_{soln.}$ may be regarded as negligible. Thus $E_{p,cell}$ will tend to zero, and $E_{p,c}$ will tend to be equal to $E_{p,a}$ (within 1–2 mV)

$$\therefore \qquad E_{p,c} = E_{p,a} = E_{corr.} \qquad \text{...(1.66)}$$

where $E_{corr.}$ is the *corrosion potential*, and from equation 1.64

$$E_{r,cell} = \eta_c + \eta_a \qquad \ldots(1.67)$$

The above considerations show that the rate of a corrosion reaction is dependent on both the thermodynamic parameter $E_{r,cell}$ and the kinetic parameters η_a and η_c. It is also apparent that (*a*) the potential actually measured when corrosion reaction occurs on a metal surface is a *mixed*, *compromise* or *corrosion* potential whose magnitude depends on $E_{r,cell}$ and on the $E_{p,c}$–I and $E_{p,a}$–I relationships, and (*b*) direct measurement of I_e is not possible when the electrodes are inseparable.

Overpotentials[19,20]

The various types of overpotentials are dealt with in more detail in Section 9.1 but it is appropriate here to outline the significant factors in relation to their importance in controlling the rate of corrosion reaction.

Activation overpotential η_A For any given electrode process under specified conditions, charge transfer at a finite rate will involve an activation overpotential η_A, which provides the activation energy required for the reactant to surmount the energy barrier that exists between the energy states of the reactant and product. Some reactions are kinetically easy (e.g. $Ag^+(aq.) + e \rightarrow Ag$) and thus require only a small activation overpotential, whilst others (e.g. $H_3O^+ + e \rightarrow \frac{1}{2}H_2$ on metals such as Hg, Pb and Zn) are kinetically difficult and high activation overpotentials are required. Most electrode processes involve more than one step; one of them is usually more sluggish than the others and is thus rate determining, and the activation energy is required, therefore, to maintain the rate of the rate-determining step (r.d.s.), since the other steps may be regarded as being at equilibrium. The activation energy $E^{\ddagger}$ is given by

$$E^{\ddagger} = zF\eta_A \qquad \ldots(1.68)$$

where $E^{\ddagger}$ is in joules per mole and z is the number of electrons involved in one act of the rate-determining step.

The activation overpotential, and hence the activation energy, varies exponentially with the rate of charge transfer per unit area of electrode surface, as defined by the well-known Tafel equation

$$\eta_A = a + b \log i \qquad \ldots(1.69)$$

where i is the current density, and a and b are the Tafel constants which vary with the nature of the electrode process and with the nature of the solution. Thus η_A will be linearly related to $\log i$ at overpotentials greater than 0·010 V, and the position and slope of the curve will be dependent on the magnitudes of a and b, which are in turn dependent on the equilibrium exchange current density i_0, the transfer coefficient α and the number of electrons z involved in one act of the rate-determining step. The Tafel equation for a cathodic process can be expressed (*see* Section 9.1) in the form

$$\eta_{A,c} = \frac{RT}{\alpha zF}\ln i_0 - \frac{RT}{\alpha zF}\ln i_c \qquad \ldots(1.70)$$

and since $2{\cdot}303\ RT/F \ln x = 0{\cdot}059 \log x$ at 25°C

$$\eta_{A,c} = \frac{0{\cdot}059}{\alpha_c z}\log i_0 - \frac{0{\cdot}059}{\alpha_c z}\log i_c \text{ at } 25°C \qquad \ldots(1.71)$$

where $\eta_{A,c}$ is the activation overpotential of the cathodic process. Similarly, the activation overpotential of an anodic process is given by

$$\eta_{A,a} = -\frac{0{\cdot}059}{\alpha_a z}\log i_0 + \frac{0{\cdot}059}{\alpha_a z}\log i_a \text{ at } 25°C \qquad \ldots(1.72)$$

It is evident from these expressions that since in the Tafel region i (the current density actually determined) must be greater than i_0 (the equilibrium exchange current density), the signs of the overpotentials will conform to equations 1.60 and 1.61, i.e. $\eta_{A,c}$ will be negative and $\eta_{A,a}$ will be positive.

Furthermore, the smaller the magnitude of i_0 the greater the magnitude of η_A and the lower the rate of the electrode process at any given polarised

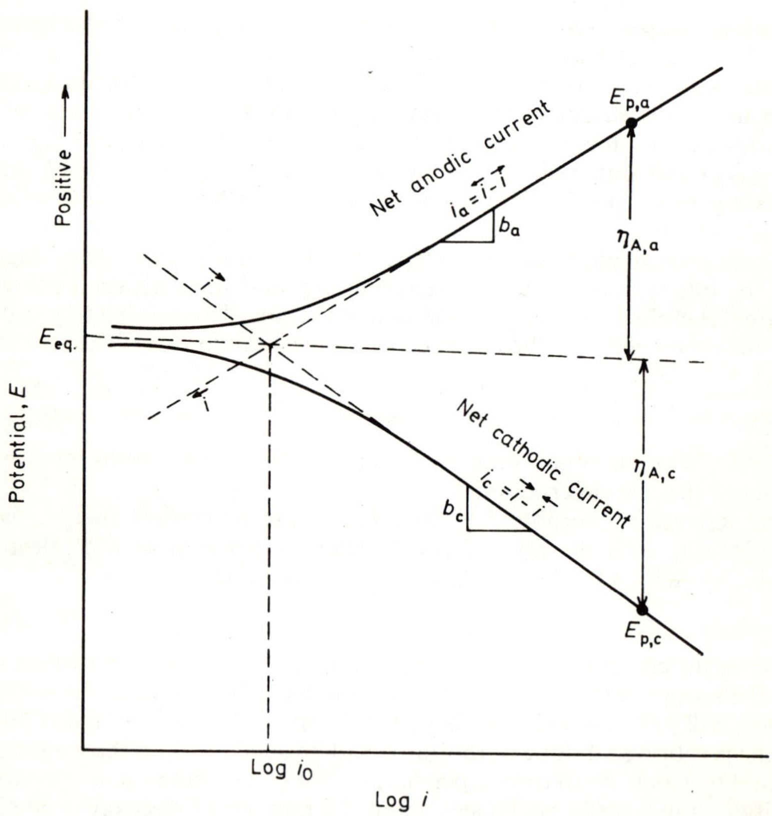

Fig. 1.24 Tafel lines for a single exchange process. The following should be noted: (*a*) linear E–$\log i$ curves are obtained only at overpotentials greater than 0·052 V (at less than 0·052 V E *vs.* i is linear); (*b*) the extrapolated anodic and cathodic E–$\log i$ curves intersect at i_0 the equilibrium exchange current density; and (*c*) i_a and i_c the anodic and cathodic current densities actually measured are the differences between $\overleftarrow{i}$ and $\overrightarrow{i}$, and $\overrightarrow{i}$ and $\overleftarrow{i}$, respectively

potential E_p. Thus the equilibrium exchange current density i_0 is the most significant parameter in controlling the rate of a corrosion process in which one (or both) of the electrode processes involve an appreciable activation energy. Figure 1.24 shows the cathodic and anodic Tafel lines for a single exchange process at an electrode, in which i_a and i_c are the anodic and cathodic current densities actually measured.

Transport (diffusion or concentration) overpotentials η_T Previous considerations have been confined to the kinetics of charge transfer but the rate of an electrode reaction will also depend on mass transfer, i.e. the rate at which the reactant is transported to the surface of the electrode and the rate at which the product is transported away from the electrode. Transport through the solution to and from the metal surface occurs by diffusion, ionic migration (transport of electrical charge through the solution) and convection, and of these diffusion through the thin static layer of solution adjacent to the metal surface, the diffusion layer δ, is usually of the greatest significance. However, this is not always the case in practical systems, particularly where dissolved oxygen is the cathodic reactant, and in certain circumstances the rate of diffusion through the bulk solution to the metal/solution interface may be rate determining.

The limiting current density (the maximum possible rate/unit area under the conditions prevailing) for a cathodic process is given by

$$i_L = \frac{DzFc}{\delta(1-n_+)} \times 10^{-3} \qquad \ldots(1.73)$$

where i_L is the limiting current density (A cm^{-2}), z is the number of electrons required for one step of the electrode process involving 1 mol of the cathode reactant, D is the diffusion coefficient, c is the concentration of the reactant (mol dm^{-3}), δ is the thickness of the diffusion layer (cm), and n_+ the transport number of the cation; the term $(1-n_+)$ can be neglected if ions other than the species involved in the electrode process are responsible for ionic migration.

The relation between transport overpotential and current density for a cathodic reaction is given by

$$\eta_T = \frac{RT}{zF}\ln\left(1-\frac{i}{i_L}\right) = \frac{0{\cdot}059}{z}\log\left(1-\frac{i}{i_L}\right) \text{ at } 25^\circ\text{C} \qquad \ldots(1.74)$$

and it is evident that the smaller i_L the greater the magnitude of the overpotential due to transport. Unlike activation overpotential, transport overpotential is not controlled by the kinetics of charge transfer, and the magnitude of η_T will be the same for any cation (providing z, D_i and c are the same) and any metal surface. Thus the rate-controlling parameter in transport overpotential is i_L, and it will be seen that any factor in equation 1.73 that causes i_L to increase will result in an increase in the corrosion rate, providing the latter is solely determined by the kinetics of the cathodic process. Figure 1.25*a* shows the relationship between η and $\log i$ when the rate is controlled solely by transport, and Fig. 1.25*b* shows the relationship when both transport and activated charge transfer are involved. It should be noted that whereas in transport overpotential z is the number of electrons involved in one act of the reaction, in activation overpotential z is the number of electrons involved in one act of the rate-determining step.

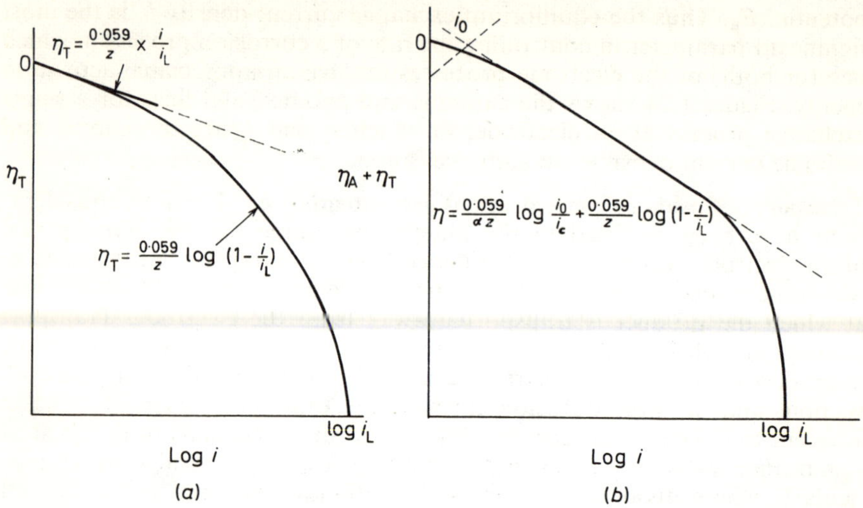

Fig. 1.25 η *vs.* $\log i$ curves for a cathodic reaction (*a*) when the rate is solely controlled by transport and (*b*) when both transport and activated charge transfer are rate determining. (Derivations of the relationships are provided in Section 1.9)

Resistance overpotential η_R Since in corrosion the resistance of the metallic path for charge transfer is negligible, resistance overpotential η_R is determined by factors associated with the solution or with the metal surface. Thus resistance overpotential may be defined as

$$\eta_R = I(R_{soln.} + R_f) \qquad \text{...(1.75)}$$

where $R_{soln.}$ is the electrical resistance of the solution, which is dependent on the electrical resistivity (Ω cm) of the solution and the geometry of the corroding system, and R_f is the resistance produced by films or coatings formed on or applied to the surface of the sites. Thus, in addition to the resistivity of the solution, any insulating film deposited either at the cathodic or anodic sites that restricts or completely blocks contact between the metal and the solution will increase the resistance overpotential, although the resistivity of the solution is unaffected. This applies particularly to the deposition of $CaCO_3$ [and $Mg(OH)_2$] at the cathodic sites during corrosion in hard waters due to the increase in pH produced by the cathodic process, and since the anodic and cathodic sites are usually close together the calcareous scale will also block the anodic sites, and thus decrease the corrosion rate.

Similar considerations also apply to the dielectric films formed on the metal surface during anodising, and, for example, in the case of the valve metals (Al, Ti, Ta, Nb, etc.) *IR* drops of hundreds of volts may be produced by the anodic oxide film formed on the metal surfaces. Paint films applied to a metal surface also exert resistance control (*see* Section 15.3).

All these types of polarisation will be present to a greater or lesser extent in most corrosion reactions, but if one is more significant than the others it will control the rate of the reaction. This leads to a classification of corrosion reactions according to whether the cathodic or anodic reaction is rate determining (cathodic control or anodic control), which can be made even more specific by including such terms as 'activation', 'transport' and 'resistance'. Thus the slow corrosion of zinc in solutions of reducing acids is controlled by the high activation energy required for the hydrogen evolution reaction (cathodic activation control), whereas the rapid corrosion of the metal in concentrated sodium hydroxide is controlled by transport of OH^- and ZnO_2^- to and away from the metal/solution interface, respectively (anodic transport control).

Graphical Methods of Expressing Corrosion Rates

The graphical method of showing how the corrosion rate $I_{corr.}$ is dependent on the extent of the polarisation of the anodic and cathodic reactions constituting the corrosion reaction was due originally to Evans[17,22] who used

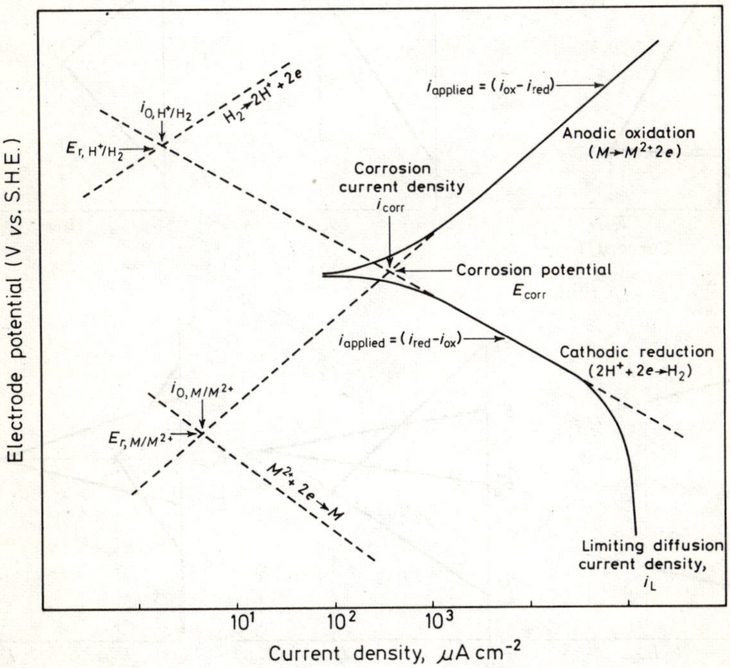

Fig. 1.26 *E vs.* log *i* curves for the corrosion of a metal in a reducing acid in which there are two exchange processes (c.f. Fig. 1.24) involving oxidation of M M^{2+} and reduction of H^+ H_2. Note that (*a*) the reverse reactions for exchange process are negligible at potentials removed from E_r, (*b*) the potential actually measured is the corrosion potential $E_{corr.}$, which is a mixed potential, and (*c*) the *E vs.* $i_{appl.}$ curves (where $i_{appl.}$ is the applied current density) when extrapolated intersect at $E_{corr.}$

the co-ordinates E and I to illustrate how the electrochemical mechanism of corrosion could be applied to a variety of corroding systems. In these 'Evans' diagrams, both the cathodic and anodic partial reactions constituting the overall corrosion reaction are presented as linear E–I curves that converge and intersect at a point, which defines the corrosion potential $E_{corr.}$ and the corrosion current $I_{corr.}$. Figure 1.26 shows the $E \log i$ curves for the two half-reactions involved in the corrosion of a metal in an acid. Comparison should be made with Fig. 1.24 for a *single* exchange process, and it should be noted that at significantly high overpotentials the reverse reaction for each half-reaction may be neglected.

A typical Evans diagram for the corrosion of a single metal is illustrated in Fig. 1.26*a* (compare with Fig. 1.23 for two separable electrodes), and it can be seen that the E_c–I and E_a–I curves are drawn as straight lines that intersect at a point that defines $E_{corr.}$ and $I_{corr.}$ (it is assumed that the resistance for the solution is negligible). $E_{corr.}$ can of course be determined by means of a reference electrode, but since the anodic and cathodic sites are inseparable direct determination of $I_{corr.}$ by means of an ammeter is not possible, and

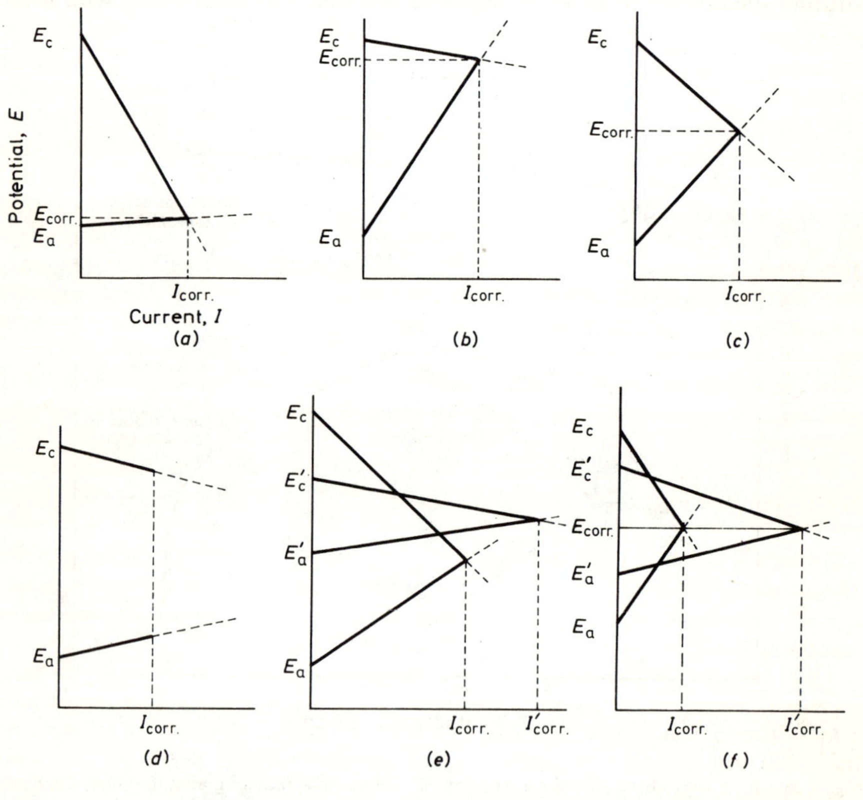

Fig. 1.27 Evans diagrams illustrating (*a*) cathodic control, (*b*) anodic control, (*c*) mixed control, (*d*) resistance control, (*e*) how a reaction with a higher thermodynamic tendency ($E_{r,cell}$) may result in a smaller corrosion rate than one with a lower thermodynamic tendency and (*f*) how $E_{corr.}$ gives no indication of the corrosion rate

indirect methods must be used (e.g. weight loss and the application of Faraday's law).

The equilibrium potentials $E_{r,c}$ and $E_{r,a}$ can be calculated from the standard electrode potentials of the H^+/H_2 and M/M^{z+} equilibria taking into account the pH and $a_{M^{z+}}$; although the pH may be determined an arbitrary value must be used for the activity of metal ions, and $a_{M^{z+}} = 1$ is not unreasonable when the metal is corroding actively, since it is the activity in the diffusion layer rather than that in the bulk solution that is significant. From these data it is possible to construct an Evans diagram for the corrosion of a single metal in an acid solution, and a similar approach may be adopted when dissolved O_2 or another oxidant is the cathode reactant.

Figures 1.27*a* to *d* show how the Evans diagram can be used to illustrate how the rate may be controlled by either the polarisation of one or both of the partial reactions (cathodic, anodic or mixed control) constituting corrosion reaction, or by the resistivity of the solution or films on the metal surface (resistance control). Figures 1.27*e* and *f* illustrate how kinetic factors may be more significant than the thermodynamic tendency ($E_{r,cell}$) and how $E_{corr.}$ provides no information on the corrosion rate.

The Evans diagram has been used for illustrating various types of corrosion phenomena ranging from the uniform corrosion of a single metal to the enhanced corrosion of one metal when it is coupled to another (bimetallic corrosion), and since the diagram can include only the predominant cathodic and anodic reactions all others are regarded as negligible. Thus if zinc is coupled to iron and the couple is immersed in an oxygenated neutral solution there are at least four possible exchange processes (eight half-reactions), but for the purpose of the Evans diagram only the reduction of dissolved oxygen on the iron surface and the oxidation of $Zn \rightarrow Zn^{2+}$ need to be considered. This tends to create the erroneous impression that each metal sustains only one electrode reaction, whereas in reality the more anodic metal may support a cathodic reaction, although it is predominantly anodic, and the converse applies to the cathodic metal.

In the Evans diagram the curves show the E_p–I relationship, whereas it is evident from previous consideration that E_p and η are functions of the current density i. In the case of a single metal $I_c = I_a$, and since $S_c = S_a$, $i_c = i_a$. However, this is not possible when the anodic and cathodic areas are not equal, and Fig. 1.28 shows how bimetallic corrosion of two dissimilar metals can be represented by an Evans-type diagram. It can be seen that although the curves intersect at $E_{corr.}$, which must be determined by placing the reference electrode at some distance from the couple, the magnitudes of i_c and i_a are different; it is also evident that the more dangerous bimetallic situation is when S_c is large and S_a is small (*see* Section 1.7).

Over the years the original Evans diagrams have been modified by various workers[23] who have replaced the linear E–I curves by curves that provide a more fundamental representation of the electrode kinetics of the anodic and cathodic processes constituting a corrosion reaction (*see* Fig. 1.26). This has been possible partly by the application of electrochemical theory and partly by the development of newer experimental techniques. Thus the cathodic curve is plotted so that it shows whether activation-controlled charge transfer (equation 1.70) or mass transfer (equation 1.74) is rate determining. In addition, the potentiostat (*see* Section 20.2) has provided a powerful tool for

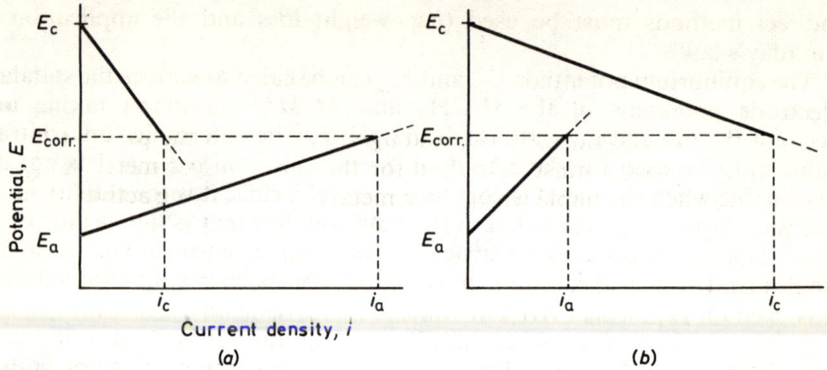

Fig. 1.28 Evans diagram illustrating a corrosion process (e.g. a bimetallic couple) in which the area of the cathode is not equal to that of the anode. (*a*) $S_c > S_a$ so that $i_c < i_a$ and (*b*) $S_a > S_c$ so that $i_a < i_c$

studying the detailed shape of the anodic curve of metals that show an active–passive transition, which has meant that the linear anodic *E–I* curve used originally has been replaced by the characteristic discontinuous potentiostatic curve. Nevertheless, all these modifications are based on the original concepts of U. R. Evans whose 'Evans diagrams' provided a major step forward in our understanding of the electrochemical mechanism of corrosion. In conclusion it is appropriate to mention that whereas in the Evans diagrams both the anodic and cathodic currents are drawn on the same side of the *E* axis (i.e. both positive) many workers (particularly Pourbaix and his co-workers) adopt the approach originally devised by Wagner and Traud[24], in which the cathodic curve is taken as negative and drawn on the left-hand side of the *E* axis whilst the converse applies to the anodic curve (Fig. 1.29).

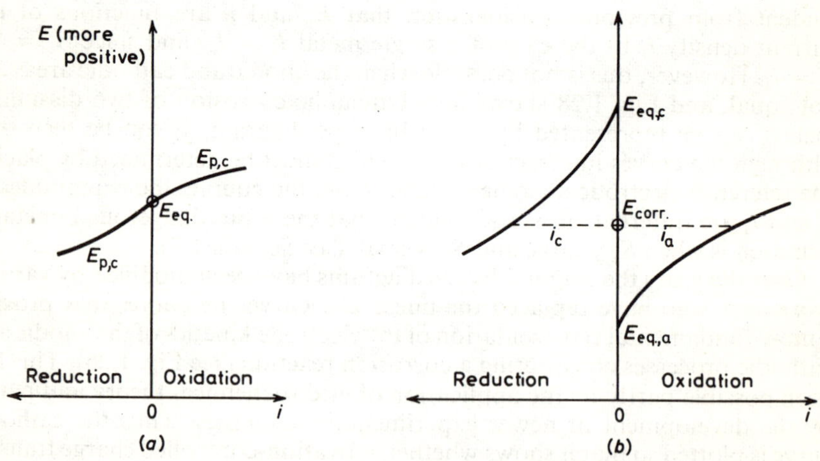

Fig. 1.29 Wagner–Traud method of representing (*a*) a single reversible reaction and (*b*) a corrosion reaction (note that $E_{corr.}$ is the potential when $i_c = i_a$)

Cathodic Reactions in Corrosion

General Considerations

It follows from the electrochemical mechanism of corrosion that the rates of the anodic and cathodic reactions are interdependent, and that either or both may control the rate of the corrosion reaction. It is also evident from thermodynamic considerations (Tables 1.9 and 1.10) that for a species in solution to act as an electron acceptor its redox potential must be more positive than that of the M^{z+}/M equilibrium or of any other equilibrium involving an oxidised form of the metal.

The hydrogen evolution reaction (h.e.r.) and the oxygen reduction reaction (equations 1.11 and 1.12) are the two most important cathodic processes in the corrosion of metals, and this is due to the fact that hydrogen ions and water molecules are invariably present in aqueous solution, and since most aqueous solutions are in contact with the atmosphere, dissolved oxygen molecules will normally be present.

In the complete absence of oxygen, or any other oxidising species, the h.e.r. will be the only cathodic process possible, and if the anodic reaction is only slightly polarised the rate will be determined by the kinetics of the h.e.r. on the particular metal under consideration (cathodic control). However, when dissolved oxygen is present both cathodic reactions will be possible, and the rate of the corrosion reaction will depend upon a variety of factors such as the reversible potential of the metal/metal ion system, the pH of the solution, the concentration of oxygen, the kinetics of the h.e.r. and the oxygen reduction reaction on the metal under consideration, temperature, etc. In general, the contribution made by the h.e.r. will increase in significance with decrease in pH, but this too will depend upon the nature of the metal and metal oxide. Thus metals like zinc and aluminium, whose oxides are amphoteric, are thermodynamically unstable in alkaline solutions (*see* Fig. 1.17) and will react with water at high pHs with consequent hydrogen evolution and formation of metal anions. In this connection it should be noted that in neutral or alkaline solutions the activity of H_3O^+ is too low for it to participate in the h.e.r., and under these circumstances the water molecule will act as the electron acceptor

$$H_3O^+ + e \rightarrow \tfrac{1}{2}H_2 + H_2O \quad \text{(acid solutions)} \quad \ldots(1.76)$$

$$H_2O + e \rightarrow \tfrac{1}{2}H_2 + OH^- \quad \text{(neutral and alkaline solutions)} \quad \ldots(1.77)$$

and for the oxygen reduction reaction

$$\tfrac{1}{2}O_2 + 2H_3O^+ + 2e \rightarrow 3H_2O \quad \text{(acid solutions)} \quad \ldots(1.78)$$

$$\tfrac{1}{2}O_2 + H_2O + 2e \rightarrow 2OH^- \quad \text{(neutral and alkaline solutions)} \quad \ldots(1.79)$$

It should also be noted that both reactions will result in an increase in pH in the diffusion layer.

The Hydrogen Evolution Reaction (H.E.R.)[19,20]

Although the h.e.r. involves transport of H_3O^+ ions (or H_2O molecules) to the metal surface by diffusion and migration, the activation energy for charge transfer is usually of the greater significance, and a corrosion reaction in which the h.e.r. is the cathodic process is frequently controlled by the activation overpotential of the latter.

If it is assumed that the transfer coefficient $\alpha = 0{\cdot}5$, and taking $z = 1$, equation 1.70 becomes

$$\eta_{A,H} = 0{\cdot}12 \log i_0 - 0{\cdot}12 \log i_c \qquad \ldots(1.80)$$

which is identical with the original Tafel equation[25], since i_0 is a constant for a given metal and for given conditions of the solution (*see* Chapter 21, Table 21.12, for values of i_0). Thus for activation-controlled transfer the significant parameter is the equilibrium exchange current density i_0, and the smaller i_0 the smaller i_c at a given overpotential. For a corrosion reaction in which both the anodic and cathodic reactions are under activation control, then

for cathodic control $\quad i_{0,c} \ll i_{0,a}$ and $E_{corr.} \rightarrow E_{r,a}$

and

for anodic control $\quad i_{0,a} \ll i_{0,c}$ and $E_{corr.} \rightarrow E_{r,c}$

Figure 1.30 gives examples of single metals corroding in a highly conducting acid in which both the anodic and cathodic reactions are assumed to be under activation control, and it can be seen that at $E_{corr.}$

$$E_{p,c} = E_{p,a} = E_{corr.} \qquad \ldots(1.81)$$

and

$$i_c = i_a = i_{corr.} \qquad \ldots(1.82)$$

but

$$\eta_c \neq \eta_c$$

In order to evaluate $i_{corr.}$ from the Tafel equation it must be expressed in terms of E_p. By definition

$$\eta = E_p - E_r$$

and from equations 1.81 and 1.82 for the cathodic reaction

$$E_{corr.} = E_{r,c} + \frac{0{\cdot}059}{\alpha_c z_c} \log \frac{i_{0,c}}{i_{corr.}} \qquad \ldots(1.83)$$

and for the anodic reaction

$$E_{corr.} = E_{r,a} + \frac{0{\cdot}059}{\alpha_a z_a} \log \frac{i_{corr.}}{i_{0,a}} \qquad \ldots(1.84)$$

By equating equations 1.83 and 1.84 $i_{corr.}$ can be evaluated, providing the Tafel parameters are known, and $E_{corr.}$ can then be calculated from either of these two equations. Alternatively, by replacing $(E_{r,c} - E_{r,a})$ by $E_{r,cell}$

$$E_{r,cell} = \frac{0{\cdot}059}{\alpha_a z_a} \log \frac{i_{corr.}}{i_{0,a}} + \frac{0{\cdot}059}{\alpha_c z_c} \log \frac{i_{corr.}}{i_{0,c}} \qquad \ldots(1.85)$$

which provides another convenient method of calculating $i_{corr.}$.

It can also be shown[19] by means of the above equations, and by assuming that $z_a = z_c = 1$, that

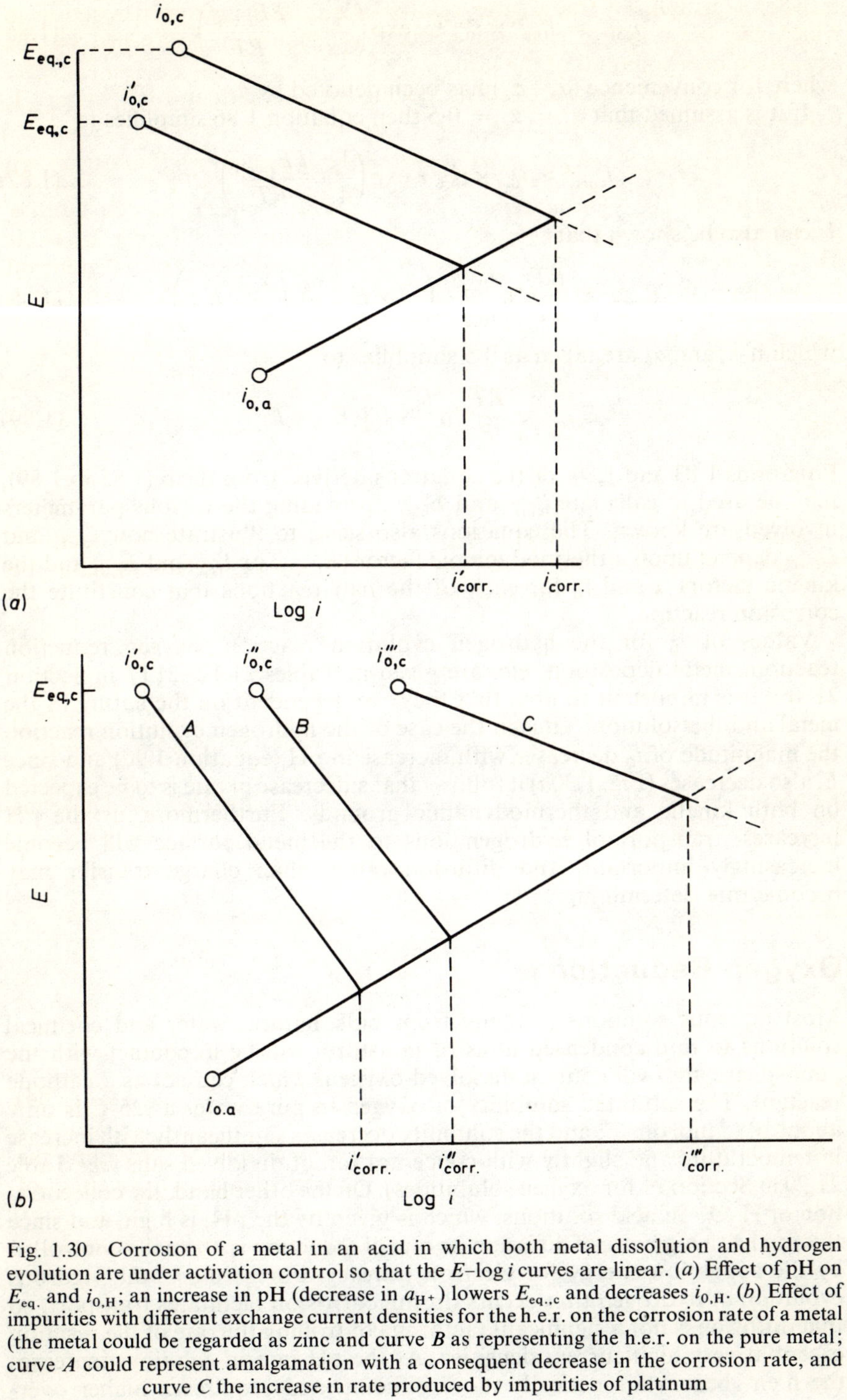

Fig. 1.30 Corrosion of a metal in an acid in which both metal dissolution and hydrogen evolution are under activation control so that the E–log i curves are linear. (*a*) Effect of pH on $E_{eq.}$ and $i_{0,H}$; an increase in pH (decrease in a_{H^+}) lowers $E_{eq.,c}$ and decreases $i_{0,H}$. (*b*) Effect of impurities with different exchange current densities for the h.e.r. on the corrosion rate of a metal (the metal could be regarded as zinc and curve *B* as representing the h.e.r. on the pure metal; curve *A* could represent amalgamation with a consequent decrease in the corrosion rate, and curve *C* the increase in rate produced by impurities of platinum

$$i_{corr.} = i_{0,a}^{\alpha_c/\alpha_t} \times i_{0,c}^{\alpha_a/\alpha_t} \times \exp\left(\frac{\alpha_a\alpha_c}{\alpha_t} \cdot \frac{FE_{r,cell}}{RT}\right) \qquad ...(1.86)$$

where for convenience $(\alpha_a + \alpha_c)$ has been denoted by α_t.

If it is assumed that $\alpha_a = \alpha_c = 0{\cdot}5$ then equation 1.86 simplifies to

$$i_{corr.} \approx i_{0,a}^{\frac{1}{2}} \times i_{0,c}^{\frac{1}{2}} \times \exp\left(\frac{1}{4} \cdot \frac{FE_{r,cell}}{RT}\right) \qquad ...(1.87)$$

It can also be shown that

$$E_{corr.} \approx \frac{RT}{\alpha_t F}\ell n \frac{i_{0,c}}{i_{0,a}} + \left(\frac{\alpha_a}{\alpha_t} \times E_{r,a}\right) + \left(\frac{\alpha_c}{\alpha_t} \times E_{r,c}\right) \qquad ...(1.88)$$

which if α_a and α_c are taken as 0·5 simplifies to

$$E_{corr.} \approx \frac{RT}{F}\ell n \frac{i_{0,c}}{i_{0,a}} + \tfrac{1}{2}(E_{r,a} + E_{r,c}) \qquad ...(1.89)$$

Equations 1.83 and 1.84, or the equations derived from them (1.85 to 1.89), may be used to calculate $i_{corr.}$ and $E_{corr.}$, providing the various parameters involved are known. The equations also serve to illustrate how $i_{corr.}$ and $E_{corr.}$ depend upon a thermodynamic factor ($E_{r,cell}$, or $E_{r,c}$ and $E_{r,a}$) and the kinetic factors α and i_0 for each of the half reactions that constitute the corrosion reaction.

Values of i_0 for the hydrogen evolution reaction, oxygen reduction reaction, metal deposition, etc. are given in Tables 21.12–21.17 in Section 21 and it is important to note that they are dependent on the nature of the metal and the solution. Thus in the case of the hydrogen evolution reaction the magnitude of i_0 decreases with increase in pH (equation 1.90) and since E_r also decreases (Fig. 1.30*a*) it follows that a decrease in rate is to be expected on both kinetic and thermodynamic grounds. Furthermore, as the pH increases, transport of hydrogen ions to the metal surface will become increasingly important, and diffusion rather than charge transfer may become rate determining.

Oxygen Reduction

Most aqueous solutions (ranging from bulk natural water and chemical solutions to thin condensed films of moisture) will be in contact with the atmosphere and will contain dissolved oxygen, which can act as a cathode reactant. The saturated solubility of oxygen in pure water at 25°C is only about 10^{-3} mol dm^{-3}, and the solubility decreases significantly with increase in temperature and slightly with concentration of dissolved salts (*see* Table 21.20 in Section 21 for oxygen solubilities). On the other hand, the concentration of H_3O^+ in acid solutions, which is given by the pH, is high, and since this ion has a high rate of diffusion its rate of reduction is normally controlled by the activation energy for electron transfer. Furthermore, the vigorous evolution of hydrogen that occurs during corrosion facilitates transport, so that diffusion is not a significant factor in controlling the rate of the reaction except at very high current densities. As the pH in acid solutions increases the h.e.r. becomes kinetically more difficult, and requires a higher overpotential. It has been shown[20] that the exchange current density decreases with increase in pH:

$$i_{0,\mathrm{H}} = i^{0}_{0,\mathrm{H}} \times c^{0\cdot5}_{\mathrm{H_3O^+}} \qquad \ldots(1.90)$$

in which $i^0_{0,\mathrm{H}}$ is the exchange current density at $c_{\mathrm{H^+}} = 1$.

The situation is different, however, in near-neutral or alkaline solutions in which the concentration of H_3O^+ will be small ($<10^{-7}\,\mathrm{mol\,dm^{-3}}$), and in these solutions the water molecule will act as the electron acceptor, and although diffusion occurs rapidly its reduction is kinetically more difficult than that of H_3O^+, and will therefore require a higher activation overpotential.

The relationship between E and pH for the H^+/H_2 and O_2/H_2O equilibria have been given in Table 1.8, and it is evident that thermodynamically dissolved oxygen is a far more powerful electron acceptor than H_3O^+ at all pHs. Thus, in the case of the corrosion of Fe → Fe^{2+} ($E^{\ominus}_{\mathrm{Fe^{2+}/Fe}} = -0\cdot44\,\mathrm{V}$) in an oxygen-free neutral solution the thermodynamic tendency for corrosion will be very small ($E_{\mathrm{reaction}} \approx 0\cdot02\,\mathrm{V}$) compared with the thermodynamic tendency in oxygenated water ($E_{\mathrm{reaction}} \approx 1\cdot2\,\mathrm{V}$). Similar consideration will not, of course, apply to the more negative metals such as magnesium ($E^{\ominus}_{\mathrm{Mg^{2+}/Mg}} = -2\cdot1\,\mathrm{V}$).

From these considerations it follows that, in general, oxygen reduction will be more significant than hydrogen evolution in near-neutral solutions, and that in the case of the former, transport of oxygen to the metal surface will be more significant than activation-controlled electron transfer. A further important factor is that in near-neutral solutions solid corrosion products will be thermodynamically stable and will affect the corrosion rate either by passivating the metal or by forming a barrier that hinders transport of oxygen to the metal surface. For these reasons corrosion rates in acid solutions are usually much higher than in neutral solution.

Transport of Oxygen

Before electron transfer can occur the oxygen in the atmosphere must be transported to the metal/solution interface, and this involves the following steps[21]:

(*a*) Transfer of oxygen across the atmosphere/solution interface.

(*b*) Transport through the solution (by diffusion and by natural and forced convection) to the diffusion layer.

(*c*) Transport across the static solution at the metal/solution interface (the diffusion layer δ) by diffusion.

As far as (*a*) is concerned the greater the surface area of the interface (as may be produced by agitating the solution) the greater the transport of oxygen to the solution. On the other hand, a restricted interface (e.g. a volumetric graduated flask filled to the graduation mark) will decrease oxygen transfer, a factor which must be taken into account in corrosion testing. In the case of (*b*) oxygen will be transported through the solution by natural convection (e.g. thermal gradients, evaporation at surfaces) or by forced convection (e.g. agitation of the solution, movement of the metal), but these are fairly rapid compared with (*c*), which is normally rate determining; however, (*b*) could be significant when a metal is immersed at depth in a static solution.

It follows from this that the limiting current density i_L is the most significant parameter in a corrosion reaction in which oxygen is the cathodic reactant, and that any factor that increases i_L will increase the corrosion rate, since at $E_{corr.}$

$$i_L = i_{corr.} \qquad \ldots(1.91)$$

For the reaction $O_2 + 2H_2O + 4e = 4OH^-$, and taking the concentration of oxygen to be $10^{-3}\,mol\,dm^{-3}$, the diffusion coefficient to be $10^{-5}\,cm^2\,s^{-1}$ and δ to be 0·05 cm in an unstirred solution, it can be calculated from equation 1.73 that $i_L \approx 80\,\mu A\,cm^{-2}$; for a vigorously stirred solution $\delta \approx 0{\cdot}001$ cm and $i_L \approx 4\,mA\,cm^{-2}$. Thus in this particular case the corrosion rate due to oxygen reduction can range from about 0·08 to $4\,mA\,cm^{-2}$, depending on the degree of agitation of the solution.

Figure 1.31*a* to *c* shows how an increase in the concentration of dissolved oxygen or an increase in velocity increases i_L and thereby increases $i_{corr.}$. It has been shown in equation 1.73 that i_L increases with the concentration of oxygen and temperature, and with decrease in thickness of the diffusion layer, and similar considerations apply to $i_{corr.}$. Thus Uhlig, Triadis and Stern[26] found that the corrosion rate of mild steel in slowly moving water at

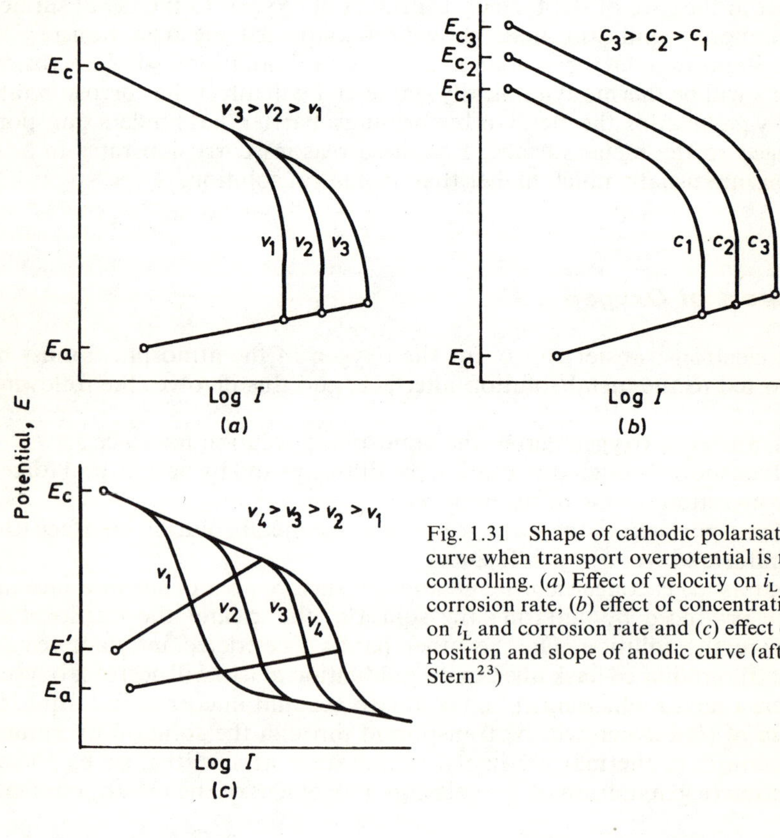

Fig. 1.31 Shape of cathodic polarisation curve when transport overpotential is rate controlling. (*a*) Effect of velocity on i_L and corrosion rate, (*b*) effect of concentration on i_L and corrosion rate and (*c*) effect of position and slope of anodic curve (after Stern[23])

25°C containing 165 p.p.m. of $CaCl_2$ increased linearly with increase in the concentration of dissolved oxygen. However, in distilled water it is possible to passivate pure iron, providing oxygen is brought rapidly and simultaneously to all parts of the metal surface; this was demonstrated first by Evans in his 'whirling ring experiment' in which a steel ring was supported on a glass rod that rotated eccentrically[27]. A similar effect was observed by Uhlig, Triadis and Stern[26] who showed that although initially the corrosion rate of iron in distilled water increased rapidly with concentration of oxygen, it started to decrease when a critical concentration of oxygen (12 mℓ/ℓ) was attained; this critical value increased with concentration of dissolved salts and temperature, and decreased with increase in velocity and pH. Thus under certain circumstances the rate of reduction of oxygen may be sufficient to exceed the critical current density required for passivation so that even mild steel in oxygenated distilled water will become passive. This is considered further in later paragraphs.

The effect of temperature is complex since there are two conflicting factors, (*a*) a decrease in the oxygen concentration which results in a decrease in $i_{corr.}$ and (*b*) an increase in the diffusion coefficient that increases about 3% per degree K rise in temperature. In a closed system from which oxygen cannot escape there is a linear increase in rate with temperature that corresponds with the increase in the diffusion coefficient. However, in an open system although the rate follows that for the closed system initially, the rate starts to decrease at about 70°C due to the decrease in oxygen solubility, which at that temperature becomes more significant than the increase in the diffusion coefficient[28] (*see* Section 2.1).

Mechanism

The mechanism of the oxygen reduction reaction[29] is by no means as fully understood as the h.e.r., and a major experimental difficulty is that in acid solutions (pH = 0) $E^{\ominus}_{O_2/H_2O} = 1{\cdot}23$, which means that oxygen will start to be reduced at potentials at which most metals anodically dissolve. For this reason accurate data on kinetics is available only for the platinum metals. In the case of an iridium electrode at which oxygen reduction is relatively rapid, a number of reaction sequences have been proposed, of which the most acceptable appear to be the following[30]:

1. $O_2 + 2M \rightleftharpoons 2MO$
2. $MO + H^+ + e \xrightarrow{\text{r.d.s.}} MOH$
3. $MOH + H^+ + e \rightleftharpoons M + H_2O$

where step 2 is the rate-determining step. The exchange current densities for oxygen reduction on the platinum metals (Pt, Rh, Ir) in acid solutions are about 10^{-10} A cm^{-2}, and it is even smaller for iron, i.e. about 10^{-14} A cm^{-2}.

It has been emphasised that the oxygen reduction reaction is diffusion controlled, and it might be thought that the nature of the metal surface is unimportant compared with the effect of concentration, velocity and temperature that all affect i_L, and hence $i_{corr.}$. However, in near-neutral solutions the surface of most metals will be coated (partially or completely) with either thin invisible films or with thick visible corrosion products, and the latter in particular will impede electron transfer; thus metals like platinum, gold, silver and copper are more efficient cathodes for oxygen reduction than metals like steel, zinc or lead (*see* Section 1.7).

Simultaneous Cathodic Reactions

Corrosion reactions involving two simultaneous cathodic processes have already been referred to, and it is now appropriate to consider the graphical method of representing the corrosion rate. It should be noted that although the simultaneous reduction of H_3O^+ and dissolved oxygen occurs frequently this does not exhaust the possibilities, and reactions such as $Fe^{3+} \rightarrow Fe^{2+}$, $Cu^{2+} \rightarrow Cu$, $Cl_2 \rightarrow Cl^-$ may accompany either or both of the above reactions. Similar considerations apply to electrolytic processes such as cathodic protection using an impressed current system in which simultaneous reduction of H_3O^+ ions and dissolved oxygen (and sometimes reduction of dissolved Cl_2 and ClO^- produced by the anodically generated Cl_2) occurs at the surface of the metal protected.

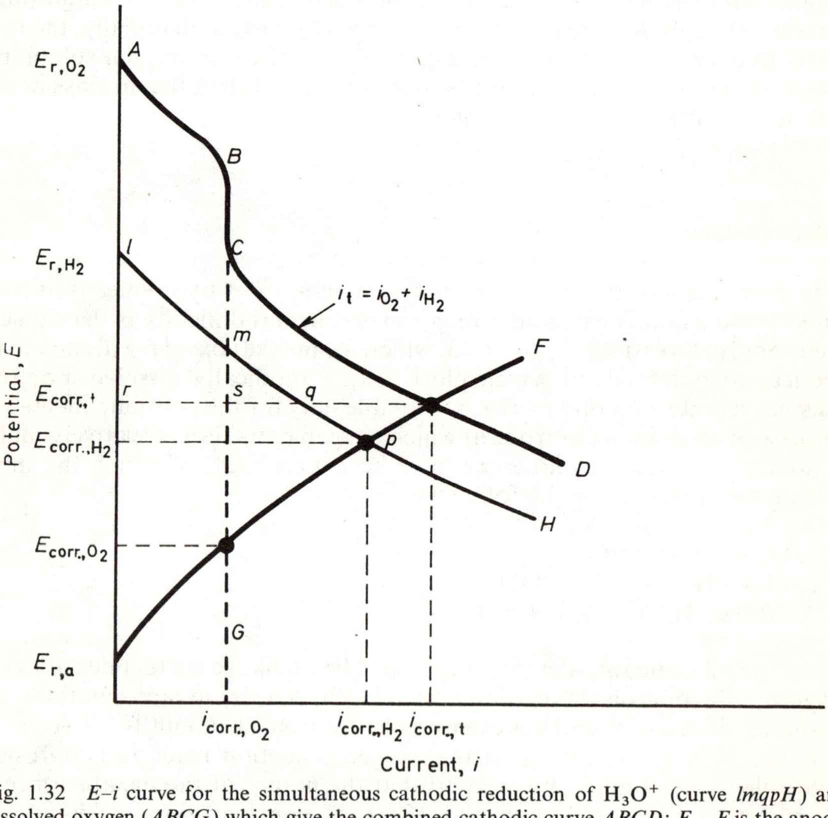

Fig. 1.32 *E–i* curve for the simultaneous cathodic reduction of H_3O^+ (curve *lmqpH*) and dissolved oxygen (*ABCG*) which give the combined cathodic curve *ABCD*; $E_{r,a}$ *F* is the anodic curve for $M \rightarrow M^{z+}$

If the anodic process is assumed to be solely $M \rightarrow M^{z+}$ then for the corrosion of a single metal

$$i_{\text{corr.}} = \Sigma i_c = i_{H_2} + i_x = i_t \qquad \ldots(1.92)$$

where i_x is the partial cathodic current density produced by an oxidant in solution and i_t is the total current density.

Figure 1.32 shows the E–i curves for a metal corroding in an acid in which both dissolved oxygen and H_3O^+ ions act as cathode reactants (note that in order to illustrate the summation of the partial currents to give the total current $i_t = i_{H_2} + i_{O_2}$, i rather than $\log i$ has been used). The total cathodic curve *ABCD* is the sum of the partial cathodic currents, and it can be seen that the corrosion potential due to the total cathodic current is more positive than either $E_{\text{corr.},H_2}$ or $E_{\text{corr.},O_2}$. However, whereas in the case of oxygen reduction the rate remains constant once i_L is attained, this does not apply to the h.e.r., which means that the presence of O_2 will decrease the rate of hydrogen evolution from p to q. At $E_{\text{corr.},t}$ the relative rate i_{O_2}/i_{H_2} will be in the ratio rs/rq with the h.e.r. dominating. The relative contributions to the corrosion rate made by dissolved oxygen and H_3O^+ ions will depend on the position of each of the curves; thus if the anodic curve passed through the cathodic curves at their point of intersection m, then $i_{\text{corr.},H_2} = i_{\text{corr.},O_2}$ and their contributions to the total corrosion rate would be equal.

The situation will of course be different if the potential is maintained constant, e.g. in cathodic protection. For example, if the potential was maintained constant at, say, $E_{\text{corr.},t}$ in an oxygen-free solution the rate due to the h.e.r. would be rq which would be unaffected if the solution was then aerated. However, since the corrosion rate would then increase, the current required to maintain the potential constant would have to be increased.

Table 1.13 shows the effect of dissolved oxygen on the corrosion rate of mild steel in oxygenated and oxygen-free dilute acids[31], and it can be seen that the ratio varies with the nature and concentration of the acid; oxygen in 1·2 N HNO_3 as would be anticipated has little effect since this acid is readily reduced cathodically. In general, the higher the concentration of the acid the lower the contribution made by dissolved oxygen (*see* results for HCl), and this appears to be due to more rapid hydrogen evolution in the more concentrated acid that tends to screen the surface of the metal from the dissolved oxygen. Uhlig[32] quotes examples, showing how traces of oxygen in dilute H_2SO_4 or substantial amounts in the more concentrated acid can act as an inhibitor. Thus zone-refined iron corroded at 680 mg dm^{-2} d^{-1} in oxygen-free 0·05 M H_2SO_4, but at only 415 mg dm^{-2} d^{-1} when the acid was aerated.

Table 1.13 Effect of dissolved oxygen on the corrosion rate (mm y^{-1}) of mild steel in acids*

Acid	*Oxygen-containing (O.C.)*	*Oxygen-free (O.F.)*	*Ratio O.C./O.F.*
6% acetic	13·97	0·15	87
6% H_2SO_4	9·14	0·76	12
4% HCl	12·19	0·79	16
0·04% HCl	9·91	0·14	71
1·2% HNO_3	46·23	39·88	12

*Data after Whitman and Russell[31].

In the case of a pure Fe–9·2 Co alloy in 0·05 M H_2SO_4, in which the corrosion rate is high, the rate was found to increase when the oxygenated acid was deoxygenated. These examples show that the role of oxygen in corrosion reactions is far more complex than would appear from the kinetic curves illustrated above.

Anodic *E–i* Curves

It is now appropriate to consider the kinetics of the anodic reaction with particular reference to the phenomenon of passivity, but since the mechanism is dealt with in detail in Section 1.5 this discussion will place the emphasis on the anodic *E–i* curves.

It has been assumed that the anodic curve for $M \rightarrow M^{z+}$(aq.) conforms with the Tafel relationship, and that *E*–log *i* is linear throughout the range of potentials under consideration. It follows, therefore, that charge transfer rather than mass transfer is rate determining, and that the linearity of the *E*–log *i* curve will be maintained until transport of metal ions away from the surface becomes significant. It is not proposed to consider metal dissolution in detail here, but it is appropriate to illustrate the complexity of the process by considering the anodic dissolution of iron:

$$Fe \rightarrow Fe^{2+}(aq.) + 2e \qquad ...(1.93)$$

This follows the Tafel relationship with a Tafel slope of $2/3\,RT/F = 0{\cdot}04$ V. Since direct transfer of two electrons as shown in equation 1.92 is highly unlikely, it would appear that the mechanism might involve a two-step process in which one step is rate determining, say

$$Fe \xrightarrow{\text{r.d.s.}} Fe^{+} + e \qquad ...(1.94)$$

$$Fe^{+} \rightleftharpoons Fe^{2+} + 2e \qquad ...(1.95)$$

However, Despic and Drazic[33] found that the reaction rate was dependent on the concentration of both Fe^{2+} and OH^-, and that

$$i_{0,\text{Fe}} = i^{0}_{0,\text{Fe}} \times c_{\text{Fe}^{2+}} \times c_{\text{OH}^-} \qquad ...(1.96)$$

and although the former dependence was expected the latter was not. This observation that the rate increased with increase in hydroxyl concentration (increase in pH), together with a number of other diagnostic criteria, has lead to the view that the most probable mechanism involves the following sequence of steps:

$$Fe + H_2O \rightleftharpoons FeOH + H^{+} + e \qquad ...(1.97)$$

$$FeOH \xrightarrow{\text{r.d.s.}} Fe(OH)^{+} + e \qquad ...(1.98)$$

$$FeOH^{+} + H^{+} \rightleftharpoons Fe^{2+} + H_2O \qquad ...(1.99)$$

It must be emphasised that although the rate of anodic dissolution of iron increases with increase in pH this will not necessarily apply to the corrosion rate which will be dependent on a number of other factors, e.g. the thermodynamics and kinetics of the cathodic reaction, film formation, etc.

Active–Passive Transitions

On the basis of previous considerations it is evident that the rate of an activation-controlled electrode reaction should increase with increase in potential according to the Tafel relationship. However, in certain metal (alloy)/electrolyte solutions the rate of corrosion decreases to a comparatively low value when the potential is raised above a critical value, and the metal is said to have become *passive*. It must be emphasised again that the passivation of a metal or alloy is dependent on its nature and on the nature of the solution; iron can be passivated only under specific environmental conditions whereas tantalum is passive in most environments.

Although passivity is normally the result of a corrosion reaction involving both a cathodic and anodic reaction, it is studied more easily by making the whole of the metal surface anodic by means of an external source of e.m.f. and a counter-electrode. In this connection it should be noted that there are a number of procedures that can be used to determine the anodic *E–i* relationship (*see* Fig. 1.39, page **1**.116).

1. Galvanostatically—the current density is maintained constant at a predetermined value and the steady-state potential is measured.
2. Potentiostatically—the potential of the metal/solution interface is maintained constant at a predetermined value and the steady-state current density is measured.

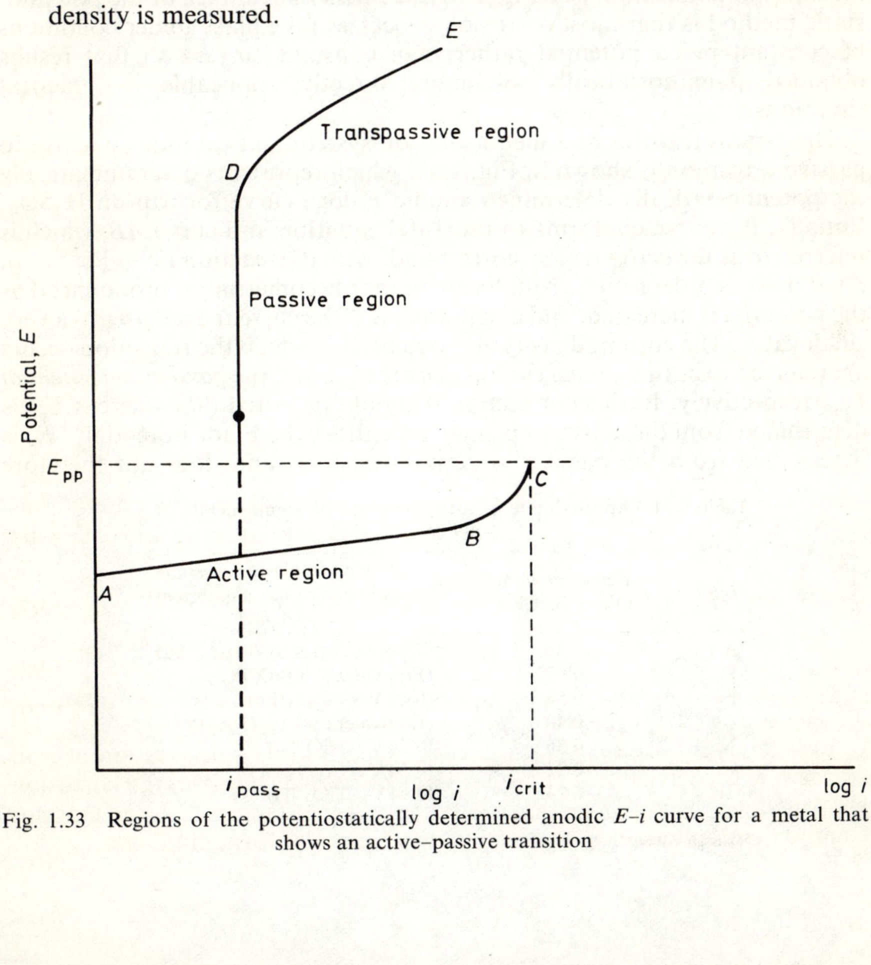

Fig. 1.33 Regions of the potentiostatically determined anodic *E–i* curve for a metal that shows an active–passive transition

The former method requires a constant-current power source whilst the latter require a potentiostat[34] (*see* Section 20.2).

Each of these two procedures can be varied by proceeding from a low to a high current density (or potential) or from a high to a low current density (or potential); the former is referred to as *forward* polarisation and the latter as *reverse* polarisation. Furthermore, there are a number of variations of the potentiostatic technique, and in the potentiokinetic method the potential of the electrode is made to vary continuously at a predetermined rate, the current being monitored on a recorder; in the pulse method the electrode is given a pulse of potential and the current transient is determined by means of an oscilloscope.

Both the galvanostatic and potentiostatic method have their own particular spheres of application, and it is not always advantageous to reject the former in favour of the latter, although there is an increasing tendency to do so. Nevertheless, the potentiostatic method does have a distinct advantage in studies of passivity, since it is capable of defining more precisely the potential and current density at which the transition from the active (charge transfer controlled $M \rightarrow M^{z+}$) to the passive state takes place; this is far more difficult to achieve with the galvanostatic method owing to the sudden decrease in current (equivalent to an increase in resistance) that accompanies the onset of passivation (*see* Fig. 1.39). A further advantage of the potentiostatic method is that most corrosion processes take place under conditions of constant redox potential rather than constant current so that results obtained potentiostatically are more directly applicable to practical situations.

The typical features of a metal/solution system that exhibits an active to passive transition is shown in Fig. 1.33, which represents diagrammatically the potentiostatically determined anodic E–$\log i$ curve for iron in H_2SO_4. Initially, the curve conforms to the Tafel equation and curve AB, which is referred to as the *active region*, corresponds with the reaction $Fe \rightarrow Fe^{2+}(aq)$. At B there is a departure from linearity that becomes more pronounced as the potential is increased, and at a potential C the current decreases to a very small value. The current density and potential at which the transition occurs are referred to as the *critical current density* $i_{crit.}$, and the *passivation potential* E_{pp}, respectively. In this connection it should be noted that whereas E_{pp} is determined from the active to passive transition, the Flade potential[35] E_F is determined from the passive to active transition and does not therefore

Table 1.14 Approximate Flade potentials of some metals at pH 0*

Metal	*Flade potential* (V, *vs.* S.H.E.)	*Comparison with reversible potential of the equilibrium specified*
Au	1·36	(Approximates to Au_2O_3/Au)
Pt	0·87	(Equivalent to PtO/Pt)
Fe	0·58	(0·62 V > Fe_2O_3/Fe)
Ag	0·40	(Equivalent to Ag_2O/Ag)
Ni	0·36	(0·24 V > NiO/Ni)
Cr	−0·22	(0·54 V > Cr_2O_3/Cr)
Ti	−0·24	(0·94 V > TiO_2/Ti)

*Data after West[37].

include extraneous factors such as the *IR* drop through the insulating film, a precursor of true passivity, or pH changes that occur in the base of the pores in the film. For these reasons E_{pp} is somewhat more positive (about 0·001–0·03 V) than E_F, although for most purposes they may be regarded as being approximately the same (*see* Section 1.5). The Flade potential E_F was found by Franck[36] to be a linear function of pH, and the general equation can be expressed in the following form:

$$E_F = E_F^{\ominus} - 0{\cdot}059\,\text{pH} \qquad \ldots(1.100)$$

in which $E_F^{\ominus}$ is the Flade potential at pH 0. This equation, which shows that E_F becomes more negative with decrease in pH, is applicable to iron, nickel and chromium–iron alloys.

The two significant parameters in passivation are E_{pp} and $i_{crit.}$, and although they are evaluated by means of the potentiostatically determined anodic *E–i* curve they are equally applicable to chemical passivation in which the redox potential and kinetics of the cathodic reaction determines the potential of the metal/solution interface. Typical values of E_F in acid solutions at pH 0 are given in Table 1.14[37], and it is evident that whereas metals like chromium and titanium, which have negative passivation potentials, may be passivated by the hydrogen ion in acid solutions ($E_{H^+/H_2}^{\ominus} = 0{\cdot}00$ V at pH 0) the other metals will require solutions of higher redox potentials. For passivation two conditions must be satisfied, (*a*) the redox potential of the solution must be more positive than E_{pp} and (*b*) the rate of the cathodic reaction must be greater than $i_{crit.}$. This is illustrated by the passivation of iron ($E_F = 0{\cdot}58$ V) in nitric acid, an oxidising acid of very high redox potential (about 1·1 V) and high limiting current density. Iron can be passivated chemically in acid solutions only by means of the powerful oxidising agent fuming nitric acid, but in reducing acids (e.g. H_2SO_4) it can be passivated by raising its potential into the passive region by means of an external e.m.f. and a counter-electrode—a process that is known as *anodic protection.* However, if iron is alloyed with metals that passivate more readily (smaller $i_{crit.}$ and more negative E_F) such as chromium and nickel, the ease of passivation of these two metals is imparted to the alloy. Table 1.15[38] shows how $i_{crit.}$, E_F and the quantity of charge required to fully passivate the alloys all decrease with increase in chromium and chromium plus nickel additions.

Table 1.15 Critical current densities ($i_{crit.}$), Flade potentials (E_F) and charge required for passivation of Fe–Cr alloys in 10% H_2SO_4*

Composition (wt. % Cr)	$i_{crit.}$ (mA cm^{-2})	E_F (V, *vs.* S.H.E.)	*Charge* (mC cm^{-2})
0	1 000	0·58	1 600
2·8	360	0·58	620
6·7	340	0·35	70
9·5	27	0·15	14
12	27	0·01	15
14	19	−0·03	9
16	12	−0·02	9
18	11	+0·10	8
18+8% Ni	2	−0·10	8

*Data after Olivier[38] (see also Ref. 37).

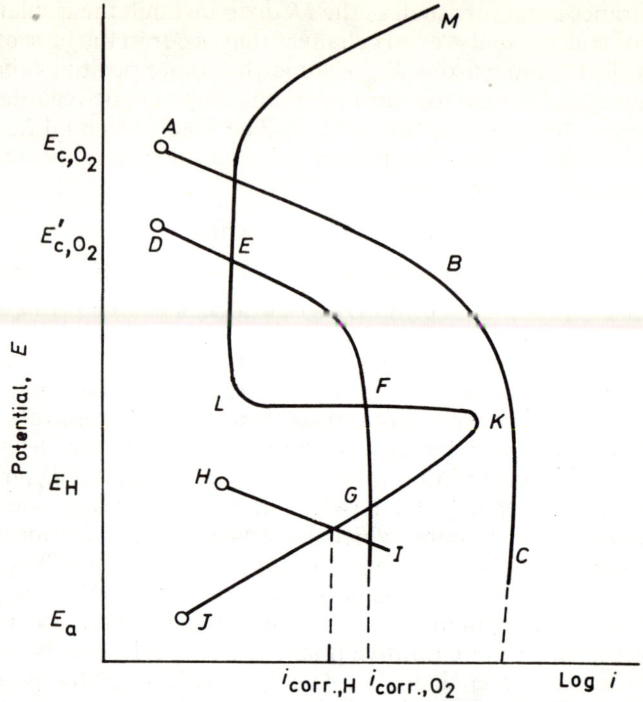

Fig. 1.34 Corrosion and passivation of Fe–18Cr–8Ni stainless steel. Potentiostatic anodic curve *JKLM*; hydrogen evolution reaction, curve *HI*; low concentration of dissolved oxygen, curve *DEFG*; high concentration of dissolved oxygen, curve *ABC*

Figure 1.34 shows diagrammatically how Fe–18Cr–8Ni stainless steel in sulphuric acid will corrode when the solution is oxygen free, since neither of the two criteria for passivation is satisfied. The presence of a small amount of oxygen in the solution (curve *DEFG*) could worsen the situation, since $i_L < i_{crit.}$. On the other hand an ample supply of oxygen could result in complete passivation. Figures 1.35[39] and 1.36[40] are experimentally determined anodic curves for a ferritic steel (Type 430, 18Cr) and for an Fe–18Cr–8Ni austenitic steel, respectively, and it can be seen how $i_{crit.}$ varies with alloy composition and with concentration of the sulphuric acid. The ferritic stainless steels in acid solutions cannot be passivated by additions of dissolved oxygen, since i_L for oxygen reduction is insufficiently large, but passivation could be achieved if another oxidant (HNO_3, Fe^{3+}, Cu^{2+}) with a high limiting current density was present in the acid.

A further possibility in the passivation of a metal in reducing acid is to change the kinetics of the h.e.r. by alloying the metal with a noble metal that has a higher exchange current density than that of the metal to be passivated. This was achieved first by Tomashov[41] who alloyed Fe–18Cr–8Ni stainless steel with Pt, Pd or Cu. Subsequently, Stern and Wissenberg[42] applied the same principle to Ti, a metal that has a low $i_{crit.}$ in hydrochloric acid as well as in sulphuric acid, and the addition of 0·1–0·2% palladium to the titanium enables passivation to take place even in boiling 10% HCl.

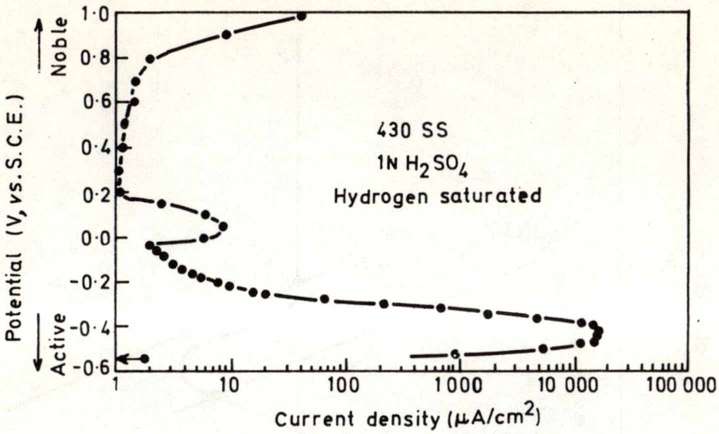

Fig. 1.35 Potentiostatic anodic curve[34] for an 18Cr ferritic stainless steel (type 430) in H_2SO_4 showing the high value of $i_{crit.}$ (after Greene[39])

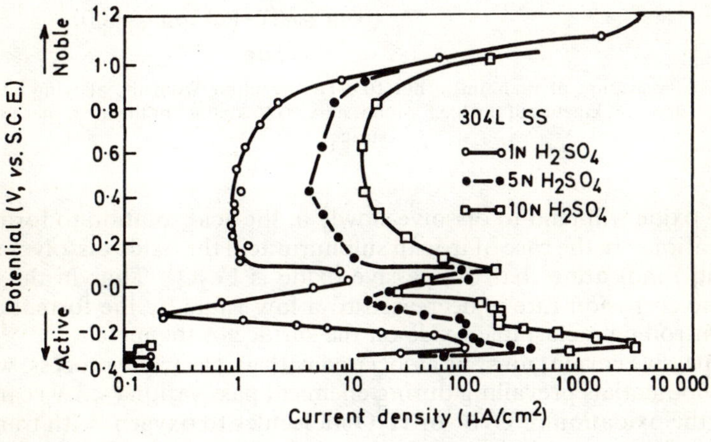

Fig. 1.36 Potentiostatic anodic curve[40] for an Fe–18Cr–8Ni austenitic stainless steel (type 304L) in various concentrations of H_2SO_4. (Note how both $i_{crit.}$ and $i_{pass.}$ increase with concentration of H_2SO_4 and how the passive region ranges from about 0·1 to 0·8 V *vs.* S.C.E.). (After Fontana and Greene[40])

Figure 1.37 shows diagrammatically how the kinetics of the h.e.r. are altered by the presence of noble-metal alloying additions.

It is not appropriate here to consider the mechanism of passivation (*see* Section 1.5), but it is apparent from Fig. 1.33 that the transition from the active to the passive state must be associated with a fundamental change in the nature of the metal surface, and it is now the generally accepted view that passivity is due to the formation of a very thin solid film of metal corrosion product, usually of oxide, on the metal surface. In this connection it should be noted that metal oxides are thermodynamically unstable in acid solutions so that the oxide formed during passivation must be regarded as a *metastable* form of the oxide that is stable at a higher pH (Fig. 1.18). It follows that the

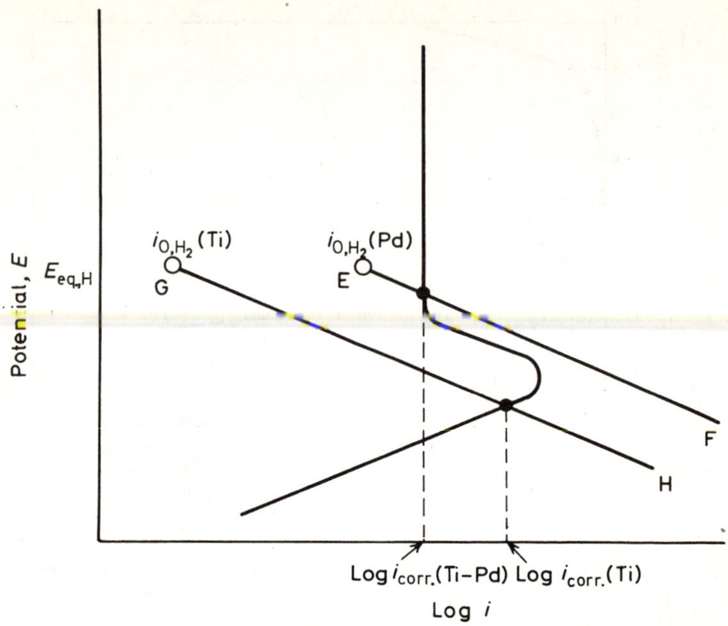

Fig. 1.37 Passivation of titanium in hot 10% HCl resulting from the alloying addition of 0·2% Pd. Curve *EF*, kinetics of the h.e.r. on Pd; curve *GH*, kinetics of the h.e.r. on the titanium surface

passive oxide will tend to dissolve slowly in the acid solution to form stable metal cations; in the case of iron in sulphuric acid the oxide dissolves to form Fe^{3+}(aq.) indicating that the passive oxide is Fe_2O_3. Thus in the passive state the corrosion rate is decreased to a low value by the formation of a rate controlling metastable oxide on the surface of the metal.

The passive current must be associated with an electrode process, which at the low potentials prevailing during chemical passivation (<1 V) cannot be due to the oxidation of OH^- or H_2O molecules to oxygen, with transfer of electrons from the solution to the metal through the intervening oxide. It follows that it must be an ionic current, and since ionic currents in oxide lattices require high fields (about $10^6\,V\,cm^{-1}$) the thickness of the oxide cannot exceed 10 nm. Since the potential is constant the field must be constant, which means that the thickness of the oxide must remain constant, and this is possible only if its rate of formation $i_{pass.}$ equals its rate of dissolution in the acid. Thus in the passive state the interface between the metal and the passivating metal oxide progresses slowly into the metal, which confirms the view expressed earlier that passivity may be regarded as a limiting case of a corrosion reaction.

As the potential is raised the passive current density $i_{pass.}$ remains reasonably constant until *D* (Fig. 1.33) when it starts to increase along curve *DE*, which is referred to as the *transpassive* region of the potentiostatic *E–i* curve. Depending on the nature of the metal/solution system and the potential, the transpassive region may be associated with gas evolution (oxygen, chlorine, etc.) or corrosion of the metal, or both reactions may proceed simultaneously.

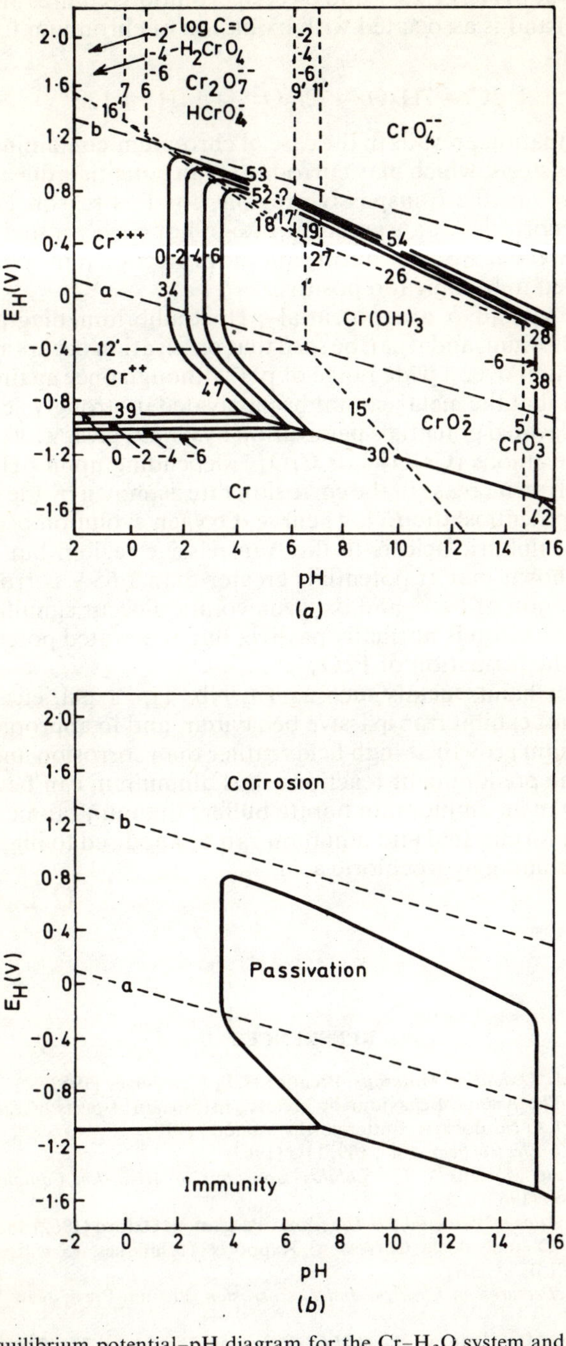

Fig. 1.38(*a*) Equilibrium potential–pH diagram for the Cr–H_2O system and (*b*) potential–pH diagram showing zones of corrosion, passivity and immunity (after Pourbaix[6])

In the case of chromium in 1 N H_2SO_4 transpassivity occurs at about 1·1 V (below the potential for oxygen evolution, since the equilibrium potential in acid solutions at pH 0 is 1·23 V and oxygen evolution requires an appreciable overpotential) and is associated with oxidation of chromium to dichromate anions:

$$2Cr + 7H_2O \rightarrow Cr_2O_7^{2-} + 14H^+ + 12e \qquad ...(1.101)$$

A similar situation prevails in the case of chromium-containing alloys such as the stainless steels, which may corrode at high potentials due to the formation of $Cr_2O_7^{2-}$ in the transpassive region; for this reason Fe–18Cr–8Ni stainless steel corrodes at a significant rate in hot concentrated HNO_3, and may corrode in reducing acids during anodic protection if the potential of the metal is allowed to become too positive.

Figures 1.38*a* and *b* are potential–pH equilibrium diagrams for the Cr–H_2O equilibrium, and it can be seen that passivation occurs at a relatively negative potential over a wide range of pH, although once again the impression is created that the metal cannot be passivated in strongly acid solutions. However, at elevated potentials passivation by Cr_2O_3 gives way to the formation of soluble anions ($Cr_2O_7^{2-}$ or CrO_4^{2-} depending upon pH) and this is associated with an increase in the corrosion rate as shown by the transpassive region of the potentiostatic E–log i curve. Oxygen evolution occurs initially when iron in sulphuric acid is in the transpassive region, but Franck and Weil[43] have shown that at potentials greater than 1·65 V corrosion of iron with the formation of Fe^{3+} and oxygen evolution occur simultaneously; in alkaline solutions iron is normally passive, but at elevated potentials it may corrode with the formation of FeO_4^{2-}.

On the other hand, metals such as Ta, Nb, Ti, Zr, Al, etc. (the 'valve' metals[44]) do not exhibit transpassive behaviour, and in appropriate electrolyte solutions film growth at high fields rather than corrosion and/or oxygen evolution is the predominant reaction; thus aluminium can be anodised to 500 V or more in an ammonium borate buffer; titanium can be anodised to about 400 V in formic acid and tantalum can be anodised to high voltages in most acids, including hydrochloric acid.

L. L. SHREIR

REFERENCES

1. Latimer, W. M., *Oxidation Materials*, Prentice Hall, New Jersey (1952)
2. Hoar, T. P., 'The Anodic Behaviour of Metals', in *Modern Aspects of Electrochemistry*, No. 2, Ed. by J. O'M. Bockris, Butterworths, London (1959)
3. Uhlig, H. H., *J. Electrochem. Soc.*, **109**, 11C (1962)
4. Sillen, L. C. and Martell, A. E., *Stability Constants of Metal-Ion Complexes*, Chemical Society, London (1962)
5. Pourbaix, M., *Atlas of Potential/pH Diagrams*, Pergamon, Oxford (1962) and CEBELCOR (Centre Belge D'etude de la Corrosion) Rapports Techniques. (*See* 'List of Reports' published by CEBELCOR)
6. Pourbaix, M., *Lectures on Electrochemical Corrosion*, Plenum Press, New York, London (1973)
7. Deltombe, E. and Pourbaix, M., *CEBELCOR Rapport Techniques*, **7** (1954) and Proceedings of 6th Meeting CITCE, Butterworths, London, 118 (1954)

8. Ashworth, V. and Boden, P. J., *Corros. Sci.*, **10**, 709 (1970)
9. Townsend, H. E., *Corros. Sci.*, **10** No. 5, 343 (1970)
10. Delahay, P., Pourbaix, M. and Van Rysselberghe, P., *Proceedings*, 2nd Meeting of CITCE, Tamburino, Milan, 118 (1950)
11. De Gromoboy, T. S. and Shreir, L. L., *Electrochim. Acta.*, **3**, 895 (1966)
12. Mattsson, E., *Electrochim. Acta.*, **3**, 279 (1961)
13. Hoar, T. P. and Booker, C. J. L., *Corros. Sci.*, **5**, 821 (1965)
14. Booker, C. J. L., *Proc. Int. Conf. Fundamental Aspects of Stress Corrosion Cracking* (ed. by R. W. Staehle), N.A.C.E., Houston, 178 (1969)
15. Hoar, T. P. and Rothwell, G. P., *Electrochim. Acta.*, **15**, 1037 (1970)
16. Hoar, T. P. and Notman, G. K., *Trans. Inst. Met. Finish.*, **39**, 166 (1962)
17. Evans, U. R., *The Corrosion and Oxidation of Metals*, Arnold, London (1961)
18. Greene, M. D. and Fontana, M. G., *Corrosion*, **15**, 41 (1959) and **15**, 55 (1959)
19. Bockris, J. O'M. and Drazic, D., *Electro-chemical Science*, Taylor and Francis, London (1972)
20. Bockris, J. O'M. and Reddy, A. K. N., *Modern Electrochemistry*, Plenum Press, 1970 (published by Macdonald and Co., London, 1970)
21. Tomashov, N. D., *Theory of Corrosion and Protection of Metals*, Collier-Macmillan Ltd., London (1966)
22. Evans, U. R., *J. Franklin Inst.*, **208**, 52 (1929)
23. Stern, M., *J. Electrochem. Soc.*, **104**, 56 (1957); ibid., *Corrosion*, **13**, 755 (1957); ibid., **14**, 329t (1958); Bonhoeffer, K. F., *Corrosion*, **11**, 304t (1955); and Edeleanu, C., *Nature*, **173**, 739 (1954)
24. Wagner, C. and Traud, W., *Z. Elektrochem.*, **44**, 391–402 (1938)
25. Tafel, J., *Z. Physik. Chem.*, **50A**, 641 (1905); Frumkin, A. N., 'Hydrogen Overvoltage and Adsorption Phenomena' in P. Delahey and C. Tobias's *Advances in Electrochemistry and Electrochemical Engineering*, Vol. 1 (1961); ibid., Vol. 3 (1964); and Vetter, K. J., *Electrochemical Kinetics*, Academic Press, Inc., New York (1967)
26. Uhlig, H., Triadis, D. and Stern, M., *J. Electrochem. Soc.*, **102**, 59 (1955)
27. See Ref. 17, p. 102–103
28. Speller, F., *Corrosion, Causes and Prevention*, McGraw-Hill, New York (1951)
29. Hoare, J. P., *The Electrochemistry of Oxygen*, Interscience, New York (1962)
30. Damjanovic, A., 'The Kinetics of the Mechanism of the Oxygen Reduction', in Bockris, J. O'M. and Conway, B. E. (editors), *Modern Aspects of Electrochemistry*, No. 5, Plenum Press, New York (1969)
31. Whitman, W. and Russell, R., *Ind. Eng. Chem.*, **17**, 348 (1925)
32. Uhlig, H., *Corrosion and Corrosion Control*, Wiley, 103 (1972)
33. Despic, A. R., *Electrochim. Acta.*, **4**, 325 (1961); Drazic, D. M., *Electrochim. Acta.*, **7**, 293 (1962). (*See* also Ref. 20, p. 1080)
34. von Fraunhofer, J. A. and Banks, C. H., *Potentiostat and its Applications*, Butterworths, London (1972)
35. Flade, F., *Z. Physik. Chem.*, **76**, 513 (1911)
36. Franck, U. F., *Z. Naturforschung*, **4A**, 378 (1949)
37. West, J. F., *Electrodeposition and Corrosion Processes*, van Nostrand Reinhold Co. Ltd., London, 90 (1970)
38. Olivier, R., *6th CITCE Meeting*, Poitiers, 1954, Butterworths, London, 314 (1955)
39. Greene, N. D., *Corrosion*, **18**, 136t (1962)
40. Fontana, M. G. and Greene, N. D., *Corrosion Engineering*, McGraw-Hill, 337 (1967)
41. Tomashov, N. D., *Corrosion*, **14**, 299t (1958)
42. Stern, M. and Wissenberg, H., *J. Electrochem. Soc.*, **106**, 759 (1959)
43. Franck, U. F. and Weil, K. G., *Z. Elektrochem.*, **56**, 814 (1952)
44. Young, L., *Anodic Oxide Films*, Academic Press, London (1961)

BIBLIOGRAPHY

Electrochemistry

Bockris, J. O'M. and Reddy, A. K. N., *Modern Electrochemistry*, Macdonald, London (1971)
Bockris, J. O'M. and Drazic, D., *Electro-chemical Science*, Taylor and Francis, London (1972)
Davies, C. W., *Electrochemistry*, Newnes, London (1967)

Delahey, P. and Tobias, C. W. (Eds.), *Advances in Electrochemistry and Electrochemical Engineering*, Vols. 1–9, John Wiley (1972)

Delahey, P., *New Instrumental Methods in Electrochemistry*, Interscience, London (1954)

Potter, E. C., *Electrochemistry, Principles and Applications*, Cleaver-Hume, London (1961)

Textbooks and reference books on corrosion

Chilton, J. P., *Principles of Metallic Corrosion*, 2nd edition, Royal Institute of Chemistry, London (1969)

Evans, U. R., *The Corrosion and Oxidation of Metals*, Arnold, London (1961); 1st Supplementary Volume (1968)

Evans, U. R., *An Introduction to Metallic Corrosion*, Arnold, London (1963)

Fontana, M. G. and Greene, N. D., *Corrosion Engineering*, McGraw-Hill, New York (1967)

Fontana, M. G. and Staehle, R. W., *Advances in Corrosion Science and Technology*, Plenum Press, New York [Vol. 1 (1970); Vol. 2 (1971); Vol. 3 (1973); Vol. 4 (1974)]

La Que, F. and Copson, H. R., *Corrosion Resistance of Metals and Alloys*, van Nostrand-Reinhold, New York (1963)

McKay, R. and Worthington, R., *Corrosion Resistance of Metals and Alloys*, 2nd edition, Reinhold, New York (1963)

Pourbaix, M., *Atlas of Electrochemical Equilibria in Aqueous Solutions*, Pergamon Press, Oxford

Pourbaix, M., *Lectures on Electrochemical Corrosion*, Plenum Press, New York–London (1973)

Rabold, E., *Corrosion Guide*, Elsevier (1968)

Scully, J. C., *The Fundamentals of Corrosion*, Pergamon Press, Oxford (1966)

Tomashov, N. D., *Theory of Corrosion and Protection of Metals*, MacMillan, New York (1966)

Uhlig, H. H. (Ed.), *Corrosion Handbook*, John Wiley, New York (1955)

Uhlig, H. H., *Corrosion and Corrosion Control*, John Wiley, New York (1967)

West, J. M., *Electrodeposition and Corrosion Processes*, van Nostrand-Reinhold, London (1971)

Books on specialised topics

Ailor, W. H., *Handbook on Corrosion Testing and Evaluation*, John Wiley, New York (1971)

Bregman, J. I., *Corrosion Inhibitors*, Collier-MacMillan Co., New York (1963)

Chapman, F. A., *Corrosion Testing Procedures*, Butler and Tanner, London (1964)

Kofsted, P., *High Temperature Oxidation of Metals*, John Wiley, New York (1965)

Kubaschewski, O. and Hopkins, B. F., *Oxidation of Metals and Alloys*, Butterworths, London (1972)

Logan, H. I., *The Stress Corrosion of Metals*, John Wiley, New York (1968)

Miller, J. D. A. (Ed.), *Microbial Aspects of Metallurgy*, Medical and Technical Publishing Co., London (1971)

Morgan, J. H., *Cathodic Protection*, Leonard Hill, London (1959)

Parker, M. E., *Pipeline Corrosion and Cathodic Protection*, Gulf Publishing Co., Houston, Texas

Rhodin, T. N. (Ed.), *Physical Metallurgy of Stress Corrosion Fracture*, Interscience, New York (1959)

Scully, J. C. (Ed.), *The Theory of Stress Corrosion Cracking in Alloys*, NATO, Brussels (1971)

Potential–pH diagrams in research

Markovic, T., 'Ein Thermodynamisches Kriterium für den Kathodischen Korrosionsschutz des Eisens mit Magnesium als Opferanode', *Corros. Sci.*, **2**, 51 (1962)

Clerbois, F. and Massart, J., 'Comportement Anodique des Metaux en Milieu Aqueux—I Utilisation des Courbes Potentiostatiques à la Determination des Mecanismes de Corrosion', *Corros. Sci.*, **2**, 1 (1962)

Nagayama, M. and Okamoto, G., 'The Anodic Behaviour of Passive Iron in Chromic Acid–Chromate Solutions', *Corros. Sci.*, **2**, 203 (1962)

Uusitalo, E. and Heinanen, J., 'Corrosion of Steel in Soft Waters', *Corros. Sci.*, **2**, 281 (1962)

Littlewood, R., 'Diagrammatic Representation of the Thermodynamics of Metal-fused Chloride Systems', *J. Electrochem. Soc.*, **109**, 525 (1962)

Bouet, J. and Brenet, J. P., 'Contribution a l'Etude du Diagramme Tension/pH du Fer en Milieux Sulfures', *Corros. Sci.*, **3**, 51 (1963)

Horvath, J. and Novak, M., 'Potential/pH Equilibrium Diagrams of Some Me–S–H_2O Ternary Systems and Their Interpretation from the Point of View of Metallic Corrosion', *Corros. Sci.*, **4**, 159 (1964)

Ruetschi, P. and Angstadt, R. T., 'Anodic Oxidation of Lead at Constant Potential', *J. Electrochem. Soc.*, **111**, 1323 (1964)

Richaud, H., 'Application des Courbes de Polarisation a l'Etude des Inhibiteurs de l'Aluminium, Notamment en presence de Chlorures', *Corros. Sci.*, **4**, 191 (1964)

Pourbaix, M., 'A Comparative Review of Electrochemical Methods of Assessing Corrosion and the Behavior in Practice of Corrodible Material', *Corros. Sci.*, **5**, 67 (1965)

Pourbaix, M. and Vandervelden, F., 'Intentiostatic and Potentiostatic Methods, Their Use to Predetermine the Circumstances for Corrosion or Non-corrosion of Metals and Alloys', *Corros. Sci.*, **5**, 81 (1965)

Horváth, J. and Hackl, L., 'Check of the Potential/pH Equilibrium Diagrams of Different Metal–Sulphur–Water Ternary Systems by Intermittent Galvanostatic Polarisation Method', *Corros. Sci.*, **5**, 525 (1965)

Johnson, H. E. and Leja, J., 'On the Potential/pH Diagrams of the $Cu–NH_3–H_2O$ and $Zn–NH_3–H_2O$ Systems', *J. Electrochem. Soc.*, **112**, 638 (1965)

Letowski, F. and Niemiec, J., 'On the Potential–pH Diagrams of the $Cu–NH_3–H_2O$ and $Zn–NH_3–H_2O$ Systems', Discussion, *J. Electrochem. Soc.*, **113**, 629 (1966)

Verink, E. D., 'Simplified Procedure for Constructing Pourbaix Diagrams', *Corros. Sci.*, **23**, 371 (1967)

Katoh, M., 'Influence of Chelating Agent (Citric Acid) and F^- on Corrosion of Aluminium', *Corros. Sci.*, **8**, 423 (1968)

Bartonicek, R. and Lukasovská, M., 'A Potential–pH Diagram for the System $Cu–NH_3–Cl^-–H_2O$', *Corros. Sci.*, **9**, 35 (1969)

Pourbaix, M., 'Recent Applications of Electrode Potential Measurements in the Thermodynamics and Kinetics of Corrosion of Metals', *Corros.*, **25**, 267 (1969)

de Nora, O., Gallone, P., Traini, C. and Meneghini, G., 'On the Mechanism of Anodic Chlorate Oxidation', *J. Electrochem. Soc.*, **116**, 147 (1969)

Lennox, T. J., 'Limitations on the Use of Pourbaix Diagrams to Predict De-alloying or Other Corrosion Characteristics', Discussion, *Corrosion*, **26**, 397 (1970)

Verink, E. D. and Parrish, P. A., 'Use of Pourbaix Diagrams in Predicting the Susceptibility to De-alloying Phenomena', *Corrosion*, **26**, 214 (1970)

Townsend, H. E., 'Potential–pH Diagrams at Elevated Temperature for the System $Fe–H_2O$', *Corros. Sci.*, **10**, 543 (1970)

Ashworth, V. and Boden, P. J., 'Potential–pH Diagrams at Elevated Temperatures', *Corros. Sci.*, **10**, 709 (1970)

Verink, E. D. and Pourbaix, M., 'Use of Electrochemical Hysteresis Techniques in Developing Alloys for Saline Exposures', *Corrosion*, **27**, 495 (1971)

Pourbaix, A., 'Characteristics of Localised Corrosion of Steel in Chloride Solutions', *Corrosion*, **27**, 449 (1971)

Cron, C. J., Payer, J. H. and Staehle, R. W., 'Dissolution Behaviour of $Fe–Fe_3C$ Structures as a Function of pH, Potential and Anion; an Electron Microscopic Study', *Corrosion*, **27**, 1 (1971)

Cowan, R. L. and Staehle, R. W., 'The Thermodynamics and Electrode Kinetic Behaviour of Nickel in Acid Solution in the Temperature Range 25–300°C', *J. Electrochem. Soc.*, **118**, 557 (1971)

Brook, P. A., 'A Computer Method of Calculating Potential–pH Diagrams', *Corros. Sci.*, **11**, 389 (1971)

Bardal, E., 'pH and Potential Measurements on Mild Steel and Cast Iron During Periodic Cathodic Polarisation at 20°C and 90°C', *Corros. Sci.*, **11**, 371 (1971)

Bartoníček, R. and Lukasovská, M., 'A Completed Potential–pH Diagram for the System $pH–NH_3–Cl^-–H_2O$', *Corros. Sci.*, **11**, 111 (1971)

Zehr, S. W., 'A Study of the Intergranular Cracking of U—7·5 wt. %, Nb—2·5 wt. % Zr (Mulberry) Alloy in Aqueous Chloride Solutions', *Corrosion*, **28**, 196 (1972)

Leidheiser, H. and Kissinger, R., 'Chemical Analysis of the Liquid Within a Propagating Stress Corrosion Crack in 70:30 Brass Immersed in Concentrated NH_4OH', *Corrosion*, **28**, 218 (1972)

Heidersbach, R. H. and Verink, E. D., 'The Dezincification of α and β-brasses', *Corrosion*, **28**, 397 (1972)

Macdonald, D. D. and Butler, P., 'The Thermodynamics of the $Al–H_2O$ System at Elevated Temperatures', *Corros. Sci.*, **13**, 259 (1973)

Misawa, T., 'The Thermodynamics Consideration for the $Fe–H_2O$ System at 25°C', *Corros. Sci.*, **13**, 659 (1973)

Cowan, R. L. and Kaznoff, A. I., 'Electrochemical Measurements of Corrosion Processes in a Boiling Water Nuclear Reactor', *Corrosion*, **29**, 123 (1973)

1.5 Passivity, Passivation, Breakdown and Pitting

A metal surface is said to be 'passive' when, although it is exposed to an environment under conditions such that reaction of the metal with the environment ought to occur with decrease in the thermodynamic free energy of the system, it remains visibly unchanged for an indefinite period. 'Passivation' of a reacting or of a potentially reactive metal surface means the creation by any means of the passive state upon it. Passivity may often be substantially complete as just defined. It may also be partial either in the sense that the entire metal surface undergoes change but at a much slower rate than does 'active' metal, or in the sense that some areas of the metal surface remain completely passive while others are active.

When iron is immersed in dilute aqueous nitric acid, vigorous reaction occurs. The reaction rate is greater in more concentrated nitric acid; but in 'fuming' acid, iron is either at once passive or becomes passivated after a few seconds of activity, as indicated by the stoppage of an initially very vigorous evolution of gas. Although the formation of ferric nitrate from the metal and the acid is even more thermodynamically favourable in the very concentrated acid than in the dilute, its rate is so small as to be immeasurable except by refined means.

Faraday suggested that passivation is caused by an invisible oxide film on the metal surface, or by an oxidised state of the surface, that prevents contact between the metal and the solution. The oxide-film theory in its modern form, due primarily to Evans[1], has been challenged by proponents of theories of change of the metallic surface involving not film formation but allotropic modification (Smits), electronic modification (Uhlig), or adsorption of ions or molecules (Uhlig, Kolotyrkin)—but it has been supported and extended, with modifications of detail, by the great weight of the experimental evidence accumulated since it was first proposed. Powerful qualitative evidence for the presence of a compact solid film (not necessarily oxide) on passivated metal surfaces is:

1. Mercury does not wet passive metal, although it quickly amalgamates with metal surfaces freed from oxide film by acid pickling or by reduction in hydrogen.
2. Thin transparent films can be detached from passivated iron, becoming visible when they leave the metal basis; this experiment, due to Evans, convinced the late W. J. Müller of the reality of the passivating film, when he performed it in Evans's laboratory in 1931.

3. The change of ellipticity of polarised light on reflection from passive surfaces is consistent with the presence of thin transparent films in the approximate thickness range 1–10 nm on passive surfaces.

This qualitative evidence has been supported by quantitative measurement of the films by many microchemical, ellipsometric and electrochemical techniques, and by electron-diffraction and electron-microscope examination of passive surfaces. The films have also been analysed chemically by electron spectroscopy.

All processes by which metals become oxidised to ionic compounds are in principle anodic processes, in that the metal cation leaves its associated electrons in the metal lattice and becomes instead associated with anions in solution or in an ionic solid; the concomitant cathodic process absorbing the released electrons can often be separated from the anodic process, so that each can be the more conveniently studied. Passivation is one such process that has been greatly clarified by studies of metal anodes under conditions producing passivity, and the following discussion makes extensive use of the phenomena of 'anodic' passivity[2]. 'Chemical' passivity, typified by iron in fuming nitric acid already mentioned and by metals in corrosion-inhibitive solutions, is a slightly more complicated set of phenomena wherein cathodic processes are involved on the same metal, as in 'natural' corrosion.

Anodic Passivity

When a metal is made an anode in an aqueous solution containing only anions with which it forms very sparingly soluble compounds, it often immediately becomes passive. Very small currents flow, but as the applied e.m.f. is increased so as to move the anode potential in the positive direction, anodic processes such as the oxidation of hydroxyl ion or water to oxygen, or of other anions to per-anions, begin. Iron and nickel in dilute sodium hydroxide solution, and lead in sodium sulphate or carbonate solution, demonstrate the effect, although passivation by lead sulphate or carbonate is incomplete and is superseded by further changes. Silver and mercury in chloride solutions are also partially passivated by the formation of films of their sparingly soluble chlorides, but these films are so porous that they grow readily and afford only a minor degree of passivation; these films grow through dissolution of the metal cations followed by their immediate precipitation by the solution anions very close to the metal surface.

When a metal is made an anode in a nearly neutral unbuffered aqueous solution containing only anions with which it forms freely soluble salts, the initial anodic product is usually hydroxide; but this is accompanied by the equivalent amount of hydrogen ion, and the pH quickly falls to the value at which the hydrated metal cations can exist in high concentration in solution, so that the metal dissolves freely. Similar free dissolution is of course observed in initially acid solutions. As the anode polarisation is increased, the current density and the rate of production of soluble cations increases; but at a certain polarisation and current density, passivation occurs. If the polarisation is applied from a galvanostatic circuit, the anode potential rises sharply at the constant current density; if it is applied from a potentiostatic circuit, the current density falls sharply at the constant potential—the 'passivating

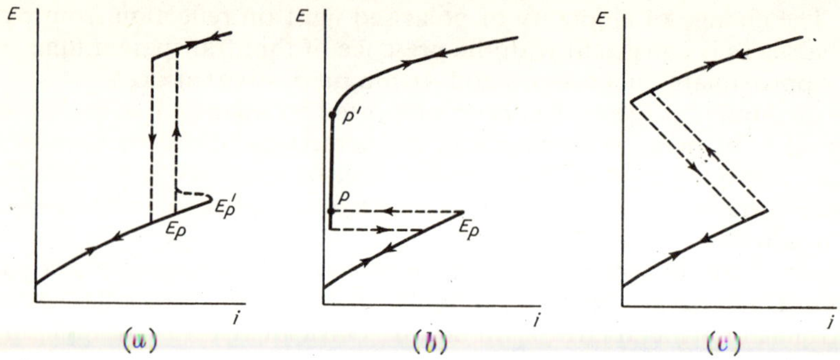

Fig. 1.39 Passivation–activation potential/current-density curves. (*a*) Galvanostatic control, (*b*) potentiostatic control and (*c*) mixed control. The 'hysteresis' between the passivating (potential becoming more noble) and the activating curves is caused by the finite time needed for the passivating film to form and (especially) to disappear. Curves traced very slowly show little hysteresis (but see Fig. 1.40)

potential' E_p (Fig. 1.39*a* and *b*); intermediate current density/potential curves are obtained if the circuit gives neither pure galvanostatic nor pure potentiostatic control (Fig. 1.39*c*).

If the increase in anode polarisation is made fairly rapidly, 'ideal' curves such as those of Fig. 1.39 are closely approached. A slower increase in polarisation, made with a potentiostatic circuit, often gives curves of the general type shown in Fig. 1.40, especially if the anode is horizontal facing upwards or is otherwise situated so as to accumulate anodic products. As was early emphasised by Müller[3], the precipitation of the freely soluble salt from its saturated solution as a magma of small crystals on the anode is often a precursor of true passivation; the effect of this magma is to make the measured, apparent current density considerably less than that flowing at the 'accessible' points of the anode, so that the current density/potential curves are distorted in the sense shown. The formation of such a precipitate has been observed microscopically[3]; in the case of iron polarised in dilute sulphuric acid, the very small crystals in constant motion produce a 'swarming' effect[4]. The onset of passivation on surfaces polarised galvanostatically from the outset with constant current density in the range E_p to E_p' (Fig. 1.39*a*) occurs sooner with larger current density according to simple laws deducible on the basis of the convection, diffusion and migration of the cations away from the anode, with the assumption of the precipitate as a precursor of passivation (Fig. 1.41)[2]. However, if an anode is polarised from the outset to a potential well above the passivating potential, passivation quickly occurs without requiring the previous formation of the magma. The passivating potential frequently corresponds to within a few millivolts to that for a metal/metal-oxide electrode in the solution; one of the higher metal oxides, in the case of metals of variable valency, is usually, but not always, the passivating oxide. Iron sometimes requires a passivating potential some 0·6 V higher than that corresponding to the formation of Fe_2O_3; probably a film containing appreciable Fe(IV) or Fe(VI) is formed[2,5]. Nickel may be passivated by one or more of its several oxides, usually generated directly from the metal[6].

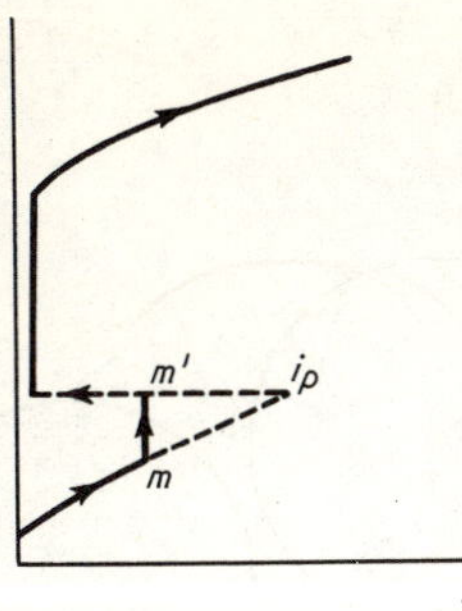

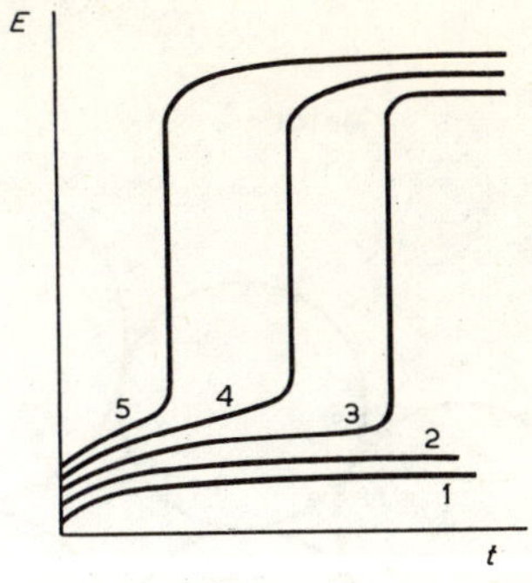

Fig. 1.40 (left) Schematic passivation curve when magma is formed as a precursor to passivation. The real passivating current density i_p is not measured because the magma forms over the range mm' and reduces the effective anode area. The effect becomes less if the curve is traced rapidly

Fig. 1.41 (right) Schematic potential/time curves for a series of constant current densities. 1 Low current density—anode remains active; 2 limiting current density—anode just remains active; and 3, 4, 5 increasing current density—anode passivates after a period τ that is shorter for greater current density. Usually[2], $(i-i_{\text{lim}})\tau^2 \approx$ constant for $\tau \gg 2$ s, and $(i-i_{\text{lim}})\tau \approx$ constant for $\tau \ll 2$ s

Studies of the amount of charge required to produce passivation give information on the thicknesses of passivating films produced on initially film-free surfaces[2]. In alkaline solutions, platinum, gold, cobalt, nickel and iron require only a monolayer of oxide for passivation, as indicated by rise of potential at constant current density, although the oxide thickens somewhat with maintenance of a potential at or above the passivating potential. Copper and silver in alkaline solutions require about four molecular layers of oxide before becoming passive; silver and lead in acid solutions require films of visible thickness. Evidently, the nucleation and growth characteristics of the film substance are vitally important in deciding how thick it must grow before becoming compact enough to produce passivation.

Mechanism of Film Nucleation and Growth

When a metal anode operates in a solution of pH low enough for the thermodynamically easiest process to be the production of aquo-cations, e.g.

$$M + x\text{H}_2\text{O} \rightarrow M^{z+}.x\text{H}_2\text{O} + ze$$

considerable polarisation is often required to make the process proceed at quite moderate rates, e.g. 50–500 mA/cm^2. If the polarisation brings the anode potential up to that for hydroxide or oxide formation at the pH concerned, either of the following processes

$$M + z\text{H}_2\text{O} \rightarrow M(\text{OH})_z + z\text{H}^+ + ze$$

and

$$M + z\text{H}_2\text{O} \rightarrow M\text{O}_{z/2} + {}^z/_2\text{H}_2\text{O} + z\text{H}^+ + ze$$

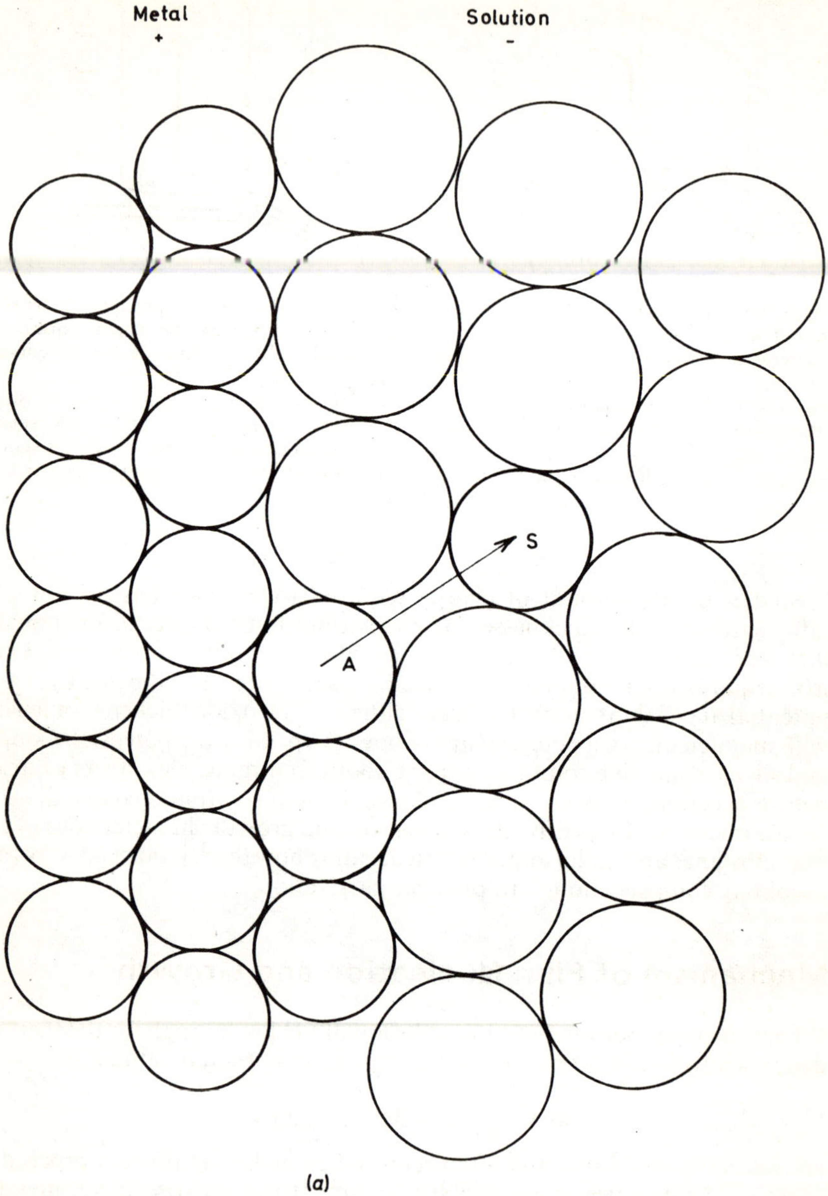

(a)

Fig. 1.42(*a*) Scheme for dissolution. A cation at kink site A becomes a hydrated cation in solution S (metal cations are represented by small circles, water molecules by large circles). (*b*) Scheme for passivation. Protons move from adsorbed water molecules to others, P → P, leaving oxide ions O^{2-} that expand laterally through mutual electrostatic repulsion to accept cations from kink site A and closer-packed sites B with almost equal ease into their film positions F. Once the oxide monolayer is established passivation may become more nearly complete as the film grows by repetition of proton loss and solid-state migration of cations

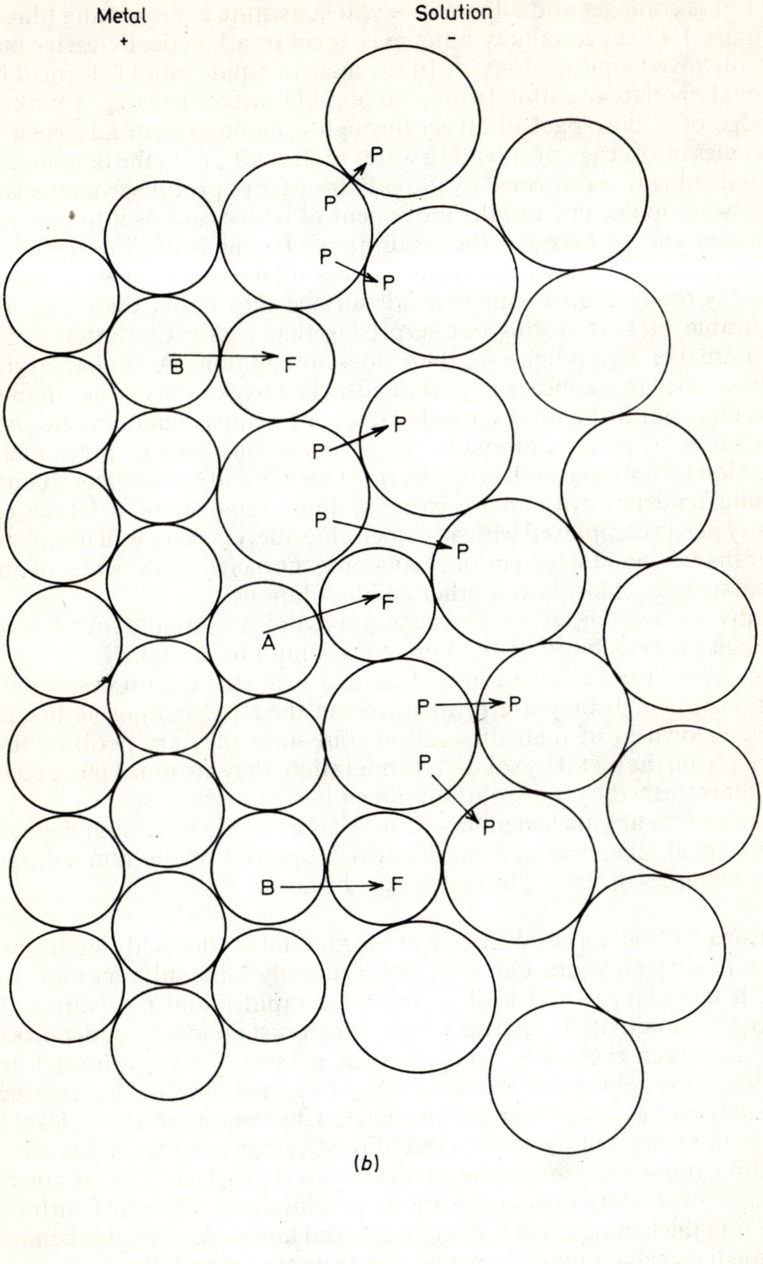

(b)

also becomes thermodynamically possible; if it is *kinetically easier*, solid hydroxide or oxide may be formed preferentially, in contact with the metal, and if it is compact and adherent, it will constitute a passivating film.

Figure 1.42 suggests how films may form in a kinetically easier manner than dissolved aquo-cations[2,7]. In (*a*) such an aquo-cation is formed by the removal of a lattice cation from a favourable surface site (e.g. a kink site in an edge of a close-packed layer) through a monolayer of adsorbed water molecules into a cage of solvating water molecules. In (*b*) the first monolayer of passivating film is formed by the easy transfer of protons from the layer to other water molecules and the movement of lattice cations into positions of minimum energy between the oxide ions. The activation energy for this second process is likely to be much smaller than that for the first, and it can probably occur almost equally at all surface sites rather than only at rare favourable sites. It is thus not surprising that passivation often occurs at much smaller overpotentials than does dissolution. A similar argument applies if the layer consists in part of adsorbed oxy-anions rather than water molecules; but if the layer consists largely of anions containing no oxygen, passivation (at least by an oxide or hydroxide film) may be prevented. The halide ions are strongly adsorbed by most metals and have a strong complex-forming tendency with many cations; thus, removal of a lattice cation already partly complexed with adsorbed chloride, etc. may well be kinetically easier than oxide-film formation, explaining the increased difficulty of obtaining passivity in chloride and other halide solutions.

Many passivating 'oxide' films contain hydrogen, usually found as OH^-, although the possibility of H_2O presence cannot be excluded[8].

The films just described are of course metastable with respect to the solution in which they are formed, owing their passivating nature to the extreme slowness of their dissolution; but since they are perforce formed directly from the metal by solid-state migration, they are much more compact than films formed by a dissolution–precipitation mechanism.

When a film has nucleated on an anode, the further behaviour depends on the electrical, chemical and mechanical properties of the film substance[9]. Some examples of these effects will now be given.

Electrical If the ion conductivity of the film (under the field conditions that exist across it) is low, the film will spread laterally while still very thin—in the limit, a monolayer—and lead to the most rapid initial passivation. If its electron conductivity is relatively high, it can sustain other anodic processes (such as oxygen evolution) on its surface, passing electrons through to the metal, and the field across the film cannot be raised to values high enough to cause substantial ion movement through it. Chromium, iron and nickel form films in this class. On the other hand, if the electron conductivity is very low, the film cannot sustain anodic processes on its surface, and, if current is forced through it by raising the anode potential, ion transport with consequent film thickening occurs. Aluminium and tantalum show this behaviour. Intermediate behaviour is shown by, for example, tin and titanium, on which oxygen evolution and film thickening can both occur, simultaneously in the case of tin and consecutively in the more complicated case of titanium. Stainless steels can grow films to interference-colour thickness in certain solutions.

A passivated anode held at potentials less noble than those at which alternative anodic processes such as oxygen evolution can occur (the region *pp'* in Fig. 1.39*b*) can be shown to pass very small currents, which may be as low as 10^{-10} A/cm^2 (platinum) or as high as 10^{-5}–10^{-4} A/cm^2 (iron under various conditions). This current is caused by very small ion flow in the passivating film; it is increased by increase of anode potential and it may diminish with time as the film thickens[10,11], or may remain nearly constant because of slow film dissolution. Where the line should be drawn between 'complete' and 'partial' passivation, as indicated by the rate of metal transformation, is an arbitrary matter; however, when currents of the order of 10^{-1} A/cm^2 are found, with anodic brightening becoming obvious after a few seconds or minutes, 'passive' is clearly not a proper description of the anode. A partially passive iron anode (in, for instance, aqueous sulphuric acid containing a little Cl^-) dissolves very slowly to mainly Fe(III), not Fe(II).

Chemical If the film substance has a relatively high solubility and rate of dissolution, and dissolves at the film/solution interface as fast as it is formed at the metal/film interface, *brightening* occurs[11–13]. The metal is not passive in any proper sense of the term; although its potential is high, and although the current density/potential relationship is *qualitatively* the same as in passivation, the high current density corresponds to rapid dissolution. Dissolution in the film surface may be assisted by the field, or may be largely a chemical effect, occurring as it does in concentrated acid solutions, where more dilute solutions of the same acid may allow passivity in the true sense. Another mode of breakdown, in part chemical, is the formation of pores in the original film formed when aluminium is passivated in acid solutions; the initial passivity quickly gives way to porous film growth, to visible thickness.

Films of substances with appreciable although sparing solubility, such as lead sulphate and silver chloride, usually grow by deposition from supersaturated solution rather than by ion transport. They are therefore coarse and porous, as compared with the compact films giving nearly perfect passivation, and give only partial passivation.

In some cases, high anodic polarisation can oxidise passivating films to higher oxides that are freely soluble. This gives rise to the *transpassivity* shown, for example, by manganese and chromium when the potentials for the formation of Mn_2O_7 and CrO_3 are reached; it is of practical importance in stainless steels. Alternatively, the higher oxide may be insoluble, and may provide *better* passivation than the lower-valent compound from which it is formed, as in the conversion of lead sulphate to lead dioxide and the production of Fe_2O_3 rather than Fe_3O_4 on iron.

Mechanical The less effective passivating films often break down during prolonged anodic treatment through mechanical cracking or flaking; the lead dioxide layers just mentioned are readily 'crumbled' off lead treated at high anode current density, with oxygen evolution. Generally, thick films are formed in compression and have a limiting thickness beyond which cracking, flaking or crumbling occurs. It is unusual, however, for mechanical breakdown to occur spontaneously in the very thin films giving the most complete passivation, because their strong adhesion to the metal substrate overcomes any compressional effect.

These considerations indicate that the most complete passivation is achieved when the film substance has:

1. Very low ion conductivity.
2. Appreciable electron conductivity.
3. Very low chemical solubility and dissolution rate.
4. A large range of potential in which it is thermodynamically stable.
5. High compressional strength and good adhesion to the metal.

Chemical Passivity

It is generally recognised that so-called chemical passivity, wherein an isolated metal surface is passive and substantially unchanged in a solution with which it has a thermodynamic tendency to react, is often if not always a manifestation of anodic passivity induced by the operation of a suitable cathodic process on the same metal surface. In a classical experiment of Evans[14], iron specimens (prepared in air, and carrying air-formed oxide films) were kept moving rapidly in aerated water, in which stationary iron corroded freely; and it was demonstrated that little corrosion occurred. Bengough and Wormwell[31] showed that similar passivity may be induced under stagnant conditions if the solution is saturated with oxygen at a pressure of several atmospheres. Evidently, a rapid rate of oxygen supply to the metal surface is the factor promoting passivity. The iron surface is originally covered with an air-formed oxide film; any weak places in this become anodes, and iron enters solution, *unless* the cathodic reduction of oxygen on the better parts of the film can occur at a sufficiently high potential and fast enough to polarise the anodes above the passivating potential with a current density beyond the passivating current density (Fig. 1.43). Although the oxygen reduction process is specifically sluggish and easily polarised, the anodic zones are so small that the much larger cathode can none the less, if plentifully supplied with oxygen, induce passivation. In the complete *absence* of oxygen or other cathodic reactant, on the other hand, salt solutions ordinarily regarded as *inhibitive* towards iron—dilute alkalis, phosphates, etc.—become corrosive; although the anions present can precipitate emerging iron cations, the partial blocking of the developing anodes is not nearly as complete by a precipitate as it is when compact iron oxides are produced anodically on the metal surface through the operation of the local cells just considered. Cohen *et al.*[16] and Mayne *et al.*[17] have indeed shown that the best inhibiting films formed in phosphate and similar solutions are in fact *oxide*, not phosphate—although it is very likely that precipitation of sparingly

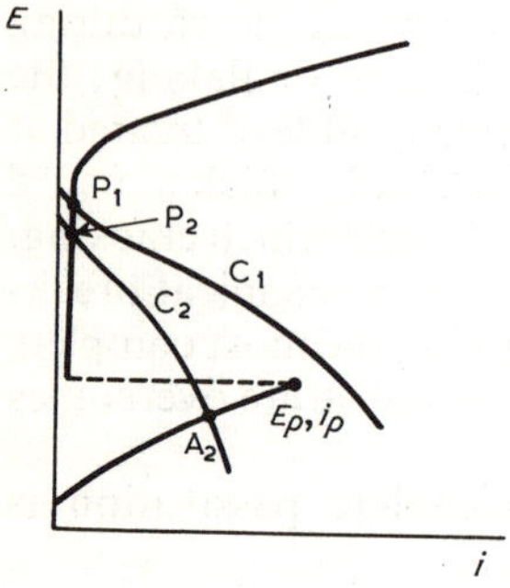

Fig. 1.43 Schematic potential/current-density curves for chemical passivation. The more active cathode C_1 provides sufficient current at a sufficiently high potential to polarise the anode beyond E_p, i_p and to lead to passivity at potential P_1. The less active cathode C_2 cannot passivate the anode but leads to corrosion at the 'active' potential A_2. It can, however, maintain a passivated anode, already polarised above E_p, i_p, at potential P_2. If the surface initially carries an oxide film, so that the anode area is much smaller than the cathode area, a specifically rather feeble cathodic reaction may none the less give enough total current to provide the passivating anode current density i_p

soluble phosphates *assists* passivation by reducing the bare area requiring to be passivated at any instant.

When passivity is induced by fuming nitric acid, it is likely that the original air-formed film is removed by initially rapid dissolution and gas evolution, and is rapidly replaced by a much more protective passivating film over the whole surface. Here, the supply of cathodic reactant is so plentiful and the reaction (reduction of nitric acid, as at the platinum electrodes of the early Grove cell) so specifically rapid at small polarisation, that the cathode can polarise the anode reaction over the whole surface of the metal to the passivating potential, even though this involves a momentarily very high anodic current. This is an instance of a 'mixed' or 'mixture' electrode[18] whereon (different) anodic and cathodic processes take place simultaneously at one and the same potential, i.e. the 'mixture' potential.

Maintenance of Passivity

A passivated surface, once produced, can often be *maintained* passive by conditions much less stringent than those required to *produce* passivation (Fig. 1.43). Any electrochemical or chemical environment that maintains the potential of the surface in the passive range while supplying the very small passive-range current suffices; the very large current demand of active anodes just *before* passivation is not required. The passive potential can be maintained by means of an external e.m.f. and an auxiliary cathode, as in *anodic protection*[19]. It may likewise be maintained by the presence in the solution next to the passive surface of any oxidising agent that, in near-equilibrium with its reduced form on the passive film surface at a potential within the passive range, provides for the very small passive-range anode current by the cathodic reaction $Ox + e \rightarrow Red$. Thus iron first passivated chemically by fuming nitric acid can be maintained passive in other solutions by the presence of small concentrations of relatively weak oxidising agents such as O_2, H_2O_2, Fe^{3+} and Cu^{2+}—provided that powerful depassivators such as halide ions are absent (see below). Stainless steels, easily passivated because both their air-formed oxide films and anodically formed oxide [containing Cr(III)] are 'stronger' and stable at more negative potentials than the corresponding films on iron, are even more easily maintained passive; and *a fortiori* titanium, with its passivating oxide stable at very negative potentials, can be readily chemically passivated and maintained passive by the relatively feeble oxidising agents H_3O^+ and water itself—especially if it is alloyed with small amounts of a noble metal such as palladium, which provides catalytically active spots for their easier cathodic reduction[20]. A similar effect occurs with stainless steels[21].

Breakdown of Passivity

Partial or complete breakdown of passivity, with the onset of corrosion, is brought about by any factor that produces partial or complete removal of the passivating film. In aqueous solutions at ordinary temperatures, such removal can be variously effected, e.g. by electrochemical reduction or oxidation, chemical dissolution, undermining by attack on the underlying metal at film imperfections, or mechanical disruption. Some instances have already been mentioned; we now consider breakdown in more detail.

Electrochemical Breakdown

The oxides of some metals are readily reduced cathodically to the metal in many solutions—copper, tin and lead afford examples; ferric oxide (dissolving extremely slowly in acid solutions) is readily reduced and dissolved to Fe^{2+}(aq.) in many acid solutions—'reductive dissolution'[22]. Cathodic treatment of passivating films in either of these classes activates the metal substrates; the film thickness can indeed be estimated from the charge required to produce the full activation associated with a sharp fall of cathode potential at constant current. An external e.m.f. is often not necessary for activation to occur, if the passivating film contains a sufficiency of 'weak' places where local anodes can develop, and if insufficient oxidising agent other than the film substance is present; then, the film itself becomes the cathodic reactant in local cells and may be completely reduced. There is a fairly sudden fall of potential, starting from a potential a little less noble than that required for passivation—the 'activation' potential, early noted by Flade[23]*.

Other metal oxides, such as those of zinc and, especially, of aluminium, are not cathodically reducible at the potentials obtainable in aqueous media; hydrogen ion is reduced instead. However, the vigorous evolution of hydrogen gas from a passivated zinc surface (the oxide is electron-conducting) usually leads to depassivation even in solutions that do not dissolve zinc oxide chemically. In the case of aluminium, thick passivating oxide films can be broken down by cathodic treatment, even though little hydrogen is at first evolved from the very feebly electron-conducting film; probably protons enter the film without discharge[24] and render it conducting, while aluminium ions may even transfer from film to metal under high field conditions—the metal/film interface is non-aqueous. Furthermore, the combination $2H \rightarrow H_2$ at the metal/film interface may provide hydrogen gas there which can push the film off under the prevailing local high-pressure conditions. The films eventually become detached, with at least partial depassivation, in many solutions.

The 'weak' places in the film leading to self-activation of a passive surface may be present naturally due to the presence in the metal surface of mechanical flaws, inclusions or alloy phases on which the oxide film is more pervious to cation egress than the main film. This is shown by the much more ready activation exhibited by passivated surfaces of abraded or worked metal, surfaces containing extraneous inclusions and surfaces containing more than one alloy phase, as compared with a passivated surface of an electropolished annealed 'clean' single-phase alloy. Generally, metals and alloys that rely on oxide films (whether deliberately produced by passivation or not) for their resistance to corrosion are most resistant when the surface is as uniform as possible. Polished austenitic stainless steel free from surface flaws, and

*Many subsequent workers have used the term 'Flade potential' indiscriminately for the passivation and activation potentials—an unnecessary confusion.

anodised super-pure aluminium, illustrate the practical application of these principles.

'Weak' places may, however, be induced or encouraged to a greater or less degree by the environment; chloride ion is among the most powerful agents leading to self-activation. Its mode of action has been much discussed and is still not unequivocally known; probably several factors come into play. Thus, an early view was that the small size of the chloride ion gives it high 'penetrating power' for oxide films[25]; another, that its chemical nature in forming a strong acid leads to anodic acidity and film undermining in oxide-film pores[26]; yet another, that its strong adsorption on metal lattices prevents repassivation of any momentarily activated surface (however caused). It has been suggested that anion exchange of solution Cl^- with film O^{2-} can occur; this might well lead to growing cation deficiency in the slowly changing film, which in turn would provide easier cation egress through vacancy diffusion, and thus promote 'active' anodic behaviour at favoured spots. Evidence for Cl^- segregation on the surface of the film (perhaps through thermal fluctuations), followed by the removal of a cation by three or four Cl^- ions, has been offered for stainless-steel film breakdown: the rate of breakdown at high potentials is proportional to $[Cl^-_{\text{bulk solution}}]^n$, where n is about 3–4, and the rate is strongly temperature- and potential-dependent[27]. In other cases, e.g. nickel and iron, entry of Cl^- into the film and its slow permeation *without* exchange with O^{2-}, seems likely.

The electrochemical breakdown of passive films by strong anodic oxidation—transpassivity, already mentioned—may be termed *oxidative* dissolution by analogy with *reductive* dissolution. It occurs whenever anode potentials that lead to oxidation of the passivating film to a freely soluble product are reached; it is of practical interest, e.g. whenever stainless steels are used under very strongly oxidising conditions that may oxidise Cr(III) to Cr(VI), CrO_3 being easily soluble. The cathodic process leading to the high anode potential can, of course, be local rather than external; bromine and concentrated nitric acid are cathodic reactants that can produce 'chemical' transpassivity in stainless steels.

Chemical Breakdown

Ordinary chemical dissolution of passivating oxide films is frequently a thermodynamically favourable process, even in the same solution in which passivation has been produced. The higher-valent oxides that are usually the best passivating films owe their protective property largely to the extreme *slowness* of their dissolution under many conditions, rather than to their thermodynamic stability. The dissolution rate of oxides such as Fe_2O_3, Cr_2O_3 and Al_2O_3 is, of course, much accelerated by increase in temperature, and consequently passivation produced in cold acid solutions by such films may often break down in hot. Another important factor in the dissolution rate of such an oxide is its defect structure. Thus ferric oxide ordinarily contains an appreciable concentration of Fe(II) in its lattice, and surface Fe^{2+} appears to provide points of initial dissolution, which is slow but not negligible at room temperature[22]. However, when ferric oxide is formed at anodic potentials well above the thermodynamically reversible value for

$2Fe + 3H_2O \rightarrow Fe_2O_3 + 6H^+ + 3e$—as it has to be if passivation is to be produced—it contains appreciable amounts of Fe(IV) or Fe(VI), and it seems that such oxide is even less readily dissolved in acid solutions than the ordinary oxygen-deficient oxide.

Metals such as copper and zinc, where a bivalent oxide is the highest stable oxide, are much more easily depassivated by chemical dissolution of the film in acid or complex-forming solutions—such oxides are much more rapidly dissolved than the M_2O_3 type. Why this should be has not been satisfactorily explained; the whole question of the 'corrosion' rate of *oxides* awaits systematic examination.

Many 'resistant' alloys show much improved passivation properties as compared with their constituent metals. A particular case is the 1:1 nickel–tin alloy produced by electrodeposition. The films on resistant alloys are often almost 'structureless' when examined by electron diffraction. Such alloys frequently contain a 'basic' metal (e.g. Fe, Ni, Cu) alloyed with one or more 'acidic' metals (e.g. Cr, Si, Al, Zu), and their oxide films are very probably glass-like, containing the glass-forming polyanions or heteropolyanions[28,29].

Breakdown by Undermining

When a passivated surface is exposed to a reagent that can attack the metal substrate rapidly although it attacks the passivating film slowly, undermining with disruption of the film can occur at any fortuitous breaks in the film. An extreme example is the breakdown of most oxide films on aluminium in the presence of mercury, which quickly dissolves any exposed metal—a hazard in the use of mercury thermometers in aluminium vessels, and a useful technique in the production of oxide-film replicas in the electron microscopy of aluminium alloys.

Mechanical Breakdown

Passivating oxide films are subject to damage by bending, stretching, impact, scratching, and other forms of mechanical force. Since they usually consist of brittle material, there is little hope of their remaining adherent, compact and non-porous by ductile extension or compression. Therefore, when a passivating film is utilised in practice for metal protection, it can be relied on only under conditions where it is chemically or electrochemically *self-healing* when the inevitable mechanical damage occurs. Fortunately, many of the well-known uses of passive metals such as stainless steel and titanium are under conditions giving self-healing by chemical processes such as those discussed earlier. There are also, however, many practical cases where 'passivated' material—e.g. stainless steel passivated by nitric acid or chromate treatment—is little better than the same material with a natural air-formed oxide film, because the use involves a corrosive environment that does not give self-healing, and slight mechanical damage soon leads to partial or complete depassivation.

Pitting

Two general forms of the pitting that follows breakdown of a passivated metal or alloy surface can be distinguished: pitting at low potential, following cathodic (or self) activation, and high-potential pitting following anodic action in the presence of film-breakdown factors. Low-potential pitting is 'crystallographic', leading to merging etch pits that can eventually extend to general attack with etching[30]. High-potential pitting takes the form of hemispherical pits that are electrobrightened[13]— they are evidently produced by anodic dissolution in the electrobrightening mode and indeed in many cases can extend and link up to give a completely electrobrightened surface. For electrobrightening a requirement is *random* dissolution independent of the metal crystal structure; this takes place through a randomly defective solid film[12,31] that is none the less a very good *ion* conductor.

Pits of either kind, once begun, lead to a form of 'occluded' corrosion[32]. The rapid production of metal ions within a pit leads to a concomitant increase of anions there[33]—often Cl^-, since this is the commonest 'breakdown' ion—and this exacerbates either pitting mode at the expense of the parallel 'passive' mode still obtaining around the pits. Furthermore, the production of most dissolved metal ions also leads to an increase of acidity within the pit[26,32,34,35]. Pit extension, both inwards and sideways, is thus to some extent autocatalytic, especially if the passivating film remains as a partial cover over the pits.

Actively pitting material can be repassivated by lowering its potential if the pitting has occurred through breakdown at high potential, but the repassivation potential is always lower than the original breakdown potential[36], no doubt because of the 'occlusion' effect. Similarly, material that is suffering from low-potential attack requires a somewhat *higher* potential to repassivate it than does a non-attacked surface.

A factor militating against indefinite pit extension is provided by any movement of the solution within the pit tending to mix it with the main solution. Large pits—beyond a certain size—tend to become washed out by solution movement, especially in vessels carrying heat transfer: aluminium hollow-ware and heat exchangers often begin to pit and then stop, owing to this fortunate effect.

An especially damaging form of 'pitting' occurs when an ordinarily resistant alloy is under tensile or shear stress while exposed to an environment in which it is passive. Slight extension or torsion of the metal cracks the oxide film. Metal at the bared surface begins to dissolve, and repassivation at the base of the micro-pit becomes progressively more difficult as the pit proceeds. Also, the stress-raising effect at the base of the pit becomes progressively greater, and finally repassivation occurs only on the *sides* of the fissure that are under very much smaller stress. The damage is now a 'stress-corrosion crack', and it proceeds by rapid dissolution at its advancing edge while suffering very little attack on the repassivated sides[37–39]. Provided that the repassivation 'chases' the advancing edge at the appropriate rate, the crack maintains acuity and hence the high stress at its edge necessary to produce sufficient metal extension to maintain bare metal and a high dissolution rate there. If the 'chaser' film is too slow to develop, sideways dissolution blunts the crack, which spreads into a wide pit; if it is very fast, it may catch up the advancing edge and repassivate it. Conditions leading to anodic stress-corrosion cracking are thus very critical; this conclusion would be welcome to the practical engineer were it not for the fact that, of all forms of corrosion failure, stress-corrosion cracking is potentially the most disastrous.

T. P. HOAR

REFERENCES

The following references have been selected from the thousands available on several grounds. Some are recent critical reviews, some are pioneer or classical papers, and almost all are easily available to readers of English. It would of course be possible to give several other equally valid lists.

1. Evans, U. R., *The Corrosion and Oxidation of Metals*, Arnold, London, Ch. 7 (1960); *First Supplementary Volume*, Arnold, London
2. Hoar, T. P., 'The Anodic Behaviour of Metals', in *Modern Aspects of Electrochemistry*, No. 2, Ed. J. O'M. Bockris, Butterworths, London (1959); *Corros. Sci.*, **7**, 341 (1967)
3. Müller, W. J., *Die Bedeckungstheorie der Passivität der Metalle und ihre experimentelle Begründung*, Verlag Chemie, Berlin (1933)
4. Stephenson, L. and Bartlett, J. H., *J. Electrochem. Soc.*, **101**, 571 (1954)
5. Nagayama, Y. and Cohen, M., *J. Electrochem. Soc.*, **109**, 781 (1962)
6. de Gromoboy, T. S. and Shreir, L. L., *Electrochim. Acta*, **11**, 895 (1966)
7. Evans, U. R., Ref. 1, Addendum, p. 1042
8. Okamoto, G., *Corros. Sci.*, **13**, 471 (1973)
9. Hoar, T. P., *Proceedings*, 3rd Meeting Int. Comm. Elect. Therm. Kin. (CITCE), Manfredi, Milan, 146 (1952)
10. Franck, U. F. and Weil, K. G., *Z. Elektrochem.*, **56**, 814 (1952)
11. Hoar, T. P. and Mears, D. C., *Proc. Roy. Soc.*, **A294**, 486 (1966)
12. Hoar, T. P. and Mowat, J. A. S., *Nature*, **165**, 64 (1950)
13. Hoar, T. P., Mears, D. C. and Rothwell, G. P., *Corros. Sci.*, **5**, 279 (1965)
14. Evans, U. R., *Industr. Engng. Chem.*, **17**, 363 (1925)
15. Bengough, G. D. and Wormwell, F., in Evans, U. R., *Metallic Corrosion, Passivity and Protection*, Arnold, London, 2nd edn., 297–298 (1946)
16. Cohen, M., *J. Phys. Chem.*, **56**, 415 (1952); Mellors, G. W., Cohen, M. and Beck, A. F., *J. Electrochem. Soc.*, **105**, 332 (1958)
17. Mayne, J. E. O. and Pryor, M. J., *J. Chem. Soc.*, 1831 (1949); Mayne, J. E. O., Menter, J. W. and Pryor, M. J., *J. Chem. Soc.*, 3229 (1950); Mayne, J. E. O. and Menter, J. W., *J. Chem. Soc.*, 103 (1954)
18. Wagner, C. and Traud, W., *Z. Elektrochem.*, **44**, 391 (1938)
19. Edeleanu, C. and Gibson, J. G., *Chem. Ind.*, 301 (1961)
20. Stern, M. and Wissenberg, H., *J. Electrochem. Soc.*, **106**, 755, 759 (1959)
21. Tomashov, N. D., *et al.*, *Problems of Corrosion and Protection of Metals, Report of the All-Union Council for the Corrosion and Protection of Metals* (Akad. Nauk. SSSR), 17, 68 (1956); summarised by Hoar, T. P., *Platinum Met. Rev.*, **2**, 117 (1958)
22. Pryor, M. J. and Evans, U. R., *J. Chem. Soc.*, 3330 (1949); ibid., 1259, 1266, 1274 (1950)
23. Flade, F., *Z. Phys. Chem.*, **76**, 513 (1911)
24. McMullen, J. J. and Pryor, M. J., *Proceedings of the First International Congress on Metallic Corrosion, London, 1961*, Butterworths, London, 52 (1962)
25. Britton, S. C. and Evans, U. R., *J. Chem. Soc.*, 1781 (1930)
26. Hoar, T. P., *Trans. Faraday Soc.*, **45**, 683 (1949)
27. Hoar, T. P. and Jacob, W. R., *Nature*, **216**, 1299 (1967)
28. Hoar, T. P., *J. Electrochem. Soc.*, **117**, 17C (1970)
29. Hoar, T. P., *Proc. 2nd South African Conf. Corrosion* (1972)
30. Pickering, H. W. and Frankenthal, R. P., *J. Electrochem. Soc.*, **112**, 761 (1965)
31. Hoar, T. P. and Farthing, T. W., *Nature*, **169**, 324 (1952)

32. Brown, B. F., Fujii, C. T. and Dahlberg, E. P., *J. Electrochem. Soc.*, **116**, 218 (1969)
33. Frank, U. F., *Proc. 1st Int. Congr. Metallic Corros.*, Butterworths, London, 197 (1961)
34. Hoar, T. P., *Trans. Faraday Soc.*, **33**, 1152 (1937)
35. Edeleanu, C. and Evans, U. R., *Trans. Faraday Soc.*, **47**, 1121 (1951)
36. Pourbaix, M., *et al.*, *Corros. Sci.*, **3**, 239 (1962)
37. Hoar, T. P., *Corrosion*, **19**, 331t (1963)
38. Hoar, T. P. and Jones, R. W., *Corros. Sci.*, **13**, 725 (1973)
39. Hoar, T. P. and Slater, J. E., *Corros. Sci.*, **14**, 415 (1974)

1.6 Localised Corrosion

The various types of localised corrosion have been enumerated in Table 1.2 in Section 1.1, and many of them are dealt with in some detail in other sections of this volume. For this reason this section will be confined to a consideration of the factors that give rise to crevice corrosion, filiform corrosion, pitting, selective leaching and erosion–corrosion and of the mechanisms of these forms of localised attack.

Corrosion can range from the highly uniform (chemical or electrochemical polishing) to the highly localised such as occurs during pitting, intergranular attack and stress-corrosion cracking. Uniform, or near-uniform, corrosion without doubt accounts for the greater proportion of metal deterioration in terms of both mass of metal converted to corrosion products and cost. However, although detrimental it is at least predictable on the basis of laboratory and field testing, so that allowance can be made for it in the design of a structure (*see* Section 10.2). In completely uniform corrosion the anodic and cathodic sites are physically inseparable, and although this type of corrosion occurs in many practical situations (e.g. chemical and electrochemical polishing, passivity) corrosion is more usually near-uniform, particularly in natural environments. Thus the appearance of the rust on steel that has been exposed to the atmosphere for some time may give the impression that attack has been quite uniform, but removal of the rust with an inhibited acid will reveal that the surface is undulating, indicating that although the whole surface has corroded the rates at different areas of the surface are not uniform. This is due to a variety of factors associated with the structure of the metal, the nature of the surface, deposits of dirt and corrosion products, etc.

Localised corrosion, on the other hand, is far less predictable, and this applies particularly to pitting in which the location of the pits on the metal surface and their distribution and size will depend upon the precise structure of the metal and environmental conditions prevailing. In pitting, the rate of corrosion into the metal is significantly greater than that parallel to the metal surface, but a precise distinction between highly localised attack and pitting is largely arbitrary. Champion[1] distinguishes between 'semi-local' corrosion and pitting by defining the latter as a pit geometry in which the ratio of average width to average depth of the attacked area is 4:1 or less; however, a ratio of 1:1 is a generally more acceptable definition of a pit. Pitting is a particularly insidious form of corrosion, since although the extent of the reaction (moles corroded) is very small the fact that the attack is predominantly into the metal results in rapid perforation when thin metal sections are involved with consequent leakage of the fluid contained in the structure. Pitting is usually regarded as more detrimental than near-uniform corrosion, but this is not always the case, and Hoar[2] has pointed out that whereas uniform corrosion of a bearing could lead to misalignment and seizure, pitting could be beneficial in providing pockets for the lubricant. Figure 1.44 shows the various types of attack[3] that can range from completely uniform to highly localised and Fig. 1.45 how a pit can initiate a stress-corrosion crack[4] thus transforming one type of localised attack into another.

Although the form of pits and their rate of propagation into the metal is far less predictable than near-uniform attack, their location is frequently determined by the metallurgical structure of the metal and the geometry of the system. Thus in certain alloys (*see* Section 1.3) attack is localised at grain boundaries leading to intergranular corrosion, in which the loss of metal may be far greater than that predicted by Faraday's law since grains of the

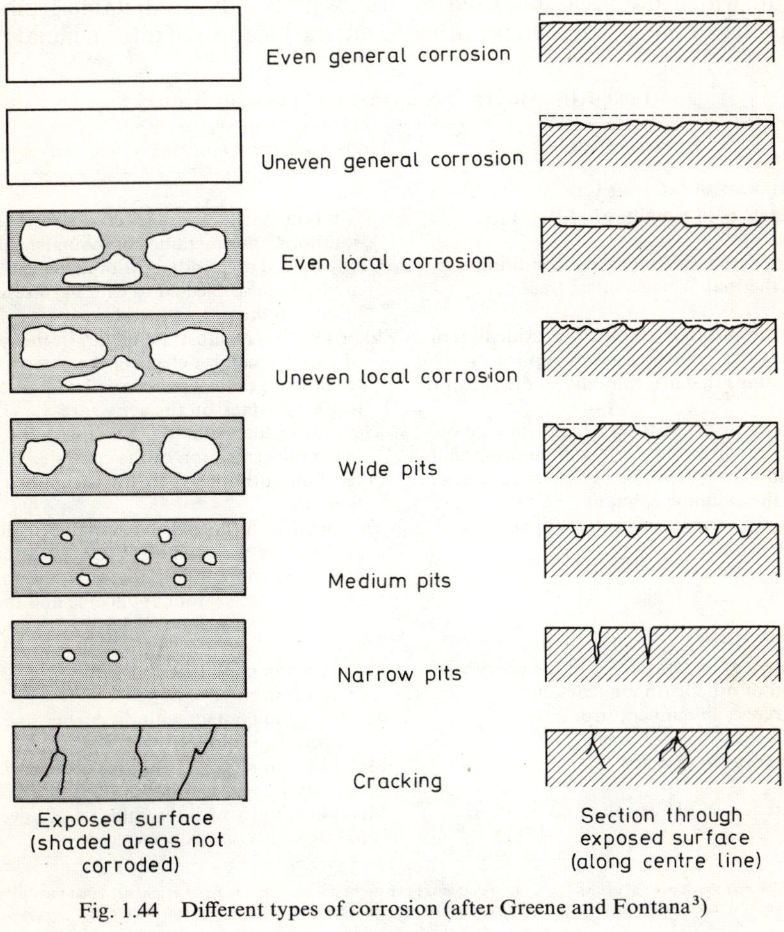

Fig. 1.44 Different types of corrosion (after Greene and Fontana[3])

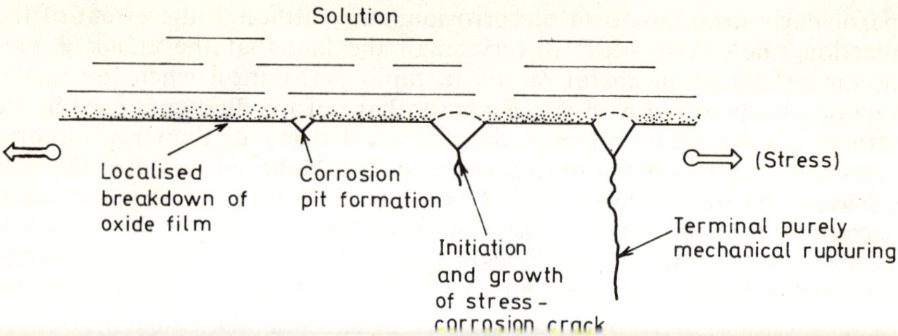

Fig. 1.45 Breakdown of oxide film leading to a pit and crack when a high-strength steel is subjected to a tensile stress in a chloride solution (after Brown[4])

metal may become detached from the substrate with a consequent rapid decrease in the cross-section leading to cracking and fracture. Similar considerations apply to localised attack resulting from dissimilar metals in contact (bimetallic corrosion), crevices, impingement of high velocity waters, etc. in which the area subjected to attack is usually predictable from the geometry of the system. On the other hand, the location of pits on metals that

Table 1.16 Heterogeneities that lead to localised attack*

Heterogeneity	*Area of metal that is predominantly anodic*
METAL AND METAL SURFACE	
Differences in metallurgical structure	Grain boundaries, more reactive phases (solid solutions, intermetallic compounds, etc.)
Differences in metallurgical condition due to thermal or mechanical treatment	Cold-worked areas adjacent to annealed areas, metal subjected to external stress anodic to unstressed metal
Discontinuities in conducting oxide film or scale or discontinuities in applied metallic coatings that are cathodic to the substrate	Exposed area of substrate metal. In the case of passive metals defects in the passive film result in an active–passive cell with intense localised attack on the active area
Crevices or deposits on a metal surface or any other geometrical configuration that results in differences in the concentration of the cathodic reactant	The area of the metal in contact with the lower concentration of the cathode reactant, although there are exceptions to this rule
Dissimilar metals in contact (bimetallic corrosion)	The metal with the more negative corrosion potential in the environmental conditions prevailing (note that the standard electrode potentials are seldom applicable and the galvanic series can be misleading)
ENVIRONMENT	
Differences in aeration or in the concentration of other cathode reactants	Metal area in contact with the lower concentration
Differences in temperature	Metal area in contact with the higher temperature solution
Differences in velocity	Metal in contact with solutions of higher velocity
Differences in pH or salt concentration	Metal in contact with the solution of lower pH or higher salt concentration

*The table provides a general indication of the area that is likely to be anodic, but it must be emphasised that there are many situations in which the heterogeneity will have no effect or where the converse to the above may occur.

depend upon passivity for their low corrosion rate is frequently unpredictable, and the pits will tend to be randomly dispersed on a metal surface that is apparently quite uniform.

Although corrosion is due to the thermodynamic instability of a metal in a specific environment, and although in many metal/environment systems attack will tend to be uniform, there are a variety of factors associated with the metal, the environment and the geometry of the system that may result in the attack being localised.

Table 1.16 gives examples of heterogeneities and geometrical factors that lead to localised attack, but it must be emphasised that they must be regarded as general guidelines, and whether or not the heterogeneity leads to localised attack will depend on the environmental conditions prevailing; stressed areas of a metal may be predominantly anodic compared with unstressed areas in a mild environment, but will have very little effect if the environment is an aggressive acid; a sensitised stainless steel will suffer enhanced intergranular attack in nitric acid, but will be unaffected if the environment is non-aggressive, e.g. milk (*see* Section 1.3).

Principles

Unfortunately, there is no general theory that will explain all the forms of localised attack that occur with the variety of metal/environment systems encountered in practice, e.g. the mechanism of the pitting of stainless steels in Cl^--containing solutions is quite different from the dezincification of brass in a fresh natural water. Nevertheless, many of the following factors play an important part in most forms of localised attack:

1. The cathode/anode area relationship.
2. Differential aeration.
3. pH changes at the cathodic and anodic sites.
4. Corrosion products (films) present initially on the metal surface and those formed during the corrosion reaction.

Anodic and Cathodic Current Densities

In Section 1.4 it was pointed out that a fundamental principle of corrosion is that the sum of the rates of the cathodic reactions must equal the sum of the rates of the anodic reactions, irrespective of whether the attack is uniform or localised:

$$\Sigma I_a = \Sigma I_c \qquad \text{...(1.102)}$$

Furthermore, if attack is *uniform*, and assuming that there is only a single predominantly anodic and cathodic reaction, then

$$\frac{I_a}{S_a} = \frac{I_c}{S_c} \quad \text{or,} \quad i_a = i_c \qquad \text{...(1.103)}$$

since the area of the cathode S_c equals the area of the anode S_a. On the other hand, if attack is localised $S_a < S_c$, and

$$i_a > i_c \quad \text{or,} \quad \frac{i_a}{i_c} > 1$$

and the larger the ratio $i_a : i_c$ the more intense the attack.

Thus localised attack usually involves a corrosion cell consisting of a large cathodic area and a small anodic area, and since I_a must equal I_c, the effect will become more pronounced the higher the rate of the cathodic process and the larger the effective area of the cathode.

This principle may be exemplified by considering the corrosion of steel in an oxygenated water, in which attack will be uniform providing the surface is free from a magnetite scale (millscale) and the geometry of the system is such that the concentration of dissolved oxygen is the same on all parts of the metal surface (Fig. 1.46*a*). If, however, a discontinuous film of magnetite Fe_3O_4 (a stable oxide of high electronic conductivity) is present on the surface, attack will be localised at the discontinuity owing to the large cathode area/small anode area relationship (Fig. 1.46*b*), and the more rapidly the oxygen arrives at the surface of the millscale the more rapid is the attack on the exposed steel (Fig. 1.46*c*).

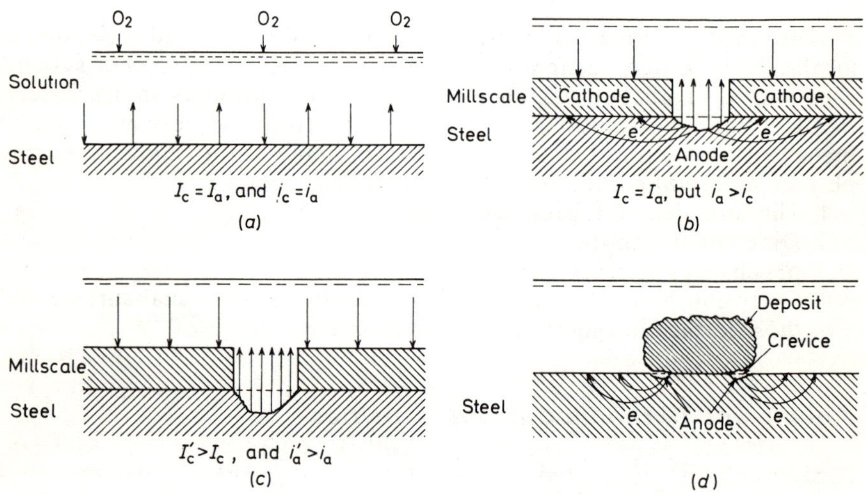

Fig. 1.46 Localised attack due to discontinuity in millscale or a deposit at a steel surface. (*a*) Uniform corrosion, (*b*) localised attack at a discontinuity in millscale, (*c*) increase in rate due to increase in the supply of oxygen and (*d*) crevice formed by deposit. (Arrows pointing downwards represent the cathodic current, and arrows pointing upwards the anodic current)

A similar effect can be produced if a crevice is present in the steel, since the geometry of the system is such that whereas oxygen can diffuse readily to the metal surface outside the crevice it can only gain access to the metal within the crevice through its very narrow mouth (Fig. 1.46*d*), and the large cathode:anode area ratio leads to localised attack of the metal within the crevice.

From these two examples, which as will be seen subsequently, present a very oversimplified picture of the actual situation, it is evident that macro-heterogeneities can lead to localised attack by forming a large cathode/small anode corrosion cell. For localised attack to proceed, an ample and continuous supply of the electron acceptor (dissolved oxygen in the example, but other species such as the H^+ ion and Cu^{2+} can act in a similar manner) must be present at the cathode surface, and the anodic reaction must not be stifled by the formation of protective films of corrosion products. In general, localised attack is more prevalent in near-neutral solutions in which dissolved oxygen is the cathode reactant; thus in a strongly acid solution the millscale would be removed by reductive dissolution (*see* Section 12.2) and attack would become uniform.

Localised attack can, however, occur on a surface of metal that is apparently uniform, and this occurs particularly with the highly passive metals that depend on a thin invisible protective film of oxide for their corrosion resistance. In such cases submicroscopic defects in the passive film may form the sites at which pits are initiated, thus giving rise to a situation similar to that shown in Fig. 1.46.

Differential Aeration

It can be seen from Table 1.16 that differences in composition (nature or concentration) of the environment can lead to localised attack, and in Section 1.4 it was shown how differences in the activity of silver ions can give rise to a reversible concentration cell in which the silver electrodes in contact with the solution containing the lower and higher concentration of Ag^+ ions are the anode and cathode, respectively. Concentration cells of this type are rare in practice, but can occur during the corrosion of copper and copper alloys.

However, most solutions are in contact with atmospheric oxygen, and geometrical situations can arise where transport of oxygen through the solution by convection (natural or forced) and diffusion to one part of the metal occurs rapidly, whereas it is slow or even negligible at another (Fig. 1.46*d*). Under these circumstances the concentration of oxygen will be higher at one part of the metal surface and lower at the other, and this situation can lead to localised attack. Evans[5,6] in his Differential Aeration Principle stated that any geometrical factor that results in a higher concentration of oxygen at one part of a metal surface and a lower concentration (or zero concentration) at another will result in the former becoming the cathode and the latter the anode of the corrosion cell, with consequent localised attack. The evidence for this principle was based on a cell divided into two components by a porous diaphragm in which two steel electrodes were immersed in KCl solution (Fig. 1.47). Oxygen was bubbled into one compartment, and it was shown from the direction of the transfer of the electrons that the steel electrode in this compartment became the cathode, whilst the steel in the non-oxygenated compartment became the anode of the differential aeration cell.

At first sight it might appear that differential aeration could be explained in terms of a reversible oxygen concentration cell, for which

$$E = \frac{0{\cdot}059}{4}\log\frac{c_{\mathrm{II}}}{c_{\mathrm{I}}} \qquad \ldots(1.104)$$

where c_{II} and c_{I} are the concentrations of oxygen at the aerated and non-aerated electrodes, respectively. However, since a reversible oxygen electrode is extremely difficult to achieve in practice, and since the potentials are irreversible corrosion potentials involving at least two electrode reactions, this explanation is not valid.

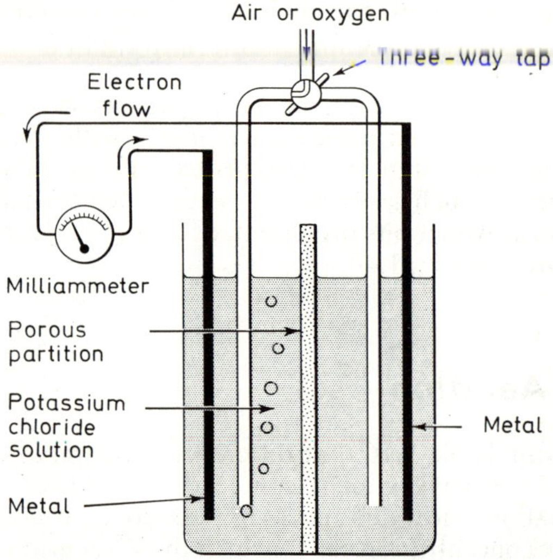

Fig. 1.47 Early form of the Evans 'differential aeration cell'. (Courtesy U. R. Evans)

In Section 1.4 (*see* Fig. 1.31*b*) it has been shown that when a corrosion reaction is controlled by the rate of oxygen diffusion, both the rate of corrosion and the corrosion potential increase with i_L, the limiting current density, i.e.

$$i_L = i_{\mathrm{corr.}} = kc_{O_2} \qquad \ldots(1.105)$$

Thus if the two pieces of steel in the differential aeration cell are not connected together the aerated steel will have a higher corrosion rate and more positive potential than the non-aerated steel. On short-circuiting the two pieces of steel, electrons will be transferred from the more negative to the more positive electrode (aerated steel), and this will result in an increase in the corrosion rate of the former and decrease in the rate of the latter, the potentials becoming equal, i.e. the potential of the aerated electrode becomes more negative and that of the non-aerated electrode more positive (*see* Fig. 1.62). On the other hand, if the two electrodes are connected through a high resistance, the non-aerated electrode will corrode at a lower rate than the aerated electrode.

It is evident from this description of the operation of the differential aeration cell that other factors must be involved if the differential aeration principle is to operate in the manner described by Evans, and this involves a consideration of the pH at the interfaces, the nature of the electrolyte, the anodic reaction and ohmic resistance effects.

Pourbaix[7,8] has shown by means of *E–i* diagrams (Wagner and Traud type) that the differential principle is applicable only at certain pH values, and the situation when the pH is the same in *both* the aerated and non-aerated zones is as follows:

1. pH$<$8; differential aeration results in a small increase in the electrode potentials of both zones with a consequent increase in both corrosion rates.
2. pH $\approx 8-10$; the differential aeration cell operates normally, and although the potential of the aerated zone increases, its corrosion rate decreases owing to passivation; the potential and corrosion rate of the non-aerated zone increases; a high current flows from the non-aerated to the aerated zone.
3. pH $\approx 10-13$; aeration causes a significant increase in the potentials of both zones, without appreciable current flow.

It is now appreciated that for the differential aeration principle to apply the aerated zone must become passive, and that whether this occurs or not will depend on the initial pH of the solution, the change in pH produced by the OH^- ions resulting from the oxygen-reduction reaction and on the potential of the electrode. Similar conclusions were drawn by Bianchi[9] during a study of the corrosion of zinc, and it was observed that differential aeration did not occur when a buffered solution was used, since the pH at the aerated zone was unable to increase to a value sufficient to cause passivation; under these circumstances attack occurred at the water line. Schaschl and Marsh[10] showed that differences in the concentration of NaCl may result in differences in oxygen concentration, the higher the salt concentration the lower the solubility of oxygen so that these areas become the anodic zones of the differential aeration cell. Irrespective of salt concentration, however, the predominantly cathodic zone was always the one where the oxygen concentration was highest, and the cell current was found to be proportional to the difference in oxygen concentration in the solutions in the two compartments. They also point out that although initially the non-aerated zone may not be completely oxygen-free, any oxygen present is rapidly depleted in oxidising the $Fe(OH)_2$ to $Fe(OH)_3$.

Many of the studies of differential aeration have been carried out in the laboratory using a differential aeration cell containing sodium chloride solution, and measuring the current flowing in the external circuit when the electrodes are short-circuited. Although the results obtained in this way are useful in studying the mechanism of differential aeration they are sometimes misleading, and in certain situations the more highly aerated zone (the cathode of the macrocouple) corrodes at a higher rate. Tomashov[11] found that the less aerated zones of steel piles driven into the sea bed corroded at a slower rate than the more aerated zones in spite of the fact that the former was the anode and the latter the cathode of the macro-cell. He points out that this situation can arise where an increase in aeration has little effect on the anodic reaction, but markedly decreases the cathode polarisation of the local-action cells. This situation applies to sea-water in which the high concentration of Cl^- ions prevents anodic passivation of the steel, and under these circumstances the more highly aerated zone will corrode at a higher rate. A further factor that must be taken into account is the electrical resistance of the metal separating the two, since if this is significant the two zones will not be short-circuited, a situation that applies in the case of the steel piling[11].

Examples of differential aeration Before considering the factors that complicate localised attack it is relevant to refer to some of the classical work of U. R. Evans[6] on differential aeration. Perhaps the simplest example, although in corrosion science nothing is as simple as it first appears, is the distribution of anodic and cathodic sites produced when a drop of an electrolyte solution (e.g. NaCl or Na_2SO_4) is placed on a horizontal steel surface. Assuming that no dissolved oxygen is present initially and that mass transfer occurs solely by diffusion, it is apparent that transport of oxygen from the atmosphere to the surface of the metal by diffusion will occur most rapidly through the thin layer of solution at the periphery of the drop and most slowly at the centre of the drop. Thus the metal surface adjacent to the periphery will become the cathode of the differential aeration cell whilst the metal in the interior will become the anode (Fig. 1.48*a* and *b*). The OH^- ions formed by the cathodic reduction of dissolved oxygen and the Fe^{2+} ions formed by anodic dissolution will be brought together by migration and diffusion and will combine to form $Fe(OH)_2$ that is rapidly oxidised by the dissolved oxygen to rust, $Fe_2O_3.H_2O$. If the electrolyte contains a small

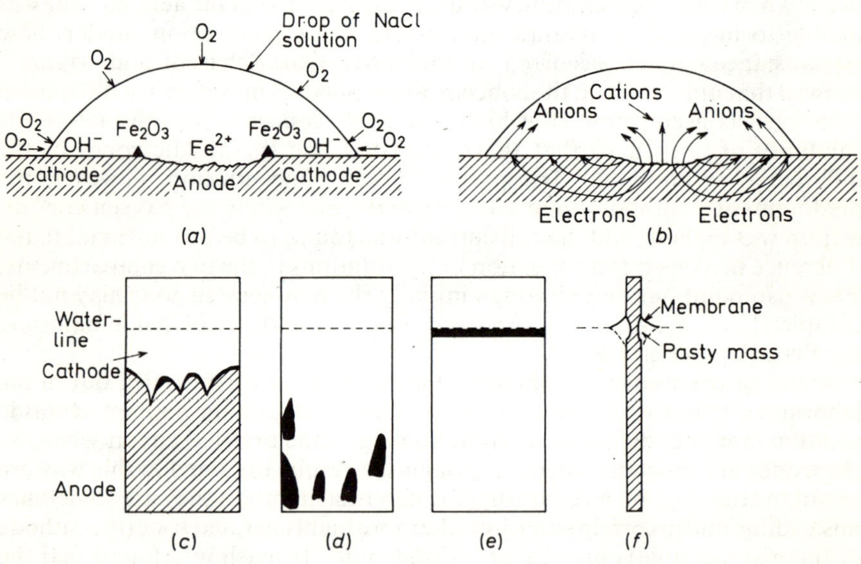

Fig. 1.48 Examples of differential aeration cells. (*a*) and (*b*) Differential aeration cells formed by the geometry of a drop of NaCl solution on a steel surface; (*c*) differential aeration cell formed by the geometry of a vertical steel plate partly immersed in a NaCl solution. Increasing concentrations of Na_2CO_3 decrease the anodic area (*d*) until at a sufficient concentration attack is confined to the water line (*e*); (*f*) shows the membrane of corrosion products formed at water line (after Evans[6])

concentration of ferroxyl indicator (potassium ferricyanide plus phenolphthalein) the Fe^{2+} ions formed at the anodic area are revealed by the formation of Prussian blue and the OH^- ions formed at the cathodic area by the pink colour of phenolphthalein in alkaline solutions.

It should be noted that since the rust is formed at a position inbetween the anodic and cathodic sites it will not influence the kinetics of the corrosion reaction.

A similar situation arises when a vertical metal plate is partly immersed in an electrolyte solution (Fig. 1.48*c*), and owing to differential aeration the upper area of the plate will become cathodic and the lower area anodic. With time the anodic area extends upwards owing to the mixing of the anolyte and catholyte by convection and by the neutralisation of the alkali by absorption of atmospheric carbon dioxide.

Severe attack frequently occurs at a water-line, which in practice can range from structural steel partly immersed in a natural water to a lacquered tin can used for containing emulsion paint. This can be illustrated by adding increasing amounts of sodium carbonate to a sodium chloride solution in which a steel plate is partly immersed (Fig. 1.48*c*, *d* and *e*). With increase in concentration of the inhibitor, attack decreases and becomes confined to the water-line. The attack at the water-line is intense and is characterised by a triangular pasty mass of corrosion products bounded on the upper surface by a dark-brown membrane that follows the contour of the water-line. The mechanism of water-line attack is not clear, but it is likely that the membrane of corrosion products results in the formation of an *occluded cell*, in which the anolyte and catholyte are prevented from mixing. These occluded cells are discussed in more detail subsequently.

pH Changes During Localised Attack

The majority of the different types of localised attack dealt with in this section occur in near-neutral solutions containing dissolved oxygen, and the pH changes that occur at the anodic and cathodic sites are of fundamental importance when the reaction products are prevented from mixing by the geometry of the system.

It has been shown in Section 1.4 that cathodic reduction of dissolved oxygen may result in an increase in pH of the solution in the vicinity of the metal

$$O_2 + H_2O + 2e = 2OH^- \quad \text{...(1.106)}$$

and a similar consideration applies if the cathodic reaction involves hydrogen evolution. It is also evident that this increase in pH and the presence of dissolved oxygen will favour passivation of the metal surface so that its potential will increase and it will become the cathode of the macro-corrosion cell.

At the anodic site the metal will form metal cations

$$M \rightarrow M^{2+} + 2e \quad \text{...(1.107)}$$

and if it is assumed that the metal hydroxide is the thermodynamically stable phase in the solution under consideration, the metal ions will be hydrolysed by water with the formation of H^+ ions

$$M^{2+} + 2H_2O \rightarrow M(OH)_2 + 2H^+ \quad \text{...(1.108)}$$

Thus if the cathodic and anodic sites are separated from one another by the geometry of the system and if the solution is relatively stagnant the pH of the anolyte will decrease whereas that of the catholyte will increase.

It should be noted that whereas a completely soluble hydroxide (e.g. NaOH) will give a solution of high pH in which the pH will increase with concentration of the hydroxide, the pH of a solution of a sparingly soluble hydroxide will depend upon the equilibrium constant for hydrolysis and the activity of metal ions.

The equilibrium between the ions of a metal M_i and its hydroxide (or hydrated oxide) M_0 due to hydrolysis can be written in the form[12]:

$$M_i + cH_2O \rightarrow M_0 + mH^+ \qquad ...(1.109)$$

and since

$$\log K = \frac{-\Sigma v_i \mu_i^\ominus}{5710}$$

then

$$\log K = \frac{-[(\mu_{M_0}^\ominus + m\mu_{H^+}^\ominus - (\mu_{M_i}^\ominus + c\mu_{H_2O}^\ominus)]}{5710} \qquad ...(1.110)$$

Thus for the equilibrium

$$Fe^{3+} + 3H_2O = Fe(OH)_3 + 3H^+ \qquad ...(1.111)$$

$$\log K = \log \frac{a_{H^+}^3}{a_{Fe^{3+}}} \qquad ...(1.112)$$

and $\log K$ can be evaluated as $-4{\cdot}84$ from the standard chemical potentials of the species involved in the equilibrium. Thus from equation 1.112

$$pH = \tfrac{1}{3}(4{\cdot}84 - \log a_{Fe^{3+}}) \qquad ...(1.113)$$

and, for example, when $a_{Fe^{3+}} = 10^{-2}$ the pH will be 2·3.

Similar considerations will apply to other metal hydroxides, and Table 1.17[13] gives the hydrolysis reactions and equilibrium pHs for metal ions that

Table 1.17 Hydrolysis reactions and equilibrium pHs for metal ions of significance in acid formation in an occluded cell in stainless steels*

Hydrolysis reaction	*Equilibrium* pH
$Fe^{2+} + 2H_2O \rightleftharpoons Fe(OH)_2 + 2H^+$	$pH = 6{\cdot}64 - \tfrac{1}{2}\log a_{Fe^{2+}}$
$Cr^{3+} + 3H_2O \rightleftharpoons Cr(OH)_3 + 3H^+$	$pH = 1{\cdot}53 - \tfrac{1}{3}\log a_{Cr^{3+}}$
$Ni^{2+} + 2H_2O \rightleftharpoons Ni(OH)_2 + 2H^+$	$pH = 6{\cdot}1 - \tfrac{1}{2}\log a_{Ni^{2+}}$
$Mo^{3+} + 2H_2O \rightleftharpoons MoO_2 + 4H^+ + e$	$pH^\dagger = (0{\cdot}311 - 0{\cdot}059\log a_{Mo^{3+}} - E^\ominus)/0{\cdot}236$
$Mn^{2+} + 2H_2O \rightleftharpoons Mn(OH)_2 + 2H^+$	$pH = 7{\cdot}66 - \tfrac{1}{2}\log a_{Mn^{2+}}$

*Data after Pourbaix[13].
†$E^\ominus = -0{\cdot}20$ V (*vs.* S.C.E.).

are of significance in acid formation during the corrosion of stainless steels, and it is evident that metal ions like Cr^{3+} and Mo^{3+} will have the greatest effect in the pH decrease. However, this treatment is valid only if the metal forms simple cations, and if complex ions are formed the pH may be even lower than that predicted from equilibria of the type given in equation 1.109.

Cells of this type, in which the anolyte and catholyte are separated from one another, are referred to as *occluded* cells, and they may be formed by (*a*) the original geometry of the system (crevices), (*b*) by a corrosion reaction that results in a pit or crack, or (*c*) by the formation of a membrane of insoluble corrosion products that form over the anodic site. In these occluded cells the pH of the solution at the cathodic sites will tend to increase and the pH at the anodic sites to decrease, thus stimulating attack at the latter.

The development of acidity within an occluded cell is by no means a new concept, and it was used by Hoar[14] as early as 1947 in his 'Acid Theory of Pitting' to explain the pitting of passive metals in solutions containing Cl^- ions. According to Hoar the Cl^- ions migrate to the anodic sites and the metal ions at these sites hydrolyse with the formation of HCl, a strong acid that inhibits the formation of a protective film of oxide or hydroxide. Edeleanu and Evans[15] followed the pH changes when aluminium was made anodic in Cl^- solutions and found that the pH decreased from 8·8 to 5·3.

However, it is only in recent years that attempts have been made to determine the actual pH within an occluded cell. Thus Brown *et al.*[16] determined the pH of the solution exuding from a propagating crack in a high-strength steel immersed in neutral NaCl (using indicator paper, an Sb/Sb_2O_3 electrode and a glass electrode) and found that it decreased to about pH 3. Wilde[17] studied hydrogen permeation through a 12% Cr martensitic steel membrane by means of the Devanathan permeation cell (*see* Sections 9.1 and 20.2) and showed that in 3% NaCl permeation occurred when the cathode side of the membrane was not cathodically polarised, indicating the formation of acid within the pits. Both these workers were concerned with stress-corrosion cracking of the alloys being studied, and both concluded that the mechanism of failure of the alloys was due to hydrogen absorption and hydrogen cracking resulting from the formation of acid.

Suzuki, Yamake and Kitamura[13] determined the pHs, chloride ion concentrations, metal ion concentrations and the potentials of artificial pits in Fe, Cr, Ni and Mo, and in three austenitic stainless steels during anodic polarisation in 0·5 N NaCl at 70°C. In the case of the pure metals the pH values were found to be lower than those calculated from the metal ion concentrations (Table 1.17), and the experimentally determined pHs were as follows:

Fe 4·71; Cr 0·09; Ni 2·93; Mo 0·020

indicating that chromium and molybdenum when present in an alloy would have the greatest effect in lowering the pH within the pit.

The results for the steels are given in Table 1.18.

Table 1.18 pH, chloride ion concentration and potential for artificial pits in stainless steels*

	Type 304L (Fe–18Cr–10Ni–0·13Mo)	*Type 316L* (Fe–17Cr–13·7Ni–2·4Mo)	Fe–18Cr–16Ni–5Mo
pH	approx. 0·6 to 0·80	approx. 0·06 to 0·17	approx. −0·13 to 0·08
Cl^- conc. (N)	3·87	6·47	6·20
Steady-state potential (V, *vs.* S.C.E.)	−0·25 to −0·021	−0·02 to −0·22	−0·18 to −0·20

*Data after Suzuki, Yamake and Kitamura[13].

The fact that the pH values of the pure metals were lower than the theoretical values was attributed to the formation of hydroxy-chloro complexes of the metal and to the high chloride ion concentration in the pit, and the results highlight the very pronounced decrease in pH that can occur in an occluded cell, particularly when the alloy contains high concentrations of chromium and molybdenum. They also showed that migration of chloride ions into the solution in the pit can result in a 7–12-fold increase in concentration, and that the potential in the pit is in the active region.

Similar results have been obtained by Bogar and Fujii[18] using an artificial crevice formed between the steel and an acrylic plastic for a range of Fe–Cr alloys immersed in 3·5% NaCl. Initially the pH corresponded with the bulk solution, but after a short time it started to decrease and attained a steady-value after a few hours. The pH decreased with increase in chromium content of the alloy and attained a minimum value of 1·8 with the Fe–25Cr alloy, and higher chromium contents produced no further decrease. It was concluded that hydrolysis of Fe^{2+} determined the pH when the chromium content was less than 1%, hydrolysis of both Fe^{3+} and Cr^{3+} was responsible in the range 1–15% chromium and that Cr^{3+} dominated when the chromium content was above 15%. These results indicate that in any occluded cell the pH will be significantly lower than that of the bulk solution and the chloride ion concentration will be significantly higher. In the case of the passive metals the potential of the metal within the occluded cell will be in the active region, whereas the freely exposed surface will be in the passive region, with a consequent difference in potential between these two areas.

Pickering and Frankenthal[18a] have determined the potentials of the bottom of a pit and the passive surface of the freely exposed metal during the pitting of pure iron and certain stainless steels, and concluded that the metal within the pit is in the active state. It was established during potentiostatic studies that the bottom of the pit could have a potential as low as about −0·2 V whereas that of the passive surface could be as high as 1·4 V, so that the *IR* drop between these two surfaces could be as much as 1·6 V. They consider that this is due to the formation of bubbles of hydrogen gas at the bottom of the pit, which constricts the current flow with a consequent high *IR* drop, and results in a high anodic current density at that area which is independent of the resistivity of the solution. On the other hand, Franck[18b] came to the conclusion that the large *IR* drop was due to a layer of solution of high resistivity at the pit/solution interface that was caused by the inability of the corrosion products to diffuse rapidly away from the surface. Pickering and Frankenthal[18a] consider that the fluctuations in the corrosion current observed during these studies was due to the movement of gas bubbles that caused a variation in the *IR* drop between the passive cathode and active anode within the pit; they also consider that a similar mechanism involving hydrogen bubbles is applicable to crevice corrosion.

The importance of occluded cells cannot be overemphasised, and Brown[19] considers that pitting, crevice corrosion, intergranular attack, filiform corrosion and hydrogen cracking are characterised by local acidification due to hydrolysis of metal ions, and that this phenomenon is of major significance in the overall mechanism.

Crevice Corrosion

Intense localised corrosion, ranging from small pits to extensive corrosion over the whole surface, can occur within narrow crevices that may be formed by:

1. The geometry of the structure, e.g. riveted plates, welded fabrications, threaded joints.
2. Contact of the metal with non-metallic solids, e.g. plastics, rubber, glass.
3. Deposits of sand, dirt or permeable corrosion products on the metal surface (a type of crevice corrosion that is referred to as *deposit attack*).

Different types of crevices are shown in Fig. 1.49, and reference should also be made to Section 10.2 for other examples. The phenomenon is referred to as *crevice* corrosion, and is characterised by a geometrical configuration in which the cathode reactant (usually dissolved oxygen) can readily gain

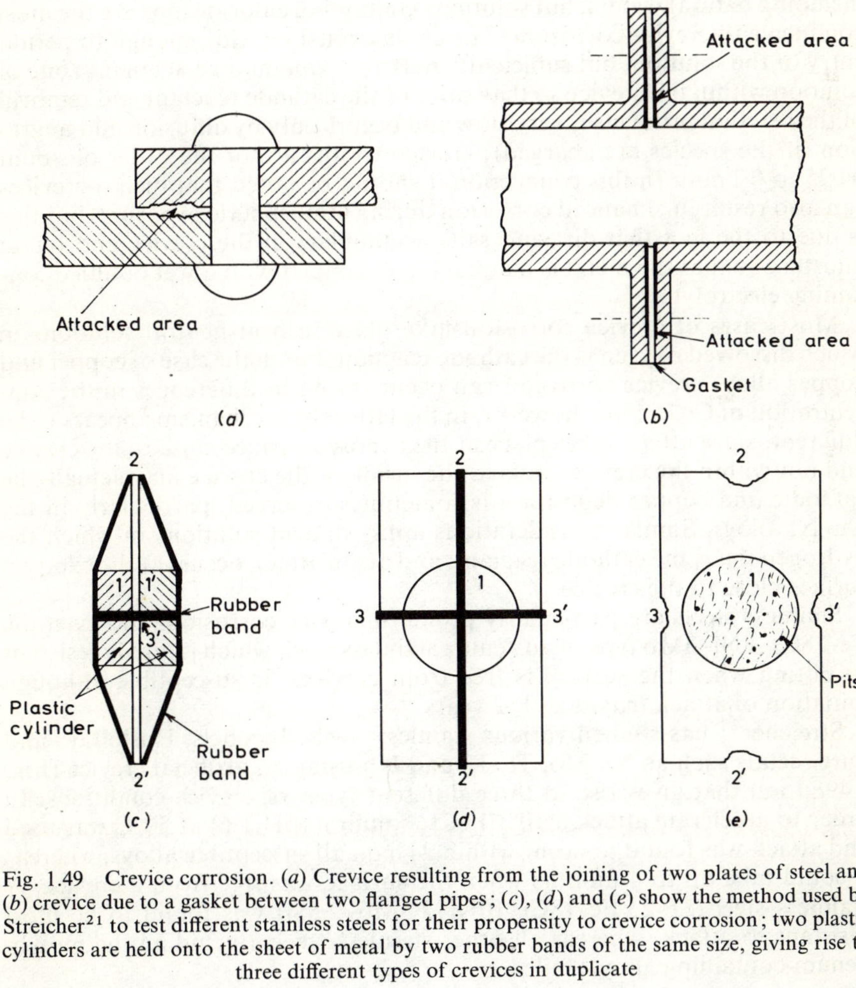

Fig. 1.49 Crevice corrosion. (*a*) Crevice resulting from the joining of two plates of steel and (*b*) crevice due to a gasket between two flanged pipes; (*c*), (*d*) and (*e*) show the method used by Streicher[21] to test different stainless steels for their propensity to crevice corrosion; two plastic cylinders are held onto the sheet of metal by two rubber bands of the same size, giving rise to three different types of crevices in duplicate

access by convection (natural and forced) and diffusion to the metal surface outside the crevice, whereas access to the layer of the stagnant solution within the crevice is far more difficult and can be achieved only by diffusion through the narrow mouth of the crevice. This leads to a difference in concentration of the cathode reactant at the two surfaces, and for this reason it is sometimes referred to as *concentration cell* corrosion. Fibrous materials (such as gaskets for joints) that draw the solution into the crevice formed between the gasket and the metal by capillary action are particularly conducive to crevice corrosion, and the pitting of stainless steels is invariably stimulated by the presence of a crevice or a deposit (dirt, sand, marine growths, etc.).

Crevice corrosion, which can range from near-uniform attack to pitting at the metal surface within the crevice, can occur with a variety of metal and alloys ranging from the noble metals silver and copper to the very electronegative metals such as aluminium and titanium, and it is particularly prone to occur with metals and alloys that rely on passivity for their corrosion resistance. The environment can be any aggressive solution (acid or neutral), including natural waters, but solutions containing chloride ions are the most conducive to crevice corrosion. The crevice must be wide enough to permit entry of the solution, but sufficiently narrow to maintain a stagnant zone of solution within the crevice so that entry of the cathode reactant and removal of the reaction products is very slow and occurs only by diffusion and migration (if the species are charged); crevice widths are of the order of about 0·025 to 0·1 mm. In this connection it should be noted that macro-crevices can also result in enhanced corrosion during atmospheric exposure, but this is due to the fact that dirt and salts accumulate in the crevice and retain moisture so that the metal is maintained in contact with a wet poultice containing electrolytes.

Most cases of crevice corrosion take place in near-neutral solutions in which dissolved oxygen is the cathode reactant, but in the case of copper and copper alloys crevice corrosion can occur owing to differences in the concentration of Cu^{2+} ions; however, in the latter the mechanism appears to be different, since attack takes place at the exposed surface close to the crevice and not within the crevice; in fact, the inside of the crevice may actually be cathodic and copper deposition is sometimes observed, particularly in the Cu–Ni alloys. Similar considerations apply in acid solutions in which the hydrogen ion is the cathode reactant, and again attack occurs at the exposed surface close to the crevice.

Stainless steels are particularly prone to crevice corrosion, and even the Fe–18Cr–8Ni–3Mo type of austenitic stainless steel, which is highly resistant to pitting when the surface is free from crevices, is susceptible although initiation of attack may take 1–2 years[20].

Streicher[21] has studied various stainless steels, Inconels, Hastelloys and pure metals such as Ni, Mo, Ti, Ta and Nb using an artificial crevice (Fig. 1.49*c* to *e*) that gives rise to three different types of crevice conditions. In order to accelerate attack a 10% $FeCl_3$ solution (pH 1·6) at 50°C was used and attack was found to occur within 24 h on all susceptible alloys, whereas it could take up to 4 months when the surface was free from a crevice. A stainless steel (SP-2; Fe–18Cr–10Ni–2·5Mo–2·5Si) was found to be most resistant to attack, whereas all other stainless steels (including the molybdenum-containing alloys) failed.

Mechanism

At first sight the mechanism of crevice corrosion appears to be simply the formation of a differential aeration cell in which the freely exposed metal outside the crevice is predominantly cathodic whilst the metal within the crevice is predominantly or solely anodic; the large cathode current acts on the small anodic area thus resulting in intense attack. However, although differential aeration plays an important rôle in the mechanism, the situation in reality is far more complex, owing to the formation of acid within the crevice.

It is appropriate to consider first the crevice corrosion of mild steel in oxygenated neutral sodium chloride, and then to consider systems in which the metal is readily passivated. Initially, the whole surface will be in contact with a solution containing oxygen so that attack, with oxygen reduction providing the cathodic process, occurs on both the freely exposed surface and the surface within the crevice (Fig. 1.50). However, whereas the freely exposed surface will be accessible to dissolved oxygen by convection and diffusion, access of oxygen to the solution within the crevice can occur only

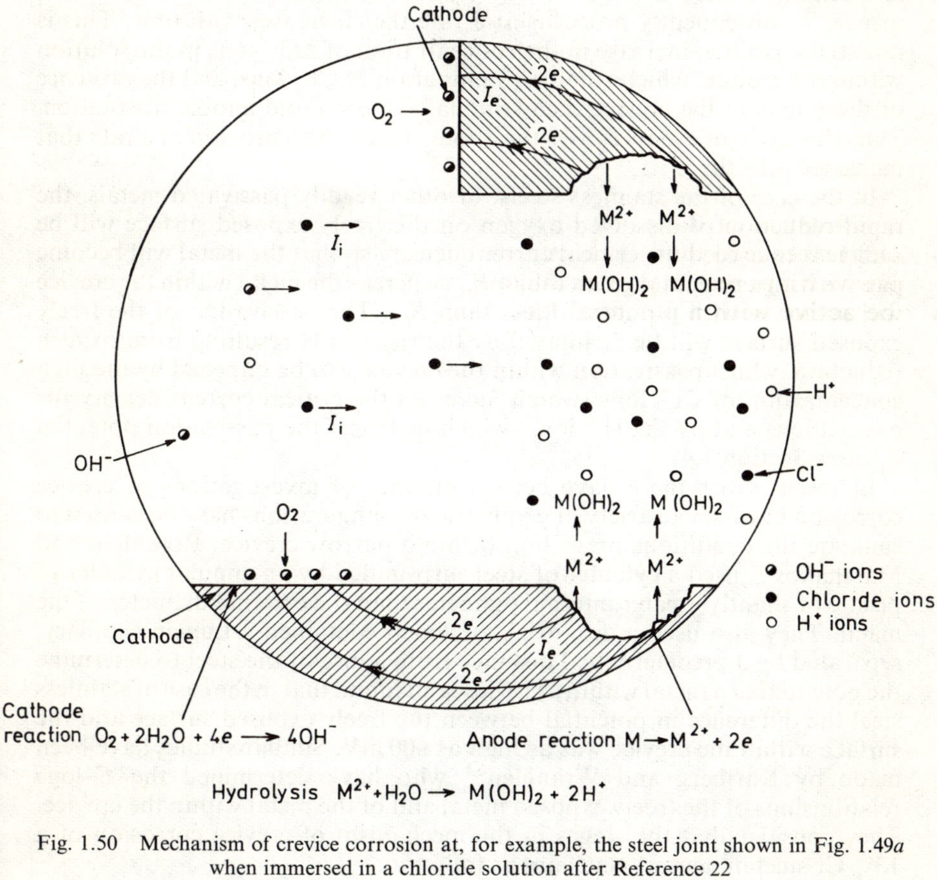

Fig. 1.50 Mechanism of crevice corrosion at, for example, the steel joint shown in Fig. 1.49*a* when immersed in a chloride solution after Reference 22

by diffusion; furthermore, it will be rapidly removed from the solution outside the crevice by cathodic reduction on the freely exposed surface and by reaction with $Fe(OH)_2$ at the mouth of the crevice. Thus under these conditions the oxygen concentration within the crevice will rapidly become negligible, and oxygen reduction within the crevice will cease.

The cathodic reduction of oxygen on the large area of freely exposed surface will result in anodic dissolution of the metal within the crevice, and the excess of positive charges (metal cations) formed within the solution will result in the migration of Cl^- and OH^- from the bulk solution to maintain charge balance (*see* Fig. 20.37); since the contribution made by the very low concentration of OH^- ions to ionic conduction is small it will not affect the pH significantly. Hydrolysis of the metal chloride then occurs with the formation of $Fe(OH)_2$, which is not protective, and H^+ ions, and the pH falls to about 3; the presence of the high chloride concentration (3–10 times that of the bulk solution) and H^+ ions prevents passivation and facilitates anodic dissolution. As the anodic dissolution becomes kinetically easier, the potential of the freely exposed surface becomes more negative, i.e. it becomes partially cathodically protected by the electrons liberated by the anodic reaction[22]. This sequence of events results in an *autocatalytic* process, and although there is frequently an incubation period before attack commences it subsequently proceeds at a rate that increases with time. This is due to the gradual increase in the concentration of M^{z+} ions in the solution within the crevice, which stimulates migration of Cl^- ions, and the presence of these ions in the crevice then stimulates more rapid anodic dissolution. Thus this cycle of events causes the anodic reaction to proceed at a rate that increases with time.

In the case of the stainless steels, or other readily passivated metals, the rapid reduction of dissolved oxygen on the freely exposed surface will be sufficient to exceed the critical current density so that the metal will become passive with a potential greater than E_{pp}, whereas the metal within the crevice will be active with a potential less than E_{pp}. The passivation of the freely exposed surface will be facilitated by the rise in pH resulting from oxygen reduction, whilst passivation within the crevice will be impeded by the high concentration of Cl^- ions (which increases the critical current density for passivation) and by the H^+ ions (which increases the passivation potential E_{pp}, *see* Section 1.4).

In recent years there have been a number of investigations of crevice corrosion in which a variety of geometrical configurations have been used to simulate the conditions prevailing within a narrow crevice. Rosenfeld and Marshakov[23] used a cylinder of steel surrounded by an annular cylinder of plastic of slightly greater internal diameter than the external diameter of the metal. They also used a disc of glass, which contained a Luggin capillary, separated by a predetermined distance from a disc of the steel to determine the potential of a metal within a crevice, and found that in the case of stainless steel the difference in potential between the freely exposed surface and the surface within the crevice was as high as 600 mV. Similar studies have been made by Karlberg and Wranglen[24] who have determined the E–$\log i$ relationships of the freely exposed metal and of the metal within the crevice. They consider that the stages in the mechanism of crevice corrosion of a 13% Cr steel in neutral NaCl are as follows:

1. All areas of the steel are passive, although i_{pass} is comparatively high.
2. Thickening of the passive film on all surfaces occurs and i_{pass} decreases; a small current now flows from the crevice to the outer surface but this is so small that the surfaces are practically equipotential.
3. The difference in oxygen concentration is now large and the potential of the metal within the crevice is now more negative than the freely exposed metal; the predominant reaction in the crevice is anodic dissolution resulting in a high concentration of Fe^{2+} and Cr^{3+} ions.
4. Hydrolysis of Fe^{2+} and Cr^{3+} results in a lowering of the pH; the cell current increases with consequent increased migration of Cl^- into the crevice.
5. Since E_{pp} of the metal within the crevice is raised by the decrease in pH, the metal is active whereas the outer surface is passive, and this represents the active propagation stage of crevice corrosion; there is a significant *IR* drop between the two zones.

In conclusion it is appropriate to refer to titanium, a highly passive metal that when first used industrially was thought to be resistant to crevice corrosion. Although this is the case at ambient temperatures it is attacked by concentrated solutions of Cl^-, I^-, Br^- and SO_4^{2-} at elevated temperatures of greater than 95°C. Figure 1.51 shows crevice corrosion at the gasket/metal interface on the flange of a titanium pipe used for hot hypochlorite solutions. Takamura[25], on the basis of studies of freely exposed surfaces of titanium and surfaces within an artificial crevice in concentrated neutral and acid chloride solutions ($NaCl$, $AlCl_3$, $ZnCl_2$, $MgCl_2$, $CaCl_2$, NH_4Cl), concluded that attack within a crevice was due to acid formation by hydrolysis reactions of the type

$$Ti \rightarrow Ti^{2+} + 2e$$

$$Ti^{2+} + 2H_2O \rightarrow TiO_2 + 4H^+ + 4e$$

Under these circumstances the metal's surface within the crevice became active and it corroded with the formation of a yellowish-white corrosion product that was identified as being mainly rutile TiO_2. On the other hand, a Ti–0·13Pd alloy was found to be immune from crevice corrosion, since the presence of the palladium facilitated passivation of the metal surfaces forming the crevice.

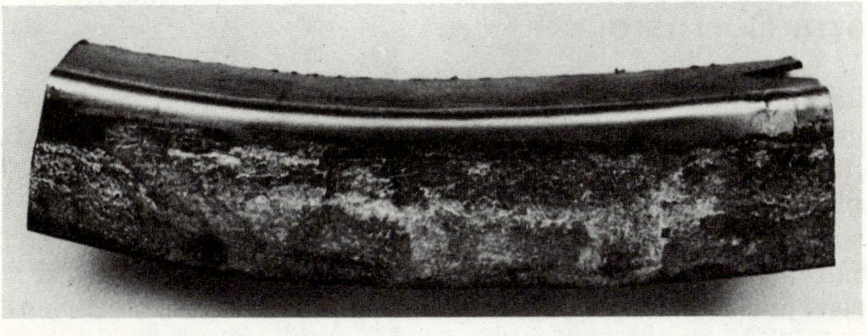

Fig. 1.51 Crevice corrosion resulting from the crevice produced between the gasket and the flange of a titanium pipe used for conveying a hot hypochlorite solution. The attacked areas are coated with a hard deposit of titanium oxides, whilst the unattacked area of metal outside the crevice remains bright

Griess[26] has observed crevice corrosion of titanium in hot concentrated solutions of Cl^-, SO_4^{2-} and I^- ions, and considers that the formation of acid within the crevice is the major factor in the mechanism. He points out that at room temperature $Ti(OH)_3$ precipitates at pH 3, and $Ti(OH)_4$ at pH 0·7, and that at elevated temperatures and at the high concentrations of Cl^- ions that prevail within a crevice the activity of hydrogen ions could be even greater than that indicated by the equilibrium pH values at ambient temperatures. Alloys that remain passive in acid solutions of the same pH as that developed within a crevice should be more immune to crevice attack than pure titanium, and this appears to be the case with alloys containing 0·2% Pd, 2% Mo or 2% Ni.

Controlling Crevice Corrosion

An outline of the methods of controlling crevice corrosion is provided here, and for further details reference should be made to Chapter 10, and particularly to Sections 10.2 and 10.5. The obvious method of controlling crevice corrosion is to avoid crevices in the design of the structure and to avoid geometrical conditions that lead to the formation of deposits on the metal surface. Thus welded butt joints should be used in preference to riveted or bolted joints. Alternatively, if crevices cannot be avoided in design they should be sealed by welding, soldering, or by the use of caulking compounds.

Solids should be removed from process liquors and vessels should be designed so that deposition of solids does not occur, e.g. by avoiding sharp corners and stagnant areas by providing facilities for complete drainage of the process liquor.

The interface between a gasket and a metal flange forms a very effective crevice, particularly if the gasket is composed of an absorbent material, but this can be minimised by using non-absorbent materials such as p.t.f.e.

Metals and alloys vary in their ability to resist crevice corrosion, and this applies particularly to those that rely on passivity for their resistance to corrosion. Titanium and high-nickel alloys such as the Inconels and Hastelloys are amongst the most resistant, but even these will be attacked under highly aggressive environmental conditions.

Filiform Corrosion

Filiform corrosion[27] is characterised by the formation of a network of threadlike filaments of corrosion products on the surface of a metal coated with a transparent lacquer or a paint film, as a result of exposure to a humid atmosphere. This phenomenon first attracted attention because of its formation on lacquered steel, and for this reason it is sometimes referred to as *underfilm* corrosion, but although it is most readily observed under a transparent lacquer it can also occur under an opaque paint film or on a bare metal surface. Filiform corrosion has been observed on steel, zinc, magnesium and aluminium coated with lacquers and paints, and with aluminium foil coated with paper. Surface treatment of the metal by phosphating or chromating lessens the tendency for filiform corrosion to occur, but it is not completely effective in preventing it. The damage caused by corrosion is very slight, and the major objection to this form of attack is its detrimental affect on the appearance of the metal, particularly when the coating is transparent.

Filiform corrosion on steel covered with a transparent lacquer, which in common with other organic finishes is permeable to oxygen and water, and takes the form of red threads with a green or blue head (Fig. 1.52). Propagation of the thread occurs via the active head, the body or tail left behind being inactive. The threads are 0·1–0·5 mm in width and propagate at 0·4 mm d^{-1} in random directions, but a feature of their propagation is that they never intersect. If a propagating head approaches an inactive tail it becomes deflected in such a way that the angle of approach equals the angle of deflection, and if two propagating heads meet propagation ceases[20].

The main factor in causing filiform corrosion is the relative humidity of the atmosphere, and if this is below 65% (the critical relative humidity for the atmospheric corrosion of most metals, *see* Section 2.2) it will not occur. As the relative humidity increases the thickness of the filaments increases; at 65–80% relative humidity they are very thin, at 80–95% relative humidity they are much wider and at approximately 95% relative humidity they broaden sufficiently to form blisters.

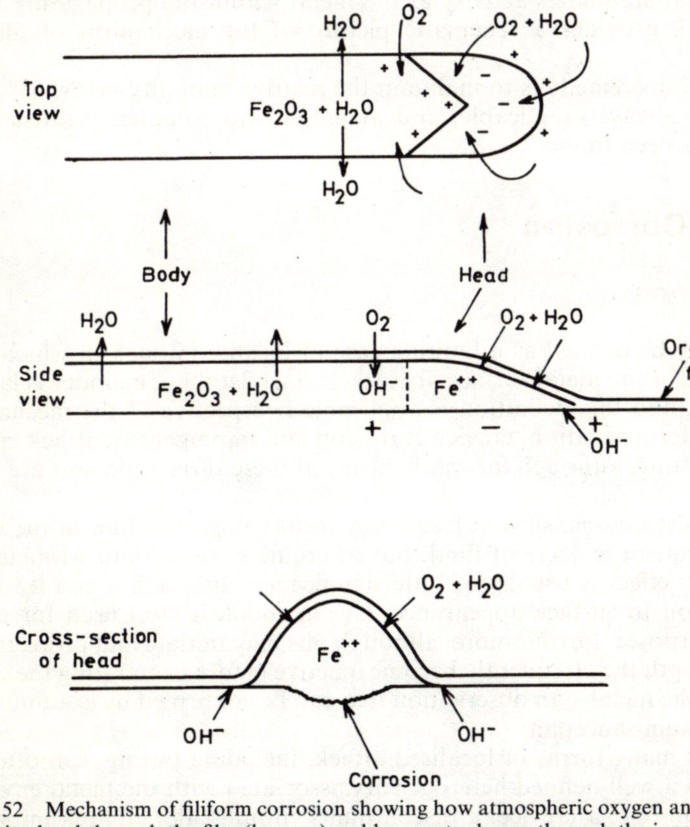

Fig. 1.52 Mechanism of filiform corrosion showing how atmospheric oxygen and water enter the active head through the film (lacquer) and how water leaves through the inactive tail. This results in a high concentration of oxygen at the 'V'-shaped interface between the tail and the head, and to a differential aeration cell (after Uhlig[20])

The mechanism of filiform corrosion[28-30] is highly complex, but it has many of the features of crevice corrosion in that differential aeration and acid formation have an important rôle in the propagation of the active head. Slabaugh and Grutheer[29] found that the blue-green active head was composed of a concentration solution of Fe^{2+}, which is supplied with moisture from the atmosphere by osmosis through the film of lacquer. In the tail the concentration of Fe^{2+} is kept low by the precipitation of $Fe_2O_3 . H_2O$ (rust) and water is removed by osmotic action, i.e. atmospheric water enters at the head and is removed at the tail, which therefore tends to dry out. Oxygen also diffuses through the film, but owing to lateral diffusion it attains a higher concentration at the 'V'-shaped interface between the head and the tail (although it is also high at the periphery of the head) and a lower concentration in the centre of the head. Thus a differential aeration cell is set up, the active head forming an occluded cell in which the metal corrodes anodically at the centre and forward portions of the head, and it is possible that owing to hydrolysis acid formation also occurs at this area. Hydroxyl ions formed at the periphery of the head play a rôle in attacking the film at the interface and thus weakening the film/metal bond. Behind the 'V'-shaped interface the metal for a portion of the length of the filament acts as a relatively large cathode that stimulates activity at the metal within the propagating head. Figure 1.52 provides a schematic picture of the mechanism of filiform corrosion[20].

The obvious remedy is to maintain the relative humidity below 65%, but this is not always practicable, and at present no completely satisfactory remedy has been found.

Pitting Corrosion

Introduction

Pitting may be defined as a limiting case of localised attack in which only small areas of the metal surface are attacked whilst the remainder is largely unaffected, and this definition is applicable irrespective of the mechanism involved; dezincification, crevice corrosion and impingement attack can all result in pitting, although the mechanisms of these three processes are quite different.

Pitting is insidious, since it frequently results in perforation of the metal with consequent leakage of fluid, but where massive sections of metal are involved its effect is usually of little significance, although it can lead to a deterioration in surface appearance, e.g. in stainless steel used for architectural purposes. Furthermore, although pits may initiate and propagate to a certain depth they frequently become inactive before penetrating the cross-section of the metal—an observation that can be confirmed by examining an old aluminium saucepan.

Whereas many forms of localised attack, including pitting, can often be ascribed to a well-defined heterogeneity associated with the metal/environment system (e.g. a crevice, a discontinuity in millscale, an area impinged upon by a high velocity solution), pitting can occur in a system that is apparently free from heterogeneities, providing the solution contains an

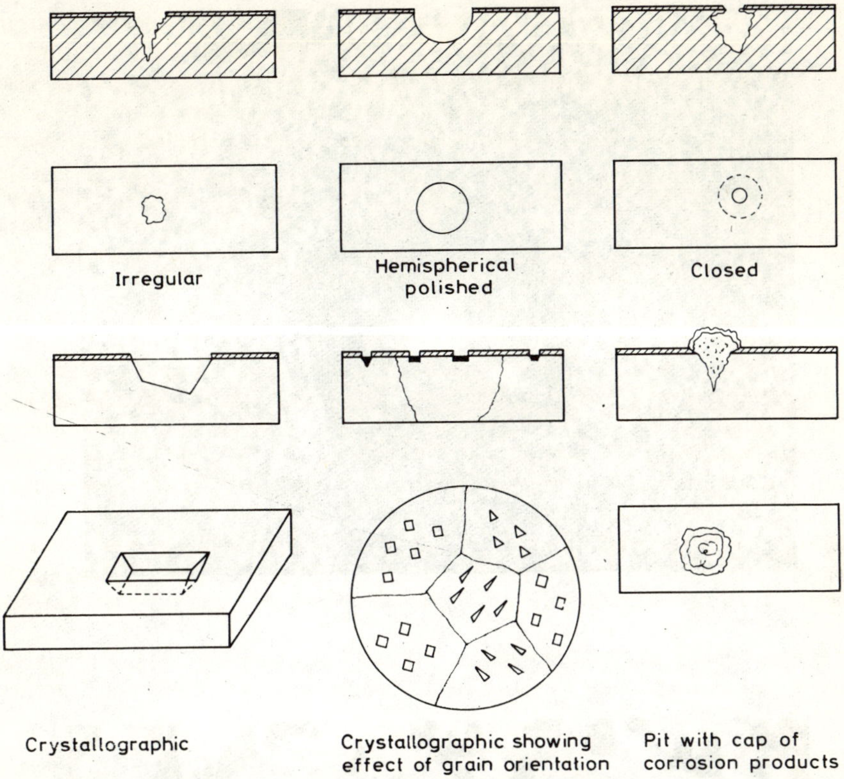

Fig. 1.53 Some different types of pits

aggressive anion such as Cl^-. It should be emphasised that whereas crevice corrosion can take place in most corrosive electrolyte solutions, pitting on a freely exposed surface will take place only if certain anions are present in the solution. Thus an 18/8 stainless steel free from crevices will pit in a stagnant solution of sodium chloride and the pit sites will be random and unpredictable; if a crevice is present the pits will be located within the crevice and the rate of pitting will be greater than at the freely exposed surface. On the other hand, the steel will not pit if immersed in a solution of, for example, sodium sulphate.

Crevice corrosion and pitting have a number of features in common, and it has been stated that pitting may be regarded as crevice corrosion in which the pit forms its own crevice; however, whereas a macroscopic heterogeneity determines the site of attack in crevice corrosion, the sites of attack in pitting are determined by microscopic or sub-microscopic features in the passive film (*see* Sections 1.3 and 1.5).

Figure 1.53 shows diagrammatically various types of pits that can range from hemispherical with a polished surface, in which crystallographic etching has been completely suppressed, to crystallographic pits whose sides are composed of the crystal planes that corrode at the slowest rate. Pits formed on Ni during anodic polarisation in an acetic acid–acetate buffer of pH 4·6 are shown in Fig. 1.54.

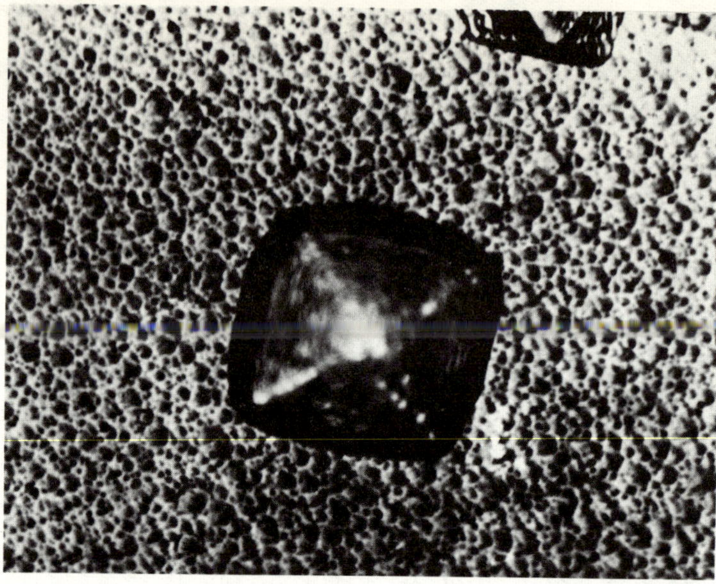

(*a*)

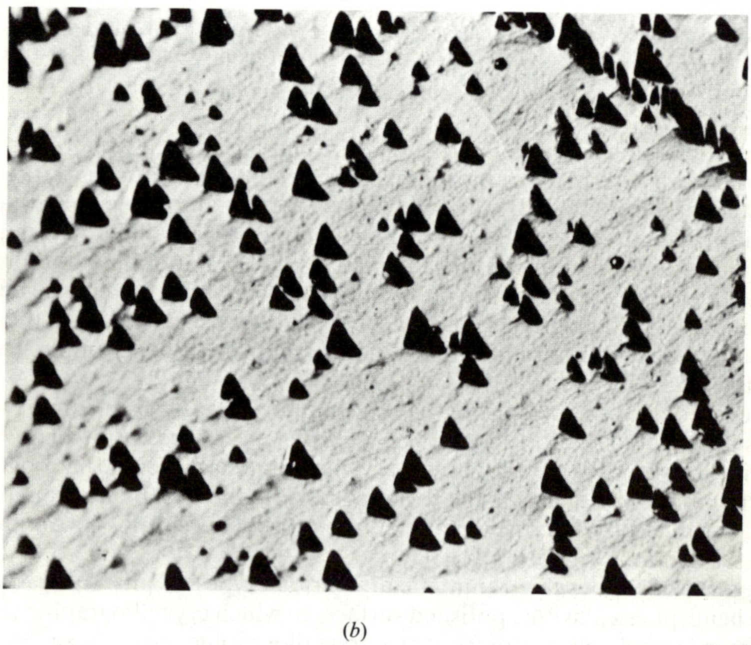

(*b*)

Fig. 1.54 Pits formed during anodic polarisation in an acetic acid–acetate buffer of pH 4·6 containing thiourea or NaCl. (*a*) General attack and formation of crystallographic pits on nickel in the buffer + 10^{-5} M thiourea (×200), (*b*) crystallographic pits formed in the buffer

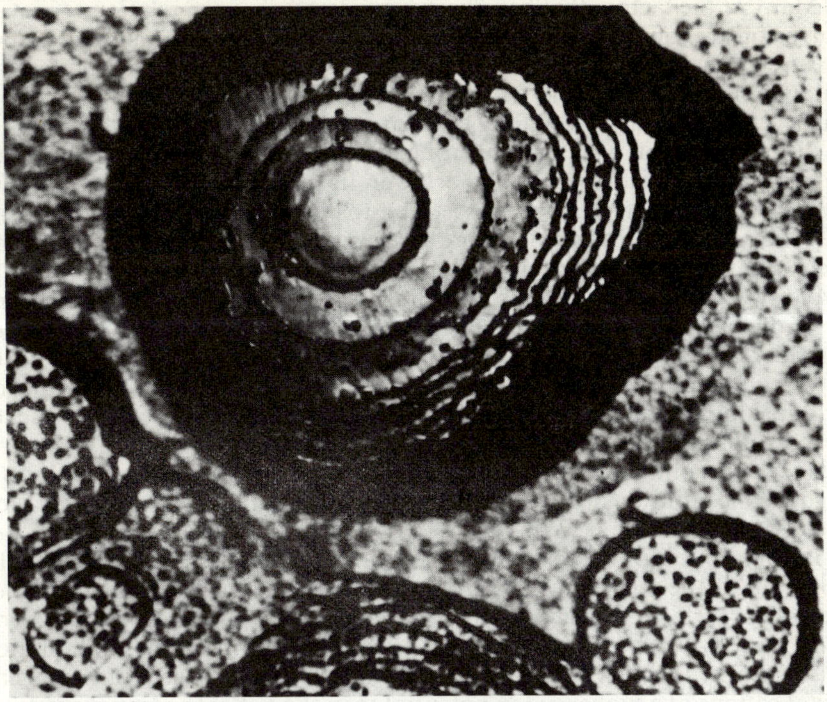

(c)

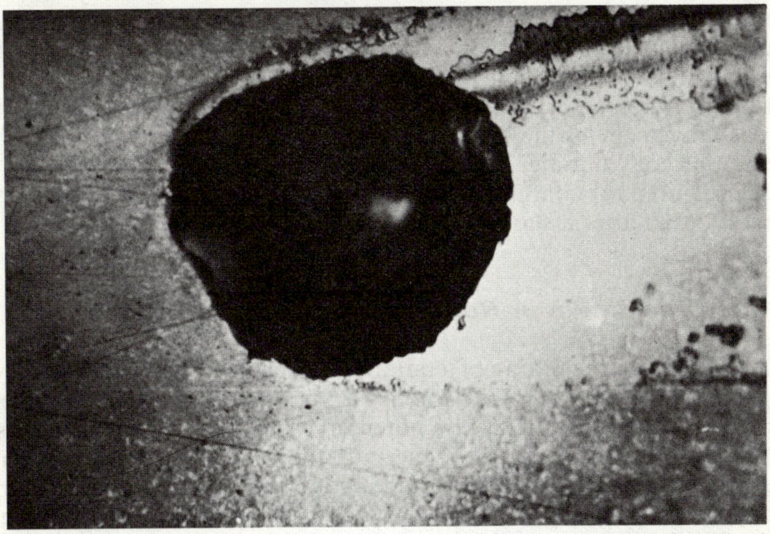

(d)

+10^{-6} M thiourea (×400), (c) hemispherical spiral-shaped pits formed on nickel in the buffer +10^{-4} M thiourea (×400) and (d) hemispherical pit formed on Fe–12Cr in the buffer+ 0·5 M NaCl (×200). (After P. Surendra, PhD thesis, University of London, 1971)

Many of the factors that have been considered above in relation to crevice corrosion are equally applicable to pitting, e.g. the large cathode/small anode area relationship, differences in concentration of the cathode reactant, autocatalytic formation of acid within the occluded cell (*see* Fig. 1.49), etc. but there are certain distinct differences that must be considered.

Pitting can occur with a number of metals and alloys, and as a rough guide the more expensive the alloy the lower its propensity for pitting! Thus Fe–17Cr stainless steel is more susceptible to pitting in chloride solutions than Fe–18Cr–8Ni, which in turn is more susceptible than the Fe–18Cr–8Ni–3Mo steel, and titanium is superior to all of them. However, it is important to avoid sweeping generalisations regarding the propensity of an alloy to pitting or of the pitting or inhibitive properties of anions, and each system must be considered individually. Thus the pitting of certain stainless steels will occur in solutions containing Cl^- ions, Br^- (but not I^- or F^-) ions, hypochlorite or thiosulphate anions, but pitting will tend to be suppressed by the presence of oxyanions such as NO_3^- or SO_4^{2-}. On the other hand, copper pits in natural fresh waters containing a very small concentration of Cl^- and a significantly higher concentration of SO_4^{2-} ions; in waters with a higher Cl^- concentration attack is far less localised[31].

Pitting can occur over a range of pHs and potentials, but in all cases the potential of the metal must be within the passive region so that the major part of the surface is passive. Furthermore, the redox potential of the solution must exceed a certain critical value for pits to initiate. In near-neutral solution this potential is achieved by the presence of dissolved oxygen, but oxidising metal cations of a higher redox potential than the critical potential are more conducive to rapid pitting than dissolved oxygen, e.g. $FeCl_3$, $CuCl_2$ and $HgCl_2$. For this reason $FeCl_3$ is widely used in testing alloys for their resistance to pitting. Stagnant conditions that prevent mixing of the alkaline catholyte and acid anolyte will favour pitting; whilst any movement of the solution will tend to suppress pitting. It follows that a crevice or the entrapment of a solution in an inactive part of a plant will be conducive to pitting, whereas movement of the solution will lower the pitting propensity. Conversely, a pit may initiate, but the local concentration of Cl^- and H^+ ions that are required for continued propagation may be removed by convection currents so that after a short time the pit becomes inactive.

Pitting, Breakdown or Rupture Potential

A characteristic of the pitting of a number of metals that rely on passivity for their corrosion resistance is the *breakdown* potential E_b (which is also referred to as the *critical pitting* or *rupture* potential), which may be regarded as the most negative potential to cause the initiation and propagation of one or more pits, or as the most positive potential that results in a decrease in current due to passivation of the entire surface. The breakdown potential for a given metal will, of course, vary with the nature of the solution, but it will also be dependent on the method used for its determination (anodic polarisation using potentiostatic, potentiokinetic or galvanostatic techniques or chemical solutions of high redox potential such as $FeCl_3$) and on the duration of the test. It cannot, therefore, be regarded as a precise parameter.

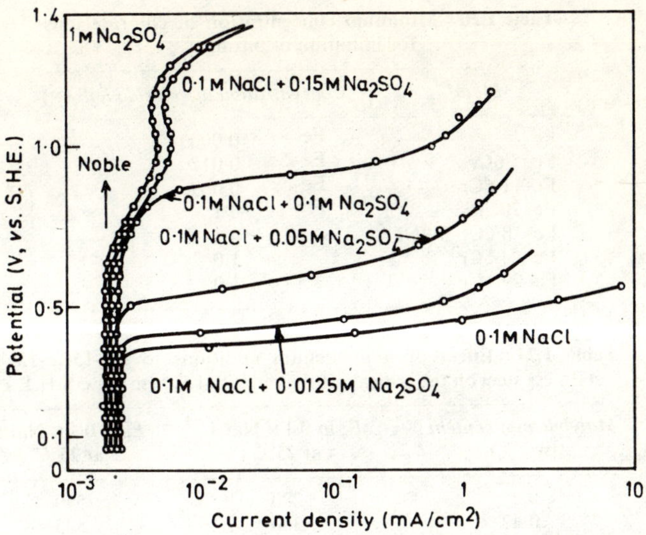

Fig. 1.55 Breakdown potentials of Fe–18Cr–8Ni stainless steel in 0·1 M NaCl plus various concentrations of SO_4^{2-} ions during potentiostatic anodic polarisation (after Leckie and Uhlig[32])

Figure 1.55 shows potentiostatic anodic curves obtained by Leckie and Uhlig[32] for Fe–18Cr–8Ni stainless steel in 0·1 M NaCl containing various concentrations of SO_4^{2-} ions, and it can be seen that in all solutions containing less than 0·15 M Na_2SO_4 there is a sudden increase in current at a certain potential, named the *critical pitting* or *breakdown* potential, within the passive region. This potential is a manifestation of pitting, and it can be seen that it becomes more positive with increasing concentration of SO_4^{2-} ions and is suppressed at concentrations at and above 0·15 M Na_2SO_4.

Table 1.19 gives some critical pitting potentials collated by Uhlig[20] for different metals in 0·1 N NaCl at 25°C, in which in most cases the potential was first increased incrementally and held constant at each potential for 5 min in order to obtain an approximate value of the critical pitting potential. This may be followed by a separate series of experiments in which fresh specimens are held at potentials close to E_b for 12 h or more and the surface

Table 1.19 Critical pitting potentials in 0·1 N NaCl at 25°C[20]

Metal	*Potential* (V, *vs.* S.H.E.)
Al	−0·37
Ni	0·28
Zr	0·46
Fe–18Cr–8Ni stainless steel	0·26
Fe–30Cr	0·62
Fe–12Cr	0·20
Cr	>1·0
Ti*	>1·0 (1 N NaCl)
	approx. 1·0 (1 N NaCl at 200°C)

*Approx. 8–12 V in sea-water.

Table 1.20 Minimum concentration of chloride ions for initiation of pitting[33]

Metal	*Concentration of chloride ions* (N)
Fe	0·0003
Fe–5·6Cr	0·017
Fe–11·6Cr	0·069
Fe–20Cr	0·1
Fe–18·6Cr–9·9Ni	0·1
Fe–24·5Cr	1·0
Fe–29·4Cr	1·0

Table 1.21 Effect of molybdenum additions to Fe–15Cr–13Ni stainless steel on the critical pitting potential E_b* (mV, *vs.* S.H.E.)

Molybdenum content (wt. %)	E_b in 0·1 N NaCl at 25°C	E_b in 0·1 N NaBr at 25°C
0·0	280	540
0·42	300	525
0·92	340	—
1·4	400	510
1·8	535	—
2·4	722	467

*Data after V. Hospadaruk and J. V. Petrocelli, *J. Electrochem. Soc.*, **113**, 878 (1962) and N. D. Tomashov, G. P. Chernova and O. N. Marcova, *Corrosion*, **20**, 166t (1964).

observed with a low-power microscope to detect pit formation. Pitting will occur in practice if the redox potential of the solution is more positive than the potentiostatically determined critical pitting potential, and providing the conditions (concentration of chloride ions, temperature, etc.) are the same. It follows from Table 1.19 that chromium and titanium will not pit in oxygenated 0·1 N NaCl at 25°C, but titanium will pit in boiling saturated $CaCl_2$ solution, owing to the fact that E_b becomes more negative with increase in temperature and chloride ion concentration; it will also pit in 0·1 N NaCl at 25°C if the potential is raised by an external source of e.m.f. to about 10 V. A measure of the tendency of metals to pit can also be obtained from the minimum concentration of chloride ions that are required to initiate pitting, and Stolica[33] has used this method to study a range of Cr–Fe alloys in 1 N H_2SO_4 containing different concentrations of chloride ions. He showed that whereas iron pitted when the concentration of chloride ions was as small as 3×10^{-5} N, concentrations as high as 1 N were required to cause pitting of Fe–29Cr alloy (Table 1.20). Table 1.21 shows how additions of Mo increase the critical potential of a 15Cr–13Ni stainless steel in chloride and bromide solutions.

Nature of Solution and Temperature

The critical pitting potential will, in general, decrease with the concentration of chloride ions, but will increase with the concentration of inhibiting oxy-anions such as OH^-, SO_4^{2-}, NO_3^-, CrO_4^{2-}, etc. Thus whereas Fe–18Cr–8Ni stainless steel will pit in 10% $FeCl_3$ in a matter of hours, pitting is completely arrested if 3% $NaNO_3$ is present in the solution owing to the fact that the presence of the NO_3^- ion increases the breakdown potential above that of the redox potential of the Fe^{3+}/Fe^{2+} equilibrium[34].

Uhlig *et al.* have established the following relationships between E_b (V, *vs.* S.H.E.) and a_{Cl^-} for Fe–18Cr–8Ni stainless steel[32,35]

$$E_b = -0{\cdot}088 \log a_{Cl^-} + 0{\cdot}168 \qquad \ldots(1.114)$$

and for aluminium[36]

$$E_b = -0{\cdot}124 \log a_{Cl^-} - 0{\cdot}0504 \qquad \ldots(1.115)$$

Uhlig *et al.* have also determined the minimum activity of inhibiting oxy-anions $a_{inh.}$ to inhibit the pitting of Fe–18Cr–8Ni stainless steel[32,35] and aluminium[36], and found that equations of the type

$$\log a_{Cl^-} = A \log a_{inh.} + B \qquad \ldots(1.116)$$

are obeyed in which A and B are constants that depend on the metal and the nature of the inhibiting anion. The order of the effectiveness of oxyanions in inhibiting pitting was found to be as follows:

for Fe–18Cr–8Ni: $OH^- > NO_3^- > \bar{A}c > SO_4^{2-} > ClO_4^-$

for aluminium: $NO_3^- > CrO_4^- > \bar{A}c >$ benzoate $> SO_4^{2-}$

where $\bar{A}c$ signifies the acetate anion.

The effect of pH appears to be controversial. Some workers find a slight increase in E_b with increase in pH in the acid region, whilst others report that there is practically no change. In the alkaline region, however, E_b becomes significantly more positive with increase in pH owing to the passivating ability of the OH^- ion.

An increase in temperature significantly decreases E_b, an observation that is frequently neglected with unfortunate consequences. An interesting example was observed recently in which Fe–18Cr–8Ni stainless-steel steam-heated pans used for the manufacture of synthetic cream containing a small concentration of sodium chloride were found to pit after 3–4 weeks. The cream was manufactured at 70°C, but the pan was heated with superheated steam, and on removal of the cream by an outlet at the bottom of the pan the residue of cream on the sides of the pan was subjected to temperatures well above 70°C, with consequent pitting as a result of the small amount of salt present in the cream. The obvious solution to this problem was to use an Fe–18Cr–10Ni–3Mo stainless steel, which is more resistant to pitting attack.

Induction Period for Pitting

It is apparent that the critical pitting potential for a given alloy depends on the concentration of chloride ions, on the concentration of inhibiting anions in the solution and on the temperature of the solution. Unfortunately, the situation is complicated further by the fact that there is an induction period for the onset of pitting, which means that the pitting propensity of an alloy cannot be assessed precisely on the basis of potentiostatic determination of short duration. The induction time τ will decrease with increase in potential and with increase in chloride ion concentration, and in connection with the latter Stolica[37] obtained the following relationship:

For pure iron $\tau = 21{\cdot}4\, c_{Cl^-}\, \text{min}^{-1}$ ...(1.117)

For Fe–5·6Cr $\tau = 1{\cdot}54\, (c_{Cl^-} - 0{\cdot}02)\text{min}^{-1}$ ...(1.118)

For Fe–11·6Cr $\tau = 2{\cdot}13\, (c_{Cl^-} - 0{\cdot}069)\text{min}^{-1}$ ...(1.119)

Mutually Protective Effect

Pits seldom form in close proximity to one another and it would appear that the area of passivated metal, which acts as the cathode for the local cell, is protected by the anodic dissolution of metal within the pit—a phenomenon that is referred to as the *mutually protective effect* (*see* Section 1.5).

Protection Potential

It has been pointed out in Section 1.3 that although the *equilibrium* potential–pH diagrams are based solely on thermodynamic data it is possible to construct practical potential–pH diagrams from experimentally determined E–i curves. Figure 1.56*a* shows the potentiokinetic E–i curves for Armco iron in chloride-free solutions of different pHs obtained by Pourbaix[38], in which general corrosion occurs below the passivation potential P and above the region of immunity, and it can be seen that it is possible to construct a *practical* potential–pH diagram from these curves showing the zones of immunity, general corrosion and passivity that will prevail under various conditions of pH and potential. However, if the same procedure is adopted with solutions of different pH containing 10^{-2} mol dm^{-3} of chloride ion, a sudden increase in current occurs when the potential is raised to a value r, the breakdown potential E_b, at which pits are initiated and give rise to an anodic current (Fig. 1.56*b*). If after attaining E_b the potential is now lowered, the curve is not retraced (electrochemical hysteresis) and will intersect the i axis ($i = 0$) at a potential p at which neither anodic oxidation nor cathodic reduction can occur, i.e. pitting is arrested. The potential p is referred to as the *protection* potential E_p against pitting, since at and below E_p the metal will not pit and the whole surface will remain passive. E_p is always more negative than E_b, and whereas pitting will occur on a pit-free surface above E_b, it will occur only in the range of potentials between E_p and E_b if the surface is already pitted, i.e. between E_p and E_b prior pits will continue to propagate, but initiation of new ones will not be possible.

Pourbaix, on the basis of the breakdown potential E_b and the protection potential E_p, distinguishes between the following states of a metal surface, which have been incorporated in the potential–pH diagram shown in Fig. 1.56*b*:

1. Perfect passivation. The potential–pH region between the passivation potential E_{pp} and the protection potential E_p, in which pits are not initiated nor do they propagate if already present owing to passivation.

2. Imperfect passivation. The potential–pH region between E_p and E_b, in which pits already present can continue to propagate.
3. Pitting region. The potential–pH region above E_b, at which pits can both initiate and propagate.

The electrochemical hysteresis method just described is now becoming widely used to characterise the pitting propensity of alloys, and it has been used by Verink and Pourbaix[39] to study the behaviour of a range of Fe–Cr alloys (0·5–24·9% Cr), Cu–10Ni and Cu–10Ni–1Fe alloys in solutions of different pH and chloride ion concentrations. However, Wilde[39b] has shown that E_p is not a unique parameter since it varies in magnitude with the amount of localised attack produced during anode polarisation. This is discussed more fully in Chapter 20 when considering testing for crevice corrosion and pitting.

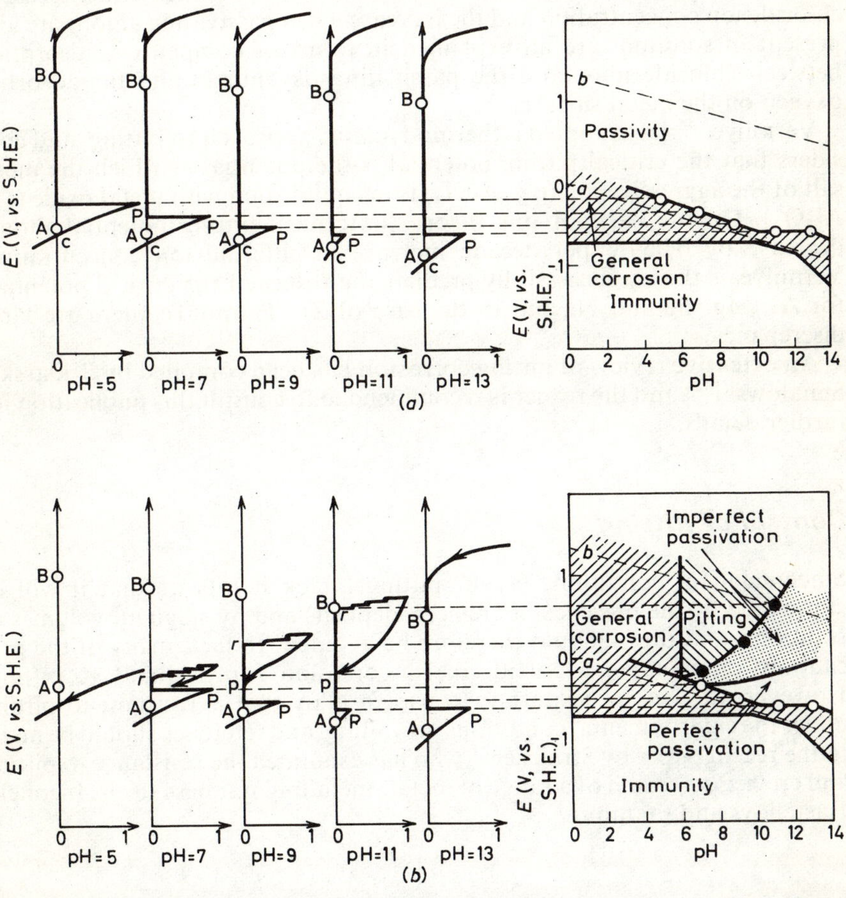

Fig. 1.56(*a*) *E–i* curves and experimental potential–pH diagram for Armco iron in chloride-free solutions of different pHs (*A* is the unpolarised potential and *P* the passivation potential) and (*b*) *E–i* curves and experimental potential–pH diagram for Armco iron in solutions of different pHs containing 10^{-2} mol dm^{-3} of chloride ion (*r* is the rupture potential and *p* the protection potential). (After Pourbaix[38,39])

Mechanism

In view of the fact that there are two opposing views on the mechanism of passivity it is not surprising that a similar situation prevails concerning the mechanism of breakdown of passivity. The solid film theory of passivity and breakdown of passivity is dealt with in some detail in Section 1.5, so that it is appropriate here to discuss briefly the views based on the adsorption theory.

Uhlig and Gilman[34], and Kolotyrkin[40] explain the breakdown of passivity in terms of competitive adsorption between chloride ions in solution and the adsorbed monolayer of oxygen on the surface of the metal. According to Uhlig[20,34] although the metal has a greater affinity for oxygen than chloride ions, adsorption of the latter will be favoured by an increase in the potential until a value is reached where the adsorbed oxygen at specific sites is replaced by chloride ions, which catalyse anodic dissolution. Thus the induction period required for pitting, the decrease in the breakdown potential with increase in chloride ion concentration and the increase when passivating anions are also present in solution, are all explained in terms of competitive adsorption between chloride ions (and the passivating oxyanions) and the adsorbed oxygen on the metal surface.

Vermilyea[41] has adopted a thermodynamic approach to pitting, and considers that the critical pitting potential is the potential at which the metal salt of the aggressive ion (e.g. $AlCl_3$) is in equilibrium with metal oxide (e.g. Al_2O_3). On the basis of this theory the critical pitting potential should decrease by 0·059 V per decade increase in chloride ion concentration. Vermilyea's theory successfully predicts the values of the critical potentials for Al, Mg, Fe and Ni, but in the case of Zr, Ti and Ta there are large discrepancies.

An extensive review of pitting corrosion has been compiled by Szklarska-Smialowski[42], and the reader is recommended to consult this publication for further details.

Control of Pitting

Since stagnant conditions favour pitting attack it follows that it will be stimulated by the presence of crevices, deposits and by stagnant volumes of solutions, and these should therefore be avoided. In fact, many of the precautions enumerated to avoid crevice corrosion apply equally to pitting. However, the best procedure is to use an alloy that is resistant to pitting under the environmental conditions prevailing, and reference should be made to the recent paper by Streicher[21] who has examined the resistance to pitting and crevice corrosion of a range of metals including stainless steels, Inconels, Hastelloys and titanium.

Pitting of Carbon Steels

At the beginning of this section a simple explanation was provided of the localised attack that can occur when steel with a discontinuous coating of

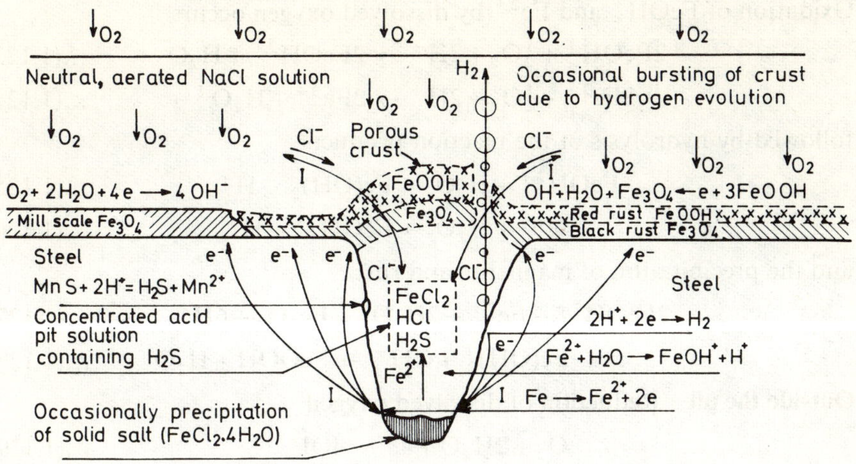

Fig. 1.57 Electrochemical reactions that occur when a pit is initiated at sulphide inclusion in a carbon steel (after Wranglen[43])

magnetite is exposed to an oxygenated water (Fig. 1.46). However, the situation is far more complex than that described, and is characterised by the formation of a hemispherical membrane of corrosion products (a tubercle, cap or mound) that inhibits diffusion of dissolved oxygen to the metal beneath, thus resulting in an occluded cell.

Wranglen[43] considers that sulphide inclusions are responsible for the initiation of attack in both carbon steels and stainless steels, and on this basis he has provided a detailed exposition of the pitting of a carbon steel at an inclusion of MnS when the steel is immersed in an oxygenated chloride solution (Fig. 1.57). The reactions of significance are given in the diagram, but certain features of the mechanism are of interest since they illustrate the complexity of the process.

Pit interior Within the pit the primary anodic reaction is

$$Fe \rightarrow Fe^{2+} + 2e \quad \text{...(1.120)}$$

which is followed by hydrolysis and the generation of H^+

$$Fe^{2+} + H_2O \rightarrow FeOH^+ + H^+ \quad \text{...(1.121)}$$

The decrease in pH results in dissolution of some of the MnS

$$MnS + 2H^+ \rightarrow H_2S + Mn^{2+} \quad \text{...(1.122)}$$

thus providing S^{2-} and HS^- that stimulate attack by decreasing the activation overpotential for the dissolution of Fe (and Ni).

The electrons released are partly accepted by dissolved oxygen at the surface millscale and partly by the H^+, with the consequent formation of H_2 gas. The concentration of chloride ions within the pit will increase owing to migration, and this too will stimulate dissolution.

Pit mouth A membrane of magnetite (Fe_3O_4) and rust (FeOOH) is formed, which prevents the intermingling of the acid anolyte and alkaline catholyte, by the following steps:

Oxidation of $FeOH^+$ and Fe^{2+} by dissolved oxygen occurs

$$2FeOH^+ + \tfrac{1}{2}O_2 + 2H^+ \rightarrow 2FeOH^{2+} + H_2O \qquad ...(1.123)$$

$$2Fe^{2+} + \tfrac{1}{2}O_2 + 2H^+ \rightarrow 2Fe^{3+} + H_2O \qquad ...(1.124)$$

followed by hydrolysis of the reaction products

$$FeOH^{2+} + H_2O \rightarrow Fe(OH)_2^+ + H^+ \qquad ...(1.125)$$

$$Fe^{3+} + H_2O \rightarrow FeOH^{2+} + H^+ \qquad ...(1.126)$$

and the precipitation of magnetite and rust

$$2FeOH^{2+} + Fe^{2+} + 2H_2O \rightarrow Fe_3O_4 + 6H^+ \qquad ...(1.127)$$

$$Fe(OH)_2^+ + OH^- \rightarrow FeOOH + H_2O \qquad ...(1.128)$$

Outside the pit Reduction of dissolved oxygen

$$O_2 + 2H_2O + 4e \rightarrow 4OH^- \qquad ...(1.129)$$

and reduction of rust to magnetite

$$3FeOOH + e \rightarrow Fe_3O_4 + H_2O + OH^- \qquad ...(1.130)$$

This area will be passivated by the increase in pH due to the cathodically produced OH^- ions, and partially cathodically protected by the electrons liberated by the anodic processes within the pit. The tubercle thus results in an occluded cell with the consequent acidification of the anodic sites. Wranglen considers that in view of the fact that crystals of $FeCl_2.4H_2O$ are sometimes observed at the bottom of a pit the solution within the pit is a saturated solution of that salt, and that this will correspond with an equilibrium pH of about 3·5.

It is also of interest to note that Wranglen considers that the decrease in the corrosion rate of steel in the atmosphere and the pitting rate in acid and neutral solution brought about by small alloying additions of copper is due to the formation of Cu_2S, which reduces the activity of the HS^- and S^{2-} ions to a very low value so that they do not catalyse anodic dissolution, and a similar mechanism was put forward by Fyfe *et al.*[44] to explain the corrosion resistance of copper-containing steels when exposed to industrial atmospheres.

Pitting of Aluminium

The pitting of aluminium in chloride-containing waters follows a similar mechanism to that of steels (Fig. 1.58), and again the characteristic feature of the process is the formation of acid within the occluded cell[45]. The passivating film of Al_2O_3 surrounding the pit acts as the cathode, but its effectiveness in reducing dissolved oxygen is significantly enhanced if copper is either deposited on the surface or enters the lattice of the Al_2O_3, and it is well known that the pitting of aluminium occurs rapidly when the water contains a trace of copper ions (*see* Section 4.1). Similar considerations apply to intermetallic phases such as $FeAl_3$ and $CuAl_2$, which can increase the kinetics of oxygen reduction.

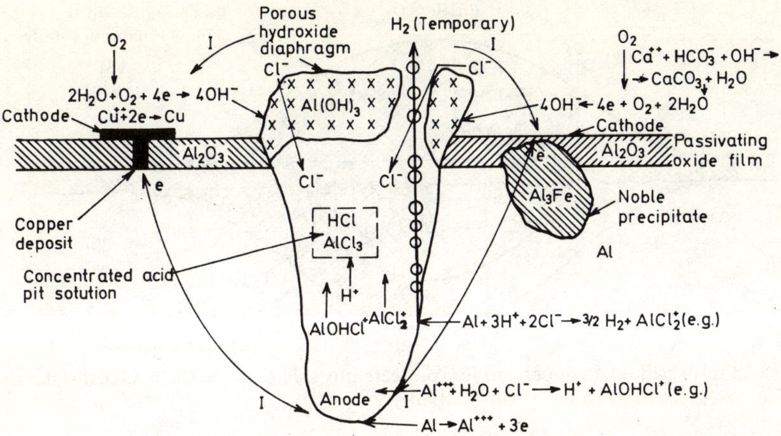

Fig. 1.58 Pit on aluminium showing how the rate of pitting may be facilitated by an intermetallic phase (Al_3Fe) or by a deposit of copper (after Wranglen[45])

Pitting of Copper

Previous considerations of pitting have been largely confined to metals and alloys that have a strong tendency to passivate, but since the pitting of copper has a number of unusual features it is appropriate to consider it in some detail. Reference to the potential–pH diagram for the Cu–H_2O (Section 4.2) system shows that in neutral solutions at the potentials encountered in oxygenated waters the stable form of copper is Cu_2O, and the corrosion resistance of copper thus depends upon whether or not the Cu_2O forms a protective film.

Copper and its alloys in certain fresh waters give rise to a form of localised attack that is referred to as *nodular* pitting in which the attacked areas are covered by small mounds or nodules composed of corrosion products and of $CaCO_3$ precipitated from the water. This is a serious problem in view of the extensive use of copper pipes and tanks for water supplies, and in aggressive water these may perforate in a relatively short time.

Figure 1.59 shows the type of pit and corrosion products[46] that form on copper pipes used for hard (or moderately hard) well waters, and this type of attack is most prevalent when the pipe is used for conveying cold water. The pit interior is almost invariably covered with solid Cu_2Cl_2, and across the mouth of the pit there is a thin membrane of Cu_2O containing one or more holes; this membrane is supported on the underside by a more substantial layer of coarsely crystalline Cu_2O formed by hydrolysis of Cu_2Cl_2. Above the Cu_2O membrane there is a roughly hemispherical mound of $CaCO_3$ containing insoluble copper salts, mainly basic carbonate and chloride. According to Campbell[47,48] it is possible to distinguish two types of pits and those described above are most prevalent in waters used in the USA, Belgium, Holland and the UK. Another type of pit occurs in certain soft water areas (mainly in Sweden and Germany), but only when the temperature is above

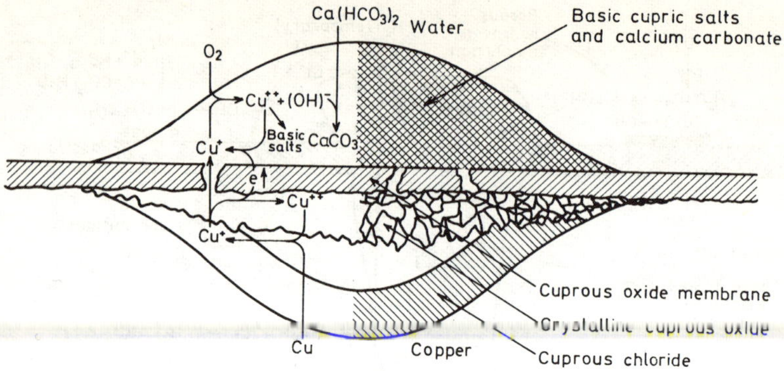

Fig. 1.59 Pit formed on a copper surface (protected by a film of Cu_2O) in a hard water (after Lucey[46])

60°C; these pits are of a smaller cross-section than those obtained in hard waters, and contain a very hard crystalline Cu_2O that may be capped by small black or greenish-black mounds of Cu_2O and basic copper sulphate, but often no mound of corrosion product is produced. Subsequent considerations will be confined to the type of pit shown in Fig. 1.59.

Campbell[49–51] during an investigation of pitting of copper tubes in the UK showed that pitting only occurred in deep well supplies of very low organic content, whereas river or lake waters that contain organic impurities did not give rise to pitting. He suggested that these non-aggressive waters contained an organic substance that facilitated the formation of a protective layer of Cu_2O, whereas in its absence the Cu_2O formed a loose, coarsely crystalline, non-protective deposit. The inhibitor has not been identified, but it has been shown to be a negatively charged colloidal substance with acidic properties, which can be detected by a whitish-blue fluorescence when the water is exposed to ultra-violet radiation; it has been suggested that it is possibly an unsaturated delta lactone.

Campbell also showed that a high proportion of copper pipes that failed in the aggressive cold waters contained a carbon deposit that was formed by the breakdown of the drawing lubricant residue during bright annealing. A glassy thin film of Cu_2O can also form during bright annealing, and initially it was considered that both the carbon and the glassy Cu_2O could act as a cathode, and thus stimulate attack at discontinuities in the film and lead to pitting.

Devroey and Depommier[52] in Belgium, and Campbell have found, however, in further investigations that glassy cuprous oxide scales formed in tubes during annealing do not cause pitting corrosion but become converted in service to a dull protective oxide. The glassy cuprous oxide found in tubes that had failed in service were shown to have been formed by the action of the water and to be associated with the presence of very thin carbon films produced in the tubes during manufacture.

The compositions of waters that give rise to pitting have been the subject of numerous investigations, and studies by Obrecht *et al.*[31] of the pitting of copper tubes in the USA have shown that it is invariably a cold water phenomenon that occurs with hard well waters. These waters contained over 5 p.p.m. of dissolved CO_2, 10–40 p.p.m. being typical, a pH in the range

7·0–7·8, 10–12 p.p.m. of dissolved O_2 and an SO_4^{2-}/Cl^- ratio of 3–4:1. Lucey[53] has analysed about 120 waters in the UK, whose behaviour in relation to the pitting of copper is well established, and has constructed a nomogram that provides a means of predicting pitting propensity from an analysis of the water. It would appear that an increase in the SO_4^{2-} or Na^+ ion concentration or an increase in the concentration of dissolved oxygen increases the pitting propensity, whereas the converse applies to the Cl^- or NO_3^- ion concentration and pH. Thus whereas the pitting of stainless steels is favoured by a high $Cl^-:SO_4^{2-}$ ratio (*see* Fig. 1.55) the pitting of copper is favoured by a high $SO_4^{2-}:Cl^-$ ratio. A high pH reduces the pitting propensity, and many surface waters contain 'humic acid' that enables more calcium bicarbonate to be held in solution than the equilibrium value, thus resulting in a high and stable pH.

Mechanism

The mechanism of pitting is highly complex, and reference should be made to the original papers for further details. However, it is of interest to consider certain views on this subject, since some of them introduce new concepts.

May[54] was the first to stress the important rôle of Cu_2Cl_2 within the pits on the mechanism, and he considered that it acted as a screen that prevented dissolved oxygen gaining access to the bottom of the pit thus preventing the formation of a protective Cu_2O film; the low solubility of Cu_2Cl_2 also maintained the activity of copper ions at a low value and thus facilitated anodic dissolution of the copper.

Pourbaix[12,55] has constructed equilibrium potential–pH diagrams for the Cu–Cl–H_2O system and for the Cu–Cl–CO_2–SO_3–H_2O system using concentrations of ions that correspond to those actually present in Brussels water. This work has shown that at the bottom of a pit the phases Cu, Cu_2O and Cu_2Cl_2 are at equilibrium in a solution containing 246 p.p.m. of Cl^- and 270 p.p.m. of Cu^{2+} only at a potential of 270 mV (*vs.* S.H.E.) and pH 3·5. The potential of the metal outside the pit in a water of pH 7·8 depends on the concentration of oxygen, the overpotential for its reduction on the passive surface film, etc. but it is normally about 300 mV, which is only 30–70 mV higher than the potential within the pits. Under these circumstances the pits do not grow appreciably. However, if a carbon film is present on the surface of the copper, a differential aeration cell will be set up, and the potential of the surface will increase—this in turn will increase the potential of the surface of the interior of the pit above the equilibrium value of 270 mV. Under these circumstances the equilibrium will be disturbed and the copper within the pit will corrode rapidly forming Cu^{2+}. This mechanism thus involves many of the conventional features of pitting, i.e. differential aeration, a large cathode:anode surface area ratio and the development of acidity within the pit by hydrolysis of Cu_2Cl_2 that prevents a protective film of Cu_2O from forming.

Lucey[46] examined a number of examples of pitting of copper pipes and tanks from hard water districts, and found that there was no more calcium carbonate scale deposited around the pits than on other parts of the metal surface. There was, however, a large amount of $CaCO_3$ in the mound immediately above the pit, and this suggested that the reduction of dissolved oxygen takes place immediately above the pit and not on the surrounding surface. He also showed that the Cu_2O membrane could act as thin bipolar electrode, the upper surface acting as a cathode and the under surface as an anode. Thus the Cu_2Cl_2 produced within the pit is anodically oxidised to Cu^{2+}, and this ion can then attack the copper within the pits by the disproportionation reaction

$$Cu^{2+} + Cu \rightarrow 2Cu^{+} \qquad ...(1.131)$$

The principal cathodic reaction on the upper surface of the membrane is the reduction of Cu^{2+} that is formed by the reaction of Cu^{+} with dissolved oxygen in the water; these Cu^{+} ions are provided partly from the diffusion through the pores in the oxide membrane from within the pit and partly from those produced by cathodic reduction (equation 1.131). Lucey's theory thus rejects the conventional large cathode: small anode relationship that is invoked to explain localised attack, and this concept of an electronically conducting membrane has also been used by Evans[56] to explain localised attack on steel due to a discontinuous film of magnetite.

Selective Leaching or De-alloying

In certain alloys and under certain environmental conditions selective removal of one metal (the most electrochemically active) can occur resulting in either localised attack, with the consequent possibility of perforation (plug type), or in a more uniform attack (layer type) that results in a weakening of the strength of the component. Although the selective removal of metals such as Al, Fe, Co, Ni and Cr from their alloys is known, the most prevalent form of de-alloying is the selective removal of zinc from the brasses—a phenomenon that is known as *dezincification*.

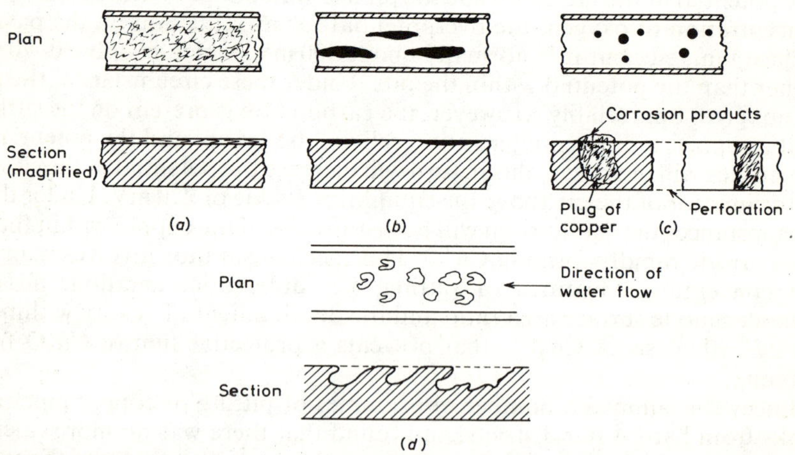

Fig. 1.60 Dezincification and impingement attack of copper-alloy tubes. (*a*) Uniform layer dezincification of a brass, (*b*) banded dezincification of a brass, (*c*) plug-type dezincification and (*d*) impingement attack

In this connection it is of interest to refer to the parting of Ag–Au alloys in nitric acid in which the silver can be completely removed from the alloy providing the ratio of Ag:Au is greater than 2·5:1; at somewhat lower ratios separation is incomplete and at low ratios no separation is possible and the alloy is unattacked by the acid. Similarly, dezincification does not occur when the zinc in a brass is less than 15%, and red brass (15% Zn) has only a slight tendency to dezincify, whereas the α-brasses (Cu–30Zn) are highly susceptible and the β-brasses (Cu–40Zn) even more susceptible; low alloying additions of Al and Mn in brasses appear to have very little effect in reducing dezincification.

Dezincification is readily apparent, since the yellow colour of the brass is replaced by the characteristic red of copper, which may take the form of small plugs or of layers that in some cases can extend over the whole of the surface (Fig. 1.60). In plug-type dezincification a mechanically weak, porous residue of copper is produced, which may remain *in situ* or become removed by the pressure of water, leading to a perforation. In the layer type the transformation of the alloy into a mechanically weak layer of copper results in loss of strength, and failure may occur by splitting when the metal is subjected to water pressure or to external stress.

Dezincification can occur over a wide range of pH, although the pH appears to affect the form of attack. Thus the layer type is favoured when the environment is acid and the brass has a high zinc content, whilst the plug type is more prevalent when the environment is neutral, slightly acid or alkaline and the zinc content is relatively low. In both cases dezincification is favoured by stagnant conditions (compare erosion–corrosion that becomes prevalent at high velocities), by the presence of chloride ions and by the formation of porous scales and deposits that lead to stagnant crevice conditions.

Dezincification of α-brass can be readily prevented by suitable alloying additions, and this was achieved first by adding 1% Sn. However, elements such as As, Sb and P are more effective, and alloying additions of 0·02–0·06% As are widely used for this purpose. Unfortunately, no alloying element has been found that prevents the dezincification of the two-phase αβ-brasses, which are more susceptible than the α-brasses, and their use must be avoided under environmental conditions that are conducive to dezincification.

Mechanism

Two theories have been proposed to explain dezincification, but since both have considerable support the precise mechanism remains unresolved. One theory proposes that the zinc is selectively leached from the alloy leaving a porous residue of metallic copper *in situ* (*cf.* parting of Ag–Au alloys), whilst the other proposes that the whole of brass dissolves and that the copper immediately redeposits at sites close to where the brass was dissolved.

There is considerable metallographic and electrochemical evidence in support of each theory and it is of interest to note that two of the most authoritative works on corrosion appear to support opposite views—Uhlig[20] favours the selective dissolution of zinc theory, whereas Fontana and Greene[22] favour the dissolution–precipitation theory.

Metallographic studies by Polushkin and Shuldener[57] revealed that the copper residue in a dezincified α-brass contained twins and residual grain boundaries that resembled those present in the parent metal, which appears to support the dissolution of zinc theory. However, other workers claim that twins found in the copper residue are similar to those found in electrolytically deposited copper. Horton[58] used time-lapse colour photomicrography to observe the changes that took place during the dezincification of an α-brass in 3% NaCl at 50°C. The brass was cold worked and annealed to produce a coarse-grained structure with a number of annealing twins, but none of these features was reproduced in the copper, indicating that the whole alloy dissolved.

Any selective dissolution mechanism must explain how the zinc within the alloy diffuses to the surface at which the reaction takes place. Pickering and Wagner[59] used X-ray and electron diffraction to study the selective dissolution of copper from Cu–Au alloys, and subsequently a similar approach was used by Pickering[60] to study the selective dissolution of zinc from ε- and γ-brasses (zinc-rich alloys) during anodic polarisation in a variety of electrolyte solutions. Pickering found that the partially dissolved alloys gave rise to new intermediate and terminal phases having a higher copper concentration than the original alloys. It was concluded that volume diffusion of zinc could occur via the vacancies (mono- and di-) created at the surface by anodic dissolution; the diffusion coefficient for di-vacancies in copper at 25°C was calculated to be $1{\cdot}3 \times 10^{-2}\,cm^2\,s^{-1}$.

Lucey[61] concludes from his electrochemical studies that dezincification involves anodic dissolution of both copper and zinc followed by the cathodic deposition of copper, and on this basis he has explained why arsenic is capable of inhibiting dezincification of α-brass but not of αβ-brass.

When dezincification occurs in service the brass dissolves anodically and this reaction is electrochemically balanced by the reduction of dissolved oxygen present in the water at the surface of the brass. Both the copper and zinc constituents of the brass dissolve, but the copper is not stable in solution at the potential of dezincifying brass and is rapidly reduced back to metallic copper. Once the attack becomes established, therefore, two cathodic sites exist—the first at the surface of the metal, at which dissolved oxygen is reduced, and a second situated close to the advancing front of the anodic attack where the copper ions produced during the anodic reaction are reduced to form the porous mass of copper which is characteristic of dezincification. The second cathodic reaction can only be sufficient to balance electrochemically the anodic dissolution of the copper of the brass, and without the support of the reduction of oxygen on the outer face (which balances dissolution of the zinc) the attack cannot continue.

The potentials of film-free α-brass and β-brass in solutions comparable to those existing inside the alloy at the advancing front of attack were found to be −0·38 V and −0·56 V (*vs.* S.H.E.), respectively. It was also established, taking into account the activities of copper ions in equilibrium with the sparingly soluble corrosion product Cu_2Cl_2, that whereas Cu^{2+} ions can be reduced to copper at −0·16 V the reduction of Cu^{+} ions is possible only at potentials more negative than −0·41 V. Thus whereas the β-phase of an αβ-brass can reduce both Cu^{2+} and Cu^{+} ions, the α-brass can reduce only the Cu^{2+} ion.

Lucey points out that although arsenic can prevent dezincification of an

α-brass it cannot prevent pitting, during which Cu_2Cl_2 is formed and is subsequently hydrolysed or oxidised to the secondary corrosion product Cu_2O. Thus during dezincification the advancing front of attack results in the formation of Cu_2Cl_2, and the Cu^+ ions can be readily reduced to metallic copper by the very negative β-phase, but not by the α-phase. The fact that α-brasses dezincify is explained by the formation of Cu^{2+} by the disproportionation of Cu^+ ions (formed from Cu_2Cl_2)

$$2Cu^+ \rightleftharpoons Cu^{2+} + Cu \qquad \ldots(1.132)$$

which are readily reduced directly to copper by the α-phase. Under these circumstances dezincification will proceed as long as Cu^{2+} ions are present at the advancing front. If, however, metallic arsenic is deposited at the advancing front in preference to copper metal, the Cu^{2+} ions are reduced to Cu^+ according to the following cycle:

$$3Cu^2 + As \rightarrow 3Cu^+ + As^{3+}$$

$$3Cu + As^{3+} \rightarrow 3Cu^+ + As$$

in which arsenic metal is regenerated. Thus dezincification of α-brass is prevented by the presence of a very small quantity of arsenic, and slow hydrolysis of Cu_2Cl_2 to Cu_2O takes place.

Erosion–Corrosion

The effect of movement of the solution or of the metal on the rate and form of corrosion is complex, and on the basis of previous considerations (*see also* Sections 1.4 and 2.1) the situation can be summarised as follows:

1. Increase in velocity may increase the rate by bringing the cathode reactant more rapidly to the surface of the metal thus decreasing cathodic polarisation, and by removing metal ions thus decreasing anodic polarisation.
2. Increase in velocity may decrease the rate by bringing the cathode reactant to the surface at a rate that exceeds $i_{crit.}$, thus causing passivation.
3. Decrease in velocity will favour all forms of localised attack in which an occluded cell is involved in the mechanism, and will also favour selective dissolution of alloys that are susceptible to this form of attack.

However, movement at appreciable rates can result in another form of attack that is brought about by the conjoint action of erosion and corrosion; hence the term *erosion–corrosion* that includes all forms of accelerated attack in which protective films, and even the metal surface itself, are removed by the abrasive action of movement of a fluid (gas or liquid) at high velocity. In general, the higher the velocity the more abrasive the solution.

Erosion–corrosion in the widest sense of the term will include impingement attack, cavitation damage and fretting corrosion, but since the latter two are dealt with in separate sections (*see* Sections 8.7 and 8.8) they will not be considered here.

The most significant effect of erosion–corrosion is the constant removal of protective films (which may range from thick visible films of corrosion products to the thin invisible passivating films) from the metal's surface, thus resulting in localised attack at the areas at which the film is removed. This can be caused by movement at high velocities, and will be particularly prone to occur if the solution contains solid particles (e.g. insoluble salts, sand and silt) that have an abrasive action. Impingement attack is a form of erosion–corrosion in which a turbulent stream of water containing entangled air bubbles and solid particles hits a metal surface, disrupts the protective film and thus results in pitting.

In addition to the mechanical damage of the protective film, velocity or movement will also bring the cathode reactant more rapidly to the metal surface thus decreasing cathode polarization.

The ability of a metal or alloy to withstand erosion–corrosion depends upon the nature of the environment which can range from a natural water to a concentrated acid, so that it is difficult to make sweeping generalisations and each system must be considered individually. In most systems the rate of attack increases with velocity, and metals and alloys that have an acceptable corrosion rate in static solutions may corrode rapidly when the solution (or the metal) is moving at a high velocity. Table 1.22[22] shows that up to about 120 cm s^{-1} the effect is small, but at about 820 cm s^{-1} there is a rapid increase in the corrosion rate. However, there are exceptions to this rule, and Fontana and Greene[22] point out that whereas the corrosion rate of aluminium in fuming nitric acid at 42°C increases with velocity, the converse applies to type 347 stainless steel owing to the different corrosion mechanisms involved. Thus aluminium is protected by films of aluminium nitrate and aluminium oxide and since the former is removed at intermediate velocities (up to 100 cm s^{-1}) and the latter at higher velocities, the corrosion rate increases

Table 1.22 Corrosion of metals by sea-water moving at different velocities ($g\,m^{-2}\,d^{-1} \times 10^2$)*

Metal	*Velocity*		
	30·5 cm s^{-1} (1 ft s^{-1})†	122 cm s^{-1} (4 ft s^{-1})‡	823 cm s^{-1} (27 ft s^{-1})§
Carbon steel	3·4	7·2	25·4
Cast iron	4·5	—	27·0
Silicon bronze	0·1	0·2	34·3
Admiralty brass	0·2	2·0	17·0
Hydraulic bronze	0·4	0·1	33·9
G bronze	0·7	0·2	28·0
Aluminium bronze (10% Al)	0·5	—	23·6
Aluminium brass	0·2	—	10·5
Cu–10Ni–0·8Fe	0·5	—	9·9
Cu–30Ni–0·05Fe	0·2	—	19·9
Cu–30Ni–0·5Fe	<0·1	<0·1	3·9
Monel	<0·1	<0·1	0·4
Stainless steel (*316*)	0·1	0	<0·1
Hastelloy *C*	<0·1	—	0·3
Titanium	0	—	0

*Data after Fontana and Greene[22]; source International Nickel Co.
†Immersed in tidal current.
‡Immersed in sea-water flume.
§Attached to immersed rotating disc.

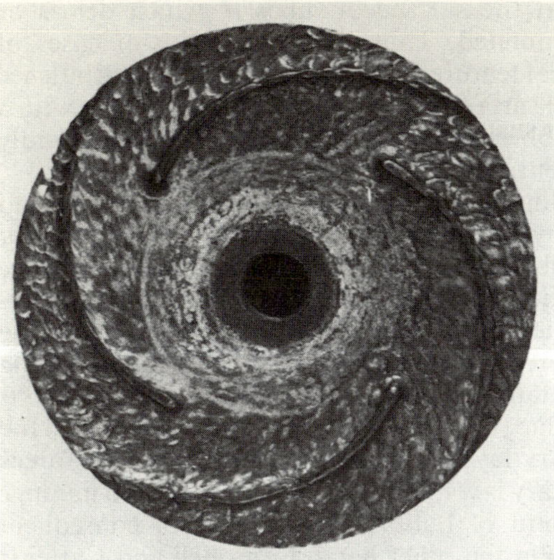

Fig. 1.61 Erosion–corrosion. Impeller fabricated from Cu–10Sn–2Zn alloy used for pumping a slurry containing 5–10% H_2SO_4 after six months' service. The blades of the impeller were extensively corroded. Note the directional form of the attack

with increase in velocity. In the case of the stainless steel in stagnant fuming nitric acid the metal is attacked autocatalytically by the nitrous acid formed by cathodic reduction of the nitric acid, and if the nitrous acid is swept away the rate decreases.

This example of aluminium illustrates the importance of the protective film, and films that are hard, dense and adherent will provide better protection than those that are loosely adherent or that are brittle and therefore crack and spall when the metal is subjected to stress. The ability of the metal to reform a protective film is highly important and metals like titanium and tantalum that are readily passivated are more resistant to erosion–corrosion than copper, brass, lead and some of the stainless steels. There is some evidence that the hardness of a metal is a significant factor in resistance to erosion–corrosion, but since alloying to increase hardness will also affect the chemical properties of the alloy it is difficult to separate these two factors. Thus although copper is highly susceptible to impingement attack its resistance increases with increase in zinc content, with a corresponding increase in hardness. However, the increase in resistance to attack is due to the formation of a more protective film rather than to an increase in hardness.

Attack by erosion–corrosion may be uniform or localised, and in the case of the latter it is characterised by the formation of grooves or rounded depressions that are smooth and free from corrosion products. Frequently the areas of localised attack follow a pattern that is indicative of the direction of movement of the metal or the environment (Fig. 1.61).

Impingement Attack

Impingement attack, as is implicit in the name, is a form of erosion–corrosion in which the solution strikes the metal surface at a high velocity—a situation that can occur at bends, tees and sudden changes in section in a water system;

at steam-turbine blades; and at the surface of tubes in which debris or corrosion products have accumulated. Thus in the majority of cases of impingement attack a geometrical feature of the system results in turbulence at one or more parts of the metal surface. Attack can occur in this way with a solution that is free from gas bubbles or suspended solids, but it is markedly accentuated when they are present.

Air bubbles entangled in cooling waters are particularly detrimental, and their importance was stressed by Bengough and May[62] in early studies of condenser-tube alloys, in which they used a 'jet impingement test' (*see* Section 20.1) to simulate conditions that prevail in practice. These workers considered that the almost static boundary layer of water adjacent to the metal surface (*see* Section 2.1) prevents the impingement of rapidly moving water. However, if the water contains air bubbles that have a diameter that is greater than a certain critical value (the thickness of the boundary layer) the bubbles on striking the boundary layer are subjected to differential forces that disrupt them and the boundary layer, thus enabling the water to impinge on the metal surface at the point of bubble disruption with consequent removal of the protective film. On the other hand, Campbell and Carter[63] consider that attack is due to highly turbulent flow at the attacked area, which can be caused by air-free water at comparatively high speeds and at lower speeds if air bubbles of a certain size are present in the water. A similar turbulent condition is produced by particles of foreign debris, by corrosion products on the tube wall and by marine crustacea lodged within the tube. The turbulent water breaks through the boundary layer and causes continuous breakdown of the protective film and removal of metal ions so that anode polarisation is small. The conducting Cu_2O film on the metal surrounding the impinged area provides a large efficient cathode for the cathodic reduction of the dissolved oxygen that is brought rapidly to its surface, and under these circumstances attack is rapid and intense. The pits formed during the impingement of water containing entangled air bubbles are of a smooth rounded shape, and frequently resemble the imprints of tiny horseshoes (as if the horse was walking up-stream) indicating the direction of the flow (Figs. 1.60*d* and 4.11).

The Role of Iron in the Resistance of Aluminium Brass to Impingement Attack

The nature of the protective film on the surface of a metal in relation to its resistance to erosion–corrosion has already been emphasised, and it is generally accepted that alloying additions that improve resistance to this form of attack do so by modifying the nature of the film. This applies to the addition of 2% Al to brass to form aluminium-brass and to the addition of 1–2% Fe to the cupro-nickels, and in both cases the alloying addition results in a pronounced increase in resistance to impingement attack. However, since impingement attack involves the characteristic large cathode/small anode relationship its control could also be effected by making the cathodic reaction more difficult, and in this connection it is of interest to consider the rôle of iron in relation to the resistance of aluminium-brass to impingement attack.

It was established in the 1920s by chemical analysis that certain aluminium-brass condenser tubes that were found to have a remarkably high resistance to impingement attack were characterised by the presence of iron in the protective film, which had entered the film owing to the presence of iron corrosion products in the water. Subsequently, when it became common practice to cathodically protect the steel boxes, deliberate injections of small amounts of $FeSO_4 . 7H_2O$ were made to the water to maintain the source of iron, and this procedure is now common practice when aluminium-brass tubes are used, particularly when the cooling water is polluted. North and Pryor[64] found that when a solution of sodium chloride containing a small concentration of Fe^{2+} ions was electrolysed between copper electrodes the anode became coated with a film of Cu_2O whilst the cathode became coated with a dark-brown film that was identified as lepidocrocite γ-FeOOH, which was formed in the alkaline catholyte by the reaction

$$2Fe^{2+} + 4OH^- + \tfrac{1}{2}O_2 \rightarrow 2FeOOH + H_2O \qquad \ldots(1.133)$$

According to these workers the FeOOH is present in the water as a positively charged colloid that migrates and adheres to the negatively charged cathode forming a non-conducting film that impedes electron transfer between the copper and the dissolved oxygen. A similar situation would prevail in practice, so that the rôle of the iron is to control the kinetics of the cathodic reaction rather than that of the anodic reaction. Studies of a 2·5% Fe–Cu alloy, in which only 0·1% of the iron is in solid solution, indicated that the iron corrodes by galvanic action and that the FeOOH is deposited initially around the iron particles where the cathodic current density is a maximum, but it subsequently spreads out over the whole surface.

Gasparini *et al.*[65] have emphasised the importance of the zeta-potentials of the filmed metal's surface and the colloidal FeOOH in relation to the mechanism, and consider that the latter has a positive zeta-potential only in demineralised water. In natural waters the zeta potential at pHs at or greater than 7·5 is negative, and since the Cu_2O film has a positive zeta potential that extends into the solution during water flow (streaming potential) the FeOOH deposits on the Cu_2O, thus forming a non-conducting film. Although North and Pryor[64], and Gasperini *et al.*[65] differ as to the nature of the charge on the FeOOH they are in accord as to the mechanism involved.

Control of Corrosion–Erosion

The most important method for controlling corrosion–erosion is the use of materials that are resistant to this form of attack, and further information can be obtained by consulting the sections that are devoted to metals and alloys (*see* particularly Section 4.2).

The geometry of the system is a further factor that enables control to be effected, and in general, reducing the velocity of the solution and ensuring that flow is laminar and not turbulent will reduce the tendency for attack by erosion–corrosion. Thus the pipe diameter should be as large as possible, consistent with other considerations; bends should have a large radius and inlet and outlets should be streamlined so that there is not a sudden change in section.

As far as the environment is concerned the main factor is to remove solid particles by settling and filtration and to remove marine crustacea by screening.

L. L. SHREIR

REFERENCES

1. Champion, F. A., *Corrosion Testing Procedures*, Chapman and Hall, London (1962)
2. Hoar, T. P., *J. Soc. Chem. Ind.*, **69**, 356 (1950)
3. Greene, N. D. and Fontana, M. G., *Corrosion*, **15**, 25t (1959)
4. Brown, B. F., in *The Theory of Stress Corrosion Cracking of Alloys* (ed. J. C. Scully), NATO, Brussels, 187 (1971)
5. Evans, U. R., *Industr. Engng. Chem.*, **17**, 363, 370 (1925)
6. Evans, U. R., *The Corrosion and Oxidation of Metals*, Arnold, London (1961)
7. Pourbaix, M., *Rapports Technique*, No. 84, CEBELCOR (1960)
8. Pourbaix, M., *J. Electrochem. Soc.*, **101**, 217C (1954)
9. Bianchi, G., *Corrosion*, **14**, 245t (1958)
10. Schaschl, E. and Marsh, G. A., *Corrosion*, **16**, 461t (1960)
11. Tomashov, N. D., *Theory of Corrosion and Protection of Metals*, Collier-Macmillan, London, 470 (1966)
12. Pourbaix, M., *Lectures on Electrochemical Corrosion*, Plenum Press, New York–London (1973)
13. Suzuki, T., Yamake, M. and Kitamura, Y., *Corrosion*, **29**, 18 (1973)
14. Hoar, T. P., *Disc. Faraday Soc.*, No. 1, 299 (1947)
15. Edeleanu, C. and Evans, U. R., *Trans. Faraday Soc.*, **47**, 1121 (1951)
16. Brown, B. F., Fujii, C. T. and Dahlberg, E. P., *J. Electrochem. Soc.*, **116**, 218 (1969) and Smith, J. A., Peterson, M. H. and Brown, B. F., *Corrosion*, **26**, 539 (1970)
17. Wilde, B. E., *Corrosion*, **27**, 32 (1971)
18. Bogar, F. D. and Fujii, C. T., N.R.L. Report 7690, Washington (1974)

18*a*. Pickering, H. W. and Frankenthal, R. P., *J. Electrochem. Soc.*, **119**, 1297 (1972)

18*b*. Franck, U. F., *Proc. 1st International Congress on Metallic Corrosion*, Butterworths, London (1962)

19. Brown, B. F., *Corrosion*, **26**, 249 (1970)
20. Uhlig, H. H., *Corrosion and Corrosion Control*, John Wiley, New York (1971)
21. Streicher, M. A., *Corrosion*, **30**, 77 (1974)
22. Fontana, M. G. and Greene, N. D., *Corrosion Engineering*, McGraw-Hill, New York (1967)
23. Rosenfeld, I. L. and Marshakov, I. K., *Corrosion*, **20**, 115t (1964)
24. Karlberg, G. and Wranglen, G., *Corros. Sci.*, **11**, 499 (1971)
25. Takamura, A., *Corrosion*, **23**, 306 (1967)
26. Griess, J. C., *Corrosion*, **24**, 96 (1968)
27. Sharmon, C., *Nature*, **153**, 621 (1944) and *Chem. Ind.*, **46**, 1126 (1952)
28. Van Loo, M., Laiderman, D. and Bruhn, R., *Corrosion*, **9**, 277 (1953)
29. Slabaugh, W. and Grutheer, M., *Ind. Eng. Chem.*, **46**, 1014 (1954)
30. Kaesche, H., *Werkst. Korros.*, **10**, 688 (1959)
31. Obrecht, M. F., Sartor, W. E. and Keyes, J. M., *Proc. 4th International Congress on Metallic Corrosion* (1969)
32. Leckie, H. and Uhlig, H. H., *J. Electrochem. Soc.*, **113**, 1262 (1966)
33. Stolica, N. D., *Corros. Sci.*, **9**, 455 (1969)
34. Uhlig, H. H. and Gilman, J., *Corrosion*, **19**, 261t (1963)
35. Uhlig, H. H. and Gilman, J., *Z. Phys. Chem.*, **226**, 127 (1964)
36. Bohni, H. and Uhlig, H. H., *J. Electrochem. Soc.*, **116**, 906 (1969)
37. Stolica, N. D., *Corros. Sci.*, **9**, 205 (1969)
38. Pourbaix, M., *Corrosion*, **26**, 431 (1970)
39. Verink, E. D. and Pourbaix, M., *Corrosion*, **27**, 495 (1971)

39*b*. Wilde, B. E., *Corrosion*, **28**, 283 (1972)

40. Kolotyrkin, Y., *Corrosion*, **19**, 261t (1963)
41. Vermilyea, D., *J. Electrochem. Soc.*, **118**, 529 (1971)
42. Szklarska-Smialowski, Z., *Corrosion*, **27**, 223 (1971)

43. Wranglen, G., *Corros. Sci.*, **14**, 331 (1974)
44. Fyfe, D., Shanahan, C. E. A. and Shreir, L. L., *Corros. Sci.*, **10**, 817 (1971)
45. Wranglen, G., *Introduction to Corrosion and Protection of Metals*, Butler and Tanner, London (1972)
46. Lucey, V. F., *Br. Corr. J.*, **2**, 175 (1967)
47. Campbell, H. S., Miscellaneous Publications MP 574, Aug. (1972) and MP 568, Feb. (1972), British Non-ferrous Research Association
48. Campbell, H. S., *Water Treatment and Examination*, **20**, 11 (1971)
49. Campbell, H. S., *J. Inst. Metals*, **77**, 345 (1950)
50. Campbell, H. S., *J. Appl. Chem.*, **4**, 633 (1954)
51. Campbell, H. S., *Proc. 2nd International Congress on Metallic Corrosion*, 237 (1966)
52. Devroey, P. and Depommier, C., *Visseries et Tritileries Reunies S.A.*, Madrelen, Belgium, Unpublished Report (1962)
53. Lucey, V. F., BNFRA Research Report No. 2, June, 1692 (1968)
54. May, R., *J. Inst. Metals*, **32**, 65 (1953)
55. Pourbaix, M., Van Muylder, J. and Van Laer, J., *CEBELCOR Rapport Technique*, No. 125 (1965) and *Corros. Sci.*, **7**, 795 (1967); Pourbaix, M., *Corrosion*, **25**, 267 (1969)
56. Evans, U. R., *Corros. Sci.*, **9**, 813 (1969)
57. Polushkin, E. and Shuldener, H., *Trans. Amer. Inst. Mining Met. Eng.*, **161**, 214 (1945)
58. Horton, R. M., *Corrosion*, **26**, 160 (1970)
59. Pickering, H. and Wagner, C., *J. Electrochem. Soc.*, **114**, 698 (1967)
60. Pickering, H., *J. Electrochem. Soc.*, **117**, 8 (1970)
61. Lucey, V. F., *Br. Corr. J.*, **1**, 9 (1965) and **1**, 52 (1965)
62. Bengough, G. D. and May, R., *J. Inst. Met.*, **32**, 81 (1924)
63. Campbell, H. S. and Carter, V. E., *J. Inst. Met.*, **90**, 362 (1961)
64. North, R. F. and Pryor, M. J., *Corros. Sci.*, **8**, 149 (1968) and **9**, 509 (1969)
65. Gasparini, R., Rocca, C. D. and Ioannilli, E., *Corros. Sci.*, **10**, 157 (1970)

BIBLIOGRAPHY*

Differential aeration

Shatalov, A. Ya. and Santsova, V. P., 'Effect of the Solution Composition of Zinc During Uneven Aeration', *Sb. Tr. Voroneshsk Otd. Vses. Khim Obshehestva*, **2**, 89 (1959); *C.A.*, **59**, 12478g

Skorchelletti, V. V. and Golubeva, N. K., 'Effect of Nonequilibrium Aeration on the Rate of Corrosion of Iron', *Zh. Prikl. Khim.*, **35**, 1570 (1962); *C.A.*, **57**, 12198a

Ruther, W. E. and Hart, R. K., 'Influence of Oxygen on High Temperature Aqueous Corrosion of Iron', *Corrosion*, **19**, 127t (1963)

Ulanovskii, I. B., Korovin, Yu. M. and Sevastyanov, V. F., 'Effect of Hydrogen Sulfide on the Electrode Potential of Carbon Steel', *Zh. Prikl. Khim.*, **37**, 1736 (1964); *C.A.*, **61**, 11607d

Golubeva, N. K. and Skorchelletti, V. V., 'The Effect of Differential Aeration on the Rate of Corrosion of Iron', *Corros. Sci.*, **5**, 203 (1965)

Gladysheva, V. P. and Shatalov, A. Ya., 'Effect of Hydrogen-ion Concentration on the Performance of Differentially Aerated Couples', *Izv. Vysshikh. Uchebn. Zavedenii, Khim. i Khim. Tekhnol.*, **9**, 48 (1966); *C.A.*, **65**, 5002a

Grubitsch, H., 'The Area of Capture Principle with Oxygen Corrosion in Electrolytes', *Werkstoffe Korrosion*, **17**, 679 (1966); *C.A.*, 19674f

Saraby-Reintjes, A., 'Differential Aeration Cell', *J. Electro. Chem. Interfacial Electrochem.*, **37**, 357 (1972); *C.A.*, **77**, 55484f

Gladysheva, V. P. and Shatalov, A. Ya., 'Electrochemical Principles of the Operation of Differential Aeration Couples', *Izv. Vyssh. Ucheb. Zavad., Khim. i Khim. Tekhnol.*, **16**, 1508 (1973); *C.A.*, **80**, 43386z

Gladysheva, V. P. and Shatalov, A. Ya., 'Corrosion of Metals During Differential Aeration', Part 2 'Work of Differential Aeration Pairs During Varied Oxygen Concentration Proportions at the Cathode and Anode', *Izv. Vyssh. Ucheb. Zaved., Khim. i Khim. Tekhnol.*, **16**, 1663 (1973); *C.A.*, **80**, 55218j

Richard, A. and Nicolaides, G., 'Differential Aeration Corrosion of Passivating Metal Under a Moist Film of Locally Variable Thickness', *J. Electrochem. Soc.*, **121**, 183 (1974)

Note: C.A. means *Chemical Abstracts.*

*Acknowledgements are made to Mrs J. Sadowska-Mazur for preparing this bibliography on localised attack.

Crevice corrosion

Anderson, W. K. and McGoff, M. J., 'Corrosion of Zircaloy in Crevices Under Nucleate Boiling Conditions', *US At. Energy Comm. Kapl.*, 2203 (1962); *C.A.*, **57**, 3173

Smirnova, T. N., Ganina, N. V., Kolotygina, V. B. and Kazakova, V. M., 'Crevice Corrosion on Objects Made of the Alloy AMr6', *Zashchita Metal ot Karrozii, Sb. Kuibyshev*, 27 (1965); *C.A.*, **65**, 1908d

Ulanovskii, I. B., 'Conditions for Cathodic Protection of Stainless Steels in Crevices', *Zashchita Metal*, **1**, 643 (1965); *C.A.*, **64**, 10751

Holmes, D. R. and Mann, G. M. W., 'A Critical Survey of Possible Factors Contributing to Internal Boiler Corrosion', *Corrosion*, **21**, 370 (1965)

Ruskol, Yu. S. and Klinov, I. Ya., 'Crevice Corrosion of Ti-alloys in Acids', *Khim. i Neft. Mashinostr.*, 28 (1966); *C.A.*, **65**, 16488a

France Jr., W. D. and Greene Jr., N. D., 'Passivation of Crevices During Anodic Protection', *Corrosion*, **24**, 247 (1968)

Swann, P. R., 'Mechanism of Corrosion Tunnelling with Special Reference to Cu_3Au', *Corrosion*, **25**, 147 (1969)

Landrum, R. J., 'Designing for Corrosion Resistance', *Chem. Eng.*, **76**, 118 (1969)

Lizlovs, E. A., 'Polarization Cell for Potentiostatic Crevice Corrosion Testing', *J. Electrochem. Soc.*, **117**, 256C, Atlantic City Meeting (1970)

Lizlovs, E. A., 'Polarization Cell for Potentiostatic Crevice Corrosion', *J. Electrochem. Soc.*, **117**, 1335 (1970)

Vermilyea, D. A. and Tedmon Jr., C. S., 'A Simple Crevice Corrosion Theory', *J. Electrochem. Soc.*, **117**, 437 (1970)

Isasi, J. A. and Metzger, M., 'Structure—dependent Corrosion of High-purity Al in HNO_3 and H_2SO_4', *Corros. Sci.*, **11**, 631 (1971)

Karlberg, G. and Wranglen, G., 'On the Mechanism of Crevice Corrosion of Stainless Cr Steels', *Corrosion Sci.*, **11**, 499 (1971)

Vincentini, B., Sinigaglia, D. and Taccani, G., 'Crevice Corrosion: Calculation of the Voltage and Current Distribution Along the Crevice', *Werks. Korros.*, **22**, 916 (1971)

Chen, C. M., Beck, F. H. and Fontana, M. G., 'Crevice Effect During Polarisation of Ti–8Al–1Mo–1V Alloy in Aqueous and Methanol Environment', *Corrosion*, **27**, 234 (1971)

Korovin, Yu. M. and Ulanovskii, I. B., 'Corrosion in the Crevice of Stainless Steel Alloyed with Copper', *Zashch. Metal*, **8**, 425 (1972); *C.A.*, **77**, 121420k

Richman, R. B., 'Device for Determining Crevice Corrosion by Polarisation Techniques', U.S./633,099(Cl.324/71C; G.01n) (1972); *C.A.*, **76**, 120896n

Covington, L. C., 'Rôle of Multivalent Metal Ions in Suppressing Crevice Corrosion of Ti', *2nd Titanium Sc. Technol. Proc. Int. Conf., 1972* (1973); *C.A.*, **76**, 73409b

Peterson, M. H. and Lennox, T. J., 'A Study of Cathodic Polarisation and pH Changes in Metal Crevices', *Corrosion*, **29**, 406 (1973)

Bates, J. F., 'Cathodic Protection to Prevent Crevice Corrosion of Stainless Steel in Halide Media', *Corrosion*, **29**, 28 (1973)

Bowers, J. M., 'Crevice Corrosion of a 70–30 Chromium Modified Cupronickel and its Relations to the Experimental Pourbaix Diagram', US Nat. Tech. Inform. Serv., AD Rep. No. 762596 (1973). Avail. NTIS from Govt. Rep. Announce. (US), **76**, 84 (1973); *C.A.*, **79**, 148936m

Crolet, J. L. and Defranoux, J. M., 'Calculation of the Incubation Period for Crevice Corrosion of Stainless Steel', *Centre Belge Étude Corros. Rapport Tech.*, **122**, 210 3/1 to 210 3/10 (1973); *C.A.*, **79**, 72821x

Korovin, Yu. M. and Ulanovskii, I. B., 'Sea Water Crevice Corrosion of Ni, Some of its Alloys and Stainless Steel', *Zashch. Metal*, **9**, 309 (1973)

Crolet, J. and Defranoux, J. M., 'Calcul du Temps d'Incubation de la Corrosion Caverneuse des Aciers Inoxydables', *Corros. Sci.*, **13**, 575 (1973)

Isaacs, H. S., 'The Behaviour of Resistive Layers in the Localised Corrosion of Stainless Steel', *J. Electrochem. Soc.*, **120**, 1456 (1973)

Degerbeck, J., 'On Accelerated Pitting and Crevice Corrosion Tests', *J. Electrochem. Soc.*, **120**, 175 (1973)

McCafferty, E., 'Electrochemical Behaviour of Iron with Crevices in Nearly Neutral Chloride Solutions', *J. Electrochem. Soc.*, **120**, 234C (1973)

Ambrose, J. R. and Kruger, J., 'Crevice Corrosion of 304 Stainless Steel in 1·0 N NaCl Solution', *J. Electrochem. Soc.*, **120**, 234C (1973)

Occluded corrosion cell—electrochemical changes at the crack tip

Barth, C. F., Steigerwald, E. A. and Troiano, A. R., 'Hydrogen Permeability and Delayed Failure of Polarised Martensitic Steels', *Corrosion*, **25**, 353 (1969)

Rhodes, P. R., 'Mechanism of Chloride Stress-corrosion Cracking of Austenitic Stainless Steel', *Corrosion*, **25**, 462 (1969)

Sandoz, G., Fujii, C. T. and Brown, B. F., 'Solution Chemistry Within Stress-corrosion Cracks in Alloy Steels', *Corros. Sci.*, **10**, 839 (1970)

Smith, J. A., Peterson, M. H. and Brown, B. F., 'Electrochemical Conditions at the Tip of an Advancing Stress-corrosion Crack in AISI 4340 Steel', *Corrosion*, **26**, 539 (1970)

Brown, B. F., 'Concept of the Occluded Corrosion Cell', *Corrosion*, **26**, 249 (1970)

Sedriks, A. J., Green, J. A. S. and Novak, D. L., 'On the Chemistry of the Solution at Tips of Stress-corrosion Cracks in Aluminium Alloys', *Corrosion*, **27**, 199 (1971)

Vermilyea, D. A., 'A Theory for the Propagation of Stress-corrosion Cracking in Metals', *J. Electrochem. Soc.*, **119**, 405 (1972)

Wilde, B. E. and Kim, C. D., 'The Rôle of Hydrogen in the Mechanism of Stress-corrosion Cracking of Austenitic Stainless Steel in Hot Chloride Media', *Corrosion*, **28**, 350 (1972)

Lin, F. and Hochman, R. F., 'Electrochemical Study of Stress-corrosion Cracking of Ti 8–1–1 Alloy and NaCl Solutions', *Corrosion*, **28**, 182 (1972)

Brignold, G. J., 'Electrochemical Aspects of Stress Corrosion of Steels in Alkaline Solutions', *Corrosion*, **28**, 307 (1972)

Gray, H. R., 'Ion and Laser Microprobes Applied to the Measurement of Corrosion-Produced Hydrogen on a Microscopic Scale', *Corrosion*, **28**, 47 (1972)

Smith, T., 'A Capillary Model for Stress-corrosion Cracking of Metals in Fluid Media', *Corros. Sci.*, **12**, 45 (1972)

Gabrielli, C., Keddam, M. and Takenouti, H., 'Progression of a Hole at the Surface of Passive Iron', *C.R. Acad. Sci. Ser. C.*, **277**, 743 (1973); *C.A.*, 43389c

Gal-Or, L., Raz, Y. and Yahalom, J., 'Mathematical Characterisation of Corrosion Currents in Local Electrolytic Cells', *J. Electrochem. Soc.*, **120**, 598 (1973)

Suzuki, T., Yambe, M. and Kitomura, Y., 'Composition of an Anolyte Within Pit Anode of Austenitic Stainless Steel in Chloride Solution', *Corrosion*, **29**, 18 (1973)

Pourbaix, M., 'Electrochemical Aspects of Stress-corrosion Cracking', *CEBELCOR Rapp. Tech.*, **118**(RT 199) (1971); *C.A.*, **76**, 12065e

Wilde, B. E., 'Mechanism of Cracking of High-strength Martensitic Stainless Steel in NaCl Solutions', *Corrosion*, **27**, 326 (1971)

Intergranular attack

Levin, I. A. and Maksimova, G. F., 'Effect of Cold Work on the Tendency Towards Intergranular Corrosion of Type 18–8 Stainless Steel', *Khim Mashinostroenie*, **5**, 35 (1961); *C.A.*, **56**, 3219

3

Gerasimov, V. V. and Alexandrova, V. N., 'Intercrystalline Corrosion of Steels of the 1 Kh 18N9T Type in Distilled Water', *Metalloved i Term. Obrabotka Metal*, **2**, 53 (1962); *C.A.*, **56**, 13900f

Kuzub, V. S., Tsinmar, A. I., Kuzub, L. G. and Dolotova, T. S., 'Intercrystalline Corrosion of Stainless Steel in Concentrated HNO_3', *Zh. Prikl. Khim.*, **35**, 2794 (1962); *C.A.*, **58**, 9873e

Aylward, J. R. and Whitener, E. M., 'Dissolution of Zr in HCl plus CH_3OH', *J. Electrochem. Soc.*, **109**, 87 (1962)

Steverding, B., 'Intergranular Sulfur Corrosion in Missile Thrust Chamber Nickel Tubes', *Corrosion*, **18**, 433t (1962)

Rowlands, J. C., 'Electrochemical Aspects of Preferential Phase Corrosion in Complex Alloys', *Corros. Sci.*, **2**, 89 (1962)

Younger, R. N., Baker, R. G. and Littlewood, R., 'The Relationship Between Microstructure and Intercrystalline Corrosion in an 18Cr–12Ni–1Nb Austenitic Steel', *Corros. Sci.*, **2**, 157 (1962)

Greenblatt, J. H., 'Initial Corrosion of Al Alloys in High Temperature Water', *Corrosion*, **19**, 295t (1963)

Parker, E. A., 'Roping and Intergranular Corrosion of 430 Stainless Steel', *Am. Soc. Testing Mater.*, Spec. Tech. Pub. No. 369, 348 (1963); *C.A.*, **63**, 14476a

Resek, F., 'Intergranular Corrosion of C–Cr, Cr–Ni, Cr–Ni–Mn and Other Steels', *Zashchita Mater.*, **11**, 379 (1963); *C.A.*, **60**, 15518e

Voeltzel, J., Henry, G., Plateau, J. and Crussard, C., 'Precipitation and Intercrystalline Corrosion in a Stainless Steel of the 18–8Ti Type', *Mem. Sci. Rev. Met.*, **60**, 879 (1963); *C.A.*, **60**, 8956c

Cihal, V., 'Explanation of Intercrystalline Corrosion of Welded, Stabilised Cr–Ni Stainless Steels', *Bergakademie*, **15**, 23 (1963); *C.A.*, **58**, 13523d

Fleitman, A. H., Romano, A. J. and Kiermut, C. J., 'Corrosion of Carbon Steel by High Temperature Liquid Mercury', *J. Electrochem. Soc.*, **110**, 964 (1963)

Haynie, F. H. and Ketcham, S. J., 'Electrochemical Behaviour of Al Alloys Susceptible to Intergranular Corrosion. Electrode Kinetics of Oxide-covered Al', *Corrosion*, **19**, 403t (1963)

Ketcham, S. J. and Haynie, F. H., 'Electrochemical Behaviour of Al Alloys Susceptible to Intergranular Corrosion. Effect of Cooling Rate on Structure and Electrochemical Behaviour in 202A Al Alloy', *Corrosion*, **19**, 242t (1963)

Tingley, I. I., 'Corrosion Resistance of Type 329 Stainless Steel in HNO_3 containing Chloride Ion', *Corrosion*, **19**, 400t (1963)

Stickler, R. and Vinckier, A., 'Electron Microscope Investigation of the Intergranular Corrosion Fracture Surfaces in a Sensitised Austenitic Stainless Steel', *Corros. Sci.*, **3**, 1 (1963)

von Schwenk, W. and Buhler, H.-E., 'Beoboshtungen an einem Kornzerfallsen falligen Austenitischen Cr–Ni–Stahl im Aktivzustand', *Corros. Sci.*, **3**, 145 (1963)

Streicher, M. A., 'Relationship of Heat Treatment and Microstructure to Corrosion Resistance in Wrought Ni–Cr–Mo Alloys', *Corrosion*, **19**, 272t (1963)

Levin, I. A. and Volikova, I. G., 'The Accelerated Method for Testing the Stability of a Single-phase Corrosion Resistant Steel to Intercrystalline Corrosion', *Zavodsk Lab.*, **30**, 816 (1964)

Hua, P.-T., Shen, H.-S., Chou, C. C. and Chung, T. T., 'Relation Between Intercrystalline Corrosion and Electrode Potential of Austenitic Stainless Steel', *K'O Hsueh T'ung Pao*, **2**, 56 (1964); *C.A.*, **60**, 15516h

Baumel, A., 'Deutung der Ursachen der Interkristallinen Korrosion von Nichtrostended Stahlen in Zusammenhang mit der Chromverormungstheorie', *Corros. Sci.*, **4**, 89 (1964)

Mieluch, J. and Smialowski, M., 'The Behaviour of Grain Boundaries in Iron During Anodic Polarisation in Ammonium Nitrate Solution', *Corros. Sci.*, **4**, 237 (1964)

Streicher, M. A., 'Effect of Heat Treatment, Composition and Microstructure on Corrosion of 18Cr–8Ni–Ti Stainless Steel in Acids', *Corrosion*, **20**, 57t (1964)

Hua, P.-T., Hsin, K.-H., Sun, C. and Hsi, S.-Y., 'An Electrochemical Method for the Estimation of Intergranular Corrosion Susceptibility of Duralumin', *Chung Kuo K'O Hsueh Yuan Ying Yung Hua Hsueh Yen Chiu So Chi K'an*, **14**, 78 (1965); *C.A.*, **64**, 9233e

Singrorell, G. and D'Angelo, M., 'Intergranular Corrosion of Cr–Ni Austenitic Steels by Potentiodynamic Polarisation Curves', *Chim. Ind.*, **47**, 1215 (1965); *C.A.*, **64**, 6216b

Nichols, H. and Rostoker, W., 'Intergranular Corrosion Penetration in an Age-hardenable Al Alloy', *J. Electrochem. Soc.*, **112**, 108 (1965)

Doshi, C. P. and Austin, W. W., 'Effect of Grain Size on Carbide Precipitation and Intergranular Corrosion in AISI Type 201 Stainless Steel', *Corrosion*, **21**, 332 (1965)

Armijo, J. S., 'Intergranular Corrosion of Nonsensitised Austenitic Stainless Steels—Part I: Environmental Variables', *Corrosion*, **21**, 235 (1965)

Lynes, W., 'Intergranular Corrosion of Alpha-brass and Some Effects of Stress', *Corrosion*, **21**, 125 (1965)

Baeumel, A. and Tramposch, O., 'Investigation of Intercrystalline Grain Boundary Corrosion of Austenitic Manganese–chromium Steels by Water and Aqueous Salt Solutions', *Werkst. Korros.*, **17**, 110 (1966); *C.A.*, **64**, 13847d

Myers, J. R., Crow, W. B., Beck, F. H. and Saxer, R. K., 'Observation on the Anodic Behaviour of Nickel and Chromium Surface Topography and Temperature Effect', *Corrosion*, **22**, 32 (1966)

Fleitman, A., Romano, J. and Klamut, C. J., 'Boiling Mercury Corrosion of Certain Refractory Metals and Stainless Steels from 593–703°C', *Corrosion*, **22**, 137 (1966)

Samans, C. H., Meyer, A. R. and Tisinai, G. F., 'Sensitisation and Thermal Stabilisation of Ni–Cr–Mo–Fe–W Alloy', *Corrosion*, **22**, 336 (1966)

Osozawa, K., Bohenkamp, K. and Engell, H. J., 'Potentiostatic Study on the Intergranular Corrosion of an Austenitic Cr–Ni Stainless-steel', *Corros. Sci.*, **6**, 421 (1966)

Coriou, M., Grall, L., Mahieu, C. and Pelas, M., 'Sensitivity to Stress Corrosion and Intergranular Attack of High-nickel Austenitic Alloys', *Corrosion*, **22**, 280 (1966)

Gellings, P. J. and de Jongh, M. A., 'Grain Boundary Oxidation and the Chromium-depletion Theory of Intercrystalline Corrosion of Austenitic Stainless Steels', *Corros. Sci.*, **7**, 413 (1967)

Armijo, J. S., 'Impurity Adsorption and Intergranular Corrosion of Austenitic Stainless Steel in Boiling HNO_3–$K_2Cr_2O_7$ Solutions', *Corros. Sci.*, **7**, 143 (1967)

Raymond, E. L., 'Mechanism of Sensitisation and Stabilisation of Incoloy Nickel–Iron–Chromium Alloy 825', *Corrosion*, **24**, 180 (1968)

Haneman, R. E. and Aust, K. T., 'Solute Clustering and Intergranular Corrosion', *Scr. Met.*, **2**, 235 (1968); *C.A.*, **67**, 63432d

Nikitin, V. I., 'Intercrystalline Corrosion by Liquid Metals by Dissolution Process', *Izv. Akad. Nauk. SSSR, Metal*, **1**, 213 (1968); *C.A.*, **67**, 98156g

Desestret, A., Epelboin, I., Froment, M. and Guiraldenq, P., 'Sur la Comparison des Facies d'Attaque Electrolytique et Thermique d'Aciers Inoxydables Austenitiques a Teneur Variable en Si', *Corros. Sci.*, **8**, 225 (1968)

France, W. D. and Greene, N. D., 'Predicting the Intergranular Corrosion of Austenitic Stainless Steels', *Corros. Sci.*, **8**, 9 (1968)

Butler, G., Ison, H. C. K. and Stretton, P., Short Communication—'Catastrophic Grain Boundary Corrosion in a Cr–Fe Alloy', *Corros. Sci.*, **8**, 281 (1968)

Bond, A. P. and Lizlovs, E. A., 'Intergranular Corrosion of Ferritic Stainless Steel', *J. Electrochem. Soc.*, **115**, 233C (1968)

Armijo, J. S., 'Intergranular Corrosion of Nonsensitised Austenitic Stainless Steel', *Corrosion*, **24**, 24 (1968)

Ward, C. T., Mathis, D. L. and Staehle, R. W., 'Intergranular Attack of Sensitised Austenitic Stainless Steel by Water Containing Fluoride Ions', *Corrosion*, **25**, 394 (1969)

Hill, B. and Trueb, L. F., 'Resistance of Explosion-bonded Stainless Steel Clads to Intergranular Corrosion and Stress Corrosion Cracking', *Corrosion*, **25**, 23 (1969)

Galvele, J. R. and De Micheli, S. M., 'Mechanism of Intergranular Corrosion of Al–Cu Alloys', *Proc. 4th Int. Congr. Metal. Corros.*, 439 (1969); *C.A.*, **77**, 134210s

Bond, A. P. and Lizlovs, E. A., 'Intergranular Corrosion of Ferritic Stainless Steels', *J. Electrochem. Soc.*, **116**, 1305 (1969)

Lev, V. Z., Klinov, I. Ya. and Shapiro, M. B., 'Potentiostatic Method for Determining the Intercrystalline Corrosion Tendency of Cast Alloy EP 375', *Tr. Mosk. Inst. Khim. Mashinostr.*, **2**, 245 (1970); *C.A.*, **77**, 82803r

Hajto, N. and Vrabely, E., 'Comparison of Tests for Intercrystalline Corrosion Tendency', *Banyasz. Kohasz. Lapok. Kohasz.*, **103**, 385 (1970); *C.A.*, **77**, 48934n

Berge, P., 'Corrosion Intergranulaire d'un Alliage du Type Inconel 600 dans la Vapeur d'Eau', *Corros. Sci.*, **10**, 185 (1970)

Galvele, J. R. and de Micheli, S. M., 'Mechanism of Intergranular Corrosion of Al–Cu Alloys', *Corros. Sci.*, **10**, 795 (1970)

Cihal, V. and Kasova, I., 'Relation Between Carbide Precipitation and Intercrystalline Corrosion of Stainless Steel', *Corros. Sci.*, **10**, 875 (1970)

France, W. D. and Greene, N. D., 'Some Effects of Experimental Procedures on Controlled Potential Corrosion Tests of Sensitised Austenitic Stainless Steel', *Corros. Sci.*, **10**, 379 (1970)

Tedmon, C. S. (Jr.), 'Intergranular Corrosion of Austenitic Stainless Steel', *J. Electrochem. Soc.*, **118**, 192 (1971)

Hodges, R. J., 'Intergranular Corrosion in High Purity Ferritic Stainless Steel. Isothermal Time–Temp. Sensitisation Measurements', *Corrosion*, **27**, 164 (1971)

Hodges, R. J., 'Intergranular Corrosion in High Purity Ferritic Stainless Steel. Effect of Cooling Rate and Alloy Composition', *Corrosion*, **27**, 119 (1971)

Tedmon, C. S. (Jr.) and Vermilyea, D. A., 'Carbide Sensitisation and Intergranular Corrosion of Ni Base Alloys', *Corrosion*, **27**, 376 (1971)

Gulyaev, A. P. and Tokareva, T. B., 'Intergranular Corrosion of Austenitic Chromium–Nickel Steels', *Inst. Khim. Mashinostr.*, **38**, 3 (1971); *C.A.*, **79**, 44784u

Tedmon, C. S. (Jr.), Vermilyea, D. A. and Broecker, D. E., 'Effect of Cold Work on Intergranular Corrosion of Sensitised Stainless Steel', *Corrosion*, **27**, 104 (1971)

Henthorne, M. and De Bold, T. A., 'Intergranular Corrosion Resistance', *Corrosion*, **27**, 255 (1971)

Streicher, M. A.—Letter to the Editor, *Corros. Sci.*, **11**, 275 (1971)—Reply to Authors to 'Some Effects of Experimental Procedures', Short Communication by France, W. D. and Greene, N. D., *Corros. Sci.*, **10**, 379 (1970)

Wilson, F. G., 'Mechanism of Intergranular Corrosion of Austenitic Stainless Steel', *Br. Corros. J.*, **6**, 100 (1971)

Gizhermo, R. and Khristo, E., 'Effect of the Deoxidation Method on the Intercrystalline-corrosion Tendency of Cr–Ni Austenitic Steels', *Metalurgiye*, **5**, 17 (1972); *C.A.*, **80**, 98898y

Joshi, A. and Stein, D. F., 'Chemistry of Grain Boundaries and its Relation to Intergranular Corrosion of Austenitic Stainless Steel', *Corrosion*, **28**, 321 (1972)

Heinz, K., 'Susceptibility of Materials to Intergranular Corrosion', *Prakt. Metallogr.*, **9**, 441 (1972); *C.A.*, **77**, 167771d

Gegelova, N. B., Mudzhiri, Y. N., Knyazheva, V. M. and Topchiashvili, L. I., 'Determination of Intercrystalline Corrosion Tendency of Stainless Steel by Potentiodynamic Method', *Zashch. Metal.*, **8**, 420 (1972); *C.A.*, **77**, 121422n

van der Horst, J. M. A., 'Grain Boundary Attack of 316L Stainless Steel in Ammonia-rich Environments', *J. Electrochem. Soc.*, **119**, 216C (1972)

Froment, M. and Vignaud, C., 'Study of Intergranular Corrosion of Aluminium Using Bicrystals', *J. Electrochem. Soc.*, **119**, 216C (1972)

Steiner, A., 'Effect of Phosphorous Level on General and Intercrystalline Corrosion in 18/8 Type Steel', *Pr. Inst. Hutn.*, **24**, 255 (1972); *C.A.*, **80**, 6185n

Anisimova, M. S. and Chikurova, A. A., 'Effects of Heat-treatment Conditions and Structure of Kh18N9TL Cast Steel on its Susceptibility to Intercrystalline Corrosion', *Optimiz Met. Protsessov*, **6**, 163 (1972)

Zehr, S. W., 'A Study of the Intergranular Cracking of U–7·5wt.%Nb–2·5wt.%Zr (Mulberry) Alloy in Aqueous Chloride Solutions', *Corrosion*, **28**, 196 (1972)

Ballard, D. B., Bennett, L. H. and Swartzendruber, L. J., 'Intergranular Embrittlement of Cu–Pd Alloys in Sea-water', *Corrosion*, **28**, 368 (1972)

Westbrook, J. H., 'The Rôle of Precipitation and Segregation at Grain Boundaries in Corrosion Failure', *J. Electrochem. Soc.*, **119**, 216C (1972)

Lewis, E. C., 'Grain Boundary Corrosion of Sensitised Type 304 Stainless Steel by a Noble-to-active Polarisation Scan', *J. Electrochem. Soc.*, **119**, 219C (1972)

Frankenthal, R. P. and Pickering, H. W., 'Intergranular Corrosion of Ferritic Stainless Steels', *J. Electrochem. Soc.*, **119**, 216C (1972)

Cihal, V. and Jezek, J., 'Corrosion of Stainless Steel in the Immediate Vicinity of the Weld Metal', *Brit. Corr. J.*, **7**, 76 (1972)

Pakhomova, N. A. and Levin, I. A., 'Effect of Grain Size on the Inter-crystalline Corrosion of Stainless Steel Type 18–8', *Zashch. Metal*, **9**, 676 (1973); *C.A.*, **80**, 86207j

Brown, M. H. and Kirchner, R. W., 'Corrosion of High Alloy Weldments: 1. Sensitisation of Wrought High Nickel Alloys', *Corrosion*, **29**, 470 (1973)

Osozawa, K., 'Metallurgical Aspects of Intergranular Corrosion of Stainless Steel', *Boshoku Gijutsu*, **22**, 267 (1973); *C.A.*, **80**, 85981V

Mikhovskii, M., Dzhambazova, L. and Marinova, I., 'Determining the Extent of Inter-crystalline Corrosion by Measuring Barkhausen Noise', *Tekh. Misul.*, **10**, 101 (1973); *C.A.*, **80**, 103010j

Levente, B., 'Effect of Stabilising Elements on the Susceptibility to Intercrystalline Corrosion of Heat and Corrosion Resistant Austenitic Steels', *Gepgyartastechnologia*, **13**, 528 (1973); *C.A.*, **80**, 102971m

Gruetzner, G., 'Sensitising a Cr–Ni–N_2 Austenitic Stainless Steel to Intergranular Corrosion by Dichromium Nitride Precipitation', *Cent. Doc. Siderurg, Circ. Inform. Tech.*, **30**, 1165 (1973); *C.A.*, **80**, 39816x

Van der Horst, J. M. A., 'Grain Boundary Attack of *3116L* Stainless Steel in Ammonia-rich Environments', *J. Electrochem. Soc.*, **120**, 512 (1973)

Frankenthal, R. P. and Pickering, W. H., 'Intergranular Corrosion of a Ferritic Stainless Steel', *J. Electrochem. Soc.*, **120**, 23 (1973)

Dzhambasova, L., Mikhovskii, M. and Ivanov, D., 'Acoustic Study of Intergranular Corrosion in Non-ferrous Alloys', *Tekh. Misul.*, **10**, 115 (1973)

Freid, M. Kh. and Suprunov, V. A., 'Effect of Secondary Phases on the Corrosion Behaviour of Cr–Ni–Ti Steel Kh 18N 10T', *Izv. Vyssh. Ucheb. Zaved. Khim., Khim. Tekhnol.*, **16**, 1037 (1973); *C.A.*, **79**, 111101j

Sandor, P., 'Resistance to Intercrystalline Corrosion in Stainless Steel Weld Metal', *Anti-corros. Methods Mater.*, **20**, 3 (1973); *C.A.*, **79**, 56661q

Savkina, L. Ya., Lazareva, N. A. and Feldgandler, E. G., 'Intercrystalline Corrosion Tendency of Low-carbon Cr–Ni Steel', *Metalloved, Term. Obrab. Metal.*, **55** (1973); *C.A.*, **79**, 34366V

Hodge, F. G., 'Effect of Ageing on the Anodic Behaviour of Ni–Cr–Mo Alloys', *Corrosion*, **29**, 375 (1973)

Cowan, R. L. and Tedmon, C. S., 'Intergranular Corrosion of Fe–Ni–Cr Alloys', in *Advances in Corrosion Science and Technology*, **3** (ed. M. G. Fontana and R. W. Staehle), Plenum Press (1973)

Streicher, M. A., 'The Rôle of Carbon, Nitrogen and Heat Treatment in the Dissolution of Fe–Cr Alloys in Acids', *Corrosion*, **29**, 337 (1973)

Selective dissolution

Desai, M. N., Talati, J. D. and Trivedi, A. K. M., 'Dezincification of Brasses', *J. Indian Chem. Soc.*, **38**, 565 (1961); *C.A.*, **56**, 11334e

Bumbulis, J. and Graydon, W. F., 'Dissolution of Brass in H_2SO_4 Solutions', *J. Electrochem. Soc.*, **109**, 1130 (1962)

Crennell, J. T. and Sawyer, L. J. E., 'Cathodic Protection Against Dezincification', *J. Appl. Chem.*, **12**, 170 (1962)

Marshakov, I. K. and Bogdanov, V. P., 'Mechanism of the Selective Corrosion of Cu–Zn Alloys', *Zh. Fiz. Khim.*, **37**, 2767 (1963); *C.A.*, **60**, 7758q

Tanabe, Z., 'The Relation Between Dealuminisation and Anodic Polarisation of Aluminium Bronze', *Sumitomo Keikinzoku Giho*, **4**, 59 (1963); *C.A.*, **62**, 223b

Gerischer, H., 'Anodic Behaviour of Noble Metal Alloys and Resistance Limits in the Corrosion of Such Alloys', *Korrosion*, **16**, 21 (1963); *C.A.*, **61**, 14180g

Joseph, G. and Arce, M. T., 'Brass Dezincification', *Rev. Indian*, **2**, 60 (1963); *C.A.*, **61**, 11693c

Hashimoto, K., Ogava, S. and Shimodaira, S., 'Dezincification of α-brass', *Trans. Japan Inst. Metals*, **4**, 42 (1963); *C.A.*, **60**, 2589g

Piatti, L. and Grour, R., 'Selective Corrosion of Cu Alloys', *Korrosion*, **14**, 551 (1963); *C.A.*, **59**, 8436g

Robinson, F. P. A. and Sholit, M., 'The Dezincification of Brass', *Corrosion Technol.*, **11**, 11 (1964); *C.A.*, **61**, 2785f

Tanabe, Z., 'Dealuminisation of Cu–Al Alloys and Polarisation Measurements', *Boshoku Gijutsu*, **14**, 56 (1965); *C.A.*, **63**, 3971d

Turner, M. E. D., 'Further Studies on the Influence of Water Composition on the Dezincification of Duplex Brass Fittings', *Proc. Soc. Water Treat. Exam.*, **14**, 81 (1965); *C.A.*, **63**, 15984e

Frade, G. and Lacombe, P., 'Influence of Intergranular Diffusion in the Kinetics of Dezincification of 70/30 Brasses Without Additions at 450–900°C', *Compt. Rend.*, **260**, 5022 (1965); *C.A.*, **63**, 5323c

Lucey, V. F., 'The Mechanism of Dezincification and the Effect of Arsenic, Part 1', *Brit. Corrosion J.*, **1**, 9 (1965–66)

Lucey, V. F., 'The Mechanism of Dezincification and the Effect of Arsenic, Part 2', *Brit. Corrosion J.*, **1**, 51 (1965–66)

Rowlands, J. C., 'Preferential Phase Corrosion of Naval Brass in Sea-water', *2nd Int. Congr. Metal Corrosion*, 1963, New York City (1966); *C.A.*, **65**, 10282h

Joseph, G. and Arce, M. T., 'Contribution to the Study of Brass Dezincification', *Corros. Sci.*, **7**, 597 (1967)

Sugawara, H. and Ebiko, H., 'Dezincification of Brass', *Corros. Sci.*, **7**, 513 (1967)

Wilde, B. E. and Teterin, G. A., 'Anodic Dissolution of Copper–Zinc Alloys in Alkaline Solutions', *Brit. Corrosion J.*, **2**, 125 (1967)

Ugiansky, G. M. and Ellinger, G. A., 'Corrosion of Monel 400 in High CO_2 Well Water', *Corrosion*, **24**, 134 (1968)

Feller, H. G., 'Untersuchungen zur Selectiven Korrosion mit der Ringscheibenelektrode im System Cu–Zn in Schwefelsaure', *Corros. Sci.*, **8**, 259 (1968)

Pickering, H. W. and Byrne, P. J., 'Partial Currents During Anodic Dissolution of Cu–Zn Alloys at Constant Potential', *J. Electrochem. Soc.*, **116**, 1492 (1968)

Heidersbach, R., 'Clarification of the Mechanism of De-alloying Phenomenon', *Corrosion*, **24**, 171 (1968)

Langenegger, E. E. and Robinson, F. P. A., 'Effect of the Polarisation Technique on Dezincification Rates and Physical Structure of Dezincified Zones', *Corrosion*, **24**, 411 (1968)

Brooks, W. B., 'Discussion of the De-alloying Phenomenon', *Corrosion*, **24**, 171 (1968)

Pickering, H. W., 'Volume Diffusion During Anodic Dissolution of a Binary Alloy', *J. Electrochem. Soc.*, **115**, 143 (1968)

Pickering, H. W., 'Preferential Anodic Dissolution in Binary Alloys', *Proc. Conf. Fundam. Aspects Stress Corros. Cracking*, 159, N.A.C.E. (1969)

Langenegger, E. E. and Robinson, F. P. A., 'A Study of Mechanism of Dezincification of Brasses', *Corrosion*, **25**, 59 (1969)

Langenegger, E. E. and Robinson, F. P. A., 'The Rôle of As in Preventing Dezincification of α-brass', *Corrosion*, **25**, 137 (1969)

Horton, R. M., 'New Metallographic Evidence for Dezincification of Brass by Redeposition of Cu', *Corrosion*, **26**, 160 (1970)
Verink, E. D. (Jr.) and Parrish, P. A., 'Use of Pourbaix Diagrams in Predicting the Susceptibility to De-alloying Phenomena', *Corrosion*, **26**, 214 (1970)
Mattsson, E., 'Mechanism of Exfoliation (Layer Corrosion) of Al–5%Zn–1%Mg', *Brit. Corrosion J.*, **6**, 73 (1971)
Gal-Or, L., Zanker, L. and Yahalom, J., 'Preferential Dissolution of Inclusions in Hastelloy C', *Corros. Sci.*, **11**, 107 (1971)
Vernik, E. D. (Jr.) and Heindersbach, R. H. (Jr.), 'Evaluation of the Tendency for Dealloying in Metal Systems', *Amer. Soc. Test. Mater., Spec. Tech. Publ.*, **516**, 303 (1972); *C.A.*, **80**, 102944e
Leidheiser, H. and Kissinger, R., 'Chemical Analysis of the Liquid Within Propagatory Stress-corrosion Cracks in 70:30 Brass Immersed in Concentrated Ammonium Hydroxide', *Corrosion*, **28**, 218 (1972)
Langenegger, E. E. and Callaghan, B. G., 'Use of an Empirical Potential Shift Technique for Predicting Dezincification Rates of $\alpha\beta$-brasses in Chloride Media', *Corrosion*, **28**, 245 (1972)
Heidersbach, R. H. and Vernik, E. D., 'The Dezincification of α and β Brasses', *Corrosion*, **28**, 397 (1972)
Revie, R. W. and Uhlig, H. H., 'Further Evidence Regarding the Dezincification Mechanism', *Corros. Sci.*, **12**, 669 (1972)
Nielsen, K. and Rislund, E., 'Comparative Study of Dezincification Tests', *Br. Corrosion J.*, **8**, 106 (1973)

Pitting (papers published in English)

Robinson, F. P. A., 'Pitting Corrosion; Cause, Effect, Detection and Prevention', *Corrosion Technol.*, **7**, 237, 266 (1960)
Kolotyrkin, Y. M., 'Electrochemical Behaviour of Metals During Anodic and Chemical Passivation in Electrolytic Solutions', *1st Intern. Congr. Metallic Corrosion*, Butterworths, London, 10 (1962)
Clarke, F. E. and Ristaino, A. J., 'New Clues in the Boiler Tube Pitting Puzzle', *1st Intern. Congr. Metallic Corrosion*, Butterworths, London, 403 (1962)
Kimmel, A. L., Holingshand, W. R. and Shea, E. P., 'Effect of Dissolved Oxygen on Pitting of Boiler Tubes', *Am. Soc. Mech. Engrs.*, Paper No. 61-WA-285
Colborne, G. F., Allen, A. R. and Thunaes, A., 'Crash Repair Program to Control Serious Pitting Corrosion on Process Tanks', *Corrosion*, **17**, 20 (1961)
Flint, G. N. and Melbourne, S. H., 'The Corrosion of Decorative Ni and Cr Coatings—a Metallographic and Potential Study', *Trans. Inst. Met. Finishing*, **38**, 35 (1961)
Menasce, S., 'Corrosion of 18–5 Stainless Steel in Dilute HCl', *Rev. Met.*, **58**, 951 (1961)
Doyle, D. P. and Godard, H. P., 'Influence of Cl^- Additions to Water on the Corrosion Behaviour of Al', *Oil Can.*, **14** No. 33, 42 (1962); *C.A.*, **57**, 8330h
Fraser, W. A. and Bloom, M. C., 'Corrosion of Mild Steel in 40% NaOH Solution at 316°C', *Corrosion*, **18**, 163t (1962)
Bloom, M. C., Fraser, W. A. and Krulfeld, M., 'Corrosion of Steel in Concentrated LiOH Solution at 316°C', *Corrosion*, **18**, 401t (1962)
Videm, K., 'Pitting Corrosion of Aluminium in Contact with Stainless Steel', *Proc. Conf. on Corrosion Reactor Mater.*, Salzburg, Austria, **1**, 391 (1962); *C.A.*, **60**, 1412g
Lyon, D. H., Salva, S. J. and Shaw, B. C., 'Etch Pits in Germanium: Detection and Effects', *J. Electrochem. Soc.*, **110**, 184c (1963)
Bertocci, U., Hulett, L. D. and Jenkins, L. H., 'On the Formation of Electrochemical Etch Pits on the (111) Face of Copper', *J. Electrochem. Soc.*, **110**, 1190 (1963)
Weed, R. D., 'Dendritic Pitting of Stellite by Alkaline Permanganate Solutions', *J. Electrochem. Soc.*, **110**, 178c (1963)
Fraser, W. A. and Bloom, M. C., 'Corrosion of Mild Steel in 15%NaOH at 316°C', *J. Electrochem. Soc.*, **110**, 177C (1963)
Pourbaix, M., Klimczak-Mathieu, Martens, C. and Meunier, J., 'Potentiokinetic and Corrosimetric Investigations of the Corrosion Behaviour of Alloy Steels', *Corros. Sci.*, **3**, 239 (1963)
Lefrancois, P. A. and Hoyt, W. B., 'Chemical Thermodynamics of High Temperature Reactions in Metal Testing Corrosion', *Corrosion*, **19**, 360t (1963)
Wesley, W. A., 'Pitting of Decorative Nickel–Chromium Coatings', *Tech. Proc. Am. Electroplaters Soc.*, **50**, 9 (1963)
Bloom, M. C., Newport, G. N. and Fraser, W. A., 'The Growth and Breakdown of Protective

Films in High-temperature Aqueous Solutions 15% NaOH at 316°C', *J. Electrochem. Soc.*, **111**, 1343 (1964)

Dugdale, I. and Cotton, J. B., 'The Anodic Polarisation of Titanium in Halide Solutions', *Corros. Sci.*, **4**, 397 (1964)

Herman, R. S., 'Radiographic–Photographic Method for Measuring Depth and Distribution of Pitting', *Corrosion*, **20**, 361 (1964)

Tomashov, N. D., Chernova, G. P. and Marcova, O. N., 'Effect of Supplementary Alloying Elements on Pitting Corrosion Susceptibility of 18Cr–14Ni Stainless Steel', *Corrosion*, **20**, 166 (1964)

Schwenk, W., 'Theory of Stainless Steel Pitting', *Corrosion*, **20**, 129 (1964)

Murray, G. A. W., 'Artificial Pits for Quantitative Studies of Corrosion of Aluminium Alloys in Natural Waters', *Corrosion*, **20**, 329 (1964)

Koehler, E. L. and Evans, S., 'Pitting and Uniform Corrosion of Aluminium by pH 3·5 Citrate Buffer Solution', *J. Electrochem. Soc.*, **111**, 17 (1964)

Uhlig, H. H. and Gilman, J. R., 'Pitting of 18–8 Stainless Steel in Ferric Chloride Inhibited by Nitrates', *Corrosion*, **20**, 289 (1964)

Fisher, A. O., 'Magnesium Anodes Control Pitting in an Inhibited Circulating Cooling Water System', *Mater. Protect.*, **3**, 64 (1964)

Pouillard, E. and Finckbohner, A., 'Removal of Surface Chromium in Relation to Pitting Corrosion of Stainless Steel', *Dev. Met.*, **62**, 721 (1965); *C.A.*, **63**, 14476b

Hoar, T. P., Mears, D. C. and Rothwell, G. P., 'The Relationships Between Anodic Passivity, Brightening and Pitting', *Corros. Sci.*, **5**, 279 (1965)

Piggott, A. R., Leckie, H. and Shreir, L. L., 'Anodic Polarisation of Titanium in HCOOH—I: Anodic Behaviour of Titanium in Relation to Anodising Conditions', *Corros. Sci.*, **5**, 165 (1965)

Heine, M. A., Keir, D. S. and Pryor, M. J., 'The Specific Effects of Chloride and Sulphate Ions on Oxide Covered Aluminium', *J. Electrochem. Soc.*, **112**, 24 (1965)

Freiman, L. I. and Kolotyrkin, Ya. M., 'Pitting Corrosion of Iron by Perchlorate Ions', *Corros. Sci.*, **5**, 199 (1965)

Greene, N. D. and Judd, G., 'Relation Between Anodic Dissolution and Resistance to Pitting Corrosion', *Corrosion*, **21**, 15 (1965)

Bond, A. P., Bolling, G. F. and Damian, H. A., 'Influence of Microsegregation on Pitting Corrosion in High-purity Aluminium', *J. Electrochem. Soc.*, **112**, 178c (1965)

Pickering, H. W. and Frankenthal, R. P., 'A Transmission Electron Microscope Study of the Breakdown of Passivity on Fe–24% Cr', *J. Electrochem. Soc.*, **112**, 761 (1965)

Board, P. W. and Elbourne, R. G. P., 'Pitting Corrosion in Plain Cans Containing Acid Foods', *Food Technol.*, **19**, 1571 (1965); *C.A.*, **64**, 2661c

Voogel, P., 'Pit Corrosion of Internally Painted Cargo and Cargo/Ballast Tanks of Ocean-going Crude Oil Tankers', *J. Oil Colour Chemists Assoc.*, **48**, 597 (1965)

White, J. H., Yaniv, A. E. and Schick, H., 'The Corrosion of Metals in the Water of the Dead Sea', *Corros. Sci.*, **6**, 447 (1966)

Novakovskii, V. M. and Sorokina, A. N., 'Model Study of Chloride Pitting in 18–8 Stainless Steel', *Corros. Sci.*, **6**, 227 (1966)

Kato, T. R. and Price, D. H., 'Effect of Environmental Conditions on the Pitting of Uranium', *AEC*, Accession No. 25486, Dept. No. NLCO-957 (1966); *C.A.*, **65**, 18069e

Davies, D. E. and Lotlikat, M. M., 'Passivation and Pitting Characteristics of Zinc', *Brit. Corrosion J.*, **1**, 149 (1966)

Bond, A. P., Bolling, G. F., Damian, H. A. and Biloni, H., 'Microsegregation and the Tendency for Pitting Corrosion in High-purity Aluminium', *J. Electrochem. Soc.*, **113**, 773 (1966)

Camp, E. K., 'Pitting Corrosion of Titanium in Aqua Regia', *J. Electrochem. Soc.*, **113**, 204c (1966)

Craig, H. L., 'Rotating Disc Study of the Pitting of Aluminium in Alkaline Chloride Solutions', *J. Electrochem. Soc.*, **113**, 203c (1966)

Hospadaruk, V. and Petrocelli, J. V., 'The Pitting Potential of Stainless Steel in Chloride Solutions', *J. Electrochem. Soc.*, **113**, 878 (1966)

Steigerwald, R. F., 'Effect of Chromium Content on Pitting Behaviour of Fe–Cr Alloys', *Corrosion*, **22**, 107 (1966)

Lebet, R. and Piotrowski, A., 'Resistance to Pitting of Types *202* and *321* Steels to Sulphuric Acid and Sodium Chloride Solutions', *Corrosion*, **22**, 257 (1966)

Spaepen, G. J. and Fevery-de Meyer, M. J., 'Electrochemical Corrosion Experiments at Temperatures About 100°C', *Corros. Sci.*, **7**, 405 (1967)

Bond, A. P. and Lizlovs, E. A., 'Anodic Polarisation of Some Ferritic Stainless Steels in Chloride Media', *J. Electrochem. Soc.*, **114**, 199c (1967)

Joseph, G. and Perret, R., 'Inhibitor Action on the Corrosion of Iron in Salt-water Solution', *Corros. Sci.*, **7**, 553 (1967)

Frazer, M. J. and Langstaff, R. D., 'Influence of Cupric Ions on the Behaviour of Surface-active Agents Towards Aluminium', *Brit. Corrosion J.*, **2**, 11 (1967)

Szklarska-Smialowska, Z. and Janik-Czachor, H., 'Pitting Corrosion of 13Cr–Fe Alloy in Na_2SO_4 Solutions Containing Chloride Ions', *Corros. Sci.*, **7**, 65 (1967)

Rosenfeld, I. L. and Danilov, I. S., 'Electrochemical Aspects of Pitting Corrosion', *Corros. Sci.*, **7**, 129 (1967)

Hoar, T. P., 'The Production and Breakdown of the Passivity of Metals', *Corros. Sci.*, **7**, 341 (1967)

Butler, G. and Beynon, J. G., 'The Corrosion of Mild Steel in Boiling Salt Solutions', *Corros. Sci.*, **7**, 385 (1967)

Beck, T. R., 'Pitting Corrosion of Titanium in Halide Solutions', *J. Electrochem. Soc.*, **114**, 201c (1967)

Frankenthal, R. P., 'The Effect of Surface Preparation on Pitting and Anodic Dissolution of Iron–Chromium Alloys', *J. Electrochem. Soc.*, **114**, 201c (1967)

Horvath, J. and Uhlig, H., 'Metallurgical Factors Affecting the Critical Potential for Pitting Corrosion of Cr–Fe–Ni Alloys', *J. Electrochem. Soc.*, **114**, 201c (1967)

Murray, G. A. W., 'Rôle of Iron in Aluminium on the Initiation of Pitting in Water', *Br. Corros. J.*, **2**, 216 (1967)

Penn, J. H. and Murray, G. A. W., 'Effect of Acidic Gelatinous Materials on Pitting Corrosion of Aluminium Hollow-ware', *Br. Corros. J.*, **2**, 193 (1967)

Lucey, V. F., 'Mechanism of Pitting Corrosion of Copper in Supply Waters', *Br. Corros. J.*, **2**, 175 (1967)

Finley, H. F., 'An Extreme-value Statistical Analysis of Maximum Pit Depths and Time to First Perforation', *Corrosion*, **23**, 83 (1967)

Mears, D. C. and Rothwell, G. P., 'Effects of Probe Position on Potentiostatic Control During the Breakdown of Passivity', *J. Electrochem. Soc.*, **115**, 36 (1968)

Frankenthal, R. P., 'The Effect of Surface Preparation and of Deformation on the Pitting and Anodic Dissolution of Fe–Cr Alloys', *Corros. Sci.*, **8**, 491 (1968)

Bond, A. P. and Lizlovs, E. A., 'Anodic Polarisation of Austenitic Stainless Steel in Chloride Media', *J. Electrochem. Soc.*, **115**, 1130 (1968)

Szummer, A., Szklarska-Smialowska, Z. and Janik-Czachor, M., 'Electron Microprobe Study of the Corrosion Pits Nucleation of Fe–16Cr Single Crystals', *Corros. Sci.*, **8**, 827 (1968)

Ijzermans, A. B. and van der Krogt, A. J., 'Pitting Corrosion of an Austenitic Cr–Ni Stainless Steel in H_2SO_4 Containing H_2S', *Corros. Sci.*, **8**, 679 (1968)

Horvath, J. and Uhlig, H. H., 'Critical Potentials for Pitting Corrosion of Ni, Cr–Ni, Cr–Fe and Related Stainless Steels', *J. Electrochem. Soc.*, **115**, 791 (1968)

Black, J. R., 'Etch Pit Formation in Silicon at Al–Si Contacts Due to the Transport of Silicon in Aluminium by Momentum Exchange with Conducting Electrons', *J. Electrochem. Soc.*, **115**, 242c (1968)

Posey, F. A., 'Pitting and Crevice Corrosion of Titanium and its Alloys', *J. Electrochem. Soc.*, **115**, 231c (1968)

Mattson, E. and Fredriksson, A. M., 'Pitting Corrosion in Copper Tubes—Cause of Corrosion and Counter Measures', *Br. Corros. J.*, **3**, 246 (1968)

Arora, G. P., Isasi, J. A. and Metzger, M., 'Protective Films and Structure Dependent Corrosion in Aluminium', *Corrosion*, **25**, 445 (1969)

Pourbaix, M., 'Recent Applications of Electrode Potential Measurements in the Thermodynamics and Kinetics of Corrosion of Metals', *Corrosion*, **25**, 267 (1969)

Lizlovs, E. A. and Bond, A. P., 'Anodic Polarisation of Some Ferritic Stainless Steels in Chloride Media', *J. Electrochem. Soc.*, **116**, 574 (1969)

Smialowski, M., Szklarska-Smialowska, Z., Rychcik, M. and Szummer, A., 'Effect of Sulphide Inclusions in a Commercial Stainless Steel on the Nucleation of Corrosion Pits', *Corros. Sci.*, **9**, 123 (1969)

Rajagopalan, K. S., Venu, K. and Viswanathan, M., 'Activation of Passivated Steel in Borax Solution', *Corros. Sci.*, **9**, 169 (1969)

Jones, D. A. and Greene, N. D., 'Electrochemical Detection of Localised Corrosion', *Corrosion*, **25**, 367 (1969)

Rossum, J. R., 'Prediction of Pitting Rates in Ferrous Metals from Soil Parameters', *J. Amer. Water Works Ass.*, **61**, 305 (1969); *C.A.*, **71**, 35404c

Szklarska-Smialowska, Z. and Czachor-Janik, M., 'Electrochemical Investigation of the Nucleation and Propagation of Pits in Iron–Chromium Alloys', *Br. Corros. J.*, **4**, 138 (1969)

Jones, R. L., 'Potter–Mann and Bloom Iron Oxide Films Grown on Mild Steel in 280C 5 N NaOH: Their Significance in Pitting', *Corrosion*, **29**, 133 (1969)

Obrecht, M. F., Sastor, W. E. and Keyes, J. M., 'Integrated Design of Field Test Panel Pilot Unit for Investigating Pitting Corrosion of Copper Water Tube by Potable Water Supplies', *Proc. 4th Int. Congr. Met. Corr.*, 1969, 576 (1972)

Joseph, G., Perret, R. and Spacek, J., 'Pitting Corrosion of Copper Tubings', *Proc. 4th Int. Congr. Metal. Corros.*, 1969, 817 (1972)

Bohni, H. and Uhlig, H. H., 'Environmental Factors Affecting the Critical Pitting Potential of Aluminium', *J. Electrochem. Soc.*, **116**, 906 (1969)

Hodge, F. G. and Wilde, B. E., 'Effect of Chloride Ion on the Anodic Dissolution Kinetics of Cr–Ni Binary Alloys in Dilute H_2SO_4', *Corrosion*, **26**, 146 (1970)

Foley, R. T., 'Rôle of the Chloride Ion in Iron Corrosion', *Corrosion*, **26**, 58 (1970)

Richardson, J. A. and Wood, G. C., 'A Study of the Pitting Corrosion of Aluminium by Scanning Electron Microscopy', *Corros. Sci.*, **10**, 313 (1970)

France, W. D. (Jr.) and Greene, N. D., 'Comparison of Chemically and Electrolytically Induced Pitting Corrosion', *Corrosion*, **26**, 1 (1970)

Brigham, R. J., 'Discussion on Chemically and Electrolytically Induced Pitting', *Corrosion*, **26**, 200 (1970)

Abdul Azim, A. A., 'Pitting Corrosion of Lead in $H_2SO_4 + HClO_4$', *Corros. Sci.*, **10**, 421 (1970)

Szklarka-Smialowska, Z., 'Electron Microprobe Study of the Effect of Sulphide Inclusions on the Nucleation of Corrosion Pits in Stainless Steels', *Br. Corros. J.*, **5**, 159 (1970)

Weinstein, M. and Speirs, K., 'Mechanisms of Chloride-activated Pitting Corrosion of Martensitic Stainless Steels', *J. Electrochem. Soc.*, **117**, 256 (1970)

Wilde, B. E. and Williams, E., 'On the Correspondence Between Electrochemical and Chemical Accelerated Pitting Corrosion Tests', *J. Electrochem. Soc.*, **117**, 775 (1970)

Bianchi, G., Cerguetti, A., Mazza, F. and Torchio, S., 'Chemical Etching and Pitting of Stainless Steels', *Corros. Sci.*, **10**, 19 (1970)

Ijzermans, A. B., 'Pitting Corrosion and Intergranular Attack of Austenitic Cr–Ni Stainless Steels in NaSCN', *Corros. Sci.*, **10**, 607 (1970)

Ashworth, V., Boden, P. J., Leach, J. S. Ll. and Nehru, A. Y., 'On the Cl^- Breakdown of Passive Films on Mild Steel', *Corros. Sci.*, **10**, 481 (1970)

Payer, J. H. and Staehle, R. W., 'Localised Attack on Metal Surfaces', *Corros. Fatigue: Chem. Mech. Microstruct.*, 1971, 211 (1972); *C.A.*, **80**, 40010

Vijh, A. K., 'A Possible Interpretation of the Influence of Chloride Ions on the Anodic Behaviour of Some Metals', *Corros. Sci.*, **11**, 161 (1971)

Forchhammer, P. and Engell, H. J., 'Investigations of the Corrosion of Stainless Steels by Potentiokinetic Measurements', *Corros. Sci.*, **11**, 49 (1971)

Jackson, R. P. and van Rooyen, D., 'Electrochemical Evaluation of Resistance of Stainless Alloys to Chloride Media', *Corrosion*, **27**, 203 (1971)

Pourbaix, A., 'Characteristics of Localised Corrosion of Steel in Chloride Solutions', *Corrosion*, **27**, 449 (1971)

Tokuda, T. and Ives, M. B., 'Pitting Corrosion of Nickel', *Corros. Sci.*, **11**, 297 (1971)

Rajagopalan, K. S. and Venu, K., 'Activation of Passivated Steel in Na_2CrO_4 Containing NaCl', *Corrosion*, **27**, 506 (1971)

Szklarska-Smialowska, Z., 'Effect of the Ratio of Cl^-/SO_4^{2-} in Solution on the Pitting Corrosion of Nickel', *Corros. Sci.*, **11**, 209 (1971)

Vermilyea, D. A., 'Concerning the Critical Pitting Potential', *J. Electrochem. Soc.*, **118**, 529 (1971)

Wilde, B. E. and Williams, E., 'The Relevance of Accelerated Electrochemical Pitting Tests to the Long Term Pitting and Crevice Corrosion Behaviour of Stainless Steels in Marine Environments', *J. Electrochem. Soc.*, **118**, 1056 (1971)

Fraker, A. C. and Ruff, A. W. (Jr.), 'Observations of Hot Saline Water Corrosion of Aluminium Alloys', *Corrosion*, **27**, 151 (1971)

Janik-Czachor, M., 'Electrochemical and Microscopic Study of Pitting Corrosion of Ultra-pure Iron', *Br. Corros. J.*, **6**, 57 (1971)

Butler, G., Ison, H. C. K. and Mercer, A. D., 'Some Important Aspects of Corrosion in Central Heating Systems', *Br. Corros. J.*, **6**, 31 (1971)

Johnson, W. K., 'Recent Development in Pitting Corrosion of Aluminium', *Br. Corros. J.*, **6**, 200 (1971)

Bond, A. P., 'Effects of Molybdenum on the Pitting Potentials of 18%Cr Ferritic Stainless Steels at Various Temperatures', *J. Electrochem. Soc.*, **118**, 208c (1971)

Mansfeld, F., 'The Effect of Water on Passivity and Pitting of Titanium in Solutions of Methanol and Hydrogen Chloride', *J. Electrochem. Soc.*, **118**, 1412 (1971)

Takuda, T. and Ives, M. B., 'Aggressive Ion Accessibility and the Morphology of Corrosion Pits', *J. Electrochem. Soc.*, **118**, 1404 (1971)

Szklarska-Smialowska, Z. and Janik-Czachor, M., 'The Analysis of Electrochemical Methods for the Determination of Characteristic Potentials of Pitting Corrosion', *Corros. Sci.*, **11**, 901 (1971)

Pryor, M. J., contribution to a discussion on 'A Study of the Pitting Corrosion of Aluminium by Scanning Electron Microscopy', by J. A. Richardson and G. C. Ward, *Corros. Sci.*, **11**, 163 (1971)

Zahavi, J. and Metzger, M., 'Electron Microscope Study of Breakdown and Repair of Anodic Films on Aluminium', *J. Electrochem. Soc.*, **119**, 1479 (1972)

Wilde, B. E., 'A Critical Appraisal of Some Popular Laboratory Electrochemical Tests for Predicting the Localised Corrosion Resistance of Stainless Alloys in Sea-water', *Corrosion*, **28**, 283 (1972)

Suzuki, T. and Kitamura, Y., 'Critical Potential for Growth of Localised Corrosion of Stainless Steel in Chloride Media', *Corrosion*, **28**, 1 (1972)

Rarey, C. R. and Aronson, A. H., 'Pitting Corrosion of Sensitised Ferritic Stainless Steel', *Corrosion*, **28**, 255 (1972)

Brigham, R. J., 'Pitting of Molybdenum Bearing Austenitic Stainless Steel', *Corrosion*, **28**, 177 (1972)

Vernik, E. D., Lee, T. S. and Cusumano, R. L., 'Influence of Prior Electrochemical History on the Propagation of Localised Corrosion', *Corrosion*, **28**, 348 (1972)

Szklarska-Smialowska, Z., 'Influence of Sulfide Inclusions on the Pitting Corrosion of Steels', *Corrosion*, **28**, 388 (1972)

Lizlovs, E. A. and Bond, A. P., 'Polarisation Behaviour of High-purity 13 and 18% Chromium Stainless Steels', *J. Electrochem. Soc.*, **119**, 219c (1972)

Bogar, F. D. and Foley, R. T., 'The Influence of Chloride Ion on the Pitting of Aluminium', *J. Electrochem. Soc.*, **119**, 462 (1972)

Beck, J. R., Mahaffey, D. W. and Olsen, J. H., 'Pitting and Deposits With an Organic Fluid by Electrolysis and by Fluid Flow', *J. Electrochem. Soc.*, **119**, 155 (1972)

Kato, M., Inoue, T. and Sato, O., 'New Microautoradiography Using Auger Electrons in the Study of the Structure and Pitting Corrosion of Aluminium', *Proc. Symp. Nucl. Tech. Basic Metal Ind.*, 1972, 541 (1973)

Seys, A. A. and van Haute, A. A., 'Pitting Potential Measurements by the Static Potential Band Method', *Philipp. Geogr. J.*, **16**, 107 (1972); *C.A.*, **79**, 99726g

Pourbaix, M., 'Significance of Protection Potential in Pitting, Intergranular Corrosion and Stress-corrosion Cracking', *J. Less-common Metals*, **28**, 51 (1972)

Abdul Azim, A. A. and Afifi, S. E., 'Electrochemical Behaviour of Lead and Some Selected Pb–Sb Alloys in $H_2SO_4+HClO_4$', *Corros. Sci.*, **12**, 603 (1972)

Isaacs, H. S. and Kissel, G., 'Surface Preparation and Pit Propagation in Stainless Steels', *J. Electrochem. Soc.*, **119**, 1628 (1972)

Pickering, H. W. and Frankenthal, R. P., I: electrochemical studies 'On the Mechanism of Localised Corrosion of Iron and Stainless Steel', *J. Electrochem. Soc.*, **119**, 1297 (1972)

Frankenthal, R. P. and Pickering, W. H., II: morphological studies 'On the Mechanism of Localised Corrosion of Iron and Stainless Steel', *J. Electrochem. Soc.*, **119**, 1304 (1972)

Zahavi, J. and Metzger, M., 'On the Rôle of Grain Boundaries in Film Breakdown and Pitting of Aluminium', *J. Electrochem. Soc.*, **119**, 218c (1972)

Szklarska-Smialowska, Z., 'The Kinetics of Pit Growth on Nickel in Solutions with Different Cl^-/SO_4^{2-} Ratios', *Corros. Sci.*, **12**, 527 (1972)

Bianchi, G., Corguetti, A., Mazza, F. and Torchio, S., 'Electronic Properties of Oxide Films and Pitting Susceptibility of Type *304* Stainless Steel', *Corros. Sci.*, **12**, 495 (1972)

Szklarska-Smialowska, Z. and Mankowski, J., 'Effect of Temperature on the Kinetics of Development of Pits in Stainless Steel in 0·5 N NaCl + 0·1 N H_2SO_4 Solution', *Corros. Sci.*, **12**, 925 (1972)

Vijh, A. K., 'The Influence of Solid State Cohesion of Metals on Their Pitting Potentials', *Corros. Sci.*, **12**, 935 (1972)

Thomas, J. G. N. and Tiller, A. K., 'Formation and Breakdown of Surface Films on Copper in $NaHCO_3$ and NaCl. Effects of Temperature and pH', *Br. Corr. J.*, **7**, 263 (1972)

Butler, G., Stretton, P. and Beynon, J. G., 'Initiation and Growth of Pits on High-purity Iron and its Alloys with Chromium and Copper in Neutral Chloride Solutions', *Br. Corr. J.*, **7**, 168 (1972)

Janik-Czachor, M., Szummer, A. and Szklarska-Smialowska, Z., 'Effect of Sulphur and Manganese on Nucleation of Corrosion Pits in Iron', *Br. Corr. J.*, **7**, 90 (1972)

Board, P. H., Holland, R. V. and Steele, R. J., 'Prediction of Pitting Corrosion in Tinplate from Capacitance Measurements', *Br. Corr. J.*, **7**, 87 (1972)

Lucey, V. F., 'Developments Leading to the Present Understanding of the Mechanism of Pitting Corrosion of Copper', *Br. Corr. J.*, **7**, 36 (1972)

Maskell, R. V., 'Pitting Potentials', AECL(Rep.) 4159, Atomic Energy Can. Ltd., 17 (1972)

Beck, T. R., 'One-dimensional Pits in Titanium', *J. Electrochem. Soc.*, **119**, 218c (1972)

Clark, M. B. and Zupancic, 'The Electrochemical Corrosion of Magnesium', *J. Electrochem. Soc.*, **119**, 218c (1972)

Richardson, J. A. and Wood, G. C., 'The Interpretation of Impedance Charges on Oxide-coated Aluminium Produced by Immersion in Inhibitive and Corrosive Aqueous Media', *J. Electrochem. Soc.*, **120**, 193 (1973)

Mao, K. W. and Hoare, J. P., 'The Anodic Dissolution of Mild Steel in Solutions Containing Both Cl^- and NO_3^- Ions', *Corros. Sci.*, **13**, 799 (1973)

Broli, A. and Holtan, H., 'Use of Potentiokinetic Methods for the Determination of Characteristic Potentials for Pitting Corrosion of Aluminium in a Deaerated Solution of 3% NaCl', *Corros. Sci.*, **13**, 237 (1973)

Rozgonyi, G. A. and Lizuka, T., 'Etch Pit Studies of GaP Liquid Phase Epitoxial Layers', *J. Electrochem. Soc.*, **120**, 673 (1973)

Bond, A. P., 'Effects of Molybdenum on the Pitting Potentials of Ferritic Stainless Steel at Various Temperatures', *J. Electrochem. Soc.*, **120**, 603 (1973)

Kato, M., Inoue, T., Goto, K., Ito, G. and Shimizu, Y., 'Effect of Dissolved Oxidising Agents and Inhibitors on Pitting Corrosion of Aluminium in Water', *Aluminium*, **49**, 289 (1973)

Richardson, J. A. and Godwin, A. W., 'Localised Corrosion of Stainless Steel During Food Processing', *Br. Corros. J.*, **8**, 259 (1973)

Bonewitz, R. A., 'An Electrochemical Evaluation of 1100, 5052 and 6063 Aluminium Alloys for Desalination', *Corrosion*, **29**, 215 (1973)

Datha, M. and Landolt, D., 'Soichiometry of Anodic Nickel Dissolution in NaCl and $NaClO_3$ Under Active and Transpassive Conditions', *Corros. Sci.*, **13**, 187 (1973)

Mansfeld, F., 'Passivity and Pitting of Al, Ni, Ti and Stainless Steel in $CH_3OH+H_2SO_4$', *J. Electrochem. Soc.*, **120**, 188 (1973)

Yahalom, J., Ives, L. K. and Kruger, J., 'On the Nature of Films Over Corrosion Pits in Stainless Steel', *J. Electrochem. Soc.*, **120**, 384 (1973)

Beck, T. R., 'Pitting of Titanium—1: Titanium Foil Experiments', *J. Electrochem. Soc.*, **120**, 1310 (1973)

Beck, T. R., 'Pitting of Titanium—2: One-dimensional Pit Experiments', *J. Electrochem. Soc.*, **120**, 1317 (1973)

Brigham, R. J. and Tozer, E. W., 'Temperature as a Pitting Criterion', *Corrosion*, **29**, 33 (1973)

Bruce, S., 'Specialist Steels Combat Corrosion by Chloride-containing Cooling Water', *Process Eng.*, **88** (1973); *C.A.*, **80**, 6176k

Cornwell, F. J., Wildsmith, G. and Gilbert, P. T., 'Pitting Corrosion in Copper Tubes in Cold Water Service', *Br. Corros. J.*, **8**, 202 (1973)

Broli, A., Holtan, H. and Midjo, M., 'Use of Potentiokinetic and Potentiostatic Methods for the Determination of Characteristic Potentials for Pitting Corrosion of an Fe–Cr Alloy', *Br. Corros. J.*, **8**, 173 (1973)

Retief, R., 'Corrosion of Condenser Tubes Due to Carbonaceous Film on the Surface', *Br. Corros. J.*, **8**, 264 (1973)

Seys, A. A. and Van Haute, A. A., 'Pitting Potential Measurements by Means of the Static Potential Band Method', *Corrosion*, **29**, 329 (1973)

Zamin, M. and Ives, M. B., 'Effect of Chloride Ion Concentration on the Anodic Dissolution Behaviour of Nickel', *Corrosion*, **29**, 319 (1973)

Uhlig, H. H., 'Distinguishing Characteristics of Pitting and Crevice Corrosion', *Mater. Prot. Performance*, **12**, 42 (1973); *C.A.*, **78**, 105248a

Herbsleb, G. and Schwenk, W., 'Electrochemical Investigations of Pitting Corrosion', *Corros. Sci.*, **13**, 739 (1973)

Vijh, A. K., 'The Pitting Potentials of Metals: the Case of Titanium', *Corros. Sci.*, **13**, 805 (1973)

Fraker, A. C., 'Effect of Solution pH on the Saline Water Corrosion of Titanium Alloys', *Proc. 2nd Int. Conf. Titanium Sci. Technol.*, 1972 (1973)

Tatsuya, K. and Shuichi, F., 'Pitting Corrosion of Titanium in High-temperature Halide Solutions', *Proc. 2nd Int. Conf. Titanium Sci. Technol.*, **4**, 2383 (1973)

Mao, K. W. and Chin, D. T., 'Anodic Behaviour of Mild Steel in $NaClO_3$ at High Current Densities', *J. Electrochem. Soc.*, **121**, 191 (1974)

Zahavi, J. and Metzger, M., 'Effect of Chloride on Growth on Anodic Film', *J. Electrochem. Soc.*, **121**, 268 (1974)

Pitting (papers published in foreign languages)

Kutzelnigg, 'Pitting, the Typical Corrosion of Nickel Coatings', *Korrosion*, **13**, 64 (1960)

Schmeken, H., 'Electrochemical Measurements on Corrosion of Zinc with Regard to Pitting Possibilities', *Korrosion*, **13**, 65 (1960)

Sverepa, O., 'Pitting Corrosion of Aluminium in Water', *Korose Ochrana Mater.*, **1** (1961); *C.A.*, **57**, 12237h

Videm, K., 'Pitting of Aluminium in Contact with Stainless Steel', *Tek. Ukeblad*, **108**, 775 (1961); *C.A.*, **57**, 10912h

Sverepa, O. and Svobodny, Z., 'Corrosion Resistance of Aluminium Storing Vessels', *Chem. Prumysl.*, **11**, 644 (1961); *C.A.*, **56**, 7030b

Coriou, H., 'Les Problemes de Corrosion Aqueuse dans le Domaine de l'Energie Nucleaire', *Corros. Sci.*, **1**, 132 (1961)

Rozenfeld, I. L. and Danilov, I. S., 'The Mechanism of Pitting Corrosion of Stainless Steel', *Dokl. Akad. Nauk. SSSR*, **147**, 1417–19 (1962); *C.A.*, **58**, 8773e

Pourbaix, M., 'Potentiokinetic and Corrosimetric Studies on the Behaviour of Alloy Steels', CEBELCOR Rappt. Tech. No. 120, **1** (1962); *C.A.*, **58**, 13431b

Sakiyama, K. and Fuyimoto, M., 'Effects of Free Chlorine on the Pitting Corrosion of Gas Cooler Tubes', *Boshoku Gijutsu*, **11**, 299 (1962); *C.A.*, **58**, 5257b

Kaesche, H., 'Uniform Dissolution and Pitting of Aluminium Electrodes', *Z. Physik. Chem.*, **34**, 87 (1962); *C.A.*, **58**, 41546

Gillman, V. A. and Kolotyrkin, Ya. M., 'The Mechanism of Zirconium Pitting in Halide Solutions', *Dokl. Akad. Nauk. SSSR*, **143**, 640 (1962); *C.A.*, **57**, 3184a

Videm, K., 'Pitting Corrosion of Aluminium in Contact with Stainless Steel', Kjeller Rept. KR-18, 16 (1962); *C.A.*, **57**, 8315a

Prazak, M. and Cihal, V., 'Die Potentiostatische Untersuchung des Einflusses Einiger Leigierungselemente auf die Electrochemischem und Korrosions-eigenschaften Nischtrostender Stahle', *Corros. Sci.*, **2**, 71 (1962)

Vanleugengaghe, C., Klimczak-Mathieu, L., Meunier, J. and Pourbaix, M., 'Influence de Chlorures et d'Oxyolants sur le Compartement d'Aciers Inoxydables en Solution Bicarbonique', *Corros. Sci.*, **2**, 29 (1962)

Prazak, M., Tousek, F. and Spanily, V., 'Rôle of Anion Adsorption During Pitting Corrosion and Corrosion Cracking of Metals', *Zashch. Metal.*, **5**, 371 (1962); *C.A.*, **71**, 97654t

Rozenfeld, I. L. and Danilov, I. S., 'Pitting Tendency of Stainless Steels', *Tr. Vses. Mezhvuz. Nauchn. Konf. po Vopr. Bor'by s Korrozici* (*Moscow: Gos. Zzd. Neft i Gorno-Toplivn Prom.*), 18 (1962); *C.A.*, **60**, 3806f

Schwenk, W., 'Diskussion uber die Ursache und Stabilitat der Lochfrasskorrosion in Zusammenlang mit dem 'Alles-oder-Nichts' Gesetz der Passivitat und Einen Kinetichen Modell', *Corros. Sci.*, **3**, 107 (1963)

Defranoux, J. M., 'Methodes Electrochimiques d'Etude du Compartement des Aciers Inoxydables en Milieu Chlorure', *Corros. Sci.*, **3**, 75 (1963)

Harzbecker, H., 'Pitting Corrosion in Steam Generators', *Energietechnic*, **13**, 502 (1963); *C.A.*, **64**, 9434a

Tomashov, N. D., Chernova, G. P. and Markova, O. N., 'Effect of Alloying Elements on the Tendency of Stainless Chromium–Nickel Steels to Pitting Corrosion', *Sa. Korroziya Metal i Splavov*, 73 (1963); *C.A.*, **60**, 3805b

Freiman, L. I. and Kolotyrkin, Ya. M., 'Pitting Corrosion of Iron by Perchlorate Ions', *Dokl. Akad. Nauk. SSSR*, **153**, 886 (1963); *C.A.*, **60**, 7732e

Roedeker, W. and Friche, W., 'An Experimental Investigation into the Pitting of Various Types of Steel Plate', *Blech*, **11**, 302 (1964); *C.A.*, **61**, 9230f

Rozenfeld, I. L. and Danilov, I. S., 'Pitting Corrosion of Passivated Stainless Steel', *Z. Physik. Chem.*, **226**, 257 (1964); *C.A.*, **61**, 13005d

Uhlig, H. H. and Gilman, J. R., 'Inhibition of Pitting Corrosion of Stainless Steel 18/8 in Iron(III) Chloride Solutions by Nitrates', *Z. Physik. Chem.*, **226**, 127 (1964); *C.A.*, **61**, 9231c

Fisher, W. R., 'Pitting Corrosion, Especially of Titanium. 1: Corrosion Studies', *Techn. Mitt. Krupp, Forschber*, **22**, 65 (1964); *C.A.*, **62**, 15859c

Kazinachei, B. Ya., Balashova, N. N. and Shuvalova, M. A., 'Control of Anti-pitting Additives in Nickel Bath', *Sb. Statei Vses. Zaochn. Politechn. Ins.*, **32**, 52 (1964); *C.A.*, **62**, 15859

Storchai, E. I. and Turkovskaya, A. V., 'The Pitting Corrosion of Aluminium Alloys', *Zashch. Metal*, **1**, 293 (1965); *C.A.*, **63**, 7999c

Schwenk, W., 'Diskussion der Eigenschaften des Lochfrass-potentials Nichtnostender Stahle in Zusemmenhang mit Seinen Bestimmungsmethoden', *Corros. Sci.*, **5**, 245 (1965)

Blauchet, 'The Pitting Attack of Magnesium and its Alloys', *Comm. Energie At. Rappt. CEA-R*, 2815 (1965); *C.A.*, **65**, 1761e

Morioka, S., Sawanda, Y. and Shisbara, K., 'Effect of SO_2 Gas on Pitting of Austenitic Stainless Steel', *Boshoku Gijutsu*, **14**, 535 (1965); *C.A.*, **64**, 19086h

Fokin, M. N., Kurtepov, M. M. and Bochkareva, E. F., 'Investigation of Pitting and Crevice Corrosion of Stainless Steels in Sea-water', *Korroziya Metal i Splavov*, **2**, 35 (1965); *C.A.*, **64**, 122820

Waight, F. H., 'Fatigue Pitting in Gears. Programme, Results and Discussion of the Results', *Erdoel Kohle*, **18**, 976 (1965) and **18**, 899 (1965); *C.A.*, **64**, 10877d

Blanchet, J. and Coriou, H., 'Electrochemical Methods for the Study of Corrosion, Experimental Techniques and Industrial Applications, Electrochemical Study of Corrosion by Pitting of Magnesium', CEBELCOR Rappt. Tech. No. 135, 5 (1965); *C.A.*, **64**, 9233a

Herbsleb, G., 'The Inhibition of Chloride Corrosion (Pitting) on Chemical-resistant Steels by Nitrate, Sulfate and Chromate Ions, and the Methods for Determining the Pit Corrosion Potential', *Werkst. Korros.*, **16**, 929 (1965); *C.A.*, **64**, 4702c

Vetter, K. J., 'Thermodynamic Theory of Corrosion Pitting', *Ber. Bunsenges. Physik. Chem.*, **69**, 683 (1965); *C.A.*, **64**, 3018f

van Muylder, J., Pourbaix, M. and van Lear, P., 'Electrochemical Characteristics of Pits from the Corrosion of Copper in the Presence of H_2O and Aqueous Chloride Solutions', CEBELCOR Rappt. Tech. No. 127, 26 (1965); *C.A.*, **64**, 3018c

Szklarska-Smialowska, Z., 'Pitting Corrosion of Iron in Ammonium Nitrate Solution Containing Chloride Ions', *Bull. Acad. Polon. Sci., Ser. Scr. Chim.*, **13**, 221 (1965); *C.A.*, **63**, 6662h

Golovina, G. V., Florianovich, G. M. and Kolotyrkin, Ya. M., 'Kinetics of the Initial Stages of Activation of Iron–Chromium Alloys by Halogen Ions', *Elektrokhimiya*, **1**, 12 (1965); *C.A.*, **63**, 302h

Van Lear, P., Van Muylden, J., de Zoubov, N. and Pourbaix, M., 'Accelerated Methods for Estimating the Risk of Pitting of Copper in the Presence of Water. Application to the Study of the Influence of a Copper Contact with Another Metal or with Graphite', CEBELCOR Rappt. Tech. No. 128, May (1965)

Tousek, J., Cihal, V. and Prazak, M., 'Die Einwirkung von Halogenionen auf die Korrosion der durch Molybdan Modifizurten Chrom–Nickel–Stahle', *Corros. Sci.*, **6**, 105 (1966)

Hubner, W., 'Pitting Corrosion of Passivated Metals', *Svensk Kem. Tidstr.*, **78**, 321 (1966); *C.A.*, **66**, 25356w

Hatch, G. B., 'Maximum Self-generated Anodic Current Density as an Inhibitor Pitting Index', Ill. State Water Surv., Circ. No. 91, 24 (1966); *C.A.*, **66**, 81814f

Herbsleb, G., 'Pitting Corrosion on Metals with Electron-conductive Passive Layers', *Werkst. Korros.*, **17**, 649 (1966); *C.A.*, **66**, 5337m

Freiman, L. I. and Kolotyrkin, Ya. M., 'Pitting Corrosion of Aluminium in Solutions of Sodium Perchlorate and Perchloric Acid', *Zashch. Metal*, **2**, 488 (1966); *C.A.*, **65**, 19674d

Novakovskii, V. M. and Sorokina, A. N., 'Comparative Electrochemistry of Stress Corrosion and Pitting of Stainless Steels in Chloride Solutions', *Zashch. Metal*, **2**, 416 (1966); *C.A.*, **65**, 18152g

Willert, H., 'Investigations of the Corrosion Resistance of a 13%Cr Steel in Aqueous Salt Solutions', *Neue Huette*, **11**, 372 (1966); *C.A.*, **65**, 14954b

Tousek, J., 'Passivity of Nickel', *Collection Czech. Chem. Commun.*, **31**, 3083 (1966); *C.A.*, **65**, 10116h

Rozenfel'd, I. L. and Danilov, I. S., 'Electrochemistry of Pitting Corrosion. 1: Pit Formation During Spontaneous Dissolution of Stainless Steels', *Zashch. Metal.*, **2**, 134 (1966); *C.A.*, **65**, 10114a

Bond, A. P., Bolling, G. F. and Domian, H. A., 'Microsegregation and the Tendency for Pitting Corrosion in High-purity Aluminium', *Ber. Bunsenges. Physik. Chem.*, **70**, 773 (1966); *C.A.*, **65**, 8510b

Kuzub, V. S. and Kachanov, V. A., 'Possibility of Anodic Protection of Stainless Steels from Chloride Pitting in Dilute Nitric Acid', *Zashch. Metal.*, **2**, 358 (1966); *C.A.*, **65**, 6716f

Rickert, H. and Holzaepfel, G., 'The Coexistence of Active and Passive Zones on an Electrode Surface', *Werkst. Korros.*, **17**, 376 (1966); *C.A.*, **65**, 3322f

Pourbaix, M., 'Effect of Chlorides on the Bahaviour of Metals and Alloys in the Presence of Aqueous Solutions', *Symp. Coupling Basic Appl. Coross. Res. Dialogue*, 67 (1966); *C.A.*, **72**, 50238r

Pourbaix, M., Van Muylder, J. and Van Laer, P., 'Sur la Tension d'Electrode du Cuivre en Presence d'Eau de Bruxelles; Influence de la Lumière et des Conditions de Circulation de l'Eau', *Corros. Sci.*, **7**, 795 (1967)

Takenori, N., Norio, S., Tatsuo, I. and Okamoto, G., 'Electrochemical Test for Pitting Corrosion in Stainless Steels', *Hakkaido Daigaku Kogakubu Kenkyu Hokoku*, **44**, 1 (1967); *C.A.*, **70**, 16534h

Izaki, T. and Arai, K., 'Pitting Corrosion of Aluminium and its Alloys in Neutral Solutions', *Nippon Kinzoku Gakkaishi*, **31**, 1023 (1967); *C.A.*, **68**, 26190r

Herbsleb, G. and Engell, H. J., 'Pitting Corrosion of Passive Iron in Sulphuric Acid Containing Chloride Ions', *Werkst. Korros.*, **17**, 365 (1966); *C.A.*, **67**, 104517y

Herbsleb, G. and Schwenk, W., 'A Pitting Corrosion Indicator Test on Chromium and Cr–Ni Steels in Solutions Containing Chloride and Bromide', *Werkst. Korros.*, **18**, 685 (1967); *C.A.*, **67**, 87126a

Defranoux, J. M., 'Sur le Compartement des Aciers Inoxydables en Presence d'Eau de Mor Froide et Chaude', *Corros. Sci.*, **8**, 245 (1968)

Garz, I., Worch, H. and Schatt, W., 'Untersuchungen uber die Anodische Auflosung und die Lochfrapkorrosion von Nickel-Einkristallelektrochen', *Corros. Sci.*, **9**, 71 (1969)

Riskin, I. V. and Turkovskaya, A. V., 'Pitting Corrosion of Kh 18N10T Steel on a Rotating Disk Electrode', *Zashch. Metal*, **5**, 443 (1969); *C.A.*, 97645r

Graefen, H., 'Electrochemical Investigation of Localised Corrosion', *Ind. Chim. Belge*, **32**, 422 (1967); *C.A.*, **70**, 111038h

Kuznetsov, V. V. and Verzhbitskaya, L. V., 'Mechanism of Pitting Corrosion of Carbon Steel in Fresh Water', *Uch. Zap. Perm. Univ.*, **229**, 131 (1970); *C.A.*, 120661g

Yoshihiro, H., 'Influence of Crystal Structure on Corrosion', *Denki Kagoku*, **9**, 691 (1970); *C.A.*, **74**, 60040a

Vetter, K. J. and Strehblow, H. H., 'Formation and Shape of Pitting Corrosion Pits in Iron and Theoretical Conclusions on Pitting Corrosion', *Ber. Bunsenges Phys. Chem.*, **74**, 1024 (1970); *C.A.*, **74**, 8865j

Uhlig, H. H., 'Modern Look at Pitting Corrosion', *Boshoku Gijutsu*, **19**, 171 (1970); *C.A.*, **74**, 8835z

Parlapanskii, M. and Mutafchiev, I., 'Antipitting Action of Some Surface-active Agents', *God. Vissh. Khimikotekhnol Inst.*, **16**, 263 (1971); *C.A.*, **79**, 37944t

von Baeckmann, W., Gisen, F. and Funk, D., 'Breaks Caused by Corrosion Pitting', *Gesammelte Ber. Betr. Forch Ruhrgas AG.*, **19**, 25 (1971); *C.A.*, **76**, 120662h

Schatt, W. and Worch, H., 'Nahere Charakterisierung der bei Nickel-Einkristallen zum Lochfrass Fuhrenden Versetzungen', *Corros. Sci.*, **11**, 623 (1971)

Tinck, M., Petitjean, J. P., Van der Poosten, P. H. and Blave, A., 'Multipotentiostat for the Rapid Determination of the Pitting Potential of Stainless Steel', *Cent. Belge Étude Corros. Rapp. Tech.*, **122**, 210 (1972); *C.A.*, **79**, 99730d

Kato, K. and Kono, T., 'Effect of Alloying Elements on Pitting Corrosion Resistance of 17%Cr–13%Ni–3%Mo (SOS316) Austenitic Stainless Steel', *Denki Seiko*, **43**, 229 (1972); *C.A.*, **79**, 34374w

Grassiani, M., 'Pit Corrosion of Austenitic Stainless Steels in Artificial Sea-water', *Trib. CEBEDEAU*, **25**, 515 (1972); *C.A.*, **79**, 8531a

Tousek, J., 'Die Aktivierende Wirkung der Nitrationen auf die Geschwiadigkeit der Metallauflosung im Passiven Zustand', *Corros. Sci.*, **12**, 799 (1972)

Tousek, J., 'Die Kinetik der Lochfrasskorrosion von Metallen', *Corros. Sci.*, **12**, 1 (1972)

Yashihiro, H., 'Electrochemistry of Pitting Corrosion', *Boshoku Gijutsu*, **21**, 503 (1972); *C.A.*, **78**, 143032h

Cherepakhova, G. L. and Shreider, A. V., 'Effect of the Ions of Cooling Waters on the Pitting Corrosion of Al–Mg Alloy', *Zh. Prik. Khim.*, **45**, 1958 (1972)

Tousek, J., 'Kinetics of the Pitting Corrosion of Iron and Nickel', *Czech. Chem. Commun.*, **37**, 1454 (1972); *C.A.*, **77**, 55490e

Tousek, J., 'Pitting Corrosion Mechanism', *Werkst. Korros.*, **23**, 109 (1972); *C.A.*, 120666n

Pourbaix, M., 'Significance of Protective Potential During Pitting Corrosion, Intergranular Corrosion, and Stress Cracking Corrosion', *Rev. Roum. Chim.*, **17**, 239 (1972); *C.A.*, **76**, 120554z

Boehni, H., 'Pit Corrosion of Metallic Materials and Detection Methods', *Schweiz. Arch. Angar. Wiss. Tech.*, **36**, 41 (1970); *C.A.*, **72**, 106498v

Tousek, J., 'Untersuchungen uber den Lochfrass von Ni and Fe der Durch Sulfationen Hervongerufen Wird', *Corros. Sci.*, **12**, 15 (1972)

Freiman, L. I., Lap Le, M. and Raskin, G. S., 'Rôle of Local Changes in Solution Composition During Pitting on Iron', *Zashch. Metal*, **9**, 680 (1973); *C.A.*, 43390w

Inoue, T. and Kato, M., 'Pitting Corrosion of Aluminium Using Macro- and Microautoradiography', *Keikinzoku*, **23**, 78 (1973); *C.A.*, **79**, 121179e

Brigham, R. J. and Tozer, E. W., 'Effect of Microstructure on Pitting of Stainless Steel', *Can. Met. Quart.*, **12**, 171 (1973); *C.A.*, **79**, 1174576

Brennert, S. and Eklund, G., 'Pitting Corrosion of Stainless Steel. Propagation Mechanism', *Scan. J. Met.*, **2**, 269 (1973); *C.A.*, **79**, 138952z

Sato, E., Tamura, T. and Okabe, T., 'Aluminium Anode for Cathodic Protection. VII: Pitting and Protection Potentials of Aluminium in NaCl Solution (Effects of pH)', *Kinzoku Hyomen Gijutsu*, **24**, 139 (1973); *C.A.*, **79**, 142225g

Ciolac, S., Vasilescu, E. and Stretcu, M., 'Pitting Corrosion of Metals. II: Characteristics of the Pitting Process and Methods of Study and Protection', *Stud. Cercet. Chim.*, **21**, 669 (1973); *C.A.*, **79**, 99678t

Ciolac, S., Vasilescu, E., Stretcu, M. and Joachimescu, O., 'Pitting Corrosion of Metals. Pitting–Passivity Correlation', *Stud. Cercet. Chim.*, **21**, 597 (1973)

Herbsleb, G. and Schwenk, W., 'Flow Dependence of the Pitting Corrosion of Cr–Ni Steel in NaCl Solution. 2: Tests with Ultrasonics', *Werkst. Korros.*, **24**, 267 (1973); *C.A.*, **79**, 56638n

El Din Shams, A. M., Bodran, M. M. and Khalil, S. E., 'Corrosion Behaviour of Manganese-containing Stainless Steel. 3: Their Susceptibility Towards Pitting Corrosion', *Werkst. Korros.*, **24**, 290 (1973); *C.A.*, **79**, 56642j

Imoi, H., Saito, Y., Kobayashi, M. and Fujiyama, S., 'Pitting-corrosion-resistant Chromium Stainless Steel', *Japan Kokai 7300*, 221 (1973); *C.A.*, **79**, 22569a

Sato, E., Tamura, T. and Okabe, T., 'Aluminium Anode for Cathodic Protection. 7: Pitting and Corrosion Potentials for Gallium in Sodium Chloride Solutions', *Kinzoku Hyomen Gijutsu*, **24**, 82 (1973); *C.A.*, **79**, 12792d

Freiman, L. I., Lap Le, M. and Kolotyrkin, J. M., 'Depassivation and Repassivation of Iron and Iron Alloys in Halide Solutions', *Z. Phys. Chem.*, **252**, 76 (1973); *C.A.*, **79**, 8530z

Freiman, L. I. and Lap Le, M., 'Limiting Potentials Which Determine the Resistance of Metals to Pitting Corrosion', *Z. Phys. Chem.*, **252**, 65 (1973); *C.A.*, **78**, 154238f

Janik-Czachor, M., 'Pitting Corrosion of Zinc Single Crystals', *Bull. Acad. Pol. Sci., Ser. Sci. Chim.*, **21**, 159 (1973); *C.A.*, **78**, 154240a

Herbsleb, G. and Schwenk, W., 'Electrochemical Study of the Pitting Corrosion of Corrosion-resistant Steels', *Werkst. Korros.*, **24**, 763 (1973); *C.A.*, **88**, 21970h

Poatsch, W., 'Optical Studies of the Mechanism of Pitting Corrosion', *Ber. Bunsenges Phys. Chem.*, **77**, 895 (1973); *C.A.*, **80**, 90201v

Zhurin, A. I., Kosmynin, A. I. and Vlasenko, O. B., 'Corrosion of Aluminium Cathodes During the Electrodeposition of Zinc', *Izv. Vyssh. Ucheb. Zaved. Tsvet. Met.*, **16**, 71 (1973); *C.A.*, **80**, 77471p

Worch, H., Garz, I. and Schott, W., 'Influence of Surface Structure and Chemical Composition on the Pitting Corrosion of Nickel Single Crystals', *Werkst. Korros.*, **24**, 872 (1973); *C.A.*, **80**, 90185

1.7 Bimetallic Corrosion

This section is concerned with the kinetics of reactions occurring when dissimilar metals are in direct electrical contact in corrosive solutions or atmospheres. Under these conditions, enhanced and aggressive corrosion of the more negative member of the bimetallic couple can be experienced together with partial or complete cathodic protection of the more positive metal*. In many cases the flow of current $I_{galv.}$ resulting from the bimetallic contact is intense and much greater than that experienced from variations of oxygen content over the surface of a single metal (differential aeration) or from variations of metal-ion concentration at different locations on a single metal (concentration-cell corrosion). Bimetallic corrosion is qualitatively well understood, but since it is a highly complex process, it has proved difficult to handle its kinetics in a quantitative fashion. Furthermore, it must be emphasised that metal surfaces are seldom unfilmed so that the kinetics of electrode reactions will be affected by the film on the metal surface.

General Theory

The factors that are of importance in the enhancement of the corrosion rate of one metal when it is in direct electronic contact with another (*cf.* cathodic protection where contact is by conducting wire) are as follows.

1. The *corrosion potentials* of the metals M_A and M_B forming the couple under the environmental conditions that prevail in practice.
2. The nature and kinetics of the cathodic reaction at the surface of the more positive metal and the nature and kinetics of the anodic reaction at the surface of the more negative metal.
3. The relative areas of M_A and M_B (*see* page **1**.79).
4. The nature and the conductivity of the electrolyte solution.

The corrosion potentials of the two metals in the environment under consideration will determine the direction of the transfer of electrons, but will provide no information on the rate of electron transfer, i.e. the magnitude of the galvanic current $I_{galv.}$. Thus if $E_{corr., A}$ is more positive than $E_{corr., B}$ the transfer of electrons will be from M_B to M_A with a consequent increase in the corrosion potential (more positive) of M_B and a decrease in that of M_A; the corrosion rate of M_B will consequently increase and the corrosion rate of M_A will decrease compared with the rates when the metals are uncoupled. However, it must be emphasised that the difference in the corrosion potentials

*This is utilised in the cathodic protection of metals using sacrificial anodes (*see* Section 11.2).

provides no information on the kinetics of bimetallic corrosion, a fact that may be illustrated by considering the couples Mg–Hg and Mg–Pt immersed in sea-water. The corrosion potentials (*vs*. S.H.E.) of the metals under consideration are Mg $\approx -1{\cdot}0$ V, Pt $\approx 0{\cdot}0$ V and Hg $\approx 0{\cdot}0$ V, so that the e.m.f.s are the same, i.e. about 1·0 V. However, whereas coupling platinum to magnesium results in a pronounced increase in the corrosion rate of the latter, the effect produced by coupling magnesium and mercury is insignificant. The reason for this apparent anomaly is that magnesium in sea-water corrodes by the overall reaction

$$Mg + 2H_2O \rightarrow Mg(OH)_2 + H_2$$

in which the cathodic reaction is reduction of water to hydrogen gas. The rate of corrosion is thus determined by the rate of the hydrogen evolution reaction which occurs with kinetic ease at a platinum surface ($i_0 \approx 10^{-3}$ Acm^{-2}), but with extreme difficulty at a mercury surface ($i_0 \approx 10^{-11}$ Acm^{-2}). Thus the platinum provides an additional cathode at which hydrogen evolution is kinetically easy, and since $I_c = I_a$ (*see* Section 1.4) the corrosion rate of the magnesium is increased.

Graphic estimation of the corrosion rate and corrosion potential of a metal immersed in a corrosive high-conductivity electrolyte, from the intersection of the polarisation curves for the appropriate anodic and cathodic reactions, has been proposed and explained by several authorities[1,2]. These polarisation curves can be further used to illustrate the effect of imposing additional anodic or cathodic potentials on to a corroding metal (*see also* Sections 1.4 and 11.1).

Figure 1.62*a* shows the effect of depressing the potential of a metal corroding in a solution of high electrolytic conductivity. As the potential is progressively depressed, the relation between potential and total cathodic current is defined by the curve *CBD*, which is a typical cathodic polarisation curve for reduction of dissolved oxygen, of which the section *CB* is the cathodic polarisation curve for local cell corrosion in the absence of an applied potential. As the potential is depressed along the curve *BD* by the increasing polarising current *yz*, *y′z′*, *y″z″*, the corrosion current (rate) of the metal is given by the progressively decreasing horizontal intercept of the anodic polarisation curve, i.e. *xy*, *x′y′*, *x″y″*, etc. When the potential of the metal is depressed below *A*, complete cathodic protection results and the original anodic sites now support the cathodic reduction of oxygen, as shown by curve *AE*. At potentials between *B* and *A*, partial cathodic protection is obtained, with the corrosion rates decreasing as the potentials approach *A*.

Figure 1.62*b* shows the result of raising the potential of a corroding metal. As the potential is raised above *B*, the current/potential relationship is defined by the line *BD*, the continuation of the local cell anodic polarisation curve, *AB*. The corrosion rate of an anodically polarised metal can very seldom be related quantitatively by Faraday's law to the *external* current flowing, $I_{galv.}$. Instead, the measured corrosion rate will usually exceed that rate calculated from the external current flow by an amount that will be designated as the local-cell corrosion rate of the anode, i.e. *xy*, *x′y′*, *x″y″*, which are the horizontal current intercepts, at the appropriate potential, of the cathodic polarisation curve *CB* (Fig. 1.62*b*). Thus the higher the potential of the anodically polarised metal, the smaller will be the local-cell corrosion rate of the anode and the more closely will the total corrosion rate of the anode approach that calculated from the flow of current in the external circuit. Since the slope of the cathodic polarisation curve is usually much greater than that of the anodic polarisation curve, it follows that the local-cell corrosion rate should be relatively little affected by variations of galvanic *current*.

When two dissimilar metals of different corrosion potentials are coupled in the same solution, the more positive metal will serve as a cathode, its

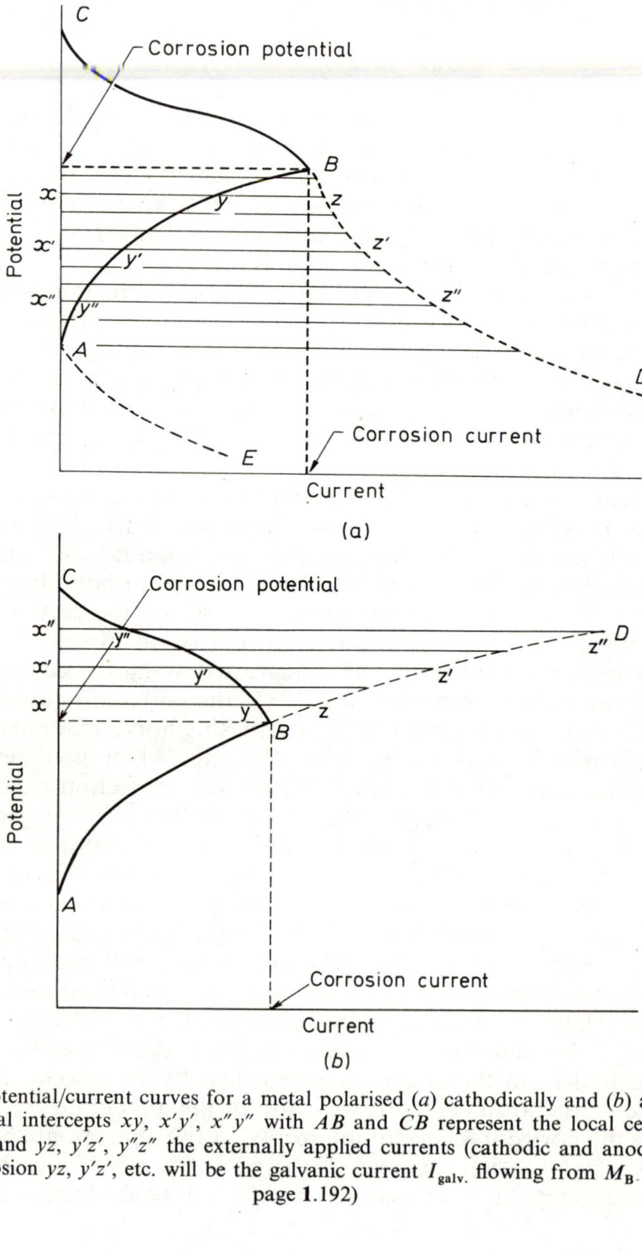

Fig. 1.62 Potential/current curves for a metal polarised (*a*) cathodically and (*b*) anodically. The horizontal intercepts *xy*, *x′y′*, *x″y″* with *AB* and *CB* represent the local cell currents respectively, and *yz*, *y′z′*, *y″z″* the externally applied currents (cathodic and anodic). In bimetallic corrosion *yz*, *y′z′*, etc. will be the galvanic current $I_{galv.}$ flowing from M_B to M_A (*see* page **1**.192)

characteristic uncoupled corrosion rate in the same solution being partly or completely retarded as shown in Fig. 1.62*a*. The more negative metal will suffer enhanced corrosion, its total corrosion rate being given by the sum of the rate due to the flow of current in the external circuit and the potential-dependent local-cell corrosion rate as shown in Fig. 1.62*b*. If the areas of the two metals are comparatively small and if the solution has a high conductivity the effect of the transfer of electrons $I_{galv.}$ will be to make the whole surface equipotential, but this will not be so if the areas are large or if the solution has a high resistivity. In this connection (*see* Section 1.4) it should be emphasised that the effective areas of the metals constituting the bimetallic couple are not necessarily the same as the geometrical areas, and in a solution of high resistivity only the areas of the metals adjacent to their interface will be significant in relation to the magnitude of $I_{galv.}$ and the anodic current density at the more negative metal of the couple.

Potentials

It is evident (*see* Appendix 1.9B) that the standard electrode potentials of M^{z+}/M equilibria, as given in the e.m.f. series of metals, are quite irrelevant in predicting the effect of coupling two dissimilar metals. The galvanic series of metals (*see* Section 21.1) in which metals are listed according to their corrosion potentials in chloride solutions, are more helpful, but these too can be misleading when other environments are involved, and when data are not available in tables or in the literature it is necessary to determine the corrosion potentials of the metals under the environmental conditions that prevail in practice. Furthermore, the magnitude of the galvanic current is dependent only in part on the difference in corrosion potentials, being also strongly influenced by the slope of the cathodic polarisation curve of the more positive metal, and/or that of the anodic polarisation curve of the less positive metal. Films of oxide or insoluble reaction products on the anode or the cathode also introduce important kinetic and resistive effects which have a major influence on the magnitude of the galvanic current flow and which

Table 1.23 Weight loss of iron for Fe/*M* bimetallic couples in 1% NaCl*

M	*Weight loss of iron* (mg)	*Weight loss of M* (mg)	*Difference in standard electrode potentials*[4] V†
Copper	183·1	0·0	+0·785
Nickel	181·1	0·2	+0·19
Tin	171·1	2·5	+0·30
Lead	183·2	3·6	+0·31
Aluminium	9·8	105·9	−1·23
Cadmium	0·4	307·9	+0·04
Zinc	0·4	688·0	−0·32
Magnesium	0·0	3104·0	−1·90

* Weight loss data after Bauer and Vogel[3].
† The + sign indicates that iron has a more negative standard electrode potential than the second metal.

are not accounted for by standard electrode potentials. Some of the foregoing factors are illustrated by Table 1.23 which contains early information developed by Bauer and Vogel[3] on the weight losses of iron and various dissimilar metals coupled together in 1% sodium chloride solution containing dissolved oxygen.

Even discounting the case of aluminium, which is usually covered by protective oxide films, it is evident from Table 1.23 that the quantitative connection between the galvanic corrosion rate of the more active member of the couple and the difference of reversible potentials of the two metals, is non-existent.

This situation has encouraged the empirical approach of using the difference in *corrosion* potentials of dissimilar metals in a single solution (usually sea-water or sodium chloride) as a means of assessing the anticipated intensity of galvanic corrosion of the less positive metal in the same solution. This approach has sometimes been carried to the extreme of setting a maximum value of 0·25 V difference in corrosion potential in sea-water as representing the safe limit for elimination of greatly aggravated corrosion of the less positive member of the couple. Such a criterion clearly has little or no technical significance, not only because it ignores polarisation effects, but also because irreversible corrosion potentials cannot be subtracted in the fashion of reversible half-cell potentials (*see* page **1**.56).

Table 1.24 Grouping of compatible metals and alloys

Group I	Magnesium and its alloys*
Group II	Cadmium, zinc and aluminium, and their alloys
Group III	Iron, lead and tin, and their alloys (except stainless steels)
Group IV	Copper, chromium, nickel, silver, gold, platinum, titanium, cobalt and rhodium, and their alloys; stainless steels; graphite

*The studies of Bothwell[25] would suggest that super-purity aluminium could be placed in Group I.

Galvanic corrosion rates can, of course, be reliably obtained by direct measurement (*see* Appendix 1.7B) or can be estimated in various electrolytes if the corrosion potentials of the dissimilar metals and the slopes of the appropriate anodic and cathodic polarisation curves are known. Unfortunately, the slopes of the anodic and cathodic polarisation curves are strongly dependent on the particular metal and on the composition of the electrolyte. Non-stable oxide or reaction-product films, which may either thin or thicken during galvanic corrosion, also impose variable resistive effects which make the practical prediction of galvanic corrosion rates uncertain. Accordingly, it is hardly surprising that a combination of practical experience and less frequent quantitative investigation dictates the choice of compatible metals in different environments. Promisel and Mustin (Table 1.24[5]) and Evans and Rance (Table 1.7A1[6]) have published reliable classifications of compatible and incompatible metals. The most commonly considered electrolyte is sea-water because highly aggravated corrosion of less noble metals commonly occurs in this environment.

Bimetallic Corrosion in Aqueous Solutions

Dissolved Oxygen

Most cases of practical bimetallic corrosion in solutions occur under conditions when the solution contains dissolved oxygen. Accordingly, the primary cathodic reaction is the reduction of dissolved oxygen

$$O_2 + 4H^+ + 4e \rightarrow 2H_2O$$

at the surface of the cathodic metal. The maximum rate at which dissolved oxygen, at a given bulk concentration and pH, can be cathodically reduced is clearly limited by the rate at which the oxygen can diffuse from the bulk electrolyte to the surface of the cathode. Accordingly, in static solutions a maximum concentration and temperature dependent diffusion rate exists, which often limits the galvanic current, and the rate of bimetallic corrosion is then under cathodic diffusion control. This situation is quite common in practical bimetallic corrosion. Although i_L should be independent of the nature of the metal (c.f. the hydrogen evolution reaction) it will be seen that the noble metals are far more efficient cathodes than metals filmed with unreducible corrosion products.

It also follows that if the solution is stirred the rate of arrival of oxygen at the cathode will be increased. This will result in a corresponding increase in the rate of bimetallic corrosion as is shown in Fig. 1.63 for the aluminium–mild steel couple in stirred 1·0 N NaCl solution[7]. The increase in galvanic

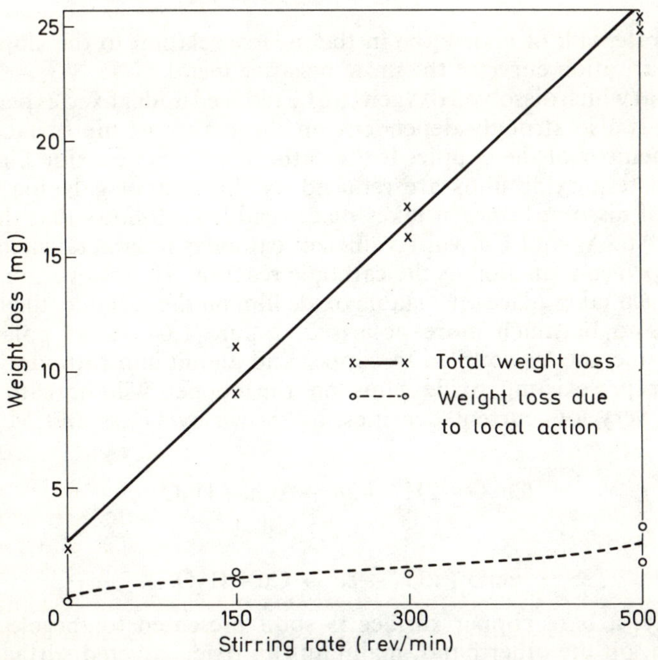

Fig. 1.63 Effect of stirring rate on the aluminium–mild steel couple in 1·0 N NaCl at 25°C. Equal areas (20 cm^2) of aluminium and steel coupled together for 24 h (after Pryor and Keir[7])

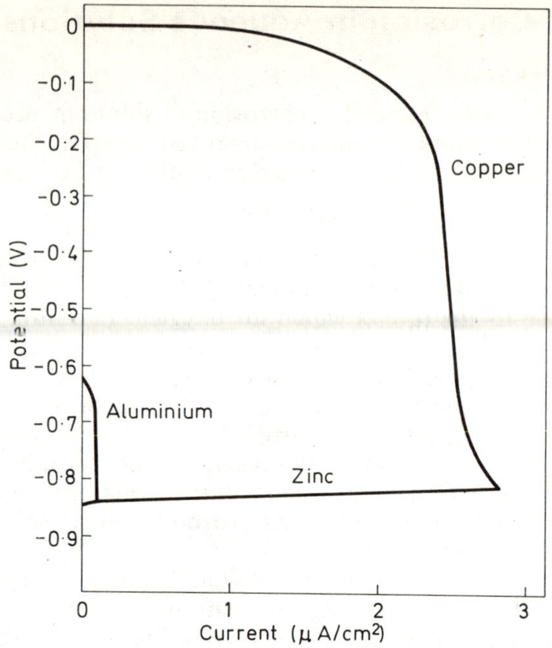

Fig. 1.64 Polarisation curves for zinc anodes coupled to aluminium and copper cathodes in 1·0 N NaCl at 25°C (potentials *vs.* S.H.E.) (after Pryor and Keir[7,9])

corrosion rate will, of course, be in the inverse relation to the slope of the anodic polarisation curve of the more negative metal.

The rate at which dissolved oxygen can be reduced under fixed experimental conditions is also strongly dependent on the nature of the surface of the cathodic member of the couple. If the cathode carries no oxide films, or if any pre-existing oxide films are reduced by the flow of galvanic current, reduction of dissolved oxygen takes place readily; it follows that the noble metals Pt, Au, Ag and Cu will be efficient cathodes in bimetallic corrosion involving oxygen reduction as the cathodic reaction. If, however, the reduction of oxygen takes place on a stable oxide film on the cathode, the reaction often tends to be much more polarised. Figure 1.64 shows polarisation curves for zinc anodes coupled to copper and aluminium cathodes in 1·0 N NaCl. Any pre-existing oxide films on the copper will be cathodically reduced at very low current densities, as shown by Dyess and Miley[8], as follows:

$$Cu_2O + 2H^+ + 2e \rightarrow 2Cu + H_2O$$

and/or

$$CuO + 2H^+ + 2e \rightarrow Cu + H_2O$$

Accordingly, a bare copper surface is soon presented to the electrolyte. Aluminium, on the other hand, maintains an oxide-covered surface under these conditions, and it is evident from Fig. 1.64, which is constructed from the work of Pryor and Keir[7,9], that the reduction of dissolved oxygen is

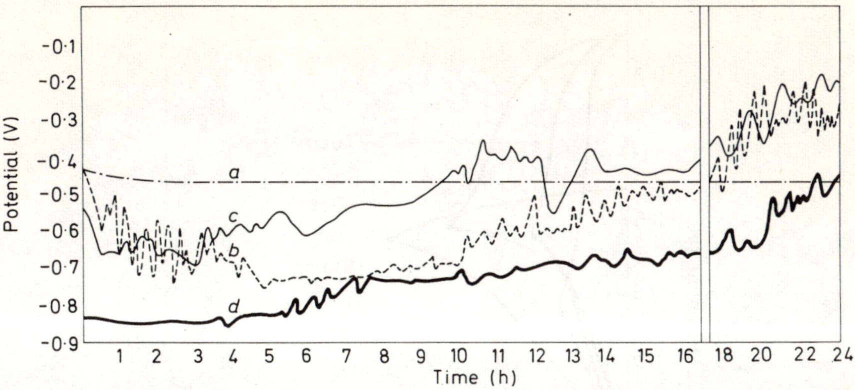

Fig. 1.65 Potential/time curves for cathodically polarised and unpolarised iron in 1·0 N NaCl at 25°C. Curve *a* unpolarised Fe; curve *b* Fe polarised cathodically at 20 μA/cm², pH = 6·2; curve *c* Fe polarised cathodically at 20 μA/cm², pH = 8·9; and curve *d* Fe polarised cathodically at 40 μA/cm², pH = 6·2 (potentials *vs.* S.H.E.) (after Pryor[10])

highly polarised and severely limits the galvanic current flow. Aluminium is an extreme example of an oxide-covered cathode because the electronic resistance of the natural oxide film is so high that it permits only small isolated areas to act as effective cathodes at low current density[9].

The existence of oxide films of varying resistance on the surface of a cathode results in a corresponding variation of galvanic current flow. This situation is far more common than is generally believed, since it is frequently observed on steel cathodes at current densities of less than 40 μA/cm². Figure 1.65 shows the variation of potential with time of mild-steel cathodes polarised cathodically at constant impressed current densities of 20 and 40 μA/cm² in 1·0 N sodium chloride solution[10]. The potential fluctuates rapidly but generally moves in the noble direction with increasing time to values of around −0·2 V (S.H.E.). This behaviour has been attributed to an overall thickening of the original air-formed oxide film due to a combination of locally increased pH at the cathode and the presence of dissolved oxygen[11]. The potential fluctuations in the active direction have been attributed to sporadic film breakdown by the large concentration of chloride ions.

Similar effects to those shown in Fig. 1.65 have been observed by Pryor and Keir[7] in certain bimetallic couples having steel cathodes. Figure 1.66 shows anodic and cathodic polarisation curves for the aluminium–mild steel couple in 1·0 N NaCl. The polarisation curves were determined by shorting different areas of mild-steel cathode to 100 cm² of aluminium for 24 h, rapidly introducing a variable and increasing external resistance into the circuit, and measuring the potential of both electrodes. Figure 1.66 shows that the transient open-circuit potentials of the steel cathodes are ennobled far above the normal value of corrosion potential (around −0·45 V) in this solution, particularly when smaller cathodes are used. In fact, maximum values of around −0·25 V are obtained, which are similar to the values obtained by impressed current methods (Fig. 1.65). Since the closed-circuit potential of the couples in Fig. 1.66 remains remarkably constant, the galvanic current at zero external resistance shows marked fluctuations

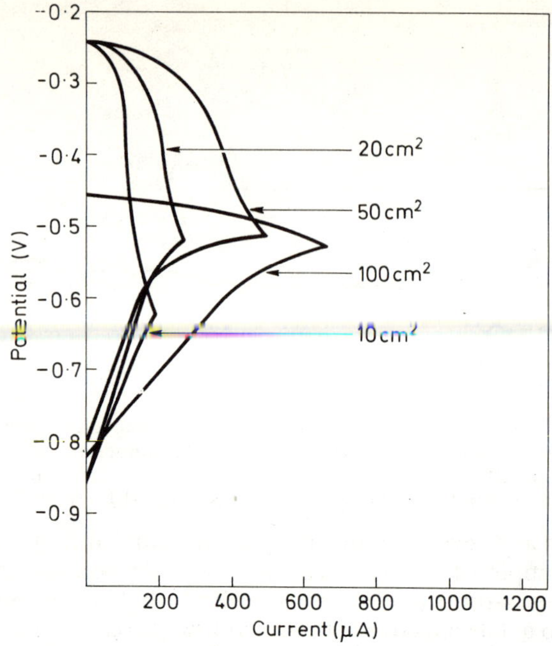

Fig. 1.66 Effect of area of steel cathode on the polarisation curves of the Al–Fe couple in 1·0 N NaCl at 25°C using 100 cm^2 of aluminium (potentials *vs.* S.H.E.) (after Pryor and Keir[7])

with time, somewhat similar to, but less extreme than, the potential fluctuations in Fig. 1.65. Comparable behaviour has also been observed with the zinc–mild steel couple in sodium chloride solution[7], but the open-circuit potentials of the cathodes are less ennobled and the current fluctuations are more gradual.

The electrochemical effects of slowly and erratically thickening oxide films on iron cathodes are, of course, eliminated when the film is destroyed by reductive dissolution[12] and the iron is maintained in the film-free condition. Such conditions are obtained when iron is coupled to uncontrolled magnesium anodes in high-conductivity electrolytes and when iron is coupled to aluminium in high-conductivity solutions of pH less than 4·0 or more than 12·0[13]. In these cases, the primary cathodic reaction (after reduction of the oxide film) is the evolution of hydrogen.

Catchment Area Principle

Practical instances of bimetallic corrosion usually involve electrolytes containing dissolved oxygen. Where the galvanic current is limited by the rate of diffusion of dissolved oxygen to the cathode, the galvanic corrosion rate will be under cathodic control. Whitman and Russell[14], in their classic investigation, showed that, under these conditions, the galvanic corrosion rate was directly proportional to the area of the cathodic metal and independent of the areas of the anodic metal. This principle was later referred to

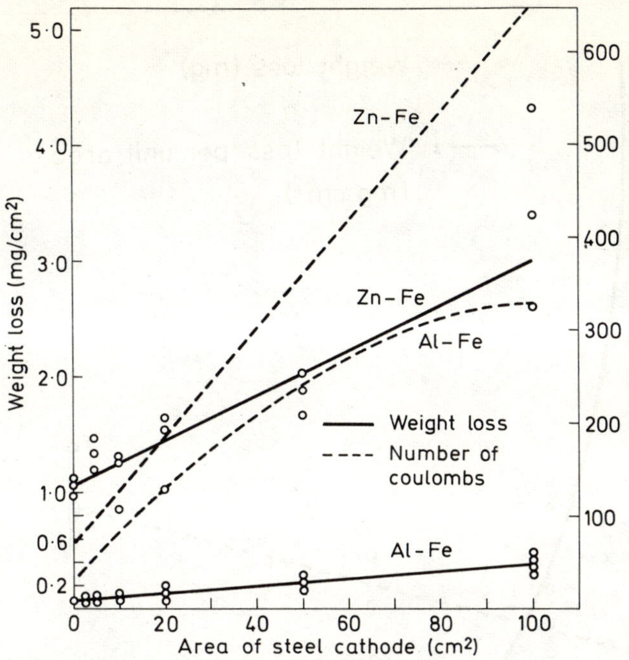

Fig. 1.67 Effect of mild-steel cathode on the weight loss of aluminium and zinc anodes (100 cm^2 area) and on the number of coulombs flowing in the aluminium–mild steel and zinc–mild steel couples over a 96 h period in 1·0 N NaCl at 25°C (after Pryor and Keir[7])

as the *catchment area* principle. It follows from the work of Whitman and Russell that the intensity (i.e. the anodic current density) of galvanic corrosion of an anode coupled to a cathode of constant area is inversely proportional to the anodic area. This principle is of major practical importance in view of its obvious use in minimising galvanic corrosion by proper selection of the relative areas of cathodic and anodic metals.

The catchment area principle is well illustrated by experiments on the Al–Fe and Zn–Fe couples shown in Figs. 1.67 and 1.68[7]. In Fig. 1.67 the area of aluminium and zinc was held constant and the area of steel cathode varied. Both the number of coulombs flowing between the two metals and the anodic weight losses increase with increasing cathodic area, and the current flow is approximately proportional to the cathodic area of each couple. Figure 1.68 shows the effect of holding the steel cathode area constant (100 cm^2) and varying the area of the aluminium anode. The weight loss of aluminium (the dotted line) increases slightly with increase in the anodic area exposed rather than remaining constant as predicted by the catchment area principle. Similar results have been obtained with zinc anodes of various area coupled to steel cathodes[7]. In part this is due to an increase in the local action corrosion with increasing anodic area[7]. Furthermore, the anodic polarisation curves for both aluminium and zinc have small but finite slopes and so the couple potential decreases slightly with increasing anodic area (decrease in polarisation of the anode metal) thereby increasing the propor-

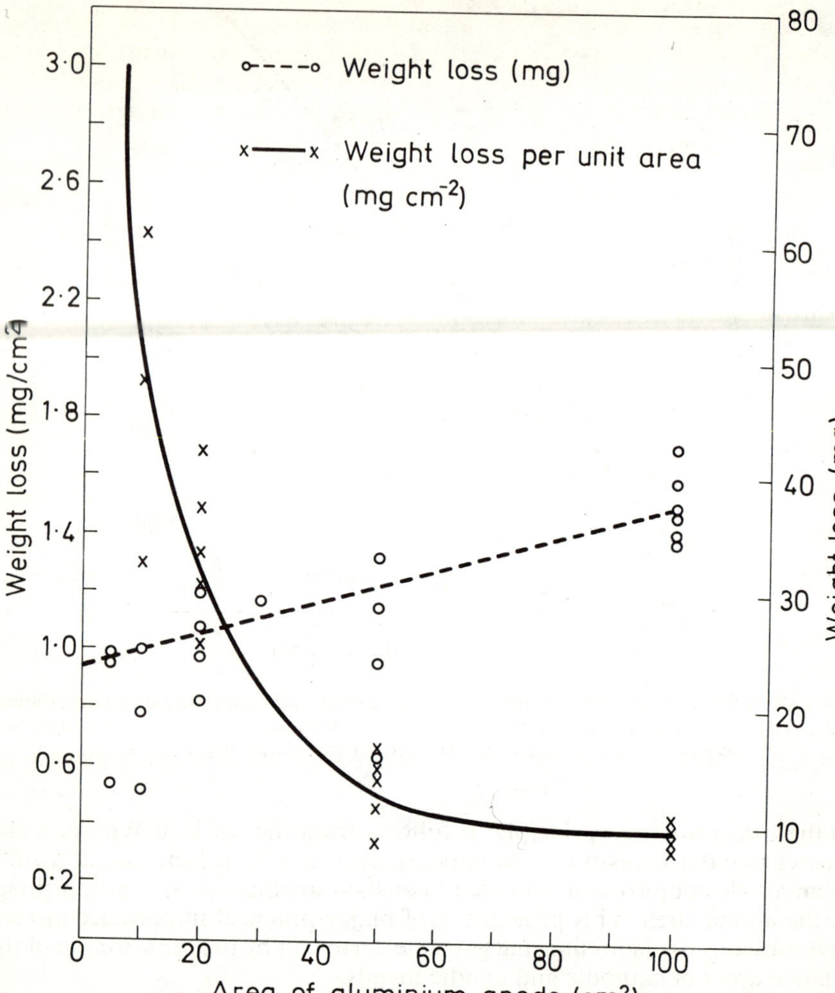

Fig. 1.68 Effect of anode area on the total corrosion and on the intensity of corrosion of aluminium coupled with mild-steel cathodes (100 cm^2 area) over a 96 h period in 1·0 N NaCl at 25°C (after Pryor and Keir[7])

tion of the weight loss due to local action (Fig. 1.62*b*). However, the *intensity* of corrosion (solid line) is very much as predicted by the catchment area principle and increases greatly with decreasing anodic area.

Hydrogen Evolution

When the predominant cathodic reaction is reduction of dissolved oxygen the magnitude of $I_{galv.}$ will be determined by the rate of diffusion of oxygen to the surface of the more positive metal constituting the bimetallic couple, and providing the metal surface is free from films or corrosion products the rate will be independent of the nature of the cathode. The position is quite different when the predominant cathodic reaction is hydrogen evolution, since under these circumstances charge transfer rather than diffusion is rate determining, and different metals will have different catalytic activities for the hydrogen evolution reaction.

If it is assumed that anodic polarisation of the more negative metal of the couple is insignificant, then it can be shown[14a] from the Tafel relationship for the hydrogen evolution reaction and from the E–$\log i$ anodic curve that

$$\log i_{\text{galv.}} = \frac{-E_{\text{corr.}} - 0{\cdot}059\,\text{pH}}{b_c} + \log \frac{S_e}{S_a} i_{0,c}$$

where $E_{\text{corr.}}$ is the corrosion potential of the couple (*vs.* S.H.E.) determined at some distance away from the couple to obtain the average value, b_c is the Tafel slope, and $i_{0,c}$ the exchange current density for the hydrogen evolution reaction at the surface of the more positive metal, and S_c and S_a are the areas of the cathode and anode metal, respectively. In this equation it is assumed that the potential of the more positive metal at which hydrogen evolution occurs is that of a reversible hydrogen electrode at 1 atm pressure, i.e. $E_H = 0{\cdot}00 - 0{\cdot}059\,\text{pH}$.

Electrolyte Composition

The composition of the electrolyte has, of course, a major influence on the magnitude and often the direction of the galvanic current flow. The majority of these effects can be rationalised from the effect of the electrolyte on the corrosion potential and, more particularly, the slopes of the appropriate

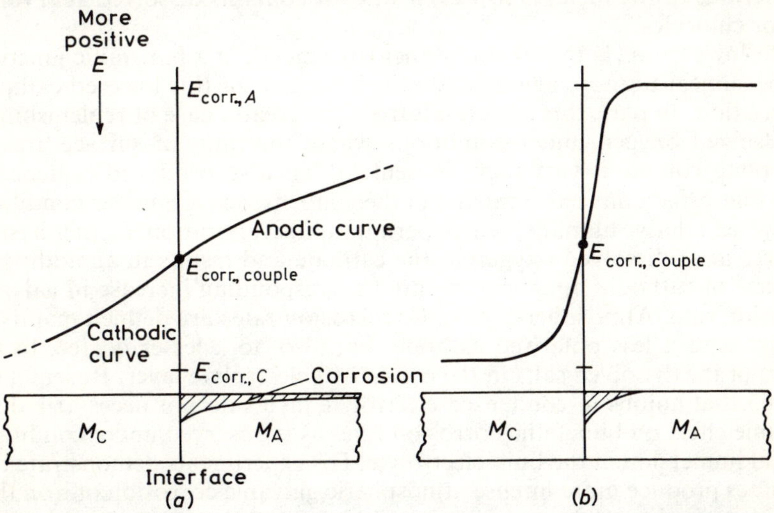

Fig. 1.69 Effect of resistivity of solution on the distribution of corrosion on the more negative metal M_A of a bimetallic couple. (*a*) Solution of very low resistivity and (*b*) solution of very high resistivity. Note that when the resistivity is low the effective areas of the cathodic and anodic metals are confined to the interface between the two metals

anodic and cathodic polarisation curves. The electrolyte does, however, have one further and important effect on the *distribution* of galvanic corrosion on the anodic metal[30]. When the electrolyte has a high conductivity, such as would be experienced in sea-water, the galvanic corrosion on the less noble metal is distributed in a comparatively uniform fashion and the total weight loss is high. As the electrolyte conductivity decreases, galvanic corrosion of the less noble metal becomes concentrated primarily around the bimetallic junction and the total weight loss of the anode decreases. The intensity of the galvanic corrosion in the interfacial region is relatively little influenced by electrolytic conductivity since it is generally controlled at near to a maximum value by polarisation effects. It is also implicit that the catchment area principle is not applicable in very low conductivity solutions. The effect of the resistivity of the electrolyte solution on the distribution of attack is illustrated in Fig. 1.69.

Bimetallic Corrosion in Corrosive Atmospheres

There are many special factors controlling atmospheric bimetallic corrosion that entitle it to separate treatment. The 'electrolyte' in atmospheric corrosion consists of a thin condensed film of moisture containing any soluble contaminants in the atmosphere such as acid fumes from industrial atmospheres and chlorides from marine atmospheres. This type of electrolyte has two characteristics which are summarised in a paper by Rosenfel'd[15].

1. The cathodic reduction of dissolved oxygen contained in a thin condensed electrolyte layer is much less severely polarised than the reduction of dissolved oxygen in the corresponding bulk electrolyte.
2. The electrolytic conductivity of the condensed layer parallel to the surface of the metal is low even when it contains dissolved acid fumes or chlorides.

It follows from (1) that the more negative metal, at a bimetallic junction, can be subject to more aggravated attack because of this lowered cathodic polarisation. In part, this may result from the greater ease of replenishment of dissolved oxygen under conditions where the ratio of surface area to electrolyte volume is very high. Rosenfel'd has also produced evidence to show that rapid convective mixing in the condensed layer, under conditions of lowered relative humidity which permit rapid evaporation, further hastens the arrival of dissolved oxygen at the cathode and results in an additional decrease of cathodic polarisation and a corresponding increase in galvanic corrosion rate. Atmospheric galvanic corrosion rates are defined mainly by contact with a less polarised cathode, but also, to a lesser degree, by the nature of the dissolved salts in the condensed electrolyte layer. Rosenfel'd[15] showed that anions in condensed electrolyte layers do not necessarily have the same effect on bimetallic corrosion rates as is observed under conditions of total immersion in the bulk electrolyte. His experiments demonstrate that sulphates produce more intense atmospheric galvanic corrosion of iron than chlorides and that the latter anion is without specific harmful effects on iron. This contention is not universally supported by atmospheric bimetallic corrosion tests on couples with iron anodes. The high electrolytic resistance of the thin condensed electrolyte layer parallel to the bimetallic surface has a controlling influence on the distribution of corrosion on the anode. The galvanic attack will normally be highly localised and qualitatively similar to that experienced under conditions of total immersion in very low conductivity electrolytes. Indeed it is quite rare to find atmospheric galvanic corrosion extending much further than 12–25 mm from the bimetallic junction. The effective cathodic area of the more noble metal tends to be limited by similar considerations (*see* Fig. 1.69*b*). It is, accordingly, evident that the catchment area principle is not applicable in the strict sense to atmospheric bimetallic corrosion since only small and approximately *equal* areas of the dissimilar metals in the immediate vicinity of a bimetallic junction interact electrochemically; thus although the geometrical area of the cathode metal may be large, its effective area is very limited. It should also be noted that whereas under immersed conditions the rate of bimetallic corrosion remains reasonably constant, it could be variable when the couple is exposed to the atmosphere owing to the variation in atmospheric conditions, i.e. periodic wetting and drying of the surface.

Variable Polarity

Instances where the normal polarity of a galvanic couple becomes reversed during service have given rise to severe practical corrosion problems and have been the subject of a great deal of investigation. Schikorr[16] was probably the first to note that the polarity of the zinc–iron couple could reverse, with iron becoming anodic to the zinc in certain supply waters when the temperature was raised above 60°C. Subsequently this problem was widely investigated by Hoxeng and Prutton[17], Kenworthy and Smith[18], Schuldener and Lehrman[19], and many others in view of its major importance relative to the severe corrosion of galvanised-steel domestic hot-water heaters[20]. It has been confirmed that the polarity reversal occurs at elevated temperature, in excess of about 60°C, and is aided by the presence of bicarbonates and nitrates in the water[17]. It has also been shown that the presence of sulphates and/or chlorides in the water decreases the probability of the polarity reversal. Schuldener and Lehrman[21] showed that under conditions of temperature and electrolyte composition which favour the zinc–steel polarity reversal, the reversal occurs more rapidly when the area of steel exposed to the electrolyte is increased. This observation suggests that galvanised steel that has been extensively pre-corroded, with zinc as the anode, may be more susceptible to subsequent polarity reversal under the appropriate conditions. Furthermore the grain size of hot-dipped galvanised coatings may well influence the rate of polarity reversal, since Caplan and Sereda[22] showed that voids of up to 0·5 mm in diameter and extending at least as far as the Fe–Zn intermetallic layers, exist in these coatings where three zinc 'spangles' meet at a point. The size of the flaws was observed to increase with increasing thickness of the base steel.

This polarity reversal has sometimes been attributed[23] to the formation, at the higher temperatures involved, of an electronically conducting zinc oxide film rather than the normally observed zinc hydroxide of poor electronic conductivity which is stable at room temperature. It is generally less well realised that the aluminium–iron couple also undergoes a polarity reversal, like the zinc–iron couple, at elevated temperatures and under somewhat similar conditions. The form of the corrosion product on aluminium also changes with increasing temperature from bayerite (β-$Al_2O_3.3H_2O$, monoclinic) to boehmite (γ-AlO.OH, orthorhombic), but since the boehmite has a very low electronic conductivity, the intensity of galvanic attack on the steel is low.

The relative polarities of iron and tin are considered in Sections 4.6 and 14.5 and will not be discussed here.

The variable polarity of the aluminium–iron couple has already been noted. It is also known that it is virtually impossible to protect commercial aluminium alloys cathodically by coupling them to magnesium in a high-conductivity electrolyte such as sea-water, since the development of a high local pH at the aluminium cathode results in cathodic corrosion. Some interesting results have been reported recently by Bothwell[25] concerning the compatibility of super-purity >99·99% Al aluminium with the magnesium alloys AZ31B (3% Al, 1% Zn, 0·4% Mn) and AZ31A (3% Al, 1% Zn, 0·4% Mn, 0·15% Ca) in sodium chloride and sea-water. Here the higher electronic resistance of the natural oxide film on super-purity aluminium appears to limit the current flow and the local rise of pH at the cathode, thereby preventing normal aggressive cathodic corrosion.

Akimov[26] was one of the earliest investigators to report that the pitting of aluminium and aluminium alloys in sea-water could be prevented by coupling to zinc. He also drew attention to the occurrence of a tenacious black film on the aluminium. Since then additional observations have been made on the beneficial effects of coupling zinc to various aluminium alloys in chloride solutions. For instance, zinc will often stop the stress-corrosion cracking of susceptible Al–Mg alloys. The protective action of zinc is, however, somewhat unreliable and strongly dependent on the method of surface preparation of the aluminium alloy[27].

Experiments carried out by Keir, van Rooyen and Pryor have clarified the variable behaviour of the Al–Zn couple in chloride solutions. Using high-purity aluminium and zinc electrodes of equal size coupled together in sodium chloride solution, it was found that zinc is initially anodic to aluminium but that within one day the polarity of the couple reverses and remains as such subsequently (Fig. 1.70). This reversal in polarity appears to be due to the accumulation of Zn^{2+} in solution. Accordingly, with decrease in distance between the electrodes, and in solution-volume: electrode-area ratio, the polarity reversal occurs much more rapidly. The accumulation of Zn^{2+} in solution depresses the potential of the aluminium from an initial value of about $-0{\cdot}5$ V to a final open-circuit value of about $-1{\cdot}0$ V (*vs.* S.H.E.). The corrosion rates of both the aluminium and the zinc electrodes are greater than in the absence of bimetallic contact, but the corrosion of the aluminium is changed from the characteristic pitting, usually observed in nearly neutral chlorides, to a desirable mild uniform attack. The polarity reversal is not detectable by potential measurements alone, since the potential of the couple remains constant, to within a few millivolts at a value of around $-0{\cdot}83$ V. When the aluminium is cathodic, polarisation curves indicate that the galvanic corrosion is under cathodic control; after the polarity reversal, the much weaker galvanic corrosion appears to be under anodic control.

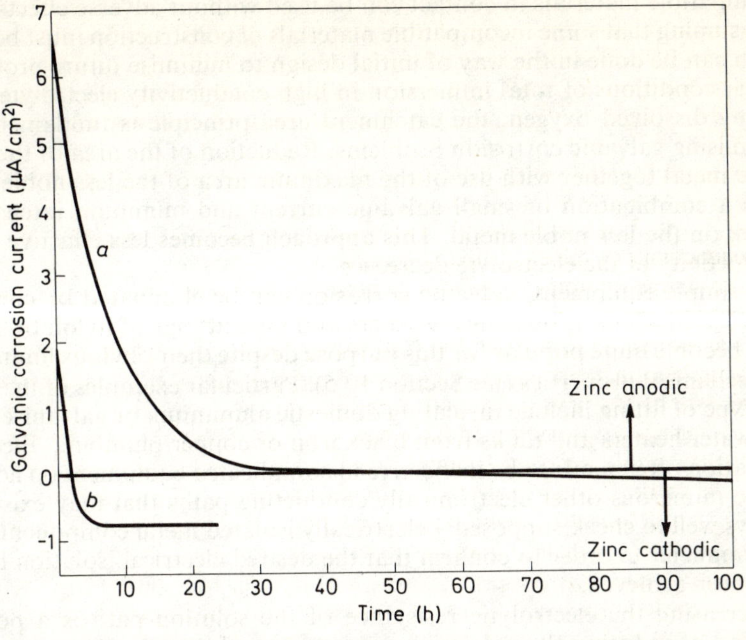

Fig. 1.70 Polarity reversal of the Al–Zu couple in 1·0 N sodium chloride at 25°C. Curve *a* aluminium and zinc electrodes 150 mm apart; 16 ml of solution per square centimetre of electrode; curve *b* aluminium and zinc electrodes 20 mm apart, 10 ml of solution per square centimetre of electrode

It remains to be determined whether the previous experiments[26,27], which have been interpreted as confirming the cathodic protection of aluminium by zinc, can be truly interpreted in this fashion or whether they are due to the accumulation of Zn^{2+} in the electrolyte. Under laboratory conditions, and under some practical conditions in stagnant solutions or in recirculating systems, the latter explanation is quite likely.

Two dissimilar metals, such as iron and aluminium, may cause aggravated corrosion effects even if they are not in electrical contact. This subject is, however, outside the scope of this section, and has been treated in detail elsewhere[28,29]. Heavy metal ions, such as copper ions, are particularly liable to produce galvanic effects by redeposition on a less noble metal; the phenomenon is discussed in Sections 4.1, 4.2 and 10.3.

Protective Measures

Under Conditions of Total Immersion

Protective measures against bimetallic corrosion should ideally start before the particular installation or equipment is built. Reference should be made to tables showing compatibility of metals, alloys and non-metallic materials (*see* Appendix 1.7A) and to the literature. However, it must be emphasised that the environment obviously plays a most important rôle in bimetallic corrosion, and that there are a number of situations in which apparently incompatible materials in contact can be used without adverse effects.

Assuming that some incompatible materials of construction must be used, much can be done in the way of initial design to minimise future problems. Under conditions of total immersion in high-conductivity electrolytes containing dissolved oxygen, the catchment area principle is fundamental in minimising galvanic corrosion problems. Reduction of the area of the more noble metal together with use of the maximum area of the less noble metal gives a combination of small galvanic current and minimum intensity of attack on the less noble metal. This approach becomes less effective as the conductivity of the electrolyte decreases.

In simple equipment, galvanic corrosion can be eliminated by complete electrical insulation. Bushings, washers and pipe fittings of nylon or Teflon have become quite popular for this purpose despite their obvious limitations in mechanical properties (*see* Section 10.5). Particular examples of the use of this type of fitting include insulating domestic aluminium or galvanised-iron hot-water heaters and tanks from black iron or copper plumbing. Electrical insulation often tends to be ineffective in complicated equipment on account of the numerous other electronically conducting paths that may exist. It is always well to check supposedly electrically isolated metal components with an ohmmeter in order to confirm that the desired electrical isolation has, in fact, been achieved.

Increasing the electrolytic resistance of the solution path is a possible method of reducing the galvanic corrosion rate. Little significant practical benefit accompanies this approach in high-conductivity electrolytes since severe galvanic corrosion can exist at locations which are several metres distant from the actual bimetallic joint. This approach, accordingly, has merit mainly in low-conductivity electrolytes such as certain supply waters, and in the case of atmospheric galvanic corrosion.

Deaeration has occasionally been used as a means of controlling bimetallic corrosion under conditions of total immersion, and this method of control can be used successfully, if physical conditions permit, provided that the less noble metal is not sufficiently electrochemically active to permit rapid evolution of hydrogen at the more noble metal, as is observed, for instance, in many bimetallic couples involving magnesium anodes.

Metallic coatings have been widely and successfully used as a means of alleviating many bimetallic corrosion problems both under conditions of total immersion and in corrosive atmospheres. If, for instance, aluminium and steel must be jointed together in sea-water, the galvanic corrosion can be largely eliminated by aluminising the steel either by hot dipping or by flame spraying, as is more popular in Europe. Both zinc and cadmium are also fairly compatible with aluminium and so the steel may be protected with thin coatings of these metals without incurring the risk of aggravated galvanic corrosion; cadmium plating has even been applied to stainless steel for this purpose. The use of dissimilar metallic coatings eliminates bimetallic corrosion only if the coating is initially free from voids and remains so in service, a circumstance seldom realised in practice. Metallic coatings on the steel that contain or develop voids still reduce the galvanic corrosion rate (because of the smaller area of steel exposed) provided that they are anodic to the substrate.

Cathodic protection with a sacrificial anode that is less noble than either member of the couple is frequently used to reduce the severity of bimetallic corrosion, particularly that resulting from the use of bronze propellers in steel ship hulls.

Paint coatings also receive extensive practical use for protecting against galvanic corrosion in atmospheres and under conditions of total immersion. The best practice, where feasible, calls for complete painting of both members of the bimetallic couple. If only one member of the couple can be painted, the *cathodic* metal should receive this treatment; since paint coatings are seldom free from holidays, painting the cathode will reduce the total cathodic area and hence the galvanic corrosion rate. Painting the anodic metal alone represents bad practice under conditions of total immersion in a high-conductivity electrolyte, because the original cathode area is undiminished, and corrosion will then take place at holidays or damaged areas in the coating on the anodic metal at a high intensity. For optimum protection against galvanic corrosion, repainting should be carried out on a regular schedule since the protection afforded by most paints can be rather limited in duration.

The use of soluble inhibitors as a means of controlling bimetallic corrosion presents many technical problems. Apart from the fact that this method is limited in applicability to recirculating systems, efficient anodic inhibitors, such as chromates, are frequently quite specific in their action and so certain bimetallic couples, such as the Al–Cu couple in chloride solutions[30], are extremely difficult to control by a single anodic inhibitor. Accordingly, other methods of treating bimetallic corrosion, as already described, are often preferred, with inhibition being relegated to special applications such as automotive cooling systems where these previous methods are either not feasible or economical. Inhibition of automotive cooling systems is generally achieved by highly complex mixtures of inhibitors, often involving combinations of borax, nitrates and organic adsorption inhibitors, and even then complete success is not always attained (*see* Section 18.2).

In Corrosive Atmospheres

Bimetallic corrosion in atmospheres is confined to the area of the less noble metal in the vicinity of the bimetallic joint, owing to the high electrolytic resistance of the condensed electrolyte film. Electrolytic resistance considerations limit the effective anodic and cathodic areas to approximately equal size and therefore prevent alleviation of atmospheric galvanic corrosion through strict application of the catchment area principle.

With this exception, many of the methods already described for protecting against bimetallic corrosion under conditions of total immersion may be similarly used for preventing atmospheric galvanic corrosion. These include selection of compatible metals, metallic coatings and painting. It is, however, more common to use the principle of increasing the resistance of the solution path for preventing galvanic corrosion. Since the solution-path resistance is already high, the additional means that are required to increase resistance further are simple and generally inexpensive. In many cases, taping the immediate joint area with mastic tapes, with or without chromate impregnation, will suffice, provided that the whole of the bimetallic contact is covered to a distance of about 25 mm from the junction and on either side. Vulcanising a rubber or Neoprene ball around small joints has also been used very satisfactorily. While these methods protect a joint against atmospheric galvanic corrosion, it can hardly be overemphasised that they are *not* applicable to protecting against bimetallic corrosion where the same joint is totally immersed in a high-conductivity electrolyte.

Some Beneficial Effects of Galvanic Coupling

So far this section has been primarily concerned with the harmful aspects of bimetallic corrosion, in which the less noble member of the couple is subjected to attack of unusual severity. It is, however, implicit that bimetallic corrosion can be beneficial in that it will reduce or prevent corrosion of the more noble metal.

The principles of bimetallic corrosion have, in addition, been used recently in an elegant fashion for the development of highly corrosion-resistant alloys.

Draley and Ruther[31] observed that commercial-purity aluminium (1100 alloy) was subject to catastrophic intergranular corrosion in distilled water above 200°C with the corrosion rate increasing very rapidly with temperature. In most cases enhanced attack occurred at grain boundaries and around second-phase stringers. Draley and Ruther showed that the rapid intergranular disintegration of the aluminium was associated with the entry of hydrogen into the metal from cathodic sites. They proposed that, if alternate cathodic sites of lower hydrogen overpotential could be provided, the hydrogen would have a much better chance of being evolved as harmless bubbles instead of entering the aluminium and causing intergranular disintegration. This hypothesis was confirmed by adding 5 p.p.m. of Ni^{2+} to the distilled water. The small amounts of nickel plating-out on the aluminium surface were sufficient to protect against catastrophic corrosion of 1100 alloy at 275°C, although the 1100 was subject to slow and uniform attack. Subsequently, 1100 alloy specimens were electroless nickel-plated and found to resist catastrophic corrosion for 80 days at 315°C. Finally Draley and Ruther alloyed small amounts of nickel (0·5% or more) with commercially pure aluminium and obtained consistent protection against catastrophic corrosion at temperatures of up to 350°C. The nickel, being largely insoluble in aluminium, exists primarily as $NiAl_3$ constituent which evidently possesses a low hydrogen overpotential and protects against catastrophic corrosion. Its action is augmented by the simultaneous presence of iron in the alloy, the improved corrosion resistance probably being due to the presence of an Al–Ni–Fe constituent[32]. A commercial aluminium alloy designated as 8001 (1% Ni, 0·6% Fe) now exists for high-temperature water service.

It has long been known that alloys such as austenitic stainless steel and metals such as titanium, while exhibiting passive behaviour in mildly or strongly oxidising solutions, often suffer active corrosion at a high rate in reducing acids. Tomashov *et al.*[33,34] were among the first to point out the possibility of improving the corrosion resistance of stainless steel, chromium and titanium by increasing the stability of the passive state with small alloying additions of noble metals such as silver, palladium and platinum. This work was extended for titanium in reducing-type acids by Stern and

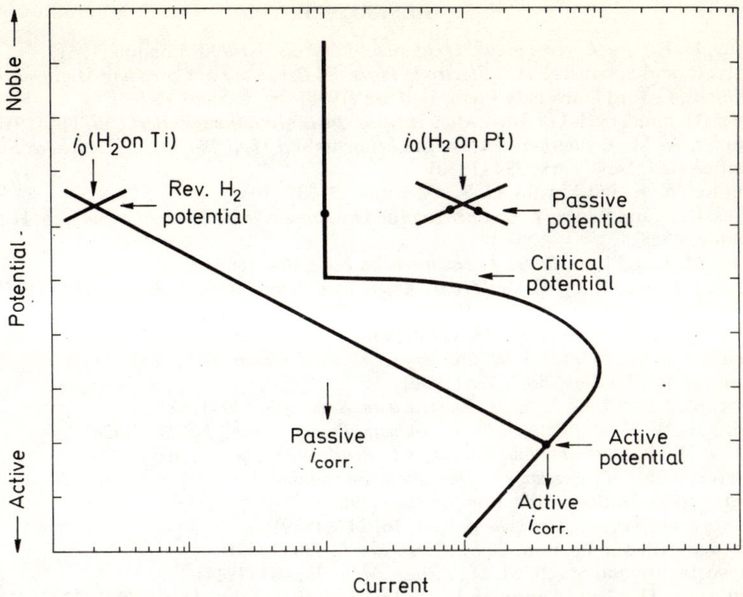

Fig. 1.71 How alloying with a noble metal produces a passive mixed potential and a marked reduction in corrosion rate (after Stern and Wissenberg[35])

Wissenberg[35] who primarily investigated the effect of platinum and palladium. The principle by which noble alloying additions are effective in improving the corrosion resistance of titanium is illustrated in Fig. 1.71 which is taken from the work of Stern and Wissenberg. In a reducing acid where hydrogen is evolved from a titanium surface, the exchange current is relatively small. It may be increased and the cathodic Tafel slope decreased by providing local noble-metal cathodes. The intercept of the cathodic and the anodic polarisation curves is shifted in the more noble direction and, if the shift is large enough to raise the mixed potential into the passive potential region in Fig. 1.71, essentially passive behaviour in reducing acids such as boiling HCl can result (*see* Section 5.4). Similar improvements in the corrosion resistance of chromium in sulphuric and hydrochloric acids have been found by Greene, Bishop and Stern[36] to accompany alloying with small amounts of rhodium, palladium or osmium. These noble-metal alloying additions have the further advantage of not significantly reducing the normally excellent corrosion resistance of chromium in oxidising acids such as nitric acid, whereas elements such as platinum, iridium and ruthenium confer excellent corrosion resistance on chromium in non-oxidising acids, but increase its corrosion rate in oxidising acids.

It is evident from the foregoing that the principles of bimetallic corrosion are being applied in a progressively more widespread and successful fashion to the development of alloys of maximum corrosion resistance.

M. J. PRYOR
(revised by L. L. Shreir)

REFERENCES

1. Evans, U. R., *The Corrosion and Oxidation of Metals*, Arnold, London (1961)
2. Gatty, O. and Spooner, E. C., *Electrode Potential Behaviour of Corroding Metals in Aqueous Solutions*, Oxford University Press, London (1938)
3. Bauer, O. and Vogel, O., *Mitt. MatPrüfAmt. Inst. Metallforsch. Berl.*, **36**, 114 (1918)
4. Latimer, W. M., *Oxidation States of the Elements and Their Potentials in Aqueous Solutions*, Prentice-Hall, New York, 294 (1950)
5. Promisel, N. E. and Mustin, G. S., *Corrosion*, **7**, 339 (1951)
6. Evans, U. R. and Rance, V. E., *Corrosion and Its Prevention at Bimetallic Contacts*, H.M.S.O., London (1958)
7. Pryor, M. J. and Keir, D. S., *J. Electrochem. Soc.*, **104**, 269 (1957)
8. Dyess, J. B. and Miley, H. A., *Trans. Amer. Inst. Min. (Metall.) Engrs.*, **133**, 239 (1939)
9. Pryor, M. J. and Keir, D. S., *J. Electrochem. Soc.*, **102**, 605 (1955)
10. Pryor, M. J., *Nature, Lond.*, **178**, 1245 (1956)
11. Mayne, J. E. O., Menter, J. W. and Pryor, M. J., *J. Chem. Soc.*, 1831 (1949)
12. Evans, U. R., *J. Chem. Soc.*, 478 (1930)
13. Pryor, M. J. and Keir, D. S., *J. Electrochem. Soc.*, **105**, 629 (1958)
14. Whitman, W. G. and Russell, R. P., *Industr. Engng. Chem.*, **16**, 276 (1924)

14*a*. Uhlig, H. H., *Corrosion and Corrosion Control*, Wiley, New York (1971)

15. Rosenfel'd, N., *Proceedings of the First International Congress on Metallic Corrosion, London, 1961*, Butterworths, London, 243 (1962)
16. Schikorr, G., *Trans. Electrochem. Soc.*, **76**, 247 (1939)
17. Hoxeng, R. B. and Prutton, C. F., *Corrosion*, **5**, 330 (1949)
18. Kenworthy, L. and Smith, M. D., *J. Inst. Met.*, **70**, 463 (1944)
19. Schuldener, H. L. and Lehrman, L., *J. American Wat. Wks. Assoc.*, **49**, 1432 (1957)
20. Cohen, M., Thomas, W. R. and Sereda, P. J., *Canad. J. Technol.*, **29**, 435 (1951)
21. Schuldener, H. L. and Lehrman, L., *Corrosion*, **14**, 545t (1958)
22. Caplan, D. and Sereda, P. J., *Canad. J. Technol.*, **31**, 172 (1953)
23. Gilbert, P. T., *J. Electrochem. Soc.*, **99**, 18 (1952)
24. Keller, J. D. and Edwards, F., *Metal Progr.*, **54**, 35, 195 (1948)
25. Bothwell, M. R., *J. Electrochem. Soc.*, **106**, 1014, 1019 (1959)
26. Akimov, G., *Korros. Metallsch.*, **6**, 84 (1930)
27. Edeleanu, C. and Evans, U. R., *Trans. Faraday Soc.*, **47**, 1121 (1951)
28. Bird, C. E. and Evans, U. R., *Corr. Technol.*, **3**, 279 (1956)
29. See for instance Evans, U. R., *The Corrosion and Oxidation of Metals*, Arnold, London, 205–6 (1960)
30. Eldridge, G. G. and Mears, R. B., *Industr. Engng. Chem.*, **37**, 736 (1945)
31. Draley, J. E. and Ruther, W. E., *Corrosion*, **12**, 480t (1956); *J. Electrochem. Soc.*, **104**, 329 (1957)
32. Phillips, H. W. L., *J. Inst. Met.*, **69**, 275 (1943)
33. Tomashov, N. D. and Chernova, G. P., *C. R. Acad. Sci. U.R.S.S.*, **89**, 121 (1953)
34. Tomashov, N. D., Altovsky, R. M. and Arakelov, A. G., *C. R. Acad. Sci. U.R.S.S.*, **121**, 885 (1958)
35. Stern, M. and Wissenberg, H., *J. Electrochem. Soc.*, **106**, 755, 759 (1959)
36. Greene, N. D., Bishop, C. R. and Stern, M., *J. Electrochem. Soc.*, **108**, 836 (1961)

BIBLIOGRAPHY

Bimetallic corrosion*

Veilleux, E. D., 'Galvanic Corrosion', *Machine Design*, **32**, 181 (1960); *C.A.*, **61**, 11609f

Valanti, V. O., 'Galvanic Corrosion of Welded Joints on Ships', *Rev. Deut. Ausg*, No. 4, 1 (1961) and No. 1, 12 (1962); *C.A.*, **60**, 8956d

Koehler, W. and Machu, W., 'Corrosion of the Aluminium–Steel System', *Werkstoffe Korrosion*, **13**, 601 (1962); *C.A.*, **58**, 1122g

Krutikov, A. N., Arest, T. V. and Kristal, M. M., 'Welding and Corrosion Resistance of the Bimetals Steel–Copper, Steel–Brass, and Steel–Bronze', *Svarchnoe Proizvodstvo*, **2**, 15 (1962); *C.A.*, **56**, 11313a

* *Note: C.A.* means *Chemical Abstracts.*

Kristal, M. M., Adugina, N. A., Krutikov, A. N. and Arest, T. V., 'Corrosion Resistance of Welded Bimetal Joints of Steel–10–Copper, Steel–10–Bronze, and Steel–10–Brass', *Tr. Vses. Nauchn-Issled i Konstrukt. Inst. Khim. Mashinostr.*, **45**, 66 (1963); *C.A.*, **61**, 14268f

Godard, H. P., 'Galvanic Corrosion Behaviour of Aluminium in the Atmosphere', *Mater. Protect.*, **2**, 40 (1963); *C.A.*, **59**, 4874

Subramanyan, N. and Shyamala, N., 'Corrosion and Electrochemical Behaviour of Copper–Mild Steel Couple Bielectrode System', *Indian J. Technol.*, **1**, 338 (1963); *C.A.*, **60**, 2543

Shimizu, K., Miyauchi, N. and Ono, M., 'Corrosion Phenomena on Galvanic Couple of Stainless Steel and Copper Alloy in Concentrated Brine', *Boshoku Gijutsu*, **13**, 109 (1964); *C.A.*, **61**, 1578h

Kopczynski, C., 'Corrosion Testing of Thermostatic Bimetals', *Prace. Inst. Elektrotech.*, **13**, 49 (1965); *C.A.*, **64**, 17136

Tajima, S., Mor, T. and Kamiya, M., 'Contact Corrosion of Bimetallic Couples of Aluminium or Anodised Aluminium and Mild Steel or Stainless Steel, in Inorganic Corrosive Media', *Boshoku Gijutsu*, **14**, 109 (1965); *C.A.*, **63**, 7897h

Latimer, K. G., 'Corrosion Behaviour of Aluminium-to-Steel Joints with Special Reference to Aluminium Superstructures on Ships', *2nd Inter. Congress Metal Corrosion, New York City, 1963*, 780 (discussion 788) (1966)

Spaehn, H. and Faessler, K., 'Theory of Contact Corrosion and its Effect on Metallic Protective Coatings', *Metalloberflaeche*, **21**, 38 (1967); *C.A.*, **66**, 68447u

Spaehn, H. and Faessler, K., 'Contact Corrosion in Machine and Equipment Construction. Some Principles of Corrosion Mechanisms and Practical Examples of Damage', *Machinenschaden*, **40**, 81 (1967); *C.A.*, **68**, 26188w

Kaesche, H., 'Rôle of Galvanic Short Circuit Cells in Electrolytic Corrosion of Metals', *Korrosion*, **19**, 9 (1965); *C.A.*, **71**, 8907n

Keir, D. S., Pryor, M. J. and Sperry, P. R., 'Galvanic Corrosion Characteristics of Aluminium Alloyed with Group IV Metals', *J. Electrochem. Soc.*, **114**, 777 (1967)

Venczel, J. and Wranglen, G., 'Corrosion and Corrosion Product Deposition in Steel–Copper Systems with Circulating Hot Water Containing Various Inhibitors', *Corros. Sci.*, **7**, 461 (1967)

Wei, M. W., 'Experience with Alclad Aluminium in Deep Sea Buoyancy Sphere', *Corrosion*, **23**, 261 (1967)

Creamer, E. L., Rozalsky, I. and Lochmann, W. J., 'Sulfuric Corrosion of Chromel–Alumel Thermocouples', *Corrosion*, **23**, 297 (1967)

Schaefer, B. A., 'The Corrosion of Copper in Ammoniacal Solutions', *Corros. Sci.*, **8**, 623 (1968)

Toy, S. M., English, W. D. and Crane, W. E., 'Studies of Galvanic Corrosion Couples in Liquid Fluorine', *Corrosion*, **24**, 418 (1968)

Wranglen, G. and Inam Khokhar, M., 'Short Communication on the Relation Between Corrosion Potential and Galvanic Corrosion of Carbon Steels in Acid Solutions', *Corros. Sci.*, **9**, 439 (1969)

Zanker, L. and Yahalom, J., 'Galvanic Behaviour of the Fe–Cd Couple', *Corros. Sci.*, **9**, 157 (1969)

Tomashov, N. D., 'Development of the Electrochemical Theory of Metallic Corrosion', *Corrosion*, **20**, 7t (1969)

Devay, J., Lengyel, B. (Jr.) and Meszaros, L., 'Application of a Potentiostat for Determining the Maximum Corrosion Current of Galvanic Couples', *Acta Chim.*, **62**, 157 (1969); *C.A.*, **72**, 38135x

Dekany, I. and Fekete, K., 'Results of Contact Corrosion Tests on Some Industrially Important Metal Pairs', *Corros. Week, Manifestation Eur. Fed. Corros.*, **41**, 1051, 1968 (1970); *C.A.*, **73**, 83041v

Kennard, E. and Waber, J. T., 'Mathematic Study of Galvanic Corrosion', *J. Electrochem. Soc.*, **117**, 880 (1970)

Greene, H. D., Bennett, D. C. and Dresty, J. E., 'Galvanic Corrosion', *J. Electrochem. Soc.*, **119**, 247C (Miami Beach Florida Mtg.) (1972)

Gouda, V. K., Shalaby, L. A. and Abdul Azim, A. A., 'Corrosion Behaviour of Lead in Salt Solutions. Lead–Steel Couple', *Br. Corr. J.*, **8**, 81 (1973)

Hiroshi, S., Kazuo, Y. and Yashiharu, K., 'Galvanic Corrosion Between Titanium and a Ti–Pd Alloy in HCl', *Kinyotu Hyomen Gijutsu*, **24**, 433 (1973); *C.A.*, **80**, 43365s

Naoji, F., Kozunobu, N. and Tadao, H., 'Corrosion Resistance of Ni–Cr Electroplate V. Contact Behaviour in Ni–Cr Electroplates', *Denki Kagaku*, **41**, 421 (1973); *C.A.*, **80**, 43372s

Storchai, E. I., Elchinova, L. N., Sokolova, A. V. and Kuznetsova, M. V., 'Corrosion Behaviour of Joints of Aluminium and its Alloys Soldered by Low Temperature Lead–Tin System Solders', *Khim. Neft. Mashinostr.*, **10**, 15 (1973); *C.A.*, **80**, 77443f

Mansfeld, F., 'Galvanic Corrosion of Aluminium Alloys Coupled to Coated PH13–8Mo Stainless Steel', *J. Electrochem. Soc.*, **120**, 231C (1973)

Mansfeld, F. and Oldham, K. B., 'Galvanic Corrosion Resulting from Rupture of a Protective Metallic Coating', *J. Appl. Electrochem.*, **2**, 183 (1972)

1.7A Appendix—Bimetallic Corrosion

Degree of Corrosion at Bimetallic Contacts

Table 1.A2 is reproduced from *Corrosion and its Prevention at Bimetallic Contacts* (H.M.S.O., London, 1958) by permission of the Director of Publications. The reader should note that it is recommended that the Table be used only in conjunction with the Introduction to the original publication.

Notes to Table 1.A2

(*a*) The exposure of iron, steel, magnesium alloys and unclad aluminium–copper alloys in an unprotected condition in corrosive environments should be avoided whenever possible even in the absence of bimetallic contact.

(*b*) The behaviour of magnesium alloys in bimetallic contacts is particularly influenced by the environment, depending especially on whether an electrolyte can collect and remain as a bridge across the contact. The behaviour indicated in the table refers to fairly severe conditions. Under conditions of total immersion or the equivalent, magnesium alloys should be electrically insulated from other metals. In less severe conditions complete insulation is not necessary, but steel, brass and copper parts should be galvanised or cadmium plated, and jointing compound (D.T.D. 369A) used during assembly. Under conditions of good ventilation and drainage, contacts classified as *D* have given satisfactory service, e.g. brass and steel push-fit and cast-in inserts in magnesium castings.

(*c*) Where contact between magnesium alloys and aluminium alloys is necessary, adverse galvanic effects will be minimised by using aluminium alloys containing little or no copper (0·1% max.).

(*d*) If in contact with thin (decorative) chromium plate, the symbol is *C*, but with thick plating (as used for wear resistance) the symbol is *B*.

(*e*) When contacts between copper or copper-rich materials and aluminium alloys cannot be avoided, a much higher degree of protection against corrosion is obtained by first plating the copper-rich material with tin or nickel and then with cadmium, than by applying a coating of cadmium of similar thickness.

(*f*) The corrosion of mild steel may sometimes be increased by coupling with cast iron, especially when the exposed area of the mild steel is small compared with the cast iron.

(*g*) Instances may arise in which corrosion of copper or brasses may be accelerated by contact with bronzes or gunmetals, e.g. the corrosion of copper sea-water-carrying pipelines may be accelerated by contact with gunmetal valves, etc.

(*h*) When magnesium corrodes in sea-water or certain other electrolytes, alkali formed at the aluminium cathode may attack the aluminium.

(*Notes continue on page 218*)

Table 1.A2 Degree of corrosi

Metal considered / *Contact metal*	1 *Gold, platinum, rhodium, silver*	2 *Monel, Inconel, nickel–molybdenum alloys*	3 *Cupronickels, silver solder, aluminium bronzes, tin bronzes, gunmetals*	4 *Copper brasses, 'nickel silvers'*	5 *Nickel*	6 *Lead, t and sof solders*
1. Gold, platinum, rhodium, silver	—	*A*	*A*	*A*	*A*	*A*
2. Monel, Inconel, nickel–molybdenum alloys	*B*	—	*A*	*A*	*A*	*A*
3. Cupronickels, silver solder, aluminium bronzes, tin bronzes, gunmetals	*C(k)*	*B* or *C*		*A*	*A*	*A*
4. Copper, brasses, 'nickel silvers'	*C(k)*	*B* or *C*	*B* or *C(g)*	—	*B* or *C*	*B* or *C(*
5. Nickel	*C*	*B*	*A*	*A*	—	*A*
6. Lead, tin and soft solders	*C*	*B* or *C(t)*	*B* or *C(q)*	*B* or *C(q)*	*B*	—
7. Steel and cast iron (*a*) (*f*) (*w*)	*C*	*C*	*C*	*C*	*C(k)*	*C(k)*
8. Cadmium (*u*)	*C*	*C*	*C*	*C*	*C*	*B*
9. Zinc (*u*)	*C*	*C*	*C*	*C*	*C*	*B*
10. Magnesium and magnesium alloys (chromated) (*b*) (*a*)	*D*	*D*	*D*	*D*	*D*	*C*
11. Stainless steel: Austenitic Fe–18Cr–8Ni	*A*	*A*	*A*	*A*	*A*	*A*
12. Stainless steel: Fe–18Cr–2Ni	*C*	*A* or *C(s)*	*A* or *C(s)*	*A* or *C(s)*	*A*	*A*
13. Stainless steel: 13% Cr	*C*	*C*	*C*	*C*	*B* or *C*	*A*
14. Chromium	*A*	*A*	*A*	*A*	*A*	*A*
15. Titanium	*A*	*A*	*A*	*A*	*A*	*A*
16. Aluminium and aluminium alloys (*n*) (*a*) (*w*)	*D*	*C*	*D(e)*	*D(e)*	*C(k)*	*B* or *C*

*Based on data provided by members of the I.S.M.R.C. Corrosion and Electrodeposition Committee and others, and arranged by Mrs. V. E. Rance.

A. The corrosion of the 'metal considered' is not increased by the 'contact metal'.

B. The corrosion of the 'metal considered' may be slightly increased by the 'contact metal'.

C. The corrosion of the 'metal considered' may be markedly increased by the 'contact metal'. (Acceleration is likely to occur only when the metal becomes wet by moisture containing an electrolyte, e.g. salt, acid, combustion products. In ships, acceleration may be expected to occur under in-board conditions, since salinity and condensation are frequently present. Under less severe conditions the acceleration may be slight or negligible.)

D. When moisture is present, this combination is inadvisable, even in mild conditions, without adequate protective measures.

at bimetallic contacts*

7	8	9	10	11	12	13	14	15	16
…teel …nd …ast …ron	*Cad-mium*	*Zinc*	*Magnesium and magnesium alloys (chromated)*	*Stainless steels* Austenitic Fe–18Cr–8Ni	Fe–18Cr–2Ni	13% Cr	*Chromium*	*Titanium*	*Aluminium and aluminium alloys*
A	*A*	*A*	*A*	*A*	*A*	*A*	*A*	*A*	*A*
A	*A*	*A*	*A*	*A*	*A*	*A*	*A*	*A*	*A*
A	*A*	*A*	*A*	*B* or *C*	*B*	*A*	*B* or *C*	*B* or *C*	*A(e)*
A	*A*	*A*	*A*	*B* or *C*	*B* or *C*	*A*	*B* or *C*	*B* or *C*	*A(e)*
A	*A*	*A*	*A*	*B* or *C*	*B* or *C*	*A*	*B* or *C*	*B* or *C*	*A*
… or *C(r)*	*A*	*A* or *C(r)*	*A*	*B* or *C*	*B* or *C*	*B* or *C*	*B* or *C*	*B* or *C*	*A*
—	*A(m)*	*A(m) (l)*	*A*	*C*	*C*	*C*	*C(k)*	*C*	*B(m)*
C	—	*A*	*A*	*C*	*C*	*C*	*C*	*C*	*B*
C	*B*	—	*A*	*C*	*C*	*C*	*C*	*C*	*C(j)*
D	*B* or *C*	*B* or *C*	—	*C*	*C*	*C*	*C*	*C*	*B* or *C(c)*
A	*A*	*A*	*A*	*(v)*	*A*	*A*	*A*	*A*	*A*
A	*A*	*A*	*A*	*A*	*(v)*	*A*	*A*	*(o)*	*A*
A	*A*	*A*	*C*	*C*	*(v)*	*C*	*C*	*C*	*A*
A	*A*	*A*	*A*	*A*	*A*	*A*	—	*A*	*A*
A	*A*	*A*	*A*	*A*	*A*	*A*	*A*	—	*A*
… or *C*	*A*	*A*	*A(c) (h)*	*B* or *C*	*B* or *C*	*B* or *C*	*B* or *C(d)*	*C*	*(v)*

Notes to Table 1.A2 (*continued from page 215*)

(*j*) When it is not practicable to use other more suitable methods of protection, e.g. spraying with aluminium, zinc may be useful for the protection of steel in contact with aluminium, despite the accelerated attack upon the coating.

(*k*) This statement should not necessarily discourage the use of the 'contact metal' as a coating for the 'metal considered', provided that continuity is good; under abrasive conditions, however, even a good coating may become discontinuous.

(*l*) In most supply waters at temperatures above about 60°C, zinc may accelerate the corrosion of steel.

(*m*) In these cases the 'contact metal' may provide an excellent protective coating for the 'metal considered', the latter usually being electrochemically protected at gaps in the coating.

(*n*) When aluminium is alloyed with appreciable amounts of copper it becomes more noble and when alloyed with appreciable amounts of zinc or magnesium it becomes less noble. These remarks apply to bimetallic contacts and not to inherent corrosion resistance. Such effects are mainly of interest when the aluminium alloys are connected with each other.

(*o*) No data available.

(*p*) In some immersed conditions, the corrosion of copper or brass may be seriously accelerated at pores or defects in tin coatings.

(*q*) In some immersed conditions there may be serious acceleration of the corrosion of soldered seams in copper or copper alloys.

(*r*) When exposed to the atmosphere in contact with steel or galvanised steel, lead can be rapidly corroded with formation of PbO at narrow crevices where the access of air is restricted.

(*s*) Serious acceleration of corrosion of Fe–18Cr–2Ni stainless steel in contact with copper or nickel alloys may occur at crevices where the oxygen supply is low.

(*t*) Normally the corrosion of lead–tin soldered seams is not significantly increased by their contact with the nickel-base alloys, but under a few immersed conditions the seams may suffer enhanced corrosion.

(*u*) The corrosion product on zinc is, in certain circumstances, more voluminous and less adherent than that on cadmium. Where this is known to be the case, it should be borne in mind in making a choice between these two metals.

(*v*) These joints are liable to corrosion in crevices where these are not filled with jointing compound (*see* para. 8 of the Introduction).

(*w*) Corrosion products from iron or steel reaching aluminium, or corrosion products from aluminium reaching iron or steel, may sometimes cause serious local corrosion through oxygen screening or in other ways, even when the total destruction of metal is finished.

1.7B Appendix—Bimetallic Corrosion

The theoretical treatment accorded to this subject in the earlier pages of this section clearly indicates the fallacy of employing potential difference, or difference in electrode potentials between two metals, as a means of quantifying or even predicting the severity of bimetallic corrosion. In the past, attempts to remedy this defect took the form of classification tables such as that of Evans and Rance (Table 1.A2), in which experience of practical behaviour is combined with some electrochemical knowledge to produce an indication of the probable acceleration of corrosion resulting from the bimetallic coupling of a range of metals and alloys. The Evans and Rance table probably constitutes the most useful practical guide to the subject that has appeared in the seventeen years between 1958 and 1975, but it suffers from the inevitable compression of a great deal of information into tabular form and, in any event, it has to be used in conjunction with modifications detailed in the extensive notes that accompany it.

The type of information in Table 1.A2 has now been further refined and extended under the guidance of a committee meeting under the auspices of the British Standards Institute. This body is producing a more substantial guide to be issued before the end of 1975. In the revised document, each metal under consideration is accorded separate and detailed treatment with respect to a wide range of metals and alloys to which it is likely to be coupled. This information is related to the influence of severe, moderate and mild environments. Additionally, there is a section describing the effects of specific complications, e.g. conductivity of the environment, metallurgical condition and area relationship. There is also a section on preventive measures. The reader is advised to contact the British Standards Institution to obtain a copy of this guide.

The realisation that the current flowing between the two members of a bimetallic couple, rather than their potential difference, provides a more realistic measure of the degree of hazard involved, is inherent from the discussion on polarisation curves already presented earlier in this chapter. Because, however, the traditional method of measurement of current flow is necessarily made in series between the coupled elements, this affects the behaviour of the couple, and valid galvanic-current-flow measurement has only become possible with the development of zero-resistance ammeters and potentiostats. Although the first zero-resistance ammeter was developed in

Table 1.B1 Total corrosion rate acceleration factors due to dissimilar metal coupling at 1:1 area ratio in flowing sea-water (mean flow rate about 1–2 m/s)

Coupled metal (wrought form)	*Uncoupled corrosion rate** (mm/y)	*Acceleration factor† due to dissimilar metal coupling with:*	
		Titanium	*Mild steel*
Zinc	0·05	4	10
SIC aluminium	0·008	30	60
Mild steel	0·15	2	1
Lead	0·01	3	3 MS
2% aluminium brass	0·01	3	3 MS
10% aluminium bronze	0·02	3	3 MS
Nickel aluminium bronze	0·015	1	3 MS
Copper	0·03	6	3 MS
90/10 cupro-nickel (1% Fe)	0·02	3	3 MS
Monel 400	0·005	2	3 MS
Stainless steel type 316	0·005	2	3 MS

*Corrosion rates relate to general corrosion only and are average rates obtained over about one year's exposure.
†Acceleration factors quoted for coupled metal corroding unless indicated by suffix (MS) for mild steel corroding.

1951[1] they did not come into more general use until about 1970[2,3]. Fully automatic balancing zero-impedance ammeters are now available commercially and their use has enabled a more accurate assessment of the degree of galvanic hazard to be made. Using this technique, a fresh assessment is being made of the galvanic hazard involved when a range of materials of construction used in marine engineering are coupled in the presence of sea-water[4]. A summary of the results of this investigation is provided in Table 1.B1, in which the degree of acceleration of corrosion is expressed by means of a numerical factor which should be compared with the uncoupled rate of corrosion of each metal or alloy in sea-water, which is obtained under the same experimental conditions. Even this sophisticated treatment does not produce a faultless assessment of the hazard in every circumstance. In practice, uncertainty is almost inevitable in systems subject to varying speeds of liquid flow, and where anyway deep crevices may exist between coupled metals. For example, with aerated sea-water, an increase in the rate of flow almost invariably brings about an increase in the rate of galvanic corrosion and in using the table the higher acceleration factors are more likely to apply under conditions of fast water movement.

In the compilation of the additional information provided in this appendix, reference has been made to various published papers which should be consulted for further elaboration. Thus, in respect of copper-base alloys, see References 5 to 9; for mild steel see 13, for titanium see 10 to 12, for sea-water velocity see 14 and for the area relationship see 15.

REFERENCES

1. Godard, H. P., 'A New Instrument for Measuring Galvanic Corrosion', *Corrosion*, **7**, 93 (1951)
2. Laver, G. and Mansfeld, F., 'Measurement of Galvanic Corrosion Current at Zero External Impedance', *Corrosion*, **26**, 504 (1970)
3. Greene, N. D., France, W. D. (Jr.) and Wilde, B. E., 'Electrode Mounting for Potentiostatic Anodic Polarisation Studies', *Corrosion*, **21**, 275 (1965)
4. Private communication; J. C. Rowlands, Admiralty Materials Laboratory, Poole; I. R. Scholes and D. J. Astley, Imperial Metals Industries Ltd., Birmingham
5. Taylor, A. H., 'The Corrosion Behaviour of Copper and Naval Brass in 0·5 M NaCl Solutions at Ambient Temperature', *J. Electrochem. Soc.*, **118**, 854 (1971)
6. Rowlands, J. C., 'Electrochemical Aspects of Preferential Phase Corrosion in Complex Alloys', *Corrosion Science*, **2**, 89 (1972)
7. Rowlands, J. C., 'Preferential Phase Corrosion of Naval Brass in Sea Water', *Proc. 2nd Int. Cong. Met. Corrosion*, N.A.C.E., Houston, 795 (1966)
8. Upton, B., 'Corrosion Resistance in Sea Water of Medium Strength Aluminium Bronzes', *Corrosion*, **19**, 204 (1963)
9. Upton, B., 'The Corrosion Resistance in Sea Water of Some Soldered and Brazed Joints with Copper', CDL Tech Memo GPJ
10. Cotton, J. B. and Downing, B. P., 'Corrosion Resistance of Titanium to Sea Water', *Trans. Inst. Mar. Eng.*, **69**, 311 (1957)
11. Schlain, D., 'Corrosion Properties of Titanium and its Alloys', Bulletin 619, Bur. of Mines (1964)
12. Shalaby, L. A., 'Galvanic Coupling of Ti with Cu and Al Alloys in Chloride Media', *Corrosion Science*, **11**, 767 (1971)
13. Johnson, K. E. and Abbott, J. S., 'Galvanic Corrosion of Mild Steel when Coupled to Other Materials', Corporate Engineering Laboratory/B.S.C., Report No. CEL/CH/23/73 (1973)
14. Danek, G. J., 'The Effect of Sea Water Velocity on the Corrosion Behaviour of Metals', *Naval Engineers Journal*, No. 763, Oct. (1966)
15. Mansfeld, F., 'Area Relationship in Galvanic Corrosion', *Corrosion*, **27**, 436 (1971)

J. B. COTTON

1.8 Lattice Defects in Metal Oxides

When a metal oxide is in contact with one of its components (metal or oxygen), the condition for thermodynamic equilibrium cannot, in general, be satisfied unless the crystal is non-stoichiometric, i.e. unless it contains an excess of one of the two components. The reason for this is that although energy must be expended in incorporating the excess component, the entropy of the system increases extremely rapidly at first, and then more slowly as the non-stoichiometry increases. Thus the equilibrium condition, namely that the free energy of the system is a minimum, is satisfied only for some finite degree of non-stoichiometry. The thermodynamic functions are shown schematically as functions of α, the degree of non-stoichiometry, in Fig. 1.72. We note at this stage that it is sufficient to discuss the equilibrium between an oxide and oxygen gas because the other case (equilibrium between the oxide and the metal) is then covered by putting the oxygen pressure equal to the dissociation pressure of the oxide.

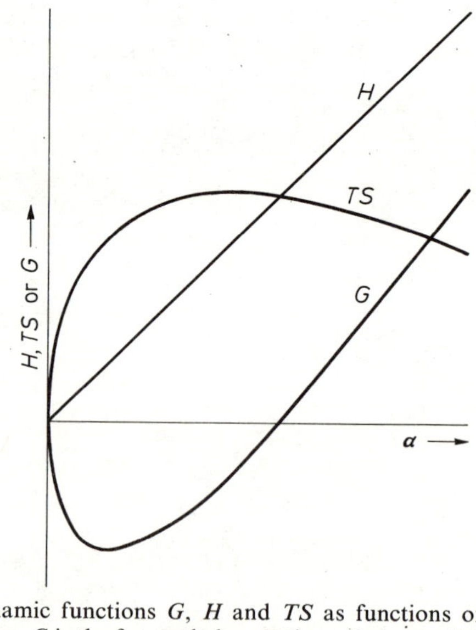

Fig. 1.72 Thermodynamic functions G, H and TS as functions of α, the degree of non-stoichiometry. G is the free enthalpy, H the enthalpy and S the entropy

There are two basic questions which can be decided only by experiments. First, we must know whether the metal or the oxygen is present in excess, and second, we must know how the excess component is incorporated in the oxide lattice. In connection with the latter question we have to remember that a non-stoichiometric crystal remains electrically neutral (except in narrow regions near the surfaces), so that if the excess component is present in the crystal as ions, lattice defects with charges of opposite sign must necessarily be present also (*see* Figs. 1.77 and 1.78). The most important defect structures will be discussed in this section.

The presence in an oxide of an excess of one component provides a mechanism for the transport of material. This transport mechanism, which is vital in understanding the formation of a continuous oxide film on a metal, is also discussed in this section. An important feature here is that an excess of one component may provide a transport mechanism, not for itself, but for the other component.

p-type Oxides

Cu_2O contains excess oxygen which is taken up in such a way as to build up new layers of the oxide. Thus, the excess oxygen is present as O^{2-} ions on their normal lattice sites. To form oxygen ions, electrons are needed, and to build up new layers of the crystal, Cu^+ ions are needed. Both species, electrons and Cu^+ ions, are supplied from the interior of the crystal. Thus excess oxygen is incorporated in Cu_2O by forming vacant cation sites ($Cu^+\square$), and vacant electron levels ($e\square$) in the crystal. The vacant electron levels are called *positive holes*, and in Cu_2O they may be pictured as Cu^{2+} ions at normal cation sites in the lattice. The cation vacancies are negative charges in the Cu_2O lattice, while the positive holes are positive charges. Consequently, they attract each other, and at low temperatures may stick together. We shall assume that the temperature is high enough for this trapping of positive holes by cation vacancies to be ignored. The chemical reaction giving non-stoichiometric Cu_2O may then be written

$$\tfrac{1}{2}O_2 \rightarrow Cu_2O + 2(Cu^+\square) + 2(e\square) \qquad \ldots(1.134)$$

four cation vacancies and four positive holes being produced by every oxygen molecule absorbed. The non-stoichiometric oxide conducts electricity with the movement of positive holes, and because of this the conductivity is said to be *p*-type (positive carriers).

If n is the concentration of defects (cation vacancies or positive holes) at equilibrium, then, applying the law of mass action to equation 1.134

$$n = p^{1/8}K^{1/4} \qquad \ldots(1.135)$$

where p is the oxygen pressure, and K the equilibrium constant for the reaction. The formula for K can be found only from statistical mechanics, and a simple calculation is instructive. The important quantity which we need to estimate is the entropy change when one molecule of oxygen is absorbed, with the formation of four defects in the lattice. This entropy change may be divided into two parts. The first is the entropy change, ΔS_1, when one molecule is absorbed, with the formation of defects at specified lattice points; the second is the entropy change, ΔS_2, when we allow for the fact that the defects may be formed anywhere in the crystal. ΔS_1 is independent of the existing defect concentration, and, in addition, is almost certainly negative because it involves the loss of translational and rotational degrees of freedom of the oxygen molecule. ΔS_2 is positive and depends on the existing defect concentration. To calculate ΔS_2 we proceed as follows. Let there be N cation sites in the crystal and n defects. The number of arrangements, P, of the cation vacancies on the lattice sites is

$$P = N!/n!(N-n)! \qquad \text{...(1.136)}$$

which contributes an entropy $S_c = k\,\ell n\, P$, where k is Boltzmann's constant. Taking logarithms in equation 1.136, using Stirling's approximation

$$(\ell n\, x! \approx x\,\ell n\, x - x \text{ as } x \to \infty)$$

and putting $n/N = \alpha$ we find

$$S_c = -Nk[(1-\alpha)\,\ell n\,(1-\alpha)+\alpha\,\ell n\,\alpha] \qquad \text{...(1.137)}$$

We note that S_c is positive and goes through a maximum as α increases. If the positive holes were localised on the cations, they would give an entropy contribution S_p exactly equal to S_c. The positive holes have, however, considerable mobility (see below), and are perhaps best treated as an ideal gas consisting of particles of effective mass m. In this case[1]

$$S_p = Nk\alpha[\tfrac{3}{2}\,\ell n\,(\phi\Omega^{2/3})+\tfrac{5}{2}-\ell n\,\alpha+\ell n\, 2] \qquad \text{...(1.138)}$$

where $\phi = 2\pi mkT/h^2$, h being Planck's constant. Ω is the volume of the oxide per metal ion, and the term $Nk\alpha\,\ell n\, 2$ takes account of electron spin. We note that S_p is positive but shows no maximum. Adding equations 1.137 and 1.138, and differentiating with respect to α, we obtain ΔS_2. Remembering that $\alpha \ll 1$, we find

$$\Delta S_2 = k[6\,\ell n\,(\phi\Omega^{2/3})+6+4\,\ell n\, 2-8\,\ell n\,\alpha]$$

showing that ΔS_2 is large and positive for small values of α.

ΔS, the entropy change in equation 1.134, is given by

$$\Delta S = \tfrac{1}{2}(\Delta S_1+\Delta S_2)$$

Since ΔS_1 is independent of α, we see that ΔS becomes positive and increases without limit as $\alpha \to 0$. Thus, although the enthalpy change ΔH in equation 1.134 may be large and positive, the equilibrium condition $\Delta G = \Delta H - T\Delta S = 0$ is satisfied for some value of α different from zero except at $T = 0$. This proves that Cu_2O is non-stoichiometric at any temperature above the absolute zero.

Regarding ΔS_1, the simplest assumption we can make is that it comes entirely from the loss of translational and rotational degrees of freedom of the oxygen molecule when it is absorbed into the crystal. A standard calculation then gives[1] (the electronic ground state of O_2 is a triplet)

$$\Delta S_1 = -k[25+(7/2)\,\ell n\,(T/298)-\ell n\,(p/\text{atm})-\ell n\, 3]$$

and equating ΔH and $T\Delta S$ to find the equilibrium condition we obtain

$$\alpha = (9{\cdot}3\times 10^{-2})(298/T)^{7/16}(48\phi^6\Omega^4)^{1/8}(p/\text{atm})^{1/8}\exp(-\Delta H/4kT)$$

Comparison with equation 1.135 would give the formula for K, but because of the approximate nature of the formula for ΔS_1 we cannot rely on the factor $(9{\cdot}3\times10^{-2})(298/T)^{7/16}$.

Other p-type oxides are NiO, FeO, CoO, Ag_2O, MnO and SnO. For NiO the reaction giving non-stoichiometry is

$$\tfrac{1}{2}O_2 \rightarrow NiO+(Ni^{2+}\square)+2(e\square)$$

and the concentration of positive holes is twice that of the cation vacancies. From the mass-action formula, the concentration of positive holes, n_p, is given by

$$n_p = 2^{1/3}p^{1/6}K^{1/3}$$

For all p-type oxides, the defect concentration, and hence the electrical conductivity, increases with the oxygen pressure.

n-type Oxides

ZnO contains excess metal which is accommodated interstitially, i.e. at positions in the lattice which are unoccupied in the perfect crystal. The process by which ZnO in oxygen gas acquires excess metal may be pictured as follows*. The outer layers of the crystal are removed, oxygen is evolved, and zinc atoms go into interstitial positions in the oxide. We represent interstitial zinc by $(Zn\bigcirc)$. However, the interstitial zinc atoms may ionise to give $(Zn^{+}\bigcirc)$ or even $(Zn^{2+}\bigcirc)$. The extra electrons produced in this way must occupy electron levels which would be vacant in the perfect crystal. We represent them by the symbol $(e\bigcirc)$, and refer to them as free electrons. They can be pictured as Zn^{+} ions at normal cation sites. We see therefore that three reactions can be written, each giving non-stoichiometric ZnO:

$$ZnO \rightarrow \tfrac{1}{2}O_2+(Zn\bigcirc) \qquad \ldots(1.139)$$

$$ZnO \rightarrow \tfrac{1}{2}O_2+(Zn^{+}\bigcirc)+(e\bigcirc) \qquad \ldots(1.140)$$

$$ZnO \rightarrow \tfrac{1}{2}O_2+(Zn^{2+}\bigcirc)+2(e\bigcirc) \qquad \ldots(1.141)$$

The exact situation with ZnO is not altogether clear. Under most experimental conditions it seems that equation 1.140 is the important reaction, but equation 1.141 cannot be ignored at high temperatures. Applying the mass-action formula to equation 1.140 we have for n, the concentration of defects (interstitial Zn^{+} ions or free electrons)

$$n = p^{-1/4}K^{1/2}$$

where K is the equilibrium constant of the reaction. The non-stoichiometric oxide conducts electricity with the movement of free electrons, and because of this the conductivity is said to be n-type (negative carriers).

Other n-type oxides are TiO_2, CdO, Al_2O_3 and V_2O_5. For TiO_2, the reaction giving the non-stoichiometry is[2]

$$\tfrac{1}{2}TiO_2 \rightarrow \tfrac{1}{2}O_2+(O^{2-}\square)+2(e\bigcirc)$$

*The mechanism is of no consequence in determining the equilibrium situation.

and the defects are anion vacancies and free electrons (Ti^{3+} ions). From the mass-action formula, the concentration of free electrons n_e is given by*

$$n_e = 2^{1/3} p^{-1/6} K^{1/3}$$

For all *n*-type oxides, the defect concentration, and hence the electrical conductivity, decreases with the oxygen pressure.

Motion of Lattice Defects

Lattice defects do not occupy fixed positions in the crystal. In Cu_2O for example, a positive hole moves about because an electron from a neighbouring Cu^+ ion can fall into the vacant level which the positive hole represents [Fig. 1.73 (top)]. Likewise, the cation vacancy moves when a neighbouring cation jumps into the vacant site [Fig. 1.73 (bottom)]. We expect the positive hole to be the more mobile because its motion is an electronic process, not an ionic one. This expectation is confirmed by measurements of the defect mobilities. The mobility, v, of a charged defect is the drift velocity acquired by the defect in an applied electric field of unit strength. It is related to the conductivity, σ, by $\sigma = nev$, where n is the concentration and e the charge of the defect, and to the diffusion coefficient D by Einstein's relation $v/D = e/kT$ (*see* Section 1.9). For Cu_2O at 1000°C, Wagner and Hammen[3] found for the hole mobility $v_p \approx 10\,\mathrm{cm^2\,s^{-1}\,V^{-1}}$, and for the mobility of cation vacancies, $v_c \approx 5 \times 10^{-4}\,\mathrm{cm^2\,s^{-1}\,V^{-1}}$. The temperature dependencies of v_p and v_c are, however, fundamentally different from one another, as with decreasing temperature v_c falls but v_p increases. At room temperature[4], v_p is about $80\,\mathrm{cm^2\,s^{-1}\,V^{-1}}$.

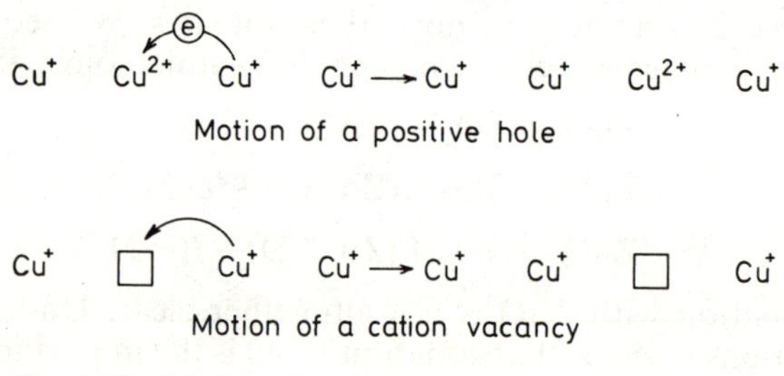

Fig. 1.73 Motion of positive holes and cation vacancies

The reason for this can be seen as follows. In a perfect crystal with the ions held fixed, a positive hole would move about like a free particle with a mass m depending on the nature of the crystal. In an applied electric field, the hole would be uniformly accelerated, and a mobility could not be defined. The existence of a mobility in a real crystal derives from the fact that the uniform acceleration is continually disturbed by deviations from a perfect lattice structure. Among such deviations, the thermal motions of the ions, and in particular, the longitudinal polarisation vibrations, are most important in obstructing the uniform acceleration of the hole. Since the amplitude of the lattice vibrations increases with temperature, we see how the mobility of a

*This assumes that anion vacancies are present to a significant extent only in the non-stoichiometric oxide, which is not in fact true for TiO_2.

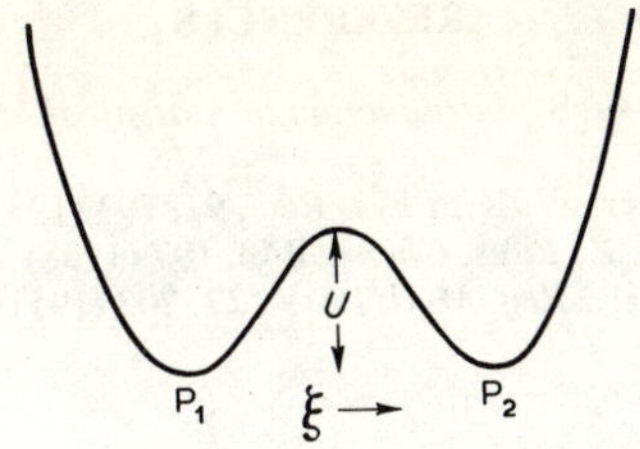

Fig. 1.74 Energy of the crystal as a function of ξ, the co-ordinate of the moving cation

hole decreases as the temperature is raised. On the other hand, a cation vacancy can move only by virtue of the thermal motions of the neighbouring cations so that the vacancy becomes more mobile as the temperature is raised. A parallel argument can of course be given for *n*-type oxides which contain free electrons and either interstitial cations or anion vacancies.

When a metal oxidises to form a continuous oxide film, the transport of material through the oxide is a necessary step in the oxidation process. From what has already been said, it is clear that the slow process will almost certainly be the transport of ions (vacancies) not electrons (positive holes). Let us therefore examine the temperature dependence of the diffusion coefficient (and hence the mobility) of a cation vacancy. Let q be the probability per second that a cation next to a vacancy jumps into the vacancy. When this happens, the vacancy moves a distance a equal to the cation–cation distance. The diffusion coefficient D_c is by definition given by $D_c = qa^2$. Hence we need a formula for q. In Fig. 1.74 we show the energy of the lattice as a function of ξ, the co-ordination of the moving cation. When the cation is at P_1, the vacancy is at P_2, and vice versa. Between P_1 and P_2 there is a symmetrical barrier of height U, and the cation can pass from P_1 to P_2 only if it acquires energy greater than U. This energy is of course supplied by the lattice vibrations. The probability that the ion has this energy is proportional to the usual Boltzmann factor, $\exp(-U/kT)$, so we can write

$$q = q_o \exp(-U/kT)$$

q_o, which has dimensions of s^{-1}, can be calculated only by a detailed analysis of the jumping process, but this elementary consideration is sufficient to establish the characteristic formulae

$$\left.\begin{aligned} D_c &= D_o \exp(-U/kT) \qquad & D_o &= q_o a^2 \\ v_c &= v_o \exp(-U/kT) \qquad & v_o &= D_o e/kT = q_o a^2 e/kT \end{aligned}\right\} \quad \ldots(1.142)$$

showing the strong temperature dependencies of D_c and v_c. The *self*-diffusion coefficient of Cu^+ in Cu_2O is, of course, $(n/N)D_c$. Naturally, we can also derive relations like equation 1.142 for interstitial ions.

Finally we notice that in the *p*-type oxides Cu_2O and NiO, the presence of excess oxygen actually provides, through the formation of cation vacancies, a transport mechanism for the metal, while in an *n*-type oxide like TiO_2, the excess metal, by forming anion vacancies, provides a transport mechanism for oxygen. With *n*-type oxides like ZnO and Al_2O_3, where the excess metal is accommodated interstitially, a transport mechanism is, of course, provided for the excess component itself.

T. B. GRIMLEY

REFERENCES

1. See for example Rushbrooke, G. S., *Introduction to Statistical Mechanics*, Oxford University Press, 135 (1949)
2. Breckenridge, R. G. and Hosler, W. R., *Phys. Rev.*, **91**, 793 (1953)
3. Wagner, C. and Hammen, H., *Z. Phys. Chem.*, **B40**, 197 (1938)
4. See for example Brattain, W. H., *Rev. Mod. Phys.*, **23**, 203 (1951)

BIBLIOGRAPHY

Kröger, T. A. and Vink, H. J., *Solid State Phys.*, **3**, 310 (1956)

Van Gool, W., *Principles of Defect Chemistry of Crystalline Solids*, Academic Press, London and New York (1966)

1.9 Continuous Oxide Films

When a solid metal is attacked by oxygen gas, the product of the reaction is the metal oxide which, if it is not volatile, builds up as a surface layer on the metal. The oxide layer may be protective or non-protective. A non-protective layer does not inhibit the continued access of oxygen to the unchanged metal; the rate of growth of such an oxide layer is independent of its thickness X and the law of growth is $\mathrm{d}X/\mathrm{d}t = k_1$. On integration this gives the linear law

$$X - X_0 = k_1 t$$

and k_1 is the *linear rate constant*. Discontinuous films are considered in Section 1.10.

A protective oxide layer forms a continuous barrier between the reactants (oxygen and metal), which inhibits the reaction. The simplest assumption that can be made about the effectiveness of this barrier is that its protecting power is directly proportional to its thickness. Mathematically, $\mathrm{d}X/\mathrm{d}t = k_2/X$, which on integration gives the parabolic law

$$X - X_0^2 = 2k_2 t$$

where k_2 is the *parabolic rate constant*. This law is obeyed for the high-temperature oxidation of many metals.

Since a protective layer separates the primary reactants, further growth of the layer involves two essential steps: (*a*) surface reactions at the metal/oxide and oxide/oxygen interfaces and (*b*) transport of material through the oxide. The rate of growth will be controlled by whichever of these two steps is the slower. If the layer is thin enough, a surface reaction must control the rate. Indeed, during the initial stages of oxide formation step (*b*) is absent. As the layer thickens, step (*b*) may be distinguished as a definite process which ultimately controls the rate. When this stage is reached, the surface reactions, by which both metal and oxygen are incorporated into the oxide, proceed to near equilibrium. The products of these reactions are then transported through the oxide by a driving force derived from the free energy change in the oxidation reaction. The first problem is to decide the nature of these reaction products.

Surface Reactions

Metal oxides are not normally stoichiometric, although the non-stoichiometry may be too small to be detected by ordinary methods of analysis (Section 1.8).

Cuprous oxide, for example, absorbs oxygen to build up new layers of the oxide with the formation of vacant cation sites ($Cu^+\square$) and positive holes ($e\square$).

The reaction of Cu_2O with oxygen may be written

$$\tfrac{1}{2}O_2 \rightarrow Cu_2O + 2(Cu^+\square) + 2(e\square) \qquad \text{...(1.143)}$$

two cation vacancies and two positive holes being formed for each oxygen atom absorbed. If $n(Cu^+\square)$ and $n(e\square)$ are the concentrations of cation vacancies and positive holes when thermodynamic equilibrium is established, the usual mass action formula is

$$n^2(Cu^+\square)n^2(e\square) = p^{1/2}K \qquad \text{...(1.144)}$$

where p is the oxygen pressure and K the equilibrium constant. Except in narrow regions near the surfaces where space charges can exist, the oxide must be electrically neutral. In the interior, therefore,

$$n(Cu^+\square) = n(e\square) = n$$

say, and

$$n = p^{1/8}K^{1/4} \qquad \text{...(1.145)}$$

The electrical conductivity is proportional to n. Equation 1.145 therefore predicts an electrical conductivity varying as $p^{1/8}$. Experimental results show proportionality to $p^{1/7}$, and this discrepancy is probably due to incomplete disorder of cation vacancies and positive holes. An effect of this sort (deviation from ideal thermodynamic behaviour) is not allowed for in the simple mass action formula of equation 1.144.

Consider now the system Cu/Cu_2O in oxygen gas at a pressure p_X (X signifies the oxide/oxygen interface in Fig. 1.75). Ignoring space charges, n_X the equilibrium concentration of cation vacancies or positive holes at the Cu_2O/O_2 interface, is given by

$$n_X = p_X{}^{1/8}K^{1/4} \qquad \text{...(1.146)}$$

and n_0 their equilibrium concentration at the Cu/Cu_2O interface, by

$$n_0 = p_0{}^{1/8}K^{1/4} \qquad \text{...(1.147)}$$

In equation 1.147 p_0 is the dissociation pressure of Cu_2O (oxygen pressure for the equilibrium $2Cu + \tfrac{1}{2}O_2 \rightleftharpoons Cu_2O$). If $p_X > p_0$, then $n_X > n_0$, and the oxidation proceeds with cation vacancies and positive holes being created at the Cu_2O/O_2 interface and moving inwards to be destroyed at the Cu/Cu_2O interface. A similar situation exists whenever a p-type oxide is formed, for example with Ni/NiO.

As an example of a different type of oxide, we may consider ZnO. This oxide evolves oxygen and forms cations in interstitial positions ($Zn^+\bigcirc$) or ($Zn^{2+}\bigcirc$), and free electrons ($e\bigcirc$). If the interstitial zinc ions are only singly charged, the reaction describing the non-stoichiometry may be written

$$ZnO \rightarrow \tfrac{1}{2}O_2 + (Zn^+\bigcirc) + (e\bigcirc)$$

Using the simple mass action formula, the equations corresponding to 1.146 and 1.147 are

$$\left.\begin{aligned} n_X &= p_X^{-1/4}K^{1/2} \\ n_O &= p_O^{-1/4}K^{1/2} \end{aligned}\right\} \qquad \text{...(1.148)}$$

with p_O equal to the dissociation pressure of ZnO. If $p_X > p_O$ then $n_X < n_O$, and the oxidation proceeds with interstitial cations and free electrons moving outwards from the Zn/ZnO interface. A similar situation exists whenever the oxidation product is an *n*-type oxide. The system Al/Al_2O_3 is another example.

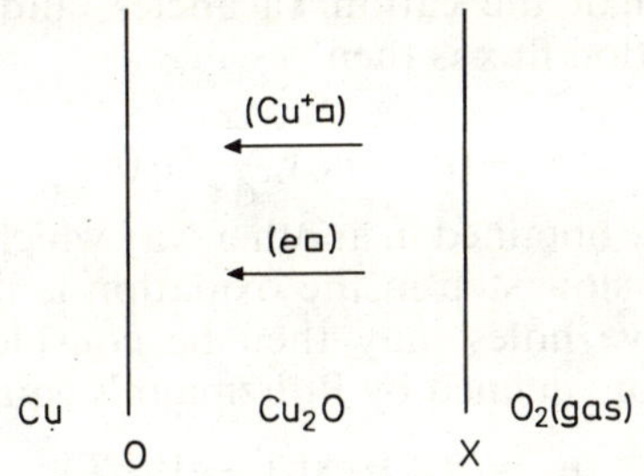

Fig. 1.75 Movement of defects in the oxidation of copper

Wagner's Theory of the Parabolic Law

For definiteness, the oxidation of copper to copper(I) oxide may be considered. Our picture of the process is that cation vacancies and positive holes formed at the Cu_2O/O_2 interface by equation 1.143 are transported to the Cu/Cu_2O interface where they are destroyed by copper dissolving in the non-stoichiometric oxide. We require an expression for the rate of oxidation.

We denote by x the distance from the metal surface, and by $n_c(x)$ and $n_p(x)$ the concentrations of cation vacancies and positive holes in the oxide. Let v_c and v_p be their mobilities, and D_c and D_p their diffusion coefficients. Let $F(x)$ be the electrostatic field in the oxide. J_c, the flux of cation vacancies (number crossing unit area per second), will be expressed by

$$J_c = -D_c\frac{dn_c}{dx} - n_c v_c F \qquad \text{...(1.149)}$$

and J_p, the flux of positive holes, by

$$J_p = -D_p\frac{dn_p}{dx} + n_p v_p F \qquad \text{...(1.150)}$$

In equation 1.149 the first term on the right-hand side is the flux due to a concentration gradient, the second is that due to the electric field. It should be noted that cation vacancies are negatively charged carriers of electricity. The terms in equation 1.150 have similar meanings; positive holes are positively charged carriers. If the system were at equilibrium we would have $J_c = J_p = 0$ and the field F would be negligible except in narrow regions near the two interfaces, but when oxidation is proceeding this is not generally the case.

In a steady state of oxidation $J_p = J_c = J$, say. Eliminating F from equations 1.149 and 1.150 and using Einstein's relation (see below) $v/D = e/kT$, where k is Boltzmann's constant and T the absolute temperature,

$$J = -(kT/e)[n_c v_c n_p v_c/(n_c v_c + n_p v_p)]\frac{d}{dx}\ell n\,(n_c n_p) \qquad ...(1.151)$$

The negative sign means that cation vacancies and positive holes move inwards, i.e. in the negative direction of x. For Cu_2O, the positive holes are much more mobile than the cation vacancies, and we can assume that $n_p v_p \gg n_c v_c$. The oxidation flux is then

$$J = -(kT/e)n_c v_c \frac{d}{dx}\ell n\,(n_c n_p) \qquad ...(1.152)$$

This equation can be obtained in another way which may be more instructive. Assume that the slow step in the oxidation is the transport of cation vacancies. The positive holes may then be considered to take up their equilibrium distribution, defined by Boltzmann's equation

$$n_p = n_p(0)\exp(-eV/kT) \qquad ...(1.153)$$

Here n_p is the concentration of positive holes at any point in the oxide where the electrical potential is V, and $n_p(0)$ is their concentration at the Cu/Cu_2O interface where $V = 0$. Differentiating equation 1.153, we obtain an equation for the electric field, namely

$$F = -\frac{dV}{dx} = (kT/e)\frac{d}{dx}\ell n\, n_p \qquad ...(1.154)$$

Substituting this in equation 1.149 and using Einstein's relation between the mobility and the diffusion coefficient of cation vacancies, we obtain equation 1.152 for the oxidation flux J_c. This derivation shows that equation 1.152 is valid if all processes involving positive holes (and therefore electrons) are so fast that the oxidation flux can be carried without significantly disturbing their equilibrium distribution. We note at this stage that to derive Einstein's relation it is only necessary to compare equation 1.153 with the equation obtained by integrating equation 1.150 for $J_p = 0$. The integration gives

$$n_p = n_p(0)\exp(-v_p V/D_p)$$

and this establishes the result $v_p/D_p = e/kT$. The corresponding result for cation vacancies is derived similarly.

Returning now to equation 1.152, we cannot in general proceed without knowing how n_c and n_p vary through the oxide. However, if the oxide layer is thick enough, the situation is simple, for we can assume that electrical neutrality is preserved in the oxide except in narrow regions near the two interfaces where there are space charges. Thus we have $n_c = n_p = n$ say, except in the space-charge regions which, however, we can neglect if the oxide layer is thick enough. Then equation 1.152 gives

$$J = -2(kT/e)v_c\frac{dn}{dx} \qquad ...(1.155)$$

However, J must be independent of x, so this equation can be integrated to give

$$JX = -2(kT/e)v_c(n_X - n_0) \qquad ...(1.156)$$

where X is the thickness of the oxide layer, and n_X and n_0 are the concentrations of cation vacancies or positive holes at the two interfaces. These concentrations are of course given by equations 1.146 and 1.147.

Our picture of the transport process in these thick oxide layers is that there is a uniform concentration gradient of defects (cation vacancies and positive holes) across the layer. But it is important to notice that the oxidation flux is exactly twice that to be expected if diffusion alone were responsible for the transport of cation vacancies. The reason for this is, of course, that the more mobile positive holes set up an electric field which assists the transport of the slower-moving cation vacancies.

If Ω is the volume of the oxide per metal atom, the rate of growth, dX/dt, is equal to $|J|\Omega$. Thus from equation 1.156 we derive the parabolic law

$$\left.\begin{aligned} dX/dt &= k_2/X \\ k_2 &= 2D_c\Omega(n_X - n_0) \end{aligned}\right\} \qquad ...(1.157)$$

This formula for k_2 can be cast into another form by using equations 1.146 and 1.147. We note first that in these latter equations $K^{1/4}$ is the concentration of defects in Cu_2O at 1 atm pressure of oxygen, so that $(K^{1/4}D_c\Omega)$ is the self-diffusion coefficient of Cu^+ in Cu_2O at this oxygen pressure. Call this self-diffusion coefficient D_c^0, then

$$k_2 = 2D_c^0(p_X^{1/8} - p_0^{1/8}) \qquad ...(1.158)$$

This equation shows that if $p_X \gg p_0$, k_2 depends upon the oxygen pressure and the temperature in the same way as the self-diffusion coefficient of Cu^+ in Cu_2O. Regarding the temperature dependence, self-diffusion coefficients in solids depend exponentially on the temperature (*see* Section 1.8), i.e.

$$D_c^0 = A\exp(-Q/kT)$$

and Q is the *activation energy* for self-diffusion. Thus for the temperature dependence of k_2 we have

$$k_2 = B\exp(-Q/kT)$$

and the activation energy for the oxidation reaction should be the same as that for self-diffusion of cations in the oxide.

The above account of the oxidation of Cu to Cu_2O is a simplified version of the more general theory developed by Wagner[1,2]. Cu_2O is a *p*-type oxide. As an example of a system where an *n*-type oxide is formed, we shall consider the oxidation of Zn to ZnO. Here Zn dissolves in the oxide at the Zn/ZnO interface to give interstitial cations and free electrons. These defects cross to the ZnO/O_2 interface and react with oxygen to build up new layers of the oxide. The slow step in the oxidation is the transport of interstitial cations, and if these are singly charged we still have equation 1.156 for the flux, except that v_c is the mobility of interstitial cations and n_X and n_0 are given by equation 1.148. We note further that J is positive in this case because defects move outwards. The expression for the parabolic rate constant corresponding to equation 1.158 is

$$k_2 = 2D_c^0(p_0^{-1/4} - p_X^{-1/4})$$

and if $p_X \gg p_0$, the oxidation rate should be *independent* of the oxygen pressure. This pressure independence should be noted since the self-diffusion coefficient of Zn in ZnO does depend on the oxygen pressure. Further, the temperature dependence of k_2 is *not* simply that of the self-diffusion coefficient, since p_0, the dissociation pressure of ZnO, is also temperature dependent. Because of this the activation energy for the oxidation reaction is less than that for self-diffusion by half the heat of formation of ZnO.

Thin Oxide Films

Wagner's theory of the parabolic law involves the following assumptions:

1. The oxide layer is compact and adherent.
2. The slow step is the transport of material through the oxide layer.
3. The layer is so thick that the space-charge regions at the two interfaces are unimportant and the oxide can be regarded as electrically neutral.

In the early stages of oxidation when the oxide layer is thin, it is clear that assumption (3) must be invalid. The limiting simple case when the layer is so thin that space charges can be neglected because they are small compared with the surface charges has been considered by Mott[3]. All other assumptions are the same as in Wagner's theory. In the oxidations of Cu, for example, we assume that electronic equilibrium is established in the system $Cu/Cu_2O/O_2$. This sets up an electrical potential difference across the oxide layer because electrons are transferred from the metal to form oxygen ions adsorbed on the outer surface of the oxide. If the surface charge formed in this way is large compared with the space charge in the oxide, the electric field is uniform and equal to V/X where V is the potential drop across the oxide. In a thin oxide layer this field may be very large. For example, a potential difference of 1 V gives a field of 10^5 V/cm in a layer 100 nm thick. Now the flux of cation vacancies (and hence the oxidation flux) is given by equation 1.149, and if F is as large as this, a significant oxidation rate is to be expected even at ordinary temperatures where the diffusion coefficient is very small. For the system Cu/Cu_2O the theory gives a cubic law of growth ($dX/dt = k_3/X^2$) at ordinary temperatures, and a parabolic law at high temperatures. The parabolic rate constant is, however, entirely different from that in Wagner's theory. When the oxidation product is an *n*-type oxide like ZnO or Al_2O_3, the law of growth is parabolic both at ordinary temperatures and at high temperatures. The two rate constants are different, and both differ from that in Wagner's theory. For further details, the original papers[3–6] should be consulted.

Very Thin Oxide Layers

Many metals oxidise rapidly at first when exposed to oxygen at sufficiently low temperatures, but after a few minutes, when a very thin oxide layer has been formed, the reaction virtually ceases. Oxide layers formed in this way are about 5 nm thick. Aluminium and chromium are well-known examples, showing this sort of behaviour at room temperature. A theory of the effect has been proposed by Mott[3,4].

For definiteness consider the system Al/Al_2O_3. It is assumed as before that electronic equilibrium is established so that there is a field in the oxide associated with the presence of oxygen ions adsorbed on the outer surface. In a very thin layer this field will be enormous (about 10^6 V/cm if $X = 10$ nm), and we cannot assume, as we did in equation 1.149, that the contribution which this field makes to the flux of cations is simply proportional to the field. An investigation of the transport process in such strong fields (see below) shows that the flux increases exponentially with the field, and because of this cations are transported through the oxide much more rapidly than would be expected on the basis of equation 1.149. It seems unlikely therefore that cation transport can be the slow step in the reaction, and the rate will be controlled instead by a surface reaction. For the system Al/Al_2O_3, the slow step is probably that by which Al^{3+} ions enter interstitial positions in the oxide at the Al/Al_2O_3 interface. The rate of this process is also influenced by the strong electric field. The potential energy diagram for an ion leaving the metal and entering the oxide is shown in Fig. 1.76. P represents an ion in the metal surface and I_1, I_2, are interstitial positions in the oxide. The height

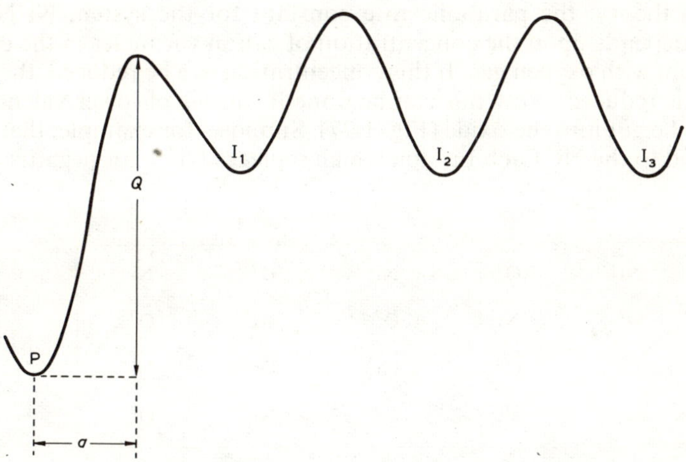

Fig. 1.76 Potential energy of an interstitial ion near the metal/oxide interface

of the barrier between P and I_1 is Q when the field is zero. The field lowers this barrier by an amount $zeaF$ where ze is the charge on the ion and a the distance between P and the top of the barrier. The probability per second that an ion jumps from P to I_1 in the absence of a field is $\nu \exp(-Q/kT)$, where ν is the vibration frequency of an ion at P. In a field F, this probability is increased by the factor $\exp(zeaF/kT)$. Hence if $\mathscr{R}$ is the rate at which ions can enter the oxide when there is no field, the rate in the field is*

$$\mathscr{R} \exp(zeaF/kT)$$

If this process determines the oxidation rate, then with $F = V/X$, the law of growth is

*An argument similar to this establishes the exponential relation between the cation flux *through* the oxide and the field F.

$$\left.\begin{aligned} dX/dt &= \Omega\mathscr{R}\exp(\lambda/X) \\ \lambda &= zeaV/kT \end{aligned}\right\} \qquad \ldots(1.159)$$

If $V = 1$ V, $a = 0{\cdot}25$ nm, and $z = 3$, $\lambda = 30$ nm at 300 K, so that for a film 1 nm thick, the field increases the rate of growth by a factor of about 10^{13}. The term in the growth law due to the field, namely $\exp(\lambda/X)$, is large only when X is small. Because of this a thin oxide film can form even at low temperatures where $\mathscr{R}$, the ordinary rate of entry of ions into the oxide, is negligible. As the film thickens, the factor $\exp(\lambda/X)$ decreases rapidly, and the rate of growth soon falls to such a low value that, for practical purposes, oxidation has ended.

Effects of Alloying

An important aspect of any theory of the oxidation of a pure metal is that it enables us to see how the protective power of the oxide layer can be altered by the introduction of alloying constituents into the metal. According to Wagner's theory, the parabolic rate constant for the system Ni/NiO for example depends upon the concentration of cation vacancies in the oxide in equilibrium with oxygen gas. If this concentration can be reduced, the oxidation rate is reduced. Now this can be done if cations of lower valency than Ni^{2+} can be got into the oxide (Fig. 1.77). Suppose, for example, that a little Li is added to the Ni. Each Li^+ ion which replaces Ni^{2+} is a negative charge

NiO

O^{2-}	Ni^{2+}	O^{2-}	Ni^{3+}	O^{2-}	Ni^{2+}
Ni^{2+}	O^{2-}	Ni^{2+}	O^{2-}	Ni^{2+}	O^{2-}
O^{2-}	□	O^{2-}	Ni^{3+}	O^{2-}	Ni^{2+}
Ni^{2+}	O^{2-}	Ni^{2+}	O^{2-}	Ni^{2+}	O^{2-}

$NiO + Li_2O$

O^{2-}	Li^{+}	O^{2-}	Ni^{3+}	O^{2-}	Ni^{2+}
Ni^{2+}	O^{2-}	Ni^{2+}	O^{2-}	Ni^{2+}	O^{2-}
O^{2-}	Li^{+}	O^{2-}	Ni^{3+}	O^{2-}	Ni^{2+}
Ni^{2+}	O^{2-}	Ni^{2+}	O^{2-}	Ni^{2+}	O^{2-}

$NiO + Cr_2O_3$

O^{2-}	Ni^{2+}	O^{2-}	Ni^{3+}	O^{2-}	Ni^{2+}
Cr^{3+}	O^{2-}	Ni^{2+}	O^{2-}	Ni^{2+}	O^{2-}
O^{2-}	□	O^{2-}	Ni^{3+}	O^{2-}	Ni^{2+}
Ni^{2+}	O^{2-}	Cr^{3+}	O^{2-}	□	O^{2-}

Fig. 1.77 Effects of Li_2O and Cr_2O_3 on the defect structure of NiO

in the NiO lattice. To preserve electrical neutrality, one positive hole ($e\square$) must be created for each Li^+ ion introduced. But the product $n(Ni^{2+}\square)$ $n^2(e\square)$ is fixed by the reaction governing the non-stoichiometry of NiO. Hence $n(Ni^{2+}\square)$ falls and the oxidation rate is reduced. By a similar argument, an alloying constituent of higher valency than Ni^{2+} (Cr^{3+} for example) which enters the oxide layer in place of Ni^{2+} increases the oxidation rate.

When the oxidation product is an *n*-type oxide like ZnO, the conditions are reversed (Fig. 1.78). If a monovalent ion like Li^+ enters the oxide layer in place of Zn^{2+}, one free electron ($e\bigcirc$) is destroyed. But the product $n(Zn^+\bigcirc)n(e\bigcirc)$ is fixed by the reaction governing the non-stoichiometry of ZnO. Hence $n(Zn^+\bigcirc)$, the concentration of interstitial Zn^+ ions, increases, and the oxidation rate, which depends upon the concentration of these ions in the oxide in equilibrium with metallic Zn, increases.

This simple account of the effect of alloying constituents is valid only if the second metal shares in the oxide formation by dissolving freely in the oxide of the basis metal. Further, the second metal should be present in the oxide layer in such low concentrations that it can be regarded as an impurity in the oxide of the basis metal. If the alloying constituent is insoluble in the oxide of the original metal, or if a new phase, for example a spinel, is formed, the discussion fails. The spinel $NiCr_2O_4$ is in fact formed in the oxidation of Ni–Cr alloys when the Cr content is high enough, and the oxidation rate is decreased, not increased as we would expect from the simple discussion above.

To examine the situation with alloys in a little more detail, the Cu–Ni alloys will first be considered. Here the mutual solubility of the two oxides

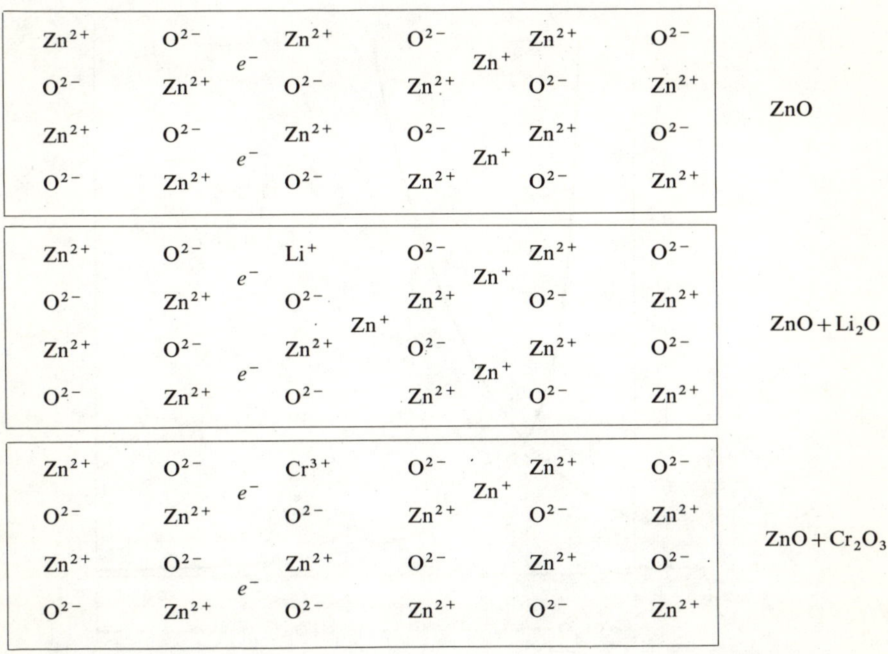

Fig. 1.78 Effects of Li_2O and Cr_2O_3 on the defect structure of ZnO

NiO and Cu_2O can probably be neglected, and these are the only two possible oxidation products. Assume for simplicity that the alloy is thermodynamically ideal, and let x_{Cu} and x_{Ni} be the mole fractions in the alloy. Consider the reactions

$$2Cu + \tfrac{1}{2}O_2 \rightarrow Cu_2O$$
$$Ni + \tfrac{1}{2}O_2 \rightarrow NiO$$

whereby Cu_2O and NiO are formed by oxidation of the *alloy*. The equilibrium conditions are

$$\left.\begin{aligned} x_{Cu}^2 p^{1/2} &= p^{1/2}(Cu_2O) \\ x_{Ni} p^{1/2} &= p^{1/2}(NiO) \end{aligned}\right\} \quad \ldots(1.160)$$

where $p(Cu_2O)$ and $p(NiO)$ are the dissociation pressures of the two oxides, and p is the effective oxygen pressure at the alloy surface. The two relations of equation 1.160 are illustrated in Fig. 1.79 by plotting p against x_{Ni}. Note that $p(NiO) < p(Cu_2O)$ and $x_{Ni} + x_{Cu} = 1$. The two curves intersect at one value of x_{Ni}, and this defines the alloy composition for which Cu_2O and NiO can co-exist on the surface. If the Ni content is higher than this critical amount, only NiO is stable; for lower Ni contents only Cu_2O is stable. If therefore the diffusion coefficients of Cu and Ni in the alloy were very large so that the composition of the alloy in the surface region did not change during oxidation, the situation would be simple, and the oxidation product would be either Cu_2O or NiO except at one critical alloy composition where both oxides would be formed. But the diffusion coefficients in the alloy are not usually large enough for this simple analysis to apply, and if we start with

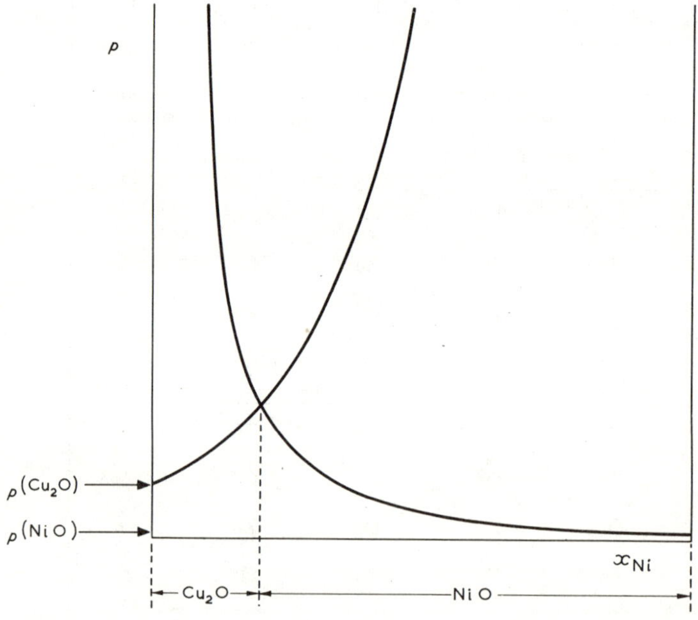

Fig. 1.79 Surface oxides on Cu–Ni alloys

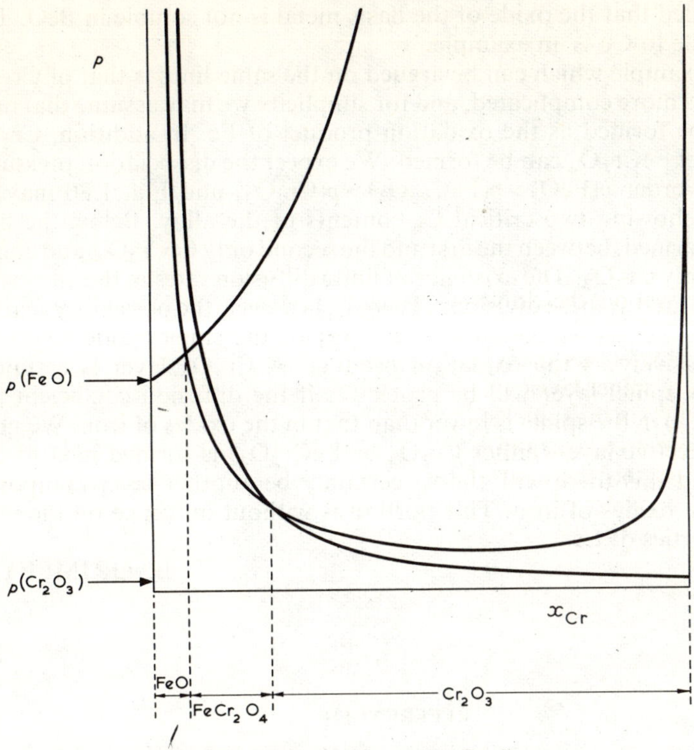

Fig. 1.80 Surface oxides on Fe–Cr alloys

a bulk alloy composition in the region where we would expect only NiO to be formed then, as oxidation proceeds, the surface region of the alloy is depleted of Ni. Thus Cu diffuses inwards and Ni has to be supplied from the interior of the alloy. The composition of the alloy at the surface necessary to maintain the supply of Ni into the oxide layer may be such that the Ni content is below the critical concentration for which only NiO is formed. If this is so, both Cu_2O and NiO are formed as oxidation products; if not, we still get only NiO. The effect of finite diffusion rates in the alloy is therefore that the critical *bulk* alloy composition for the exclusive formation of NiO is pushed to a higher Ni content than we would expect from elementary consideration. Similarly, the minimum Cu content for the exclusive formation of Cu_2O is pushed higher. There is therefore a *range* of compositions of the bulk alloy in which both Cu_2O and NiO are formed together.

This analysis shows that if the oxides of the two components of a binary alloy are mutually insoluble, and if one of the components has a much greater affinity for oxygen than the other, then the oxide of the baser metal will be formed exclusively even though it is present in the alloy in only a small amount. It seems that the importance of beryllium as an alloying constituent can be explained in this way. It has a high affinity for oxygen [$p(\text{BeO}) \approx 10^{-30}$ atm at 1000°C] and also forms a highly protective oxide layer. The oxidation resistance of metals more stable to oxygen than Be, but which normally oxidise faster, should therefore be improved by the addition of Be, provided that the oxide of the basis metal is not soluble in BeO. The addition of Be to Cu is an example.

Another example which can be argued on the same lines is that of Cr–Fe alloys. This is more complicated, and for simplicity we may assume that only FeO could be formed as the oxidation product of Fe. In addition, Cr_2O_3 and the spinel $FeCr_2O_4$ can be formed. We expect the dissociation pressures to be in the order $p(FeO) > p(FeCr_2O_4) > p(Cr_2O_3)$ and Fig. 1.80 may be constructed showing two critical Cr contents of the alloy. Below the first only FeO is formed, between the first and the second only $FeCr_2O_4$, and above the second only Cr_2O_3. The existence of finite diffusion rates in the alloy will, of course, smear out these divisions. There is, however, the possibility that by adding Cr to Fe, either the spinel $FeCr_2O_4$ or the single oxide Cr_2O_3 is formed exclusively as the oxidation product. A Cr_2O_3 layer is certainly protective. A spinel layer will be protective if the diffusion coefficient for Fe^{2+} (or Fe^{3+}) in the spinel is lower than that in the oxides of iron. We note that the protective layer (either Cr_2O_3 or $FeCr_2O_4$) is formed next to the alloy. Beyond this there will almost certainly be another layer composed mainly of the oxides of iron. This portion is without influence on the protective properties of Cr.

T. B. GRIMLEY

REFERENCES

1. Wagner, C., *Z. Phys. Chem. B.*, **21**, 25 (1933)
2. Wagner, C., *Z. Phys. Chem. B.*, **32**, 47 (1936)
3. Mott, N. F., *J. Chim. Phys.*, **44**, 172 (1947)
4. Cabrera, N. and Mott, N. F., *Rep. Progr. Phys.*, **12**, 163 (1948–49)
5. Grimley, T. B. and Trapnell, B. M. W., *Proc. Roy. Soc. A.*, **234**, 405 (1956)
6. Hauffe, K. and Schattky, W., *Halbleiterprobleme*, **5**, 203 (1960)

BIBLIOGRAPHY

Kubaschewski, O. and Hopkins, B. E., *Oxidation of Metals and Alloys*, Butterworths, London (1962)

Hauffe, K., *Oxydation von Metallen und Metallegierungen*, Springer, Berlin (1956)

Garner, W. E. (Ed.), *Chemistry of the Solid State*, Butterworths, London (1955)

Bénard, J. (Ed.), *L'Oxydation des Méteaux*, Vol. I, Gauthier–Villars et Cie, Paris (1962)

1.10 Discontinuous Oxide Films

The Applicability of Rate Laws

Section 1.9 showed that as long as an oxide layer remains adherent and continuous it can be expected to increase in thickness in conformity with one of a number of possible rate laws. This qualification of continuity is most important; the direct access of oxidant to the metal by way of pores and cracks inevitably means an increase in oxidation rate, and often in a manner in which the lower rate is not regained. In common with other phase change reactions the volume of the solid phase alters during the course of oxidation; it is the manner in which this change is accommodated which frequently determines whether the oxide will develop discontinuities. It is found, for example, that oxidation behaviour depends not only on time and temperature but also on specimen geometry, oxide strength and plasticity or even on specific environmental interactions such as volatilisation or dissolution.

The models derived for continuous oxide layers remain valuable when porous oxides are formed; they provide a frame of reference against which deviations may be examined and give a basis for understanding the factors governing the location of new oxide. In many cases, however, the experimentally derived rate 'laws' no longer have a unique interpretation. For example, the linear rate law relating the thickness of oxide, x, to the time, t

$$x = k_l t \qquad \ldots(1.161)$$

can describe a situation in which an oxide regularly fails when it reaches a critical thickness[1], or one in which the oxide volatilises[2] at a uniform rate. Similarly, the logarithmic rate law

$$x = k_a \log(1 + k_b t) \qquad \ldots(1.162)$$

has been shown to describe many differing situations, including rate control by chemisorption[3], by oxide nucleation[4], or by cavity production[5].

In certain circumstances even the parabolic rate law may be observed under conditions in which the oxide is porous and permeated by the oxidising environment[6]. In these cases it has been shown that it is diffusion of one or other of the reactants through the fluid phase which is rate controlling. More usually however the porous oxide is thought to grow on the surface of a lower oxide which is itself growing at a parabolic rate. The overall rate of growth is then said to be *paralinear*[7,8] and may be described by the sum of linear and parabolic relationships (*see* equations 1.174 and 1.175).

It will be seen that although each system must be described on its merits there are general principles governing mass transport which can be used as a guide.

The Volume Change on Oxidation of a Metal

The formation of discontinuities, particularly the grosser forms of pores and cracks, in an oxide layer is often attributable to the mass flows and volume changes occasioned by oxidation. As can be seen from Table 1.25 it is usual

Table 1.25 Metals which form porous oxides in dry oxygen

Linear oxidation Type 1 Curve (Fig. 1.89)		*Paralinear oxidation Type 2 Curve (Fig. 1.89)*	
Metal	*Oxide/metal volume ratio* (r)	*Metal*	*Oxide/metal volume ratio* (r)
Mg	0·81	Ce	1·22
Ca	0·64	La	1·1
Nb (400°)	2·49	Mo	3·24
		W	3·35
		U	1·94
		Nb (450°)	2·49
		Ta	2·54
		Ti	1·73
		Zr	1·56
		Hf	1·60
		Th	1·35

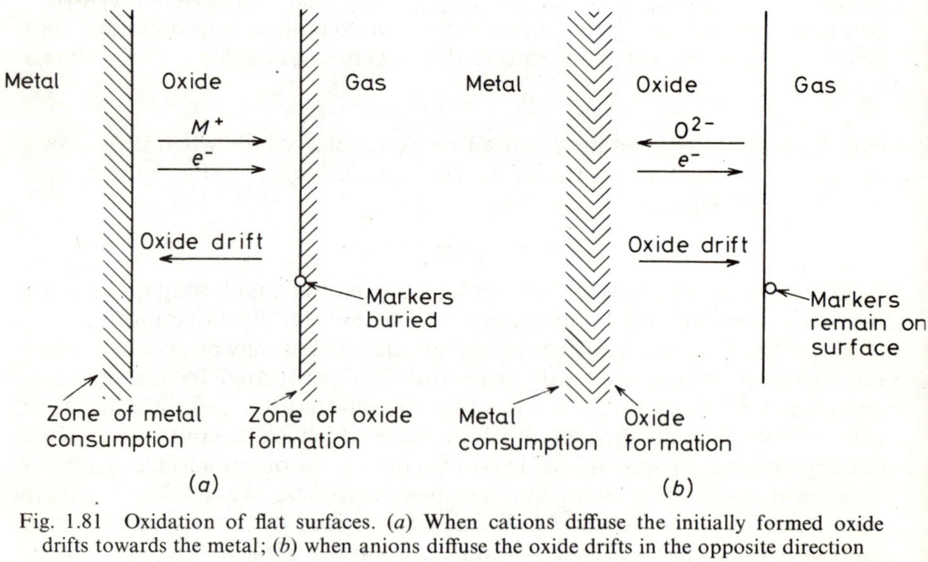

Fig. 1.81 Oxidation of flat surfaces. (*a*) When cations diffuse the initially formed oxide drifts towards the metal; (*b*) when anions diffuse the oxide drifts in the opposite direction

for the volume of an oxide to differ greatly from that of the equivalent amount of metal. It is in fact remarkable that such a phase change can occur in so many cases of oxidation without oxide disruption. Moreover, oxidation reactions are rarely topotactic but take place by transport of one of the reactants to a plane of reaction. It is thus not only the net volume change which must be considered, but also the local volume changes due to oxide creation in some zones and metal consumption in others[9].

Fortunately the oxidation of many metals takes place by the diffusion of the metal cation[10]. This flux is outwards through the oxide layer, and the work of adhesion[11] enables the loss of metal to be compensated for by a drift of the oxide towards the metal (Fig. 1.81). Thus the stresses set up in the maintenance of oxide/metal contact are compressive and, as such, can be more readily withstood by most oxides. Nevertheless, it is these general movements of the oxide scale which are ultimately responsible for discontinuities in the majority of cases and it is appropriate to discuss transport-induced flows before proceeding any further.

Mass Transport in Growing Oxide Layers

Oxide Drift

In the very early stages of oxidation the oxide layer is discontinuous; both kinetic[12] and electron microscope[13–15] studies have shown that oxidation commences by the lateral extension of discrete oxide nuclei. It is only once these interlace that the direction of mass transport becomes of importance. In the majority of cases the metal then diffuses across the oxide layer in the form of cations and electrons (cationic diffusion), or as with the heavy metal oxides, oxygen may diffuse as O^{2-} ions with a flow of electrons in the reverse direction (anionic diffusion). The number of metals oxidising by both cationic and anionic diffusion is believed to be small, since a favourable energy of activation for one ion generally means an unfavourable value for the other[10].

A consequence of single-ion diffusion is that the mass movement must be compensated for by an opposing drift (relative to a fixed point deep in the metal) of the existing oxide layer if oxidation is not to be stifled by lack of one of the reactants. The effect may be illustrated by reference to a metal surface of infinite extent (Fig. 1.81).

When cations are the diffusing species (Fig. 1.81*a*), metal is consumed either by solution in the oxide as interstitial cations and electrons

$$M = M^{+} + e^{-} \qquad \ldots(1.163)$$

or by reaction with metal ion vacancies in the oxide

$$M + M^{+}\square + \oplus = \text{null} \qquad \ldots(1.164)$$

where as is conventional the right-hand side of the equation represents the excess or deficiency of ions in the stoichiometric oxide, and $M^{+}\square$ and $\oplus$ represent the vacant sites in the ionic lattice and positive holes in the full band, respectively. Since oxide is formed at the gas/oxide interface, oxide–metal contact is only maintained by translation of the initial oxide through the volume occupied by that part of the metal which has been oxidised.

In systems in which anionic diffusion prevails (Fig. 1.81*b*), metal is consumed by direct reaction to form the diffusing oxygen species

$$M \rightarrow MO + O^{2-}\square + 2e^{-} \qquad \ldots(1.165)$$

The oxygen vacancies then diffuse to the gas interface where they are annihilated by reaction with adsorbed oxygen. The important point, however, is that metal is consumed and oxide formed in the same reaction zone. The oxide drift has thus only to accommodate the *net* volume difference between the metal and its equivalent amount of oxide. In theory this net volume change could represent an increase or a decrease in the volume of the system, but in practice all metal oxides in which anionic diffusion predominates have a lower metal density than that of the original metal. There is thus a net expansion and the oxide drift is away from the metal.

Oxide movements are determined by the positioning of 'inert' markers on the surface of the oxide[16–18]. At various intervals of time their position can be observed relative to, say, the centreline of the metal as seen in metallographic cross-section. In the case of cation diffusion the metal–interface–marker distance remains constant and the marker moves towards the centreline; when the anion diffuses, the marker moves away from both the metal–oxide interface and the centreline of the metal. In the more usual observation the position of the marker is determined relative to the oxide/gas interface. It can be appreciated from Fig. 1.81 that when anions diffuse the marker remains on the surface, but when cations move the marker translates at a rate equivalent to the total amount of new oxide formed. Bruckman[19] recently has re-emphasised the care that is necessary in the interpretation of marker movements in the oxidation of lower to higher oxides.

Oxidation of Non-planar Surfaces

Oxide movements on plane surfaces, such as those just described, do not create stress; stress will arise however, when the oxide movement is constrained by the presence of a corner, or when the metal is curved, so that there is a progressive strain on the lateral dimensions of the oxide. Since oxides are brittle the appearance of tensional stress can be expected to lead to brittle failure; examples are given in Figs. 1.82 and 1.83.

When convex surfaces oxidise by cationic diffusion compressive stresses occur within the oxide along planes parallel to the metal surfaces and reach a maximum at the metal/oxide interface. Frequently oxides are able to flow plastically under compression at moderate temperatures and the stress is relieved. Hales[20], for example, has proved this to occur by the density of dislocations present in a nickel oxide layer. Sometimes however, relief is abrupt, giving rise to oxide distortions of the type illustrated in Fig. 1.82. An example of oxide buckling observed during the oxidation at 500° of a Cu–10Ni alloy is shown in Fig. 1.84; features of this kind are not unusual and have been followed in the oxidation of pure iron using a hot stage and oxygen blanket in the scanning electron microscope[21].

The compressive forces may be contained by adhesion and cohesion in the oxide–metal system; experiments designed to reveal the presence and

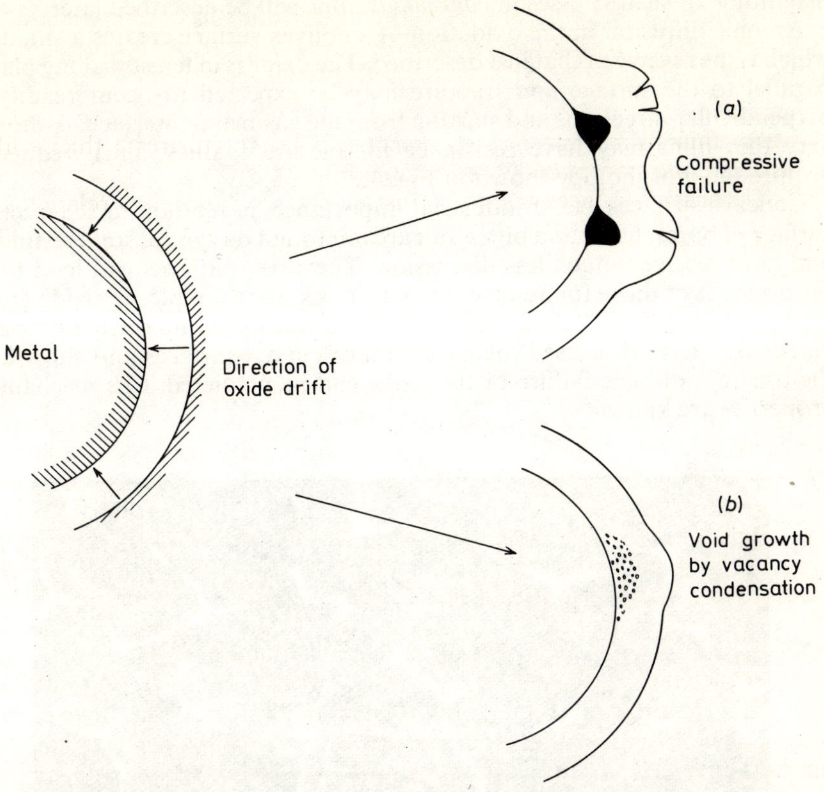

Fig. 1.82 Oxidation of a convex surface by cation diffusion; the compressive stress in the initially formed oxide may lead to (*a*) failure by buckling or to (*b*) void precipitation

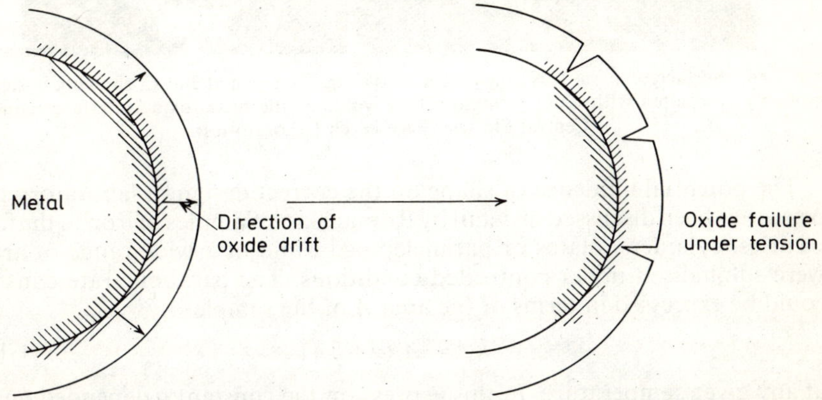

Fig. 1.83 Oxidation of a convex surface by anion diffusion; the outward translation of the oxide gives tensile cracking in the initially formed oxide

magnitude of such stresses in *continuous* films will be described later.

Anionic diffusion in the oxidation of a convex surface creates a situation which is the reverse of that just described. The oxide is in tension along planes parallel to the surface and fracture may be expected to occur readily in perpendicular directions and starting from the gas/metal interface. Although very thin films may have resistance to fracture[22], thick films frequently acquire the morphology shown in Fig. 1.83.

Concave surfaces are of industrial importance, in relation to the internal surface of bores, holes and pipes, but are not found on typical solid testpieces and have received much less discussion. The stress patterns will tend to be the opposite of those found on convex surfaces; for example, an oxide growing by cation diffusion should be in tension at the metal interface. Bruce and Hancock[23] have discussed the oxidation of curved surfaces and show how the time to adhesive failure of the oxide can be predicted if its mechanical properties are known.

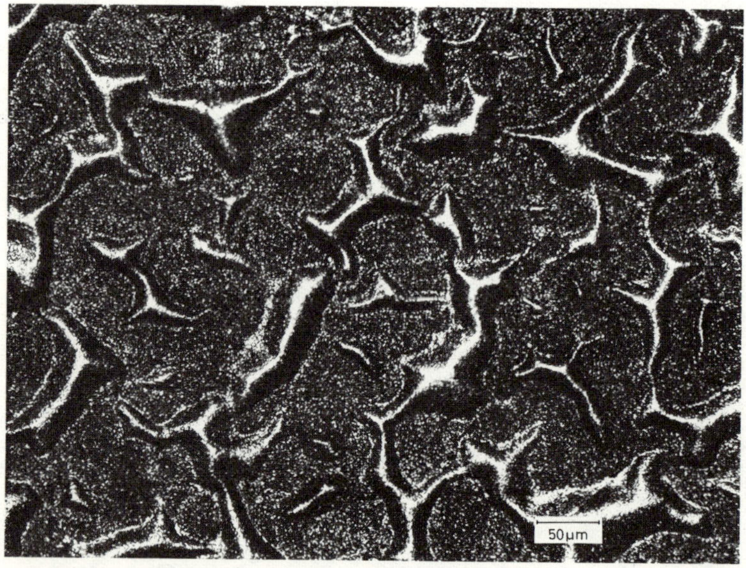

Fig. 1.84 Surface of a Cu–10Ni alloy after oxidation in oxygen at 500°C, showing blistering, probably associated with CuO formation over voids at the metal/oxide interface (courtesy Central Electricity Research Laboratories)

The potential influence of shape on the correct design of laboratory testpieces has been discussed in detail by Romanski[24]. Samples of iron in the form of discs, cylinders, plates or parallelepipeds, and of a wide range of areas, were sulphidised under controlled conditions. The parabolic rate constant could be expressed in terms of the area A of the samples by

$$k_p = A/(a+bA) \qquad \ldots(1.166)$$

at any given temperature. In this expression the constant a depended on the geometry of the specimen and hence on scale rigidity, and b on specimen purity, and thus probably on scale plasticity. The form of this equation was confirmed with the systems Ni–S, Cu–S and Cu–O. Romanski concluded that plate or disc specimens with an area of 30–40 cm^2 were required to obtain rates approaching the maximum for infinite surfaces.

Oxide Drift and Pore Formation

Clearly geometric constraints have an important influence on the formation of oxide layers. There are, however, many reported examples[25–28] of the development of porosity at the oxide–metal interface on plane surfaces well away from corners. This phenomenon is generally ascribed to the suitability of the interface as a sink for lattice vacancies generated during oxidation; the point was lucidly discussed by Vermilyea[29]. Voids of this type may range in size from a few nanometres, as described by Boggs[26] for the oxidation of tin, through the crypt-like cavities in the beautiful micrographs of Howes[30], to the complete separation witnessed in the oxidation of iron by Pheil[31] and described in a paper which also recorded the first use of inert markers. Birchenall has also given interesting observations on the behaviour of large cavities[32]. Such porosity is more commonly associated with cation diffusion although Pemsler[33] has suggested that condensation of *oxygen* vacancies at the metal–oxide interface may be the cause of the voids observed during the oxidation of zirconium. The work of Cohen and his group at Ottawa concerning the oxidation of pure irons[34–39] and more recently, of nickel[40] has made a notable contribution to this field. In particular, they have shown how the dislocation network of the underlying metal is effective in removing vacancies from the metal–oxide interface and thus preventing void formation. The effect is less marked in the case of nickel where the more important effect of cold work in the metal is to yield an oxide structure which is consistent with high diffusion rates. It seems likely that the major difference between the oxidation of these two metals lies in the plasticity of their oxide layers. Bruce and Hancock[41] have detected repetitive oxide cracking during the oxidation of iron by use of a vibrating testpiece, whereas oxide cracking did not occur when nickel testpieces were used. This is in good accord with the behaviour of dislocations in nickel *oxide* during oxidation; Hales[20] has shown that these become optimised to permit maximum creep rates and moreover that the preferred orientation and columnar nature of the oxide also derives from this selection principle. The work is related very nicely to the analysis by Harris and Masters[11] of the work of metal/oxide adhesion.

When voids do form the first formed oxide may lose contact with the retreating metal surface; the way is then open for the metal consumption zone[9] to be filled with a secondary growth which is usually fine grained and porous[42]. Often the inner layer appears to have formed from the inception of oxidation since the inner–outer layer interface retains the shape and dimensions of the original metal surface[43–45]. More often, however, the balance between inner and outer layer formation alters from point to point, with the inner layer being favoured at corners and edges. On alloys the inner layer may consist of particles of the more noble metal oxide which, as in the case of the copper–nickel alloys[45], provide a spaceframe supporting the outer layer. Duplex layers of this nature are sometimes seen on pure metals, the classic example being the case of nickel. Sartell and Li[46] have shown from diffraction measurements that the two layers on nickel differ in their lattice constants, the outer being in compression. They envisaged the outer oxide to grow by anionic diffusion and the inner by cation movement. It now seems likely that the inner layer grows in oxygen formed by decomposition of the outer layer; the case has been well argued by Bruckman[9] and supported by the work of Douglass[42]. What is not so clear is how the outer layer on pure metals continues to grow since it demands the movement of both metal and oxygen across the inner layer when they already have sufficient chemical potential to react. The problem does not arise when alloys oxidise since the inner layer can be traversed at potentials below those at which the species react. Mrowec and Webber[43] have made numerous observations on systems of this kind and neatly summarise their findings in a discussion of a paper by Kofstad and Hed[47] on cobalt–chromium alloys.

There are a number of factors which will trigger void formation during oxidation: Cohen and his co-workers[34–39] have shown that annealing of the metal (to remove dislocations) will do so, whilst Douglass[42], and Wulf, Carter and Wallwork[48] have shown that the presence of a continuous film of a more noble alloying element is just as effective. Tuck, *et al.*[49] consider that the extra adhesion of the oxide on iron when this metal is heated at 950°C in oxygen containing water vapour is due to the beneficial effect of hydrogen on oxide plasticity. There are many similar observations in the literature.

The Influence of Voids on Oxidation Kinetics

Voids as Diffusion Barriers

It was Evans[5,50] who first suggested that cavities could act as diffusion barriers and derived the logarithmic rate law from the progressive nucleation of voids. Boggs[51], and several co-workers[26,27] proved the Evans mechanism to apply to the oxidation of tin over the temperature range 150–220°C in a series of papers illustrating the combined use of electron microscopy and kinetic measurements. The logarithmic rate law also describes the oxidation of a number of other metals in the thin-film range, e.g. Mg[52], Cu[53] and Ni[54], but there is as yet no evidence that the Evans mechanism applies to these cases. Possibly pores have not been sought with the same rigour as that used by Boggs. However, Douglass[42] used a rearrangement of a method first used by Tylecote[55] to show that void condensation causes a three-fold reduction in the parabolic rate constant for a dilute chromium–nickel alloy at 800°C; he compared the rates of oxidation of two cylinders, one of them being plugged at either end to exclude the atmosphere from the interior. Voids did not form at the metal–oxide interface on the plugged cylinder since the interior then acted as a sink for diffusing vacancies.

Probably the most comprehensive measurements of the effect of voids on rates are those of Cohen[34–39] and his school. They have published data on the oxidation of pure irons for a wide temperature range and for oxygen pressures ranging from $1{\cdot}3 \times 10^{-4}\,\mathrm{N/m^2}$ to $100\,\mathrm{kN/m^2}$. The interactions between void formation and oxygen uptake are complex but only at pressures below $1{\cdot}3 \times 10^{-2}\,\mathrm{N/m^2}$ do voids have no effect. Some of their results are summarised in Fig. 1.85; over the pressure range $1{\cdot}3 \times 10^{-2}\,\mathrm{N/m^2}$ to

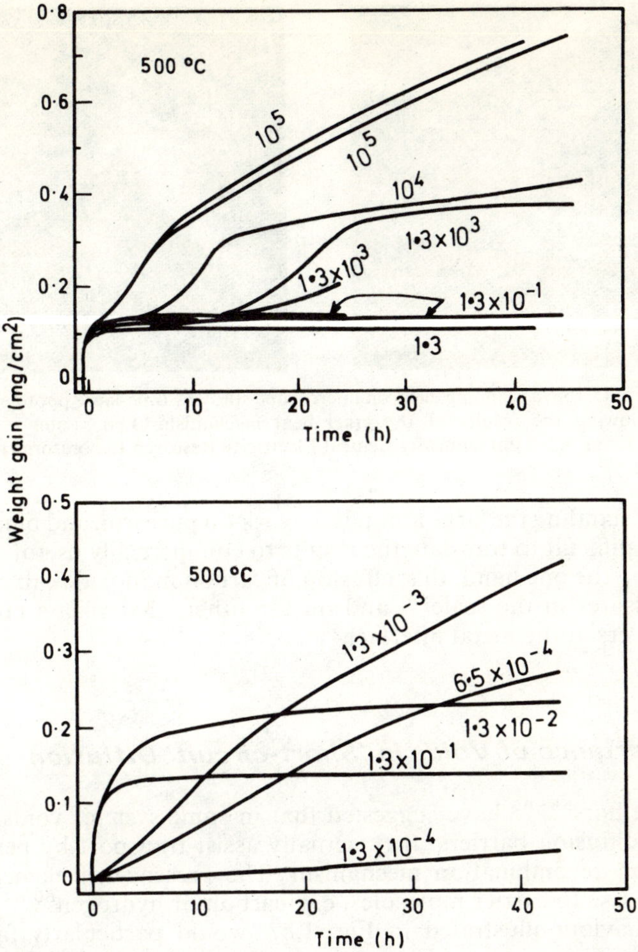

Fig. 1.85 Oxidation of high-purity iron in oxygen at differing pressures. All figures on curves are in $N m^{-2}$. At $1{\cdot}3\times10^{-4}$ to $1{\cdot}3\times10^{-3}$ N/m^2 torr the rate is controlled by the impact of molecular oxygen; at $1{\cdot}3\times10^{-2}$ N/m^2 torr the initial rate of oxidation is sufficiently high to give void precipitation and the rate decreases with pressure increasing to $1{\cdot}3$ N/m^2; at pressures greater than this the crack-heal mechanism becomes operative and the rate again increases with pressure (after Hussey and Cohen[38])

$1{\cdot}3$ N/m^2 the rate decreases as vacancies start arriving at the interface more rapidly than they can be assimilated. Hussey and Cohen[37,38] have evidence that the voids form in the Fe_3O_4 layer before the outer Fe_2O_3 is nucleated. However, this rapidly takes place once the supply of iron is stifled by the voids, oxide blisters and cracks, and oxygen ingress causes the rate to accelerate again. The curves indicate that blistering is a repetitive process and this is confirmed by micrographs from the hot-stage scanning electron microscope (Fig. 1.86). It is a salutary fact that the material showing the most protective behaviour (Fig. 1.85) has an oxide which is already partly separated from the metal.

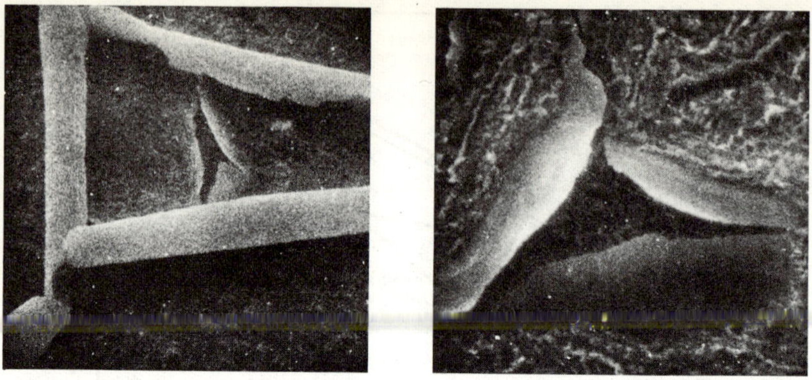

Fig. 1.86 'Stills' from a scanning electron microscope study by time-lapse photography of iron oxidation showing the results of the crack-heal mechanism. Left, 1 mm ≡ 1 μm; right, 1 mm ≡ 0·5 μm (courtesy Central Electricity Research Laboratories)

Notwithstanding the large amount of work on pure iron and binary alloys, it remains difficult to translate the results to commercially useful steels. It is believed, on the one hand, that effusion of carbon monoxide can cause non-healing fissures in the scale[56], and on the other, that silicon creates self-healing layers at the metal interface[57].

The Importance of Voids in 'Short-circuit' Diffusion

Several authors[58–63] have suggested that in some systems voids, far from acting as diffusion barriers, may actually assist transport by permitting a dissociation–recombination mechanism. The presence of elements which could give rise to carrier molecules, e.g. carbon or hydrogen[60,62], and thus to the behaviour illustrated in Fig. 1.87, would particularly favour this mechanism. The oxidant side of the pore functions as a sink for vacancies diffusing from the oxide/gas interface by a reaction which yields gas of sufficiently high chemical potential to oxidise the metal side of the pore. The vacancies created by this reaction then travel to the metal/oxide interface where they are accommodated by plastic flow, or they may form additional voids by the mechanisms already discussed. The reaction sequence at the various interfaces (Fig. 1.87*b*) for the oxidation of iron (prior to the formation of Fe_2O_3) would be

at *A* $$2O_2 = Fe_3O_4 + Fe^{\frac{8}{3}+}\square + 8e \quad \text{...(1.167)}$$

at *B* $$Fe_3O_4 + 3Fe^{\frac{8}{3}+}\square + 8e + 4CO = 4CO_2 \quad \text{...(1.168)}$$

at *C* $$4CO_2 = Fe_3O_4 + 3Fe^{\frac{8}{3}+}\square + 8e + 4CO \quad \text{...(1.169)}$$

at *D* $$3Fe^{\frac{8}{3}+}\square + 8e + 3Fe_{metal} = \text{null} \quad \text{...(1.170)}$$

Notice that oxide is utilised by the reaction at interface *B* at the same rate as it is formed at *A*, so that the void effectively moves through the growing oxide with the distance *AB* remaining constant. It may be recalled that a truly inert marker placed at the point B in the *continuous* oxide would be expected to remain fixed with reference to the point D (the metal–oxide interface). The distinction is important since voids involved in the mechanism in this way remain in a string across the oxide and can thus co-operate to move the oxidant towards the metal at a high rate[64]. What is not so easy to explain is why, when this mechanism is thought to apply, the oxide/metal and metal/oxide interfaces remain flat; the implication is that the abnormally high flux carried by the porous regions redistributes, at the gas/oxide interface, to the benefit of the whole surface. Gibbs[65] has shown that as long as surface diffusion is fast this mechanism gives rise to an altered parabolic rate law. Smeltzer[66], however, has argued that short-circuited routes may be progressively lost as the oxide increases in thickness; thus there is a transition from a short-circuit diffusion process at short times to the usual parabolic dependence at long times.

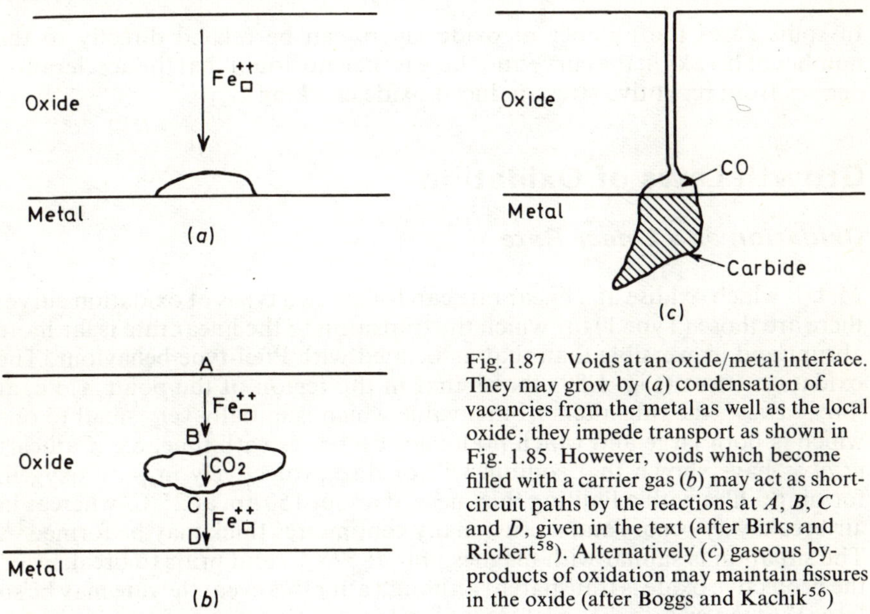

Fig. 1.87 Voids at an oxide/metal interface. They may grow by (*a*) condensation of vacancies from the metal as well as the local oxide; they impede transport as shown in Fig. 1.85. However, voids which become filled with a carrier gas (*b*) may act as short-circuit paths by the reactions at *A*, *B*, *C* and *D*, given in the text (after Birks and Rickert[58]). Alternatively (*c*) gaseous by-products of oxidation may maintain fissures in the oxide (after Boggs and Kachik[56])

All voids described so far have been formed at, or released from, the metal–oxide interface. Birchenal[67] has discussed the formation and growth of voids within the oxide scale by condensation of vacancies. For this to occur anions must be removed in some way and he has suggested creep and slip in the anion lattice as possible mechanisms. More recently, Cox[64] has used a sensitive porosimeter as well as special metallographic techniques to study strings of voids formed in zirconium oxide as a result of recrystallisation—the work is discussed in more detail below. Although Birchenal's model predicted greater deviations from the parabolic rate law than were in fact found, the phenomenon of pore growth within the scale seems to be real and it is as well to keep the implications for anion movement in mind (*see* page **1**.260 *et seq*.).

Oxide Cracking and Void Cavities

As the area of individual cavities at the metal–oxide interface increases there is an increased probability that the oxide will crack and permit oxygen access to a large area of unprotected metal. Such cracking is then reflected in the overall kinetics and the rate curve takes the form shown by the curves in Fig. 1.85 for pressures greater than 1·3 kN/m^2; it is known to occur with copper[68,69], iron[38] and iron–chromium alloys[70,71] over certain temperature ranges. The oxide formed has become known as a Pfeil-type porous oxide[31]. When failure occurs rather infrequently the rate curve appears as a succession of parabolas, each having as origin the critical values of time and thickness (t_c and x_c) at which the previous film failed. The net rate of oxidation is thus close to that given by the linear rate law with the value of the constant k_l equal to x_c/t_c, or

$$k_l = (x_c/k_p)^{\frac{1}{2}} \qquad \text{...(1.171)}$$

In some cases the number of oxide layers can be related directly to the number of breaks in the curve and there is then no doubt that the acceleration derives from repetitive stress-induced oxide cracking.

Growth Laws of Oxidation

Oxidation at a Linear Rate

Metals which oxidise at a linear rate can follow two types of oxidation curve; there are those (Type 1) for which the transition to the linear rate is far more abrupt and irreversible than that associated with Pfeil-type behaviour. The oxidation curve (Fig. 1.88) shows that in the region of the point A, i.e. at 'breakaway', the rate alters from a value which is initially very small to one which is both large and which no longer decreases with time, e.g. Cathcart *et al.*[72] have shown that sodium will oxidise protectively in pure oxygen, forming a film with a limiting thickness of about 150 nm at 25°C, whereas in air breakaway occurs and a crust many centimetres thick may be formed[73]. The situation is similar with magnesium; at 575°C, and prior to breakaway, the protective oxide is transparent although after this event the rate may be so large that the metal takes fire[1]. Not all rate transitions are as large in magnitude; that of steel in the CO_2 coolant of nuclear reactors leads to a linear rate of about 25 μm/y[74], and this, because of the inaccessibility of the steel, has necessitated temperature reductions in operating reactors, consequently with a considerable loss in power output.

The second type of behaviour (Fig. 1.89) is much closer to that which one might predict from the regular cracking of successive oxide layers, i.e. the rate decreases to a constant value. Often the oxide–metal volume ratio (Table 1.25) is much greater than unity, and oxidation occurs by oxygen transport in the continuous oxide; in some examples the data can be fitted by the paralinear rate law, which is considered later. Destructive oxidation of this type is shown by many metals such as molybdenum, tungsten and tantalum which would otherwise have excellent properties for use at high temperatures.

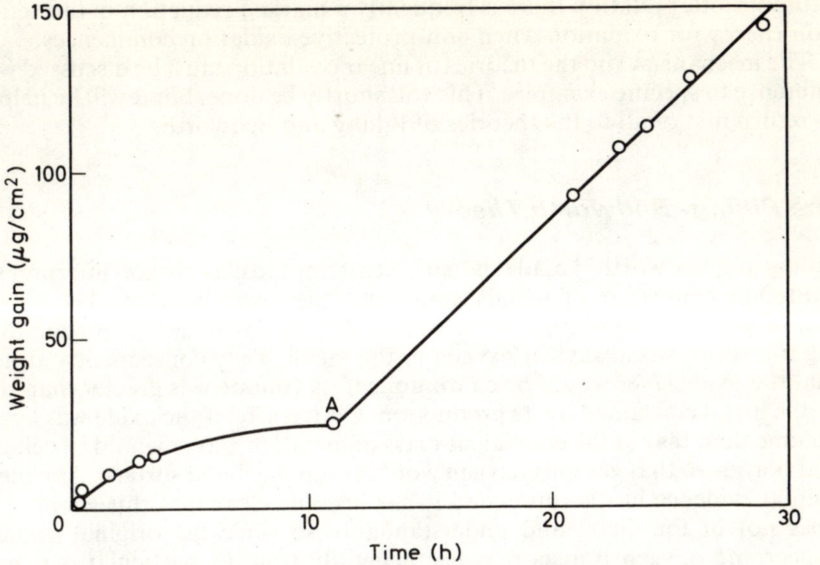

Fig. 1.88 Early stages of oxidation of magnesium at 525°C, but at a lower pressure of 13 kN/m^2 than the example in Fig. 1.89 illustrating breakaway

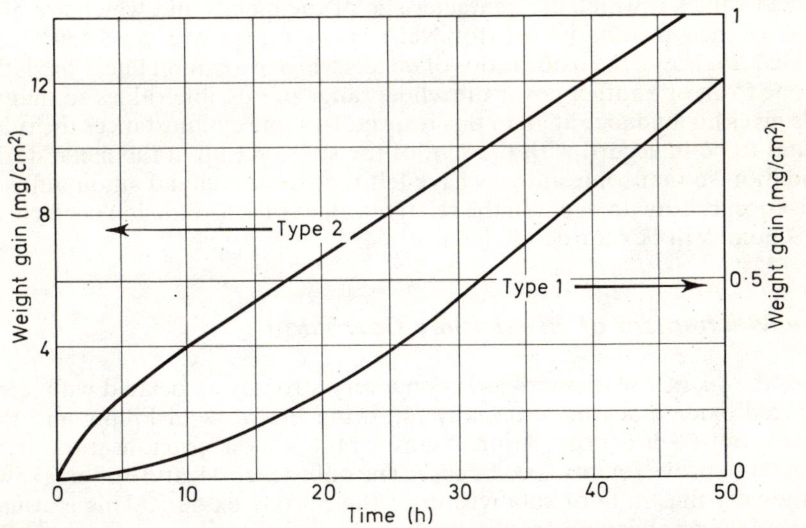

Fig. 1.89 Oxidation of magnesium at 500°C illustrating the increase in rate to the constant value (Type 1) (after Gregg and Jepson[1]) and the oxidation of tungsten at 700°C (after Webb, Norton and Wagner[88]) illustrating the decrease in rate to the constant value (Type 2)

Probably the only feature common to the mechanism of oxidation of the two groups is that, because of crack or pore formation in the continuous oxide, the rate of transport of oxygen in a molecular form has increased to the point where a phase-boundary reaction has assumed rate control. In accord with this interpretation there is frequently a marked reduction in the activation energy for oxidation when non-protective oxidation commences.

The mechanism and the theories of linear oxidation must be discussed with reference to specific examples. This will shortly be done, but it will be helpful to return first of all to the theories of Pilling and Bedworth.

The Pilling–Bedworth Theory

Pilling and Bedworth[68] made the earliest attempt to classify and interpret the oxidation behaviour of metals, and since they were working before any marker experiments had been carried out, they assumed that the predominating transport was always of oxygen to the metal. They consequently argued that the oxide layer would be continuous if its volume was greater than that of the metal consumed by its production. Conversely, if the oxide was less in volume than that of the equivalent mass of metal, the layer would be cellular and porous so that gaseous oxygen would reach the metal surface. As experimental evidence has accumulated it has become clear that this simple rule does not fit the facts; and understandably so since the original premise concerning oxygen transport is not generally true. In particular it is now known that all the metals with oxide: metal volume ratios of less than unity, e.g. Na_2O:Na = 0·55[72], MgO:Mg = 0·81[52] and CaO:Ca = 0·64[75], form oxides which remain thin and continuous indefinitely at temperatures below critical values t_c which are characteristic of the metals and which are often close to their melting points (for Na, t_c = 48°C; for Mg, t_c = 550°C). As Table 1.25 shows, the oxide ratios of metals which do exhibit linear oxidation of one form or another cover the whole range of possible values so that the rule gives no guidance at all in this respect. In some circumstances the rule is found to be in accord with the sign of the stress set up in the metal during oxidation so that some authors have felt that some residual anion diffusion must occur, notwithstanding the relative values of self-diffusion coefficients. This point will be returned to later.

The Mechanism of Breakaway Corrosion

Over the years, breakaway has become very strongly associated with stress-induced oxide cracking, especially following the work of Pilling and Bedworth, but such a proposition is unwarranted as a generalisation and is difficult to prove for any specific case. The only general feature of breakaway is the very fine state of subdivision of the porous oxide[76]. This is usually beyond the resolving power of even the scanning electron microscope[77]. The coarse cracks which are seen in many optical micrographs are almost certainly secondary features associated with the proximity of the reaction zone to the metal–oxide interface and the subsequent drift of the oxide scale. The factors suggesting that the development of fine-grained porosity in oxides is not merely a stress-assisted reticulation type of failure are (*a*) the long induction periods, when stresses should be relieved rather than developed; (*b*) the major rôle of gas phase impurities (examples are water in the case of sodium[72] and beryllium[78], hydrocarbons with magnesium[52] and CO with steel in CO_2[44]); (*c*) the promotion of breakaway of steel[79] in high-temperature water or CO_2[44], and of zirconium[80] by layers of platinum; and (*d*) the curious fact that metals which happily formed a protective oxide when they first contacted oxygen refused to do so when the process was repeated.

Much of the difficulty in demonstrating the mechanism of breakaway in a particular case arises from the thinness of the reaction zone and its location at the metal–oxide interface. Workers must consider (*a*) whether the oxide is cracked or merely recrystallised[81]; (*b*) whether the oxide now results from direct molecular reaction, or whether a barrier layer remains[82]; (*c*) whether the inception of a side reaction (e.g. $2CO \rightarrow CO_2 + C$)[44] caused failure; or (*d*) whether a new transport process, chemical transport or volatilisation[83], has become possible. In developing these mechanisms both arguments and experimental technique require considerable sophistication. As a few examples one may cite the use of density and specific surface-area measurements as routine[75]; of porosimetry by a variety of methods[64]; of optical microscopy[84,85], electron microscopy[21] and x-ray diffraction[86] at reaction temperature; of tracer[44], electric field[87] and stress measurements. Excellent metallographic sectioning is taken for granted in this field of research.

As has been intimated certain breakaway reactions are of great technological importance and a correspondingly large amount of research has been carried out on these, but as yet no consensus has obtained for the mechanism of linear oxidation of Type 1 (Fig. 1.89) for any one of the metals. The papers of Cox[64,87] on the oxidation of zirconium and its alloys are, however, well worth study; the work included the development of a mercury porosimeter sensitive to pores of about 10 nm in size and the investigation of electron transport in the pre-breakaway oxide layer. Cox concludes that it is electron transport which is rate controlling in the early stages and that breakaway is the recrystallisation of the oxide, induced by a tensional stress, which creates continuous porosity[81] by void condensation at grain boundaries. Intermetallic particles, which in zirconium alloys are associated with easy routes for electron transport in the pre-breakaway film, appear to be located at the base of pores after breakaway. The porosimetry measurements indicate that the pores have a diameter of less than 10 nm and this correlates with the size of pore-like features seen in replicas by transmission electron microscopy. The model received additional support from hot-stage x-ray diffraction[86] which showed recrystallisation to a columnar structure to be concurrent with the rate transition. Bradhurst and Heuer[85] prefer to turn the mechanism around, arguing that recrystallisation occurs as a result of cracking in the traditional way; they have confirmed that there is stress relief in the oxide–metal system at the time of breakaway and have evidence from the hot stage microscope of crack-like features moving across the surface. The viewpoints of both schools are summarised in letters to the *Journal of Nuclear Materials*[81,85].

The discovery by Fiegna and Weisgerber[80] that noble metals are able to catalyse the breakaway corrosion of zirconium has not been built into either of the main theories. Antill, *et al.*[44] have also found it difficult to explain their similar observation for the oxidation of steel by CO_2. Reactor grade CO_2 contains both water and CO as impurities; CO is also produced by the reaction

$$3Fe + 4CO_2 = Fe_3O_4 + 4CO \qquad ...(1.172)$$

and Antill's thesis is that it is the disproportionation of the carbon monoxide (the Boudouard reaction)

$$2CO = CO_2 + C \qquad ...(1.173)$$

which causes fragmentation of the protective oxide. However, in this system the action of the noble metal appears to be associated with the water impurity.

Water as an impurity is known to promote the breakaway corrosion of a number of metals; in addition to iron in CO_2 the effect has been reported for magnesium (hydrocarbons have more effect on the oxidation of this metal), beryllium, zirconium and sodium. In the latter case water is known to convert the oxide to deliquescent NaOH but acceleration of beryllium oxidation probably results from hydride formation and mechanical damage to the oxide.

The Mechanisms of Paralinear Oxidation

Some metals oxidise at a rate which decreases, rather than increases (Type 2 in Fig. 1.89). Cerium behaves in this fashion at temperatures between 40°C and 130°C, and Loriers[7,8] has suggested that the curve derives from the competition between the two oxides Ce_2O_3 and CeO_2. It was proposed that the inner layer Ce_2O_3 was continuous and grew under diffusion control but transformed at a constant rate to an outer layer of CeO_2. That is, if we write y and z as the thickness of the inner and outer layers respectively, then

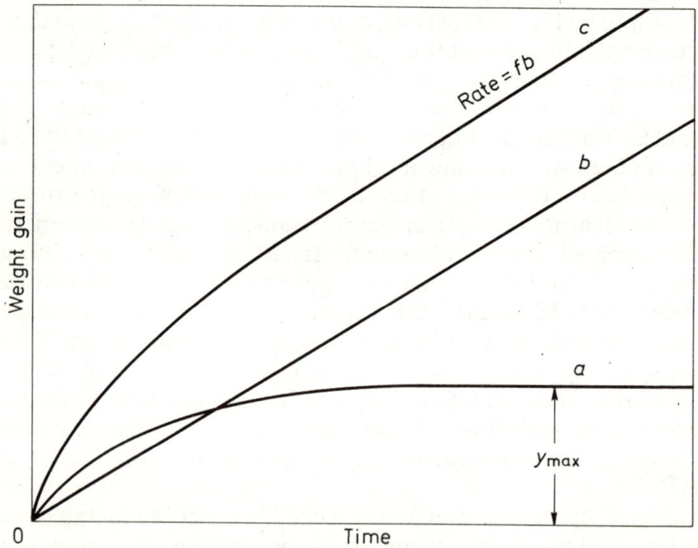

Fig. 1.90 Kinetic interpretation of paralinear oxidation. Curves *a* and *b* correspond to the growth of the inner compact layer and the outer porous layer, respectively; curve *c* represents the total weight and is the algebraic sum of curves *a* and *b*. Note that as oxidation proceeds, y tends to a limiting value $y_{max.}$ (curve *a*) and the overall rate of oxidation tends to a constant value fb

$$dy/dt = (a/y)-b \qquad ...(1.174)$$

and

$$dz/dt = fb \qquad ...(1.175)$$

where f is the ratio of the oxygen content per gramme atom in the outer layer to that in the inner layer, a is one half the parabolic rate constant for diffusion in the inner layer, and fb is the linear rate constant.

The integrated forms of equations 1.174 and 1.175 are illustrated in Fig. 1.90 together with that for the total weight gain w which is given by the sum of y and z. It is evident that as oxidation proceeds the rate of thickening of the inner layer progressively decreases and its thickness tends to a limiting value, i.e.

$$y_{max.} = a/b \qquad ...(1.176)$$

so that the overall rate of oxidation tends to the constant value

$$dw/dt = fb \qquad ...(1.177)$$

The particular interest in this form of oxidation stems from the fact that the important group of metals Nb, Ta, Mo and W show similar behaviour although it is only with tungsten that the kinetics are strictly paralinear. The model was first applied to the oxidation of tungsten by Webb, Norton and Wagner[88] for the temperature range 700–1000°C. They were able to demonstrate the presence of a barrier film (probably a metastable modification of one of the intermediate oxide phases of tungsten such as W_4O_{11}) between the metal and the outer layer of porous tungstic oxide. Subsequently Jepson and Aylmore[89] measured the amount of porous oxide formed on tungsten oxidised at temperatures in the range 750–800°C by krypton sorption and by metallographic methods. They discovered that the porous oxide did not form at a constant rate as required by the paralinear model; instead its rate of formation decreased with time with kinetics similar to those of weight gain *vs.* time. They concluded that the weight of combined oxygen in the barrier layer must be very small. Kellet and Rodgers[90] have since thrown even more doubt on the concept of a barrier layer by their finding that the black oxide 'barrier layer' and the yellow porous oxide have the same chemical composition.

Oxidation in the Presence of Subscales

The aforementioned inconsistencies between the paralinear model and actual observations point to the possibility that there is a different mechanism altogether. The common feature of these metals, and their distinction from cerium, is their facility for dissolving oxygen. The relationship between this process and an oxidation rate which changes from parabolic to a linear value was first established by Wallwork and Jenkins[91] from work on the oxidation of titanium. These authors were able to determine the oxygen distribution in the metal phase by microhardness traverses across metallographic sections; comparison of the results with the oxidation kinetics showed that the rate became linear when the metal surface reached oxygen saturation, at a composition of about Ti_3O. It was thought that the porous layer of TiO_2 was formed by exfoliation of metal layers but remained in good contact since (*a*) when the oxygen demand of the metal fell, cations diffused into, and discoloured, the oxide and (*b*) if the oxidising atmosphere was removed, the oxide redissolved.

The general thesis of this work was supported by the work of Osthagen and Kofstad[92] with zirconium in oxygen at 800°C, who found that the rate became linear at a surface composition of Zr_6O, and by that of Smeltzer, *et al.*[93]. The former authors related the rate change to the poor influence of the suboxide (which is believed to be volatile) on the interface adhesion. Pemsler[94], in an important series of papers, has developed the theme of oxidation of prior-formed zirconium–oxygen alloys but, curiously, was unable to reproduce the linear portion of the oxidation curve. He found the rate to remain parabolic at all temperatures up to 1300°C and with a rational rate constant which was independent of the degree of oxygen saturation. Pemsler developed a sensitive technique for the determination of the oxygen profile in the metal and showed ordered alloys to exist at the compositions ZrO_x, where x takes, in turn, the values 0·16, 0·21, 0·26 and 0·32.

Kofstad[95] had earlier suggested that the initial protective period in the oxidation of niobium (this metal belongs to the group which shows Type 1 behaviour) is to be interpreted in terms of the dissolution of oxygen and the formation of a suboxide. Breakaway then corresponds to the nucleation of the pentoxide Nb_2O_5 on the surface. Cathcart[96] and his co-workers observed blister-like cracks in the oxide at breakaway, which is to be expected since Nb_2O_5 grows by anion diffusion and has an oxide:metal volume ratio of 2·49:1. Later work has confirmed this and it thus seems possible that there is no barrier layer, merely an interaction with the underlying metal[97], at least for certain conditions of temperature and oxygen pressure.

Theories of the oxidation of tantalum in the presence of suboxide have been developed by Stringer[98]. By means of single-crystal studies he has been able to show that a rate anisotropy stems from the orientation of the suboxide which is precipitated in the form of thin plates. Their influence on the oxidation rate is least when they lie parallel to the metal interface, since the stresses set up by their oxidation to the pentoxide are most easily accommodated. By contrast, when the plates are at 45° to the surface, complex stresses are established which create characteristic chevron markings and cracks in the oxide. The cracks in this case follow lines of pores generated by oxidation of the plates. This behaviour is also found with niobium, but surprisingly, these pores are not formed when Ta–Nb alloys are oxidised[99], and the rate anisotropy disappears. However, the rate remains linear; it seems that this is another case in which molecular oxygen travels by submicroscopic routes.

The Rôle of Metal Dissolution or Volatilisation in the Formation of Porous Oxides

As we have seen, a consequence of the formation of porous oxide is that the rate-controlling step reverts to that of a phase boundary reaction and therefore becomes independent of the oxide thickness. When these circumstances are such that the porous oxide becomes unusually thick, or if the oxidising medium is unusually dense, a form of the parabolic rate law may be re-established. In this case the relevant diffusion coefficient is that of the transported component in the fluid phase permeating the oxide layer. Perhaps the best known system in which the apparently paradoxical association of porous oxide formation with parabolic kinetics is observed is that in which iron or mild steel reacts with water or alkaline solutions at temperatures within the region of 300°C. At this temperature water acts as an oxidising agent, even in the absence of electrochemical coupling, and very hard compact layers of oxide are formed. This oxide scale is found to consist of two layers, both magnetite, and marker experiments have been used to establish the important fact that the inner layer occupies exactly the volume of the metal consumed by oxidation[100]. It follows from these observations that the inner layer of oxide supports counter diffusing and nearly equivalent fluxes of iron and oxygen and, by inference from the observed parabolic rate law, one of these is rate controlling. Potter[101] considered the inner layer to be continuous and to grow by oxygen ion diffusion; however the rate of oxygen transport during corrosion is 10^8 times greater than would be expected to occur by diffusion in a continuous oxide film, and it is difficult to account for the equivalence of the ion fluxes over a wide range of temperature and solution composition[102].

The fact that the outer layer was porous was known to Potter and Mann, but the possibility that the inner layer was porous was first discussed by Field, *et al.*[103] who adduced evidence that the oxide had only 90% of its bulk theoretical density. It was subsequently shown that the inner layer consists of individual crystallites of 0·1–0·2 μm in size[104] and that the porosity between them was interconnected[6]. The knowledge that molecular water penetrated to the reaction zone close to the metal surface enabled Castle and Masterson[6] to construct a model in which the dissolution of the metal matrix represented a competing reaction with the growth of oxide nuclei. When dissolution is sufficiently fast the growing oxide nuclei can be undermined before a continuous stable film of oxide is formed. Since dissolution will also expose new sites for oxide nucleation the process can be repeated indefinitely so long as the resolved iron is removed from the vicinity of the metal surface. It is this efflux of iron in soluble form from the metal surface which becomes the rate-controlling transport reaction during the oxidation of steel in high-temperature water. The model is useful since it correctly predicts the dependence of the reaction rate on the solution pH[6,105]. This model, or variants[106] of it, is also able to explain the behaviour of aluminium in high-temperature water[107], of steels in molten salts[108] and of non-ferrous metals and alloys in low-temperature aqueous solutions[109,110]. There is also evidence that the transport of metals in volatile form, across similarly porous oxide, may be an important feature of oxidation in steam[83] and in special circumstances where the vapour pressure of the metal is high[82].

The more important cases of oxide volatilisation occur in the platinum metals[111] and with the refractory metals[2] at high temperatures. In these systems, unlike the aforementioned, it is the higher valence oxide which is the more volatile so that at sufficiently high temperature the metal may be oxide free. Gulbransen[2] has shown that the rate of oxidation is then constant and agrees well with that derived from the kinetic theory of gases and the relevant thermochemical data.

Stresses in Oxide Layers

Much of the earlier part of this chapter has dealt with the observable effects of stress in an oxide layer. Oxide buckling or even failure, recrystallisation, the promotion of columnar grain growth or whiskers, and, above all else, the creation of non-protective oxide, have all been attributed to stress. It has been shown (page **1**.243) how mass transport in oxides leads to oxide drift and thus, on finite and non-planar surfaces, to a change of shape with consequent generation of stress. Other mechanisms of stress generation are: epitaxial strain, i.e. the development of oxide layers with an orientation which permits some correspondence between the lattice parameter of the metal sub-lattice in the oxide and that of the metal; oxide formation in cracks or grain boundaries; the formation of higher oxides; oxygen solution in the metal, with or without subscale formation; and shrinkage of the metal by assimilation of vacancies from the metal oxide interface. The experimental evidence relating to these forms of stress generation has been reviewed by Stringer[112] in an article which draws together work from wide-ranging fields.

In order to understand the more catastrophic effects of oxide stress its magnitude must be determined in oxide layers which are still continuous. Several methods have been evolved to permit this measurement including low-energy electron diffraction (L.E.E.D.)[113] for the very early stages of oxide growth, x-ray diffraction-line broadening[114] and techniques in which the strain developed by the oxide layer when grown under conditions which permit stress relief is measured by bending or extension of the metal substrate[112,115,116]. This latter class encompasses a variety of geometrically designed metal testpieces. The method has become known as the *flexure* or Stoney[117] method and is worthy of special comment particularly since it is the method most widely adopted for measurements on thick film.

The Flexure Method

The method makes use of the tendency of a metal foil to bend when it is oxidised unilaterally, i.e. on one side only, and has been developed from the method used by Stoney to measure stresses in electrodeposits, as long ago as 1909. There are important problems in the translation of the technique to oxidation studies. Firstly, it is difficult to completely arrest oxidation on the 'inert' side of the testpiece. The problems of diffusion across a barrier layer of electrodeposited or vapour-deposited metal for a long time restricted its use to low temperatures, e.g. Ta at 350–550°C[118], Nb at 425°C[118] and Cu at 200–400°C, but recently Pawel and Cathcart have reported the use of Al–3Au evaporated layers which, on uranium alloys, will enable temperatures of up to 800°C to be used[119]. Secondly, there is an interpretive difficulty since although there are two stress systems, acting at right angles in the plane of the oxide, a strip (or even a disc) generally bends in a plane perpendicular to only one of them[120].

Because of the experimental difficulty other workers have circumvented the problem in various ways. In one of the first demonstrations of stress, Evans[121] examined the flexure of detached oxides at room temperature whilst soon after this Dankov and Churaev[122] examined the stresses present at reaction temperature by oxidation of layers of the metals evaporated onto mica substrates. Their technique has been criticised[29] on the grounds that the major stresses would refer to oxide formation in the pores of the evaporated layer—and also that the thin films oxidised may unduly reflect epitaxial stresses. Engell and Wever[123] oxidised both sides of a spiral of iron at 700°C and relied on the different lengths of the inward- and outward-facing surfaces of the helix to provide a differential force. This technique would be unduly influenced by rate differences on either side of the helix—especially as derived from the curvature itself[112]. Bradhurst and Heuer[124] also oxidised both sides of the testpiece and obtained their results from the change in curvature at room temperature when the oxide was removed from one face. The method was used to determine stresses in the oxide on Zircalloy 2 and zirconium at 500–700°C and is successful only if the oxide does not spall on cooling and if the correction for differential thermal expansion can be experimentally determined. Bradhurst and Heuer were successful and showed that stresses steadily increased until relieved on the Zircalloy 2 at the point of breakaway. Yet another technique has been utilised by Appleby and Tylecote[125] who protected one side of disc-shaped specimens by reducing atmospheres while the upper side was oxidised at temperatures up to 950°C.

The interpretive difficulty has been discussed in detail by Pawel[120] and by Morton[126] in papers which appeared almost simultaneously. Both authors use arguments to show that the simple formula of bending-beam theory utilised by Stoney

$$\sigma = \frac{Et^2}{6rd} \qquad \text{...(1.178)}$$

where E is Young's modulus, t is the thickness of metal, r the radius of curvature, d the oxide thickness and σ the stress, is suitable only as a first approximation.

Since there is an isotropic growth stress in the plane of the oxide it is necessary to consider the two principal stresses σ_x and σ_y given by

$$\sigma_x = \frac{E(\varepsilon_x + \nu\varepsilon_y)}{1-\nu^2} \qquad \text{...(1.179)}$$

for biaxial plane stress in a plate, and

$$\sigma_y = \frac{E(\nu\varepsilon_x + \varepsilon_y)}{1-\nu^2} \qquad \text{...(1.180)}$$

where ε_x and ε_y are the strains, x is the long axis and y the short axis of the testpiece, respectively, and ν is the Poisson's ratio. If the testpiece is free to bend in both directions under these stresses then $\varepsilon_x = \varepsilon_y$ and

$$\sigma_x = \frac{Et^2}{(1-\nu)\,6rd} \qquad \text{...(1.181)}$$

More usually however the initiation of bending on a plane perpendicular to y (say) and through x will render the testpiece sufficiently rigid to preclude bending in the orthogonal plane through y. As Morton points out this mode of deformation must inevitably occur in the helical testpieces used in the spiral contractometer. An appropriate formula is then $\varepsilon_y \approx 0$ and

$$\sigma_x = \frac{Et^2}{(1-v^2)\,6rd} \qquad \text{...(1.182)}$$

although Timoshenko's[127] early analysis shows that it is this relationship which reverts to the Stoney formula for suitably shaped specimens.

The rigidity of the y axis prevents the development of spherical surfaces for all but very small displacements. Morton suggests that the limit is reached when the displacement is equal to the metal thickness. This condition was satisfied in the high-temperature studies of Appleby and Tylecote[125] and spherical doming of the disc specimen occurred. When the oxide is not very thin compared with the metal both the moduli for oxide and metal must be considered. Stringer[112], in his excellent review of stress generation and relief in oxide layers, quotes a corrected formula, originally due to Brenner and Senderoff[128]

$$\sigma = E_{\text{ox.}}\,t\,\frac{(t+d)}{6rd} - \frac{(E_{\text{ox.}}-E_{\text{met.}})t^3}{6rd(t+d)} \qquad \text{...(1.183)}$$

(which omits the Poisson's ratio correction) and quotes data for magnesium which show that the simple formula would be in error by more than 50% when the oxide layer had a thickness of 10% of that of the metal. It is perhaps apposite to remind the reader that these corrections are pertinent to films which are very thick compared with those formed at breakaway in the Type 1 oxidation. Magnesium undergoes breakaway when the oxide has maintained a stable thickness of about 5×10^{-7} m for several hundred hours. Either on the Pilling–Bedworth model or the oxide drift argument the film would contain a biaxial stress of the order of only about 2 MN/m^2 tension or compression even if there has been no stress relief by plastic flow. The oxide should be well able to support this. In fact it is occasionally possible to evaporate magnesium leaving a near-perfect box of transparent oxide which shows no sign whatsoever of having been in strained conditions—yet eventually such magnesium will go into breakaway.

As reliable stress measurements become extended to greater temperatures the extent to which growth stresses are dissipated by plastic flow of the oxide becomes more apparent—the values of around 700 MN/m^2 measured for the oxidation of tantalum[118], niobium[118,129] and Zircalloy 2[124] must be con trasted with the value of 70 MN/m^2 (no breakaway) found for the oxidation of zirconium at 700–900°C[124] and 869 kN/m^2 for copper at this same temperature[125]. The stress measured for the formation of cuprous oxide at 900°C was only 371 kN/m^2[125]. The very low stress calculated for the oxidation of copper at 700°C is, however, surprising since the crack-heal mechanism of oxidation is known to be operative at slightly lower temperatures, and also in view of the direct relation between stress in the oxide scale and the formation of whiskers[130]. They are, however, in reasonable accord with the x-ray strain values measured by Homma and Issike[131] for the oxidation of copper at 500 and 600°C and the very high values are undoubtedly associated with oxygen solution and oxide wedging in the metal.

Stress Relief by Metal Creep

The fact that uniaxially oxidised metals bend suggests that thin foils should stretch when biaxially oxidised. This has been frequently observed[112] and moreover thin-walled tubing may both decrease in diameter[132] and increase in length[133,134] under the compressive loading of oxidation. Such a loading on the metal may seriously reduce its effective tensile strength[135] without relieving much of the compressive stress in the oxide, and introduces a dependence of creep behaviour on the oxygen potential of an environment[136]. Although outside the scope of this section, the paper on anodic oxidation of loaded aluminium wires by Leach and Neufeld[137] gives an indication of the probable depth of this field, in which little work has yet been carried out.

J. E. CASTLE

REFERENCES

1. Gregg, S. J. and Jepson, W. B., *J. Inst. Met.*, **87**, 187 (1958–59)
2. Gulbransen, E. A., *Corrosion*, **26**, 1 (1970)
3. Landsberg, P. T., *J. Chem. Phys.*, **23**, 1079 (1955)
4. Vernon, W., Akeroyd, E. and Stroud, E., *J. Inst. Met.*, **65**, 301 (1939)
5. Evans, U. R., *Trans. Electrochem. Soc.*, **91**, 547 (1947)
6. Castle, J. E. and Masterson, H. G., *Corros. Sci.*, **6**, 93 (1966)
7. Lorriers, J., *C.R. Acad. Sci., Paris.*, **229**, 547 (1949)
8. Lorriers, J., *C.R. Acad. Sci., Paris*, **231**, 522 (1950)
9. Bruckman, A., *Corros. Sci.*, **7**, 51 (1967)
10. Harrop, P. J., *J. Mats. Sci.*, **3**, 206 (1968)
11. Masters, B. C. and Harris, J. E., *Proc. Roy. Soc.*, **292A**, 240 (1966)
12. Boggs, W. E., Kachik, R. H. and Pellissier, G. E., *J. Electrochem. Soc.*, **112**, 539 (1965)
13. Taylor, M. E., Holmes, E. and Boden, P. J., *Corros. Sci.*, **9**, 683 (1969)
14. Pignocco, A. J. and Pellissier, G. E., *J. Electrochem. Soc.*, **112**, 1188 (1965)
15. Uhlig, H. H., *Corros. Sci.*, **7**, 325 (1967)
16. Sacks, K., *Metallurgia*, **54**, 11 (1956)
17. Mrowec, S. and Weber, T., *Acta Met.*, **8**, 819 (1960)
18. Jorgenson, P. J., *J. Chem. Phys.*, **37**, 874 (1962)
19. Bruckman, A. and Simkovich, G., *Corros. Sci.*, **12**, 595 (1972)
20. Hales, R., *Corros. Sci.*, **12**, 555 (1972)
21. Castle, J. E. and Hunt, M. R., to be published in *Corros. Sci.*
22. Pashley, D. W., *Advances Phys.*, **5**, 173 (1956)
23. Bruce, D. and Hancock, P., *J. Iron and Steel Inst.*, **208**, 1021 (1970)
24. Romanski, J., *Corros. Sci.*, **8**, 67 (2 papers) (1968)
25. Tylecote, R. F., *J. Iron Steel Inst.*, **195**, 380 (1960)
26. Boggs, W. E., Kachick, R. H. and Pellisier, G. E., *J. Electrochem. Soc.*, 1086 (1961)
27. Boggs, W. E., Trozzo, P. S. and Pellisier, G. E., ibid., 13 (1961)
28. Caplan, D. and Cohen, M., *Corrosion*, **15**, 141t (1969)
29. Vermilyea, D. A., *Acta Met.*, **5**, 492 (1957)
30. Howes, V. R., *Corros. Sci.*, **10**, 99 (1970)
31. Pfeil, L. B., *J. Iron and Steel Inst.*, **119**, 501 (1929)
32. Juenker, D. W., Meussener, R. A. and Birchenall, C. E., *Corrosion*, **14**, 57 (1958)
33. Pemsler, J., *J. Electrochem. Soc.*, **112**, 477 (1965)
34. Caplan, D. and Cohen, M., *Corros. Sci.*, **6**, 521 (1966)
35. Caplan, D. and Cohen, M., *Corros. Sci.*, **7**, 725 (1967)
36. Caplan, D., Graham, M. J. and Cohen, M., *Corros. Sci.*, **10**, 1 (1970)
37. Hussey, R. J. and Cohen, M. J., *Corros. Sci.*, **11**, 699 (1971)
38. Hussey, R. J. and Cohen, M. J., *Corros. Sci.*, **11**, 713 (1971)

39. Caplan, D., Sproule, G. I. and Hussey, R. J., *Corros. Sci.*, **10**, 9 (1971)
40. Caplan, D., Graham, M. J. and Cohen, M., *J. Electrochem. Soc.*, **119**, 1205 (1972)
41. Bruce, D. and Hancock, P., *J. Inst. Met.*, **97**, 140 (1969)
42. Douglass, D. L., *Corros. Sci.*, **8**, 665 (1968)
43. Mrowec, S. and Weber, T., *J. Electrochem. Soc.*, **117**, 1531 (1970)
44. Antill, J. E., Peakall, K. A. and Warburton, J. B., *Corros. Sci.*, **8**, 689 (1968)
45. Whittle, D. P. and Wood, G. G., *Corros. Sci.*, **8**, 295 (1968)
46. Sartell, J. A. and Li, C. H., *J. Inst. Met.*, **90**, 92 (1961)
47. Kofstad, P. N. and Hed, A. Z., *J. Electrochem. Soc.*, **116**, 1542 (1969)
48. Wulf, G. L., Carter, J. J. and Wallwork, G. R., *Corros. Sci.*, **9**, 689 (1969)
49. Tuck, C. W., Odgers, M. and Sacks, K., *Corros. Sci.*, **9**, 271 (1969)
50. Evans, U. R., *The Corrosion and Oxidation of Metals*, Arnold, London (1961)
51. Boggs, W. E., *J. Electrochem. Soc.*, **108**, 124 (1961)
52. Castle, J. E., Gregg, S. J. and Jepson, W. B., *J. Electrochem. Soc.*, **109**, 1018 (1962)
53. Uhlig, H. H., *Acta Met.*, **4**, 541 (1956)
54. Uhlig, H. H., Pickett, J. and MacNairn, J., *Acta Met.*, **7**, 111 (1959)
55. Tylecote, R. F. and Michett, T. E., *J. Iron and Steel Inst.*, **196**, 445 (1960)
56. Boggs, W. E. and Kachik, R. H., *J. Electrochem. Soc.*, **116**, 424 (1969)
57. Wood, G. C., Richardson, J. A., Hobby, M. C. and Boustead, J., *Corros. Sci.*, **9**, 655 (1969)
58. Birks, N. and Rickert, H., *J. Inst. Met.*, **91**, 30 (1962)
59. Wood, G. C., Wright, J. G. and Fergusson, J. M., *Corros. Sci.*, **5**, 645 (1965)
60. Fuji, C. T. and Meussner, R. A., *J. Electrochem. Soc.*, **114**, 435 (1967)
61. Mrowec, S., *Corros. Sci.*, **7**, 563 (1967)
62. Antill, J. E., Peakall, K. A. and Warburton, J. B., *Corros. Sci.*, **8**, 689 (1968)
63. Birks, N., *British Corros. J.*, **3**, 56 (1968)
64. Cox, B., *J. Nuc. Mats.*, **27**, 1 (1968)
65. Gibbs, G. B., *Corrosion Sci.*, **7**, 165 (1967)
66. Smeltzer, W. W., Haering, R. R. and Kirkaldy, J. S., *Acta. Met.*, **9**, 880 (1961)
67. Birchenal, C. E., *J. Electrochem. Soc.*, **103**, 619 (1956)
68. Pilling, N. B. and Bedworth, R. E., *J. Inst. Met.*, **29**, 529 (1923)
69. Tylecote, R. F., *J. Inst. Met.*, **81**, 681 (1953)
70. Caplan, D. and Cohen, M., *J. Metals, Trans. Amer. Inst. Min. (Metall.) Engrs.*, **203**, 336 (1955)
71. Mortimer, D. and Post, M. L., *Corros. Sci.*, **8**, 499 (1968)
72. Cathcart, J. V., Hall, C. C. and Smith, G. P., *Acta. Met.*, **5**, 249 (1957)
73. Howland, W. H. and Epstein, L. F., *Ind. Eng. Chem.*, **49**, 1931 (1957)
74. Antill, J. E., Campbell, C. B., Goodison, D., Jepson, W. B. and Stevens, C. G., *Proc. 3rd Conf. Peaceful Uses of Atomic Energy*, **9**, 523 (1964)
75. Gregg, S. J. and Jepson, W. B., *J. Chem. Soc.*, 712 (1960)
76. Aylmore, D. W., Gregg, J. J. and Jepson, W. B., *J. Electrochem. Soc.*, **106**, 1010 (1959)
77. Castle, J. E. and Wood, C. G., *Scanning Electron Microscopy* (Ed. O. Johari), I.I.T. Research Inst., Chicago, 39 (1968)
78. Aylmore, D. W., Gregg, S. J. and Jepson, W. B., *J. Nuc. Mats.*, **3**, 190 (1961)
79. Castle, J. E. and Mann, G. M. W., *Corros. Sci.*, **6** (1966)
80. Fiegna, A. and Weisgerber, P., *J. Electrochem. Soc.*, **115**, 369 (1968)
81. Cox, B., *J. Nuc. Mats.*, **41**, 96 (1971)
82. Dupre, B. and Shreiff, R., *J. Nuc. Mats.*, **42**, 260 (1972)
83. Surman, P. L. and Castle, J. E., *Corros. Sci.*, **9**, 771 (1969)
84. Fern, F. H. and Antill, J. E., *Corros. Sci.*, **10**, 649 (1970)
85. Bradhurst, D. H. and Heuer, P. M., *J. Nuc. Mats.*, **41**, 101 (1971)
86. Roy, C. and David, G., *J. Nuc. Mats.*, **37**, 71 (1970)
87. Cox, B., *J. Nuc. Mats.*, **31**, 48 (1969)
88. Webb, W. W., Norton, J. T. and Wagner, C., *J. Electrochem. Soc.*, **103**, 107 (1956)
89. Jepson, W. B. and Aylmore, D. W., *J. Electrochem. Soc.*, **108**, 942 (1961)
90. Kellett, E. A. and Rodgers, S. E., *J. Electrochem. Soc.*, **110**, 503 (1965)
91. Wallwork, G. R. and Jenkins, N. E., *J. Electrochem. Soc.*, **106**, 10 (1959)
92. Osthagen, K. and Kofstad, P., *J. Electrochem. Soc.*, **109**, 204 (1962)
93. Smeltzer, W. W., *Can. Met. Quart.*, **11**, 41 (1962)
94. Pemsler, J., *J. Electrochem. Soc.*, **111**, 383 (1964)
95. Kofstad, P., *Proc. 1st Int. Cong. Met. Corrosion, London, 1961*, Butterworths, London, 181 (1962)

96. Cathcart, J. V., Campbell, J. J. and Smith, G. P., *J. Electrochem. Soc.*, **105**, 442 (1958)
97. Weirich, L. J. and Larsen, W. L., *J. Electrochem. Soc.*, **119**, 465 (1972)
98. Stringer, J., *J. Less Common Metals*, **12**, 301 (1967)
99. Dooley, R. B. and Stringer, J., *J. Less Common Metals*, **24**, 139 (1971)
100. Potter, E. C. and Mann, G. M. W., *Proc. 1st Int. Cong. Met. Corrn.*, London, 1961, Butterworths, London, 417 (1962)
101. Potter, E. C., *Mitt. V.G.B.*, **76**, 19 (1962)
102. Castle, J. E. and Surman, P. L., *J. Phys. Chem.*, **71**, 4255 (1967)
103. Field, L. M., Stanley, R. C., Adams, A. M. and Holmes, D. R., *Proc. 2nd Int. Cong. Met. Corrosion*, N.A.C.E., New York (1963)
104. Harrison, P. L., Holmes, D. R. and Teore, P., V.G.B. Conference on Feed-water Treatment, Essen (1965)
105. Potter, E. C. and Mann, G. M. W., *B. Corrosion J.*, **1**, 26 (1965)
106. Bignold, G. J., Garnsey, R. and Mann, G. M. W., *Corros. Sci.*, **12**, 325 (1972)
107. Castle, J. E., unpublished work
108. Holmes, D. R., Discussion of Paper by A. U. Seybolt, *4th Int. Cong. on Met. Corros.*, Amsterdam, 560 (1969)
109. Vermilyea, D. A. and Vedder, W., *Trans. Farad. Soc.*, **66**, 2644 (1970)
110. Green, J. A. S., Mengelberg, H. D. and Yolhen, H. T., *J. Electrochem. Soc.*, **117**, 433 (1970)
111. Betteridge, W. and Rhys, D. W., *1st Int. Cong. Met. Corros.*, Butterworths, London, 186 (1962)
112. Stringer, J., *Corros. Sci.*, **10**, 513 (1970)
113. Sickafus, E. N. and Bonzel, H. P., *Recent Progress in Surface Science IV*, A.P., New York and London (1971)
114. Swank, T. F. and Lawless, K. R., *Advances in X-ray Analysis*, **10**, Plenum Press, New York, 234 (1966)
115. Pawel, R. E., Cathcart, J. V. and Campbell, J. J., *J. Electrochem. Soc.*, **110**, 551 (1963)
116. Jaenicke, W., Leistikow, S. and Städder, J., *J. Electrochem. Soc.*, **111**, 1031 (1964)
117. Stoney, G. C., *Proc. Roy. Soc., Lond.*, **A82**, 172 (1909)
118. Pawel, R. E. and Campbell, J. J., *Acta Met.*, **14**, 1827 (1966)
119. Pawel, R. E. and Cathcart, J. V., *J. Electrochem. Soc.*, **118**, 1776 (1971)
120. Pawel, R. E., *J. Electrochem. Soc.*, **116**, 1144 (1969)
121. Evans, U. R., *Inst. Met. Symp. on Stresses in Metals*, 219 (1947)
122. Dankov, D. D. and Churaev, P. V., *Dokl. Akad. Nauk. SSSR*, **73**, 1221 (1950)
123. Engell, H. and Wever, F., *Acta Metall.*, **5**, 695 (1957)
124. Bradhurst, D. H. and Heuer, P. M., *J. Nuc. Mats.*, **37**, 35 (1970)
125. Appleby, W. K. and Tylecote, R. F., *Corros. Sci.*, **10**, 325 (1970)
126. Morton, V. M., *Corros. Sci.*, **9**, 261 (1969)
127. Timoshenko, S., *Mech. Engrs.*, **45**, 259 (1923)
128. Brenner, A. and Senderoff, S., *J. Res. Natn. Bur. Stand.*, **42**, 105 (1949)
129. Weirich, L. J. and Larsen, W. L., *J. Electrochem. Soc.*, **119**, 465 (1972)
130. Sartell, J. A., Stokes, R. J., Bendel, S. H., Johnson, T. L. and Li, C. H., *Trans. A.I.M.M.E.*, **215**, 211 (1959)
131. Homma, T. and Issiki, S., *Acta Met.*, **12**, 1092 (1964)
132. Moore, W. J., *J. Chem. Phys.*, **21**, 1117 (1953)
133. Hoden, J. D., Knight, C. J. and Thomas, M. W., *Br. Corros. J.*, **3**, 47 (1968)
134. Buresch, F. L. and Bollenrath, F., *J. Nuc. Mats.*, **248**, 270 (1967)
135. Weirich, L. J. and Larsen, W. L., *J. Electrochem. Soc.*, **119**, 472 (1972)
136. Knights, C. E. and Perkins, R., *J. Nuc. Mats.*, **36**, 180 (1970)
137. Leach, J. S. L. and Neufeld, P., *Proc. Br. Ceramic Soc.*, 49 (1966)

2 ENVIRONMENTS

2.1 Effect of Concentration, Velocity and Temperature

In Chapter 1 the electrochemical mechanism of corrosion was considered in detail and it was shown how the kinetics of cathodic and anodic partial reactions control the rate of overall corrosion reaction. In this section the effects of environmental factors such as concentration, velocity and temperature will be considered on the assumption that either the anodic or cathodic reaction, but not both, is rate controlling. Thus if a metal is corroding under cathodic control it is apparent that the velocity of the solution will be more significant when diffusion of the cathodic reactant is rate controlling, although temperature may still have an effect. On the other hand if the cathodic process requires a high activation energy, temperature will have the most significant effect.

The effects of concentration, velocity and temperature are complex and it will become evident that these factors can frequently outweigh the thermodynamic and kinetic considerations detailed in Section 1.4. Thus it has been demonstrated in Chapter 1 that an increase in hydrogen ion concentration will raise the redox potential of the aqueous solution with a consequent increase in rate. On the other hand, an increase in the rate of the cathodic process may cause a decrease in rate when the metal shows an active/passive transition. However, in complex environmental situations these considerations do not always apply, particularly when the metals are subjected to certain conditions of high velocity and temperature.

The Effect of Anion Concentration on the Rate of Corrosion

The numerous metals and alloys used in practice show such a wide variation in response to various anions in acid and alkaline solutions that common features are difficult to discern and a basis for predicting corrosion behaviour is not very apparent.

Although Pourbaix (potential–pH) diagrams have led to a greater understanding of the changes in the corrosion behaviour of a metal due to a change in pH, they are less instructive about the behaviour of alloys and about the maxima in the corrosion rate that frequently occurs with increasing concentration of acid. For instance, Fe–18Cr–8Ni stainless steel has a low rate of corrosion in dilute sulphuric acid which increases with increase in concentration, to a maximum, followed by a decreasing rate on further increase of concentration (Fig. 2.1). Recently thermochemical data has become available so that potential–pH diagrams[1–4] have been calculated for anions such as SO_4^{2-}, Cl^-, citrate, S^{2-}, etc. However, there is no method of calculating the

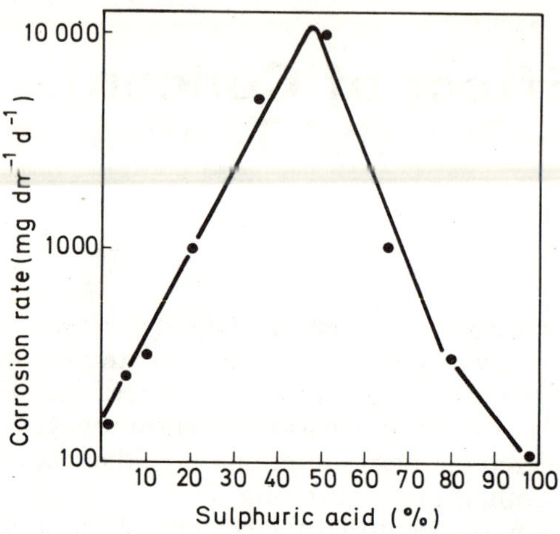

Fig. 2.1 Corrosion rate of Fe–18Cr–8Ni as a function of sulphuric acid concentration (20°C)

relationships for alloys, although present studies on alloy dissolution[5,6] suggest that for some simple alloy systems this may be possible since dissolution of each phase appears to be a simple additive mechanism.

Anodic Dissolution under Film-free Conditions

With regard to the anodic dissolution under film-free conditions in which the metal does not exhibit passivity, and neglecting the accompanying cathodic process, it is now generally accepted that the mechanism of active dissolution for many metals results from hydroxyl ion adsorption[7–9], and the sequence of steps for iron are as follows:

$$Fe + H_2O \rightleftharpoons Fe(OH)^-_{ads.} + H^+$$
$$Fe(OH)^-_{ads.} \rightleftharpoons Fe(OH)_{ads.} + e$$
$$Fe(OH)_{ads.} \rightleftharpoons Fe^{2+} + OH^- + e$$

where $Fe(OH)^-_{ads.}$ signifies adsorption of OH^- on the Fe surface.

The most important outcome of this theory is that the rate of dissolution should be potentially greater as the pH increases, which is in conflict with simple concepts of corrosion kinetics. However, the theory has been proved to be applicable to many systems, and Bonhoeffer and Heusler[8] found that iron in sulphuric acid corroded at a greater rate with increase in pH, whilst Kabanov *et al.*[9] found that it corroded faster in alkaline solution than in acid solution for the same electrode potential.

Since the hydroxyl *anion* is involved in the mechanism given before, the implication is that other anions may also take part in the dissolution process, and that the effect of various chemicals may be interpreted in the light of the effect of each anion species. Most studies have been in solutions of sulphuric and hydrochloric acid and typically the reaction postulated for active dissolution in the presence of sulphuric acid is:

$$Fe(OH)_{ads.} + SO_4^{2-} \rightarrow FeSO_4 + OH^- + e,$$

followed by dissociation to Fe^{2+} and SO_4^{2-} aquo ions.

Studies of other metals in sulphuric acid[10], hydrochloric acid plus sodium chloride[11] and formic acid strongly support direct anion participation in the dissolution process. In general, it appears that whilst OH^- ions (and water molecules) have the largest accelerating effect on the rate of corrosion, other anions are also effective and this explains why some strong acids are more aggressive than others, in that they have different abilities to compete with hydroxyl ions in the dissolution process. This effect of different anions in increasing the rate of dissolution manifests itself as an increase in the exchange current density, as shown by Bockris *et al.*[7] who gave the following series of anions in order of acceleration of dissolution:

$$NO_3^- < CH_3COO^- < Cl^- < SO_4^{2-} < ClO_4^-$$

However, in the pH range 1–4, the effect of the OH^- ion predominates to such an extent that corrosion rates are similar in the presence of many other anions at concentrations less than 0·1M. Since an adsorption process is involved in the mechanism, the corrosion rate in the pH range 1–4 may be represented by the Freundlich equation:

$$\text{Corrosion rate} = KC^n_{OH^-}$$

where C_{OH^-} is the concentration of OH^- ions and n is a small integer, often $=2$. At higher pHs and concentrations of anions the rate of corrosion can be markedly reduced by either (*a*) precipitation and crystallisation on the surface of corrosion products, or (*b*) adsorption of specific anions that cover the surface and decrease adsorption of OH^-, i.e. competitive adsorption.

An important example of (*a*) is mild steel which may be used for containing concentrated sulphuric acid (greater than 70% H_2SO_4) because of the process of sulphation[12] or in the case of 10M phosphoric acid[13] because of phosphatisation by ferrous phosphate; in each case the salt crystallises on the metal surface forming a mechanical barrier. Under these circumstances, and providing the salt layer is not disturbed by mechanical scraping or by flow of the solution, the corrosion rate will decrease to a low level. An example of (*b*) is the decrease in the corrosion rate of iron in dilute sulphuric acid caused by halide ions, e.g. I^- ions lower the rate by 95% in 1N H_2SO_4, perhaps as a consequence of the high polarisability of halide anions[14].

An increase followed by a decrease in corrosion rate at a certain critical concentration is a commonly observed phenomenon for many metals and alloys. If the anion concentration at which the decrease takes place is high, then the anion species is deemed to be aggressive, but if low the anion is referred to as inhibitive. A considerable amount of experimental work in relation to the effect of specific ions on corrosion has been carried out on mild steel[15–17] and zinc[18].

Film-forming Conditions

The corrosion rate of many important metals and alloys is controlled by the formation of a passive film, and the thermodynamics and kinetics of their formation and breakdown are dealt with in Section 1.2.

The dissolution of passive films, and hence the corrosion rate, is controlled by a chemical activation step. In contrast to the enhancement of the rate of dissolution by OH^- ions under film-free conditions, the rate of dissolution of the passive film is increased by increasing the H^+ ion concentration, and the rate of corrosion in film-forming conditions such as near-neutral solutions follows the empirical Freundlich adsorption isotherm:

$$\text{Corrosion rate} = KC^n_{H^+}$$

where n is an integer and C_{H^+} is the concentration of hydrogen ions. It has been observed that in general, rates are controlled mainly by this equation, but the nature and concentration of anions do have an effect. Many anions such as Cl^- appear to be capable of causing pitting and breakdown of the film as the concentration increases. As the concentration in the bulk solution increases, corrosion products precipitate in the pits and blocking occurs with a subsequent reduction in dissolution rate.

For both film-free and film-forming conditions a decrease in corrosion rate is observed as the concentration of the anion increases. For some anions the maximum in the corrosion rate may be attained at low concentrations depending on the species and concentration (Fig. 2.2). One form of inhibition

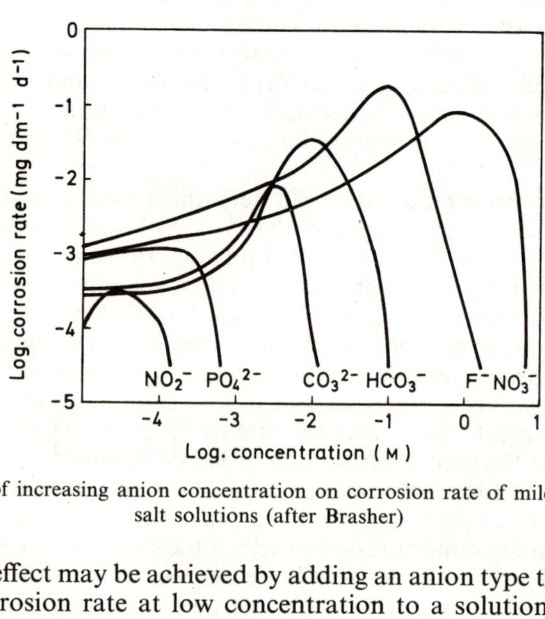

Fig. 2.2 Effect of increasing anion concentration on corrosion rate of mild steel in sodium salt solutions (after Brasher)

based on this effect may be achieved by adding an anion type that reaches its maximum corrosion rate at low concentration to a solution containing a more aggressive anion. Legault *et al.*[19] have studied this subject in detail and show that a Freundlich-type equation may still apply:

$$\text{Corrosion rate} = a + b\,\frac{C_{\text{agg.}}}{C_{\text{ihb.}}}$$

where a and b are constants, $C_{\text{agg.}}$ is the concentration of anions which stimulate corrosion and $C_{\text{ihb.}}$ is the concentration of anion type which precipitates and blocks the surface.

Effect of Solution Velocity on the Rate of Dissolution

Since the effect of concentration has been shown to lead usefully to salt passivation in many metal/anion systems the flow rate will markedly affect precipitation and subsequent corrosion behaviour. There have been many investigations of the influence of flow on the rate of corrosion, but lack of an awareness of hydrodynamic parameters has led to many experiments of questionable validity. In these circumstances it is not surprising that results from service failures do not correlate with laboratory tests. Some of these difficulties arise from the incomplete understanding of the theory of mass transport that still exists, especially in concentrated solution, although methods are now becoming available that are leading to more accurate prediction of corrosion rates in flowing systems. The most successful application of hydrodynamic theory to date has been for metals dissolving under essentially film-free conditions for a process that is unambiguously controlled by the arrival of the reactants at the surface, i.e. when the activation process is very fast compared to diffusion. For this reason it is customary to decide for each corrosion process which of the four main processes predominates and controls the corrosion rate. These four processes arise from the two electrochemical reactions, anodic and cathodic, and whether activation or concentration overpotential is the dominant process. Thus, of the four possibilities the rate of mass transport will be involved in at least three of them as follows:

Anode:	act.	act.	conc.	conc.
Cathode:	act.	conc.	act.	conc.

In contrast, the temperature may only become important in the case of perhaps one combination since this parameter will have the greatest influence on activation-controlled processes. This arises because corrosion processes controlled by concentration overpotential have a limiting diffusion current, which in many cases imposes a maximum value on the corrosion rate even when activation polarisation is decreased (see Section 9.1). When concentration overpotential predominates then the limiting current will give a good estimate of the corrosion rate. Since the limiting current is determined by the flow-rate it should be possible to predict the changes in corrosion rate if the relationship between flow and concentration overpotential is known. If the limiting current density of the corrosion cell exceeds the critical current density for passivation then increased flow rate could lead to a marked decrease in dissolution.

Concentration Overpotential at an Anode

There is an important difference between anode and cathode concentration overpotential. In the former, where anions directly participate in the dissolution process, ions accumulate at the surface and the rate is governed by metal ions moving out to the bulk solution; in the latter the rate is controlled by cathodic reactants (hydrogen ions, dissolved oxygen) moving towards the surface. In the case of cathode processes the limiting diffusion current is due to the depletion of ions as a result of the high rate of reaction, but in anodic processes no such limit is possible. There is, however, a type of rate-limiting behaviour when the solution next to the surface becomes saturated and crystallisation on the surface occurs. Solution flow will stimulate both electrode reactions by providing fresh solution with more ions for the cathodic process and with more water molecules to dilute the saturated solution formed at the anode. The rate of corrosion when concentration overpotential is controlling is governed by the diffusion of ions and the length of the path between concentration of ions in the bulk solution and at the surface. Fick's first law* can be applied to this situation and the rate/unit area, in terms of the current density i, can be described by:

$$i = \frac{zFD(a_s - a_b)}{\delta(1-t)}$$

where z = number of electrons in the electrode process,
F = Faraday constant (10^5 C),
D = diffusion coefficient (m^2/s),
a_s = activity of ions at the surface (mol/m^3),
a_b = activity of ions in the bulk solution (mol/m^3),
δ = path length of diffusing ions (m) and
t = transport number of all the ions involved in the electrode process; in high concentrations of electrolytes, NaCl, Na_2SO_4, HCl, etc. the charge is carried by ions not involved in the corrosion process and may be assumed to be unity.

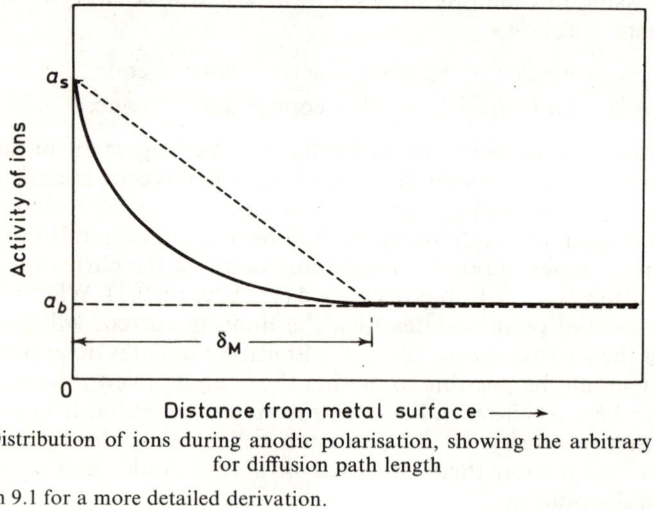

Fig. 2.3 Distribution of ions during anodic polarisation, showing the arbitrary value used for diffusion path length

*See Section 9.1 for a more detailed derivation.

The rate is a maximum when $a_s = 0$ for a cathodic process and when it reaches saturation for an anodic process. For isothermal conditions D is a constant, and at the limiting current $(a_s - a_b)$ is a constant, and in this case the controlling factor is δ the path length for diffusion. Thus the variation of δ with solution flow rate determines the corrosion rate.

The change in activity along the diffusion path length is unlikely to be linear and from diffusion theory the distribution of ions is most likely to be that given in Fig. 2.3. The distribution shows the difficulty of choosing a value for δ and a compromise value is used, δ_M, which is the effective diffusion layer thickness when the diffusion gradient is assumed to be linear. The effect of changing velocity on corrosion processes can best be understood through the factors that change δ_M. Fortunately, there are many practical examples of corrosion where the rate is wholly controlled by either the anode or cathode concentration overpotential so that the parameters that control the effective layer thickness should be known. This is best done by the use of hydrodynamic theory.

Application of Hydrodynamics

It is important at the outset to define more closely the effective film thickness δ_M. In any electrolyte solution in contact with a metal surface there is a static layer of solution next to the surface whose thickness will decrease as the solution velocity increases. The way in which this velocity changes the hydrodynamic thickness (δ_H) is complex and depends on such factors as viscosity, geometry, temperature and surface roughness*.

It is also necessary to separate *laminar* flow when a stagnant layer of well defined thickness δ_H is formed, from *turbulent* flow when values of δ_H are very low and when flow towards and away from the surface is complex. The analogy of mass transport with heat transfer[20,21] has led to successful methods of regarding the mass transport interaction with fluid flow, since the behaviour of heat is in many ways similar to mass transport depending as it does on a driving force, i.e. the heat gradient may be regarded as analogous to the concentration gradient.

Relationships between δ_H and δ_M have been established for certain geometries, e.g. for a rotating disc Levich[21] has found that for laminar flow $\delta_H \approx 5\delta_M$. The mathematical proofs of these relationships are not appropriate here, but a useful non-mathematical account of the application of hydrodynamic theory to mass transport has been given by King[22]. The most important variables are the main stream solution velocity U, the characteristic length L (diameter in the case of a rotating cylinder) and the kinematic viscosity ν (m^2/s). For application to mass transport such as cathodic reduction or anodic dissolution, which are dependent on ion or molecular (dissolved oxygen) transport, the variables are the diffusion coefficient, D, and the activity difference between surface and main solution, Δ_a (mol/m^3).

Using the mathematical technique of dimensionless group analysis, the rate of mass transport (R_M) in terms of mass per unit area per unit time can be shown to be a function of these variables, which when grouped together

*Reference 20 gives definitions pertaining to the various diffusion layers for the special case of electrochemical mass transfer.

can be related to the rate by a power term. For many systems under laminar flow conditions it has been shown that the following relationship holds:

$$\frac{R_M L}{D\Delta_a} = K\left(\frac{UL}{\nu}\right)^x \left(\frac{\nu}{D}\right)^y \qquad \ldots(2.1)$$

where K is a constant, and x and y are exponents that are very often $\frac{1}{2}$ and $\frac{1}{3}$, respectively. The dimensionless groups in equation 2.1 are referred to as follows:

$$\frac{UL}{\nu} = (Re) \text{ (Reynolds No.)}; \quad \frac{\nu}{D} = (Sc) \text{ (Schmidt No.)};$$

$$\frac{R_M L}{D\Delta_a} = (Sh) \text{ (Sherwood No.)};$$

where U is the main stream velocity (m/s), D is the diffusion coefficient (m^2/s), ν is the viscosity (m^2/s) and L is the characteristic length (m). The application of this equation is only useful if:

1. The relationship between δ_H and δ_M is known.
2. Concentration has a power exponent of unity, i.e. conforms to a first order reaction.
3. Dissolution is uniform (etching); otherwise, for rough surfaces such as pitting, turbulent flow regimes may occur even at low solution velocities.

The rate R_M can be converted to the limiting current density i_L by Faraday's law, so that from physical measurements on the solution it is possible to calculate $i_{corr.}$, since it is of the same order as i_L.

It has not been possible to calculate constant K or exponents x and y directly from theory, except in onc case, so that they have to be determined by experiment. The geometry of a system, i.e. flat plates, stirrers, pipes, etc. have a large effect in determining the magnitude of the constants and by using reactions whose parameters are well established for one geometry it has been possible to gather data for many other systems, by using the known reactions in other geometries.

In the case of the one system that can be predicted from theory, i.e. the rotating-disc electrode, this is proving to be a useful tool for understanding the effects of flow on corrosion reactions. The equation can be rearranged to give the limiting current i_L, and the characteristic length × velocity UL term can be interpreted as the angular velocity ω. The equation developed by Levich[21] by substituting in equation 2.1 is:

$$i_L = 0{\cdot}62zF\left(\frac{UL}{\nu}\right)^{\frac{1}{2}} \left(\frac{\nu}{D}\right)^{\frac{1}{3}} D\Delta_a \qquad \ldots(2.2)$$

Substituting for UL by ω and for the limiting case for the cathodic reduction process when $a_s = 0$, the activity term Δ_a is then equal to a_b (the concentration of ions in the bulk solution), and then

$$i_L = 0{\cdot}62zF\omega^{\frac{1}{2}}\nu^{-\frac{1}{6}}D^{\frac{2}{3}}a_b \qquad \ldots(2.3)$$

Zembura[23] has made specific use of the rotating disc for investigation of the effect of flow on corrosion reactions. This work has shown that it is possible to determine the type of control (activation or concentration polarisation) of zinc dissolving in 0·1N Na_2SO_4 (de-aerated), which followed closely the predicted increase in hydrogen ion reduction as the flow rate increased, and proved that in this example

$$i_{\text{corr.}} \approx i_L$$

Although the rotating disc is useful in understanding the mechanism of corrosion it is necessary to evaluate flow rates in the turbulent region which characterises the effect of velocity in real systems. King[22] collected data for rotating cylinders, which give turbulent flow regimes*, even at low rotation speeds. For many systems he found that several data fitted the relationship:

$$i_L = 1{\cdot}7zFa_bU\left(\frac{D}{\nu}\right)^{0{\cdot}83}$$

where i_L is in A/m², a_b is the activity in the bulk solution and the equation assumes that the reaction is under cathodic control.

The corrosion rates in pipes and annuli are very important, but for these geometries the effect must be evaluated experimentally, since the theory cannot predict the relationships with any accuracy. Although the importance of the interaction of flow rate and corrosion has long been appreciated, there have been very few studies where hydrodynamic parameters have been considered or even measured and controlled[24,25]. However, there has been much work where these factors have been taken into consideration for

Table 2.1 Corrosion rates from hydrodynamic parameters for pipes and annuli

Flow regime	*Appropriate equation*	*Ref. No.*
Pipes 1. Laminar flow $(Re) > 2\,000$	$\frac{i_L}{zFD\Delta_a} = 1{\cdot}64\left[(Re)(Sc)\frac{\text{diameter}}{L}\right]^{0{\cdot}33}$ $\left[\text{only if } (Re)(Sc)\frac{d}{L} < 8\right]$	25
2. Turbulent flow $(Re) < 5\,000$	$\frac{i_L}{zFD\Delta_a} = 0{\cdot}276(Re)^{0{\cdot}58}\left[(Sc)\frac{\text{diameter}}{L}\right]^{0{\cdot}33}$	25
3. Turbulent flow $(Re) > 5\,000$	$\frac{i_L}{zFD\Delta_a} = 0{\cdot}023(Re)^{-0{\cdot}2}(Sc)^{0{\cdot}66}\left(\frac{UD}{\text{diameter}}\right)$	27
Annuli 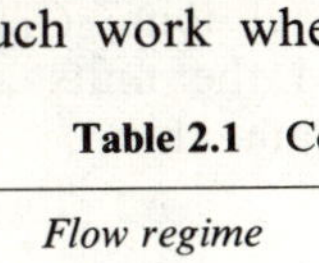 1. Laminar flow $(Re) > 2\,000$	$\frac{i_L}{zFD\Delta_a} = K\left[(Re)(Sc)\frac{d_e}{L}\right]^{0{\cdot}33}$, radius ratio $= \frac{\text{core radius}}{\text{outer radius}}$ When radius ratio = 0·25, K = 1·8 and when radius ratio = 0·125, K = 2·03; d_e = annular equivalent diameter	27
2. Turbulent flow	for radius ratio 0·5: $\frac{i_L}{zFD\Delta_a} = 0{\cdot}276U(Sc)^{-0{\cdot}66}(Re)^{-0{\cdot}42}\left(\frac{d_e}{L}\right)^{0{\cdot}33}$	27

*For rotating cylinders the exponent x for Reynolds number is very often unity for turbulent flow, and therefore L may be included in the constant term for a particular geometry of cylinder.

corroding systems[26,27], and the work of Ross and Wragg[27] contains a valuable review of previous work. A great deal of work has been done in the general field of electrochemistry which may be applicable to corroding electrodes when concentration polarisation of one of the half reactions predominates. Table 2.1 gives some useful equations for pipes and annuli.

Effect of Flow During Conditions that Lead to Film Formation

For many corroding systems the rate is controlled by the presence of a film of corrosion products. This may range from the relatively thin films on metals such as aluminium and magnesium to massive deposits found on corroding iron pipes. Both situations have been considered from the viewpoint of hydrodynamic theory. In the case of magnesium corroding in hydrochloric acid, Marangozis[28] has derived a relationship that includes chemical dissolution of $Mg(OH)_2$ as a controlling factor, the rate being controlled by diffusion of OH^- ions. For rotating cylinders the rate of dissolution followed the relationship

$$R_{Mg^{2+}} \doteq 2{\cdot}10^7 (K_{OH^-})(C_{Mg^{2+}})$$

where K_{OH^-} is the mass transport coefficient of hydroxyl ions, $C_{Mg^{2+}}$ is derived from the solubility product of $Mg(OH)_2$ and the units are $gm^{-2}d^{-1}$. K_{OH^-} can be calculated from purely physical data, thus

$$K_{OH^-} = K^0_{OH^-}\left[\frac{(Sc)^{0{\cdot}33}_{H^+}}{(Sc)^{0{\cdot}33}_{OH^-}} + \frac{(Sc)^{0{\cdot}66}_{OH^-}}{(Sc)^{0{\cdot}66}_{H^+}}\right]\frac{(C_{H^+})}{(C_{OH^-})}$$

where (Sc) = Schmidt number (ν/D),
(C_{H^+}) = concentration (or activity) in main solution,
(C_{OH^-}) = concentration (or activity) at the surface which may be obtained from the solubility product of $Mg(OH)_2$, and
$K^0_{OH^-}$ = the mass transport coefficient of hydroxyl ions *without* chemical reaction and is obtained from:

$$K^0_{OH^-} = \left[0{\cdot}143(Sc)^{0.33}_{OH^-}\left(\frac{Uh}{\nu}\right)^{0{\cdot}7}\left(\frac{D}{r+h}\right)\right],$$

where U = peripheral velocity of the rotating cylinder,
h = distance between cylinder and vessel wall,
r = radius of cylinder, and
D = diffusion coefficient of OH^- ions.

The equations are valid up to about 0·1M HCl, but at high rates surface roughening gives errors as the flow regime changes to turbulent flow. The analysis of the hydrodynamic situation when a porous film, several micrometres in thickness, is formed by precipitation of corrosion products has been made by Mahato *et al.*[29]. The importance of this work is that it takes into account the varying rate as this layer begins to thicken. The system they considered was the internal rusting of steel pipes by natural water. The presence of corrosion product gave rise to turbulent flow regime, even at low flow and the characteristic length L was the internal pipe diameter which decreased with time as corrosion product accumulated. Thus the dimensional group theory was modified to accommodate the 'unsteady state' mass transport.

The details of this calculation are given by Mahato, but the final result was:

$$\text{Weight loss} = 52{\cdot}1(Re)^{0{\cdot}54}[(0{\cdot}96t+0{\cdot}31)^{0{\cdot}5}-0{\cdot}56]$$

where the weight loss is in $gm^{-2}d^{-1}$ and the time t is in hours.

An important assumption was that the solution was dilute (in this case natural water of approximately 100 p.p.m. total dissolved solids) since there are difficulties in applying mass transport equations for certain situations in concentrated electrolyte solution, where a knowledge of activities is uncertain and leads to large errors.

The above work is important, since many practical corrosion systems involve a thick but porous film of corrosion products, e.g. rusting, sulphatising, tuberculation and atmospheric corrosion, and the approach may lead to a more valid corrosion testing technique for these situations.

Some Effects of Temperature on Corrosion Reactions

In contrast to the influence of velocity, whose primary effect is to increase the corrosion rates of electrode processes that are controlled by the diffusion of reactants, temperature changes have the greatest effect when the rate determining step is the activation process. In general, if diffusion rates are doubled for a certain increase in temperature, activation processes may be increased by 10–100 times, depending on the magnitude of the activation energy.

Bearing in mind the importance of the rate determining process and because of the complex situation in corrosion processes of having two electrode processes, the effect of temperature is best illustrated by reference to specific situations.

Cathode Reactions

Hydrogen evolution process In de-aerated solutions when this process is under activation control the main effect of increasing the temperature is to increase the exchange current. Typical examples of the magnitude of this change have been given by Conway *et al.*[30] who found that for nickel the exchange current increased from approximately $10^{-2}\,A/m^2$ to $1{\cdot}0\,A/m^2$ when the temperature changed from 10 to 75°C and the activation energy was about 59 kJ. Thus the rate of corrosion would be increased by at least 100 times if the anode process was unaffected by the temperature increase, whilst for control by concentration polarisation, the diffusion coefficient for hydrogen ions would increase perhaps only twice over the same temperature range.

Dissolved oxygen reduction process Corrosion processes governed by this cathode reaction might be expected to be wholly controlled by concentration polarisation because of the low solubility of oxygen, especially in concentrated salt solution. Temperature increase is complex in that the diffusivity of oxygen molecules increases, but solubility decreases. Data are scarce for these effects but the net mass transport of oxygen should increase with temperature[31] until a maximum is reached (estimated at about 80°C) when the concentration falls as the boiling point is approached. Thus the corrosion rate should attain a maximum at 80°C and then decrease with further increase in temperature.

A striking example of the interaction of solution velocity and concentration is given by Zembura[23] who found that for copper in aerated 0·1N H_2SO_4, the controlling process was the oxygen reduction reaction and that up to 50°C, the 'slow step' is the activation process for that reaction. At 75°C the process is now controlled by diffusion, and increasing solution velocity has a large effect on the corrosion rate (Fig. 2.4), but little effect at

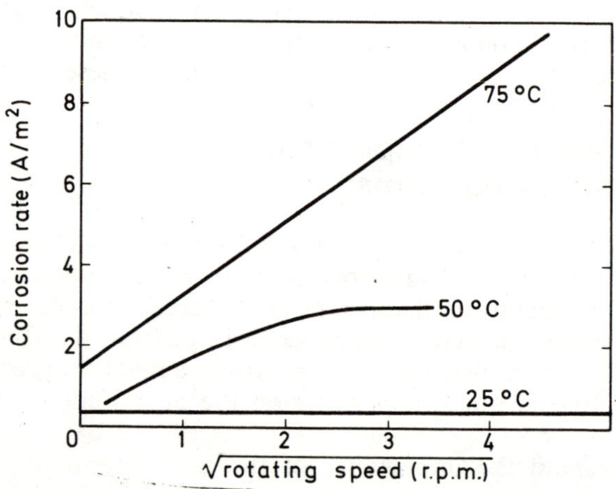

Fig. 2.4 Corrosion rate of a copper rotating disc in air-saturated 0·1N H_2SO_4 at various temperatures (after Zembura[23])

temperatures below 50°C. This study shows how unwise it is to separate these various parameters of concentration, velocity and temperature until the dominating electrode process is determined.

Anode Reactions

It is convenient to consider three stages of anode polarisation with regard to temperature effects, (*a*) under film-free conditions, (*b*) under film-forming conditions and (*c*) at the active–passive transition.

Film-free conditions It has been observed for many metals that the magnitude of $i_{crit.}$ (see Section 1.4) increases with temperature and that the activation energy for dissolution is low, suggestive of a diffusion-limited anode process when the migration of corrosion products away from the surface is rate controlling. Some examples of the value of the activation energy for this process are given in Table 2.2.

These data have been obtained by anodic polarisation work and might therefore be more relevant when cathodic reduction of oxygen takes place that can increase the corrosion potential to high positive values.

Table 2.2 Activation energies for critical currents ($i_{crit.}$) for passivation

Temperature range	*Activation energy* (kJ/mol)	*System*	*Reference No.*
25–98°C	−19·2	Ni in 0·05N H_2SO_4	32
25–75°C	−23·4	Ni in 0·12N H_2SO_4	32
25–100°C	−37·6	Cr–Ni–Mn steels in H_2SO_4	33
25–98°C	−46·0	Ti in HCl	34
4–30°C	−41·8	Fe in 1N H_2SO_4	35

The relationship between $i_{crit.}$ and temperature can be described by an Arrhenius type of equation:

$$\log i_{crit.} = A - \frac{E^{+}}{RT} \qquad \ldots(2.4)$$

where A is a constant and E^{+} is the activation energy. A more useful form is:

$$\log i_{crit.} = \log i^{T}_{crit.} - \frac{E^{+}}{RT} + K$$

where $\log i^{T}_{crit.}$ is the value at, say, 25°C.

For many metals the critical current density for passivation ($i_{crit.}$) increases with increasing pH of the solution:

$$\log i_{crit.} = \log i_{crit.o} + K\,\text{pH} \qquad \ldots(2.5)$$

where K is a constant depending on the metal or alloy and $\log i_{crit.o}$ is the value when pH = 0.

This relationship is in accordance with the hydroxyl adsorption theory where polarisation is decreased with increasing hydroxyl concentration (unless film formation interferes). Equations 2.4 and 2.5 may be combined to determine the magnitude of $i_{crit.}$ at various temperatures and pH. Since control is due mainly to the diffusion of anodic products from the surface, changes in flow would be expected to have a large effect, and should be considered when designing anodic protections systems.

Film formation In a few instances the temperature dependency of i_p, the passive current, has been observed and high activation energies (−46 to −84 kJ/mol) have been obtained, indicating a large increase in rate as the temperature increases (see Table 2.3).

Thus the rate of change of i_p under activation control is much faster than $i_{crit.}$ which is under diffusion control, and for the same condition of solution velocity the two rates could become equal at some common temperature, i.e. $i_{crit.} = i_p$, and there is *no* active–passive transition. For many of the systems given in the table this temperature is about 100°C. Above this temperature the measured activation energy is lower and diffusion control is established.

In practice the danger of aerated systems becomes apparent when the temperature is above a certain minimum, for there is no passive film formation, and it is clear that anodic protection cannot be effective in these circumstances.

Table 2.3 Activation energies of passivation current densities, i_p

Temperature range	*Activation energy* (kJ/mol)	*System*	*Reference No.*
25–60°C	−60	Sn in 1N NaOH	37
25–60°C	−49	Sn in 0·1N NaOH	37
4–25°C	−54	Fe in 1N H_2SO_4	35
25–100°C	−78	Ni in 1N H_2SO_4	36
25–75°C	−56	Cu in 0·1N H_2SO_4	23
60–100°C	−50	Inconel 825	38
	−46	Corronel 230	38
25–60°C	−48	Cr–Mn–Ni steels	33
10–25°C	−75	Ti in 10N HCl	39
25–70°C	−92	Ti in 10N HCl	39

The dissolution of passive films is, in the main, controlled by a chemical activation step in contrast to film-free conditions at $i_{crit.}$. Many protective anodic films are oxides and hydroxides whose dissolution depends upon the hydrogen ion concentration, and the rate follows a Freundlich adsorption equation:

$$\log i_p = \log k + n \log C^{H^+} \quad \ldots(2.6)$$

where k is constant and n is related to the valency of the cations; $n = 0{\cdot}5$, 0·33 and 0·25 for monovalent, divalent and trivalent ions, respectively. On rearranging equation 2.6 a more useful form is obtained:

$$\log i_p = \log i_p^0 - n\text{pH}^*$$

where $\log i_p^0$ is the current density (at constant temperature) when pH = 0. Combining this equation with the temperature relationship, and assuming the pH is constant:

$$\log i_p = \log i_{pT} - \frac{-E^{\ddagger}}{2{\cdot}3RT} + K$$

where i_{pT} is the passivation current at some standard temperature, e.g. 25°C.

Hence $$\log i_p = \tfrac{1}{2}\left\{\log i_p^0 + \log i_{pT} - \left[n\text{pH} + \frac{E^{\ddagger}}{2{\cdot}3R}\left(\frac{1}{T}\right)K\right]\right\}$$

and $$\log i_p = \tfrac{1}{2}\left[K_1 - n\text{pH} + K_2\left(\frac{1}{T}\right)\right] \quad \ldots(2.7)$$

This assumes that the pH does not change over the temperature range considered. However, corrections are small for pH values of from 0 to 5, and vary in the range 6–9 as shown in Fig. 2.5.

Active–passive transition It has been shown that i_p, the current required to maintain a passive film, increases with temperature at a much greater rate than the critical current for passivation as a result of an activation-controlled

*Note that the pH dependency is opposite to that for $i_{crit.}$ which increases as pH increases.

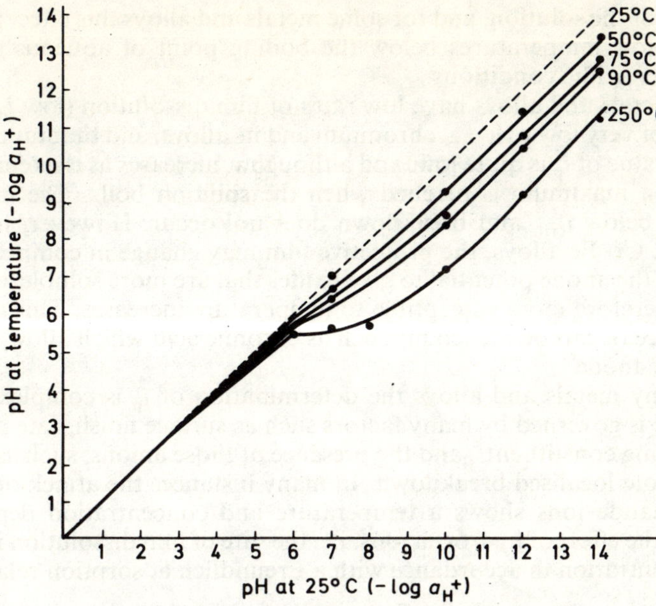

Fig. 2.5 Change in $-\log a_{H^+}$ with temperature[54]

process. At some temperature i_p will exceed $i_{crit.}$ and *no* active–passive transition will be observed, and more important no protection by a passive film is possible because of the high rate of dissolution. At this stage the slow process becomes the diffusion of reactants and control of the rate is sensitive to solution flow changes. The temperature at which this occurs has only been determined in a few cases and an example is given in Fig. 2.6. It must be emphasised that the rate of dissolution of the passive film is also determined in many cases

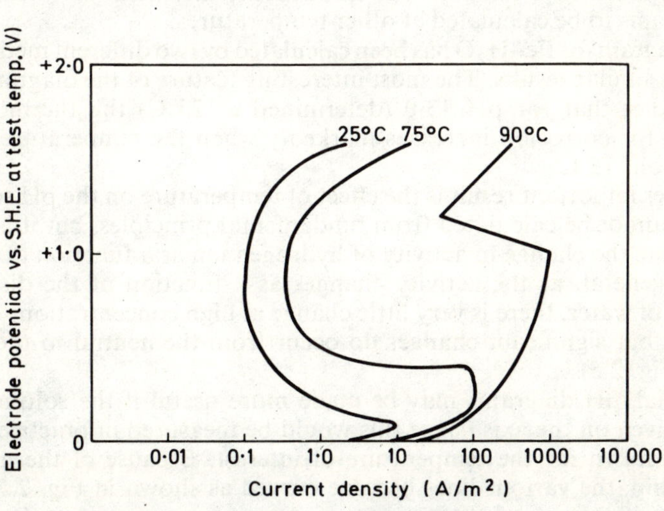

Fig. 2.6 Anodic polarisation of nickel at various temperatures, in 0·05N H_2SO_4 + 0·05N K_2SO_4, pH 1·3 (after Cowan and Staehle[32])

by the pH of the solution, and for some metals and alloys this effect probably occurs at low temperatures below the boiling point of aqueous solutions when under acidic conditions.

Some metals and alloys have low rates of film dissolution (low i_p) even in solutions of very low pH, e.g. chromium and its alloys, and titanium. In these cases the value of i_p is quite low, and although it increases as the temperature increases, a maximum is reached when the solution boils. The maximum current is below $i_{crit.}$ and breakdown does not occur. However, in certain alloys, e.g. Cr–Fe alloys, the protective film may change in composition on increasing the anode potential to give oxides that are more soluble at low pH and are therefore more susceptible to temperature increases. This occurs in the presence of cathode reactants such as chromic acid which allow polarisation of the anode.

For many metals and alloys the determination of i_p is complex, and its magnitude is governed by many factors such as surface finish, rate of formation, alloying constituents, and the presence of those anions, such as halides, that promote localised breakdown. In many instances the attack on passive films by halide ions shows a temperature and concentration dependence similar to the effect of hydrogen ions, i.e. the rate of film dissolution increases with concentration in accordance with a Freundlich adsorption relationship

$$\text{Rate} = KC^n$$

where C is the concentration of halide, and the effect of temperature indicates a diffusion-controlled process.

Thermodynamic Considerations

Potential–pH diagrams have been calculated for most metals and many non-metal systems by Pourbaix[40] and others for temperatures at 25°C. In addition, theoretical relationships have been established[41, 42] which enable the diagrams to be calculated at other temperatures.

The diagram for Fe–H_2O has been calculated by two different methods[43, 44] and gives similar results. The most interesting feature of the diagrams is that they predict that for pH 13·0 (determined at 25°C) the thermodynamic tendency for corrosion increases markedly when the temperature is raised above about 75°C.

Another important result is the effect of temperature on the pH scale. This change cannot be calculated from fundamental principles, but it is possible to calculate the change in activity of hydrogen ion as a function of temperature. In general, as the activity changes as a function of the dissociation constant of water, there is very little change at high concentrations of hydrogen ions, but significant changes do occur from the neutral to the alkaline region (Fig. 2.5).

Potential–pH diagrams may be made more useful if the solution pH at 25°C is given on the axis (since this would be measured in practice) and the diagram drawn for the temperature of interest. Because of the non-linear relationship, the various lines become curved as shown in Fig. 2.7. Experimental verifications of these types of diagram are few, but Cowan and Staehle[34] have studied the behaviour of nickel in sulphate solutions at pH values in the acidic region. Their main conclusions were that at 300°C in the pH range 5–10, the corrosion of nickel should be thermodynamically impossible, and in the range pH 2–5 and 10–13 the driving force for corrosion should be small in oxygen-free solutions. These observations confirm subsequent experimental tests.

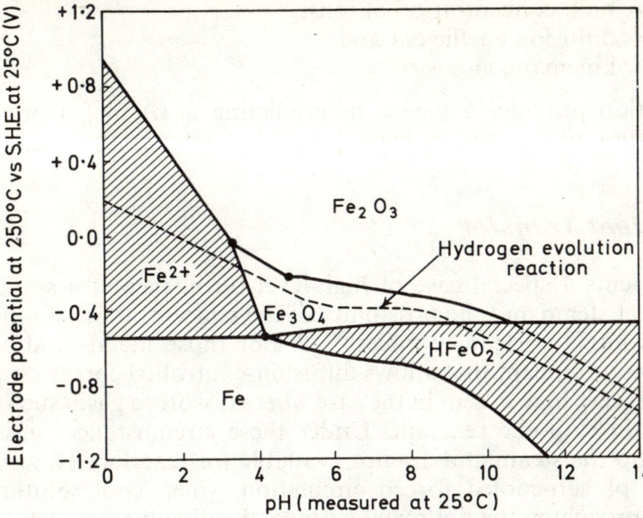

Fig. 2.7 Potential–pH diagram calculated for Fe–H_2O system at 250°C. The pH scale refers to the solution measured at 25°C and then raised to 250°C (after Ashworth[55])

The calculations of these diagrams are complicated, but computer methods are now available[45], and using these techniques the effect of temperature on potential–pH diagrams has been calculated for many metals[46]. Their usefulness, however, awaits further experimental evidence to confirm the predictions.

Heat Transfer

The discussion to date has assumed isothermal conditions, but in many practical situations corrosion reactions have to be considered when the electrode is acting as a cooler or a heater.

From various studies[47, 48] it is becoming clear that in spite of a heat flux, the overriding parameter is the temperature at the interface between the metal electrode and the solution, which has an effect on diffusion coefficients and viscosity. If the variations of these parameters with temperature are known, then i_L (and $i_{corr.}$) can be calculated from the hydrodynamic equations.

When film-forming reactions occur and activation control is the rate-determining factor then interfacial temperature again will determine the extent of corrosion.

The work of Porter *et al.*[48] has shown that for copper in phosphoric acid the interfacial temperature was the main factor, and furthermore this was the case for positive or negative heat flux. Activation energies were determined for this system; they indicated that concentration polarisation was the rate-determining process, and by adjustment of the diffusion coefficient and viscosity for the temperature at the interface and the application of dimensional group analysis it was found that:

$$i_L = 43{\cdot}3(C_S - C_B)D^{\frac{2}{3}}\nu^{-\frac{1}{6}}$$

where C_S = surface concentration of ions,
C_B = bulk concentration of ions,
D = diffusion coefficient and
ν = kinematic viscosity.

This equation provides a means of predicting i_L (A/m^2), from which an approximation of $i_{corr.}$ can be made from physical measurements alone.

Boiling Heat Transfer

This represents a special case of high-level turbulence at a surface by the formation of steam and the possibility of the concentration of ions as water evaporates into the steam bubbles[49, 50]. For those metals and alloys in a particular environment that allows diffusion-controlled corrosion processes, rates will be very high except in the case where dissolved gases such as oxygen are the main cathodic reactant. Under these circumstances gases will be expelled into the steam and are not available for reaction. However, under conditions of sub-cooled forced circulation, when cool solution is continually approaching the hot metal surface, the dissolved oxygen appears to be effective[51] and cathodic processes are stimulated. When the activation step is rate controlling the boiling temperature represents a maximum in the rate. Only small changes (of about 10°C) are possible for metal temperatures to exceed the boiling point because of film boiling, when steam effectively

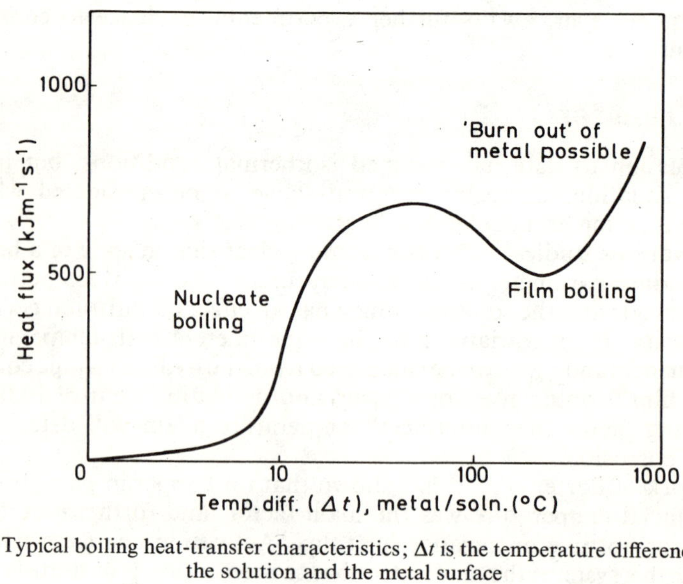

Fig. 2.8 Typical boiling heat-transfer characteristics; Δt is the temperature difference between the solution and the metal surface

covers the whole surface and corrosion rates become negligible. There is a danger of metal damage by a rapid rise in temperature (Fig. 2.8) when the 'cooling' action of evaporation at the surface is prevented, a situation that is obviously to be avoided in the design. When activation processes are in control the small temperature rise allowable above the equilibrium boiling point may increase rates of dissolution by two orders of magnitude[51].

Butler and Ison[49] have suggested that variation in corrosion rate can be influenced by surface roughness, which allows a large number of nuclei for steam bubble formation. In these circumstances they have suggested that concentration of ions in solution next to the surface will be greater, and their observations on corrosion damage indicate that the steam bubbles may provide crevices or at least enhanced conditions for dissolution at the triple interface (solution/metal/steam).

Thermogalvanic Corrosion

It is impossible to design heat exchangers where all surfaces are isothermal and in many cases such differences are required by the design. For instance, a steam cooler may have a de-superheating zone, a condensing zone and a liquid cooling zone on the same metal tube, but at different positions along its length (Fig. 2.9). The question arises as to whether such temperature differences on the same metal surface in contact with the same electrolyte

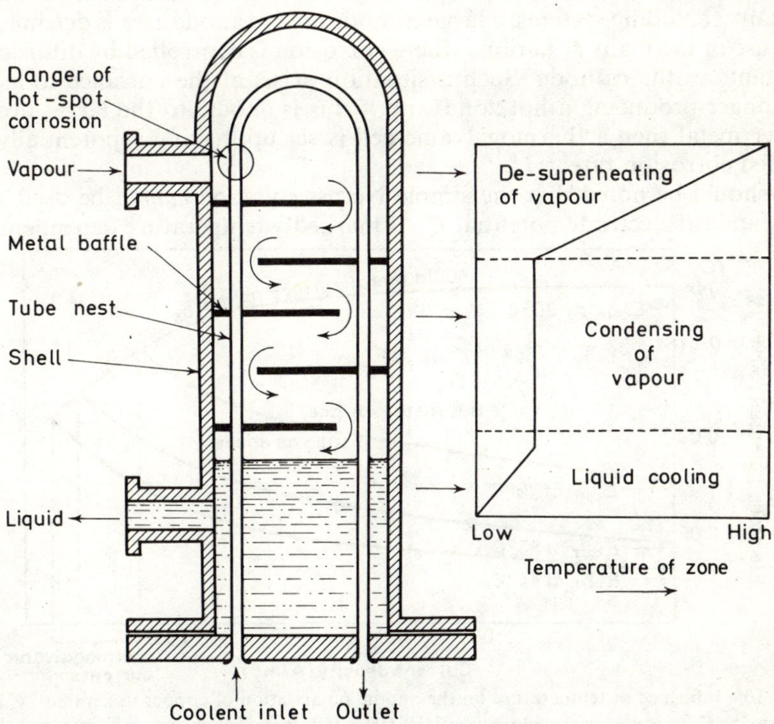

Fig. 2.9 Temperature distribution in a typical heat exchanger

solution (on the cooling side), can have sufficient electrode potential differences to give rise to a galvanic cell, i.e. a thermal galvanic cell. Electrode potentials change with temperature, but as shown previously, temperature changes may also affect the kinetics of dissolution, especially activation-controlled processes. The main rôle of thermogalvanic cells is in polarising existing electrode processes, which, depending on other aspects of the environment, may accelerate or decelerate corrosion.

Origin and Magnitude of Thermogalvanic Potentials

The e.m.f. of a thermogalvanic cell is the result of four main effects[42]: (*a*) electrode temperature, (*b*) thermal liquid junction potential, (*c*) metallic thermocouple and (*d*) thermal diffusion gradient or Soret.

The driving force of a thermogalvanic corrosion cell is therefore the e.m.f. attributable to these four effects, but modified by anodic and cathodic polarisation of the metal electrodes as a result of local action corrosion processes.

In practical systems, (*c*) and (*d*) are often very small especially on the same metal surface when solution flow occurs by convection or forced circulation. In neutral solutions, (*b*) may be small but is somewhat larger in acid solution. On this basis several workers have determined effect (*a*) as a guide to the subsequent behaviour of a thermogalvanic cell. The main usefulness of such a calculation is to decide whether a hot anode or a hot cathode is produced. In many corroding systems, a large cathode area to anode area is detrimental, because of the many situations where corrosion is controlled by diffusion of reactants to the cathode. Such a situation exists at the entrance to a heat exchanger producing a hot zone, and if this is anodic to the larger area of cooler metal then a thermogalvanic cell is set up, having a potentially enhanced corrosion rate[52, 53].

It should be noted that the simple Nernst equation *cannot* be used since the standard electrode potential $E^{\ominus}$ is markedly temperature dependent. By

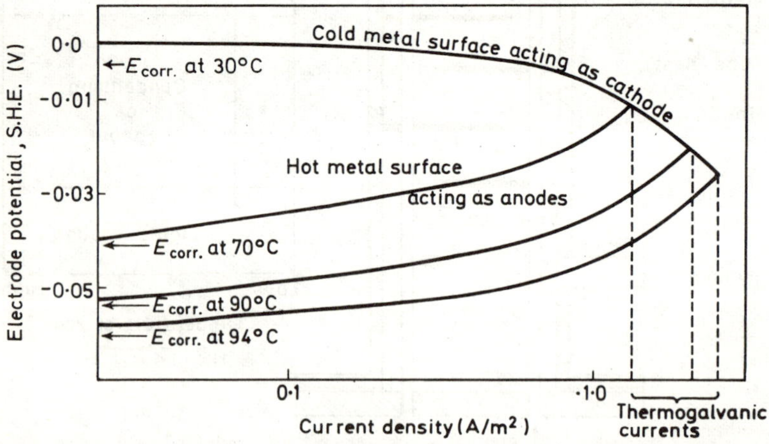

Fig. 2.10 Influence of temperature on the anodic polarisation of copper in aerated 3% NaCl solution[51]. $E_{corr.}$ is the corrosion potential of the hot metal when not in contact with the cold metal

means of irreversible thermodynamics[42] equations have been computed to calculate these potentials and are in good agreement with experimentally determined results.

In general, temperature coefficients of electrode potential are in the range ±0·1 to 2 mV/°C and in many practical systems temperature differences rarely exceed 75°C, so that the driving force for thermogalvanic corrosion is small and would be subject mainly to resistance control. However, in many instances the temperature change also decreases polarisation (see Fig. 2.10) so that if the resistance of the solution is not high severe attack can ensue. Because of the resistance effect attack is confined to a small area of largest temperature gradient, leading to deep notches at the edge of the heated zone, i.e. the dangerous situation of a small anode and large cathode.

Active–passive Transitions

Whilst temperature coefficients suggest modest potential differences, these calculations do not take into account the large potential changes that can occur when thermal effects allow transition from active to passive states. Potentials become more positive (a hot cathode) as a result of thickening of the passive film. Such changes have been observed[54] (Fig. 2.11), and moreover, in the presence of aggressive anions, when the thermal effects allow the change to passivity, then overall general corrosion is changed to deep pitting on the electrode surface. Thermogalvanic coupling would enhance this effect.

These observations show that kinetic factors can outweigh thermodynamic effects and the situation of the mutual polarisation of two electrodes in a corrosion cell leads to either negative or positive temperature coefficients.

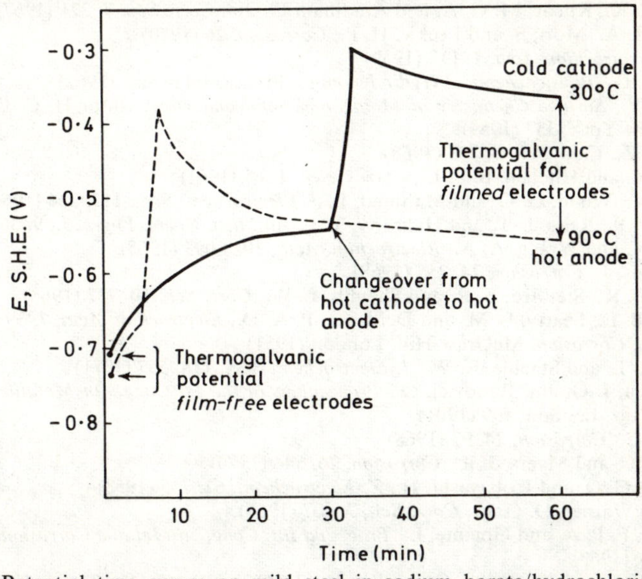

Fig. 2.11 Potential–time curves on mild steel in sodium borate/hydrochloric acid buffer solutions, pH 7·60, oxygen-saturated solution (after Ashworth[55])

Prediction of thermogalvanic action from consideration of the anode process alone can therefore be misleading. Thermogalvanic corrosion rates may be low in most circumstances but they are persistent for long periods, existing as long as temperature differences exist, i.e. the operating period of plant. They represent a dormant situation that can accelerate corrosion if the environment changes, e.g. high conductivity, and increase in aggressive ion concentration.

P. J. BODEN

REFERENCES

1. Pourbaix, M., *Electrochim. Acta*, **12**, 184 (1967) and Ness, P., *Electrochim. Acta*, **12**, 161 (1967)
2. Pourbaix, M., *Corrosion*, **25**, 267 (1967)
3. Mattson, E., *Electrochim. Acta*, **3**, 279 (1961)
4. Horvath, J. and Hackl, L., *Corr. Sci.*, **5**, 528–538 (1965)
5. Greene, N. D., *J. Electrochem. Soc.*, **107**, 457 (1960) and France, W. D. and Greene, N. D., *Corrosion*, **24**, 403 (1968)
6. Cron, C. J., Payer, J. H. and Staehle, R. W., *Corrosion*, **27**, 1 (1971)
7. Bockris, J. O'M., Drazic, D. and Despic, A. R., *Electrochim. Acta*, **4**, 325 (1961)
8. Bonhoeffer, K. F. and Heusler, K. E., *Z. Phys. Chem.*, N.F., **8**, 390 (1956)
9. Kabanov, B. N. and Leikis, D. I., *Dokl. Akad. Nauk. SSSR*, **58**, 1 685 (1947)
10. Florianovich, G. M., Sokolova, L. A. and Kolotyrkin, Ya. M., *Electrochim. Acta*, **12**, 897 (1967)
11. Oakes, G. and West, J. M., *Brit. Corr. J.*, **4**, 66 (1969)
12. Hines, J. G., *Electrochim. Acta*, **10**, 225 (1965)
13. Florianovich, G. M., Kolotyrkin, Ya. M. and Kononova, M. D., *Proceedings of the 4th International Congress on Metallic Corrosion*, Amsterdam, N.A.C.E. (1969)
14. Ammar, I., Darwish, S. and Etman, M., *Electrochim. Acta*, **12**, 485 (1967)
15. Brasher, D. M. and Mercer, A. D., *Brit. Corr. J.*, **3**, 120 (1968)
16. Brasher, D. M., Reichenberg, D. and Mercer, A. D., *Brit. Corr. J.*, **3**, 144 (1968)
17. Brasher, D. M., *Brit. Corr. J.*, **4**, 122 (1969)
18. Gouda, V. K., Khedr, M. G. A. and Am Sham El Din, *Corr. Sci.*, **7**, 221 (1967)
19. Legault, R. A., Mori, S. and Leckie, H. P., *Corrosion*, **26** (1970)
20. Ibl, N., *Electrochim. Acta*, **1**, 117 (1959)
21. Levich, V. G., *Physicochemical Hydrodynamics*, Prentice-Hall Inc. (1962)
22. King, C. V., *Surface Chemistry of Metals and Semiconductors*, editor H. C. Gatos, John Wiley, New York, 357 (1959)
23. Zembura, Z., *Corr. Sci.*, **8**, 703 (1968)
24. Ross, T. K. and Hitchen, B. P. L., *Corros. Sci.*, **1**, 65 (1961)
25. Ross, T. K., Wood, G. C. and Mahmud, I., *J. Electrochem. Soc.*, **113**, 334 (1966)
26. Van Shaw, P., Reiss, L. P. and Hanratty, T. J., *Am. Inst. Chem. Engrs. J.*, **9**, 362 (1963)
27. Ross, T. K. and Wragg, A. A., *Electrochim. Acta*, **10**, 1 093 (1965)
28. Marangozis, J., *Corrosion*, **24**, 255 (1968)
29. Mahato, B. K., Steward, F. R. and Shemilt, L. W., *Corr. Sci.*, **10**, 737 (1968)
30. Conway, B. E., Beatty, E. M. and DeMaine, P. A. D., *Electrochim. Acta*, **7**, 39 (1962)
31. Speller, F., *Corrosion*, McGraw-Hill, London (1951)
32. Cowan, R. L. and Staehle, R. W., *J. Electrochem. Soc.*, **118**, 557 (1971)
33. Murgulescu, I. G. and Radovici, O., *Proceedings of the 1st Cong. On Metallic Corrosion*, Butterworths, London, 109 (1961)
34. Griess, J. C., *Corrosion*, **24**, 97 (1968)
35. Finley, T. C. and Myers, J. R., *Corrosion*, **26**, 544 (1970)
36. Okomoto, G. O. and Kobayashi, H., *Z. Electrochem.*, **62**, 755 (1958)
37. Pugh, M., Warner, D. Gabe, *Corr. Sci.*, **7**, 807 (1967)
38. Robinson, F. P. A. and Golante, L., *Proc. 2nd Int. Cong. on Metallic Corrosion*, N.A.C.E., New York (1963)
39. Kucera, V., Novak, P., Franz, F. and Koritta, J., *Korrozija i Zaščita Titana*, G.N.T.I.M.L., Moscow (1964)

40. Pourbaix, M., *Atlas of Electrochemical Equilibria in Aqueous Solutions*, Pergamon Press, Oxford (1966)
41. Criss, C. M. and Cobble, J. W., *J. Am. Chem. Soc.*, **86**, 5 390 (1964)
42. de Bethune, A. J., Licht, T. S. and Swendeman, N., *J. Electrochem. Soc.*, **106**, 616 (1959)
43. Townsend, H. E., *Corr. Sci.*, **10**, 343 (1970)
44. Ashworth, V. and Boden, P. J., *Corr. Sci.*, **10**, 709 (1970)
45. Brook, P. A., *Corr. Sci.*, **11**, 389 (1971)
46. Brook, P. A., *Corr. Sci.*, **12** (1972)
47. Ross, T. K., *Brit. Corr. J.*, **2**, 131 (1967)*
48. Porter, D. T., Donimirska, M. and Wall, R., *Corr. Sci.*, **8**, 833 (1966)
49. Butler, G. and Ison, H. C. K., *Proc. 1st Int. Cong. on Metallic Corrosion*, Butterworths, London (1961)
50. Freeborn, J. and Lewis, D., *J. Mech. Eng. Soc.*, **4**, 46 (1962)
51. Boden, P. J., *Corr. Sci.*, **11**, 353 (1971)
52. Breckon, G. and Gilbert, P. T., *Proc. 1st Int. Cong. on Metallic Corrosion*, Butterworths, London (1961)
53. Bem, R. S. and Campbell, H. S., ibid.
54. Ashworth, V. and Boden, P. J., *J. Electrochem. Soc.*, **119**, 6 (1972)
55. Ashworth, V., to be published

*Detailed review presented at symposium on 'Corrosion under Heat Transfer in Liquid Media', reported in *Br. Corr. J.*, **2** (1967)

2.2 The Atmosphere

Metals are more frequently exposed to the atmosphere than to any other corrosive environment. Atmospheric corrosion is also the oldest corrosion problem known to mankind, yet even today it is not fully understood. The principal reason for this paradox lies in the complexity of the variables which determine the kinetics of the corrosion reactions. Thus, corrosion rates vary from place to place, from hour to hour and from season to season. Equally important, this complexity makes meaningful results from laboratory experiments very difficult to obtain.

However, the object of this section is to outline the principles which govern atmospheric corrosion, and the emphasis is placed on metals whose atmospheric corrosion is of economic importance. These include iron and steel, zinc, copper, lead, aluminium and chromium.

Classification of Atmospheric Corrosion

Atmospheric corrosion can be conveniently classified as follows:

(*a*) Dry oxidation.
(*b*) Damp corrosion.
(*c*) Wet corrosion.

Dry Oxidation

This takes place in the atmosphere with all metals that have a negative free energy of oxide formation. Gold does not oxidise and this property is utilised in the coating of electronic components where even the thinnest layers of corrosion product cannot be tolerated. For metals forming non-porous oxides (alkali metals are an exception) the films rapidly reach a limiting thickness since ion diffusion through the oxide lattice is extremely slow at ambient temperatures, and at the limiting thickness, the oxide films on metals are invisible. For example, those on iron are typically 30 Å thick. For certain metals and alloys these films are so fault-free or rapidly self-healing that they confer remarkable protection on the substrate, e.g. stainless steel, titanium and chromium.

The tarnishing of copper and silver in dry air containing traces of hydrogen sulphide (Table 2.4) is another example of film growth by lattice diffusion at ambient temperatures. In these cases defects in the sulphide lattice enable the films to grow to visible thicknesses with the consequent formation of tarnish films which are aesthetically objectionable and may have a significant effect on the behaviour of the metals in particular applications, e.g. electrical contacts.

Table 2.4 Typical concentrations of atmospheric impurities

Impurity	*Typical concentrations* (μg/m³)
Sulphur dioxide*[1]	Industrial region: winter 350, summer 100 Rural region: winter 100, summer 40
Sulphur trioxide	Approximately 1% of the sulphur dioxide content
Hydrogen sulphide[2]	Industrial region: 1·5–90 Urban region: 0·5–1·7 Rural region: 0·15–0·45 } values measured in the spring
Ammonia[3]	Industrial region: 4·8 Rural region: 2·1
Chloride[3] (air sampled)	Industrial inland: winter 8·2, summer 2·7 Rural coastal: annual average 5·4
Chloride[3] (rainfall sampled)	Industrial inland: winter 7·9, summer 5·3 Rural coastal: winter 57, summer 18 (these values in mg/l)
Smoke particles[1]	Industrial region: winter 250, summer 100 Rural region: winter 60, summer 15

*There are two methods that are commonly used for estimating sulphur dioxide:
(*a*) Lead peroxide 'candle' method. The weight gain, caused by lead sulphate formation as sulphur dioxide reacts with a specified surface area of lead peroxide paste, is measured.
(*b*) Hydrogen peroxide titrimetric method. A known volume of air is pumped through a weak hydrogen peroxide solution in which the sulphur dioxide is oxidised to sulphuric acid. The acid content is estimated by titration. In Ref. 1 the second method was used, the air first being filtered to yield an estimate of particulate matter.

However, in this section emphasis is placed upon *damp* and *wet* atmospheric corrosion which are characterised by the presence of a thin, invisible film of electrolyte solution on the metal surface (damp type) or by visible deposits of dew, rain, sea-spray, etc. (wet type). In these categories may be placed the rusting of iron and steel (both types involved), 'white rusting' of zinc (wet type) and the formation of patinae on copper and its alloys (both types).

The corrosion products may be soluble or insoluble. If insoluble, they usually reduce the rate of corrosion by isolating the substrate from the corrosive environment. Less commonly, they may stimulate corrosion by offering little physical protection while retaining moisture in contact with the metal surface for longer periods.

Soluble corrosion products may increase corrosion rates in two ways. Firstly, they may increase the conductivity of the electrolyte solution and thereby decrease 'internal resistance' of the corrosion cells. Secondly, they may act hygroscopically to form solutions at humidities at and above that in equilibrium with the saturated solution (Table 2.5). The 'fogging' of nickel in SO_2-containing atmospheres, due to the formation of hygroscopic nickel sulphate, exemplifies this type of behaviour. However, whether the corrosion

Table 2.5 Relative humidities of air in equilibrium with saturated salt solutions at 20°C[4–7]

Salt in solution	r.h. (%)	*Salt in solution*	r.h. (%)
$CuSO_4 \cdot 5H_2O$	98	$NaCl$	76
K_2SO_4	98	$CuCl_2 \cdot 2H_2O$	68
Na_2SO_4	93	$FeCl_2$	56
$Na_2CO_3 \cdot 10H_2O$	92	$NiCl_2$	54
$FeSO_4 \cdot 7H_2O$	92	$K_2CO_3 \cdot 2H_2O$	44
$ZnSO_4 \cdot 7H_2O$	90	$MgCl_2 \cdot 6H_2O$	34
$3CdSO_4 \cdot 8H_2O$	89	$CaCl_2 \cdot 6H_2O$	32
KCl	86	$ZnCl_2 \cdot xH_2O$	10
$(NH_4)_2SO_4$	81	NH_4Cl	80

products are soluble or insoluble, protective or non-protective, the corrosive atmosphere experienced by the substrate (often referred to as the 'micro-environment') is modified from the macro-environment experienced by a bare substrate. For this reason, corrosion rates are rarely constant for extended periods of atmospheric exposure.

Composition of the Atmosphere

Nominal Composition

The composition given in Table 2.6 is global and, for most components, is reasonably constant for all locations, but the water vapour content will obviously vary according to the climatic region, season of the year, time of the day, etc. However, only oxygen, carbon dioxide and water vapour need to be considered in the context of atmospheric corrosion.

Carbon dioxide was once thought essential for the rusting of ferrous metals (viz. the carbonic acid theory of rusting) but is now considered of relatively minor importance[8, 9]. However, basic zinc carbonate is frequently found in the corrosion products of zinc and small amounts of siderite ($FeCO_3$) are found in ferrous rusts.

Table 2.6 Approximate constitution[10] of the atmosphere at 10°C and 100 kN/m^2 (excluding impurities)

Constituents	g/m^3	*Weight* (%)	*Constituents*	mg/m^3	*p.p.m. by weight*
Air	1 172	100	Neon	14	12
Nitrogen	879	75	Krypton	4	3
Oxygen	269	23	Helium	0·8	0·7
Argon	15	1·26	Xenon	0·5	0·4
Water vapour	8	0·70	Hydrogen	0·05	0·04
Carbon dioxide	0·5	0·04			

Water vapour is essential to the formation of an electrolyte solution which will support the electrochemical corrosion reactions, and its concentration in the atmosphere is usually expressed in terms of the *relative humidity* (r.h.).

This is defined as the percentage ratio of the water vapour pressure in the atmosphere compared to that which would saturate the atmosphere at the same temperature. Alternatively, the difference in temperature between the ambient atmosphere and that to which it would have to be cooled before moisture condensed from it, is also used as a measure of moisture content. This difference in temperature is called the *dew point depression*. The actual temperature at which condensation takes place is known as the *dew point*. The relative humidity is then expressed as:

$$\text{r.h.} = \frac{\text{Saturated vapour pressure of } H_2O \text{ at the dew point}}{\text{Saturated vapour pressure of } H_2O \text{ at ambient temp.}} \times 100\%.$$

Oxygen from the atmosphere, dissolved in the electrolyte solution provides the cathode reactant in the corrosion process. Since the electrolyte solution is in the form of thin films or droplets, diffusion of oxygen from the atmosphere/electrolyte solution interface to the solution/metal interface is rapid. Moreover, convection currents within these thin films of solution may play a part in further decreasing concentration polarisation of this cathodic process[11]. Oxygen may also oxidise soluble corrosion products to less soluble ones which form more or less protective barriers to further corrosion, e.g. the oxidation of ferrous species to the less soluble ferric forms in the rusting of iron and steel.

Atmospheric Contaminants

In a sense this subdivision of the composition of the atmosphere is arbitrary since some of the so-called contaminants are derived partly or wholly from natural sources. However, in that their concentrations vary appreciably within very narrow geographical limits, they may be distinguished from the contents of Table 2.6 (with the possible exception of water vapour). Table 2.4 lists those contaminants which are important from a corrosion standpoint. Excluded are contaminants found only in very specific locations, e.g. in the vicinity of a chemical works. The concentrations given are intended only to indicate general levels in the usual classification of environments and not to define a particular environment.

Sulphur oxides These (SO_2 is the most frequently encountered oxide) are powerful stimulators of atmospheric corrosion, and for steel and particularly zinc the correlation between the level of SO_2 pollution and corrosion rates is good[12–14]. However, in severe marine environments, notably in the case of zinc, the chloride contamination may have a higher correlation coefficient than SO_2.

The SO_2 in the atmosphere is derived from two sources. Firstly, from the aerial oxidation of H_2S produced naturally (see later) and secondly from the combustion of sulphur-containing fuels. In industrialised countries the second source predominates, but on a global scale only about one-fifth of the total sulphur pollution is derived from human activity. In 1969, the total sulphur emission, expressed in terms of SO_2, from burnt fuel in the UK was $6{\cdot}06 \times 10^6$ tons. In densely populated countries sulphur pollution levels are very much related to the domestic heating cycle, and in the UK maximum pollution levels are reached in January/February and the minimum usually occurs in August[1]. This cyclic pattern is closely reflected by corrosion rate variations[15,16], corrosion being heaviest in the winter months despite lower average temperatures.

A more detailed consideration of the rôle of SO_2 as a corrosion stimulator will be given later.

Hydrogen sulphide This is produced by the putrefaction of organic sulphur compounds or by the action of sulphate-reducing bacteria in anaerobic conditions (e.g. in polluted river estuaries). It is fairly rapidly oxidised to SO_2 and concentrations are considerably lower than those of SO_2[2] (Table 2.4). Nevertheless it is responsible for the tarnishing of copper and silver at normal atmospheric concentrations.

Nitrogen compounds These also arise from both natural and synthetic sources. Thus ammonia is formed in the atmosphere during electrical storms, but increases in the ammonium ion concentration in rainfall over Europe in recent years are attributed to increased use of artificial fertilisers. Ammonium compounds in solution may increase the wettability of a metal[17] and the action of ammonia and its compounds in causing 'season cracking', a type of stress-corrosion cracking of cold-worked brass, is well documented.

Saline particles These are of two main types. The first is ammonium sulphate formed in heavily industrialised areas where appreciable concentrations of ammonia and SO_3 or of H_2SO_4 aerosol co-exist. It is a strong stimulator of the initiation of corrosion, being hygroscopic and acidic. The second is marine salt, mainly sodium chloride but quite appreciable quantities of potassium, magnesium and calcium ions are analysed in rainfall[3]. Chlorides are also produced in industrial areas and for the UK the fall-off in concentration of marine salt with distance from the sea is partially masked by chloride produced by the industrial regions in the centre of the country[3]. Chlorides are also hygroscopic and the chloride ion is highly aggressive to some metals, e.g. stainless steel.

Other airborne particles These are also divisible into two groups. Firstly, the inert non-absorbent particles, usually siliceous, which can only affect corrosion by facilitating differential aeration processes at points of contact.

Secondly, absorbent particles such as charcoal and soot are intrinsically inert but have surfaces or infrastructures that adsorb SO_2, and by either co-adsorption of water vapour or condensation of water within the structure, catalyse the formation of a corrosive acid electrolyte solution. 'Dirt' with soot assists the formation of patinae on copper and its alloys by retaining soluble corrosion products long enough for them to be converted to protective, insoluble basic salts.

Other Atmospheric Variables

Temperature This may be more or less of an important factor, depending on the metal considered. For example, while zinc is characterised by a very low positive temperature coefficient of corrosion rate[19], steel has a high positive coefficient[15,19]. The rate of drying of electrolyte solution from the metal surface, directly into the atmosphere or through layers of corrosion product, is strongly temperature dependent. In these contexts the metal *surface* temperature is probably more important than ambient temperature although the latter obviously strongly influences the former. However, many other factors will affect the metal temperature, including the thermal capacity of the metal structure, its orientation with respect to the sun, the intensity of sunlight, the reflectivity of the metal surface or its corrosion products, wind velocity and direction, the thermal insulating properties of insoluble corrosion products, and so on. The prevailing wind direction is also an important factor in relation to increases in corrosion rates to be expected from the proximity of large industrial plants producing appreciable concentrations of potentially corrosive pollutants.

Electrolyte Solution Formation

Wetness of a metal surface The *time of wetness* of the metal surface is an exceedingly complex, composite variable. It determines the duration of the electrochemical corrosion process. Firstly it involves a consideration of all the means by which an electrolyte solution can form in contact with the metal surface. Secondly, the conditions under which this solution is stable with respect to the ambient atmosphere must be considered, and finally the rate of evaporation of the solution when atmospheric conditions change to make its existence unstable. Attempts have been made to measure directly the time of wetness[18], but these have tended to use metals forming non-bulky corrosion products (see Section 20.1). The literature is very sparse on the rôle of insoluble corrosion products in extending the time of wetness, but considerable differences in moisture desorption rates are found for rusted steels of slightly differing alloy content, e.g. mild steel and Cor-Ten.

Critical relative humidity The primary value of the critical relative humidity denotes that humidity below which no corrosion of the metal in question takes place. However, it is important to know whether this refers to a clean metal surface or one covered with corrosion products. In the latter case a secondary critical humidity is usually found at which the rate of corrosion increases markedly[8]. This is attributed to the hygroscopic nature of the corrosion product (see later). In the case of iron and steel it appears that there may even be a tertiary critical humidity[15]. Thus at about 60% r.h. rusting commences at a very slow rate (primary value)[20]; at 75–80% r.h. there is a sharp increase in corrosion rate probably attributable to capillary condensation of moisture within the rust[8, 21]. At 90% r.h. there is a further increase in rusting rate[15] corresponding to the vapour pressure of saturated ferrous sulphate solution[5], ferrous sulphate being identifiable in rust as crystalline agglomerates[16]. The primary critical r.h. for uncorroded metal surfaces seems to be virtually the same for all metals, but the secondary values vary quite widely.

Moisture precipitation Apart from wetting by sea-spray, moisture may either be deposited on a surface by rainfall or dew formation. For a known ambient humidity the dew point can be calculated, using the expression given previously, from standard tables giving the saturated vapour pressure of water at various temperatures (e.g. Handbook of the Chemical Rubber Company). However, Fig. 2.12 sets out these relationships graphically and

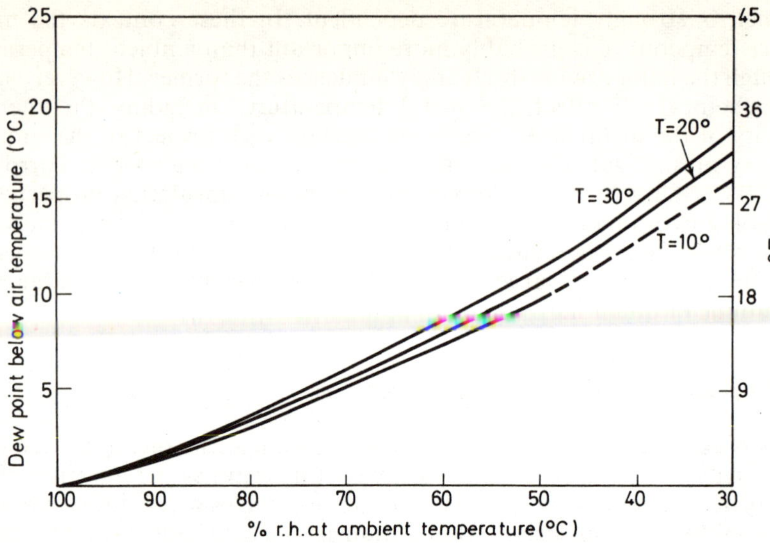

Fig. 2.12 Dew point depression below ambient temperature as a function of the relative humidity of the ambient atmosphere over a range of temperature

from a knowledge of the ambient relative humidity and ambient temperature, the dew point depression may be read off.

Since gaseous pollution, particularly of SO_2, tends to be concentrated near ground level, dew can be considerably more acid than rain which forms at higher altitudes. Moreover, dew can, unlike rain, wet completely sheltered surfaces. Thin sheets of metal which closely follow changes in ambient temperature are more likely to have dew formed on them than more massive sections of higher thermal capacity which will cool more slowly. 'White rusting' of galvanised sheeting is usually attributable to dew formation in poorly ventilated conditions.

Rainfall, besides wetting the metal surface, can be beneficial in leaching otherwise deleterious soluble species and this can result in marked decreases in corrosion rate[15,16]. A recent survey of rainfall analyses for Europe has shown that, with the exception of the UK, the acidity and sulphate content of rainfall markedly increased in the period 1956 to 1966, pH values having fallen by 0·05 to 0·10 units per annum[22]. The exception of the UK may be due to anti-pollution measures introduced in this period. However, even in the UK a pH of 4 is not uncommon for rainfall in industrial areas. The significance of electrolyte solution pH will be discussed in the context of corrosion mechanisms. The remaining cases of electrolyte formation are those in which it exists in equilibrium with air at a relative humidity below 100%.

Capillary condensation The vapour pressure above a concave meniscus of water is less than that in equilibrium with a plane water surface. It is therefore possible for moisture to condense in narrow capillaries from an atmosphere of less than 100% r.h.

The relative lowering of the saturated vapour pressure of water is described by the Thomson equation:

$$p = p_0 \, e^{-2\sigma M/dRTr}$$

where p and p_0 are the saturated vapour pressures above a concave meniscus of radius r, and a plane surface, respectively; σ is the surface tension of the liquid at an absolute temperature T, d its density and M its molecular weight; and R is the gas constant. Thus, as the value of r decreases (r can be approximately equated to the radius of the capillary concerned) so the relative humidity at which condensation takes place within the capillary also decreases (Table 2.7).

Table 2.7 Capillary radii for condensation at given humidities

Capillary radius (Å)	*Relative humidity for condensation*
360	98
94	90
47	80
30	70
21	60
15	50

This concept may be invoked to account for electrolyte formation in microcracks in a metal surface or in the re-entrant angle formed by a dust particle and the metal surface. More importantly, it can also explain electrolyte formation in the pores of corrosion product and hence the secondary critical humidity discussed earlier. Ferric oxide gel is known to exhibit capillary condensation characteristics[23] and pore sizes deduced from measurements of its adsorptive capacity[23] are of the right order of magnitude to explain a secondary critical relative humidity $\approx 70\%$ for rusted steel[20].

Chemical condensation This occurs when soluble corrosion products or atmospheric contaminants are present on the metal surface. When the humidity exceeds that in equilibrium with a saturated solution of the soluble species, a solution, initially saturated, is formed until equilibrium is established with the ambient humidity. The contaminants have already been detailed and of the corrosion products, obviously sulphates, chlorides and carbonates are most important in this context. However, in some cases there is a lack of reliable data on the vapour pressure exerted by saturated solutions of likely corrosion products. The useful data was summarised in Table 2.5.

In practice, however, the soluble components are often contained in a matrix of insoluble product and formation of electrolyte by both capillary and chemical condensation may occur in the same humidity range.

Adsorbed electrolyte layers In this case the water molecules are bound to the metal surface by Van der Waals' forces. It is estimated that at 55% r.h. the film on polished iron is about 15 molecular layers thick, increasing to 90 molecular layers at just below 100% r.h.[4]. Such films are capable of supporting electrochemical corrosion processes and these have been studied. As the humidity is reduced below 100% and the moisture layers become thinner, polarisation of the cathodic and particularly the anodic process rapidly becomes enormous and corrosion virtually ceases below about 60% r.h.[4].

The Rôle of Sulphur Dioxide in Atmospheric Corrosion

Sulphur dioxide plays such an important rôle in the corrosion of metals in the atmospheres of industrialised countries that detailed consideration of its action seems justified. For all metals SO_2 appears to be selectively adsorbed from the atmosphere, less so for aluminium than for other metals, and for rusty steel it is almost quantitatively adsorbed even from dry air at 0°C[15]. Under humid conditions sulphuric acid is formed, the oxidation of SO_2 to SO_3 being catalysed by metals and by metallic oxides.

For some non-ferrous metals (copper, lead, nickel) the attack by sulphuric acid is probably direct with the formation of sulphates. Lead sulphate is barely soluble and gives good protection. Nickel and copper sulphates are deliquescent but are gradually converted (if not leached away) into insoluble basic sulphates[15], e.g. $Cu\{Cu(OH)_2\}_3SO_4$, and the metals are thus protected after a period of active corrosion. For zinc and cadmium the sulphur acids probably act by dissolution of the protective basic carbonate film[15]. This reforms, consuming metal in the process, redissolves, and so on. Zinc and cadmium sulphates are formed in polluted winter conditions whereas in the purer atmospheres of the summer the corrosion products include considerable amounts of oxide and basic carbonate[15].

Thus for non-ferrous metals, SO_2 is consumed in the corrosion reactions whereas in the rusting of iron and steel it is believed[15,24] that ferrous sulphate is hydrolysed to form oxides and that the sulphuric acid is regenerated. Sulphur dioxide thus acts as a catalyst such that one SO_4^{2-} ion can catalyse the dissolution of more than 100 atoms of iron before it is removed by leaching, spalling of rust or the formation of basic sulphate[24]. These reactions can be summarised as follows:

$$Fe + SO_2 + O_2 \rightleftharpoons FeSO_4$$
$$4FeSO_4 + O_2 + 6H_2O \rightleftharpoons 4FeOOH + 4H_2SO_4$$
$$4H_2SO_4 + 4Fe + O_2 \rightleftharpoons 4FeSO_4 + 4H_2O, \textit{ et seq.}$$

Rosenfel'd[11,25] considers that SO_2 can act as a depolariser of the cathodic process. However, this effect has only been demonstrated with much higher levels of SO_2 (0·5%) than are found in the atmosphere (Table 2.4) and the importance of this action of SO_2 has yet to be proved for practical environments. However, SO_2 is 1 300 times more soluble than O_2 in water[11] and therefore its concentration in solution may be considerably greater than would be expected from partial pressure considerations. This high solubility would make it a more effective cathode reactant than dissolved oxygen even though its concentration in the atmosphere is comparatively small.

Electrochemistry of Atmospheric Corrosion

This has already been touched upon in several of the previous paragraphs.

Russian workers have extensively examined the electrochemistry of corrosion under thin moisture films and the reader is referred to the work of Rosenfel'd, Tomashov, Klark and co-workers for fuller details[4,11,25,26]. It has been found that the corrosion rate reaches a maximum when the moisture film is around 150 μm thick.

The cathodic process in atmospheric corrosion is often stated to be oxygen reduction, and indeed in many cases the evidence is that this is so[27,28], i.e.

$$O_2+2H_2O+4e^- \rightleftharpoons 4OH^-$$

Kaesche[9] considers that proton reduction may also play a rôle in polluted environments where the pH of the electrolyte is likely to be low. This would be particularly likely in the case of iron if the Schikorr mechanism, involving the presence of sulphuric acid, did in fact operate. However, Russian work[4,11] has shown that oxygen depolarisation is many times more efficient in thin moisture films than in bulk solutions and therefore proton reduction may not be important in affecting corrosion rates.

In the rusting of iron and steel, Evans[29] considers that the anodic reaction of

$$Fe \rightleftharpoons Fe^{2+}+2e^-$$

is balanced by the cathodic reduction of ferric rust to magnetite under wet conditions when access of oxygen is limited:

$$4Fe_2O_3+Fe^{2+}+2e^- \rightleftharpoons 3Fe_3O_4$$

As the rust dries and is permeated by oxygen, magnetite is reoxidised to rust with a net gain of $0{\cdot}5Fe_2O_3$:

$$3Fe_3O_4+0{\cdot}75O_2 \rightleftharpoons 4{\cdot}5Fe_2O_3$$

There is considerable evidence that under certain conditions this mechanism may be operative.

Effect of Corrosion Products on Corrosion Rates

The change in corrosion rate with time varies markedly for different metals due to the differing degrees of protection conferred by the corrosion products.

Lead, aluminium and copper corrode initially but eventually form completely protective films[4]. Nickel in urban atmospheres does not form a completely protective film, the corrosion/time curve being nearly parabolic[4]. The corrosion rate of zinc appears to become linear after an initial period of decreasing corrosion rate[4].

The behaviour of steel depends very much on the alloying elements present for any given environment. Thus the decrease in corrosion rate with time for mild steel is very much slower than for a low-alloy steel. This can be attributed to the much more compact nature of the rust formed on the latter type of steel and this is clearly illustrated in Figs. 2.13(*a*) and (*b*).

Weather conditions at the time of initial exposure of zinc and steel have a large influence on the protective nature of the initial corrosion products[15]. This can still be detected some months after initial exposure. Finally, rust on

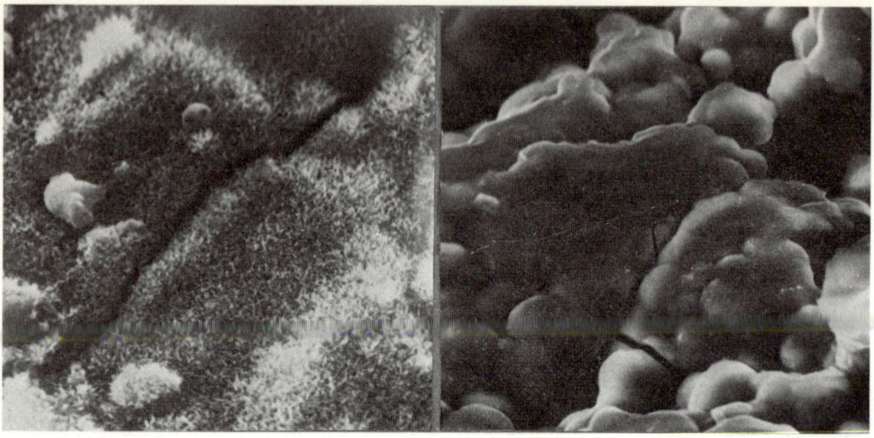

(a) (b)

Fig. 2.13 Surface textures of rust on (a) mild steel and (b) Cor-Ten steel exposed in an industrial atmosphere for 2·5 years [(a) ×1 260 and (b) ×1 320]

steel contains a proportion of ferrous sulphate which increases with increase in SO_2 pollution of the atmosphere. The effect of this on corrosion rate is so strong that mild steel transferred from an industrial atmosphere to a rural one corrodes for some months as though it was still exposed to the industrial environment[24].

Conclusion

In a section of this brevity it is impossible to cover all aspects of the 'atmosphere'. There are therefore gaps concerning such topics as corrosion at sub-zero temperatures, effect of surface orientation and inclination on corrosion rates and the influence of organic vapours on metallic corrosion. Neither has it been possible to describe the intensive efforts now being made to monitor continuously atmospheric variables such as 'time of wetness' with a view to predicting the long-term corrosion behaviour of metals in a particular area without resorting to long-term trials.

D. FYFE

REFERENCES

1. *Investigation of Air Pollution, April 1968 to March 1969—National Survey, Smoke and Sulphur Dioxide*, Min. Tech., Warren Spring Laboratory, Stevenage, Herts.
2. Smith, A. F. *et al.*, *J. Appl. Chem.*, **11**, 317 (1961)
3. Stevenson, C. M., *Q. J. Roy. Meteorol. Soc.*, **94**, 56 (1968)
4. Tomashov, N. D., *Theory of Corrosion and Protection of Metals*, Macmillan, New York (1966)
5. Schikorr, G., *Werk. Korr.*, **18**, 514 (1967)
6. Young, J. F., *J. Appl. Chem.*, **17**, 241 (1967)
7. O'Brien, F. E. M., *J. Sci. Instrum.*, **25**, 73 (1948)
8. Vernon, W. H. J., *Trans. Far. Soc.*, **31**, 1 668 (1935)
9. Kaesche, H., *Werk. Kørr.*, **15**, 379 (1964) and B.I.S.I.T.S. No. 5 271

10. Meetham, A. R., *Atmospheric Pollution: its Origins and Prevention*, Pergamon, London (1956) and Stern, A. C., (Ed.), *Air Pollution*, Academic Press, New York, 2nd edn. (1968)
11. Rosenfel'd, I. L., *Proc. 1st Int. Corros. Cong. (London)*, Butterworths (1962)
12. Hudson, J. C. and Stanners, J. F., *J. Appl. Chem.*, **3**, 86 (1953)
13. Chandler, K. A. and Kilcullen, M. B., *Br. Corros. J.*, **3**, 87 (1968)
14. Haynie, F. H. and Upham, J. B., *Mats. Prot.*, **9** No. 8, 35 (1970)
15. Schikorr, G., *Werk. Korr.*, **15**, 457 (1964) and B.I.S.I.T.S. No. 3 947
16. Schwarz, H., ibid., **16**, 93 (1965) and B.I.S.I.T.S. No. 4 269
17. Ross, T. K. and Callaghan, B. G., *Nature*, **211**, 25 (1966)
18. Sereda, P. J., *Ind. Engng. Chem.*, **53**, 157 (1960)
19. Stanners, J. F., *Br. Corros. J.*, **5**, 117 (1970)
20. Skorchelletti, V. V. and Tukachinsky, S. E., *J. Appl. Chem. (USSR)*, **28**, 615 (1955)
21. Barton, K. and Bartonova, Z., *Werk. Korr.*, **21** No. 2, 85 (1970) and B.I.S.I.T.S. No. 8 349
22. Persson, G., *Acidity and Concentration of Sulphate in Precipitation Over Europe*, Report, Swedish National Nature Conservancy Office, December (1968)
23. Broad, D. W. and Foster, A. G., *J. Chem. Soc.*, 446 (1946)
24. Schikorr, G., *Werk. Korr.*, **14**, 69 (1963)
25. Rosenfel'd, I. L. and Zhigalova, K., *Corrosion of Metals and Alloys* (Ed. by C. Booker), Metallurgizdat, Moscow
26. Klark, G. B., *et al.*, ibid.
27. Brokskii, A. I., *Zhur. Fiz. Khim.*, **30**, 676 (1956)
28. Roikh, I. L., ibid., **32**, 1137 (1958)
29. Evans, U. R., *Trans. Inst. Met. Fin.*, **37**, 1 (1960)

2.3 Natural Waters

Introduction

Metals immersed or partly immersed in water tend to corrode because of their thermodynamic instability. Natural waters contain dissolved solids and gases and sometimes colloidal or suspended matter; all these may affect the corrosive properties of the water in relation to the metals with which it is in contact. The effect may be either one of stimulation or one of suppression, and it may affect either the cathodic or the anodic reaction; more rarely there may be a general blanketing effect. Some metals form a natural protective film in water and the corrosiveness of the water to these metals depends on whether or not the dissolved materials it contains assist in the maintenance of a self-healing film.

The metals most commonly used for water systems are iron and steel. These metals often have some sort of applied protective coating; galvanised steel, for example, relies on a thin layer of zinc, which is anodic to the steel except at high temperatures. Many systems, however, contain a wide variety of other metals and the effect of various water constituents on these must be considered. The more usual are copper, brasses, bronzes, lead, aluminium, stainless steel and solder.

The passage of a natural water through a pipe may modify the composition of the water and hence its corrosive properties. Consumption of constituents which in the given circumstances may be corrosion-inducing—e.g. oxygen or carbon dioxide, may reduce the water's corrosive properties. Dissolution of a metal into water may, on the other hand, make it more corrosive. An example of this is the attack of some waters on copper and the subsequent increased pitting corrosion of less noble metals such as iron, galvanised steel and aluminium. It has been suggested that this enhanced pitting is caused by the redeposition of minute quantities of copper on the less noble metal thus setting up numerous bimetallic corrosion cells[1]. An example of this is the attack of some waters on copper and the redeposition of minute quantities of this metal on galvanising or aluminium.

Failure of the metal is the main effect of a corrosive water, but other effects may arise from small concentrations of metallic ion produced by corrosion. A natural water passed through a lead pipe may contain a toxic concentration of that metal; with copper there is a greater tolerance from the toxicity point of view but staining of fabrics and sanitary fittings may be objectionable.

With iron, similarly, discoloration of the water may be unpleasant and may cause damage to materials being processed.

Constituents or Impurities of Water

The concentrations of various substances in water in dissolved, colloidal or suspended form are relatively low but vary considerably; for example, a hardness of 300–400 p.p.m. (as $CaCO_3$) is sometimes tolerated in public supplies, whereas dissolved iron to the extent of 1 mg/litre would be unacceptable.

In treated water for high-pressure boilers or where radiation effects are important, as in some nuclear energy projects, impurities are measured in very small units (e.g. μg/litre or p.p. 10^9), but for most purposes it is convenient to express results in mg/litre. In water analysis, determinations (except occasionally for dissolved gases) are made on a weight/volume basis but some analysts still express results in terms of parts per million (p.p.m.). The difference between mg/litre and p.p.m. is small and for practical purposes the two units are interchangeable. For some calculations, the use of milli-equivalents per litre or equivalents per million (e.p.m.) has advantages but has not found much application. Hardness, whatever the constituent salts, is usually expressed as p.p.m. $CaCO_3$ (see Table 2.8).

Table 2.8 Units of measurement and of hardness

Units of Measurement—Conversion Factors

Milligrams per litre (mg/l) = parts per million (p.p.m.).
Part per 100 000 = 10 mg/litre.
Grains per Imperial gallon = 14·25 mg/litre.
Grains per US gallon = 17·1 mg/litre.

Hardness Units

Parts per million (mg/l) as $CaCO_3$.
Degree French = parts per 100 000 as $CaCO_3$ (=10 mg/litre $CaCO_3$).
Degree Clark, English or British = grains per Imperial gallon or p.p. 70 000 as $CaCO_3$ (=14·3 mg/litre $CaCO_3$).
Degree German = parts per 100 000 as CaO (=17·8 mg/litre $CaCO_3$).

Water analysis for drinking-water supplies is concerned mainly with pollution and bacteriological tests. For industrial supplies a mineral analysis is of more interest. Table 2.9 includes a typical selection and gives some indication of the wide range that can be found.

The important constituents can be classified as follows:

1. Dissolved gases (oxygen, nitrogen, carbon dioxide, ammonia, sulphurous gases).
2. Mineral constituents, including hardness salts, sodium salts (chloride, sulphate, nitrate, bicarbonate, etc.), salts of heavy metals, and silica.
3. Organic matter, including that of both animal and vegetable origin, oil, trade (including agricultural) waste constituents and synthetic detergents.
4. Microbiological forms, including various types of algae and slime-forming bacteria.

Table 2.9 Typical water analyses (results in mg/litre)

	Very soft lake water	*Moderately soft surface water*	*Slightly hard river water*	*Moderately hard river water*	*Hard borehole water (chalk formation)*	*Slightly hard borehole water containing sodium bicarbonate*	*Very hard underground water*
pH value	6·3	6·8	7·4	7·5	7·1	8·3	7·1
Alkalinity to methyl orange ($CaCO_3$)	2	38	90	180	250	278	470
Total hardness ($CaCO_3$)	10	53	120	230	340	70	559
Calcium hardness ($CaCO_3$)	5	36	85	210	298	40	451
Sulphate (SO_4)	6	20	39	50	17	109	463
Chloride (Cl)	5	11	24	21	4	94	149
Silica (SiO_2)	trace	0·3	3	4	7	12	6
Dissolved solids	33	88	185	332	400 (also nitrate 51 mg/litre NO_3)	620	1 670

Dissolved Gases

Of the dissolved gases occurring in water, oxygen occupies a special position as it stimulates the corrosion reaction. Carbon dioxide is scarcely less important; this constituent must, however, be considered in relation to other constituents, especially calcium hardness.

Nitrogen is present with oxygen, although the ratio is not the same as in air. It has little importance in connection with corrosion, but can be a nuisance if changes in physical conditions bring about its release from solution.

Other gases which are occasionally present usually arise from pollution. Ammonia, which in various forms may be present in waste waters, attacks copper and copper alloys; its presence in estuarine waters is one of the main causes of condenser-tube corrosion.

Hydrogen sulphide and sulphur dioxide are also usually the result of pollution; sometimes they are produced by the interaction of two contaminants, but sometimes bacterial action may be contributory. Both gases may initiate or accelerate corrosion of most metals.

The significance of small concentrations of these and other impurities in high-pressure steam-boiler feed water is discussed in Section 18.4.

Oxygen Dissolved oxygen is probably the most significant constituent affecting corrosion, its importance lying in the fact that it is the most important cathodic depolariser in neutral solutions. Other depolarisers also occur, but as oxygen is an almost universal constituent of natural waters its importance will readily be understood.

In surface waters, the oxygen concentration approximates to saturation, but in the presence of green algae supersaturation may occur. Underground waters are more variable in oxygen content and some waters containing ferrous bicarbonate are oxygen-free. Contact with air, however, usually gives rise to an oxygen concentration similar to the figures in Table 2.10, which are for distilled water. The solubility is slightly less in the presence of dissolved solids, but this effect is not very significant in natural waters containing less than 1 000 p.p.m. dissolved solids.

Table 2.10 Solubility of oxygen in distilled water

Temperature (°C)	*Oxygen content of air-saturated water*	
	mg/litre	ml/litre
0	14·16	9·90
5	12·37	8·65
10	10·92	7·64
15	9·76	6·83
20	8·84	6·18
25	8·11	5·67

For some applications, notably feed-water treatment for high-pressure boilers, removal of oxygen is essential. For most industrial purposes, however, de-aeration is not applicable, since the water used is in continuous contact with air, from which it would rapidly take up more oxygen. Attention must therefore be given to creating conditions under which oxygen will stifle rather than stimulate corrosion.

It has been shown that pure distilled water is least corrosive when fully aerated and that some inhibitors function better in the presence of oxygen[2]. In these cases oxygen acts as a passivator of the anodic areas of the corrosion cells. These facts do not, however, modify the foregoing statements on the significance of oxygen in waters as used in practice.

A major difficulty in applying corrosion inhibitors is that the oxygen content of the water may not be the same at all points. For example, in a thin layer of water between a flake of scale (or almost any other foreign body) and the metal on which it is lying the oxygen can be depleted; the difference in oxygen content between the body of water and the stagnant water will then set up a corrosion current which is difficult to suppress. Rather similar conditions occur at the water line of a vessel containing water with air above it. Even if the water is conditioned to prevent corrosion under submerged conditions, the protection may not extend to the water line, especially if the water has a high dissolved-solids content. Fluctuation in water level extends the area of localised attack.

Carbon dioxide and calcium carbonate The effect of carbon dioxide is closely linked with the bicarbonate content. Normal carbonates are rarely found in natural waters but sodium bicarbonate is found in some underground supplies. Calcium bicarbonate is the most important, but magnesium bicarbonate may be present in smaller quantities; in general, it may be regarded as having properties similar to those of the calcium compound except that on decomposition by heat it deposits magnesium hydroxide whereas calcium bicarbonate precipitates the carbonate.

Calcium bicarbonate requires excess carbon dioxide in solution to stabilise it; the necessary concentration depends on the other constituents of the water and the temperature.

Quantities of carbon dioxide in water can be classified as follows:

1. The amount required to produce carbonate.
2. The amount required to convert carbonate to bicarbonate.
3. The amount required to keep the calcium bicarbonate in solution.
4. Any excess over that accounted for in types 1, 2 and 3.

With insufficient carbon dioxide of type 3 (and none of type 4) the water will be supersaturated with calcium carbonate and a slight increase in pH (at the local cathodes) will tend to cause its precipitation. If the deposit is complete and adherent the metal surface will be isolated from the water and hence protected from corrosion. Under these conditions the water will usually be non-corrosive, but occasionally the deposit may be incomplete or not properly adherent, and corrosion may then occur. If type 4 carbon dioxide is present there can be no deposition of calcium carbonate and old deposits will be dissolved; there cannot therefore be any protection by calcium carbonate scale.

The mathematical relationship between carbon dioxide, calcium bicarbonate and calcium carbonate has been studied by several workers, including Langelier[3,4]. The simpler form of his equation is

$$pH_s = pCa + pAlk + (pK_2 - pK_s) \text{ at constant temperature}$$

where pH_s = saturation pH value,

pCa = negative logarithm of the calcium concentration expressed as p.p.m. $CaCO_3$,

$pAlk$ = negative logarithm of the alkalinity to methyl orange expressed in p.p.m. of equivalent $CaCO_3$,

pK_2 = ionisation constant of HCO_3^-, $\left[\frac{[H^+][CO_3^{2-}]}{[HCO_3^-]}\right]$ and

pK_s = solubility product of $CaCO_3$.

This simple form does not apply to pH values over 9·0, and high salinities affect its accuracy. The term $(pK_2 - pK_s)$ is a function of temperature and ionic strength (dissolved solids). In an analysis of a given water at a constant temperature much useful information can be obtained from the equation.

The saturation index of a water (S.I.) is defined as:

$$S.I. = pH - pH_s$$

where pH is the actual pH of the water. If the saturation index is positive the water will be supersaturated with calcium carbonate whereas if it is negative the water will be aggressive to calcium carbonate. Graphical forms of the expression are of most practical value and that devised by Powell, Bacon and Lill[5] is shown in Fig. 2.14. These authors have also developed the method so that it can be applied to provide a water which has a constant saturation index over a fair temperature range; this is mainly of interest to operators of industrial cooling systems[6]. The corresponding formula for magnesium hydroxide equilibrium.

$$pH_s(Mg) = \tfrac{1}{2}(pMg - pK_{s(Mg)}) + pK_w$$

is incorporated in the Langelier diagram[4].

A distinction must be made between a thick layer of deposit—whether of calcium carbonate or of other material—and a protective layer. The ideal

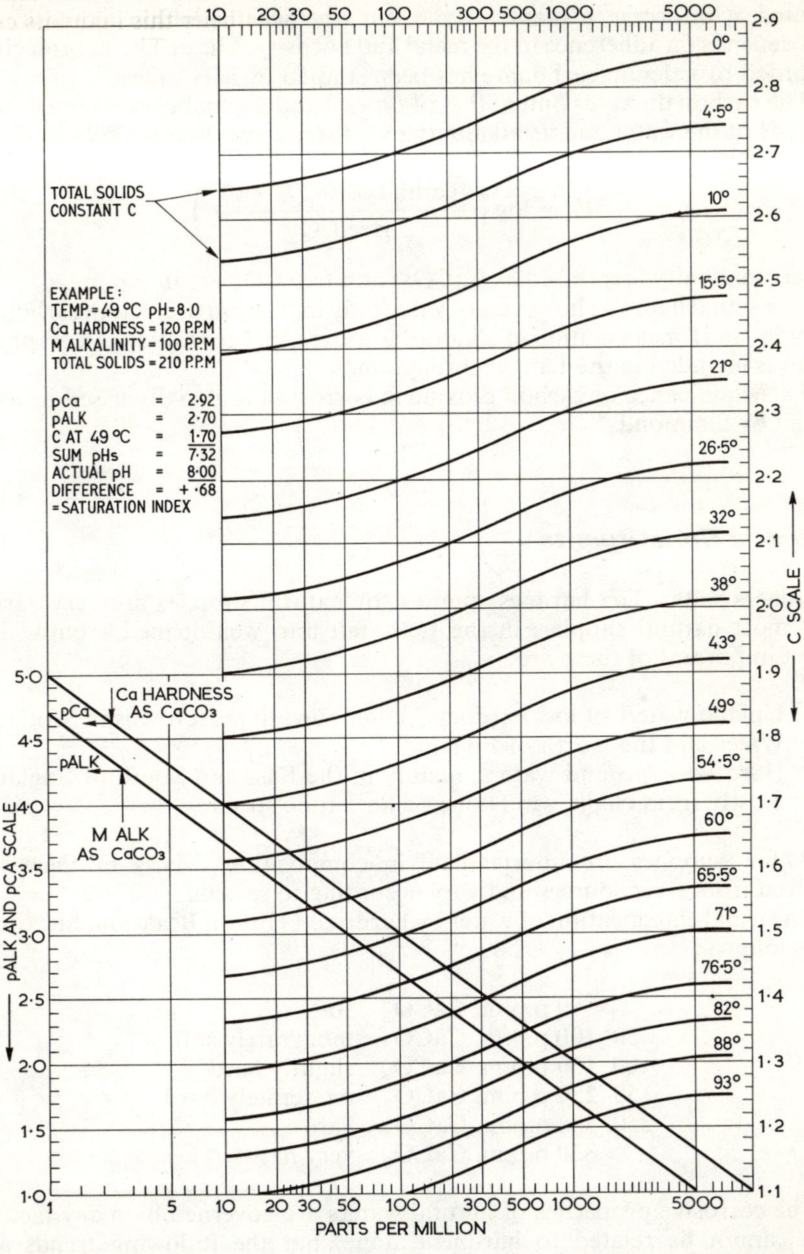

Fig. 2.14 Langelier saturation index chart (after the American Chemical Society)

protection in fact consists of layers of negligible thickness which do not impede water or heat flow and which are self-healing. This is difficult to achieve with natural waters. A water which is exactly in equilibrium in respect of calcium carbonate is normally corrosive to steel (unless it contains natural inhibitors of other types) because it has no power to form a calcium carbonate deposit. Supersaturated waters on the other hand, unless suitably treated, will form a substantial scale, but whether this inhibits corrosion or not depends on adherence to the metal and porosity. The degree of protection afforded by calcium carbonate has been studied by McCauley[7].

The carbon dioxide content (i.e. of types 3 and 4) can be ascertained from the pH of the water and its alkalinity by a formula devised by Tillmans[8]

$$\text{pH} = \log\left(\frac{\text{alkalinity} \times 0{\cdot}203 \times 10^7}{\text{free } CO_2}\right)$$

where alkalinity (expressed as $CaCO_3$) and free CO_2 are in p.p.m.

The actual figure is, however, of value only in relation to calcium carbonate content and for calculation of alkali additions for pH corrections. A graphical form is included in the Langelier diagram[4].

The significance of carbon dioxide in corrosion is also discussed in some detail by Simmonds[9].

Mineral Constituents

Hardness salts The hardness figures for natural supplies are very varied but most natural supplies in the U.K. fall into well-defined groups. The most important of these are:

1. Upland waters of low hardness, as supplied to most towns in Scotland, Wales and the North of England.
2. Hard underground waters, mainly in the East and South of England, mostly from chalk, sandstone or limestone strata.

A few supplies are intermediate in composition. Many of them are derived from river sources and vary according to season.

The usual classification of water by hardness (Thresh, Beale and Suckling) is as follows:

<50 p.p.m. $CaCO_3$	soft
50–100 p.p.m. $CaCO_3$	moderately soft
100–150 p.p.m. $CaCO_3$	slightly hard
150–250 p.p.m. $CaCO_3$	moderately hard
250–350 p.p.m. $CaCO_3$	hard
>350 p.p.m. $CaCO_3$	very hard.

The corrosive properties of natural waters are governed by many factors and cannot be related to hardness alone, but the following trends are apparent:

1. Soft upland waters are aggressive to most metals, their behaviour depending to some extent on pH values as discussed on p. **2**.48. They are inevitably unsaturated with respect to calcium carbonate and it is not usually practicable to modify the carbonate equilibrium to make them non-aggressive.
2. Very hard waters are usually not very aggressive provided that they are supersaturated with calcium carbonate. Underground waters with a low pH value and high carbon dioxide content are, however, aggressive unless corrective treatment is applied.
3. Waters of intermediate hardness frequently contain fair amounts of other constituents and there is often a tendency for the scale to be loosely attached, permitting corrosion to occur irregularly underneath. In most waters the bicarbonate content is less than the hardness, but a few natural waters are known where the reverse is the case. These waters have been partially softened by the zeolite process which occurs underground, and then contain sodium bicarbonate which, together with the high concentration of chloride and other minerals, may accelerate attack.

Dissolved mineral salts The principal ions found in water are calcium, magnesium, sodium, bicarbonate, sulphate, chloride and nitrate. A few parts per million of iron or manganese may sometimes be present and there may be traces of potassium salts, whose behaviour is very similar to that of sodium salts. From the corrosion point of view the small quantities of other acid radicals present, e.g. nitrite, phosphate, iodide, bromide and fluoride, have little significance. Larger concentrations of some of these ions, notably nitrite and phosphate, may act as corrosion inhibitors, but the small quantities present in natural waters will have little effect. Some of the minor constituents have other beneficial or harmful effects, e.g. there is an optimum concentration of fluoride for control of dental caries and very low iodide or high nitrate concentrations are objectionable on medical grounds.

Chlorides have probably received the most study in relation to their effect on corrosion. Like other ions, they increase the electrical conductivity of the water so that the flow of corrosion currents will be facilitated. They also reduce the effectiveness of natural protective films, which may be permeable to small ions; the effect of chloride on stainless steel is an extreme example but a similar effect is noted to a lesser degree with other metals. Turner[10] has observed that the meringue dezincification of duplex brasses is affected by the chloride/bicarbonate hardness ratio.

Nitrate is very similar in its effects to chloride but is usually present in much smaller concentrations.

Sulphate in general appears to behave very similarly; Hatch and Rice have shown that small concentrations in distilled water increase corrosion more than similar concentrations of chloride[11]. In practice, high-sulphate waters may attack concrete, and the performance of some inhibitors appears to be adversely affected by the presence of sulphate. Sulphates have also a special rôle in bacterial corrosion under anaerobic conditions. Both sulphates and nitrates are acceptable in low-pressure boiler feed water as they are believed to be of value in controlling caustic cracking.

Conventional combinations Salts of strong acids and alkalis are, of course, almost completely ionised in dilute solutions. For some purposes, however, it is convenient to regard the ions as being in combination, and various systems of 'conventional combinations' have been developed. In Britain, the system most used takes the metals and acid radicals in the order shown in Table 2.11 (after Thresh, Beale and Suckling). For example, if the amount of bicarbonate ion is more than the sum of the equivalents of ferrous iron and calcium the presence of magnesium bicarbonate, and possibly sodium bicarbonate, is postulated. If, however, the bicarbonate content is less than the calcium equivalent the water is assumed to contain calcium sulphate.

Table 2.11 Conventional combinations*†

Ferrous carbonate	56 Fe	≡	60 CO_3	≡	116 $FeCO_3$
Calcium carbonate	40 Ca	≡	60 CO_3	≡	100 $CaCO_3$
Calcium sulphate	40 Ca	≡	96 SO_4	≡	136 $CaSO_4$
Calcium chloride‡	40 Ca	≡	71 Cl	≡	111 $CaCl_2$
Calcium nitrate‡	40 Ca	≡	124 NO_3	≡	164 $Ca(NO_3)_2$
Magnesium carbonate	24 Mg	≡	60 CO_3	≡	84 $MgCO_3$
Magnesium sulphate	24 Mg	≡	96 SO_4	≡	120 $MgSO_4$
Magnesium chloride‡	24 Mg	≡	71 Cl	≡	95 $MgCl_2$
Magnesium nitrate‡	24 Mg	≡	124 NO_3	≡	148 $Mg(NO_3)_2$
Sodium carbonate	46 Na	≡	60 CO_3	≡	106 Na_2CO_3
Sodium sulphate	46 Na	≡	96 SO_4	≡	142 Na_2SO_4
Sodium chloride‡	46 Na	≡	71 Cl	≡	117 NaCl
Sodium nitrate‡	46 Na	≡	124 NO_3	≡	170 $NaNO_3$
Sodium silicate§	46 Na	≡	136 Si_2O_5	≡	182 $Na_2Si_2O_5$

*If potassium is present in significant quantities and is determined it is usually inserted in the conventional combination table after magnesium and before sodium (78 K = 138 K_2CO_3 ≡ 174 K_2SO_4 ≡ 149 KCl ≡ 202 KNO_3).

†The figures incorporated in the table are 'equivalent to $CaCO_3$' or double the chemical equivalent weights. If it is desired to express analytical figures as milli-equivalents per litre (e.p.m.), the concentrations in mg/litre (p.p.m.) must be divided by half the table figure (e.g. 150 p.p.m. $CaCO_3$ ≡ 3 e.p.m.; 96 p.p.m. Mg ≡ 8·0 e.p.m.). If this procedure is adopted for both anions and cations, the totals of each should be identical.

‡Sometimes chlorides and nitrates are taken in reversed order throughout but this is rarely of much significance.

§The sodium silicates present are variable in composition; the formula given above must be regarded as a 'rough average'.

The significance of conventional combinations arises largely in two classes of supply: (*a*) those in which the method indicates sodium bicarbonate to be present and (*b*) those similarly found to contain magnesium chloride or calcium chloride.

Waters containing sodium bicarbonate are derived from underground sources where zeolitic materials are present. They occur in various parts of Britain but the best-known group comprises those in central London and other parts of the Thames valley where the water, originating in the chalk, is modified by passage through zeolitic rock. Although some softening occurs, the hardness may still be appreciable; the waters also have high chloride concentrations and usually a fair amount of carbon dioxide. (A typical analysis is given in Table 2.1.) They are among the most difficult corrosive waters to deal with unless the bicarbonates and excess carbon dioxide are removed. Alkali additions are rarely effective and conditions are far from ideal for the effective use of inhibitors.

The importance of magnesium chloride has probably been exaggerated. There is little doubt that it can act as a catalyst in corrosion reactions by hydrolysing to form hydrochloric acid, being then regenerated by reaction between ferrous chloride and magnesium hydroxide. There is, however, little evidence that this reaction takes place in cold- or hot-water systems, and it is probably confined to steam boilers where it might be a cause of corrosive attack underneath scale deposits; it does not constitute a problem in a properly conditioned boiler water.

Brief reference has already been made to iron- and manganese-bearing waters. The small amount of deposit formed from these waters is not likely to have much effect on corrosion although there is always a possibility that attack will occur under sludge deposits. Most iron-bearing waters contain substantial amounts of carbon dioxide which may be troublesome. Manganese-bearing waters may be of a similar type but they sometimes contain complex organic compounds of manganese for which special treatment may be needed.

Another mineral constituent of water is silica, present both as a colloidal suspension and dissolved in the form of silicates. The concentration varies very widely and, as silicates are sometimes applied as corrosion inhibitors, it might be thought that the silica content would affect the corrosive properties of a water. In general, the effect appears to be trivial; the fact that silicate inhibitors are used in waters with a high initial silica content suggests that the form in which silica is present is important.

Organic Matter

The types of organic matter in water supplies are very diverse and may be present in suspension, or in colloidal or true solution. It is largely decaying vegetable matter but there are many other possible sources including run-off from fields and domestic and industrial wastes. An increasing volume of literature is appearing on organic pollution but the significance of this in relation to corrosion receives little, if any, attention. A comprehensive account of all the possible constituents is beyond the scope of this section but it may be useful to consider the effects of some types of organic matter.

In the first place, there is the masking effect of deposits which may result from suspended matter thrown down on to hot surfaces or at areas where velocity is reduced. They may also form from material coming out of true or colloidal solution. A partially covered surface is always liable to attack. In an aerated water, the distribution of oxygen will be uneven so that corrosion currents will be set up by the cells so produced, corrosion normally occurring at the points where the oxygen content is lower. In waters free from oxygen, other 'differentials' may result in corrosion cells. Another aspect, especially in systems where de-aerated water might be used, is that deposits may lead to over-heating and failures of a different kind, e.g. bursting of boiler tubes.

Among contaminants one of the most objectionable is oil, especially in systems where water is strongly heated. A relatively small amount of oil on a heating surface can produce very rapid failures. An indirect effect of oil, or other contaminants which form films on the water surface, is that the film isolates the water from air so that in polluted water anaerobic conditions may develop with the encouragement of objectionable bacterial activity.

Some of the worst corrosive effects in soft waters are attributed to a rather wide group of organic acids abstracted from peat and mosses, sometimes called *peaty* acids. Such waters have low pH values and are often discoloured. They affect ferrous metals appreciably and also attack lead and copper. Corrosion control, either of steel or of copper, is rarely achieved solely by pH correction of such waters.

Organic Growths

Natural waters may contain organic growths of various kinds, including algae and slime-forming bacteria, which may have a direct or indirect effect on corrosion.

The effects may be of two main types: (*a*) the masking effect of living or dead organisms which is little different from that of other materials, and (*b*) the effect of alterations in composition brought about by the organisms. Algae, for example, may remove carbon dioxide and produce oxygen, while other organisms often consume oxygen. Under anaerobic conditions the sulphate-reducing bacteria produce sulphide, and hence hydrogen sulphide, from dissolved sulphate, with disastrous effects. This mechanism is often responsible for external attack on cast iron mains in waterlogged soil, but is not unknown in hot-water systems where the temperature in stagnant branches may promote the development of such organisms (see Sections 2.5 and 2.6).

The voluminous deposits associated with iron bacteria, although objectionable in other ways, rarely have much effect on corrosion as they form only over a long period and the alteration in water composition is negligible.

pH of Water

Reference has previously been made to pH in connection with calcium carbonate, but it has also a more general significance. The pH of natural waters is, in fact, rarely outside the fairly narrow range of 4·5 to 8·5. High values, at which corrosion of steel may be suppressed, and low values, at which gaseous hydrogen evolution occurs, are not often found in natural waters.

According to weight-loss measurements, steel corrodes at approximately the same rate throughout the range of pH found in natural waters. The form which the corrosion takes is, however, affected by the pH. At values between 7·5 and 9·0 there is a tendency for the corrosion products to adhere in a hard crusty deposit[12]. Sometimes there are separate ‘tubercles’, but these are more usually joined up to form a more or less continuous layer. Attack under the deposit is, however, usually irregular. At lower pH values, adherent corrosion products are not so evident although a very hard form of deposit is sometimes seen in pipes which have been in service for some years. Loss of head due to scaling of a pipe is more commonly found in the higher pH range; at lower pH values, ‘red water’ complaints, arising from corrosion products in suspension, are more common. Inhibitors may reduce the amount of corrosion, but if inhibition is not complete the type of attack is unaltered.

For this reason, it is difficult to prevent corrosion in the tuberculating range as a small amount of attack produces an adherent corrosion product which puts a barrier between the inhibitor and the metal.

Cast iron behaves in a manner similar to steel at alkaline pH values but at low pH values it is subject to graphitisation.

Copper is affected to a marked extent by pH value. In acid waters, slight corrosion occurs and the small amount of copper in solution causes green staining of fabrics and sanitary ware. In addition redeposition of copper on aluminium or galvanised surfaces sets up corrosion cells resulting in pitting of these metals. In most waters the critical pH value is about 7·0 but in soft water containing organic acids it may be higher. The 'pitting' corrosion of copper is independent of the general nature of the water but occurs only when (*a*) certain carbonaceous or oxide films occur on the metal surface and (*b*) when the water does not contain a natural organic inhibitor* which is, in fact, present in many supplies[13,14] (see also Sections 1.6 and 4.2).

Lead is affected by carbonate content, pH value and mineral constituents. The simplest method of control is usually to increase the pH value by adding alkali, and few waters are plumbo-solvent if the pH value is above 7·0.

Zinc coatings on steel (galvanised) are attacked in the same way as iron, but usually more slowly. Very alkaline waters are usually aggressive to zinc and will often remove galvanised coating; the corrosion products consist of basic zinc carbonate or other basic compounds and may take the form of a thick creamy deposit or hard abrasive particles.

Rates of Flow

The effect of deposits has been referred to in relation to organic matter. Oxygen depletion can, of course, also occur under other types of deposit. Water velocity plays some part here, as with a good flow deposition is less likely.

Apart from this effect, increased velocity usually increases corrosion rates by removing corrosion products which otherwise might stifle the anodic reaction, and by providing more oxygen, may stimulate the cathodic reaction[15].

Temperature

The effect of temperature is complex. At very high temperatures such effects as reversal of polarities, as in the Zn/Fe couples of a galvanised surface, may be produced, and, where there are temperature gradients, corrosion cells may be set up. The more general effects may, however, be summarised as follows: (*a*) the velocity of corrosion reactions is greater at increased temperatures, (*b*) temperature changes may affect solubility of corrosion products or shift the position of such equilibria as that existing between calcium carbonate and carbon dioxide, (*c*) gases are less soluble at increased temperature, an effect which is, however, partly offset by greater diffusion rates

*The chemical nature of the inhibitor is not known.

and (*d*) modification of pH value. This last effect is bound up with the previous two and is mainly of importance in affecting the form and distribution of corrosion products.

The overall effect is that corrosion is usually more rapid at higher temperatures, the corrosion products being often more objectionable in nature. There are, however, exceptions to this generalisation and the increased rate of reaction at high temperature can sometimes be put to good advantage when corrective measures are being applied.

C. W. DRANE

REFERENCES

1. Kenworthy, L., *J. Inst. Met.*, **69**, 67 (1943)
2. Uhlig, H. H., Triadis, D. N. and Stern, M., *J. Electrochem. Soc.*, **102**, 59 (1955)
3. Langelier, W. F., *J. Amer. Wat. Wks. Ass.*, **28**. 1 500 (1936)
4. Langelier, W. F., *J. Amer. Wat. Wks. Ass.*, **38**, 169 (1946)
5. Powell, S. T., Bacon, H. W. and Lill, J. R., *Industr. Engng. Chem.*, **37**, 842 (1945)
6. Powell, S. T., Bacon, H. W. and Lill, S. R., *Industr. Engng. Chem.*, **40**, 435 (1948)
7. McCauley, R. F. and Abdullah, M. O., *J. Amer. Wat. Wks. Ass.*, **50**, 1 419 (1958)
8. Tillmans, J., *Die Chemische Untersuchung von Wasser und Abwasser*, Verlage von Wilhelm Knapp, Saale (1936)
9. Simmonds, M. A., 'Carbon Dioxide in Domestic Water Supplies', *Proc. Soc. Water Treatment and Examination*, **12**, 4, 197 (1963) and **13**, 1, 40 (1964)
10. Turner, M. E. D., ibid., **10**, 2, 162 (1961) and **14**, 2, 81 (1965)
11. Hatch, G. B. and Rice, Owen, *J. Amer. Wat. Wks. Ass.*, **51**, 719 (1959)
12. Hatch, G. B. and Rice, Owen, *Industr. Engng. Chem.*, **37**, 710 (1945)
13. Gilbert, P. T., 'Dissolution by Fresh Waters of Copper from Copper Pipes', *Proc. Soc. Water Treatment and Examination*, **15**, 3, 165 (1966)
14. Lucey, V. F., *Br. Corros. J.*, **2**, 175 (1967)
15. Eliassen, R., Perada, C., Romeo, A. S. and Skrinde, R. T., *J. Amer. Wat. Wks. Ass.*, **48**, 1 005 (1956)

BIBLIOGRAPHY

Holden, W. S. (Ed.), *Water Treatment and Examination*, Churchill, London (1970)

Tarzwell, C. M., Proc. 29th International Water Conference, Engineers' Society of Western Pennsylvania, 1 (1968)

Corrosion of Iron and Steel by Industrial Waters and its Prevention, Special Report No. 41, I.S.I., London (1949)

Manual of British Water Engineering Practice, Vols. I to III. Published for the Institution of Water Engineers, Heffer, Cambridge, 4th edn. (1969)

Manual on Industrial Water and Industrial Waste Water, American Society for Testing Materials, Philadelphia (1962)

Water Quality and Treatment, American Water Works Association, New York, 3rd edn. (1971)

Water Quality Criteria, Resources Agency of California, State Water Quality Control Board, 2nd edn. (1963)

Water Quality Criteria, American Society for Testing and Materials, S.T.P. No. 416 (1967)

Water—1968, *Chemical Engineering Progress*, Symposium Series, American Institute of Chemical Engineers (1968)

2.4 Sea Water

Sea water is the only electrolyte containing a relatively high concentration of salts that occurs commonly in nature, covering as it does over two-thirds of the earth's surface. It is both the most familiar and one of the most severe of natural corrosive agents.

Chemical Composition

Ocean sea water is roughly equivalent in strength to a $3\frac{1}{2}\%$ w/v solution of sodium chloride, but it has a much more complex composition, embodying a number of major constituents, and traces at least of almost all naturally occurring elements. For convenience, however, the concentration of salts in any sample of sea water is expressed in terms of the chloride content, either as *chlorinity* or as *salinity*. Both these units are again subject to arbitrary definition and do not conform simply to the chemical composition.

Chlorinity When a sample of sea water is titrated with silver nitrate, bromides and iodides, as well as chlorides are precipitated. In calculating the chlorinity (Cl), the entire halogen content is taken as chloride, and chlorinity is defined as the weight in grams of silver required for precipitation of total halogen content per kilogram of sea water, multiplied by 0·328 533. (Chlorinity is always expressed as parts per thousand, using the symbol ‰.)

Salinity This term is intended to denote the total proportion of dissolved salts in sea water. As it is inconvenient to determine directly, it is normally derived from the chlorinity, defined and determined as above, using the empirical relationship:

$$\text{Salinity} = 0{\cdot}03 + (1{\cdot}805 \times \text{Chlorinity})$$

Like chlorinity, it is expressed in parts per thousand.

Constancy of composition The validity of these arbitrary conversions depends on the constancy of the ratios of the various dissolved salts. It is a remarkable and important fact that, except where there is gross dilution or contamination, the relative proportions of the major constituents of sea water are practically constant all over the world.

Table 2.12[1] gives the composition of sea water of 19 parts per thousand chlorinity.

Table 2.12 Major constituents of sea water (parts per thousand) (Chlorinity = $19‰$, density at 20°C = 1·024 3)

Constituent	Parts per thousand
Chloride (Cl^-)	18·979 9
Sulphate (SO_4^{2-})	2·648 6
Bicarbonate (HCO_3^-)	0·139 7
Bromide (Br^-)	0·064 6
Fluoride (F^-)	0·001 3
Boric acid (H_3BO_3)	0·026 0
Sodium (Na^+)	10·556 1
Magnesium (Mg^{2+})	1·272 0
Calcium (Ca^{2+})	0·400 1
Potassium (K^+)	0·380 0
Strontium (Sr^{2+})	0·013 3

(Data courtesy Prentice-Hall, Inc., USA.)

Variations of salinity In the major oceans the salinity of sea water does not vary widely, lying in general between 33 and 37 parts per thousand; a figure of 35 parts per thousand, equivalent to 19·4 parts per thousand chlorinity, is commonly taken as the average for 'open-sea' water.

Local conditions may modify this profoundly in special areas. In the Arctic and Antarctic, and where there is dilution by large rivers, the salinity may be considerably less, and it may vary greatly according to season. Salinity is well below normal in the Baltic, and may fall nearly to zero at the head of the Gulf of Bothnia. In enclosed seas like the Mediterranean, Black Sea and Red Sea, on the other hand, where there is rapid evaporation, salinity may reach 40 parts per thousand.

Minor constituents Sea water contains a multitude of organic and inorganic molecules some of which form metallic complexes which even in trace amounts can significantly affect the corrosion mechanism[2]. Trace metallic complexes also play an important rôle in determining the physiology of biological organisms whose presence in sea water can exert considerable control over corrosion reactions. The presence of such complexing agents in sea water could explain the difficulty of simulating the natural product for corrosion research investigations in the laboratory. (See Section 20.1 for compositions of artificial sea waters.)

Physical Properties

Density The density of sea water is, of course, related to its salinity (or chlorinity). If ρ_0 is the density of sea water at 0°C in g/ml, σ_0 is defined as $(\rho_0 - 1)1\,000$ and the relationship between density and chlorinity is given by the equation

$$\sigma_0 = -0{\cdot}069 + 1{\cdot}470\,8\ \mathrm{Cl} - 0{\cdot}001\,570\ \mathrm{Cl}^2 + 0{\cdot}000\,039\,8\ \mathrm{Cl}^3$$

Since, however, density is affected to a considerable degree by temperature, and since its accurate measurement demands special apparatus and great care, it is not a reliable measure of the 'strength' of sea water.

Electrical conductivity This is often a convenient and accurate measurement of salinity or chlorinity. Here, too, there is considerable variation with temperature, so that simultaneous observation of temperature is essential. Figure 2.15[3] shows the relationship between conductivity and chlorinity at various temperatures.

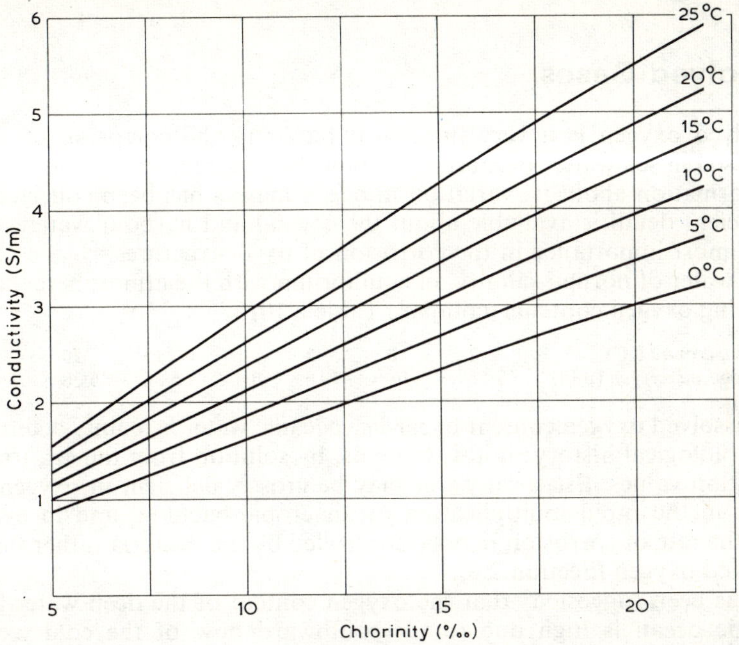

Fig. 2.15 Relationship between conductivity and chlorinity of sea water (after Prentice-Hall, Inc., USA). Note that $S = \Omega^{-1}$

Temperature The surface temperature of sea water ranges between about -2°C and 35°C, while the temperature of a shallow surface layer may run even higher. A general picture of the variation with geographical location is given by Table 2.13[4].

Table 2.13 Average surface temperature of the oceans between parallels of latitude (°C)

North latitude	*Atlantic ocean*	*Indian ocean*	*Pacific ocean*	*South latitude*	*Atlantic ocean*	*Indian ocean*	*Pacific ocean*
70°–60°	5·60	—	—	70°–60°	−1·30	−1·50	−1·30
60°–50°	8·66	—	5·74	60°–50°	1·76	1·63	5·00
50°–40°	13·16	—	9·99	50°–40°	8·68	8·67	11·16
40°–30°	20·40	—	18·62	40°–30°	16·90	17·00	16·98
30°–20°	24·16	26·14	23·38	30°–20°	21·20	22·53	21·53
20°–10°	25·81	27·23	26·42	20°–10°	23·16	25·85	25·11
10°– 0°	26·66	27·88	27·20	10°– 0°	25·18	27·41	26·01

(Data courtesy Prentice-Hall, Inc., USA.)

Seasonal variations are associated not only directly with the elevation of the sun, but also with changes of surface currents depending on the prevailing winds. The annual variation is generally quite small in the tropics and greatest in the temperate zones, where it may amount to about 10°C.

In general, water at great depths in the oceans is not subject to temperature fluctuations and even in the tropics seldom exceeds 10°C.

The 'freezing point' of sea water, defined as the temperature at which ice crystals begin to form, is −2°C.

Dissolved Gases

Dissolved oxygen is a very important factor in the corrosion of metals immersed in sea water. Because of its biological significance, a vast amount of information about its variation in ocean masses has been collected, but insufficient detail is available about the coastal and harbour waters which are of most importance in the corrosion of fixed structures.

Sea water of normal salinity, in equilibrium with the atmosphere, has the following oxygen contents (compare Table 2.10):

Temperature (°C)	−2	0	5	10	15	20	30
Dissolved oxygen (ml/l)	8·52	8·08	7·16	6·44	5·86	5·38	5·42

The dissolved oxygen content of *surface* oceanic water is mainly determined by its biological history; it always tends, by solution from the air, towards saturation values. Estuarial water may be grossly deficient in oxygen; this results in the rapid multiplication of anaerobic bacteria, and in extreme cases the rate of corrosion may be controlled by the bacteria rather than by dissolved oxygen (Section 2.6).

It has been suggested[2] that the oxygen content of the deep water in the Atlantic ocean is high due to the southward flow of the cold oxygen-saturated water through the funnel of the north Atlantic, but in the Pacific ocean the oxygen content decreases with depth due to negligible water flow through the Bering Strait.

pH

Natural sea water is well buffered and normally lies between 8·1 and 8·3 but may fall to 7·0 in stagnant basins with the formation of hydrogen sulphide produced by anaerobic bacteria.

Potentials of Metals in Sea Water

An important factor in the corrosion of a metal in sea water is its electrical potential. This is of course especially the case when two or more electrically connected metals are immersed in a single system.

The 'open-circuit' potential of most metals in sea water is not a constant and varies with the oxygen content, water velocity, temperature and metallurgical and surface condition of the metal.

In static air-free sea water the potential of iron or steel reaches a steady-state value of −0·75 V (*vs*. S.C.E., $E = 0{\cdot}246$ V) which should be compared with the more noble potential of −0·61 V observed under conditions of high velocity and aeration (Table 2.14). This potential of −0·75 V for iron in sea water is important in the practice of cathodic protection.

Table 2.14 Potentials of metals in aerated moving sea water (Potentials are *negative* to the S.C.E., E = 0·246 V)

Metal	*Potential* (V)	*Reference from which figures taken*
Magnesium	1·5	5
Zinc	1·03	6
Aluminium	0·79	6
Cadmium	0·7	5
Steel	0·61	6
Lead	0·5	5
Solder (50/50)	0·45	5
Tin	0·42	5
Naval brass	0·30	5
Copper	0·28	5
Aluminium brass	0·27	5
Gun metal	0·26	5
Cupro-nickel 90/10	0·26	5
Cupro-nickel 80/20	0·25	5
Cupro-nickel 70/30	0·25	5
Nickel	0·14	5
Silver	0·13	6
Titanium	0·10	7
Stainless steel 18/8 (passive)	0·08	6
Stainless steel 18/8 (active)	0·53	6

The values in Table 2.14 show how the potentials obtained under service conditions differ from the standard electrode potentials which are frequently calculated from thermodynamic data. Thus aluminium, which is normally coated with an oxide film, has a more noble value than the equilibrium potential $E^{\ominus}_{Al^{3+}/Al} = -1{\cdot}66$ V *vs.* S.H.E. and similar considerations apply to 'passive' stainless steel (see Chapter 21).

Although Table 2.14 shows which metal of a couple will be the anode and will thus corrode more rapidly, little information regarding the corrosion current, and hence the corrosion rate, can be obtained from the e.m.f. of the cell. The kinetics of the corrosion reaction will be determined by the rates of the electrode processes and the corrosion rate of the anode of the couple will depend on the rate of reduction of hydrogen ions or dissolved oxygen at the cathode metal (Section 1.4).

Corrosion Rates

Ferrous Metals

Ferrous metals, of which steel is technically the most important, have a remarkably steady rate of corrosion when fully immersed in sea water. The corrosion of mild steel is very rapid initially but falls off gradually over several months to a fairly steady rate (Table 2.15).

Table 2.15 Effect of exposure period on corrosion rate of mild steel, copper and aluminium[5]

Exposure time (months)	*Average corrosion rate for period* (mm/y)		
	Steel	*Copper*	*Aluminium*
1	0·33	—	—
2	0·25	—	—
3	0·19	—	—
6	0·15	—	—
12	0·13	0·034	0·004 3
24	0·11	0·019	0·002 1
48	0·11	0·018	0·001 7

In extended exposure periods of up to 16 years in tropical sea water, Southwell and Alexander[8] obtained an average corrosion rate for steel of 0·18 mm/y in the first year, falling off to a constant rate after 4 years at 0·025 mm/y. They also quote pitting rates as 1 mm/y in the first year falling off dramatically over the second to fourth years and ultimately continuing at a rate comparable with the average rate of penetration giving an average rate for exposure for 16 years of 0·08 mm/y. However, the pitting rate is generally quoted as several orders of magnitude greater than the average rate of penetration, with values of 0·25 to 0·4 mm/y[9].

A comprehensive table of corrosion rates in sea water has been compiled by LaQue[10]. This appears to show no obvious dependence of corrosion rates on the geographical location of the testing site, and few of the rates depart widely from an average of 0·11 mm/y. It is suggested that a figure of 0·13 mm/y may be taken as a reasonable estimate of the expected rate of corrosion of steel or iron continuously immersed in sea water under natural conditions, in any part of the world.

The theory has been advanced that the rapid growth of marine fouling in the tropics may provide a protective shield which counteracts the effect of the greater activity of the hotter water, and LaQue[11] has pointed out that in flowing sea water, when no fouling organisms became attached to small fully immersed specimens, corrosion of steel at 11°C proceeded at 0·18 mm/y compared with 0·36 mm/y at 21°C. This increase corresponds with what would be expected from chemical kinetics, where the rate of reaction is approximately doubled for a rise of 10°C.

It is significant that most of the data from which a remarkable uniformity of attack is deduced are derived from small isolated panels. This is the most convenient form of specimen for measurements of corrosion rates by loss of weight; but it eliminates the important effect of galvanic currents passing between remote parts of a large structure. It is believed that the experience of civil engineers and other users would not support the conclusion suggested by panel tests that corrosion is no faster in tropical than in temperate waters.

Ambler and Bain[12] found that isolated panels exposed in half-tide conditions are normally more rapidly corroded than those fully immersed, a factor of 2 to 4 being not unusual, but in commercial ports the presence of oil contamination may greatly reduce half-tide corrosion by filming the metal surface.

Humble[13] investigated the corrosion of coupled and uncoupled steel plates distributed in a vertical line extending above high-water and below low-water levels and gives a diagram showing the corrosion profile of steel piling in sea water, based on five years' exposure at Kure Beach. This shows two maxima, one in the 'splash zone' above high-tide level, and the other just below low-tide level. In the tidal zone, between these, there is a minimum corrosion rate.

The explanation of this pattern is that the well-aerated areas in the tidal zone become strongly cathodic while the metal just below water becomes anodic. This distribution is in striking contrast to the results quoted by Ambler and Bain[12].

It is generally agreed that steel composition within the range practical for ship plate has little influence on the corrosion rate in sea water[14–17].

Non-ferrous Metals

Many of the common non-ferrous metals corrode relatively slowly in still or slowly-moving sea water. Typical figures are given in Table 2.16.

Table 2.16 Corrosion rates of non-ferrous metals and alloys in sea water

Material	*Corrosion rate* (mm/y)
Copper (*a*) full immersion	0·003 8
(*b*) half-tide	0·002 5
Brass (Cu–10 to 35Zn)	0·004 5
Aluminium brass (Cu–22Zn–2Al)	0·002 0
Admiralty brass (Cu–29Zn–1Sn)	0·004 6
Gun-metal (Cu–10Sn–2Zn)	0·002 5
Phosphor bronze	0·002 5
Aluminium bronze (95Cu–5Al)	0·003 8
Copper-nickel-iron (Cu–5Ni–1Fe)	0·003 8
Cupro-nickel (70Cu–30Ni)	0·001 3
Nickel	0·002 5
Monel	0·002 5
Aluminium (99·8%)	0·000 38
(98%)	0·000 76
NS. 6. (5 Mg)	0·000 30
Lead	0·001 0
Zinc	0·001 8

The effect of exposure time on the corrosion of copper and aluminium is illustrated in Table 2.15. The results quoted by Southwell, Hummer and Alexander[18–19] for corrosion rates of copper and aluminium in tropical waters compared with those obtained around the British Isles suggests that the corrosion rate increases by a factor of two for every 10°C rise in temperature.

The effect of alloying additions on the marine corrosion properties of non-ferrous metals can be very significant, and for copper-based alloys has been comprehensively reviewed by Bradley[20].

Crevice Corrosion

Stainless steel, and aluminium and its alloys, derive their excellent corrosion resistance from the self-repairing protective oxide film which renders them passive. Repair of the film depends on access of oxygen, and in crevices this is often inadequate, with the result that the metal in a crevice becomes 'active'. As the fully exposed areas, usually relatively large, act as a cathode, rapid and sometimes disastrous corrosion may result. The use of stainless steel under sea water needs the greatest care to avoid this trouble, and as a rule one of the resistant copper alloys is the better choice; the danger from built-in crevices may be foreseen and avoided by careful design, but crevice corrosion also occurs behind marine fouling organisms or other deposits which it may not be possible to prevent. (See Section 1.6.)

Effect of Depth

Little scientific examination of the deterioration of materials at depth has been undertaken except that by the US Naval Civil Engineering Laboratory and Naval Research Laboratory. The results of this work were reported by Reinhart in 1966 and more recently the work has been reviewed by Kirk[21]. Typical corrosion data for a selection of metals exposed in the Pacific Ocean at several sites and for different times are shown in Table 2.17 and are compared with results obtained in surface waters at Wrightsville Beach by International Nickel Inc.

Table 2.17 Effect of depth on the corrosion rate of some metals and alloys

Material	Rate of metal penetration at various exposure depths (mm/y)					Form of attack
	0 m	704 m	1 600 m	1 700 m	2 050 m	
Zinc	0·015	0·058	0·018	0·091	0·150	General
Mild steel	0·127	0·043	0·023	0·020	0·058	General
Aluminium alloy 5052*	—	—	—	>0·576	—	Pitting
'G' bronze†	0·008	0·005	0·008	0·018	0·008	General
Cu–10Ni	0·008	0·020	0·018	0·015	0·015	General
Cu–30Ni	0·005	0·023	0·015	0·025	0·030	General
Stainless steel (type 410)	—	1·270	1·270	1·270	1·270	Pitting
Incoloy 825	—	Slight	Slight	0	Slight	Pitting
Stainless steel (type 316)	—	0	0·025	0	Slight	Pitting
Monel 400	—	0·033	>0·035	0·038	>0·092	Pitting
Exposure conditions						
Temperature (°C)	5–30	7·2	2·5	2·3	2·7	
Oxygen concentration (p.p.m.)	5–10	0·6	1·8	2·1	1·7	

* Al–4Mg alloy. † Admiralty Gunmetal (Cu–10Sn–2Zn, BS 1400 G1).

The general indication of the results in this table is that the corrosion rates of non-ferrous metals increase with depth in spite of lower temperatures and lower oxygen concentrations than at the surface. It was noted in the paper by

Kirk[21] that the results at depth were typical of the variation of performance of these materials experienced on numerous occasions in surface sea water. A notable exception was for aluminium alloys of the 5000 (Al–Mg) and 6000 (Al–Mg–Si) series which had good resistance to corrosion in shallow waters, but were found to suffer very severe crevice corrosion in deep sea water. Interpreting these data into practical terms it would seem that there is little justification for expecting lower corrosion rates with increased depth in spite of changes in the sea-water chemistry, and corrosion rates would be expected to show the normal variations encountered in surface waters.

Although in the deep-ocean corrosion tests the oxygen concentrations were considerably lower than at the surface, it is perhaps not too surprising that this does not lead to reduction in corrosion since water movement brings a fresh supply of oxygen to the corroding metal surface. In desalination investigations it has been shown that even one part per million of dissolved oxygen in sea water can sustain a corrosion reaction on some materials[22]. In terms of applications for deep-water engineering it is probably the deterioration of materials by the combined action of mechanical loading and corrosion such as stress corrosion and corrosion fatigue which is of major concern.

Drisko and Brouillett[23] have examined a number of protective coatings on mild steel and compared their performance at 2·1 km with that at shallow depth. They concluded that with thick coatings of over 0·3 mm there was negligible difference, but with thinner coatings there was some loss of protection at holidays under deep immersion conditions. An exception was with soft coatings such as asphalt and coal tar which performed better at depth owing to their susceptibility to damage by certain marine fouling organisms such as barnacles at the surface, whereas such species were not encountered at depth.

Effect of Water Speed

Hardly any quantitative results on the effect of movement on corrosion of steel are available. Water movement can markedly affect the corrosion process in controlling the rate of transport of reactants to the corrosion site, and the removal of the corrosion reaction products.

A curve is given by LaQue[10] which indicates that the corrosion rates are approximately as follows:

Water speed (mm/s)	0	1·5	3	4·6	6·1	7·6
Corrosion rate (mm/y)	0·013	0·05	0·074	0·084	0·089	0·089

This is presumably an estimated average curve, as no numerical data are quoted, and it may be assumed to refer to bare steel. This conclusion is not supported by the results of Volkening[24], whose main interest was in the effect of chlorination and who shows that although corrosion increased with velocity of chlorinated sea water, when plain sea water was used velocity had little effect. There can be no doubt that painting will very much reduce the effect of water speed, as also will marine fouling or slime.

Excessive corrosion rates are commonly observed on those parts of a ship's hull which are exposed to high and turbulent flow of water, e.g. leading edges of rudders and shaft-brackets. The pitting found in these places is stimulated by selective local damage to paint films and possibly also by the proximity of a bronze propeller. The contribution of high water speed to this accelerated corrosion cannot be separately assessed. An indirect relation between water speed and corrosion arises from the fact that marine fouling organisms (in particular shell-fouling) do not settle if the water speed is more than about 1·5–2 m/s. Fouling may somewhat restrict general corrosion by its shielding effect, but may also cause crevice corrosion.

Impingement Attack

This form of attack, especially as affecting copper alloys in sea water, has been widely studied since the pioneer work of Bengough and May[25]. Impingement attack of sea water pipe and heat exchanger systems is considered in Sections 1.6 and 4.2. In such engineering systems the water flow is invariably turbulent and the thickness of the laminar boundary layer is an important factor in controlling localised corrosion.

At very high water speeds *cavitation-damage* (Section 8.8) is sustained by any metal; high-speed bronze propellers, for instance, may suffer seriously. This form of attack is mainly mechanical, although an element of true corrosion may be present, and is not specifically associated with sea water.

J. C. ROWLANDS

REFERENCES

1. Sverdrup, H. V., Johnson, M. W. and Fleming, R. H., *The Oceans*, Prentice-Hall, N.Y., 173 (1942)
2. Compton, K. G., *Corrosion*, **26**, 448 (1970)
3. As Reference 1, but p. 72
4. As Reference 1, but p. 127
5. Central Dockyard Laboratory, Portsmouth (unpublished)
6. LaQue, F. L., *Proc. Amer. Soc. Test. Mater.*, **51**, 541 (1951)
7. Cotton, J. B. and Downing, B. P., *Trans. Inst. Mar. Engrs.*, **69**, 311 (1959)
8. Southwell C. R. and Alexander A. L., *Materials Protection*, **9**, 4 (1970)
9. Fink, F. W., *Corrosion of Metals in Sea Water*, Battelle Memorial Inst., PB 171 344 (1960)
10. LaQue, F. L., *The Corrosion Handbook*, Wiley, New York; Chapman and Hall, London, 384 (1948)
11. LaQue, F. L., *Corrosion*, **6**, 162 (1958)
12. Ambler, H. R. and Bain, A. J., *J. Appl. Chem.*, **5**, 437 (1955)
13. Humble, H. A., *Corrosion*, **5**, 292 (1949)
14. Hudson, J. C., *J. Iron Steel Inst.*, **166**, 123 (1950)
15. Boudot, H. and Chaudron, G., *Rev. Métall.*, **43**, 1 (1946)
16. Forgeson B. W., Southwell, C. R. and Alexander, A. L., *Corrosion*, **16**, 105t (1960)
17. Evans, U. R. and Rance, V., *Corrosion and its Prevention at Bimetallic Contacts*, H.M.S.O., 3rd edn. (1963)
18. Southwell, C. R., Hummer, C. W. and Alexander, A. L., *Materials Protection*, **4**, 30 (1965)
19. Southwell, C. R., Alexander, A. L. and Hummer, C. W., *Materials Protection*, **7**, 41 (1968)
20. Bradley, J. N., *Inst. Metals Metallurgical Review* (1971)
21. Kirk, W. W., *Proc. Workshop Conf. on High Pressure Aquarian Systems*, N.A.C.E. (1971)
22. Schreiber, C. F., Osborn, O. and Coley, F. H., *Mat. Prot.*, **7**, 24 (1968)
23. Driakt, R. W. and Brouillett, C. V., *Mat. Prot.*, **32** (1966)
24. Volkening, V. B., *Corrosion*, **6**, 123 (1950)
25. Bengough, G. D. and May, R., *J. Inst. Met.*, **32**, 204 (1924)

2.5 Soil in the Corrosion Process

Introduction

Soil has been defined in many ways, often depending upon the particular interests of the person proposing the definition. In discussion of the soil as an environmental factor in corrosion, no strict definitions or limitations will be applied; rather, the complex interaction of all earthen materials will come within the scope of the discussion. It is obvious only a general approach to the topic can be given, and no attempt will be made to give full and detailed information on any single facet of the topic.

Soil is distinguished by the complex nature of its composition and of its interaction with other environmental factors. No two soils are exactly alike, and extremes of structure, composition and corrosive activity are found in different soils. Climatic factors of rainfall, temperature, air movement and sunlight can cause marked alterations in soil properties which relate directly to the rates at which corrosion will take place on metals buried in these soils.

Soil Genesis

The condition of any soil represents a stage in the changing process of soil evolution. Soils develop, mature and change with the passage of time. Whereas the time required for a true soil to develop from the parent rock of the earth may be thousands of years, rapid changes can result in a few years when soils are cultivated, irrigated, or otherwise subjected to man's manipulation. The type of soil that develops from the parent material will depend upon the various physical, chemical and biological factors of the environment.

The weathering process which eventually reduces the rock of the parent material to the inorganic constituents of soil comprises both physical and chemical changes. Size reduction from rocks to the colloidal state depends not only upon the mechanical action of natural forces but also on chemical solubilisation of certain minerals, action of plant roots, and the effects of organic substances formed by biological activity.

Interrelated with change in particle size and changes in type and kind of soil minerals present, organic matter is formed and accumulates as an integral part of the soil. Organic-matter content varies from practically none in sands to almost 100%, as exemplified by peat formations. The amount of organic matter present thus reflects the interaction of all environmental factors influencing chemical and biological activity. Whether the percentage of organic matter increases or decreases depends then upon the relation between the rate at which it is being formed by growth, death and accumulation of plant material, and that at which the microbiological activities within the soil are causing the decomposition of the complex organic molecules.

Moisture must be considered of primary importance in soil formation, in weathering, and in all of the changes taking place within the soil. The types of soil that form depend to a great extent upon the rainfall situation. Too little rainfall will prevent development of plant and animal life with their soil-building action. Too much moisture has a similar effect in preventing normal soil formation.

Closely associated with rainfall and climate is the acid or basic reaction which develops as a soil matures. When rainfall is high, water percolates through the soil, dissolving the soluble components, and leaching out alkaline minerals of the weathering rock. This happens whether a soil is developing from a naturally acidic or a naturally alkaline parent material. The end result is a shift in reaction to an acid condition. The degree to which this acidity develops depends upon many factors such as the parent minerals, biological activity, and temperature, related to the moisture situation. Should the loss of water from a soil be mainly by surface evaporation (as in arid regions), the dissolved salts tend to accumulate near the surface and alkaline conditions usually develop.

Although conditions of high rainfall and moderate to warm temperatures usually lead to an overall decrease in organic matter (particularly in cultivated soils), exceptions occur when the amount of water is great enough to prevent the adequate aeration necessary for maximum microbial activity. Swampy areas with peat and muck soils are the result. In a parallel manner, low temperatures of sub-polar regions slow down decomposition of organic materials and again highly organic soils develop.

The Corrosion Process in Soil

Although the soil as a corrosive environment is probably of greater complexity than any other environment, it is possible to make some generalisations regarding soil types and corrosion. It is necessary to emphasise that corrosion in soils is extremely variable and can range from the rapid to the negligible. This can be illustrated by the fact that buried pipes have become perforated within one year, while archaeological specimens of ancient iron have probably remained in the soil for hundreds of years without significant attack.

Corrosion in soil is aqueous, and the mechanism is electrochemical (see Section 1.4), but the conditions in the soil can range from 'atmospheric' to completely immersed (Sections 2.2 and 2.3). Which conditions prevail depends on the compactness of the soil and the water or moisture content. Moisture retained within a soil under field dry conditions is largely held within the capillaries and pores of the soil. Soil moisture is extremely significant in this connection, and a dry sandy soil will, in general, be less corrosive than a wet clay.

Although the mechanism will be essentially electrochemical, there are many characteristic features of soil as a corrosive environment which will be considered subsequently; it can, however, be stated here that the actual corrosiveness of a soil will depend upon an interaction between rainfall, climate and soil reaction.

A characteristic feature of the soil is its heterogeneity. Thus variation in soil composition or structure can result in different environments acting on different parts of the same metal surface, and this can give rise to differing electrical potentials at the metal/soil interface. This will result in the establishment of predominantly cathodic or predominantly anodic areas, and the consequent passage of charge through the metal and through the soil. Differences in oxygen concentration (differential aeration), or differences in acidity or salt concentrations may thus give rise to corrosion cells. The distance of the separation of the anodic and cathodic areas can range from very small to miles ('long-line' corrosion).

The conductivity of the soil is important as it is evident from the electrochemical mechanism of corrosion that this can be rate-controlling; a high conductivity will be conducive of a high corrosion rate. In addition, the conductivity of the soil is important for 'stray-current corrosion' (see Section 11.5), and for cathodic protection (Chapter 11).

Properties of Soils Related to Corrosion

Soil Texture and Structure

Soils are commonly named and classified according to the general size range of their particulate matter. Thus sandy, silt and clay types derive their names from the predominant size range of inorganic constituents. Particles between 0·07 and about 2 mm are classed as sands. Silt particles range from 0·005 mm to 0·07, and clay particle size ranges from 0·005 mm mean diameter down to colloidal matter.

The proportion of the three size groups will determine many of the properties of the soil. Although a number of systems have been used to classify soils as to texture, the one shown in Fig. 2.16 represents commonly used terminology for various proportions of sand, silt and clay.

Since soils contain organic matter, moisture, gases and living organisms as well as mineral particles, it is apparent that the relative size range does not determine the whole nature of the soil structure. In fact most soils consist of aggregates of particles within a matrix of organic and inorganic colloidal matter rather than separate individual particles. This aggregation gives a crumb-like structure to the soil, and leads to friability, more ready penetration of moisture, greater aeration, less erosion by water and wind, and generally greater biological activity. The loss of the aggregated structure can occur as the result of mechanical action, or by chemical alteration such as excess alkali accumulation. Destruction of the structure or 'puddling' greatly alters the physical nature of the soil.

Mention should be made of the soil profile (section through soil showing various layers) because it is important to recognise that the soil's surface

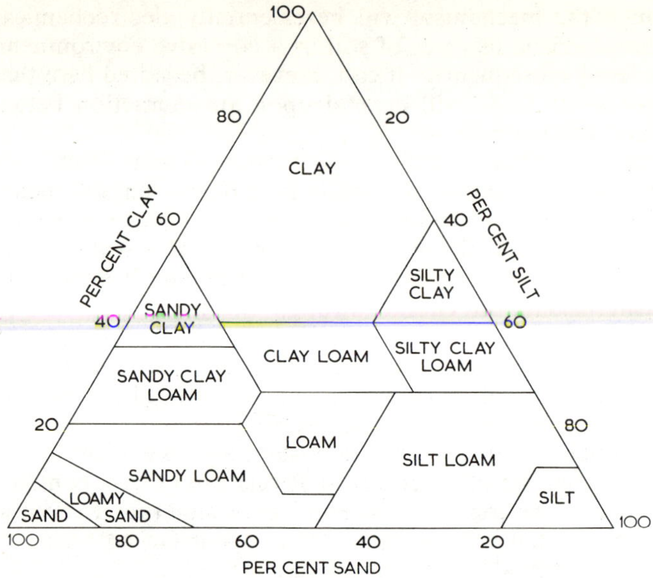

Fig. 2.16 Proportions of sand, silt and clay making up the various groups of soils classified on the basis of particle size

gives a very poor indication of the underlying strata. Pipe-lines are buried several feet below surface soils and corrosion surveys based on surface observations give little information as to the actual environs of the pipe when buried.

The Clay Fraction

Clays make up the most important inorganic constituents of soil. They consist of various minerals depending on the mineral composition of the parent material, and on the type and degree of weathering. Often clays may be grouped in a family series, depending upon the weathered condition, as, for example, montmorillonite → illite → kaolinite. Weathering of montmorillonite causes loss of potassium and magnesium which alters the crystalline structure, and eventually kaolinite results. In this example (and also for other clay mineral groups) marked changes occur in the physical properties of a soil as clay minerals undergo the weathering process.

Montmorillonite clays absorb water readily, swell greatly and confer highly plastic properties to a soil. Thus soil stress (Section 15.8) occurs most frequently in these soils and less commonly in predominantly kaolinitic types. Similarly, a soil high in bentonite will show more aggressive corrosion than a soil with a comparable percentage of kaolinite. A chalky soil usually shows low corrosion rates. Clay mineralogy and the relation of clays to corrosion deserves attention from corrosion engineers. Many important relationships are not fully understood and there is need for extensive research in this area.

Aeration and Oxygen Diffusion

The pore space of a soil may contain either water or a gaseous atmosphere. Thus the aeration of a soil is directly related to the amount of pore space present and to the water content. Soils of fine texture due to a high clay content contain more closely packed particles and have less pore capacity for gaseous diffusion than an open-type soil such as sand.

Oxygen content of soil atmosphere is of special interest in corrosion. It is generally assumed that the gases of the upper layers of soil are similar in composition to the atmosphere above the soil, except for a higher carbon dioxide content. Relatively few data are available showing oxygen content of soils at depths of interest to the corrosion engineer. Judging by the fact that plant roots require oxygen to penetrate a soil, however, it may be assumed that soil gases at depths of 6 m or more contain significant amounts of oxygen.

Diffusion of gases into soil is enhanced by a number of climatic factors. Temperature changes from day to night conditions cause expansion and contraction of the surface-soil gases. Variation in barometric pressure has a bellows-like effect on gaseous diffusion. To illustrate the magnitude of this diffusion rate on a large scale, it may be recalled that air within the more than 43 km of underground passages of the Carlsbad Caverns in New Mexico undergoes a complete change each day, despite the fact that the single opening of these caverns to the surface is only a metre or so in diameter.

Biological activity within the soil tends to decrease the oxygen content and replace the oxygen with gases from metabolic activity, such as carbon dioxide. Most biological activity occurs in the upper 150 mm of soil, and it is in this region that diffusion would be most rapid. Factors which tend to increase microbial respiration, such as the addition of large amounts of readily decomposed organic matter, or factors which decrease diffusion rates (water saturation) will lead to development of anaerobic conditions within the soil. The significant microbiological relationships to corrosion under both aerobic and anaerobic situations are discussed in Section 2.6.

Water Relations

No corrosion occurs in a completely dry environment. In soil, water is needed for ionisation of the oxidised state at the metal surface. Water is also needed for ionisation of soil electrolytes, thus completing the circuit for flow of a current maintaining corrosive activity. Apart from its participation in the fundamental corrosion process, water markedly influences most of the other factors relating to corrosion in soils. Its rôle in weathering and soil genesis has already been mentioned.

Types of Soil Moisture

1. *Free ground water*. At some depth below the surface, water is constantly present. This distance to the water table may vary from a few metres to hundreds of metres, depending upon the geological formations present.

Only a small amount of the metal used in underground service is present in the ground water zone. Such structures as well casings and under-river pipelines are surrounded by ground water. The corrosion conditions in such a situation are essentially those of an aqueous environment.

2. *Gravitational water*. Water entering soil at the surface from rainfall or some other source moves downward. This gravitational water will flow at a rate governed largely by the physical structure regulating the pore space at various zones in the soil profile. An impervious layer of clay, a 'puddled' soil, or other layers of material resistant to water passage may act as an effective barrier to the gravitational water and cause zones of water accumulation and saturation. This is often the situation in highland swamp and bog formation. Usually gravitational water percolates rapidly to the level of the permanent ground water.

3. *Capillary water*. Most soils contain considerable amounts of water held in the capillary spaces of the silt and clay particles. The actual amount present depends upon the soil type and weather conditions. Capillary moisture represents the important reservoir of water in soil which supplies the needs of plants and animals living in or on the soil. Only a portion of capillary water is available to plants. 'Moisture-holding capacity' of a soil is a term applied to the ability of a soil to hold water present in the form of capillary water. It is obvious that the moisture-holding capacity of a clay is much greater than that of a sandy type soil. Likewise, the degree of corrosion occurring in soil will be related to its moisture-holding capacity, although the complexities of the relationships do not allow any quantitative or predictive applications of the present state of knowledge.

Significance of Fluctuations in Water Content

Except for zones below the level of permanent ground water where the environment is water-saturated, and for zones of dry surface sand, continual variation may be expected to occur in the water content of soils. This is usually dependent on rainfall, snow, flooding and such climatic influences, though irrigation practices in many agricultural areas influence water content and hence the corrosion rates.

Water losses from the soil represent the sum of downward movement of gravitational water and surface losses by evaporation. Man's activities, other than drainage procedures or long-term water use from pumps in industrial areas, do not usually influence the downward movement of water. On the other hand, agricultural practices have a great effect on surface evaporation losses.

As mentioned earlier, there is an inverse relationship between water volumes and oxygen concentration in soil. As soils dry, conditions become more aerobic and oxygen diffusion rates become higher. The wet–dry or anaerobic–aerobic alternation, either temporal or spatial, leads to higher corrosion rates than would be obtained within a constant environment. Oxygen-concentration-cell formation is enhanced. This same fluctuation in water and air relations also leads to greater variation in biological activity within the soil.

Chemical Properties of Soils

Soil reaction (pH) The relationship between the environment and development of acid or alkaline conditions in soil has been discussed with respect to formation of soils from the parent rock materials. Soil acidity comes in part by the formation of carbonic acid from carbon dioxide of biological origin and water. Other acidic development may come from acid residues of weathering, shifts in mineral types, loss of alkaline or basic earth elements by leaching, formation of organic or inorganic acids by microbial activity, plant root secretions, and man-made pollution of the soil, especially by industrial wastes.

As with other factors, no direct statements can be made relating the reaction of a soil to its corrosive properties. Extremely acid soils (pH 4·0 and lower) can cause rapid corrosion of bare metals of most types. This degree of acidity is not common, being limited to certain bog soils and soils made acid by large accumulations of acidic plant materials such as needles in a coniferous forest. Most soils range from pH5·0 to pH8·0, and corrosion rates are apt to depend on many other environmental factors rather than soil reaction *per se*. The 45-year study of underground corrosion conducted by the United States Bureau of Standards[1] included study of the effect of soils of varying pH on different metals, and extensive data were reported.

Soluble salts of the soil Water in the soil should most properly be considered as the solvent for salts of the soil; the result being the soil solution. In temperate climates and moderate rainfall areas, the soil solution is relatively dilute, with total dissolved salts ranging from 80 to 1 500 p.p.m.[2]. Regions of extensive rainfall show lower concentrations of soluble salts as the result of leaching action. Conversely, soils in arid regions are usually quite high in salts as these salts are carried to the surface layers of the soil by water movement due to surface evaporation.

Generally, the most common cations in the soil solution are potassium, sodium, magnesium and calcium. Alkali soils are high in sodium and potassium, while calcareous soils contain predominantly magnesium and calcium. Salts of all four of these elements tend to accelerate metallic corrosion by the mechanisms mentioned. The alkaline earth elements, calcium and magnesium, however, tend to form insoluble oxides and carbonates in non-acid conditions. These insoluble precipitates may result in a protective layer on the metal surface and reduced corrosive activity.

The anionic portions of the soil solution play a rôle of equal importance to the cations. The anions function in the manner outlined for cations in conductivity and concentration-cell action, and have an additional action if they react with the metal cation and form insoluble salts. Thus, if the metal is lead and the predominant anion is sulphate, a layer of insoluble lead sulphate may precipitate on the metal surface and form an effective barrier against further loss of metal.

Another important relationship between the salts of the soil and corrosion has to do with biological activity. Since the growth of plants and micro-organisms depends upon the proper inorganic mineral nutrients, the action of these forms of life varies with the mineral content of the soil. While many of the possible indirect effects, such as the rôle of various nitrogenous materials in bacterial growth and corrosion, have not as yet been studied, one well-documented situation is known. This relationship of sulphur and sulphates to bacterial activity in corrosion is fully discussed in Section 2.6.

The salts content of soils may be markedly altered by man's activities. The effect of cathodic protection will be discussed later in this section. Fertiliser use, particularly the heavy doses used in lawn care, introduces many chemicals into the soil. Industrial wastes, salt brines from petroleum production, thawing salts on walks and roads, weed-killing salts at the base of metal structures, and many other situations could be cited as examples of alteration of the soil solution. In tidal areas or in soils near extensive salt deposits, depletion of fresh ground-water supplies has resulted in a flow of brackish or salty sea water into these soils, causing increased corrosion.

The Environment of the Pipe-line Ditch

A comprehensive study of the soil and microbial situation in the backfilled zone of pipe-line ditches has shown a number of significant facts[3-6]. The results of over a thousand bell-hole studies along operating oil and gas pipe-lines in widely separated geographical areas of the United States has led to the conclusion that the pipe-line ditch represents a marked disturbance of the

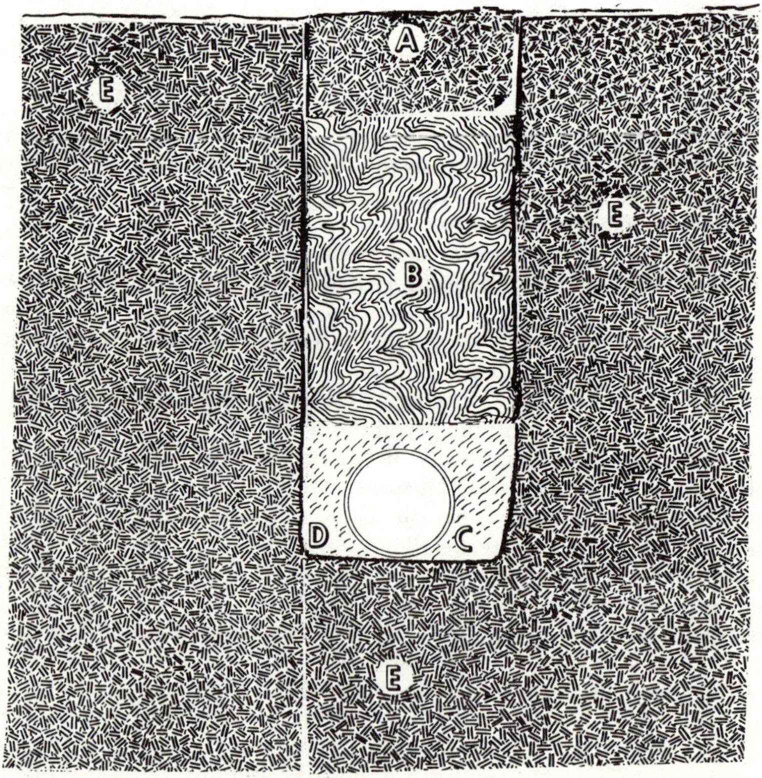

Fig. 2.17 Cross-section of soil and backfill areas surrounding underground pipe-lines

natural soil situation. Figure 2.17 indicates the general zones of interest. The operation of ditching and back-filling has resulted in a zone above the pipe (*B*) which never settles but remains less compact than the undisturbed soil. In this zone, water may penetrate and leave more rapidly. Aeration is more efficient, as shown by the presence of strictly aerobic bacteria in abundant population.

The bottom of the pipe-line ditch (*C* and *D*, Fig. 2.17) has a higher moisture content than undisturbed soil at comparable depths. Many instances of free water at the ditch bottom were reported. Differences at the surface between backfill (*A*) and soil of the right-of-way (*E*) were less and tended to decrease with passage of time. Conclusive evidence was obtained indicating a greatly increased activity of bacteria in the backfill zone. Some of this may have resulted from the mixing of surface and sub-surface soils during ditching and back-filling. High populations of bacteria adjacent to the organic matter of coating or coating and wrapped-in systems on the external surface of the pipe indicated that these organic compounds served as an available food supply. The presence of hydrocarbon-utilising bacteria was a common finding particularly when the soil was in contact with asphalt protective coatings.

These concepts of the altered soil situation in the pipe-line ditch have important implications to the corrosion process. The increased aeration, the high moisture of the ditch bottom, presence of organic matter in coatings, and high microbial populations all lead to greatly increased possibilities for the development of heterogeneity and the formation of zones differing in oxidation–reduction potentials. The action of moisture and micro-organisms on asphaltic coatings, unbonding of the coating[6], and formation of cathodic and anodic areas on metal surfaces are all directly related to the disturbed environment of the pipe-line ditch.

Cathodic Protection and Soil Properties

The modern procedure to minimise corrosion losses on underground structures is to use protective coatings between the metal and soil and to apply cathodic protection to the metal structure (see Chapter 11). In this situation, soils influence the operation in a somewhat different manner than is the case with unprotected bare metal. A soil with moderately high salts content (low resistivity) is desirable for the location of the anodes. If the impressed potential is from a sacrificial metal, the effective potential and current available will depend upon soil properties such as pH, soluble salts and moisture present. When rectifiers are used as the source of the cathodic potential, soils of low electrical resistance are desirable for the location of the anode beds. A protective coating free from holidays and of uniformly high insulation value causes the electrical conducting properties of the soil to become of less significance in relation to corrosion rates (Section 15.8).

Effect of cathodic protection on soils Long-term application of an electrical potential to the metal structure with resulting flow of electrical current through the soil has two noticeable effects, the magnitude of which will be in proportion to the time and amount of current passing through the soil.

The most commonly observed effect of current flow is the development of alkaline conditions at the cathode. On bare metal this alkaline zone may exist only at the metal surface and may often reach pH values of 10 to 12. When the soil solution contains appreciable calcium or magnesium these cations usually form a layer of carbonate or hydroxide at the cathodic area. On coated lines the cations usually move to holidays or breaks in the coating. On failing asphalt or asphalt mastic type coatings, masses of precipitated calcium and magnesium often form nodules or tubercles several centimetres in diameter.

The existence of an electrical potential causes not only cation and anion movement but also migration of moisture toward the cathode. This movement of water (electroendosmosis) is due to the asymmetrical nature of the polar groups of the water molecule. In arid regions water leaving the anode area may cause the soil surrounding the anodes to become so dry that proper current densities cannot be maintained along the line. To alleviate this, some pipe-line companies have had to transport water into desert areas to re-moisten anode beds.

Moisture films are frequently found under unbonded protective coatings of asphalt and plastic tapes. The nature and origin of this water is still unknown but is of great interest because of its relationship to bond failure, microbial utilisation of asphalt and hydrocarbons, and efficiency of cathodic protection[6].

Long-line currents As the result of the use of protective coatings and cathodic protection, present-day pipe-lines are usually constructed of welded joints and the line forms a continuous conductor rather than a series of insulated sections. This situation led to the finding of the so-called long-line currents. Often low currents of medium to high voltage have been observed. The cause and significance of this phenomenon is not known. Theories as to the origin of these currents are:

1. *Pick-up of stray current* (*a.c. or d.c.*) (Section 11.5). Decreased use of d.c. in many areas has led to less possibilities of pick-up of direct current from utilities, mines, etc. The importance of grounded a.c. systems has been discounted, but Waters[7] has shown that alternating currents can accelerate corrosion. Furthermore the rectifying effects of oxide films, clay minerals and other soil factors are not understood.
2. *Current induction due to earth's rotation*. The long lines act as conductors, and variations in magnetic flux could cause induced currents. A few studies have shown long-line current activity to be greater at high activity of the aurora borealis, which is known to be related to earth's magnetism. The existence of extremely long-length waves of electromagnetic radiation[8] gives another possibility.
3. *Atmospheric lightning*. Discharges of static electricity in the various forms of lightning represent high potentials of extremely short duration. The dissipation of this potential through the earth's crust may well be the origin of the long-line currents.
4. *Differences in soil potential*. Since pipe-lines pass through zones of aerated and unaerated soil and possibilities for electrolytic-cell formation are great, the observed currents may have resulted from soil dissimilarities.

Those interested in long-line currents are referred to the publications of Gish[9], Gill and Rogers[10] and Mudd[11].

Methods Proposed for Evaluation of Soil Corrosivity

Because corrosion rates of metals buried in soil show extreme variation, a test procedure to indicate the expected corrosion activity of a given soil would be extremely valuable. The discussion on the heterogeneity of soils, however, indicates the complex nature of the situation and thus also suggests that the likelihood of success of a single survey-type procedure would be slight. Many types of tests have been suggested. Certain ones are in use by corrosion engineers, and others remain to receive further study. The various types of survey vary from the observations of Denison and Ewing[12] that corrosivity of Ohio soils could be related to colour and texture, to complex laboratory testing equipment. It is obvious that a useful test procedure should be relatively rapid and capable of field use, show small changes in environmental relations, and give results predicting relative corrosion rates. In the paragraphs which follow, only the general nature of the test will be discussed, and the reader is referred to Sections 11.4 and 11.7.

Soil resistivity The rôle of soil in the electrical circuitry of corrosion is now apparent. Thus the conductivity of the soil represents an important parameter. Soil resistivity has probably been more widely used than any other test procedure. Opinions of experts vary somewhat as to the actual values in terms of ohm centimetres which relate to metal-loss rates. The extended study of the US Bureau of Standards[1] presents a mass of data with soil-resistivity values given. A weakness of the resistivity procedure is that it neither indicates variations in aeration and pH of the soil, nor microbial activity in terms of coating deterioration or corrosion under anaerobic conditions. Furthermore, as shown by Costanzo[13], rainfall fluctuations markedly affect readings. Despite its shortcomings, however, this procedure represents a valuable survey method. Scott[14] points out the value of multiple data and the statistical nature of the resistivity readings as related to corrosion rates (see also Chapter 11).

Oxidation–reduction potential Because of the interest in bacterial corrosion under anaerobic conditions, the oxidation–reduction situation in the soil was suggested as an indication of expected corrosion rates. The work of Starkey and Wight[15], McVey[16], and others led to the development and testing of the so-called *redox* probe. The probe with platinum electrodes and copper sulphate reference cells has been described as difficult to clean. Hence, results are difficult to reproduce. At the present time this procedure does not seem adapted to use in field tests. Of more importance is the fact that the data obtained by the redox method simply indicate anaerobic situations in the soil. Such data would be effective in predicting anaerobic corrosion by sulphate-reducing bacteria, but would fail to give any information regarding other types of corrosion.

Electrolytic method This procedure is also known as the Williams Corfield test[17]. It is based on loss of metal from iron electrodes buried in a water-saturated soil through which current from a 6-V battery is passed. It does not reflect field conditions and depends upon soil conductance under saturated conditions.

Polarisation-curve procedures The Denison[18] method is to measure the current at various degrees of polarisation of metal in soil in a special cell. While this test is considered quite accurate, it has the disadvantage that the measurements are made in the laboratory and cannot be made in the field.

Combination electrical methods Tomashov and Mikhailovsky[19] describe a method developed in the Soviet Union. This test is essentially a combination of resistivity measurement and polarisation rates on iron electrodes in soil *in situ*. The usefulness and value of this procedure has not as yet been determined by practical application by corrosion engineers. The development of this combination test does, however, represent an attempt to integrate some of the complex factors controlling corrosion rates in soil. Much more research on these factors and methods of measurement should in the future enable the corrosion engineer to evaluate soil properties with respect to application of corrosion-alleviating operations.

Polarisation-resistance method The polarisation-resistance method (see Section 20.1) has been used for determining corrosion rates of metals buried underground.

J. O. HARRIS

REFERENCES

1. Romanoff, M., *Underground Corrosion*, Nat. Bur. Stand., Circular No. 579, Washington (1957)
2. Russell, E. J., *Soil Conditions and Plant Growth*, Longmans, London (1932)
3. Harris, J. O., *Kansas Agric. Exp. Sta. Tech. Bull.*, No. 102, Manhattan (1959)
4. Harris, J. O., *Corrosion*, **16**, 149 (1960)
5. Harris, J. O., *Proceedings*, Sixth Ann. Appalachian Underground Corrosion Short Course, West Virginia Univ., Morgantown, 198 (1961)
6. Harris, J. O., *Proc. Pacific Cst. Gas Ass.*, **52**, 109 (1961)
7. Waters, F. O., *Mater. Protect.*, **1** No. 3, 26 (1962)
8. Heirtzler, J. R., *Sci. Amer.*, **206**, 128 (1962)
9. Gish, O. H., *Sci. Mon., N.Y.*, **32**, 5 (1930)
10. Gill, S. and Rogers, W., *Physics*, **1**, 194 (1931)
11. Mudd, O. C., *Oil Gas J.*, **38**, 48 (1939)
12. Denison, I. A. and Ewing, S. P., *Soil Sci.*, **40**, 287 (1935)
13. Costanzo, F. E., *Corrosion*, **14**, 363 (1958)
14. Scott, G. N., *Corrosion*, **14**, 396 (1958)
15. Starkey, R. L. and Wight, K. M., *Anaerobic Corrosion of Iron in Soil*, Amer. Gas Assoc., New York (1945)
16. McVey, R. E., Proceedings of the Fifth Ann. Appalachian Underground Corrosion Short Course, West Virginia Univ., Morgantown, 23 (1960)
17. Corfield, G., *West. Gas.*, **7**, 123 (1930)
18. Denison, I. A., *Nat. Bur. Stand. J.*, **17**, 363 (1936)
19. Tomashov, N. D. and Mikhailovsky, Y. N., *Corrosion*, **15**, 77 (1959)

2.6 The Microbiology of Corrosion

The rôle of microbes in the corrosion of metals is due to the chemical activities (metabolism) associated with microbial growth and reproduction[1]. Under favourable conditions doubling times of 10–60 min are obtained. Thus the concomitant chemical reactions accelerate exponentially and finally become massive and catastrophic[2]. By reason of such rapid growth the onset of such changes may be sudden, and even when apparently suppressed by mechanical or chemical cleaning often return because a residual low number of living organisms rapidly grow again when favourable conditions are restored. These characteristics are typical of widespread biodeterioration caused by microbes in all industries of which corrosion is a special case. With a few exceptions such as synthetic polymers, all materials including natural products such as cotton, wood, rubber and oils, and man-made materials such as concrete, complex organic chemicals and metals, can be attacked. Often a single microbial species is involved, but usually biodeterioration, including corrosion, results from an association of a number of different microbes. For instance, a rapid growth of an aerobic organism may so deplete oxygen that strictly anaerobic sulphate reducers associated with cathodic depolarisation then appear, and metallic corrosion results. In recent years the complicated associations in such microbial ecosystems have become increasingly recognised. Their corrosive effects on metals can be attributed to:

(*a*) Direct chemical action of metabolic products such as sulphuric acid, inorganic or organic sulphides and chelating agents such as organic acids.
(*b*) Cathodic depolarisation associated with anaerobic growth.
(*c*) Changes in oxygen potential, salt concentration, pH, etc. which establish local electrochemical cells.
(*d*) Removal of corrosion inhibitors (oxidation of nitrite or amines) or protective coatings (bitumen on buried pipes).
(*e*) The presence of the biomass itself or residues of biomass such as hygroscopic salt deposits from cells burnt-on in annealing.

It must be stressed that the identification of the causative organism may be extremely difficult since it often depends on the quantitative determination of numbers of each microbial type in a complex mixture together with an assessment of its chemical and physical activities in that particular environment.

Although bacteria may predominate, moulds, yeasts and protozoa may be associated with bacteria, or, under some conditions, may either cause corrosion by themselves, or modify it drastically.

Methods of protecting materials against microbial corrosion include:

(*a*) Coatings, particularly of resistant synthetic polymers or paints containing inhibiting salts (e.g. Cu^{2+}, Cr^{3+} and Zn^{2+}).
(*b*) Controlled dosing with appropriate biocides.
(*c*) Change in environmental conditions unfavourable to microbial growth, e.g. removal of water from lubricating or fuel oils, good industrial housekeeping, temperature changes.
(*d*) Designs based on fundamental knowledge of microbial ecology. This implies co-operation between engineers and biologists aimed at reducing infection and maintaining unfavourable conditions for microbial growth.

Although many of the effects of microbes on metal are associated with growth this is not necessarily so because a biomass once established may cease to increase but continue its chemical activities often at an accelerated rate, once the controls on growth are released.

Acid Corrosion

Massive and rapid corrosion of metal, concrete and limestones is usually caused by the action of sulphuric acid formed by the oxidation of sulphur or sulphide by members of the genus *Thiobacillus*[3]. The majority of this genus grow by assimilating CO_2 at the expenditure of energy produced by the oxidation of sulphur, sulphite, thiocyanate and tri- or tetrathionate; some strains are sensitive to low concentrations of H_2S. They thus grow as chemolithotrophs* in contrast to the chemo-organotrophs whose energy supplies come from organic materials, or phototrophs who utilise light energy. The oxidation of sulphur may produce a concentration of up to 10–12% H_2SO_4 and it is this that produces corrosion; these organisms are also exploited for ore leaching[4] and in the biological treatment of coke oven effluents[5]. *Thiobacillus thio-oxidans* commonly occurs in soil and water and is to be suspected where corrosion occurs associated with very low pH in the immediate vicinity[6]. It may be isolated on acid media and reliably estimated by plate counting on a solid medium.

Acid production and corrosion associated with pyritic deposits is caused by *Ferrobacillus ferro-oxidans*[7].

Detection *T. thio-oxidans* is best detected by the strongly acid conditions it generates in a mineral salt solution on which sulphur is floating.

Prepare and sterilise by steaming a medium of $(NH_4)_2SO_4$, 0·2 to 0·4 g; KH_2PO_4, 3 to 4 g; $CaCl_2$, 0·25 g; $MgSO_4$, 0·5 g; $FeSO_4$, 10 mg; sulphur,

*The commonly used term *autotrophy* describes the ability to grow on CO_2.

10 g; tap water to 1 000 ml; pH $5 \pm 0{\cdot}3$ (some authorities recommend a trace metal mixture in place of the ferrous salt). Add 1 ml or 1 g of material to be tested to 100 ml of medium in a conical flask and incubate in air at 30°C. From 4 days to 2 weeks the pH of samples should be measured at intervals. An abrupt drop to 2·5 or lower indicates growth of *T. thio-oxidans*. Little turbidity appears in the medium; under a microscope the sulphur particles are seen to be surrounded by motile stubby Gram-negative rods.

Estimation[8] The above medium is reinforced with 10 g/l of thiocyanate, sulphur is omitted and it is prepared as pour plates by the addition of 3% agar. Organisms other than Thiobacilli will grow from spread samples, but the Thiobacilli are easily distinguished by sulphur haloes (see Fig. 2.18).

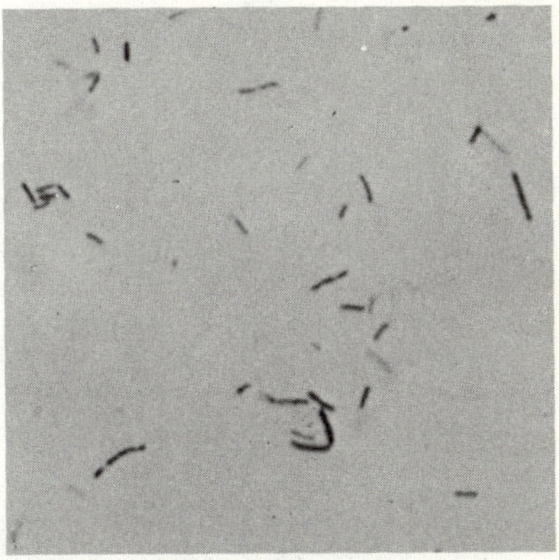

Fig. 2.18 *Thiobacillus thio-oxidans* (NCIB 8 342). Usually stubby rods, but a few elongated forms can be seen (these are most common in old cultures), × 260 (Crown copyright courtesy Microbiological Research Establishment)

Examples Parker[9] described the rôle of Thiobacilli in the corrosion of concrete sewer pipes: evolution of hydrogen sulphide from the sewage leads to the deposition of inorganic sulphur compounds on the roof of the pipe, and these are oxidised to sulphuric acid by the bacteria, causing a characteristic corrosion pattern in which the roof of the pipe becomes decayed. Iron pipes carrying polluted effluents, and concrete manhole covers examined at the National Chemical Laboratory have been corroded for similar reasons; corrosion of Mouchel cooling towers[10] had a like origin, as had corrosion of buried iron gas mains in south London[11]. Corrosion of statues in France has been partly attributed to Thiobacilli[12]; corrosion of buildings and stonework probably has a similar origin, the source of sulphur for the bacteria being atmospheric pollution. Occasionally, in the experience of the National Chemical Laboratory, ornamental cements containing sulphur have been used for facing buildings, and these form ideal substrates for Thiobacilli. Materials containing sulphur have been used for jointing water mains and Thiobacilli may cause corrosion by forming acid from them[13]. Instances in which vulcanised rubber has been corroded by these bacteria are known[14]. Corrosion by Thiobacilli is probably more widespread than the published examples would indicate.

Prevention Methods of prevention already summarised may be used singly or in combination. Elimination of sulphur and certain of its compounds are most effective but more recently more resistant materials such as polythene or asbestos are used to replace iron and concrete where acid corrosion is severe.

Corrosion by Ferrobacillus

F. ferro-oxidans[15] is capable of accelerating the oxidation of pyritic (FeS_2) deposits at acid pH values. It is usually found in association with Thiobacillus and was known as *Thiobacillus ferroxidans* before the distinction between the two organisms was appreciated. It is responsible for pollution problems arising from acid waters in gold and bituminous coal mines[16,17]; such waters are corrosive to pumping machinery[18] and mining installations (see Fig. 2.19).

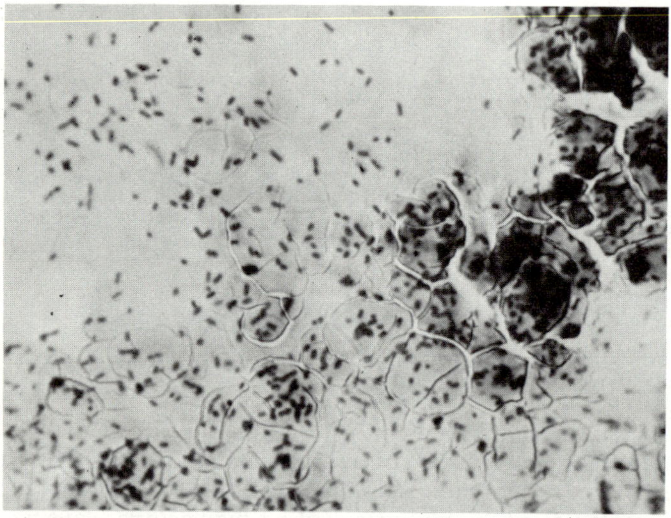

Fig. 2.19 *Ferrobacillus ferro-oxidans* (NCIB 8 451), bacteria and encrustations of ferric oxides; the proportion of bacteria was much increased by filtering and centrifugation, ×260 (Crown copyright courtesy Microbiological Research Establishment)

Detection Corrosive waters formed by these bacteria have a pH in the region of 2 to 3, show a brown deposit of basic ferric sulphate, and contain free sulphuric acid.

Prepare and sterilise by heating a medium of $(NH_4)_2SO_4$, 0·15 g; KCl, 0·05 g; $MgSO_4{\cdot}7H_2O$, 0·5 g; KH_2PO_4, 0·05 g; $Ca(NO_3)_2$, 0·01 g in 1 000 ml of tap water. Prepare a 10% solution of $FeSO_4{\cdot}7H_2O$ and sterilise by filtration. Add 1 ml $FeSO_4$ solution to 100 ml of medium in a 250 ml conical flask, inoculate with the material being examined, check that the pH is approximately 3·5, and incubate in air at room temperature. Growth of *F. ferro-oxidans* is indicated by formation of a brown precipitate as compared with slow, slight browning in an uninoculated culture; the organism grows very slowly and may take up to a month to show unequivocal results; the responsible bacteria are straight rods, often difficult to see microscopically amid the debris.

Prevention Neutralisation of acid waters with lime has been recommended but the resulting sulphate-laden water may present a disposal problem. Use of acid-resistant machinery and pipes is more satisfactory.

Mechanism of sulphur oxidation Apart from its intrinsic interest the economic importance of acid corrosion and more lately interest in ore leaching, has stimulated considerable work on the oxidation of sulphur, Fe^{2+} and Mn^{2+}. It must be stressed that the Thiobacilli are obligate aerobes, i.e. that depend on molecular oxygen as a terminal electron acceptor. Possible reactions for the oxidation of sulphur are[19]:

$$4S^- \rightarrow 2S_2O_3^- \rightarrow S_4O_6^- \rightarrow SO_3^- + S_3O_6 \rightarrow 4SO_3^- \rightarrow 4SO_4^{2-}$$

The central rôle of SO_3^- is apparent.

Reactions leading to SO_4^{2-} formation from pyrites are possibly:

$$2FeS_2 + 7O_2 + 2H_2O \rightarrow 2FeSO_4 + 2H_2O \quad \ldots(2.8)$$

$$4FeSO_4 + O_2 + 2H_2SO_4 \rightarrow 2Fe_2(SO_4)_3 + 2H_2O \quad \ldots(2.9)$$

$$FeS_2 + Fe(SO_4)_3 \rightarrow 3FeSO_4 + 2S \quad \ldots(2.10)$$

$$2S + 3O_2 \rightarrow 2H_2O + H_2SO_4 \quad \ldots(2.11)$$

some of these reactions, e.g. 2.8 and 2.10, take place slowly in the absence of bacteria but are accelerated in their presence.

Other Acid Corrosion

Less well studied than the effects of the Thiobacilli are corrosion reactions due to the formation of acids from the oxidation of organic materials. These may include the products of microbial attack on protective coatings such as hessian sacking and bitumen coatings used for iron pipes initiated by the cellulose-decomposing bacteria. Paper and synthetic rubber coatings for insulation cables may also be attacked. Under strongly aerobic conditions CO_2 is the end product of the oxidation of the organic material, and lead carbonate has been detected as a corrosion product of lead-coated underground cables. Under semi-anaerobic conditions organic acids accumulate and these may lead to simple acid corrosion or alternatively may accelerate corrosion by chelation of passive layers on metal. Besides bacteria, moulds and yeasts may accumulate organic acids even under aerobic conditions and in some cases may synthesise complex 'secondary metabolites' some of which, although only weakly acid, are powerful chelating agents. These may be of special significance when microbial slimes accumulate on metal surfaces, as relatively high concentrations of potentially corrosive products may be trapped in them, and corrosion pits result. It is possible that massive pitting in aluminium fuel tanks in aircraft may originate in this way[20].

Microbial-accelerated Cathodic Depolarisation of Ferrous Metals

Corrosion of iron and steel, especially in anaerobic conditions such as waterlogged soils, is usually caused by sulphate-reducing bacteria of which the genus *Desulphovibrio* is the most commonly occurring. The presence of organic material often stimulates these organisms' reducing power whereby sulphate is reduced to sulphide, but some at least of this genus appear to grow as essentially as chemolithotrophs, and reduce sulphate as follows:

$$SO_4^{2-} + 4H_2 \rightarrow S^{2-} + 4H_2O \qquad ...(2.12)$$

Simple corrosion by H_2S should yield exclusively FeS:

$$Fe + H_2S \rightarrow FeS + H_2 \qquad ...(2.13)$$

whereas both hydroxide and sulphide would be expected if the cathodic reaction was due to the mechanism proposed by von Wolzogen Kühr and van der Vlugt[21].

$$4Fe + 4H_2O + SO_4^{2-} \rightarrow 3Fe(OH)_2 + FeS + 2OH^- \qquad ...(2.14)$$

It can be seen from equation 2.14 that the ratio of iron corroded to iron in the form of sulphide should be 4:1, but values from 0·9 to 48 are commonly obtained experimentally. Subsequently it was shown by Booth[22] and his co-workers that the ratios of the corrosion products were dependent on the particular strain of *Desulphovibrio* and on their rates of growth. Later the activity of the enzyme hydrogenase which brings about the reaction:

$$H_2 \rightleftharpoons 2e^- + 2H^+ \qquad ...(2.15)$$

was correlated directly to the ratios of corrosion products. Further work, especially with methods of continuous or semi-continuous growth of pure cultures coupled with appropriate enzymic and electrochemical measurement, largely confirms the important rôle of hydrogenase in cathodic depolarisation but also suggests rates are affected by precipitated ferrous sulphide which can, in the presence of excess ferrous ions, form films on the metal with the possibility of setting up local concentration cells. The rôle of ferrous sulphide and ferrous salts is as yet unexplained, but the concentration of ferrous ions appears to have an effect even on the anaerobic corrosion of buried pipes in the field as well as in the laboratory. The proposed mechanisms emphasise the rôle of environmental conditions on anaerobic microbial corrosion. In considering this, two aspects concern engineers, firstly in any new engineering venture involving buried iron or steel structures, the soil conditions must be evaluated. The term *aggressive* is often used to imply some approximately quantitative estimate of the likelihood of corrosion and depends on measuring factors such as soil water (resistivity), pH, redox potential, salt concentrations and bacterial populations in order to establish criteria for the prediction of corrosion rates[22,23]. Similar measurements for predicting corrosion in rivers and bottom deposits have been described. Confirmation of these prognostic tests was made on buried or immersed metal plates with periodic measurements of metal/soil or metal/water potentials with reference to copper/copper sulphate electrodes; weight losses were determined after burial for up to 3 years. Aggressive soils are characterised by a resistivity of less than 2000 Ω cm or a redox potential of less than 400 mV (see Table 2.17 for evaluation of E_h); when a soil is 'borderline' and its resistivity and redox potential approximate to the values above, the water content provides a further criterion and it is regarded as aggressive when the water content is greater than 20%. These criteria have been confirmed to be valid for mild steel with a few exceptions, and have been found to be fairly satisfactory for lead and zinc but invalid for aluminium. The metal/soil potential gave little information of value.

Detection of Anaerobic Corrosion

This is immediately recognised in smooth pitting with a black corrosion product and smell of hydrogen sulphide when the pipe is first exposed; cast iron shows graphitisation. The iron sulphide corrosion product oxidises rapidly on exposure to air and should be examined quickly: in doubtful cases acidification on exposure is confirmatory. Isolation and counting the bacteria depends on establishing strict anaerobiosis. Methods for examining soil and water are given below.

Sampling

Soil samples from the levels in which structures or pipes are to be laid are filled to the top of screw capped bottles, and bacteriological tests are made within 24 h.

Isolation and Enumeration

Methods have recently been evaluated by Mara and Williams[24] and for most purposes the modified ISA medium is suitable: iron sulphite agar (oxoid) 23 g/l; $FeSO_4 \cdot 7H_2O$, 0·5 g/l; 7% sodium lactate 5·0 g/l; $MgSO_4 \cdot 7H_2O$, 2·0 g/l. The medium is adjusted to pH 7·5 with sterile NaOH after autoclaving at 121°C for 15 min. For liquid cultures, screw-cap test tubes are filled to the brim, for enumeration sterile test tubes (150 × 16 mm) are filled to within 5 mm of the top with inoculated agar medium, covered with a cap of sterile 1·5% w/v agar and closed with a polypropylene cap. Incubation is at 30°C and colonies are counted until the maximum number has developed. These methods are prone to give false positives and large errors,

Table 2.18 Redox potentials* of soils in relation to corrosiveness[23]

Range of E_h	*Classification of corrosiveness*
100 mV	Severe
100–200 mV	Moderate
200–400 mV	Slight
400 mV	Non-corrosive

*The redox potential is determined with a probe consisting of a platinum electrode and a Hg/Hg_2Cl_2—Cl^- or $Ag/AgCl$—Cl^- reference electrode. If E_p is the potential of the platinum probe, E_r the potential of the reference electrode and E_h the redox potential of the soil (in mV on the hydrogen scale) then

$$E_h = E_p + E_r + 60\,(\text{pH} - 7)$$

where pH is the pH of the soil.

and should only be attempted where adequate microbiological backing is available.

Sulphate-reducing bacteria are present in virtually all soils, and the qualitative procedure is valuable because it works only when relatively large numbers are present. Hence a positive result with this test is a rough indication of a particularly aggressive soil, though a negative result does not necessarily mean that the soil is innocuous. Quantitative procedures for enumerating sulphate-reducing bacteria in soils and water have been available only for the last few years and data on populations in aggressive and non-aggressive soils are therefore scanty. Highly polluted waters and soil are known to contain 10^5 to 10^6 viable sulphate-reducing bacteria per millilitre or gram; waters with less than 10^2 bacteria per millilitre are usually innocuous from the pollution point of view. It is likely that the aggressiveness of soils would follow broadly the population of viable sulphate-reducing bacteria within these limits.

As already stated, aggressives will also depend on soil E_h according to the scale given in Table 2.18. Temperature is also a controlling factor and both pscrophylic (cold) and thermophilic (hot) forms are known, e.g. in electrical transformers, hot water systems.

Hydrogenase Determination

This enzyme is of wide occurrence in bacteria where it is concerned with the reduction of nitrate and CO_2 as well as sulphur. Methods for its estimation depend on measuring some activity of hydrogenase by (*a*) dye reduction (benzyl viologen or methylene blue), (*b*) isotopic exchange and (*c*) evolution of molecular hydrogen. Interpretation of quantitative results is difficult due to the complex relationship between the enzyme cell structure and the particular method selected[25].

Corrosion Due to Other Microbes

In recent years it has become apparent that widespread microbial infections of materials in the manufacturing industries can lead to corrosion for the reasons briefly outlined above. Examples include the rusting of machined parts, corrosion of machine tools, aircraft fuel tanks, hydraulic systems, strip steel, etc.

The precise rôle of any specific organism in these instances is difficult to determine and will probably remain the province of specialists, largely because of the ubiquitous occurrence of microbes. However, a number of simplified tests have recently been devised to assist engineers and chemists in diagnosing whether or not a particular corrosion is biological in origin. These are based on (1) direct microscopy, (2) measurements of microbial metabolism (oxygen uptake, dye reduction, extracellular enzyme activity) and (3) direct enumeration of specific species of selective media. These are briefly discussed below.

1. *Microscopy*[26]. A binocular phase-contrast microscope equipped with

low-power lens and an achromatic ×40 objective and ×10 compensating eyepiece. An oil-immersion lens (×100) is required for stained material. Dark ground illumination is also probably worthwhile. Size measurements can be made with an image splitting eyepiece (Vickers Instruments, York). The significance of microscopal examination depends on familiarity with the material and long experience of its microbial populations.

2. Two simple metabolic tests may be used for evaluating infections by aerobic and anaerobic organisms.

(*a*) *Red spot test*[27]. This relies on the reduction of a soluble colourless tetrazolium dye to an insoluble coloured formazan. This can be done on plates (Oxoid Ltd.) or ampoules (BDH Ltd.). The test can be calibrated by estimating by eye the extent of red formazan deposition and correlating it with elaborate counting methods or merely by taking account of any particular situation. Anaerobic bacteria give rapid responses, slower or little response is given by most moulds and yeasts[27]. Anaerobes[28] can be enumerated in the solidifying media for *Desulphovibrio* (p. **2**.79).

(*b*) A test based on an enzyme released by microbial growth is illustrated by the Avtur Test (BDH Ltd.)[29] for jet fuel infections. This is based on an approximate assay of the enzyme acid phosphatase.

3. The medium already described for enumerating Thiobacilli (p. **2**.74) is a typical selective media and depends on selecting for a mixed population an organism with a specific growth requirement sufficiently different from most of its fellows in that environment. Further selectivity can be designed into a medium by the addition of suitable growth inhibitors, e.g. bile salts (to select coliforms) penicillin (to select Gram negatives), etc. Temperature and other physical parameters can also be used for selection pressure on a mixed community. The use of media enriched with biocide is particularly important for estimating the likelihood of resistant species occurring during treatment of an industrial plant to stop microbial corrosion[30]. Selection media are available for isolating and enumerating fungi (Corn meal agar; Oxoid Ltd.) and yeasts (Tryptone soya broth (Oxoid Ltd.) containing 10% w/v sucrose, which inhibits the growth of most bacteria and fungi[31]). Sampling can be very important especially if false positive results are to be avoided. The usual practice is to collect directly into sterilised bottles, e.g. plastic (Sterilin Ltd.), from appropriate sections of a plant assuming that, for instance, in multi-phase systems representative samples of each phase are collected, e.g. oil- and water-lubricating lines. Both liquids should be run to waste for at least 30 s before collecting. Samples are assayed as quickly as possible or stored at 2–4°C in the dark. After testing, samples are disposed by autoclaving or immersion in a suitable disinfectant overnight.

To conclude it must be stressed that recent work has directed attention to the interplay between different microbial species in most of the corrosion effects described. Considerable care is needed therefore in interpreting the results of the comparatively simple tests described, especially if costly remedial action is necessary.

D. E. HUGHES

REFERENCES

1. Hughes, D. E. and Hill, E. C., Mining and Metallurgical Congress Paper No. 24 (1969)
2. Hill, E. C., in *Microbial Aspects of Corrosion*, Ed. Miller, J. D. A., Medical and Technical Publishing, Aylesbury (1971)
3. Miller, J. D. A. and Tiller, A., ibid.
4. Le Roux, N. W., ibid.
5. Stafford, D. A. and Callely, A. G., Symposium on Effluent Treatment, B.C.R.A., Chesterfield, to be published.
6. Butlin, K. R. and Postgate, J. R., in 'Autotrophic Micro-organisms', Symp. Soc. Gen. Microbiol., Cambridge University Press (1954)
7. Fletcher, A. W., in *Microbial Aspects of Corrosion*, Ed. Miller, J. D. A., Medical and Technical Publishing, Aylesbury (1971)
8. Williams, A. R., Stafford, D. A., Callely, A. G. and Hughes, D. E., *J. Bact.*, **33**, 656 (1970)
9. Parker, C. D., *Aust. J. Exp. Biol. Med. Sci.*, **23** No. 81, 91 (1945)
10. Taylor, C. B. and Hutchinson, G. H., *J. Soc. Chem. Ind.*, London, **66**, 54 (1947)
11. *Chemistry Research 1956*, H.M.S.O., London, 16, 61 (1956)
12. Pochon, J., Coppier, O. and Tchan, Y. T., *Chim. et Industr.*, **65**, 496 (1951)
13. Frederick, L. R. and Starkey, R. L., *J. Amer. Wat. Wks. Ass.*, **40**, 729 (1948)
14. Thaysen, A. C., Bunker, H. J. and Adams, M. E., *Nature*, London, **155**, 322 (1945)
15. Leathen, W. W., Kinsel, N. A. and Braley, S. R., *J. Bact.*, **72**, 700 (1956)
16. Braley, S. A., *Min. Engng.*, N.Y., **8**, 314 (1956)
17. Braley, S. A., *Min. Engng.*, N.Y., **9**, 76 (1957)
18. Butlin, K. R. and Vernon, W. H. J., *J. Inst. Water Engineers*, **3**, 627–637 (1949)
19. Trudinger, P. A., *Advances in Microbial Physiology*, **3**, 111 (1969)
20. Elphick, J., in *Microbial Aspects of Metallurgy*, Ed. Miller, J. D. A., Medical and Technical Publishing, Aylesbury (1970)
21. von Wolzogen Kühr, C. A. H. and van der Vlugt, L. S., *Water*, Den Haag, **18**, 147 (1934)
22. Booth, G. H., *Microbiological Corrosion*, M & B Monographs, Mills and Booth Ltd., London (1971)
23. Starkey, R. L. and Wight, K. M., *Anaerobic Corrosion of Iron in Soil*, Amer. Gas. Ass., New York (1945)
24. Mara, D. D. and Williams, D. J. A., *J. Appl. Bact.*, **33**, 543 (1970)
25. San Pietro, A., *Methods in Enzymology*, Vol. II, Academic Press, N.Y., 861
26. Barer, R., *The Microscope*, Blackwell, Oxford (1959)
27. Hill, E. C. and Penberthy, I., *Metals and Materials*, **2**, 359 (1968)
28. Hill, E. C. and Gibbon, O., *Metal Finishing*, **15**, 395 (1969)
29. Hill, E. C., *Aircraft Engineering*, July, 24 (1970)
30. Callely, A. G., *Process Biochemistry*, **3**, 11 (1967)
31. Private communication

2.7 Chemicals

For the purposes of this section a *chemical* may be defined as a substance useful technologically and containing over 95% of the principal chemical.

In general most textbooks on corrosion control, including this one, give data on the physical and mechanical properties of a particular metal or alloy and then outline its suitability in various environments, rather than considering a particular chemical in relation to its effect on various materials of construction. This approach may be difficult to appreciate, since the design engineer will be given, at first, the environment he has to work with and then is required to choose the most suitable materials. To date this information is grossly inadequate because of the enormity of collating such a large amount of data. For example, if some 400 chemicals are identified as being handled and processed on a large scale and there are 10 suitable materials, then 4 000 systems would have to be considered. Since temperature, concentration and solution velocity are important in determining corrosion rate, and if only five levels of each of the three variables are considered, then the number of experiments to be carried out would be $4\,000 \times 5^3 = 500\,000$.

There are in addition several other factors that accelerate corrosion and must be taken into account; these include crevices, galvanic coupling, tensile stress, aeration, presence of impurities, surface finish, etc. If these were also taken into consideration then several million experiments would have to be performed to compile such data. There are many instances where two or more chemicals exert a marked synergistic action such that low dissolution rates obtained in either environment become much greater in the presence of both. Further, the corrosiveness of a chemical will be affected by the presence of certain impurities, which may act as either accelerators or inhibitors. To take all these factors into account would add to an already impossible task and as Evans has remarked[1], 'There are not enough trained investigators in the world to obtain the empirical information to cover all combinations of conditions likely to arise'. Unfortunately corrosion science has not yet reached the stage where prediction, based on a few well established laws, allows selection of materials to be made without recourse to a vast amount of data.

Sources of Information

In the chemical industry, many processes are required to contain and handle solutions of complex composition containing many aggressive ingredients, but there exists a readily available supply of materials that can be used in the most aggressive environments. The purpose of this section is to indicate the procedure that an engineer could adopt if he was called upon to design plant items for chemicals for which he had little or no prior knowledge or experience. This is a situation that exists within many engineering industries that do not utilise metallurgists or materials scientists and are generally unaware of the existence of the specialist corrosion engineer or scientist and his importance at the planning stage. The most important sources are:

1. The 'ideal' source book for designers, which is the one in which the individual chemicals are listed together with the corrosion rates for a variety of materials under different conditions of temperature, pressure, velocity, etc.
2. Corrosion resistance data lists on specific materials offered for sale. However the engineer must also consider mechanical and physical properties and, last but not least, the cost of the material, its fabrication and protection. Thus reference must be made to books, journals and data that provide this information.
3. Information based on experience, which includes national standards, specifications and codes of practice, and also the technical and scientific literature reviews and reports. Corrosion engineers, technologists and scientists employed by specialist organisations or as consultants form the most important source of information, and their advice should be sought wherever possible.

Selection Based on Chemical Environment

The design engineer who requires full information will be disappointed because such books, tables and monographs do not exist. In the present situation the engineer will turn to official standards, etc. However, they are relatively few in number and, though an important source of information, suffer from the lack of reliable data for many popular materials in a wide range of environments that may lead to pitting, stress-corrosion cracking, crevice corrosion and corrosion fatigue.

For the engineer who cannot find a suitable specification for a particular chemical environment various sources of information exist. Perhaps the most comprehensive text presently available is the *Corrosion Guide*[2]. In this book a great deal of information has been gleaned from the world's literature in an attempt to fulfil the engineers' requirements. Because of the magnitude of the problem, three or four levels of corrosivity are given, but only a small but detailed amount of data are reported for certain chemicals with regard to temperature, concentration and velocity. The data given may be useful for making an initial survey of materials prior to further detailed planning or testing. Many of the 800 or so chemicals mentioned do not corrode all of the materials (metals, ceramics and plastics) and this book could be considered as the most useful source of preliminary information. The guide gives corrosive effects of a large number of individual inorganic and organic chemicals and also some important groupings such as fatty acids, cements, concrete, mortar, plaster, oil and varnishes. However, it must again be emphasised that the Corrosion Guide provides no information other than corrosion behaviour, and that this alone is insufficient in a rational approach to materials selection. The corrosion resistance of mild steel is poor in most environments, but this is frequently outweighed by the fact that it has good mechanical properties, and is readily available and cheap.

When the information is available, the Corrosion Guide provides detailed corrosion data on the preparation of various chemicals. For example, in the section on sulphuric acid, the corrosion rates for several alloys are given when used at various stages of an actual process involving that acid. In the section on phosphoric acid, cognisance of the method of production shows also the influence of the minor constituents as well as the major chemical on the corrosion of various materials. Nevertheless, it must be emphasised that even a book as comprehensive as the Corrosion Guide can only cover a limited number of all the possible chemicals used in practice.

Where a large collection of data exists then it may be effectively condensed in the form of diagrams. A popular method is the use of 'iso-corrosion rates'

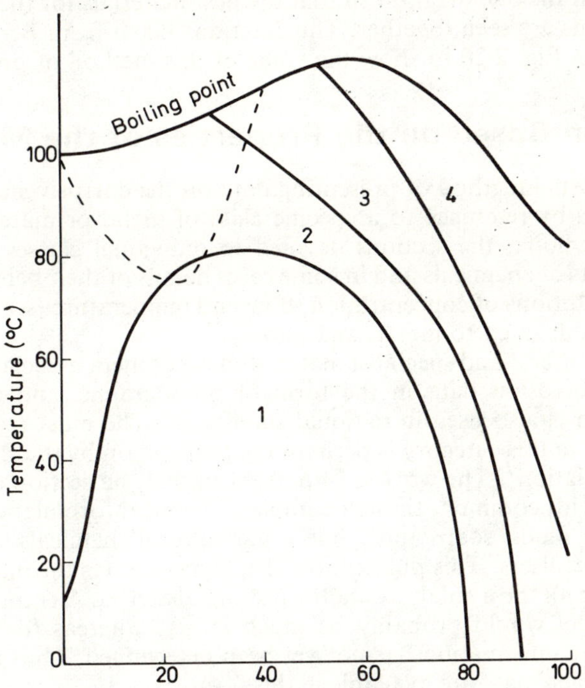

Fig. 2.20 Safe operating conditions for various steels in nitric acid solutions. Iso-corrosion lines at 0·1 $gm^{-2}h^{-1}$

Area 1: Fe–13Cr
Areas 1 + 2: Fe–17Cr
Areas 1 + 2 + 3: Fe–18Cr–8Ni
Areas 1 + 2 + 3 + 4, but excluding area within broken line: Fe–15Si

(after Berg[3])

plotted on co-ordinates of temperature and concentration for one material and one chemical. Because of the large amount of data on the common acids there are many examples of this type of diagram, e.g. the work of Berg[3] who has chosen metals and alloys that are readily available. He has excluded many metals and alloys on the grounds that they are either 'Non-resistant or can be substituted by cheaper materials. . . .'

From these diagrams an initial selection can be made for materials to be used for the chemicals considered, and although temperature and concentration are included, the effects of flow and the effect of crevices, galvanic coupling or stresses are not indicated. Thus for sulphuric acid, 'chemical lead' and 'silicon iron' appear to cover a wide range of temperature and concentration, however, the mechanical weakness of the former and the brittleness and difficulties in fabrication of the latter, might mitigate against their use in say pumps, valves and unsupported pipes. For these purposes the more expensive nickel-base alloys have also been listed which may be preferred on various grounds which might include ease of fabrication and tolerance to thermal cycling.

An interesting interpretation of these temperature–concentration diagrams has been given by Nelson[4]. He has combined information of all the materials on the one diagram so that the best materials for the whole range of conditions are seen together. The diagrams taken from Berg[3] have been combined in Fig. 2.20 to show the value of this method of presentation.

Selection Based on the Properties of the Material

The more usual method of presenting data on the corrosiveness of various chemicals is by reference to a specific class of metal or material. Thus in the present book, the sections devoted to individual classes of materials contains lists of chemicals and in some cases details of their behaviour under various conditions of concentration, flow and temperature (see in particular the sections devoted to metals and alloys).

Manufacturers[5] and specialist materials development associations publish extensive corrosion data in the form of monographs, and this form of presentation is also used in national standards[6]. The most recent comprehensive text in this category is perhaps the publication by the Zinc Development Association[7]. The work is important in that the section on chemicals also deals with common, though complex, chemical formulations, e.g. fire-extinguisher fluids, soaps and syndets, agricultural chemicals such as pesticides and fertilisers. This publication also demonstrates the mammoth task of recording all the available data for just one material. A comparable book for mild steel would probably be much larger, whereas for many other materials the information has not yet been determined. Thus at best, only very incomplete data are available in this form.

Selection Based on Experience

The scientific and technical literature[8] abounds with much information, adding slowly to the massive factual data required for the design engineer who relies solely on his own literature search.

The material development associations and manufacturers have, by their own research and development, accumulated a great deal of information about their own product and this is transmitted directly to potential users of their materials.

Selection Based on Scientific Principles

Ideally the design engineer requires an equation which condenses all this information and from which he can calculate the effect of a particular chemical upon a range of materials, and the limiting conditions of say temperature, concentration and velocity. To achieve this objective he needs to know which of the properties of the chemical and the material are the most important in determining the interaction leading to corrosion.

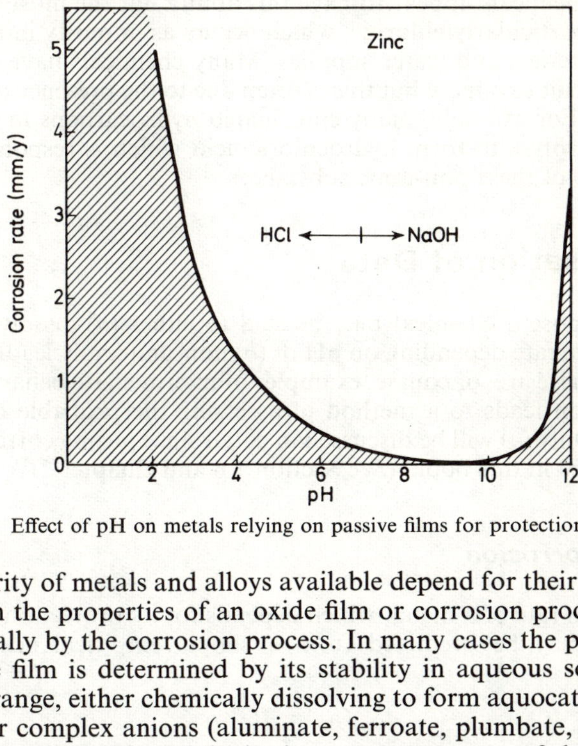

Fig. 2.21 Effect of pH on metals relying on passive films for protection, e.g. zinc

The majority of metals and alloys available depend for their resistance to corrosion on the properties of an oxide film or corrosion product which is formed initially by the corrosion process. In many cases the protectiveness of the oxide film is determined by its stability in aqueous solutions in a specific pH range, either chemically dissolving to form aquocations at lower pH values or complex anions (aluminate, ferroate, plumbate, zincate, etc.) at higher pH values (Fig. 2.21). An important property of the chemical is therefore the pH value that it develops when dissolved in water. For many materials and many chemicals this is the overriding factor and in many cases the oxide film is found to dissolve uniformly by interaction with hydrogen ions, the rate varying with concentration in accordance with equations such as:

$$\text{Rate} = \text{constant} \times (\text{conc. of H}^+ \text{ ions})^n$$

As the film dissolves more oxide film is formed (the metal/oxide interface progresses into the metal) and the overall rate may be low enough to be acceptable for a particular process. In other cases, the corrosion products precipitate on the surface of the oxide and either accelerate the overall rate by enhancing diffusion of ions through the porous outer layers or, when less porous layers are formed, access of hydrogen ions to the inner oxide surface is reduced thus decreasing the rate.

The ability to form a second barrier film when the conditions are such that the oxide film is dissolving rapidly, is another important property of the chemical, i.e. its ability to form highly insoluble corrosion products which would allow the physical blockage of the surface. There are, however, a certain number of chemicals that have the ability to cause non-uniform dissolution of oxide films and the more serious form of pitting corrosion occurs. Certain anions have the ability to attack the oxide film in small areas and allow corrosion at high rates confined to these areas. This often leads to deep penetration of the metal although overall weight-loss is very low. Many chemicals appear to have this ability but the most insidious are the halides, particularly chloride, which occurs as impurity in many chemicals, raw materials and water supplies. Many chemicals have a reputation of being difficult to handle but this is often due to the presence of these types of impurity. For example, many chlorinated hydrocarbons in the presence of water hydrolyse to form hydrochloric acid which is responsible for the corrosiveness of these non-ionic substances.

Rationalisation of Data

For this purpose a chemical may be said to either (*a*) dissolve a material uniformly, the rate depending on pH or (*b*) non-uniformly leading to pitting corrosion. There are, of course, examples of intermediate behaviour but this simple division leads to a method of assessing the probable behaviour of a chemical. Only (*a*) will be discussed in detail since pitting corrosion is dealt with elsewhere in this book. (See Section 1.6 and Chapter 21.)

Uniform Corrosion

The corrosion rate of a metal, which depends for its protection on a passive oxide film, may be predicted from a simple empirical adsorption law (Freundlich):

$$C_R = AC^n$$

where C_R = the corrosion rate,
A = the specific rate constant,
C = the concentration of hydrogen ions and
n = an integer,

which may be modified by taking logarithms and substituting pH for C, when the equation becomes

$$\log C_R = K - n(\text{pH})$$

where $K = \log A$.

Reliable pH data and activities of ions in strong electrolytes are not readily available. For this reason calculation of corrosion rate has been made using weight-loss data (of which a great deal is available in the literature) and concentration of the chemical in solution, expressed as a percentage on a weight of chemical/volume of solution basis. Because the concentration instead of the activity has been used, the equations are empirical; nevertheless useful predictions of corrosion rate may be made using the equations.

In strong acid solutions many common structural materials dissolve uniformly and this assumption is reasonable in many real situations. The data given in the monograph by Berg[3] are used in order to demonstrate the universal application of the technique. Four main types of behaviour may be identified for metals and alloys in various acids at different temperatures and concentrations.

Type 1. *Increasing corrosion rate with increasing concentration and temperature* In this case the equation obeyed is

$$\log C_R = n \log C - \frac{A}{T} + K,$$

where A, K and n are constants specific to a particular material and chemical. This relationship predicts that corrosion rate C_R increases continuously with temperature and concentration.

Example 1. Hard lead (antimoniacal) can be used for sulphuric acid to quite high concentration but it displays an increasing corrosion rate with increasing temperature and concentration. Relationships are complex, but the general form of the equation may be used:

$$\log C_R = 4 \log C - \frac{2\,500}{T} - 1{\cdot}25$$

The prediction is not very accurate, but in Fig. 2.22 comparison between the actual and calculated value shows the effectiveness of the empirical equations especially at higher temperatures. The iso-corrosion line obtained by substituting $0{\cdot}1\ \text{gm}^{-2}\text{h}^{-1}$ in the equation gives a reasonable guide to the temperature and concentration limitations of this material. The activation energy, which may be obtained from the temperature coefficient, indicates an activation-controlled reaction, so that flowing solutions should have a small effect on corrosion rates. Providing that the designer allows for the low mechanical strength of lead, the material could be considered for situations where high flow is involved, but abrasion of the protective film of corrosion products must be avoided.

Type 2. *Decreasing corrosion rate with increasing concentration and temperature* Metals and alloys that show this behaviour are important because they can be used for more concentrated acids. Mild steel is a good example in that it resists concentrated H_2SO_4 and HNO_3. This behaviour is associated with true passivity in the case of HNO_3, but in H_2SO_4 there is also a precipitation of corrosion products which appear to block the surface and prevent further attack. For higher temperatures and less concentrated

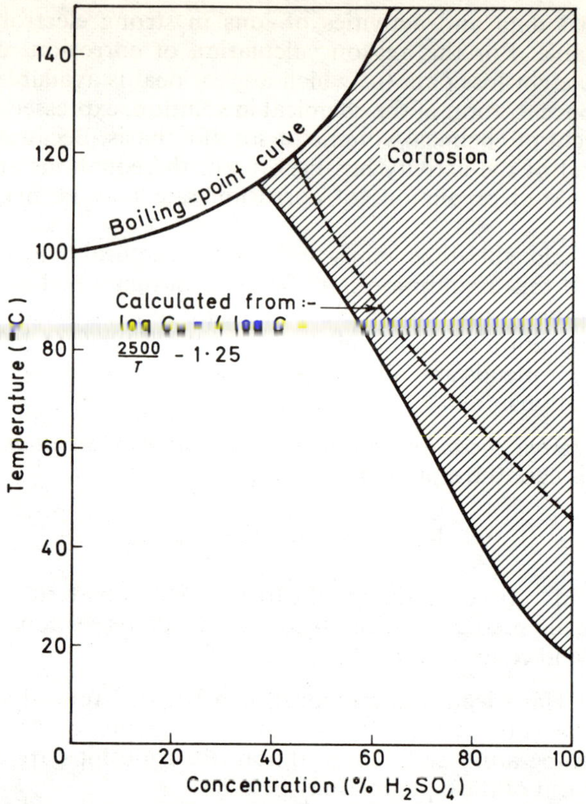

Fig. 2.22 Type 1 behaviour; increasing corrosion rate with increasing temperature and concentration, e.g. lead in sulphuric acid. Iso-corrosion lines at $0{\cdot}1\ \mathrm{gm^{-2}h^{-1}}$

solutions, mild steel is not resistant but other ferrous alloys have been developed that retain this characteristic over a much wider range of temperature and concentration. The empirical equations may be applied, and although an increase in temperature increases the rate, the form of the equation must be changed to allow for the decrease in corrosion rate with increasing concentration:

$$\log C_{\mathrm{R}} = K - n\log C - \frac{A}{T}$$

From the temperature factor A, an activation energy may again be calculated which gives a useful indication of the influence of flow-rate on corrosion rate.

Example 2. The corrosion rate of silicon iron (Fe–15Si) in static H_2SO_4 may be predicted from:

$$\log C_{\mathrm{R}} = 6{\cdot}54 - 1{\cdot}71 \log C - \frac{1\,540}{T}$$

The calculated and experimental iso-corrosion lines are given in Fig. 2.23 and show reasonable agreement with experimental values when the corrosion rate was $0{\cdot}1\ \mathrm{gm^{-2}h^{-1}}$. However, at a concentration of 50% there is a sudden decrease in rate and the material is useful at all temperatures beyond this concentration. The activation energy has been calculated as approximately 30 kJ/m and suggests that the corrosion rate will increase with increasing flow-rate, which could increase considerably the corrosion rates for static solutions quoted above. A diffusion-controlled reaction would be in accord with a salt passivation or blocking-type mechanism on a surface otherwise protected by SiO_2. The material is, however, used for pumps, stirrers, etc. in high concentrations of sulphuric acid where a thick anodic film is formed.

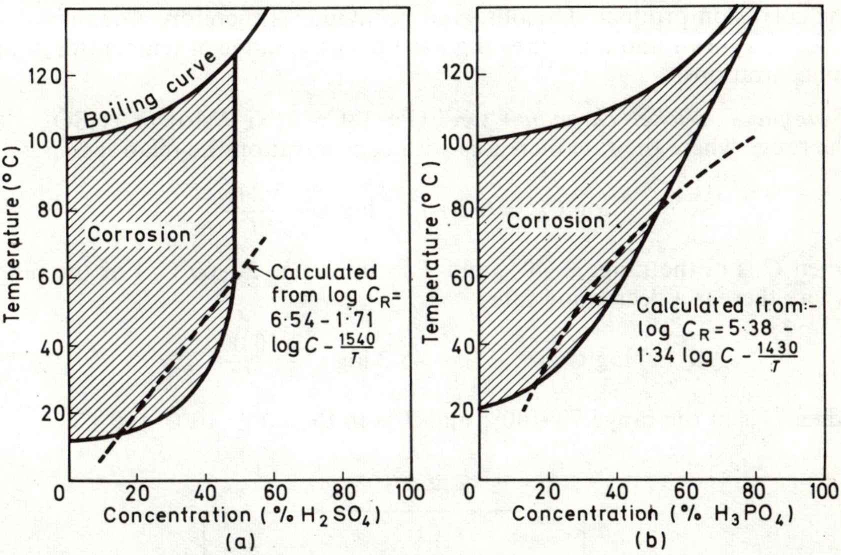

Fig. 2.23 Type 2 behaviour; decreasing corrosion with increasing temperature and concentration, e.g. (*a*) Fe–15Si in H_2SO_4 and (*b*) aluminium bronzes in H_3PO_4. Iso-corrosion lines at 0·1 $gm^{-2} h^{-1}$

Example 3. Aluminium bronze (Cu–7Al–3Fe) in H_3PO_4 follows the equation:

$$\log C_R = 5{\cdot}38 - 1{\cdot}34 \log C - \frac{1430}{T}$$

This alloy is suitable for high concentrations of de-aerated phosphoric acid providing the concentration is above a certain maximum, which varies with temperature in accordance with the equation. The calculated activation energy suggests that diffusion control may be operating and that flowing solutions should influence corrosion rate to a marked extent. The corrosion rate is influenced by aeration which is perhaps responsible for the observed diffusion control.

Type 3. *Chemicals that show a maximum rate at a certain concentration* A great many metals and alloys are unattacked in dilute solution but the rate increases with increase in concentration up to a maximum and then decreases with further increase in concentration. Thus the rates in the very dilute and highly concentrated acid may be acceptable. This behaviour has been observed for many metals and alloys in acid, alkali and neutral solutions of many salts. Brasher[9] *et al.* have shown that for many chemicals the maximum rate occurs at an acceptably low concentration of chemical followed by a low rate with further increase in concentration. Therefore, chemicals may be classed as 'aggressive' or 'inhibitive', depending on the position of the maximum corrosion rate in relation to anion concentration. The falling off in corrosion rate arises from the formation of an anodic film that is not affected by an increasing concentration of hydrogen ions. This has often been identified as a crystalline metal salt or oxide associated with the low solubility of the corrosion product. The observed behaviour is therefore a combination of both Types 1 and 2 manifesting itself over the range of temperatures and concentrations.

Example 4. Nickel–chromium steel (Fe–18Cr–10Ni–2Mo) *in* H_2SO_4. In the range where the rate increases with concentration the relationship is:

$$\log C_R = 8{\cdot}2 + 1{\cdot}67 \log C - \frac{3\,245}{T}$$

when C is in the range 0–40% and T in the range 20–60°C. For the range where there is a decreasing rate:

$$\log C_R = 117{\cdot}5 - 45{\cdot}2 \log C - \frac{10\,000}{T}$$

when C is in the range 70–100% and T is in the range 10°C–100°C.

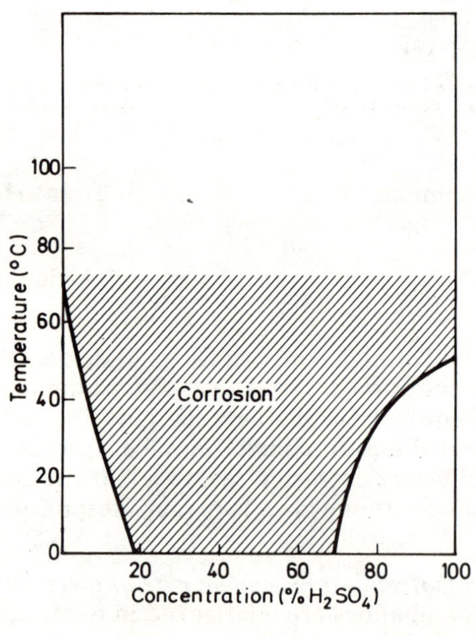

Fig. 2.24 Type 3 behaviour; maximum corrosion rate with change in concentration, e.g. nickel and chromium steel in H_2SO_4. Iso-corrosion lines at $0{\cdot}1\ \mathrm{g\,m^{-2}\,h^{-1}}$

In the region of ascending rate the reaction is clearly governed by a diffusion process and is susceptible to flow-rate, whereas at high concentrations temperature is of greater importance than solution velocity which would have very little effect on the corrosion rate. At the high rates of corrosion (above about 10 $gm^{-2}h^{-1}$) these relationships do not apply between 40 and 60% H_2SO_4 because of the transition between the two types of behaviour. The use of the equations is most effective when some acceptably low rate is chosen, say 0·1 $gm^{-2}h^{-1}$ and the iso-corrosion line is calculated over a range of temperatures and concentrations (see Fig. 2.24).

Type 4. *Decrease in corrosion rate with increase in temperature* An increase in temperature generally increases reaction rates, and the previous examples show that this applies to corrosion rates. However, at certain high rates of corrosion a decrease in rate can occur when the solubility of certain anodic products is exceeded, owing to surface coverage by such films. Because an increase in temperature leads to an increase in corrosion then it should be possible to reduce corrosion when the dissolution is greater than a certain high rate. This type of behaviour has been found and gives an insight into the seemingly uncharacteristic behaviour of some materials in chemical processing plant.

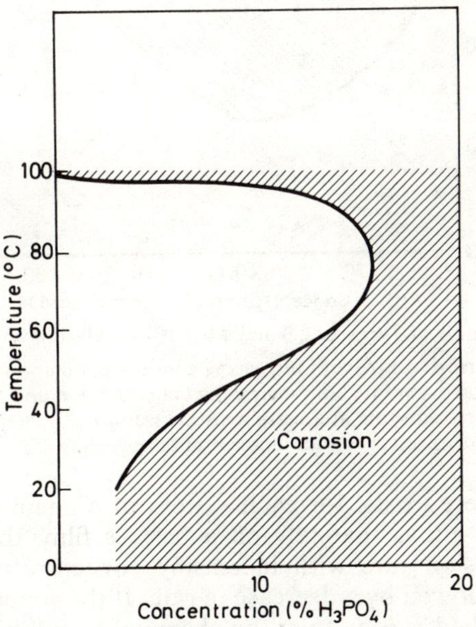

Fig. 2.25 Type 4 behaviour; decreasing corrosion rate with increase in temperature, e.g. Fe–17Cr in H_3PO_4. Iso-corrosion lines at 0·1 $gm^{-2}h^{-1}$

Example 5. Chromium steel (Fe–17Cr) in phosphoric acid at low concentrations shows a decreasing rate with increasing temperature (see Fig. 2.25) presumably due to surface coverage by metal phosphates.

It is apparent that each metal/chemical system should conform with one or more of the four types of behaviour already mentioned, and the behaviour should therefore be capable of prediction by means of one of the following:

(*i*) $\log C_R = K + n\log C - \frac{A}{T}$ (*iii*) $\log C_R = K + n\log C + \frac{A}{T}$

(*ii*) $\log C_R = K - n\log C - \frac{A}{T}$ (*iv*) $\log C_R = K - n\log C + \frac{A}{T}$

Example 6. Hastalloy *B* (Ni–26Mo–4Fe) in formic acid shows all the previous types of behaviour and includes the falling off in rate with increase in temperature (see Fig. 2.26).

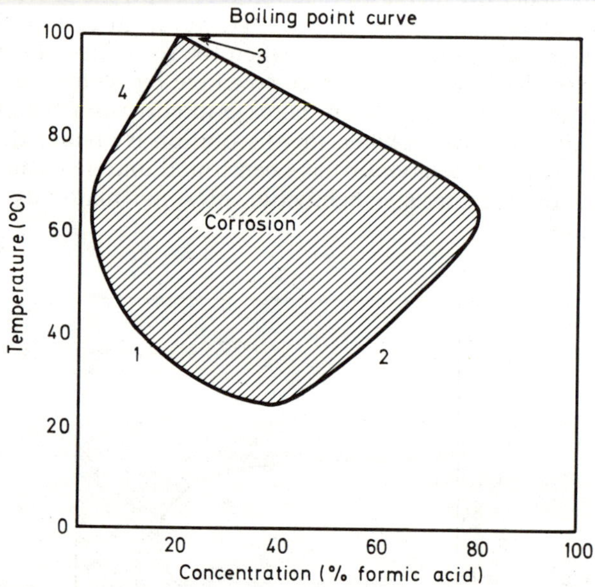

Fig. 2.26 Example of Types 1, 2, 3 and 4 behaviour. Hastalloy *B* in formic acid

1. Increasing corrosion rate with increasing concentration and temperature
2. Decreasing corrosion rate with increasing concentration and temperature
3. Maximum corrosion rate at a certain concentration
4. Decrease in corrosion rate with increase in temperature

In the main there exists, for each system of a chemical in contact with those metals and alloys that rely on a passive film, the possibility of an increase in corrosion rate with increasing concentration but reaching a maximum and followed by a decrease in rate. If the concentration when this maximum is reached is low, then the chemical is 'inhibitive'. The effect of temperature on corrosion is dependent on the position of the maximum concentration. For many chemical/metal systems this maximum may be at a temperature beyond the boiling point of the solution, in which case only part of the behaviour described in Types 1, 2 and 3 is observed.

The boiling point line is an artificial barrier, since constant temperature prevails only at constant pressure and concentration. However, many chemical processes are carried out under pressure and the onset of the behaviour described for Type 4 will manifest itself for many other metals and alloys. Thus serious corrosion at higher temperatures resulting from higher pressures may not necessarily occur.

Non-uniform Corrosion

The arbitrary division of behaviour has been made because of the extreme behaviour of some chemicals that initiate small areas of attack on a well-passivated metal surface. The form of attack may manifest itself as stress-corrosion cracking, crevice attack or pitting. At certain temperatures and pressures, minute quantities of certain chemicals can result in this form of attack. Chloride ions, in particular, are responsible for many of the failures observed, and it can be present as an impurity in a large number of raw materials. This has led to the development of metals and alloys that can withstand pitting and crevice corrosion, but on the whole these are comparatively expensive. It has become important, therefore, to be able to predict the conditions where more conventional materials may be used. The effect of an increase in concentration on pitting corrosion follows a similar relationship to the Freundlich equation where

$$C_R = KC^n,$$

but concentrations below which pitting does not occur are generally very low. The effect of temperature may again be represented by the Arrhenius equation:

$$C_R = B \exp(-A/T)$$

where A and B are constants. The corrosion rate C_R is meaningful when expressed as $1/\gamma$, where γ is the time required for the onset of pitting. Conditions may therefore be chosen when γ is large.

In other sections of this book the anions and other chemicals which enhance pitting, crevice and stress corrosion are discussed in greater detail.

P. J. BODEN

REFERENCES

1. Evans, U. R., *The Corrosion and Oxidation of Metals*, Arnold, London (1960)
2. Rabald, E., *Corrosion Guide*, Elsevier, London, 2nd edn (1968)
3. Berg, F. F., *Corrosion Diagrams*, V.D.I.-Verlag, G.m.b.H., Düsseldorf (1965)
4. Nelson, G. A., *Corrosion Data Survey*, Shell Development Co., Emeryville, Calif. (1960)
5. Examples are: *Wiggin Alloys 100*, Henry Wiggin Co. Ltd., Hereford; *Copper in Chemical Plant*, Copper Development Association, London (1960); *Corrosion Resistance of Stainless Steel*, Publication 112/1, Firth Vickers Stainless Steel Ltd., Sheffield; *Corrosion Resistance of Titanium*, Imperial Metals Industries (Kynoch) Ltd., Birmingham
6. British Standard Code of Practice C.D. 3 003: *Linings of Vessels and Equipment for Chemical Processes*. Part 1: Rubber, Part 2: Glass Enamel, Part 3: Lead, Part 4: P.V.C., Part 5: Epoxy Resins, Part 6: Phenolic Resin, Part 7: Corrosion and Heat Resistant Materials, Part 8: Precious Metals, Part 9: Titanium and Part 10: Brick and Tile
7. Slunder, C. J. and Boyd, W. K., *Zinc: Its Corrosion Resistance*, Zinc Development Association, London (1971)
8. Examples are: *Conference on Materials Selection*, Gen. Ed. Verink, E. D., Jr., Gordon Beach, Met. Soc., **40**, London (1966). See particularly, Gachenbach, J. E., 'Material Selection in Chemical Industry' and Koelbl, H. and Schulze, J., *Selection of Materials for Chemical Industry*, Proceedings 4th International Conference on Metallic Corrosion, Amsterdam, N.A.C.E., New York (1972)
9. Brasher, D. M., Reichenberg, D. and Mercer, A. D., *Brit. Corr. J.*, **3**, 144 (1968)

2.8 Corrosion by Foodstuffs

Foodstuffs, like other chemical substances, are responsible in all phases of processing (including packaging) for corrosive effects of different kinds on constructional materials. These effects are, in addition, influenced by environmental conditions of processing, i.e. by the temperatures involved, by the rates of flow (in particular, erosion is associated with very high rates of flow), and by alternating stresses which may be present in component parts of process machinery (this effect is specifically called *stress* corrosion). The presence or absence of oxygen is generally important in relation to the extent of corrosion produced.

The corrosive effects to be considered (mainly simple corrosion of metals) are, as would be expected from the edible nature of foodstuffs which are not excessively either acidic or basic but which may contain sulphur, less severe than those often encountered with inedible materials containing reactive substances. The importance of corrosive effects where foodstuffs are concerned lies not so much in the action of the foodstuffs on the metal involved as in the resultant metal contamination of the foodstuff itself, which may give rise to off-flavours, in the acceleration of other undesirable changes (by the Maillard reaction[1] for example), and in the possible formation of toxic metallic salts. Metal ions generally have threshold values of content for incipient taste effect in different liquid foodstuffs. Except in the case of the manufacture of fruit juices and pickles, process plant failure through corrosion must be rare. Nevertheless all foodstuffs, particularly liquid ones, should be regarded as potentially corrosive and capable of metal pick-up which may be undesirable.

Construction Materials

The most suitable construction material should usually be selected on the basis of prior experience or direct tests. An exact analysis of the constituents of foodstuffs is not always possible, since they are often of complex composition, but it is usually the organic acids present which are the important corrosive agents. Corrosion rates not in excess of 10–100 $gm^{-2} d^{-1}$ are generally to be sought in food process plant. It is also necessary to bear in mind that in many cases the plant will have to be regularly disinfected by cleaning and sterilising solutions. These should either be non-corrosive or contain inhibiting additives. As always, the choice of construction material is determined in practice in relation to ease of fabrication and cost, which depend upon complexity of construction (cf. jacketed pans with a h.t.s.t. pasteuriser). Electrolytic couples should be avoided. Since bacteria may be pathogenic or, like certain metals, cause off-flavours, the constructional material for foodstuffs plant should be of such a nature that its surfaces can easily be kept clear and free of bacterial lodgment.

In selection of construction materials, the tendency is therefore to play safe; stainless steel is on this account generally specified[2] and is always used in cases of doubt. The suitability of other materials should not, however, be overlooked[3]. Prior to the general advent of stainless steel in 1924, tinned copper was much in vogue. Today titanium, although expensive, represents a possible alternative[4].

The precise effect of different quantities of various trace metals should be considered in relation to the specific type and variety of foodstuff in question. Differing quantities of inhibiting substances may be present in different varieties of the same foodstuff. Thus for example different quantity levels of trace metals are reported as causing off-flavours in various fruit juices, and again the quantity level of trace metals causing rancidity in edible oils depends upon whether the oils are crude or refined. The possible toxicity of these trace metals in different foodstuffs has been investigated, recommendations[4] by the Food Standards Committee of the Ministry of Agriculture, Fisheries and Food are made from time to time, and there are statutory regulations in force in different countries[5] for some foodstuffs. To state the position briefly, aluminium and tin in the quantities normally encountered are considered as non-toxic. Tin, however, in excess is considered undesirable, and may be responsible for gastro-intestinal upsets, so that the accepted limit for canned foods is taken at 250 p.p.m.

Particular groups of foodstuffs which create corrosive environments, and the processing of these foodstuffs, will now be briefly discussed. Indications will also be given of construction materials recommended to meet these situations.

Foodstuffs

Liquid Foodstuffs

As these demand a considerable degree of processing, they are probably the most important group. Within this group, there are certain foodstuffs which are slightly acid in reaction, such as instant coffee extracts and the fruit juices (for example, lemon juice at a pH of 4, and blackcurrant at 2·5). The processing[6,7] of these fruits involves milling, pressing, concentration by evaporation as necessary (temperatures are of the order of 40 to 50°C), pasteurisation (h.t.s.t.), subsequent storage and canning. The common metals, together with aluminium, will be attacked; the degree to which they are susceptible to corrosion is determined by oxygen content and the presence of hydrogen acceptors. Some published information on rates of attack is available[8-10], e.g. boiling lemon juice and tomato juice give figures of 1 400 and 180 $gm^{-2}d^{-1}$ respectively on aluminium. The use of sulphur dioxide for the protection of fruit juices against moulds and bacteria presents an additional corrosion hazard.

The important commercial feature of these juices, especially significant with blackcurrant and tomato juices, is their ascorbic acid (or vitamin C) content, of which loss by oxidation is known to be accelerated both by heat and by metal (particularly copper) contamination. The effect of copper has been carefully investigated for pure ascorbic acid[11], and more recently ascorbic acid in blackcurrant juice and model systems[12]. There are, however, oxidation inhibitors of different kinds (which may themselves be heat-sensitive) present in various fruits, which give differing results. The presence of metals will also affect flavours[13], may cause discoloration, and may give rise to clouding effects, as in apple juice[14].

The first choice of construction material for processing fruit juices is stainless steel (BS En 58A or En 58B—the B type being similar to the A type but providing in addition for resistance to weld decay—or En 58J). Where sulphur dioxide preservation is conducted, high-molybdenum stainless steel (type En 58J) is used. Both mild steel and copper should be avoided. There are, however, other materials available, e.g. tinned copper (provided that the coating is satisfactory, and that it is shown to be otherwise suitable), enamelled metal (less popular than formerly on account of the danger of chipping), and Pyrex glass for piping and linings. Gun-metal and aluminium bronze are useful metals for various parts of pressing equipment. Plastic materials are suitable for storage-tank linings[15], provided that the surfaces can be sterilised. Nickel and Monel have certain applications; Monel for example offers good resistance to erosion. Aluminium, though known to give fair resistance to organic acids, is not generally used for normal-temperature work; certain alloys may be found more suitable than the pure metal.

For canning, conventional tin cans are available[16, 17], but it is generally recommended that the coating be protected by acid-resisting lacquers, and, when sulphur dioxide has been used[18], by an acid- and sulphur-resisting lacquer. The effect of nitrates in water causing detinning is of further importance.

Milk is subjected to the process operations[6] of pasteurisation (h.t.s.t., 71·1–72·2°C for 15 s), evaporation (temperatures of the order of 40–50°C), homogenisation, sterilisation and drying. In addition, milk is processed into other dairy products such as butter, cream and cheese by what are essentially normal-temperature operations. Milk is approximately neutral in reaction, although lactic acid is present; the lactic acid content is increased by natural souring or by the artificial souring necessary for cheese and butter manufacture. This is perhaps the only constituent of milk which is responsible for any metal attack. Protective films of precipitated colloids may be formed by heating. Milk is sensitive in flavour to the presence of such metals as copper and iron, the chief result being a 'castor oil' or 'tallowy' taste which probably arises from fat or lecithin turning rancid. Tin is regarded as not affecting the flavour of milk. Since the flavour of milk is also influenced to a great extent by bacterial action, cleansing and sterilising solutions are regularly used in milk process plant.

Stainless steel is now generally recommended for all phases of processing (type En 58J is not normally necessary except in cases where the process liquid is specially acid)[19]. Aluminium is widely used but suffers from the defect of pitting[20] at higher temperatures. Tinned copper is suitable, where the construction permits its use. Nickel is satisfactory for cold milk and for milk that is being heated up, but is stated to be unsuitable for milk which is cooling down. Glass or enamelled steel is suitable for storage materials and for pipe-lines. Where rapid temperature fluctuations in process operations are involved, however, glass is at a disadvantage. Plastics are used for lining milk-storage vessels and for piping up to temperatures of say 50°C. There are certain reservations; with p.v.c., certain plasticisers used may impart an objectionable taste, and certain brands become opaque. Alkaline cleansing agents (up to strengths of 5% caustic soda solution equivalent) do not affect p.v.c., whereas phenolformaldehyde linings are affected. There are, however, other agents available for such linings.

The edible oils and the margarine emulsions derived from them are large tonnage foodstuffs. High- (i.e. 190°C in deodorisation) and medium-temperature operation, and low-temperature storage are involved. The refining of crude oils is carried out in the presence of caustic soda at 60°C. Refined oils would not be expected to attack the common metals, but in fact edible oils, particularly in the crude state, contain varying amounts of fatty acids, and margarine emulsions contain salt, both of which are capable of attacking metals. Here again, the effect of the metal contaminant on the foodstuff is more important than the corrosive effect of the foodstuff on the metal. It is now common knowledge that copper and iron impurities act as catalysts in the oxidation of oils, to cause slight or marked rancidity. Various oils, however, contain different inhibiting substances. Stainless steel is the prescribed material (En 58A, En 58B; or En 58J for emulsions containing salt). Deodorising plant handling refined oils may be fabricated in mild steel. Glass-lined mild-steel tanks are available for storage.

The manufacture by fermentation and/or distillation of alcoholic beverages containing a wide range of organic materials including acids, whose effect it would be difficult to assess, is traditionally carried out in copper plant, and this has found full consumer acceptance.

Sugar Products

The foodstuff sugar is relatively inert as a chemical, though when it is processsed as an aqueous solution slightly acid conditions may be present and boiling temperatures may be involved. There are no deleterious effects due to trace metals. Mild steel is generally recommended for sugar processing and for the handling of aqueous solutions in allied industries. The processing of fruits into jams and purées, which consists essentially in boiling in open pans, is closely related. The environmental factors here are the natural acidity of the fruits and the possible presence of sulphur dioxide from stored fruit pulps. Traditionally, copper plant, clad on cast iron, uncoated, tinned or even silvered, is used, but stainless steel is now widely adopted.

In the manufacture of sugar confectionery, including chocolate, the main ingredients are sugar and glucose, milk (including condensed milk), and cocoa fat, and the essential operations those of boiling and compounding. Cocoa fat, like other edible fats, is liable to oxidative rancidity. The modern choice of constructional material is again stainless steel.

Vegetable Processing

In pickle and sauce manufacture, vinegar (an approximately 5% solution of acetic acid), and salt at temperatures up to boiling point are the important corrosive agents. Sodium chloride, particularly in the presence of oxygen and acids, is known to cause rapid attack. Metal contamination (particularly iron) is likely to cause darkening by reaction with tannin present, which is either leached from wooden vessels or derived from certain spices. In pickle and sauce manufacture, stainless steel (En 58J) is recommended; it is interesting to note, however, that it is claimed that the flavour of piccalilli liquor is more readily brought out in the presence of iron[21]. This belief finds expression in the practice of using cast iron process vessels.

Meat and Fish Products

Despite attention to hygiene and cleanliness, corrosion of base metals by meat juices and deterioration of meat and fish owing to metal contamination is liable to occur. Stainless steel is recommended in soup and paste manufacture, and aluminium has a certain application. Mild steel is however used in the corned beef industry, for meat pre-cooking.

Other Food Products

Many foodstuffs are in the form of solids or processed powders, and do not offer serious corrosion problems, though mild steel equipment in infrequent use, or after washing down, can develop slightly rusted surfaces. This material is usually undesirable if it finds its way into food products. Scouring batches of dry foodstuffs is one solution to the problem if stainless steel is not used or affordable. Hygiene and cleanliness are, however, dominant factors.

Meat and fish are very liable to bacterial putrefaction; in this connection an interesting innovation is the increasing use of easily cleaned aluminium fish boxes. It is also possible that copper should be avoided in contact with herrings, which have a high fat content.

Mention should also be made of glutamic acid and invert sugar which are used in foodstuffs and demand the use of hydrochloric acid-resistant material in manufacture, and of the essential flavouring oils which should preferably be stored and prepared in stainless steel and aluminium equipment.

R. J. CLARKE

REFERENCES

1. Reynolds, T. M., 'The Chemistry of Non-enzymic Browning', *Advances in Food Science*, **14**, 168 (1965)
2. Gilroy, P. E., *Food*, **21**, 255 (1952)
3. Various authors, *Ann. Falsif.*, **213** (1950)
4. O'Keefe, J., *Bell's Food and Drugs*, Butterworths, London, 14th edn. (1968)
5. Reith, J. F., *Ann. Bromatologia*, **8**, 145 (1956)
6. Clarke, R. J., *Process Engineering in the Food Industries*, Butterworths, London (1957)

7. Tressler, D. K. and Joslyn, M. A., *Fruit and Vegetable Juice Processing Technology*, Avi. Westport (1968)
8. Blount, A. L. and Bailey, H. S., *Trans. Amer. Inst. Chem. Eng.*, **18**, 139 (1926)
9. Gilroy, P. E. and Champion, F. A., *J. Soc. Chem. Ind.*, **67**, 407 (1948)
10. McKay, R. J. and Worthington, R., *Corrosion Resistance of Metals and Alloys*, Reinhold, New York (1936)
11. Joslyn, M. A. and Miller, J., *Food Res.*, **14**, 325 (1949)
12. Timberlake, C. F., *J.S.F.A.*, **8**, 159 (1957) and **2**, 268 (1960)
13. Schrader, J. H. and Johnson, A. H., *Industr. Eng. Chem. (Industr.)*, **26**, 179 (1934)
14. Kieser, M. E. and Timberlake, C. F., *J.S.F.A.*, **8**, 151 (1957)
15. Docherty, A. C. and Hughes, H., *Chem. Ind.*, 1 171 (1959)
16. Hartwell, R. R., *Advances in Food Research*, **3**, 327 (1950)
17. Alderson, M. G., *Food Manuf.*, **67** (1970)
18. Board, P. W., Holland, R. V. and Elborne, R. G., *Journal Sci. Food. Agric.*, **18**, 232 (1967)
19. Bottom, G. H., *J. Soc. Dairy Tech.*, **6**, 179 (1953)
20. Aziz, P. H. and Goddard, H. P., *Industr. Eng. Chem. (Industr.)*, **44**, 1 791 (1952)
21. Smith, E., *Corrosion Resisting Steels—Application in the Pickle and Sauces Industry*, B.F.M.I.R.A., Scientific and Technical Survey, December (1950)

2.9 Mechanisms of Liquid-metal Corrosion*

The corrosion of metals and alloys by liquid metals generally follows the pattern of the formation of metallic alloys, i.e. solution and intermetallic compound formation and the corrosion process is often one of simple dissolution in the liquid metal. In some special cases electron-transfer processes—involving reducible impurities in the liquid metal—may modify or even override the simple dissolution process. This is especially true when, as with liquid alkali metals, the solubility of structural metals is very low. Very often under isothermal conditions equilibrium between an alloy and liquid can be approached. Continued corrosive attack is then possible only if the equilibrium is disturbed by removing in some way the dissolved corrosion product from the system. Thus the nature of liquid-metal corrosion varies depending on whether the fluid is static or is moving relative to its container, on whether the temperature is constant or varying throughout the system, and on whether the container is a single metal or a composite of two or more metals. All the processes, however, involve solution as a first step.

It is curious and indeed lamentable that no adequate theory is yet available to explain the variation in the solubilities of metals in molten metals. Both Kerridge[1] and Strachan and Harris[2] noted that plotting the solubilities of metals in a number of solvent metals showed a periodic variation with the *solute* and not the solvent (see Fig. 2.27), i.e. a given metal such as manganese showed a consistently high solubility in molten magnesium, tin, bismuth and copper, compared with iron or chromium. Kerridge correlated this variation with the solute lattice energy and hence with the latent heat of fusion. This correlation is useful in making qualitative predictions.

In the more practical sense solution may be uniform or localised.

Preferential solution can take two forms:

1. Leaching—one component of an alloy is preferentially dissolved, an example being nickel which is leached from stainless steels by molten lithium or bismuth, sometimes to such an extent that voids are left in the steel.
2. Intergranular attack—the liquid penetrates along the grain boundaries, owing either to the accumulation of soluble impurities in the boundaries or to the development at the junction of a grain boundary with the metal surface of a low dihedral angle to satisfy surface-energy relationships.

*Testing procedures for liquid-metal corrosion are given in Chapter 20.

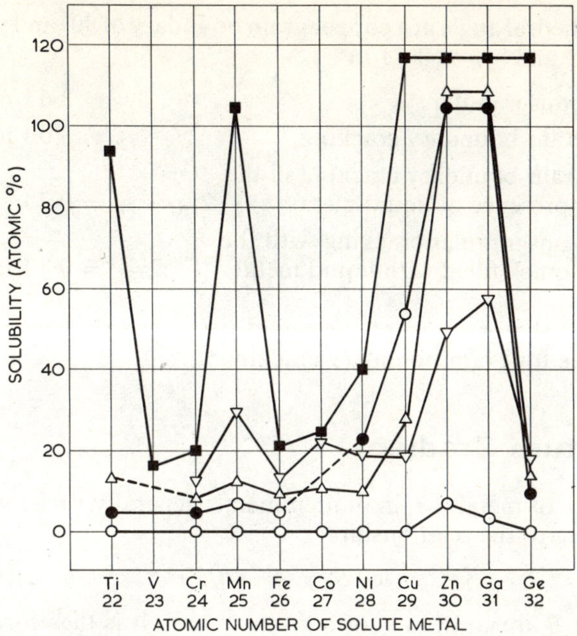

Fig. 2.27 Solubilities of the first row of transition elements in five liquid metals. (To avoid overlap of points, the graphs have been set at different positions on the solubility axis)

	Solvent metal	*Temperature* (°C)	*For true solubility, deduct from axis reading:*
○	Hg	20	—
●	Mg	700	4%
△	Sn	450	8%
▽	Bi	400	12%
■	Cu	1 200	16%

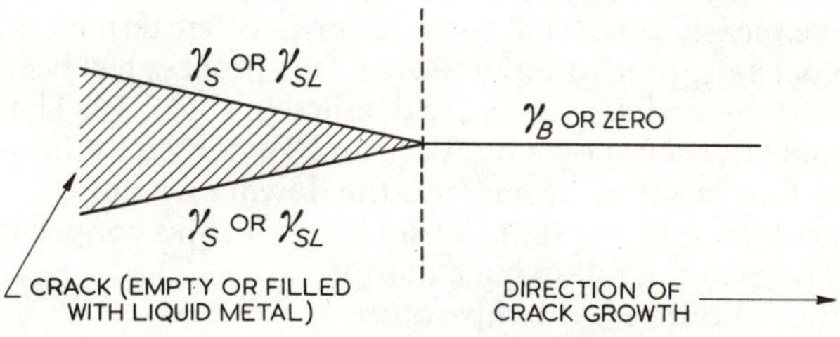

Fig. 2.28 Energies involved in the growth of a crack

When this process is accompanied by stress, catastrophic failure can occur, a classical example being the action of mercury on brass.

The situation may be described in terms of the surface-energy changes when a crack propagates through a solid metal as shown in Fig. 2.28 where γ_S is the solid–gas interfacial energy, γ_{SL} is the solid–liquid interfacial energy and γ_B is the grain-boundary energy.

Tabulated below are the energy changes involved for different cracking modes, with numerical values for the case of copper in contact with liquid lead with a dihedral angle at a copper grain boundary of 90° and $\gamma_S = 1{\cdot}8\ \mathrm{J/m^2}$, $\gamma_B = 0{\cdot}6\ \mathrm{J/m^2}$ and $\gamma_{SL} = 0{\cdot}4\ \mathrm{J/m^2}$:

Transgranular:	$2\gamma_S = 3{\cdot}6\ \mathrm{J/m^2}$
Grain-boundary cracking:	$2\gamma_S - \gamma_B = 3{\cdot}0\ \mathrm{J/m^2}$
Grain-boundary cracking in the presence of liquid metal:	$2\gamma_{SL} - \gamma_B = 0{\cdot}2\ \mathrm{J/m^2}$
Transgranular cracking with the crack filled with liquid metal:	$2\gamma_{SL} = 0{\cdot}8\ \mathrm{J/m^2}$

It is seen that the presence of the liquid metal greatly lowers the surface-energy change for grain-boundary cracking[3].

Temperature Gradient

The solubility of metals, S, in molten metals generally varies with temperature according to the relationship

$$\log S = A - (B/T)$$

where A and B are constants for a given system. It is therefore possible for more material to dissolve from a container at its highest-temperature end than at the low-temperature end, and if the melt flows round the container by natural or forced convection the liquid arriving at the cold region will be supersaturated and will precipitate solute until equilibrium is attained. If it is then recycled to the hot end it dissolves more metal until saturated and then returns to the cold end to precipitate this excess. This process is termed 'thermal gradient mass transfer'. It can best be illustrated by circulating a corrosive metal such as bismuth round a thermal convection loop of the type shown in Fig. 2.29. After a prolonged period of operation, thinning of the inner wall of the hot section and a precipitate on the wall of the cold section can be clearly observed. (The latter is often termed a *plug* since it eventually blocks the pipe to liquid flow.) This process has been analysed in some detail and the various stages are detailed in Fig. 2.30. The overall rate-controlling step has been shown by Ward and Taylor[4] to be diffusion through the boundary film of solute atoms into the flowing stream.

To be precise, they found that the solution of solid copper in liquid lead and bismuth obeyed the following equations.

Under static conditions, at temperature T

$$n_t = n_0[1 - \exp(-KS/V)]$$

where n_t = concentration of solute after time t, n_0 = saturation concentration of solute, S = surface area of solid exposed to liquid of volume V, $K = K_0 \exp[-\Delta E^{+}/RT]$ (ΔE^{+} = activation energy for solution).

Under flowing conditions

$$\frac{\mathrm{d}n_t}{\mathrm{d}t} = K(S/V)(n_0 - n_t)$$

If, therefore, the solute atoms can be prevented from entering the boundary

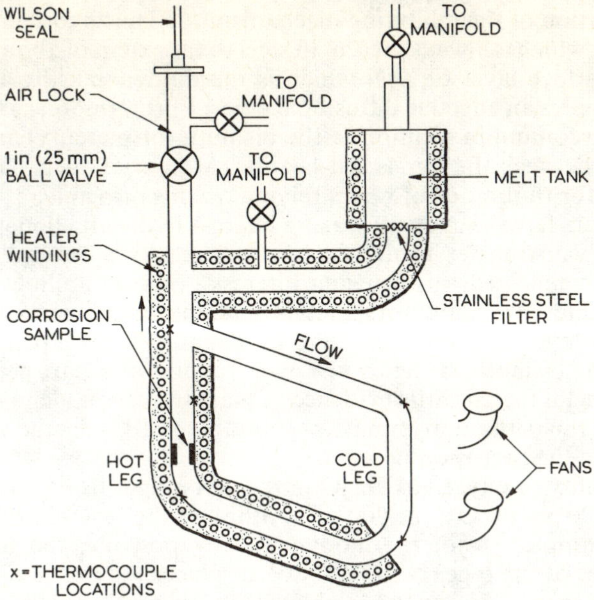

Fig. 2.29 Thermal convection loop

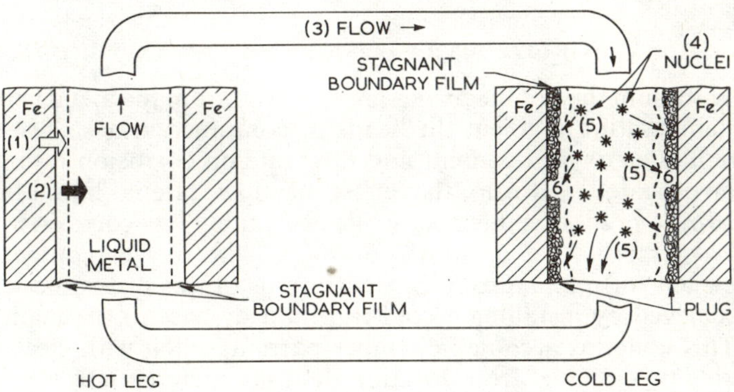

Fig. 2.30 Stages in thermal gradient mass transfer

1. Solution
2. Diffusion
3. Transport of dissolved metal
4. Nucleation
5. Transport of crystallites
6. Crystal growth and sintering (plug formation)

(after Brookhaven National Laboratory)

film from the solid the process will be halted. A method for doing this was discovered by workers at the U.S. General Electric Company some time ago when developing a mercury boiler for electric power generation[5]. They found that small quantities of dissolved titanium, zirconium, chromium, nickel and aluminium were effective as inhibitors of the corrosion of steels by hot mercury, the first two being particularly so. Later interest in the use of liquid bismuth as a carrier of uranium in a liquid-metal-fuelled reactor led to the extension of the use of zirconium inhibitor to bismuth in steel circuits and to an elucidation of the inhibiting mechanism[6, 7]. The zirconium reacts with the nitrogen, which is always present in steel to the extent of about 100 p.p.m., to form a surface layer of ZrN which is thermodynamically a very stable compound and is an effective diffusion barrier. Furthermore, as long as there is residual zirconium in solution in the bismuth (or mercury) and dissolved nitrogen in the steel, the film is self-healing. Mercury boilers have operated successfully for thousands of hours relying on this principle.

Recent years have seen an increasing interest in the alkali metals, notably sodium, as heat-transfer media for nuclear reactors. Several major conferences have reflected this growing interest*. The solubilities of possible structural materials—refractory metals and the components of iron and nickel base alloys—are much less than in the more noble liquid metals such as mercury or bismuth, being as low as a fraction of a part per million. In pure alkali metal there is little evidence of mass transfer under a temperature gradient. In this situation impurities—notably oxide—in the liquid metal can dominate the corrosion process.

The solubility of oxygen as Na_2O in sodium ranges from 3 p.p.m. O^{2-} at 150°C to 1000 p.p.m. O^{2-} at 500°C[9]. Although the reaction of Na_2O with iron, for example, is not thermodynamically possible, the formation of double oxides of the type $(Na_2O)_2FeO$[10] displaces the reaction in favour of attack of the iron. Addison[11] has shown that many transition metals form double oxides with Na_2O, sometimes accompanied by a change in valency state, e.g.

$$Nb_2O_5 + Na_2O \rightarrow Na_2O{\cdot}Nb_2O_3 + O_2$$

In flowing sodium the double oxides are removed from the metal surface and under a temperature gradient this leads to continued attack. The effect of such variables as oxygen content and flow rate on corrosion rate has been widely investigated[12]. From the effect of flow rate a liquid-boundary controlled-rate process is inferred, while acceptably low corrosion rates are achieved at temperatures up to 650 or 700°C with oxygen contents of less than a few tens of p.p.m. Oxide concentrations of this order and below are readily achieved by installing a cold-trap in a by-pass to the main sodium circuit. This contains a cooled chamber partially filled with steel wool on which oxide precipitated from solution deposits preferentially. Periodically the trap is isolated and the oxide removed.

Chemical Gradient

In a container or a 'loop' in which a molten metal is flowing but where there is not necessarily a temperature gradient, a type of corrosion can occur by transfer of material along an activity gradient ('activity' being used here in the thermodynamic sense—see Section 9.1). Again it is convenient to discuss this phenomenon under two headings:

Metal Transfer

This is often referred to as 'isothermal mass transfer' since it does not require

*See Bibliography p. 2:109.

a thermal gradient. When two different metals are separated by a third, liquid, metal and one or both solid metals dissolve in the liquid, an alloying action may be observed. The liquid metal itself may form an intermetallic compound on the surface of one or both of the solid metals, in which case a diffusion barrier will build up and the reaction will gradually slow down. If, however, this does not occur, an alloy of the two solid metals will be formed either in the melt or on the surface of one of the solids. The most common effect is the formation of a layer at the surface of the less soluble of the two solid metals, this layer being made up of all or some of the compounds which are formed between the two solids. The rate-controlling step is the diffusion of the less soluble metal through this surface layer to react with the other species of solute atom. Some good examples have been provided by the studies by Covington and Woolf[8], who have investigated, among other systems, aluminium and molybdenum in molten lead, tin and bismuth. In bismuth, three of the four intermetallic compounds were identified in the layer on the molybdenum in the following order from the molybdenum surface: Al_3Mo, Al_5Mo and $Al_{12}Mo$. In tin only Al_3Mo was observed, while in lead two compounds were formed, the controlling process probably being the rate of diffusion through the layers of the different compounds and the rate of solution of the more soluble metal (aluminium) in the liquid metal.

Non-metal Transfer

The elements of primary importance in this context are oxygen, nitrogen, carbon and hydrogen. In the technology of the liquid alkali metals they play a predominant rôle. Their origin is not, in general, the massive element but rather leakages in the circuit or impurities remaining after construction. It is convenient to discuss these four elements separately.

Oxygen Even in cold-trapped sodium containing only a few p.p.m. of oxide, refractory metals such as Nb, V, Ta and Zr suffer severe oxidation. Their behaviour is much the same as in oxidising gases at high temperature, forming non-adherent oxide films. Oxygen diffuses into the metal to produce a substitutional solid solution with consequent embrittlement. To prevent this occurring the oxygen level in the sodium is brought down below the p.p.m. level in a 'hot-trap', where in a by-pass circuit oxygen is gettered by sacrificial foils or turnings of zirconium heated to 600°C or above.

Nitrogen Nitrogen is virtually insoluble in pure sodium (10^{-4} p.p.m. at 500°C and 1 MN/m^2) and it can react with circuit materials only when physically entrained as bubbles. It is, however, quite soluble in molten lithium (2 000 p.p.m. at 300°C and 100 kN/m^2) and considerably accelerates the already corrosive action of lithium, particularly on steels. Zirconium and yttrium getters have been used with some success.

Carbon The solubility of carbon in sodium has been measured; it is considered lower than the corresponding value for oxygen (2 p.p.m. of carbon at 500°C) but is sufficiently high to give rise to undesirable effects. Carburisation of refractory metals and of austenitic stainless steels has been observed in sodium contaminated with carbon, the source of which can be either decomposed organic material, e.g. diffusion-pump oil, or a ferritic steel of low or zero alloy content. The latter is an example of chemical-gradient transfer since the activity of carbon in Fe–18 Cr, possibly stabilised with titanium or niobium, is clearly lower than that in a plain carbon steel and there is, therefore, a driving force for transfer. A deterioration in properties of both steels is the consequence of using austenitic and ferritic steels in one system, the austenitic being embrittled and the ferritic softened. The effect can be minimised if the carbon activity in the ferritic steel is reduced to that in the stainless steel by the incorporation in the steel of a 'carbon stabiliser' such as titanium. Hot trapping with zirconium removes carbon as well as residual oxygen, but generally carbon sources should be kept from a sodium circuit containing readily carburised metals.

Hydrogen Sodium and the other alkali metals react with moisture to produce the oxide or hydroxide plus hydrogen. The latter dissolves to some extent and may transfer to refractory metals which dissolve hydrogen, e.g. zirconium, thereby causing embrittlement. It is probable that this will be serious only in systems running at fairly low temperatures, since hydrides are generally unstable at high temperatures when hydrogen diffuses very rapidly.

We can summarise this discussion of non-metal transfer by saying that chemical thermodynamics is a useful guide to probable behaviour. The transfer of a non-metal, X, dissolved in a molten metal, M′, to another metal, M″, will depend on the relative free energies of formation of M′X and M″X (see Section 7.6). Thus sodium will give up oxygen to Zr, Nb, Ti and U, as the free energy of oxide formation of these metals is greater than that for sodium; on the other hand, sodium will remove oxygen from Fe, Mo and Cu. Alternatively, double oxides may be formed.

Conclusion

Liquid-metal corrosion may be summarised by the general statement that two factors are predominant—solution and alloying behaviour—but that non-metallic contaminants may produce oxidation or carburisation effects similar to those observed in gaseous corrosion.

G. LONG

REFERENCES

1. Kerridge, D. H., *Reactor Technol.*, **1**, 215 (1961)
2. Strachan, J. F. and Harris, N. L., *J. Inst. Met.*, **85**, 17 (1956–7)
3. Rostoker, W., McCaughey, J. M. and Markus, H., *Embrittlement by Liquid Metals*, Reinhold, New York (1960)
4. Ward, A. G. and Taylor, J. W., *J. Inst. Met.*, **85**, 145 (1956–7) and **86**, 36 (1957–8)
5. Reid, R. C., Amer. Soc. Mech. Engrs., Tech. Paper No. 51–S–13 (1951)
6. Frost, B. R. T. *et al.*, *Proceedings of the Second Conference on the Peaceful Uses of Atomic Energy*, United Nations, Geneva, **7**, Paper P.270 (1958)
7. Klamut, C. J. *et al.*, ibid., paper P.2406 (1958)
8. Covington, A. K. and Woolf, A. A., *Reactor Technol.*, **1**, 35 (1959) and **3** (1961)

9. Eichelberger, R. L., *Atomics International/U.S.A.E.C. Report*, No. AI–AEC–12685 (1968)
10. Horsley, G. W., *J. Iron St. Inst.*, **182**, 43 (1956)
11. Addison, C. C. *et al.*, *International Symposium on the Alkali Metals*, Nottingham, 1966, Spec. Pub. No. 22, The Chemical Society, London, 460 (1967)
12. Thorley, A. W. and Tyzack, G., *Symposium on Alkali Metal Coolants*, Vienna, 1966, International Atomic Energy Agency, Vienna, 97 (1967)

BIBLIOGRAPHY

Various authors, *International Symposium on the Alkali Metals, Nottingham, 1966*. Special Publication No. 22, The Chemical Society, London (1967)
Various authors, *Symposium on Alkali Metal Coolants, Vienna, 1966*. International Atomic Energy Agency, Vienna (1967)
Various authors, *International Conference on Sodium Technology and Large Fast Reactor Design*, Argonne, Ill., U.S.A. ANL 7520, Pts. 1 and 2, N.B.S., U.S. Dept. of Commerce, Springfield, Va. (1968)
Various authors, *International Conference on Liquid Alkali Metals*, British Nuclear Energy Soc., Nottingham Univ., April (1973)

2.10 Corrosion in Fused Salts

Interest in the use of fused salts in industrial processes is continually increasing and these media are gradually becoming accepted as a normal field of chemical engineering. The change is being accelerated by the increasing demand for the newer refractory metals—often produced by processes involving fused salts—and also by the novel chemical engineering techniques which have been developed in the nuclear-energy industry. For example, a nuclear reactor using molten fluorides as a fluid fuel has operated, and this has involved the use of pumps, heat exchangers and similar equipment to circulate the high-temperature melt[1].

In certain applications it has not always been easy to find suitable metallic container materials, particularly in the nuclear-energy industry, where, for certain applications, corrosion resistance of the same order as that required by the fine chemical industry has to be achieved in order to prevent contamination of the process stream. Such difficulties have stimulated the study of corrosion in fused salts and have led to a fairly high degree of understanding of corrosion reactions in these media.

The subject is also closely related to fuel-ash corrosion which in most cases is caused by a layer of fused salts such as sulphates and chlorides[2, 3]. Attention has been focused on the electrochemistry of this type of corrosion[4, 5] and the relevant thermodynamic data summarised in the form of diagrams[6–8]. Fluxing and descaling reactions also resemble in some respects reactions occurring during the corrosion of metals in fused salts. A review of some of the more basic concepts underlying corrosion by fused salts (such as acid-base concepts and corrosion diagrams) has appeared[8].

Scope of the Present Treatment

There are two cases in which a metal can be attacked by a salt melt: if it is soluble in the melt, or if it is oxidised to metal ions. In the first case, attack occurs by direct dissolution without oxidation of the metal and the mechanism is likely to be closely similar to attack by liquid metals (Section 2.9). If the solubility is appreciable, excessive corrosion can be expected, but with few exceptions metals appear to be appreciably soluble only in their own salts. Most of the metals of the first and second groups of the periodic table are soluble in their own halides, and in certain cases there is complete miscibility at high temperatures[9–11].

The present survey will be confined to corrosion arising as a result of oxidation of the metal to ions, since little information on corrosion involving only metal-solubility effects is available.

General Principles

Electrochemically, the system metal/molten salt is somewhat similar to the system metal/aqueous solution, although there are important differences, arising largely from differences in temperature and in electrical conductivity. Most fused salts are predominantly ionic, but contain a proportion of molecular constituents, while pure water is predominantly molecular, containing very low activities of hydrogen and hydroxyl ions. Since the aqueous system has been extensively studied, it may be instructive to point out some analogues in fused-salt systems.

Setting aside for the moment effects due to passivity, a metal can be oxidised (corroded) only if it is in contact with something it can reduce; in an aqueous electrolyte, either hydrogen ions are displaced to form hydrogen gas, or some other solute (often dissolved oxygen) is reduced; in a salt melt, either one of the metal cations (corresponding to hydrogen ions in the aqueous solution) is displaced to produce metal, or some molecular species (such as dissolved chlorine or oxygen) or an ion (such as nitrate or ferric ion) is reduced. In aqueous electrolytes, therefore, metals which are too noble to displace hydrogen corrode only if dissolved oxygen or some other reducible substance is present, and when the reducible material has been used up, corrosion ceases; the metal has become 'immune', i.e. has reached electrochemical equilibrium with its environment. The position is similar in fused salt corrosion, and the relative nobilities of the salt melt and the metal are important. A noble metal in contact with a pure melt of a base-metal cation can react only to a very limited extent provided the anion is not reducible (but see References 4–8); for example, nickel cannot displace sodium from a sodium chloride melt to any appreciable extent and hence nickel will not corrode in molten sodium chloride unless some other reducible impurity is present.

The displacement reaction between a metal, M, and a salt melt, AB, can be written:

$$\underset{\text{metal}}{\mathrm{M}} + \underset{\substack{\text{salt}\\\text{melt}}}{\mathrm{AB}} \rightleftharpoons \underset{\substack{\text{metal}\\\text{corrosion}\\\text{product}}}{\mathrm{MB}} + \underset{\substack{\text{displaced}\\\text{metal}}}{\mathrm{A}} \qquad \ldots(2.16)$$

for which there is an equilibrium constant*

$$K = \frac{a_{\mathrm{MB}}a_{\mathrm{A}}}{a_{\mathrm{M}}a_{\mathrm{AB}}} \qquad \ldots(2.17)$$

The metal M and the salt AB, being pure substances, have an activity of one, while the corrosion product MB and the displaced metal A will in most cases

*See Sections 1.4 and 9.1.

be in solution in the melt at an activity of less than one. Where the metal M is noble and the cation A of the salt is base, the equilibrium constant (2.17) has a very small value, so that the activities of displaced metal A and corrosion product MB remain small. Provided there are no other factors present which might remove or combine with MB or A, the system will reach equilibrium after a small amount of reaction (corrosion) has occurred and the system will then be 'immune' against further corrosion. Distillation of the displaced metal A out of the melt is one of the factors which can prevent equilibrium being reached, and under such conditions corrosion can continue indefinitely. Such factors have been considered in some detail for chloride melts[12].

It is usual to choose a container metal for fused salts sufficiently noble for the displacement reaction (2.16) to be negligible, and the most important aspects of corrosion are, as in aqueous solutions, those which involve reducible impurities, although in a salt melt there is also the additional possibility of a reducible anion (see above). All such factors can be described as controlling the 'oxidising power' of the melt, which can be defined in terms of a 'redox potential' just as in aqueous solutions[13]. The redox potential is expressed by relationships of the form

$$E = E^{\ominus} + \frac{RT}{zF} \ln \frac{a_{ox}}{a_{red}} \qquad \ldots(2.18)$$

For example, the redox potential of a system containing oxide ions and oxygen* could be expressed as:

$$E = E^{\ominus}_{O_2/O^{2-}} + \frac{RT}{2F} \ln \frac{a^{\frac{1}{2}}_{O_2}}{a_{O^{2-}}} \qquad \ldots(2.19)$$

Similar equations can be formulated for other equilibria, e.g. NO_3^-/NO_2^-, Na^+/Na, etc.

The redox potential of a melt is a measure of its aggressiveness towards metals, and a metal in contact with the melt will react with it until its potential becomes the same as the redox potential of the melt[14]. The potential of the metal will depend on the activity of its ions in solution in the melt

$$E = E^{\circ}_{M} + \frac{RT}{zF} \ln a_{M^{z+}} \qquad \ldots(2.20)$$

so that the higher the redox potential (oxidising power) of the melt, the more will the metal have to oxidise (corrode)—giving M^{z+} ions—before equilibrium can be reached. (Oxidation of the metal will also result in reduction of the oxidising species present in the melt, which under certain conditions can cause the redox potential of the melt to fall during the course of the reaction, in which case equilibrium will occur at a potential rather lower than the initial redox potential of the melt[14].)

*Peroxide and superoxide ions have recently been identified in molten systems. These ions are formed by the oxidation of oxide ions by oxygen or oxy-anions such as nitrate (see Reference 11).

Conditions Under Which Corrosion Occurs in Practice

It follows from what has been said that the conditions under which metals are exposed to fused salts will in practice be of three types:

1. Conditions under which equilibrium between metal and salt melt can be reached and maintained.
2. Conditions under which equilibrium cannot be reached.
3. Conditions under which passivity is possible.

Item 3 has not been considered yet, so its importance in melts will be discussed first.

In metal/melt systems, just as in aqueous systems, one possible way of ensuring adequate corrosion resistance is to choose conditions such that the metal is passive, which requires that it should become covered with an adherent, compact, insoluble film or deposit, preventing direct contact of the metal with its environment. Any melt which reacts with a metal to give a corrosion product insoluble in the melt is in principle capable of passivating the metal, e.g. passivity can be expected to occur in oxidising salts in which metal oxides are sparingly soluble. Thus, iron is highly resistant to alkali nitrate melts because it becomes passive, and passivity has also been observed by electrode potential measurements of an iron electrode in chloride melts containing nitrates[14], although in this case the oxide corrosion product is not particularly protective. In general, fused salts are 'good' solvents for inorganic compounds so that passivity is not likely to be a widely encountered phenomenon.

In cases where passivity is impossible, corrosion can be prevented if the metal can reach equilibrium with the melt (case 1). The system usually undergoes some corrosion initially, when traces of oxidising impurities are reduced and the redox potential of the melt falls (Fig. 2.31). Finally, after a certain amount of corrosion has occurred, the metal becomes immune and corrosion ceases. In Fig. 2.31 complete equilibrium between metal and melt was still not quite reached even after several hundred hours exposure.

The equilibrium of metals with molten sodium hydroxide[15] and with fused alkali chlorides[14] has been studied in detail. Williams, Grand and Miller[15] studied the reaction of molten sodium hydroxide with nickel,

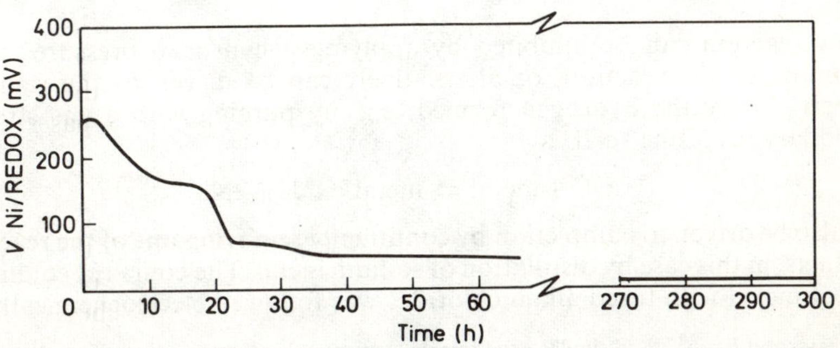

Fig. 2.31 Potential difference V between a redox electrode and a nickel electrode immersed in an alkali chloride melt; 700°C, argon atmosphere[14]

copper, Monel, gold, chromium, titanium, tantalum, beryllium, manganese, silver, cobalt, iron and ferrous alloys. The primary reaction can be represented as:

$$NaOH + metal \rightleftharpoons Na_2O + metal\ oxide + H_2$$

and for each metal at each temperature, equilibrium was reached at some characteristic and reproducible pressure. Fig. 2.32 shows the rate of approach to equilibrium for nickel at 800, 900 and 1 000°C*.

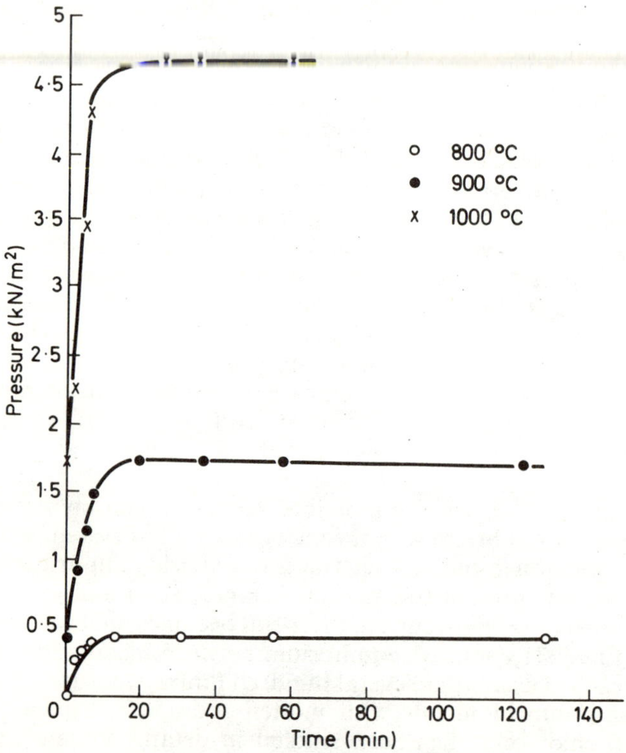

Fig. 2.32 Rate of pressure development over NaOH/Ni system[15]

The reaction can be inhibited by applying a hydrogen pressure at the beginning of the reaction, or alternatively can be driven to the right by sweeping away the hydrogen formed, e.g. by purging with a gas stream. Secondary reactions such as

$$Na_2O + metal \rightleftharpoons metal\ oxide + 2Na$$

can also be driven to completion by continually removing one of the reaction products, in this case by distillation of sodium metal. The course of equilibration of nickel with fused alkali chlorides was followed electrochemically by

*Mathews and Kruh[15(a)], in similar experiments, were unable to obtain steady-state pressures, but this was probably due to the larger size of vessel used, to temperature gradients, and to difficulties due to 'creeping' of the fused sodium hydroxide up the vessel walls.

Littlewood and Argent[14], who showed the importance of geometric factors. The time taken for the metal to come to equilibrium with the melt depends on the amount of oxidising impurities present and on the ratio 'surface area of metal/volume of melt', being longer for a small surface area of metal in contact with a large volume of melt. Littlewood and Argent suggested that the kinetics were determined largely by diffusion processes in the melt. Equilibration was hindered when the chloride melts were in contact with glass containers which are not inert at high temperature.

Under certain conditions, it will be impossible for the metal and the melt to come to equilibrium and continuous corrosion will occur (case 2); this is often the case when metals are in contact with molten salts in practice. There are two main possibilities: first, the redox potential of the melt may be prevented from falling, either because it is in contact with an external oxidising environment (such as an air atmosphere) or because the conditions cause the products of its reduction to be continually removed (e.g. distillation of metallic sodium and condensation on to a colder part of the system); second, the electrode potential of the metal may be prevented from rising (for instance, if the corrosion product of the metal is volatile). In addition, equilibrium may not be possible when there is a temperature gradient in the system or when alloys are involved, but these cases will be considered in detail later. Rates of corrosion under conditions where equilibrium cannot be reached are controlled by diffusion and interphase mass transfer of oxidising species and/or corrosion products; geometry of the system will be a determining factor.

Special Features of Fused-salt Corrosion

'Electrochemical' corrosion of metals Since the aggressiveness of salt melts is governed by redox equilibria, and is often controlled by composition of the external atmosphere[16], effects analogous to 'electrochemical' or 'oxygen-concentration' corrosion in aqueous systems can occur in salt melts. Tomashov and Tugarinov[17] determined cathodic polarisation curves in fused chlorides and concluded that the cathodic reactions of impurities could be represented as:

$$O_2+4e \rightarrow 2O^{2-}$$
$$CO_2 \rightleftharpoons CO+O;\ O+2e \rightarrow O^{2-}$$
$$H_2O \rightleftharpoons H^+ +OH^-;\ H^+ +e \rightarrow \tfrac{1}{2}H_2$$
$$SO_4^{2-}+8e \rightarrow S^{2-}+4O^{2-}$$
$$2NaNO_3+6e \rightarrow Na_2O+2NO+3O^{2-}$$
$$2NaNO_2+2e \rightarrow Na_2O+2NO+O^{2-}$$

The anodic reaction was:

$$M-e \rightarrow M^+$$

Since metals are electronic conductors, the anodic and cathodic reactions will not necessarily occur at the same site, and 'anodic' and 'cathodic' areas can develop as in aqueous solutions. For example, 'wash-line' attack is often a feature of corrosion by fused salts in contact with air.

Thermal mass transfer (see also Section 2.9) When a temperature gradient exists in a system containing metal in contact with molten salt, thermal potentials are set up, causing removal of metal at high-temperature points and deposition of metal at cooler places[18, 19]. This mass transfer is essentially different in nature from that met in liquid-metal corrosion, which is simply a temperature–solubility effect. In fused salts, both the corrosion and deposition reactions are electrolytic, and it has been shown that an electrical path is necessary between the hot and cold regions of the metal. Edeleanu and Gibson suggest that this type of mass transfer be called '*Faradaic mass transfer*' to indicate that it requires an electrolytic current[19a].

Mass-transfer deposits can lead to blockages in non-isothermal circulating systems, as in the case of liquid-metal corrosion. In fused salts, the effect can be reduced by keeping contamination of the melt by metal ions to a minimum; e.g. by eliminating oxidising impurities or by maintaining reducing conditions over the melt[19].

Selective attack Corrosion of alloys at high temperatures is complicated by effects due to diffusion, particularly where the alloy components have different affinities for the environment, and corrosion of an alloy in a fused salt at high temperature often exhibits features similar to those of internal oxidation. Selective removal of the less noble component occurs, and as it diffuses outwards, vacancies move inwards and segregate to form visible voids (Kirkendall effect). Since diffusion rates are faster at grain boundaries than in the grains, voids tend to form at the grain boundaries and specimens often have the appearance of having undergone ordinary intercrystalline corrosion. More careful examination has shown, however, that in the case of Fe–18Cr–8Ni corroding in a fused 50–50 NaCl/KCl melt at 800°C in the presence of air, the attack is not continuous at the boundaries, and the voids formed are not in communication with each other[20]. In high-nickel alloys, a greater proportion of voids are formed within the grains and the appearance of intercrystalline attack is less marked[20–24]. When Inconel is exposed to fused sodium hydroxide, a two-phase corrosion-product layer is formed, resulting from growth of the reaction product—a mixture of oxides and oxysalts—into the network of channels[22].

Selective removal of the less noble constituent has been demonstrated by chemical analysis in the case of nickel-rich alloys in fused caustic soda[21–23] or fused fluorides[24], and by etching effects and X-ray microanalysis for Fe–18Cr–8Ni steels in fused alkali chlorides[20]. This type of excessive damage can occur with quite small total amounts of corrosion, and in this sense its effect on the mechanical properties of the alloy is comparable with the notorious effect of intercrystalline disintegration in the stainless steels.

Practical Determinations of 'Corrosion Rates'

From what has been said already, it is clear that determinations of 'corrosion rates' from small-scale experiments must be treated with great caution. If the metal cannot passivate, it will corrode until it becomes immune, at which point the corrosion rate will fall to zero; between initial exposure and the attainment of immunity the corrosion rate will be continually changing. If, on the other hand, it is impossible for the metal to come to equilibrium with the melt, then the rate of corrosion, although probably constant, will be primarily controlled by diffusion and interphase mass-transfer rates, and the geometry of the system will be an overriding factor. For this reason, it is not always possible to correlate the results of different workers under apparently similar conditions, nor can such results be expected to correspond particularly closely to the amount of corrosion encountered in larger-scale apparatus.

It is not worth while, therefore, to give a digest of experimentally determined 'corrosion rates', but Table 2.19 indicates some sources of such data and their nature. (Some references to data on compatibility of fused salts with non-metallic materials have been included for the sake of completeness.) It should be remembered that in the case of alloys, failure usually arises from selective attack which causes porosity of the container, even though the wall may appear on casual inspection to be quite sound[26, 27].

Table 2.19 Corrosion data in fused salts

Melt	*Metal*	*Remarks*	*Reference*
NaOH	Ni, Cu, Monel, Au, Ag, Cr, Fe, Ti, Be, Ta, Mn	System allowed to reach equilibrium	15
NaOH	Inconel		22
NaOH	High-Ni alloys		21
NaOH	Ni	Including thermal mass transfer	19
NaOH, LiOH, KOH	Ni, Cu, Fe, stainless steel		32
Fused alkalis	Various		25
$LiNO_3$, $NaNO_3$, KNO_3	Ni, Cu, Duralumin		30
$BaCl_2$/KCl/NaCl	Fe–Ni–Cr alloys and mild steel	With reference to salt-bath heat treatment pots	26, 27
NaCl, KCl, $CaCl_2$	Fe		17
KCl, NaCl, LiCl	Ni, Cu and 3 steels		28
NaCl/KCl and LiCl/KCl mixtures	Fe		29
$BaCl_2$/KCl, $CaCl_2$/NaCl and $MgCl_2$/KCl eutectic mixtures	Fe		31
Carnallite ($MgCl_2{\cdot}KCl{\cdot}6H_2O$)	Fe	Especially effect of dehydration	33
KCl, NaCl and K_2TaF_7 mixture	Inconel		23
Fluoride mixtures	High-Ni alloys		24
Various	Various	Includes glasses and ceramics	34
Various	Various	Includes glasses	35

One interesting feature of comparative experiments with a series of salts having a common anion is that the aggressiveness of the salts towards metals is dependent on the nature of the cation. The aggressiveness of chloride melts in contact with air is in the order[17, 28]:

$$LiCl \sim MgCl_2 \sim CaCl_2 \gg NaCl > KCl$$

In the case of $CaCl_2$ and NaCl, the order corresponds with the corrosion behaviour expected from cathodic polarisation curves[17]. The order of aggressiveness of chlorides can also be explained on the basis of redox potentials of the melts, calculated on thermodynamic grounds from the free energies of formation of the appropriate oxides and chlorides[36]. The order of aggressiveness of nitrates is complicated by passivity effects[30], while that of alkalis in contact with air is [32]:

$$KOH > NaOH > LiOH$$

This is the reverse order to the aggressiveness of chlorides and indicates that the mechanism of corrosion in the two systems is different, i.e. in the latter case it involves the discharge of hydrogen as in acid aqueous solutions.

Measures to Reduce Corrosion

The obvious method of reducing corrosion in fused salts is to choose a system in which either the metal can come to equilibrium with the melt, or else truly protective passivity can be attained. In most cases in industry neither of these alternatives is used. In fact, fused salt baths are usually operated in air atmosphere, and the problem is the prevention of *excessive* corrosion. This can be done in two ways, (*a*) by reducing rates of ingress of oxidising species (mainly O_2 and H_2O) from the atmosphere, and rates of their diffusion in the melts, and (*b*) by keeping the oxidising power (redox potential) of the melt low by making periodic additions to the bath.

Access of air to the melt can be reduced by fitting a cover over it, or by floating a layer of powdered graphite on the melt surface. Diffusion rates in the melt can be reduced only by lowering the temperature, and this invariably reduces the amount of corrosion (except in certain cases in nitrate melts where passivity effects occur[30]).

Additions made to salt baths to lower the oxidising power of the melt are known as 'regenerators'. In 'neutral' chloride salt baths (i.e. those causing neither carburisation nor decarburisation of steel), regenerators fall into two classes, *viz.* alkalis and reactive metals. Alkali additions raise the O^{2-} ion activity in the melt, and since the oxygen partial pressure over the melt is constant, this results in a fall in redox potential (equation 2.19). Reactive metals combine with oxidising impurities in the melt, and in general the more base the metal the greater its effectiveness. The baser metals, such as sodium or potassium, are the most reactive, but would be inconvenient in practice. Magnesium is sometimes used, but slightly more noble metals with volatile chlorides, such as silicon, are commoner. Evaporation of the $SiCl_4$ as it is formed drives the reaction to completion. Besides reducing the aggressiveness of salt baths, certain regenerators also affect the carburising tendency of salt baths, but the subject is outside the scope of this discussion.

D. INMAN

REFERENCES

1. Bettis, E. S., *et al.*, *Nuclear Sci. Engng.*, **2**, 804, 841 (1957)
2. Shirley, H. T., *J. Iron St. Inst.*, **182**, 144 (1956)

3. Jarvis, W. R., *Chem. & Ind.* (*Rev.*), 1 687 (1961)
4. Burrows, B. W. and Hills, G. J., *J. Inst. Fuel*, **39**, 168 (1966)
5. Cutler, A. J. B., Hart, A. B., Fountain, M. J. and Holland, N. H., A.S.M.E. Publication 67-WA/CD-4 (1967)
6. Rahmel, A., *Electrochim. Acta*, **13**, 495 (1968)
7. Bombara, G., Baude, G. and Tamba, A., *Corrosion Sci.*, **8**, 393 (1968)
8. Inman, D. and Wrench, N. S., *Brit. Corros. J.*, **1**, 246 (1966)
9. Blander, M. (Ed.), *Molten Salt Chemistry*, Interscience, New York (1964)
10. Sundheim, B. R. (Ed.), *Fused Salts*, McGraw-Hill, New York (1964)
11. Mamantov, G. (Ed.), *Molten Salts*, Marcel Dekker, New York (1969)
12. Edeleanu, C. and Littlewood, R. *Electrochim. Acta*, **3**, 195 (1960)
13. Littlewood, R. and Edeleanu, C., *Silicates Industr.*, **26**, 447 (1961)
14. Littlewood, R. and Argent, E. J., *Electrochim. Acta*, **4**, 155 (1961)
15. Williams, D. D., Grand, J. A. and Miller, R. R., *J. Amer. Chem. Soc.*, **78**, 5 150 (1956)
15a. Mathews, D. M. and Kruh, R. F., *Industr. Engng. Chem.*, **49**, 55 (1957)
16. Littlewood, R. and Argent, E. J., *Electrochim. Acta*, **4**, 114 (1961)
17. Tomashov, N. D. and Tugarinov, N. I., *Zh. Prikl. Khim.*, **30**, 1 681 (1957)
18. Smith, G. P., Steidlitz, M. E. and Hoffman, E. E., *Corrosion*, **13**, 561t (1957)
19. Lad, R. A. and Simon, S. L., *Corrosion*, **10**, 435 (1954)
19a. Edeleanu, C. and Gibson, J. G., *J. Inst. Met.*, **88**, 321 (1960)
20. Edeleanu, C., Gibson, J. G. and Meredith, J. E., *J. Iron St. Inst.*, **196**, 59 (1960)
21. Smith, G. P. and Hoffman, E. E., *Corrosion*, **13**, 627t (1957)
22. Smith, G. P., Steidlitz, M. E. and Hoffman, E. E., *Corrosion*, **14**, 47t (1958)
23. Bakish, R. and Kern, F., *Corrosion*, **16**, 533t (1960)
24. Manly, W. D. *et al.*, *Proceedings of the Second Conference on the Peaceful Uses of Atomic Energy*, United Nations, Geneva, 1958, Pub. No. A/CONF/15/P/1990
25. Evans, U. R., *The Corrosion and Oxidation of Metals*, Arnold, London, 356–357 (1960)
26. Jackson, J. H., *Alloy Cast. Bull.*, No. 16, 1 (1952)
27. Jackson, J. H. and LaChance, M. H., *Trans. Amer. Soc. Metals*, **46**, 157 (1954)
28. Gurovich, E. I., *Zh. Prikl. Khim.*, **27**, 395 (1954)
29. Kochergin, V. P. and Stolyarova, G. I., *Zh. Prikl. Khim.*, **29**, 730 (1956)
30. Gurovich, E. I., *Zh. Prikl. Khim.*, **29**, 1 358 (1956)
31. Kochergin, V. P., Garpinenko, M. S., Skornyakova, O. N. and Minullina, M. S., *Zh. Prikl. Khim.*, **29**, 621 (1956)
32. Gurovich, E. I., *Zh. Prikl. Khim.*, **32**, 836 (1959)
33. Kochergin, V. P., Kabirov, A. V. and Skornyakova, O. N., *Zh. Prikl. Khim.*, **27**, 883 (1954)
34. Mackenzie, J. D., In *Physicochemical Measurements at High Temperatures*, Eds. J. O'M. Bockris, J. L. White and J. D. Mackenzie, Butterworths, London, 334 (1959)
35. Lunden, A., *Corrosion Sci.*, **1**, 62 (1961)
36. Littlewood, R., *J. Electrochem. Soc.*, **109**, 525 (1962)

2.11 Corrosion in Lubricant Systems

Lubricants are not generally regarded as being corrosive, and in order to appreciate how corrosion can occur in lubricant systems it is necessary to understand something of the nature of lubricants. Once, lubricants were almost exclusively animal or vegetable oils or fats, but modern requirements in the way of volume and special properties have made petroleum the main source of supply. In volume, lubricants now represent about 2% of all petroleum products; in value, considerably more.

There are many hundreds of different varieties of lubricants, many of them tailored to meet particular requirements. Lubricating greases are solid or semi-solid lubricants made by thickening lubricating oils with soaps, clays, silica gel or other thickening agents. Synthetic lubricants, which will operate over a very wide range of temperature, have been developed mainly for aviation gas-turbine engines. These are generally carboxylic esters and are very expensive products.

The main function of most lubricants is to reduce friction and wear between moving surfaces and to abstract heat. They also have to remove debris from the contact area, e.g. combustion products in an engine cylinder, swarf in metal-cutting operations. Sometimes they have to protect the lubricated or adjacent parts against corrosion, but this is not a prime function of most lubricants. On the other hand, many lubricants do contain corrosion inhibitors and some lubricating oils, greases, mineral fluids and compounds are specially formulated to prevent the corrosion of machinery or machine parts, particularly when these components are in storage or transit. These temporary protectives are described in Section 17.3.

The Manufacture of Lubricating Oils

Mineral lubricants may be distillates or residues derived from the vacuum distillation of a primary distillate with a boiling point range above that of gas oil[1,2,3]. They are mixtures of hydrocarbons containing more than about 20 carbon atoms per molecule, and range from thin, easily flowing 'spindle' oils to thick 'cylinder' oils. For hydrocarbons having the same number of carbon atoms per molecule, the higher the proportion of carbon to hydrogen, the more viscous the oil and the lower the viscosity index.

Distillate lubricating oils can be conveniently divided into three groups—low viscosity index oils (LVI oils), medium viscosity index oils (MVI oils) and high viscosity index oils (HVI oils). LVI oils are made from naphthenic distillates, with low wax contents so that costly dewaxing is not required. MVI oils are produced from both naphthenic and paraffinic distillates; the paraffinic distillates have to be dewaxed. HVI oils are prepared by the solvent extraction and dewaxing of paraffinic distillates. Solvent extraction is a physical process which removes the undesirable constituents, thereby improving viscosity index and the oxidation and colour stability. White oils are obtained by the more drastic refining of low viscosity lubricating oil distillates to remove unsaturated compounds and constituents that impart colour, odour and taste. They are usually solvent extracted and then repeatedly treated with strong sulphuric acid or oleum and alkali, and finally 'clay'-treated to remove surface-active compounds. Acid and clay treating is expensive and is being superseded by hydrofinishing, a catalytic hydrogenation treatment.

The residues from the vacuum distillation can also be refined to provide very viscous lubricants. The residues from paraffinic base oils are generally solvent extracted and dewaxed. The main use of these products (bright stocks) is as blending components for heavy lubricants. Thus residues from naphthenic base oils, which are also used as blending components for heavy lubricants, are normally not extracted.

The performance characteristics of a lubricating oil depend on its origin and on the refining processes employed, and in order to ensure consistent properties these are varied as little as possible. Some aero-engine builders insist on a complete re-evaluation of a lubricant, costing many thousands of pounds, whenever there is a change of source (crude) or refining process.

Corrosion and Corrosion Prevention in Lubricant Systems

Deterioration of lubricants in use Lubricating oils deteriorate in service in two ways, *viz.* they become contaminated and they undergo physical and chemical changes due to oxidation[4]. In engines the common contaminants are airborne dust and wear products, unburnt fuel, fuel combustion products and water. The oxidation products are mainly acidic materials and asphaltenes. Asphaltenes in association with fuel contaminants and water form sludges and lacquers. The acidic materials resulting from oxidation are generally weak organic acids although in extreme cases strong mineral acids may be produced. However, contamination by fuel combustion products is the source of almost all strong-acid contamination in lubricants[5]. Thus, distillate diesel fuels can contain up to 1% sulphur (by weight) and residual diesel fuels up to 3% sulphur. This sulphur is oxidised to sulphur acids, and sulphuric acid condensate may be encountered on the cooler surfaces[6]. In gasolines the sulphur content is very low but halogen compounds are added to the gasolines as scavengers for the lead-based anti-knock compounds. These halogen compounds can give rise to halogen acids[7].

Lubricants with a high aromatic content (LVI and some MVI oils) tend to oxidise to give sludge-forming compounds, although some naphthenic oils give organic acids. Paraffinic oils (HVI and some MVI oils) oxidise more slowly to give weak acids. In a plentiful supply of oxygen, oxidation proceeds at a significant rate at temperatures above about 60–130°C depending on the composition of the lubricant[8]. This oxidation is an extremely complicated process which involves the formation of organic peroxides as intermediates[9]. It is catalysed by the presence of metals, particularly copper, iron and lead[10].

The rôle of additives The progressive development of engines and general machinery resulting in more arduous operating conditions, and particularly the use of longer oil-change periods, means that neither straight mineral oils nor compounded oils (mineral oils to which a proportion of an animal or vegetable oil has been added) are adequate for modern service requirements.

Despite the introduction of new, improved methods of refining it has been necessary to enhance the performance of lubricants by the use of additives, either to reinforce existing qualities or to confer additional properties. Once additives were regarded with some suspicion—an oil that needed an additive was necessarily an inferior oil; today they are an accepted feature of lubricants. Almost all quality lubricants on sale today contain one or more additives. An enormous range of additives are available for use in lubricants[1,11–13], some produced by the oil companies and others provided by specialist manufacturers. Additives are usually named after their particular function, but many additives are multifunctional. Thus, an anti-wear additive may also protect a surface against corrosion. The main types of additives that can enhance the anti-corrosion behaviour of lubricants are listed in Table 2.20.

Extreme pressure additives Many additives, essential to the performance of the lubricant, provide no corrosion protection and some additives may become corrosive in certain circumstances. Thus EP (Extreme Pressure) additives contain chemical groups which are designed to react chemically with metal surfaces when normal lubrication fails, forming easily sheared layers of metal oxides, sulphides, chlorides or phosphates, thereby preventing catastrophic wear and seizure. Reaction between EP compounds and metal surfaces should only occur at local hot spots and the layers formed are extremely thin. However, if the operating conditions are very severe these layers are continually generated and removed as they fulfil their anti-wear function. A process of this nature is sometimes called 'chemical' wear, and if sliding surfaces operate continually under these conditions loss of metal from the rubbing surfaces can ultimately result in failure. Alternatively, all the EP additive may be used up (depleted) and then failure by seizure will occur. EP agents are intended to cater for the occasional overload condition and it must be emphasised that machinery should be designed so that it does not require the continual action of EP agents to function satisfactorily.

Obviously the selection of an EP additive requires great care; if it is too active it may give rise to excessive metal removal under normal operating conditions (see the section on corrosion by sulphur additives). Also, if a component is prone to fatigue pitting in service the presence of an overactive EP agent may result in corrosion fatigue[14].

Table 2.20 Additives that can enhance the anti-corrosion behaviour of lubricants

Additive	*Function*	*Typical chemical types*
Oxidation inhibitor (anti-oxidant)	To increase oxidation resistance of lubricants by interfering with the chain reactions which give rise to acid and asphaltene formation	Oil-soluble amine and phenol derivatives for temperatures up to about 120°C. For higher temperatures dialkyldithiophosphates and other compounds listed below as metal deactivators; also P_2S_5-olefin and -terpene reaction products
Metal deactivator	To form inactive protective films on metal surfaces which otherwise might catalyse oxidation and corrosion reactions	Trialkyl and triaryl phosphites, organic dihydroxyphosphines, some active sulphur compounds, diamines; in lubricating greases, mercaptobenzothiazole and phosphites
Corrosion inhibitor	To protect metal surfaces, particularly bearing surfaces (copper, silver, lead) against corrosion	Zinc dialkyldithiophosphates, sulphurised olefins, sulphurised terpenes
Rust inhibitor	To eliminate rusting in presence of moisture	Polar compounds, such as metallic soaps, esters and derivatives of dibasic acids; barium and calcium sulphonates
Water repellant	To impart water-resistant properties, particularly in greases	Aliphatic amines, hydroxy fatty acids and some organic silicone polymers
Basic compound	To neutralise strong mineral acids (from fuels)	Barium or basic calcium salts of sulphonic acids and alkylsalicylic acid. 'Superbasic' and 'hyperbasic' additives are basic salts with a large excess of base
Dispersants	To keep insoluble combustion and oxidation products in suspension and dispersed	Salts of phenolic derivatives; polymers containing barium, sulphur and phosphorus; calcium or barium soaps of petroleum sulphonic acids

Additive interaction Modern high-performance lubricants contain a number of additives, each with a particular, special function. Thus a turbine lubricant may contain an oxidation inhibitor, a rust inhibitor, an extreme pressure (EP) agent and an anti-foam compound. A high-grade diesel-engine lubricant may contain a VI improver, a dispersant, an anti-oxidant, a corrosion inhibitor, a basic compound, a pour-point depressant and an anti-foam compound. Sometimes these additives may have undesirable side effects or interact adversely[1]; in a turbine oil the rust inhibitor may act as an emulsifier interfering with demulsification; in a diesel lubricant the dispersant may promote oil oxidation. Frequently anti-corrosion additives may not be able to exert their maximum effect because they are competing for sites on metal surfaces. The development of a successful new lubricating oil requires much skill and experience and always necessitates considerable laboratory and field testing in order to strike the right balance between the various additives.

Additives in greases Oil additives may be incorporated in the oil phase in a grease. However, with greases there is also the opportunity to employ oil-insoluble anti-corrosion agents, since these can be incorporated in the grease along with other thickening agents, e.g. sodium nitrite is incorporated in greases in this way.

Additive depletion Additives are consumed or 'depleted' as they fulfil their respective functions. It is for this reason, more than any other, that it is necessary to change engine oils regularly.

Corrosion by sulphur additives Sulphur compounds occur naturally in most lubricants and many oil additives contain sulphur. In a properly formulated lubricant these sulphur compounds should be inactive at ambient temperature. At elevated temperatures they may decompose to give more active materials which can stain and corrode metals, particularly silver and copper. However, these same sulphur compounds have many beneficial qualities[1,4,10]; this is why they are not removed completely in refining and why they are used as additives.

Thus, sulphur compounds in lubricants generally act as anti-oxidants, preventing acid and sludge formation. They also form very thin films on metal surfaces protecting them from acid or peroxide attack. In addition, sulphur compounds are often used as EP agents. The oil chemist must try to strike a balance—the activity of the sulphur must be high enough for it to exert a beneficial effect and yet not so high as to stimulate corrosion.

All too frequently lubricants containing sulphur are exposed to more severe operating conditions than intended, and staining and corrosion results. This has given sulphur compounds and sulphur-containing additives, particularly those of the dithiosphosphate type, a bad name. Silver bearings are still used in certain diesel and aero-engines and if the lubricants for these engines contain sulphur compounds with too much chemical activity, severe corrosion ensues[5]. A more widespread problem is the corrosion of phosphor-bronze alloys[5,15] (containing about 10% tin) particularly in little-end bushes in diesel engines where temperature can exceed 200°C. Some engine builders and operators hold sulphur additives entirely responsible, but this opinion cannot be substantiated, for corrosion can occur with lubricants containing only natural sulphur compounds[5]. Two important metallurgical factors affecting the corrosion resistance of phosphor-bronze alloys are the amount of alloying element in solution in the copper-rich phase and the porosity of the alloy. For example, if the amount of tin in solution can be increased by special casting techniques or heat treatments the corrosion resistance is greatly increased. Similarly, zinc or silicon in solution also increases the resistance of copper to sulphur corrosion. If the alloy is porous the lubricant is drawn into the pores where it stagnates, and, at high temperatures, becomes very corrosive, e.g. copper catalyses oil oxidation with the consequent formation of corrosive sulphur compounds.

The most satisfactory solution to this problem is to employ a corrosion-resistant alloy, and alloys of the gun-metal type, containing 2–4% zinc, have proved completely satisfactory. The substitution of zinc for phosphorus gives sounder castings and improves the corrosion resistance of the copper-rich matrix.

Corrosion in Specific Lubricant Systems

Lubricants are specially formulated to meet all manner of requirements and the major corrosion problems associated with important classes of lubricants will now be outlined.

Engine lubricants Engine lubricants are exposed to severe operating conditions, being subjected to high temperatures, the products of combustion and a plentiful supply of oxygen. Consequently, unless the oil is changed at appropriate intervals strong mineral acids and weak organic acids may accumulate. In addition, droplets of water may be formed and these can contain strong mineral acids derived from the fuel combustion gases. These droplets sometimes give rise to emulsions which deposit in the colder portions of an engine, e.g. on the rocker-box covers; ferrous surfaces are most affected by this condensed moisture. A special dynamic corrosion test has been developed to study corrosion in these two-phase (water-in-oil) systems[16]. Problems associated with the retention of water in engine lubricants are likely to become more acute as anti-pollution devices are fitted to engines. The harmful effects are best countered by anti-rust and basic additives, and in diesel engines burning high-sulphur fuels, e.g. marine diesel engines, very high levels of basicity are required.

Cast or sintered copper-lead or lead-bronze alloys are widely used for engine bearings. The lead phase in such bearings is readily attacked by weak organic acids and almost all the lead can be leached out unless preventive measures are employed[17]. The lead may be protected by a precision electrodeposited overlay of a lead-tin or lead-indium alloy[18]. About 3% tin or 5% indium in lead will render the lead resistant to attack by oil-oxidation acids[5,19]. One reason why leaded bearings are protected by an overlay, and not by incorporating the protective alloying elements in the underlying lead, is that both tin and indium dissolve preferentially in copper. In a cast or sintered bearing, therefore, any tin or indium will be found in solution in the copper-rich phase, leaving the lead-rich phase susceptible to attack.

In overlay bearings operating above about 140°C, the tin or indium in the overlay diffuses towards, and alloys with, the underlying copper, depleting the overlay and reducing its resistance to corrosion[5]. This depletion by diffusion can be combatted by the use of a diffusion barrier or 'dam', e.g. a nickel-rich layer between the bearing alloy and the overlay[20].

In gasoline engines lead halides accumulate in the lubricant; occasionally these give rise to the corrosion of aluminium-alloy pistons[7] and very rarely to corrosion on aluminium-tin bearings[5].

Steam-turbine lubricants Lubricants in steam turbines are not exposed to such arduous conditions as those in engines. The main requirement is for high oxidation stability. However, they may be exposed to aqueous condensate or, in the case of marine installations, to sea water contamination, so they have to be able to separate from water easily and to form a rust-preventing film on ferrous surfaces, and it is usual to employ rust inhibitors. The problem of tin oxide formation on white-metal bearings is associated with the presence of electrically conducting water in lubricants and can be overcome by keeping the lubricant dry[21,22].

Gear lubricants In addition to the usual oxidation and corrosion inhibitors, lubricants for heavily loaded gears almost always contain EP additives containing sulphur, chlorine or phosphorus. In order to function, these additives must react locally with the metal surfaces, and yet the extent of the reaction should not be such that it could be described as corrosive, or promote fatigue pitting[1]. These EP additives may be quite safe with ferrous metal surfaces, but may cause severe corrosion on copper alloys, e.g. on bronze worm wheels[12], if for any reason excessive temperatures arise.

Some turbine lubricants have to lubricate the turbine gears as well as the turbine; in these circumstances any EP additives employed should not be corrosive in the presence of moisture[12].

Metal-working lubricants Metal-working lubricants can be divided into two categories, *viz.* cutting lubricants and forming lubricants. Three types of cutting fluids are widely used: soluble oils, water-base fluids and straight cutting oils. Soluble oils are low-viscosity mineral oils which when added to water form stable oil-in-water emulsions. Their main function is to cool rather than to lubricate, but they should contain rust inhibitors to prevent corrosion of both the workpiece and the machine. Water-base fluids contain no oil and employ the normal corrosion inhibitors used in coolants, e.g. sodium nitrite, sodium benzoate and triethanolamine phosphate. Straight cutting oils are used in severe operations where good lubrication is essential, and generally contain EP agents. These EP agents must be carefully selected so that they do not stain the workpiece or the machine tool[12]. For example, very active sulphur compounds should not be employed with copper alloys[23]. Chlorinated additives should be stable in moist air to avoid the risk of hydrochloric acid formation.

Forming lubricants include lubricants for rolling, drawing, extruding and forging. A vast range of compounds including fatty oils and compounded oils are used in these operations, and a major requirement is that they should not stain the workpiece during the forming operation or during subsequent annealing or in storage. Consequently, all oils and additives employed should be completely volatile, in addition to affording protection against rusting.

Lubricants for other applications Specially formulated lubricants are required for steam engines, compressors and exhausters, refrigerators, hydraulic equipment, textile machinery, transformers and switchgear, nuclear power plants and many other diverse applications. Most of these lubricants will contain a carefully balanced set of additives, including some to prevent corrosion either directly by protecting the metal surface, or indirectly by preventing deterioration of the lubricant and combatting the action of contaminants. When corrosion is encountered with a specialised lubricant, the cause is not likely to be any weakness on the part of the lubricant; it is probable that the lubricant is being exposed to operating conditions far more extreme than those for which it was designed.

R. W. WILSON

REFERENCES

1. O'Connor, J. J. (Ed.), *Standard Handbook of Lubrication Engineering*, McGraw-Hill Book Company, New York (1968)
2. Cameron, A. (Ed.), *Principles of Lubrication*, Longmans Green, London (1966)
3. *Modern Petroleum Technology*, Institute of Petroleum, London, 3rd edn (1962)
4. Zuidema, H. H., *The Performance of Lubricating Oils*, Reinhold Publishing Corporation, New York, Chapman and Hall, London, 2nd edn. (1959)
5. Wilson, R. W. and Shone, E. B., *Joint Course on Tribology*, Paper 4, Institution of Metallurgists, London (1968)
6. Broeze, J. J. and Wilson, A., *Gas and Oil Power*, **44**, 386 (1949)
7. Graham, R., *Shell Aviation News*, June, 18 (1950)
8. Fowle, T. I., *Conference on Lubrication and Wear*, Paper 19, Institution of Mechanical Engineers, London, 568 (1968)
9. Garner, F. H. and Wilson, B. S., *J. Inst. Petrol*, **37**, 225 (1951)
10. Denison, G. H., *Ind. Eng. Chem.*, **36**, 477 (1944)
11. Ford, J. F., *J. Inst. Petrol*, **54**, 198 (1968)
12. Braithwaite, E. R. (Ed.), *Lubrication and Lubricants*, Elsevier Publishing Company, London (1967)
13. *Lubrication*, **54**, 85 (1968)
14. Galvin, G. D. and Naylor, H., *Lubrication and Wear*, Third Convention, Paper 6, Inst. Mech. Engrs., 56 (1965)
15. Quayle, J. P., *Copper*, **3** No. 5, 12 (1969)
16. Hughes, R. I., *Corrosion Science*, **9**, 535 (1969)
17. *Lubrication*, **8**, 145 (1953)
18. Forrester, P. G., *Met. Rev.*, **5**, 507 (1960)
19. Wilson, R. W. and Shone, E. B., *Anti-Corrosion Methods and Materials*, **17** No. 8, 8 (1970)
20. Forrester, P. G., *Modern Materials*, **4**, Academic Press, New York and London, 173 (1964)
21. Bryce, J. B. and Roehner, T. G., *Trans. Inst. Mar. Engrs.*, **73**, 377 (1961)
22. Lloyd, K. A. and Wilson, R. W., *Lubrication and Wear*, Seventh Convention, Paper 10, Inst. Mech. Engrs., 76 (1969)
23. *Lubrication*, **55** No. 7, 65 (1969)

3 FERROUS METALS AND ALLOYS

3.1 Iron and Steel

Introduction

Bare iron and steel are liable to rust in most environments but the extent of the corrosion depends upon a number of factors, the most important of which are the composition and surface condition of the metal, the corrosive medium itself and the local conditions.

With regard to the effect of composition, ferrous metals fall into three broad categories:

1. The ordinary cast irons, wrought irons and steels, to which no alloying elements are added, and which are vulnerable to corrosion.
2. Low-alloy steels, which contain about 2–3% of alloying elements, commonly copper, chromium and nickel. These steels still rust, but under certain conditions in the atmosphere, the rust formed becomes adherent and protective so that the corrosion rate becomes several times less rapid than with the ordinary steels mentioned above. These steels are often termed *weathering* steels.
3. Stainless steels, which contain high percentages of alloying elements, e.g. 18% chromium, 8% nickel and 3% molybdenum. Steels of this type are practically non-corrodible in appropriate circumstances.

The discussion that follows will be concerned mainly with materials in the first of these categories. Moreover, since the corrosion of cast iron is discussed elsewhere (see Section 3.6), and since little wrought iron is produced nowadays, the subject matter will virtually resolve itself into the corrosion of ordinary carbon steels, as used in mass for general purposes. The corrosion of low-alloy steels and that of stainless steels are considered in Sections 3.2 and 3.3 respectively.

The treatment will begin with a brief consideration of the mechanism of rusting and of the influence of variations in the steel itself. It will be completed by short surveys of present knowledge of the rusting of ordinary mild steel in the three natural media: air, water and soil.

Mechanism of Rusting

In pure dry air at normal temperatures a thin protective oxide film forms on the surface of polished mild steel. Unlike that formed on stainless steels it is not protective in the presence of electrolytes and usually breaks down in air, water and soil. The anodic reaction is:

$$Fe \rightarrow Fe^{2+} + 2e^-$$

In de-aerated solutions, the cathodic reaction is

$$2H^+ + 2e^- \rightarrow H_2$$

This occurs fairly rapidly in acids but very slowly in alkaline and neutral solutions. In the presence of oxygen the following reaction occurs in slightly alkaline and neutral solutions:

$$O_2 + 2H_2O + 4e^- \rightarrow 4OH^-$$

This is the common form of cathodic reaction in most environments. The OH^- ions react with Fe^{2+} ions to form ferrous hydroxide:

$$Fe^{2+} + 2OH^- \rightarrow Fe(OH)_2$$

This is oxidised to ferric hydroxide $Fe(OH)_3$, which is a simple form of rust. The final product is the familiar reddish brown rust $Fe_2O_3 \cdot H_2O$, of which there are a number of varieties, the most common being the α form (goethite) and the γ form (lepidocrocite). In situations where the supply of oxygen is restricted, Fe_3O_4 (magnetite) or γ Fe_2O_3 may be formed.

This is a simplified treatment but it serves to illustrate the electrochemical nature of rusting and the essential parts played by moisture and oxygen. The kinetics of the process are influenced by a number of factors, which will be discussed later. Although the presence of oxygen is usually essential, severe corrosion may occur under anaerobic conditions in the presence of sulphate-reducing bacteria (*Desulphovibrio desulphuricans*) which are present in soils and water. The anodic reaction is the same, i.e. the formation of ferrous ions. The cathodic reaction is complex but it results in the reduction of inorganic sulphates to sulphides and the eventual formation of rust and ferrous sulphide (FeS).

Effect of Variations in the Steel

Ordinary steels are essentially alloys of iron and carbon with small additions of elements such as manganese and silicon added to provide the requisite mechanical properties. The steels are manufactured from a mixture of pig iron and scrap, which is treated in the molten state to remove excess carbon and other impurities. The steel may be continuously cast into strands or cast into individual ingots. The final product is then produced by rolling, drawing or forging. During hot rolling and forging the steel surface is oxidised by the air and the scale produced, usually termed *millscale*, may have an important influence on the corrosion of the steel, as will be discussed later.

The structure of millscale consists of three superimposed layers of iron oxides in progressively higher states of oxidation from the metal side outwards, viz. ferrous oxide (FeO) on the inside, magnetite (Fe_3O_4) in the middle and ferric oxide (Fe_2O_3) on the outside. The relative portions of the three oxides vary with the rolling temperatures. A typical millscale on 9·5 mm mild steel plate would be about 50 μm thick, and contain approximately 70% FeO, 20% Fe_3O_4 and 10% Fe_2O_3.

If millscale was perfectly adherent, continuous and impermeable, it would form a good protective coating, but in practice millscale soon cracks and flakes off in places. In air, the presence of millscale on the steel may reduce the corrosion rate over comparatively short periods, but over longer periods the rate tends to rise. In water, severe pitting of the steel may occur if large amounts of millscale are present on the surface.

Generally, the process of manufacture has no appreciable effect on the corrosion characteristics of steel. Slight variations in composition that inevitably occur from batch to batch in steels of the same quality have little effect with the exception of copper, the effect of which is discussed more fully in Section 3.2.

Briefly, the addition of about 0·2% of copper results in a two to three-fold reduction in the corrosion rate in air compared with a copper-free steel[1,2].

Variations in the other elements in ordinary steels affect the corrosion rate to a marginal degree, the tendency being for the rate to decrease with increasing content of carbon, manganese and silicon. For example, in the open air a steel containing 0·2% of silicon rusts about 10% less rapidly than an otherwise similar steel containing 0·02% of silicon.

Wrought iron, which has been replaced by mild steel, contained appreciable quantities of slag and corroded at a rate about 30% less than mild steel when exposed in air. However, artificially produced Aston-Byers iron corrodes at about the same rate as mild steel.

The corrosion rates of wrought iron and mild steel when immersed in sea-water or buried in soil are not significantly different when the copper contents are similar.

Rusting in Air

Controlling Factors

The rusting of bare steel in the atmosphere is controlled by the climatic conditions at the exposure site. The main factors are the availability of moisture, and the extent to which the air is polluted, but other less important ones, such as temperature, must also be considered. (See also Section 2.2.)

Moisture: the critical humidity Moisture can reach a steel surface directly in liquid form as a result of precipitation processes, i.e. rain and dew, but the water vapour that is always present in air can also, under certain conditions, cause steel to rust at relative humidities well below saturation.

This important fact was first demonstrated by Vernon[3] in a series of classical experiments, some of which are summarised graphically in Fig. 3.1. He showed that rusting is minimal in pure air of less than 100% relative humidity but that in the presence of minute concentrations of impurities, such as sulphur dioxide, serious rusting can occur without visible precipitation of moisture once the relative humidity of the air rises above a critical and comparatively low value. This value depends to some extent upon the nature of the atmospheric pollution, but, when sulphur dioxide is present, it is in the region of 70–80%. Below the critical humidity, rusting is inappreciable, even in polluted air.

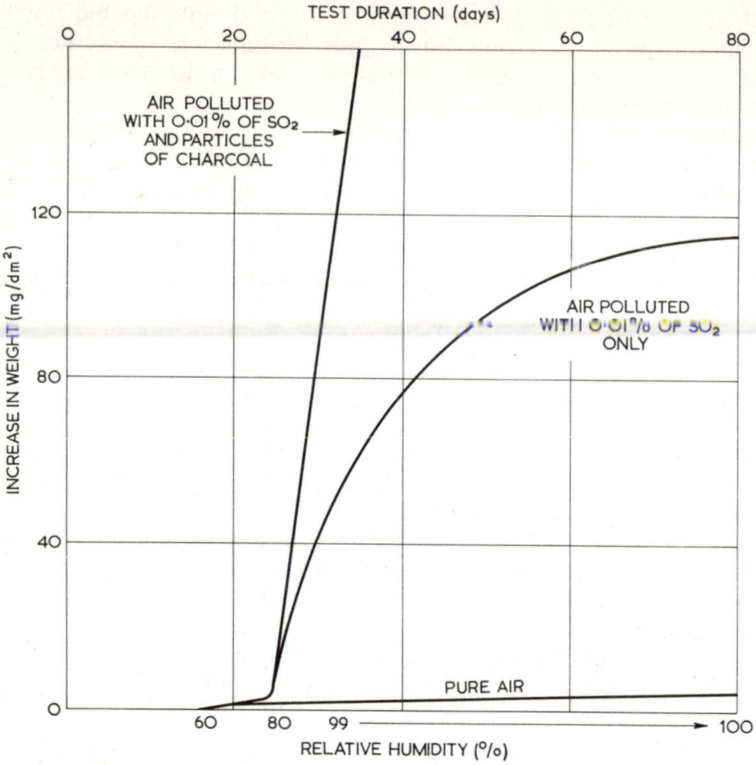

Fig. 3.1 Effect of relative humidity and atmospheric pollution on the rusting of iron (after Vernon[3])

The presence of moisture on steel above the critical humidity but below the saturation point may be caused by an adsorption mechanism or by the presence of particles of deliquescent salts on the surface. Once rusting has begun, the composition of the rust already formed will influence the relative humidity at which further rusting will occur, because rusts formed in polluted atmospheres contain hygroscopic salts. The method by which moisture reaches the surface is probably less important, however, than the length of time during which the steel remains wet.

Pollution Although humidity plays a vital part, the impurities in the air are decisive in determining the rate of rusting in atmospheres of the requisite humidity; in their absence rusting is not serious even in highly humid air. The most important impurity in industrial atmospheres is sulphur dioxide, although chlorides and ammonium salts[4] also may have an effect. Generally, near the coast chlorides have the most pronounced effect. In the presence of chlorides, rusting can continue at humidities as low as 40%[5]. As a rule, the chloride concentration in the air falls off rapidly with distance inland, but steel rusts at almost incredible rates on surf beaches in the tropics where it is exposed to a continuous spray of sea salts in the surf. This is shown by the results of some tests made in Nigeria that are given in Table 3.1.

Table 3.1 Effect of sea salts on the rate of rusting*
(Tests made in Nigeria on behalf of BISRA by the former Tropical Testing Establishment, Ministry of Supply)

Distance from surf	*Salt content of air*†	*Rate of rusting* (mm/y)
50 yd	11·1	0·950
200 yd	3·1	0·380
400 yd	0·8	0·055
1 300 yd	0·2	0·040
25 miles	—	0·048

*The specimens were of ingot iron and were exposed for one year.
†The salt content of the air was determined by exposing wet cloths, and is expressed as mg NaCl d^{-1} $(100\ cm^2)^{-1}$.

In most districts, however, sulphur dioxide and dust particles are the main corrosive pollutants. It has been demonstrated that there is a direct relationship between sulphur dioxide in the atmosphere and the corrosion of steel exposed to it (see Fig. 3.2). In a series of tests carried out in the Sheffield area, sulphur dioxide accounted for about 50% of the variations in corrosion rate at the different sites[6].

Sulphur dioxide in the air originates from the combustion of fuel and influences rusting in a number of ways. For example, Russian workers consider that it acts as a cathodic depolariser[7], which is far more effective than dissolved oxygen in stimulating the corrosion rate. However, it is the series of anodic reactions culminating in the formation of ferrous sulphate that are generally considered to be of particular importance. Sulphur dioxide in the air is oxidised to sulphur trioxide, which reacts with moisture to form sulphuric acid, and this in turn reacts with the steel to form ferrous sulphate. Examination of rust films formed in industrial atmospheres have shown that 5% or more of the rust is present in the form of iron sulphates[8] and $FeSO_4 \cdot 4H_2O$ has been identified in shallow pits[9].

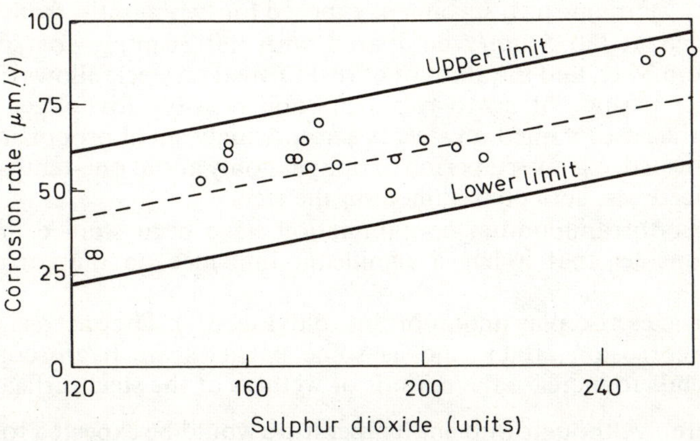

Fig. 3.2 Relationship between sulphur dioxide and corrosion (after Chandler and Kilcullen[6]). Note that 1 unit is the average daily reading in μg of SO_2 per m^3 of air

Workers who have studied atmospheric corrosion processes are generally agreed that the loss of iron as a sulphate accounts in only a small measure for the effects of sulphur dioxide[10–13]. There is not complete agreement on the detailed mechanism but once ferrous sulphate has been formed it is able to promote further rusting. This has been demonstrated by allowing steel to rust in an atmosphere containing sulphur dioxide and then transferring it to a clean atmosphere, where the corrosion is continued at an enhanced rate, at least for a time[14].

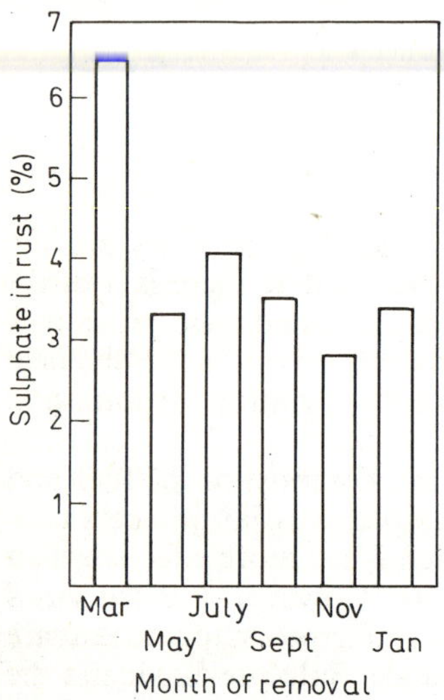

Fig. 3.3 Sulphate in adherent rust on steel exposed at Battersea in January (after Chandler and Stanners[8])

There is a cyclic variation in the amount of sulphate found in rust formed on steel exposed at different times of the year (Fig. 3.3), and the amount depends on the month of the year rather than on the period of exposure, at least for periods of up to two years[8]. Consequently, the month of exposure will have an important influence on the corrosion rate up to a year or so (see Table 3.2). In one test, specimens exposed for two months from September corroded at 0·035 mm/y compared with 0·0136 mm/y for specimens exposed from May, and the amount of rust formed on steel follows this cyclic pattern over periods of up to two years and possibly for longer periods. However, the rust retained on steel is a surprisingly small proportion of the total rust formed, e.g. over a period of one year only about one-tenth of all the rust produced was actually retained on the steel.

The effects of ammonium contamination have been studied and some workers consider that it has a significant influence on the corrosion of steel[15,16].

Dust particles also play an important rôle (Fig. 3.1). They act as nuclei for the initial corrosion attack and as some particles are hygroscopic their presence tends to increase the periods of wetness of the steel surface.

Temperature Although ambient temperature would be expected to have an influence on the rate of rusting, its effect is not clearly defined despite the efforts of workers to establish a relationship[17,18]. It has an effect on relative

Table 3.2 Variations in rate of rusting of steel exposed at different months of the year

Month test begun	*Rate of rusting* (mm/y)
January	0·081
March	0·064
May	0·076
July	0·070
September	0·094
November	0·089

humidity and consequently an indirect effect on corrosion. However, fluctuations in temperature may be more important than average temperatures because they influence condensation and the rate of drying of moisture in contact with steel.

Rates of rusting in different climates Some representative figures to show how the rate of rusting of bare steel in the open air varies in different parts of the world are given in Table 3.3. They relate to $100 \times 50 \times 3{\cdot}2$ mm (4 in × 2 in × $\frac{1}{8}$ in) pieces of ingot iron freely exposed in a vertical position for one year.

Table 3.3 Rates of rusting of mild steel outdoors in different climates (BISRA)

Type of atmosphere	*Site*	*Corrosion rate* (mm/y)
	Great Britain	
Rural or suburban	Godalming	0·048
	Llanwrtyd Wells	0·069
	Teddington	0·070
Marine	Brixham	0·053
	Calshot	0·079
Industrial	Motherwell	0·095
	Woolwich	0·102
	Sheffield	0·135
	Frodingham	0·160
	Derby	0·170
	Overseas	
Rural or suburban	Khartoum	0·003
	Abisko, North Sweden	0·005
	Delhi	0·008
	Basrah	0·015
	State College, Pa, U.S.A.	0·043
	Berlin-Dahlem	0·053
Marine	Singapore	0·015
	Apapa, Nigeria	0·028
	Sandy Hook, N.J., U.S.A.	0·084
Marine/industrial	Congella, South Africa	0·114
Industrial	Pittsburgh, Pa, U.S.A.	0·108
Marine, surf beach	Lagos	0·615

The results follow the expected pattern, the highest rate being that observed on a surf beach at Lagos. The magnitude of this rate (0·6 mm/y) will be better appreciated by the statement that an ingot iron specimen as above, 3·2 mm thick, would be completely destroyed by rusting from both sides in about $2\frac{1}{2}$ years. As previously stated, rusting in Great Britain is roughly proportional to the atmospheric pollution; in accordance with this, industrial sites are several times more corrosive than rural, suburban or marine ones.

Table 3.4 Effect of the surface condition at the time of exposure on the atmospheric corrosion of mild steel (BISRA) (5 years outdoor exposure at Sheffield)

Surface condition	*Average penetration* (mm)
As-rolled	0·545
Pickled	0·545
Sandblasted	0·532
Machined	0·534
Polished	0·532

The low rates of rusting observed at Khartoum, Abisko, Delhi, Basrah and Singapore are primarily associated with the absence of serious pollution. Moreover, at most of them the relative humidity is low, e.g. at Khartoum the relative humidity lies below the critical value for rusting throughout the whole year.

Effect of the exposure conditions The absolute values for the rate of rusting given in Table 3.3 would be affected by the mass of the specimen itself and by other factors such as the orientation of the steel, the climatic conditions prevailing at the time of exposure, and the duration of exposure.

The *orientation* of the steel influences the rate of rusting through its effect on the amounts of moisture and pollutants that can reach the surface. The groundward side of a horizontal surface is protected from rain but it is also shielded from the drying action of the sun and often of the wind, so that dew tends to remain in contact with the steel there for longer periods; moreover, harmful solid particles and soluble salts are not leached away. Consequently, the groundward side may corrode more rapidly than the skyward side. The same considerations apply to steel exposed obliquely. The relative corrosion of the opposite faces of a vertical steel plate will largely depend on the direction of the prevailing wind.

In agreement with these considerations, the results of tests at Derby[19] have shown that specimens exposed at 45° corroded 10–20% more than vertical specimens, and that 54% of the total loss was on the underside. In American tests[20] on specimens exposed at 30° to the horizontal, 62% of the loss was on the underside.

Effect of mass The rate of rusting of steel in the atmosphere is affected to some extent by the mass of the part concerned, because this determines the speed at which the surface temperature adjusts itself to fluctuations in the ambient temperature, the amount of condensation during humid periods, and the time during which dew or rain remains in contact with the steel.

For example, in a test over 12 months at the National Chemical Laboratory under sheltered conditions outdoors, thick steel plates rusted more than thin ones as is shown below.

Plate thickness (mm)	55	28	12·5	5
Average general penetration (mm)	0·038	0·033	0·031	0·030

Evidently, therefore, correct design can play an important part in reducing the corrosion of steel structures through lessening the danger of local attack due to mass metal effects and condensation.

Effect of surface condition As previously noted, millscale on steel may decrease the corrosion rate over short periods. However, over longer periods the surface condition is not usually a determining factor as can be seen from Table 3.4 which shows the results of 5-year tests at Sheffield.

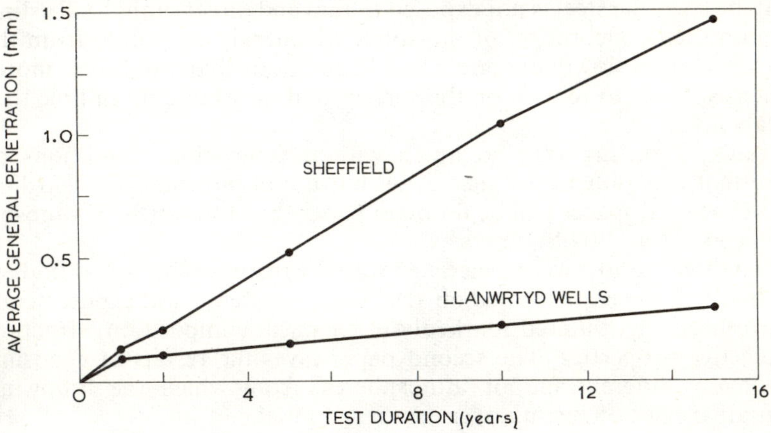

Fig. 3.4 Rusting of mild steel (0·02% Cu) in the open air at Sheffield and Llanwrtyd Wells (after Reference 2)

It is important to know whether the rate of rusting of steel outdoors increases, remains constant or decreases with increasing *duration of exposure*. The position over a long period is adequately illustrated by the curves drawn In Fig.3.4 showing the results of tests carried out for 15 years in a rural environment at Llanwrtyd Wells and in an industrial atmosphere at Sheffield[2]. At both sites the rate of rusting fell considerably after the first year or two, but then remained fairly constant for the rest of the test period. In actual figures the average rate of rusting in mm/y over the first two years was 0·058 for Llanwrtyd Wells and 0·118 for Sheffield; the corresponding rates over the last 10 years were 0·015 and 0·025, respectively.

Similar observations have been made in the United States, where a marked decrease in the rate of rusting has generally been observed after the early stages. For example, in one test, the total corrosion over the first two years was about 0·1 mm but the additional corrosion during the next 10 years was only about 0·06 mm. These are slow rates of rusting, however, and it is possible that the rust formed was more protective than usual, because of the relatively non-aggressive test environment.

The corrosion curves in Fig. 3.4 were obtained some years ago. Corrosion is markedly influenced by the pattern of pollution, which is changing in the United Kingdom, and consequently the long-term corrosion rates may change. There is some evidence based on more recent tests to indicate that in many industrial environments the corrosion rate of steel over periods of 15 years will drop to a greater extent than is shown in Fig. 3.4.

It should be made clear that all the rates of rusting in the atmosphere just quoted, relate to average general penetration and take no account of pitting. Serious pitting of steel exposed to atmospheric corrosion is uncommon on simple test plates, but it may be necessary to allow for this in some practical cases, where local attack may be occasioned by faulty design and other factors.

Rusting indoors and in enclosed spaces Although much of what has been written above is of general application, the experimental data apply specifically to exposure outdoors. The conditions of exposure indoors are somewhat different, because the steel is not exposed to rain and direct sunlight. Rusting depends on the condensation of moisture, which may evaporate from the surface much more slowly under enclosed conditions than outdoors; moreover, the rust tends to remain on the surface and may build up in time to a thick scaly layer.

The rates of rusting vary so much with the particular conditions of exposure that it is difficult to generalise, but useful information regarding rusting in enclosed spaces will be found in papers by Holzworth, Thompson and Boegehold[21] and by Stanners[22].

The first three authors were concerned with the rusting of motorcar bodies. They found that the rusts formed on steel under sheltered and exposed conditions, respectively, differed markedly in chemical composition, structure and protective properties. The second paper gives the results of exposure tests in many different indoor atmospheres, from which the following representative rates of rusting over one year are taken.

Type of atmosphere	*Rate of rusting* (mm/y)
Domestic kitchens and bathrooms	0·0025–0·010
Laundry	0·0075
Bleach house	0·043
Sulphuric acid plant	0·048
Paper mill	0·068
Locomotive shed	0·080
Steelworks' pickling plant	>0·45

Rusting in Water

Introduction

The corrosion of steel by natural and industrial waters is a complicated and many-sided phenomenon, which cannot be dealt with completely within the narrow compass available here. The difficulty arises from the fact that,

Table 3.5 Rates of rusting of mild steels in sea-water (BISRA)
(total immersion for 203 days at Plymouth)

*Type of steel**	*Analysis (%)*				*Average general penetration*
	C	Mn	P	S	(mm/y)
1. Basic Bessemer, rimming					
Ordinary	0·05	0·64	0·06	0·02	0·143
High phosphorus	0·03	0·31	0·14	0·04	0·143
High phosphorus and sulphur	0·03	0·30	0·10	0·07	0·148
2. Open-hearth, rimming					
Ordinary	0·13	0·33	0·03	0·03	0·143
From haematite pig	0·06	0·32	0·01	0·03	0·140
3. Open-hearth, killed					
Ordinary	0·10	0·35	0·03	0·02	0·140
From haematite pig	0·11	0·34	0·01	0·03	0·136
4. Open-hearth, killed					
Ordinary	0·22	0·71	0·03	0·03	0·143
From haematite pig	0·21	0·58	0·02	0·03	0·158

*The copper contents of the steels, which were supplied through the courtesy of l'Office Technique pour l'Utilisation de l'Acier (France), varied from 0·03 to 0·11%. The killed steels contained about 0·04% Al and 0·1% Si.

of the three main considerations involved—the composition and surface condition of the steel, the quality of the water, and the operating conditions—the last is generally the most important. Moreover, in industry the operating conditions vary widely, and many of them need individual study. Consequently, all that will be attempted here is to state a few general facts and principles. For more detailed treatment the reader is referred to the recognised textbooks such as those of Evans[10], Hasse[23] and Butler and Ison[24]. A brochure published by the Iron and Steel Institute[19] may also prove helpful.

Effect of the Metal Composition

All ordinary ferrous structural materials, mild steels, low-alloy steels and wrought irons corrode at virtually the same rate when totally immersed in natural waters. Wrought iron may be slightly more resistant than mild steel; in a test in sea-water at Gosport, Scottish wrought-iron specimens lost about 15% less weight after 12 months immersion than specimens of ordinary mild steel. As shown in Table 3.5, the process of manufacture and the composition of mild steel do not affect its corrosion rate appreciably[25].

Although small copper additions generally have no influence on the corrosion rate of steel immersed in water, under certain boiler conditions small amounts of copper of up to 0·3% may have a marked effect on reducing the depth of pitting. This is attributed to the catalytic action of metallic copper on the formation of a protective layer of magnetite which diverts the attack sideways producing general rather than pitting corrosion[23].

Generally, over 3% of alloying additions such as chromium are necessary to obtain any marked improvement in the corrosion-resistance of steel in waters.

Effect of the Surface Condition

The surface condition of the steel at the time of exposure is of great importance however. This is because many natural waters are good electrolytes, so that there is ample opportunity for electrolytic corrosion when steel is permanently in contact with them. The presence of millscale on the surface is more dangerous, for example, when steel is immersed in sea-water than when it is exposed to air, for the galvanic cell formed by millscale and bare steel can operate much more freely under the former conditions. This may lead to rapid pitting; pits up to 1·25 mm deep were found on as-rolled steel specimens after six months immersion in sea-water at Gosport[26].

It follows that for many practical purposes where steel is exposed to water without a protective coating—boiler tubes are a good example—it is desirable to remove millscale before putting the parts into use.

Effects of Welds

Serious pitting may occur in the area of welds, particularly in sea-water. Corrosion rates of up to 10 mm/y have been reported in weld joints of ice-breakers. The severe corrosion has been attributed to galvanic effects between the weld metal and the steel plate. The use of more noble electrodes for welding are reported to overcome this problem[27].

Effect of the Water Composition

Saline and acid waters are particularly aggressive to mild steel, so the composition of the water is clearly important in determining the rate of rusting of steel exposed to it. Some of the main factors here are the nature and amount of the dissolved solids, which influence the electrical conductivity, pH value and hardness of the water, the carbon dioxide and oxygen contents, and the presence of organic matter.

Dissolved solids The effect of dissolved solids is complex. The presence of inorganic salts, notably of chlorides and sulphates, should promote corrosion, because they increase the conductivity of the water, thereby facilitating the electrochemical rusting process; moreover, chlorides at least may be detrimental to the development of protective films (see below). Alkaline waters tend to be less aggressive than acid or neutral waters, and rusting can be repressed entirely by making the water strongly alkaline. Unfortunately, at pH values just insufficient to give complete passivation, there is a grave danger of severe pitting, even though the total corrosion is reduced, and this for many purposes is a greater evil.

The most important property of the dissolved solids in fresh waters is whether or not they are such as to lead to the deposition of a protective film on the steel that will impede rusting. This is determined mainly by the amount of carbon dioxide dissolved in the water, so that the equilibrium between calcium carbonate, calcium bicarbonate and carbon dioxide, which has been studied by Tillmans and Heublein[28] and others, is of fundamental significance.

Since hard waters are more likely to deposit a protective calcareous scale than soft waters, they tend as a class to be less aggressive than these; indeed, soft waters can often be rendered less corrosive by the simple expedient of treating them with lime (Section 2.3).

Dissolved gases Oxygen and carbon dioxide are the most important dissolved gases in water. Oxygen is an effective cathodic depolariser and the cathodic reaction in water is generally oxygen reduction (see Sections 1.4 and 9.1). At ordinary temperatures in neutral or near neutral water, dissolved oxygen is necessary for any appreciable corrosion of steel. Increasing the oxygen concentration results in an acceleration of the corrosion of steel up to a certain concentration but beyond this the rate of corrosion is reduced. In slowly-moving distilled water, it has been found[29] that the critical concentration is 12 m/l. This value increases with temperature and in the presence of certain dissolved salts; it decreases with high velocities.

Carbon dioxide affects the acidity of the water and, as already noted, influences the formation of protective carbonate scales.

Presence of organic matter Another important factor is that most natural waters are far from being sterile. They contain greater or lesser amounts of organic matter, both living and dead. Some of the dead organic matter, e.g. peat residues, may render the water corrosive by making it acid, but in most cases the living organisms probably exert the greater influence. In natural sea-water fouling occurs, and in fresh waters algae may grow. Moreover, there are a number of strains of bacteria, such as the sulphate-reducing bacteria (see Section 2.6), that can influence the rusting process under immersed conditions.

Effect of the operating conditions The operating conditions have an important influence. Generally the factors involved are complex; they include temperature, rate of flow, design features and stray currents.

The temperature of the water affects the rate of rusting in several ways. First, the corrosion process shares the general tendency of chemical reactions to increase in speed with rising temperature. More important, however, are the effects of temperature on the nature and solubility of the corrosion products. For example, a rise in temperature will often throw down a carbonate scale; moreover, it increases the rate of diffusion of oxygen through water but decreases the solubility of this gas. Some of these effects are conflicting, with the result that under certain laboratory conditions at least the rate of rusting/temperature curve for steel immersed in water passes through a maximum before the boiling point is reached—at about 80°C in experiments made by Friend[30].

The rate of water flow is also most important. This determines the supply of oxygen to the rusting surface, and may remove corrosion products that would otherwise stifle further rusting. A plentiful oxygen supply to the cathodic areas will stimulate corrosion, but so may smaller supplies at a slow rate of flow, if this leads to the formation of differential aeration cells (see Section 1.6).

At sufficiently high rates of flow in natural waters enough oxygen may reach the surface to cause partial passivity, in which case the corrosion rate may decrease[31]. In sea-water, owing to the high concentration of chloride ions, the corrosion rate increases with velocity. In one series of tests, corrosion under static conditions was 0·125 mm/y, 0·50 mm/y at 5 ft/s and 0·83 mm/y at 15 ft/s.

Design must also be considered. Sharp changes in the direction of flow, as in a badly designed water box, may lead to severe local damage by impingement attack. Severe galvanic corrosion can result from the injudicious juxtaposition of steel with non-ferrous metals, such as copper or bronze. Even when the ferrous and non-ferrous metals are not in direct contact, local corrosion cells can be set up round small particles of non-ferrous metal that have been dissolved by the water and redeposited on to the steel. Corrosion cells can also be formed when steel is in contact with solutions of different saline contents; for example, 'long line' currents have been observed in a lock gate that was in contact with a layer of fresh river water flowing out over a layer of sea-water.

Table 3.6 Rates of rusting of mild steel in natural waters *(total immersion)*

Type of water	*Test authority*	*Test site*	*Test duration* (years)	*Average general penetration* (mm/y)
Sea-water	Institution of Civil Engineers[32]	Halifax, Nova Scotia	15	0·108
		Plymouth	15	0·065
	BISRA[33]	Emsworth	5	0·065
Fresh water	Institution of Civil Engineers[32]	Plymouth: fresh-water reservoir	15	0·043
River water	Office Technique pour l'Utilisation de l'Acier[34]	La Cadène: granite bed, very pure water	5	0·068
		Dôle: highly calcareous water	5	0·010

Rates of rusting in natural waters It will be evident from the preceding remarks that the results of laboratory experiments or field tests on corrosion by natural waters can be applied to practical cases only with considerable reserve. Yet the reader may well wish to gain some general idea of the rates of rusting involved. For this purpose a few representative experimental results are given in Table 3.6; they apply to ordinary low-carbon structural steel tested under the conditions stated. The figures are for the average general penetration over the whole test areas. As an indication of the rate of pitting, it may be noted that in the sea-water tests of the Institution of Civil Engineers[32] the maximum depth of pitting for descaled mild steel after 15 years immersion was about 2·3 mm; when the steel had been immersed in the as-rolled condition with its millscale, a figure as high as 7·6 mm was observed. Under half tide immersion conditions the corrosion rate of steel may be increased by a factor of 2 to 5 compared with the results for total immersion quoted above.

Rusting in Soil

Introduction

The practical importance of soil corrosion to Great Britain was well brought out some years ago by a report of a Departmental Committee of the Ministry of Health[35] which will repay study. It is even more difficult to generalise about this problem than about rusting in water. With regard to the bare metals, the problem is of little more than academic interest, because in good practice iron and steel are not buried in corrosive soils without an adequate protective coating, often supplemented by an effective scheme of cathodic protection. The operating conditions are again of extreme importance, e.g. corrosive long-line currents may originate where sections of a pipeline run through adjacent dissimilar soils, or, more frequently, stray currents from adjacent electrical installations may enter the pipe; in both cases severe corrosion damage may be caused where the currents leave the pipe. The presence of stones, tree roots or attack by rodents may perforate the protective coating.

Although all these complicating factors cannot be reproduced in small-scale tests, it will be of value to summarise the main knowledge that has been gained from long-period burial trials conducted in the United States[36] and in Great Britain[37]. The subject will be considered under two heads: effect of metal composition and effect of the soil.

Effect of the Metal Composition

The corrosion of iron and steel in soil is generally electrochemical in character but the conditions are such that the corrosion products usually remain in contact with the metal. Moreover, the rate of oxygen supply is often low in comparison with that in air or in water. This is probably the main reason why, considered broadly, there are no major differences in the general corrosion rates of ordinary steels and cast irons when buried in most types of soil. These rates of general attack are comparatively low, say 0·038 mm/y over a 10-year period for burial in a clay soil, as compared with 0·1–0·13 mm/y for complete exposure outdoors in an industrial atmosphere.

Local corrosion or pitting is more important for practical purposes than the rate of general corrosion, and may proceed 10 times or so more rapidly than this. Inasmuch as certain types of cast iron are liable to suffer *graphitic corrosion**, whereas steel does not, steel might theoretically be expected to show to some advantage when used for buried pipelines. In practice, however, a cast-iron pipe has to be of stouter wall than a steel pipe for equal strength, and it is doubtful whether any distinction between the rust resistance of the two materials in the soil is justified.

It is also doubtful whether the surface condition of the bare metal at the time of burial has much significance. Some authorities consider that the casting skin on cast iron is protective, but the evidence on this point is conflicting. On the other hand, it is desirable to remove the millscale from steel

*Graphitic corrosion is associated with the presence of graphite flakes in the iron matrix, and results in the local replacement of iron by 'plugs' of graphite and corrosion products.

when, as is usual, a protective coating is applied. If the millscale was left on, it might cause the coating to spall.

The effect of stray currents arising from a d.c. source or from cathodic protection of an adjacent structure are considered in Sections 11.5 and 11.6.

Effect of the Soil

Soils vary greatly in corrosiveness, and the type of soil affects the corrosion rate much more than any variation in the ferrous material or in its method of manufacture. Although it is difficult to assess the corrosiveness of a particular soil beforehand, much useful information can be obtained from a well-conducted soil survey on the site.

In general, dry, sandy or calcareous soils, with a high electrical resistance, are the least corrosive. At the other end of the scale are the heavy clays and the highly saline soils, whose electrical conductivity is high. The depth of the water table is also important; much depends on whether the buried iron or steel is permanently above or below this, or even more perhaps on whether it is alternately 'wet' or 'dry'. The variation in corrosion rate with depth of burial is illustrated by the results given in Table 3.7, which also serve to indicate the rates of average general penetration in typical British soils. It will be noted that the depth of burial had no consistent effect, which is not surprising since the average depth of the water table and the seasonal fluctuations in this varied from one site to another.

Table 3.7 Effect of depth of burial on the rusting of mild steel flats (BISRA)

(test duration 5 years)

Site	*Type of soil*	*Average general penetration* (mm/y)	
		1·37 m	0·61 m
Benfleet	London clay	0·0185	0·0361
Gotham	Keuper marl	0·0132	0·0094
Pitsea	Alluvium	0·0353	0·0284
Rothamsted	Clay with flints	0·0201	0·0213

The maximum general corrosion rate reported in tests carried out by the U.S. National Bureau of Standards[36, 38] is 0·068 mm/y, the maximum rates obtained in tests carried out in the United Kingdom by BISRA[39] and the National Physical Laboratory[40] are 0·035 mm/y and 0·050 mm/y. However, the pitting rate was much greater, maximum pits of 0·25 mm/y have been reported from American and 0·30 mm/y from British tests.

Bacterial activity often plays a major part in determining the corrosion of buried steel. This is particularly so in waterlogged clays and similar soils, where no atmospheric oxygen is present as such. If these soils contain sulphates, e.g. gypsum and the necessary traces of nutrients, corrosion can occur under anaerobic conditions in the presence of sulphate-reducing bacteria. One of the final products is iron sulphide, and the presence of this is characteristic of attack by sulphate-reducing bacteria, which are frequently present (see Section 2.6).

Effect of the Duration of Burial

Finally, it should be added that the extensive field tests made in the United States[38] indicate that buried steel rusts less and less rapidly as time goes on, both as regards general attack and pitting. This can be illustrated by the

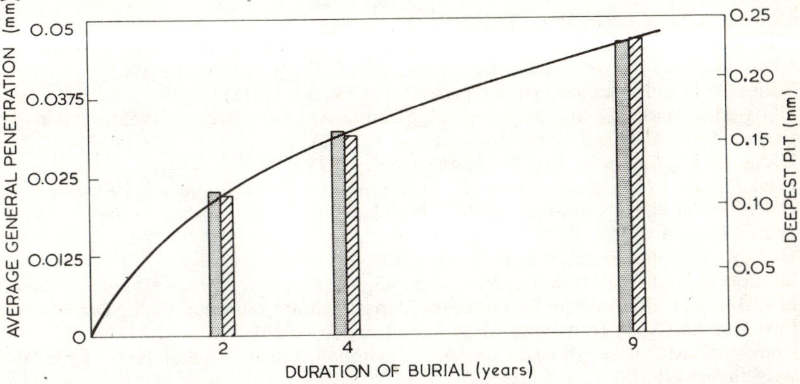

Fig. 3.5 Effect of duration of burial on the corrosion of mild steel (after Romanoff[38]). *Left:* average general penetration deduced from loss in weight, *right:* deepest pit. Figures are averages for 16 soils

typical results shown in Fig. 3.5. Field tests made in British soils by BISRA have not, however, exhibited the same tendency; in these rusting has been roughly proportional to the duration of burial.

K. A. CHANDLER
J. C. HUDSON

REFERENCES

1. Larrabee, C. P. and Coburn, S. K., *Proc 1st Int. Congr. Met. Corros.*, 1961, 276, London, Butterworths (1962)
2. Sixth Report of the Corrosion Committee, Spec. Rep. No. 66, Iron and Steel Institute, London (1959)
3. Vernon, W. H. J., *Trans. Faraday Soc.*, **31**, 1 668 (1935)
4. Ross, T. K. and Callaghan, B. G., *Corr. Sci.*, **6**, 337 (1966)
5. Chandler, K. A., *Br. Corros. J.*, **1**, July (1966)
6. Chandler, K. A. and Kilcullen, M. B., *Br. Corr. J.*, **3**, March (1968)
7. Rosenfel'd, I. L., *Proc. 1st Congr. Met. Corros.*, 1961, 243, London, Butterworths (1962)
8. Chandler, K. A. and Stanners, J. F., *Proc. 2nd Int. Congr. Met. Corr.* (N.A.C.E. Houston), 325 (1963)
9. Tanner, A. R., *Chemistry & Industry*, 1 027 (1964)
10. Evans, U. R., *The Corrosion & Oxidation of Metals*, Edward Arnold, London (1960)
11. Schikorr, G., *Werkstoffe Korros.*, **14** No. 2, 69 (1963)
12. Schwartz, H., *Werkstoffe Korros.*, **16** No. 2, 93 (1965)
13. Schwartz, H., *Werkstoffe Korros.*, **16** No. 3, 208 (1965)
14. Schikorr, G., *Korrosion und Metallschutz*, **17**, 305–313 (1941)
15. Harrison, J. B. and Tickle, T. C. K., *J. Oil Colour Chem. Assoc.*, **45** No. 8, 571 (1962)

16. Ross, T. K. and Callaghan, B. G., *Nature*, London, **211**, 25 (1966)
17. Golubev, A. Y. and Kadyrov, M. Kh., Repr. No. 3–67–487/13, Gosinti, Moscow (1967)
18. Sereda, P. J., *Ind. Eng. Chem.*, **52**, 157 (1960)
19. Dearden, J., *J.I.S.I.*, **159**, 241 (1948)
20. Larrabee, C. P., *Trans. Electrochem. Soc.*, **85**, 297 (1944)
21. Holzworth, J., Thompson, R. F. and Boegehold, A. L., *Trans. Soc. Auto. Engrs.*, **64**, 221 (1956)
22. Stanners, J. F., *J. Appl. Chem.*, **10**, 461 (1960)
23. Haase, L. W., *Werkstoffzerstorüng und Schutzschich bildung Im Wasserfach*, Verlag Chemie Weinheim Bergstr. (1951)
24. Butler, G. and Ison, H. C. K., *Corrosion and its Prevention in Waters*, Leonard Hill, London, 74 (1966)
25. Hudson, J. C., *J.I.S.I.*, **166**, 123 (1950)
26. Hudson, J. C., The Corrosion of Iron & Steel, Chapman and Hall, London, 61 (1940)
27. Uusitalo, *Proc. 2nd Int. Congr. Met. Corr.*, N.A.C.E., Houston, 812 (1963)
28. Tillmans, J. and Heublein, O., *Gesundheitsing*, **35**, 669 (1912)
29. Uhlig, H., Triadis, D. and Stern, M., *J. Electrochem. Soc.*, **102**, 59 (1955)
30. Friend, J. N., *Carnegie Schol. Mem.*, **11**, 113 (1922)
31. Uhlig, H. H., *Corrosion Handbook*, John Wiley, 391 (1948)
32. Friend, J. N., *18th Report of the Committee of the Institution of Civil Engineers on the Deterioration of Structures of Timber, Metal and Concrete exposed to the Action of Sea Water*, London (1940)
33. Hudson, J. C. and Stanners J. F., *J.I.S.I.*, **180**, 271 (1955)
34. Baudot, H. and Chaudron, G., *Rev. Met.*, **43**, 1 (1946)
35. Ministry of Health, Interim Report of the Departmental Committee on the Deterioration of Cast Iron and Spun Iron Pipes, H.M.S.O., London (1950)
36. Romanoff, M., *Underground Corrosion*, National Bureau of Standards, Circular 579, Washington (1957)
37. Hudson, J. C. and Acock, J. P., *Symposium on the Corrosion of Buried Metals*, The Iron & Steel Inst., Special Report No. 45, London (1952)
38. Romanoff M., *J. Res. Nat. Bur. Stand.*, **660**, 223–224 (1962)
39. Hudson, J. C. and Watkins, K. O., BISRA Open Report No. MG/B/3/68
40. Booth, C. N. *et al.*, *Br. Cor. J.*, **2** No. 3, 104–118 (1967)

3.2 Low-Alloy Steels

Physical and Mechanical Properties

The mechanical properties of low- or medium-carbon structural steels can be improved considerably by small alloy additions; for example, 1% of chromium will raise the yield point of 0·2% carbon steel from about 280 MN/m^2 to 390 MN/m^2. This has led to the development of a range of so-called 'low-alloy steels' with high tensile properties; a typical material, which originated in America, contains about 0·8% Cr, 0·4% Cu, 0·5% Si and 0·15% P. Other elements used in low-alloy steels include manganese, molybdenum and nickel.

Although, originally at least, the main object was to increase the strength of the steel, improvements in the mechanical properties of unalloyed steels have resulted in a considerable overlap in properties between the two classes. In some cases, though by no means all, low-alloy additions, besides making further improvements in properties possible, may also enhance resistance. It is to such steels that this chapter specifically refers. As a class they are by no means uncorrodible but under favourable conditions such as when they are freely exposed outdoors, some of them rust several times less rapidly than unalloyed mild steel. Methods of fabrication differ little in principle from those already described. To maintain rust resistance and uniform appearance, matching welding rods made from suitable low-alloy steels have to be used.

The most common alloying element is copper, but it is doubtful whether plain copper steels should be classed as low alloy, since, when copper is progressively added to steel, there is little further improvement in rust resistance once about 0·2% copper has been reached. As copper contents of 0·1% or more are unavoidably introduced into steel at many works through the scrap used in the steelmaking process, the point of distinction between ordinary steels and copper steels is ill-defined. Copper steels will, however, be discussed here, because their behaviour helps to explain the reasons for the better rust resistance of slow-rusting steels generally.

Corrosion Behaviour in Aqueous Environments

Theoretical Considerations

The improvement in rust resistance achieved through low-alloy additions obviously depends on the nature and amounts of the alloying elements—

incidentally their effects are not additive—and to an even greater degree on the nature of the corrosive environment. To make a broad generalisation, slow-rusting steels show to maximum advantage when they are freely exposed to the open air in industrial environments; it is doubtful whether, from the corrosion aspect, their use is worthwhile under sheltered conditions of atmospheric exposure, or where immersion in natural waters, or burial in the soil is involved. Data to illustrate these points, taken mainly from British and American sources, are given in the following paragraphs. No fully satisfactory theory has yet been advanced to cover all aspects of the corrosion properties of the low-alloy steels. The effect of copper has been the most fully studied. Copson's proposal[1] that the copper in a copper steel consolidates the rust by forming basic sulphates that are less soluble than those of iron, may partially account for the effect of copper and nickel but cannot be extended to other beneficial elements. Copper has long been known[2] to suppress the adverse effect of sulphur in the steel on its corrosion resistance, although the manganese normally present in steels may also fulfil this rôle.

Porosity in the rusts appears to be a factor; several observers have reported a compact, dense rust layer next to the steel on slow-rusting low-alloy steels that is not present on mild steel. Reduced water uptake observed at humidities below saturation[3] may be the result of lower porosity.

Cathodic additions (such as copper and chromium) to low-alloy steels may also influence the rate of rusting by raising the potential of the surface to more noble values and so encouraging passivation[4]. Electrochemical measurements certainly seem to bear this out and they have recently been used in attempts to develop improved compositions[5,6].

Characteristic Features of Corrosion Behaviour

When low-alloy steels are exposed outdoors, the rust formed on them is generally darker in colour and much finer in grain than that formed on ordinary steel. Moreover, the slowing down in rusting rate with time (cf. Section 3.1, p. **3**:11) seems to be more marked for low-alloy steels than for ordinary steels. This can be illustrated by the BISRA[7] figures given in Table 3.8.

Table 3.8 Variation of rate of rusting with time (BISRA)[7]

Steel	*Rate of rusting* (mm/y) A *1st and 2nd years*	*Rate of rusting* (mm/y) B *6th to 15th year*	*Ratio,* B/A
Ordinary mild steel (Cu 0·02%)	0·129	0·094	0·73
Low-alloy steel (Cr 1·0%; Cu 0·6%)	0·077	0·025	0·33

The distinguishing feature of the behaviour of the slow-rusting low-alloy steels is the formation of this protective rust layer. Corrosion in conditions where it cannot form is little different from that of unalloyed steel, although the particular alloying elements present will have some influence on the actual rate at which corrosion occurs.

Statements of this kind can give only a general idea of the improvements to be gained by the use of low-alloy steels in the atmosphere, because so much depends upon the conditions of exposure. In particular, the beneficial effects observed in the open air do not generally extend to conditions where the steel is enclosed and sheltered from the rain. Thus the rates of rusting of bare steels, exposed for five years in Dove Holes Tunnel, England, were as follows[8]:

Copper content of steel (%)	0·02	0·2	0·5
Rate of rusting (mm/y)	0·061	0·060	0·059

As a further example of the ineffectiveness of low-alloy additions in slowing down rusting under sheltered conditions, tests by BISRA in indoor atmospheres have failed to reveal any substantial difference in the rusting of a chromium–copper steel and of an ordinary mild steel in most of them[9]. The test sites covered a wide range of domestic and industrial conditions, from bathrooms to locomotive sheds.

Corrosion in Natural Environments

In the Atmosphere

The general effect of low-alloy additions on the rusting of structural steel in the open air is illustrated in Fig. 3.6, which shows the results of tests conducted by BISRA for 15 years in an industrial district of Sheffield[7]. Figure 3.7 refers to similar trials of 9 years duration at Rotherham[10]. Similar curves based on American tests[11], reproduced in Fig. 3.8, show substantially the same features. The American specimens were exposed obliquely, so that the rates of rusting of the skyward and groundward faces differed; the figures given are averages for both. The fact that the rates of rusting are markedly slower in the American than in the British tests is mainly due to differences in the corrosiveness of the test atmospheres.

The effects of the various alloying elements are not additive. Bearing this in mind, the practical effect of individual elements can be summarised as follows.

(i) *Copper* additions up to about 0·2% give a marked improvement, but further additions make little difference.

(ii) *Phosphorus*, at least when combined with copper, is also highly beneficial. However, in practice, levels above about 0·10% adversely affect mechanical properties.

(iii) *Chromium*, in fractional percentages, has a significant influence on corrosion rates. While it appears to be beneficial, some conflicting results have been reported and its contribution to the reduced corrosion of complex low-alloy steels containing copper and phosphorus is not large.

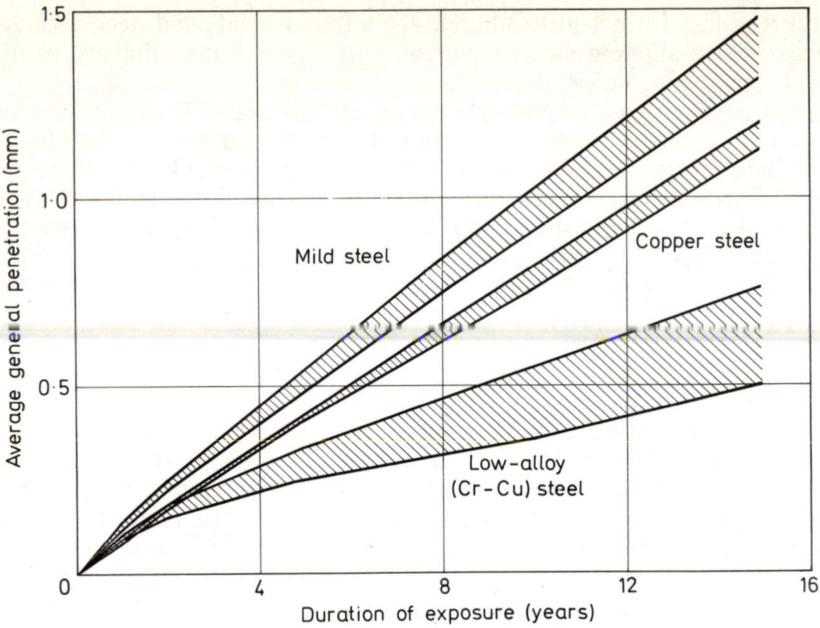

Fig. 3.6 Effect of low-alloy additions on the corrosion of steel outdoors[7]

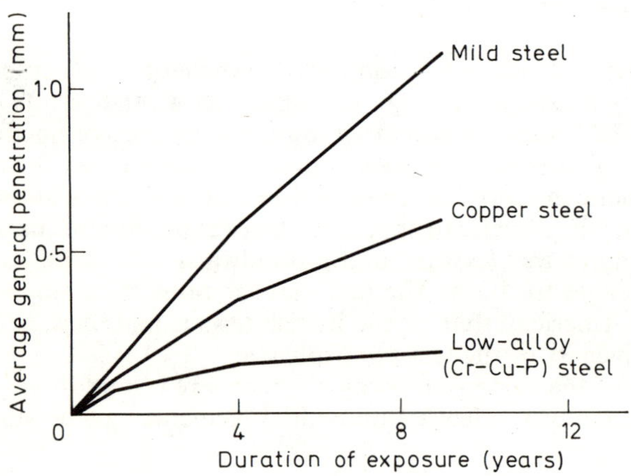

Fig. 3.7 Effect of low-alloy steels on the corrosion of steel outdoors (after Edwards[10])

(iv) *Nickel*, while reducing corrosion rates a little, is not as important in its effect as the aforementioned elements.

(v) *Manganese* may have a particular value in chloride-contaminated environments, but its contribution is little understood.

(vi) *Silicon* is in a similar position to manganese, with conflicting evidence as to its value.

(vii) *Molybdenum* had been little used in low-alloy steels, but may be as effective as copper and is worthy of further study.

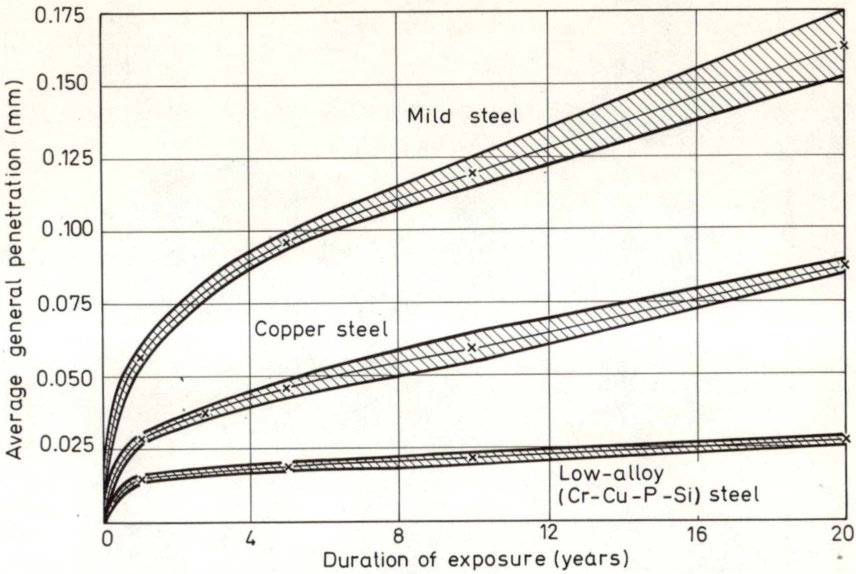

Fig. 3.8 Effect of low-alloy additions on the corrosion of steel outdoors at Kearny, N.J. (after Larrabee and Coburn[11])

Recently, the effects of these and other alloying additions and of their combinations have been extensively reviewed[12] (over 180 references).

The data in Table 3.9, while not entirely consistent, give some indication of the extent of the variations that can occur.

Immersed

Of the elements mentioned in Table 3.9 only chromium has been found to reduce the rate of rusting of steel appreciably when immersed in sea-water. For example, the following data were obtained in British tests[13]:

Chromium content of steel (%)	0	1	1·5	2	2·5
Rate of rusting (mm/y)	0·066	0·056	0·053	0·048	0·041

An addition of 3% of chromium might be expected to halve the corrosion rate in this medium. Such steels are rather outside the range of low-alloy steels, because they are relatively expensive and may need heat treatment in certain applications. They have, however, been used for special purposes, such as bolts for lock gates.

The effect of copper and chromium–copper additions on the corrosion rate of steel in natural waters, both fresh and salt, has also been studied in France by the Office Technique pour l'Utilisation de l'Acier[14]. The results obtained were not entirely concordant, and were markedly affected by the composition of the water. In general, chromium–copper steels failed to show any advantage, particularly in sea-water, but there were indications that copper steels with a high phosphorus content resisted rusting somewhat better than ordinary mild steels. The magnitude of the difference should not be exaggerated, however, because at all sites the loss in weight of the most and least

Table 3.9 Effect of individual low-alloy additions on the rate of rusting of low-carbon steels in the open air

Basis steel composition (%)	Alloying elements, %* and rusting rates†														
	Al	*Rusting rate*	Cr	*Rusting rate*	Cu	*Rusting rate*	Mn	*Rusting rate*	Mo	*Rusting rate*	Ni	*Rusting rate*	P	*Rusting rate*	Si	*Rusting rate*
Tests at Kearny, N.J.[11]‡																
0·01–0·02 Cu 0·25–0·40 Mn ⩽0·01 P, ⩽0·10 Si					0·24	21							0·10	26	0·61	49
0·01–0·02 Cu 0·25–0·40 Mn 0·06 P, 0·2 Si			1·3	100	0·21	52					1·0	44			0·53	100
0·05 Cu 0·25–0·40 Mn ⩽0·01 P, ⩽0·10 Si					0·24	69							0·10	59	0·56	76
0·05 Cu 0·25–0·40 Mn 0·06 P, 0·2 Si			1·3	91	0·21	77					1·0	64			0·50	92
Tests at Sheffield[13]§																
0·01–0·02 Cu 0·5 –0·6 Mn 0·02–0·06 P 0·1 –0·2 Si			2·6	38	0·23	52	1·8	79	0·55	60	3·1	30				
0·06 Cu 0·5 Mn 0·05–0·06 P 0·2 Si	1·6	72	1·0	67	0·23	72							0·16	87	0·81	93

*Total amount, including that in basis steel. †Taking the rate of rusting of the unalloyed basis as 100. ‡Duration of test 15·5 years. §Average values for 1, 2 and 5 years.

corroded steels lay within plus or minus 25% of the average value for the whole set.

On the whole, it is improbable that low-alloy steels other than those of relatively high chromium content mentioned above will give much better results than ordinary steel when immersed in natural waters. This conclusion is supported by Forgeson, Southwell and Alexander, who concluded from extensive tests in fresh and salt waters of the Panama Canal Zone that: 'Proprietary low-alloy steels were not in general more resistant to underwater corrosion than the mild unalloyed carbon steel'[15].

Underground

In tests by BISRA, made over three years in a heavy clay soil at Binfield, additions of chromium and copper had no beneficial effect on the rusting of buried mild steel, as is shown in the following figures[16]:

Alloying elements (%)	Nil	Cu 0·5	Cr 1·0	Cr 0·6, Cu 0·5
Rate of rusting (mm/y)	0·026	0·033	0·031	0·026

The apparent differences have little significance since the experimental error was in the region of 10%.

Additional evidence is available from tests made by the National Bureau of Standards on 10 varieties of steel, which were buried in 15 typical American soils from 1937 to 1950[17]. The results showed that, with few exceptions, the corrosion of low-alloy steels containing copper, nickel and molybdenum in various combinations did not differ by more than 20% from that of ordinary open-hearth steel. Additions of 2 or 5% of chromium did increase the corrosion resistance, however, as is indicated in Fig. 3.9*.

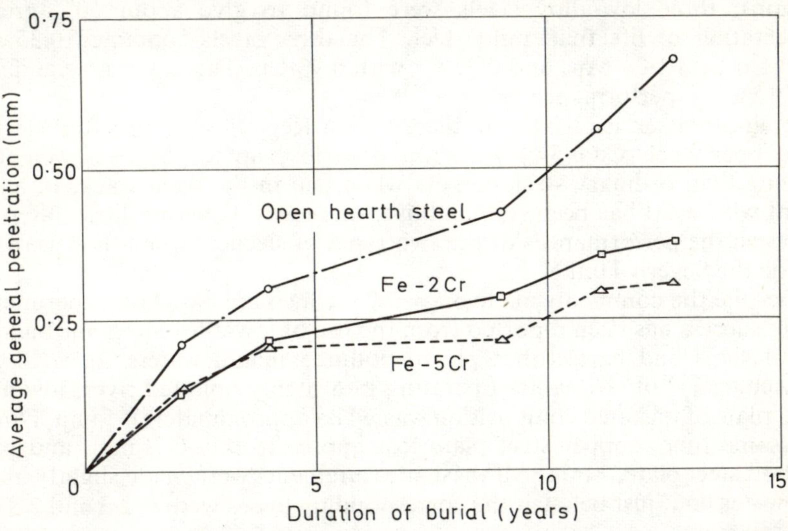

Fig. 3.9 Effect of chromium content on the corrosion of buried steel (after Romanoff[17])

*The points shown in these graphs have been calculated by the present authors from the original data; they are averages for the three soils Nos. 58, 63 and 64 in the American paper. There were only single samples of open-hearth steel and of 2% Cr steel, but there were three 5% Cr steels, the results of which have been averaged. Some of the chromium steels also contained about 0·5% Mo.

Applications in Industry

Low-alloy steels suitably selected for the protective nature of the rust formed on them in the atmosphere may be used in the bare condition. An important feature of this application is that they are less susceptible than unalloyed steel to increased pollution.

Extra care must be taken with the geometrical design and in detailing to avoid moisture traps and to prevent unsightly run-off of rust stain on to light-coloured surfaces such as concrete. The currently available steels should not be used in severe marine and industrial environments, nor do they develop a satisfactory appearance indoors[18]. Technical brochures giving full advice on the use of these materials are available[19].

Low-alloy steels show to most advantage in open-air structures that are deliberately built of bare steel or on which it is impracticable to renew the protective coating after they have entered into service. Their use for structures such as motorway bridges has now become widespread and, in many countries, they are also in use in buildings that feature the successful application of their novel architectural appearance[18].

Such publicised uses tend to mask their wide use in other applications over many years. Railway wagons are a good example, and low-alloy steels have long been used in the United States for their construction. The results of trials on coal gondolas in service on an American railroad[20] indicate that sheets made from a commercial low-alloy steel (0·14 P, 0·02 S, 0·8 Si, 0·4 Cu, 1·0 Cr) would last one and a half to two times as long as copper steel (0·3% Cu) sheets, if undue local attack at laps and ledges (due to the accumulation of wet coal dust) was prevented*. Similar results have been obtained in England. In a service trial[21] of 15 years duration on floor-plates of coal wagons, three low-alloy steels were found to give about 50% greater expectation of life than mild steel. The three steels contained 0·35 Cu; 0·44 Cu with 1·34 Mn; and 0·33 Cr with 0·90 Cu. There was no significant difference in performance between them.

In another service trial[22] on the Western Region of British Rail sleepers of copper steel (Cu 0·5%) were found to lose up to 50% less weight by rusting than ordinary steel sleepers when laid in the open track. In agreement with what has been stated above, there was, however, little difference between the performances of the two types of sleepers when they were laid inside the Severn Tunnel.

Despite the comments made on p. 3.25, which are based on experimental data, success has been reported from the use of low-alloy steels for building small ships and barges that ply in polluted inland waters. In a 10-year service trial[23] on a steamer operating in a highly polluted river, low-alloy steel plate of unstated composition wasted by approximately 0·5 mm. During the same time, copper steel plate lost approximately 0·75 mm, and plain carbon steel plate, 1 mm. All these measurements were made slightly above the water line; just below it the corresponding losses were 1, 2·3 and 2·8 mm

*This proviso draws attention to the fact already mentioned that the enhanced corrosion resistance of low-alloy steels is not usually obtained in wet conditions. Presumably the copper steel sheets, in turn, would have outlasted sheets of ordinary steel.

respectively, so that the superiority of the low-alloy steel was even more marked under conditions of complete immersion than in air.

Other obvious applications of low-alloy steels are for roofing sheets, wire ropes and chain-link fencing. The advantages here, though appreciable, are not so great as would appear at first sight, because a high proportion of such products is at least galvanised, if not plastics-coated as well, and the effect of the low-alloy additions begins to operate only after the zinc coating has failed.

Most structural steelwork that is exposed to the atmosphere is given a protective coating of some kind. If this coating is continuously maintained in perfect condition, so that no rusting of the steel takes place, there is no advantage from the corrosion aspect in using a low-alloy steel instead of mild steel. If, on the other hand, it is probable that the protective coating will be damaged or allowed to deteriorate, the use of a low-alloy steel should be considered. The more compact rust film formed on these steels will be less likely to cause the coating adjacent to the corroded areas to spall off, and the rate at which breakdown of the coating spreads will be reduced. Several investigators have reported better performance and durability of painting schemes on low-alloy steels than on ordinary steel. For example, Copson and Larrabee have written[24]: 'Both field tests and service experience have shown that paint coatings are more durable on high-strength low-alloy steels than on carbon steel or on copper steel. Any rust which forms at breaks or holidays or underneath the paint film is less voluminous on the low-alloy steels. Owing to the smaller volume of rust there is less rupturing of the paint film and, hence, less moisture reaches the steel to promote further corrosion.' In conformity with this, low-alloy steels could be used with advantage for such purposes as the production of agricultural machinery, on which the coating is liable to get knocked about, and which is frequently left in the open fields for long periods.

J. C. HUDSON
J. F. STANNERS

REFERENCES

1. Copson, H. R., *Proc. Amer. Soc. Test. Mater.*, **45**, 554 (1945)
2. Buck, D. M., *Proc. A.S.T.M.*, **19** No. 2, 224 (1919)
3. Skorchelletti, V. V. and Tukachinsky, S. E., *Zh. Prikl. Khim.*, Leningr., **26** No. 1, 30 (1953)
4. Tomashov, N. D. and Lokotilov, A. A., *Annals-Korriziya i Zaschita Stallei*, 171, Mashgiz, Moscow (1959)
5. Becker, G., *Arch. Eisenhüttenwes*, **36**, 489 (1965)
6. Pourbaix, M., Rapport Technique No. 160, Cebelcor, Brussels (1969)
7. Sixth Report of the Corrosion Committee, The Iron and Steel Institute, Special Report No. 66, London (1959)
8. Fourth Report of the Corrosion Committee, The Iron and Steel Institute, Special Report No. 13, London (1936)
9. Stanners, J. F., *J. Appl. Chem.*, **10**, 461 (1960)
10. Edwards, A. M., Proc. Symp. on Developments in Methods of Prevention and Control of Corrosion in Buildings, British Iron and Steel Federation, London (1966)
11. Larrabee, C. P. and Coburn, S. K., Proceedings of the First International Congress on Metallic Corrosion, London, 1961, Butterworths, London, 276 (1962)

12. Chandler, K. A. and Kilcullen, M. B., *Brit. Corros. J.*, **5** No. 1, 24 (1970)
13. Hudson, J. C. and Stanners, J. F., *J. Iron St. Inst.*, **180**, 271 (1955)
14. Baudot, H. and Chaudron, G., *Rev. Métall.*, **43**, 1 (1946)
15. Forgeson, B. W., Southwell, C. R. and Alexander, A. L., *Corrosion*, **16**, 105t (1960)
16. Hudson, J. C., Banfield, T. A. and Holden, H. A., *J. Iron St. Inst.*, **146**, 107 (1942)
17. Romanoff, M., *Underground Corrosion*, National Bureau of Standards Circular 579, US Government Printing Office, Washington (1957)
18. Colbridge, G. B. and Chandler, K. A., *Construction Steelwork*, 28–32, Jan./Feb. (1970)
19. *Weathering Steel in Bridgework*, British Steel Corporation, London (1971)
20. Kelly, B. J., *Corrosion*, **7**, 196 (1951)
21. Hudson, J. C., *J. Iron St. Inst.*, **194**, 45 (1960)
22. Hudson, J. C., *J. Iron St. Inst.*, **169**, 13 (1951)
23. *Mat. and Meth.*, **24**, 117, Manual No. 102, Feb. (1954)
24. Copson, H. R. and Larrabee, C. P., *Bull. Amer. Soc. Test. Mater.*, No. 242, 68 (1959)

3.3 Stainless Steels

Introduction

The scope of the term *stainless* steel has not been precisely defined, but for general purposes it may be considered to include alloys whose main constituent is iron but which also contain not less than 10% Cr. As with low-alloy steels, a distinction between low or medium carbon grades and high carbon grades must also be drawn, the latter being more in the nature of alloy cast irons. These are used mainly for oxidation resistance at high temperatures and for applications where abrasion resistance allied to a certain amount of corrosion resistance is required, and will not be considered in this section.

The term *stainless* steel is not, of course, a strictly accurate description, but the difference in behaviour between chromium-bearing steels and carbon steels in many environments is so marked as to justify the adoption of the name.

H. Brearley was the first to see clearly the commercial possibilities of the 12–13% Cr martensitic steel, introduced to the public in the form of stainless cutlery over the period 1912–1915. As a result, he has been very widely considered the 'inventor' of stainless steel. In fact, a remarkable amount of study had been made of such alloys over the previous decade. For example, Guillet (1902–1906) and Portevin (1909–1911) in France, and Giesen (1907–1909) in England had worked on the metallurgy and physical properties. They noted etching difficulties, but made no special study of corrosion resistance. On the other hand, Monnartz in Germany had, in 1908–1909, given detailed consideration to this aspect. He showed a remarkable appreciation of passivity and its relation to oxidation and potential, the rôle of chromium content in resistance to nitric acid, and the effect of carbon. He discussed carbide stabilisation by chromium and the use of molybdenum additions for improved corrosion resistance. He also carried out tests in the atmosphere, tap water and sea-water. In spite of this, it was left to Brearley to visualise the full practical possibilities of such a steel, given the optimum combination of composition, heat treatment and surface finish.

The austenitic steels with the well known 18Cr–8Ni basis came into use in the 1920s.

As with the chromium–iron alloys, the study of the records of early development reveal a complicated story, again leading to the pioneering work of Guillet in France and Giesen in England during the first decade of this century. Maurer and Strauss, working for Krupp in Germany, were most prominent in the development over the next 10 years, but the earlier Krupp steel was of higher chromium and carbon content, as indicated by Strauss, who in 1924 gave 0·25C–20Cr–7Ni as typical (A.S.T.M. 1924). Practical development of this steel in Sheffield led to lowering of the chromium and carbon contents and by 1924 the familiar 18–8 composition with 0·10–0·15% C was established.

The Krupp work had shown interesting improvements in acid resistance resulting from molybdenum and copper additions, and the use of 2–3% Mo for more difficult acid conditions was soon established. Other early additions were made to overcome susceptibility to intercrystalline corrosion, culminating in the general use, by the early thirties, of titanium additions for carbide stabilisation, followed shortly after by the alternative use of niobium.

Classification of Stainless Steels

The various requirements of not only the designer and user but also the fabricator and the steel producer have led to considerable development in alloys. Not only have new grades been devised, but the traditional grades have in some instances been slightly modified too, sometimes repeatedly, to facilitate developing production by modern methods. This has resulted in a situation which may appear bewildering to many, with numerous specifications and proprietary grades, the objectives of which may not be immediately clear. Fortunately the situation has been clarified to some extent by the adoption of new sets of British Standards which cover the major tonnage of the steels used. These are BS 970; Part 4:1970 which covers blooms, billets, bars and forgings and BS 1449; Part 4:1967 which covers plate, sheet and strip. The specifications are summarised in Tables 3.10–3.12, the classification being according to structure (martensitic, ferritic, austenitic) which establishes the basic physical properties. The alloying elements shown are those which are of significance in establishing important properties. Other elements are present as incidental impurities or as deliberate additions made to facilitate production (e.g. silicon and manganese added as de-oxidants) but have no significant effect on properties. For details, the specifications should be consulted. (See Chapter 21.)

Martensitic Steels

A plain carbon steel, when heated to elevated temperature has an f.c.c. structure (austenite) but when subsequently cooled to ambient temperature transforms either to a b.c.c. form (ferrite) with iron carbides or to an acicular phase (martensite), depending on the cooling rate; with a simple Fe–C alloy very rapid cooling rates are necessary to produce the martensitic structure. While ferrite is soft, martensite is hard, the actual hardness depending on the carbon content, and it is the production and heat treatment of this structure which allows development of strengths higher than those of the mild steels of essentially ferritic structure.

The addition of chromium, even in relatively small quantity, allows the martensitic structure to form on cooling at relatively slow rates and so the grades shown in Table 3.10 are air hardening, i.e. they form martensitic structures on cooling in air without resort to quenching, even in quite large sections. Thus, as opposed to the simple carbon steels, special treatments must be adopted to soften these grades rather than to harden them. The treatment adopted is to 'temper' the martensite by reheating to some temperature below that at which austenite reforms. The degree of softening depends on the temperature and time of tempering and, during the treatment, carbides are precipitated while the structure changes to 'tempered martensite' or 'ferrite'. A detailed description of the tempering processes is given in Reference 1.

Reference to Fig. 3.10 shows why the chromium content of the martensitic steels is limited, despite the fact that increasing chromium content gives increasing corrosion resistance. This diagram, which is a modification of

Table 3.10 Martensitic steels

Steel number	*British Standard specification conformed with*	C	Cr	Mn	Ni	Mo	Al	S	Se
403S17	BS 970 and BS 1449 (sheet and strip)	0·08 max.	12·0–14·0						
405S17	BS 1449 (plate)	0·08 max.	12·0–14·0				0·30 max.		
410S21	BS 970 and BS 1449 (plate)	0·09–0·15	11·5–13·5						
420S29	BS 970	0·14–0·20	11·5–13·5						
420S37	BS 970	0·20–0·28	12·0–14·0						
420S45	BS 970	0·28–0·36	12·0–14·0						
416S21	BS 970	0·09–0·15	11·5–13·5	1·50 max.		0·60 max.		0·15–0·30	
416S41	BS 970	0·09–0·15	11·5–13·5	1·50 max.		0·60 max.			0·15–0·30
416S29	BS 970	0·14–0·20	11·5–13·5	1·50 max.		0·60 max.		0·15–0·30	
416S37	BS 970	0·20–0·28	12·0–14·0	1·50 max.		0·60 max.		0·15–0·30	
431S29	BS 970	0·12–0·20	15·0–18·0		2·0–3·0				
441S29	BS 970	0·12–0·20	15·0–18·0	1·50 max.	2·0–3·0	0·60 max.		0·15–0·30	
441S49	BS 970	0·12–0·20	15·0–18·0	1·50 max.	2·0–3·0	0·60 max.			0·15–0·30

Table 3.11 Ferritic steels

Steel number	*British Standard specification conformed with*	C	Cr	Mo
430S15	BS 970 and BS 1449	0·10 max.	16·0–18·0	
434S19	BS 1449 (sheet and strip)	0·10 max.	16·0–18·0	0·90–1·30
442S19	BS 1449 (sheet and strip)	0·10 max.	18·0–22·0	

Table 3.12 Austenitic steels

Steel number	*B.S.S. conformed with*	C	Cr	Mn	Ni	Mo	Ti	Nb	S	Se
301S21	BS 1449 (sheet and strip)	0·12 max.	16·0–18·0		6·0–8·0					
302S25	BS 970 and BS 1449 (sheet and strip)	0·12 max.	17·0–19·0		8·0–11·0					
303S21	BS 970	0·12 max.	17·0–19·0	1·00–2·00	8·0–11·0				0·15–0·30	
303S41	BS 970	0·12 max.	17·0–19·0	1·00–2·00	8·0–11·0					0·15–0·30
304S12	BS 970 and BS 1449	0·03 max.	17·5–19·0		9·0–12·0					
304S15	BS 970 and BS 1449 (plate)	0·06 max.	17·5–19·0		8·0–11·0					
304S16	BS 1449 (sheet and strip)	0·06 max.	17·5–19·0		9·0–11·0					
305S19	BS 1449 (sheet and strip)	0·10 max.	17·0–19·0		11·0–13·0					
315S16	BS 970 and BS 1449 (sheet and strip)	0·07 max.	16·5–18·5		9·0–11·0	1·25–1·75				
316S12	BS 970 and BS 1449	0·03 max.	16·5–18·5		11·0–14·0	2·25–3·0				
316S16	BS 970 and BS 1449	0·06 max.	16·5–18·5		10·0–13·0	2·25–3·0				
317S12	BS 970	0·03 max.	17·5–19·5		14·0–17·0	3·0–4·0				
317S16	BS 970 and BS 1449	0·06 max.	17·5–19·5		12·0–15·0	3·0–4·0				
320S17	BS 970 and BS 1449	0·08 max.	16·5–18·5		11·0–14·0	2·25–3·0	4 × C–0·60			
321S12	BS 970 and BS 1449	0·08 max.	17·0–19·0		9·0–12·0		5 × C–0·70			
321S20	BS 970	0·12 max.	17·0–19·0		8·0–11·0		5 × C–0·90			
325S21	BS 970	0·12 max.	17·0–19·0	1·00–2·00	8·0–11·0		5 × C–0·90		0·15–0·30	
326S36	BS 970	0·12 max.	16·5–18·5	1·00–2·00	10·0–13·0	2·25–3·00				0·15–0·30
347S17	BS 970 and BS 1449	0·08 max.	17·0–19·0		9·0–12·0			10 × C–1·0		

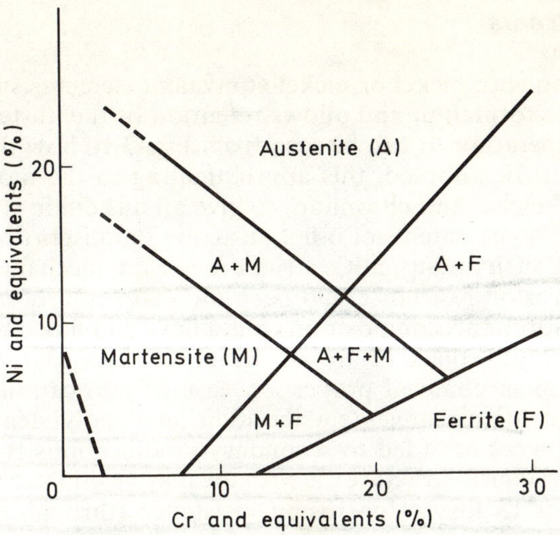

Fig. 3.10 Structure of steels after cooling from elevated temperature as determined by composition. *A* austenite, *M* martensite and *F* δ-ferrite (after Schaeffler)

that first proposed by Schaeffler, indicates the effect of composition on the structure of material after rapid cooling to room temperature from a temperature of 1050°C. The other alloying elements used in steel manufacture exhibit the same effect as either chromium or nickel, although to different extents. These elements, when present in significant quantity, must be taken into account when making practical use of such a diagram. For further details, References 1 and 2 should be consulted; in this instance the diagram is shown to illustrate the principle involved. It can be seen that too high a chromium content in a low nickel-equivalent steel leads to the presence of δ-ferrite in the martensitic matrix. This is termed *δ-ferrite* to distinguish it from the transformation product of austenite or martensite which is termed *α-ferrite*. δ-ferrite contains a higher proportion of chromium than the matrix; this excess chromium, being segregated in a second phase, is of no advantage to the bulk corrosion resistance of the alloy. In addition, the presence of δ-ferrite in a martensitic steel can have adverse effects on mechanical properties and hot workability. In 431S29 and its variants, a higher chromium content is utilised and a nickel addition is used to limit the δ-ferrite content. Nevertheless, some δ-ferrite is often present in this grade and its effect must be considered. The physical metallurgy of stainless steels is considered in Sections 1.3 and 9.2.

Ferritic Steels

It can be seen by reference to Fig. 3.10 that by having a sufficiently high chromium content and limiting nickel-equivalent elements, a fully δ-ferritic structure can be obtained in which the chromium is evenly distributed. This has therefore improved corrosion resistance over the martensitic types. The ferritic group of steels (Table 3.11) are of this type. However, the ferritic structure introduces limitations in mechanical properties and fabrication.

Austenitic Steels

Suitable alloying with nickel or nickel-equivalent elements suppresses martensite or ferrite formation and allows retention of the austenitic structure at ambient temperature. It can be seen from Fig. 3.10 how the '18–8' composition came to be adopted, this approximating to the most economical combination of nickel and chromium to give an austenitic structure, in the presence of tolerable contents of other austenite stabilisers (mainly carbon). Major assets of such an austenitic structure are the mechanical properties, but even if corrosion resistance had been the main consideration in establishing composition, this composition would have had much to commend it. The steels listed in Table 3.12 are all variants of this base, changes being in the interest of mechanical properties, ease of fabrication or basic corrosion resistance. Most changes (e.g. the addition of molybdenum to increase corrosion resistance) have led to secondary modifications to ensure retention of a fully austenitic structure; as with the martensitic steels, the presence of δ-ferrite leads to loss of corrosion resistance (through segregation of chromium and molybdenum in ferrite) and can affect mechanical properties and hot workability.

Free Cutting Grades

Several of the martensitic and austenitic variants listed have additions of sulphur or selenium with increased manganese and possibly molybdenum. These additions are made in the interests of improved machinability and use of such grades can facilitate the machining of complex parts or allow increased production rate. They do, however, adversely affect some mechanical properties and corrosion resistance although these are still perfectly acceptable for many applications. However, full discussion of these grades cannot be attempted here.

Precipitation Hardening Martensitic and Semi-austenitic Grades

A relatively recent development is the precipitation hardening types[3] which may be subdivided into martensitic and semi-austenitic. These were developed because of the need for certain projects for steels which could be

Table 3.13 Compositions of typical precipitation hardening and typical 'semi-austenitic' steels

	C	Si	Mn	Cr	Ni	Mo	Cu	Nb	Ti
Precipitation hardening	0·04	0·4	0·7	14·0	5·5	1·6	1·8	0·25	—
Semi-austenitic	0·05	0·4	1·3	15·8	5·5	1·6	2·0	—	0·08

hardened to high strength levels but which would also exhibit better corrosion resistance and weldability than the 13% chromium group. The need for improved fabrication characteristics also led to the development of the semi-austenitic types which can be fabricated in the austenitic state and then transformed by heat treatment to a strong martensitic state. The metallurgy of these steels is complex and for further information References 2 and 4 should be consulted. While included in some material specifications, these steels are known by their proprietary names and typical examples are given in Table 3.13.

Physical Properties

Typical physical properties for a selection of steels are given in Table 3.14.

Mechanical Properties

Martensitic Steels

These steels are normally used in the hardened and tempered conditions and the effect of tempering temperature on the strength of two grades is shown in Fig. 3.11. The lower carbon grades are most frequently used in a relatively soft state and typical properties are shown in Table 3.15. The toughness indicated is excellent, but it should be remembered that these steels show a sharp ductile/brittle transition at a critical testing temperature that varies according to strength and other features (such as grain size). Low toughness values at ambient temperature must be expected with treatments giving U.T.S. values above approx. 950 MN/m^2. Special proprietary grades with superior transition temperatures at higher strength levels have been developed (Table 3.15). Typical curves showing how toughness is related to temperature are given in Fig. 3.12 for 13% Cr steel and a modified type.

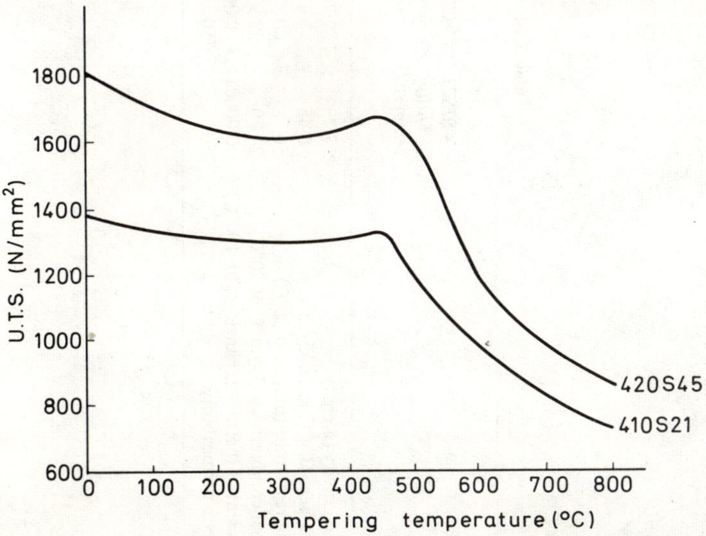

Fig. 3.11 Strength (U.T.S.) of two 13% Cr, martensitic steels as affected by tempering at various temperatures for 1 h following air cooling from 980°C

Table 3.14 Physical properties of some stainless steels

	410S21 (fully tempered)	*420S45 (cutlery temper)*	*431S39 (fully tempered)*	*Precipitation hardening (see Table 3.13, 560°C overaged)*	*430S15 (softened)*	*304S15 (softened)*	*321S12 (softened)*	*347S17 (softened)*	*316S16 (softened)*
Specific gravity (g/cm^3)	7·73	7·74	7·70	7·70	7·70	7·90	7·90	7·93	7·96
Specific heat (J/g°C at 20°C)	0·48	—	0·48	—	—	0·50	0·50	0·50	0·50
Specific resistance (μΩ cm at 20°C)	56	60	72	77	62	73	71	72	73
Thermal conductivity (W/m °C at 20°C)	24·70	20·93	18·84	17·17	21·77	15·91	15·91	15·91	16·33
Coefficient of thermal expansion (20–100°C)	0·000011	0·000011	0·000010	0·000011	0·000010	0·000016	0·000016	0·000017	0·000016
Magnetic permeability	750	750	190	110	290	1·005–2·03	1·5–2·0	1·005–1·03	1·05

Table 3.15 Mechanical properties of some martensitic steels

	410S21 (750°C temper)	*420S29 (750°C temper)*	*420S45 (cutlery temper, 200°C)*	*Alloyed 12% Cr steel* (650°C temper)*
0·2% P.S. (MN/m^2)	350	480	1 380	825
U.T.S. (MN/m^2)	605	710	1 530	1 020
Elongation (% $4\sqrt{A}$)	33	27	17	22
Reduction of area (%)	70	60	47	61
Fatigue limit (10^8 cycles, MN/m^2)	±230	±310	—	—
Impact strength, Izod (J)	133	108	22	79

*Composition (%): C 0·12, Cr 11·5, Ni 2·3, Mo 1·4, V 0·15 and Nb 0·3.

Table 3.16 Mechanical properties of typical precipitation hardening and semi-austenitic stainless steels

	*Precipitation hardening**			*Semi-austenitic*†*	
	Fully overaged[1]	*550°C overaged*[2]	*Peak hardened*[3]	*Softened*[4]	*Peak hardened*[5]
0·2% P.S. (MN/m^2)	550	990	1 130	310	1 080
U.T.S. (MN/m^2)	930	1 040	1 210	800	1 160
Elongation (% $4\sqrt{A}$)	27	23	22	18	10
(% 50 mm)	—	—	—	30	30
Reduction of area (%)	66	65	59	—	—
Fatigue limit (10^8 cycles, MN/m^2)	±650	±590	±475	—	—
Impact strength, Izod (J)	115	100	68	—	—

Notes: * See Table 3.12
† Sheet specimens
1. A.C. 1 050°C A.C. + 2 h 750°C + 2 h 620°C
2. A.C. 1 050°C A.C. + 2 h 750°C + 2 h 550°C
3. A.C. 1 050°C A.C. + 2 h 850°C + 4 h 450°C
4. A.C. 1 050°C A.C.
5. A.C. 1 050°C A.C. + 2 h 750°C + 4 h 450°C

Table 3.17 Mechanical properties of austenitic steels

0·2% P.S. (MN/m^2) (varies according to section and method of working)	150–240
U.T.S. (MN/m^2) (varies according to section and method of working)	500–650
Elongation (% $4\sqrt{A}$) (varies according to grade)	45– 70
Reduction of area (%) (varies according to grade)	55– 80
Izod (J) (varies according to grade)	110–180
Fatigue limit (10^8 cycles, MN/m^2)	±275

Where high hardness values are required with steels such as those listed in Table 3.10, it is customary to use low temperature tempering in order to obtain optimum corrosion resistance. Such treatments are usually confined to lighter sections as the toughness is limited. Typical properties of 420S45 are given in Table 3.15.

With the precipitation hardening types, high strengths can be obtained with good toughness. A feature of these steels is that there is no sharp

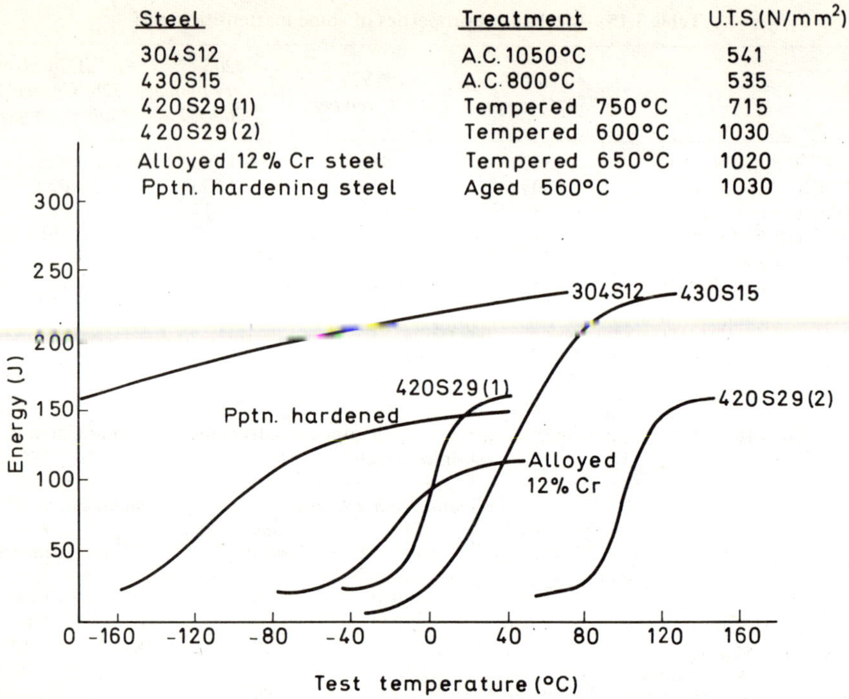

Fig. 3.12 Impact strengths (Charpy 'V'-notch) at various temperatures for several stainless steels

ductile/brittle transition, although lower impact values are obtained at lower temperatures. Properties for a typical example are shown in Tables 3.16 and Fig. 3.12.

Ferritic Steels

The ferritic steels also show a sharp ductile/brittle transition, but the transition temperature is normally above ambient for all conditions of heat treatment (Fig. 3.12). Hence these steels are normally used in small section or flat product form. They cannot be usefully hardened by heat treatment and correctly treated steel has a typical yield point of 400 MN/m^2 and a U.T.S. of 850 MN/m^2.

Austenitic Steels

The austenitic steels have excellent toughness, even at very low temperatures (Fig. 3.12). They cannot be hardened by heat treatment and characteristically have a low limit of proportionality, but because of a high work hardening rate, they have a relatively high U.T.S. value and high uniform elongation. These characteristics are very useful in cold forming operations. The properties to be expected from the whole group except for 301S21 are shown in Table 3.17.

The austenitic structure can be strengthened by cold work, and suitable forms (i.e. strip, sheet, wire) can be worked to very high strengths[5]. Plate can be strengthened to a lesser degree by warm rolling[6] while the proof strengths of certain types can be increased significantly by alloying with nitrogen.

Methods of Fabrication

All the standard stainless steels may be hot forged by hammering, pressing, drop stamping or extrusion, without difficulty although they are 'stiffer', especially the nickel-bearing types, than the low alloy and carbon steels. Working ranges of 1 100–900°C and 1 200–900°C are common for the iron–chromium and iron–chromium–nickel types, respectively. It is normal to heat treat after forging in order to develop optimum mechanical properties and, in some instances, corrosion resistance. With the martensitic types, normalising and tempering (air cooling from the austenitic range followed by reheating to some temperature below the austenitic range), annealing (furnace cooling from the austenitic range) or simply tempering, may be adopted. For the ferritic steels, heating at 750–800°C and air cooling is common, while for the austenitics, heating to 1 000–1 100°C followed by air cooling or quenching depending on steel and section, is normally carried out. In larger sections, severe thermal shock must be avoided, both when heating and cooling with the ferritic types and with the martensitic types in the hardened condition, or cracking may occur. The austenitic types are very crack resistant but distortion can occur if severe thermal gradients are caused.

Shaping of sheets, bars, plates, tubes, etc. may be carried out hot or cold in most instances. The ductility of all the steels in the softened state is good in the cold, and it is normally only necessary to heat if insufficient power is available to allow cold forming. As low a temperature as possible is normally adopted to minimise oxidation, but the metallurgical considerations already noted or to be discussed later must be borne in mind when selecting the temperature. While ductility is high in all cases, the rates of work hardening of the softened martensitics and ferritics on the one hand, and the austenitics on the other are quite different. With the former it is low and so these steels are ideal for cold heading and deep drawing, but can only be stretch formed to a moderate degree. The austenitics work harden rapidly and while this is a drawback for some forming methods, it gives the steels a very high uniform elongation value in tension and they are admirably suited to stretch forming.

All the steels can be joined by mechanical methods, but the possibility of and the consequences of crevice corrosion must be considered. The various methods can be utilised successfully in many circumstances, although hot rivetting should not be adopted as oxidation between surfaces makes crevice corrosion very probable. Welding by all the normal processes is feasible with the austenitic types, which, being non-hardening and tough, require few precautions to prevent weld cracking. Flame welding using hydrocarbons is not ideal because of the possibility of carburisation. The 'heat treatment' caused locally by the welding heat can reduce corrosion resistance (i.e. lead to 'weld decay') in some circumstances, but this phenomenon is well understood and the standard grades given in Table 3.12 are either very resistant or virtually immune. The martensitic steels, being air hardening, require special care with pre- and post-welding heating to avoid weld or heat-affected zone cracking, and the problem is greater with higher carbon levels. The precipitation hardening types are much easier to weld. Ferritic steels are also liable to cracking except in thin section.

The need for heat treatment after forming or welding is a complex topic and can only be mentioned briefly here. Generally speaking, however, unless the forming operation has been very severe it is not necessary to heat treat to restore mechanical properties, although in special cases it may be advisable to do so to relieve stresses. With the martensitic steels it is imperative to heat treat after welding to soften the weld and with the ferritics to restore corrosion resistance.

All the stainless steels can be machined in the softened states, but they may present some problems unless the correct techniques are adopted. This is especially so with the austenitic grades where the extreme ductility minimises chip breaking and the work hardening may cause difficulties unless modest cuts are made. Detailed machining instructions are readily available from steel suppliers.

Corrosion Behaviour

Theoretical Considerations

The useful corrosion resistance of a stainless steel is due almost entirely to the fact that it exhibits passivity in a wide range of environments. The nature of passivity has been discussed extensively in Sections 1.4 and 1.5 and so will not be considered in detail here. The working hypothesis adopted by many concerned with stainless steel is that the passive film is essentially an oxide, and this has proved a reasonable assumption for practical purposes.

The general form of the anodic polarisation curve of the stainless steels in acid solutions as determined potentiostatically or potentiodynamically[7] is shown in Fig. 3.13, curve *ABCDE*. If the cathodic curve of the system *PQ* intersects this curve at *P* between *B* and *C* only, the steel is passive and the film should heal even if damaged. This, then, represents a condition in which the steel can be used with safety. If, however, the cathodic curve *P'Q'* also intersects *ED* the passivity is unstable and any break in the film would lead

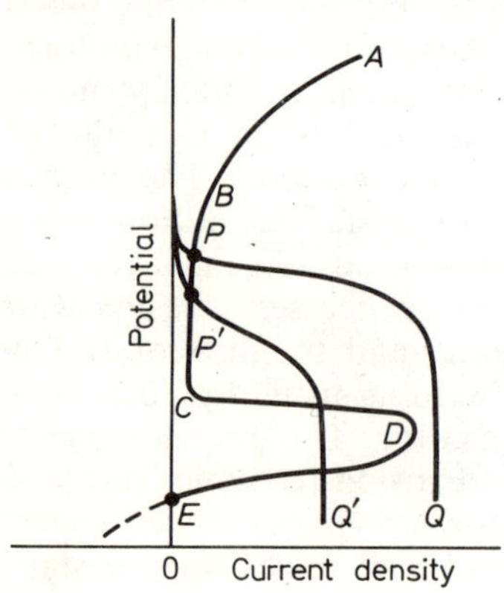

Fig. 3.13 Idealised form of a potentiostatic anodic polarisation curve *ABCDE* for stainless steels as determined in sulphuric acid solution. *PQ* and *P'Q'* are two cathodic polarisation curves that lead to passivity and corrosion, respectively

Table 3.18 Some critical values from anodic polarisation curves determined potentiodynamically in 20% sulphuric acid at 27°C (see Fig. 3.13)

Steel	*Potential* C* (V)	*Potential* E* (V)	*Critical current density D* (mA/cm²)
410S21	+0·17	−0·42	50
431S29	−0·02	−0·42	12
430S15	−0·20	−0·55	100
434S19	−0·25	−0·55	15
304S16	−0·25	−0·37	0·15
316S16	No breakdown under test conditions used		

*Versus S.C.E.

to rapid metal solution, since the potential is now in the active region and the intersection at Q' gives the stable corrosion potential and corrosion current.

Reducing acids or acid mixtures tend to give cathodic curves at low potentials, thus favouring corrosion, but oxidising solutions give curves of high potential and so favour passivity unless the curve is at such a high level that it intersects the anodic curve at *BA*, the region of transpassivity where the film is rendered unstable by oxidation. Besides providing an insight into the passivity breakdown mechanism, the determination of curves of this type provides a convenient method for studying the effects of steel composition, heat treatment and microstructure on corrosion resistance. The actual values of points *A*, *B*, *C* and *D* depend, of course, on the solution and temperature used, but standardisation of test technique allows useful comparisons to be made. The relative effects of some of the alloying elements utilised may be gauged from the figures for several standard steels shown in Table 3.18. Increasing chromium content depresses potential *C*, therefore usefully extending the range of stable passivity, but it also increases the critical current at point *D* so that higher corrosion rates are obtained in the absence of passivity. Nickel also depresses potential *C* but markedly reduces the critical current. Molybdenum also has beneficial effects on both breakdown potential and critical current. Other alloying elements also have beneficial effects (while some are harmful), but, with the exception of copper and silicon which are utilised in some proprietary alloys, these are the only alloying elements which are used with the objective of improving the basic corrosion resistance of the steel.

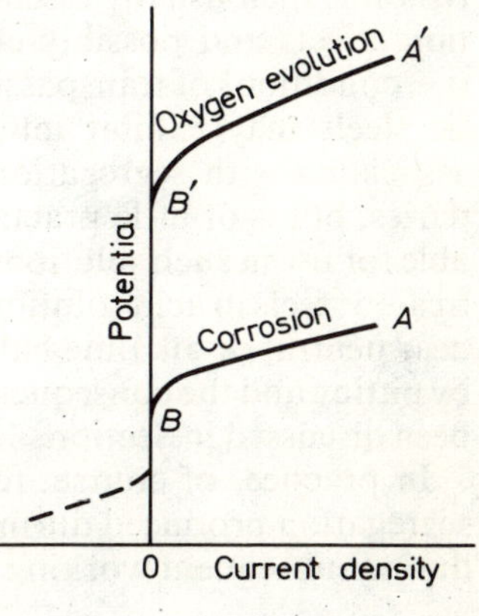

Fig. 3.14 Idealised form of a potentiostatic anodic polarisation curve for stainless steels as determined in a neutral solution containing chloride

Table 3.19 Breakdown potentials for various stainless steels determined potentiodynamically in 0·1 M NaCl + 0·1 M $NaHCO_3$ solution (Point *B* in Fig. 3.14)

Steel	*Breakdown potential* (V, *vs.* S.C.E.)
410S21	−0·22
430S25	−0·12
431S29	−0·12
434S19	−0·08
Precipitation hardening (Table 3.13)	−0·08
304S16	−0·07
315S16	0·00
316S16	+0·14
317S16	+0·30

In near neutral or alkaline salt solutions, anodic polarisation curves of the type shown in Fig. 3.14 are obtained, in which the active loop *CDE* of Fig. 3.13 is no longer apparent. At high potentials curve *B′A′* will represent oxygen evolution, but at low potentials (curve *BA*) in the presence of halide ions it will be indicative of pitting corrosion. This form of attack represents a considerable hazard for some service conditions, and a measure of the breakdown potentials under controlled conditions can give useful indications of the relative resistances of alloys (see Section 1.6).

Breakdown potentials for various alloys are given in Table 3.19, and it can be seen that increasing chromium, nickel and molybdenum contents all lead to improved resistance to pitting, which is indicated by the magnitude of the breakdown potential. As with acid corrosion resistance, other elements could also be beneficial, but have not proved economical or desirable from other viewpoints.

Metallurgical Considerations and Forms of Corrosion

The corrosion resistance of any of the stainless steels is at its best when the material is single phase and in a homogeneous state. In acid solutions corrosion is then usually essentially uniform in nature except for grain orientation effects and possibly etch pitting. In very strongly oxidising solutions (i.e. conditions of transpassivity) however, apparently homogeneous austenitic steels may exhibit intercrystalline corrosion, which is believed to be associated with segregation of certain interstitial elements at grain boundaries, but is of little practical importance since stainless steels are not suitable for use in such solutions. Pitting or localised corrosion is rare in correctly treated steels in acid solutions, but can be obtained in some circumstances. In near neutral or alkaline halide-bearing solutions, however, any corrosion is by pitting and the consequences may be serious. The mechanism of pitting has been discussed in Sections 1.5 and 1.6 and will not be considered further here.

In practice, of course, few alloys are homogeneous in composition, the segregation produced during solidification being reduced but not eliminated during subsequent working and heat treatment. This is the case with stainless steels, and while microsegregation of the major alloying elements does not usually lead to any significant effect on overall corrosion resistance, some differential corrosion rates and etching effects of various surfaces can be noted under environmental conditions causing significant corrosion.

The presence of second phases also means partitioning of the elements with consequent alloying element segregation. The most likely second phase in austenitic stainless steel is δ-ferrite which is higher in chromium and molybdenum (if present in the alloy) and lower in nickel content than austenite and therefore has different corrosion resistance characteristics. From a consideration of the idealised polarisation curves and data it can be shown that in some circumstances the ferrite could corrode at a more rapid rate while in others the reverse could be the case. Both phases have good resistance to corrosion and in very many environments no attack occurs on either phase. δ-ferrite is rarely present in more than trace amounts in the standard wrought steel, but it is often intentionally present in weld or cast structures when progressive selective attack may be a hazard in some aggressive media if the content and distribution of the ferrite is such as to provide a continuous network. Other phases theoretically possible in austenitic steels are sigma (σ) or chi (χ), although in the current standard grades they are usually only formed by prolonged heat treatments in the region 650–900°C. The phases themselves are very resistant to corrosion, but during formation they can deplete adjacent austenite of chromium and molybdenum so marginally affecting corrosion resistance in aggressive media. The only really serious effect of these phases is when they form in δ-ferrite pools especially in molybdenum-bearing steels (in this instance formation is rapid and can occur on air cooling of moderate or large sections). This can result in the ferrite being seriously depleted in alloying element content. Again this is only really a hazard with certain castings and weld metal as wrought material normally has a low ferrite content.

Precipitation of Carbides

The most marked structural effect on corrosion resistance is that associated with chromium-bearing carbides, which can occur in all three groups of steels. The martensitic types of hypo-eutectic composition have essentially all carbon in solid solution in the as-transformed (i.e. the hardened) condition. Tempering reduces strength as shown in Fig. 3.11 and causes the precipitation of carbides; at lower temperatures the carbides are essentially of iron, but at higher temperatures they are chromium-rich. The diffusion rate of carbon which is an interstitial element is much greater than that of chromium, and as a result chromium gradients are set up adjacent to the growing carbide particles. If the carbide particle distribution is such that the chromium-reduced regions can overlap, a continuous low-chromium path can be formed which may lead to selective attack. Such attack can occur along prior austenite boundaries and also along martensite lath boundaries. The treatments giving susceptibility to such attack are indicated in Fig. 3.15; at low temperatures there is insufficient chromium in the carbides to have an effect, while at higher temperatures diffusion of chromium is rapid enough to prevent severe gradients, although the overall effective chromium content is reduced somewhat, as is the corrosion resistance as a

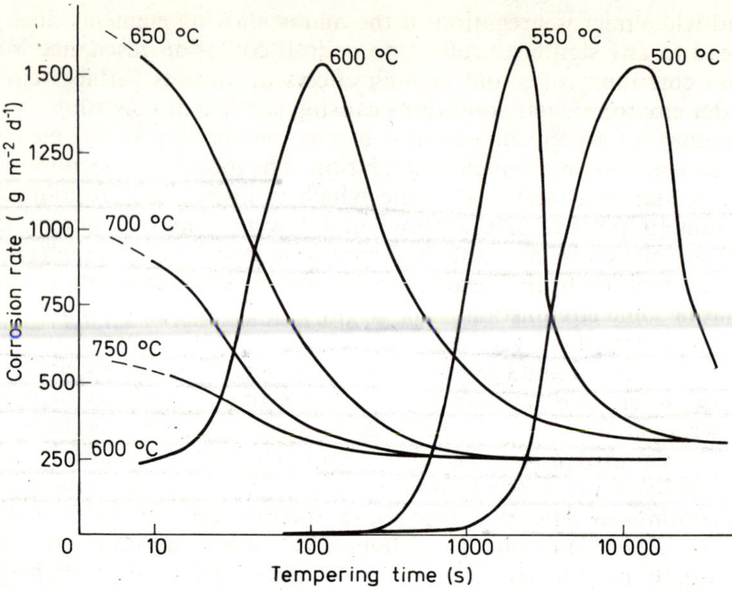

Fig. 3.15 Effect of tempering treatment on corrosion resistance of 420 S45 (air cooled, 980°C). Corrosion tests in 10% nitric acid solution at 20°C

result. Obviously the effects are more pronounced with higher carbon steels. As the effects are associated with a precipitation reaction, the time of tempering has an effect as well as temperature*. This is illustrated in Fig. 3.15. For discussion of the tempering processes see Reference 1.

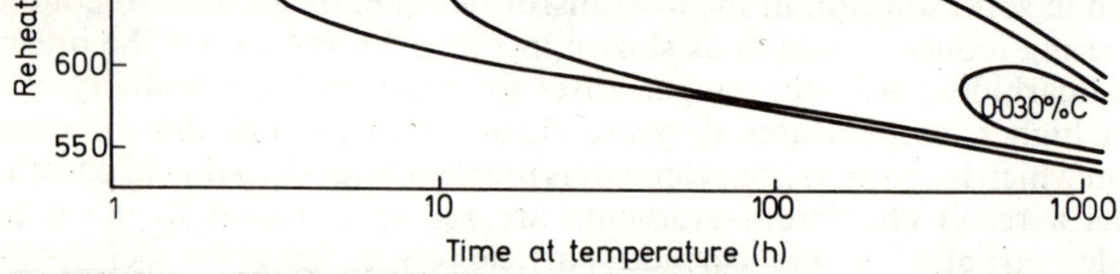

Fig. 3.16 Temperature–time–sensitisation diagrams for three austenitic Cr–Ni steels solution treated at 1 050°C. The curves enclose the treatments causing susceptibility to intercrystalline corrosion in a boiling $CuSO_4 + H_2SO_4$ test reagent

*Martensitic stainless steels are usually used in the softened (tempered at or above 650°C) or in the fully hardened condition (tempered at or below 250°C) so that there is no substantial reduction in corrosion resistance resulting from carbide precipitation. However, the hard soldering of knife blades can result in carbide precipitation and pitting of the blade at the area adjacent to the handle, and care must be taken in the soldering process to avoid this danger.

Austenitic steels of the 302S25 type are normally heat treated at 1 050°C and cooled at a fairly rapid rate to remove the effects of cold or hot working, and in this state much of the carbon is in supersaturated solid solution. Reheating to temperatures below the solution treatment temperature leads to the formation of chromium-rich $M_{23}C_6$ precipitates predominantly at the grain boundaries with the production of chromium gradients and reduced corrosion resistance as is the case with the martensitic steels. Any attack is intergranular in this instance. The effect of temperature and the time of reheating on establishing whether steels are susceptible to this form of attack or not, is shown in Fig. 3.16. The corrosive medium used was that specified in BS 970 and it should be noted that the precise position and area of the curves are relevant to this test medium only, although the general form is similar for other test media. Much has been written regarding the relative merits of various accelerated tests, but it is generally accepted in this country that the copper sulphate-sulphuric test gives an adequate guide to steel condition for most purposes. It should also be noted that not all corrosive media produce intergranular attack on 'susceptible' steels.

The metallurgical condition leading to susceptibility to intercrystalline corrosion can be caused by cooling from the solution treatment temperature at a sufficiently slow rate as well as by isothermal treatment. Large sections in certain grades may thus require quenching in oil or water to ensure optimum corrosion resistance. The heating introduced by welding can also produce the undesirable metallurgical condition in bands a short distance on each side of the weld. This may lead to the localised attack known as *weld decay* (see Section 10.5).

Since the susceptibility to intercrystalline attack is due to carbide formation it follows that low carbon steels are more resistant to the phenomenon, as illustrated in Fig. 3.16. Thus the alloy with a carbon content suitable for the thermal treatment to be encountered in fabrication should be selected, i.e. 302S25, 304S15 or 304S12. An alternative method for ensuring resistance to intercrystalline corrosion is to 'stabilise' the carbon, that is to restrict its solubility at heat treatment temperature by incorporating a strong carbide-forming element in the steel, and titanium or niobium are utilised for this purpose (321S12, 321S20 and 347S17). Whether low-carbon grades or stabilised grades are selected for a given application where intercrystalline corrosion may be a hazard depends on a variety of factors. The stabilised grades are especially useful where service may involve periodic heating into the critical temperature range followed by contact with some aqueous corrodent. As shown in Fig. 3.16, even the low carbon grades can be sensitised in time. Low carbon grades may be preferred where welding is followed by heating into the critical range for a relatively short time for the purpose of stress relief. The zone immediately adjacent to the weld is subjected to a very high temperature in welding which causes solution of the titanium or niobium carbides so that chromium-rich carbides may form on reheating. Thus a narrow band immediately adjacent to the weld may corrode in service. This is known as *knife-line* attack to distinguish it from *weld decay* where the corrosion zones are some slight distance ($\simeq$ 2 mm) from the weld.

The ferritic steels, 430S15 and 434S19 may also undergo intercrystalline corrosion as a result of boundary carbide formation. In the softened condition ($\simeq$ 800°C) these grades will withstand reheating at lower temperatures for considerable periods without sensitisation. High temperature heat treatment, such as that that occurs immediately adjacent to a weld, followed by even rapid cooling causes susceptibility. This difference in response to the austenitic types may be attributed to a very low solubility of carbon in ferrite at softening temperature but increased solubility at higher temperatures, together with the formation of some austenite with relatively high carbon solubility. Sensitisation following solution treatment is more rapid because of the high diffusion rates in ferrite. The attacked zones with welded material are immediately adjacent to the welds. All welded components in these steels should be given a post welding heat treatment (800°C). Stabilised grades can be produced but are not included in the British Standards.

At least some forms of stress corrosion cracking may be considered a form of selective corrosion. With all three types of stainless steel, in adverse conditions of heat treatment, as described in the preceding paragraphs, the application of a critical stress in the presence of a corrosive environment can lead to rapid intergranular cracking instead of general intercrystalline penetration at a much slower rate in the absence of the critical stress. The term *stress-accelerated intercrystalline corrosion* rather than *stress corrosion* may be preferred for this type of attack, but the possibility of turning superficial attack into rapid cracking must be considered. Transgranular stress-corrosion cracking to which the austenitic grades are susceptible may also be a form of selective corrosion; this phenomenon is discussed in detail in Section 8.3. The corrodents causing such failure are quite specific and the most common are chlorides. For failure to occur, critical combination of stress and chloride strength are required and, in practice, failures are most prevalent when the metal is hot. All the austenitic steels listed in Table 3.12 are susceptible to cracking to a generally similar degree. The ferritic steels given in Table 3.11 however, appear to be immune, but it has recently been noted that alloying with nickel, copper or cobalt can induce susceptibility in a ferritic structure. The martensitic group are not susceptible to transgranular cracking in the softened states but are prone to failure under stress in the harder conditions, susceptibility generally increasing with hardness. Opinions differ as to whether this is truly selective corrosion or predominantly a form of brittle fracture (although to the engineer this is perhaps of little consequence). The corrodents which may cause failure are not as specific as with the austenitic types. Cowan and Tedman have compiled a comprehensive review of the intergranular corrosion of Fe–Ni–Cr alloys[8(a)].

Corrosion in Natural Environments

Atmospheric Corrosion

One of the major assets of the stainless steels has proved to be their resistance to discoloration in the atmosphere, and even the least corrosion resistant steels have been used with success indoors. Knives, for instance, have to be made from the 13% chromium types to achieve the hardness necessary and yet to retain their bright, attractive appearance without cleaning or special attention. Many other articles such as domestic utensils, kitchen equipment and decorative articles have been made from the ferritic and the austenitic types. Outdoor service is more arduous, however, and steels of the martensitic group are not used, at least where appearance is important. The ferritic steels, especially 430S15, and to an increasing extent 434S19, are used to manufacture motorcar trim and accessories, primarily because they retain their bright appearance. In one way this is surprising since simple exposure tests in industrial atmospheres show the 430S15 type in particular to be prone to rust staining and pitting, and it must be presumed therefore that thin films of wax or grease applied in the normal course of car cleaning have a beneficial effect. Stainless steels are normally used for external architectural purposes because of their bright appearance (although they can also be functional in nature), and often this must be retained without the necessity for periodic cleaning. In the UK at least, type 430S15 is not really suitable, suffering rust staining caused by very fine pitting corrosion. Type 432S19 and the molybdenum-free austenitic types are better and have been used with success, but again these may show rust staining unless cleaned regularly.

Table 3.20 Atmospheric exposure tests (Sheffield)

Material	*1938/1943 series*	*1962/1967 series*
Carbon steel	3 700	
410S21	270	
430S25	135	45
434S19	—	15·5
18Cr–8Mn	165	
12Cr–12Ni	160	
304S16	85	4·5
25Cr–20Ni	80	
321S20	70	
30% Cr	70	
316S16	1	0·4
Al	165	
Cu	185	
Phosphor bronze	200	
Zn	385	
54Cu–44Ni	440	

Note. The results are average losses in weight (g/m^2) for multiple samples exposed for 5 y and not cleaned during test.

For this reason the molybdenum-bearing 316S12 steel is being used, which for most practical purposes is unaffected by polluted atmospheres.

The mechanism of atmospheric corrosion of the stainless steels has not been widely explored, but it is generally accepted that pollution of marine, industrial and domestic origin is a major influence. With marine atmospheres it may reasonably be assumed that chlorides are a major factor. With industrial atmospheres it has been suggested that sulphurous gases represent a major factor, although chlorides again could well be of considerable importance. Any attack is in the form of pitting, although with the 13% chromium types the pits become so numerous in time that they can overlap and the steel can appear to be suffering general attack. With all types the pits are small and do not represent the same hazard of penetration that occurs in chloride solutions.

Table 3.21 Effect of periodic washing on the behaviour of several stainless steels exposed to an industrial atmosphere (Sheffield)

	Area corroded (%)			
	430S25	*434S19*	*304S16*	*316S16*
Specimen washed weekly	23·8	3·66	3·60	<0·01
Specimen washed monthly	27·9	4·20	4·19	<0·01
Specimen washed quarterly	47·3	4·74	4·59	<0·01
Specimen not washed	26·5	3·60	2·0	<0·01

Note. Exposure period 1962–1968

Weight loss values for samples exposed to the Sheffield atmosphere for 5-year periods are shown in Table 3.20. Two series of tests are represented, and an interesting feature is that the corrosivity of the atmosphere to stainless steels appears to have decreased significantly over the 20 years separating the two sets. Exposure tests at other sites in the UK including marine and industrial/marine are described in Reference 8.

Where retention of appearance is of prime importance the 316S12 grade is now used almost exclusively. There are various opinions expressed as to whether regular cleaning is necessary to obtain the best behaviour from stainless grades. The results of the tests given in Table 3.20 show that this is certainly not necessary with this grade, although cleaning may be desirable to remove dirt films in order to preserve the appearance. The arguments in favour of cleaning arise from experience, largely pre-war, with non-molybdenum grades where rust staining became apparent unless the surface was washed periodically. It was argued that dirt particles set up crevice conditions and helped retain corrodents. Some recent tests (some results from which are given in Table 3.21) indicate that cleaning has no significant effect however, at least on steel fully exposed to rain, and the practical experience may simply be that regular cleaning removes corrosion products while the corrosion itself is too superficial to be noticed.

Corrosion in Natural Waters

One of the main factors which establishes the corrosivity of a water to stainless steel is the chloride content. Also significant are oxygen content and pH, and it is also probable that other features such as hardness, and the nature and content of other anions and of cations has an effect. Temperature and velocity of the water can also be important.

Any corrosion normally takes the form of pitting, but where crevices are operative, breakdown tends to be more general in nature within the crevice.

The 13% Cr martensitic types are unsuitable for use in sea-water. Type 431S29 and the precipitation hardening grades on the other hand, have proved successful for certain applications but it must be recognised that corrosion can occur with these grades in some circumstances. Pollution giving reduced pH, contact with low velocity or stagnant water, high temperatures and crevices (whether as a result of engineering design or deposition of foreign matter) all favour attack. The same reservations apply to the ferritic steels, but in fact these are rarely used in contact with sea-water.

The austenitic grades have superior resistance to breakdown, especially the molybdenum-bearing grades, and have been used with success for a variety of purposes. It must again be recognised, however, that even the best of this group is not immune to attack under all circumstances, and the same principles mentioned in connection with the martensitic steels apply. Particularly dangerous are marine organism growths which can occur on prolonged contact with low velocity waters, especially if they are warm.

Contact of brass, bronze, copper or the more resistant stainless steels with the 13% Cr steels in sea-water can lead to accelerated corrosion of the latter. Galvanic contact effects on metals coupled to the austenitic types are only slight with brass, bronze and copper, but with cadmium, zinc, aluminium and magnesium alloys, insulation or protective measures are necessary to avoid serious attack on the non-ferrous material. Mild steel and the 13% chromium types are also liable to accelerated attack from contact with the chromium-nickel grades. The austenitic materials do not themselves suffer anodic attack in sea-water from contact with any of the usual materials of construction.

With river waters and other less saline waters the possibility of corrosion is reduced, but the same general principles apply.

Potable waters when cold naturally represent less of a hazard than many of the natural waters, but on the other hand they are often used at higher temperatures. The admirable service over many years of kitchen and catering utensils and cutlery show how resistant all the grades are, but prolonged contact at elevated temperatures may introduce special problems. For relatively high chloride waters (say >200 p.p.m.) only the more corrosion resistant types should be used for hot water. Heat transfer from steel to water can lead to corrosion at very high heat fluxes and also at lower heat fluxes with hard waters due to the concentration of chlorides underneath water scale deposits. Similar concentrations can occur in crevices in heat transfer surfaces, and can, in both instances, lead either to pitting, or, in extreme cases with the austenitic types, to stress-corrosion cracking.

Soil Corrosion

Stainless steels have not been widely used in applications where they are buried in soil, but some applications have involved underground service. Various stainless steels from the 13% Cr to the molybdenum-bearing austenitic types were included in the comprehensive series of tests in a variety of soils reported by Romanoff[9]. High-chloride poorly-aerated soils proved most aggressive, but even here the austenitic types proved superior to the other metals commonly used unprotected. Of special interest is the fact that though corrosion was by pitting there was little or no increase in pit depth after the first few years.

Corrosion in Chemical Environments

Acids

The value of stainless steels may be illustrated by consideration of some of the stronger acids.

It should be noted that the data presented were obtained under laboratory conditions in still solutions of pure acids.

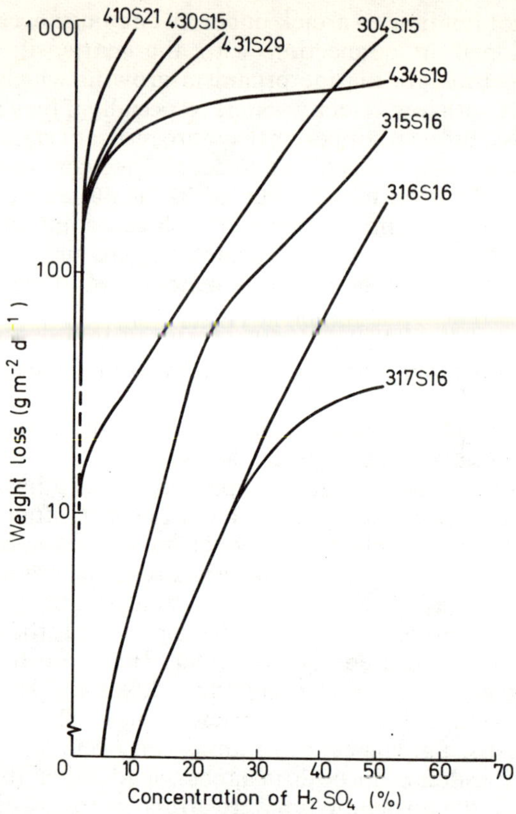

Fig. 3.17 Corrosion rates of various stainless steels in different concentrations of sulphuric acid at 20°C

Sulphuric acid Stable passivity is possible under a range of conditions for all the stainless types, the transition to the active state being favoured by an increase in acid concentration and temperature. With the 13% Cr types, passivity is only possible in very dilute solutions and corrosion in the active region is very rapid; the range of acid strengths over which steels are passive is extended somewhat with the higher chromium ferritics. The nickel, and in some instances, molybdenum content in the austenitic grades and in the precipitation-hardening martensitic types, not only enhances passivity but also retards corrosion in the active state so that in some instances materials can give a useful life while corroding. Titanium and niobium present in some grades do have a slightly beneficial effect, but are not added primarily for this reason, and for practical purposes types 321S12, 321S20 and 347S17 may be considered similar to 304S16. Corrosion rates for various grades determined in static solutions, not deliberately aerated, at 20°C are shown in Fig. 3.17. The effect of acid temperature on the behaviour of the more resistant grades is illustrated in Fig. 3.18.

The presence of reducing agents or oxidising agents in the acid, even in small quantity, can reduce or extend the range of passivity by depressing or raising the cathodic curve in the diagram illustrated in Fig. 3.13. Practical

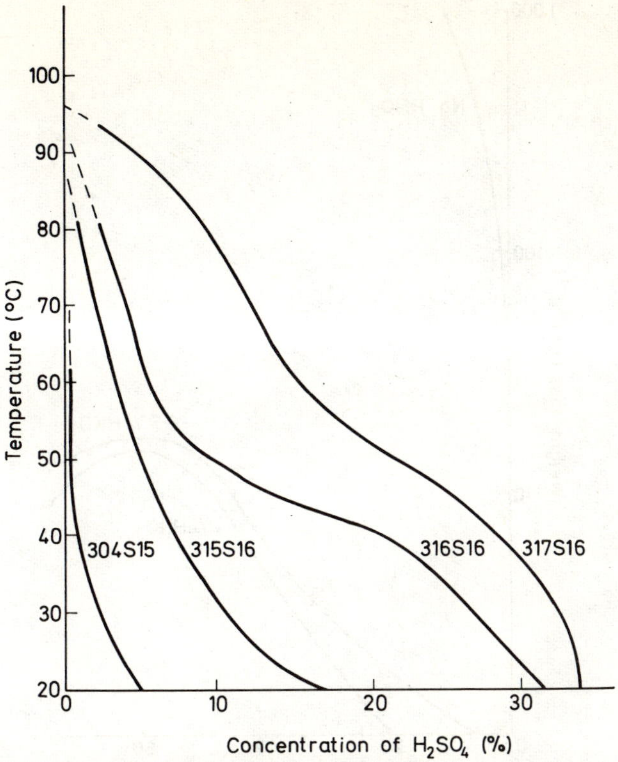

Fig. 3.18 Strength and temperature of sulphuric acid solutions and type of stainless steel to give a corrosion rate of $2{\cdot}5\,g\,m^{-2}\,d^{-1}$

use may be made of this by utilising small additions of an oxidant as a cathodic reactant. Copper sulphate or nitric acid are often used for this purpose; the marked effect of the latter is illustrated in Fig. 3.19.

Polarisation from an external source may also affect the range of passivity. Cathodic polarisation may depress the potential from the passive to the active region (see Fig. 3.13) and thus care should be taken to avoid contact with any other corroding metal. Anodic polarisation, on the other hand, can stabilise passivity provided that the potential is not increased into the range of transpassivity (see Fig. 3.13) and anodic protection is quite feasible.

In certain applications involving sea-water (e.g. pumps) stainless steel is connected to mild steel, and the latter has to be protected by cathodic protection. Under these circumstances the stainless steel will also be protected by the cathodic current, but it is unusual to protect cathodically stainless steel on its own.

Hydrochloric acid Most of the general remarks in the last section dealing with sulphuric acid are also applicable to hydrochloric acid, but the chloride ion makes this acid a very aggressive one and the application of stainless steels is confined to low concentrations. Corrosion behaviour is illustrated in Figs. 3.20 and 3.21. While passivity is possible with this acid as was noted for sulphuric acid, anodic protection either by impressed current or addition of an oxidant is not practicable and, indeed, could be harmful.

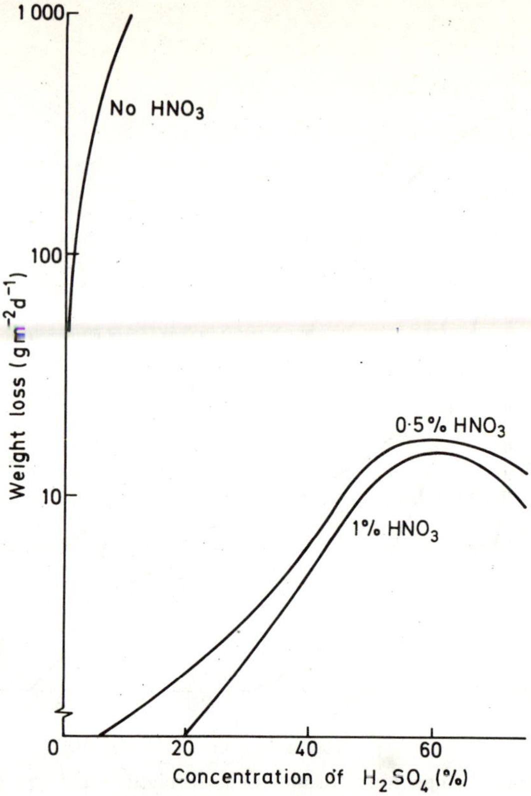

Fig. 3.19 Inhibiting effect of nitric acid on corrosion of 304 S15 steel in sulphuric acid at 100°C

Nitric acid Of the alloying elements utilised, chromium and nickel are beneficial in reducing attack in nitric acid. The effect of acid strength on corrosion rates of three types of steels is shown in Fig. 3.22. At the 13% Cr level, enhanced attack occurs at both low and high acid strength but at 17% Cr the attack is less at lower acid strengths. Nitric acid is especially likely to produce more rapid attack at grain boundaries and thus for long term service, especially if welded plant is involved, the extra-low-carbon austenitic grades (<0·03% C) or the stabilised grades are preferred. Of the latter, the niobium-bearing variety withstands the more onerous conditions better than the titanium-bearing type. From laboratory and plant experience, the top limits of concentration tolerable appears to be those summarised in Table 3.22. It should be noted that some corrosion occurs even below these limits and some contamination of acid must be expected. High concentrations of nitric acid raise the potential of the steel into the transpassive region with a consequent increase in the corrosion rate.

Phosphoric acid The austenitic grades are resistant to all strengths up to 80°C but are limited to 30–40% concentration at boiling point, the molybdenum-bearing types having the best corrosion resistance. Some test data for various types is shown in Fig. 3.23. Industrially, this acid is often encountered in an impure state with appreciable amounts of sulphuric and

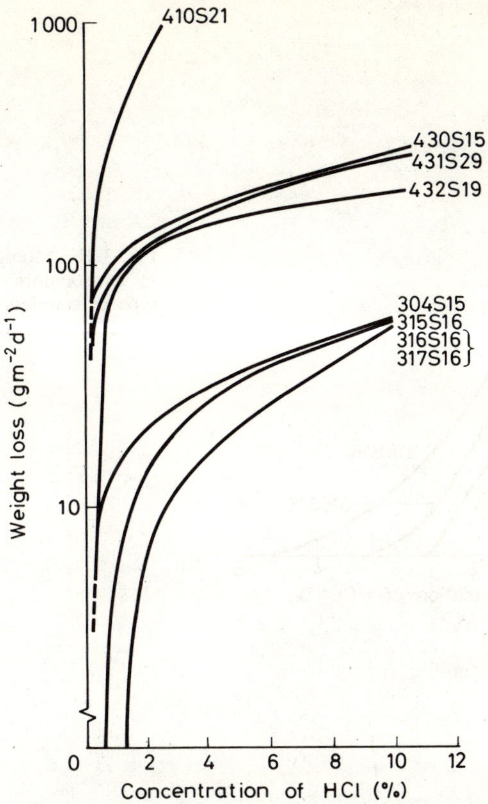

Fig. 3.20 Corrosion rates of various stainless steels in different concentrations of hydrochloric acid at 20°C

hydrofluoric acid present so that process testing is likely to be particularly necessary.

Acetic acid Acetic acid may be taken as being typical of the stronger organic acids. Considerable variability between tests is shown in the 'borderline' region at boiling point, and attack on plain austenitic types may rise to approximately 1 mm/y in the more concentrated solutions. 316S16 steel is superior to non-molybdenum steels, with losses of up to 0·1 mm/y with boiling 80% acid.

With stainless steels optimum resistance is dependent on aeration, and due account must be taken of this in considering operating conditions where previous experience has been with non-ferrous alloys which require an absence of oxygen for full resistance. There is therefore no need to run off the early distillate from a batch to avoid impurity during initial de-aeration. On the other hand, complete exclusion of oxygen for the benefit of non-ferrous metals in a mixed plant have very serious consequences for the stainless steel, and this must be kept in mind at the design stage.

Other acids Formic acid is more liable to cause pitting than is acetic acid, and it must be considered with particular care at higher temperatures. Citric,

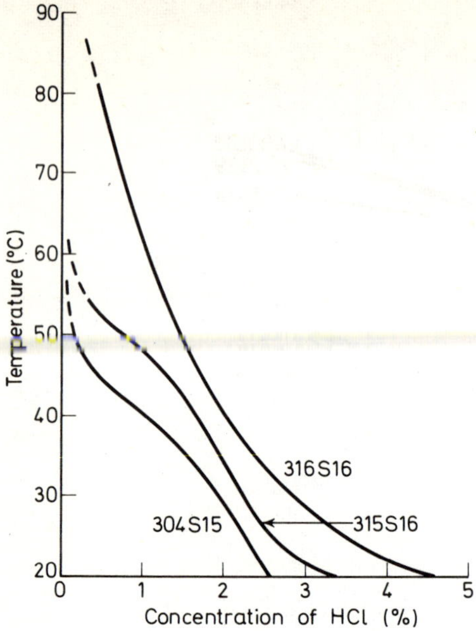

Fig. 3.21 Strength and temperature of hydrochloric acid solutions and type of stainless steel to give a corrosion rate of 2·5 g m^{-2} d^{-1}

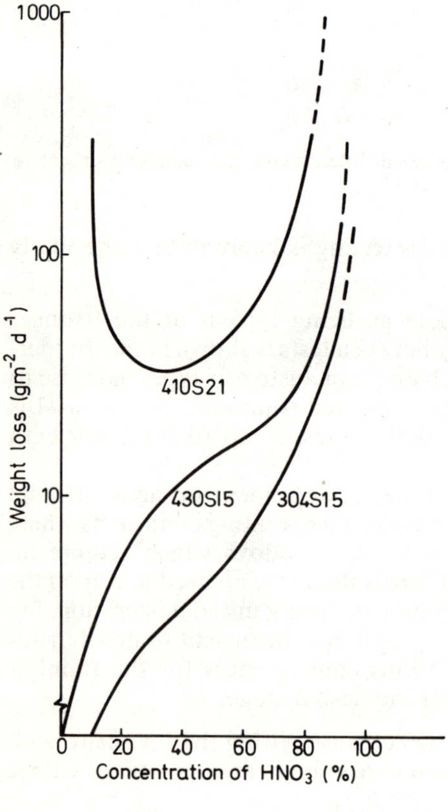

Fig. 3.22 Corrosion rates of various stainless steels in nitric acid solutions at boiling point

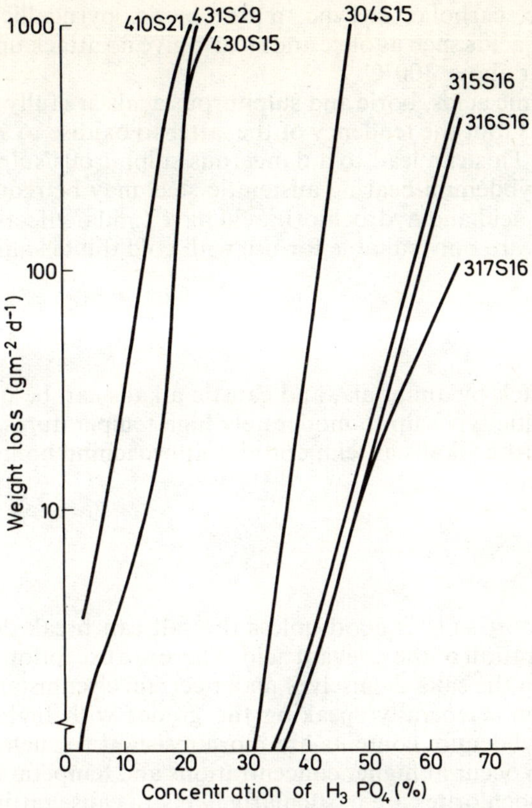

Fig. 3.23 Corrosion rates of various stainless steels in phosphoric acid solutions at boiling point

Table 3.22 Maximum nitric acid concentration that can be tolerated by austenitic stainless steels

Temperature (°C)	*Stabilising element*	*Nitric acid concentration* (%)
20°C	Ti	70
	Nb*	99
40°C	Ti	70
	Nb*	95
60°C	Ti	70
	Nb*	80
80°C	Ti	70
	Nb	70

*Or extra low carbon (<0·03%) grade.

oxalic, lactic and sebacic acids are also liable to cause appreciable attack in strong solutions at the higher temperatures, and for the borderline conditions molybdenum is a useful addition to the steel. On the other hand, many organic acids are less corrosive than acetic, and non-molybdenum austenitic steels may be used with them without trouble even at 100°C. Such acids include benzoic, carbolic, cresylic, malic, picric, pyrogallic and tartaric. The higher fatty acids such as oleic and stearic give no attack up to processing temperatures far above 100°C.

Of the inorganic acids, boric and sulphurous acids are fully resisted under many conditions, but the tendency of the latter to oxidise to sulphuric must be appreciated. This can lead to a dangerous sulphurous/sulphuric mixture so that the molybdenum-bearing austenitic steel may be required.

Hydrofluoric acid and hydrochloric acid have similar effects, and even the austenitic steels are only suitable for use with cold dilute solutions.

Alkalis

There is no attack by ammonia, and caustic alkalis can be used in contact with the austenitic types up to moderately high temperatures, but in strong solutions of caustic alkalis at temperatures approaching boiling point some attack will occur.

Salts

Resistance to most salts is good unless the salt can break down to give a critical concentration of the relevant acid. There are exceptions in the case of the halides when the salts themselves under certain circumstances can cause pitting corrosion. Generally speaking the grades with higher chromium, nickel and molybdenum contents are more resistant to such attack, which is more likely to occur at higher concentrations and temperatures and lower pH values. Hypochlorites are particularly likely to cause pitting and should not be used under conditions of prolonged contact.

Halogens

In the presence of moisture all the halogens are corrosive to stainless steel, although dilute iodine solutions can be used, e.g. under hospital conditions, for contact for short duration.

Non-acid Organic Compounds

Resistance to non-acid organic materials is usually very good. It is necessary to bear in mind that some halogen derivatives are liable to either break down or to hydrolyse in contact with moisture, giving the corresponding halogen acid, which may cause attack.

Domestic and Miscellaneous Situations

Resistance to most foodstuffs is excellent. In cooking it is necessary to avoid prolonged contact with hot strong sodium chloride solutions and, as far as possible, with acid chloride solutions at any temperature. Foodstuffs may, however, contain natural inhibitors which reduce the rate of attack.

In sterilising it is necessary to avoid prolonged contact with hypochlorite sterilisers.

Conjoint Action of Stress and Corrosion

With the martensitic grades, stress-corrosion cracking is only likely in the high strength conditions. As steels in such conditions are usually only used if strength is required, stressing in service must be expected. If this is compressive and of a transient nature there is little danger unless there is also a residual tensile stress, and care must be taken in the heat treatment and fabrication to minimise such residual stresses if a corrosive environment is to be encountered. Consideration of the environment is even more important if service involves sustained loading of a high strength martensitic steel[10].

With the austenitic steels, failure by stress-corrosion cracking may occur in specific environments, e.g. warm strong solutions of chloride and caustic substances. The stresses involved in most failures have proved to be incidental rather than functional and in some cases could have been avoided by suitable modifications in fabrication and stress relief. Stress relieving treatments, however, are inconvenient as the temperatures required are high (usually >900°C) and descaling is necessary afterwards. Moreover, many incidental stresses cannot be avoided and are the result of thermal expansion effects. The best preventative procedure is to avoid the presence of the corrodent in critical concentrations. In most cases where stress-corrosion cracking has occurred in practice, the corrodent was proved to have formed locally on the metal surface by evaporation, and ways in which this can occur should be 'designed out'[11].

Stainless steels are subject to fatigue failure, like all metallic materials. In the absence of significant corrosion all the types exhibit a true fatigue limit which is of the order of one half of the U.T.S. value, unless this is very high when the fraction can be lower. The presence of a corrodent removes the limit so that the number of cycles to failure becomes a function of the cyclic stress at all stress levels. The stress against the logarithm of the number of cycles to failure plot is also displaced to lower stress values. The interrelation of steel composition, strength, corrodent and fatigue behaviour is too complex to discuss in detail, but in general terms the fatigue endurance limit (in 3% NaCl solution for 10^8 cycles) of the softened martensitics is of the order of ± 120 MN/m^2 and of the softened austenitics ± 200 MN/m^2.

Applications

Martensitic Types

By far the greatest use has been for engineering purposes where the corrosion-resisting requirements are not high. Typical applications are steam and water valves, pump, and turbine and compressor components and shafting. Where the simple 13% Cr steel has proved inadequate in corrosion resistance, the higher chromium type 431S29 has been used and, more latterly, the precipitation hardening types. In the hardened state, the steels are used where abrasion resistance is required. 13% Cr steels are widely used for steam turbine blading and, with the increasing size of turbines, higher strength, tough versions have been developed.

A major use of the 13% Cr higher carbon type is for the manufacture of corrosion-resistant cutting edges. Type 420S45 has been the standard for domestic knives for many years.

Ferritic Types

The major use of the ferritic steels is on motor vehicles as trim and in domestic equipment such as cutlery and hollow ware, but it has also been used in refrigerators, washing machines and on sinks and similar fittings. Some types would no doubt find much wider application in the chemical field and other fields where their superior corrosion resistance would be a considerable advantage if it was not for the fact that the austenitic types have advantages (sometimes considerable) in fabrication.

Austenitic Types

Architectural Considerable use has been made of austenitic steels for decorative aspects such as shop fronts, fittings, door and window frames and, in some cases, extensive wall cladding. Many of these applications may be termed functional as well as decorative since although initial expenditure may be more than for competitive materials, the absence of the necessity for maintenance is a considerable attraction. In the UK, at least, use of the molybdenum-bearing 316S16 type is now almost universal although molybdenum-free grades have been used with success in the past. The use of steels for fully functional purposes is also growing and the molybdenum-bearing grade may not be necessary. Roofing in stainless steel (even 430S15) is being used to an increasing extent on the continent, while load-bearing members such as masonry fixings are being made in the austenitic grades. An important recent additional use of the austentitic types is as water tubing.

Marine engineering Austenitic steels have been extensively used for marine propeller shafts. Where tropical conditions or effluents increase the risk of attack, the molybdenum-bearing types should be used. This also applies to yacht rigging, deck fittings and other items exposed to spray, for which the highest possible resistance is required.

Domestic and catering applications Austenitic steels are ideal for most food preparation, and in this connection are used for table tops, sinks and cooking vessels, including steam-heated pans for hotels and canteens. Other household uses include washing machines, steam irons, mixing machines and cutlery. The importance of a polished non-porous surface in promoting ease of cleaning and general hygiene is obvious, and this type of finish is readily available with these steels.

Food and drink production and distribution

Dairy work Applications include vessels for milk storage and sterilisation, cooling units, cream separators and cheese and butter-making equipment, as well as general dairy fittings, bottle-washing machinery and tankers for bulk milk transport. Extensive use is made of these steels in equipment used in production of ice cream and dried milk.

Brewing Austenitic steels are used in fermentation vessels, storage tanks and attemperators, in yeast processing, and for transport tanks and barrels.

Preserves and fruit juices Much use is made of sulphur dioxide to preserve fruit pulps and juices, and for these applications molybdenum-bearing steels should be specified.

Soups and sauces These products may involve highly corrosive mixtures, in this case from combined chlorides and acidity. For this reason it is often necessary to use the molybdenum-bearing type of steel.

Baking Ease of cleaning is again important, and austenitic steels are very suitable for making items such as mixing machinery and working tables.

Surgery, hospital and antibiotic uses Surgical instruments are often made from the non-molybdenum types of 18 Cr–8 Ni steel, but for work such as bone implants where it is essential to have maximum resistance, the molybdenum-bearing type is necessary.

For general hospital work such as instrument vessels, sterilising equipment and similar uses, the non-molybdenum steels are frequently satisfactory, but where conditions are onerous and likely to involve exposure to acid chlorides, prolonged boiling or hypochlorite sterilisation, it is safer to use molybdenum-bearing steels.

Photographic processing Developing gives no trouble with the Cr-Ni steels, but for contact with acid fixing solutions 316S16 steel should be specified.

Textile applications Much use has been made of non-molybdenum Cr-Ni steels in dyeing. On the other hand, many dyeing solutions contain chlorides or other corrosive substances such as formic acid, and here it is often wiser to use type 316.

For bleaching with alkaline peroxide, 304S16 is satisfactory but with moist sulphur dioxide or hypochlorites 316S16 should be used. In the case of hypochlorites, contact should be short or confined to alkaline sodium hypochlorite.

Chemical engineering

Acids One of the most extensive uses of austenitic steels has been for synthetic nitric acid plant, including oxidation coolers, absorption towers, storage vessels and pipelines, as well as transport tanks. For storage of cold 98% acid or for near-boiling solutions above 60% strength, 347S17 or 304S12 should be used.

Phosphoric acid processing is usually associated with onerous conditions, involving hot strong acid, often with appreciable contents of impurities such as sulphuric or hydrofluoric acids. Limited use is made of the austenitic steels, especially 316S16.

Sulphurous acid occurs in bisulphite manufacture, as well as in various processes such as cellulose pulp digestion and oil refining. Since some oxidation to sulphuric acid, which makes the conditions much more corrosive, is often involved, the use of molybdenum-bearing types is generally preferred.

316S16 steel is used for vinegar distillation and other high-temperature acetic acid applications, the non-molybdenum types being used at lower temperatures. With the higher fatty acids such as oleic and stearic, the non-molybdenum types are resistant even up to boiling point. Lactic, citric and oxalic acids are used with these steels over more limited ranges, the 316S16 type being necessary in many cases.

Processes involving less corrosive organic acids such as carbolic, cresylic, salicylic and many others, are satisfactorily catered for by the non-molybdenum steels. This also applies to the less corrosive inorganic acids such as boric, nitrous, carbonic and silicic.

Fertilisers For nitrates, non-molybdenum austenitic steels are satisfactory, but in the manufacture of ammonium sulphate some free acid is often present, so that evaporators and centrifugal dryer baskets in this case are generally made from molybdenum-bearing steels. For super-phosphates this has limited application.

Cellulose products There is little scope for the use of austenitic steels in the alkali processes for the digestion of wood pulp, but both molybdenum-free and molybdenum-bearing steels are extensively used for sulphite pulp digestion, choice depending on concentration, temperature and working experience.

The presence of free sulphuric acid in rayon-spinning baths limits application of the austenitic steels, but they are used for acetylation of cellulose in the acetate process. They are also used for dissolving and spinning solutions in the cuprammonium processes.

The austenitic steels are extensively used for nitrating cellulose, owing to their good resistance to many sulphuric-nitric acid mixtures.

Fats and soaps Their excellent resistance to the higher fatty acids makes the austenitic steels valuable constructional materials for plant dealing with hydrogenation or other treatment of oleic, stearic, and similar acids.

In soap-making some use is made of the molybdenum-bearing type for dealing with glycerine recovery from soap liquors containing sodium chloride.

Oil refining and by-products Various acid conditions, e.g. sulphuric acid and sulphur dioxide, are encountered in refining, and for such conditions 316S16 steel has proved useful, with the most resistant higher-nickel Cu–Mo types for the more difficult applications.

Fine chemicals There are many applications of the austenitic steels in the manufacture and storage of fine chemicals and pharmaceutical products. These include storage tanks, pipelines, valves, stills, steam-jacketed pans, mixing vessels, filters and tableting machinery. Considerable use has been made of these steels in penicillin production.

High-pressure synthesis Here austenitic steels are applied to autoclaves and pressure-vessel linings, pipelines, valves and other equipment dealing with a variety of products, often at quite high temperatures.

Atomic power production The necessity for avoiding contamination of operative liquids, together with other requirements which must be met in selecting constructional materials in this highly specialised field has resulted in the choice of austenitic steels for applications in heat exchangers, pressure vessels, pipelines and fuel processing.

J. E. TRUMAN

REFERENCES

1. Irvine, K. J., Crowe, D. J. and Pickering, F. B., *J.I.S.I.*, **195**, 386, Aug. (1960)
2. Truman, J. E., *I.S.I. Special Report No. 86*, 84 (1964)
3. Morley, J. I., *I.S.I. Special Report No. 86*, 92 (1964)
4. Irvine, K. J. and Llewellyn, D. T., *J.I.S.I.*, **192**, 218 (1959)
5. Llewellyn, D. T. and Murray, J. D., *I.S.I. Special Report No. 86*, 197 (1964)
6. Kemper, K. J., Morley, J. I., McWilliam, J. A. and Slater, D., *Proc. Inst. Mech. Eng.*, **181 Pt. 1** (1966/67)
7. Edeleanu, C., *J.I.S.I.*, **188**, 122 (1958)
8. Chandler, K. A., *I.S.I. Publication No. 117*, 128 (1969)

8(a). Cowan Jr, R. L. and Tedman Jr, C. S., 'Intergranular Corrosion of Iron–Nickel–Chromium Alloys', in *Advances in Corrosion Science and Technology*, Vol. 3 (Ed. Fontana, M. G. and Staehle, R. W.), Plenum (1973)

9. Romanoff, M., Nat. Bureau of Stds., Circular 579, 222, April (1957)
10. Truman, J. E., Perry, R. and Chapman, G., *J.I.S.I.*, **202**, 746 (1964)
11. Truman, J. E., *I.S.I. Publication No. 117*, 102 (1969)

BIBLIOGRAPHY

Colombier, L. and Hochmann, J., *Stainless and Heat Resisting Steels*, Arnold (1967)

Keating, F. H., *Chromium-Nickel Austenitic Steels*, Butterworths, London (1956)

Monypenny, J. H. G., *Stainless Iron and Steel*, Chapman and Hall, Vol. I (1951) and Vol. II (1954)

Uhlig, H. H., (ed.), *The Corrosion Handbook*, Wiley, New York (1948)

3.4 Corrosion Resistance of Maraging Steels

Composition

Maraging steels are a class of high strength steels of very low carbon content. Strengthening is achieved by the use of substitutional elements to produce age hardening in the martensitic iron–nickel matrix. The term *maraging* was thus coined from the words 'martensite' and 'age hardening'.

The development of maraging steels began in the late 1950s, on steels containing 20% and 25% Ni using a combination of aluminium, titanium and niobium as age-hardening elements. The characteristics of these steels are contained in Reference 1. Later work[2] revealed the important synergistic age-hardening effect of cobalt plus molybdenum and led to the development of the 18% Ni maraging steels. Using titanium as a supplementary hardening element, and with appropriate balancing of cobalt and molybdenum, nominal yield strengths in the range of 1 370–2 400 MNm^{-2} can be achieved. Table 3.23 lists the nominal composition of the 18% Ni maraging steels. Other types of maraging steel include an alloy (17% Ni) developed for use as a casting, and a 12Ni–5Cr–3Mo alloy. Stainless-type alloys have also been developed but are outside the scope of this section.

Table 3.23 Nominal composition of 18% Ni maraging steels

Maraging steel type	*Nominal composition (%)*					*Nominal yield strength*
	Ni	Co	Mo	Ti	Fe	($MN\ m^{-2}$)
18% Ni 200	18	8·5	3	0·2	Balance	1 380
18% Ni 250	18	8	5	0·4	Balance	1 720
18% Ni 300	18	9	5	0·6	Balance	2 050
18% Ni 350	17·5	12·5	3·75	1·7	Balance	2 390
Cast alloy	17	10	4·6	0·3	Balance	1 580

Data for the cast and the Cr-bearing alloys are contained in References 36 and 5 respectively. Discussion in this article is restricted to the 18% Ni maraging steels.

Structural Features

On cooling to room temperature after annealing, maraging steels transform completely to martensite. The as-annealed structure consists of packets of parallel lath-like martensite platelets arranged within a network of prior-austenite grain boundaries. The platelets have a high dislocation density but are not twinned.

On heat treating at 485°C a very rapid age-hardening reaction takes place and greatly strengthens the material. Although the nature of the precipitates formed is still uncertain, the consensus of opinion is that ageing for several hours at 485°C results in a Ni_3Mo phase while longer times produce an Fe_2Mo phase. There may also be a titanium precipitate, η-Ni_3Ti or $Ni_3(Mo, Ti)$[3]. Ageing at higher temperatures or longer times results in some reversion to austenite, which may be stable at room temperature (depending on time and temperature of ageing), and a lower strength.

Physical and Mechanical Properties

Table 3.24 Summary of physical properties for the 18% Ni 200 to 18% Ni 350 alloys

Density 8·0–8·1 g/cm^3
Crystal structure Martensite (body centred cubic); austenite (face centred cubic)
Lattice parameter Martensite 2·856–2·862 Å at room temperature; austenite (retained) 3·58 Å
Thermal conductivity 19·68–20·93 kW/m^2 °C (20–100°C)
Electrical resistivity 60–70 μΩ cm when solution annealed at 815°C; 35–50 μΩ cm when maraged at 485°C for 3 h
Melting temperature 1 430–1 445°C
Transformation temperature M_s 145–200°C; M_f 77–145°C; A_s 445°C
Nominal length change −0·06% to −0·10% during maraging

Table 3.25 Summary of nominal mechanical properties for the 18% Ni maraging alloys

Property	18% Ni 200	18% Ni 250	18% Ni 300	18% Ni 350
Yield strength—0·2% offset (MNm^{-2})	1 310–1 550	1 650–1 820	1 780–2 060	2 270–2 480
Ultimate tensile strength (MNm^{-2})	1 340–1 580	1 680–1 860	1 820–2 100	2 300–2 510
Elongation in 25·4 mm (1 in), (%)	11–15	10–12	7–11	6–10
Reduction in area (%)	35–67	35–60	30–50	25–45
Modulus of elasticity *E* (kNm^{-2})	$18{\cdot}0 \times 10^7$	$18{\cdot}5 \times 10^7$	$18{\cdot}9 \times 10^7$	$19{\cdot}3 \times 10^7$
Hardness (Rockwell *C*)	44–48	48–50	51–55	56–59
Impact Charpy V-notch (J)	35–68	24–45	16–26	7–14
Notch tensile				
0·0128 m bar K_t = 10 (MNm^{-2})	2 390	2 350–2 510	2 900–3 100	—
0·00762 m bar K_t = 10 (MNm^{-2})	—	2 560–2 660	—	1 360–1 490
Fracture toughness (K_{Ic}) ($MNm^{-2}\sqrt{m}$)	101–176	98–165	88–143	44–82
Endurance limit (MNm^{-2})				
Smooth bar 10^8 cycles	620–795	620–760	760–900	690
Notched bar 10^8 cycles (K_t = 2·2)	275–345	275–380	275–415	—
(K_t = 2·8)	—	—	—	352

Note. In all cases treatment was for 1 h at 815°C plus 3 h at 485°C.

Summaries of the physical and mechanical properties of the 18% Ni maraging steels are given in Tables 3.24 and 3.25. The mechanical properties are highlighted by good ductility, toughness and a lack of notch sensitivity. The plane strain fracture toughness of maraging steels is superior to other alloys at comparable strength levels (Table 3.26).

Table 3.26 Comparison of toughness of maraging steels and other high strength alloys*

Alloy	*Yield strength* (MNm^{-2})	K_{Ic}($MNm^{-2}\sqrt{m}$)
18% Ni 200	1 380	110–176
18% Ni 250	1 720	98–165
18% Ni 300	1 930	88–143
D6 AC	1 380	88–99
H–11	1 790	66–71
AISI 4340	1 790	61–66
AMS 6430	1 510	61–71
Ti–16V–2·5Al	1 170	49–55
Aluminium 7075.T6	415–485	39–66

*Data after Reference 9.
Note. K_{Ic} is the plane strain fracture toughness (see Section 8.9).

Fabrication

Maraging steels have been produced both by air and vacuum melting. Small amounts of impurities can decrease toughness significantly, sulphur in particular is detrimental and should be kept as low as possible. Silicon and manganese also have a detrimental effect on toughness and should be maintained below a combined level of 0·20%. Such elements as C, P, Bi, O_2, N_2 and H_2 are kept at the lowest levels practicable.

The maraging steels are readily hot worked by conventional rolling and forging operations. A preliminary homogenisation at 1 210°C–1 260°C is normally used prior to hot working at that temperature. During subsequent hot working, extended times at, or slow cooling through, temperatures from 760°C–1 100°C should be avoided since they produce embrittlement[4]. Maraging steels can be cold worked up to 85% before requiring intermediate annealing (because of low work-hardening characteristics) but are usually annealed after smaller reductions.

Heat treatments are relatively simple and normally consist of annealing for 1 h at 815°C followed by ageing for 3 h at 485°C. Recently, double annealing treatments have grown in favour. Machining or fabrication is easily performed in the as-annealed condition. Subsequent age hardening generally introduces small and predictable dimensional changes.

Theoretical Considerations and Characteristics of Corrosion Behaviour

Corrosion in Natural Environments

In atmospheric exposure 18% Ni maraging steel corrodes in a uniform manner[5], and becomes completely rust covered. Pit depths tend to be more shallow than for the low-alloy high-strength steels[6]. Atmospheric corrosion rates[7] in industrial (Bayonne, New Jersey) and marine (Kure Beach, North Carolina) atmospheres are compared with those for low-alloy steel in Figs. 3.24, 3.25 and 3.26. The corrosion rates drop substantially after the first year or two and in all cases the rates for maraging steel are about half the corrosion rate for HY80 and AISI 4340 steels.

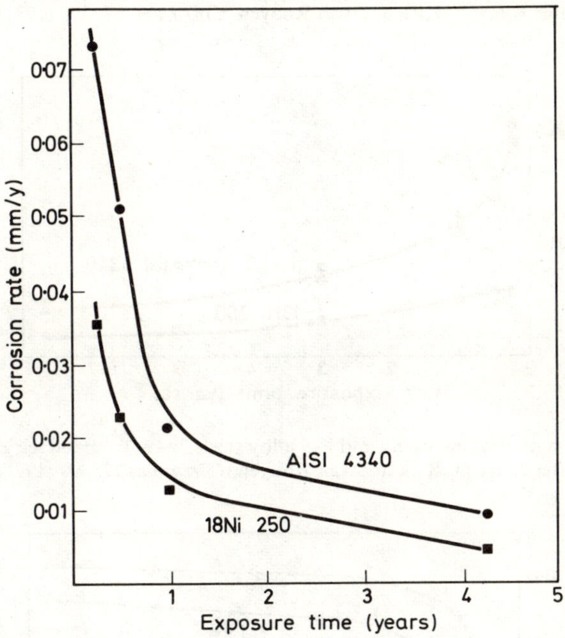

Fig. 3.24 Corrosion rates of maraging and low alloy steels in an industrial atmosphere at Bayonne, N.J. (after Kenyon, Kirk and van Rooyen, *Corrosion*, **27**, 390 (1971)[7])

The corrosion rates for both maraging steel and the low alloy steels in sea-water are similar initially, but from about 1 year onwards the maraging steels tend to corrode more slowly as indicated in Fig. 3.27[7]. The corrosion rates for both low alloy and maraging steel increase with water velocity[5]. During sea-water exposure the initial attack was confined to local anodic areas, whereas other areas (cathodic) remained almost free from attack; the latter were covered with a calcareous deposit typical of cathodic areas in sea-water exposure. In time, the anodic rust areas covered the entire surface[6].

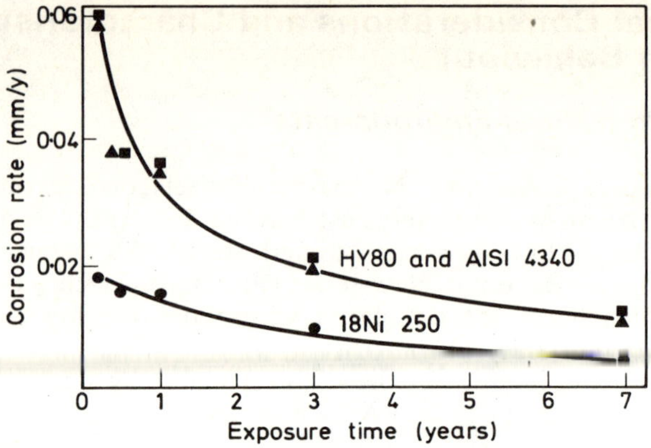

Fig. 3.25 Corrosion rates of maraging and low alloy steels 24·4 m from the sea at Kure Beach, N.C. (after Kenyon, Kirk and van Rooyen, *Corrosion*, **27**, 390 (1971)[7])

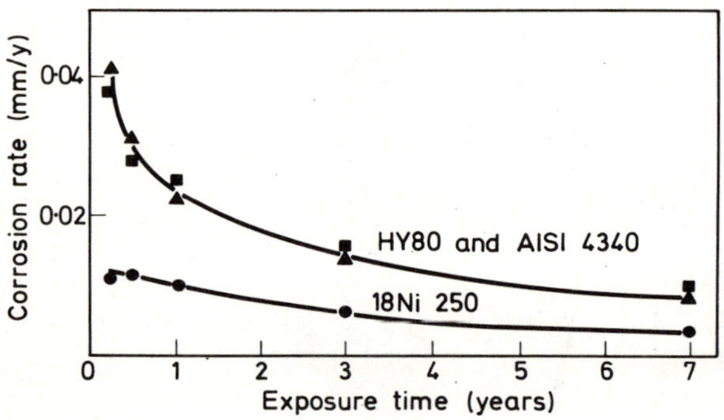

Fig. 3.26 Corrosion rates of maraging and low alloy steels 244 m from the sea at Kure Beach, N.C. (after Kenyon, Kirk and van Rooyen, *Corrosion*, **27**, 390 (1971)[7])

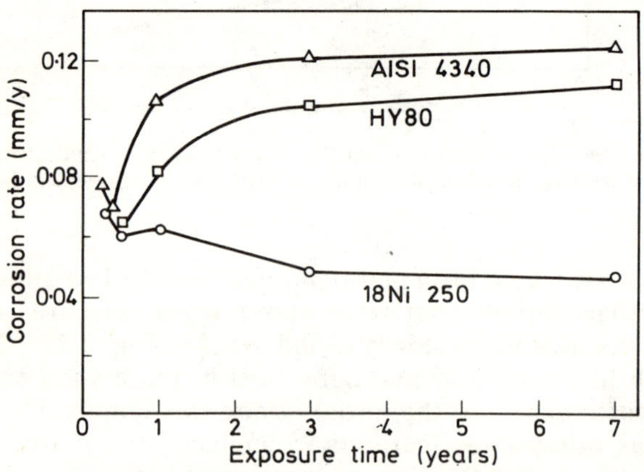

Fig. 3.27 Corrosion rates of maraging and low alloy steels in sea-water flowing at 0·6 m/s (after Kenyon, Kirk and van Rooyen, *Corrosion*, **27**, 390 (1971)[7])

Polarisation tests[8] indicate that maraging steel does not exhibit passive behaviour in 3% NaCl, and that the polarisation curves are unaffected by changes in heat treatment.

Corrosion in Chemicals

The corrosion rates of maraging steels in acid solutions such as sulphuric, hydrochloric, formic and stearic acids are substantial although lower than those of the low-alloy high-strength steels[9]. Polarisation studies[10] indicate that maraging steels exhibit active–passive behaviour in 1N and 0·1N sulphuric acid. The corrosion potential, critical current density, primary passivation potential and passive current density are all affected by variations in ageing treatment. The critical and passive current densities increase as the structure is varied from fully annealed to fully aged. The normal heat treatment produces critical and passive current densities of $0{\cdot}4\,mA/cm^2$ and $0{\cdot}2\,mA/cm^2$ respectively ($0{\cdot}1\,mA/cm^2 \approx 1{\cdot}2\,mm/y$[11]). Stress corrosion of maraging steel may also occur in acid environments.

Stress Corrosion Cracking

The 18% Ni maraging steels do not display passivity and normally undergo uniform surface attack in the common environments. Of more serious consequence, however, for all high strength steels, is the degree of susceptibility to stress corrosion cracking (s.c.c.).

For high strength steels in general, the s.c.c. process in aqueous media is characterised by delayed failure, which consists of an incubation period followed by slow and, at times, intermittent crack growth. Failure can occur on loading to some fraction of the yield stress, or through the action of residual stresses, often in environments as mild as humid air[12]. The degree of susceptibility depends on the mode of loading and is highest in cases of plane strain loading (triaxial stresses), with tensile loading and plane stress bending, in that order, representing less severe loading conditions[13] (tensile stresses of course are present in all the above modes of loading). In general, susceptibility to s.c.c. increases with increasing yield strength. Different alloy types however, vary in their degree of resistance, and at comparable strength levels maraging steels compare favourably in cracking resistance to other high strength steels.

There has been some controversy as to whether s.c.c. occurs by active path corrosion or by hydrogen embrittlement. Lack of space does not permit a full treatment of this subject here. References 14 and 15 are recent reviews on the s.c.c. of high strength steels and deal with the mechanism of cracking (see also Section 8.4). It is appropriate to discuss briefly some of the latest work which appears to provide pertinent information on the cracking mechanism. It should be noted, however, that cracking in all alloy systems may not be by the same mechanism, and that evidence from one alloy system need not constitute valid support for the same cracking mechanism in another.

Recently-developed techniques[16,17] have enabled studies to be made of the electrochemical conditions near the tip of a growing stress corrosion crack. It was found that under freely corroding conditions the pH of the solution near the crack tip was about 3·8 for all eleven steels (including maraging steels) studied[18]. Furthermore, potential measurements indicated that thermodynamic conditions were satisfied for hydrogen ion reduction. Further potential–pH measurements were made on AISI 4340 steel exposed to 0·6N NaCl solutions of different pH and polarised to potentials both negative and positive with respect to the corrosion potential[17]. It was found that the pH of the solution in the crack was determined solely by electrochemical reactions at the crack tip irrespective of starting pH. It was also apparent that regardless of the impressed potential, the electrochemical conditions in the crack satisfied the thermodynamic requirements for the production of hydrogen. These results indicate that it is not necessary to invoke an active path mechanism and that hydrogen is available even during anodic polarisation of AISI 4340 steel.

Activation energy measurements for s.c.c. of H-11 (38 kJ/g atom)[19] and AISI 4340 (36 kJ/g atom)[20] steels in water and moist air suggest that the cracking rate in these alloys is controlled by the diffusion of hydrogen, since these activation energies are in close agreement with the value of 39 kJ/g atom obtained for the diffusion of hydrogen through AISI 4340 membranes[21]. Activation energies have not been determined for cracking in maraging steels, nor have they been determined for any steel in situations of anodic polarisation.

There is little question that at strongly negative potentials cracking occurs by hydrogen embrittlement. Also, the activation energy studies of H-11 and AISI 4340 strongly suggest that hydrogen embrittlement is the controlling mechanism in these cases. For all other cases there does not appear to be a strong basis for favouring one mechanism over the other or to discard the possibility that each mechanism shares in the control of the cracking process. The latter possibility for maraging steel has been supported by a recent study of cracking response to polarisation, fractographic studies and pH determinations of the corrodent at the tip of the crack[22].

Testing for Stress Corrosion Cracking

The stress corrosion resistance of maraging steel has been evaluated both by the use of smooth specimens loaded to some fraction of the yield strength and taking the time to failure as an indication of resistance, and by the fracture mechanics approach[23] which involves the use of specimens with a pre-existing crack. Using the latter approach it is possible to obtain crack propagation rates at known stress intensity factors (K) and to determine critical stress intensity factors (K_{ISCC}) below which a crack will not propagate (see Section 8.9).

Any test (several such tests are used) in which time to failure of smooth specimens is determined is an overall measure of the incubation period to initiate a crack, the ability to resist the propagation of a stress corrosion crack and the ability to resist final mechanical fracture. Since this test does not indicate the relative merits of an alloy in each individual aspect of the cracking process it is probably of less benefit to a design engineer. The use of precracked specimens in the fracture mechanics approach follows from the philosophy that structures are likely to contain crack-like defects. The use of precracked specimens promotes a rapid change in chemistry of the solution at the crack tip and shortens or eliminates the initiation time for crack propagation. Data from both types of tests are available and are reviewed here.

Cracking Resistance in Smooth Materials

Maraging steel in the strength range 1 240–1 720 MNm^{-2} tested as 'U' bends in sea-water, displayed good resistance, as it did not fracture in periods up to 2–3 years although there was considerable general corrosion and fouling. However, microcracks were observed after 6 months. Similar behaviour[5,6] of 'U'-bends and bent beam specimens can be expected in industrial or marine atmospheres although general corrosion is less severe. By comparison AISI 4340 at strength levels of 1 660 MNm^{-2} failed in about 1 week in both sea-water and atmospheric tests[5,6]. Maraging steel of yield strength at or above 2 060 MNm^{-2} was not resistant and failed rapidly.

Maraging steel welds are somewhat less resistant than base plate. 'U'-bend exposure of 1 240 MNm^{-2} strength welds survived for up to 2 years in sea-water while at 1 380 MNm^{-2} failures occurred in 2–18 months[7].

It is possible to provide cathodic protection to base plate up to 1 720 MNm^{-2} yield strength, by coupling to mild steel or possibly to zinc[5,6], but zinc and metals more active than zinc tend to induce hydrogen embrittlement. Welds up to 1 380 MNm^{-2} may be cathodically protected by zinc, but at impressed potentials of −1·25 V (S.C.E.) both 1 240 and 1 380 MNm^{-2} welds fail rapidly due to hydrogen embrittlement[7]. Neither mild steel nor zinc couples protect AISI 4340 steel[5].

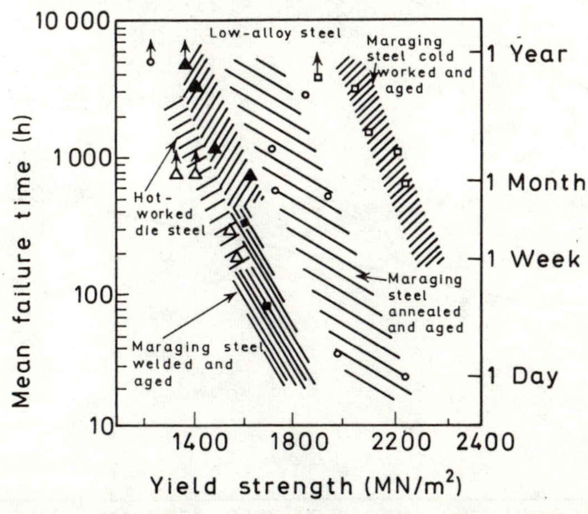

Fig. 3.28 Bent-beam test results in aerated distilled water. These specimens were exposed to the environment at a stress of 70% of yield (after Setterlund, *Materials Protection*, **4** No. 12, 27 (1965)[24])

Although tests on smooth specimens indicate that cathodic protection of maraging steel is possible, tests on specimens with pre-existing cracks indicate a greater sensitivity to hydrogen embrittlement during cathodic polarisation[7]. The use of cathodic protection on actual structures must therefore be applied with caution, and the application of less negative potentials than are indicated to be feasible in smooth specimen tests is to be recommended if it is assumed that structures contain crack-like defects.

Further evidence of the relative resistance of maraging steel is reproduced in Fig. 3.28 from Reference 24. Maraging steel is shown to be superior to a die steel and low alloy steel (both unidentified) in bent beam tests stressed at 75% of the yield strength in distilled water. Also shown is the beneficial effect in smooth surface tests of cold rolling. Shot peening has a similar beneficial effect[7].

Resistance to Propagation of Pre-existing Cracks

Critical Stress Intensity Factor It has become common to use K_{ISCC}, the critical stress intensity factor, as a measure of the resistance of an alloy to s.c.c. Tests are performed on specimens which are precracked by a fatigue machine and must be of sufficient dimensions to ensure plane strain conditions. Recommendations on precracking and dimensions are given elsewhere[25,26].

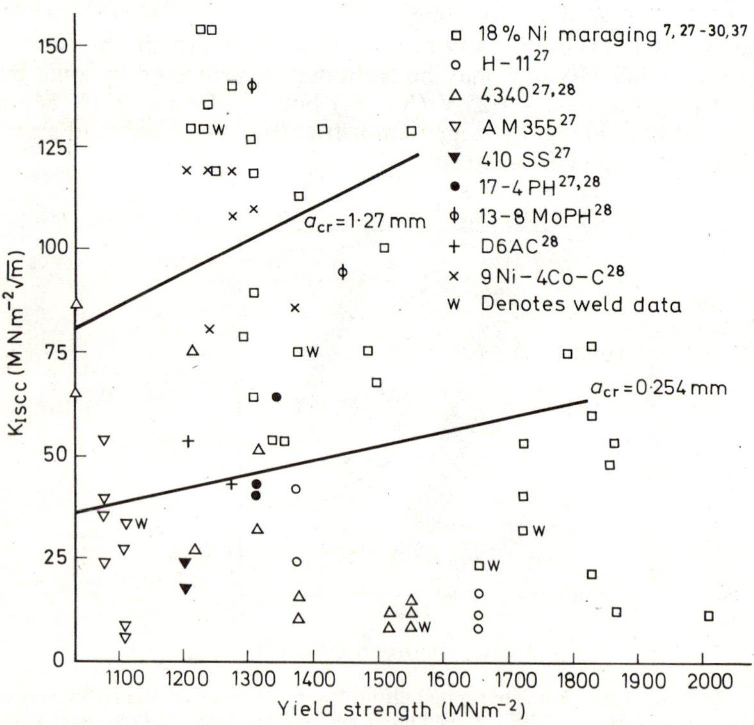

Fig. 3.29 Comparison of K_{ISCC} as a function of yield strength of 18% Ni maraging and other high strength steels

Figure 3.29 presents a summary comparing K_{ISCC} values of maraging steel, with values for H-11, AISI 4340, AM 355*, 410 stainless steel, 17–4 PH†, 13–8 Mo PH†, D6AC‡ and 9Ni–4Co–C§ steels. The data have been taken from a number of sources[7,27–30,37] and are for exposures in aqueous environments, with and without NaCl. No attempt has been made to distinguish between different environments since they do not affect the results appreciably. Plotted points marked *W* refer to data for welds. Also included in Fig. 3.29 are lines of critical crack depth a_{cr}. The region below a line of specified critical crack depth corresponds to combinations of strength and K_{ISCC} for which a long crack of the specified depth will propagate when stressed to the yield point, whereas the region above the line corresponds to the strength and K_{ISCC} combinations for which the crack will not propagate. The critical crack depth (assuming yield point stresses) for cracks whose length greatly exceeds their depth is given by

$$a_{cr} = 0{\cdot}2\left(\frac{K_{ISCC}}{\sigma_y}\right)^2$$

where σ_y is the yield stress[23].

In general it is clear that maraging steels compare favourably with other high strength steels, and offer comparatively high K_{ISCC} values over a wide range of strength. It is also clear that maraging steels can withstand a greater crack depth without crack propagation.

Table 3.27 Crack propagation rates for a number of high strength steels*

Alloy	*Yield strength* (MNm^{-2})	*Crack velocity* (mm/s)
Maraging 250	1 570	$1{\cdot}31 \times 10^{-5}$
Maraging 250	1 690	$1{\cdot}40 \times 10^{-5}$
Maraging 300	1 950	$2{\cdot}75 \times 10^{-4}$
Modified maraging 300†	2 180	$1{\cdot}18 \times 10^{-4}$
Modified maraging 300† (underaged)	1 810	$0{\cdot}72 \times 10^{-3}$
Modified maraging 300† (overaged)	1 760	$2{\cdot}88 \times 10^{-5}$
4340	1 430	$1{\cdot}65 \times 10^{-3}$
D6AC	1 540	$1{\cdot}74 \times 10^{-4}$
H-11	1 420	$1{\cdot}19 \times 10^{-5}$
HP 9-4-25	1 330	$2{\cdot}54 \times 10^{-5}$
HP 9-4-45 (bainitic)	1 460	$1{\cdot}06 \times 10^{-5}$

*Data after Carter[31].
†No details on the modification were available.

A further estimation of the corrosion resistance of maraging steel can be obtained from data on the rate of crack propagation. Although the rate of crack propagation has been found to be a function of stress intensity in some alloys, for many alloys and heat treatments there is a range of stress

*Registered Trademark of Allegheny Ludlum Industries
†Registered Trademark of Armco Steel Corporation
‡Registered Trademark of Vasco-Teledyne Company
§Registered Trademark of Republic Steel Corporation

intensity above K_{ISCC} and approaching K_{IC} where the rate of crack propagation is independent of stress intensity[31]. Table 3.27, summarised from Reference 31, presents a list of crack growth rates for a number of high strength steels whose cracking rates were independent of stress intensity. The cracking rate for maraging steel is seen to be slower than for 4340 and D6AC and equivalent to H-11 and HP 9Ni–Co–C steels, at about 10^{-5} mm/s. The fact that cracks propagate very slowly in maraging steels has an important consequence related to their stress corrosion testing. In determining K_{ISCC} by dead-weight-loaded cantilever-beam tests, it has been recommended[32] that maraging steel should withstand 1 000 h without failure to ensure that the applied stress intensity is at, or lower than, K_{ISCC}. 100 h was considered adequate for the more rapidly cracking low-alloy steels[32].

Effect of Metallurgical Variables on s.c.c.

It is notable that while it is possible to produce maraging steels with consistently uniform mechanical properties, the stress corrosion properties are subject to scatter, as indicated in Fig. 3.29. To a large extent this scatter is an indication of the greater sensitivity of s.c.c. resistance to metallurgical variables. Although the variation in cracking resistance is not well understood*, and the reaction to certain treatments not always consistent, certain observations may be used to indicate guidelines for improved properties.

Cracking has generally been observed to be intergranular[6,33–35] with isolated cases of transgranular cracking[29]. Both Ti_2S[29] and Ti(C, N)[29,35] on prior austenite boundaries havc been suggested to be related to greater susceptibility. Since prolonged time in the temperature range 760–1 100°C favours the precipitation of Ti_2S and Ti(C, N) on prior austenite boundaries, such prolonged exposures should be avoided both in processing and in annealing. It is preferable that annealing should be done at the minimum temperatures required to develop desired properties.

Studies of the effect of ageing temperature on cracking behaviour have shown rather marked effects. Underageing at temperatures of 455°C or lower was found to increase greatly the rate of crack propagation without affecting the mode of crack propagation (intergranular) or K_{ISCC}[31,33,34]. Table 3.27 lists crack growth rates for a modified 18% Ni 300 steel as a function of ageing treatment. Other work[29] found that underageing not only increased the crack growth rate, but also decreased the resistance to cracking (K_{ISCC}) and resulted in a change from a transgranular cracking mode to an intergranular mode. Electron microscopy was unable to detect any grain boundary phases. No satisfactory explanation for this behaviour was apparent. Overageing appears to offer slightly slower crack growth rates (Table 3.27), however it did not significantly improve K_{ISCC}[29]. The best combination of properties is obtained with the normal ageing treatment at 485°C.

Studies on welds showed that cracking resistance as indicated in 'U'-bend tests[7] was substantially increased by a post weld anneal (1 h at 815°C) prior

*This aspect and the general stress corrosion and hydrogen embrittlement behaviour are considered in detail in Reference 37.

to the normal ageing treatment. Material aged in the as-welded condition was less resistant. The most significant structural difference resulting from the two heat treatments was a finer dispersion of austenite ribbons in material annealed before ageing.

High Temperature Corrosion

Little data are available on hot corrosion behaviour. Figure 3.30 indicates maraging steel to have better resistance to air exposure at 535°C than a 5% Cr tool steel[9]. Metallographic examination indicates that exposure to

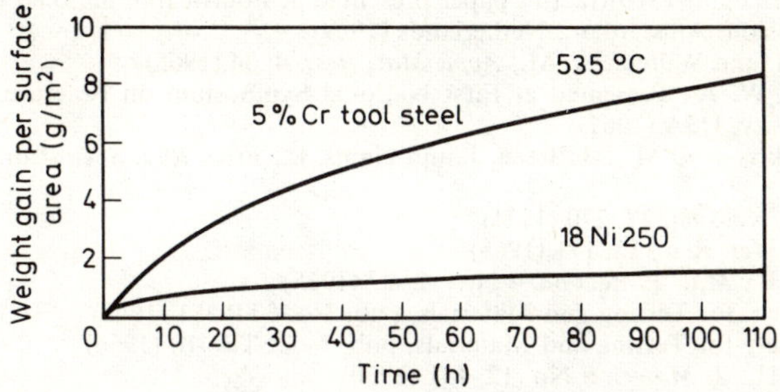

Fig. 3.30 Oxidation rate at 535°C of 18Ni250 maraging steel compared with a generally-available tool steel. These tests were performed on $\frac{1}{4}$in (6·35 mm) cubes placed in refractory cubicles and exposed to still air for total times of 5, 25 and 100 h. The weight gain includes the scale formed during heating and cooling (after Reference 9)

air at elevated temperatures results in reaction with both oxygen and nitrogen, forming both oxides and titanium carbo-nitrides. Under some conditions of reduced oxygen partial pressure, selective sub-surface oxidation of iron can occur.

Applications

Maraging steels have found varied uses in the aerospace and aircraft industries. These uses have included rocket motor cases, landing gear components, aircraft forgings and fasteners. Other areas of usage include machine tool and die applications, and extrusion hardware. Marine uses include hydrofoil foil systems and aircraft arrestor hooks.

D. P. DAUTOVICH

REFERENCES

1. Bieber, C. G., *Metal Progress*, **78**, 99 (1960)
2. Decker, R. F., Eash, J. T. and Goldman, A. J., *Trans. Quart. A.S.M.*, **55**, 58 (1962)
3. Floreen, S., *Met. Rev.*, **13**, 115 (1968)
4. Novak, C. J., D.M.I.C. Memo 196 (1964)
5. Kirk, W. W., Covert, R. A. and May, T. P., *Met. Eng. Quart.*, **8**, 31 (1968)
6. Dean, S. W. and Copson, H. R., *Corrosion*, **21**, 95 (1965)

7. Kenyon, N., Kirk, W. W. and van Rooyen, D., *Corrosion*, **27**, 390 (1971)
8. Stavros, A. J. and Paxton, H. W., Paper presented at NACE Meeting, Chicago, Illinois (1971)
9. *Data Bulletin on 18% Ni Maraging Steels*, The International Nickel Company, Inc. (1964)
10. Nam Bui, Pieraggi, B. and Dabosi, F., *Mem. Sci. Rev. Met.*, **68**, 223 (1971)
11. France, W. D., Jr. and Mazzatenta, E. D., *Materials Engineering*, **62**, Sept. (1970)
12. Johnson, H. H., *Proceedings of Conference of Fundamental Aspects of Stress Corrosion Cracking*, National Association of Corrosion Engineers, 439 (1969)
13. Hayden, H. W. and Floreen, S., *Corrosion*, **27**, 429 (1971)
14. Phelps, E. H., *Proceedings of the Conference of Fundamental Aspects of Stress Corrosion Cracking*, National Association of Corrosion Engineers, 398 (1969)
15. Kennedy, J. W. and Whittaker, J. A., *Corrosion Science*, **8**, 359 (1968)
16. Brown, B. F., Fujii, C. T. and Dalhberg, E. P., *J. Electrochem. Soc.*, **116**, 218 (1969)
17. Smith, J. A., Peterson, M. H. and Brown, B. F., *Corrosion*, **26**, 539 (1970)
18. Brown, B. F., Extended Abstract of paper presented at Fourth International Congress on Metallic Corrosion, Amsterdam, Netherlands (1969)
19. Johnson, H. H. and Willner, A. M., *Appl. Mat. Res.*, **4**, 34 (1965)
20. van der Sluys, W. A., Presented at First National Symposium on Fracture Mechanics, Lehigh University, USA (1967)
21. Beck, W., Bockris, J. O'M., McBreen, J. and Nanis, L., *Proc. Roy. Soc.*, (London), **A290**, 221 (1966).
22. Syrett, B. C., *Corrosion*, **27**, 270 (1971)
23. Brown, B. F., *Met. Rev.*, **13**, 171 (1968)
24. Setterlund, R. B., *Mat. Protection*, **4** No. 12, 27 (1965)
25. American Society for Testing and Materials, pub. No. STP 381 (1965)
26. American Society for Testing and Materials, pub. No. STP 410 (1966)
27. Freedman, A. H., *J. Mater.*, **5** No. 12, 431 (1970)
28. Brown, B. F., Naval Research Laboratory Report 7168 (1970)
29. Dautovich, D. P., The International Nickel Company, Inc., unpublished work
30. Steigerwald, E. A., presented at ASM Conference on Stress Corrosion Cracking, Philadelphia, Pa., Aug. (1970. Original work by Carter, C. S., Boeing Co. Report D6–19770, Nov. (1967)
31. Carter, C. S., *Corrosion*, **27**, 471 (1971)
32. Brown, B. F., paper presented at ASM Conference on Fracture Control, Philadelphia, Pa., Jan. (1970). Referred to by Carter, C. S., (Ref. 33)
33. Carter, C. S., *Met. Trans.*, **1**, 1 551 (1970)
34. Stavros, A. J. and Paxton, H. W., *Met. Trans.*, **1**, 3 049 (1970)
35. Parkins, R. N. and Haney, E. G., *Trans. Met. Soc.*, **242**, 1 943 (1968)
36. Sadowski, E. P. and Koppi, W. A., *Trans. Am. Foundrymen's Soc.*, **75**, 294 (1967)
37. Dautovich, D. P. and Floreen, S., Paper presented at the International Conference on Stress Corrosion Cracking and Hydrogen Embrittlement of Iron Base Alloys, Unieux-Firminy, France, June (1973)

3.5 Nickel–Iron Alloys

Use of Nickel–Iron Alloys

Nickel–iron alloys have a number of important applications that are derived from such special physical properties as their unique magnetic characteristics in the regions of 35, 50 and 80% nickel and from their abnormally low thermal expansion in the region of 36–50% nickel. Although not specifically used as corrosion-resistant materials, their high resistance to attack from many common environments is of benefit in their specialised applications.

Electrochemical Characteristics

The potentiodynamically-derived polarisation curves of Beauchamp[1] (Fig. 3.31) demonstrate the effect of increasing nickel content on the anodic behaviour of iron–nickel alloys in 1N H_2SO_4. The maximum current in the active region is reduced and the potentials moved to more noble values; the current in the passive region is increased and some evidence of secondary passivity appears. The greater nobility with increasing nickel content in the active region is of importance in acid environments where hydrogen evaluation is the major cathodic reaction, and results in significantly lower rates of corrosion. In neutral environments, the protection provided by a layer of insoluble corrosion products is of greater significance.

Atmospheric Corrosion

The addition of small amounts of nickel to iron improves its resistance to corrosion in industrial atmospheres due to the formation of a protective layer of corrosion products. Larger additions of nickel, e.g. 36% or 42%, are not quite so beneficial with respect to overall corrosion since the rust formed is powdery, loose and non-protective, leading to a linear rate of attack as measured by weight loss. Figure 3.32 of Pettibone[2] illustrates the results obtained.

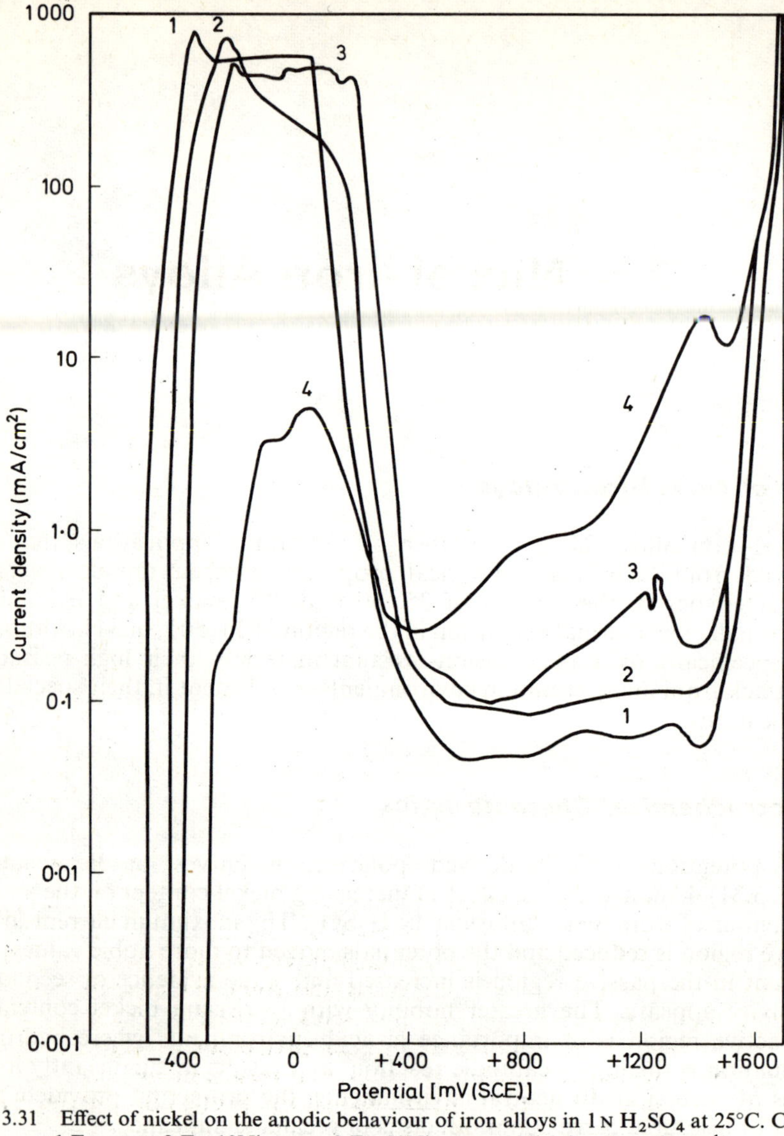

Fig. 3.31 Effect of nickel on the anodic behaviour of iron alloys in 1 N H_2SO_4 at 25°C. Curve 1 Fe; curve 2 Fe–10Ni; curve 3 Fe–20Ni; curve 4 Ni (after Beauchamp[1])

With respect to resistance to pitting corrosion, there is an increasing advantage to be obtained by increasing the nickel content up to 50%. There is little distinction between the Fe–50Ni alloy and pure nickel. Data on the corrosion of Fe–36Ni alloy at an industrial site in the USA are reported by La Que and Copson[3] and at a European site by Evans[4].

In marine atmospheres the overall rates of corrosion are reduced progressively with increase in nickel content up to about 35%, but with small improvement thereafter. The rates of corrosion at various sites, reported by

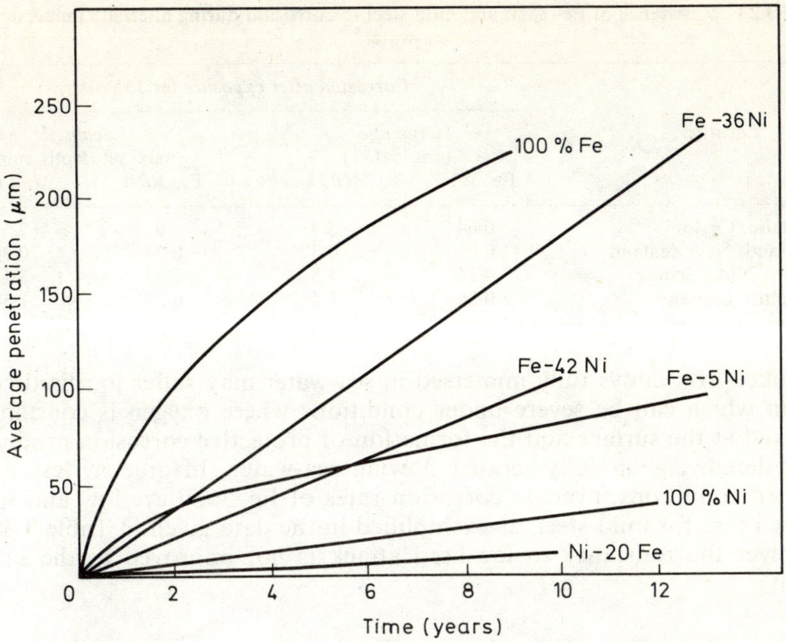

Fig. 3.32 Resistance of nickel–iron alloys to corrosion by an industrial atmosphere (Bayonne N.J., USA) (after Pettibone[2])

Table 3.28 Resistance of Fe–36Ni and mild steel to corrosion in marine atmospheres[5]

Location	*Corrosion after exposure for 15 years*			
	Average ($gm^{-2} d^{-1}$)		*Localised* (max. pit depth, mm)	
	Fe–36Ni	*Mild steel*	Fe–36Ni	*Mild steel*
Colombo, Ceylon	0·08	5·5	0	—
Auckland, New Zealand	0·02	1·5	0	2·43
Halifax, Nova Scotia	0·03	0·65	0·1	1·64
Plymouth, England	0·08	3·5	0·19	1·09

Friend[5], show the superiority of Fe–36Ni over mild steel with respect to both average and localised corrosion (Table 3.28).

Sea-water

The average rates of corrosion of Fe–36Ni alloy exposed to alternate immersion in sea-water are appreciably greater than those that occur when the alloy is exposed to marine atmospheres. Although the rates of corrosion are significantly below those observed for mild steel (Table 3.29) the superiority over mild steel is not so great with respect to pitting attack.

Table 3.29 Resistance of Fe–36Ni and mild steel to corrosion during alternate immersion in sea-water[5]

Location	Corrosion after exposure for 15 years			
	Average ($gm^{-2}d^{-1}$)		Localised (max. pit depth, mm)	
	Fe–36Ni	*Mild steel*	Fe–36Ni	*Mild steel*
Colombo, Ceylon	0·64	3·4	1·0	2·55
Auckland, New Zealand	0·09	0·32	0·24	0·36
Halifax, Nova Scotia	0·24	1·5	2·59	2·15
Plymouth, England	0·36	1·4	0·25	1·58

Nickel–iron alloys fully immersed in sea-water may suffer localised corrosion which can be severe under conditions where oxygen is constantly renewed at the surface and the formation of protective corrosion products is hindered, e.g. in fully-aerated flowing sea-water. In quieter, less oxygenated conditions, average corrosion rates of Fe–36Ni are low and well below those for mild steel, as exemplified in the data given in Table 3.30[5]. However the resistance to localised attack is not improved to the same extent.

Table 3.30 Resistance of Fe–36Ni alloy and mild steel to corrosion in sea-water[5]

Location	Corrosion after exposure for 15 years			
	Average ($gm^{-2}d^{-1}$)		Localised (max. pit depth, mm)	
	Fe–36Ni	*Mild steel*	Fe–36Ni	*Mild steel*
Colombo, Ceylon	0·8	2·0	2·5	6·5
Auckland, New Zealand	0·5	2·0	1·08	2·59
Halifax, Nova Scotia	1·3	2·2	3·49	1·23
Plymouth, England	0·8	1·5	1·82	2·75

Fresh Water

Nickel–iron alloys suffer significantly less corrosion than mild steel when exposed to a soft, fresh water but Friend[5] found that the resistance to pitting is only slightly greater (Table 3.31).

Table 3.31 Resistance of Fe–36Ni alloy and mild steel to corrosion in fresh water[5]

Corrosion after exposure for 15 years			
Average ($gm^{-2}d^{-1}$)		Localised (max. pit depth, mm)	
Fe–36 Ni	*Mild steel*	Fe–36 Ni	*Mild steel*
0·07	0·94	2·00	2·20

Resistance to Acids

Much of the information available on resistance of nickel–iron alloys to corrosion by mineral acids is summarised by Marsh[6]. In general, corrosion rates decrease sharply as the nickel content is increased from 0 to 30–40%, with little further improvement above this level. The value of the nickel addition is most pronounced in conditions where hydrogen evolution is the major cathodic reaction, i.e. under conditions of low aeration and agitation. Results reported by Hatfield[7,8] show that the rates of attack of Fe–25Ni alloy in sulphuric and hydrochloric acid solutions, although much lower than those of mild steel, are still appreciable (Tables 3.32 and 3.33). In solutions of nitric acid, nickel–iron alloys show very high rates of corrosion.

Table 3.32 Resistance of Fe–25Ni and carbon steel to corrosion by sulphuric acid

Alloy	*Corrosion rate* (mm/year)					
	5% H_2SO_4		25% H_2SO_4		50% H_2SO_4	
	15°C	40°C	15°C	40°C	15°C	40°C
Carbon steel	62	183	93	378	0·9	2·8
Fe–25Ni	0·45	1·4	0·45	1·8	0·9	1·8

Table 3.33 Resistance of Fe–25Ni and carbon steel to corrosion by hydrochloric acid

Alloy	*Corrosion rate* (mm/year)					
	5% HCl			25% HCl		
	15°C	40°C	60°C	15°C	40°C	60°C
Carbon steel	23	40	41	63	188	185
Fe–25Ni	0·45	2·3	5·4	0·9	7·7	19

Resistance to Alkalis

Addition of nickel improves the resistance of iron and steel to corrosion by alkaline solutions. The beneficial effect is most pronounced in hot, strong caustic solutions as illustrated by the results[9] on nickel cast irons in Table 3.34.

Table 3.34 Resistance of nickel cast irons to corrosion by hot caustic soda*

% Nickel in cast iron	*Corrosion rate* (mm/year)	*% Nickel in cast iron*	*Corrosion rate* (mm/year)
0	1·8 to 2·3	15	0·8
3·5	1·2	20	0·09
5	1·2	30	0·01

*Specimens immersed for 54 d in evaporator concentrating caustic soda from 50 to 60% under 88 kN/m² vacuum.

Resistance to Salt Solutions

Nickel–iron alloys are more resistant than iron to attack by solutions of various salts. In alternate immersion tests in 5% sodium chloride solution Fink and De Croly[10] determined values of 2·8, 0·25 and 0·5 $gm^{-2} d^{-1}$ for alloys containing 37, 80 and 100% nickel compared with 46 $gm^{-2} d^{-1}$ for iron. Corrosion rates of about 0·4 $gm^{-2} d^{-1}$ are reported by Hatfield[7] for Fe–30Ni alloy exposed to solutions containing respectively 5% magnesium sulphate, 10% magnesium chloride and 10% sodium sulphate; the same alloy corroded at a rate of about 1·2 $gm^{-2} d^{-1}$ in 5% ammonium chloride.

In a study of the corrosion of ten binary nickel–iron alloys in 3% sodium chloride solution, Schwerdtfeger[11] found the average corrosion rate to decrease from 1·4–1·6 $gm^{-2} d^{-1}$ for alloys containing 0–16% nickel alloy to 0·1 $gm^{-2} d^{-1}$ for a 57% nickel–iron alloy. There was little further reduction in rate of weight loss for the higher nickel alloys. However, the alloys showed increasing tendency to suffer from pitting and crevice corrosion with increasing nickel content.

Stress Corrosion

In tests lasting for 14 days, Copson[12] found that the susceptibility of steel to stress-corrosion cracking in hot caustic soda solutions increased with increase in nickel content up to at least 8·5%. Alloys containing 28% and more of nickel did not fail in this period. In boiling 42% magnesium chloride the 9% nickel–iron alloy was the most susceptible of those tested to cracking (Table 3.35). Alloys containing 28 and 42% nickel did not fail within 7 days.

Table 3.35 Resistance of iron–nickel alloys to stress corrosion cracking in boiling 42% magnesium chloride[12]

Composition of alloy Ni	C	Mn	Si	Fe	*Hardness*	*Time to cracking* (days)	*Comments*
nil	0·19	1·65	0·20	bal.	89 Rb	No cracking after 11 days	
2·02	0·19	0·46	0·18	bal.	77 Rb	7	Few shallow cracks
4·96	0·15	0·51	nil	bal.	96 Rb	<3	Profuse deep cracks
8·67	0·10	0·76	0·23	bal.	24 Rc	<3	Cracked in two
27·88	0·03	0·18	0·06	bal.	81 Rb	No cracking after 7 days	
41·79	0·02	0·18	0·08	bal.	77 Rb	No cracking after 7 days	
99·41	0·10	0·24	0·02	0·13	20 Rc	No cracking after 7 days	

Couper[13] reports cracking of an Fe–36 Ni alloy in 10–55 days in this medium. Radd *et al.*[14] have noted cracking of Fe–36 Ni alloys at ambient temperatures in an unspecified environment, but this possibly may have been residual traces of acid copper chloride etching solution.

Bimetallic Corrosion

Bimetallic corrosion of nickel–iron alloys may be of significance in welding operations. Ni–45 Fe alloys are used as filler materials in the welding of cast irons but the favourable area relationship of weld metal to base plate mitigates the effect of the more noble characteristics of the nickel–iron alloy. Thus their application in corrosive environments is rarely of concern.

Of more serious practical significance is iron contamination of nickel-clad steel welds. Tables 3.36 and 3.37 show the increase in corrosion of various nickel–iron alloys which may occur when coupled to nickel in calcium

Table 3.36 Bimetallic corrosion between nickel and nickel–iron alloys in 16% calcium chloride solution[2]

Alloy	*Corrosion rate** (mm/y)	
	Coupled	*Uncoupled*
100 Ni	—	0·02
Ni–5 Fe	0·045	0·045
Ni–10 Fe	0·7	0·055
Ni–20 Fe	1·0	0·055

*Room temperature test of duration 120 days, solution agitated. Cathode:anode area = 100:1.

Table 3.37 Bimetallic corrosion between nickel–iron alloys in sodium hydroxide[2]

Galvanic couple	*Corrosion rate** (mm/y)				
	23% NaOH		50% NaOH	75% NaOH	
	Coupled	*Uncoupled*	*Coupled*	*Coupled*	*Uncoupled*
Ni–5 Fe	0·04	0·035	0·06	0·025	0·02
Nickel	0·015	0·01	0·02	0·04	0·04
Ni–10 Fe	0·07	0·03	0·48	0·38	0·035
Nickel	0·01	0·01	0·015	0·02	0·04
Ni–20 Fe	0·095		0·04	0·05	0·015
Nickel	0·015		0·01	0·03	0·04
Ni–30 Fe	0·04	—	0·2	0·31	—
Nickel	0·02	0·01	0·01	0·02	0·04
Ni–40 Fe	0·04	—	0·21	0·34	—
Nickel	0·02	0·01	0·01	0·02	0·04

*Area ratio Ni:Ni–Fe = 10:1.

chloride or sodium hydroxide solutions. It is evident that in the calcium chloride solution, 5% iron contamination of weld metal can be tolerated whilst in the sodium hydroxide solution bimetallic corrosion would not become significant until contamination had exceeded 20%.

G. N. FLINT

REFERENCES

1. Beauchamp, R. L., Dissertion, Ohio State University, Figure 27 (1966)
2. La Que, F. L. and Copson, H. R., (Ed.), *Corrosion Resistance of Metals and Alloys*, Reinhold Publishing Corporation, New York, Van Nostrand Reinhold Ltd, London, 2nd edn., 458 (1963)
3. *Symposium on Atmospheric Corrosion of Non-Ferrous Metals*, Amer. Soc. Test. Mat., 58th Annual Meeting, June 29 (1955). Spec. Tech. Publn. No. 175, 141–158

4. Evans, T. E., *4th International Congress on Corrosion*, Amsterdam (1969)
5. Friend, J. N., *Deterioration of Structures of Timber, Metal and Concrete Exposed to the Action of Sea Water*, London, 18th Report of the Committee of the Institution of Civil Engineers (1940)
6. Marsh, J. S., *The Alloys of Iron and Nickel: Vol. 1 Special-purpose Alloys*, McGraw-Hill, New York and London, 495 et seq. (1938)
7. Hatfield, W. H., *Engineer*, **134**, 639 (1922)
8. Hatfield, W. H., *J. Iron and Steel Institute*, **108**, 103 (1923)
9. *Corrosion Engineering Bulletin No. 2*, International Nickel Co.
10. Fink, C. G. and De Croly, C. M., *Trans. Amer. Electrochem. Soc.*, **56**, 239 (1929)
11. Schwerdtfeger, W. J., *J. of Research*, **70C**, 187 (1966)
12. Rhodin, T. H., (Ed.), *Physical Metallurgy of Stress Corrosion Fracture*, Interscience, New York, 259–262 (1959)
13. Couper, A. S., *Materials Production*, **8** No. 10, 17 (1969)
14. Radd, F. J., Wolfe, L. H. and Crowder, L. H., The World Petroleum Conference (1967)

3.6 Cast Iron

Composition and Structure

Cast iron is the term applied to a wide range of ferrous alloys, whose principal distinguishing feature is a carbon content in excess of 1·7%. The relatively low melting point of these alloys compared with that of steels and their tendency to expand slightly on solidification, which make them admirably suited for the production of components by casting, result from this feature of their composition.

Four main types of iron are commonly encountered, viz. *white* iron in which all the carbon is in solid solution; *grey* iron in which most of the carbon is present as graphite flakes; *nodular* or *ductile* iron in which most of the carbon is present as graphite nodules, produced during solidification of the casting; and finally *malleable* iron in which most of the carbon is present as graphite nodules, produced subsequent to solidification by heat-treatment of the casting. These classes can be further sub-divided by consideration of their matrices. White iron has a matrix of pearlite containing amounts of free carbide which depend on the carbon content of the iron.

Table 3.38 Composition ranges of cast iron alloys

Type of iron	*Microstructure*		*Total carbon* (%)	C.C.† (%)	Si (%)	Mn (%)	S (%)	P (%)
	Graphite form	*Matrix*						
White iron	None	Pearlite + carbide	1·7–3·0	All	0·8–1·3	0·4	<0·15	<0·5
Grey iron	Flake	Pearlite	2·7–4·0	<0·9	0·5–3·3	0·3–1·0	<0·15	<1·4
Nodular graphite iron*	Nodules	Pearlite or ferrite	3·3–3·9	<0·9	1·6–2·5	0·4	<0·01	<0·1
Blackheart malleable	Nodules	Ferrite	2·0–2·7	None	0·8–1·2	0·1–0·6	<0·15	<0·2
Whiteheart malleable	Nodules	Pearlite	3·3–3·9	0–1·2	0·3–0·8	0·1–0·5	<0·4	<0·1

*Nickel 0–1·5%, magnesium 0·04–0·10%.
†Combined carbon.

Grey iron usually has a pearlitic matrix; this is the form usually meant by the term *cast* iron and is the most common alloy of the family. Except for 'metal-spun' pipes, which are produced by casting into a spinning metal mould and which solidify with a white iron structure, necessitating a subsequent annealing treatment, flake graphite irons with a predominantly ferritic structure are rare. Nodular graphite irons solidify with a pearlitic matrix, but in order to develop the full ductility of the iron, the castings are often subsequently annealed to give a ferritic matrix. Malleable irons are produced by two different processes which result in either ferritic or pearlitic matrices depending on the process adopted, but even the pearlitic iron is usually produced with a surface layer of ferrite.

The figures quoted in Table 3.38, while not authoritative in indicating upper and lower limits, give some idea of the range of analysis to be expected for each type of iron. Because of this variation in composition, cast irons are usually specified in terms of their mechanical properties rather than on an analytical basis.

Effect of Structure and Composition on Corrosion Resistance

Structure

An essential difference may be observed between the behaviours of steel and cast iron components immersed in an environment in which rust is precipitated at some distance from the corroding surface. The steel will waste away at a steady rate and its overall dimensions will steadily diminish, whereas cursory examination of the cast iron may suggest that it has not corroded at all, since its dimensions appear to be substantially unchanged. This difference arises from the fact that the cast iron contains in its microstructure several more or less corrosion-resistant components which are largely or completely absent from the microstructure of a steel. The most important of these corrosion-resistant micro-constituents are graphite, phosphide eutectic, and, to a lesser extent, carbide. When the cast iron corrodes in such a way that the corrosion product is deposited at some distance from the corroding surface, a skeleton is left behind comprising graphite flakes stiffened, in the case of phosphoric irons, by phosphide–eutectic cells and plugged by the carbonaceous debris resulting from the decomposition of the pearlite, silicic acid derived from the oxidation of the silicon dissolved in the iron, and whatever rust is precipitated relatively close to the metal (see Fig. 3.33). This skeleton, although much inferior in strength to the original iron, nevertheless retains sufficient strength to resist moderate erosion and may preserve the original contour of the component quite well.

The amount of graphitic residue retained on a corroding surface depends partly on the morphology of the graphite and partly on the corrosivity of the medium. In general, a coarse flake graphite tends to give a more permeable and less strong residue than a finer graphite, while nodules produce an even weaker residue. However, these differences are only clear at very high corrosion rates. For example, while flake graphite iron retains virtually all the graphitic residue when corroded in 0·5% sulphuric acid, the residue from a nodular graphite iron is largely detached; in 0·05% sulphuric acid, however, there is little difference in the amounts retained by the two irons[1]. The effect of the graphite in the iron on the corrosion process depends on

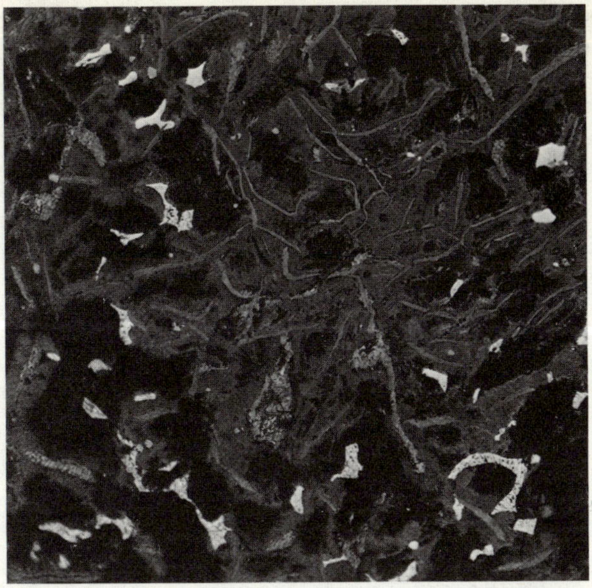

Fig. 3.33 Typical graphitic residue, containing graphite flakes and some unattacked phosphide eutectic

the residue thickness—thus in 0·5% acid, graphite stimulates attack on the nodular graphite iron because of its ability to act as an efficient cathode for hydrogen evolution, but in 0·05% acid, once the residue thickness reaches about 0·1 mm the attack tends to become stifled[1]. An interesting sidelight on this aspect is that flake graphite irons exposed in hot 70% sulphuric acid often do not evolve hydrogen, which surprisingly is apparently adsorbed by the graphite, presumably because of the high surface tension forces involved in bubble formation in this medium.

Composition

Small variations in the composition of cast irons, or even the addition of small amounts of alloying elements, generally have little effect on the corrosion resistance.

Graham *et al.*[2], however, working on the corrosive wear of automobile cylinders and piston rings exposed to high sulphur fuels, showed that irons exposed to 70% sulphuric acid at 130°C are attacked at rates dependent on the silicon content of the iron, the rate being relatively low at below 1% Si but rising to a peak at about 2% Si (Fig. 3.34). The effect seems to be associated with the development of a corrosion stifling film on the lower silicon irons. The laboratory work was broadly substantiated by piston-ring wear results and tests in boiler-combustion simulation rigs.

Whittaker and Brandes[3] followed up this work, carrying out tests on irons containing small amounts of copper and nickel. The addition of 0·6% Cu to irons containing 2% Si gave a significant improvement in corrosion resistance, but the addition to irons containing less than 1·5% Si decreased the corrosion resistance, which nevertheless remained greater than that for the higher silicon irons; nickel additions produced a smaller effect.

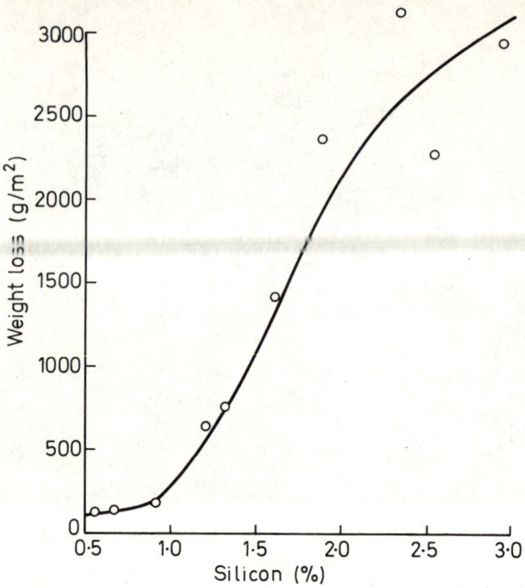

Fig. 3.34 Weight loss for cast iron exposed in 70% H_2SO_4 at 130°C for 16 h (after Graham *et al.*[2])

Copper additions appear to have the particular effect of reducing the corrosion stimulating effect of the sulphur content of an iron exposed to acid and the effect is thus less marked in low sulphur irons. Because sulphur can stimulate corrosion in acidic environments, it is usually kept as low as possible in irons to be used under these conditions. A low sulphur content is in any case metallurgically desirable.

Nickel has often been added to castings to be used with alkalis, either molten or in solution. There is, however, no conclusive evidence to support any belief in its efficacy.

Corrosion in Natural Environments

Atmospheric Corrosion

Because cast iron components are normally very heavy in section, the relatively low rates of attack associated with atmospheric corrosion do not constitute a problem and little work has been carried out on the phenomenon. A summary of some of the data available is given in Table 3.39. The most extensive work in this field was initiated by the A.S.T.M.[4] in 1958 and some of the results produced by these studies are quoted in Table 3.40. It will be noted that there is a marked fall in corrosion rate with time for all the metals tested.

Table 3.39 Corrosion rates of steels and irons in the atmosphere ($g\,m^{-2}\,d^{-1}$)

Environment	*Rural*		*Urban*				*Industrial*			*Marine*	
Source	A.S.T.M.[4]	Roll[5]	Friend[6]	Nekrytyi[7]	Roll[5]	A.S.T.M.[4]	Dearden[8]	Friend[6]	Roll[5]	LaQue[9]	A.S.T.M.[4]
Exposure period (years)	12	2	6	1	2	12	1	6	2	—	12
Material:											
Steel	0·23			1·2		0·27	3·4	2·4–3·2		3·6	1·38
Grey iron			1·4–2·1				3·2	1·1–1·2		0·6	2·0
White iron			0·1–0·3				1·3				
Malleable iron											
(a) <0·1% S	0·11			2·1		0·14			3·3		0·43
(b) >0·1% S	0·15	3·3			4·9–6·0	0·20			3·6		0·75
Nodular iron											
(a) Pearlitic	0·10					0·15				0·9	0·37
(b) Ferritic	0·11					0·17					0·72

Table 3.40 A.S.T.M. atmospheric corrosion data ($g\,m^{-2}\,d^{-1}$)

Location		*State College, Pa. (rural)*			*Kure Beach, N.C. (marine)*			*Newark, N.J. (industrial)*		
Duration (years)		1	3	12	1	3	12	1	3	12
Metal	*Condition*									
Ferritic ductile iron	As cast	0·90	0·36	0·11	1·51	0·85	0·72	1·29	0·51	0·17
	Machined	0·56	0·31	0·09	0·90	0·63	0·60	0·88	0·36	0·12
Pearlitic ductile iron	As cast	0·62	0·30	0·10	0·96	0·53	0·37	1·15	0·43	0·15
	Machined	0·50	0·22	0·07	0·82	0·47	0·27	0·70	0·30	0·10
Malleable iron >0·1% S	As cast	0·75	0·40	0·15	1·41	1·11	0·75	1·53	0·70	0·20
Mild steel	Rolled	0·97	0·52	0·23	3·02	2·01	1·38	1·75	0·81	0·27

Corrosion by Waters

The corrosivity of a natural water depends on the concentration and type of impurity dissolved in it and especially on its oxygen content. Waters of similar oxygen content have generally similar corrosivities, e.g. well-aerated quiescent sea-water corrodes cast iron at rates of 0·05–0·1 mm/y while most well-aerated quiescent 'fresh' waters corrode iron at 0·01–0·1 mm/y.

Those waters in which the carbon dioxide content is in excess of that required as bicarbonate ion to balance the bases present are among the most dangerous of the 'fresh' waters. Hard waters usually, though not invariably, deposit a carbonate scale and are generally not appreciably corrosive to cast iron, corrosion rates of less than 0·02 mm/y being frequently encountered. Water-softening processes do not increase the corrosivity of the water provided that the process does not result in the development of an excess of dissolved carbon dioxide.

Waters of pH less than 6 may be expected to be corrosive, but, because any weak acids present in the solution may not be fully ionised, it does not follow that water of pH greater than 7 will not be corrosive. Mine waters are particularly corrosive to cast iron, often to such an extent as to preclude its use with them, because of their relatively high acid content, derived from the hydrolysis of ferric salts of the strong acids, mainly sulphate, and because the ferric ion can act as a powerful cathodic depolariser.

While well-aerated near-neutral waters are normally much more corrosive than less-well-aerated waters, waters with near zero oxygen contents may cause high rates of corrosion if active sulphate-reducing bacteria, which can act as very efficient depolarising agents, are present. A corrosion rate of 1·5 mm/y has been observed on cast iron exposed to such a water.

Table 3.41 Increase in sulphur content of graphitic corrosion residue due to the presence of active sulphate-reducing bacteria

	Total carbon (%)	Si (%)	P (%)	S (%)
Metal	3·39	2·31	1·17	0·10
Graphitic corrosion residue (G.C.R.)	10·3	7·4	3·1	4·4
Ratio G.C.R./metal	3·0	3·2	2·7	44·0

The presence of active sulphate-reducing bacteria usually results in 'graphitic corrosion' and this has led to a useful method of diagnosing this cause of corrosion. The leaching out of iron from the graphitic residue which is responsible for the characteristic appearance of this type of corrosion leads to an enriched carbon, silicon and phosphorus content in the residue as compared with the original content of these elements in the cast iron. Sulphur is usually lost to some extent but when active sulphate-reducing bacteria are present, this loss is offset by the accumulation of ferrous sulphide in the residue with a consequent increase in the sulphur content of the residue out of all proportion to the content of the other elements present. The figures quoted in Table 3.41 illustrate the point.

Velocity of water As the velocity of flow of a water over cast iron increases, the supply of oxygen to the corroding surface is increased. Eventually, the supply of oxygen to the surface will reach a level sufficient to permit the formation of a strongly adherent, impermeable rust scale on the corroding surface. When this level is reached, the corrosion rate either ceases to change with increasing water velocity (Fig. 3.35[10]) or decreases, sometimes to negligible values.

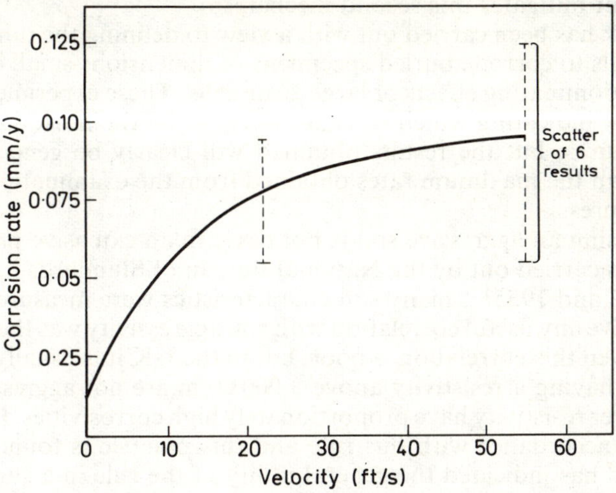

Fig. 3.35 Effect of velocity upon cast iron in sea-water; Rotor test at a water temperature of 25°C (after Higgins[10])

Very rapid and highly localised pitting is sometimes observed on components exposed to very turbulent flow conditions leading to cavitation in the stream. In general, these conditions appear to induce corrosion rather than erosion on cast iron surfaces, in contradistinction to what usually happens with other metals, apparently because the erosive component of the liquid flow scours away corrosion-stifling films and allows the development of very active electrochemical cells on the exposed metal surfaces[11].

Soil Corrosion

Since one of the major uses of cast iron is in the manufacture of pipes, the problem of the corrosion of buried cast iron structures is very important. It is however also very complex and only relatively general observations can be made on the subject.

The corrosion observed on a pipe buried in soil is the result of two separate effects, viz:

1. Interaction of the metal with the soil electrolyte—the 'intrinsic' corrosivity of the soil.
2. Development of a very-large-scale galvanic cell, due, for example, to variations in salt concentration or oxygen availability from point to point along the pipeline or to the presence of stray electrical currents.

In general, the former is the component measured during corrosion testing of isolated small lengths of pipe, but the latter is potentially the most dangerous to buried pipelines since the corrosion rates which can be supported by the mechanism suggested can be very much greater than those due to the 'intrinsic' corrosivity of the soil itself. However, the use of flexible joints in the pipeline introduces resistances large enough to break the electrical continuity in view of the relatively low potential differences in question, and this to some extent mitigates this second mechanism.

Much work has been carried out with a view to defining the inherent tendencies of soils to corrode buried specimens of dimensions small enough to preclude the dominating effects of large-scale cells. These experiments are of some value in indicating which soils are likely to be the most corrosive to buried structures, but the results obtained will clearly be generally more optimistic than the maximum rates obtained from the examination of large buried structures.

Characterising an aggressive soil is not easy. In an extensive programme of burial tests carried out by the National Bureau of Standards in the USA between 1922 and 1955[12], many soil characteristics were measured but the only one to give any useful correlation with soil aggressivity was the electrical resistivity. Even this correlation is poor, but in the UK it is usually assumed that all soils having a resistivity above 3 000 Ω cm are not aggressive while soils with lower resistivity have proportionately high corrosivities. Protection is applied in accordance with this rule and this practice is found to work well. Booth[13] has indicated the general utility of the rule in a survey of 60 soils in the UK, although he has suggested that a better criterion is:

	Aggressive soils	*Non-aggressive soils*
Resistivity and/or	<2 000 Ω cm	>2 000 Ω cm
Redox potential at pH 7	<0·40 V (NHE) or <0·43 V if in clay	>0·40 V (NHE) or >0·43 V if in clay
Borderline cases resolved by water content	>20% w/w	<20% w/w

It is common practice to report data from soil burial trials in terms of weight loss per unit area and depth of deepest pit. Fuller and Collins[14] have suggested that a better method is to express the data as a cumulative probability function indicating the chance of finding any given depth of penetration on the corroded surface. Comparisons between different materials can then be made on the basis of penetrations at given probability levels. The function is derived by plotting the cumulative frequency of penetrations derived from a large random sample of measurements. The method has been shown to give a good representation of the corroded profile. Some corrosion rates for different proportions of the corroded surface determined in this way on specimens of spun-grey and ductile-iron pipes buried in a clay soil of 500–900 Ω cm are given in Table 3.42, together with the rates derived from the weight loss and deepest pit depths for comparison.

Corrosion results depend so much on the precise soil characteristics that it is difficult to summarise the available data accurately and readers are referred to the classic work of Romanoff[12,17] and Hudson and Watkins[15]. A summary of some of the work on buried nodular graphite irons has been

Table 3.42 Corrosion rates for buried uncoated pipes (mm/y)

Metal	Exposure (years)	% Specimen area attacked 10	1	0·1	0·01	0·001	Deepest pit	Weight loss
Ductile iron	2	0·14	0·23	0·33	0·38	0·41	0·55	0·09
	5	0·12	0·20	0·25	0·31	0·36	0·40	0·11
Grey iron	2	0·14	0·28	0·41	0·47	0·51	0·75	0·05
	5	0·13	0·30	0·44	0·49	0·55	0·77	0·10

given by Collins, Fuller and Harrison[16]. Some results reported by Collins, *et al.*[16] for spun-grey and ductile-iron pipes buried in various aggressive soils are quoted in Table 3.43.

It is a common observation that the extent of corrosion damage, however measured, can be expressed as a function of the exposure period by the expression

$$C = at^b$$

where C is the corrosion parameter, t the test duration and a and b are constants. However, when attempts are made to determine values for the constant a and b terms from burial trial results, the data are found to be liable to considerable scatter, and no broad generalisations can be made.

Table 3.43 Deepest pit depths (mm) for buried uncoated pipes exposed for 2 and 5 years (each value is the deepest pit found on a set of 6 similar specimens)

Soil type	Resistivity (Ω cm)	Ductile iron 2 years	Ductile iron 5 years	Grey iron 2 years	Grey iron 5 years
Heavy clay	500–900	1·1	2·0	1·5	3·9
Anaerobic peaty clay	900	1·6	2·0	3·9	4·8
Sandy clay liable to tidal immersion	110	0·7	3·3	3·5	7·5*

*Sample perforated: low result.

Corrosion in Industrial Environments

Corrosion by Acids

In general, unalloyed grey or white cast irons possess no useful resistance to dilute mineral acids. In very dilute acids the presence of air, or other oxidising agents such as ferric salts, appreciably increases the corrosion rate. If corrosion rates are to be held below 0·25 mm/y in moderately aerated solutions, it is unwise to exceed a total acid concentration of 0·001N, irrespective of the acid concerned.

Mineral acids Unalloyed cast iron possesses no useful resistance to hydrochloric acid at any concentration or temperature. Dilute sulphuric, nitric and phosphoric acids are also very aggressive, corrosion rates amounting to several centimetres per year in some cases. Owing to the insolubility of surface films of ferrous sulphate in strong sulphuric acid, however, concentrations of this acid in excess of 65% can generally be withstood by cast iron at room temperature, corrosion rates being less than 0·12 mm/y. Concentrations of sulphuric acid greater than 97% at temperatures up to 300°C do not corrode cast iron at more than 0·12 mm/y. Free SO_3 (oleum) does not increase the corrosion rate. Any influence operating against the formation or retention of the insoluble sulphate film, for example high liquid velocities or abrasion of the metal surfaces, may lead to excessive corrosion rates, even in acid concentrations greater than 65%. Collins[18] has described some grey iron castings which had suffered deterioration when handling oleum or sulphur trioxide due to penetration of the sulphur trioxide into the iron, apparently along

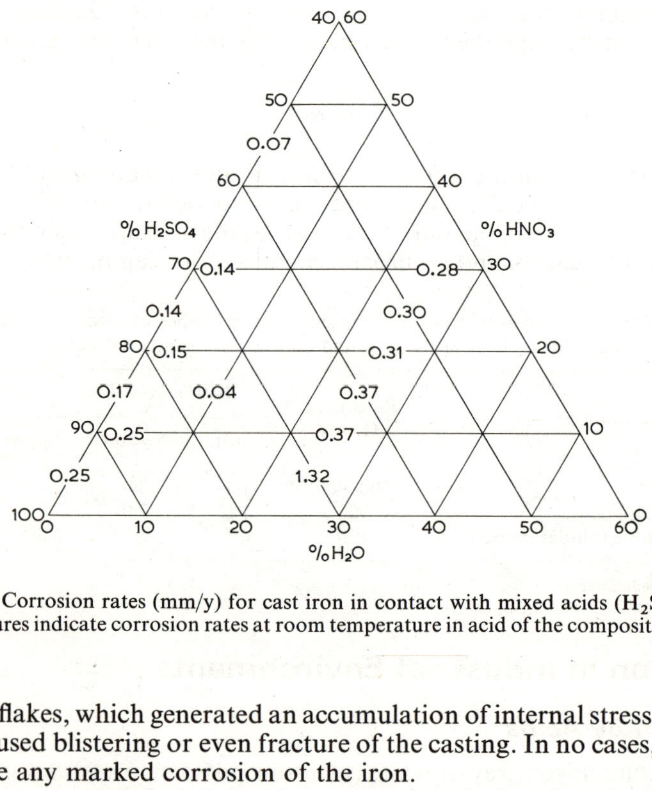

Fig. 3.36 Corrosion rates (mm/y) for cast iron in contact with mixed acids (H_2SO_4, HNO_3, H_2O). Figures indicate corrosion rates at room temperature in acid of the composition indicated

graphite flakes, which generated an accumulation of internal stress and eventually caused blistering or even fracture of the casting. In no cases, however, was there any marked corrosion of the iron.

While concentrated nitric acid passivates steel, the phenomenon is too unreliable to permit cast irons to be used with confidence, even for strong nitric acids. The evidence available[19,20] in relation to mixed nitrating acids, however, suggests that the corrosion rates may be lower here than in solutions of pure nitric or sulphuric acids of comparable dilution. Typical results are shown in Fig. 3.36.

Organic acids Dilute solutions of organic acids, especially if well aerated, attack cast iron at uneconomical rates. Temperature and velocity are also accelerating factors.

Corrosion by Alkalis

Dilute alkali solutions do not corrode cast iron at any temperature, but hot solutions exceeding about 30% concentration will attack it, with an accompanying evolution of hydrogen, to form a ferrite. Broadly speaking, if corrosion rates are to be held below 0·2 mm/y the temperature should not exceed 80°C. Corrosion rates may range as high as 1·25–2·5 mm/y in boiling solutions of more than 50% concentration. Molten caustic soda (650°C) may attack cast iron initially at rates around 20 mm/y, but it is probable that this figure decreases somewhat after a few days. In spite of these high corrosion rates, caustic-concentration and fusion pots are made from cast iron, since the material is relatively cheap, and the thick wall necessary for mechanical strength also gives the pot long life. It is of the utmost importance to ensure that concentration and fusion pots are cast in sound metal, as any unsoundness, particularly at the bottom of the pan, will lead to pitting attack and premature failure. The addition of 1 to 3% nickel to metal for these castings is said to be beneficial.

Corrosion by Salt Solutions

The corrosivity of a salt solution depends upon the nature of the ions present in the solution. Those salts which give an alkaline reaction will retard the corrosion of the iron as compared with the action of pure water, and those which give a neutral reaction will not normally accelerate the corrosion rate appreciably except in so far as the increased conductivity of the solution in comparison with water permits galvanic effects to assume greater importance. Chlorides are dangerous because of the ability of the anions to penetrate otherwise impervious barriers of corrosion products.

Those salts which hydrolyse to give an acid solution, e.g. the strong acid salts of aluminium, iron and, to a lesser extent, calcium, give solutions which may be very corrosive to cast iron, particularly if they are well aerated. When oxidising salts are also present in these acid solutions, a particularly dangerous system may be created. It is owing to this combination of oxidising and acidic character that mine waters are so corrosive.

If oxidising salts are present in neutral solutions they may reduce corrosion of the iron by the establishment of thin protective films on the metal surface. Their effectiveness is considerably diminished if much chloride ion is also present in solution.

Ammonium salts are more corrosive than those of the alkaline metals, the difference increasing directly with concentration and temperature. If free ammonia is also present, however, the pH value may be raised to a point at which little corrosion can occur. If ammonium salt solutions are exposed to locally high temperatures, it is possible that ammonia may be lost from the salts with a corresponding increase in the free acid content of the solution, and the development of a more corrosive character.

Corrosion under Stress

As far as is known, cast iron is not subject to stress corrosion cracking, although it has been suggested that this can happen with ductile irons exposed to strong caustic alkali solutions[21]. Irons exposed under conditions of cyclic stress are, however, liable to corrosion fatigue. Some data on this effect have been given by Collins and Smith[22], but a more extensive study has been reported by Palmer[23]. By its nature, it is difficult to give absolute values for resistance to corrosion fatigue since this will be affected by the duration and frequency of cycling and the corrosivity of the environment. Palmer exposed specimens in a Wöhler fatigue machine to sprays of demineralised water, 3% sodium chloride solution and demineralised water containing various inhibitors, using a cycle frequency of about 3 000 cycles/min and a duration of 50 000 000 or 100 000 000 cycles. The results for the uninhibited solutions are summarised in Table 3.44. Palmer found that corrosion fatigue due to water spray could be eliminated or mitigated by some of the inhibitor systems examined, but it was apparently easier to inhibit

Table 3.44 Corrosion fatigue limiting strengths (MN/m^2)

Environment	*Air**	*Water**	3% NaCl *Solution†*
Pearlitic grey iron	126	100	39
Ferritic grey iron	93	77	23
Pearlitic ductile iron	270	224	46
Ferritic ductile iron	208	178	46

*Fatigue strength based on 50×10^6 cycles.
†Fatigue strength based on 100×10^6 cycles.

damage on grey irons than that on ductile irons which could only be inhibited by 0·25% potassium chromate solution. He suggested this was due to an inability of the other systems to maintain a continuous passive film on the iron; grey iron is less sensitive to a notch effect in fatigue so that the presence of local sites of attack would be less important to this metal provided that the overall corrosivity of the solution were depressed.

H. H. COLLINS

REFERENCES

1. Collins, H. H., *B.C.I.R.A. Journal*, **10**, 543, Sept. (1962)
2. Graham, R., Prado, O. S., Collins, M. H., Brandes, E. A. and Farmery, H. K., *Proc. Instn. Mech. Engs.*, **174**, 617 (1960)
3. Whittaker, J. A. and Brandes, E. A., *Foundry*, 70, April (1962)
4. Mannweiler, G. B., *Proc. Amer. Soc. Testing Mats.*, **72**, Appendix 1, 42 (1972)
5. Roll, F., *Werkstoffe u. Korrosion*, **12**, 209, April (1962)
6. Friend, J. N., *Carnegie Scholarship Memoirs*, **18**, Iron and Steel Institute, 61 (1929)
7. Nekrytyi, S. S. and Karpov, V. T., *Russian Castings Production*, 369, August (1961)
8. Dearden, J. and Swindale, J. D., *J. Iron St. Inst.*, **185**, 227 (1957)
9. LaQue, F. L., *Corrosion*, **14**, 485t (1958)
10. Higgins, R. I., unpublished data
11. Collins, H. H., *Pitting of Diesel Cylinder Liners*, Diesel Engineers' and Users' Association, July (1961)
12. Romanoff, M., *Underground Corrosion*, US Nat. Bur. Stnd. Washington, D.C., 227 (1957)
13. Booth, G. H., *Brit. Corr. J.*, **2**, 109, May (1967)

14. Fuller, A. G. and Collins, H. H., *The Evaluation of Corrosion on a Non-Uniformly Corroded Surface*, Proceedings of 1968 Conference on Corrosion and Protection of Pipelines, London, 119 (1968)
15. Hudson, J. H. and Watkins, K. O., *Tests on the Corrosion of Buried Cast Iron and Mild Steel Pipes*, B.I.S.R.A. Open Report, June (1968)
16. Collins, H. H., Fuller, A. G. and Harrison, J. T., 12th World Gas Conf., Nice, Report 19U/D, 14–73 (1973)
17. Romanoff, M., *J. Amer. Water Works Assoc.*, **60**, 645 (1968)
18. Collins, H. H., *Metallurgia*, **71**, 177, April (1965)
19. Bate, S. C., *Chem. Age.*, London, **15**, 419 (1926)
20. Chapman, F. F., *Trans. Amer. Inst. Chem. Engs.*, **18**, 7 (1926)
21. Smith, S. H., Discussion to Begg, G. A. J., *B.C.I.R.A. Journal of Research and Deterioration*, **6**, 623, April (1957)
22. Collins, W. L. and Smith, J. O., *Proc. A.S.T.M.*, **42**, 639 (1942)
23. Palmer, K. B., *Proceedings of Conference on Engineering Properties and Performance of Modern Iron Castings*, B.C.I.R.A., 110–115 (1972)

3.7 High-Nickel Cast Irons

Composition and Properties

The addition of about 20% nickel to cast iron produces materials with a stable austenitic structure; these materials are sometimes known as austenitic cast irons but are more often referred to commercially as *Ni-Resist* cast irons. The austenitic matrix of these irons gives rise to very different mechanical and physical properties to those obtained with the nickel-free grey cast irons. The austenitic matrix is more noble than the matrix of unalloyed grey irons and it was shown in the early work of Vanick and Merica[1] that the corrosion resistance of cast iron increases with increasing nickel content up to about 20% (Fig. 3.37).

Although the Ni-Resist irons, due to their austenitic matrix, are tougher and more shock resistant than the nickel-free grey irons, those in which the carbon is present in the flake graphite form (F.G. irons) still exhibit certain disadvantages due to the graphite structure. Much better strength and impact properties can be obtained by treating the iron with a small quantity of magnesium sufficient to give a residual content of 0·05–0·1%, which converts the graphite to a spheroidal form (S.G. irons). The Ni-Resist irons are available in both flake and spheroidal graphite forms and typical structures, consisting of flake or spheroidal particles dispersed throughout the austenitic matrix, are shown in Figs. 3.38 and 3.39. The matrix also contains small amounts of carbides, the amounts of which increase with increasing chromium content.

The first alloys in the Ni-Resist series, containing about 20% nickel, were introduced in the 1930s and soon became established in both corrosion and heat resistance applications. The range of alloys has been extended over the years and a total of twenty grades of austenitic irons have been developed with nickel contents varying from 13 to 35%. Each material has somewhat different characteristics so that the most appropriate grade must be selected to obtain the most advantageous properties for any particular application. The compositions and mechanical properties of the principal grades of austenitic cast irons are summarised in Table 3.45. There are six basic grades of flake graphite austenitic iron and five basic grades of spheroidal graphite austenitic iron. There is no spheroidal austenitic iron corresponding to Type 1 Ni-Resist since it is difficult to obtain a good spheroidal graphite structure in an austenitic iron containing more than 2% copper. A considerable number of modified grades also exist which differ in composition and properties from these basic grades. Specifications for eight commonly used grades of austenitic cast irons are given in BS 3468:1962.

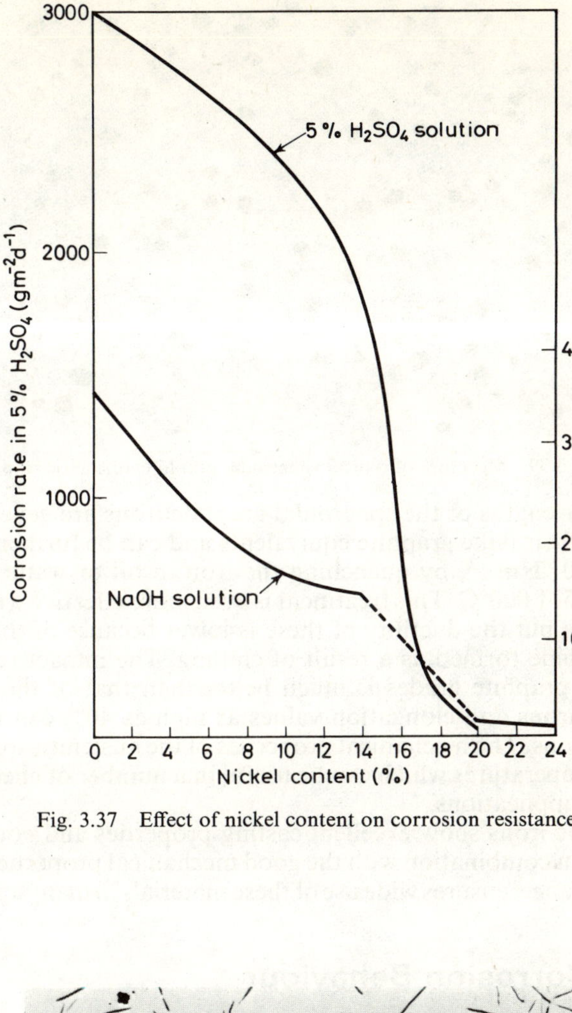

Fig. 3.37 Effect of nickel content on corrosion resistance

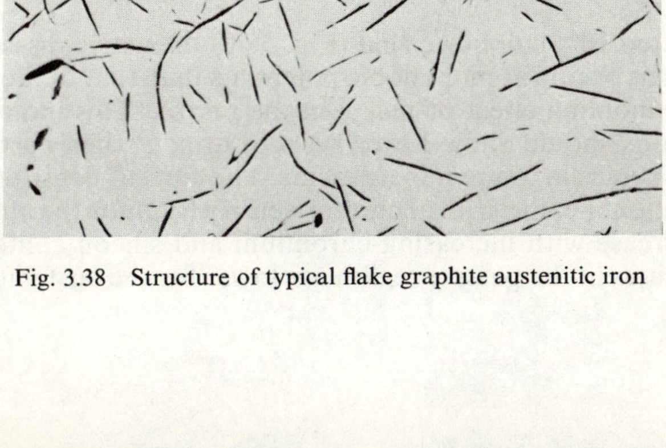

Fig. 3.38 Structure of typical flake graphite austenitic iron

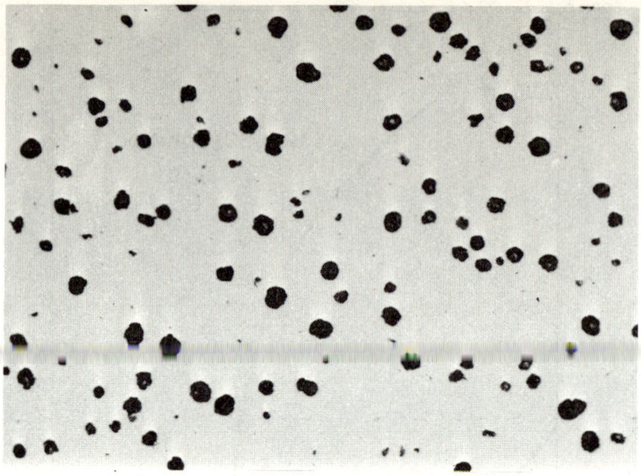

Fig. 3.39 Structure of typical spheroidal graphite austenitic iron

The tensile strengths of the spheroidal graphite irons are generally about twice those of their flake graphite equivalents and can be further improved, by about 8×10^7 Nm^{-2}, by quenching the iron in oil or water from temperatures of 925–1 000°C. This treatment is even more effective when applied to chill castings but the ductility of these is lower because of the increased amount of carbide formed as a result of chilling. The impact resistance of the spheroidal graphite grades is much better than that of the equivalent flake graphite irons and elongation values as high as 40% can be attained with the S.G. irons. The mechanical properties of the austenitic irons are also good at low temperatures which can be useful in a number of chemical plant and cryogenic applications.

The austenitic irons show excellent casting properties and good machinability, which, in combination with the good mechanical properties and good corrosion resistance, ensures wide use of these materials in many applications.

Aqueous Corrosion Behaviour

The austenitic cast irons show better corrosion resistance than the ferritic irons primarily due to the nickel content of the austenitic matrix.

Potential–current density (E–i) curves, which have been determined[2] for a number of the austenitic cast irons and also for the nickel-free ferritic irons, indicate that in general the austenitic cast irons show more favourable corrosion characteristics than the ferritic irons in both the active and passive states.

In de-aerated 10% sulphuric acid (Fig. 3.40) the active dissolution of the austenitic irons occurs at more noble potentials than that of the ferritic irons due to the ennobling effect of nickel in the matrix. This indicates that the austenitic irons should show lower rates of attack when corroding in the active state such as in dilute mineral acids. The current density maximum in the active region, i.e. the critical current density (i_{crit}) for the austenitic irons tends to decrease with increasing chromium and silicon content. Also the current densities in the passive region are lower for the austenitic irons than

Table 3.45 Compositions and properties of principal grades of flake and spheroidal graphite austenitic cast irons

BS 3468 designation	*Ni-Resist type*	*Composition* (wt. %)						*Minimum tensile strength* ($\times 10^7$ N/m^2)	*Brinell hardness*
		C	Si	Mn	Ni	Cr	Cu		
AUS101A	Type 1	3·0	1–2·8	1–1·5	13·5–17·5	1·75–2·5	5·5–7·5	17	130–170
AUS102A	Type 2	3·0	1–2·8	0·8–1·5	18–22	1·75–2·5	0·5 max.	17	125–170
AUS105	Type 3	2·6	1–2	0·4–0·8	28–32	2·5 –3.5	0·5 max.	17	120–160
—	Type 4	2·6	5–6	0·4–0·8	29–32	4·5 –5·5	0·5 max.	17	150–210
—	Type 5	2·4	1–2	0·4–0·8	34–36	0·1 max.	0·5 max.	14	100–125
AUS104	—	1·6–2·2	4·5–5·5	1–1·5	18–22	1·8 –4·5	0·5 max.	19	248 max.
AUS202A	Type D–2	3·0	1·75–3	0·7–1	18–22	1·75–2·5	—	37	140–200
AUS205	Type D–3	2·6	1·5 –2·8	0·5 max.	28–32	2·5 –3·5	—	37	140–200
—	Type D–4	2·6	5–6	0·5 max.	29–32	4·5 –5·5	—	42	170–240
—	Type D–5	2·4	1·5 –2·8	0·5 max.	34–36	0·1 max.	—	37	130–180
AUS204	—	3·0	4·5 –5·5	1–1·5	18–22	1–2·5	—	37	230 max.

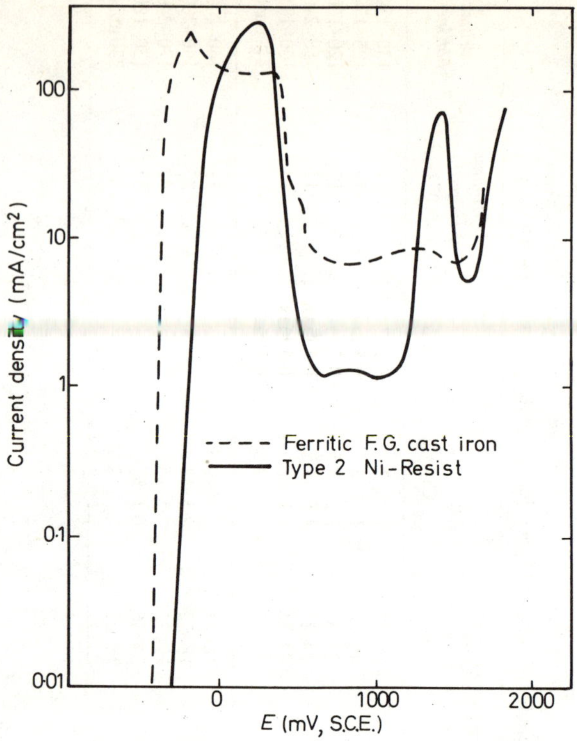

Fig. 3.40 Potential–current density curves in 10% sulphuric acid solution at 25°C

for the ferritic. These observations indicate that the austenitic irons, particularly those of higher chromium and silicon content, should show superior passivating properties to the ferritic irons.

Similar curves determined in 50% sodium hydroxide solution at 60°C show (Fig. 3.41) that the austenitic irons exhibit more noble active dissolution and also lower current densities in the active and passive regions than the ferritic irons; the current densities in both regions decrease markedly with increasing nickel content (Fig. 3.42).

In 3% sodium chloride solution at 60°C the austenitic irons again show superior characteristics to the ferritic. The breakdown potentials determined in this environment, which provide a relative measure of the resistance to attack in neutral chloride solutions, are generally more noble for the austenitic irons than for the ferritic (Table 3.46). This indicates that the austenitic irons should show better corrosion resistance in such environments.

The more favourable electrochemical characteristics exhibited by the austenitic irons in this range of environments are reflected in the corrosion behaviour of the alloys discussed below.

One of the outstanding properties of the austenitic irons is their resistance to graphitic corrosion or 'graphitisation'. In some environments ferritic cast irons corrode in such a manner that the surface becomes covered with a layer of graphite. This compact graphite layer, being more noble than the matrix, markedly increases the rate of attack. The austenitic irons rarely

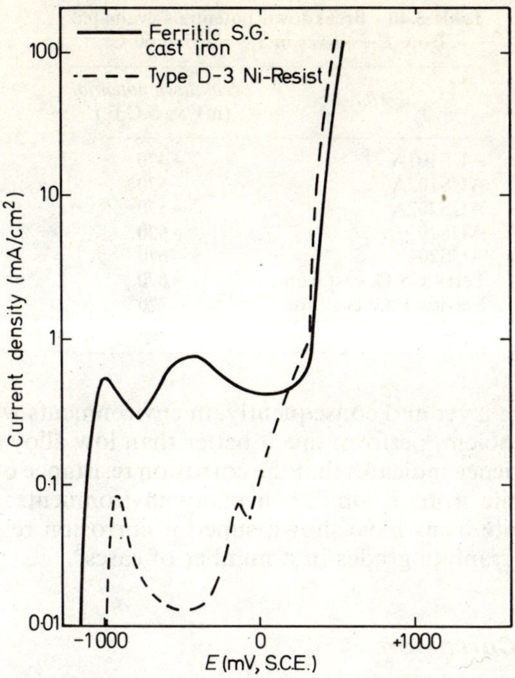

Fig. 3.41 Potential–current density curves in 50% sodium hydroxide solution at 60°C

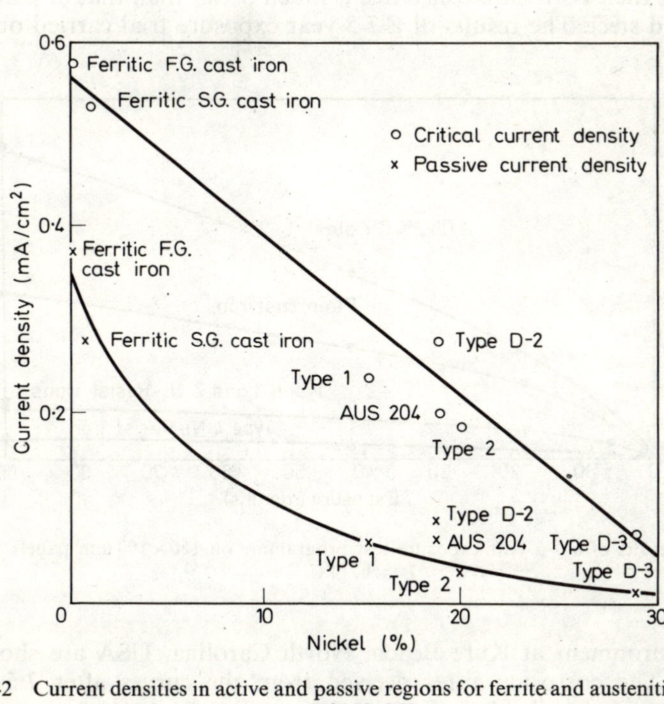

Fig. 3.42 Current densities in active and passive regions for ferrite and austenitic cast irons in 50% NaOH

Table 3.46 Breakdown potentials evaluated from *E–i* curves in 3% NaCl at 60°C

Alloy	*Breakdown potential* (mV vs S.C.E.)
AUS101A	−470
AUS102A	−520
AUS202A	−570
AUS205	−620
AUS204	−620
Ferritic S.G. cast iron	−670
Ferritic F.G. cast iron	−720
Mild steel	620

form this graphite layer and consequently, in environments where graphitic corrosion is a problem, perform much better than low alloy cast irons.

Practical experience indicates that the corrosion resistance of the flake and spheroidal graphite irons is similar in many environments; however, the spheroidal graphite irons have shown superior corrosion resistance to the equivalent flake graphite grades in a number of cases[3].

Atmospheric Corrosion

Although the Ni-Resist irons will not remain rust-free when exposed to the atmosphere their corrosion resistance is much better than that of plain cast iron or mild steel. The results of a 7·5 year exposure trial carried out in a

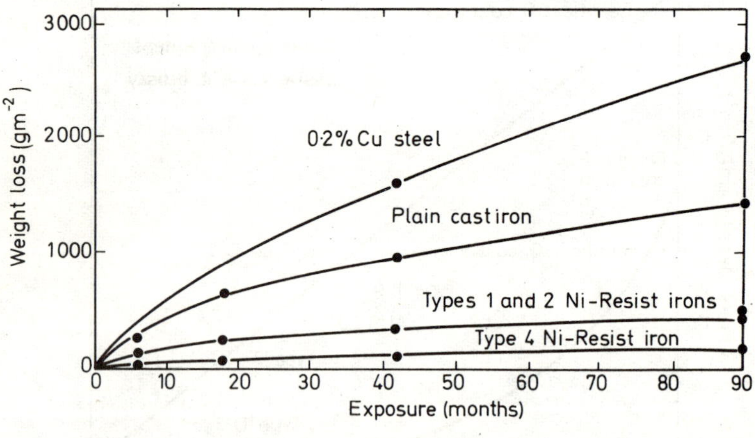

Fig. 3.43 Results of a 7.5 year exposure test programme on 150 × 100 mm panels at Kure Beach, N.C.

marine environment at Kure Beach, North Carolina, USA are shown in Fig. 3.43. The corrosion rates derived from the curves after 7·5 years exposure are given in Table 3.47.

Table 3.47 Corrosion rates after exposure for 7·5 years at Kure beach

Alloy	*Corrosion rate* ($mm\ y^{-1}$)
0·2% Cu steel	0·020
Cast iron	0·010
Types 1 and 2 Ni-Resist	<0·003
Type 4 Ni-Resist	<0·003

Natural Waters

Water which is used for cooling purposes in refineries and chemical plant can cause severe problems of corrosion and erosion. Ordinary cast irons usually fail in this type of environment due to graphitic corrosion or corrosion/erosion. Ni-Resist irons however show better corrosion resistance, due to the nobility of the austenitic matrix, and are preferred for use in the more aggressive environments such as those containing appreciable amounts of carbon dioxide or polluted with chemical wastes or sea-water.

The austenitic irons have also been shown to exhibit better corrosion resistance than the ferritic irons in sea-water. Tests over long periods of time have shown that Ni-Resist irons of Types 1, 2 and 3 corrode at rates of 0·020 to 0·058 mmy^{-1} in relatively quiet sea-water. Under similar conditions low alloy cast irons have shown corrosion rates ranging from 0·066 to 0·53 mmy^{-1}[(4)]. The Ni-Resist irons maintain this superiority over a wide variety of conditions (Figs. 3.44 and 3.45) both in stationary and flowing sea-water. In a test lasting 740 days in sea-water moving at 1·5 m/s low

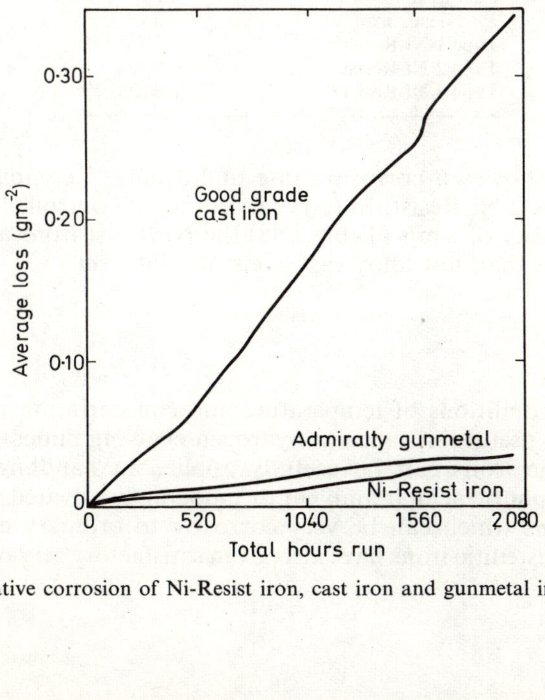

Fig. 3.44 Relative corrosion of Ni-Resist iron, cast iron and gunmetal in aerated sea-water

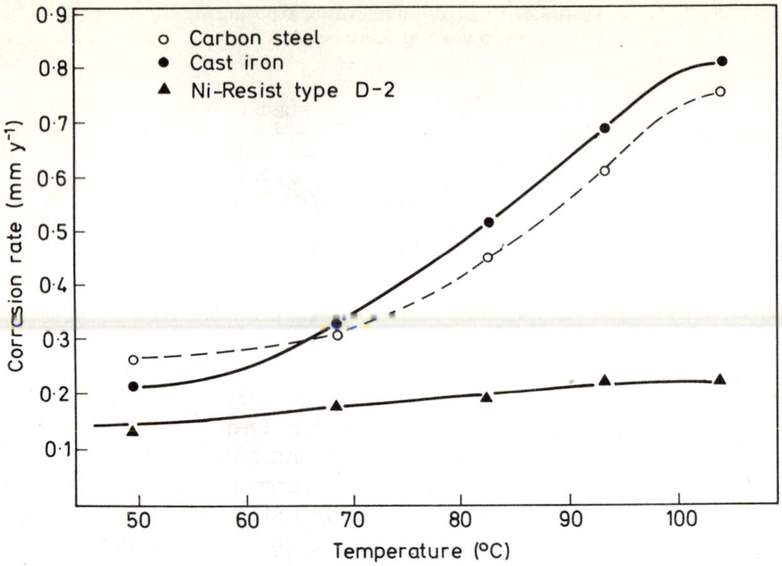

Fig. 3.45 Corrosion rate versus temperature in de-aerated sea-water; continuous test exposure for 156 d

Table 3.48 Sea-water corrosion/erosion test carried out at 8 m/s at 28°C for 60 d

Alloy	*Average corrosion rate* (mm y^{-1})
Cast iron	6·9
2% Ni cast iron	6·1
Type 1 Ni-Resist	0·74
Type 2 Ni-Resist	0·79
Type 3 Ni-Resist	0·53

alloy cast iron showed a corrosion rate of 1·3 mmy^{-1} compared to 0·050 mmy^{-1} for Type 2 Ni-Resist. In tests carried out at controlled temperature at a higher velocity of 8 m/s (Table 3.48) the Ni-Resist irons again showed better properties than low alloy cast irons or mild steel.

Acids

Under certain conditions of temperature and concentration the austenitic cast irons show useful resistance to hydrogen-evolving mineral acids.

The austenitic irons can be usefully applied in handling very dilute solutions of sulphuric acid at ambient or moderately elevated temperatures under conditions which can be very corrosive to ordinary cast iron and carbon steel. Austenitic irons have also given satisfactory service in handling concentrated sulphuric acids, but although they show low corrosion rates in such environments they are not markedly superior to the unalloyed cast irons. Type 1 Ni-Resist and the high silicon grades AUS104 and AUS204 are the types most generally used in sulphuric acid environments.

The austenitic irons are superior to ordinary cast iron in their resistance to corrosion by a wide range of concentrations of hydrochloric acid at room temperature (Table 3.49). However, for practical uses where such factors as velocity, aeration and elevated temperatures have to be considered, the austenitic irons are mostly used in environments where the hydrochloric acid concentration is less than 0·5%. Such environments occur in process streams encountered in the production and handling of chlorinated hydrocarbons, organic chlorides and chlorinated rubbers.

Table 3.49 Corrosion of Type 1 Ni-Resist, cast iron and carbon steel in unaerated hydrochloric acid solutions at room temperature

Acid concentration (%)	*Corrosion rate* (mmy^{-1})		
	Ni-Resist	*Cast iron*	*Carbon steel*
1·8	0·13	23	15
3·6	0·38	30	36
5·0	0·46	38	46
10·0	0·41	30	48
20·0	1·1	32	69
27·0	3·0	30	60
36·0	9·4	28	30

The austenitic irons are also useful in some circumstances for handling organic acids such as dilute acetic, formic and oxalic acids, fatty acids and tar acids. They are more resistant to organic acids than unalloyed cast irons, e.g. in acetic acid the austenitic irons show corrosion rates 20–40 times lower than the ferritic irons (Table 3.50).

Table 3.50 Corrosion of Type 1 Ni-Resist and ferritic cast iron in acetic acid in laboratory tests at 15°C

Acid concentration (%)	*Corrosion rate* (mmy^{-1})	
	Cast iron	*Ni-Resist*
5	17	1·0
10	22	0·5
25	20	0·5
50	16	2·0

The austenitic irons show poor resistance to solutions of nitric acid even when dilute and at low temperatures.

Alkalis

Austenitic cast irons show particularly good corrosion resistance in alkaline environments, even better than that shown by low alloy cast irons. The resistance to corrosion improves with increasing nickel content (Fig. 3.46),

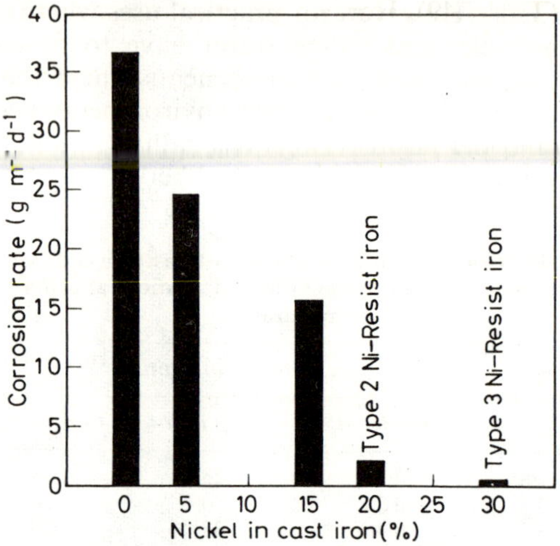

Fig. 3.46 Effect of nickel additions to cast iron in reducing corrosion by caustic alkalis

Table 3.51 Corrosion of Ni-Resist irons, cast iron and carbon steel in caustic soda solutions

Alloy	*Corrosion rate* (mmy^{-1})	
	14% NaOH (88°C)	74% NaOH (127°C)
Cast iron	0·2	1·93
Carbon steel	0·2	0·38
Ni-Resist Type 1	0·07	—
Ni-Resist Type 2	—	0·15
Ni-Resist Type 3	—	0·06
Ni-Resist Type D–2	—	0·13

and the irons containing about 30% nickel, such as Type 3 Ni-Resist, show the best resistance. This high corrosion resistance enables Ni-Resist cast irons to be used in caustic soda processes where both high temperature and concentrations are encountered (Table 3.51). One of the most important uses for the austenitic irons has for many years been in caustic soda production plant where the materials find use in many components. Molten caustic soda is very much more aggressive than aqueous solutions and a corrosion rate of 6·6 mmy^{-1} has been recorded[5] for austenitic iron in molten sodium hydroxide at 670°C.

The austenitic irons also show good corrosion resistance in caustic alkalis containing sulphides and mercaptans and have therefore proved useful materials for the construction of pumps, valves and piping in caustic soda regenerators in oil refineries.

Salts

The corrosion rates of the Ni-Resist irons in salt solutions depend upon the chemistry of the salt. Solutions which are alkaline or neutral in reaction are not generally corrosive to high nickel irons and even brines containing calcium and magnesium chloride can be safely handled by the austenitic irons (Table 3.52). Those salts which hydrolyse to give an acidic solution are more corrosive to high-nickel irons although the corrosion rate is still less than for unalloyed cast iron in the same medium. The corrosion rates of Type 1 Ni-Resist and cast iron given in Table 3.53 demonstrate that the austenitic iron shows better resistance than the nickel-free cast iron in a wide range of salt solutions.

Table 3.52 Corrosion of Type 1 Ni-Resist and cast iron in brine solutions

Environment	Temperature (°C)	Corrosion rate (mmy^{-1})	
		Ni-Resist	*Cast iron*
14% $NaCl + 16{\cdot}7\% \, CaCl_2 + 3{\cdot}4\% \, MgCl_2$. pH ≈ 6	69	0·08	0·53
Saturated NaCl	93	0·12	1·85

Table 3.53 Corrosion of Type 1 Ni-Resist and cast iron in various inorganic salt solutions

Salt solution	Concentration (%)	Temperature (°C)	Corrosion rate (mmy^{-1})	
			Ni-Resist	*Cast iron*
Aluminium sulphate	5	16	0·41	1·0
Ammonium chloride	5	16	0·08	1·3
Ammonium chloride	5	93	0·15	4·8
Ammonium sulphate	5	16	0·15	0·76
Ammonium citrate	30	Room	1·5	550
Ammonium thiocyanate	50	27	0·38	3·3
Potassium aluminium sulphate	5	16	0·25	0·76
Zinc chloride	30	Boiling	2·0	16
Ammonium nitrate	5	Room	0·23	0·76
Manganese chloride	10	77	0·038	0·79
Ammonium chloride	20	93	0·25	5·8

Applications

The individual characteristics and uses of the basic grades of the austenitic irons are given in Table 3.54. The major uses for these materials occur in the handling of fluids in the chemical and petroleum industries and also in the power industry and in many marine applications. The austenitic irons are also used in the food, soap and plastics industries where low corrosion rates are essential in order to avoid contamination of the product. Ni-Resist grades Type 2, 3 or 4 are generally used for such applications but the highly alloyed Type 4 Ni-Resist is preferred where low product contamination is of prime importance.

Table 3.54 Characteristics and uses of basic grades of austenitic cast irons

BS 3468 Designation	*Ni-Resist type*	*Characteristics*	*Uses*
AUS101A	Type 1	Least expensive austenitic iron; good corrosion resistance particularly in acidic media	Pumps, valves, furnace components
AUS102A AUS202A	Type 2 Type D-2	Good corrosion resistance; better than Type 1 in alkaline environments	As for Type 1 but preferable for alkaline solutions; used in soap and plastic industries
AUS105 AUS205	Type 3 Type D-3	Good thermal shock resistance; high resistance to erosion particularly in alkaline media	Pumps, valves, pressure vessels, filter parts, exhaust gas manifolds
— —	Type 4 Type D-4	Best corrosion resistance and erosion resistance of the austenitic irons	Castings for industrial furnaces; used in food industry for low contamination of product
— —	Type 5 Type D-5	Very low thermal expansion; good dimensional stability	Scientific instruments, glass moulds
AUS104 AUS204	— —	Good resistance to high temperature oxidation; good corrosion resistance in sulphuric acid	Pumps and valves

The modified grades of Ni-Resist are often sufficiently different to the basic grades to be used in additional applications. Ni-Resist Types 1B, 2B and D-2B have a higher chromium content than the corresponding basic grades which increases the erosion resistance of the materials; these grades are less expensive than the other grades with good erosion resistance, i.e. Ni-Resist Types 3 and 4. A chromium-free grade of austenitic iron, Ni-Resist Type D-2C, has particularly good ductility and shows good mechanical properties down to $-100°C$. Even better low temperature properties are, however, obtained with Ni-Resist Type D-2M, a chromium-free 4% manganese grade, which was specially developed for cryogenic applications such as the separation of aromatic hydrocarbons and the production of ethylene. Ni-Resist Type D-4A is a recently developed grade of austenitic iron which has particularly good resistance to high temperature oxidation and better ductility than the standard Type D-4 grade[6].

A. C. HART

REFERENCES

1. Vanick, J. S. and Merica, P. D., *Trans. Amer. Soc. Steel Treat.*, **18**, 923 (1930)
2. Evans, T. E. and Hart, A. C., to be published
3. Swales, G. L., *The Industrial Chemist*, Jan.–March (1963)
4. Uhlig, H. H., (Editor), *The Corrosion Handbook*, John Wiley, New York, 385–386 (1948)
5. La Que, F. L., *Cast Iron in the Chemical and Process Industries*, Grey Iron Founders' Society Inc., Cleveland, Ohio (1945)
6. Cox, G. J., *British Foundryman*, 1–14, Jan. (1970)

3.8 High-Chromium Cast Irons

Composition

There is no clear demarcation between high-chromium steels and high-chromium cast irons other than the fact that components are fabricated from the steels, and cast in the irons. In practice, however, the irons are usually found to have carbon contents of between 0·6 and 3%, while most of the steels contain less than 0·3% carbon.

Of the high-chromium irons, those used for components requiring a high degree of corrosion resistance normally contain 25–35% chromium, although it has been suggested by Küttner[1] that much higher contents may sometimes be necessary to give adequate resistance in certain environments. It is commonly agreed[2] that a useful formula for the minimum chromium content of a corrosion-resistant iron is:

$$\%\,Cr = (\%\,C \times 10) + 12$$

i.e., a 1·5% carbon alloy should contain not less than 27% chromium. This is supported by work carried out by Küttner[1] with various solutions, although his own interpretation of his results is regarded as faulty.

The limitations imposed by this formula, together with the fact that the alloys in practice rarely contain more than 35% chromium, suggest that the maximum carbon content of the irons should be 2·3%, and in fact the irons normally contain between 1·0% and 2·0% carbon, unless some property other than corrosion resistance is the most important.

Silicon may be present in high-chromium irons in amounts varying between 0·5 and 2·5%. Its effect is to increase fluidity in the foundry and improve the surface quality of castings. Further effects are to refine the eutectic carbides in the iron, to produce a more uniform structure and to raise the temperature at which the matrix transforms from ferrite to austenite with consequent dimensional changes. Additions above 2·5%[3,4] have an embrittling effect.

Structure

Irons of the compositions indicated above all have structures similar to that shown in Fig. 3.47, that is, a uniform dispersion of chromium–iron complex carbides in a matrix of chromium-containing ferrite. The chromium content of the ferrite is not known, although it is assumed to be about 10–13%. The

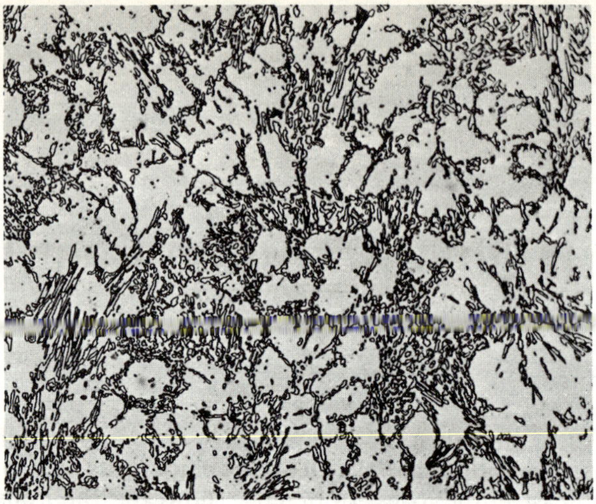

Fig. 3.47 Microstructure of 30% chromium iron. Analysis: total C 1·6, Si 1·8, Cr 31. Etched in Murakami's reagent, total magnification ×150

carbides are probably mixtures of the types Cr_7C_3 and $Cr_{23}C_6$, in which some of the chromium has been replaced by iron[5].

Mechanical Properties

Broadly speaking, the high-chromium irons are hard but not completely unmachinable. Typical properties for irons of the compositions described above in the as-cast state are:

Brinell hardness	320 HB
Transverse strength	695 MN/m^2
Tensile strength	463 MN/m^2

The hardness of the alloys makes them particularly useful in environments where abrasion or wear resistance may be important.

Production

As already indicated, these irons are used for the production of components by casting.

The principal difficulties in the production of castings in this alloy are its high shrinkage, which entails some tendency to the development of porosity, and the ready formation of oxide skin, which may cause cold laps in the casting. Castings must in consequence be produced by methods similar to those employed for steel castings and care must be taken to avoid the introduction of oxide into the mould.

Corrosion Resistance

The high-chromium irons undoubtedly owe their corrosion-resistant properties to the development on the surface of the alloys of an impervious and highly tenacious film, probably consisting of a complex mixture of chromium and iron oxides. Since the chromium oxide will be derived from the chromium present in the matrix and not from that combined with the carbide, it follows that a stainless iron will be produced only when an adequate excess (probably not less than 12%[1,2]) of chromium over the amount required to form carbides is present. It is commonly held, and with some theoretical backing, that carbon combines with ten times its own weight of chromium to produce carbides[1]. It has been said that an increase in the silicon content increases the corrosion resistance of the iron[1]; this result is probably achieved because the silicon refines the carbides and so aids the development of a more continuous oxide film over the metal surface. It seems likely that the addition of molybdenum has a similar effect, although it is possible that the molybdenum displaces some chromium from combination with the carbon and therefore increases the chromium content of the ferrite.

The irons are most useful in environments containing a plentiful supply of oxygen or oxidising agents; anaerobic or reducing conditions may lead to rapid corrosion. Physical effects such as abrasion or sudden dimensional changes induced by temperature fluctuations may rupture the film and allow corrosion to take place. The iron will also be subject to corrosion by solutions containing anions, such as those of the halides, which can penetrate surface films relatively readily.

Atmospheric Corrosion

Provided there is a suitable excess of chromium over carbon in the alloy, the irons will not rust when exposed to the atmosphere in the as-cast state. Alloys which have been found to tarnish in the as-cast state because of an inadequate excess of chromium may be found to be completely stainless in the machined and polished state, presumably because a thin film is more likely to be continuous on a smooth surface than on a rough one.

Natural and Industrial Waters

Because of its mechanical properties and the difficulties associated with its production, high-chromium iron is mostly used in environments which are particularly aggressive to other cast alloys. It is most useful for handling acid waters containing oxidising agents, for example mine waters and industrial effluents. Because many of these waters tend to contain solid matter in suspension, which can lead to abrasion of metals exposed to them, the very hard high-chromium iron is often the most suitable material for pumps handling these solutions. There is always a possibility that abrasive slurries may damage components made from this material by breaking down the oxide film and allowing corrosion to take place, but provided there is a plentiful supply of oxygen or oxidising agent at the metal surface this danger is reduced by the rapid healing of the film.

Acids

The most comprehensive data available about the corrosion behaviour of high-chromium irons in a wide variety of acid, basic and saline solutions are those issued by Bergische Stahl-Industrie of Remscheid, Germany[6], and this has formed the basis of all the following comments. Additional data given by Küttner[1] and Houdremont and Wasmuht[7] are generally, but not completely, in agreement with this.

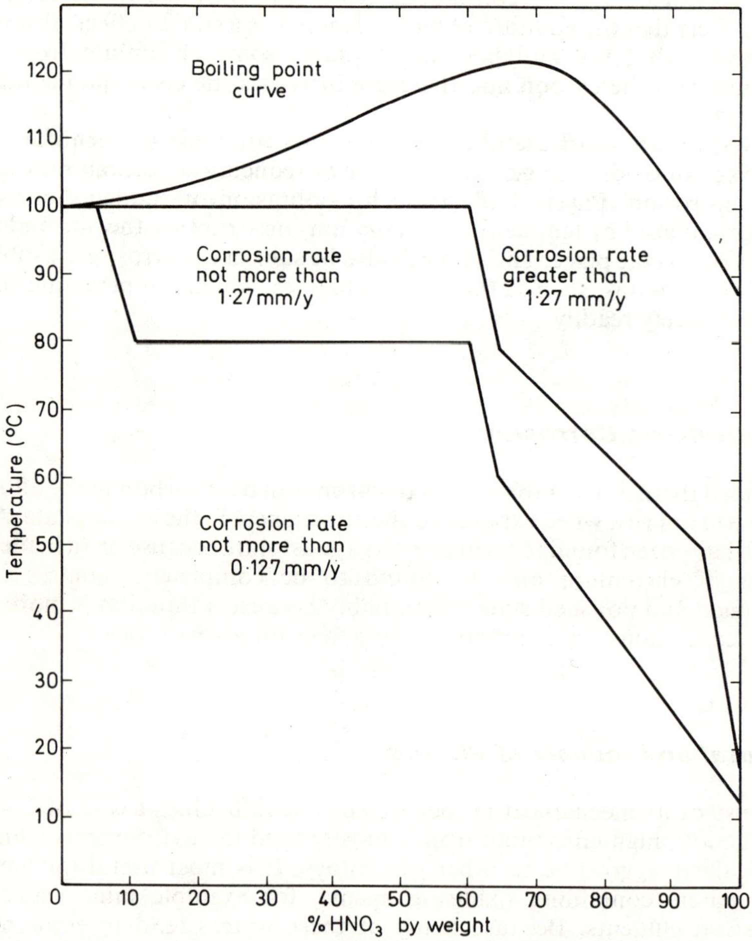

Fig. 3.48 Resistance of high-chromium iron to nitric acid solutions

Figure 3.48 has been derived from data[6] for corrosion by nitric acid solutions of an alloy nominally containing 29% chromium and 0·8% carbon.

It is interesting to note that the very useful resistance afforded by this alloy in these solutions is roughly complementary to that of the high-silicon irons, the high-chromium iron being more suitable for dilute solutions and the high-silicon iron for concentrated solutions. Because of the lack of information about the method of obtaining the results on which this and the subsequent diagram are based, it must be understood that the diagram gives only an indication of corrosion rates likely to be encountered and should not be considered as authoritative.

The data available[6] suggest that high-chromium irons have no useful resistance to sulphuric acid of more than 10% concentration at any temperature. At temperatures above 20°C corrosion rates in excess of 1·27 mm/y are probable even for acid of less than 10% concentration. The addition of 2% molybdenum appears to produce an appreciable increase in the resistance to this acid at very low and very high concentrations (Fig. 3.49).

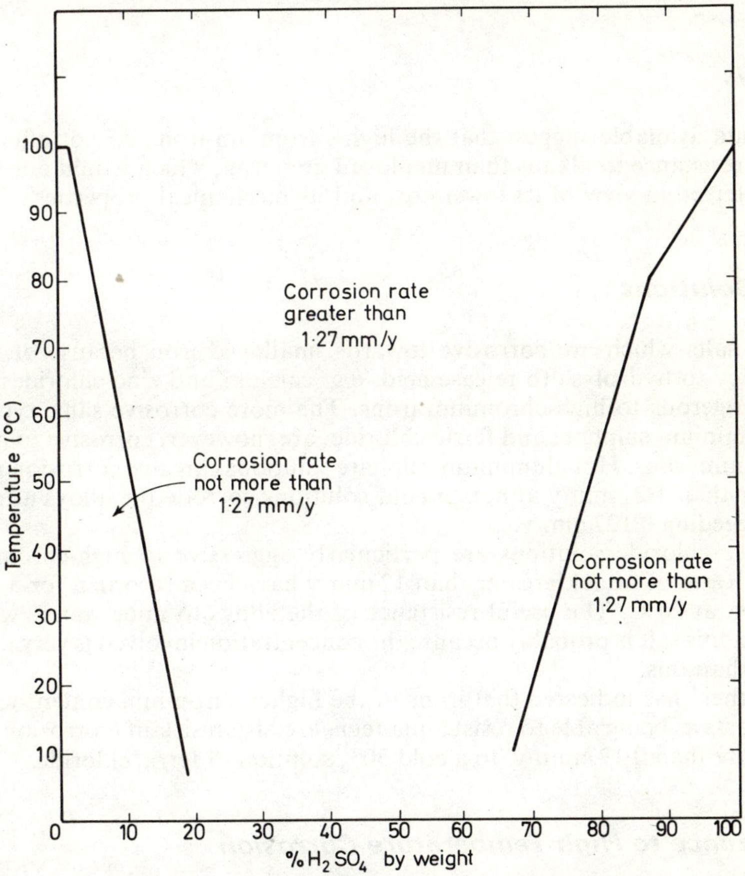

Fig. 3.49 Resistance of high-chromium iron containing 2% molybdenum to sulphuric acid solutions

It is doubtful whether the irons have any useful resistance to hydrochloric acid solutions at any concentration or temperature. It has, however, been claimed that the molybdenum-containing alloy is attacked by 1% acid at

20°C at rates not exceeding 1·27 mm/y; 2% acid, however, attacks this alloy at rates greater than 1·27 mm/y.

Although the irons have such poor resistance to sulphuric acid solutions, they are reported to be much more resistant to sulphuric acid/nitric acid mixtures which rarely cause corrosion rates of more than 1·27 mm/y. Aqua regia is corrosive to the alloys, although Küttner[1] has reported that an increase in the chromium content of the alloys, apparently according to the formula

$$\%\,\mathrm{Cr} = (\%\,\mathrm{C} \times 5) + 36$$

is effective in producing a resistance to this solution.

Other acids are completely resisted by the irons at room temperature, although high corrosion rates can sometimes develop at elevated temperatures.

Alkalis

The data available suggest that the high-chromium irons do not offer any better resistance to alkalis than unalloyed grey iron, which would normally be preferred in view of its lower cost and its mechanical properties.

Salt Solutions

Many salts which are corrosive towards unalloyed iron because of their tendency to hydrolyse to release acid, e.g. calcium and zinc chlorides, are not dangerous to high-chromium irons. The more corrosive salts, typified by aluminium sulphate and ferric chloride, are, however, corrosive to high-chromium irons. Hot aluminium sulphate solutions can give corrosion rates greater than 1·27 mm/y although cold solutions corrode the alloys at rates not exceeding 0·127 mm/y.

Ferric chloride solutions are particularly aggressive to high-chromium irons. Rates of attack greater than 12 mm/y have been recorded for a 25% solution at 20°C. The useful resistance of the alloys to mine waters which contain this salt is probably because the concentration involved is very much lower than this.

Küttner[1] has indicated that irons of the higher chromium content which he reports as being able to resist aqua regia are also resistant (corrosion rate not more than 0·127 mm/y, to a cold 30% solution of ferric chloride.

Resistance to High Temperature Corrosion

Irons containing around 12% chromium[8] may be used for service up to 900°C whilst those in the range 30–35% chromium, 1–2% carbon and 1–2% silicon have a scaling resistance up to 1 050°C[9,10], being employed for furnace hearths, kiln furniture, heat exchangers, etc. The composition range represents a compromise between ease of founding, avoiding ferrite/austenite changes on thermal cycling and maximum scaling resistance consistent with the avoidance of the long term embrittling effect of σ-phase formation in the temperature range 600–800°C[11].

J. DODD

REFERENCES

1. Küttner, C., *Tech. Mitt. Krup.*, No. 1, March 17 (1933)
2. Kinzel, A. B. and Franks, R., *Alloys of Iron and Chromium*, Vol. **2**: *High Chromium Alloys*, McGraw-Hill, New York (1940). See especially pp. 228–260
3. Valenta, E., *I.S.I. Carnegie Schol. Mem.*, **19**, 79–165 (1930)
4. Valenta, E. and Poboril, F., *Chemie et Industrie*, June, 633–648 (1933)
5. Jackson, R. S., *J.I.S.I.*, **208**, 163–167 (1970)
6. *Corrodur Bergit*, Bergische Stahl-Industrie, Remscheid, Germany (1957). See especially pp. 19–34
7. Houdremont, E. and Wasmuht, R., *Krupp. Mh.*, No. 12, 331 (1931). See also *Metals and Alloys*, **4**, 13, Feb. (1933)
8. Hallett, M. M., *J.I.S.I.*, **170**, 321–329 (1952)
9. Colton, W. J., *Brit. Foundryman*, **56**, 237–261 (1963)
10. Dixon, R. H. T. and Cumberland, J., *Foundry Trade Journal*, **116**, 721–726 and 785–791 (1964)
11. Boyes, J., *B.C.I.R.A.*, **11**, 715–731 and 461–475 (1963)

3.9 Silicon–Iron Alloys

Composition and Structure of Silicon Irons

The normal silicon content of cast iron (up to about 3%) has little effect upon the corrosion resistance of the alloys. In alloys containing much greater amounts of silicon, however, the silicon is responsible for the development of a marked increase in chemical resistance. These alloys can be divided into two types: those containing 4 to 10% silicon, which are used in applications requiring an iron with good resistance to oxidation at high temperatures, and those containing 14 to 18% silicon, which are used in applications requiring an iron with very high resistance to acid attack. These last are commonly referred to as the 'high-silicon irons'.

The alloys containing less than 11% silicon have resistances to low-temperature corrosion not substantially different from those of low-silicon irons containing similar amounts of other alloying elements, and will not be further discussed in this section.

Table 3.55 Analysis of typical silicon–iron alloys

Name	*Total carbon* (%)	Si (%)	Mn (%)	S (%)	P (%)	Ni (%)	Cr (%)	Mo (%)
Grey cast iron	3·5	2·0	0·5	0·1	0·1	—	—	—
Hypersilid 14/16	0·65	14·5	0·5	0·02	0·15	—	—	—
Duriron	0·85	14·5	0·6	<0·05	<0·1	—	—	—
Durichlor	0·85	14·5	0·6	<0·05	<0·1	—	—	3·0
Hypersilid 16/18	0·35	17·0	0·5	0·02	0·1	—	—	—

Table 3.56 Mechanical properties of silicon–iron alloys

Name	*Ultimate tensile strength* (MN/m^2)	*Elongation* 2 in (50 mm) *gauge length*	*Brinell hardness no.*
Grey cast iron	23	2%	180
14·5% silicon iron	139	Nil	450
Durichlor	108	Nil	520

The analyses of typical high-silicon irons are given in Table 3.55 and their typical mechanical properties in Table 3.56.

All these alloys are characterised by high hardness values and low resistance to impact. In this they are probably more similar to stoneware than to other metals but they are superior to stoneware in thermal conductivity and in their resistance to thermal shock, which, however, is poor compared with that of other metals. Moreover, it is usually easier to make castings of silicon iron than to fabricate required parts from stoneware.

The microstructure of the high-silicon irons containing less than 15·2% silicon consists of a matrix of α silico-ferrite[1] in which the majority of the carbon present in the alloy is distributed as fine graphite flakes. A typical microstructure of Fe–14·5 Si is shown in Fig. 3.50. In irons containing more

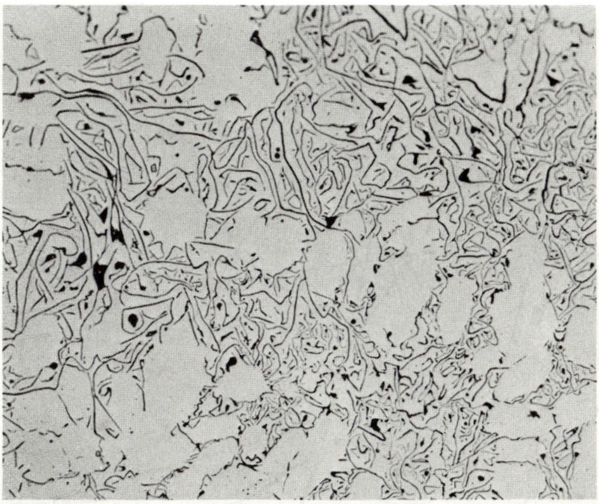

Fig. 3.50 Typical structure of 14·5% silicon iron

than 15·2% silicon some η phase is also present[1]. The hardness and brittleness of the silicon iron is due to the nature of the silico-ferrite. An attempt[2] has been made to produce high-silicon iron with a nodular graphite structure, with the intention of improving the mechanical properties of the alloy, but because the low strength of these alloys is due to the matrix rather than to the graphite form, the nodular graphite irons have proved to be little, if at all, superior in strength to the conventional alloys.

All the components made in silicon irons are produced by casting. The high-silicon irons are best melted in crucible or electric furnaces and their production tends to be concentrated in the hands of a few highly specialised foundries. Because the mechanical properties of these high-silicon irons preclude any machining other than grinding, it is necessary to design the castings made from these alloys in such a way that the necessity for subsequent treatment is kept to the minimum; for this reason the closest liaison between designer and foundry is essential.

In view of the poor mechanical properties of the high-silicon irons, the development of any stresses in the castings during solidification is very dangerous, since they may cause the casting to crack in subsequent service. To overcome this risk, it is usual to strip the castings from the moulds while they are still red hot and to anneal them at 800°C for 4–5 h, followed by slow cooling[3].

General Corrosion Behaviour

The resistance to corrosion exhibited by the high-silicon alloys is believed to be due to the development of a corrosion-resistant film containing a large proportion of silica. The full protective value of the film does not develop until at least 14·25% silicon is present in the alloy (Fig. 3.51). Increase in the content of silicon above 14·5% does not have a dramatic effect upon the corrosion resistance of the alloy (Fig. 3.51), although the further increase in film density is of service if the casting is to be exposed to solutions containing halide ions, especially hydrochloric acid.

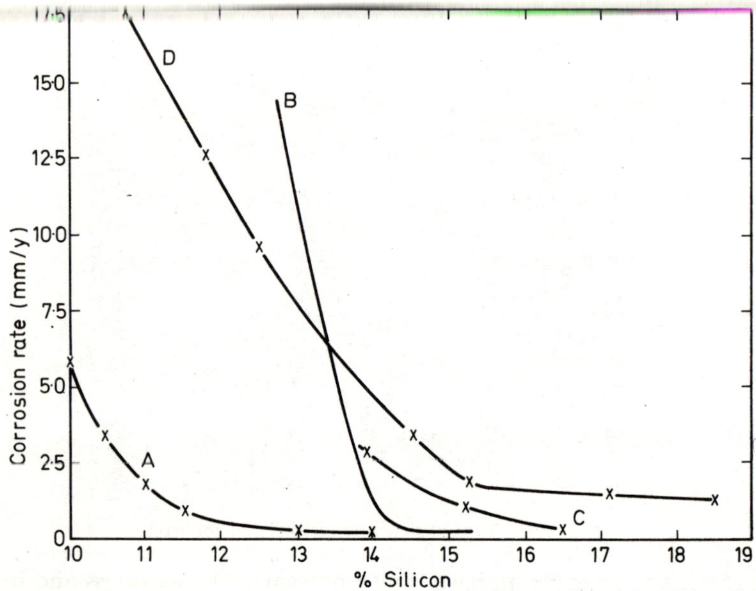

Fig. 3.51 Dependence of corrosion resistance of high-silicon iron on silicon content. *A* 70% HNO_3, *B* 20% H_2SO_4 at b.p., *C* 10% HNO_3 at b.p., *D* 10% HCl at 80°C

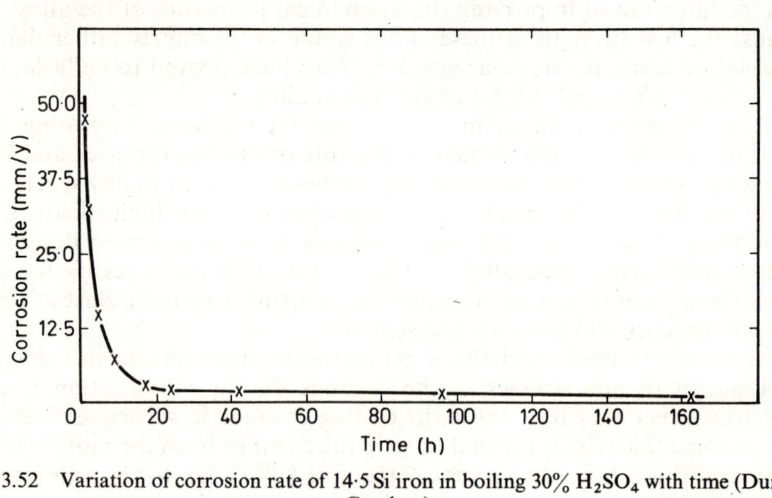

Fig. 3.52 Variation of corrosion rate of 14·5 Si iron in boiling 30% H_2SO_4 with time (Duriron Co. data)

Since the formation of the silica film does not depend on any particular property of the corrosive environment, high-silicon irons can resist attack by a very wide range of environments. Solutions which are capable of dissolving silica, even in a small degree, are, however, inimical to silicon irons, and there are also a few ions capable of penetrating the silica film, which can cause relatively serious corrosion of the metal.

Before the silica film can form, some corrosion of the metal must necessarily take place, and it follows that initial corrosion rates are high. Fig. 3.52 illustrates this point and suggests that uniform rates of corrosion are not reached until at least 100 h after the onset of the attack. As a result, useful data on the corrosion of high silicon irons can be obtained only from tests of at least this duration.

Corrosion in Natural Environments

Although the high-silicon irons are often used in circumstances which expose them to atmospheric, water or soil corrosion, they are rarely installed specifically to resist these agencies. Their corrosion resistance is such, however, that in fact no normally occurring environment ever causes serious attack. This is not to say that these irons can be regarded as stainless, and in fact alloys containing less than 14·7% silicon have been reported as becoming rusty in a moist atmosphere[3].

Corrosion in Industrial Environments

Acids

Since the corrosion resistance of the high-silicon alloys depends upon the permanence and impermeability of a thin silica film on the surface of the metal, it is obvious that any reagent which can damage the film will cause accelerated corrosion of the metal. For this reason all solutions containing hydrofluoric acid must be regarded as incompatible with the alloys.

As silica is not attacked by any acid other than hydrofluoric it might be expected to act as an effective barrier to attack by any other acid solutions, but in fact, while the high-silicon iron is resistant to attack by most acids, it is corroded relatively severely by hydrochloric, hydrobromic and sulphurous acids. The aggressive character of the two halogen acids may be ascribed to the readiness with which their relatively small anions can penetrate a passive film.

Attempts made to produce an alloy more resistant to hydrochloric acid have resulted in alloys containing 17–18% silicon or 14·5% silicon and 3% molybdenum. The first is produced in Britain, and the second in the United States. As far as is known, the molybdenum-containing alloy has no clear advantages over Fe–18 Si, in relation to corrosion resistance, strength, manufacturing difficulty or cost. Both the alloys are known to be more fragile than Fe–14·5 Si, a decrease in strength of some 30% being involved.

The reason for the increase in resistance to hydrochloric acid of the Fe–18 Si alloy is thought to lie primarily in the increased density of the silica-rich film left on the metal by initial corrosion. The addition of 3–3·5%

molybdenum to Fe–14·5 Si causes the formation of extremely stable complex carbides with the consequent complete elimination of graphite[1]. It seems possible that this removal of graphite flakes from the microstructure, and the resulting elimination of the planes of weakness in the silica film which might be expected to result from their presence may be the cause of the increased resistance of molybdenum-containing irons to hydrochloric acid.

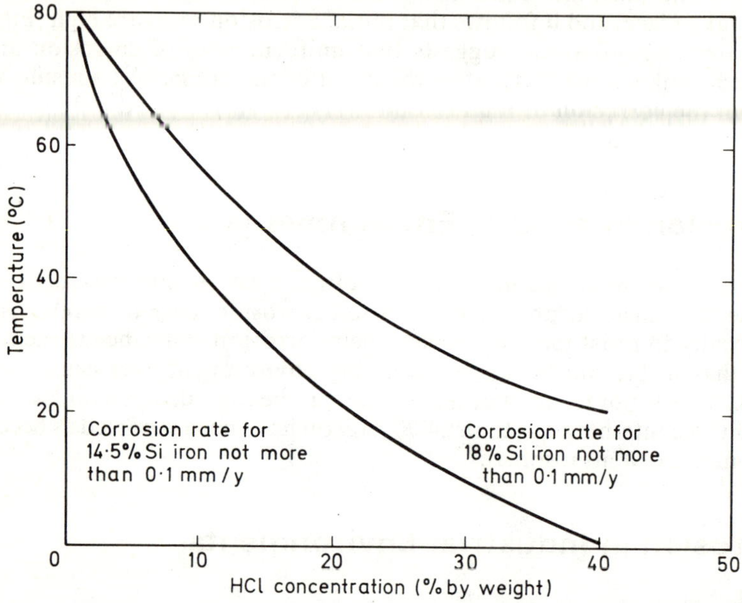

Fig. 3.53 Limits of use of silicon irons for handling hydrochloric acid solutions (former Ministry of Supply data) for Fe–18Si and Fe–14·5Si alloys

Figure 3.53 was prepared by a Silicon Iron Advisory Panel to the former Ministry of Supply between 1940 and 1945 as a guide to the ranges of hydrochloric acid concentration and temperature in which the two types of high-silicon iron could be safely used. The limits to the ranges are set by those conditions outside which the corrosion rate exceeds 0·1 mm/y. It should be emphasised that these curves give a guide to operating conditions only, and are not always in complete agreement with other published work. Borderline cases should always be checked by corrosion tests, and in the interpretation of these the large variations in corrosion rate which may arise because of subtle differences between laboratory and plant conditions must always be borne in mind.

The evidence at present available concerning the corrosion of high-silicon irons by sulphurous acid is insufficient to allow the formation of any theory about the mechanism by which the silica film barrier is broken down in the presence of this acid. As far as is known, this acid is corrosive to all types of high-silicon iron.

Probably the most useful characteristic of the high-silicon irons is their ability to withstand sulphuric acid at all temperatures and concentrations. The maximum rate of corrosion which can develop has been reported to be

0·482 mm/y in 30% sulphuric acid at boiling point[4], and this falls to a minimum rate of 0·025 mm/y when the acid concentration exceeds 60% and the temperature is at boiling point (Fig. 3.54). The former Ministry of Supply

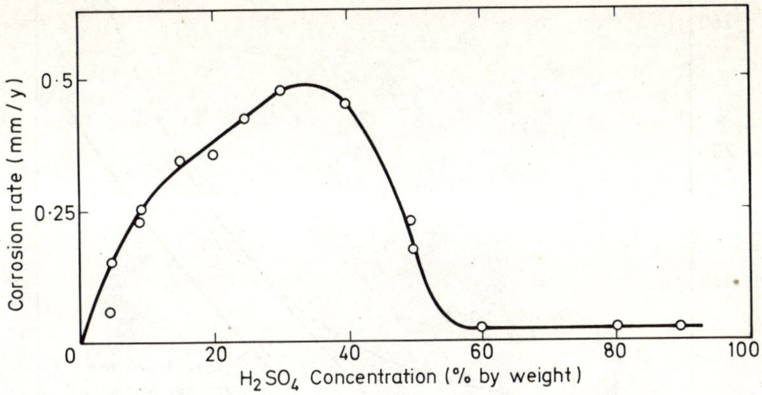

Fig. 3.54 Variation of corrosion rate with concentration of boiling sulphuric acid (Duriron Co. data)

Advisory Panel prepared a chart to indicate the regions of temperature and acid concentration in which the corrosion rate of Fe–14·5 Si was less than 0·1 mm/y (Fig. 3.55). This curve is at variance with the results from which Fig. 3.54 is derived, which suggest that the dotted curve in Fig. 3.55 may represent the true state of affairs. At best, Fig. 3.55 can be regarded only as a guide and the remarks already made about the interpretation of Fig. 3.53 are also applicable here.

It has been reported[5] that oleum is corrosive to high-silicon iron and it is not normally recommended for withstanding this acid.

Nitric acid is also withstood by high-silicon iron. The concentrated acid is believed to reinforce the silica film by the formation of a passive iron oxide film, and this assumption is supported by the fact that the highest rates of attack are associated with hot dilute solutions. The curve prepared by the former Ministry of Supply Advisory Panel to indicate the regions of acid concentration and temperature in which the corrosion rate of Fe–14·5 Si is less than 0·1 mm/y is shown in Fig. 3.56. Fontana[6] has published a similar curve which is at variance with that produced in the UK for concentrations of less than 30%, as shown by the dotted line in Fig. 3.56.

The results published in the literature suggest that solutions of orthophosphoric acid at all concentrations and temperatures are not corrosive to high-silicon irons. Kosting and Heins[7] quote figures given by Duriron Co. together with figures obtained in their own tests of 24 hours duration; these are given in Table 3.57.

It will be observed that (probably mainly because of the short test period they used) their results indicate significantly higher corrosion rates than those given by Duriron Co. but neither source suggests that the rates are dangerously high. It seems probable that solutions of pyrophosphoric or tetraphosphoric acid are liable to give rise to excessive corrosion, although there is evidence to suggest that pyrophosphoric acid may be handled safely in the cold and even

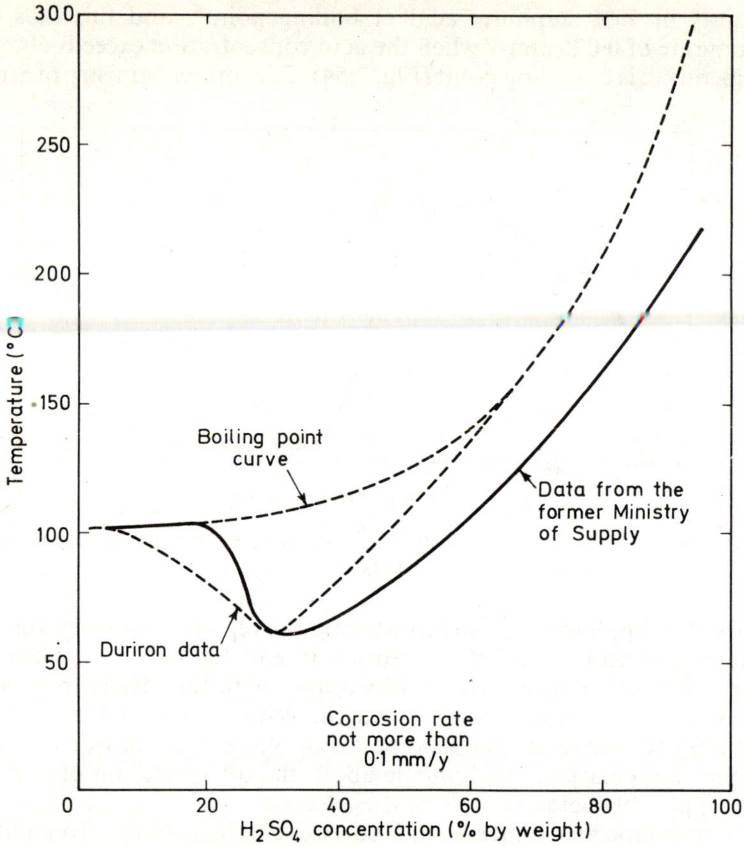

Fig. 3.55 Limits of use of silicon iron for handling sulphuric acid solutions

possibly up to 80 to 90°C. Most crude phosphoric acid contains appreciable amounts of fluoride which lead to excessive corrosion rates.

Table 3.57 Corrosion of high-silicon irons by ortho-phosphoric solutions

Acid concentration (%)	*Temperature* (°C)	*Corrosion rate* (mm/y)	*Origin of data*
10	82–88	0·007	Duriron Co.
	98	0·147	Kosting & Heins[7]
25	82–88	0·010	Duriron Co.
	98	0·038	Kosting & Heins[7]
50	98	0·185	Kosting & Heins[7]
87	82–88	0·010	Duriron Co.

High-silicon iron offers excellent resistance to attack by all concentrations of nitric–sulphuric acid mixtures. The mixed acid corrodes the iron at rates

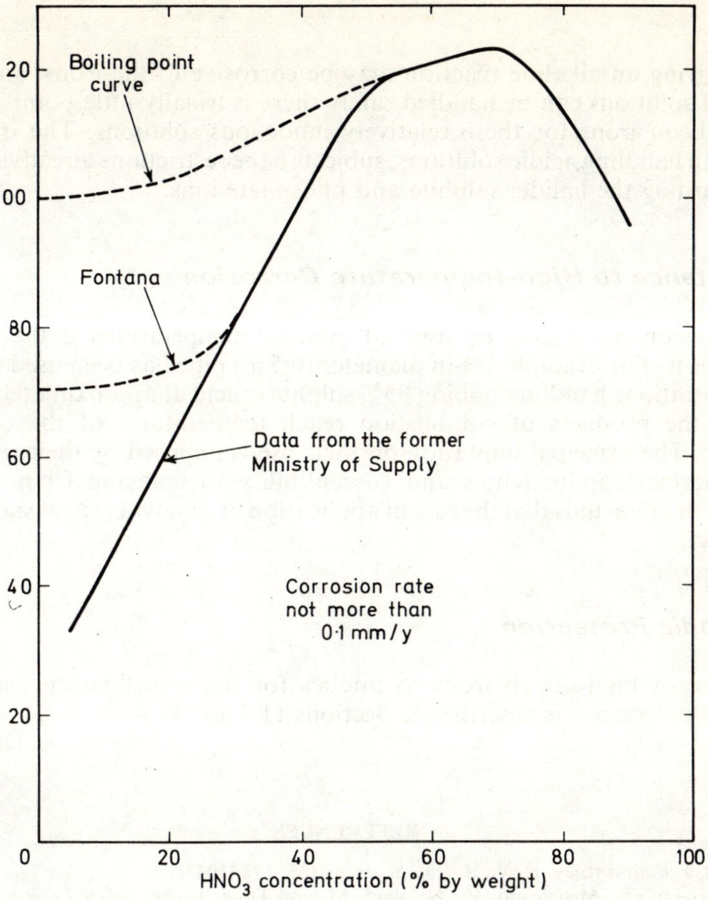

Fig. 3.56 Limits of use of silicon iron for handling nitric acid solutions

never greater and often lower than the individual acids of comparable concentration.

All organic and other weak acids commonly encountered in industry are satisfactorily handled by high-silicon irons.

Alkalis

High-silicon irons are inferior to unalloyed grey irons in their resistance to attack by alkalis. For example, boiling 20% caustic soda solution or 50% caustic soda at 80°C will attack high-silicon irons at rates of the order of 1·27 mm/y (grey iron would be attacked at rates not greater than 0·25 mm/y).

The irons are not recommended even for so weak a base as ammonium hydroxide, if the liquid temperature is greater than 20°C.

Salts

Salts giving an alkaline reaction may be corrosive to the irons, and while neutral solutions can be handled safely there is usually little point in using high-silicon irons for these relatively innocuous solutions. The irons are useful in handling acidic solutions, subject to the restrictions already referred to regarding the halide, sulphite and phosphate ions.

Resistance to High-temperature Corrosion

High-silicon irons may be used at elevated temperatures if the process requires it. For example, 20-in diameter (0·5 m) pipe has been used for acid concentrations handling boiling 95% sulphuric acid at approximately 285°C where the products of combustion reach temperatures of the order of 590°C[8]. The principal limitation on their use is imposed by their relatively low thermal conductivities and susceptibility to cracking from thermal shock; this demands that the rate of application or removal of heat should not be rapid.

Cathodic Protection

The use of high-silicon irons as anodes for impressed-current cathodic-protection systems is described in Sections 11.3 and 11.4.

J. DODD

REFERENCES

1. Hurst, J. E. and Riley, R. V., *J. Iron St. Inst.*, **155**, 172 (1947)
2. Dumitrescu, T., Medeleanu, V., Nicolaid, M. and Dinu, I., *Rev. Métall. (Roumania)*, **3** No. 2, 19 (1958)
3. Hurst, J. E., *Proc. Inst. Brit. Foundrym.*, **37**, B46 (1943–44)
4. Duriron Co. (published literature)
5. Luce, W. A., *Chem. Engng.*, **61**, 246 (1954)
6. Fontana, M. G., *Industr. Engng. Chem. (Industr.)*, **45**, 91A, 92A, 94A, May (1953)
7. Kosting, P. R. and Heins, C., *Industr. Engng. Chem. (Industr.)*, **23**, 140–150, Feb. (1931)
8. Bryan, W. T. and Uhlig, H. H. (Ed.), *The Corrosion Handbook*, Wiley, New York; Chapman and Hall, London, 201–207 (1948)

4 NON-FERROUS METALS AND ALLOYS

4.1 Aluminium and Aluminium Alloys

Aluminium and the aluminium alloys lend themselves to many engineering applications because of their combination of lightness with strength, their high corrosion resistance, their thermal and electrical conductivity and heat and light reflectivity, and their hygienic and non-toxic qualities. The variety of forms in which they are available also enhances their utility.

Composition and Mechanical Properties

Pure aluminium has good working and forming properties, high resistance to corrosion, low mechanical strength, and high ductility. The diverse and exacting technical demands made on aluminium alloys in different applications are met by the considerable range of alloys available for general and specific engineering purposes (BS 1470–75, 1490), each of which has been designed and tested to provide various combinations of useful properties. These include strength/weight ratio, corrosion resistance, workability, castability, or high-temperature properties, to mention but a few. The compositions and properties of these standard alloys are given in Tables 4.1 to 4.4.

More specialised alloys are covered by the *DTD* and *L* series for aircraft applications and include the high strength Al–Zn–Mg alloys. The medium strength weldable Al–Zn–Mg compositions are finding increasing utility in engineering and a national specification may be anticipated in the near future.

Where free machining characteristics are required, this may be achieved by additions of cadmium, antimony, tin or lead (e.g. BS 4300/5). Materials for electrical use are of special composition (BS 2627, 3988), while bearings are manufactured from Al–Sn alloys.

Composites of aluminium alloy with a thin cladding on one or both surfaces of a more anodic aluminium alloy or pure aluminium, enable sheet, plate and tube to be produced with special combinations of strength and corrosion resistance appropriate to service conditions. Although originally applied to high strength aircraft alloys, this principle of cladding is now utilised in several important industrial applications.

The I.S.O. designations may be correlated directly with the British Standard General Engineering series and partially with the American Aluminium Association designations. The nearest equivalents for the three systems are given in Table 4.5, although differences in alloying practice in

Table 4.1 Some wrought British Standard aluminium alloys for general engineering purposes (non-heat treatable alloys)

Material designation					*Suitability for*					
							Fusion welding[3]			
BS 1470–75	I.S.O.	*Major alloying constituents* (nominal %)	*Tensile strength range** (N/mm^2)	*Resistance to atmospheric attack*	*Cold forming*[2]	*Machining*	*Oxy-gas*	*Inert gas shielded arc*	*Resistance spot welding*	*Protective anodising*
1	Al 99·99	99·99Al	100 max.	V	V	P	V	V	F	V
1A	Al 99·8	99·8Al	125 max.	V	V	P	V	V	F	V
1B	Al 99·5	99·5Al	55–135	V	V	F	V	V	G	V
1C	Al 99·0	99Al	60–140	V	V	F	V	V	G	V
N3	Al Mn1	1·25Mg	90–175	V	V	F	V	V	V	G
N4	Al Mg2	2·25Mg	160–225	V	G	G	G	G	V	V
N5	Al Mg3·5	3·5Mg	215–275	V	G	G	F	G	V	V
N6[1]	Al Mg5	5·0Mg	250 min.	G	G	G	F	G	V	G
N8	Al Mg4·5 Mn	4·5Mg	275–345	V	G	G	F	G	V	G

Notes: 1. Rivet and screw stock only.
2. Ratings are for material in the optimum condition for forming.
3. Ratings are given for correct technique and filler rod and take into account the properties of material after welding.
V = very good, G = good, F = fair, P = poor and U = unsuitable.
* Where strength varies with temper the specified minima are quoted. Similarly where properties are influenced by the fabrication process, the lowest minima are given.

Table 4.2 Heat treatable alloys in the solution treated condition, but naturally aged where applicable

Material designation					*Suitability for*					
							Fusion welding[3]			
BS 1470–75	I.S.O.	*Major alloying constituents* (nominal %)	*Tensile strength** (N/mm^2)	*Resistance to atmospheric attack*	*Cold forming*[2]	*Machining*	*Oxy-gas*	*Inert-gas-shielded arc*	*Resistance spot welding*	*Protective anodising*
H9	Al Mg Si	0·7Mg, 0·5Si	125 min	V	G	G	G	G	G	V
H12	Al Cu2 Ni1 Mg Fe Si	2·3Cu, 1·0Ni, 0·9Mg, 0·9Fe, 0·9 Si	310 min	F	F	G	U	G	U	F
H15	Al Cu4 Si Mg	4·5Cu, 0·8Si, 0·5Mg, 0·8Mn	370 min	F	F	G	U	G	V	F
HC15	Al Cu4 Si Mg	As H15, clad with 1B	375 min	G	F	G	U	F	V	V
H20	Al Mg1 Si Cu	1·0Mg, 0·6Si, 0·25Cu (0·5Mn or 0·25Cr)	215 min	G	G	G	F	G	G	G
H30	Al Si Mg Mn	1·0Si, 0·8Mg, 0·7Mn	200 min	G	G	G	F	G	G	G

Notes: 2. Ratings are for material in the optimum condition for forming.
3. Ratings are given for correct technique and filler rod and take into account the properties of material after welding.
* Where strength varies with temper the specified minima are quoted. Similarly where properties are influenced by the fabrication process, the lowest minima are given.

Table 4.3 Heat treatable alloys in the fully heat treated condition

Material designation		Major alloying constituents (nominal %)	Tensile strength* (N/mm²)	Resistance to atmospheric attack	Suitability for					
					Cold forming(2)	Machining	Fusion welding(3)		Resistance spot welding	Protective anodising
BS 1470–75	I.S.O.						Oxy-gas	Inert-gas-shielded arc		
H9	Al Mg Si	0·7Mg, 0·5Si	150 min	G	G	V	G	G	G	V
H12	Al Cu2 Ni1 Mg Fe Si	2·3Cu, 1·0Ni, 0·9Mg, 0·9Fe, 0·9Si	385 min	F	P	V	U	G	U	F
H15	Al Cu4 Si Mg	4·5Cu, 0·8Si, 0·5Mg, 0·8Mn	400 min	P	P	V	U	G	V	F
HC15	Al Cu4 Si Mg	As H15, clad with 1B	400 min	G	P	G	U	F	V	V
H16	Al Cu2 Mg1·5 Fe1 Ni1	2·3Cu, 1·5Mg, 1·2Fe, 1·1Ni	430 min	P	P	V	U	G	U	F
H20	Al Mg1 Si Cu	1·0Mg, 0·6Si, 0·25Cu (0·5Mn or 0·25Cr)	280 min	G	F	V	F	G	G	G
H30	Al Si Mg Mn	1·0Si, 0·8Mg, 0·7Mn	280 min	G	F	V	F	G	G	G

Notes: 2. Ratings are for material in the optimum condition for forming.
3. Ratings are given for correct technique and filler rod and take into account the properties of material after welding.
* Where strength varies with temper the specified minima are quoted. Similarly where properties are influenced by the fabrication process, the lowest minima are given.

Table 4.4 Cast British Standard aluminium alloys for general and special purposes

BS 1490	*Major alloying constituents* (nominal %)	*Condi-tion*	*Min. tensile strength** (N/mm²) *Sand cast*	*Chill cast*	*Fluidity*	*Resistance to hot tearing*	*Pressure tightness*	*Machin-ability*	*Resistance to corrosion*	*Protective anodising*	*Comments*
					GENERAL PURPOSE ALLOYS						
LM2	10Si, 1·5Cu	M	—	150	G	V	V	G	G	P	General purpose die casting, particularly for thin sections
LM4	5Si, 3Cu	M TF	140 230	160 280	G	G	V	G	F	P	Widely used, general purpose alloy
LM6	12Si	M	160	190	V	V	V	F	V	P	Excellent castability for thin sections and intricate shapes, good corrosion resistance
LM20	12Si	M	—	190	V	V	V	G	G	P	Excellent castability, but less corrosion resistant than LM6
LM24	8·5Si, 3·5Cu	M	—	180	G	V	V	G	F	P	General purpose die casting
LM25	7Si, 0·3Mg	M TF	130 230	160 280	G	V	V	G	V	P	High strength with good corrosion resistance
LM27	7Si, 2Cu, 0·4Mn	M	140	160	G	V	V	G	G	P	General purpose alloy

Note: * Where strength varies with temper the specified minima are quoted. Similarly where properties are influenced by the fabrication process, the lowest minima are given.

Table 4.4 (continued)

SPECIAL PURPOSE ALLOYS

LM0	99·5% Al min.	M	—	—	F	F	F	F	V	V	Suitable for corrosive environments and electrical applications
LM5	4·5Mg, 0·4Mn	M	140	170	F	F	P	G	V	V	Suitable for corrosive environments and for decorative applications
LM9	12Si, 0·4Mg, 0·4Mn	M	— 240	190 295	G	V	G	F	G	P	Good combination of castability, corrosion resistance and strength
LM10	10Mg	TB	280	310	F	G	P	G	V	F	Combines toughness with shock resistance
LM12	10Cu, 0·3Mg	M	—	170	F	G	G	V	P	P	Specially suitable for hydraulic equipment
LM13	11Si, 1Cu, 1Mg	TF TF7	170 140	280 200	G	V	F	F	F	P	Used for pistons
LM16	5Si, 1·3Cu, 0·5Mg	TB TF	170 230	230 280	G	G	G	G	F	P	Elevated temperature applications
LM18	5Si	M	120	140	G	V	V	F	V	P	Used for food handling equipment
LM21	6Si, 4Cu, 0·4Mn, 0·2Mg	M	150	170	G	G	G	G	P	P	High strength with good engineering characteristics
LM22	5Si, 3Cu, 0·4Mn	TB	—	245	G	G	G	G	P	P	High shock resistance, used for structural applications
LM26	9·5Si, 3Cu, 1Mg	TE	—	210	G	V	F	F	P	P	Used for pistons
LM28	18Si, 1·5Cu, 1Mg, 1Ni	TE TF	— 120	170 190	F	G	F	P	G	P	Special piston applications
LM29	23Si, 1Cu, 1Mg, 1Ni	TE TF	120 120	190 190	F	G	F	P	G	P	Special piston applications
LM30	17Si, 4·5Cu, 0·5Mg	M TS	— —	150 160	G	F	F	F	P	P	Automobile cylinder blocks

Note: V = very good, G = good, F = fair and P = poor.

Table 4.5 British Standard, I.S.O. and comparable aluminium association designations

BS 1470–75	I.S.O.	*Nearest AA designation*	*Major alloy constituents* (nominal %)
1	Al 99·99	1099	99·99Al
1A	Al 99·8	1080	99·8Al
1B	Al 99·5	1050	99·5Al
1C	Al 99·0	1100	99Al
N3	Al Mn1	3003	1·25Mn
N4	Al Mg2	5052	2·25Mg
N5	Al Mg3·5	5154	3·5Mg
N6	Al Mg5	5456	5·0Mg
N8	Al Mg4·5 Mn	5083	4·5Mg
H9	Al Mg Si	6063	0·7Mg, 0·5Si
H12	Al Cu2 Ni1 Mg Fe Si	—	2·3Cu, 1·0Ni, 0·9Mg, 0·9Fe, 0·9Si
H15	Al Cu4 Si Mg	2014	4·5Cu, 0·8Si, 0·5Mg, 0·8Mn
HC15	Al Cu4 Si Mg	Al clad 2024	As H15, clad with 1B
H16	Al Cu2 Mg1·5 Fe1 Ni1	—	2·3Cu, 1·5Mg, 1·2Fe, 1·1Ni
H20	Al Mg1 Si Cu	6061	1·0Mg, 0·6Si, 0·25Cu (0·5Mn or 0·25Cr)
H30	Al Si Mg Mn	6351	1·0Si, 0·8Mg, 0·7Mn

America make a direct correlation impossible in several cases. Notably, where European practice utilises a minor addition of manganese, a similar effect is frequently achieved by a chromium addition in American practice; also, there is a greater tendency to use small copper additions in American alloys than in European alloys.

The British Standard alloys use a systematic letter notation to indicate the form and heat treatment of the material. Details are given in Table 4.6. For example, the strongest condition is H8 for a non-heat-treatable alloy, TB or TD for a single heat-treatment alloy, and TE, TF and TH for a double heat-treatment alloy.

The non-heat-treatable alloys (prefixed *N*) are hardened by cold work and attain the desired properties by a combination of annealing and cold work. The hard material has markedly increased strength with only slightly reduced corrosion resistance.

The heat-treatable alloys (prefixed *H*)—notably the Al–Cu–Mg and the Al–Mg–Si types—can be heated at 480–535°C for a period between 20 min and some hours to obtain solution of the alloying elements, and then rapidly quenched. This solution treatment gives increased strength, and may also give slightly increased corrosion resistance. Further strengthening of certain alloys is achieved by an additional lower temperature heat-treatment for longer periods (1–20 h or more, according to the alloy) which promotes precipitation of the alloying elements within the metal crystal structure. With some alloys this ageing treatment takes place at room temperature. The ageing or precipitation treatment slightly reduces the corrosion resistance of most alloys.

Comprehensive details of alloy properties and characteristics are provided in the publications of the major aluminium companies and independent organisations[1].

Table 4.6 Temper and heat treatment symbols for aluminium alloys–suffixes

British Standard designation	*Meaning*
	WROUGHT MATERIALS
M	As manufactured. Material which acquires some temper from shaping processes in which there is no special control over thermal treatment or amount of strain hardening
O	Annealed. Material which is fully annealed to obtain the lowest-strength condition
H1, H2 H3, H4 H5, H6 H7, H8	Strain hardened. Material subjected to the application of cold work after annealing (or hot forming) or to a combination of cold work and partial annealing/stabilising in order to secure the specified mechanical properties. The designations 1–8 are in ascending order of tensile strength
TB	Solution heat treated and naturally aged. Material which receives no cold work after solution heat treatment except as may be required to flatten or straighten it. Properties of some alloys in this temper are unstable
TD	Solution heat treated, cold worked and naturally aged
TE	Cooled from an elevated temperature shaping process and precipitation treated
TF	Solution heat treated and precipitation treated
TH	Solution heat treated, cold worked and then precipitation treated
	CAST MATERIALS
M	As cast
TS	Stress relieved only
TE	Precipitation treated
TB	Solution treated
TB7	Solution treated and stabilised
TF	Solution treated and precipitation treated
TF7	Full heat treatment plus stabilisation

Physical Properties

Some of the more useful physical and mechanical properties of aluminium are given in Tables 4.7 and 4.8. The common wrought forms are rolled plate (prefixed P), clad plate (PC), sheet and strip (S), clad sheet and strip (C), bars, rods, and sections (E), extruded round tube and hollow sections (V), drawn tubes (T), wire (G), rivet stock (R), bolt and screw stock (B), and forgings and forging stock (F). Castings are made in sand moulds or in metal moulds known as dies; the most widely used methods involve casting either under gravity or under pressure. Aluminium and aluminium alloys are fabricated into products such as rolled plate, sheet, extruded sections, drawn tube, etc. by all the familiar processes, with modifications appropriate to the temper or condition of the material. Joining may be carried out by mechanical methods (such as riveting and bolting), brazing, soldering, adhesive bonding, or welding. The argon-shielded arc welding methods (mig and tig) are particularly appropriate where corrosion resistance of welded joints is of importance[2].

Table 4.7 Physical properties of aluminium

Atomic number	13
Atomic volume	10·0
Atomic weight	26·97
Valency	3
Crystal structure	Face-centred cubic
Interatomic distance	2·863Å
Electrochemical equivalent (g/Ah)	0·3354
Density at 293 K (kg/m^3)	2 700

Thermal						Electrical			
m.p. (K)	Sp. heat at 293 K (J/kg K)	Mean sp. heat (293–931 K) (J/kg K)	Latent heat of fusion (kJ/kg)	Coeff. of linear exp. (293–393 K) (m/K)	Thermal conductivity at 273 K (W/m K)	Elec. vol. resistivity at 293 K (μΩ cm)	Elec. vol. conductivity at 293 K (% I.A.C.S.)	Temp. coeff. of elec. resistance per K for 293 K	Thermo-electric power vs platinum (mV/100 K)
931	896	1 047	387	$0{\cdot}61 \times 10^{-6}$	214	2·7–3·0	63–57	0·0041	+0·41

Table 4.8 Mechanical properties of aluminium

Young's modulus (MN/m^2)	*Torsion modulus* (MN/m^2)	*Poisson's ratio*	*Compressibility* ($dv/v_0\, dp$)	
			at 293 K	at 400 K
59×10^3	24×10^3	0·34	$1{\cdot}45 \times 10^{-6}$	$1{\cdot}70 \times 10^{-6}$

Selection of Purity or Alloy Type

It will be noted that the materials covered by the BS specifications fall into several distinct groups, sometimes with apparently small differences within the group. Characteristics which could influence the selection of the most appropriate material for a specific application are tabulated in Table 4.1 for wrought products, but some elaboration is desirable since the successful utilisation of aluminium begins with the selection of alloy. Additionally, mention should be made of materials not covered by the BS General Engineering series.

Pure Aluminium

Within the BS series the corrosion resistance of unalloyed aluminium increases with increasing metal purity.

The use of the 99·8% and 99·99% grades is usually confined to those applications where very high corrosion resistance or ductility is required. The chemical industry can advantageously use these purities for handling some products, but because of their low mechanical strength they are sometimes used as a cladding material for a stronger substrate.

Decreasing the purity results in increased strength for the 99% and 99·5% grades, which still retain a high resistance to corrosion. The 99% pure metal may be considered the more useful general purpose metal for lightly stressed applications such as panelling and cooking utensils.

Aluminium-Manganese Alloy

The alloy N3 is the sole BS alloy of this binary system. In sheet form, the combination of good corrosion resistance with adequate mechanical strength results in large tonnages being used in building, cooking utensils and sundry general applications.

Aluminium-Magnesium Alloys

For general use, the Al–Mg system is represented by N4, N5 and N8 with increasing magnesium content respectively. The corrosion resistance of all these alloys is extremely good, while the level of mechanical properties obtainable makes them ideally suited for structural use in aggressive conditions.

The characteristics of these alloys make them ideal for boat and ship-building, for which a long history of satisfactory performance is on record for the higher magnesium alloys. Where strength is less critical the lower magnesium alloys may be used with similar success and are recommended for aqueous conditions.

Elevated temperatures should be avoided with N5 and N8, since the precipitation of Mg_2Al_3 over a period of time can lead to serious structural corrosion. In case of doubt regarding this aspect, the manufacturer should be consulted.

Aluminium-Magnesium-Silicon Alloys

The heat-treatable Al–Mg–Si alloys H9, H20 and H30 are predominantly structural materials, all of which have a high resistance to corrosion.

The low Mg+Si content of H9 facilitates the production of complex extrusions with a good surface finish making H9 a natural choice for glazing sections and other architectural features. Higher mechanical properties are obtainable with the H20 and H30 compositions, which are therefore more suitable for load bearing structures.

The corrosion resistance of the Al–Mg–Si alloys is slightly inferior to that of the Al–Mg alloys, but where maximum obtainable strength is required then a fully heat-treated Al–Mg–Si alloy would generally be preferable to an Al–Mg alloy with comparable properties obtained by cold working.

Aluminium-Copper Alloys

The composition of these alloys extends beyond the binary system and they may be categorised as the Duralumin type H15 and the complex types H12 and H16.

The mechanical properties and characteristics of H15 cause it to be used for those applications where high strength is the prime criterion, outweighing its poor resistance to corrosion. Protection by anodising or painting is desirable, when satisfactory performance may be expected except in the most severe conditions. Alternatively, the HC15 clad version has a corrosion resistance similar to its pure cladding, provided that repeated heat treatments have not caused excessive copper diffusion into the pure cladding. The use of H15 for machined components is fairly common, but cannot be recommended where the service conditions will be aggressive.

Where the retention of strength at elevated temperatures is required, then the alloys H12 and H16 should be considered. Because of their copper content the corrosion resistance is mediocre and for service in aggressive environments the Al–1Zn clad version to DTD 5070 would generally be preferred to the unclad metal.

Aluminium-Zinc-Magnesium Alloys

The Al–Zn–Mg alloy system provides a range of commercial compositions, primarily for those areas where strength is a major consideration. None of these alloys are yet included in the BS General Engineering series although their use in Europe and America is quite well established. Essentially the range of compositions may be conveniently divided into two categories.

The high strength alloys contain a Zn + Mg content well in excess of 6% and are used in specialist structures such as aircraft. The risk of stress corrosion cracking in these alloys may be accentuated by incorrect heat treatment or composition and they cannot be recommended for general use.

The other group of alloys are those with a Zn + Mg content not exceeding 6%. These have been used for general engineering, when natural ageing after welding can be utilised to permit the fabrication of strong welded structures. In particular, these medium strength Al–Zn–Mg alloys have been successfully used for transport applications and it seems probable that this will increase in the near future. With correct manufacturing procedures the risk of stress corrosion with these alloys is negligible and the resistance to unstressed corrosion is only slightly inferior to the Al–Mg–Si structural alloys.

Corrosion Behaviour in Aqueous Environments

Theoretical Considerations of Corrosion Behaviour

Aluminium is a very reactive metal with a high affinity for oxygen. The metal is nevertheless highly resistant to most atmospheres and to a great variety of chemical agents. This resistance is due to the inert and protective character of the aluminium oxide film which forms on the metal surface. In most environments, therefore, the rate of corrosion of aluminium decreases rapidly with time. In only a few cases, e.g. in caustic soda, does the corrosion rate approximate to the linear. A corrosion rate increasing with time is rarely encountered with aluminium, except in aqueous solutions at high temperatures and pressures.

The corrosion resistance of aluminium and its alloys is largely due to the protective oxide film which within seconds attains a thickness of about 10Å on freshly exposed metal; continuation of growth is markedly influenced by the environment, being accelerated by increasing temperature and humidity. Immersion in water results in rapid oxide thickening. The behaviour of the oxide may be modified by impurities or alloying additions. In aluminium–magnesium alloys the presence of magnesia in the oxide imparts a characteristic bloom to metal stored under humid conditions. The possible effects of minor impurities or additions is well illustrated in the case of tin, whose modifying effect upon the oxide[3] is utilised in obtaining a highly electronegative potential in aluminium sacrificial anodes (see Section 11.2).

The oxidation of aluminium at room temperature is reported to conform to an inverse logarithmic equation for growth periods up to 5 years duration[4]. At elevated temperatures, oxidation studies over shorter periods illustrate conformity to parabolic, linear and logarithmic relationships according to time and temperature. These kinetic variations are attributed to different mechanisms of film formation[5,6].

The various equilibria of the Al–H_2O system have been collated by Pourbaix *et al.* in a potential versus pH diagram (Fig. 4.1). This diagram

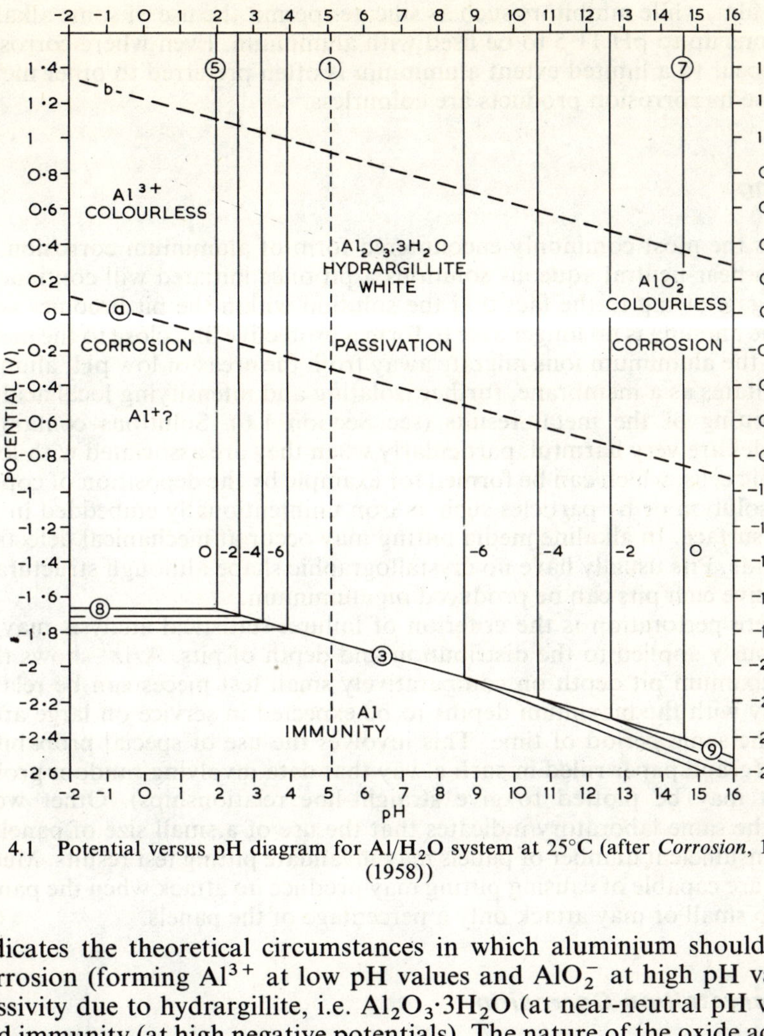

Fig. 4.1 Potential versus pH diagram for Al/H_2O system at 25°C (after *Corrosion*, **14**, 496t (1958))

indicates the theoretical circumstances in which aluminium should show corrosion (forming Al^{3+} at low pH values and AlO_2^- at high pH values), passivity due to hydrargillite, i.e. $Al_2O_3 \cdot 3H_2O$ (at near-neutral pH values) and immunity (at high negative potentials). The nature of the oxide actually varies according to temperature, and above about 75°C, boehmite ($Al_2O_3 \cdot H_2O$) is the stable form. It should be noted that the potential–pH diagram does not indicate one of the most important properties of aluminium, i.e. its ability to become passive in strongly acid solutions of high redox potential such as concentrated nitric acid (see also Section 1.4).

Characteristic Features of Corrosion Behaviour

General Dissolution

This occurs in strongly acid or strongly alkaline solutions, but there are specific exceptions. Thus in concentrated nitric acid the metal is passive and the kinetics of the process are controlled by ionic transport through the oxide film, while inhibitors such as silicates permit the use of some alkaline solutions up to pH 11·5 to be used with aluminium. Even where corrosion may occur to a limited extent aluminium is often preferred to other metals because its corrosion products are colourless.

Pitting

This is the most commonly encountered form of aluminium corrosion. In certain near-neutral aqueous solutions a pit once initiated will continue to propagate owing to the fact that the solution within the pit becomes acid, and the alumina is no longer able to form a protective film close to the metal. When the aluminium ions migrate away from the areas of low pH, alumina precipitates as a membrane, further isolating and intensifying local acidity, and pitting of the metal results (see Section 1.6). Solutions containing chlorides are very harmful, particularly when they are associated with local galvanic cells, which can be formed for example by the deposition of copper from solution or by particles such as iron unintentionally embedded in the metal surface. In alkaline media pitting may occur at mechanical defects in the oxide. Pits usually have no crystallographic shape although structurally indicative etch pits can be produced on aluminium.

Where perforation is the criterion of failure, statistical analysis may be judiciously applied to the distribution and depth of pits. Aziz[7] shows that the maximum pit depth on comparatively small test pieces can be related linearly with the maximum depths to be expected in service on large areas over the same period of time. This involves the use of special probability paper (graph paper ruled in such a way that data involving random probabilities may be plotted to give straight-line relationships). Other work from the same laboratory indicates that the use of a small size of panel or of an insufficient number of panels may invalidate pitting test results. Media which are capable of causing pitting may produce no attack when the panels are too small or may attack only a percentage of the panels.

Intercrystalline Corrosion

This is also electrochemical in nature, the galvanic cell being formed because of some heterogeneity in the alloy structure, which may arise from major or trace alloying additions or from minor elements present. In the aluminium–copper type alloys, precipitation of $CuAl_2$ particles at the grain boundaries leaves the adjacent solid solution anodic and prone to corrosion[8]. With aluminium–magnesium alloys the opposite situation occurs, since the precipitated phase Mg_2Al_3 is less noble than the solid solution. However, serious intercrystalline attack in these two alloys is not usual, provided that correct manufacturing and heat treatment conditions are observed.

In the case of the aluminium–magnesium system, most commercial alloys are usually supersaturated, so that elevated service temperatures and inexpert heat treatment are inadvisable, since any resultant grain boundary precipitation may induce susceptibility to intercrystalline attack. The extent of this susceptibility may be approximately deduced from the continuity of

Mg_2Al_3 at the boundaries, continuous or nearly continuous films being extremely detrimental and discrete widely spaced particles being relatively harmless.

Trace elements which adversely affect intercrystalline attack are normally controlled at a safe level. Copper is particularly pertinent in this respect since relatively small additions can cause a marked increase in intercrystalline attack in some alloy systems.

Stress Corrosion

This form of corrosion is of limited occurrence with only a few aluminium alloys[9], in particular the higher strength materials such as the Al–Zn–Mg–Cu type and some of the Al–Mg alloys, wrought and cast, with the higher magnesium contents, notably after specific low-temperature heat treatments such as occur during stove enamelling. Stress corrosion is intergranular on aluminium alloys (see Section 8.5).

Filiform Corrosion

This appears as a random non-branching white tunnel of corrosion product either on the surface of non-protected metal or beneath thin surface coatings. It is a structurally insensitive form of corrosion which is more often detrimental to appearance than strength, although thin foil may be perforated and attack of thin clad sheet (as used in aircraft construction) may expose the less corrosion resistant aluminium alloy core. Filiform corrosion is not commonly experienced with aluminium, as reflected by the insignificance afforded it in reviews on the phenomena[10].

Layer Corrosion

This may occur on material which has a marked fibrous structure caused by rolling or extrusion. The attack is rapid and very selective, forming partly detached layers of relatively uncorroded material. It is regarded by some authorities as a form of stress corrosion, the stress being either inherent in the metal or produced through the pressure of the larger volume of the corrosion product. It is rare, occurring mainly in copper-bearing alloys, but can occur in a number of environments, including some regarded as only mildly corrosive. Suitable adjustments of ageing treatments and copper content may largely overcome the effect in the higher-strength Al–Cu type alloys[11].

Effect of Composition

Few general statements can be made regarding the effect on corrosion resistance of alloying elements or impurities. A useful summary of the information has been prepared by Whitaker[12]. Copper is usually harmful causing increased susceptibility to intercrystalline or general attack, so that alloys containing copper should be regarded as less corrosion resistant than copper-free materials. There are however exceptions to this generalisation, such as an improved stress corrosion resistance in Al–Zn–Mg alloys obtained by a small copper addition[13, 14]. Alternatively, the presence of copper may be utilised to delay perforation at the expense of increased general corrosion.

With increasing purity of aluminium, greater resistance to corrosion is developed. On high-purity materials, however, any pits which develop are likely to be deeper though fewer in number than those formed in more impure metal. In some special applications, notably in contact with ammonia solutions or pure water at elevated temperatures and pressures, the iron and silicon present in commercial-purity metal are beneficial and retard corrosion. Up to about 5% magnesium improves the corrosion resistance to sea-water.

Bimetallic Corrosion

Aluminium is anodic to many other metals and when it is joined to them in a suitable electrolyte—which may even be a damp porous solid—the potential difference causes a current to flow and considerable corrosion can result. Corrosion is most severe when the resistance of the electrolyte is low, e.g. sea-water. In some cases surface moisture on structures exposed to an aggressive atmosphere can give rise to galvanic corrosion. In practice, copper, brasses, and bronzes in marine conditions cause the most trouble. The danger from copper and its alloys is enhanced by the slight solubility of copper in many solutions and its subsequent redeposition on the aluminium to set up active local cells. This can occur even when the copper and aluminium are not originally in contact, e.g. when water running over cuprous surfaces subsequently comes into contact with aluminium. Similarly, water washings from lead can cause pitting of aluminium. The controlling factor with lead and cuprous washings is the solvency of the water, so that soft waters are the most damaging in this respect. The successful utilisation of these metals in close proximity to aluminium, e.g. in plumbing and roofing, therefore requires careful design to avoid the transfer of a harmful solute to the aluminium.

Contact with steel, though less harmful, may accelerate attack on aluminium, but in some natural waters and other special cases aluminium can be protected at the expense of ferrous materials. Stainless steels may increase attack on aluminium, notably in sea-water or marine atmospheres, but the high electrical resistance of the two surface oxide films minimises bimetallic effects in less aggressive environments. Titanium appears to behave in a similar manner to steel. Aluminium–zinc alloys are used as sacrificial anodes for steel structures, usually with trace additions of tin, indium or mercury to enhance dissolution characteristics and render the operating potential more electronegative.

Aluminium may accelerate attack on zinc alloys; this is particularly noticeable when there is an unfavourable area ratio, as with galvanised fittings in aluminium sheets. In alkaline solutions, however, the aluminium may be preferentially attacked.

The copper-bearing aluminium alloys are more noble than most other aluminium alloys and can accelerate attack on these, notably in sea-water. Mercury and all the precious metals are harmful to aluminium.

Mechanical and Design Factors

Stress below the proof stress does not normally affect corrosion rates. Cyclic stresses in combination with a corrosive environment (corrosion fatigue) can produce failure at below the ordinary fatigue limit. Alloys susceptible to intergranular attack may corrode faster when stressed (see Section 8.5).

Soldered or brazed joints will usually have lower corrosion resistance than the parent metal, but sound welded joints with resistance to attack equal to that of the parent metal can be obtained in most alloys[2]. Many assemblies contain angles, pockets or crevices which attract moisture originating either from external sources or from condensation. The corrosion so caused could often be avoided by slight redesign of the assembly, the provision of drain holes of at least 5 mm dia., and the avoidance of horizontal surfaces being among the more important features. Crevices may be filled with jointing compounds. In static assemblies these compounds may be of the setting variety, but in assemblies subject to vibration or movement, as on ships, it is essential that the mastic used should not become too rigid as it might crack in service. It is advisable to incorporate chromates in jointing compounds to inhibit attack by any moisture that may penetrate.

Corrosion in Natural Environments

Atmospheric

The aluminium alloys as a group weather outdoors to a pleasant grey colour, which deepens to black in industrial atmospheres. Superficial pitting occurs initially but gradually ceases, being least marked on high-purity aluminium. With some alloys, including the copper-bearing alloys and the medium-strength Al–Zn–Mg alloys, additional protection, e.g. painting, is desirable in the more aggressive atmospheres to avoid any risk of intercrystalline corrosion.

Gases such as hydrogen sulphide and carbon dioxide do not increase the corrosivity of the atmosphere towards aluminium[15]. Service experience extends over 70 years and includes such well-known examples as Eros, Piccadilly Circus, London, which is in excellent condition, although cast in a low purity (98%) aluminium, and a cupola of San Gioacchino, a church in Rome which was covered in 1897 with sheet 1·25 mm thick and now shows attack to a depth of less than 0·13 mm. Twenty-year tests at selected marine, industrial and rural sites in the U.S.A.[16] have shown that the greater part of the attack takes place in the first year or two and that thereafter the rate of attack maintains a low value. Results from typical environments are shown in Fig. 4.2, and it is apparent that clad alloys give the best results. The relatively high percentage strength losses are due to the extremely thin test specimens. After 20 years the average measured depth of attack for an aluminium–copper alloy at a sea coast test site did not exceed 0·15 mm. The falling-off in rate of pitting with time is in sharp contrast to the behaviour of the older-established structural metals which have a fairly uniform corrosion rate throughout their life, and indicates that the relative merit of aluminium increases with scheduled life.

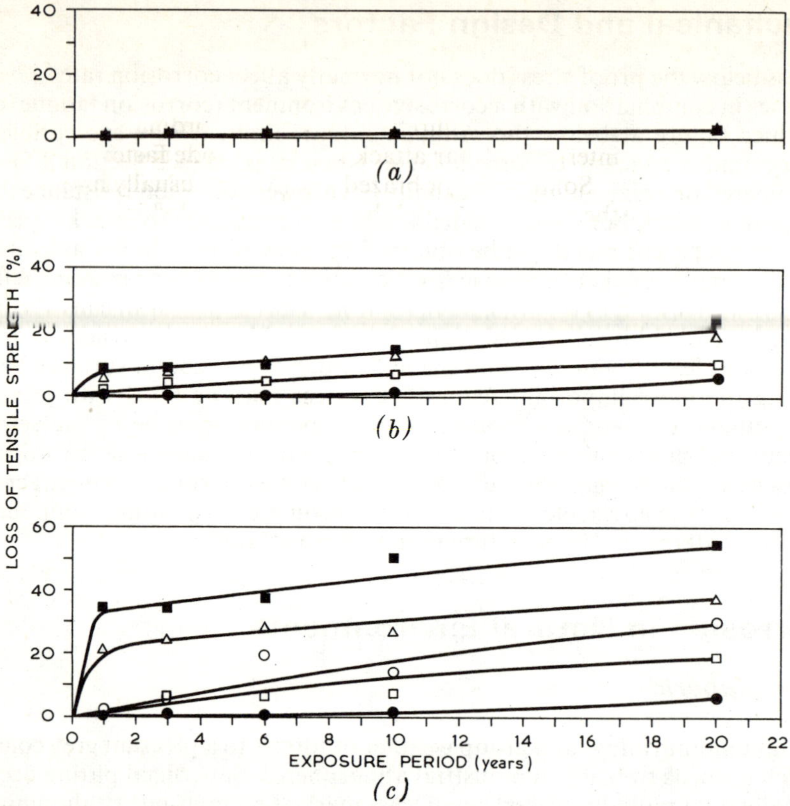

Fig. 4.2 A.S.T.M. 20-year corrosion tests:

- ■ 2017–T3
- □ 3003–H14
- △ 6051–T4
- ○ 1100–H14
- ● Alclad 2017–T3

(*a*) State College, Pa. (rural). Premachined tension specimens 0·89 mm thick. Curves for 1100–H14, 3003–H14 and Alclad 2017–T3 fall below curve shown; (*b*) New York, N.J. (industrial); (*c*) La Jolla, Calif. (seacoat) (after A.S.T.M. Symposium on Atmospheric Corrosion of Non-ferrous Metals, 27 (1955))

Aggressive environments include marine conditions and particularly industrial atmospheres containing high concentrations of acid gases such as sulphur dioxide; rain washing is beneficial in both environments, while dampness and condensation alone can accentuate the rate of attack in the presence of chlorides and acidic sulphates.

The relative aggressivity of industrial, marine and rural conditions has been clearly demonstrated by the results of seven year tests in the U.S.A. and British Isles[17], and in this work the benefit from rain washing was especially manifest for the industrial sites in the British Isles (Fig. 4.3).

The combination of acidic sulphates and condensation experienced in some industrial conditions, can cause a particularly voluminous loose corrosion product on some alloys, such as NS3. Where this is likely to be troublesome, cladding with high purity metal is recommended and has been successfully employed, for example on the underside of aluminium-roofed industrial buildings.

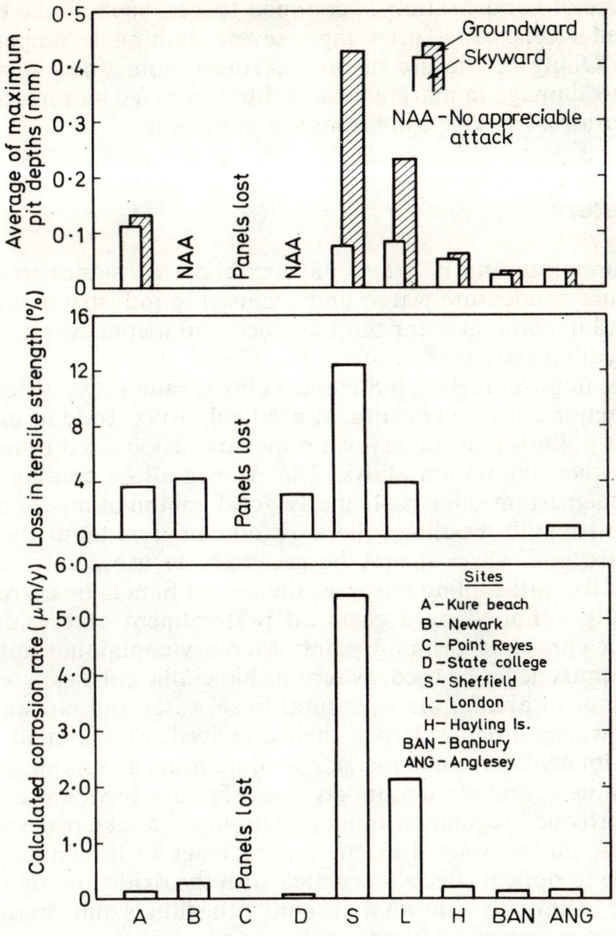

Fig. 4.3 Comparison of weathering effects at United States and English sites—1199–H18 (and S1–H) alloy—7 year data (after *Metal Corrosion in the Atmosphere*, A.S.T.M. Pub. STP435, 151 (1967))

While the continual removal of atmospheric pollution by rain washing is beneficial, the removal of the protective corrosion product is obviously undesirable. The retention of the weathered surface is therefore usually preferred unless aesthetic considerations are of major importance, in which case abrasive or specialist chemical cleaning are effective.

In urban areas, atmospheric fall-out of carbon from partially burned fuel can cause severe localised pitting by galvanic action, although this is not commonly encountered.

Indoors, aluminium retains its appearance well, and even after prolonged periods may show no more than slight dulling or on aluminium–magnesium alloys a slight bloom. This superficial deterioration can be accelerated by the presence of moist conditions and condensation which in extreme cases may lead to staining.

The presence of condensation in confined spaces, such as the turns of a coil or stacked sheets, can cause a more severe staining accompanied by a thick bloom. Oiling or the use of interleaving is sometimes successful in preventing this damage in marginal cases, but improved storage conditions or the elimination of crevice conditions are preferable.

Natural Waters

Immersed aluminium and its alloys have excellent resistance to attack by distilled or pure condensate water, and are used in industry in condensing equipment and in containers for both distilled and deionised water, as well as in steam-heating systems[18].

Of the more important British Standard alloys, only those which contain copper as a major alloying constituent are likely to corrode in unpolluted sea-water, but pollution of the sea-water may cause localised pitting attack to occur on other aluminium alloys. The Al–Mg alloys containing up to about 4·5% magnesium offer particularly good combinations of corrosion resistance and strength. Fouling collects readily on aluminium alloys, as on other materials, and where it may be necessary to use paints containing cuprous oxide for anti-fouling purposes the risk of bimetallic corrosion can be substantially inhibited by a chemical pretreatment of the aluminium followed by a chromate priming paint. Mercury-containing anti-fouling compositions must never be used, as serious bimetallic corrosion will result.

The behaviour of aluminium in natural fresh water and tap waters may vary as these waters differ widely in their dissolved solid content. No corrosion occurs immediately on immersion of aluminium and its alloys in these near-neutral waters, and aluminium gives satisfactory service with all types of tap water provided regular cleaning and drying can take place, as occurs with aluminium hollow-ware. In some waters, black or brown stains which are largely due to optical effects associated with the oxide film on the metal surface, occur. Although somewhat unsightly, the film is quite harmless and can be removed by simple methods such as boiling of fruit (e.g. rhubarb). Alternatively, preliminary boiling with pure water provides some protection against the staining, but can hardly be considered justifiable in most cases.

The combination of carbonate, chloride and copper is more damaging than if they are present singly or if one of them is absent[19, 20], so that some supply waters are naturally more aggressive than others. The role of copper is of particular relevance, since as little as 0·02 parts per million can initiate pitting in hard waters[21], although more is required in soft waters which are otherwise less aggressive. In this context however it must be remembered that soft waters are inherently more cupro-solvent than hard waters; consequently the conjoint use of aluminium and copper fittings is rarely advisable irrespective of the necessity for avoiding galvanic interaction when the two metals are in direct contact.

Once pitting has started it may continue in solutions which would themselves be incapable of initiating corrosion. In waters of all types, the rate of increase in the depth of pitting falls off rapidly with time. Water movement (of the order of 0·3 m/s or more) will reduce pitting or prevent its initiation. A rise in temperature tends to lead to higher corrosion rates at existing pits, but even with the most aggressive hard waters, above about 50°C the oxide-forming mechanisms act to prevent the initiation of pitting, as shown by the long and satisfactory service given by aluminium hollow-ware which is assisted in some waters by scale formation.

Where aluminium is to be used in direct contact with cold natural waters with no possibility of regular cleaning, clad aluminium alloys are the preferred materials. An Al–1·2Mn alloy clad with Al–1·2Zn is suitable. The cladding is anodic to the core and corrosion is therefore restricted to the surface cladding, thus obviating the risk of perforation. Cladding with super-purity aluminium is preferable where it is important to have the minimal degree of total corrosion, but in this case the potential relationship with the core is more critical and in some circumstances the cladding can actually become cathodic. Sacrificial protection may also be obtained from sprayed coatings of appropriate composition which can be applied to extrusions and castings as well as to sheet, rod, plate and tubes. In practice, unclad aluminium–manganese alloys have been used for piping soft waters in this country and, more widely, in the USA.

Underground Corrosion by Soils

This is largely related to the presence of moisture which can leach out soluble constituents from the soil. As is the case with natural waters, the nature of the corrosive environment is a more important factor than the alloy used, provided that copper-bearing alloys are avoided. At the present time it is impossible to produce a satisfactory classification of soils in respect of their aggressive action on aluminium alloys. Made-up ground, particularly when it includes cinders, is usually extremely corrosive, while neutral clays are often least corrosive. It is desirable that protection should be given to all aluminium materials buried in soils[22,23] except where there is previous experience of satisfactory service from aluminium in a given soil. Pipe wrappings based upon bitumen or chromates are effective, while for cable sheathing a continuous plastic coating provides both electrical and corrosion protection. Cathodic protection has been utilised for pipelines[24,25] but is not widely practised; close control is necessary since over-protection can result in alkali attack. Potentials in the region of −1·0 V vs saturated $Cu/CuSO_4$ are favoured, although some divergence of opinion exists in this respect.

Corrosion in Chemical Environments

Detailed information about behaviour of specific chemicals is given in several works of reference[25–30].

Acids

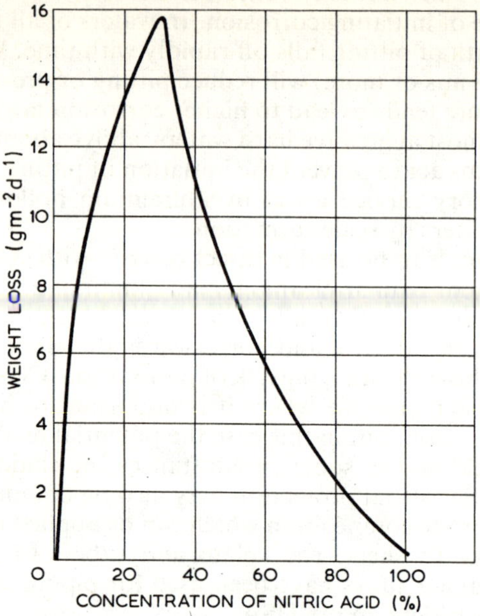

Fig. 4.4 Action of nitric acid of various concentrations on commercial-purity aluminium at 20°C (after Reference 26)

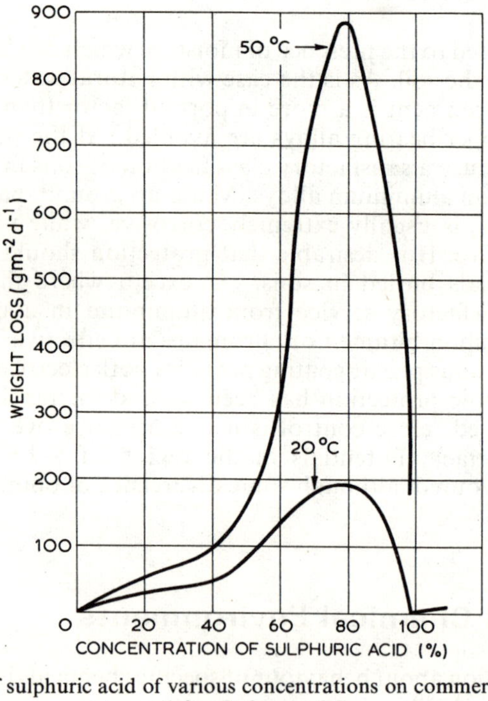

Fig. 4.5 Action of sulphuric acid of various concentrations on commercial-purity aluminium (after Reference 26)

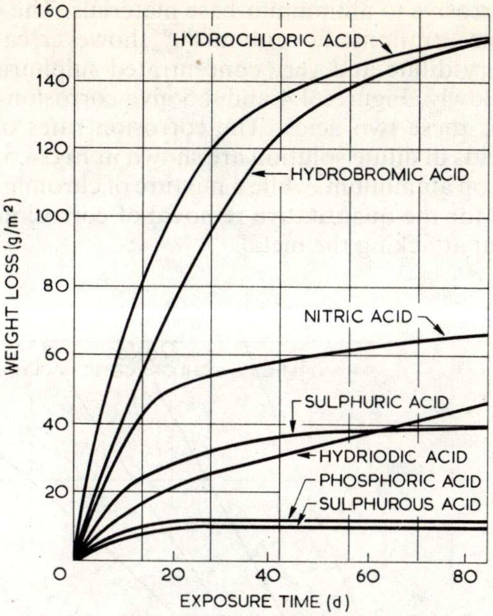

Fig. 4.6 Action of dilute (0·1 N) solutions of inorganic acids on commercial-purity aluminium at 25°C (after Reference 26)

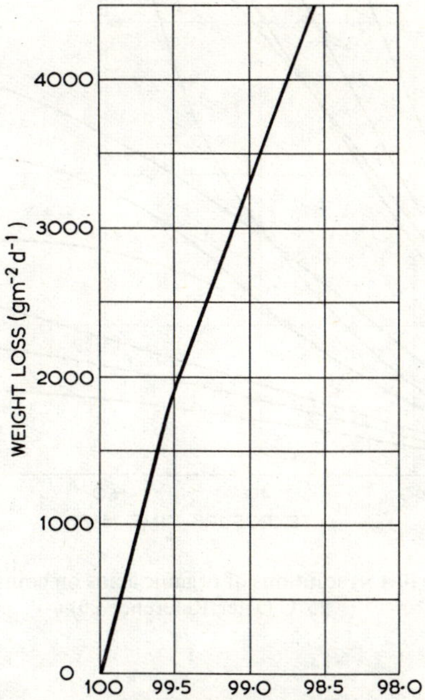

Fig. 4.7 Action of 40% hydrochloric acid on aluminium of various purities at 20°C (after Reference 26)

Most acids are corrosive to aluminium-base materials. The oxidising action of nitric acid at concentrations above about 80%, however, causes passivation of aluminium. Very dilute and very concentrated sulphuric acid dissolves aluminium only slowly. Figures 4.4 and 4.5 give corrosion data at various concentrations for these two acids. The corrosion rates of aluminium in other inorganic acids in dilute solution are shown in Fig. 4.6. Boric acid also exerts little attack on aluminium, while a mixture of chromic and phosphoric acids can be used for the quantitative removal of corrosion products from aluminium without attacking the metal.

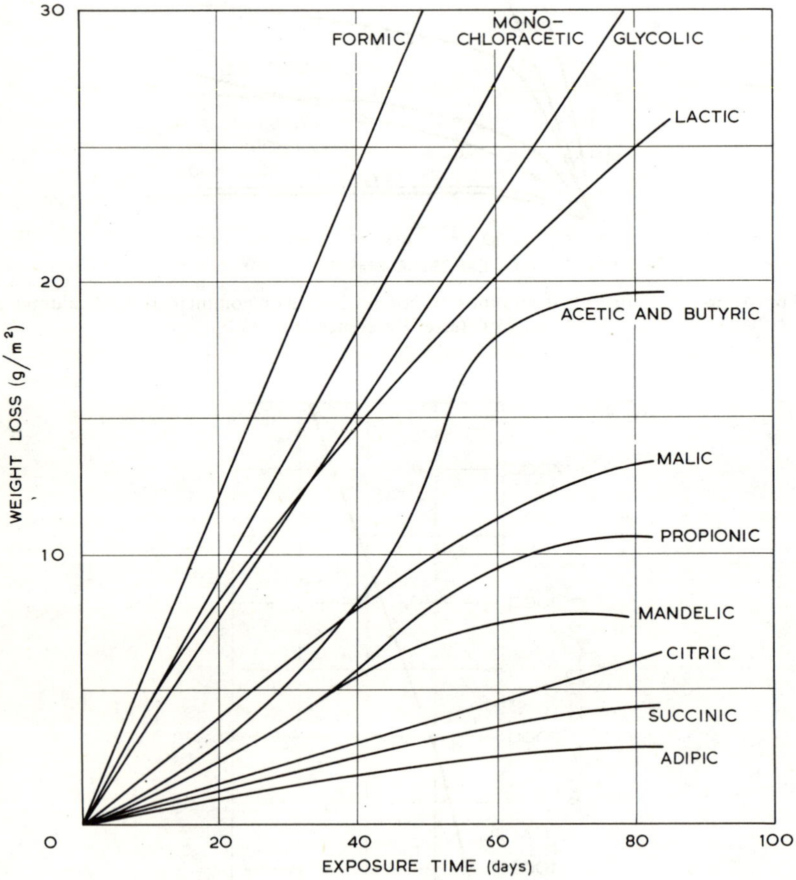

Fig. 4.8 Action of dilute (0·1 N) solutions of organic acids on commercial-purity aluminium at 25°C (after Reference 26)

The effect of commercial metal purity (impurities mainly iron and silicon) on corrosion by 40% hydrochloric acid is shown in Fig. 4.7. This curve is typical of that obtained with many acids.

Organic acids usually have low rates of attack on aluminium, notable exceptions to this generalisation being formic acid, oxalic acid and some chloride-containing acids such as trichloroacetic acid. Corrosion rates for dilute organic acid solutions are given in Fig. 4.8. Glacial acetic acid (pH 3) has no significant corrosive effect on aluminium but the rate of attack increases rapidly with decreasing concentration or in the absence of the traces of water normally present. The rate of corrosion in an acid solution rises rapidly with temperature, often doubling or more with a 10°C rise.

Alkalis

Alkalis are generally corrosive to aluminium; caustic soda is in fact used for chemical milling of aluminium. 99·0% aluminium is, however, resistant to ammonium hydroxide, even at pH 13, while the action of more dilute caustic alkalis can be inhibited by the use of silicates. Mild alkalis such as sodium carbonate are moderately corrosive and are not recommended for washing aluminium hollow-ware. Synthetic detergents, in general, give satisfactory service in cleaning aluminium, but those containing uninhibited sodium carbonate may give some surface roughening. Inhibitors such as silicates can prevent attack by the more dilute solutions.

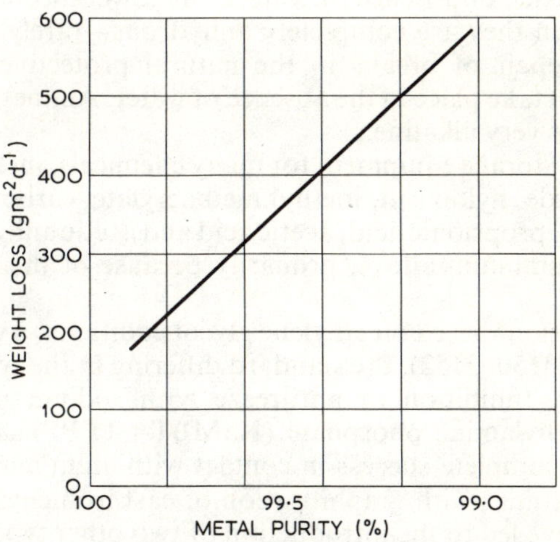

Fig. 4.9 Action of 5·6% potassium hydroxide solution on aluminium of various purities at 20°C (after Reference 26)

Alloys of aluminium with magnesium or magnesium and silicon are generally more resistant than other alloys to alkaline media. The corrosion rate in potassium and sodium hydroxide solutions decreases with increasing purity of the metal (Fig. 4.9), but with ammonium hydroxide the reverse occurs.

Inorganic Salts

Most simple inorganic salt solutions cause virtually no attack on aluminium-base alloys, unless they possess the qualities required for pitting corrosion, which have been considered previously, or hydrolyse in solution to give acid or alkaline reactions, as do, for example, aluminium, ferric and zinc chlorides. With salts of heavy metals—notably copper, silver, and gold—the heavy metal deposits on to the aluminium, where it subsequently causes serious bimetallic corrosion.

Some salts, notably chromates, dichromates, silicates, borates and cinnamates, have marked inhibitive power and are very effective in closed circuit water systems. Care must be taken to ensure that a sufficient quantity of such anodic inhibitors as chromates is added, as otherwise attack, though occurring at fewer points, may be more severe at these points. Chromates and dichromates have little inhibitive power in strongly acid solutions.

Aluminium is used in hydrogen peroxide (H.T.P.) processing and storage equipment partly because of its high corrosion resistance but also because it does not cause degradation of the peroxide.

Organic Compounds

With many organic compounds, aluminium shows high corrosion resistance either in the presence or absence of water. The lower alcohols and phenols are corrosive when they are completely anhydrous—rarely encountered in practice—since repair of breaks in the natural protective oxide film on aluminium cannot take place in the absence of water. Amines generally cause little attack unless very alkaline.

Processing and storage equipment for many chemicals, including acetaldehyde, formaldehyde, nylon salt, methyl methacrylate, carbon tetrachloride, glycerol, triacetin, proprionic acid, acetic acid and acetic anhydride, is manufactured from aluminium alloys, primarily because of their excellent corrosion resistance.

Antifreeze solutions based on ethylene glycol additions to water have been standardised (BS 3150–3152), the standard differing in the type of corrosion inhibitor present. Inhibition of antifreeze with sodium mercaptobenzothiazole and triethylamine phosphate (NaMBT+TEP) has been used for many years with complete success in contact with aluminium, e.g. in aero-engines, but difficulties with graphitisation of cast iron engine components in the solution have led to the introduction of two other types of inhibitors: (*a*) benzoate plus nitrite, and (*b*) borax, usually with soluble oils. Service experience has indicated that corrosion of aluminium components in these inhibited solutions occasionally takes place, though most trials give satisfactory results.

In refrigerating systems, halogen derivatives of methane and ethane marketed under the trade names of Arctons and Freons are without action on pure aluminium and its copper-free alloys in dry conditions, but in wet conditions monochlorodi-, dichloromono- and trichloromonofluoromethanes can hydrolyse to produce slight attack on the aluminium.

Aluminium has good resistance to petroleum products, and an Al–2Mg alloy is used for tank heating coils in crude-oil carriers. Caked-on deposits must be removed from the coils by hot sea-water cleaning in order to maintain effective heat transfer and prevent corrosion. Aluminium is also used in the petroleum industry for sheathing for towers, heat exchangers, transport and storage tanks and scrubbers. Many industries use aluminium alloys for heat exchangers, clad alloys being used where pitting corrosion is liable to be initiated by one of the contacting materials. Heat exchangers in the gas industry have utilised duplex tubes, with aluminium on the water side and steel on the gas side in cases where aluminium is unsuitable owing to the presence of catechol which can attack it.

Aluminium does not become brittle at low temperatures and for this reason (and because of its corrosion resistance) it has been adopted for the carriage and storage of liquefied methane.

High-temperature Corrosion

Dry Atmospheres

When exposed at high temperatures in dry atmospheres aluminium is highly resistant to corrosion by most of the common gases, other than the halogens or their compounds.

High-temperature Aqueous Systems

When aluminium corrodes at temperatures below 90°C in aqueous systems, attack is usually by pitting. At temperatures between 90 and 250°C (for the attainment of which considerable pressures are needed) uniform attack is the commonest form of aqueous corrosion. Above about 250°C, uniform attack is merely the prelude to highly destructive intergranular attack. The corrosion products from the uniform attack form a film which includes a barrier layer and a bulk film analogous to those formed during anodising (Section 16.1); it is the bulk film which controls the corrosion rate, which is not significantly affected by most common dissolved ions[31]. The onset of intergranular attack occurs at about the same time as the crystallisation of the amorphous barrier layer oxide. Kinetic studies indicate that over the temperature range from 100 to 363°C the oxidation rate law is successively inverse logarithmic, parabolic and linear[32].

The requirements of nuclear energy application fostered an interest in special alloys for service in high temperature aqueous environments, but their utilisation has not been widespread. Encouraging results have been reported for alloys of 2Ni–0·5Fe[33] and 1·2Ni–1·8Fe[34].

It has been suggested that the role of nickel (as $NiAl_3$) is to provide sites of low hydrogen overvoltage, where cathodically liberated hydrogen may be liberated without disrupting the protective oxide[35]. The distribution of such sites is apparently critical however, since high corrosion resistance is associated with a fine dispersion of the second phase, while the electronic conductivity of the film is probably also important[36].

Steam forms a protective white film at temperatures up to about 250°C, but above this temperature steam can, under some conditions, react with aluminium progressively to form aluminium oxide and hydrogen. Sintered aluminium powder (S.A.P.) has relatively good resistance to steam at 500°C, but at about 300°C an addition of 1% nickel to the S.A.P. is needed to prevent rapid disintegration.

Molten Salts and Metals

Aluminium-base alloys resist the action of many molten salts which are nearly neutral in reaction. Molten sodium nitrate or mixtures of sodium nitrate and potassium nitrate are used for salt bath heat treatment of some aluminium alloys.

Molten metals generally attack aluminium alloys. Both zinc and tin form alloys by dissolution of aluminium, although the aluminium does not melt. Molten lead is inert to aluminium, and molten lead baths can be used for heating aluminium alloys. Mercury, molten at room temperature, amalgamates readily with aluminium alloys if their naturally formed oxide films are temporarily removed by scratching, and rapid corrosion occurs on subsequent exposure to moist air or water. Under stressed conditions cracking will frequently result, since mercury penetrates into the aluminium alloy selectively at grain boundaries. Contact of aluminium with mercury is extremely dangerous and severe corrosion can occur with a very small amount of mercury.

Aluminium in Contact with other Materials

The aluminium alloys recommended for building purposes (not including the high strength alloys containing copper) have good resistance to concretes, mortars, plasters and asbestos cement products. When freshly mixed, some of these materials release traces of alkaline products which may be sufficient to stain aluminium or to etch it slightly. As soon as the mixture is set, however, the attack ceases and even after many years' service, attack on embedded aluminium is found to be negligible[37, 38].

With cement and concrete under continually wet conditions, there may be some surface attack. This decreases rapidly with time, and the strength of components is not significantly affected. Under embedment conditions, bituminous protection is advisable, to avoid risk of cracking of the concrete due to stresses set up owing to the bulk of the corrosion product. Plasters are generally even less aggressive than Portland cement. In damp environments, some corrosion of aluminium may arise in contact with the more open-grained building stones and brickwork, but the hard stones, such as granite, are inert. With building stone and brickwork, as with soils (Section 10.3), it is the nature of the products which can be leached out which will determine whether aluminium corrodes. Unprotected aluminium, in the form of nails for example, can be satisfactorily used in contact with precast concrete blocks, which are usually non-corrosive to aluminium. Magnesium oxychloride compositions (used for flooring), on the other hand, stimulate corrosion of aluminium under moist conditions, as will many insulation materials based on magnesium and calcium silicates.

Plastics are generally without action on aluminium and are widely used to provide insulation between other metals and aluminium, while the use of aluminium/plastic laminates is increasing. Rubber has no effect upon aluminium.

A few acid woods, such as oak, chestnut and Western red cedar, accelerate surface weathering of aluminium, but do not usually give rise to serious attack[39]. Timber preservatives containing soluble copper compounds should be avoided; creosote and zinc napthenate are satisfactory preservatives for wood in contact with aluminium.

Common packaging materials are a potential source of aggressive substances[40], and careful selection is recommended to avoid surface deterioration. Where paper is in contact with aluminium, the chloride content should be below 0·05%, sulphate content below 0·25%, copper content below 0·01% and the pH of aqueous extracts in the range pH 5·5–7·5, in order to avoid corrosion in damp conditions. Papers and felts used in building applications should also conform to this specification as a minimum requirement and be of the highest quality, since metallic copper found in materials of inferior origin can result in severe local galvanic attack of aluminium.

Tarpaulins are sometimes treated with copper-containing preservatives and water leached from these has been found to cause corrosion of underlying aluminium sheets.

Fibreglass insulation produced from soda glass can cause pitting in conditions where leaching of alkali occurs, for example, by condensation: the use of Fibreglass produced from Pyrexglass is therefore preferred. Common putties of whiting/linseed oil composition do not attack aluminium; adhesion is obtained by allowing the metal surface to weather, or by applying an etch primer treatment to the metal. Both thermosetting adhesives (e.g. the phenolic types) and thermoplastic adhesives (such as paraffin and microcrystalline waxes, or bitumen) are non-corrosive to aluminium. In general, adhesives applied to aluminium should not contain chlorides in excess of 0·05% (as NaCl) of the solid content, and should be free from copper- or mercury-containing anti-fungicides. The presence in the adhesive of borax or sodium silicate is beneficial when one of the adhesive components is of an acid character.

J. C. BAILEY
F. C. PORTER
A. W. PEARSON

REFERENCES

1. *The Properties of Aluminium and its Alloys*, 6th edn, The Aluminium Federation, Birmingham (1968)
2. Blewett, R. V. and Skerry, E. W., *Metallurgia*, **71**, 73 (1965)
3. Keir, D. S., Pryor, M. J. and Sperrey, P. R., *J. Electrochem. Soc.*, **114**, 777 (1967)
4. Godard, H. P., *J. Electrochem. Soc.*, **114**, 354 (1967)
5. Aylmore, D. W., Gregg, S. J. and Jepson, W. B., *J. Inst. Met.*, **88**, 205 (1959–60)
6. Bartlett, R. W., *J. Electrochem. Soc.*, **111**, 903 (1964)
7. Aziz, P. M. and Godard, H. P., *Corrosion*, **12**, 495t (1956)
8. Hunter, M. S., Frank, G. R. and Robinson, D. L., *2nd International Congress on Metallic Corrosion*, 66 (1963)
9. Champion, F. A., *J. Inst. Met.*, **83**, 385 (1954–55)
10. Barton, J. F., *Paint Manufacture*, Nov., 53 (1964) and Dec., 47 (1964)

11. Bell, W. A. and Campbell, H. S., *J. Inst. Met.*, **89**, 464 (1960–61)
12. Whitaker, M. E., *Metal Ind.*, 80, 183, 207, 227, 247, 263, 288, 303, 331, 346, 387 (1952)
13. Chadwick, R., Muir, N. B. and Grainger, H. B., *J. Inst. Met.*, **85**, 161 (1956–57)
14. Busby, J., Cleave, J. F. and Cudd, R. L., *J. Inst. Met.*, **99**, 41 (1971)
15. Aziz, P. M. and Godard, H. P., *Corrosion*, **15**, 529t (1959)
16. *Symposium on Atmospheric Corrosion of Non-Ferrous Metals*, Amer. Soc. Test. Mat., Special Technical Publication No. 175 (1956)
17. *Metal Corrosion in the Atmosphere*, Amer. Soc. Test. Mat., Special Technical Publication No. 435 (1968). Papers by McGeary, *et al.*, p. 141 and Ailor, J. R., p. 285
18. Symposium on Corrosion by High Purity Water, *Corrosion*, **13**, 151t (1957)
19. Davies, D. E., *J. Appl. Chem.*, **9**, 651 (1959)
20. Rowe, L. C. and Walker, M. S., *Corrosion*, **17**, 353t (1961)
21. Porter, F. C. and Hadden, S. E., *J. Appl. Chem.*, **3**, 385 (1953).
22. Gilbert, P. T. and Porter, F. C., *Iron and Steel Inst.*, Special Report No. 43, 33–74 (1951)
23. Raine, P. A., *Chem. and Ind.* (Rev.), 1102, 1196 (1956)
24. Sprowls, D. O. and Carlisle, M. E., *Corrosion*, **17**, 125t (1961)
25. *Das Chemische Verhalten von Aluminium*, Aluminium-Verlag GmbH, Dusseldorf (1955)
26. *Aluminium in the Chemical and Food Industries*, British Aluminium Co. Ltd., London (1959)
27. *Aluminium with Food and Chemicals*, Alcan Booth Industries Ltd. (1966)
28. Ritter, F., *Korrosionstabellen Metallischer Werkstoffe*, Springer-Verlag, Vienna (1944)
29. *Aluminium with Food and Chemicals*, The Aluminium Association, New York (1967)
30. *Process Industries Applications of Alcoa Aluminium*, Alcoa, Pittsburgh, U.S.A.
31. Troutner, V. H., *Corrosion*, **15**, 9t (1959)
32. Dillon, R. L., *Corrosion*, **15**, 13t (1959)
33. Perryman, E. C. W., *J. Inst. Met.*, **88**, 62 (1959)
34. Dillon, R. L. and Bowen, H. C., *Corrosion*, **18**, 406t (1962)
35. Draley, J. E. and Ruther, W. E., *Corrosion*, **12**, 480t (1956)
36. Greenblatt, J. H. and Macmillan, A. F., *Corrosion*, **19**, 146t (1963)
37. Porter, F. C., *Metallurgia*, **65**, 65 (1962)
38. Jones, F. E. and Tarleton, R. D., *Effect of Embedding Aluminium and Aluminium Alloys in Building Materials*, National Building Studies Research Paper No. 36, H.M.S.O., London
39. Farmer, R. H. and Porter, F. C., *Metallurgia*, **68**, 161 (1963)
40. Scott, D. J. and Skerrey, E. W., *Br. Corros. J.*, **5**, 239 (1970)

4.2 Copper and Copper Alloys

Copper and copper alloys are amongst the earliest metals known to man, having been used from prehistoric times, and their present-day importance is greater than ever before. Their widespread use depends on a combination of good corrosion resistance in a variety of environments, excellent workability, high thermal and electrical conductivities, and attractive mechanical properties at low, normal and moderately elevated temperatures.

A wide range of cast and wrought alloys is available. For detailed expositions of properties and uses the reader is referred to publications on many specialised aspects obtainable from the Copper Development Association's offices in various countries. Relevant publications of the British Standards Institution include BS 1400, *Copper Alloy Ingots and Castings*[1] and BS 2870–5, *Copper and Copper Alloy Wrought Products*[2]. All standards of the American Society for Testing and Materials relating to copper and copper alloys are included in a volume published annually[3].

Composition and Properties

The mechanical properties of wrought alloys[4] depend on composition and metallurgical condition. At the extremes, annealed pure copper has a tensile strength of 180 MN m^{-2} and a hardness of 40 HV, and heat-treated beryllium copper can have a tensile strength of 1 300 MN m^{-2} and a hardness of 390 HV. Summaries of typical properties of some of the more important wrought and cast copper alloys are given in Tables 4.9 and 4.10.

Coppers The purest grade of copper commercially available, and that with the highest electrical conductivity, is oxygen-free high-conductivity copper. The minimum copper content required by some specifications is 99·99%, and the method of manufacture is such that no residual deoxidant is present. Oxygen itself has very little effect on conductivity, and the 'tough pitch' coppers (either electrolytic or fire-refined), containing about 0·04% oxygen, are high-conductivity materials.

One disadvantage of tough pitch coppers is the embrittlement that is liable to occur when they are heated in atmospheres containing hydrogen. For many purposes, therefore, and particularly where fabrication is involved, deoxidised coppers are preferred. The usual deoxidising agent is phosphorus, and specifications require residual phosphorus contents of between 0·004

Table 4.9 Typical properties of wrought alloys

Alloy	*Density* (g/cm^3)	*Melting pt.* (*liquidus*, °C)	*Coefficient of expansion* × 10^6	*Electrical conductivity* % I.A.C.S.	*Thermal conductivity* (W/m°K)	*Tensile strength* (MN/m^2)	*Elongation* %	*Vickers hardness*
H.C. copper	8·94	1 083	18	103	390	180–340	10–60	40–110
Deoxidised non-arsenical copper	8·93	1 082	18	80	340	180–340	10–55	40–120
Arsenical copper	8·93	1 080	17	45	175	220–350	10–55	40–125
Tellurium copper	8·93	1 075	18	96	360	230–320	10–50	40–110
Beryllium copper	8·2	955	18	23	85	500–1 300	2–40	110–390
85/15 brass	8·74	1 025	19	35	155	280–540	8–70	65–170
70/30 brass	8·53	955	20	27	125	280–600	5–75	55–180
60/40 brass	8·38	905	21	29	125	370–600	5–45	75–180
Aluminium–brass	8·33	980	19	23	100	320–700	6–75	70–200
Naval brass	8·41	890	21	25	110	370–620	5–45	75–180
H.T. brass	8·35	890	21	23	105	520–770	8–35	90–220
15% nickel silver	8·69	1 060	16	7	35	350–700	4–55	70–220
30% nickel silver	8·87	1 190	17	5	20	390–700	4–50	90–220
5% tin–bronze	8·89	1 050	18	17	80	340–740	5–70	70–220
12% tin–bronze	8·70	995	19	8	50	460–830	5–65	110–250
Silicon–bronze	8·52	1 030	18	8	40	260–630	5–75	60–200
7% aluminium–bronze	7·95	1 050	18	15	80	430–770	4–65	80–210
90/10 cupro–nickel	8·91	1 150	16	10	50	310–620	8–55	90–200
70/30 cupro–nickel	8·94	1 240	16	5	30	370–700	5–55	95–210

Table 4.10 Properties of cast alloys

Alloy	*Min. tensile strength* (MN/m²)	*Min. elongation* (%)
10Sn–0·2P	230	7
10Sn–10Pb	190	5
10Sn–2Zn	260	15
5Sn–5Zn–5Pb	200	15
9·5Al–2Fe	490	20
Silicon–bronze	310	15
30Zn–2Pb	190	12
Naval brass	310	20
H.T. brass (up to 2·5Al)	460	20
H.T. brass (up to 5·0Al)	590	15
H.T. β brass	740	12

and 0·05%. Phosphorus-deoxidised coppers commonly have electrical conductivities about 80% of those of pure copper.

Arsenical coppers containing about 0·4% arsenic (tough pitch or deoxidised) are used where increased strength at elevated temperatures is required. Additions of cadmium (1·0%), chromium (0·5%), and silver (0·1%) also give improved high-temperature properties, but without any serious loss of electrical conductivity. Tellurium (1·0%) gives improved machinability. An addition of about 2% beryllium gives a heat-treatable alloy that can develop extremely high strength.

Brasses Brasses are basically alloys of copper and zinc, containing between about 10 and 45% Zn, but many other additions are made and the resulting alloys are the most complicated of all the copper-base series. The single-phase (α) brasses, containing up to about 37% Zn in the binary alloys, may have additions of 1% Sn (Admiralty brass), 2% Al (aluminium–brass), or 1–2% Pb for ease of machining. Duplex (α–β) brasses containing more than 37% Zn, may have additions of 1% Sn (Naval brass), or 1–3% Pb to assist machining. Both α and α–β brasses, with and without lead, are used in the cast as well as the wrought form.

High-tensile brasses are α–β (or, occasionally, β) alloys containing up to 5% Al and 1–2% of one or more of the following: Sn, Pb, Fe, Mn. These alloys also are used in both wrought and cast form.

Copper/nickel alloys Alloys containing 5–30% Ni, used mostly in the wrought condition[5], have a very good combination of properties. For optimum corrosion resistance, additions of 0·5–2·0% each of Fe and Mn are made.

Tin-bronzes and gunmetals Alloys containing 3·0–12·5% Sn and 0·02–0·4% P, known as *phosphor-bronzes*, are widely employed. Cast as well as wrought alloys are used, and cast leaded bronzes are also available.

Gunmetals are alloys of copper, tin and zinc, with or without lead, used in the cast condition. Commonly used alloys are (*a*) 10Sn–2Zn, and (*b*) 5Sn–5Zn–5Pb.

Aluminium-bronzes Aluminium-bronzes usually contain 5–10% Al, the structure being duplex when more than about 8% Al is present. Plain Cu–Al alloys are sometimes used, but wrought (single-phase) alloys may have additions of about 0·25–2% of one or more of the following: Ni, Fe, Mn, Ag, Sn, As. Cast alloys of high strength and complex structure usually contain about 10% Al and additions of Fe, Mn and Ni.

Silicon-bronzes Silicon-bronzes usually contain 1·5–3% Si and 0·5–1% Mn[6]. They are used in wrought or cast form, though the cast alloys may also contain some Zn and Fe.

Nickel silvers These wrought alloys consist essentially of Cu, Zn and Ni, with Ni in the range 10–30%. Leaded nickel brasses are also used, usually where some machining is involved.

General Considerations of Corrosion Behaviour

Copper is the first member of Group IB of the periodic table, having atomic number 29 and electronic configuration 2.8.18.1. Loss of the outermost electron gives the cuprous ion Cu^+, and a second electron may be lost in the formation of the cupric ion Cu^{2+}.

Copper occurs in the uncombined state in nature and is relatively easily obtained by the reduction of its compounds. It is not very active chemically and oxidises only very slowly in air at ordinary temperatures.

In the electrochemical series of elements, copper is near the noble end and will not normally displace hydrogen, even from acid solutions. Indeed, if hydrogen is bubbled through a solution of copper salts, copper is slowly deposited (more rapidly if the process is carried out under pressure). (See Section 1.2 for thermodynamic considerations.)

As copper is not an inherently reactive element, it is not surprising that the rate of corrosion, even if unhindered by films of insoluble corrosion products, is usually low. Nevertheless, although the breakdown of a protective oxide film on copper is not likely to lead to such rapid attack as with a more reactive metal such as, say, aluminium, in practice the good behaviour of copper (and more particularly of some of its alloys) often depends to a considerable extent on the maintenance of a protective film of oxide or other insoluble corrosion product.

Many of the alloys of copper are more resistant to corrosion than is copper itself, owing to the incorporation either of relatively corrosion-resistant metals such as nickel or tin, or of metals such as aluminium or beryllium that would be expected to assist in the formation of protective oxide films. Several of the copper alloys are liable to undergo a selective type of corrosion in certain circumstances, the most notable example being the *dezincification* of brasses. Some alloys again are liable to suffer stress corrosion by the combined effects of internal or applied stresses and the corrosive effects of certain specific environments. The most widely known example of this is the *season cracking* of brasses. In general brasses are the least corrosion-resistant of the commonly used copper-base alloys.

The various grades of copper available do not differ to any marked extent in their corrosion resistance, and a choice is usually based on other grounds. Subsequent references to the corrosion behaviour of copper may therefore be taken to apply broadly to all types of copper.

The choice of alloy for any particular application is determined by the desired physical, mechanical and metallurgical properties. Within these limits, however, a range of materials is usually available. It is essential that at the very earliest stage the choice of materials and the details of design of the installation should be considered from the point of view of corrosion, if the best performance is to be obtained in service. This is particularly true of copper alloys, where protective measures are not normally applied.

Several books contain general summaries of the corrosion behaviour of copper and its alloys[7–15] and the formation of copper corrosion products and methods for their identification have been described in a number of papers[16].

Electrode Potential Relationships

The standard potentials for the equilibria

$$Cu^{2+} + 2e \rightleftharpoons Cu \qquad \ldots(4.1)$$

$$Cu^{+} + e \rightleftharpoons Cu \qquad \ldots(4.2)$$

$$Cu^{2+} + e \rightleftharpoons Cu^{+} \qquad \ldots(4.3)$$

are $+0{\cdot}34$ V, $+0{\cdot}52$ V and $+0{\cdot}17$ V respectively, based on values in the book by W. M. Latimer[17]. For the equilibrium

$$2Cu^{+} \rightleftharpoons Cu^{2+} + Cu$$

$$K = a_{Cu^{2+}}/(a_{Cu^{+}})^2 \qquad \ldots(4.4)$$

K has the value of about 1×10^6 at 298 K, and in solutions of copper ions in equilibrium with metallic copper, cupric ions therefore greatly predominate (except in very dilute solutions) over cuprous ions. Cupric ions are therefore normally stable and become unstable only when the cuprous ion concentration is very low. A very low concentration of cuprous ions may be produced, in the presence of a suitable anion, by the formation of either an insoluble cuprous salt or a very stable complex cuprous ion. Cuprous salts can therefore exist in contact with water only if they are very sparingly soluble (e.g. cuprous chloride) or are combined in a complex, e.g. $[Cu(CN)_2]^-$, $[Cu(NH_3)_2]^+$. Cuprous sulphate can be prepared in non-aqueous conditions, but because it is not sparingly soluble in water it is immediately decomposed by water to copper and cupric sulphate.

The equilibrium between copper and cuprous and cupric ions is disturbed by the presence of oxygen in solution, since the reaction shown in equation 4.3 is facilitated, the oxygen acting as an electron acceptor.

Behaviour of Copper Electrodes

The electrode potential behaviour of copper in various solutions has been investigated and discussed in considerable detail by Gatty and Spooner[18]. According to these workers a large part of the surface of copper electrodes in aerated aqueous solutions is normally covered with a film of cuprous oxide and the electrode potential is usually close to the potential of these film-covered areas. The filmed metal simulates a reversible oxygen electrode at the existing oxygen concentration and pH, less an overvoltage determined by the existing current density. The principal factors that affect the electrode potential are thus the nature of the solution and the way in which this influences the area of oxide film, and the supply of oxygen to the metal surfaces.

In solutions containing chloride there is a tendency for the establishment of the Cu/CuCl/Cl^- electrode potential, so that the activity of chloride ions is an important factor in determining the electrode behaviour. From a knowledge of the solubility products of cuprous chloride and cuprous oxide it is possible to predict under what conditions chloride or hydroxyl ions are the potential-determining ions. According to Gatty and Spooner, chloride determines the potential if $a_{OH^-} < 10^{-8.1} \times a_{Cl^-}$ and hydroxyl if $a_{OH^-} > 10^{-8.1} \times a_{Cl^-}$. This will not hold in concentrated solutions, however, since complex $[CuCl_2]^-$ ions as well as simple ions will be present. A further factor to be considered is the ready formation of insoluble basic compounds. In solutions not containing chloride (e.g. sulphate or nitrate solutions), corrosion rates are usually lower and the electrode potential is more steady over

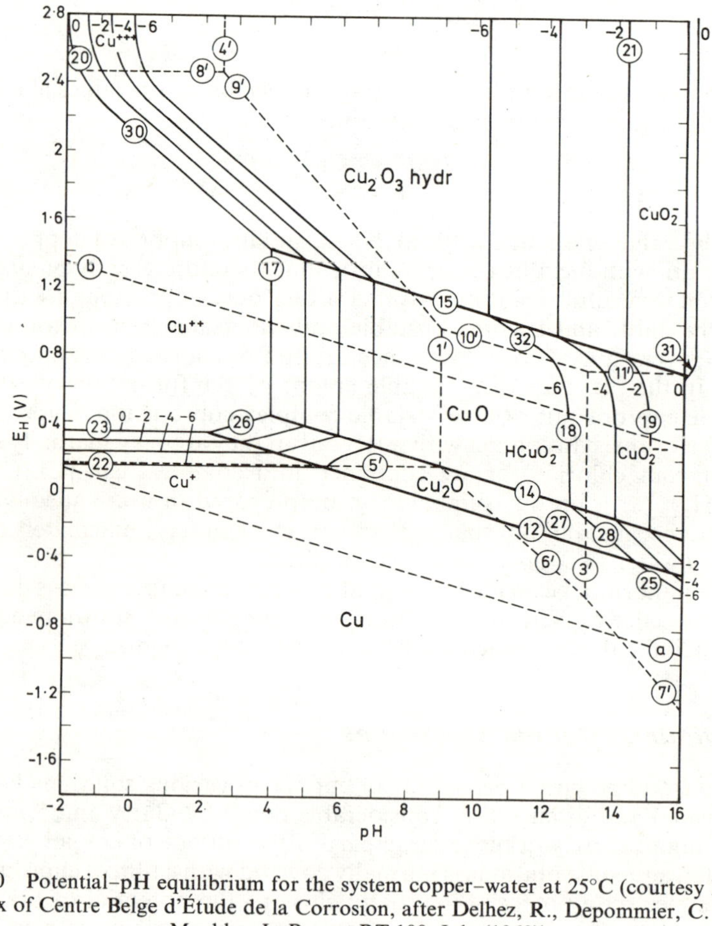

Fig. 4.10 Potential–pH equilibrium for the system copper–water at 25°C (courtesy M. J. N. Pourbaix of Centre Belge d'Étude de la Corrosion, after Delhez, R., Depommier, C. and van Muylder, J., Report RT 100, July (1962))

a wide range of conditions. In this case Gatty and Spooner consider that the rate of corrosion is probably determined by the rate at which metal ions can escape through pores in the protective oxide film, and this is supported by the results of experiments on the anodic and cathodic polarisation of copper.

The potential–pH equilibrium diagrams devised by Pourbaix[19,20] are considered in detail in Section 1.4, and one of the diagrams for the $Cu–H^2O$ system is shown in Fig. 4.10. Such diagrams are of considerable assistance in discussing many problems related to the chemistry, electrochemistry, electrodeposition and corrosion of copper. It is well recognised, of course, that the thermodynamic approach has limitations, the most important of which is that though predictions can be made about the possibility of a given reaction proceeding in certain circumstances no information can be gained about the rate at which it will proceed.

A method of representing the behaviour of copper in dilute aqueous solutions by means of corrosion–current/pH diagrams has been given by Rubinić and Marković[21].

A study of the behaviour of copper when anodically polarised has been made by Hickling and Taylor[22], using an oscillographic method that records the variation of potential with quantity of electricity passed. In alkaline solutions the main stages of polarisation were (*a*) the charging of a double layer, and (*b*) the formation of a film of cuprous oxide which was almost at once oxidised to cupric oxide. In 0·1N NaOH the film was about four molecules thick when oxygen evolution first commenced. In buffer solutions of decreasing pH, the formation of sparingly soluble salts preceded or accompanied the formation of the oxide film and in acid solutions giving soluble copper salts no passivity developed, the anodic process being merely dissolution of copper. Other workers have also studied the anodic behaviour of copper or copper alloys in alkaline[23] and in acid[24] solutions.

Atmospheric Corrosion

Copper has a high degree of resistance to atmospheric corrosion, and is widely used for roofing sheets, flashings, gutters and conductor wires, as well as for statues and plaques. The resistance of copper and its alloys is due to the development of protective layers of corrosion products, which reduce the rate of attack. The formation, in the course of time, of the typical green 'patina' gives copper roofs a pleasing appearance; indeed efforts have been made to produce it artificially or to accelerate its formation[25,26]. The nature of the corrosion products formed on copper exposed to the atmosphere was exhaustively studied by Vernon and Whitby[25,27]. In the early periods of exposure the deposit contains sulphide, oxide and soot. By the action of sulphuric acid and by the oxidation of sulphide, copper sulphate is formed, and this hydrolyses and forms a coherent and adherent basic copper sulphate. This approximates initially to $CuSO_4{\cdot}Cu(OH)_2$, but gradually increases in basicity until after 70 years or so it becomes $CuSO_4{\cdot}3Cu(OH)_2$ and is identical with the mineral brochantite. In some cases small quantities of basic carbonate, $CuCO_3{\cdot}Cu(OH)_2$ (malachite), are also present, and near the sea coast basic chloride $CuCl_2{\cdot}3Cu(OH)_2$ (atacamite) is produced. Even very near the sea coast, however, sulphate usually predominates over chloride. The presence of atmospheric pollution is thus an essential factor in the development of green patina.

In laboratory tests Vernon[28] showed that the relative humidity and the presence of sulphur dioxide have a profound effect on the rate of corrosion of copper, as of many other metals. When the relative humidity was less than 63%, there was little attack even in the presence of much sulphur dioxide, but when the relative humidity was raised to 75%, corrosion became severe and increased with the concentration of sulphur dioxide present.

By exposing specimens to the atmosphere at different times of the year, Vernon found that the rate of attack on copper was determined by the conditions prevailing at the time of first exposure. For specimens first exposed in winter there was a linear relationship between increase in weight and time of exposure, indicating that the layer of corrosion product formed under these conditions was non-protective. For specimens first exposed in summer the square of the increase in weight was proportional to the time of exposure, indicating that the coating formed in summer, when the atmospheric pollution was relatively low, was protective. The parabolic law holds when the corrosion product layer obstructs the access of the corrosive agent to the metal, the rate of attack then being inversely proportional to the thickness of the layer. The protective character of the layer persisted through subsequent periods when the pollution was relatively high.

Copper tarnishes rapidly when exposed to atmospheres containing hydrogen sulphide[29] and the reaction is not dependent on the presence of moisture.

Atmospheric corrosion tests on copper and several copper alloys were carried out by Hudson[30] at a number of sites in Great Britain. Corrosion damage was assessed by one or more of the following methods: gain in weight, loss of weight after cleaning, loss of electrical conductivity, and loss of tensile strength. Hudson found that the resistance to atmospheric corrosion was high, and that the rate of attack tended to decrease with time of exposure. Little difference was found between the behaviour of arsenical copper and high-conductivity copper, and most of the alloys tested behaved very similarly except for the brasses, which deteriorated more rapidly owing to dezincification.

Several series of atmospheric exposure tests have been carried out since Hudson's work, and the loss in weight data obtained in five of the most important investigations are summarised in Table 4.11. In all cases, losses in tensile strength were also determined, and the results from the two methods were, in general, in good agreement. However, for alloys suffering selective attack (such as dezincification of brasses), change in mechanical properties usually provided a more reliable indication of deterioration than weight loss. Some other findings common to all the tests were that (*a*) corrosion rates decreased with time, (*b*) least attack occurred at rural sites and most in urban and industrial atmospheres, (*c*) corrosion was uniform and with few exceptions there was no significant pitting.

Tracy, Thompson and Freeman[31] exposed specimens of 11 different grades of copper in the form of sheet and wire to rural, marine and industrial atmospheres in the USA for periods up to 20 years. The differences in the behaviour of the materials were small and of little, if any, practical significance. Very similar results for various types of copper were found by

Mattsson and Holm[34] in Sweden and Scholes and Jacob[35] in the UK (see Table 4.11).

The results of tests on copper alloys have been given by Tracy[32], Thompson[33], Mattsson and Holm[34] and Scholes and Jacob[35], the first two of these investigations being made under the aegis of the American Society for Testing and Materials. The tests of Tracy, and Scholes and Jacob were both for periods up to 20 years; in those of Thompson, and Mattsson and Holm specimens have been removed after 2 years and 7 years and further specimens remain exposed for removal after 20 years. The numbers of materials tested are given in Table 4.11; they included brasses, nickel silvers, cupro-nickels, beryllium coppers and various bronzes. Mattsson and Holm tested 14 alloys in the form of rod in addition to the sheet materials, the results for which are given in Table 4.11.

In the tests described by Tracy, a high-tensile brass suffered severe dezincification (Table 4.11). The loss in tensile strength for this material was 100% and for a non-arsenical 70/30 brass 54%; no other material lost more than 23% during 20 years' exposure. In Mattsson and Holm's tests the highest corrosion rates were shown by some of the brasses. Dezincification caused losses of tensile strength of up to 32% for a β brass and up to 12% for some of the α–β brasses; no other materials lost more than 5% in 7 years. Dezincification, but to a lesser degree, occurred also in the α brasses tested, even in a material with as high a copper content as 92%. Incorporation of arsenic in the α brasses consistently prevented dezincification only in marine atmospheres.

In the tests described by Thompson, the alloy showing the lowest rate of attack at all sites was a bronze containing 7Al–2Si. Relatively high corrosion rates were shown by Cu–5Sn–0·2P at a marine site, and Cu–2·5Co–0·5Be in the industrial environment. The beryllium–copper alloys were the only materials to show measurable pitting, the deepest attack being 0·06 mm after 7 years. In Scholes and Jacob's tests some pitting, intergranular or transgranular penetration, or selective attack occurred on some of the alloys. The maximum depth of attack exceeded 0·2 mm in 20 years on 6 of the 21 materials (three brasses, two nickel silvers and Cu–20Ni–20Mn), but exceeded 0·5 mm in 20 years only on Cu–20Ni–20Mn and 60/40 brass. These two latter alloys lost up to 73% and 13% respectively of their tensile strength; no other alloys lost more than 10% in 20 years.

From the work described and other investigations[199], it is evident that copper and most copper alloys are highly resistant to atmospheric corrosion. In general, copper itself is as good as, or better than, any of the alloys. Some of the brasses are liable to suffer rather severe dezincification and it is unwise to expose these to the more corrosive atmospheres without applying some protection.

When unusually rapid corrosion of copper and its alloys occurs during atmospheric exposure, it is likely to be for one of the following reasons:

1. Extreme local pollution by products of combustion.
2. Bad design or construction, e.g. the presence of crevices where moisture may lodge for long periods.
3. Constant dripping of rain water contaminated by atmospheric pollution (e.g. from near-by chimney stacks) or by organic acids from lichens, etc.

Table 4.11 Atmospheric corrosion tests on copper and copper alloys

	No. of types of copper	*No. of different alloys*	*No. of sites*	*Period of exposure* (years)	*Average rates of attack from weight losses* (mm/y × 10^4)		
					Rural sites	*Marine sites*	*Urban/industrial sites*
Tracy, Thompson and Freeman[31]	11	—	4	20	3·6–4·3	6[illegible]–9·4	8·6–12
Tracy[32]	2	9	7	20	0·5–7·6	1[illegible]–23*	13–30*
Thompson[33]	1	17	4	7	3·3–10	4[illegible]–25	13–27
Mattsson and Holm[34]	4	—	3	7	5–6	[illegible]–8	10–12
	—	18	3	7	2–5	[illegible]–11	9–22
Scholes and Jacob[35]	4	—	2	20	—	[illegible]–10	11–20
	—	17	2	20	—	[illegible]–26	14–38

*Rates of attack for a high-tensile brass were 45×10^{-4} to 115×10^{-4} mm/y.

4. Corrosion fatigue due to inadequate allowance for expansion and contraction with consequent buckling as the temperature fluctuates.

Most of these can be avoided by attention to design.

Corrosion in tropical environments has been the subject of several papers[36], some of which deal with corrosion at bimetallic contacts[37].

Soil Corrosion

Several extensive series of soil-corrosion tests have been carried out by the National Bureau of Standards in the United States, and the results have been summarised by Romanoff[38]. In one series two types of copper and ten copper alloys were exposed in fourteen different soils for periods up to 14 years. The results for the copper specimens are summarised in Table 4.12.

The behaviour of the phosphorus-deoxidised and tough pitch coppers was in general very similar. At the less corrosive sites, copper was, with few exceptions, the best material, but most of the alloys lost not more than about twice as much weight, with maximum depths of attack usually not more than two or three times as great as with copper. At the other sites copper was also usually rather better than the alloys, but some of the alloys were occasionally superior.

The three most corrosive sites were rifle peat (pH 2·6), cinders (pH 7·6) and tidal marsh (pH 6·9). Corrosion of some of the alloys was particularly severe in the cinders. The behaviour of the brasses tested, particularly those high in zinc, was rather different from that of the other materials. In most cases dezincification occurred and the brasses were the worst materials; in

Table 4.12 Soil-corrosion tests on copper by National Bureau of Standards and British Non-ferrous Metals Research Association

	Period of exposure (years)	*Average rate of attack from loss in weight* ($mm/y \times 10^4$)	*Maximum rate of pitting* (mm/y)
BNFMRA 1st series: 5 least corrosive soils	10	0·5–2·5	Nil
BNFMRA 2nd series: 4 least corrosive soils	5	5·0–25	0·040
Nat. Bur. Standards: 9 least corrosive soils	14	4·0–25	0·043
Nat. Bur. Standards: 2 next most corrosive soils	14	25–130	0·033
BNFMRA 1st series: acid clay and acid peat	10	53–66	0·046
BNFMRA 2nd series: cinders	5	66	0·32
Nat. Bur. Standards: 3 most corrosive soils:—rifle peat, cinders, tidal marsh	14	160–355	0·115

the cinders, for instance, several brass specimens were completely destroyed by dezincification. In some of the soils rich in sulphides, however, the brasses were the best materials.

The British Non-Ferrous Metals Research Association carried out two series of tests, the results of which have been given by Gilbert[39] and Gilbert and Porter[40]; these are summarised in Table 4.12. In the first series[37] tough pitch copper tubes were exposed at seven sites for periods of up to 10 years. The two most corrosive soils were a wet acid peat (pH 4·2) and a moist acid clay (pH 4·6). In these two soils there was no evidence that the rate of corrosion was decreasing with duration of exposure. In the second series[40] phosphorus-deoxidised copper tube and sheet was exposed at five sites for five years. Severe corrosion occurred only in cinders (pH 7·1). In these tests sulphides were found in the corrosion products on some specimens and the presence of sulphate-reducing bacteria at some sites was proved. It is not clear, however, to what extent the activity of these bacteria is a factor accelerating corrosion of copper.

Cinders and acid peaty soils are obviously among the soils most corrosive toward copper. There is, however, no direct relationship between the rate of corrosion and any single feature of the soil composition or constitution[41]. For instance, in the American tests corrosion in several soils with either low pH or high conductivity was not particularly severe, while the British tests show that high chloride or sulphate contents are not necessarily harmful.

The above-mentioned tests show that bare copper can safely be buried in a wide range of soils without fear of excessive corrosion. Experience of the behaviour of copper water service pipes, which are used widely, confirms this. Trouble is confined to 'made-up' ground containing cinders, etc. and a few other aggressive soils, and in these circumstances it is necessary to apply protection such as bitumen-impregnated wrappings or plastic coatings. Tin coatings cannot be recommended since experience shows that accelerated attack is liable to occur at pores and scratches in the coating, leading to premature failure. Copper water pipes have been known to fail by the action of stray electric currents but this is not a common cause of trouble.

There is also agreement between the soil-corrosion tests carried out by the National Bureau of Standards and practical experience of the behaviour of hot-pressed brass fittings used for joining copper water service pipes. These duplex-structure brass fittings are liable to suffer attack by dezincification in many soils in which copper behaves satisfactorily, and for burial underground fittings of copper or gun metal are to be preferred. In general, it may be said that unless there is some special reason for using a copper alloy, it is preferable to choose copper for applications involving service underground.

Copper and Copper Alloys in Natural Waters

Copper and copper alloy pipes and tubes are used in large quantities both for conveying fresh and salt waters and in condensers and heat exchangers where fresh or salt waters are used for cooling. Pumps, screens, valves and other ancillary equipment may also be largely constructed of copper alloys. Large tonnages of these materials are therefore used in power stations, on board ship, in sugar factories and in oil refineries, as well as in hot- and cold-water circuits and heating and cooling systems in hospitals, hotels, factories and homes.

Corrosion problems that arise are frequently discussed under the headings (*a*) sea-water, and (*b*) fresh waters, but there is, in fact, no sharp dividing line, since some harbour, estuarine and brackish well waters are mixtures of sea-water and fresh water and are often variable in composition. In the past, corrosion problems were serious, particularly in sea-water service, but resistant alloys have been developed and although trouble still occasionally arises this is more frequently due to poor design or operation rather than to lack of materials suitable for the application.

There are several distinctive types of corrosion that copper and copper alloys may suffer, particularly in sea-water, but also on occasion in fresh waters. The more important of these are discussed briefly below.

Impingement attack When moving water flows over copper or copper alloys the turbulence may be sufficient to cause breakdown of the surface film. This is particularly likely to happen if air bubbles entrained in the water break as they hit the metal surface. The resulting corrosion is characteristic, producing clean-swept pits, often of a 'horseshoe' shape as shown in Fig. 4.11. This type of attack was first described by Bengough and May[42,43]. The action can be very rapid, the local anodes being depolarised by the continuous removal of metal ions and corrosion product, and the local cathodes by the dissolved oxygen in the rapidly moving well-aerated water stream.

Factors that tend to increase the severity of impingement attack are increase of water speed and particularly of local turbulence, pollution of the water, and, within certain limits, increase in the size and the amount of entangled air bubbles (see also Section 1.6).

A laboratory test designed to simulate the conditions occurring in condenser tubes in practice was devised by May[43] and newer versions of this 'jet impingement apparatus' have been described[44], as has some of the testing equipment in use in the USA[45]. Use of the jet impingement apparatus has been an important factor in the development of alloys resistant to impingement attack, but it has to be borne in mind that the results obtained when the water is recirculated may be different from those obtained when it is passed once through the apparatus, as shown by Gilbert and LaQue[46].

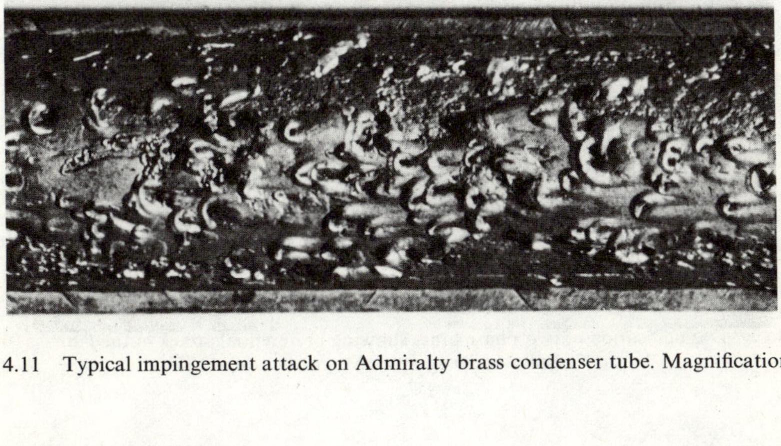

Fig. 4.11 Typical impingement attack on Admiralty brass condenser tube. Magnification ×2

Details of jet impingement tests will be found in Section 20.1. Alloys resistant to impingement attack will be considered subsequently.

Dezincification of brasses When dezincification occurs, regions of the brass become replaced by a porous mass of copper which, though retaining the shape of the original article, has virtually no strength. There has long been discussion as to whether there is selective corrosion of the zinc in the brass, which leaves the copper behind, or whether complete dissolution of the brass occurs, followed by re-deposition of copper. Possibly both processes occur in different circumstances. The mechanism has been investigated and discussed by Evans[7], Fink[47], Lucey[48], Feller[49] and Heidersbach[50], and is referred to in many other papers[51].

With a single-phase brass the whole of the metal in the corroded areas is affected. Dezincification may proceed fairly uniformly over the surface, and this 'layer type' takes much longer to cause perforation than the localised 'plug type' that more often occurs[52]. With a two-phase brass the zinc-rich β phase is preferentially attacked as shown in Fig. 4.12. Eventually the α phase may be attacked as well. The zinc corrosion products that accompany dezincification may be swept away, or in some conditions may form voluminous deposits on the surface which may lead to blockages, e.g. in fittings.

In general, the rate of dezincification increases as the zinc content rises, and great care needs to be exercised in making brazed joints with copper/zinc brazing alloys, particularly if they are to be exposed to sea-water. Under these conditions, a properly designed capillary joint may last for some time, but it is preferable to use corrosion-resistant jointing alloys such as silver solders (e.g. BS 1845, Type *AG1* or *AG5*)[53].

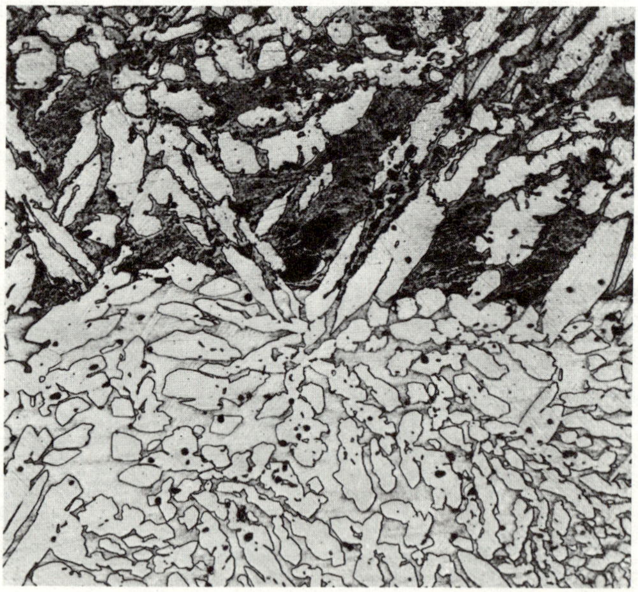

Fig. 4.12 Dezincification of two-phase brass showing preferential attack of the β phase (upper half of photomicrograph). Magnification $\times 133$

Factors that cause increased rates of dezincification are high temperature, high chloride content of water and low water speed. Dezincification is also likely to occur preferentially beneath deposits of, for instance, sand or silt on the metal surface, or in crevices where there is a low degree of aeration.

Addition of about 0·04% arsenic will inhibit dezincification of α brasses[42] in most circumstances and arsenical α brasses can be considered immune to dezincification for most practical purposes[54]. There are conditions of exposure in which dezincification of these materials has been observed, e.g. when exposed outdoors well away from the sea[34], or when immersed in pure water at high temperature and pressure, but trouble of this type rarely arises in practice. In other conditions, e.g. in polluted sea-water, corrosion can occur with copper redeposition away from the site of initial attack, but this is not truly dezincification, which, by definition, requires the metallic copper to be produced *in situ*. The work of Lucey[48] goes far in explaining the mechanism by which arsenic prevents dezincification in α brasses, but not in α–β brasses (see also Section 1.6). An interesting observation is that the presence of a small impurity content of magnesium will prevent arsenic in α brass from having its usual inhibiting effect[55].

Additions of antimony or phosphorus, in amounts similar to arsenic, are claimed to be also capable of preventing dezincification of α brasses. Most manufacturers use arsenic, however, and it certainly appears desirable to avoid phosphorus, since Bem[56] has shown that this element can, in some circumstances, lead to an undesirable susceptibility to intercrystalline corrosion. The same appears to be true of excessive amounts of arsenic (over about 0·05%).

No reliable method of inhibiting dezincification of two-phase brasses has been discovered. Various additions, including arsenic, have been advocated from time to time, but nothing is known that will render α–β brasses immune to dezincification under all conditions of exposure. The addition of 1% tin can markedly reduce the rate of dezincification, and naval brass (61Cu–38Zn–1Sn) is attacked considerably more slowly than 60/40 brass in sea-water, though there may be virtually no difference in most fresh waters. Some of the cast complex high-strength two-phase brasses containing tin, aluminium, iron and manganese appear to have relatively good resistance to dezincification, but they are by no means immune to it.

Selective attack in other alloys Selective attack analogous to dezincification can occur in other copper alloys, particularly aluminium bronzes and less frequently tin bronzes[57], cupro-nickels[58], etc. In recent years de-aluminification of aluminium bronzes has been studied extensively[59] and the results indicate that whilst α-phase alloys suffer such attack comparatively rarely, alloys of higher aluminium content can be more susceptible. The electrochemical relationships are such that preferential corrosion of the second phase is liable to occur in α–γ_2 alloys, but α–β alloys are relatively resistant to attack. Retention of β phase is favoured by rapid cooling after casting or after high-temperature heat treatment, and also by incorporating manganese in the alloy.

Deposit attack and pitting When water speeds are low and deposits settle on the surface (particularly at water speeds below about 1 m/s), pitting of copper and copper alloys is liable to occur by differential aeration effects. In sea-water systems such attack may occur under dead barnacles or shell-fish, the decomposing organic matter assisting corrosion. Pitting is most likely to occur in polluted in-shore waters, particularly when hydrogen sulphide is present. In such contaminated waters non-protective sulphide scales are formed and these tend to stimulate attack.

Corrosion of condenser tubes and related equipment There have been many surveys[60] of the problems of corrosion of condenser and heat-exchanger tubes and related components in marine service and others[61] dealing with oil refinery service.

Corrosion of condenser tubes was a problem of great magnitude during the first quarter of this century. Its solution was based on research carried out for the Institute of Metals by Bengough *et al.*[42, 52, 62], one of whom, May[43, 54], remained associated with the research when it was transferred to the auspices of the British Non-Ferrous Metals Research Association in 1930.

In early times 70/30 brass condenser tubes failed by dezincification and Admiralty brass (70Cu–29Zn–1Sn) was brought into use. This proved little better, but some time later the addition of arsenic was found to inhibit dezincification. Failures of Admiralty brass by impingement attack became a serious problem, particularly as cooling water speeds increased with the development of the steam turbine. The introduction of alloys resistant to this type of attack was a great step forward and immediately reduced the incidences of failure.

The alloys in most common use today are aluminium brass (76Cu–22Zn–2Al–0·04As) and cupro-nickels containing appropriate iron and manganese additions[15, 63]. Three cupro-nickel alloys are in widespread use containing (approximately) (i) 30Ni–0·7Fe–0·7Mn, (ii) 30Ni–2Fe–2Mn and (iii) 10Ni–1·5Fe–1Mn. These materials are extensively and successfully used in ships, power stations, oil refineries, etc., in condensers and heat exchangers, with nominal water speeds through the tubes of up to about 3 m/s, sometimes with much entangled air present. At the highest water speeds there is a rather greater factor of safety with 70/30 cupro-nickel, and this alloy is also usually to be preferred under most polluted water conditions, although occasionally other alloys are as good or even better. There is evidence to indicate that when the operating conditions involve relatively high temperatures, aluminium–brass or 90/10 cupro-nickel is to be preferred to 70/30 cupro-nickel[58]. Admiralty brass is no longer considered a suitable alloy for sea-water service, except possibly where water speeds are very low, i.e. not more than about 1 m/s. In some oil refineries, for instance, Admiralty brass is preferred because it has good resistance to corrosion by oil products, and in these installations the heat exchangers are designed to have low cooling-water speeds, so that corrosion from the sea-water side is kept within reasonable bounds.

Tin–bronze containing about 12% Sn has been shown to have good resistance to impingement attack[64] to attack by acid cooling waters, and to abrasion in cooling waters containing suspended solid particles, but the alloy has so far only been used on a limited scale. The alloy most commonly used when suspended abrasive matter is a problem is the special 30% Ni alloy containing 2% each of Fe and Mn. Aluminium–bronze tubes have sometimes given good results, but their use has been limited because of their susceptibility to pitting attack.

The occasional failures that still occur in condenser tubes are usually due to one (or sometimes several) of the following factors:

1. Localised attack or pitting in badly contaminated waters.
2. Pitting under decaying barnacles, shell fish or other deposits.
3. Impingement attack due to high local water velocities, e.g. at partial obstructions in a tube such as pieces of coke, shell fish, etc.
4. Erosion due to sand or other abrasive particles suspended in the water[65].
5. Use of tubes of the wrong alloy, or of incorrect composition, or containing manufacturing defects.

A difficult condenser-tube corrosion problem arises from the use of polluted cooling waters from harbours and estuaries that may be severely contaminated. All condenser-tube materials are liable to suffer corrosion in these circumstances, and the choice of materials is made difficult by the fact that different orders of merit apply at different locations and even at the same location at different times. The state of the water when the tubes first enter service may well determine whether or not a satisfactory life will be obtained[66]. The most corrosive waters are those containing free hydrogen sulphide produced by the action of sulphate-reducing bacteria. Waters may also be rendered abnormally corrosive by the presence of small amounts of organic sulphur compounds produced by bacterial action, as shown by Rogers[67].

Corrosion of power-station condenser tubes by polluted waters[68] has been particularly troublesome in Japan and efforts have been made to study the problem by electrochemical methods[69] and by exposing model condensers at a variety of power station sites[70]. Improved results have been reported using tin brasses[71] or special tin bronzes[72]. Pretreatment with sodium dimethyldithiocarbamate is reported to give protective films that will withstand the action of polluted waters[73], though the method would be economic only in special circumstances.

Electrochemical studies, including the determination of polarisation curves, have been carried out in recent years by many authors[74] in endeavours to understand more fully the mechanism of protective film formation on copper alloys in sea-water. Other authors have described experiences with condenser tubes in fresh or brackish waters[75]. Methods of maintaining tube cleanliness include chlorination[76], use of high molecular weight water-soluble polymers[77] and use of the Taprogge system of circulating sponge-rubber balls[78, 79].

Condenser tube-plates of Naval brass usually undergo some dezincification in sea-water service, but this is normally not serious in view of the thickness of metal involved. Attack can, however, be more serious with 60/40 brass (Muntz metal) and such plates may have to be renewed during the life of the condenser. Increasing use is now being made of tube-plates of more resistant materials such as aluminium-brass, silicon-bronze, aluminium-bronze or cupro-nickel. Plates that are too large to be rolled in one piece can be fabricated by welding together two or more pieces. In some special applications the tubes are fusion welded or explosively welded to the tube-plates. Fusion welding operations are rather more difficult with brasses than with other copper alloys (because of evolution of zinc fume from brasses).

Condenser water-boxes were hitherto usually made of unprotected (or poorly protected) cast iron and these afforded a measure of cathodic protection to the tube-plates and tube ends. This beneficial effect has been lost with the general adoption of water-boxes completely coated with rubber or some other impervious layer, or of water-boxes made from resistant materials such as gunmetal, aluminium-bronze or cupro-nickel, or steel clad with cupro-nickel or Monel. To prevent attack on tube-plates and tube-ends in these circumstances, it is highly desirable to instal either a suitable applied-current cathodic-protection system[80], or sacrificial soft-iron or mild-steel anodes. Ferrous wastage plates have the additional advantage that the iron corrosion products introduced into the cooling water assist in the development of good protective films throughout the length of the tubes. This is particularly important in the case of aluminium-brass tubes; indeed, with such tubes it may be desirable, as an additional preventive measure, to add a suitable soluble iron salt (such as ferrous sulphate) regularly to the cooling water. Cases of the success of such treatment in power station condensers have been described by Bostwick[81] and Lockhart[82], and other workers[79, 83] have since studied the effects of ferrous sulphate treatment on tube behaviour.

As it has become increasingly necessary to supplement natural sources of fresh water in various parts of the world, processes for producing fresh water from sea-water have been intensively studied and the literature dealing with the subject is now very extensive. Distillation is currently the process most widely used and during the past decade increasing numbers of multi-stage flash-distillation plants have been installed in various countries, many of the larger units being capable of producing several millions of gallons of fresh water per day. In these plants, sea-water passes through horizontally-disposed tubes and steam 'flashed' from the brine condenses on the outside. In some parts of the plant the conditions are similar to those in steam condensers, but in other parts the sea-water has been treated to remove dissolved gases and is at much higher temperatures. In another distillation process receiving considerable study, films of sea-water fall down the inner walls of comparatively large-diameter vertically-disposed tubes, usually of fluted configuration, and evaporation takes place due to the heating effect of steam condensing externally. Copper-alloy tubes are used almost exclusively for the heat-exchange units in distillation plants, mainly aluminium-brass and the various cupro-nickel alloys, and the factors affecting choice of materials have been considered in several papers[84].

For ships' cooling-water trunking and salt-water services in the engine room and elsewhere, including fire mains, plumbing and air-conditioning systems, more resistant alloys are taking the place of copper or galvanised steel, which were formerly extensively used, but which do not have adequate resistance to attack by sea-water. Both aluminium-brass and the Cu–10Ni–1·5Fe alloy are widely used and, being highly resistant to impingement attack, normally give excellent service. In some special naval applications pipelines of 70/30 cupro-nickel are used. It is important that correct fabrication and installation techniques are used. Carbonaceous residues from fillers used in bending operations must be avoided or pitting corrosion may occur in service. Jointing materials of low corrosion resistance should not be used, silver brazing or appropriate welding methods being the correct techniques. Residual stresses, if present, can cause stress-corrosion failures of aluminium-brass pipelines in service.

Copper alloys in wrought or cast form are used for other purposes in ships and other marine installations, such as for propellers[85], bearings, valves and pumps. One widespread application of aluminium-brass is its use for heating coils in tankers carrying crude oil or petroleum products. Some corrosion problems encountered in this and other applications on board ship have been described by Gilbert and Jenner[86].

Fresh Waters

Fresh waters are, in general, less corrosive towards copper than is sea-water, and copper is widely and satisfactorily used for distributing cold and hot waters in domestic and industrial installations[15,87]. Copper and copper alloys are used for pipes, hot-water cylinders, fire-back boilers, ball floats, ball valves, taps, fittings, heater sheaths, etc. In condensers and heat exchangers using fresh water for cooling, tubes of 70/30 brass or Admiralty brass are usually used, and corrosion is rarely a problem.

Joints in copper components may be a source of trouble. Copper/zinc brazing alloys may dezincify and consequently give rise to leaks[90]. In some waters, soft solders are preferentially attacked unless in a proper capillary joint. Copper/phosphorus, copper/silver/phosphorus, and silver brazing alloys are normally satisfactory jointing materials. Excessive corrosion of copper is sometimes produced by condensates containing dissolved oxygen and carbon dioxide. Rather severe corrosion sometimes occurs on the fire side of fire-back boilers and on electric heater element sheaths under scales deposited from hard waters[91].

Dezincification of brasses This may occur, particularly in stagnant or slowly-moving warm or hot waters relatively high in chloride and containing little carbonate hardness[88]. Dezincification of α brasses is inhibited by the usual arsenic addition (see Fig. 4.12), but two-phase brasses are liable to severe attack in some waters[89]. In such waters the use of duplex-structure brass fittings should be avoided.

Impingement attack Copper may occasionally suffer this form of attack in systems where the speed of water flow is unusually high and the water is one that does not form a protective scale, e.g. a soft water containing appreciable quantities of free carbon dioxide[92]. Ball valve seatings may also suffer an erosive type of attack. The corrosion of ball valves, including the effect of chlorination of the water, has been studied by several workers[93].

Dissolution Some waters continuously dissolve appreciable amounts of copper[94]. Factors that favour this action are high free carbon dioxide, chloride and sulphate contents, low hardness, and increase of temperature. The trouble is therefore most prevalent in hot, soft, acid waters. The corrosion is general and the resulting thinning is so slight that the useful life of the pipe or component is virtually unaffected (unless impingement attack occurs). Trouble is usually confined to (*a*) stimulation of the corrosion of components of zinc-coated steel[95], light alloys[96], and sometimes bare steel with which the water subsequently comes into contact; and (*b*) the formation of green stains in baths, sinks, etc. owing to the combination of copper with soaps.

In de-aerated conditions, for instance in most central heating systems, little if any attack on copper occurs[92, 97]. As far as drinking waters are concerned, the World Health Organisation states that copper should not be classified as a toxic substance or hazardous to health. To avoid any difficulties due to unpalatability, the maximum continuous copper content should not exceed 1·0 p.p.m., with a limit of 3 p.p.m. in water after standing overnight in copper pipes. A review of the subject by Grunau[98] makes reference to 394 published papers.

Pitting Occasionally copper water pipes fail prematurely by pitting. This most often occurs in cold waters originating from deep wells and boreholes and has been shown by Campbell[99] to be associated with residues of carbon produced in the bore of the tubes during bright annealing, as a result of decomposition of residual drawing lubricant. It is therefore necessary for manufacturers to take steps to avoid these harmful residues. This trouble has occurred in many different countries[100].

Failures of this type are confined to certain districts, and Campbell has shown that in many supply waters in Great Britain pitting cannot proceed, even in tubes containing dangerous cathodic films, owing to the presence in the water of small amounts of a naturally-occurring inhibitor, probably an organic colloidal material, that stifles pitting of copper. Trouble therefore only occurs in waters that contain little or no inhibitor.

Pitting failures also occasionally occur in copper water cylinders[101] and as a result of a study of this problem Lucey[102] has made suggestions about the mechanism of pitting of copper in supply waters.

In hot-water pipes, failures sometimes occur in certain areas supplied with soft waters from moorland gathering grounds. The waters concerned contain a few hundredths of a part per million of manganese, and in the course of several years' exposure, a deposit rich in manganese dioxide is laid down in the hottest parts of the system. This may cause pitting and eventually lead to failure. Hot-water pitting of another type is sometimes experienced in soft waters having a high sulphate content in relation to the carbonate hardness and a relatively low pH value[103].

Behaviour in Chemical Environments

Detailed information on the action of a wide range of chemicals on copper and copper alloys is given in a number of publications, particularly those listed under References 7–12, 104 and 105.

When contemplating the use of copper-base materials for industrial purposes it is necessary to bear in mind that even though a satisfactory life of the component may be obtained, trouble can arise from other causes:

1. Copper compounds can be tolerated only in small amounts in potable waters or substances that are to be consumed.

2. Copper compounds are highly coloured, and a very small amount of corrosion may lead to staining and discoloration of products.
3. Stimulation of the corrosion of vital parts made of more anodic metals may occur if they are connected to copper.
4. Very small amounts of copper taken into solution may cause considerable corrosion of more anodic metals elsewhere in the system, particularly zinc[95], aluminium[96], and sometimes steel[106]. Small particles of copper deposited from solution set up local cells that cause rapid pitting.

Despite these qualifications copper and its alloys are used extensively and successfully in much chemical equipment. Uses include condensers and evaporators, pipelines, pumps, fans, vacuum pans, fractionating columns, etc. Tin-bronzes, aluminium-bronzes and silicon-bronzes are used in some circumstances because they present better corrosion resistance than copper or brasses.

Acid Solutions

Copper does not normally displace hydrogen even from acid solutions, and it is therefore virtually unattacked in non-oxidising conditions. Most solutions that have to be handled contain dissolved air, however, and this will cause cathodic depolarisation and enable some corrosion to take place[107]. It is difficult, therefore, to lay down any general recommendations for the use of copper in acid solutions, since the rate of attack depends so greatly on the particular circumstances. Under fairly mild conditions copper or copper alloys are successfully used for handling solutions of hydrofluoric[104, 108], hydrochloric[8–10, 104, 109], sulphuric[8–10, 110], phosphoric[8, 9, 104, 111] and acetic and other fatty acids[8–10, 104, 112]. Rates of corrosion, in general, increase with concentration of acid, temperature, amount of aeration[113] and speed of flow[114]. Tin-bronzes[8], aluminium-bronzes[7, 115], silicon-bronzes[7, 8] and cupro-nickels[8] are among the copper alloys most resistant to acids. Brasses should not normally be used. All copper-base materials are attacked rapidly by oxidising acids such as nitric, strong sulphuric, etc.

The dissolution of copper and of brasses in acid solutions has been studied by several authors[116]. Various substances have inhibiting effects on the rate of attack of copper or brasses in nitric acid[117] and in hydrochloric acid[118].

Neutral and alkaline solutions Copper-base materials are resistant to alkaline solutions[8, 9, 119] over a wide range of conditions but may be appreciably attacked by strong solutions, particularly if hot. Copper/nickel alloys usually give the best results in alkaline solutions. Copper and copper alloys should be avoided if ammonia[8, 9, 120] is present, owing to the danger of both general corrosion and, if components are under stress, stress corrosion.

Copper is satisfactory for handling solutions of most neutral salts[7–9, 104, 121] unless aeration and turbulence are excessive. An exception is provided by those salts that form complexes with copper, such as cyanides, and solutions containing oxidising agents, such as ferric or stannic compounds[8].

Other chemicals Copper and copper alloys are unsuitable for handling hydrogen peroxide[104, 122] or molten sulphur[104, 123]. Hydrogen sulphide accelerates corrosion of most copper-base materials. In its presence brasses high in zinc are usually found to behave better than other copper alloys[8].

Halogens have little action on copper at room temperature when dry, but are corrosive when wet. Hypochlorite solutions are corrosive[124]. Most organic compounds are without appreciable action[8, 104]. Copper and copper alloys are extensively and successfully used in refrigeration systems employing organic refrigerants such as CCl_2F_2. Attack can, however, occur if halogenated compounds hydrolyse in the presence of moisture to give traces of hydrochloric acid. Copper alloys are widely used for handling hydrocarbon oils, though if sulphur compounds are present attack can be serious[7]. The effects of synthetic detergents on copper have been investigated[125], and several authors[9, 104, 126] have discussed various aspects of the behaviour of copper and its alloys in the food-processing industry.

Oxidation and Scaling

Several authors[7, 8, 127–129] have reviewed the literature on the oxidation and scaling of metals, including copper.

Copper

The volume ratio (see Section 1.9) for cuprous oxide on copper is 1·7, so that an initially protective film is to be expected. Such a film must grow by a diffusion process and should obey a parabolic law. This has been found to apply for copper in many conditions, but other relationships have been noted. Thus in the very early stages of oxidation a linear growth law has been observed (e.g. at 1 000°C)[130].

At 180–290°C it was found[131] that the parabolic law first applied but subsequently changed to a logarithmic relationship of the type

$$y = K \log B(t + 1/B)$$

B being a constant. Other workers have reported a cubic relationship under some conditions.

Evans[132] has shown how the effect of internal stresses in growing films may have various effects that can lead to any one of the first three growth laws referred to above.

At medium and high temperatures[133, 134] copper ultimately follows the parabolic law[128, 135]. It has been shown[136] using radioactive tracers that the diffusion of copper ions in cuprous oxide is the rate-determining step at 800–1 000°C, and there is considerable evidence in favour of the view that metal moves outwards through the film by means of vacant sites in the oxide lattice[133].

When oxidation is a diffusion process the oxidation rate should be related to the temperature by the Arrhenius equation

$$K = A \exp[-Q/RT]$$

where K is the rate constant, A a constant, R the gas constant, T the absolute temperature, and Q the activation energy. Values that have been obtained for A and Q are summarised by Tylecote[128], Pilling and Bedworth[137], Feitknecht[138] and others give values of Q of about 0·17 MJ for temperatures of 700–1 000°C, while at lower temperatures (up to 500°C) Vernon[139] and others obtained values of about 85 kJ. These values are in agreement with calculations by Valensi[140] based on the assumption that at the high temperatures the oxidation proceeds by the reaction of oxygen with metallic copper to produce cuprous oxide while at lower temperatures the rate is determined by the reaction between oxygen and cuprous oxide to form cupric oxide.

At low temperatures (e.g. 300°C) the film consists almost entirely of CuO. As the temperature increases the film consists of a layer of Cu_2O beneath a layer of CuO, the proportion of Cu_2O increasing until at high temperature the film is almost entirely Cu_2O. The precise composition of the film depends, however, on a number of factors, including temperature, time, oxygen concentration in the atmosphere, etc. Tylecote[141] has investigated the composition, properties and adherence of scales formed on various types of copper at temperatures between 400 and 900°C. At the higher temperatures the scales formed on coppers containing phosphorus were more brittle and less adherent than those formed on coppers containing no phosphorus.

Studies have been carried out of the effects at high temperatures of sulphur[142] and of atmospheres containing hydrogen sulphide[143], steam[144, 145], sulphur dioxide[145] and hydrogen chloride[145].

Copper Alloys

With copper alloys containing more noble metals the oxide will be substantially pure copper oxide since the oxides of the noble metals have higher dissociation pressures than the copper oxides. With alloys containing baser metals, however, the alloying element will appear as an oxide in the scale, often in greater concentration than in the alloy itself, and sometimes to the exclusion of copper oxides. The dissociation pressures of many oxides have been calculated by Lustman[146].

Whether the rate of oxidation of an alloy of copper with a baser metal is less or more than that of copper will depend on the concentration of the alloying element and the relative diffusion velocities of metal atoms or ions in the oxide layers. There is extensive literature on the oxidation behaviour of copper alloys[128, 129, 147]. According to Wagner's theory[148] the rate of oxidation will be largely influenced by the electrical conductivity of the film, and the theory is therefore supported by the fact that the alloying elements giving maximum oxidation resistance, i.e. beryllium, aluminium and magnesium, form oxides having very low conductivities, as shown by Price and Thomas[149]. Wagner calculated that when sufficient aluminium was present in copper to cause the formation of an alumina film the oxidation rate should be decreased by a factor of more than 80 000. Experiment showed a factor of only 36, but when Price and Thomas carried out initial oxidation under very slightly oxidising conditions, producing only a film of alumina, the oxidation rate on subsequent exposure to full oxidising conditions was decreased by a factor of about 240 000. Hallowes and Voce[145] found that selective oxidation of a 95Cu–5Al alloy by this method gave protection from atmospheric oxidation up to 800°C unless the film was scratched or otherwise damaged, or the atmosphere contained sulphur dioxide or hydrogen chloride.

The effects on oxidation resistance of copper as a result of adding varying amounts of one or more of aluminium, beryllium, chromium, manganese, silicon, zirconium are described in a number of papers[147]. Other authors have investigated the oxidation of copper–zinc[150,151] and copper–nickel alloys[151,152], the oxidation of copper and copper–gold alloys in carbon dioxide at 1 000°C[153] and the internal oxidation of various alloys[154].

Copper alloys have been used extensively in high-pressure feed-water heaters in power generating plant. Experience has shown that when such heaters are operated intermittently, 70Cu–30Ni or 80Cu–20Ni alloy tubes suffer fairly rapid and severe steam-side (external) oxidation with the formation of exfoliating scales. This corrosion, which may be associated with ingress of air during shutdowns, has been the subject of several published papers[155]. The behaviour of other alloys for feed-water heater service has also been discussed[156].

Stress Corrosion

Failure of copper alloys may occur by cracking due to the combined influence of tensile stress and exposure to a corrosive environment. When the stresses are produced in components during manufacture the trouble is usually known as *season cracking* and failures of brass components due to this form of stress corrosion have been known for many years[157–160].

Only certain specific environments appear to produce stress corrosion of copper alloys, notably ammonia or ammonium compounds or related compounds such as amines. Mercury or solutions of mercury salts (which cause deposition of mercury) or other molten metals will also cause cracking, but the mechanism is undoubtedly different[161]. Cracks produced by mercury are always intercrystalline, but ammonia may produce cracks that are transcrystalline or intercrystalline, or a mixture of both, according to circumstances. As an illustration of this, Edmunds[162] found that mercury would not produce cracking in a stressed single crystal of brass, but ammonia did.

Stress Corrosion of Brasses

Alloys containing only a few per cent of zinc may fail if the stresses are high and the environment sufficiently corrosive. Most types of brass, besides the plain copper/zinc alloys, appear to be susceptible to stress corrosion. An extensive investigation of the effect of additions to 70/30 brass was carried out by Wilson, Edmunds, Anderson and Peirce[163], who found that about 1% Si was markedly beneficial. Other additions were beneficial under some circumstances and none of the 36 additions tested accelerated stress-corrosion cracking. Further results are given in later papers[164].

In general, the susceptibility to stress corrosion appears to increase with increase in zinc content, but in some circumstances alloys containing 64–65% Cu were found to be rather more affected than those containing 60% Cu[165].

Many workers[157,159,166] have investigated the residual stresses introduced by different working processes in brasses of various compositions and the annealing treatments necessary to remove these stresses or reduce them to a safe level. A 'stress-relief anneal' at about 300°C will usually lower internal stresses to comparatively small values without much effect on the hardness of the material.

Specifications for brass products customarily include provision for carrying out a mercurous nitrate test[157] to ensure that unduly high residual tensile stresses are not present, but a satisfactory result in this test does not guarantee freedom from cracking in environments containing ammonia. More searching tests involving exposure to ammonia have therefore been devised. The standardisation of stress-corrosion cracking tests and their correlation with service experience have been described in several papers[167]. Other authors[168] have described practical cases of stress-corrosion cracking, usually involving tensile stresses applied in service. Two possible preventive measures are the use of coatings[169] or inhibitors[170].

The behaviour of a wide range of α, α–β and β brasses in various corrosive environments was studied by Voce and Bailey and the results summarised by Whitaker[171]. Penetration by mercury and by molten solder was intercrystalline in all three types of brass. In moist ammoniacal atmospheres the penetration of unstressed brasses of all types was intercrystalline. Internal or applied stresses accelerated the intercrystalline penetration of α brasses and initiated some transcrystalline cracking, and also caused severe transcrystalline cracking of β alloys and transcrystalline cracking across the β regions in the two-phase brasses. Immersion in ammonia solution, however, caused intercrystalline cracking of stressed β brasses.

β brasses containing 3% or more aluminium failed with an intercrystalline fracture when stressed at about the 0·1% proof stress in air. The behaviour of alloys of this type was subsequently studied by Perryman[172], and by Bailey[173], who has shown that cracking in air occurs only when moisture is present. It has been confirmed that β brasses are prone to crack in service[174].

High-strength α–β brasses containing up to about 5% Al (with small amounts of Fe, Mn, Sn, etc.) used for propellers, parts of pumps, nuts and bolts, etc. usually give good service but occasionally suffer intercrystalline failure, for instance in contact with sea-water. Examination of such failures usually reveals thin dezincified layers along the cracks, but it is difficult to decide whether the crack or the dezincification occurred first.

The theoretical aspects of stress-corrosion cracking have attracted much attention in recent years. Amongst the copper alloys, the behaviour of brasses in ammoniacal environments has been most studied. Whilst cracking has been shown to be possible in contact with some other corrosive agents, ammonia has the most powerful effect. Evans[132] suggests that this is because ammonia is a feeble corrodent that produces little attack except at regions such as grain boundaries or other lattice imperfections and because it prevents accumulation of copper ions in the crevices formed owing to the formation of stable complexes, $Cu[(NH_3)_4]^{2+}$. The mode of cracking (intercrystalline or transcrystalline) can be affected by changes in composition of the brass or by changes in the nature of the environment[175]. Mattsson[176] found that on immersion in ammoniacal solutions of different pH values, stressed brasses cracked most rapidly at pH 7·1–7·3 and that in this range black surface films formed on the metal. The rôle of tarnish films has been further studied subsequently[177]. Many authors have studied electrochemical[178] or metallurgical[179] aspects of the stress corrosion of copper alloys and discussed theories of the mechanism. Papers on the subject have been included in several symposia or conferences on stress corrosion of metals[159, 180, 181]. The stress cracking of brasses was reviewed by Bailey[182] and subsequent reviews of stress corrosion[183] contain references to the subject.

Stress Corrosion of Other Copper Alloys

Evidence indicates that pure copper is not liable to undergo stress corrosion[184–186] but instances are known of the failure by stress corrosion of copper containing about 0·4% As[187] or 0·02% P[184]. Failure can also occur with copper–beryllium[188], copper–manganese[189], aluminium-bronzes[190], tin-bronzes, silicon-bronzes, nickel silvers[191] and cupro-nickels[191, 192]. Most of these alloys are much less susceptible to cracking than brasses[185, 193]. Under some conditions, however, aluminium-bronzes can be very prone to cracking[194]. In ammoniacal environments the cracks tend to be transcrystalline, and in steam atmospheres intercrystalline[195]. Additions of 0·35% Sn or Ag are claimed to be effective in preventing intercrystalline cracking[196, 197]. An aluminium–bronze containing 2% Ni and 0·5–0·75% Si is claimed to have good resistance to stress corrosion[196, 198].

Thompson and Tracy[184] carried out tests in a moist ammoniacal atmosphere on stressed binary copper alloys containing zinc, phosphorus, arsenic, antimony, silicon, nickel or aluminium. All these elements gave alloys susceptible to stress corrosion. In the case of zinc the breaking time decreased steadily with increase of zinc content, but with most of the other elements there was a minimum in the curve of content of alloying elements against breaking time. In tests carried out at almost 70 MN/m^2 these minima occurred with about 0·2% P, 0·2% As, 1% Si, 5% Ni and 1% Al. In most cases cracks were intercrystalline.

Protective Measures

The good behaviour of copper and copper alloys is dependent upon correct choice of material, good design of equipment, and proper methods of operation. If proper attention is given to these factors there will usually be no need for protective measures. In special cases, however, e.g. to prevent the dissolution of small amounts of copper or to maintain a high-grade finish, metallic coatings of one or more of the following metals may be applied: tin, lead, nickel, silver, chromium, rhodium, gold. In other cases painting, varnishing or lacquering may be desirable, or if the conditions are very severe, as in some corrosive soils, heavier protection such as bituminous or plastic coatings may be necessary. Brasses that are liable to suffer dezincification or stress corrosion may need protection where other copper alloys would be satisfactory unprotected. Sometimes use is made of the principles of cathodic protection, e.g. steel 'protector blocks' in condenser water-boxes.

In some circumstances, use of inhibitors may be a desirable remedial measure. For instance, benzotriazole has been found of considerable value for preventing staining and tarnishing of copper products[200]. Sodium diethyldithiocarbamate also has useful inhibiting properties[201]. Other types of inhibitors can be of value in condensate systems[202] and in acid solutions[117,118]. Reviews have been given of corrosion inhibitors for copper[203] and brasses[204].

The danger of accelerated attack on copper-base materials due to coupling with other metals is small since copper is usually the cathodic member of the couple, but precautions are often necessary to prevent excessive corrosion of the anodic member. Surveys of the behaviour of couples involving copper and copper alloys have been given[11,205]. One material that has been found capable of accelerating attack on copper in practice is graphite; hence graphitic paints are undesirable. Occasionally the action between different copper-base materials may be appreciable, e.g. gunmetal may stimulate the corrosion of copper or brass in sea-water.

P. T. GILBERT

Acknowledgement This section is based on the article 'Chemical Properties and Corrosion Resistance of Copper and Copper Alloys' which formed Chapter XVIII of the American Chemical Society Monograph No. 122, *Copper: The Science and Technology of the Metal its Alloys and Compounds*, edited by Professor Allison Butts, and published by the Reinhold Publishing Corporation, New York, in 1954. Acknowledgement is hereby made to the Reinhold Publishing Corporation for permission to use the above-mentioned section as a basis for the present chapter.

REFERENCES

1. *Copper Alloy Ingots and Copper and Copper Alloy Castings*, BS 1400 (1969)
2. Specifications for Copper and Copper Alloys:
   BS 2870, *Sheet, Strip and Foil* (1968); BS 2871, *Tubes* (1971); BS 2872, *Forging Stock and Forgings* (1969); BS 2873, *Wire* (1969); BS 2874, *Rods and Sections* (1969); BS 2875, *Plate* (1969)
3. *Copper and Copper Alloys*, Part 5, American Society for Testing and Materials, Philadelphia (1970)
4. *Standard Handbook—Copper, Brass, Bronze, Wrought Mill Products*, 6th edn, Copper Development Association Inc., New York (1968); 'Properties and Applications of Wrought Coppers and Copper Alloys', *Metal Prog.*, **98**, 85 (1970)
5. 'Wrought Cupro-Nickels', *Mater. in Des. Engng.*, **49**, 127 (1959); Katz, W., *Werkst. u. Korrosion*, **15**, 977 (1964)
6. 'Silicon Bronzes', *Mater. in Des. Engng.*, **50**, 112 (1959)
7. Evans, U. R., *Metallic Corrosion Passivity and Protection*, Edward Arnold, London (1946); Evans, U. R., *The Corrosion and Oxidation of Metals*, Edward Arnold, London (1960)
8. Uhlig, H. H. (Ed.), *Corrosion Handbook*, Wiley, New York and Chapman and Hall, London (1948)
9. LaQue, F. L. and Copson, H. R., *Corrosion Resistance of Metals and Alloys*, 2nd edn, Reinhold, New York (1963)
10. Speller, F. N., *Corrosion, Causes and Prevention*, McGraw-Hill, New York (1951)
11. Burghoff, H. L., *Corrosion of Metals*, Amer. Soc. Metals Monograph, Cleveland, Ohio, 100–130 (1946)
12. *Metals Handbook*, Amer. Soc. Metals, Metals Park, Ohio, **1**, 983–1005 (1961)
13. Rogers, T. Howard, *Marine Corrosion Handbook*, McGraw-Hill Co. of Canada Ltd., Toronto (1960)

14. Rogers, T. Howard, *Marine Corrosion*, Butterworths, London (1968)
15. Cairns, J. H. and Gilbert, P. T., *The Technology of Heavy Non-Ferrous Metals and Alloys*, Butterworths, London (1967)
16. Anon., *Corrosion*, **15**, 199t (1959); **16**, 131t (1960); Lasko, W. R. and Tice, W. K., *Corrosion*, **18**, 116t (1962); Erdös, E., *Werkst. u. Korrosion*, **19**, 385 (1968)
17. Latimer, W. M., *Oxidation Potentials*, Prentice-Hall, New Jersey (1952)
18. Gatty, O. and Spooner, E. C. R., *Electrode Potential Behaviour of Corroding Metals in Aqueous Solutions*, Clarendon Press, Oxford (1938)
19. Pourbaix, M. J. N., *Thermodynamics of Dilute Aqueous Solutions* (Trans. by J. N. Agar), Edward Arnold, London (1949)
20. Pourbaix, M. J. N., *Atlas of Electrochemical Equilibria in Aqueous Solutions*, Pergamon, Oxford (1966)
21. Rubinić, L. and Marković, T., *Werkst. u. Korrosion*, **10**, 666 (1959)
22. Hickling, A. and Taylor, D., *Trans. Faraday Soc.*, **44**, 262 (1948)
23. Feitknecht, W. and Lehnel, H. W., *Helv. Chim. Acta.*, **27**, 775 (1944); Wilde, B. E. and Teterin, G. A., *Brit. Corrosion J.*, **2**, 125 (1967)
24. Bonhoeffer, K. F. and Gerischer, J., *Z. Elektrochem.*, **52**, 149 (1948); Stolica, N. D. and Uhlig, H. H., *J. Electrochem. Soc.*, **110**, 1215 (1963); Mansfield, F. and Uhlig, H. H., *J. Electrochem. Soc.*, **117**, 427 (1970); Varenko, E. S. *et al.*, *Zashchita Metallov.*, **6**, 103 (1970)
25. Vernon, W. H. J., *J. Inst. Met.*, **49**, 153 (1932)
26. Freeman, J. R. and Kirby, P. H., *Metals and Alloys*, **5**, 67 (1934)
27. Vernon, W. H. J. and Whitby, L., *J. Inst. Met.*, **42**, 181 (1929); **44**, 389 (1930); **49**, 153 (1932); Vernon, W. H. J., *J. Chem. Soc.*, 1853 (1934)
28. Vernon, W. H. J., *Trans. Faraday Soc.*, **27**, 255, 582 (1931)
29. Evans, U. R., *Trans. Electrochem. Soc.*, **46**, 247 (1924)
30. Hudson, J. C., *Trans. Faraday Soc.*, **25**, 177 (1929); *J. Inst. Met.*, **44**, 409 (1930); *J. Birmingham Met. Soc.*, **14**, 331 (1934); *Metal Ind.*, **44**, 415 (1934); *J. Inst. Met.*, **56**, 91 (1935)
31. Tracy, A. W., Thompson, D. H. and Freeman, J. R., Special Technical Publication No. 175, A.S.T.M., 77–87 (1955)
32. Tracy, A. W., STP No. 175, A.S.T.M., 67–76 (1955)
33. Thompson, D. H., *Metal Corrosion in the Atmosphere*, STP 435, A.S.T.M., 129 (1968)
34. Mattsson, E. and Holm, R., *Metal Corrosion in the Atmosphere*, STP 435, A.S.T.M., 187 (1968)
35. Scholes, I. R. and Jacob, W. R., *J. Inst. Met.*, **98**, 272 (1970)
36. Compton, K. G., *Trans. Electrochem. Soc.*, **91**, 705 (1947); Ambler, H. R. and Bain, A. A. J., *J. Appl. Chem.*, **5**, 437 (1955); Forgeson, B. W. *et al.*, *Corrosion*, **74**, 73t (1958); Hummer, C. W., Jr., Scuthwell, C. R. and Alexander, A. L., *Mater. Protection*, **7** No. 1, 41 (1968)
37. Cole, H. G., R.A.E. Report MET82 (1954); Compton, K. G., Mendizza, A. and Bradley, W. W., *Corrosion*, **11**, 383t (1955)
38. Romanoff, M., *Underground Corrosion*, Nat. Bur. Stand. Circ. 579, Supt. of Documents, Washington, D.C. (1957)
39. Gilbert, P. T., *J. Inst. Met.*, **73**, 139 (1947)
40. Gilbert, P. T. and Porter, F. C., Iron and Steel Inst. Special Report No. 45, 55–74, 127–134 (1952)
41. Marković, T., Sevdić, M. and Rubinić, L., *Werkst. u. Korrosion*, **11**, 87 (1960)
42. Bengough, G. D. and May, R., *J. Inst. Met.*, **32**, 81 (1924)
43. May, R., *J. Inst. Met.*, **40**, 141 (1928)
44. May, R. and de V. Stacpoole, R. W., *J. Inst. Met.*, **77**, 331 (1950)
45. LaQue, F. L. and Stewart, W. C., *Métaux et Corros.*, **23**, 147 (1948); LaQue, F. L., *Proc. Amer. Soc. Test. Mater.*, **52**, 1 (1952)
46. Gilbert, P. T. and LaQue, F. L., *J. Electrochem. Soc.*, **101**, 448 (1954)
47. Fink, F. W., *Trans. Electrochem. Soc.*, **75**, 441 (1939); Evans, U. R., *ibid.*, 446
48. Lucey, V. F., *Brit. Corrosion J.*, **1**, 9 and 53 (1965)
49. Feller, H-G., *Z. Metallkunde*, **58**, 875 (1967)
50. Heidersbach, R., *Corrosion*, **24**, 38 (1968)
51. Polushkin, E. P. and Shuldener, H. L., *Trans. Amer. Inst. Min. (Metall.) Engrs.*, **161**, 214 (1945); Kleinberger, R., Okuzumi, H. and Perio, P., *Métaux (Corrosion-Ind.)*, No. 413, 40–43 (1960); Kenworthy, L. and O'Driscoll, W. G., *Corrosion Tech.*, **2**, 247 (1955); Piatti, L. and Grauer, R., *Werkst. u. Korrosion*, **14**, 551 (1963); Hashimoto, K., Ogawa, S. and Shimodaira, S., *Trans. Japan Inst. Met.*, **4**, 42 (1963); Robinson, F. P. A. and Shalit, M.,

Corrosion Tech., **11**, 11 (1964); Rowlands, J. C., *Proc. 2nd Internat. Cong. Met. Corrosion*, New York, 1963, N.A.C.E., Houston, 795 (1966); Frade, G. and Lacombe, P., *Mém. Sci. Rev. Métall.*, **63**, 649 (1966); Sugawara, H. and Ebiko, H., *Corrosion Sci.*, **7**, 513 (1967); Joseph, G. and Arce, M. T., *Corrosion Sci.*, **7**, 597 (1967); Langenegger, E. E. and Robinson, F. P. A., *Corrosion*, **24**, 411 (1968) and **25**, 59 (1969); Horton, Ralph M., *Corrosion*, **26**, 160 (1970); Rothenbacher, P., *Corrosion Sci.*, **10**, 391 (1970); Pötzl, R. and Lieser, K. H., *Z. Metallkunde*, **61**, 527 (1970)

52. Bengough, G. D., Jones, R. M. and Pirret, R., *J. Inst. Met.*, **23**, 65 (1920)
53. Upton, B., *Brit. Corros. J.*, **1**, 134 (1966)
54. May, R., *Trans. Inst. Mar. Engrs.*, **49**, 171 (1937); Sherborne, H. F. (p. 76), Bailey, G. L. (p. 78). Discussion of Bradbury, E. J. and Johnson, L. W., *Trans. Inst. Mar. Engrs.*, **63**, 59 (1951)
55. Breckon, C. and Gilbert, P. T., *Chem. and Ind.*, Jan. 4, 35 (1964)
56. Bem, R. S., *The Engineer*, **206**, 756 (1958)
57. Clark, W. D., *J. Inst. Met.*, **73**, 263 (1947)
58. Breckon, C. and Gilbert, P. T., *Proceedings of the First International Congress on Metallic Corrosion*, London, 1961. Butterworths, London, 624 (1962)
59. Gleekman, L. W. and Swanby, R. K., *Corrosion*, **17**, 144t (1961); Rowlands, J. C., *Corrosion Sci.*, **2**, 89 (1962); Smith, A. A., *Corrosion Prev. and Control*, **10**, 29 (1963); Upton, B., *Corrosion*, **19**, 204t (1963); Piatti, L. and Grauer, R., *Werkst. u. Korrosion*, **14**, 551 (1963); Neiderberger, R. B., *Modern Castings*, **45** No. 3, 115 (1964); Gaillard, F. and Weill, A. R., *Mem. Sci. Rev. Met.*, **61**, 437 (1964); Maersch, R. E. and Ciesleiwicz, J. M., *Mater. Protection*, **3** No. 7, 54 (1964); Shinoda, G. and Amano, Y., *Trans. Japan Inst. Met.*, **4**, 231 (1963); Tanabe, Z., *Corrosion Sci.*, **4**, 413 (1964); Arnaud, D. *et al.*, *Fonderie*, No. 226, 403 (1964); Arnaud, D., *Fonderie*, No. 275, 88 and No. 281, 355 (1969); Shibard, P. R. and Balachandra, J., *Anti Corrosion Methods Mater.*, **14** No. 2, 10 (1967)
60. Gilbert, P. T. and May, R., *Trans. Inst. Mar. Engrs.*, **62**, 291 (1950); Gilbert, P. T., *Trans. Inst. Mar. Engrs.*, **66**, 1 (1954); Breckon, C. and Baines, J. R. T., *Trans. Inst. Mar. Engrs.*, **67**, 1 (1955); Bradbury, E. J. and Johnson, L. W., *Trans. Inst. Mar. Engrs.*, **63**, 59 (1951); Slater, I. G., Kenworthy, L. and May, R., *J. Inst. Met.*, **77**, 309 (1950); Bethon, H. E., *Corrosion*, **4**, 457 (1948); Eichhorn, K., *Werkst. u. Korrosion*, **8**, 657 (1957) and **21**, 535 (1970); Nothing, F. W., *Metall.*, **10**, 520 (1956) and **16**, 1089, 1196 (1962); Maurin, A. J., *Corros. and Anti-Corros.*, **5**, 275, 383 (1957) and **6**, 15 (1958); Todhunter, H. A., *Corrosion*, **11**, 221t (1955) and *Power*, **100**, 85 (1956); Gilbert, P. T., *Chem. and Ind.*, July 11, 888 (1959); Nowlan, N. V., *Corrosion Tech.*, **7**, 397 (1960); Otsu, T., *Sumitomo Light Metal Tech. Rep.*, **1** No. 1, 62 (1960); Gilbert, P. T., *Inst. Mar. Engrs. Materials Section Symposium*, March, 14 (1968); Gilbert, P. T., *Trans. Inst. Mar. Engrs.*, **82** No. 7, 6 (1970); Kingerley, D. G., *Brit. Chem. Engng.*, **6** No. 1, 20 (1961); Sisson, A. B., *Corrosion*, **17**, 18 (1961); Hall, B. N., *Corrosion Prev. and Control*, **10**, 49 (1963); Malcolm, R. R., *Australasian Corros. Engng.*, **7** No. 3, 17 and **7** No. 10, 25 (1963); Serre, J. and Laureys, J., *Corrosion et Anti-Corrosion*, **11**, 305, 360 (1963) and *Corrosion Sci.*, **5**, 135 (1965); Kenworthy, L., *Trans. Inst. Mar. Engrs.*, **77**, 149 (1965); Page, G. C., *Anti-Corrosion Methods Mater.*, **14** No. 5, 13 (1967); Yandushkin, K. N., *Zashchita Metallov.*, **6**, 46 (1970)
61. Tracy, A. W., *Corrosion*, **1**, 103 (1945); Mitchell, N. W., *Corrosion*, **3**, 243 (1947); Van der Baan Sj., *Corrosion*, **6**, 14 (1950); Mason, J. F., *Corrosion*, **12**, 199t (1956); Rust, A. D., *Corrosion Tech.*, **3**, 185 (1956); Gilbert, P. T., *Soc. Chem. Ind. Monograph*, No. 10, 111–120 (1960); Bird, D. B. and Moore, K. L., *Mater. Protection*, **1** No. 10, 70 (1962)
62. Bengough, G. D., *J. Inst. Met.*, **5**, 28 (1911); Bengough, G. D. and Jones, R. M., *J. Inst. Met.*, **10**, 13 (1913); Gibbs, W. E., Smith, R. H. and Bengough, G. D., *J. Inst. Met.*, **15**, 37 (1916); Bengough, G. D. and Hudson, O. F., *J. Inst. Met.*, **21**, 167 (1919)
63. Bailey, G. L., *J. Inst. Met.*, **79**, 243 (1951); Tracy, A. W. and Hungerford, R. L., *Proc. Amer. Soc. Test. Mater.*, **45**, 591 (1945); LaQue, F. L. and Mason, J. F., *Proc. Amer. Petrol. Inst.*, **30M III**, 103 (1950); Todhunter, H. A., *Corrosion*, **16**, 226t (1960) and *Mater. Protection*, **6** No. 7, 45 (1967); LaQue, F. L. and Stewart, W. C., *Corrosion*, **8**, 259 (1952); Krafack, K. and Franke, E., *Werkst. u. Korrosion*, **4**, 310 (1953); Se Ui Yu and Turkovsaya, A. V., *Tsvetnaya Metall.*, **4**, 145 (1961); May, T. P. and Weldon, B. A., *Rev. Nickel*, No. 3, 183 and No. 4, 219 (1966)
64. Chapman, J. and Cuthbertson, J. W., *J. Soc. Chem. Ind.*, **58**, 100, 330 (1939); Cuthbertson, J. W., *J. Inst. Met.*, **72**, 317 (1946)
65. Tanabe, Z., *Sumitomo Light Metal Tech. Rep.*, **9** No. 3, 167 (1968)
66. Baker, L., *Trans. Inst. Mar. Engrs.*, **65**, 26 (1953)

67. Rogers, T. Howard, *J. Inst. Met.*, **75**, 19 (1948–49)
68. Sato, S., *Sumitomo Light Metal Tech. Rep.*, **6** No. 1, 42 (1965); Tanaka, R., *Sumitomo Light Metal Tech. Rep.*, **3** No. 3, 55 and **3** No. 4, 1 (1962) and **6** No. 3, 152 (1965); Changarnier, J., *Corros. Anti-Corros.*, **1**, 8 (1953)
69. Shimodaira, S., Sugawara, H. and Sato, S., *Sumitomo Light Metal Tech. Rep.*, **4** No. 1, 31 (1963); Tanabe, Z., *Sumitomo Light Metal Tech. Rep.*, **5** No. 1, 16 (1964)
70. Otsu, T. and Sato, S., *Sumitomo Light Metal Tech. Rep.*, **2** No. 2, 23 and **2** No. 4, 27 (1961); Otsu, T. and Okawa, M., *Sumitomo Light Metal Tech. Rep.*, **3** No. 3, 35 (1962); Otsu, T., Sato, S. and Watanabe, T., *Sumitomo Light Metal Tech. Rep.*, **4** No. 2, 21 (1963); Tanaka, R. and Tanabe, Z., *Sumitomo Light Metal Tech. Rep.*, **5** No. 1, 9 (1964); Tanaka, R., *Sumitomo Light Metal Tech. Rep.*, **5** No. 3, 188 (1964) and **6** No. 1, 71 (1965); Sato, S. and Sagiska, K., *Sumitomo Light Metal Tech. Rep.*, **11** No. 2, 1 (1970)
71. Tanabe, Z., *Sumitomo Light Metal Tech. Rep.*, **6** No. 2, 119 (1965)
72. Sato, S., 4th International Congress Met. Corros., Amsterdam (1969). Nat. Assoc. Eng., Houston, 795 (1972)
73. Rowlands, J. C., *J. Appl. Chem.*, **15**, 57 (1965)
74. Grubitsch, H., Hilbert, F. and Sammer, R., *Werkst. u. Korrosion*, **17**, 760 (1966); Tanabe, Z., *Sumitomo Light Metal Tech. Rep.*, **8** No. 1, 10 (1967); Meany, J. J., Jr., *Mater. Protection*, **8** No. 10, 27 (1969); Mor, E., Scotto, V. and Trevis, A., *Corrosion (Paris)*, **18** No. 2, 67 (1970); Baudo, G. and Giuliani, L., *Werkst. u. Korrosion*, **21**, 332 (1970); North, R. F. and Pryor, M. J., *Corrosion Sci.*, **9**, 509 (1969) and **10**, 297 (1970); Giuliani, L. and Bombara, G., *Brit. Corrosion J.*, **5**, 179 (1970)
75. McAllister, R. A., *et al.*, *Corrosion*, **17**, 579t (1961); Erdös, E., *Schweizer Archiv.*, **30**, 251 (1964); Mifflin, R. C. and Bird, D. B., *Mater. Protection*, **8** No. 9, 72 (1969)
76. Sato, S., *Sumitomo Light Metal Tech. Rep.*, **3** No. 3, 106 (1962)
77. Sherry, A. and Gill, E. R., *Chem. and Ind.*, Jan. 18, 102 (1964); Edwards, B. C., *Corrosion Sci.*, **9**, 395 (1969)
78. Gilbert, P. T., *Chem. and Ind.*, July 11, 888 (1959)
79. Sato, S., Nagata, K. and Ogiso, A., *Sumitomo Light Metal Tech. Rep.*, **11** No. 3, 1 (1970)
80. Peplow, D. B., *Brit. Power Engineering*, **1** No. 5, 51 (1960); Attwood, P. G. and Richards, N. G., *Corrosion*, **17**, 8t (1960); Crennel, J. T. and Sawyer, L. J. E., *J. Appl. Chem.*, **12**, 170 (1962); Page, G. G., *Proc. 2nd Internat. Cong. Met. Corros.*, New York (1963). *Nat. Assoc. Corros. Eng.*, Houston, 275 (1966)
81. Bostwick, T. W., *Corrosion*, **17**, 12 (1961)
82. Lockhart, A. M., *Proc. Inst. Mech. Engrs.*, **179**, 495 (1964–65)
83. North, R. F., *Corrosion Sci.*, **8**, 149 (1968); Gasparini, R., Della Rocca, C. and Ioannilli, E., *Corrosion Sci.*, **10**, 157 (1970)
84. *Proceedings of Conference on the Role of Copper and its Alloys in Desalination Equipment*, London, December (1966), Copper Development Association. See Stewart, J. M., p. 21, Gilbert, P. T., p. 31, Weldon, B. A. and Tuthill, A. H., p. 39; Tuthill, A. H. and Sudrabin, D. A., *Metals Engng. Quart.*, **7** No. 3, 10 (1967); Fink, F. W., Tech. Rep. 704/6, Copper Development Association, New York (1966) and *Mater. Protection*, **6** No. 5, 40 (1967); Schoraten, A., *Metall.*, **22**, 1153 (1968); Cohen, A. and Rice, L., *Mater. Protection*, **8** No. 12, 67 (1969) and **9** No. 11, 29 (1970); Bom, P. R., *Brit. Corrosion J.*, **5**, 258 (1970)
85. Campbell, H. S. and Carter, V. E., *J. Inst. Metals*, **90**, 362 (1962); Murphy, T. J. and Jack, J. B., *Shipping World and Shipbuilder*, Jan. 21, 282 (1965)
86. Gilbert, P. T. and Jenner, B. J., *Inst. Marine Engrs. International Marine and Shipping Conference*, London, June (1969)
87. Campbell, H. S., *Chem. and Ind.*, 692 (1955); Hatch, G. B., *J. Amer. Waterworks Assoc.*, **53**, 1417 (1961); N.A.C.E. Tech. Rep. 60–11 and *Corrosion*, **16**, 453t (1960); Schafer, G. J., *New Zealand J. Sci.*, **5**, 475 (1962)
88. Turner, M. E. D., *Proc. Soc. Water Treatm. Exam.*, **10**, 162 (1961) and **14**, 81 (1965)
89. Baldwin, A. B. and Campbell, H. S., *Brit. Waterworks Assoc. J.*, **43**, 13 (1961); Schafer, G. J. and Dall, R. A., *Australasian Corrosion Engng.*, **10** No. 3, 9 (1966); Ladeburg, H., *Metall.*, **20**, 33 (1966); Simmonds, M. A. and Huxley, W. G. S., *Australasian Corrosion Engng.*, **11** No. 11, 9 (1967)
90. Schafer, G. J., Foster, P. K. and Marshall, T., *New Zealand J. Sci.*, **4**, 194 (1961)
91. Schafer, G. J. and Dall, R. A., *Brit. Corrosion J.*, **3**, 12 (1968); Harrison, P. S., *Electrical Times*, **153**, 219 (1968)
92. Obrecht, M. F., *Corrosion*, **18**, 189t (1962)
93. Ingleson, H., Sage, A. M. and Wilkinson, R., *J. Inst. Wat. Engrs.*, **3**, 81 (1949); Wormwell,

F. and Nurse, T. J., *J. Appl. Chem.*, **2**, 685 (1952); Solelev, A., *J. Inst. Wat. Engrs.*, **9**, 208 (1955)
94. Tronstad, L. and Veimo, R., *J. Inst. Met.*, **66**, 17 (1940); Kenworthy, L., *J. Instn. Heat. Vent. Engrs.*, **8**, 15 (1940); Gilbert, P. T., *Proc. Soc. Water Treatm. Exam.*, **15**, 165 (1966)
95. Kenworthy, L., *J. Inst. Met.*, **69**, 67 (1943)
96. Porter, F. C. and Hadden, S. E., *J. Appl. Chem.*, **3**, 385 (1953)
97. Davenport, W. H., Nole, V. F. and Robertson, W. D., *J. Electrochem. Soc.*, **106**, 1005 (1959); Ives, D. J. G. and Rawson, A. E., *J. Electrochem. Soc.*, **109**, 447 (1962); Obrecht, M. F. and Pourbaix, M., *J. Amer. Waterworks Assoc.*, **59**, 977 (1967)
98. Grunau, E. B., *Städtehygiene*, No. 7, 153 (1967)
99. Campbell, H. S., *J. Inst. Met.*, **77**, 345 (1950); *J. Appl. Chem.*, **4**, 633 (1954); *Proc. Soc. Wat. Treatm. Exam.*, **3**, 100 (1954) and *Proc. 2nd. Internat. Cong. Met. Corrosion*, New York (1963), N.A.C.E., Houston, 237 (1966)
100. Schafer, G. J., *Australasian Corrosion Engng.*, **6** No. 8, 15 (1962) and *New Zealand Plumbing Rev.*, **1** No. 9, 10 (1964); Rambow, C. A. and Holmgren, R. S., Jr., *J. Amer. Waterworks Assoc.*, **58**, 347 (1966); Pourbaix, M., Van Muylder, J. and Van Laer, P., *Corrosion Sci.*, **7**, 795 (1967); von Franqué, O., *Werkst. u. Korrosion*, **19**, 377 (1968); Lihl, F. and Klamet, H., *Werkst. u. Korrosion*, **20**, 108 (1969); Walker, I. K. and Page, G. G., *Australasian Corrosion Engng.*, **13** No. 4, 13 (1969); Kennett, A., *Australasian Corrosion Engng.*, **13** No. 4, 5 (1969); Gilbert, P. T., *Australasian Corrosion Engng.*, **13** No. 5, 13 (1969)
101. Schafer, G. J. and Dall, R. A., *Australasian Corrosion Engng.*, **7** No. 10, 33 (1963)
102. Lucey, V. F., *Brit. Corrosion J.*, **2**, 175 (1967)
103. Mattsson, E. and Fredriksson, A-M., *Brit. Corrosion J.*, **3**, 246 (1968)
104. Lee, J. A., *Materials of Construction for Chemical Process Industries*, McGraw-Hill, New York (1950)
105. Rabald, E., *Corrosion Guide*, Elsevier Publishing Co., New York (1951); Carmenisch, K. P., *Pro-Metal*, **13** No. 73, 288 (1960); Ritter, F., *Korrosionstabellen Metallischer Werkstoffe*, Springer-Verlag, Vienna (1958); LaQue, F. L., *Corrosion*, **10**, 391 (1954); Heim, A. T., *Industr. Engng. Chem.*, **49**, 63A, 64A, 66A (1957); Baker, S., *Corrosion Tech.*, **8**, 8 (1961); Tracy, A. W., *Chem. Engng.*, **69**, 130, 152 (1962); Anon., *Industr. Engng. Chem.*, **40**, 1827 (1948), **43**, 2218 (1951) and **49**, 63A (1957)
106. Gould, A. J. and Evans, U. R., *J. Iron St. Inst.*, **155**, 195 (1947)
107. Lacan, M., Markovic, T. and Rubinic, L., *Werkst. u. Korrosion*, **10**, 767 (1959)
108. Holmberg, M. E. and Prange, F. A., *Industr. Engng. Chem.*, **37**, 1030 (1945); Anon., *Industr. Chem.*, **30**, 609 (1954); Lingnau, E., *Werkst. u. Korrosion*, **8**, 216 (1957)
109. Fontana, M. G., *Industr. Engng. Chem.*, **42**, 69A (1950)
110. Groth, V. J. and Hafsten, R. J., *Corrosion*, **10**, 368 (1954)
111. Bulow, C. L., *Chem. Engng.*, **53**, 210 (1946)
112. Friend, W. Z. and Mason, J. F., *Corrosion*, **5**, 355 (1949), N.A.C.E. Report and *Corrosion*, **13**, 757t (1957)
113. Russell, R. P. and White, A., *Industr. Engng. Chem.*, **19**, 116 (1927); Damon, G. H. and Cross, R. C., *Industr. Engng. Chem.*, **28**, 231 (1936)
114. Cornet, I., Barrington, E. A. and Behrsing, G. U., *J. Electrochem. Soc.*, **108**, 947 (1961)
115. Caney, R. J. T., *Aust. Engr.*, **64**, 54 (1954) and U.K. Pat 718,987; Zitter, H. and Kraxner, G., *Werkst. u. Korrosion*, **14**, 80 (1963); Piatti, L. and Fot, E., *Werkst. u. Korrosion*, **15**, 27 (1964)
116. Gregory, D. P. and Riddiford, A. C., *J. Electrochem. Soc.*, **107**, 950 (1960); Talati, J. D., Desai, M. N. and Trivedi, A. M., *Werkst. u. Korrosion*, **12**, 422 (1961); Bumbulis, J. and Graydon, W. F., *J. Electrochem. Soc.*, **109**, 1130 (1962); Kagetsu, T. J. and Graydon, W. F., *J. Electrochem. Soc.*, **110**, 856 (1963); Feller, H-G., *Corrosion Sci.*, **8**, 259 (1968); Otsuka, R. and Uda, M., *Corrosion Sci.*, **9**, 703 (1969)
117. Rana, S. S. and Desai, M. N., *Indian J. Technol.*, **5**, 393 (1967); Desai, M. N. and Shah, Y. C., *Anti-Corrosion Methods Mater.*, **15** No. 12, 9 (1968); Desai, M. N., Shah, Y. C. and Gandhi, M. H., *Corrosion Sci.*, **9**, 65 (1969); Padma, D. K. and Rama Char, T. L., *Anti-Corrosion Methods Mater.*, **16** No. 4 (1969); Desai, M. N., Shah, Y. C. and Punjani, B. K., *Brit. Corrosion J.*, **4**, 309 (1969)
118. Ammar, I. A. and Riad, S., *Corrosion Sci.*, **9**, 423 (1969)
119. Anon., *Proc. Amer. Soc. Test. Mater.*, **35**, 161 (1935); Desai, M. N. and Rana, S. S., *Werkst. u. Korrosion*, **17**, 870 (1966)
120. Radley, J. A., Stanley, J. S. and Moss, G. E., *Corrosion Tech.*, **6**, 229 (1959); Schaefer, B. A., *Corrosion Sci.*, **8**, 623 (1968); Bartoníček, R., Holinka, M. and Lukašovská, M., *Werkst. u.*

Korrosion, **19**, 1032 (1968); Green, J. A. S., Mengelberg, H. D. and Yolken, H. T., *J. Electrochem. Soc.*, **117**, 433 (1970); Jenkins, L. H. and Durham, R. B., *J. Electrochem. Soc.*, **117**, 768 (1970)

121. Dubrisay, R. and Chesse, G., *Compt. Rend. Acad. Sci. Paris*, **220**, 707 (1945)
122. Reichert, J. S. and Pete, R. H., *Chem. Engng.*, **54**, 218 (1947)
123. West, J. R., *Chem. Engng.*, **58**, 281 (1951)
124. Botham, G. H. and Dummett, G. A., *J. Dairy Res.*, **16**, 23 (1949)
125. Holness, H. and Ross, T. K., *J. Appl. Chem.*, **1**, 158 (1951); Bukowiecki, A., *Schweizer Archiv. Angew. Wiss.*, **24**, 355 (1958)
126. Mason, J. F., *Corrosion*, **4**, 305 (1948); Inglesent, H. and Storrow, J. A., *J. Soc. Chem. Ind.*, **64**, 233 (1945); Clendenning, K. A., *Canad. J. Res. F (Technology)*, **26**, 277 (1948)
127. *Review of Oxidation and Scaling of Heated Metal Solids*, D.S.I.R., H.M.S.O., London (1935); Vernon, W. H. J., *Chem. and Ind. (Rev.)*, **59**, 87 (1940)
128. Tylecote, R. F., *J. Inst. Met.*, **78**, 259 (1950–51); Cabrera, N. and Mott, N. F., *Rep. Progr. Phys.*, **12**, 163 (1948–49); Rönnquist, A. and Fischmeister, H., *J. Inst. Met.*, **89**, 65 (1960–61)
129. Kubaschewski, O. and Hopkins, B. E., *Oxidation of Metals and Alloys*, Butterworths, London (1953); Hauffe, K., *Oxydation von Metallen und Legierungen*, Springer-Verlag, Berlin (1956)
130. Wagner, C. and Grunewald, K., *Z. Phys. Chem.*, **40**, 455 (1938B)
131. Dighton, A. L. and Miley, H. A., *Trans. Electrochem. Soc.*, **81**, 321 (1942)
132. Evans, U. R., *Symposium on Internal Stresses in Metals and Alloys*, Inst. Metals, London, 291–310 (1947), *Trans. Electrochem. Soc.*, **91**, 547 (1947) and *Research*, London, **6**, 130 (1953)
133. Mott, N. F., *Trans. Faraday Soc.*, **35**, 1175 (1939); **36**, 472 (1940); **43**, 429 (1947); *Nature*, London, **145**, 996 (1940); *J. Chim. Phys.*, **44**, 172 (1947); *Research*, London, **2**, 162 (1949); Price, L. E., *Chem. and Ind. (Rev.)*, **56**, 769 (1937)
134. de Carli, F. and Collari, N., *Chim. et Industr.*, **33**, 77 (1951); McKewan, W. and Fassell, W. M., *J. Metals. N.Y.*, **51**, 1127 (1953); Paidassi, J., *Acta Metallurgica*, **6**, 216 (1958); Lohberg, K. and Wolstein, F., *Z. Metallk.*, **46**, 734 (1955); Baur, J. P., Bridges, D. W. and Fassell, W. M., *J. Electrochem. Soc.*, **103**, 273 (1956); Gulbrausen, E. A., Copan, T. P. and Andrew, K. F., *J. Electrochem. Soc.*, **108**, 119 (1961); Rönnquist, A., *J. Inst. Met.*, **91**, 89 (1962); Yoda, E. and Siegel, B. M., *J. Appl. Phys.*, **34**, 1512 (1963); Wallwork, G. R. and Smeltzer, W. W., *Corrosion Sci.*, **9**, 561 (1969)
135. Tylecote, R. F., *J. Inst. Met.*, **78**, 327 (1950–51) and **81**, 681 (1952–53)
136. Bardeen, J., Brattain, W. H. and Shockley, W., *J. Chem. Phys.*, **14**, 714 (1946); Castellan, G. W. and Moore, W. J., *J. Chem. Phys.*, **17**, 41 (1949)
137. Pilling, N. B. and Bedworth, R. E., *J. Inst. Met.*, **29**, 529 (1923)
138. Feitknecht, W., *Z. Elektrochem.*, **35**, 142, 500 (1929)
139. Vernon, W. H. J., *J. Chem. Soc.*, 2273 (1926)
140. Valensi, G., *Pittsburgh International Conference on Surface Reactions*, Corrosion Publishing Co., Pittsburgh, 156–165 (1948)
141. Tylecote, R. F., *J. Inst. Met.*, **78**, 301 (1950–51)
142. Oudar, J., *Métaux*, **35**, 397, 445 (1960)
143. Dyess, J. B. and Miley, H. A., *Trans. Amer. Inst. Min. (Metall.) Engrs.*, **133**, 239 (1939); Vernon, W. H. J., *Trans. Faraday Soc.*, **19**, 839 (1924)
144. Preston, G. D. and Bircumshaw, L. L., *Phil. Mag.*, **20**, 706 (1935)
145. Hallowes, A. P. C. and Voce, E., *Metallurgia, Manchr.*, **34**, 95 (1946)
146. Lustman, B., *Metal. Prog.*, **50**, 850 (1946)
147. Dennison, J. P. and Preece, A., *J. Inst. Met.*, **81**, 229 (1952–53); Spinedi, P., *Metallurg. Ital.*, **45**, 457 (1953); Collari, N. and Spinedi, P., *Metallurg. Ital.*, **46**, 403 (1954); Blade, J. C. and Preece, A., *J. Inst. Met.*, **88**, 427 (1959–60); Maak, F. and Wagner, C., *Werkst. u. Korrosion*, **12**, 273 (1961); Wallbaum, H. J., *Werkst. u. Korrosion*, **12**, 417 (1961); Maak, F., *Z. Metallkunde*, **52**, 538 (1961); Zwicker, U., *Metall.*, **16**, 1110 (1962); Kapteijn, J., Couperus, S. A. and Meijering, J. L., *Acta Metall.*, **17**, 1311 (1969); Sanderson, M. D. and Scully, J. C., *Corrosion Sci.*, **10**, 165 (1970)
148. Dunwald, H. and Wagner, C., *Z. Phys. Chem.*, **22**, 212 (1933B); Wagner, C., *Z. Phys. Chem.*, **21**, 25 (1933B), Pittsburgh International Conference on Surface Reactions, Corrosion Publishing Co., Pittsburgh, 77–82 (1948); Hoar, T. P. and Price, L. E., *Trans. Faraday Soc.*, **34**, 867 (1938)
149. Price, L. E. and Thomas, G. J., *J. Inst. Med.*, **63**, 21 (1938)
150. Schückher, F. and Lampe, V., *Pro-Metal.*, No. 105, 192 (1965)

151. Wood, G. C. and Chattopadhyay, B., *J. Inst. Met.*, **98**, 117 (1970)
152. Whittle, D. P. and Wood, G. C., *J. Inst. Met.*, **96**, 115 (1968) and *Corrosion Sci.*, **8**, 295 (1968)
153. Swaroop, B. and Wagner, J. B., Jr., *J. Electrochem. Soc.*, **114**, 685 (1967)
154. Rhines, F. N., *Corros. Mat. Prot.*, **4**, 15 (1947); Ashby, M. F. and Smith, G. C., *J. Inst. Met.*, **91**, 182 (1963); Bolsaitis, P. and Kahlweit, M., *Acta Metall.*, **15**, 765 (1967); Pötschke, J., Mathew, P. M. and Frohberg, M. G., *Z. Metallkunde*, **61**, 152 (1970)
155. Moore, C. and Bindley, D., *Proc. 2nd Internat. Cong. Met. Corros.*, New York (1963), Nat. Assoc. Corros. Eng., Houston, 391 (1966); Castle, J. E., Harrison, J. T. and Masterson, H. G., *ibid*, 822; Hopkinson, B. E., *A.S.M.E.*, Paper No. 62-WA-274 (1962); Wiedersum, G. C. and Tice, E. A., *A.S.M.E.*, Paper No. 64-WA/CT-3 (1964); Castle, J. E., Harrison, J. T. and Masterson, H. G., *Brit. Corros. J.*, **1**, 143 (1966)
156. Otsu, T. and Sato, S., *Trans. Jap. Inst. Met.*, **2**, 153 (1961); Sato, S., *Sumitomo Light Metal Tech. Rep.*, **5** No. 1, 2 (1964); **5** No. 2, 27 (1964); **5** No. 3, 231 (1964); **5** No. 4, 290 (1964); Brush, E. G. and Pearl, W. L., *Corrosion*, **25**, 99 (1969)
157. Moore, H., Beckinsale, S. and Mallinson, C. E., *J. Inst. Met.*, **25**, 35 (1921); Moore, H. and Beckinsale, S., *J. Inst. Met.*, **23**, 225 (1920)
158. *Bibliography on Season Cracking*, Proc. Amer. Soc. Test. Mater., **41**, 918 (1941)
159. *Symposium on Stress-corrosion Cracking in Metals*, Amer. Soc. Test. Mater. - Amer. Inst. Min. (Metall.) Engrs., Philadelphia (1944)
160. Nelson, G. A., *Bull. Amer. Soc. Test. Mat.*, No. 240, 39 (1959)
161. Robertson, W. D., *Trans. Amer. Inst. Min. (Metall.) Engrs.*, **191**, 1190 (1951)
162. Edmunds, G., Ref. 159, pp. 67–89
163. Wilson, T. C., Edmunds, G., Anderson, E. A. and Peirce, W. M., Ref. 159, pp. 173–193
164. de Jager, W. G. R., *Metall.*, **4**, 138, 185 (1950); Steinle, H., *Metall.*, **9**, 492 (1955); Kamath, K. V., *Freiberger Forschungshefte*, B56 (1961); Lihl, F. and Hutter, H., *Metall.*, **21**, 884 (1967); Sato, S. and Nosetani, T., *Sumitomo Light Metal Tech. Rep.*, **10** No. 2, 83 (1969); Syrett, B. C. and Parkins, R. N., *Corrosion Sci.*, **10**, 197 (1970)
165. Ref. 8, pp. 78–79
166. Sato, S., *Sumitomo Light Metal Tech. Rep.*, **1** No. 3, 45 (1960); Kamath, K. V. and Erdmann-Jesnitzer, F., *Metall.*, **14**, 1061 (1960); Thompson, D. H., *Chem. Engng.*, **68** No. 3, 130 (1961); Laub, H., *Metall.*, **20**, 1174 (1966); Adamson, K., *Corrosion Sci.*, **7**, 537 (1967); Erdmann-Jesnitzer, F. and Kaeslingk, N., *Werkst. u. Korrosion*, **20**, 493 (1969); Sabbadini, L., *Metallurgia Italiana*, **62**, 228 (1970)
167. Edmunds, G., Anderson, E. A. and Waring, R. K., Ref. 159, pp. 7–18; Bulow, C. L., Ref. 159, pp. 19–35; Jamieson, A. L. and Rosenthal, H., Ref. 159, pp. 36–46; Hellsing, S., Lissner, O., Rask, S. and Ström, B., *Werkst. u. Korrosion*, **8**, 569 (1957); Aebi, F., *Z. Metallk.*, **49**, 63 (1958); Thompson, D. H., *Mater. Res. Standards*, **1**, 108 (1961); Szabo, E., *Werkst. u. Korrosion*, **14**, 162 (1963); Mattsson, E., Lindgren, S., Rask, S. and Wennström, G., *Current Corrosion Research in Scandinavia*, Kemian Keskusliitto, Helsinki, 171 (1965)
168. Breckon, C. and Gilbert, P. T., *Metal Ind.*, **93**, 89, 114 (1958); Sinclair, N. A. and Albert, H. J., *Mater. Protection*, **1** No. 3, 35 (1962); Fox, D. K., *Modern Castings*, **42** No. 6, 51 (1962); Baumann, G., *Werkst. u. Korrosion*, **13**, 737 (1962); Sato, S., *Sumitomo Light Metal Tech. Rep.*, **4** No. 1, 48 (1963); Uhlig, H. H. and Sansone, J., *Mater. Protection*, **3** No. 2, 21 (1964); Peters, B. F., Carson, J. A. H. and Barer, R. D., *Mater. Protection*, **4** No. 5, 24 (1965); Logan, H. L. and Ugiansky, G. M., *Mater. Protection*, **4** No. 5, 79 (1965); Serre, J., *Corrosion et Anti-corrosion*, **14** No. 1, 9 (1966)
169. Laub, H., *Metall.*, **20**, 597 (1966) and **21**, 173 (1967); Laub, H., *Metalloberfläche*, **20**, 413, 453, 493 (1966)
170. Sato, S. and Nosetani, T., *Sumitomo Light Metal Tech. Rep.*, **10** No. 3, 175 (1969)
171. Whitaker, M. E., *Metallurgia*, Manchr., **39**, 21, 66 (1948)
172. Perryman, E. C. W., *J. Inst. Met.*, **83**, 369 (1954–55); Perryman, E. C. W. and Goodwin, R. J., *J. Inst. Met.*, **83**, 378 (1954–55)
173. Bailey, A. R., *Metal Ind.*, **80**, 519 (1952) and *J. Inst. Met.*, **87**, 380 (1959)
174. Sheehan, T. L. and Dickerman, H. E., *J. Amer. Soc. Nav. Engrs.*, **58**, 586 (1946)
175. Pugh, E. N., Craig, J. V. and Montague, W. G., *A.S.M. Trans. Quart.*, **61**, 468 (1968)
176. Mattsson, E., *Electrochimica Acta*, **3**, 279 (1961)
177. Forty, A. J. and Humble, P., *Phil. Mag.*, **8** No. 86, 247 (1963); Proc. 2nd Internat. Cong. Met. Corrosion, New York (1963), N.A.C.E., Houston, 80 (1966); McEvily, A. J., Jr. and Bond, A. P., *J. Electrochem. Soc.*, **112**, 131 (1965)

178. Dix, E. H., Jr., *Proc. Amer. Soc. Test. Mater.*, **41**, 928 (1941); Read, T. A., Reed, J. B. and Rosenthal, H., Ref. 159, 90–110; Chaston, J. C., *Sheet Metal Ind.*, **24**, 1395 (1947); Zürrer, T., *Pro-Metal.*, **13**, 307 (1960); Forty, A. J., *Metal Progr.*, **75**, 154 (1959); Graf, L. and Budki, J., *Z. Metallk.*, **46**, 378 (1955); Graf, L., *Corros. Anti-Corros.*, **6**, 151 (1958); Graf, L. and Lacour, H. R., *Z. Metallk.*, **51**, 152 (1960) and **53**, 764 (1962); Graf, L. and Richter, W., *Z. Metallk.*, **52**, 834 (1961); Aebi, F., *Z. Metallk.*, **46**, 547 (1955) and **47**, 421 (1956); Logan, H. L., *J. Res. Nat. Bur. Stand.*, **48**, 99 (1952) and **56**, 159 (1956); Bakish, R. and Robertson, W. D., *J. Electrochem. Soc.*, **103**, 320 (1956); Edeleanu, C. and Forty, A. J., *Phil. Mag.*, **5** No. 58, 1029 (1960); Graf, L., *Proc. 2nd Internat. Cong. Met. Corrosion*, New York (1963), N.A.C.E., Houston, 89 (1966) and *Metall.*, **18**, 1163, 1287 (1964); Lynes, W., *Corrosion*, **21**, 125 (1965); Ohtani, N. and Dodd, R. A., *Corrosion*, **21**, 161 (1965); Pugh, E. N. and Westwood, A. R. C., *Phil. Mag.*, **13**, 167 (1966); Pugh, E. N., Montague, W. G. and Westwood, A. R. C., *A.S.M. Trans. Quart.*, **58**, 665 (1965); Hoar, T. P. and Booker, C. J. L., *Corrosion Sci.*, **5**, 821 (1965); Fairman, L., *Corrosion Sci.*, **6**, 37 (1966); Tanabe, Z., *Sumitomo Light Metal Tech. Rep.*, **7** No. 3, 137 (1966); Takano, M. and Shimodaira, S., *Trans. Japan Inst. Met.*, **8**, 239 (1967) and *Corrosion Sci.*, **8**, 55 (1968); Murakami, Y. and Ikai, Y., *Trans. Japan Inst. Met.*, **8**, 246 (1967); Lahiri, A. K., *Brit. Corrosion J.*, **3**, 289 (1968); Lahiri, A. K. and Banerjee, T., *Corrosion Sci.*, **8**, 895 (1968); Hoar, T. P. and Rothwell, G. P., *Electrochimica Acta.*, **15**, 1037 (1970)

179. Swann, P. R. and Nutting, J., *J. Inst. Met.*, **88**, 478 (1960); Swann, P. R., *Corrosion*, **19**, 102t (1963); Swann, P. R. and Pickering, H. W., *Corrosion*, **19**, 369t, 373t (1963); Tromans, D. and Nutting, J., *Corrosion*, **21**, 143 (1965); Rönnquist, A., *Jernkontorets Ann.*, **149**, 604 (1965); Brown, B. F., *Met. Mater.*, **2** No. 12, 171 (1968); Graf, L., *Werkst. u. Korrosion*, **20**, 408 (1969)

180. *Symposium on Internal Stresses in Metals and Alloys* (1947), Institute of Metals, London (1948)

181. *Stress Corrosion Cracking and Embrittlement* (Electrochem. Soc. Symposium), Ed. Robertson, W. D., Wiley, New York (1956); *Physical Metallurgy of Stress-Corrosion Fracture* (A.I.M.E. Symposium), Ed. Rhodin, T. N., Interscience, New York (1959); *Conference on Fundamental Aspects of Stress Corrosion Cracking*, Ohio State Univ. (1967), N.A.C.E., Houston (1969)

182. Bailey, A. R., *Metall. Rev.*, **6** No. 21, 101 (1961)

183. Logan, H. L., *Met. Engng. Quart.*, **5**, 32 (1965); Parkins, R. N., *Metall. Rev.*, **9** No. 35, 201 (1964); Engell, H-J. and Speidal, M. O., *Werkst. u. Korrosion*, **20**, 281 (1969)

184. Thompson, D. H. and Tracy, A. W., *J. Metals*, N.Y., **1**, 100 (1949)

185. Cook, M., Ref. 180, p. 73

186. Pugh, E. N., Montague, W. G. and Westwood, A. R. C., *Corrosion Sci.*, **6**, 345 (1966); Uhlig, H. H. and Duqette, D. J., *Corrosion Sci.*, **9**, 557 (1969)

187. White, L. F. and Blazey, C., *Metal Ind.*, **75**, 92 (1949)

188. Sylwestrowicz, W. D., *Corrosion*, **25**, 168, 405 (1969) and **26**, 160 (1970)

189. Lahiri, A. K. and Banerjee, T., *Corrosion Sci.*, **5**, 731 (1965); Chatterjee, U. K., Sircar, S. C. and Banerjee, T., *Corrosion*, **26**, 141 (1970); Chatterjee, U. K. and Sircar, S. C., *Brit. Corrosion J.*, **5**, 128 (1970)

190. Blackwood, A. W. and Stoloff, N. S., *A.S.M. Trans. Quart.*, **62**, 677 (1969)

191. Helliwell, B. J. and Williams, K. J., *Metallurgia*, **81**, 131 (1970)

192. Graf, L., *et al.*, *Z. Metallk.*, **54**, 406 (1963)

193. Thompson, D. H., *Corrosion*, **15**, 433t (1959)

194. Marshall, T. and Hugill, A. J., *Corrosion*, **13**, 329t (1957)

195. Klement, J. F., Maersch, R. E. and Tully, P. A., *Corrosion*, **15**, 295t (1959)

196. Norden, R. B., *Chem. Engng.*, **65**, 194, 196 (1958)

197. Klement, J. F., Maersch, R. E. and Tully, P. A., *Metal Prog.*, **75**, 82 (1959), *Corrosion*, **16**, 519t (1960) and U.S. Pat. 2 829 972

198. Robertson, W. D., Grenier, E. G., Davenport, W. H. and Nole, V. F., *Metal Prog.*, **75**, 152 (1959) and U.K. Pat. 802 044

199. Wiederholt, W., *Werkst. u. Korrosion*, **15**, 633 (1964); Laub, H., *Metall.*, **22**, 1116 (1968)

200. Dugdale, I. and Cotton, J. B., *Corrosion Sci.*, **3**, 69 (1963); Cotton, J. B., *Proc. 2nd Internat. Cong. Met. Corrosion*, New York (1963), N.A.C.E., Houston, 590 (1966); Walker, R., *Anti-Corrosion Methods and Mater.*, **17** No. 9, 9 (1970); Cotton, J. B. and Scholes, I. R., *Conf. on the Protection of Metal in Storage and in Transit*, Brintex Exhibitions Ltd., London (1970); Poling, G. W., *Corrosion Sci.*, **10**, 359 (1970)

201. Bhatt, I. M., Soni, K. P. and Trivedi, A. M., *Werkst. u. Korrosion*, **18**, 968 (1967)

202. Tinley, W. H., *Chem. and Ind.*, Dec. 12, 2036 (1964); Obrecht, M. F., *Proc. 2nd Internat. Cong. Met. Corrosion*, New York (1963), N.A.C.E., Houston, 624 (1966)
203. Desai, M. N., Rana, S. S. and Gandhi, M. H., *Anti-Corrosion Methods Mater.*, **17** No. 6, 17 (1970)
204. Desai, M. N., Shah, Y. C. and Gandhi, M. H., *Australasian Corros. Engng.*, **12** No. 3, 3 (1968); Desai, M. N. and Shah, Y. C., *Werkst. u. Korrosion*, **21**, 712 (1970)
205. Rance, V. E. and Evans, U. R., *Corrosion and its Prevention at Bimetallic Contacts*, H.M.S.O., London (1956)

4.3 Lead and Lead Alloys

Lead forms a series of relatively insoluble compounds which, so long as they are produced and maintained in close contact with the metal surface, often prevent or at least severely restrict further reaction when the metal is exposed to corrosive environments. Thus the general good corrosion resistance of lead results from the formation of relatively thick adherent films of corrosion product.

Composition and Mechanical Properties

Lead

While lead of purity in excess of 99·99% is commercially available, it is put to direct use only in special cases, and more frequently small amounts of other metals are introduced to improve mechanical properties.

About nine varieties of lead are in common use in the UK; their compositions and impurity limits are listed in Table 4.13. Of the impurities, zinc and bismuth aggravate corrosion in most circumstances, while copper, tellurium, antimony, nickel, silver, tin and arsenic (which are often introduced into the alloys to improve some specific mechanical property) may actually improve corrosion resistance or at any rate reduce it only slightly in the particular environment in which the lead containing them is designed to serve.

BS 334:1934, *Chemical Lead* (*Types A and B*), defines the levels of various impurities and also gives details of two empirical tests for corrosion resistance: the flash test* and the aqua regia test, which is seldom used.

BS 334 Type A lead, on account of its comparatively high purity, has better all-round resistance to a wide range of chemical reagents than any of the alloys.

While the softness of lead is often an advantage, it does mean that the metal must be suitably supported to prevent movement and mechanical failure; in particular, care must be taken with lead sheathing in chemical plant to prevent buckling caused by the difference between the coefficients of expansion of the lead and the material protected by it[1], and pipework must always be adequately supported, without excessive rigidity.

Unalloyed lead is characterised by coarse grain size, which frequently results in proneness to intercrystalline cracking under certain conditions of

*Test for evaluating resistance of lead to H_2SO_4. The specimen is placed in 95–96 w% H_2SO_4 and temperature is raised from ambient to 300°C in 7 min. The 'flashing' is due to the formation of $PbSO_4$, and chemical lead should show no sign of flashing below 285°C or in special cases at 300°C.

mechanical stress. The recrystallisation temperature of Type A lead is about 0°C, so that it does not remain work-hardened.

Lead Alloys

Alloys are frequently used in preference to unalloyed lead on account of the net improvement due to increased strength and finer grain size. Occasionally the presence of an alloying element, while improving mechanical properties, renders the material more susceptible to corrosion, and therefore alloying has to be carefully controlled.

The alloys of lead in common use may be divided into two classes, those in which the additions are present in quantities less than 0·25%, and those containing greater quantities. The various grades of lead specified in Table 4.13 can form the basis for alloy preparation.

The first category includes the 0·07% Cu alloy prepared from Type A lead, which is widely used in the chemical industry because the finer grain size resulting from alloying renders it not very susceptible to intercrystalline cracking. Its mechanical properties are generally similar to those of Type A lead, and it has good resistance to sulphuric acid but is more sensitive to traces of nitrates. The same may broadly be said of the 0·06Cu–0·04Te alloy, which work hardens, and of the 0·004Ag–0·004Cu (BS 1085) which has slightly superior creep characteristics. The following specifications have been laid down for lead alloys containing small amounts of protective elements: BS 334, BS 602 (lead pipes for water, gas and effluent services), BS 1178 (lead sheet for domestic purposes).

In the second category, where large quantities of alloying elements are added, usually to increase the hardness, the most common addition is antimony. Alloys of 6 to >12% Sb are known as Regulus Metal and are covered by BS 335. The alloys, although not possessing the outstanding corrosion resistance of Type A Lead, find widespread application in sulphuric acid works for pump bodies, valves, etc. In the lead acid battery the lead grids, basically an antimonial alloy, are proof against the comparatively weak acid used and offer good resistance to the attack of oxygen evolved during charging.

Intermediate between the two classes described above are the specialised alloys used in the cable industry which contain less than 1% total of antimony, tin, copper and cadmium (see BS 801). These alloys exhibit desirable mechanical features and, although their corrosion resistance is reduced by these additions, are adequate for underground environments which are less aggressive than those found in the chemical industry.

It has been shown that the flash points of lead alloys rise slightly with increasing copper and tellurium contents and decrease with the contents of bismuth, zinc, tin and antimony[2].

Corrosion Behaviour

The standard electrode potential, $E^{\ominus}_{Pb^{2+}/Pb} = -0{\cdot}126$ V, shows that lead is thermodynamically unstable in acid solutions but stable in neutral solutions. The exchange current for hydrogen reaction on lead is very small

Table 4.13 Composition of types of lead and lead alloys commonly used in the United Kingdom

Specification	Use	Alloying elements or impurities (%)*										Pb
		Sn	Sb	Ag	Cu	Ni	Fe	Bi	Cd	Zn	Te	
BS 334 Type A	Specialised chemical applications	Trace	0·002	0·002	0·003	0·001	0·003	0·005	Trace	0·002	—	99·99 min.
BS 334 Type B	General chemical applications	—	—	—	0·06 to 0·07	—	—	—	—	—	—	Balance
BS 334 Type B	Specialised chemical applications	—	—	—	0·06 to 0·07	—	—	—	—	—	0·04 to 0·05	Balance
BS 334 Type B and BS 1085	Specialised chemical applications	0·002	0·005	0·003 to 0·005	0·003 to 0·005	—	—	0·01	—	0·001	—	Balance
	Specialised domestic applications											

*A single figure indicates maximum permissible impurity and a double entry denotes deliberate addition.

Table 4.13 *(continued)*

Specification	*Use*	*Alloying elements or impurities (%)**										Pb
		Sn	Sb	Ag	Cu	Ni	Fe	Bi	Cd	Zn	Te	
BS 602, No. 1	Pipes	0·075	0·02	—	0·03	—	—	0·05	—	0·005	—	99·80
BS 1178	Sheet	—	0·01	—	—	—	—	—	—	0·005	—	99·90
BS 801	Cable sheathing	—	—	—	—	—	—	0·05	—	0·002	—	99·9 min.
BS 801 Alloy B	Cable sheathing	—	0·8 to 0·9	—	—	—	—	—	—	—	—	Balance
BS 801 Alloy E	Cable sheathing	0·35 to 0·45	0·15 to 0·25	—	—	—	—	—	—	—	—	Balance

*A single figure indicates maximum permissible impurity and a double entry denotes deliberate addition.

($\approx 10^{-13}$–10^{-11} A/cm^2), but control of corrosion is usually due to mechanical passivation of the local anodes of the corrosion cells as the majority of lead salts are insoluble and frequently form protective films or coatings.

Anodic behaviour Lead is characterised by a series of anodic corrosion products which give a film or coating that effectively insulates the metal mechanically from the electrolyte [e.g. $PbSO_4$, $PbCl_2$, Pb_3O_4, $PbCrO_4$, PbO, $2PbCO_3 \cdot Pb(OH)_2$]; under certain circumstances this film may be converted to lead dioxide, which has a high electronic conductivity, so that subsequent passage of charge results in the evolution of oxygen (sulphuric acid[3], chromates or alkaline solutions[4]) or chlorine (certain chloride-containing solutions[5]). The exact nature of the mechanisms occurring when the mechanically passivating films are partially or fully converted to lead dioxide is not fully understood. Quite often only a relatively small amount of lead peroxide is formed and is followed by discharge of gas, and further reaction of lead is slight.

The conditions for the formation of anodic oxides of lead have been studied in detail[6], and it has been suggested that the intermediate oxides Pb_3O_4 and PbO are not formed by the direct electrochemical oxidation of lead but from oxidising radicals in solution.

Lead dioxide exists as two polymorphs, α-PbO_2 (orthorhombic) and β-PbO_2 (tetragonal), the type formed being dependent upon potential, pH, and nature of ions present. Both have a resistivity of $1–4 \times 10^{-4}\,\Omega$ cm which is due to their non-stoichiometry (actual composition, $PbO_{1\cdot 94-1\cdot 96}$) and to the free electrons in the lattice. Lead dioxide has a low overpotential for oxygen evolution from sulphuric acid[7] and potassium hydroxide[4] solutions, the magnitude of the overpotential being influenced by the polymorphic form[8]. It also discharges chlorine at low overpotentials[9].

Cathodic behaviour If high cathodic currents are applied, lead tends to corrode with the formation of a grey cloud of fine particles of metal. The cause of this has not yet been established[10]. In cathodic protection of lead cables the current densities used are insufficient to produce this effect.

Thermodynamics of the Pb–H_2O, Pb–H_2O–X systems Pourbaix *et al.*[11] have studied the Pb–H_2O, Pb–H_2O–X systems where X is a non-metal, and have established the domains of thermodynamic stability of lead, lead cations and anions, and insoluble compounds of lead. Figure 4.13 shows the Pb–H_2O system; it can be seen that in the region of high and low pH, corrosion occurs owing to the amphoteric nature of lead (cf. Zn, Al, Sn). This is a significant factor in the behaviour of lead in actual environments.

In contrast with the Pb–H_2O system, it can be seen (Fig. 4.14) that in the presence of SO_4^{2-} the corrosion zone in the region of low pH no longer exists, owing to the thermodynamic stability of $PbSO_4$. The Pb–H_2O–CO_2 system has been expressed in a similar pH/potential diagram[11] in which account has been taken of insoluble carbonates and basic carbonates of lead.

The influence of pH on the corrosion of lead is indicated in Fig. 4.15, and it is apparent that the predictions of the pH/potential diagram are fulfilled. However, in very concentrated acid solutions, attack may diminish, owing to the relative insolubility of the relevant salt in the acid. Thus, lead nitrate although soluble in water has (owing to common ion effect) only slight solubility in concentrated nitric acid, and the corrosion rate is reduced. Similarly, lead chloride is less soluble in moderately concentrated hydrochloric acid than it is in water. In concentrated hydrochloric acid, however, the converse is the case, since in high chloride-ion concentrations lead forms soluble complex anions.

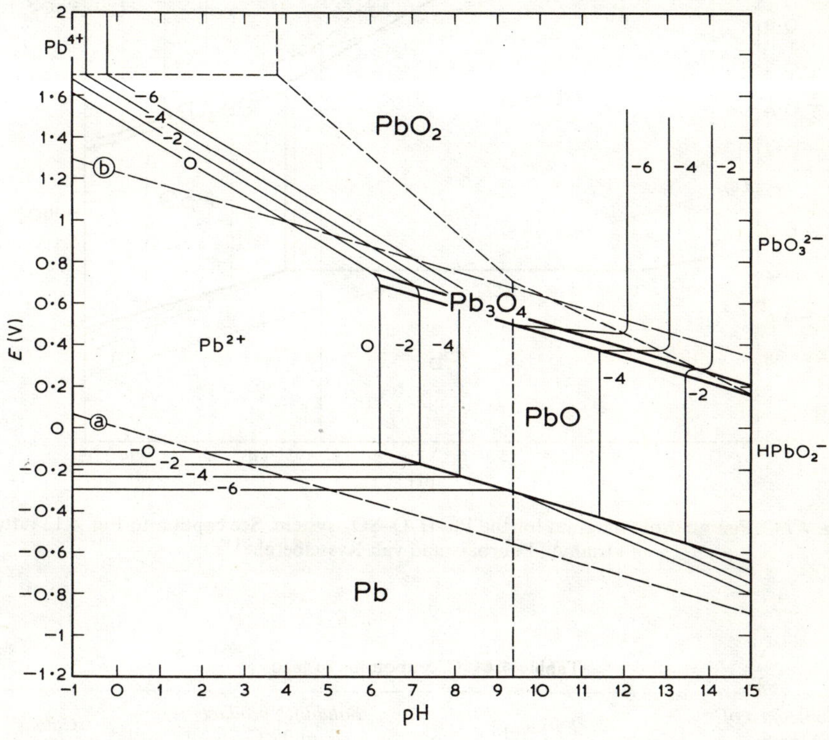

Fig. 4.13 Potential/pH diagram for the Pb–H_2O system. The area between ⓐ and ⓑ corresponds to the thermodynamic stability of water. Light lines represent equilibrium conditions between a solid phase and an ion at activities 1, 10^{-2}, 10^{-4} and 10^{-6}. Heavy lines represent equilibrium conditions between two solid phases. Broken lines represent equilibrium conditions between two ions for a ratio of these ions equal to unity (after Delahay, Pourbaix and van Rysselberghe[11])

Inspection of products of corrosion and correlation with thermodynamic data frequently gives an indication of the cause of corrosion. Thus from Fig. 4.13 it will be seen that a potential of about 1 V (neglecting the influence of other ions which would be unlikely to decrease this value) is required for the formation of lead dioxide in neutral solution. This is somewhat higher than that likely to be generated by galvanic action between dissimilar metals commonly used in civil engineering, and the presence of lead dioxide among the corrosion products is usually taken as an indication that an impressed current is responsible for corrosion.

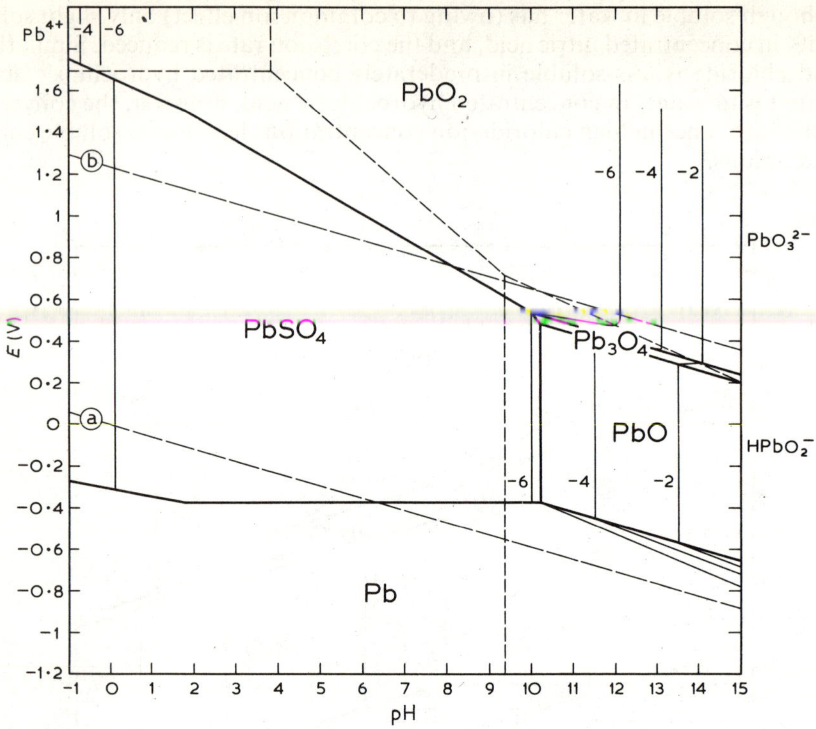

Fig. 4.14 Potential/pH diagram for the Pb–H_2O–SO_4 system. See caption to Fig. 4.13 (after Delahay, Pourbaix and van Rysselberghe[11])

Table 4.14 Compounds of lead

Compound of lead	*Formula*	*Solubility product at 25°C*	*Colour*
Acetate	$(CH_3COO)_2Pb$	soluble (55 g/100 ml)	white
Bromide	$PbBr_2$	$5{\cdot}7 \times 10^{-6}$	white
Carbonate	$PbCO_3$	$3{\cdot}3 \times 10^{-14}$	white
Basic carbonate	$2PbCO_3{\cdot}Pb(OH)_2$	as above	white
Chloride	$PbCl_2$	1×10^{-4}	white
Chromate	$PbCrO_4$	$1{\cdot}8 \times 10^{-14}$	orange
Dioxide	PbO_2	insoluble	black or dark brown
Fluoride	PbF_2	$3{\cdot}7 \times 10^{-8}$	white
Hydroxide	$Pb(OH)_2$	4×10^{-15}	white
Iodide	PbI_2	$1{\cdot}4 \times 10^{-8}$	yellow
Monoxide	PbO	insoluble	yellow-red
Nitrate	$Pb(NO_3)_2$	soluble (60 g/100 ml)	white
Phosphate	$Pb_3(PO_4)_2$	10^{-55}	white
Sulphate	$PbSO_4$	1×10^{-8}	white
Sulphide	PbS	$3{\cdot}4 \times 10^{-28}$	black
Triplumbic-tetroxide	Pb_3O_4	insoluble	red

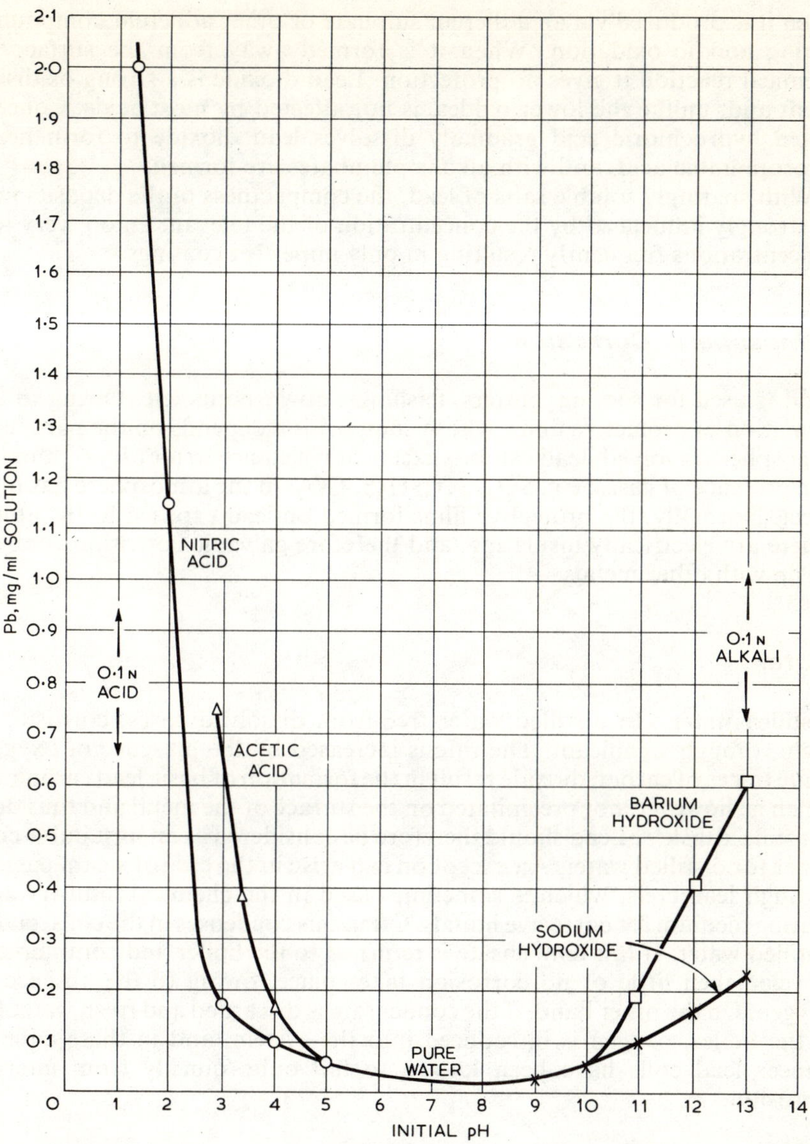

Fig. 4.15 Corrosion of lead in contact with various acid and alkali solutions (after Anderegg and Achatz[12])

From an inspection of the more common compounds of lead, it will be seen that, in many environments, the corrosion product will be relatively insoluble (Table 4.14). Often, however, vitiating circumstances prevail which prevent compact protective films from forming on the surface of the metal. The nature of the film is influenced by the mode of crystallisation, and in the case of the lower oxides, for example, frequently little protection is afforded. The higher oxide (lead dioxide) often forms a good adherent film, especially when it is produced via an adherent sulphate or other adherent compounds during anodic oxidation. When it is formed away from the surface by chemical reaction it gives no protection. Lead dioxide is a strong oxidising agent and, unlike the lower oxides, is not affected by most acids. Concentrated hydrochloric acid gradually dissolves lead dioxide to form hexachloroplumbic acid, and with alkalis plumbates are formed.

With sparingly soluble salts of lead, the compactness of the deposits may be strongly influenced by the concentration of the relevant anion, very low concentrations frequently resulting in only imperfect coatings.

Atmospheric Corrosion

Lead is used for roofing, gutters, flashings, downspouts, etc. Owing to the formation of protective films, whose composition depends on the particular atmosphere involved, lead exhibits excellent resistance to air (dry or humid). The presence of gases, e.g. SO_2, SO_3, H_2S, CO_2, in the atmosphere has little effect. Normally, the protective films formed on lead exposed to the atmosphere are electrically insulating, and therefore galvanic corrosion is rarely set up with other metals.

Water

Distilled water In distilled water free from dissolved gases, corrosion is slight, though significant. The rate is increased by the presence of oxygen, while traces of carbon dioxide result in the formation of basic lead carbonate, which is, however, not precipitated on the surface of the metal and thus does not stifle attack[13]. Lead should therefore be considered as an unsuitable container for distilled water. An exception can arise in the case of steam passing through lead coils, which is sometimes used in the chemical industry as a heating medium for corrosive liquids. Often this condenses in the coils, giving distilled water. If this condensate is returned to the boiler and continuously recycled then little or no corrosion takes place, owing to the absence of oxygen. On the other hand, if the condensate is discarded and fresh water fed to the boiler, oxygen is introduced into the system, and in these circumstances lead coils have been known to fail quite quickly from internal corrosion.

Natural waters Most natural waters show a few degrees of hardness, and the carbonate, sulphate, or silicate content is often sufficient to permit their being conveyed through lead pipes (see Section 2.3). Because of the poisonous nature of lead compounds, slight attack may be dangerous (the maximum content permissible being between 0·05 and 1 p.p.m. of lead)[14]. Rain water is frequently plumbosolvent, and stagnant waters containing perhaps nitrates, carbon dioxide and other aggressive agents are often corrosive to lead.

Waters drawn from peaty moorlands, are, owing to their organic acid content, often plumbosolvent. Treatment of aggressive waters is complicated since a complete analysis is required in order to estimate the effect of individual constituents which may not be neutralised or precipitated by conventional methods. These methods consist of adding milk of lime or whiting or percolating the water through a bed of limestone; this neutralises acidity and also provides $Ca(HCO_3)_2$ which is conducive to the formation of a protective carbonate scale on the lead[15,16].

Lead usually has excellent resistance to sea-water, the carbonate, present resulting in the formation of a protective film, which should be compared with its behaviour in solutions of alkali chlorides (see *Salts*, p. **4**.82).

Underground Corrosion

Stray-current Corrosion

Stray currents are a source of damage to buried metal structures (see Section 11.5), and lead pipes and cable sheaths are particularly susceptible to it, possibly because their shape lends itself to the passage of electricity over considerable distances. In systems where both metallic and non-metallic pipes are used, electrical continuity is broken and the corroding areas are necessarily more numerous but frequently less intense than they would be in a continuous installation passing from one environment to another. In addition, corrosion occurring as a result of the influence of a near-by electrical installation may be localised to that portion of the lead which is closest to the electrical system. Lead which becomes the anode under conditions of electrolytic attack is corroded with the formation of oxides, carbonates, hydroxides and chlorides[17], and the corrosion product may be in the form of glassy watery crystals in which the compounds $PbCl_2 \cdot Pb(OH)_2$ and $PbCl_2 \cdot 6PbO \cdot 2H_2O$ have been identified. Lead can also corrode under cathodic (alkaline) conditions if, for example, a cable network or pipeline is affected by a cathodic-protection system in the vicinity. The use of lead pipes as earths for alternating currents has also resulted in serious corrosion[18].

Prevention of corrosion by stray currents is difficult, since none of the protective coatings are entirely effective. Bituminous coatings give some protection, greases are useful in concrete ducts, and insulating material of tar and resin gives good protection. Concrete coatings are of value in some cases, but they are expensive and the alkaline content of some cements may cause chemical attack[19].

Electrochemical Corrosion

Failures of lead pipes and cable sheaths will, in the absence of incorrect mechanical support or construction, be due to chemical or electrochemical attack. Corrosion cells are established as a result of inhomogeneity of the surrounding environment, and variations in soil composition can result in the creation of large corrosion currents. Lead passing through soils of different character give rise to e.m.f.s as high as 1·5 V as a result of the creation of 'geological cells'. The soils may differ in degree of aeration, water content, or presence of various chemicals and bacteria. The worst combination of soils is wet clay and cinders. The carbon in the cinder acts as an efficient cathode, and the portion of the pipe in the clay becomes the anode and corrodes badly. The clay holds moisture and thus permits the passage of relatively high currents. Packing sand around pipes and laying them in trenches of broken chalk prevents corrosion due to this cause. Galvanic cells may also be set up owing to the presence of metallic impurities in the lead, but severe corrosion due to this cause is not common. A galvanic cell may also be established when a section of new cable (clean sheath) is joined to an old cable (dirty or encrusted sheath). The new sheath becomes anodic and can corrode. Similar circumstances can arise when cables are placed on steel racks in underground ducts, or a dissimilar metal, such as a copper bonding ribbon, is used.

Many long-term tests have been carried out on lead under various soil conditions[20], and it has been found that corrosion or passivity depends not only upon the nature of the constituents present but also upon their concentration. A coarsely textured soil which permits fairly free circulation of air produces conditions which closely resemble atmospheric exposure, and corrosion is prevented by the formation of protective complex films, but corrosion may be accelerated if the access is restricted, and the oxygen depolarises cathodic areas or results in the creation of oxygen concentration cells. These damaging circumstances can also be found in clay and silts.

High sulphate-ion concentration assists in the formation of passivating films. Silicates and carbonates assist in passivating the lead, and in some cases the corrosiveness of certain media has been reduced by the addition of small quantities of sodium silicate. On the other hand if these inhibiting radicals are not evenly distributed and afford protection in certain areas only, galvanic cells may be set up, with the consequences described earlier.

Nitrates generally have a corrosive action, especially when present in acid soil waters. This is due both to the oxidising nature of this anion which causes cathodic depolarisation and to the formation of soluble nitrates. Alkaline soils can seriously corrode lead, with the formation of alkali plumbites which decompose to give (red) lead monoxide. Soil containing rotting vegetable matter or farmyard manure, has a strong corrosive action: this is the basis of the 'stack' or 'Old Dutch' process of making white lead by the action of organic acids and carbon dioxide. It is probably the explanation of so-called 'phenol corrosion' (see later).

Free lime, present in concrete, is frequently responsible for the corrosion of lead, which is covered with, supported by, or merely placed in the vicinity of this material (see Section 10.3), and very weak alternating or direct currents, as well as currents induced by contact with other metals (e.g. bonding straps for cathodic protection), may accelerate the corrosive influence. Frequently the corrosion product is lead monoxide[21,22]. This type of corrosion is often experienced with telephone cables (bare lead sheath) laid in a duct of asbestos cement.

To prevent underground corrosion, lead is frequently mechanically protected with various coatings of tar, bitumen, resins, etc. which are of course effective only if they completely insulate the metal from corrosive agents and stray currents. Occasionally, cable sheaths protected by a layer of coal tar or bitumen and followed by paper, jute or hessian layers impregnated with these materials are attacked by a process known as 'phenol corrosion' (modern theories suggest that phenol, once thought to be the active agent, plays little or no part in the action). The corrosion product is a voluminous white powdery deposit (basic carbonate and carbonate), and, as well as surface attack, intercrystalline corrosion and internal cavities may result. In order that corrosion may proceed, air and moisture are essential, and it is evident that jute and hessian cannot resist indefinitely the ingress of water[23]. This moisture can reach the lead notwithstanding the coating of tar or bitumen, but it has been suggested that the ingredients for corrosion are initially present in the components of the cable[24]. There is evidence that attack is initiated by bacterial action on the jute or hessian with the formation of volatile organic acids which promote corrosion[25].

The most successful method of preventing underground corrosion of covered or bare lead is cathodic protection, which, if correctly applied, can protect indefinitely (see Chapter 11). There are, however, complications which should be borne in mind.

The potential of the lead surface must be suitably monitored, since too negative a value can result in an increase in pH, and the resultant alkalinity can present a hazard to the lead structure, and should the current be interrupted the protected unit will be attacked. Serious corrosion has been observed as a result of the formation of caustic soda from the electrolysis of salt solution used on streets, railways, etc.[26].

Under certain sewerage conditions, lead in the presence of H_2S or other sulphides forms a coating of electrically conducting lead sulphide. This possesses a more noble potential than bare and carbonate-covered lead, and between such portions a large corrosion current can flow[27].

Chemicals

Corrosive data reported as weight losses can be misleading because of lead's high density; volume losses or yearly penetration figures are to be preferred for this metal. It should also be remembered that in chemical applications the thicknesses of lead are usually greater than those of other metals, and higher corrosive rates, by themselves, are therefore not so serious. Since lead is protected by relatively thick films of corrosion products, short term tests can be misleading, and once the film has formed there will be a significant decrease in the corrosion rate.

Gases

Lead is resistant to chlorine wet or dry up to about 100°C, but will resist dry bromine only at lower temperatures. It is fairly resistant to wet or dry sulphur dioxide and trioxide. The coating of lead fluoride that normally protects a lead surface against corrosion by hydrofluoric acid does not afford any appreciable protection against corrosion by hydrogen fluoride, and therefore lead should not be used to contain or transport this gas.

Acids

Mineral acid

Sulphuric acid In industry, lead is frequently employed in the manufacture, storage and conveyance of this acid. Very little trouble is experienced in these applications so long as the sulphate film is not broken in circumstances which prevent or hinder its repair. Thus, corrosion will take place if the film is eroded by liquids at high velocities, or even by gases containing acid spray. The film may be ruptured by the movement of the underlying metal caused by thermal cycling. Strong sulphuric acid will dissolve lead sulphate, and it is generally held that acid of over 85% concentration is aggressive. There have, however, been instances where cold and quiescent acid at concentrations of over 90% has caused no damage. The relation between temperature and corrosion of lead by sulphuric acid at various concentrations is shown in Fig. 4.16.

Quite often in sulphuric acid lead-chamber plants operating on the intensive system, corrosion of the lead results from the action of nitrosylsulphuric acid and from nitrosyl chloride which is formed as a result of the chloride content of the water. The injurious effects of these impurities cannot generally be prevented by alloying[29], and certain alloying elements (e.g. copper) present in the lead can increase the corrosion rate.

Nitric acid Lead is readily attacked by nitric acid (Fig. 4.15), and should not be used for handling nitrate or nitrite radicals except at extreme dilutions and, for preference, in the presence of a reagent, such as a sulphate, which will accord a degree of protection. An example of the latter is the wash water from cellulose nitrate units.

Hydrochloric acid It is difficult to predict the service behaviour of lead exposed to hydrochloric acid. Generally, hydrochloric acid should be regarded as an aggressive reagent and lead should not be used for its conveyance. There have, however, been instances in which a satisfactory life has been obtained, and this is presumably associated with the formation of a closely adherent film of lead chloride which is only slightly soluble in

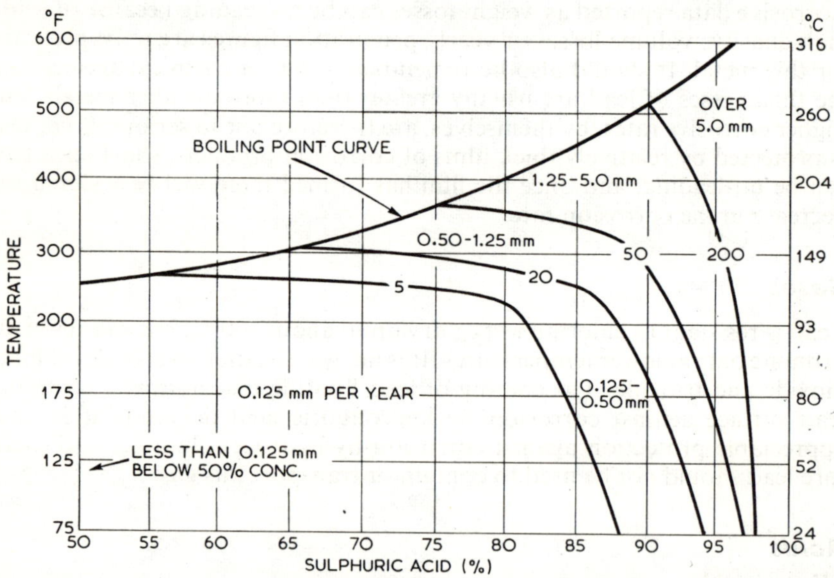

Fig. 4.16 Corrosion of lead by sulphuric acid as a function of temperature. Concentrations below 50% are not shown because resistance of lead is very good even at temperatures including boiling (after Fontana[28])

moderately concentrated hydrochloric acid. In addition lead has a high overpotential for hydrogen evolution, so that in reducing acids corrosion will be slow. With mixed hydrochloric and hydrofluoric acids, used in the pickling of steel, the behaviour of lead is uncertain. If, however, the hydrofluoric acid is introduced into the lead tank before addition of the hydrochloric acid, a protective film of lead fluoride is formed and the life of the tank is prolonged. The presence of aluminium fluoride in solution can prevent the formation of a protective film and severe corrosion may result[30].

Phosphoric acid, chromic acid Lead has good corrosion resistance to phosphoric and chromic acids; its resistance to chrome-plating solutions is discussed later.

Organic acids

Acetic acid Acetic acid and acetates rapidly attack lead (Fig. 4.15). The protective film of lead oxide is very soluble both in acetic acid and in acetates, yielding basic salts which become carbonated in the atmosphere to form basic lead carbonate. In these circumstances, the latter compound does not form in intimate contact with the lead surface and affords no protection against further attack. In the absence of oxygen the corrosion rate in weak acid solutions (0·01M or less) is slight[31].

Lead is resistant to oxalic, tartaric and fatty acids only in the absence of oxygen. Formic acid in the presence of oxygen corrodes lead rapidly.

Dilute (0·1 N–0·0001 N) acetic, propionic, butyric, succinic and lactic acids all corrode lead to about the same extent. Pyruvic acid appears to inhibit corrosion after a short period of attack. In most cases the corrosion products are $x\text{PbCO}_3.y\text{Pb(OH)}_2$, the ratio of x to y being 2:1. Corrosion is intergranular rather than general attack[32].

The use of lead in buildings where corrosion occurs owing to the formation of certain organic materials is considered in Section 10.3 and the effect of wood on lead in Section 19.10.

Fuel oil Organic acids are also thought to be responsible for the corrosion occurring when fuel oil is left in contact with lead. Formerly this attack was ascribed to the action of mercaptans (responsible for many troubles in refineries), but recently it is believed that the corrosive agents are naphthenic acids.

Lubricating oils The failure of bearings is not normally due to corrosion, but occasionally bearings are attacked in this way. It has been shown that the corrosion of lead-base bearings is associated with the acidity of white oils, the peroxide content and the presence of air. Peroxides are the controlling factor, but in the absence of air, corrosion is reduced. The corrosion product consists of a basic salt of lead of two or more organic acids[33] (see Section 2.11).

Alkalis

Lead is not attacked by strong alkalis as rapidly as it is by aggressive acids, and in some cases the corrosive action of sodium hydroxide and potassium hydroxide (up to 30% at 25°C and up to 10% at 90°C) can be tolerated. The rapid attack of lime solutions is associated with dissolved oxygen in the solution (see Section 10.3) and the absence of carbonate facilitates anions.

Salts

Lead is not generally attacked rapidly by salt solutions (especially the salts of the acids to which it is resistant). The action of nitrates and salts such as potassium and sodium chloride may be rapid. In potassium chloride the corrosion rate increases with concentration to a maximum in 0·05M solution, decreases with higher concentration, and increases again in 2M solution. Only loosely adherent deposits are formed. In potassium bromide adherent deposits are formed, and the corrosion rate increases with concentration. The attack in potassium iodide is slow in concentrations up to 0·1M but in concentrated solutions rapid attack occurs, probably owing to the formation of soluble $KPbI_3$. In dilute potassium nitrate solutions (0·001M and below) the corrosion product is yellow and is probably a mixture of $Pb(OH)_2$ and PbO, which is poorly adherent. At higher concentrations the corrosion product is more adherent and corrosion is somewhat reduced[34]. Details of the corrosion behaviour of lead in various solutions of salts are given in Figs. 4.17 and 4.18.

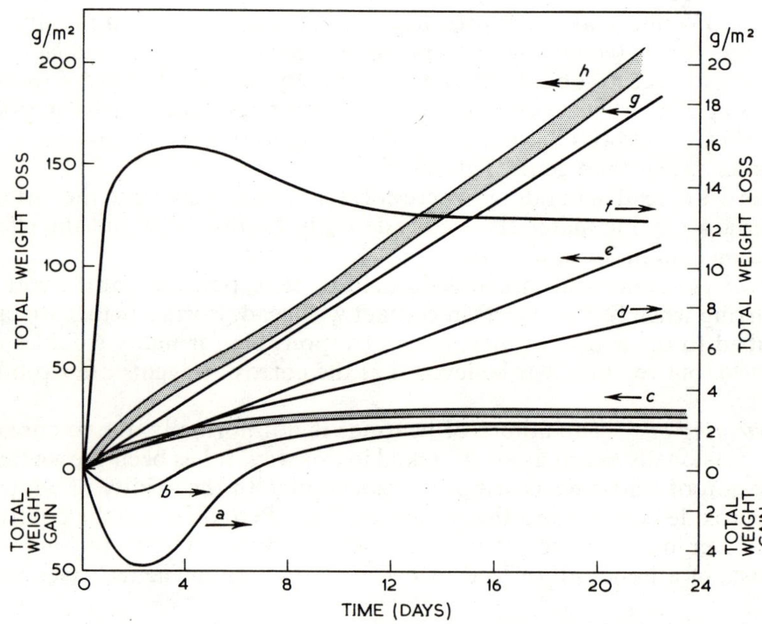

Fig. 4.17 Relationship between weight loss or weight gain and time for lead immersed in various environments (selected from Reference 35).

LEGEND

a 0·5N $(NH_4)_2SO_4$; pH 2·9
b 0·5N Na_2SO_4; pH 5·4
c 0·01N NaOH; pH 12·8
0·01N $Ba(OH)_2$; pH 11·9
d 0·5N NaCl; pH 4·9
e 0·5N CH_3COONH_4+NaOH; pH 8·9
f 0·25N NaF; pH 6·3
g 0·5N CH_3COONH_4+CH_3COOH; pH 4·6
h 0·04N $Ca(OH)_2$; pH 13
0·5N CH_3COONa; pH 16
0·5N $Ba(OH)_2$; pH > 13

Chrome-plating Solutions

Lead is the favoured material for the construction of anodes, tanks and pipelines to contain chrome-plating solutions despite occasional, and sometimes spectacular, failures. The electrolyte is usually chromic acid and sulphuric acid, but in the case of mixed catalyst baths, fluosilicic acid is also present.

In order to combat corrosion, the tank is frequently made anodic for the first few hours of its life, thus building up on its inner surface a coating of lead dioxide. It has, however, been found that if the oxide film is damaged severe local action can occur. Flame treatment of the lead, resulting in a uniform cast structure of the surface, has been advocated. Recently, it has been shown that the sulphate concentration of the bath is the factor responsible for corrosion. Tests on chemical lead, tellurium lead (0·04Te–0·06Cu), 7Sn/Pb and 8Sb/Pb, have shown that relatively little attack occurs with mixed catalyst plating solutions, whereas with simple sulphate-catalyst baths,

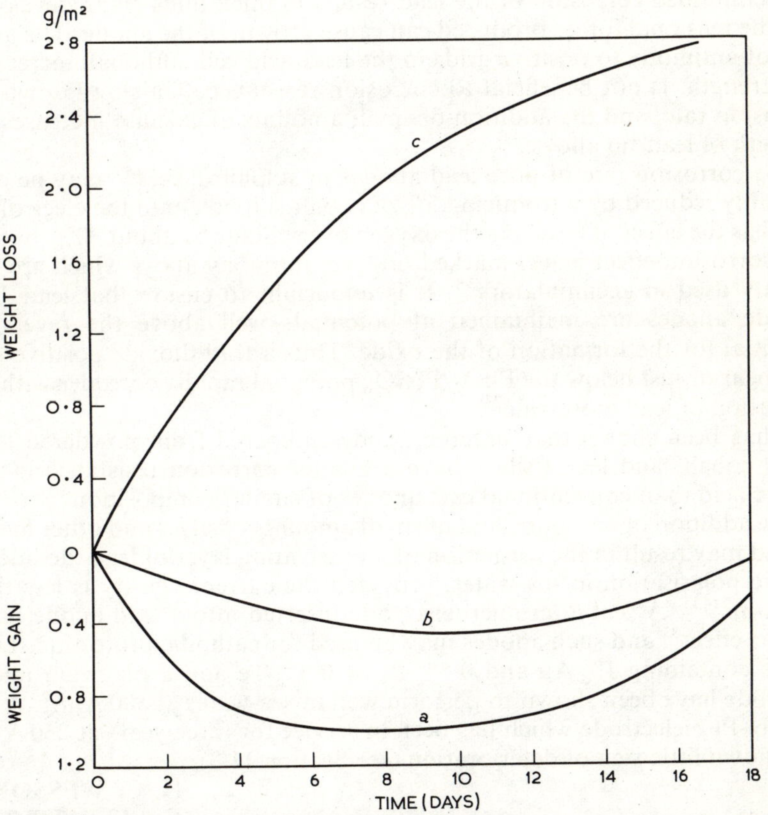

Fig. 4.18 Relationship between weight loss or weight gain and time for lead immersed in carbonate solutions (selected from results given in Reference 35).

LEGEND
a 0·5 N $NaHCO_3$; pH 8·7
b 0·01 N Na_2CO_3; pH 10·8
c 1 N Na_2CO_3; pH 11·3

tin-lead and tellurium-lead are attacked at room temperature and at 40°C, and chemical lead and antimonial lead are attacked at room temperature only. The critical sulphate concentration for maximum corrosion is in the range 4–9 g/l and depends upon temperature and whether the temperature remains constant or is cycled. It has been found that the addition of 0·5 g/l of magnesium fluosilicate to the bath completely suppresses attack and does not interfere with the plating process[36].

Lead Anodes

Lead and lead alloy anodes are frequently used for electroplating (copper, zinc, chromium), and for electrolysis of brine solution, as well as being employed in the lead/acid accumulator. The properties of the anodic film formed (lead dioxide) have been discussed. Despite the fact that once a continuous film of this oxide has formed the lead is chemically passivated, there is generally a very slow but continuous transition of lead into lead dioxide. This continued corrosion of the lead results in thick films being developed and the tensional forces produced can cause growth of the anode. The addition of antimony to positive grids in the lead/acid cell, although increasing the strength, is not beneficial to corrosion resistance. Tin slows down the corrosion rate, and the addition of small amounts of calcium increases the strength of lead/tin alloys.

The corrosion rate of pure lead anodes in sulphuric acid[37] may be considerably reduced by introducing 5% of cobalt sulphate into the electrolyte. This has the effect of reducing the oxygen overvoltage by about 30%, but the anti-corrosion effect is less marked on lead/antimony alloys which are frequently used in accumulators[38]. It is important to ensure that lead–lead dioxide anodes are maintained at potentials well above the reversible potential for the formation of the oxide. Thus a lead dioxide positive grid held near or just below the $PbO_2/PbSO_4$ potential rapidly corrodes with the formation of lead monoxide[39].

It has been shown that sintered anodes prepared from powdered lead, lead + cobalt, and lead + silver have a greater corrosion resistance in sulphuric acid than conventional cast anodes of similar composition[40].

The addition of antimony and of small amounts of silver and other metals to lead may result in the formation of a passivating layer of lead dioxide on anodic polarisation in sea-water, provided the current density is less than 120 A/m^2[41]. A platinum microelectrode inserted into a lead anode has a similar effect[5], and such anodes may be used for cathodic protection. Lead alloys containing 1% Ag and 0·5% Bi or 0·5% Te and a platinum microelectrode have been shown to perform well in sea-water at 500 A/m^2. A 1% Ag/Pb–Pt bielectrode which has been in service for three years at 250 A/m^2 has shown little sign of deterioration (see Section 11.3).

H. C. WESSON
E. LITTAUER

REFERENCES

1. Mantius, E. and Freiher, H. F., *Industr. Engng. Chem.*, **29**, 373 (1937)
2. Hanabusa, M. and Kawabata, R., *Nippon Kinz. Gakk.*, **21**, 35 (1957)
3. Thirsk, H. R. and Wynne-Jones, W. F. K., *Trans. Inst. Metal. Finish.*, **29**, 260 (1953)
4. Jones, P., Thirsk, H. R. and Wynne-Jones, W. F. K., *Trans. Faraday Soc.*, **52**, 1003 (1956)

5. Littauer, E. L. and Shreir, L. L., *Proceedings of the First International Congress on Metallic Corrosion, London, 1961*, Butterworths, London, 374 (1962); Shreir, L. L., *Corrosion*, **17**, 118t (1961)
6. Burbank, J., *J. Electrochem. Soc.*, **106**, 369 (1959)
7. Jones, P., Lind, R. and Wynne-Jones, W. F. K., *Trans. Faraday Soc.*, **50**, 972 (1954)
8. Ruetschi, P. and Cahan, B. D., *J. Electrochem. Soc.*, **105**, 369 (1958)
9. Littauer, E. L., Ph.D. Thesis, University of London (1961)
10. Van Muylder, J., Cebelcor *Rapport Technique*, **12** (1954)
11. Delahay, P., Pourbaix, M. and Van Rysselberghe, P., *J. Electrochem. Soc.*, **98**, 57 (1951)
12. Anderegg, F. O. and Achatz, R. V., *Purdue Univ. Engng. Exper. Station Bull.*, **18** (1924)
13. Evans, U. R., *Metallic Corrosion Passivity and Protection*, Arnold, London, 353 (1948)
14. *International Standards for Drinking Water*, World Health Organisation, Geneva (1971)
15. Hawkes, C. A., *Chem. Ind.*, 264 (1944)
16. Miles, G., *J. Soc. Chem. Ind.*, **67**, 10 (1948)
17. Glander, F. and Glander, W., *Z. Metallk.*, **44**, 97 (1953)
18. Amy, L. and Moujnos, C., *Rev. Gén. Elect.*, **66**, 187 (1957)
19. Forest, E. D., *Mod. Pwr. Engng.*, **46**, 63, 122 (1954)
20. Denison, I. A. and Romanoff, M., *J. Res. Nat. Bur. Stand.*, **44**, 259 (1950)
21. Dodero, M., *Métaux et Corros.*, **24**, 50 (1949)
22. Wolff, E. F. and Bonilla, C. F., *Trans. Electrochem. Soc.*, **79**, 307 (1941)
23. Radley, W. G. and Richards, C. E., *J. Inst. Elect. Engng.*, **35**, 685 (1939)
24. Senez, J. C. and Pichinoty, F., *Corros. Anticorros.*, **5**, 203 (1957)
25. Cole, E. L. and Davies, R. L., *Chem. and Ind. (Rev.)*, **39**, 1030 (1956)
26. Andrews, A. G., *Corrosion*, **4**, 93 (1948)
27. Schmeling, E. L. and Röschenbleck, B., *Werkst. u. Korrosion, Mannheim*, **9**, 529 (1958)
28. Fontana, M. G., *Industr. Engng. Chem.*, **43** No. 9, 105A (1951)
29. Wickert, K., *Korros. Metallsch. Beih.*, **20**, 147 (1944)
30. Camuil, J., *Sci. Industr. Res.*, **20**, 114 (1947)
31. Turnbull, D. and Frey, D. R., *J. Phys. Colloid Chem.*, **51**, 681 (1947)
32. Coles, E. L., Gibson, J. G. and Hinde, R. M., *J. Appl. Chem.*, **8**, 341 (1958)
33. Wilson, B. S. and Garner, F. H., *J. Inst. Petrol.*, **37**, 225 (1951)
34. Vaivads, A. and Liepina, L., *Latv. P.S.R. Zinat. Akad. Vestis.*, **8**, 119 (1954)
35. Katz, W., *Metalloberfläche*, **11**, A.161 (1953)
36. Carter, V. E. and Campbell, H. S., *J. Met. Finish.*, **8**, 103 (1962)
37. Lander, J. J., *J. Electrochem. Soc.*, **99**, 467 (1954)
38. Koch, D. F. A., *Electrochimica Acta.*, **1**, 32 (1959)
39. Lander, J. J., *J. Electrochem. Soc.*, **103**, 1 (1956)
40. Kir'yakov, G. Z. and Dunaev, Yu. D., *Izvest. Akad. Nauk. Kazakh. S.S.R., Ser. Khim.*, **2**, 32 (1957)
41. Morgan, J., *Corrosion Technol.*, **5**, 347 (1958)

BIBLIOGRAPHY

Jones, D. de G. and Masterson, H. G., 'Enhanced Corrosion of Lead Sheathed Power Cables due to Adjacent Cathodic Protection Schemes', 3rd Congress of the European Corrosion Federation, Brussels, 1963

Galimberti, C. E., 'Corrosion of Lead by Alternating Current', *Corrosion*, **20**(5), 150t (1964)

Kawabata, R., Miyase, S. and Tagaya, M., 'Effect of Alloying Elements on the "Flash Points" of Lead', *Trans. Japan Inst. Metals*, **5**(2), 85 (1964). (*Bull. Brit. Non-Ferrous Metals Res. Assoc.*, No. 427, Abstract p. 42 (1965))

Pelzel, E., 'Corrosion Behaviour of Lead Alloys in Sulphuric Acid', *Metall.*, **19**(8), 818 (1965)

Ipolyi, K., 'Clarification of the Question of "Phenol Corrosion" (Part V)', *Tech. Mitt. Post-Telegraphen Telephon-Verwaltung* (Switzerland), **63**(11), 463 (1965)

Johnson, J. W., Wu, C. K. and James, W. J., 'Anodic Dissolution of Pb in Aqueous Solutions', *Corros. Sci.*, **8**(5), 309 (1968)

Parlov, D. and Togatschev, T., 'Influence of Silver and Thallium on the Anodic Corrosion of Lead Alloys in Sulphuric Acid', *Werkst. Korros.*, **19**(8), 677 (1968)

Obrecht, M. F. and Pourbaix, M., 'Corrosion of Metals in Potable Water Systems', *J. Am. Water Assoc.*, **59**, 977 (1967)

4.4 Magnesium and Magnesium Alloys

Magnesium is a divalent metal and is silvery white in appearance. It is the eighth most abundant element and sixth most abundant metal. The atomic weight is 24·32 and the specific gravity of the pure metal 1·738 at 20°C. The structure is close packed hexagonal. The melting point is 650°C and the boiling point 1 107°C. The specific heat at 20°C is 1·030 kJ/kg °C and the thermal conductivity at 20°C is 157·5 W/m °C; the electrochemical equivalent is 0·126 mg/C. The standard electrode potential $E^{\circ}_{Mg^{2+}/Mg} = -2{\cdot}37$ V, but in 3% sodium chloride the electrode potential is $-1{\cdot}63$ V (vs S.C.E.), i.e. $-1{\cdot}38$ V (vs S.H.E.).

For engineering purposes magnesium is rarely used in the unalloyed condition. Small percentages of aluminium, zinc, etc. as indicated in Table 4.15 are added to improve mechanical and other properties. Magnesium is itself used for alloying with other metals.

Applications of magnesium alloys are many and varied but their light weight—about two-thirds that of the aluminium alloys—and high strength-to-weight ratio have made them of particular interest to the aircraft and guided-weapons industries. In the transport industry too, their light weight is attractive. Other features for which they are noted are their high stiffness-to-weight ratio, their great ease of machinability, good casting qualities and high damping capacity.

In the nuclear engineering field special magnesium-base alloys are extensively used as canning materials for uranium in gas-cooled reactors.

The compositions of the more common magnesium-rich alloys used in Great Britain are given in Table 4.15. Similar alloys are in use in the USA and elsewhere and their American designation together with the equivalent British alloys are given in Table 4.16. Many of the casting alloys are given various simple heat treatments to improve their properties, while the wrought alloys can be obtained in a number of tempers.

As with other electronegative metals, the absence of serious corrosion of these alloys in ordinary industrial atmospheres is largely a result of the formation of protective films which inhibit further attack. Similarly, when serious corrosion does occur, or when it occurs after a period of successful use, it can usually be traced to a change in conditions of such a nature that protective films already formed have suffered dissolution or break-down. No alloying ingredients are known which effect any substantial improvement

Table 4.15 Nominal composition of magnesium-rich alloys

Designation		% *Composition (remainder magnesium)*								
		Al	Zn	Zr	Th	*Rare earth metals*	*Fractionated rare earth metals*	Be	Mn	Ag
Casting alloys	Z5Z	—	4·5	0·7	—	—	—	—	—	—
	RZ5	—	4·0	0·7	—	1·2	—	—	—	—
	TZ6	—	5·5	0·7	1·8	—	—	—	—	—
	MSR A	—	—	0·6	—	—	1·7	—	—	2·5
	MSR B	—	—	0·6	—	—	2·5	—	—	2·5
	ZRE1	—	2·2	0·6	—	2·7	—	—	—	—
	ZT1	—	2·2	0·7	3·0	—	—	—	—	—
	MTZ	—	—	0·7	3·0	—	—	—	—	—
	A8	8·0	0·5	—	—	—	—	—	0·3	—
	AZ91	9·5	0·5	—	—	—	—	—	0·3	—
	AZ91X	9·4	0·4	—	—	—	—	0·0015	0·3	—
	C	7·5–9·5	0·3–1·5	—	—	—	—	—	0·15 (min)	—
	AZG	6·0	3·0	—	—	—	—	—	0·3	—
	ZE63	—	6·0	0·6	—	2·5	—	—	—	—
Wrought alloys	ZW3	—	3·0	0·6	—	—	—	—	—	—
	ZW1	—	1·3	0·6	—	—	—	—	—	—
	ZW6	—	5·5	0·6	—	—	—	—	—	—
	ZTY	—	0·5	0·6	0·75	—	—	—	—	—
	AM503	—	—	—	—	—	—	—	1·5	—
	AZ31	3·0	1·0	—	—	—	—	—	0·3	—
	AZM	6·0	1·0	—	—	—	—	—	0·3	—
	AZ855	8·0	0·4	—	—	—	—	—	0·3	—
	ZM21	—	2·0	—	—	—	—	—	1·0	—
	ZM61	—	6·0	—	—	—	—	—	1·0	—

Table 4.16 Equivalent British and American designations for magnesium alloys

Wrought alloys		*Cast alloys*	
British designation	*American designation (ASTM)*	*British designation*	*American designation (ASTM)*
ZW6	ZK60A	Z5Z	ZK51A
AM503	M1A	RZ5	ZE41A
AZM	AZ61A	TZ6	ZH62A
AZ855	AZ80A	MSR	QE22A
ZM21	ZM21	ZRE1	EZ33A
ZM61	ZM61	ZT1	HZ32A
		MTZ	HK31A
		A8	AZ81A
		AZ91X	AZ91
		ZE63	ZE63

in the general corrosion behaviour of the magnesium alloys; though manganese is usually regarded as beneficial this is probably because it offsets the deleterious effects of iron and other cathodic metals which may be present rather than because it makes any positive improvement in the resistance of the magnesium itself. All metals added to produce the various alloys are intended simply for the improvement of the physical and mechanical properties and not for any effect they may have on the corrosion behaviour. On the other hand, none of these additions has any very marked deleterious effect on the corrosion resistance of the alloys.

Since corrosion resistance depends on film formation, it follows that the behaviour of the alloys will vary considerably with the medium to which they are exposed. The corrosion of the metal is governed largely by the solubility and other characteristics of the film. Thus, magnesium fluoride is very insoluble in hydrofluoric acid and as a consequence magnesium does not dissolve in this acid. Initial attack forms a film of magnesium fluoride and even though the film is not impervious to other corrosive influences it effectively seals the metal against further reaction. In dilute aqueous hydrofluoric acid, attack may take place, and if so it will be of a pitting type similar to that which takes place in tap water. In fact the corrosion is due to the water itself and not to the acid. Magnesium sulphate on the other hand is readily soluble in dilute sulphuric acid and no protective film is formed when magnesium reacts with this acid; attack is rapid and continuous with evolution of hydrogen. It should be noted, however, that magnesium sulphate is only slightly soluble in *concentrated* sulphuric acid. When, therefore, magnesium is immersed in strong sulphuric acid, initial attack produces a film of magnesium sulphate which quickly saturates the acid at the interface, and the reaction is reduced to a vanishingly low rate. As long as water is excluded no further attack takes place.

In considering the corrosion behaviour of magnesium alloys, therefore, it is of the utmost importance to know the nature of the medium to which the metal is to be exposed. In general, atmospheric attack in damp conditions is largely superficial; aqueous solutions bring about attack which varies not only with the solute but with the volume, movement and temperature of the

Table 4.17 Standard electrode potentials (S.H.E.)

Metal	*Potential* V
Magnesium	−2·37
Aluminium	−1·66
Zinc	−0·76
Iron	−0·44

Table 4.18 Steady-state potential of several metals in 1M solutions of different electrolytes (vs 0·1N Calomel, $E_H = 0{\cdot}336$ V)

Metal	*Potential* (V)								
	Sodium chloride	*Sodium sulphate*	*Sodium chromate*	*Hydrochloric acid*	*Nitric acid*	*Sodium hydroxide*	*Ammonium hydroxide*	*Calcium hydroxide* (sat'd)	*Barium hydroxide* (sat'd)
Magnesium	−1·72	−1·75	−0·96	−1·68	−1·49	−1·47	−1·43	−0·95	−0·88
Aluminium	−0·86	−0·50	−0·71	−0·80	−0·49	−1·50	−0·80	−1·54	−1·53
Zinc	−1·15	−1·19	−0·67	−1·14	−1·06	−1·51	−1·50	−1·40	−1·49
Iron	−0·72	−0·76	−0·16	−0·66	−0·58	−0·22	−0·18	−0·30	−0·25

liquid. Many organic liquids are quite inert to magnesium, but some of those which contain reactive polar groups, as might be expected, are reactive in some degree towards the metal.

Galvanic corrosion of magnesium, i.e. the enhanced corrosion to which the anodic member of a pair of metals in contact is subject when both are in contact with a common electrolyte, is of considerable practical importance, since magnesium is anodic to all other structural metals in most electrolytes.

The standard electrode potential of magnesium is given, along with the potentials of other metals, in Table 4.17 and the steady-state potentials of magnesium in various solutions are listed in Table 4.18[1].

Corrosion of Magnesium Alloys in Atmospheric Conditions

In clean, dry atmospheres (with r.h. below about 60%) uncontaminated magnesium alloys retain a lustrous surface almost indefinitely. If the atmosphere is clean but not dry and the humidity approaches 100% a scattered pattern of corrosion spots appears after a period, but considerable areas of unaffected surface remain for a very long time. (This effect is probably attributable to galvanic corrosion on a micro-scale, each spot representing corrosion of the magnesium adjoining a cathodic particle contained in the surface as an impurity.) It is quite otherwise if the surface has been contaminated either by corrosive dusts or by cathodic particles introduced by abrasive treatments of various kinds. For example, if a piece of magnesium which has been shot blasted is exposed to damp conditions, the whole surface rapidly becomes covered with a greyish layer of corrosion product. In drier atmospheres the layer may resemble a patina and develop very slowly.

In many ways the corrosion of magnesium alloys in normal atmospheric conditions is a close approximation to the *initial* formation of rust on mild

Table 4.19 Corrosion rates of magnesium-rich alloys ($gm^{-2} d^{-1}$) in three different environments

	Type of test		
Alloy	Immersion for 30 days in 3% NaCl solution	Sea-water spray 3 times per day for 6 months	Atmospheric exposure for 2 years
A8	3·5–133·0 av. 72·7	1·5–2·5 av. 2·0	0·22–0·26 av. 0·24
AZ91	1·3–144·0 av. 52·6	0·5–2·3 av. 1·7	0·19–0·26 av. 0·23
Z5Z	5·5–9·3 av. 8·0	0·7–0·9 av. 0·8	0·34–0·35 av. 0·34
ZRE1	20·5–32·7 av. 26·6	1·2–2·0 av. 1·6	0·36–0·37 av. 0·36
RZ5	12·0–40·2 av. of 26 tests 32·6	1·1–1·6 av. of 12 tests 1·3	—
TZ6	26·8–67·6 av. of 19 tests 53·4	0·9–1·2 av. of 4 tests 1·1	0·40–0·44 av. of 4 tests 0·41
High-purity A8	3·0	0·4	0·28

steel when similarly exposed and is just as superficial. Study of the cross-sections of a number of test bars of different alloys which were exposed to the weather for three years at Clifton Junction, Manchester, showed that the loss of section was remarkably small and the deterioration of mechanical properties was not greater than could be attributed to this and to the roughening of the surface which had occurred.

Corrosion rates in normal industrial atmospheres measured as loss of weight over a period are extremely uniform among the various alloys. Table 4.19, last column, gives the corrosion rates (in $gm^{-2} d^{-1}$) for a number of alloys determined at Clifton Junction in recent years. The highest value recorded ($0{\cdot}4\ gm^{-2} d^{-1}$) is equivalent to a rate of penetration of 0·076 mm/y, which is appreciably less than that of mild steel.

Composition of Corrosion Product

The compound produced by the interaction of magnesium with moist air appears to be essentially magnesium hydroxide ($S_p = 10^{-11{\cdot}05}$). Magnesium oxide reacts slowly with water to form hydroxide so that films of oxide formed at high temperatures, during hot-forming operations for example, eventually become hydrated. The final composition of the corrosion product undoubtedly depends upon the nature of the atmosphere and the gases it contains. Thus the natural air-formed corrosion product in industrial areas always consists largely of carbonate and sulphate in addition to hydroxide.

In marine atmospheres magnesium chloride is formed and eventually oxychloride by reaction with magnesium hydroxide formed at the same time. Since the chloride is hygroscopic, moisture is attracted and the corrosive effect is hence much worse than that of water alone.

Marine Atmospheres

If by a 'marine atmosphere' one means the conditions at a site within a few metres of high-water mark in an otherwise clean country atmosphere, then corrosion of unprotected magnesium alloys is remarkably small. The presence of actual liquid spray is exceptional, and during periods of high wind, when such spray is generally produced, the humidity is usually low and evaporation of the droplets is rapid. This explains why crystals of salt may be found on exposed metal yet little evidence of severe corrosion may be seen. Salt solution is corrosive but dry salt particles are almost without effect. Two other factors which operate in such regions are: (*a*) the washing effect of clean rain followed by rapid evaporation; and (*b*) the scouring by blown sand particles. Unpainted magnesium items exposed near the sea coast invariably exhibit a dull, slightly rough surface, with no corrosion product evident. This appearance is in sharp contrast to that of metal which has been exposed to conditions of persistent dampness; in such cases the corrosion product, considerably more voluminous than the metal from which it is formed, is retained on the surface as a greyish white powder. The glossy surface of painted specimens and the polish of newly prepared bare metal specimens exposed near the sea coast is also quickly lost. Corrosion rates of painted specimens which have been exposed so long that the paint has perished are often higher than those of bare metal. The cause of this is probably the retention of water by the sponge-like texture of the deteriorated paint*. In a similar way corrosion may be most in evidence at junctions and narrow gaps. This is due not so much to true 'crevice corrosion' as to the retention of moisture. Surface moisture evaporates quickly, and the incipient corrosion to which it gives rise quickly comes to an end. In a narrow recess, however, trapped moisture, unable to evaporate, continues the corrosion process over long periods. True crevice corrosion on the other hand is a phenomenon which owes its nature to the development of anodic areas within the crevice, caused by exclusion of oxygen for example, and does not appear to occur with the magnesium alloys.

In conditions where sea-water spray may be deposited regularly on magnesium articles with no alleviating mechanism for its removal, or where breaking waves may drench the components, the effect is quite different. Corrosion of bare metal will be heavy and will be intensified at junctions with other more noble metals. Unless magnesium alloys can be adequately protected in such combinations it is better to avoid their use. This matter is dealt with under the section on protection.

In an experiment similar to that referred to on p. **4**.91, tensile test bars were exposed at Clifton Junction, Manchester, for six months, during which time they were sprayed three times daily with sea-water. Whereas exposure to industrial atmosphere alone had little effect, bars of the same alloys were much more heavily attacked by sea-water spray.

High-purity Alloys and Galvanic Corrosion

In addition to the alloying ingredients which are added, certain other metals are usually present in small amounts. In the alloys which contain aluminium, for example, iron usually amounts to about 0·02–0·05%. By special techniques and care in melting this can be reduced to about one-tenth of the above figure. Many workers have shown that such high-purity alloys have a markedly better resistance to salt water than those of normal purity, but their behaviour towards industrial atmospheres is not greatly different. Furthermore, the practical value of the higher resistance to corrosion is largely offset when components are used in electrical contact with other more cathodic metals. The effect of a steel bolt for example, even when it has been zinc or cadmium plated, is much greater at the point of contact than that of the excess of local cathodes in the impure alloys. Galvanic corrosion at joints with other metals therefore is not markedly less in the case of the high-purity alloys. Nevertheless, such alloys have their place, and when they can be used without other metal attachments provide better intrinsic resistance to corrosion by sea-water than the alloys of normal purity.

Alloys containing zirconium as a grain-refining agent have the iron content automatically reduced to about 0·004% by settling out of impurities during the alloying procedure.

*Private communication from A. P. Fenn, Esq., Birmingham Aluminium Casting (1903) Co. Ltd.

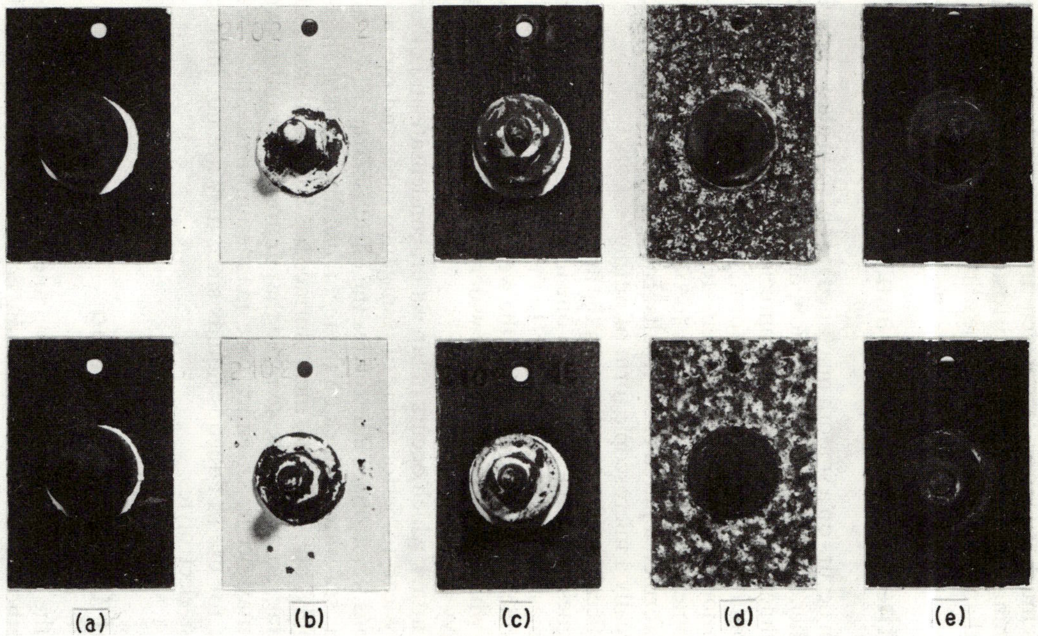

Fig. 4.19 Samples of high purity AZ31 (upper photographs) and ZW3 (lower photographs) magnesium-base alloys, fitted with mild-steel nuts and bolts and exposed to a variety of corrosion conditions. (*a*) 4·5 hours' immersion in 3% salt soln., (*b*) 180 days' immersion in distilled water, (*c*) 4 days' immersion in borehole water, (*d*) 180 days' in humidity cabinet sea-water spray and (*e*) 180 days' atmospheric exposure

Reference to Table 4.19 will show that greatly superior corrosion rates of the high-purity alloys are only in evidence in the more severe conditions of test by immersion in salt water, and that in less drastic conditions, and especially in industrial atmospheric exposure, there is little to choose between the alloys.

Figure 4.19 illustrates the corrosion occurring on high-purity AZ31 and ZW3 in contact with steel bolts. Tested alone in sea-water, the corrosion rate of the former is much the lower. It is evident from the illustration, however, that the governing factor in galvanic corrosion is the type of electrolyte present rather than the composition of the alloy.

Nature of the Corroded Surface

When corrosion occurs on a smooth machined magnesium alloy surface, this surface is roughened by the chemical action, and after the initial attack the degree of roughness does not change appreciably. In the usual industrial atmospheric conditions the attack is uniform, but in immersed conditions, including corrosion under pools of condensate, attack may be, and usually is, irregular; some areas become anodic to other areas and, as corrosion proceeds at the former, a pitted effect results. Even in atmospheric attack the roughening is really a microscopic form of pitting.

There is a noticeable difference between the appearance of the aluminium-containing magnesium-rich alloys on the one hand and the zinc/zirconium-containing magnesium alloys on the other. In the former the microscopic pits in the surface which has been exposed to the weather tend to be narrow and relatively deep, while in the latter they are wider and tend to overlap, leading to a slightly wavy appearance.

The unequal attack which occurs in tap water, condensate and other mild electrolytes may lead to perforations of thin-gauge sheet and even to deep pitting of castings. In stronger electrolytes the effect is variable. In chloride solutions such as sea-water, attack on the metal usually results in the pitting of some areas only, but where the metal surface has been rendered reactive, as by shot blasting, attack may be so rapid that uniform dissolution over the whole surface may occur. In either case magnesium-base alloys are not usually suitable for use in aqueous liquids since they are not intrinsically resistant to these electrolytes.

Methods of Corrosion Testing

In considering the corrosion of magnesium and its alloys it is important to examine the methods available for assessing corrosion tendencies and particularly those known as accelerated tests. Tests carried out by immersion in salt water or by spraying specimens regularly with sea-water are worthless as a means of determining the resistance of magnesium alloys under any other than the particular test conditions. Extrapolation to less corrosive conditions is not valid and even the assessment of the value of protective measures by such means is hardly possible. The reason is to be found in the fact that corrosion behaviour is directly related to the formation of insoluble films. In chloride solutions there is no stable insoluble film formed from the solution itself and no previously formed film (by chemical reaction) is impermeable to the chloride ion. Even existing protective films are penetrated relatively easily by chloride ions, and organic films of paint or varnish are subjected to osmosis and to swelling in conditions which are quite unlike normal experience. Except for the specific purpose of determining the behaviour of the materials in dilute chloride solutions, accelerated methods of test of this nature are inadmissible and the results misleading. Corrosion in the atmosphere is usually a continuous, relatively gentle, but persistent, process. Since it is continuous the time scale cannot be shortened and to attempt to obtain results merely by increasing the severity of the conditions is illogical. Such tests can only give information about the behaviour in the chosen conditions. The only true indication of behaviour is obtained by exposing samples to the conditions to be experienced, if these are known or can be assessed. If the severity is of an intermittent nature then it is usually permissible to accelerate the time basis by omitting or reducing the intervals of less severe exposure. Even in this case, however, the effects of recovery by drying out, etc. should not be overlooked. In particular, the samples exposed should take account of the effects of joints with both similar and dissimilar materials; acid vapours from wood and plastics and the electric stress introduced by coupling to other metals in the presence of an electrolyte can vitiate completely any deductions concerning the protective value of paints based on tests on isolated pieces of metal.

Intergranular Corrosion

True intergranular corrosion of magnesium alloys does not occur for the reason that the grain-boundary constituent is invariably cathodic to the grain body. It follows, therefore, that corrosion will be principally concentrated on the grains and the grain-boundary constituent will not only be more resistant to attack but will in some measure receive cathodic protection from the corrosion of the neighbouring grain.

Examples are occasionally quoted which purport to show attack at the grain boundaries, but this is not intergranular corrosion properly so called; indeed it is the opposite and might better be called granular attack, for it is the grain and not the boundary which is preferentially attacked. Because the grain boundary is cathodic to the grain proper, attack is concentrated on the area of the anodic grain adjoining the boundary until eventually the grain may be undercut and fall out of the matrix. The important difference from true intergranular corrosion is quite clearly that attack can proceed only grain by grain and cannot make its way through the body of the material following a grain-boundary path.

As usually cleaned for microscopical examination, a corroded specimen has invariably lost the delicate tracery of intergranular material in the cleaning process and thus may present something of the appearance associated with intergranular attack. If special steps are taken, however, it is possible to mount a corroded specimen of magnesium alloy with the grain boundaries still intact, showing where some grains have entirely disappeared while others are in the process of dissolution round the edges where the cathodic

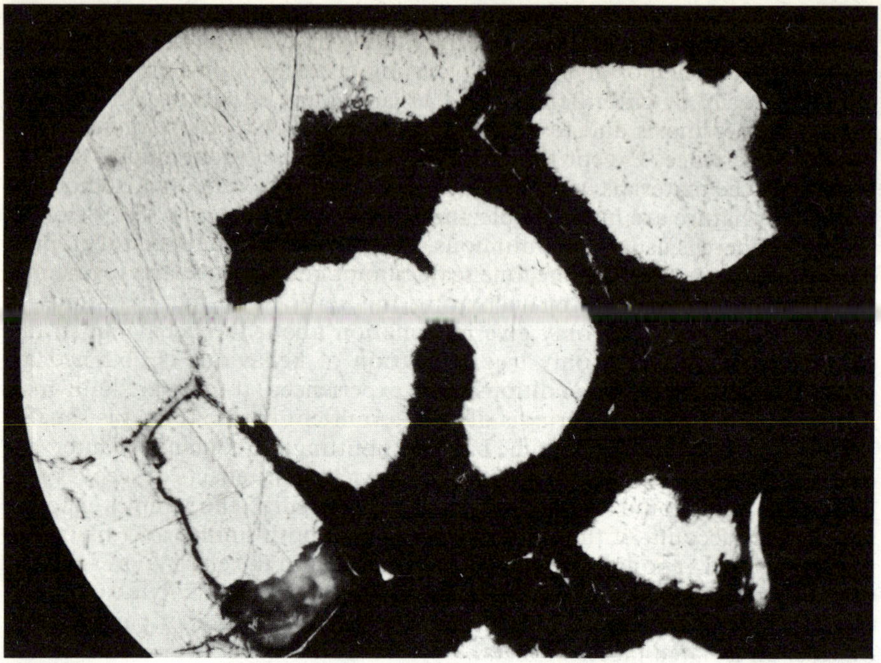

Fig. 4.20 Corroded grain in ZRE1 showing the grain boundaries still intact. Attack occurs on the periphery of the grains and thus is not intergranular

effect of the grain-boundary material is strongest. This is illustrated in Fig. 4.20 which shows a sample of corroded ZRE1 with the intergranular constituent still intact in many areas and with one grain almost etched free from the containing network.

Protection of Magnesium Alloys

The proneness or otherwise to corrosion is essentially the same in all the magnesium-base alloys and it is important to note that the requirements of protection therefore do not vary for the magnesium alloy under consideration. If conditions are such that any one of the alloys can be used satisfactorily without protection, then any other of the alloys can be so used. On the other hand, if a given protective scheme is found necessary for a particular alloy, then the same protective scheme will be found necessary (and will be equally effective) with any other magnesium-base alloy.

It will be realised that since the tendency of magnesium to corrode is governed by the nature of the environment to which it is exposed the degree of protection necessary is also controlled by the same factor.

The methods of protection available are of two basic kinds: chemical or electrochemical treatments which oxidise the metal and produce a film which is more stable than the metal itself, and coating methods which rely upon the application of some extraneous material to provide a more or less impervious coating and thus restrict access of corrosive influences. Generally the two methods are complementary and for best effects, or for protection under the more drastic conditions, they are used in conjunction with each other.

The first class referred to above includes various simple treatments in which the metal is dipped in acidic or near-neutral solutions, usually containing chromate in the hexavalent condition, which has a strong passivating action on many reactive metals. The resulting chemical film which is formed on the magnesium-rich alloys consists essentially of magnesium oxide but also contains chromium compounds. This class of treatment also includes methods of anodic oxidation achieved either by galvanic or electrical means. In general the films resulting from such electrical methods are thicker and often harder than those produced by the simple immersion methods, and usually form more effective barriers against corrosion. Both types, however, should be regarded as suitable for withstanding normal atmospheric conditions for relatively short periods only, since they are all porous or at least permeable to water to some degree. They are, in fact, to be considered as foundation treatments for coating with more impermeable organic materials such as paint and enamel. (See also Section 16.3.)

The second class of protective measures includes all those processes generally known as painting, as well as 'temporary' protective treatments such as greasing and oiling. It is conventional to apply paint films over one or other of the chromate-containing oxide films (produced in one of the immersion-type baths), but paints can be applied equally well to the electrolytically developed coatings. It is not usually wise to apply paint directly to the bare metal. The reason for this is twofold. In the first place, a chromated surface, especially a freshly chromated surface, is not as likely to have deteriorated in storage as a bare metal surface and consequently will provide a better basis for paint. Secondly, and more important, the natural surface of magnesium alloys in contact with damp air is alkaline, because of the presence of the naturally formed oxide and hydroxide, and this may lead to the rapid deterioration of paint films. This is especially true of the oil-based and some of the synthetic air-drying paints which are sensitive to alkali.

In all cases a priming coat of paint containing some form of hexavalent chromium such as zinc or strontium chromate should be used.

Table 4.20 lists a number of the better-known processes for producing protective films on magnesium alloys by chemical and electrochemical processes.

High Temperature Stoved Epoxy Resins

In recent years use has been made of the strong adhesion, toughness and water impermeability of some of the epoxy resins to secure greatly improved surface protection of magnesium alloys. By this means it has been possible to employ these alloys even in situations where they are drenched repeatedly with sea-water.

Not all the epoxy resins are equally efficacious and all have to be stoved at a relatively high temperature (180–220°C) in order to develop the requisite

Table 4.20 The more usual cleaning and chromating processes for magnesium-base alloys*

Process	*D.T.D. Spec.*	*United States Spec.*	*Purpose*	*Composition (% by wt.)*	*Method of use*	*Resulting appearance*	*Suitable container*	*Notes*
Pickling and cleaning								
Nitric acid pickle	911C, Sec. 2.1.1	MIL-M-3171C, Sec. 3.2.4	Cleaning rough castings	5–10% conc. nitric acid in water	Repeated brief dips until clean, followed by thorough rinse, preferably with hose	Clean and bright but may be loose black smut on Mg–Zr–Zn alloys	Glass, polythene, rubber and earthenware	Metal removal
Sulphuric acid pickle	911C, Sec. 2.1.1	MIL-M-3171C, Sec. 3.2.4	Cleaning rough castings	2–5% conc. sulphuric acid in water	Repeated brief dips until clean, followed by thorough rinse, preferably with hose	Clean and bright but may be loose black smut on Mg–Zr–Zn alloys	Glass, polythene, rubber and earthenware	Metal removal
Fluoride anodising treatment	911C, Sec. 2.1.3	MIL-M-3171C, Sec. Type VII, 3.9	Super cleaning for maximum corrosion resistance	10–30% ammonium bifluoride in water	Application of 120 V a.c. for 5–30 min	Smooth white matt	Rubber, ebonite and polythene	No dimensional change M.E.L.B. Pat. 721 445
Caustic soda clean	—	MIL-M-3171C, Sec. 3.2.3	Cleaning finished parts or greasy components generally Removal of paint	2–5% caustic soda in water	Immerse and boil for 15–30 min; wash thoroughly	Clean, showing no water 'break' when wetted with cold water	Steel	No dimensional change Proprietary brands of alkaline cleaners may be used provided these do not attack metal Soda ash with washing soda and soap are suitable for preparing alkaline cleaning baths
Chromic acid bath	911C, Sec. 2.1.4	MIL-M-3171C, Sec. 3.4.1	Removal of old chromate and fluoride films; also removal of corrosion product and oxide films	10–15% chromic anhydride (CrO_3) in water	Immerse and boil for 15 s to 30 min or as necessary; wash thoroughly	Clean grey appearance	Glass and steel	No dimensional change unless much impurity is present For small delicate parts 0·1–0·5% silver chromate may be added to reduce attack caused by impurities such as chloride in corrosion product

Chromating								
Acid chromate bath (I.G. bath) (chrome pickle)	911C, App. 11, Bath iv	MIL-M-3171C, Type I, Sec. 3.5	Protection in storage Paint foundation Repair of chromate film	15% sodium or potassium dichromate and 20–25% nitric acid (s.g. 1·42)	Immerse for 10 s to 2 min, drain for 5 s, wash in cold or warm water	Golden bronze often with iridescent colours	Glass, earthenware, slate, aluminium and stainless steel	Metal removal Cannot be used on parts to fine tolerances Useful for rough castings
Hot half-hour bath (R.A.E. bath) (black bath)	911C, App. 11, Bath iii	MIL-M-3171C, Sec. 6.4.1	Good paint foundation Protection in storage	3% ammonium sulphate, 1·5% potassium or sodium dichromate and 0·5–0·75% 0·880 ammonia solution	Immerse in bath and simmer for 30 min; wash	Usually black; light brown on D.T.D. 118	Glass, steel and aluminium	No dimensional change Used chiefly for finished work
Chrome–manganese bath (M.E.L. black bath)	911C, App. 11, Bath v	MIL-M-3171C, sec. 6.4.1	Good paint foundation Protection in storage	10% sodium or potassium dichromate, 5% magnesium sulphate and 5% manganese sulphate	Immerse in bath for 2 h at 20°C; proportionately less if bath is heated	Usually black or dark brown; light brown on D.T.D. 118	Glass, steel and aluminium	No dimensional change Used chiefly for finished work where heating may be undesirable where inserts are present
Special surface treatments								
H.A.E. process	—	MIL-M-45202B (ORD)	Corrosion and abrasion resistance	12% potassium hydroxide, 1% aluminium (high purity) 3·5% trisodium phosphate (crystals), 3·5% potassium fluoride (anhydrous) and 2·2% potassium manganate	Application of up to 90 V a.c. for up to 2 h	Dark brown, ceramic-like coating, brittle	Glass, rubber and steel	Increase in dimensions of 0·025 to 0·050 mm
Dow 17 process	—	MIL-M-45202B	Corrosion and abrasion resistance	24% ammonium bifluoride, 10% sodium dichromate and 8·6% orthophosphoric acid (85%)	Application of up to 110 V a.c. at 50 to 500 A/m^2; temp. 70–80°C; time 2–30 min	Grey-green to dark green dependent on thickness	Steel, rubber and vinyl-based materials	Increase in dimensions of 0·005 to 0·038 mm

*Table reproduced by permission of Magnesium Elektron Ltd.

adhesion to the metal and the necessary water resistance in the resin. Furthermore, a technique of baking the metal before application of the resin, followed by dipping of the still warm metal, is essential in order to fill subcutaneous blemishes. In the absence of such a procedure trouble can still arise from 'pockets' which retain moisture and which are only bridged by a normally applied resin film.

The Effect of Surface Finishing on the Corrosion Behaviour of Magnesium Alloys

It is probable that all corrosion of magnesium alloy surfaces exposed to a damp atmosphere, or still more immersed in an electrolyte, is largely galvanic in origin and much influenced by the presence of exposed cathodic particles. Some of these are present in the alloys as unavoidable impurities, and nearly all foreign metallic particles not in solution are cathodic to magnesium. Various methods of chemical pickling or etching as well as mechanical means of metal treatment may remove a proportion of these, but unfortunately they will usually expose a further number in the lower layers. Furthermore, some methods of mechanical abrasion may increase the number of foreign cathodic particles in the surface by entrapment, and even pickling solutions are not exempt from causing the deposition of more noble metals in solution by displacement. Partly used pickling baths which have been in use for some time, in particular, become enriched in cations of other metals and may redeposit these metals by displacement. In practice, pickling baths based on nitric acid are less likely to give rise to this effect though they may not be able to remove foreign particles already *in situ*.

Chief among the processes which bring about harmful effects on the corrosion resistance of magnesium alloy surfaces are shot and grit blasting and the use of emery cloths and papers. Various blasting operations are used for the removal of adhering foundry sand from sand castings but it should be recognised that they lower very considerably the natural resistance of the surface and reduce its ability to form protective films. Metal particles either of the abrasive itself or of material scoured from the equipment, are lodged in the magnesium surfaces by blasting operations, while splinters of emery (which form quite effective cathodes) are picked up by the use of such materials. On the other hand, glass papers are usually harmless; the splinters of glass are non-conductive and therefore incapable of acting as cathodes. Despite the fact that silicon carbide is a conductor, the use of Carborundum paper and belts does not usually lead to any serious deterioration in the corrosion resistance of magnesium alloys.

By the use of many commercial abrasive processes, the corrosion resistance of magnesium alloys can be reduced to such an extent that samples of metal that may lie quiescent in salt water for many hours will, after shot blasting, evolve hydrogen vigorously, and the corrosion rate, as measured by loss of weight, will be found to have increased many hundred-fold. The effect in normal atmospheres is naturally much less, yet the activation of the surface is an added hazard and is the opposite of passivation which is essential if later-applied paint finishes are to have proper durability.

The use of chromating baths and acid pickles is powerless to remove all the evil effects of such treatments, but one of the electrolytic processes, namely that of Fluoride Anodising at a high voltage in a solution of ammonium bifluoride, is very effective in removing cathodic foreign metals. In this process the magnesium surface itself is quickly converted to insoluble and non-conductive magnesium fluoride and this reaction thereafter terminates. The current is thereupon automatically directed to and concentrated on the local metal cathodes which are conductive, and these are either dissolved or dislodged from the surface. Carbon, in the form of graphite resulting from the use of die lubricants in forming and in pressure diecasting, may also be an active cathode. It is not dissolved by the electrolysis but the film is undercut and insulated from the metal surface by a layer of magnesium fluoride. In this condition it is less harmful than when in direct contact with the metal; furthermore, it can more readily be removed by treatment in chromic acid or in hot caustic soda solution, processes which, in the absence of prior fluoride anodising, are not completely effective.

K. G. ADAMSON

REFERENCES

1. Mears, R. B. and Brown, C. D., *Corrosion*, **1**, 113 (1945); Hanawalt, J. D., Nelson, C. E. and Peloubet, J. A., *Trans. Amer. Inst. Min. (Metall.) Engrs.*, **147**, 275 (1942)

BIBLIOGRAPHY

Metals Handbook, **1**, 8th edn, American Society for Metals, Chicago (1961)

Pearlstein, F. and Teilell, L., 'Corrosion and Corrosion Prevention of Light-Metal Alloys' Paper No. 114, *Corrosion*, 73, Anaheim, March (1973)

Emley, E. F., *Principles of Magnesium Technology*, Pergamon Press, Chapter xx (1966)

Adamson, K. G., King, J. F. and Unsworth, W., *Evaluation of the Dow 17 Treatment for Magnesium Alloys*, Ministry of Defence D.Mat. Report No. 192, February (1973)

King, J. F., Adamson, K. G. and Unsworth, W., *Impregnation of Anodic Films for the Protection of Magnesium Alloys*, Ministry of Defence D.Mat. Report No. 193, February (1973)

Adamson, K. G., King, J. F. and Unsworth, W., *Evaluation of High Temperature Resistant Coatings for the Protection of Magnesium Alloys*, Ministry of Defence D.Mat. Report No. 196, July (1973)

4.5 Nickel and Nickel Alloys

Physical and Mechanical Properties

Composition of Metal and Alloys

Commercially pure nickel has good mechanical properties and good resistance to many corrosive environments and therefore finds application where this combination of properties is required. Of more importance, however, is the fact that nickel forms a wide range of alloys having desirable engineering and corrosion-resistant properties. With regard to corrosion resistance to aqueous solutions, among the most important of these alloying elements are Cr, Fe, Cu, Mo and Si. Since the range of corrosion-resistant nickel alloys includes some that owe their corrosion resistance to passivity and others that are resistant because they are sufficiently noble not to displace hydrogen from acidic solutions, the corrosive environments in which nickel alloys can be successfully used are very varied, embracing acids, salts and alkalis (both oxidising and non-oxidising in character) sea-water, natural waters and the atmosphere.

In addition to nickel alloys, nickel also forms an important alloying element in stainless steels and in cast irons, in both of which it confers additional corrosion resistance and improved mechanical and engineering properties, and in Fe–Ni alloys for obtaining controlled physical and magnetic properties (see Chapter 3). With non-ferrous metals nickel also forms important types of alloys, especially with copper, i.e. cupro-nickels and nickel silvers; these are dealt with in Section 4.2.

Nickel is also widely used as an electrodeposited underlay to chromium on 'chromium-plated' articles, reinforcing the protection against corrosion provided by the thin chromium surface layer. In recent years the production of articles of complex shape to close dimensional tolerances in nickel by electroforming—a high-speed electrodeposition process— has attracted considerable interest. Electrodeposition of nickel and the properties of electrodeposited coatings containing nickel are dealt with in greater detail in Section 14.7.

The nominal compositions of commercially pure wrought nickel and some nickel alloys that are used for their corrosion resistance are given in Table 4.21; there are, however, many other proprietary variants of the main alloy types given in this table. Applications of nickel alloys are not confined to those where corrosion resistance to aqueous solutions is a prime requirement, and the complete range of nickel alloys that are available commercially

Table 4.21 Nominal compositions of some typical corrosion-resistant nickel alloys

Alloy	Type	Composition (%)						
		C	Cr	Fe	Cu	Mo	Ni	Other
Nickel 200	Ni	0·15 max.					99 min.	
Nickel 201	Ni	0·02 max.					99 min.	
Corronel 230	Ni–Cr	0·08 max.	35	5 max.			Bal.	
Inconel 600	Ni–Cr–Fe	0·15 max.	15	8			Bal.	
Incoloy 800	Ni–Cr–Fe	0·10 max.	20	Bal.			32	0·15–0·6Ti; 0·15–0·6Al
Hastelloy F	Ni–Cr–Fe–Mo	0·05 max.	22	20		6·5	Bal.	2·1Nb+Ta
Illium R	Ni–Cr–Fe–Mo–Cu	0·05	22	6	2·5	5	Bal.	
Illium G	Ni–Cr–Fe–Mo–Cu	0·2	22·5	6·5	6·5	6·5	Bal.	0·6Si; 1·2Mn
Illium 98	Ni–Cr–Fe–Mo–Cu	0·05	28	1	5·5	8·5	Bal.	0·7Si; 1·25Mn
Illium B	Ni–Cr–Fe–Mo–Cu	0·05	28	2·0–3·5	5·5	8·5	Bal.	2·5–6·5Si; 1·25Mn 0·05–0·55B
Hastelloy G	Ni–Cr–Fe–Mo–Cu	0·05 max.	22·2	20	2	6·5	Bal.	2·1Nb+Ta
Incoloy 825*	Ni–Cr–Fe–Mo–Cu	0·05 max.	21	Bal.	2	3	40	0·6–1·2Ti
Hastelloy C276	Ni–Cr–Mo	0·02 max.	15·5	5·5		16	Bal.	4W; 0·1Si
Hastelloy C4	Ni–Cr–Mo	0·02 max.	15·0–16·75	1 max.		14·0–15·75	Bal.	0·5W max.; 0·05Si max.; 0·35–0·65Ti
Hastelloy N	Ni–Mo–Cr	0·06	7	5 max.		16·5	Bal.	0·8Mn max.
Hastelloy B	Ni–Mo	0·05 max.		5		28	Bal.	
Monel 400	Ni–Cu	0·30 max.		2·5 max.	30		Bal.	1Mn
Monel K500	Ni–Cu				30		Bal.	3Al; 0·5Ti
Hastelloy D	Ni–Si–Cu	0·12 max.		2 max.	3		Bal.	9Si

*Formerly 'Ni-O-nel'.

N.B.: Corronel, Inconel, Incoloy, Ni-O-nel, Monel, Hastelloy and Illium are trade marks.

'Bal.' indicates that the element indicated constitutes substantially the balance of the alloy composition, although other elements may also be present in small amounts.

for other specialised uses, notably those involving service at high temperatures, is therefore much greater than indicated by Table 4.21. The corrosion and oxidation resistance of nickel alloys at elevated temperatures is described in Section 7.5.

In general, the alloys listed in Table 4.21 are confined to those in which nickel is the principal alloying element, but it should be noted that highly alloyed stainless steels containing 20–30% Cr, and 20–30% Ni with additions of molybdenum and copper have some features in common with the Ni–Cr–Fe–Mo–Cu alloys given in the table.

In addition to the alloys in Table 4.21, Ni–Sn and Ni–Ti alloys also possess useful corrosion resistance. Ni–Sn alloys are extremely brittle and, because of this, are used only as electrodeposited coatings. Ni–Ti alloys over a wide range of compositions have been studied, of which perhaps the intermetallic compound NiTi (55·06Ni–44·94Ti) has attracted the most interest.

Structural Features and Physical and Mechanical Properties

Nickel normally crystallises in the f.c.c. structure; it undergoes a magnetic transformation at 357°C and is ferromagnetic below that temperature. In all the alloys shown in Table 4.21 the f.c.c. (austenitic) structure is substantially retained, and in consequence most of the alloys possess the combination of properties required of materials for widespread industrial acceptability, i.e. tensile strength, ductility, impact strength, hardness, hot and cold workability, machinability and fabrication.

Table 4.22 gives the physical properties for nickel and a range of nickel alloys; Table 4.23 shows the mechanical properties. The data given in these tables are those published by manufacturers. It is seen that, compared with nickel, the alloys have considerably lower thermal conductivity and much higher electrical resistivity. As with nickel, some of the alloys undergo magnetic transformation; e.g. the Ni–Cu alloy 'Monel' 400 has a transformation temperature close to 0°C. In all cases alloying considerably increases the proof stress and tensile strength. The elongation values of wrought alloys are generally only slightly below those of nickel, although those of cast alloys, viz. Illium G, Illium 98, Illium B and Hastelloy D are much lower. The hardness values of wrought alloys are generally below 200 HV for annealed material, whereas with cast alloys the hardness increases rapidly with increasing silicon content.

As with stainless steels, some nickel alloys have a propensity to form intergranular precipitates of carbides and intermetallic phases during heat treatment and sometimes during welding. The presence of such intergranular precipitates may render the materials susceptible to intergranular attack in certain corrosive environments. To minimise this possibility, the content of carbon and, in some cases, other alloying elements is carefully controlled. The subject of intergranular corrosion of specific nickel alloys and of methods of avoiding it is dealt with in greater detail subsequently.

Methods of Fabrication

Nickel and wrought nickel alloys may be fabricated by welding or, less commonly, by brazing or silver soldering. In order to minimise the deleterious

Table 4.22 Physical properties of some typical corrosion-resistant nickel alloys

Alloy	*Density** (g/cm^3)	*Melting range* (°C)	*Specific heat** (J/kg K)	*Mean coefficient of thermal expansion* ($\times 10^{-6}$/K; 20–100°C)	*Thermal conductivity** (W/m K)	*Electrical resistivity** (μΩ cm)	*Modulus of elasticity** (kN/mm^2)
Nickel	8·89	1 435–1 445	456	13.3	74·9	9	207
Corronel 230	8·20	1 350		11·7		133	
Inconel 600	8·43	1 355–1 412	444	13·3	14·8	103	207
Incoloy 800	8·02	1 357–1 385	502†	14·2	11·5	99	197
Hastelloy F	8·17	1 288	431	14·6	<15	112	
Illium R	8·58	1 328–1 370	461	12·0†	13·0	120	214
Illium G	8·58	1 255–1 338	440	12·2†	12·1	124	168
Hastelloy G	8·30	≈1 315			<15		
Incoloy 825	8·14	1 370–1 400		14·0	11·1	113	195
Hastelloy C	8·94	1 265–1 343	385	11·9	8·8	130	
Hastelloy N	8·77	1 302–1 400	398	11·2	10·2	139	
Hastelloy B	9·25	1 302–1 368	381	9·5	10·4	135	
Monel 400	8·83	1 300–1 350	419	14·1	21·8	51	179
Hastelloy D	7·80	1 130–1 215	456	11·0	20·9	113	

*at 20–25°C.
†0–100°C.

Table 4.23 Mechanical properties of some typical corrosion-resistant nickel alloys

Alloy	*Form*	0·2% *proof stress* (N/mm²)	*Tensile strength* (N/mm²)	*Elongation* (%)	*Hardness* (HV)
Nickel 200	Annealed sheet	157	450	44	
Nickel 201	Hot-rolled annealed plate	83–241	345–483	60–40	
	Cold-drawn annealed tube	69–193	345–483	60–40	112 max.
Corronel 230	Hot-rolled annealed bar	318	729	53	
Inconel 600	Hot-rolled annealed plate	207–345	552–724	55–35	118–163
	Cold-drawn annealed tube	172–345	552–690	55–35	175 max.
Incoloy 800	Hot-rolled annealed plate	210–410	520–730	50–30	121–184
	Cold-drawn annealed tube	210–410	520–680	50–30	179 max.
Hastelloy F	Wrought sheet	310	710	46	167
Illium R	Solution heat-treated	359	778	46	167
Illium G	Cast	268	469	7·5	173
Illium 98	Cast		372	18	165
Illium B	Cast, 2·5–4·5% Si		421–462	1·0–4·5	205–245
	Cast, 4·5–5·7% Si		400–448	0·5–1·5	245–295
	Cast, 5·7–6·1% Si		352–393	0·5 max.	295–340
	Cast, 6·1–6·3% Si		310–352	0·5 max.	340–380
Hastelloy G	Wrought sheet	386	710	48	170
Incoloy 825	Cold-rolled annealed sheet	316	672	42	
Hastelloy C	Wrought sheet	400	834	47	188
Hastelloy N	Wrought sheet	310	793	51	213
Hastelloy B	Wrought sheet	393	834	63	192
Monel 400	Hot-rolled annealed plate	193–345	483–586	50–35	110–140
	Cold-drawn annealed tube	172–310	483–586	50–35	140 max.
Hastelloy D	Sand cast	593	858	2–3	340

effects that may result from intergranular precipitation, either low-heat-input welding procedures employing flux-coated electrodes, or the MIG, TIG or plasma arc procedures, are recommended. Thick sections may be welded using the submerged arc process and a relatively restricted heat input. Oxyacetylene welding is rarely used because of the high heat input and the danger of carbon transfer into the metal.

Corrosion Behaviour in Aqueous Environments

Theoretical Considerations

Nickel occupies an intermediate position in the electrochemical series; $E^{\ominus}_{Ni^{2+}/Ni} = -0{\cdot}25$ V, so that it is more noble than Zn and Fe but less noble than Sn, Pb and Cu. Figure 4.21 shows a simplified potential–pH equilibrium (Pourbaix) diagram[1] for the Ni–H_2O system at 25°C. From this it is seen that:

1. Nickel is thermodynamically stable in neutral and moderately alkaline solutions although not in acidic or strongly alkaline solutions.
2. The metal would be expected to dissolve in acidic solutions forming Ni^{2+} ions with liberation of H_2.
3. The metal should be capable of passivation by forming a surface layer of NiO, $Ni(OH)_2$ or higher oxides of nickel in neutral and moderately alkaline solutions.
4. The metal may be unstable in strongly alkaline solutions, dissolving to form $HNiO_2^-$ ions.
5. In oxidising alkaline conditions passivation should be possible through formation of films of higher oxides.

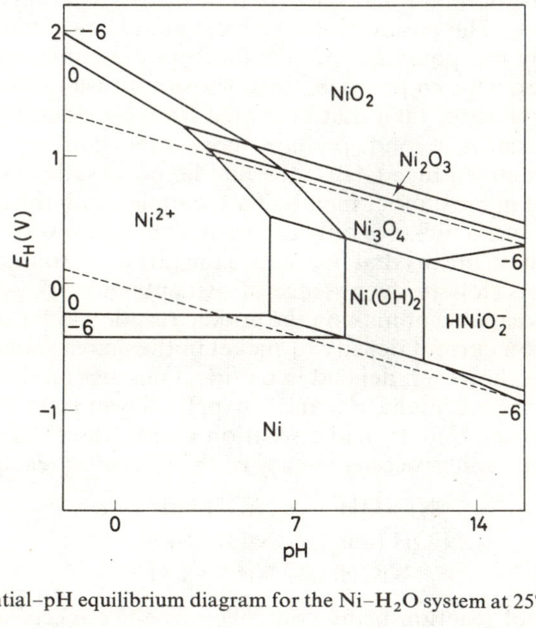

Fig. 4.21 Potential–pH equilibrium diagram for the Ni–H_2O system at 25°C (after Pourbaix[1])

On the basis of these data, nickel is considered to be a slightly noble metal[1], although in practice, as will be seen below, it is considerably more corrosion resistant in both acidic and alkaline solutions than would be predicted from Fig. 4.21.

Several complications are involved in the calculation of potential–pH equilibrium diagrams for temperatures other than 25°C[2, 3, 3(a)], including the fact that the pH scale itself varies with temperature; thus, diagrams in which the pH scale refers to the temperature for which the equilibria are calculated are probably preferable for most purposes[3(a)]. The most notable consequence of increasing temperature on the equilibria appears to be a widening of the pH range within which the oxides of Ni are thermodynamically stable.

Anodic Behaviour of Nickel

Many investigators have studied the anodic behaviour of nickel. A complete discussion of the reactions occurring during anodic dissolution and passivation of the metal is outside the scope of this section; what follows is a summary of recently published views on these topics.

Anodic E–i curves for nickel obtained by potentiostatic, potentiokinetic or, in earlier days, galvanostatic techniques, have been published by many workers. Unfortunately, good agreement is not always found between data from different sources. The principal reasons for the discrepancies appear to lie in the nature and amount of impurities in the metal[4,5] or in the solution[6], both of which may have a profound effect on the shape of the curve, and in variations in experimental procedure[7–9,9(a),9(b)].

Figure 4.22 shows a curve for nickel in 0·5M H_2SO_4[10] which illustrates the main features of the anodic behaviour of the metal that are of interest with regard to its corrosion resistance. It is seen that in acidic solution nickel is capable of passivation and that the extent of the passive range (*DE*) is considerable, ≈0·5 V. The passivation of nickel in acidic solution is a feature not predicted by the potential–pH equilibrium diagram (see Section 1.4) and is one reason why, in practice, the corrosion resistance of the metal in acidic solutions is better than that indicated from consideration of thermodynamic equilibria. A second, perhaps more important, reason lies in the fact that in the active range (*ABC*) the anodic polarisation is considerably greater than that of most other metals. This, coupled with the fact that in the electrochemical series nickel is only moderately negative with respect to the H^+/H_2 equilibrium, means that in practice the rate of dissolution of nickel in acidic solutions is slow in the absence of oxidants more powerful than H^+ or of substances capable of making the anodic reaction kinetically easy. The anodic dissolution current density of nickel in the active state as a function of potential does, however, depend in a critical manner on the rate at which the measurements are made[9,11] and on pH[9]. To explain this, Sato and Okamoto[9] proposed that in acidic solution anodic dissolution of nickel is catalysed by OH^- and proceeds by way of the following reaction sequence:

$$\mathrm{Ni} + \mathrm{OH^-} \rightarrow \mathrm{NiOH\,(ads.)} + e^-$$
$$\mathrm{NiOH\,(ads.)} \rightarrow \mathrm{NiOH^+} + e^-$$
$$\mathrm{NiOH^+} \rightarrow \mathrm{Ni^{2+}} + \mathrm{OH^-}$$

the overall rate of reaction being controlled by the concentration of OH^- ions. This mechanism appears to provide a basis for explaining the sluggish anodic dissolution of nickel in acidic solution (low concentration of OH^- ions) and also accounts in part for the variations in the anodic behaviour reported from different sources.

The anodic dissolution of nickel is also dependent on the amount of cold work in the metal[11,12], and in the active region the anodic current density of cold worked material at a given potential is up to one order of magnitude greater than that of annealed material.

In the passive region (*DE*, Fig. 4.22) the current density of nickel is greater than that of most other passive metals, viz. Al, Ti, Cr, and is also slightly greater than that of Fe, but in practice this is of little significance as the rate is sufficiently low to be acceptable for most applications. At high potentials

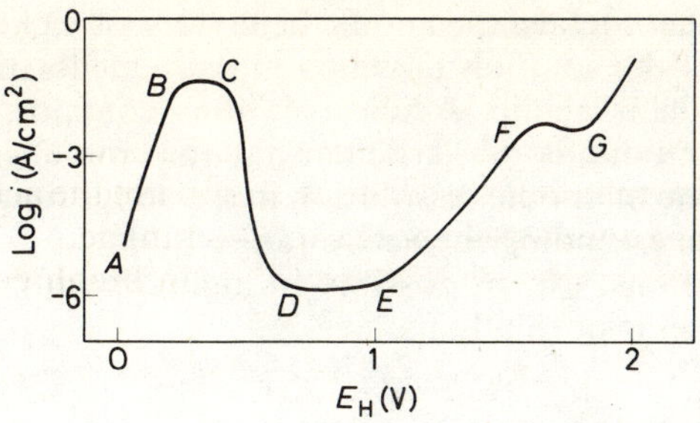

Fig. 4.22 Potentiostatic E–log i curve for nickel, anodically polarised in 0·05 M H_2SO_4 saturated with N_2 at 25°C (after Sato and Okamoto[10])

nickel becomes transpassive (EF), and in this region corrosion occurs preferentially at grain boundaries[12], as with stainless steels. In the passive and transpassive states anodic dissolution results in the formation of Ni^{2+} ions in solution[12]. At still higher potentials nickel exhibits secondary passivity (FG), and although the anodic current is several orders of magnitude greater than in the passive region (DE) it is not localised at grain boundaries[12]. At potentials above the range of secondary passivity the anodic current density increases and dissolution of nickel through the oxide film, mainly as Ni^{3+}, is accompanied by evolution of O_2. In this region grain boundaries are preferentially attacked again[12]. The corrosion behaviour of nickel in acidic solutions in the regions of transpassivity, secondary passivity and beyond, is of limited practical significance, since these potentials are beyond the range of the redox potentials of most aqueous solutions.

The influence of temperature on the anodic behaviour of nickel has been studied[2], and in acidic and neutral solutions the active–passive transition is not observed at temperatures greater than about 100°C (Fig. 4.23).

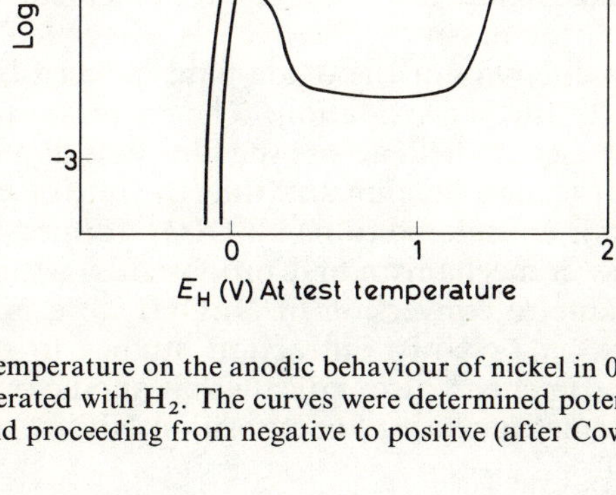

Fig. 4.23 Effect of temperature on the anodic behaviour of nickel in 0·025 M H_2SO_4 + 0·025 M K_2SO_4 (pH 1·3) de-aerated with H_2. The curves were determined potentiokinetically at a scan rate of 2 V/h and proceeding from negative to positive (after Cowan and Staehle[2])

As with most other metals, the anodic behaviour of nickel is influenced by the composition of the solution in which measurements are made, particularly if the solution is acidic. Acidic solutions containing Cl^- ions[13] or certain sulphur compounds[6] in particular have a pronounced influence both in increasing the rate of anodic dissolution in the active range and in preventing passivation, i.e. extending the active range.

In alkaline solutions, except possibly in high concentrations at elevated temperatures, nickel is normally passive.

Passivity of Nickel

In many aqueous solutions nickel has the ability to become passive over a wide range of pH values. The mechanism of passivation of nickel and the properties of passive nickel have been studied extensively—perhaps more widely than for any other element, except possibly iron. Despite this, or perhaps because of it, there is by no means agreement either on the mechanism of passivation of nickel or on the composition and thickness of the passive layer. Early studies on the passivation of nickel were stimulated by the use of nickel anodes in alkaline batteries and in consequence were conducted in the main in alkaline media. More recently, however, attention has been directed to the passivation of nickel in acidic solutions.

Most authorities nowadays probably accept the view that passivity of nickel, as of most other metals, is due to the formation of a protective, self-repairing film of oxide or hydrated oxide a few nanometres thick, although others[12], whilst acknowledging the existence of a surface oxide film, consider passivity to be due to a chemisorbed layer of O_2. Ellipsometric measurements, both in alkaline solution[14] and in acidic solution[15], support the existence of surface oxide films on passive nickel several nanometres thick, although impedance measurements[16] suggest that, in acidic solutions at least, the passive layer is electrically complex and is not an ideal dielectric.

In alkaline solutions, galvanostatic measurements[17] suggest that passivation of nickel is due to formation of a monolayer of $Ni(OH)_2$. This probably forms by a solid state process involving nucleation and growth, according to the general model for such growth proposed by Armstrong, Harrison and Thirsk[18]. In some alkaline conditions, particularly concentrated solutions at higher temperatures, thicker films are undoubtedly formed.

In acidic solutions, an interesting view of the mechanism of passivation of nickel, advanced by Bockris, Reddy and Rao[15], is that although formation of a continuous surface oxide film is a necessary precursor, it is not, in itself, a sufficient condition for passivity. Passivity is associated with an increase in the electronic conductivity of the oxide film, induced by a change in its stoichiometry, i.e. its further oxidation. The increase in electronic conductivity reduces the electric field across the film; this diminishes the rate of metal ion transport through the film and thus the rate of metal dissolution. The pre-passive film is considered to be $Ni(OH)_2$, formed by a dissolution–precipitation process, a mechanism first proposed by Müller[19] many years ago. Passivation is due to conversion of $Ni(OH)_2$ to a non-stoichiometric higher oxide, $NiO_{1\cdot5-1\cdot7}$. X-ray diffraction studies in conjunction with electrochemical experiments[20] also indicate that anodic oxidation of nickel yields a non-stoichiometric oxide of composition $NiO_{1\cdot7}$.

A different view, derived from potential decay measurements on passive nickel electrodes[10] and also from potentiostatic anodic polarisation measurements[6], is that oxidation (or reduction) proceeds through well-defined stages, corresponding to the formation of oxides of stoichiometric composition, viz.

$$Ni \rightarrow NiO \rightarrow Ni_3O_4 \rightarrow Ni_2O_3$$

at potentials close to those predicted from thermodynamic considerations. Passivation is considered to accompany the $NiO \rightarrow Ni_3O_4$ transition[10].

In certain concentrated acidic solutions, e.g. H_2SO_4, nickel, whilst not truly passive, may exhibit 'pseudo-passivity' owing to crystallisation of a layer of nickel salt (in conc. H_2SO_4 probably β-$NiSO_4 \cdot 6H_2O$) on the surface[21].

Influence of Alloying on Anodic Behaviour of Nickel

During recent years a considerable amount of information has been published on the anodic behaviour of nickel alloys. The data include studies both of binary alloy systems in which nickel forms the major alloying component and of more complex commercially produced nickel alloys. The data are sufficiently numerous to permit a rational and fairly complete interpretation of many of the corrosion-resistant properties of nickel alloys on the basis of their anodic behaviour.

Potential/anodic current density curves illustrating the influence of binary alloying additions to nickel are shown as follows: Cr, Fig. 4.24; Fe, Fig. 4.25; Cu, Fig. 4.28; Mo, Fig. 4.30 (curve for Hastelloy B); Si, Fig. 4.31;

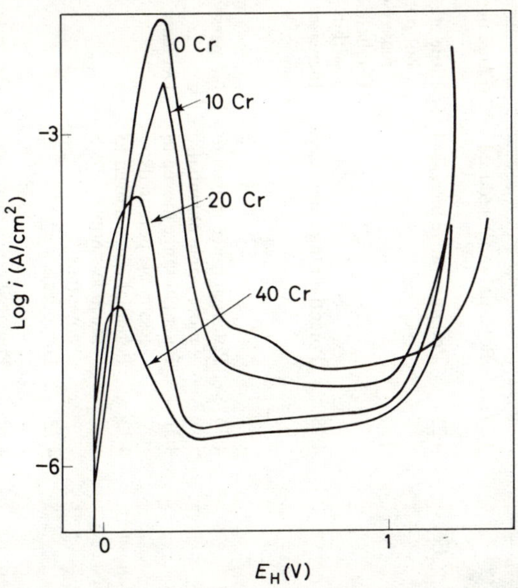

Fig. 4.24 Effect of chromium content on the anodic behaviour of Ni–Cr alloys in 0·5 M H_2SO_4 (de-aerated with H_2) at 25°C; the potential was increased incrementally by 0·025 V every 3 min (after Hodge and Wilde[22])

Table 4.24 Influence of alloying on anodic behaviour of nickel

Alloy addition (%)	Fig. no.	*Influence on anodic behaviour*				
		Active region	*Max. c.d. prior to passivation, $i_{crit.}$*	*Potential of passivation, E_p*	*c.d. in passive region, i_p*	*Passive region*
0–40Cr	4·24	Potential range reduced	Large decrease	Less noble	Considerable decrease	Potential range increased
0–70 at. % Fe	4·25	Potential range increased	Little effect on magnitude, but potential range of max. c.d. increases. Oscillations often observed within max. c.d. range	More noble	Little effect	Potential range reduced
0–70Cu	4·28	Potential range increased	Increase	More noble, no passivation above ≈50% Cu	Large increase	Potential range reduced, eliminated above ≈50% Cu
0–28Mo	4·30	Potential range moved to more noble potentials	No passivation	No passivation	No passivation	No passivation
0–16·5Si	4·31	Potential range moved to less noble potentials	Decrease	Less noble	Large decrease	Potential range increased
Electrodeposited Sn–35Ni	4·32	Potential range moved to much less noble potentials		Much less noble		Potential range much increased
0–100Ti	4·33	Disappears		Much less noble?	Large decrease	Potential range much increased
0–10Al	4·34	Potential range increased	Decrease	Less noble	0–7% Al considerable decrease	Potential range increased
0–25Ge	4·35	Potential range moved to slightly more noble potentials	>10% Ge, decrease	Less noble	Large increase	Potential range increased

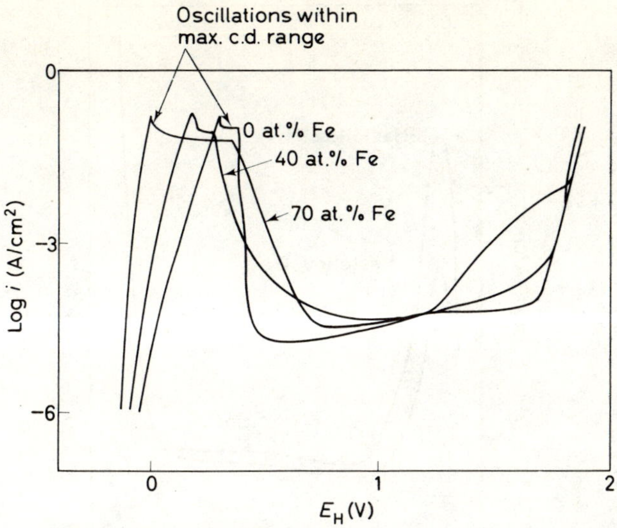

Fig. 4.25 E–log i relationship for the anodic behaviour of Ni–Fe alloys in 0·5 M H_2SO_4 (de-aerated with H_2) at 25°C (after Economy, *et al.*[23]; see also References 24–26

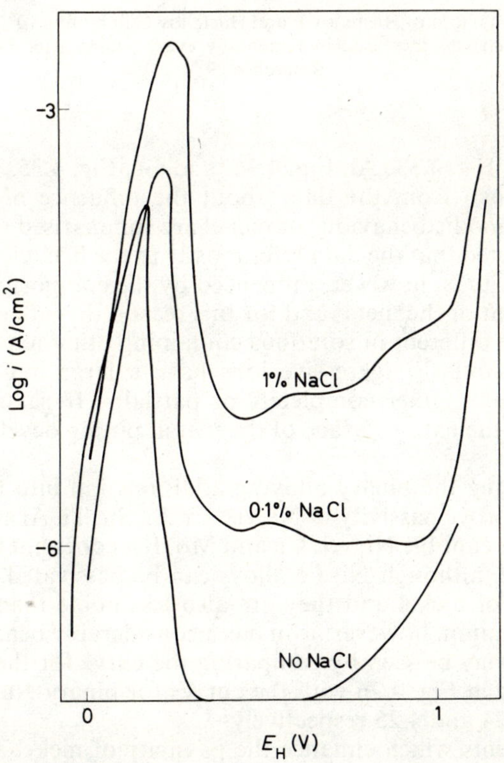

Fig. 4.26 Anodic behaviour of Inconel 600 in 0·5 M H_2SO_4 (de-aerated with N_2) at 24°C containing different concentrations of Cl^- ions (after Piron, *et al.*[27])

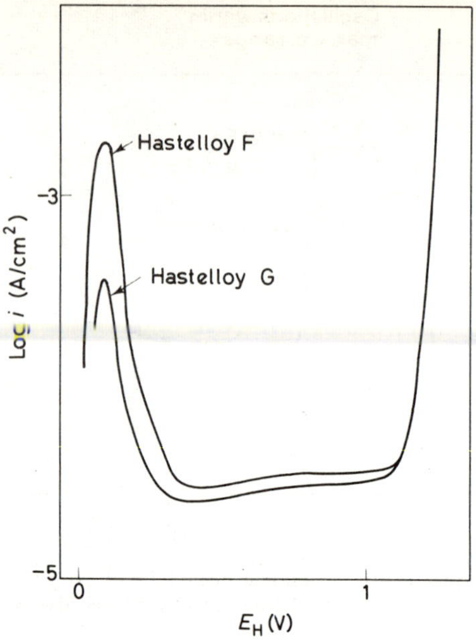

Fig. 4.27 Anodic behaviour of Hastelloy F and Hastelloy G in boiling 10% H_2SO_4 de-aerated with H_2 (the potential was increased incrementally every 3 min; after Leonard[28]; see also Reference 29)

Sn, Fig. 4.32; Ti, Fig. 4.33; Al, Fig. 4.34; and Ge, Fig. 4.35. The deductions that may be drawn from the data about the influence of these alloying elements on the anodic behaviour of nickel are summarised in Table 4.24.

It should be noted that the data refer mostly to the behaviour of the alloys in H_2SO_4. Passivity is, however, influenced by the composition of the solution as well as that of the metal and for this reason the influence of alloying additions may be different in solutions containing other ions. In particular, Cl^- and other similarly aggressive ions have a large influence and may prevent passivation, either completely or partially. If passivity cannot be maintained over the entire surface of the metal, pitting develops, and this is considered later.

Broadly speaking the binary alloying additions fall into two categories: (1) those that improve passivity of Ni, viz. Cr, Si, Sn, Ti, Al and perhaps Ge, and (2) those that ennoble Ni, viz. Cu and Mo. Iron does not belong in either category because, although Ni–Fe alloys can be passivated, their passivity is less than that of nickel and they are also less noble than nickel. In the presence of chromium, however, iron has a considerably beneficial influence on passivity, as may be seen by comparing the curve for the Ni–15Cr–8Fe alloy Inconel 600 in Fig. 4.26 with the curves for binary Ni–Cr and Ni–Fe alloys in Figs. 4.24 and 4.25 respectively.

Alloying elements which enhance the passivity of nickel are expected to improve the corrosion resistance to oxidising media, in particular acidic solutions containing oxidants. Generally, this is found to be so in practice,

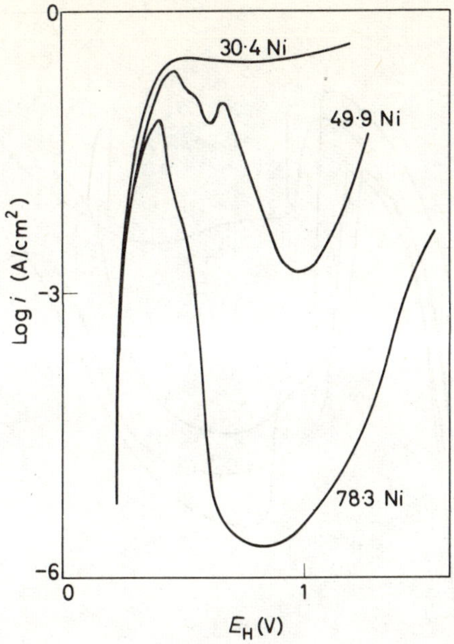

Fig. 4.28 Anodic behaviour of Ni–Cu alloys in 0·5 M H_2SO_4 (de-aerated with N_2) at 25°C; the curve was determined potentiokinetically at 0·4 V/h for the 78·3 and 49·9% Ni alloys and at 3 V/h for the 30·4% Ni alloy proceeding from more positive to more negative (after Osterwald and Uhlig[30])

although it should be noted that strongly oxidising acids, e.g. HNO_3 and H_2CrO_4, or other acidic solutions containing powerful oxidants may render such alloys transpassive, in which condition the corrosion resistance may be impaired. In less oxidising media, particularly in acidic solutions where hydrogen evolution is the cathodic process, not all alloying elements which improve passivity are beneficial, although some are. In these circumstances chromium, silicon and probably aluminium and germanium might also be expected to confer little benefit, because the passivation potential, although displaced to slightly more negative values, is not displaced sufficiently to permit passivity to develop in hydrogen-evolving acidic solutions. In contrast, alloying additions of titanium and tin (in the electrodeposited Sn–35Ni alloy) are undoubtedly beneficial because the active/passive transition is displaced to sufficiently negative potentials that enable passivity to be maintained in non-oxidising acidic solutions.

In fact silicon is also often beneficial, especially in H_2SO_4. In dilute H_2SO_4, Ni–Si alloys containing about 10% Si do not passivate spontaneously, but the rate of anodic dissolution rapidly falls to a low value owing to the formation of a silicon-rich surface layer. In concentrated H_2SO_4 such Ni–Si alloys are passive, whilst in H_2SO_4 of intermediate concentration the corrosion behaviour is complex, being governed by the nature of the cathodic process, which changes as corrosion proceeds[33].

The alloying elements molybdenum and copper do not, by themselves, enhance passivity of nickel in acid solutions, but instead ennoble the metal.

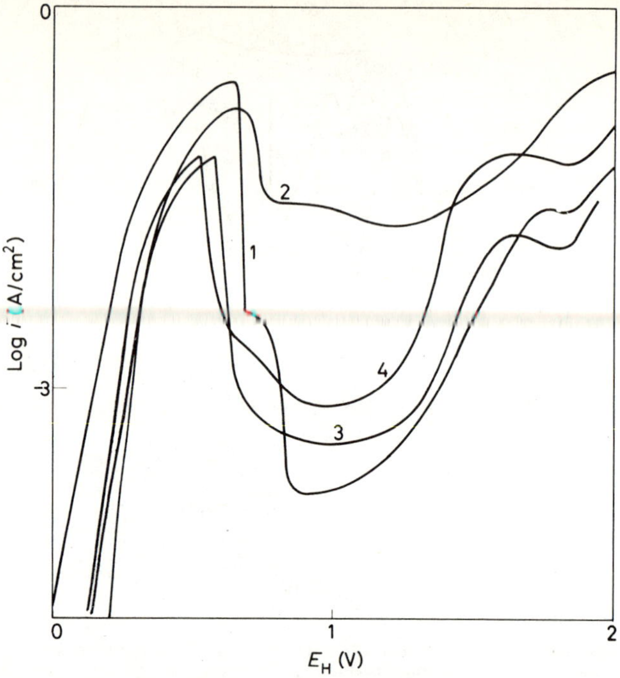

Fig. 4.29 Anodic behaviour of Monel alloys in 10% H_2SO_4 at ambient temperature. 1, Ni; 2, Monel 400; 3, Monel K500° solution treated; 4, Monel K500 aged (after Flint and Barker[31])

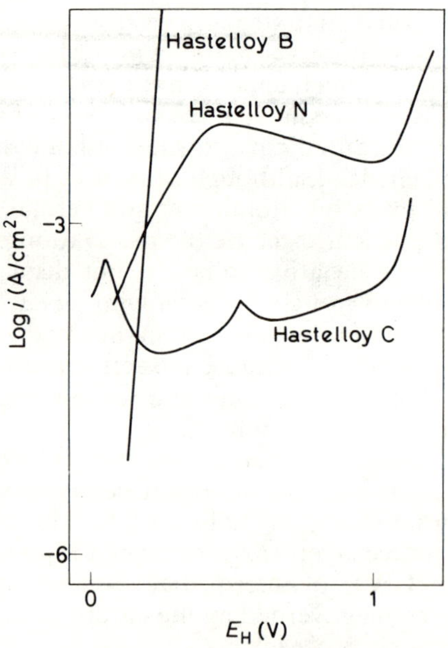

Fig. 4.30 Anodic behaviour of Hastelloy B, C and N in boiling 10% H_2SO_4 de-aerated with H_2; the potential was increased incrementally (after Leonard[28])

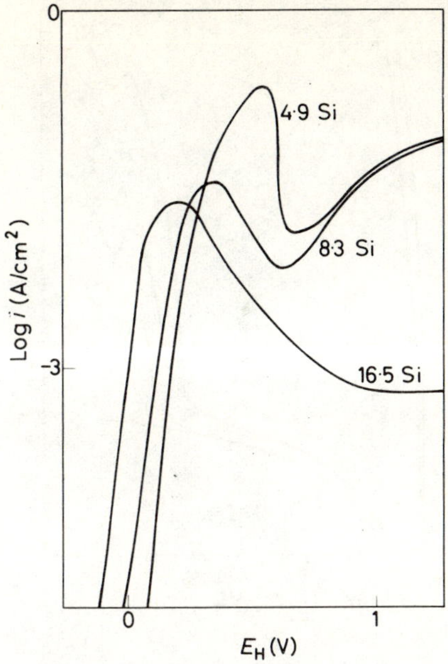

Fig. 4.31 Anodic behaviour of Ni–Si alloys in 25% H_2SO_4 (de-aerated with N_2) at ambient temperature (after Barker and Evans[32])

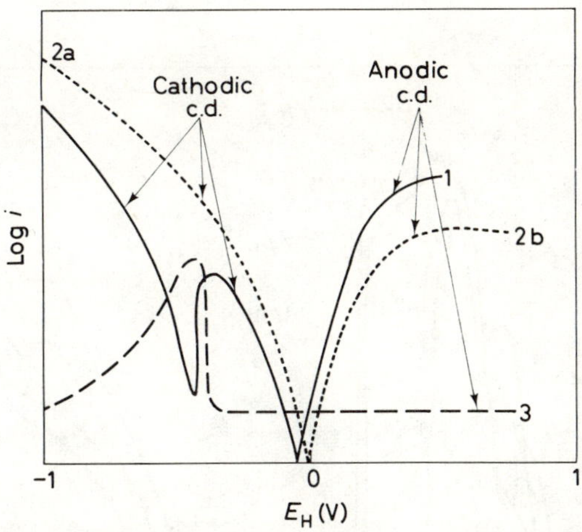

Fig. 4.32 Suggested anodic behaviour of electrodeposited Sn–35Ni alloy; 1, 'observed' curve; 2a, H_2 evolution; 2b, H_2 oxidation; 3, 'true' anodic curve (after Clarke and Elbourne[34])

This means that, in practice, these alloying elements confer benefit in precisely those circumstances where chromium does not, viz. hydrogen-evolving acidic solutions, by reducing the rate of anodic dissolution. In more oxidising media the anodic activity increases, and, since binary Ni–Mo and Ni–Cu alloys do not passivate in acidic solutions, they are generally unsuitable in such media.

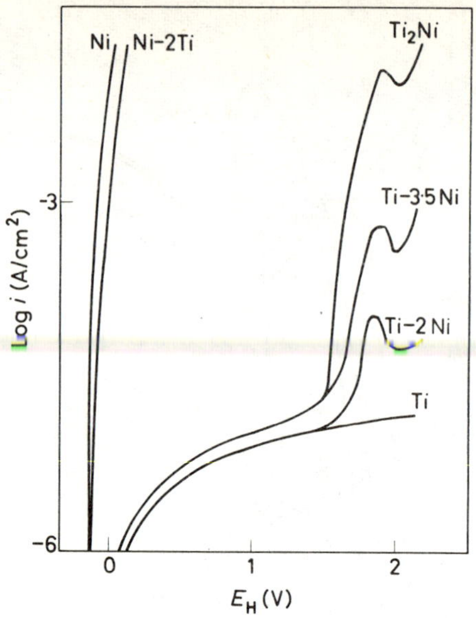

Fig. 4.33 Anodic behaviour of Ni–Ti alloys in HCl + 3·5% NaCl (pH 1), de-aerated with argon, at 22·2°C; the potential was increased by 0·02 V every minute (after Sedriks, *et al.*[35])

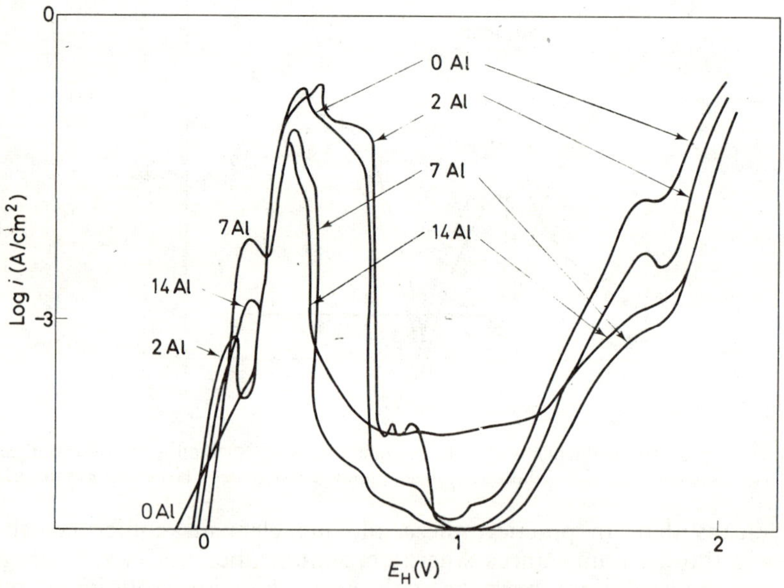

Fig. 4.34 Anodic behaviour of Ni–Al alloys in 0·5 M H_2SO_4, de-aerated with H_2, at 22°C; the potential was increased by 0·01 or 0·02 V every 3 min in the active range and by 0·04 V in the passive range (after Crow, *et al.*[36])

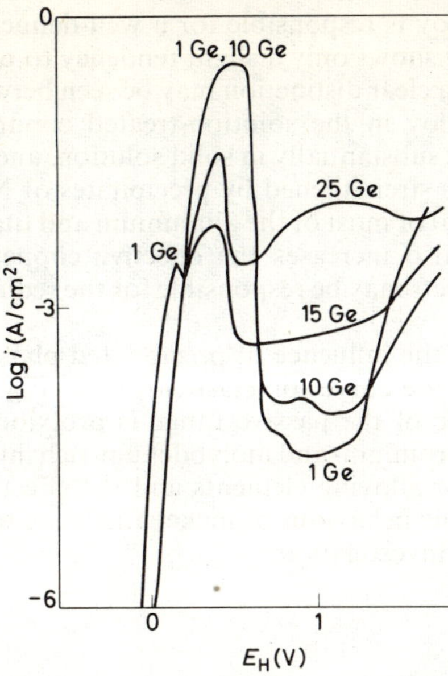

Fig. 4.35 Anodic behaviour of Ni–Ge alloys in 1 M H_2SO_4 de-aerated with N_2, at room temperature; the scan rate from negative to positive was 0·02 V/min (after Hart[37])

Relatively small amounts of molybdenum in Ni–Cr–Fe alloys, as in stainless steels, render passivation much easier and it may be seen from Fig. 4.27 that the further addition of 2% Cu enhances the effect. The major effect is to reduce the maximum current density prior to passivation, $i_{crit.}$, although the current density in the passive range, i_p, is also reduced. Potential/anodic current density curves of Ni–Cr–Fe–Mo and Ni–Cr–Fe–Mo–Cu alloys plotted in the conventional way do not show these effects clearly, but they may be illustrated by employing fast scan rates[29] or elevated temperatures (see Fig. 4.27). Because of the effect on $i_{crit.}$, Ni–Cr–Fe–Mo and Ni–Cr–Fe–Mo–Cu alloys have good corrosion resistance to acidic solutions both in oxidising conditions and when corrosion is accompanied by hydrogen evolution.

The addition of chromium to Ni–Mo alloys containing about 15% Mo confers passivity, as may be seen by comparing the curves in Fig. 4.30 for Hastelloy B (Ni–28Mo), Hastelloy N (Ni–16·5Mo–7Cr) and Hastelloy C (Ni–16Mo–15·5Cr). Chromium, however, displaces the active region in these alloys to more negative potentials, so that whilst the chromium-containing alloys are more corrosion resistant than the chromium-free alloy in oxidising acidic media, they are less resistant in most hydrogen-evolving acidic solutions.

An interesting illustration of the effect that quite small alloying additions may sometimes have on anodic behaviour is seen in Fig. 4.29[31] from a comparison of the Ni–30Cu alloy Monel 400 with its age-hardening variant Monel K500, which contains 3% Al and 0·5% Ti. The presence of these elements in the latter alloy is responsible for a well-defined passive region, whereas the former alloy shows only a slight tendency to passivate in acidic solutions. Furthermore, a clear distinction may be seen between the passivity of the age-hardening alloy in the solution-treated condition, where aluminium and titanium are substantially in solid solution, and in the aged condition, where the alloy is strengthened by precipitates of Ni_3Al and Ni_3Ti. In addition to the removal of most of the aluminium and titanium from solid solution, precipitation also increases the effective copper content of the matrix. Both of these effects may be responsible for the reduction in passivity of the aged material.

Another indication of the influence of precipitated phases on anodic behaviour may be seen in the curve for Hastelloy C in Fig. 4.30, where the small peak in the middle of the passive range is probably attributable to anodic dissolution of a chromium- and molybdenum-rich intermetallic phase.

The influence of minor alloying elements and the effect of formation of other phases on the anodic behaviour of nickel alloys are worthwhile topics for further experimental investigation.

Pitting

Pitting of nickel and nickel alloys, as of other metals and alloys, occurs when passivity breaks down at local points on the surface exposed to the corrosive environment, at which points anodic dissolution then proceeds whilst the major part of the surface remains passive. Since most (sometimes all) of the cathodic reaction accompanying corrosion is distributed over the passive surface, it follows that the fewer the number of sites of breakdown the more intense is the anodic dissolution at each site, i.e. the fewer the pits the faster they grow, at least in the early stages. Pitting of nickel has been shown to develop preferentially near structural features in the metal, such as grain boundaries, and also at imperfections in the surface, such as scratches[38]. Electropolishing appears to be helpful in reducing the tendency of pits to develop at surface imperfections, but not necessarily at sites associated with structural features of the metal.

In practice, pitting of nickel and nickel alloys may be encountered if the corrosive environment contains chloride or other aggressive ions and is more liable to develop in acidic than in neutral or alkaline solutions. In acidic solutions containing high concentrations of chloride, however, passivity is likely to break down completely and corrosion to proceed more or less uniformly over the surface. For this reason nickel and those nickel alloys which rely on passivity for their corrosion resistance are not resistant to HCl.

Figure 4.36 illustrates, by means of potential/anodic current density curves, the influence of pH and Cl^- ions on the pitting of nickel[13]. The tendency to pit is associated with the potential at which a sudden increase in anodic current density is observed within the normally passive range (E_B on Curve 1 in Fig. 4.36). It can be seen that in neutral 0·05M Na_2SO_4 containing 0·02MCl^- (Curve 1) E_B has a value of approximately 0·4 V E_H. When pitting develops, the solution in the pits becomes acidic owing to hydrolysis of the corrosion product (see Section 1.6) and when this occurs the anodic current density increases by at least two orders of magnitude and tends to follow the

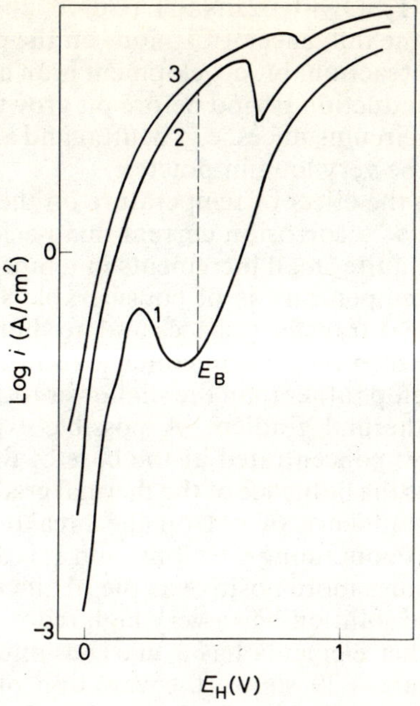

Fig. 4.36 Influence of pH and Cl^- ions on the anodic behaviour of nickel in $SO_4^{2-} + Cl^-$ ion solutions at 20°C (potentiokinetic polarisation at 0·05 V/min). 1, 0·05 M Na_2SO_4 + 0·02 M NaCl; 2, 0·05 M H_2SO_4 + 0·02 M NaCl; 3, 0·05 M H_2SO_4 + 0·05 M NaCl (after Szklarska-Smialowska[13])

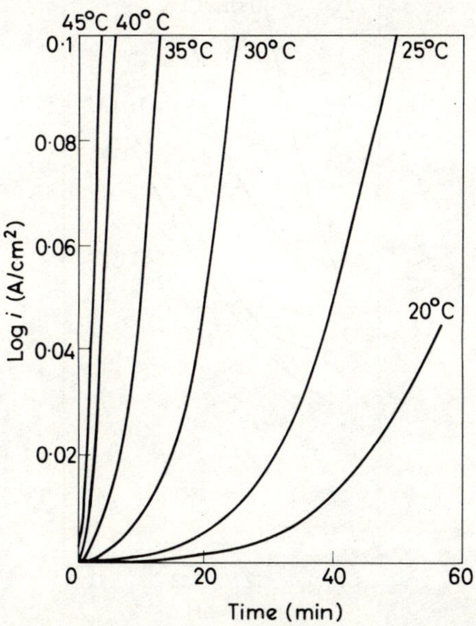

Fig. 4.37 Influence of temperature on breakdown of passivity of nickel in $H_2SO_4 + Na_2SO_4$ solution (pH 0·4) containing 0·05 M Cl^- (after Gressmann[39])

curve obtained in 0·05MH_2SO_4+0·02MNaCl (Curve 2). Comparison of Curves 2 and 3 illustrates the influence of Cl^- ions on the pitting process.

Owing to the hydrolysis reaction, pit development is an autocatalytic process and often there is an induction period before pit growth attains observable proportions. In some circumstances, e.g. neutral and alkaline solutions, the induction period may be very long in practice.

Figure 4.37[39] illustrates the effect of temperature on the rate of development of pitting, measured as a corrosion current in an acidic solution containing Cl^-; it is seen that quite small increments in temperature have large effects. The influence of temperature is of considerable significance when metals and alloys act as heat transfer surfaces and are hotter than the corrosive environment with which they are in contact. In these circumstances, deep pointed pits may develop rather than the shallower rounded pits usually found when there is no thermal gradient. A possible explanation is that anodic dissolution becomes concentrated at the base of the growing pit in preference to its sides under the influence of the thermal gradient in the metal.

Figure 4.38 shows the influence of pH on the breakdown potential of nickel in alkaline solutions containing Cl^- ions, and it is apparent that the breakdown potential becomes more positive as the pH increases, i.e. breakdown is unlikely unless the solution has a very high redox potential.

Alloying nickel with other elements has a marked influence on the susceptibility to pitting. Figure 4.39 shows the variation of the breakdown potential with chromium concentration for binary Ni–Cr alloys[41], and it is seen that breakdown becomes significantly less probable as the chromium increases above 10%. Alloying with iron in addition to chromium yields a further improvement, as may be seen from Fig. 4.26, which shows that the

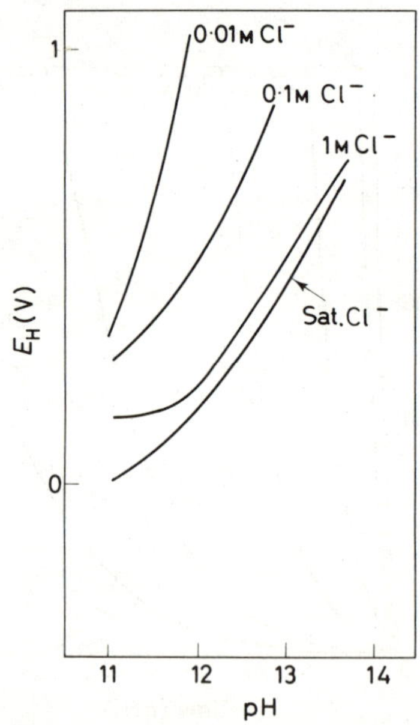

Fig. 4.38 Influence of pH and Cl^- ion on the breakdown potential of commercial nickel in alkaline solutions (0·001–5 M NaOH) de-aerated with N_2 (after Postlethwaite[40])

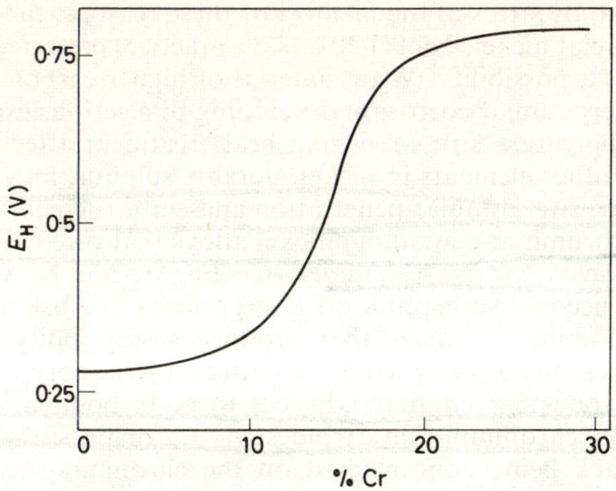

Fig. 4.39 Influence of the chromium content of Ni–Cr alloys on the breakdown potential in 0·1 M NaCl at 25°C de-aerated with N_2 (after Horvath and Uhlig[41])

Ni–15Cr–8Fe alloy Inconel 600 exhibits little tendency to breakdown even in an acidic solution containing 1% NaCl. In practice, Ni–Cr–Fe alloys exhibit a high degree of pitting resistance and, as with stainless steels, the addition of a few per cent molybdenum improves their resistance even further.

Nickel alloys which rely on nobility for their corrosion resistance, viz. Ni–Cu and Ni–Mo alloys in acidic solution, do not usually pit in these circumstances. It should be noted, however, that the Ni–Cu alloy Monel 400 normally forms a protective oxide film in neutral and alkaline solutions, and this is of particular significance with regard to its corrosion resistance to sea-water. In circumstances where the supply of O_2 is insufficient to maintain the film in good repair, as in stagnant conditions, pitting may develop.

Intergranular Corrosion

As with most other metal and alloys systems, nickel and certain of its alloys may suffer intergranular corrosion in some circumstances. In practice, intergranular corrosion of nickel alloys is usually confined to the vicinity of welds as a result of the effects produced by the welding operation on the structure of the material in those regions. Alloys that are subjected to other similarly unfavourable thermal treatments may also become susceptible. The compositions of most commercial nickel alloys that are marketed today are, however, carefully controlled to minimise the possibility of intergranular corrosion developing in welded material during service.

Intergranular corrosion of nickel and its alloys is nearly always associated with grain boundary precipitates. In certain commercial grades of nickel, which contain carbon as an impurity, lengthy exposure to high temperatures may result in the formation of a grain boundary film of graphite which in some circumstances renders the material susceptible to intergranular corrosion on subsequent exposure in an environment to which the material is otherwise well suited, viz. caustic alkalis; with nickel this form of corrosion may be intensified by stress in the metal. For these reasons, the low-carbon grade of commercial nickel, Nickel 201, is, in practice, preferred where this form of attack is a possibility. With material of higher carbon content the possibility of intergranular corrosion developing to a serious extent may be minimised by applying a stress-relieving heat treatment after fabrication. The presence of other elements in nickel, notably sulphur, may also render the metal liable to intergranular penetration and embrittlement.

The types of chromium-containing nickel alloys that owe their corrosion resistance to passivity, viz. Ni–Cr–Fe, Ni–Cr–Fe–Mo and Ni–Cr–Fe–Mo–Cu alloys, may become susceptible to intergranular corrosion in circumstances broadly similar to those that produce susceptibility in stainless steels[42,43]. In these materials, preferential attack by the corrosive environment occurs at zones immediately adjacent to grain boundaries at which precipitates of the chromium-rich carbides $M_{23}C_6$ or possibly M_7C_3 have formed, the attack being concentrated on the chromium-depleted zones adjacent to the precipitate, since these zones cannot become passivated[44,45]. As with stainless steels, the appropriate preventative measures are to minimise carbide formation by controlling the carbon content of the material to levels as low as practicable—nowadays 0·03% C max. is attainable—and to add elements such as titanium and niobium to form carbides more stable than $M_{23}C_6$ with the residual carbon and thus prevent chromium-depletion. It should be noted, however, that owing to the higher activity of carbon in nickel-rich alloys than in stainless steels, a greater proportion of stabilising element such as titanium is needed in the former materials than in the latter[46]. Intergranular corrosion of stainless steels and Ni–Cr–Fe alloys has been observed to occur in the absence of grain boundary carbide precipitates in the alloy during laboratory tests in highly oxidising acidic solutions such as HNO_3 containing chromates or dichromates[47], although the phenomenon is probably of limited significance in practice. A modern review of intergranular corrosion of alloys in the Fe–Ni–Cr system, including stainless steels and nickel alloys, is available[47].

Another type of nickel alloy with which problems of intergranular corrosion may be encountered is that based on Ni–Cr–Mo containing about 15% Cr and 15% Mo. In this type of alloy the nature of the grain boundary precipitation responsible for the phenomenon is more complex than in Ni–Cr–Fe alloys, and the precipitates that may form during unfavourable heat treatment are not confined to carbides but include at least one intermetallic phase in addition. The phenomenon has been extensively studied in recent years[48–53], though with conflicting results concerning the nature of the grain-boundary precipitates responsible for rendering the materials susceptible. It seems reasonably certain that the major cause of susceptibility to intergranular attack in this class of alloy is the existence of zones adjacent to the grain boundaries depleted in molybdenum and possibly in chromium also, and the presence of molybdenum- (and chromium-) rich phases at the grain boundaries. The major molybdenum-rich phase precipitated is an intermetallic compound, probably possessing a crystal structure similar to Fe_7MO_6[51,53], but molybdenum-rich carbide of the M_6C type may also form. It is also probable that two different mechanisms of intergranular corrosion operate in this type of alloy, one involving attack on the depleted regions being observed in HCl (and perhaps other hydrogen-evolving acidic solutions), the other, in which the intermetallic precipitate is preferentially attacked, being observed in more highly oxidising acidic media. An observation of significance made some years ago was that limiting the silicon content of this type of alloy to very low levels reduced the tendency for formation of the intermetallic phase during welding[49] and this led to the introduction of improved commercial alloys, e.g. Hastelloy C276. More recently a composition possessing even greater thermal stability, i.e. Hastelloy C4, has been introduced, in which iron and tungsten present in the earlier alloys have been largely replaced by nickel[53] (see Table 4.21).

Ni–Mo alloys containing about 28% Mo are a third category of nickel alloy liable to intergranular corrosion in the welded condition. In these alloys preferential corrosion may develop at zones adjacent to welds exposed to HCl and other hydrogen-evolving acids in which this type of alloy is used. Corrosion is preferentially concentrated on molybdenum-depleted zones adjacent to grain boundaries in which molybdenum-rich M_6C carbide has precipitated. The susceptibility of this type of alloy to intergranular corrosion is reduced by controlling the carbon and iron content to levels as low as is practicable and also by addition of about 2% V[54] or 3·5–5% W[55]. Niobium may also be a beneficial addition[54], but titanium and zirconium accelerate intergranular corrosion of this type of alloy[55].

Bimetallic Corrosion

Owing to their intermediate position in the galvanic series, nickel and nickel alloys may stimulate corrosion of metals less noble to themselves when in bimetallic contact and thus receive cathodic protection or suffer intensified corrosion from contact with more noble metals and graphite. In general, in mild environments such as unpolluted atmospheric conditions, nickel and nickel alloys are compatible with a fairly wide range of other metals and alloys, but in strong electrolytes such as sea-water and marine atmospheres the range of compatible couples is less. Table 4.25 gives guidance in very general terms, but should not be assumed to apply in every circumstance, since other factors may influence the issue. The relative surface areas of the two metals in contact plays a large part in determining whether bimetallic corrosion is serious or not, and the combination of a small area of the more negative (less noble) metal or alloy in contact with a large area of the more noble material is usually the most dangerous situation (see Section 1.7). Protection of the less noble metal by painting or other means, if properly carried out, is usually effective in minimising bimetallic corrosion. In aggressive environments nickel and the different types of nickel alloy are not necessarily wholly compatible one with another.

Corrosion in Natural Environments

The Atmosphere

Nickel and nickel alloys possess a high degree of resistance to corrosion when exposed to the atmosphere, much higher than carbon and low-alloy steels, although not as high as stainless steels. Corrosion by the atmosphere

Table 4.25 Bimetallic corrosion effects of nickel and nickel alloys
(General guidance only; other factors, including relative surface areas, often exert an important influence)

Corrosive environment	*Corrosion of nickel or nickel alloy is stimulated by bimetallic contact with:*	*Bimetallic contact with nickel or nickel alloy has little or uncertain influence*:*	*Bimetallic contact with nickel or nickel alloy stimulates corrosion of:*
Most atmospheric conditions except marine atmospheres	Au Pt	Rh Pd C (graphite) Ti Cu and Cu alloys Stainless steels Cr plate C steel† Al and Al alloys† Mg and Mg alloys†	Pb Sn Soft solders Cd Zn Galvanised steel Al clad Carbon steel‡ Al and Al alloys‡ Mg and Mg alloys‡
Sea-water and marine atmospheres	C (graphite) Graphitised cast iron Au Pt	Ag and Ag brazing alloys Cu and Cu alloys Pb Sn Soft solders Other Ni alloys	Austenitic cast iron§ Low-alloy steels Cast iron (ungraphitised) Wrought iron Carbon steel Cd Al and Al alloys Zn Mg and Mg alloys

*Little effect in most atmospheres, except marine. Effects in sea-water and marine atmospheres depend on surface area relationships.
†If properly painted. ‡If unpainted or improperly painted. §Contact with small area of Monel 400 has little effect.

is, therefore, rarely if ever a factor limiting the life of nickel and nickel alloy structures when exposed to that environment.

The appearance of bright nickel is, however, impaired by exposure to moist, polluted atmospheres owing to the phenomenon known as *fogging*. Vernon showed more than 40 years ago that for 'fogging' of nickel to occur, a high humidity—greater than about 70% r.h.—and the presence of SO_2 were both necessary[56]. Fogging is due to the catalytic oxidation of SO_2 in polluted atmospheres by the nickel surface and subsequent corrosion of the nickel by the liquid film of H_2SO_4 thus formed on the surface, the corrosion product —a basic nickel sulphate—being responsible for fogging. In the early stages the film can be readily removed by wiping with a cloth, but once the surface has become fogged the bright appearance cannot be restored merely by wiping and mild abrasion is needed. Some nickel alloys, viz. the Ni–Cu alloy Monel 400 also undergo fogging, but alloys containing 15% Cr or more do not exhibit this phenomenon. Fogging is prevented by a very thin film of chromium deposited on the surface—a fact which forms the basis for the bright appearance of decorative chromium–nickel plate (see Sections 14.7 and 14.8).

Nickel and nickel alloys do not form thick layers of corrosion products when freely exposed to outdoor atmospheres in circumstances where the surface is periodically washed by rain, but such deposits may form on sheltered surfaces. Quantitative data on the rate of loss of metal and of pitting of nickel and nickel alloys exposed to outdoor atmospheres are available[57–60]. Figure 4.40 shows results obtained at three sites in the USA over a 7 year period[59] and Fig. 4.41 gives results from a 10 year test at Birmingham[60]. In both series of tests, Ni–Cr–Fe alloys gave lower weight losses than nickel itself or Ni–Cu alloys and the American results bring out the point that over long periods the corrosion rate of Ni–Cr–Fe alloys in the atmosphere declines to a very low value whilst that of nickel and Ni–Cu alloys remains approximately linear. Comparison of the data in Figs. 4.40 and 4.41 shows that corrosion at the UK site was several times greater than that at the most aggressive American site. This has also been observed with stainless steels in a test programme where a direct comparison was made between identical test samples[61].

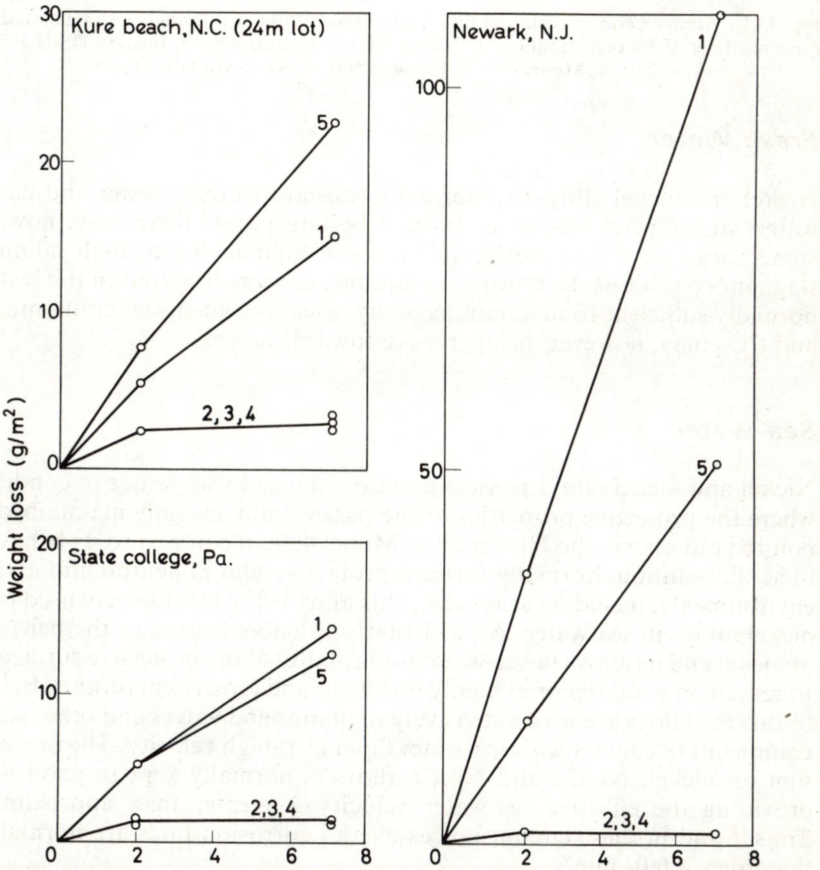

Fig. 4.40 Atmospheric corrosion of nickel and nickel alloys during exposure tests at sites in the USA. 1, Nickel 200; 2, Inconel 600; 3, Incoloy 800; 4, Incoloy 825; 5, Monel 400 (after van Rooyen and Copson[59])

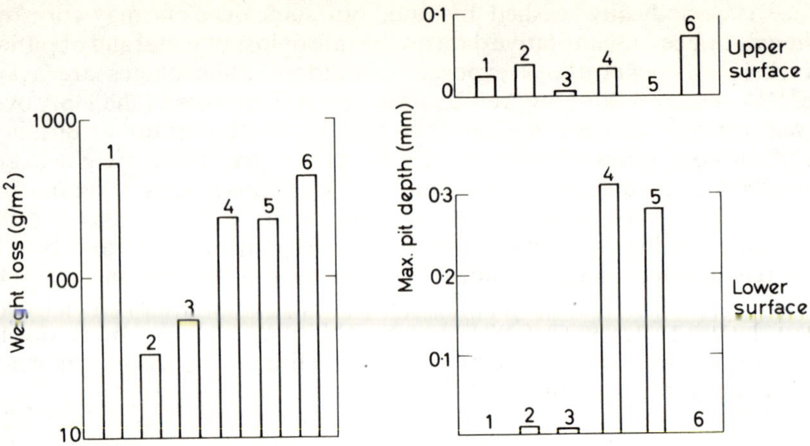

Fig. 4.41 Atmospheric corrosion of nickel and nickel alloys at Birmingham, England, during exposure tests of 10 year duration. 1, Nickel 200; 2, Inconel 600; 3, Incoloy DS (Fe–37Ni–18Cr–2Si); 4, Monel 400; 5, Monel K500; 6, Ni–28Mo (after Evans[60])

Fresh Water

Nickel and nickel alloys are normally resistant to fresh water and natural waters at temperatures up to normal boiling point; there may, however, sometimes be a risk of pitting in waters of high acidity or high salinity in stagnant conditions. In flowing conditions, oxygen dissolved in the water is normally sufficient to maintain passivity. Steam condensates containing O_2 and CO_2 may, however, be aggressive towards nickel.

Sea-water

Nickel and nickel alloys possess good resistance to sea-water in conditions where the protective properties of the passive film are fully maintained. As pointed out above, the Ni–Cu alloy Monel 400, in contrast to its behaviour in acidic solution, normally forms a protective film in neutral and alkaline environments, including sea-water; this alloy is the most widely used of the nickel alloys in sea-water. A particularly valuable feature of the behaviour of nickel and its alloys in sea-water is the ability of the protective surface film to remain in good repair in highly turbulent and erosive conditions. Because of this the alloys are used extensively in pumps and valves and other similar equipment in contact with sea-water flowing at high velocity. The protective film on nickel, Ni–Cu and Ni–Cr alloys is normally kept in good repair providing the effective sea-water velocity is greater than approximately 2 m/s[62] and in these circumstances overall corrosion rates are normally of the order of 0·01 mm/y.

In sea-water flowing at slower velocities and more especially in stagnant conditions, pitting may develop, particularly beneath deposits and marine growths at the surface of the metal. Some data for the Ni–Cu alloy Monel 400[63] are shown in Fig. 4.42; the corrosion was mostly pitting.

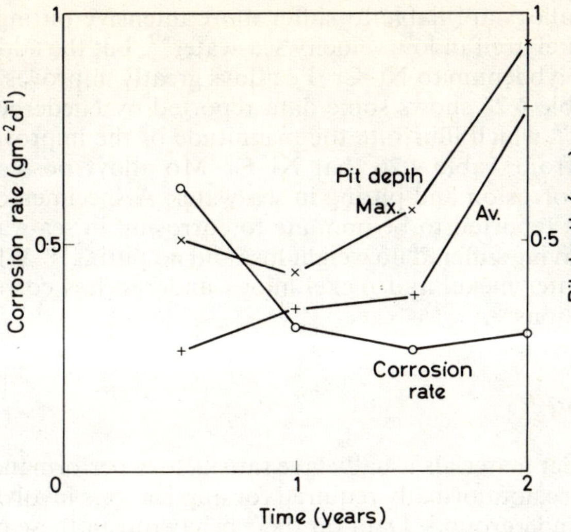

Fig. 4.42 Corrosion of Monel 400 in sea-water at Port Hueneme Harbour, Cal., USA (after Brouillette[63])

Table 4.26 Corrosion of nickel alloys in quiet and in slow moving sea-water (after Niederberger, *et al*[64])

Alloy composition	*Quiet sea-water*			*Slow moving sea-water*		
	Wt. loss (g)	*Range of pit depth* (mm)		*Wt. loss* (g)	*Range of pit depth* (mm)	
		Exposed	*Crevice*		*Exposed*	*Crevice*
Ni–35Cu	20·40	0–0·5	0·75–0·8	21·10	0·25–0·38	0·6–0·68
Ni–30Cu–3Al	19·50	0·55–0·65	0·55–0·58	24·40	0·8–1·0	0·5–0·7
Ni–16Cr–7Fe	11·85	3·25	3·25	12·55	3·25	0·73–3·25
Ni–35Cr–2Fe	9·32	3·0	0·15–0·7	7·62	3·0	0·38–1·65
Ni–47Fe–20Cr	15·72	3·4	3·63	24·50	3·63	3·63
Ni–27Mo	42·80	0·5–0·53	0·33–0·35	54·80	0·25–0·35	0·2–0·4
Ni–16Mo–7Cr–4Fe	0·50	Nil	0–0·05	1·45	Nil	0·08–0·1
Ni–30Fe–21Cr–3Mo	0·25	0–0·03	0–0·23	0·20	0·08–0·15	0·03–0·65
Ni–22Cr–9Mo–2Fe	Nil	Nil	Nil	0·25	Nil	Nil
Ni–20Cr–5Mo–6Fe	0·20	Nil	0·35–0·63	0·15	0–0·05	Nil
Ni–16Cr–16Mo–4W	Nil	Nil	Nil	0·10	Nil	Nil

Conditions of exposure: Quiet sea-water—suspended from raft.
Slow moving sea-water—velocity 0·3–0·6 m/s.
Location: Harbor Island, Wrightsville Beach, N.C., USA.
Duration of test: 2 years; panel size 12 × 3 in (0·305 × 0·076 m).
crevice area 1 × 1 in (0·025 × 0·025 m).

Ni–Cr–Fe alloys are liable to suffer more intensive pitting than Ni–Cu alloys and nickel itself in low velocity sea-water[62], but the addition of a few per cent of molybdenum to Ni–Cr–Fe alloys greatly improves the resistance to pitting. Table 4.26 shows some data reported by Niederberger, Ferrara and Plummer[64] which illustrate the magnitude of the improvement. It will also be seen from Table 4.26 that Ni–Cr–Mo alloys possess even better resistance to corrosion and pitting in sea-water. A specimen of Ni–Cr–Mo alloy has been reported to be immune to corrosion in sea-water over a 10 year period, having suffered no weight loss and no pitting[65]. When immersed in deep sea-water nickel and nickel alloys undergo less corrosion than in shallow conditions[66].

Underground

As other cheaper materials usually give satisfactory performance, nickel and nickel alloys are not normally required for applications involving resistance to corrosion underground. Data on their behaviour in these circumstances are therefore sparse; in particular, whether micro-organisms responsible for the accelerated corrosion of ferrous and other metals in certain anaerobic soils have any influence on nickel and its alloys, is uncertain.

Corrosion in Chemical Environments

Acids

The wide range of corrosion-resistant nickel alloys that are produced commercially is capable in practice of handling most types of acid. Since the nickel-alloy range includes some that are corrosion resistant by virtue of their relative nobility and others that owe their resistance to passivity, alloys suitable both for hydrogen-evolving acids and for more oxidising acids are available. Table 4.27 contains a summary of data mainly derived from laboratory corrosion tests to illustrate the behaviour of individual alloys in some common mineral and organic acids.

The data in Table 4.27 refer to solutions of pure acids; in practice the presence of impurities often has a large influence and modifies the corrosion resistance to a greater or lesser extent. Oxygen from the air stimulates corrosion of alloys of the relatively noble type, including nickel itself, Ni–Cu and Ni–Mo alloys, but may be helpful in maintaining passivity of the other type, viz. alloys containing 15% Cr or more. Other oxidants such as Fe^{3+} or Cu^{2+} (which may sometimes in practice be present through corrosion of ferrous and copper-base alloys also in contact with the acidic environment) usually have a similar influence. The presence of halide ions, especially Cl^- and F^-, in H_2SO_4, H_3PO_4 and HNO_3 is usually highly detrimental to the corrosion resistance of both the noble and passive types of alloys. It should be noted, however, that the Ni–35Cr alloy possesses sufficient resistance to HNO_3–HF mixed acids to be practically useful and that a Co–20Cr–15W–10Ni alloy is similarly resistant to certain HNO_3–HCl mixtures.

Table 4.27 Corrosion resistance of nickel and nickel alloys to acidic solutions

Alloy	H_2SO_4	H_3PO_4	HNO_3	HCl	HF	*Other acids*
Nickel 200 and 201	*A*, 0–20%[1], RT *B*, 0–50%[1], 70°C	*A*, 0–85%[1], RT	*U*	*A*, 0–10%[1]	*B*, 0–70%[1], RT *A*, 0–90%[1], RT	Acetic, *B*, 0–100%, BP Formic, *A*, 0–90%, RT
Monel 400	*A*, 0–85%[1], 30°C *B*, 0–60%[1], 95°C *B*, 25–85%[2], 30°C	*A*, 0–90%[1], 100°C	*U*	*A*, 0–8%[1], RT	*B*, 0–60%[1], RT *A*, >90%[1], RT	Acetic, *A*, 0–100%[1], RT *B*, 0–40%[2], RT *B*, 65–100%[2], RT
Hastelloy B	*A*, 0–96%[1], 65°C *A*, 0–50%, BP	*A*, 0–85%[1], 65°C *A*, 0–50%, BP	*U*	*A*, 0–37%[1], RT *B*, 0–25%[1], 65°C *B*, 0–2%, BP	*A*, 0–45%[1], RT	Acetic, *A*, 0–100%, BP Formic, *A*, 0–90%, RT, BP *B*, 0–60%, 65°C
Hastelloy N		*A*, 0–85%, RT	*B*, 0–20%, RT	*A*, 0–37%, RT	*A*, 0–45%, RT	
Hastelloy C	*A*, 0–96%, 65°C *B*, 0–10%, BP	*A*, 0–85%, 65°C *A*, 0–50%, BP	*A*, 0–70%, RT *A*, 0–30%, 65°C *B*, 30–70%, 65°C *B*, 0–10%, BP	*A*, 0–37%, RT *A*, 0–2%, 65°C *B*, 2–37%, 65°C *A*, 0–1%, BP	*A*, 0–5%, RT *B*, 5–45%, RT	Acetic, *A*, 0–100%, BP Formic, *A*, 0–90%, 65°C *B*, 0–90%, BP
Hastelloy F	*A*, 0–96%, RT *A*, 0–10%, 65°C *B*, 10–25%, 65°C	*A*, 0–85%, 65°C *A*, 0–10%, BP *B*, 10–50%, BP	*A*, 0–70%, 65°C *A*, 0–40%, BP *B*, 40–70%, BP	*A*, 0–2%, RT *B*, 2–25%, RT *A*, 25–37%, RT *B*, 0–1%, BP	*A*, 0–5%, RT *B*, 5–45%, RT	Acetic, *A*, 0–100%, BP Formic, *A*, 0–90%, 65°C
Hastelloy G	*B*, 0–10%, BP	*A*, 0–30%, BP *B*, 30–85%, BP			*A*, 0–45%, RT	
Illium R	*A*, 0–3%, BP *B*, 3–60%, BP	*A*, 0–75%, BP	*A*, 0–80%, BP	*U*		
Illium G	*A*, 0–60%, 80°C *A*, 60–98%, 40°C *B*, 0–98%, 70°C	*S*	*S*	*U*		

(continued overleaf)

Table 4.27 (continued)

Alloy	H_2SO_4	H_3PO_4	HNO_3	HCl	HF	*Other acids*
Illium 98	*A*, 0–98%, 65°C *B*, 0–65%, 100°C *B*, 90–98%, 100°C	*S*	*S*	*U*		
Incoloy 825	*A*, 0–5%, 80°C *B*, 0–45%, BP *B*, 40–80%, 100°C	*A*, 0–20%, BP *B*, 20–80%, BP *A*, 0–80%, 75°C	*A*, 0–30%, BP *B*, 30–70%, BP *A*, 0–70%, 75°C		*B*, 38–70%, 60°C	Acetic, *A*, 0–100%, BP Formic, *A*, 0–100%, BP Oxalic, *A*, 0–10%, BP *B*, 10–50%, BP
Inconel 600	*A*, 0–70%, RT(3)	*A*, 0–80%, RT(3)		*A*, 0–2%, RT *B*, 2–15%, RT(3)	*S*	Acetic, *A*, 0–100%, RT
Corronel 230	*U*		*A*, 0–70%, BP	*U*		
Hastelloy D	*A*, 0–96%, 65°C *B*, 0–50%, BP	*A*, 0–85%, 65°C *A*, 0–10%, BP *B*, 10–50%, BP	*U*	*A*, 0–10%, RT *B*, 10–20%, RT		Acetic, *A*, 0–100%, RT *B*, 0–100%, 65°C

The data show corrosion resistance as a function of acid composition, concentration and temperature. Since the data are mostly derived from laboratory corrosion tests in pure solutions, they should not be taken as a firm indication of performance in service.

A = <0·1 mm/year
B = 0·1–0·5 mm/year
S = no data, but often suitable in service
U = unsuitable

% = concentration w/w
RT = room temperature
BP = boiling point

(1) = air-free solutions; aeration increases corrosion
(2) = saturated with air
(3) = not resistant at high temperatures

In addition to impurities, other factors such as fluid flow and heat transfer often exert an important influence in practice. Fluid flow accentuates the effects of impurities by increasing their rate of transport to the corroding surface and may in some cases hinder the formation of (or even remove) protective films, e.g. nickel in HF. In conditions of heat transfer the rate of corrosion is more likely to be governed by the effective temperature of the metal surface than by that of the solution. When the metal is hotter than the acidic solution corrosion is likely to be greater than that experienced by a similar combination under isothermal conditions. The increase in corrosion that may arise through the heat transfer effect can be particularly serious with any metal or alloy that owes its corrosion resistance to passivity, since it appears that passivity breaks down rather suddenly above a critical temperature, which, however, in turn depends on the composition and concentration of the acid. If the breakdown of passivity is only partial, pitting may develop or corrosion may become localised at hot spots; if, however, passivity fails completely, more or less uniform corrosion is likely to occur.

Table 4.27 provides a basis for selecting the nickel alloy type likely to be suitable for service in particular acids. The Ni–Cr–Fe–Mo–Cu and Ni–Cr–Fe–Mo alloys, both wrought and cast, are the types most often selected for H_2SO_4, and they possess the additional advantage that their resistance is not greatly affected by the presence of SO_2. Cast Ni–Si alloys containing 9% Si or more and, preferably, alloying additions of copper, titanium and molybdenum[67, 68] are also used for H_2SO_4. Most nickel alloys have good resistance to pure H_3PO_4, but the presence of halide ion impurities reduces the resistance in the higher concentration range. The Ni–35Cr alloy possesses good resistance to HNO_3 and is one of the few metals able to withstand the combination of HNO_3 plus HF: it is not, however, resistant to mixtures of HNO_3 and HCl. Among metallic materials the Ni–28Mo is one of the most suitable for handling HCl, particularly in the absence of air and other oxidants; in oxidising conditions Ni–Cr–Mo alloys are usually more suitable. Nickel itself and Ni–30Cu both possess good resistance to HF; in practice the Ni–30Cu alloy is used for aqueous and anhydrous HF, but precautions are necessary against stress-corrosion cracking. In practice, nickel and nickel alloys have good resistance to most organic acids.

Alkalis

Nickel and its alloys are among the most resistant metallic materials to caustic alkalis. Nickel itself possesses outstanding resistance to NaOH and KOH and is used to contain these substances over the entire concentration and temperature ranges that are of practical interest, viz. 0–100% and up to 350°C. At the higher concentrations and temperatures KOH is significantly more corrosive than NaOH towards nickel and in these circumstances the metal is sometimes cathodically protected.

At temperatures above 300°C, low-carbon nickel (0·02% C) is preferred to avoid the possibility of intergranular attack developing after long exposure; if material of higher carbon is employed it should be annealed after fabrication and before exposure to caustic alkalis to prevent stress-assisted intergranular corrosion.

The corrosion resistance of nickel to caustic alkalis is adversely affected by the presence of small amounts of oxidisable alkaline sulphur-containing salts, e.g. Na_2SO_3, $Na_2S_2O_3$, Na_2S and, at high temperatures, by alkaline oxidising agents, viz. $NaClO_3$ and Na_2O_2. In the former circumstance Ni–15Cr–8Fe is more resistant than nickel, but not in the latter. When Ni–15Cr–8Fe is used for service in caustic alkalis, it should be stress relieved after fabrication to minimise the possibility of stress-corrosion cracking.

Other nickel alloys, notably Ni–30Cu, also possess good resistance to caustic alkalis.

Salts

Nickel and nickel alloys generally possess good corrosion resistance to acidic, neutral and alkaline salts, including halides, that are not oxidising in character. Oxidising salts are usually corrosive towards Ni, Ni–Cu and Ni–Mo alloys, but not to Ni–Cr and Ni–Cr–Fe–Mo–Cu alloys unless they contain appreciable quantities of both oxidiser and halide ions, e.g. $FeCl_3$, $CuCl_2$, NaOCl. Ni–Cr–Mo alloys are among the few metallic materials that are resistant to oxidising halide salts.

Wet and Dry Gases

Nickel and its alloys are usually resistant to dry gases, including NH_3, SO_2, F_2, Cl_2, HCl and HF up to high temperatures, and in many cases are the preferred materials for handling such gases. Nickel itself is used in service at temperatures up to about 540°C with dry Cl_2, HCl, F_2 and HF and up to about 300°C with dry SO_2. The Ni–15Cr–8Fe alloy is resistant to dry SO_2 up to 800°C and to dry NH_3 up to at least 600°C. When moist or under dew-point conditions these gases are in many instances appreciably more corrosive towards nickel and most nickel alloys, except Ni–Cr–Mo alloys. Ni–Cr–Fe–Mo–Cu alloys do, however, possess useful resistance to condensates containing SO_2 at temperatures well in excess of 100°C, and also to solutions containing NH_3 and its salts. Ni–Cr–Mo alloys are among the most resistant metallic materials to moist halogens.

Organic Compounds

Nickel and nickel alloys are resistant to many organic compounds and are often suitable for handling organic acids, alcohols and halogenated hydrocarbons. It should be borne in mind, however, that halogenated organic compounds may undergo hydrolysis in the presence of water or steam and release appreciable quantities of the corresponding halogen hydracid, and this will often dictate the choice of the alloy. Detailed information should be sought concerning the suitability of alloys for particular circumstances.

Water and Steam at High Temperatures

The corrosion rates of nickel and nickel alloys in pure water and steam at elevated temperatures are generally extremely low, typically of the order of 1 μm/year. The metal and its alloys are therefore often selected for service in these environments in circumstances where contamination of the water by metal ions is to be avoided. It should be noted, however, that the possibility of stress corrosion may need to be taken into account in certain circumstances.

Conjoint Action of Stress and Corrosion

As with alloys of other metals, nickel alloys may suffer stress-corrosion cracking in certain corrosive environments, although the number of alloy–environment combinations in which nickel alloys have been reported to undergo cracking is relatively small. In addition, intergranular attack due to grain boundary precipitates may be intensified by tensile stress in the metal in certain environments and develop into cracking. Table 4.28 lists the major circumstances in which stress corrosion or stress-assisted corrosion of nickel and its alloys have been recorded in service and also shows the preventive and remedial measures that have been adopted, usually with success, in each case.

Table 4.28 Stress corrosion of nickel and nickel alloys

Alloy type	*Environment*	*Type of cracking*	*Preventive or remedial measures*
Ni ($C > 0{\cdot}02\%$)	Caustic alkalis, high concentrations and high temperature	IG	1. Stress-relieve after fabrication[69] 2. Use low-carbon nickel (0·02% C max.)
Ni–15Cr–8Fe Ni–30Cu	Caustic alkalis, high concentrations and high temperature	Usually IG	Anneal after fabrication
Ni–Cr–Fe and Ni–Cr–Fe–Mo–Cu (Ni approx. <40%)	Chloride solutions at elevated temperature	TG	1. Use alloys with higher nickel content[70] 2. Control amounts of minor alloying elements and impurities[70] 3. Remove or reduce Cl^- in environment if possible
Ni–15Cr–8Fe and stainless steels	High temperature water	Usually IG	1. Stress relieve after fabrication 2. Control water purity[71,72]
Fe–32Ni–21Cr	Polythionic acids	IG	Stabilise material against intergranular corrosion[73]
Ni–30Cu	Hydrofluoric acid	Usually IG	1. Stress relieve after fabrication[74] 2. Control composition of weld electrode/filler[75] 3. Control residual O_2 and other oxidising agents[74,75]

IG = Intergranular; TG = Transgranular.

With regard to stress-corrosion cracking in the Ni–Cr–Fe system, including both nickel-base alloys and stainless steels, a vast number of papers has been published. A detailed review of work published before 1969 is available[70] and the authors have since published additional data[76]. With regard to behaviour in high-temperature water, Ni–Cr–Fe alloys in the stressed condition have been reported to undergo stress corrosion during laboratory autoclave tests in water at temperatures in the region of 300°C[71,72,77]. Cracking is usually, although not always, intergranular and is influenced by the nature and amount of impurities present in the autoclave and by the stress level of the specimens. Ni–Cr–Fe alloys have, however, given good service in high-temperature water environments in nuclear heat exchangers, where they are the preferred materials of construction.

High-temperature Corrosion and Oxidation

Nickel and Ni–Cr alloys are among the most resistant metallic materials to corrosion and oxidation at high temperatures and are widely used to resist corrosion by gases and molten salts at elevated temperatures (see Sections 7.1 and 7.5).

Applications in Industry

As will have become apparent, nickel and corrosion-resistant nickel alloys have wide ranges of application, particularly in industries where strongly acidic, strongly alkaline or strongly saline environments are encountered. Table 4.29 lists some of the more important applications in those industries where these conditions most frequently arise, i.e. in the chemical, petrochemical, nuclear and conventional power generating, textile, paper, marine, desalination and food processing industries. The list is by no means exhaustive and there are many other applications of a similar nature in these and other industries. The table should, nevertheless, serve to indicate the wide range of corrosive conditions in which nickel and its alloys are used in practice.

A major feature to note is that in many applications the requirement is not only for a corrosion-resistant material but also for one that will not contaminate the product. This is particularly so in the food processing industry, where freedom from metallic contamination is important both from the point of view of preserving the product in good condition and also to avoid rendering it toxic. Nickel is advantageous in both respects. Similar considerations of product purity arise in the production of certain chemical intermediates, notably phenol that is to be used for production of synthetic fibres. Another application where, in addition to a very high degree of integrity of the material, minimum release of corrosion product is required is in the primary coolant circuit of water-cooled nuclear reactors. The Ni–Cr–Fe alloy Inconel 600 has been widely used over recent years in the heat exchangers of pressurised water-reactors with highly satisfactory results. In the relatively few instances of failure that have occurred, these have been associated with

Table 4.29 Applications of Ni and corrosion-resistant nickel alloys

Alloy type	*Industry*						
	Chemical	*Petrochemical*	*Nuclear and power*	*Marine and desalination*	*Textile and paper*	*Food*	*Other*
Ni	Production of caustic alkalis. Production of phenol. Salt evaporators	Chlorination processes. Oxy-chlorination processes. Production of phenolic resins. Production of synthetic rubber, plastics. Production of chlorinated hydrocarbons	Fluorination processes		Production of viscose rayon	Processes where freedom from metallic contamination is necessary	Production of fine chemicals
Ni–Cr			Fuel element processing in HNO_3 + fluorides				
Ni–Cr–Fe (approx. 75% Ni)	Production of caustic alkalis especially in the presence of S	Production of halogenated hydrocarbons and phenolic resins	Heat exchangers in water-cooled reactors. Fluorination processes		Paper pulp production (alkaline processes). Steam-heated dryers. Textile rolls	Processes and storage where freedom from contamination is necessary	
Ni–Cr–Fe (approx. 32% Ni)	HNO_3 condensers	Reforming processes	Superheater tubes				
Ni–Cr–Fe–Mo Ni–Cr–Fe–Mo–Cu	Production of fertilisers. Sulphonation processes	Handling acid sludges. Hydrocarbon processes involving H_3PO_4	Fuel element processing		Wood pulp digesters. SO_2 strippers. Synthetic fibre manufacture	Processes where freedom from stress corrosion is necessary	Pickling of steel. Handling waste gases containing SO_2

(continued overleaf)

Table 4.29 (continued)

Alloy type	Industry						
	Chemical	*Petrochemical*	*Nuclear and power*	*Marine and desalination*	*Textile and paper*	*Food*	*Other*
Ni–Cr–Mo	Chlorination processes			Applications where no pitting and no loss or reflectivity are necessary	Bleaching operations		
Ni–Mo	Processes involving HCl and non-oxidising acidic chlorides	Distillation columns containing acidic chlorides					
Ni–Cu	Production and recovery of HF	HF alkylation processes. Sulphur stripping columns. Distillation columns containing acidic chlorides. Handling acid sludges	Fluorination processes	Valves, impellers, propeller shafts, fasteners. Demisters in desalination plants		Handling refrigerating brines. Salt production. Evaporators	Pickling of steel
Ni–Si	Processes involving conc. H_2SO_4						

inadequate control over water purity, sometimes coupled with adverse features involving crevices.

Nickel and most nickel alloys are available in the usual wrought forms—plate, sheet, bar, tube, etc.—and also in some cases as clad steel plate. Sheet material may be employed as corrosion-resistant liners in process vessels and some of the alloys may also be used as weld overlays to provide a corrosion-resistant surface. In applications where a higher strength than that normally available is required, high-strength variants of some of the materials, i.e. Ni, Ni–Cu and Ni–Cr alloys, are available. These materials are strengthened by precipitation hardening, and heat treatment is necessary to develop the full strength. Nickel and most of the nickel alloy types are available in cast form, and it should be noted that the Ni–Si alloy and some of the Ni–Cr–Fe–Mo–Cu alloys with higher than normal silicon are only available as castings. These materials are used primarily for pumps and valves. Certain wrought nickel, Ni–Cr–Fe and Ni–Cu alloys can be supplied and fabricated to meet the statutory requirements for pressure-vessel construction.

T. E. EVANS

REFERENCES

1. Pourbaix, M., *Atlas of Electrochemical Equilibria in Aqueous Solutions*, Pergamon Press, 333 (1966)
2. Cowan, R. L. and Staehle, R. W., *J. Electrochem. Soc.*, **118**, 557 (1971)
3. Brook, P. A., *Corr. Sci.*, **12**, 297 (1972)
3a. Ashworth, V. and Boden, P. J., *J. Electrochem. Soc.*, **119**, 720 (1972)
4. DiBari, G. A. and Petrocelli, J. V., *J. Electrochem. Soc.*, **112**, 99 (1965)
5. Hart, A. C., *Metal Fin. J.*, **19**, 332 (1973)
6. de Gromoboy, T. S. and Shreir, L. L., *Electrochim. Acta.*, **11**, 895 (1966)
7. Chatfield, C. J. and Shreir, L. L., *Corr. Sci.*, **12**, 563 (1972)
8. Ijzermans, A. B., *Corr. Sci.*, **10**, 113 (1970)
9. Sato, N. and Okamoto, G., *J. Electrochem. Soc.*, **111**, 897 (1964)
9a. Pigeaud, A., *J. Electrochem. Soc.*, **122**, 80 (1975)
9b. Kesten, M., *J. Electrochem. Soc.*, **119**, 722 (1972)
10. Sato, N. and Okamoto, G., *J. Electrochem. Soc.*, **110**, 605 (1963)
11. Hoar, T. P., *Trans. Inst. Metal Fin.*, **39**, 166 (1962)
12. Kunze, E. and Schwabe, K., *Corr. Sci.*, **4**, 109 (1964)
13. Szklarska-Smialowska, Z., *Corr. Sci.*, **12**, 527 (1972)
14. Hayfield, P., in *Advances in Corrosion Science and Technology*, Ed. by M. G. Fontana and R. W. Staehle, **2**, Plenum Press, 96 (1972); (quoting results of Tronstad)
15. Bockris, J. O'M., Reddy, A. K. N. and Rao, B., *J. Electrochem. Soc.*, **113**, 1133 (1966)
16. Lovrecek, B. and Lipanovic, S., *Corr. Sci.*, **10**, 865 (1970)
17. Davies, D. E. and Barker, W., *Corrosion*, **20**, 47t (1964)
18. Armstrong, R. D., Harrison, J. A. and Thirsk, H. R., *Corr. Sci.*, **10**, 679 (1970)
19. Müller, W. J., *Monatsh. Chem.*, **48**, 426 (1927)
20. Briggs, G. W. D. and Wynne-Jones, W. F. K., *Electrochim. Acta.*, **7**, 241 (1962)
21. Gilli, G., Borea, P., Zucchi, F. and Trabanelli, G., *Corr. Sci.*, **9**, 673 (1969)
22. Hodge, F. G. and Wilde, B. E., *Corrosion*, **26**, 146 (1970)
23. Economy, G., Speiser, R., Beck, F. H. and Fontana, M. G., *J. Electrochem. Soc.*, **108**, 337 (1961)
24. Shiobara, K., Sawada, Y. and Morioka, *Trans. Inst. Met. Japan*, **5**, 97 (1965)
25. Sayano, R. R. and Ken Nobe, *Corrosion*, **25**, 260 (1969)
26. Condit, D. O., *Corr. Sci.*, **12**, 451 (1972)
27. Piron, D. L., Koutsoukos, E. P. and Ken Nobe, *Corrosion*, **25**, 151 (1969)
28. Leonard, R. B., *Corrosion*, **24**, 301 (1968)
29. Morris, P. E. and Scarberry, R. C., *Corrosion*, **26**, 169 (1970)

30. Osterwald, J. and Uhlig, H. H., *J. Electrochem. Soc.*, **108**, 515 (1961)
31. Flint, G. N. and Barker, W., *Proc. Chem. Engng. Group (S.C.I.)*, **43**, FB1 (1964)
32. Barker, W. and Evans, T. E., unpublished data
33. Evans, T. E. and Hart, A. C., *Electrochim. Acta.*, **16**, 1955 (1971)
34. Clarke, M. and Elbourne, R. G. P., *Corr. Sci.*, **8**, 29 (1968)
35. Sedriks, A. J., Green, J. A. S. and Novak, D. L., *Corrosion*, **28**, 137 (1972)
36. Crow, W. B., Myers, J. R. and Marvin, B. D., *Corrosion*, **27**, 459 (1971)
37. Hart, A. C., *Corr. Sci.*, **12**, 137 (1972)
38. Tokuda, T. and Ives, M. B., *Corr. Sci.*, **11**, 297 (1971)
39. Gressmann, R., *Corr. Sci.*, **8**, 325 (1968)
40. Postlethwaite, J., *Electrochim. Acta.*, **12**, 333 (1967)
41. Horvath, J. and Uhlig, H. H., *J. Electrochem. Soc.*, **115**, 791 (1968)
42. Henthorne, M. and DeBold, T. A., *Corrosion*, **27**, 255 (1971)
43. Brown, M. H., *Corrosion*, **25**, 438 (1969)
44. Tedmon, C. S. and Vermilyea, D. A., *Corrosion*, **27**, 377 (1971)
45. Raymond, E. L., *Corrosion*, **24**, 180 (1968)
46. Copson, H. R., Hopkinson, B. E. and Lang, F. S., *Proc. A.S.T.M.*, **61**, 879 (1961)
47. Cowan, R. L. and Tedmon, C. S., in *Advances in Corrosion Science and Technology*, Ed. by M. G. Fontana and R. W. Staehle, **2**, Plenum Press (reviewing results of several authors), 293 (1973)
48. Gräfen, H. and Böhm, G., *Z. Metallkunde*, **51**, 245 (1960)
49. Class, I., Gräfen, H. and Scheil, E., *Z. Metallkunde*, **53**, 283 (1962)
50. Streicher, M. A., *Corrosion*, **19**, 272t (1963)
51. Samans, C. H., Meyer, A. R. and Tisinai, G. F., *Corrosion*, **22**, 336 (1966)
52. Leonard, R. B., *Corrosion*, **25**, 222 (1969)
53. Hodge, F. G., *Corrosion*, **29**, 375 (1973)
54. Flint, G. N., *J. Inst. Met.*, **87**, 303 (1959)
55. Pavlov, S. S. and Svistunova, T. V., *Metallov. Term. Obrab. Metallov.*, **10**, 20 (1970)
56. Vernon, W. H. J., *J. Inst. Met.*, **48**, 121 (1932)
57. Copson, H. R., *A.S.T.M. Spec. Tech. Publn. No. 175*, 141 (1955)
58. Copson, H. R. and Tice, E. A., *Werkst. u. Korrosion*, **15**, 645 (1964)
59. van Rooyen, D. and Copson, H. R., *A.S.T.M. Spec. Tech. Publn. No. 435*, 175 (1968)
60. Evans, T. E., *Proc. 4th Int. Congr. on Met. Corr.*, 408 (1969)
61. Evans, T. E., *Werkst. u. Korrosion*, **15**, 797 (1964)
62. Tuthill, A. H. and Schillmoller, C. M., paper presented at 'The Ocean Science and Engineering Conference', Washington D.C. (1965)
63. Brouillette, C. V., *Corrosion*, **14**, 352t (1958)
64. Niederberger, R. B., Ferrara, R. J. and Plummer, F. A., *Mats. Perf.*, **9**, 18 (1970)
65. Weisert, E. D., *Corrosion*, **13**, 659t (1957)
66. Reinhart, F. M., *U.S. Naval Civil Engng. Laboratory Tech. Note N-1023*
67. Barker, W., Evans, T. E. and Williams, K. J., *Brit. Corr. J.*, **5**, 76 (1970)
68. Barker, W., Williams, K. J. and Evans, T. E., *Chem. Proc. Engng.*, **51**, 57 (1970)
69. Fraser, O. B. J., *Symposium on Stress Corrosion Cracking of Metals*, ASTM/AIME, 458 (1944)
70. Latinision, R. M. and Staehle, R. W., *Proc. Conf. Fundamental Aspects of Stress Corrosion Cracking*, 214, NACE (1969)
71. Copson, H. R. and Dean, S. W., *Corrosion*, **21**, 1 (1965)
72. Copson, H. R. and Economy, G., *Corrosion*, **24**, 55 (1968)
73. Stephens, C. D. and Scarberry, R. C., NACE Conf., Houston, Texas, Preprint Paper No. 10 (1969)
74. Copson, H. R. and Cheng, C. F., *Corrosion*, **12**, 647t (1956)
75. Edwards, W. T., *Proc. 1st Int. Congr. on Met. Corr.*, 437 (1961)
76. Staehle, R. W., Royuela, J. J., Raredon, T. L., Serrate, E., Morin, C. R. and Farrar, R. V., *Corrosion*, **26**, 451 (1970)
77. Coriou, H., Grall, L., Mahieu, C. and Pelas, M., *Corrosion*, **22**, 280 (1966)

4.6 Tin and Tin Alloys

The most important forms in which tin is used[1] are as follows:

1. Coatings for other metals.
2. Tin of more than 99% purity.
3. Tin hardened by additions of 1–2% Cu or Sb.
4. Pewter with 90–95% Sn, 4–8% Sb and 1–2% Cu.
5. Soft solder with tin and lead in all proportions.
6. Bearing metal with a large range of proportions of tin, antimony, copper and lead, or with tin (5–30%) in aluminium.
7. Diecasting alloys containing 70–80% Sn with antimony, copper and lead, either singly or combined.

Corrosion behaviour of the above-listed forms of tin can be considered as being basically similar, except in the case of solders and bearing metals, where the wide composition range and special duties of the materials necessitate additional comment. The behaviour of massive tin is a guide to the performance of coatings (Section 14.5).

Physical Properties

The use of unalloyed tin is restricted by its low melting point (232°C) and by its low tensile strength (15 MN/m^2). On the other hand, its melting point, and its ability to wet other metals, facilitates its use as solder and as a coating, while its softness and high ductility make it suitable for cold working and for bearing applications.

Of the elements normally present in tin-rich alloys, lead forms a simple eutectic system with a eutectic composition at 63% Sn, and copper and antimony have a small solid solubility and form the intermetallic compounds Cu_6Sn_5 and SbSn respectively[1].

Tin recrystallises readily at room temperatures so that effects of mechanical working are slight, arising from differing grain size and not from stress. Corrosion in sodium carbonate solution was found[2, 3] to be less for finer grained material, but there is some doubt as to the general application of this finding.

The normal crystal form of tin is body-centred tetragonal, but a low-temperature allotrope, 'grey tin', is cubic. The transformation temperature is

13·2°C, but, unless there is inoculation with the low-temperature form, the change rarely occurs even at very low temperatures, as the quantities of impurity normally present in commercial tin act as inhibitors. Since the transformation produces 'warts' of grey product, its effects are not distinguishable by the casual observer from those of local corrosion; usually, products suspected to be grey tin prove on investigation to be corrosion product.

The impurities likely to be present in nominally pure tin are unlikely to affect its corrosion resistance, except for minor effects on the rate of oxidation in air. Small aluminium contents, however, may result in a severely embrittling intercrystalline attack by water. The addition of antimony counteracts this effect. Although 0·1% magnesium appears to be tolerable, larger amounts produce effects similar to those of aluminium.

Theoretical Consideration of Corrosion Behaviour

The Pourbaix diagram[4] for tin (Fig. 4.43) refers only to solutions in which formation of soluble tin complexes or protective layers of insoluble salts does not occur. There are few instances of the formation of protective layers other than oxide on tin, and although the formation of soluble complexes is more common, the diagram provides a useful general indication of the conditions

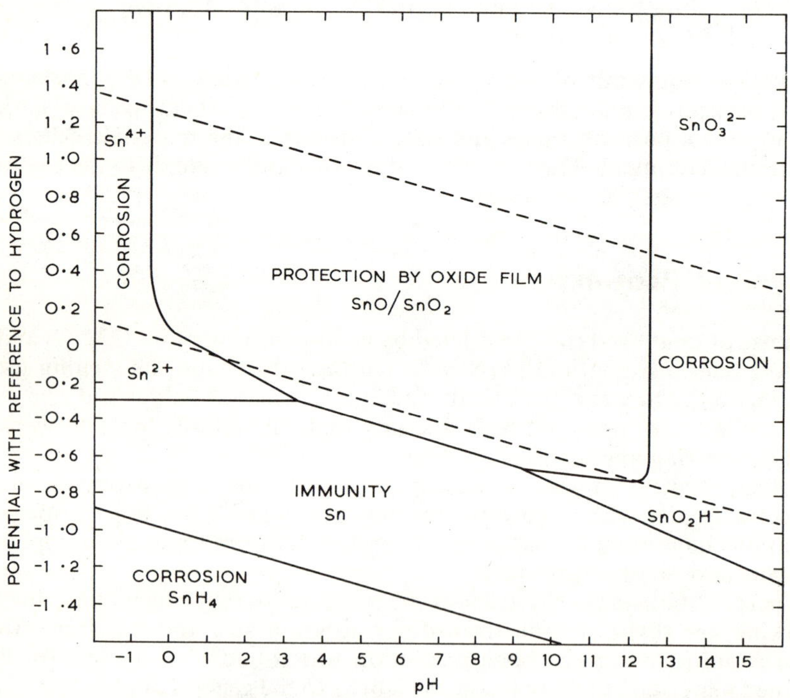

Fig. 4.43 Potential/pH diagram for tin. The full lines enclose the zones where the stability of various compounds or ions makes the indicated action possible. The broken lines indicate the limits of stability of water at a pressure of 1 atm (after Reference 4 and Cebelcor)

in which the corrosion of tin is possible. The wide field of stability of the two oxides, some of it extending below the field of stability of water, indicates easy passivation over a large range of pH, but, in either strongly acid or strongly alkaline solutions, tin may be dissolved. The region at depressed potentials where the possibility of the formation of SnH_4 exists has no known practical importance, but the action has been observed in the laboratory[5].

In zones where, according to the diagram, the dissolution of tin is possible, the rate of corrosion may be very slow. The hydrogen overpotential of tin is high. (In a range of concentrations of H_2SO_4[(6)], the Tafel constant a was 1·29 and b was 0·118, the exchange current for the hydrogen evolution reaction was 10^{-11} A cm^{-2} with a transfer coefficient of 0·50; in KOH solutions[7], the Tafel constant a varied between 1·36 and 1·53 with concentration, b was 0·120 and the exchange current was $0{\cdot}3 \times 10^{-6}$ A cm^{-2}.) Consequently, in alkaline or moderately acid solutions free from oxygen or oxidising agents, corrosion of pure tin may be barely detectable, unless the tin is in contact with another metal of low hydrogen overpotential. Also, tin forms complex ions with many acids including those commonly found in fruits (citric, oxalic, malic). This action has important effects on the galvanic behaviour of tin. The normal electrode potential of tin is −0·136 V, but the reduction of stannous ion activity resulting from formation of complexes may depress the corrosion potential to make tin anodic to iron[8,8(a)].

The stability of tin over the middle pH range (approximately 3·5–9), its solubility in acids or alkalis (modified by the high hydrogen overpotential), and the formation of complex ions are the basis of its general corrosion behaviour. Other properties which have influenced the selection of tin for particular purposes are the non-toxicity of tin salts and the absence of catalytic promotion of oxidation processes that may cause changes in oils or other neutral media affecting their quality or producing corrosive acids.

Atmospheric Corrosion

Oxidation of tin in dry air is slow; the metal remains bright and interference colours are not developed below about 180°C. On a rolled tin surface heated in air[9,10], the thickness of the oxide grew according to a logarithmic law at temperatures up to about 160°C and according to a parabolic law at higher temperatures, for which the oxide was identified as SnO. For electropolished tin heated in oxygen at pressures of 130 N/m^2 and above, three stages of oxidation were observed at temperatures up to 220°C[11–13]:

1. An initial sigmoid growth curve during nucleation.
2. A logarithmic growth curve when cavities acted as diffusion barriers.
3. Erratic behaviour caused by random film rupture.

The oxide formed was identified as α-SnO at temperatures down to 75°C. However, a mixture of SnO and SnO_2 is formed in dry air at lower temperatures and, in humid air, at temperatures up to at least 100°C[39].

Small additions, e.g. 0·1%, of indium, zinc or phosphorus, reduce the rate of oxidation[9] and an addition of antimony, thallium or bismuth accelerates it[14].

Humidity of the air, in the absence of polluting gases or dusts, increases the rate of oxidation[14(a)] and may cause yellow discoloration. In an ordinary uncleaned atmosphere, some corrosion product may be formed in due course; tin products are not hygroscopic and tin is not attacked at relative humidities below 100% unless the dust falling on the surface is hygroscopic, or impurities in the metal are able to form a hygroscopic product. Indoors, in a laboratory atmosphere without special pollution, a grey film, increasing in weight linearly with time ($0.004\,gm^{-2}d^{-1}$), is formed[15]. The reflectivity of the surface is slowly lost if it is left untouched, but may be preserved by regular washing; in one experiment, when the surface was washed at intervals of three weeks, a water wash was adequate for six weeks and although the use of soap was necessary thereafter almost complete preservation of reflectivity was achieved[16]. In a Stevenson Screen out of doors, i.e. with exposure to all atmospheric influences except rain, the corrosion rate declined somewhat with time[15].

The impurities ordinarily present in the atmosphere do not appreciably affect the character of corrosion. No tarnishing effect is exerted by hydrogen sulphide, sulphur dioxide and other acids in low concentrations, including formic, acetic and other organic acids which when evolved from wood or insulating materials are so often destructive to metals other than tin in the confined spaces of electrical equipment or of packages. Chlorides accelerate corrosion and tend to form a white corrosion product containing oxychloride. The presence of some impurities, notably zinc, in the metal may cause tarnishing and loss of brightness in atmospheres containing SO_2[(17)].

The atmospheric pollution prevailing in special industrial or laboratory locations may induce more severe corrosion, e.g. the vapours from concentrated hydrochloric or acetic acid will etch tin, and moist sulphur dioxide will produce a sulphide tarnish, as will hydrogen sulphide at temperatures above about 100°C. The halogens attack tin readily. The commonly used volatile corrosion inhibitors are without adverse action although the benefit derived from their use is doubtful.

When tin is fully exposed out of doors, corrosion is uniform, and the rate falls only slightly with time. The metal becomes dull and accumulates a compact layer of pale grey product, mainly stannous oxide. Rates observed during exposures in the USA for periods of up to 20 years were as follows[18]:

Industrial	0·0013–0·0018 mm/y
Sea coast	0·0018–0·0028 mm/y
Rural	0·005 mm/y

Corrosion in Near-neutral Aqueous Media

Pure tin is completely resistant to distilled water, hot or cold. Local corrosion occurs in salt solutions which do not form insoluble compounds with stannous ions (e.g. chloride, bromide, sulphate, nitrate) but is unlikely in solutions giving stable precipitates (e.g. borate, mono-hydrogen phosphate, bicarbonate, iodide)[19]. In all solutions, oxide film growth occurs and the potential of the metal rises. Any local dissolution may not begin for several days but, once it has begun, it will continue, its presence being manifested at first by small black spots and later by small pits. Movement of the solution may prevent pitting; stagnation, especially in crevices where the tin touches another solid surface, favours its progress. Contact with a more noble metal such as copper or nickel increases the number and intensity of pits; contact with metals such as aluminium and zinc gives cathodic protection.

As indicated above, the bicarbonate ion inhibits the process, which does not occur, therefore, in many supply waters; attack is most likely in waters which by nature or as a result of treatment have a low bicarbonate content and relatively high chloride, sulphate or nitrate content. The number of points of attack increases with the concentration of aggressive anions and ultimately slow general corrosion may occur. During exposure of 99·75% tin to sea-water for 4 years, a corrosion rate of 0·0023 mm/y was observed[20]. Corrosion in soil usually produces slow general corrosion with the production of crusts of oxides and basic salts; this has no industrial importance but is occasionally of interest in archaeological work.

Corrosion by Acids

Since the high overvoltage restricts hydrogen evolution, corrosion in organic acids or dilute non-oxidising mineral acids is controlled by the rate of supply of oxygen. In solutions of acid open to air, with specimens of size 50 × 20 mm completely immersed, corrosion rates were, in a range of 0·1N organic acids, 400–500 $gm^{-2}d^{-1}$ and, in 0·1N hydrochloric or sulphuric acids, 600 $gm^{-2}d^{-1}$. In absence of oxygen, the rates in the two mineral acids were 100–150 $gm^{-2}d^{-1}$ and in the organic acids, negligible. Phosphoric acid forms a protective layer and, even in the presence of air, the corrosion rate is only about 20 $gm^{-2}d^{-1}$. Nitric acid by reason of its oxidising action attacks tin freely even in the absence of oxygen, but chromic acid forms a protective film. This film, which contains chromic oxide and tin oxides, will, after withdrawal from the acid, give some degree of protection against mildly corrosive conditions; hot solutions of chromic acid, alone or mixed with phosphoric acid, are used as passivating solutions.

The compounds that inhibit acid attack upon steel exert little or no restraint on acid attack upon tin.

Corrosion by Alkalis

The Pourbaix diagram indicates the possibility of attack by solutions of pH values above about 9·5, but the position of this limit is influenced by temperature, by the constitution of the solution, and by the surface condition of the metal. Corrosion will ensue if the surface oxide can be dissolved; this will invariably take place if the pH exceeds 12, and may occur even at pH values between 10 and 12.

Once corrosion begins, its rate is governed by the oxygen supply and temperature and is not greatly affected by the character of the alkali. Rates of attack for specimens completely immersed in still solutions open to air are about 600 $gm^{-2}d^{-1}$ at 30°C and 1 000 $gm^{-2}d^{-1}$ at 70°C. In intermittent immersion such as is experienced in the cleaning of tinned ware by alkaline detergents, however, the rate of corrosion is affected by the nature and concentration of the solution, since these affect the time required for removal of the oxide film at each fresh immersion[21].

Saturated ammonia solutions do not attack tin, possibly because of the negligible oxygen content, but more dilute solutions behave like those of other alkalis of comparable pH.

The removal of oxygen from an alkaline solution, as by the addition of sodium sulphite, can prevent corrosion unless the tin is in contact with another metal, such as steel, from which hydrogen can be evolved. Additions of oxidising agents in small amounts stimulate corrosion but sufficiently large additions produce passivity. Alkaline chromate solutions in the passivating range produce a film containing chromium oxide, which has some protective value[9, 22].

Corrosion by Other Liquid Media

Sulphide solutions, sulphurous acid and some foodstuffs containing organic sulphur compounds, produce stains of sulphide, but the rate of loss of metal is low. Milk and milk products are usually without action, although local corrosion has been known to occur in dairy equipment.

Beer initially dissolves a trace of tin and this may be sufficient to cause a haze in the liquor, but the action is usually checked after a period of contact.

In general, near-neutral aqueous products are without action except for possible sulphide staining or, when there are dissolved salts present, some local corrosion. The slight acidity which may develop in solutions of some organic compounds such as formaldehyde or alcohols can be tolerated.

Many organic liquids, including oils (essential, animal, vegetable or mineral), alcohols, fatty acids, chlorinated hydrocarbons and aliphatic esters, are without action. The absence of any catalytic action of tin on oxidative changes is helpful in this respect. When, however, mineral acidity can arise, as with the chlorinated hydrocarbons containing water, there may be some corrosion, especially at elevated temperature.

Tin as an Anode

Tin when made anodic shows passive behaviour as surface films are built up but slow dissolution of tin may persist in some solutions and transpassive dissolution may occur in strongly alkaline solutions. Some details have been published for phosphoric acid[23] with readily obtained passivity, and sulphuric acid[24] for which activity is more persistent, but most interest has been shown in the effects in alkaline solutions. For galvanostatic polarisation in sodium borate[25] and in sodium carbonate solutions[26] at 1×10^{-6}–50×10^{-6} A/cm^2, simultaneous dissolution of tin as stannite ions and formation of a layer of SnO occurs until a critical potential is reached, at which a different oxide or hydroxide (possibly SnO_2) is formed and dissolution ceases. Finally oxygen is evolved from the passive metal. The nature of the surface films formed in KOH solutions[27] up to 7M and other alkaline solutions[28] has also been examined.

If the applied current density is reduced when a tin anode has been made passive in alkaline solution with the formation of a brown film and evolution of oxygen, the surface film changes to one of yellow colour and dissolution of tin as stannite ions proceeds freely[29]. This effect is exploited in the electrodeposition of tin from sodium or potassium stannate solutions.

It is possible to obtain conditions[30] in which the anodic film continues to grow to form a blue or black layer, and this, although not exceptionally protective, has uses in the treatment of baking pans. A typical anodising solution contains 100 g/l $Na_2HPO_4 \cdot 12H_2O$ and 20 ml/l phosphoric acid, and is used at 350 A/m^2 at 60–90°C for about 10 min.

Solders

By the nature of their use as jointing material, solders are usually presented to a corrosive environment as a small area within a much larger area of another metal. Thus, if the solder is anodic to the metal it joins, and if the corroding medium has good electrical conductivity, damaging corrosion is possible.

Solders are anodic to copper, but soldered joints in copper pipes are widely used without trouble for cold supply waters; possibly corrosion is restricted by the deposition of cathodic carbonate scales and the formation of insoluble lead compounds. Hot supply waters tend to be more aggressive and, where these are involved, it is wise to tin any copper which has a soldered joint. Electrolytes of high conductivity such as sea-water will also attack a soldered joints in copper.

Soldered brass seldom gives trouble. In radiators, antifreeze solutions have been alleged to cause corrosion, possibly because materials such as ethylene glycol sometimes detach protective deposits. Sodium nitrite, valuable as a corrosion inhibitor for other metals in a radiator, tends to attack solders, but sodium benzoate is safe and, in addition, protects the soldered joint against the action of nitrites[31]. In an investigation[32] of other inhibitors in ethylene glycol solutions, 1% borax, either alone or in combination with 0·1% mercaptobenzothiazole, appeared to be satisfactory.

Solders are cathodic to steel, zinc and cadmium, and anodic to Monel metal. Although tin or tin-coated metals may be used safely in contact with aluminium when they are not fused with it, a joint in aluminium made with a tin–lead solder is liable to destructive corrosion. The formation, on fusion, of the grain-boundary state, which, as already mentioned, makes aluminium so dangerous an impurity in tin, is responsible. Tin–zinc solders may be used; the zinc gives a useful degree of protection.

For environments in which tin is less readily corroded than lead, corrosion resistance of the alloy decreases as the lead content increases; the decrease may, in some circumstances, be sharp at a particular composition. In the more corrosive media, such as nitrite solution, a sharp increase of corrosion rate is observed as the lead content increases beyond 30%. In waters with low contents of dissolved salts, the corrosion rate increases slowly with lead content up to about 70% and then rises more steeply, but in the general run of supply waters the ability of lead to form protective insoluble anodic products is helpful to the durability of solder. Selective dissolution of tin has been reported to occur in prolonged contact of solders with solutions of anionic surface active agents[33]. This is consistent with some observations of the behaviour of pure tin in such solutions[34].

Soldered joints, especially those to be used in a static environment, are, if insufficient care is taken, liable to corrosion by residues of flux, which by their nature as oxide removers are potentially corrosive. It is, however, possible to select fluxes which are active when hot but give non-corrosive residues when cold.

If it is necessary to use more vigorous materials, such as zinc chloride, any residues must be rinsed away; owing to the tendency of zinc chloride to yield a screen of basic chloride in a neutral rinse, a preliminary rinse in water acidified with hydrochloric acid is advisable.

Bearing Metals

There are three main types of Babbitt alloy for bearings: high-tin alloys (substantially lead-free), intermediate-tin alloys containing some lead, and high-lead alloys. The high- and intermediate-tin alloys have the resistance of tin to the weak acids and sulphur compounds which may be present in used oils; the lead-rich alloys are rather less resistant[35]. The anti-oxidant action of tin is useful in limiting acidity. Rare instances of corrosion of tin-rich bearings have occurred, with the formation of hard crusts of tin oxide in which the intermetallic compounds of tin with copper and with antimony may remain embedded. Although the causes are not wholly established, some access of water to the bearing is an essential factor and accumulation of breakdown products of constituents of the lubricating oil may contribute[36]. The nobility of intermetallic compounds of tin relative to tin itself has been demonstrated[37].

When free access of salt water to a bearing is possible, the Babbitt alloys are not suitable since they are cathodic to steel shafts. For underwater bearings, alloys with 70% Sn, 1·5% Cu and the balance Zn, are used; the possible dissolution of zinc gives cathodic protection to the shaft, although the more easily replaced bearing suffers some corrosion.

Uses of Tin

Apart from the special uses in solders and bearings metal referred to already, and as coatings, tin and its alloys find employment where advantage can be taken of their physical properties and their fair resistance to tarnish and corrosion in near-neutral environments. Tin pipe is used to condense steam for high-purity distilled water, as a conveyor of beer and soft drinks, especially in coils through cooling media, and, in a larger size, as organ pipes. Some pharmaceutical and food products are packed in tin collapsible tubes and tinfoil coverings are used on cork wads for jar and bottle closures. Pewter is most valued for the decorative forms into which it is easily worked or cast, but it is also used for drinking vessels and dishes.

S. C. BRITTON

REFERENCES

1. Hedges, E. S. *et al.*, *Tin and its Alloys*, Edward Arnold, London (1960)
2. Derge, G., *Trans. Electrochem. Soc.*, **75**, 449 (1939)
3. Derge, G. and Marcus, H., *Trans. Amer. Inst. Min. (Metall.) Engrs.*, **143**, 198 (1941)
4. Deltombe, E., de Zoubov, N. and Pourbaix, M., Technical Report No. 25, Centre Belge d'Étude de la Corrosion (1955)
5. Salzburg, H. W. and Mies, F., *J. Electrochem. Soc.*, **105**, 64 (1958)
6. Quintin, M. and Hagymas, G., *J. Chim. Physique*, 541 (1964)
7. Ross, T. K. and Firoiu, C., *Electrochim. Acta.*, **8**, 877 (1963)
8. Hoar, T. P., *Trans. Faraday Soc.*, **30**, 472 (1934)
8a. Willey, A. R., *Br. Corros. J.*, **7**, 29 (1972)
9. Britton, S. C. and Bright, K., *Metallurgia, Manchr.*, **56**, 163 (1957)
10. Trillat, J. J., Tertian, L. and Britton, S. C., *Métaux-Corros.-Ind.*, **32**, 475 (1957)
11. Boggs, W. E., Kachik, R. H. and Pellissier, G. E., *J. Electrochem. Soc.*, **108**, 6 (1961)
12. Boggs, W. E., Trozzo, P. S. and Pellissier, G. E., *J. Electrochem. Soc.*, **108**, 13 (1961)
13. Boggs, W. E., *J. Electrochem. Soc.*, **108**, 124 (1961)
14. Boggs, W. E., Kachik, R. H. and Pellissier, G. E., *J. Electrochem. Soc.*, **110**, 4 (1963)
14a. Britton, S. C. and Sherlock, J. C., *Br. Corros. J.*, **9**, 96 (1974)
15. Kenworthy, L., *Trans. Faraday Soc.*, **31**, 1331 (1935)
16. Kenworthy, L. and Waldram, J. M., *J. Inst. Met.*, **55**, 247 (1934)
17. Britton, S. C. and Clarke, M., *Trans. I.M.F.*, **42**, 195 (1964)
18. Hiers, G. O. and Minarcik, E. J., Symposium on the Atmospheric Corrosion of Nonferrous Metals, Spec. Tech. Pub. No. 175, Amer. Soc. Test. Mat., 135 (1956)
19. Hoar, T. P., *Trans. Faraday Soc.*, **33**, 1152 (1937)
20. Friend, J. N., *J. Inst. Met.*, **39**, 111 (1928)
21. Britton, S. C. and Michael, D. G., *J. Appl. Chem.*, **5**, 402 (1955)
22. Britton, S. C. and Angles, R. M., *J. Appl. Chem.*, **4**, 351 (1954)
23. Ragheb, A. and Kamel, L. A., *Corrosion*, **18**, 153t (1962)
24. Machu, W., Azzam, A. R. and Halashi, G. M., *Metallöberflache*, **9A**, 53 and 73 (1955)
25. Shah, S. N. and Davies, D. E., *Electrochim. Acta.*, **8**, 663 (1963)
26. Davies, D. E. and Shah, S. N., *Electrochim. Acta.*, **8**, 703 (1963)
27. Hampson, N. A. and Spencer, N. E., *Br. Corros. J.*, **3**, 1 (1968)
28. El Wakkad, S. E. S., El Din, A. M. and Sayed, J. K., *J. Chem. Soc.*, 3103 (1954); El Din, A. M. and El Wahab, F. M. A., *Electrochim. Acta.*, **9**, 883 (1964); Barbulescu, F. and Roller, L., *Rev. Roumaine Chim.*, **10**, 491 (1965)
29. Kerr, R., *J. Soc. Chem. Ind.*, **57**, 405 (1938); Bianchi, G., *Chim. e Ind.*, **29**, 295 (1947)
30. Kerr, R. and Macnaughtan, D. J., *J. Electrodep. Tech. Soc.*, **12**, 19 (1937)
31. Wormwell, F. and Mercer, A. D., *J. Appl. Chem.*, **3**, 22 (1953)
32. Levy, M., *Ind. Engng. Chem.*, **51**, 209 (1959)
33. Watts, C., *Engineering Materials and Design*, **4**, 740 (1961)
34. Ross, T. K., *J. Appl. Chem.*, **5**, 10 (1955); Ross, T. K. and Harris, W., *J. Appl. Chem.*, **10**, 24 (1960)
35. Symposium on Lead-base Babbitt Alloys, *Metal Progr.*, **69**, 174 (1956)
36. Bryce, J. B. and Roehner, T. G., *Trans. Inst. Mar. Engrs.*, **73**, 377 (1961)
37. Clarke, M. and Britton, S. C., *Corrosion Sci.*, **3**, 207 (1963)

4.7 Zinc and Zinc Alloys

The excellent resistance of zinc to corrosion under natural conditions is largely responsible for the many and varied applications of the metal. In fact nearly half the world consumption of zinc is in the form of coatings for the prevention of corrosion of steel fabrications exposed to the atmosphere and to water. For its varied applications zinc is obtainable in a number of grades. Ordinary commercial (G.O.B.) zinc contains up to about 1·5% total of lead, cadmium and iron. Electrolytic zinc has a minimum zinc content of 99·95% and contains small amounts of the same impurities. Special high-purity zinc has a minimum of 99·99% zinc. Even purer zincs are commercially available.

The special high-purity zinc (99·99%) is used mainly for the production of diecasting alloys containing 4% aluminium and 0·04% magnesium and sometimes 1% copper, as shown in Table 4.30, which gives the composition of the two alloys laid down by BS 1004: 1972, and of some newer zinc alloys.

Physical and Mechanical Properties

The bluish-white form of the unalloyed metal is familiar. It is moderately hard and quite malleable at normal temperatures. The tensile strength and impact resistance of the unalloyed metal are low, so zinc is not to be regarded as a structural metal.

The physical properties of zinc and the diecasting alloys are given in Table 4.31.

The good mechanical properties of the diecasting alloys depend on the fine-grain structure resulting from rapid drilling in the water-cooled die. The impact strength of the pressure-diecast alloys at normal temperatures is higher than that of cast aluminium alloys and grey cast iron, and although cooling below normal temperatures produces a sudden drop in the impact strength, this is still within the range given by many other cast metals at ordinary temperatures. (Variation of mechanical properties of the alloys with temperatures is given in Table 4.32). Other factors, such as the design of castings, also influence mechanical properties.

Its malleability and ductility make it possible to produce zinc in sheet, strip and plate form by rolling and as rod and wire by extrusion; continuous casting is also used. Rolled zinc in sheet and strip form is a well-established

Table 4.30 Composition of ingot zinc alloys

Alloy	*Alloying elements* (%)					*Impurities* % (max.)*					Zn
	Al	Cu	Mg	Ti	Other	Fe	Cu	Pb	Cd	Sn	
BS 1004A	3·9–4·3	—	0·04–0·06	—	—	0·075	0·03	0·003	0·003	0·001	Balance
BS 1004B	3·9–4·3	0·75–1·25	0·04–0·06	—	—	0·075	—	0·003	0·003	0·001	Balance
No. 2†	3·9–4·3	2·7–3·5	0·03–0·06	—	—	0·03	—	0·003	0·003	0·001	Balance
No. 7†	3·5–4·3	—	0·005–0·020	—	0·005–0·020Ni	0·075	0·25	0·003	0·002	0·001	Balance
Ilzro 12†	11·0–13.0	0·5–1·25	0·01–0·03	—	—	0·075	—	0·003	0·003	0·001	Balance
Ilzro 14†	0·01–0·03	1·0–1·5	—	0·25–0·30	—	0·075	—	0·003	0·003	0·001	Balance
Ilzro 16†	0·01–0·04	1·0–1·5	—	0·15–0·25	0·10–0·20Cr 0·30–0·40Ti + Cr	0·075	—	0·003	0·003	0·001	Balance
Alzen 305†	30	5	—	—	—	0·03	—	0·003	0·003	0·001	Balance

*The limits for impurities are not all formally specified, but have been taken as for BS 1004 alloys to indicate the need for using special high-purity zinc in making zinc casting alloys even when the aluminium content is very low. In the latter case, the precaution is necessary to avoid dangerous contamination of the BS 1004 type alloys if mixing should occur.

†Ilzro 12 is primarily an alloy for gravity casting in metallic or non-metallic moulds. Ilzro 14 and 16 are for pressure-die-casting components which need better creep resistance than the BS 1004 alloys provide; No. 16 is recommended for high-temperature applications. Alloy No. 2 can be pressure die cast and is sometimes chosen for sliding components because of its hardness. It is also used for casting sheet-metal-forming dies and plastic moulds, but cannot be employed for moulding chlorine compounds (e.g. p.v.c.), some rubbers or polystyrene expanded with steam, because of its corrosion under such conditions. Alloy No. 7 is for pressure die casting and is claimed to display improved fluidity because of the reduced magnesium content. Alzen 305 is a bearing alloy used mainly as gravity cast shells, sand castings and continuously cast or extruded rod. It should not be used in contact with salt or alkaline solutions nor at temperatures above 120°C.

Table 4.31 Physical properties of zinc and cast zinc alloys

Material	*Atomic weight*	*Specific gravity* (g/cm^3)	*Melting point* (°C)	*Electrical conductivity* (Cu = 100)	*Thermal conductivity* (Cu = 100)	*Coefficient of expansion* per °C
Zinc	65·39	7·14	419·4	28·3	29·3	$3{\cdot}95 \times 10^{-5}$
Alloy A	—	6·7	382–387	26·6	30	$2{\cdot}7 \times 10^{-5}$
Alloy B	—	6·7	379–388	25·9	29	$2{\cdot}7 \times 10^{-5}$
Alzen 305	—	4·8	480–580	31	28	$2{\cdot}2 \times 10^{-6}$

Table 4.32 Variation with temperature of mechanical properties of diecast test bars of alloys *A* and *B*, BS 1004

Temperature (°C)	*Tensile strength* (N/mm^2)		*Elongation % on*		*Impact strength*	
	A	*B*	*A*	*B*	*A*	*B*
95	200	240	30	23	58	—
40	250	290	16	13	57	62
20	280	350	11	8	57	60
10	—	—	—	—	42	56
0	295	370	9	8	10	53
−10	—	—	—	—	5	24
−20	—	—	—	—	3·5	5
−40	320	370	4·5	3	3	3·5

roofing material, particularly in parts of France and northern Europe. Diecastings can be made readily on account of the low melting points and the good flow properties of the diecasting alloys. The other major uses are in brass manufacture, coatings for steel and in sacrificial anodes.

Corrosion Properties

Both zinc and zinc alloys have excellent resistance to corrosion in the atmosphere and in most natural waters. The property which gives zinc this valuable corrosion resistance is its ability to form a protective layer consisting of zinc oxide and hydroxide, or of various basic salts, depending on the nature of the environment. When the protective layers have formed and completely cover the surface of the metal, the corrosion proceeds at a greatly reduced rate.

Alloys containing up to 1% copper and about 0·1% titanium have been used in strip form for roofing on a large scale in France, Germany, the UK and elsewhere since about 1960, but no details are available on how their corrosion resistance compares with that of the traditional unalloyed rolled zinc. Initially they are lighter in colour and they remain bright a little longer.

Composition of the Protective Films

In dry air, a film of zinc oxide is initially formed by the influence of the atmospheric oxygen, but this is soon converted to zinc hydroxide, basic zinc carbonate and other basic zinc salts by water, carbon dioxide and chemical impurities present in the atmosphere.

Below about 200°C, the film grows very slowly and is invisible and very adherent. It is often said that the first layer formed governs the corrosion resistance of the zinc throughout its life. If the film becomes too thick, it is liable to break away or become porous, when, of course, it ceases to provide protection. Moreover, zinc oxide occupies a larger volume than the zinc from which it originated and, as the layer thickens, strains are set up which lead to the production of fissures.

Vernon[1] claims that in outdoor atmospheres the corrosion product consists largely of zinc oxide, hydroxide and combined water, but also contains zinc sulphide, zinc sulphate and carbonate. The following table[2] gives the composition of typical films formed in an industrial atmosphere.

Table 4.33 Composition of films formed in industrial atmosphere

Compound	%
ZnO	37·0
Included water	20·0
Sulphates:	
$ZnSO_4$	5·2
$PbSO_4$	9·0
Na_2SO_4	1·8
$ZnCl_2$	0·2
Carbonates:	
$CaCO_3$	1·1
$MgCO_3$	0·4
$ZnCO_3$	17·4
Fe_2O_3	2·5
Sand	4·5

The formation of the protective layers is governed largely by the pH of the environment. Since zinc forms an amphoteric oxide, both acid and alkaline conditions adversely affect its corrosion behaviour by interfering with the formation of the protective layers. Figure 4.44 shows how the corrosion rate of zinc varies with the pH[3], and it will be seen from this that the attack is most severe at pH values below 6 and above 12·5, while within this range the corrosion is very slow.

The electrochemical properties of zinc also have a large bearing on its corrosion behaviour. Zinc* is negative to $E^{\ominus}_{H^+/H_2}$ and, magnesium and aluminium excepted, to most other metals commonly encountered, including those found in the less pure forms of zinc. This means that when zinc is in contact with these metals sacrificial electrochemical action can take place, with zinc forming the anode. Contact with other metals and impurities can

* $E^{\ominus}_{Zn^{2+}/Zn} = -0{\cdot}76$ V and the corrosion potential of Zn approximates to this value in sea water, but is more positive than it in fresh waters, particularly at elevated temperatures.

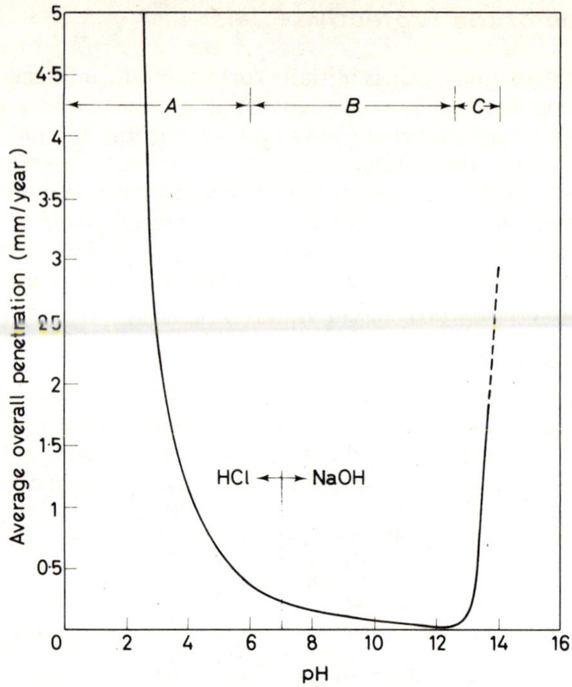

Fig. 4.44 Effect of pH on the rate of corrosion of zinc. *A*, rapid corrosion; *B*, stable film—low corrosion rate; *C*, rapid corrosion (after Roetheli, *et al.*[3])

therefore cause zinc to corrode rapidly, and it is important to avoid intermetallic contacts with metals such as copper, tin or nickel, under immersion in water. On the other hand the overvoltage for hydrogen evolution on zinc is high, and this can, in some circumstances, lead to a reduction in the corrosion rate, although under normal atmospheric conditions oxygen is generally present in sufficient concentration to act as a cathode reactant.

The Corrosion of Zinc in the Atmosphere

In dry air the stability of zinc is remarkable. Once the protective layer of zinc oxide formed initially is complete, the attack virtually ceases. Even under normal urban conditions, such as those in London, zinc sheet 0·8 mm thick has been found to have an effective life of 40 years or more when used as a roof covering and no repair has been needed except for mechanical damage. The presence of water does, of course, increase the rate of corrosion; when water is present the initial corrosion product is zinc hydroxide, which is then converted by the action of carbon dioxide to a basic zinc carbonate, probably of composition similar to $ZnCO_3 \cdot 3Zn(OH)_2$[(4)]. In very damp conditions unprotected zinc sometimes forms a loose and more conspicuous form of corrosion product known as 'wet storage stain' or 'white rust' (see p. **4**.156).

Industrial atmospheres usually accelerate the corrosion of zinc. When heavy mists and dews occur in these areas, they are contaminated with considerable amounts of acid substances such as sulphur dioxide, and the film of moisture covering the metal can be quite acid and can have a pH as low as 3. Under these conditions the zinc is dissolved but, as the corrosion proceeds, the pH rises, and when it has reached a sufficiently high level basic salts are once more formed and provide further protection for the metal. These are usually the basic carbonate but may sometimes be a basic sulphate. As soon as the pH of the moisture film falls again, owing to the solution of acid gases, the protective film dissolves and renewed attack on the metal occurs. Hudson and Stanners[5] conducted tests at various locations in order to determine the effect of atmospheric pollution on the rate of corrosion of steel and zinc. Their figures for zinc are given in Table 4.34 and clearly show the effect which industrial contamination has on the corrosion rate.

In areas near the sea coast the rates of corrosion may be increased somewhat by the sea spray containing soluble chlorides, but the rates are still much lower than those prevailing in heavily polluted industrial areas. The white corrosion product which is sometimes found under these conditions probably consists of the basic chloride Zn_2OCl_2[(4)].

Anderson[6], reporting on a series of 20-year exposure tests carried out by the ASTM, quotes the average rates of corrosion for various types of environment which are given in Table 4.35.

Table 4.34 Atmospheric corrosion of zinc in various parts of the UK

Site	*Degree of pollution*	*Standard pollution**	*Corrosion rate* (mm/y)	
			Zinc	Steel
Godalming	Slight	0·23	0·00107	0·00425
Teddington		0·82	0·00211	0·00633
Hornchurch	Moderate	0·92	0·00324	0·00627
London (Victoria Park)		1·28	0·00445	0·00756
Barking (Greatfield Park)	Heavy	1·60	0·00483	0·00750
Salford (Ladywell)		2·31	0·00843	0·0105
Sheffield (Hunshelf Bank)	Severe	3·58	0·0129	0·0107
Billingham (Council Offices)		5·24	0·0125	0·0180

*Calculated on weight (mg) of SO_3 absorbed per day by 1 dm^2 of lead dioxide.

Table 4.35 Further examples of atmospheric corrosion rates for zinc[6]

Atmosphere	*Corrosion rate* (mm/y)
Industrial	0·0064
Sea coast	0·0015
Rural	0·0011
Arid	0·00018

These results clearly demonstrate (*a*) the effect of moisture in increasing the corrosion rate six times in the rural areas compared with the arid areas and (*b*) the considerable increase of attack in the industrial areas. Anderson considers the principal features which control the rate of corrosion of zinc are (*a*) the frequency of rainfall and dew fall, (*b*) the acidity of the atmosphere and (*c*) the rate of drying. In situations where the drying of zinc is retarded corrosion is found to be most severe, for example the bottom strand of a galvanised wire fence which is shielded from the sun by grass and weeds.

Ambler[7] has attempted to find a relationship between the corrosion of zinc and iron and atmospheric salinity in the UK. This followed previous tests in Nigeria[8], when it was concluded that the governing factor in the corrosion of steel and zinc was airborne salt and that there was a relationship between corrosion and the distance from the sea. In the UK, however, no such relationship was found to exist, and the governing factor in the corrosion of zinc in the atmosphere is confirmed to be the amount of industrial pollution.

White rust If a fresh zinc surface is allowed to stand with large drops of dew on it, as may easily happen if it is stored in a closed place in which the temperature varies periodically, it is attacked by the oxygen dissolved in the water, owing to differential aeration between the edges and the centres of the drops. A porous form of zinc oxide builds up away from the surface and quickly takes up carbon dioxide from the air to form the basic carbonate known as *white rust* or *wet storage stain*.

This type of corrosion can take place on any new surface of zinc and is best prevented by storing the metal in a dry, airy place until a protective layer has been formed. Zinc which has been properly aged in this way is safe against white-rust formation. Various methods are employed to prevent white rust. A chromate treatment is widely used for zinc-plated articles and for galvanised sheet, and occasionally for zinc die castings. Fatty substances, such as oils or lanolin, are sometimes used to protect larger items.

Corrosion of Zinc in Water

When zinc is immersed in distilled water containing dissolved oxygen, a protective film, probably consisting of zinc hydroxide, is slowly formed, and this extends over nearly the whole surface. At certain points the metal seems to remain uncovered and local attack continues, resulting in pitting. This may be quite serious and can lead to the rapid penetration of a zinc sheet. The attack does not spread over the whole surface of the zinc but is confined to these small local areas, although similar pits are liable to appear at points in a vertical line directly below the seat of the original corrosion. It has been suggested that the areas attacked occur where particles of corrosion product have fallen and are resting on the surface, shielding it from oxygen. The point thus shielded becomes anodic and suffers corrosion. If oxygen replenishment is made uniform over the whole surface, e.g. by 'whirling'[9] a zinc plate, pitting is avoided and the protective film of zinc hydroxide is found over the entire surface.

Natural waters all contain dissolved salts to a certain extent and these tend to form a scale on the surface of the metal. Since such scales have a protective effect it is to be expected that corrosion will be less severe in hard water. This is in fact true, and it is often found that distilled water is more corrosive than natural waters.

The effect of pH on the corrosion of zinc has already been mentioned (p. **4**.154). In the range of pH values from 6 to 12·5, zinc is quite stable, and since most natural waters come within this range little difficulty is encountered in respect of pH. The pH does, however, affect the scale-forming properties of hard water (see Section 2.3 for a discussion of the Langelier index). If the pH is below the value at which the water is in equilibrium with calcium carbonate, the calcium carbonate will tend to dissolve rather than form a scale. The same effect is produced in the presence of considerable amounts of carbon dioxide, which also favours the dissolution of calcium carbonate. In addition, it is important to note that small amounts of metallic impurities (particularly copper) in the water can cause quite severe corrosion, and as little as 0·05 p.p.m. of copper in a domestic water system can be a source of considerable trouble with galvanised tanks and pipes.

Sea-water contains considerable amounts of soluble salts, particularly sodium chloride, which is present in concentrations from 1 to 25%. The North Sea, for example, contains about 3% sodium chloride, 0·47% magnesium sulphate, 0·2% magnesium chloride and 0·1% calcium chloride. The carbon dioxide content is about 0·005 to 0·01% and the pH is between 7·6 and 8·1. The high chloride content would tend to increase the rate of corrosion, and this usually takes the form of pitting under these conditions. The corrosive influence of the chloride ions is, however, inhibited by the presence of magnesium and calcium ions by virtue of the formation of a protective layer of magnesium and calcium salts (calcareous scale).

The effect of the magnesium salts has been clearly demonstrated by Schikorr[10]. Zinc immersed in a solution containing 30 g/l of sodium chloride showed a weight loss of 198 g/m^2 after 14 days. When the solution contained in addition 12 g/l of magnesium chloride, the loss in weight was only 4 g/m^2 after the same period. Artificial sea-water was also tested and gave a weight loss of 5 g/m^2.

Zinc roofs are quite satisfactory at the coast, where they receive a large amount of salt-water spray, and many British piers have been covered with sheet zinc which has lasted 50 years and more. Most of the zinc actually immersed in sea-water is in the form of zinc coatings, the behaviour of which is discussed in a later section of this book (see Section 14.4). Experience with these coatings has proved the superiority of zinc in sea-water over most other metals in this environment[11].

The effect of temperature on the corrosion of zinc in water It is found that the temperature has a marked effect on the rate at which zinc corrodes in water. The corrosion rate in distilled water reaches a maximum in the temperature range 65–75°C. This variation in the corrosion rate with temperature is attributed to changes in the nature of the protective film. At lower temperatures the film is found to be very adherent and gelatinous, while at temperatures around 70°C it becomes distinctly granular in character and much less adherent. Above 75°C it again tends to become more adherent and assumes a very compact and dense form. It is believed that the granular

Table 4.36 Dependence of corrosion rate of zinc on temperature

Temperature (°C)	*Corrosion rate* (mm/y)
20	0·020
50	0·071
55	0·38
65	3·05
75	2·34
95	0·31
100	0·12

coating formed at temperatures around 70°C is more porous than the others and permits greater access of the dissolved oxygen to the metal.

Experiments in which rolled high-grade zinc was immersed in distilled water for 15 days at various temperatures are described by Cox[12]. The figures which he obtained are given in Table 4.36.

Soil Corrosion

The factors influencing the corrosion of metals in soil are more numerous than those prevailing in air or water, and the electrochemical effects are more pronounced. Moreover, soils vary widely in their composition and behaviour even over very short distances. It is difficult therefore to obtain reliable data. It is evident, however, that zinc has considerable resistance to corrosion when buried, and the greatest attack is caused by soils which are acid or contain large amounts of soluble salts.

In tests carried out by the National Bureau of Standards in the USA[13] specimens of copper alloys, lead, zinc and zinc alloys were buried at a number of different sites for periods varying from 11 to 14 years. The soils tested covered a pH range from 2·6 to 9·4 and resistivities ranged from 62 to 17 800 Ω cm. The weight losses and maximum depths of pitting were recorded, and the results indicated that the most severe corrosion occurred in soils of poor aeration having high acid and soluble-salt contents.

Although there was no significant difference in weight loss between rolled zinc and zinc alloys, the maximum penetration of the alloys definitely appeared to be greater than that of rolled zinc.

BISRA tests on galvanised steel pipe buried for five years at five different sites are described by Hudson and Acock[14]. The galvanised pipes resisted corrosion rather better than steel at all sites. Galvanised pipes of small diameter are frequently used to provide underground water services in farms and similar establishments, and little trouble is experienced.

Corrosion Resistance of Zinc in Chemical Environments

Acids and alkalis Zinc dissolves in liquids whose pH is below about 5 and above 12·5. Table 4.37 shows the approximate pH value of 0·01 and 0·1N solutions of some common acids and alkalis, from which it is seen that all of the acids included would attack zinc, while only the stronger solutions of most of the alkalis would cause any corrosion. Zinc dissolves in caustic soda solution to form sodium zincate, i.e. Na_2ZnO_2.

Table 4.37 pH values of common acids and alkalis

Acids	*Concentration*	pH
Hydrochloric	0·1N	1·1
Hydrochloric	0·01N	2·0
Sulphuric	0·1N	1·2
Sulphuric	0·01N	2·1
Oxalic	0·1N	1·6
Formic	0·1N	2·3
Acetic	0·1N	2·9
Acetic	0·01N	3·4
Bases		
Sodium hydroxide	0·1N	13·0
Sodium hydroxide	0·01N	12·0
Potassium hydroxide	0·1N	13·0
Potassium hydroxide	0·01N	12·0
Sodium metasilicate	0·1N	12·6
Trisodium phosphate	0·1N	12·0
Sodium carbonate	0·1N	11·6
Borax	0·1N	9·2
Sodium bicarbonate	0·1N	8·4

Owing to the high overpotential for hydrogen evolution pure zinc is attacked only very slowly by dilute sulphuric and hydrochloric acids, but impure zinc dissolves quite readily, evolving hydrogen. Thus when ordinary commercial zinc is placed in a dilute acid it immediately starts to corrode slowly. As corrosion proceeds the impurities present are reprecipitated as a black metallic sponge which may further accelerate the reaction owing to the increased surface area of low overpotential cathodic sites, e.g. copper. Some metallic impurities, notably aluminium, retard the corrosion of zinc by acids. This is due to the production of highly protective films. Other metals, such as tin and lead, retard the corrosion initially, but when after a few hours they are deposited as a metallic sponge the corrosion is as fast as with ordinary zinc.

Salt solutions When a zinc sheet is immersed in a solution of a salt, such as potassium chloride or potassium sulphate, corrosion usually starts at a number of points on the surface of the metal, probably where there are defects or impurities present. From these it spreads downwards in streams, if the plate is vertical. Corrosion will start at a scratch or abrasion made on the surface but it is observed that it does not necessarily occur at all such places. In the case of potassium chloride (or sodium chloride) the corrosion spreads downwards and outwards to cover a parabolic area. Evans[15] explains this in terms of the dissolution of the protective layer of zinc oxide by zinc chloride to form a basic zinc chloride which remains in solution.

Feitknecht[16] has examined the corrosion products of zinc in sodium chloride solutions in detail. The compound on the inactive areas was found to be mainly zinc oxide. When the concentration of sodium chloride was greater than 0·1M, basic zinc chlorides were found on the corroded parts. At lower concentrations a loose powdery form of a crystalline zinc hydroxide appeared. A close examination of the corroded areas revealed craters which appeared to contain alternate layers and concentric rings of basic chlorides and hydroxides. Two basic zinc chlorides were identified, namely $6Zn(OH)_2 \cdot ZnCl_2$ and $4Zn(OH)_2 \cdot ZnCl_2$. These basic salts, and the crystalline zinc hydroxides, were found to have layer structures similar in general to the layer structure attributed to the basic zinc carbonate which forms dense adherent films and appears to play such an important role in the corrosion resistance of zinc against the atmosphere. The presence of different reaction products in the actual corroded areas leads to the view that, in addition to action between the major anodic and cathodic areas as a whole, there is also a local interaction between smaller anodic and cathodic elements.

Organic chemicals Most organic liquids, other than the stronger acids, only attack zinc slowly. Zinc is, therefore, suitable for storage tanks for liquid hydrocarbons such as motor fuels, for phenols and for trichlorethylene degreasers. Zinc or zinc-coated vessels are not recommended for use in contact with acid foodstuffs, but are regularly used for dry foods. Zinc in small quantities is beneficial in the human diet.

Cathodic Protection by Zinc Anodes

Zinc should give a potential of −1·05 V vs. $Cu/CuSO_4$ and should have a driving potential of about −0·25 V with respect to cathodically protected steel. Zinc is therefore sufficiently negative to act as a sacrificial anode, and its first use for such purposes was on the copper-sheathed hulls of warships more than a century ago. The first attempts to fit zinc anodes to steel hulls, however, were a complete failure, for the sole reason that it had not been realised that the purity of the zinc was of paramount importance. The presence of even small amounts of certain impurities leads to the formation of dense adherent films, which cause the anodes to become inactive.

The major harmful impurity is iron, and by keeping the iron content to less than 15 p.p.m. it became possible to produce perfectly satisfactory anodes of zinc[17,18]. Alternatively the effect of the iron can be neutralised by alloying the zinc with certain metals, among which aluminium and silicon or cadmium have been found to be particularly effective. The presence of cadmium causes the corrosion product to fall away evenly, leaving an active surface (see Section 11.2).

Zinc alloy anodes are generally very efficient, owing to their non-polarising characteristics and to the absence of parasitic reactions when buried or immersed. As zinc's 'self-corrosion' is, therefore, very small, the anode's high efficiency of 85–95% holds throughout the current-density range, whereas the efficiency of magnesium, about 50–55% at high current densities, may fall to only 30% at low current densities. The use of these zinc alloys for reference electrodes to monitor some hull-protection schemes or as permanent reference cells located along pipelines, underlines their non-polarising characteristics. The electrochemical equivalent of zinc indicates that 10·5 kg should give 1 A/year, but in practice this amount of charge is given by 11–12 kg of zinc compared with 7·7–9 kg of magnesium. As the cost of zinc is low, zinc anodes are found to be very economical. Moreover, since zinc is fairly dense, the volume consumption (1·6 litre A^{-1} $year^{-1}$) is much lower than that for magnesium (5 litre A^{-1} $year^{-1}$) and the dimensions of the zinc anode are correspondingly smaller.

The driving potential of zinc to cathodically protected steel is 200–250 mV and this is considerably lower than the 700 mV given by magnesium. While this value is ideal in sea-water or other electrolytes of low resistivity, the use of zinc is not always practicable in environments of higher electrical resistance. For example, it is not likely to be of much use for protecting large underground systems in a high resistivity soil, but on the other hand it proves to be of value for smaller underground units, such as storage tanks, situated in soils of resistivity below about 3 000 Ω cm. Olive[19], for instance, in the USA, discussed the application of zinc anodes for protecting underground equipment at petrol-filling stations. Larger installations which employ a considerable number of zinc anodes are the protective schemes for the steel gas mains in Houston and New Orleans[20]. Of a total of some 1 200 galvanic anodes in New Orleans about 1 000 are of zinc. This is a good example of a case where a system of zinc anodes can be used to protect a large underground installation under the right soil conditions. Zinc is used quite widely for the protection of bare service pipes of small diameter, and is receiving increased acceptance for the protection of large-diameter coated pipes in built-up areas in order to minimise interference on adjacent mains. It is also used for protecting galvanised cold-water tanks.

Where the use of zinc anodes is practicable, the low driving potential is a great advantage since the resistance of the steel to be cathodically protected is the controlling resistance, and the current output of the anode varies with the requirements of the cathode. Thus it can be said that zinc anodes are largely self-governing.

By far the largest use of zinc anodes is in sea-water for the protection of ships' hulls and North Sea pipelines and drilling rigs. The high conductivity of the sea-water and the excellent natural resistance of zinc to corrosion make it very effective in this application. Many examples of trials with zinc anodes on ships can be quoted, and much work has been done in this respect by the US Navy. One example is a paper by Carson[16], describing zinc anodes attached directly to the hull and found to have a life of 8 to 10 years. This paper also gives the results of tests carried out at the Pacific Naval Laboratory, showing the effect of various impurities in the zinc. Carson looks into the economics of zinc alloy anodes and concludes that for small and medium hull sizes, e.g. 1 400 m^2 wetted hull area, zinc alloy anodes are more economical than the most competitive alternative systems available and have the additional advantages of more even current distribution, minimum risk of paint-stripping, and maintenance-free long life. For larger hull areas zinc alloy anodes are at least as economic as many alternative systems.

Zinc anodes are made in a variety of shapes and sizes, ranging in weight from 2·25 to 110 kg and in shape from cylindrical rods to rectangular bars.

When used underground they are usually placed in a backfill, consisting of gypsum, sodium sulphate and clay, which may be added loose, shipped in a bag around the anode, or obtained in cast form.

A. R. L. CHIVERS

REFERENCES

1. Vernon, W. H. J., Second Experimental Report to the Atmospheric Corrosion Committee, B.N.F.M.R.A., *Trans. Faraday Soc.*, **22**, 113 (1927)
2. Bablik, H., *Galvanising Hot Dip*, Spon, London, 333 (1950)
3. Roetheli, B. E., Cox, G. L. and Littreal, W. B., *Metals and Alloys*, **3**, 73 (1932)
4. Morriset, P., *Zinc et Alliages*, **20**, 15 (1959)
5. Hudson, J. C. and Stanners, J. F., *J. Appl. Chem.*, **3**, 86 (1953)
6. Anderson, E. A., Amer. Soc. Test. Mater. Special Publication No. 175, June (1955)
7. Ambler, H. R., *J. Appl. Chem.*, **10**, 213 (1960)
8. Ambler, H. R. and Bain, A. A., *J. Appl. Chem.*, **5**, 437 (1955)
9. Evans, U. R. and Davies, D. E., *J. Chem. Soc.*, 2607 (1951)
10. Schikorr, G., *Z. Metallk.*, **32**, 314 (1940)
11. Hudson, J. C. and Banfield, T. A., *J. Iron St. Inst.*, **154**, 229 (1946)
12. Cox, G. L., *Ind. Engng. Chem.*, **23**, 902 (1931)
13. Denison, I. A. and Romanoff, M., *J. Res. Nat. Bur. Stand.*, **44**, 259 (1950)
14. Hudson, J. C. and Acock, G. P., *Tests on the Corrosion of Buried Iron and Steel Pipes*, Iron and Steel Inst. Special Report No. 45 (1953)
15. Evans, U. R., *The Corrosion and Oxidation of Metals*, Edward Arnold, London (1960)
16. Feitknecht, W., *Chem. and Ind. (Rev.)*, 1102 (1959)
17. Crennell, J. J. and Wheeler, W. C. G., *J. Appl. Chem.*, **8**, 571 (1958); **6**, 415 (1956)
18. Carson, J. A. H., *Corrosion*, **16**, 99 (1960)
19. Olive, M. J., *Corrosion*, **16**, 9 (1960)
20. Trouard, S. E., *Corrosion*, **13**, 21 (1957)

BIBLIOGRAPHY

Much detailed information on the corrosion behaviour of zinc surfaces is to be found in the following books:

Metal Corrosion in the Atmosphere—a Report on a Symposium Held in June, 1967, American Society for Testing and Materials, STP 435, Philadelphia (1968)

Schikorr, G., *Atmospheric Corrosion Resistance of Zinc*, English edition, Zinc Development Association, London and Zinc Institute, New York (1965)

Wiederholt, W., Verhalten von Zinc in Wässern, Metall-Verlag GmbH, Berlin (1965)

Slunder, C. J. and Boyd, W. K., *Zinc: It's Corrosion Resistance*, Zinc Development Association, London (1971)

5 RARER METALS

5.1 Beryllium

Introduction

As a light, strong metal, beryllium holds considerable promise as a useful engineering material, but because of an inherent directional brittleness, a really significant commercial use, e.g. in the aircraft industry, has not proved possible. It has been used to a limited extent in aerospace applications, and it was employed as heat shields for the Project Mercury space capsule. It has also found use in precision guidance systems when fairly pure environmental conditions can be assured.

In common with magnesium and zirconium the metal has little tendency to capture neutrons, and there was promise that a significant industrial use would arise in nuclear engineering, but after extensive trial in gas cooled reactors, its ultimate commercial employment in that context was also deemed inappropriate.

For many years, however, beryllium has been used as a minor alloying addition to other metals, particularly copper, and although its main attribute lies in improving mechanical properties, it generally also improves the corrosion resistance of the parent metal. For example, in the case of copper, beryllium addition has been found to increase oxidation resistance[1], as well as the wet corrosion resistance, particularly in corrosion fatigue[2] where the strengthening effect is advantageous. By ensuring conditions under which preferential oxidation of the beryllium constituent occurs—the principle of 'selective oxidation'[3]—the oxidation resistance of copper-beryllium can be much improved relative to a normal surface. A similar effect can be produced on silver, where deposition of beryllium oxide, e.g. by cathodic deposition from ammoniacal beryllium sulphate or nitrate, effects very considerable improvement in the resistance to tarnishing. Beryllium has been used for the purpose of improving oxidation resistance in magnesium, to produce the Magnox series of alloys used for fuel sheathing in carbon-dioxide-cooled reactors[4]. The particular composition used in the Calder Hall reactor, for example, contains about 0·01% beryllium and 0·8% aluminium. In accord with the improved resistance to oxidation, the incendivity of Magnox alloys is also reduced relative to that of pure magnesium. This had been found with magnesium some years earlier[5] when as little as 0·001% beryllium raised the air ignition temperature of a magnesium–aluminium–zinc alloy from 580°C to over 800°C.

Metallurgy

General

Beryllium is a light metal (s.g. 1·85) with a hexagonal close-packed structure (axial ratio 1·568). The most notable of its mechanical properties is its low ductility at room temperature. Deformation at room temperature is restricted to slip on the basal plane, which takes place only to a very limited extent. Consequently, at room temperature beryllium is by normal standards a brittle metal, exhibiting only about 2 to 4% tensile elongation. Mechanical deformation increases this by the development of preferred orientation, but only in the direction of working and at the expense of ductility in other directions. Ductility also increases very markedly at temperatures above about 300°C with alternative slip on the $10\bar{1}0$ prismatic planes. In consequence, all mechanical working of beryllium is carried out at elevated temperatures. It has not yet been resolved whether the brittleness of beryllium is fundamental or results from small amounts of impurities. Beryllium is a very poor solvent for other metals and, to date, it has not been possible to overcome the brittleness problem by alloying.

As already indicated, the brittleness of beryllium has so far been the main determining feature in the technology, and because of the mechanical anisotropy, the most widely practised method of fabrication is via powder metallurgy.

Two other factors are noteworthy: the deleterious effects on chemical and mechanical properties of small amounts of impurities residual from extraction of the metal, and its toxicity. The first of these factors is obviated by vacuum melting the raw metal (for purification) as an essential prerequisite to further processing. The toxicity of beryllium is essentially a pulmonary problem and great care must be taken in handling the finely divided metal or its compounds. In practice, this type of activity is usually carried out under well-ventilated conditions. Certain tolerance levels for atmospheric beryllium are now internationally accepted[6] and merit careful study before work on beryllium is embarked upon.

Extraction and Fabrication

Beryllium is extracted from the main source mineral, the alumino-silicate *beryl*, by conversion to the hydroxide and then through either the fluoride or the chloride to the final metal. If the fluoride is used, it is reduced to beryllium by magnesium by a Kroll-type reaction. The raw metal takes the form of 'pebble' and contains much residual halides and magnesium. With the chloride on the other hand, the pure metal is extracted by electrolysis of a mixture of fused beryllium chloride and sodium chloride. The raw beryllium is now dendritic in character, but still contains residual chloride.

Before further processing, the raw metal must be purified. Various methods of leaching have been tried on a laboratory scale, but in practice the method usually adopted is vacuum induction melting. The ingot so produced is then converted to powder by swarfing, followed by grinding. On a laboratory scale the grinding is generally done by ball milling, while in production, milling is carried out between beryllium-faced plates.

The particle size of powder most often used for consolidation is −200 mesh (74μm sieve aperture), and the most widely practised method of consolidation is hot pressing *in vacuo*.

Setting aside the necessity for hot working and the toxicity problem, the fabrication of consolidated beryllium generally follows normal lines; rolling, extrusion, drawing, forging, etc. have all been successfully carried out. It is interesting to note that because of the high chemical activity of beryllium, allied to the method of consolidation from powder, the usual grade of metal contains about 1 to 2% of beryllium oxide. It could therefore be considered almost a mild cermet rather than a conventional pure metal. Specifications for chemical composition are detailed in Table 5.1.

Table 5.1 Chemical composition of various forms of beryllium

Form	*Analysis*								*Notes*
	Be *Assay* % min.	BeO % max.	Fe % max.	Al % max.	C % max.	Mg % max.	Si % max.	Ni % max.	
Pebble	96·0	—	0·15	0·14	—	—	0·1	0·04	Remainder slag
Ingot	99·0	1·0	0·15	0·14	0·14	0·08	0·1	0·04	
Powder	98·0	2·0	0·18	0·16	0·12	0·08	0·12	—	Any other metallic impurity 0·04% max.

Corrosion Behaviour

Aqueous Environments

In its general corrosion behaviour, beryllium exhibits characteristics very similar to those of aluminium. Like aluminium, the film-free metal is highly active and readily attacked in many environments. Beryllium oxide, however, like alumina, is a very stable compound (standard free energy of formation = −579 kJ/mol), with a bulk density of 3·025 g/cm^3 as compared with 1·85 g/cm^3 for the pure metal, and with a high electronic resistivity of about 10^{10} Ω cm at 0°C. In fact, when formed, the oxide confers the same type of 'spurious nobility' on beryllium as is found, for example, with aluminium, titanium and zirconium.

It is often difficult to correlate or categorise the corrosion behaviour of beryllium because of past inconsistencies in the quality of metal studied and differences between the chemical composition and metallurgical history of current material and of past grades. For example, although vacuum melting now produces a fairly consistent grade of beryllium (often designated QMV), this method of purification was not introduced until the late 1940s. Prior to this, powder was produced by grinding 'pebble' without purification and hence small amounts of residual halide slag frequently persisted through to the fabricated product.

Furthermore, even though a consistent quality of beryllium is now produced, the chemical composition falls far short of the standards found for instance in aluminium; generally, the main impurities consist of about 1% of beryllia at grain boundaries, about 0·15% of iron and 0·05–0·1% of other elements such as silicon, aluminium and carbon.

Beryllium is readily attacked by most acids and, being amphoteric, is slowly attacked by caustic alkalis with the evolution of hydrogen. As might be anticipated, in view of the controlling influence of the surface film of beryllia on corrosion behaviour, concentrated nitric acid has little effect on beryllium[7], while the dilute acid results in slow attack. Hot acid is much more reactive. Nitric acid is in fact often used to pickle-off residual mild steel from hot-extruded clad beryllium.

Anything which will affect the characteristics of the surface film of beryllia can be expected to affect the corrosion resistance of the metal. For example, dissolved fluoride or chloride ions give rise to increased pitting attack, analogous to the attack on aluminium in similar circumstances. At beryllium carbide inclusions in the metal, hydrated oxide or beryllium hydroxide is formed in moist air by hydrolysis of the carbide[8]; depending on the size of the inclusion, it would be expected that pitting of the underlying metal might ensue. The presence of cations which will deposit heavy metals and set up local cathodic areas has a pronounced adverse effect on corrosion resistance. Cupric ions in particular, at a concentration of less than 1 p.p.m., have been found to cause serious pitting in 0·005M hydrogen peroxide aqueous solution at 85°C; ferric ions also appear to increase the rate of corrosion, although apparently not as seriously as copper.

Conversely to the above, any factor which tends to maintain the protective character of the beryllia film will obviously increase corrosion resistance, and, in this respect, the presence of anodic inhibitors such as sodium dichromate, up to about 40 p.p.m., will effectively suppress pitting of beryllium in water.

Most of the controlled corrosion studies on beryllium have been carried out in the USA in simulated reactor coolants. The latter have usually been water, aerated and de-aerated, containing small amounts of hydrogen peroxide and at temperatures up to 300–350°C. Many variables have been examined, covering surface condition, chemical composition, temperature, pH, galvanic effects and mechanical stress[8].

Surface condition Machined, abraded and pickled surfaces all exhibit much the same behaviour in water, and after exposure of up to about one year at temperatures less than 100°C average attack measures 0·0025–0·0050 mm/y. Almost always, however, corrosion of beryllium in water is accompanied by pitting and, on machined surfaces, pits of as much as 0·25 mm have been observed in 0·0005M hydrogen peroxide at 85°C. Under similar conditions, annealed material has been found to be somewhat less resistant to attack than either machined or pickled surfaces.

Chemical composition The effect of carbide inclusions in aggravating attack in moist air has already been referred to, but in testing in 0·0005M hydrogen peroxide, pH 6, at 85°C, the presence of carbide up to 0·26% by weight, iron up to 0·4% by weight, silicon up to 0·2% by weight, and aluminium up to 1·05% by weight, in both vacuum-cast and hot-pressed beryllium, did not appear to affect average corrosion rate; after tests of 13 months, the corrosion rate measured only 0·0025 mm/y. Pitting in these tests appeared to be worst in the aluminium-containing specimens, but even in those it did not exceed 0·050 mm.

Temperature Results of tests carried out in the temperature range 250–350°C are not consistent. In one series for example, on extruded beryllium for seven days in distilled water at 270°C, average attack varied between 0·015 and 0·53 mm/y. In contrast, other work on hot-pressed material gave an average attack of less than 0·0025 mm/y.

Variation of pH Between 4 and 8, variation of pH does not appear to have a very significant effect on corrosion rate in de-aerated water.

Galvanic effects Galvanic effects would be expected with such metals as copper, but even with stainless steel (US grade 347) tests in water containing small amounts of hydrogen peroxide at 85°C showed definite evidence of accelerated attack. Extruded beryllium was used in both static and dynamic conditions, and attack on the beryllium test specimens was some 3 to 5 times greater than that on uncoupled controls coupled to the steel. Galvanic effects with various aluminium alloys, however, were not so definite. This is perhaps not surprising in view of the close similarities between the two metals.

Effect of stress The first reported work on the stress-corrosion cracking of beryllium related to its use in water containing 0·005M hydrogen peroxide at pH 6–6·5 and at about 90°C[9]. Although in these circumstances some pitting occurred, there was no evidence of stress corrosion even though the tests employed extruded metal stressed at up to 90% of the yield stress. Subsequently, stress corrosion has been reported on exposure to synthetic seawater[10]. In this environment stress-corrosion failure was reported in a 40 h test at 70% of the 0·2% yield strength, the crack habit being transgranular in character.

Gaseous Environments

The possible employment of beryllium in nuclear engineering and in the aircraft industry has encouraged considerable investigation into its oxidation characteristics. In particular, behaviour in carbon dioxide up to temperatures of 1 000°C has been extensively studied [11–17], and it has been shown that up to a temperature of 600°C the formation of beryllium oxide follows a parabolic law but with continued exposure 'break-away' oxidation occurs in a similar fashion to that described for zirconium. The presence of moisture in the carbon dioxide enhances the 'break-away' reaction[15–16]. It has been suggested that film growth proceeds by cation diffusion and that oxidation takes place at the oxide/air interface[18].

It is reported that beryllium powder, of unspecified particle size, will burn in air at 1 200°C and react with nitrogen at 500°C[5]. Fluorine appears to attack beryllium at room temperature, and the other halogens, nitrogen dioxide and hydrogen sulphide are said to attack it at elevated temperatures[5].

Dry hydrogen chloride gas readily attacks solid beryllium above about 500°C with the formation of volatile beryllium chloride. Beryllium carbide and nitride are similarly attacked, but not beryllium oxide; this behaviour is of use in one method for the determination of beryllium oxide in metallic beryllium.

There is a reaction between beryllium and nitrogen that starts at about 750°C and is appreciable at 850°C, beryllium nitride being formed[11]. The reaction with oxygen is less sluggish and at 900°C in oxygen oxidation proceeds at about twice the rate of nitride formation. Thus when beryllium is heated in air, beryllium nitride forms only a small proportion of the total scale—about 0·75% after 1 h at 1 000°C.

Protective Measures

From what has already been indicated, it will be apparent that use of beryllium in almost any commercial environment involving moisture may require some form of surface protection and it should be recognised that apart from the hazard of pitting corrosion, precautions must usually be taken against adverse galvanic coupling with heavy metals.

Protective measures range from chemical conversion coatings and anodising to the application of more substantial protective layers, e.g. enamels. For a more detailed treatment of the subject, the reader is advised to consult References 19, 20, 21 and 22.

J. B. COTTON

REFERENCES

1. Fröhlich, K. W., *Z. Metallk.*, **28**, 368 (1936)
2. Gough, H. J. and Sopwith, D. G., *J. Inst. Met.*, **60**, 143 (1937)
3. Price, L. E. and Thomas, G. J., *J. Inst. Met.*, **63**, 21 (1938)
4. Brooks, P. E. *et al.*, *J. Inst. Met.*, **88**, 500 (1959–60)
5. Kroll, W. J., in *The Corrosion Handbook* (Ed. Uhlig, H. H.), Wiley, New York; Chapman and Hall, London (1948)
6. Eisenbud, M., in *The Metal Beryllium* (Eds. White, D. W., Jr. and Burke, J. E.), Amer. Soc. Metals, Cleveland, Ohio, 620 (1955)
7. Kjellgren, B. R. F., *Rare Metals Handbook*, Reinhold, New York, 31 (1954)
8. English, J. L., in *The Metal Beryllium* (Eds. White, D. W., Jr. and Burke, J. E.), Amer. Soc. Metals, Cleveland, Ohio, 530 (1955)
9. Logan, H. L. and Hessing, H., *Summarising Report on Stress Corrosion of Beryllium*, NBS–6, Nat. Bur. Stand., Washington, Dec. (1955)
10. Miller, R. A. *et al.*, *Corrosion*, **23**, 11–14 (1967)
11. Gulbransen, E. A. and Andrew, K. F., *J. Electrochem. Soc.*, **97**, 383 (1950)
12. Williams, J. and Munro, W., A.E.R.E. Report M/M108 (1956)
13. Livey, D. T. and Williams, J., *Proceedings of the Second Conference on the Peaceful Uses of Atomic Energy*, United Nations, Geneva (1958) Paper A/CONF/15/P/319
14. Werner, W. J. and Inouje, H., Conference on the Metallurgy of Beryllium, Institute of Metals, Paper 32, Oct. (1961)
15. Pennah, P. J. *et al.*, Conference on the Metallurgy of Beryllium, Institute of Metals, Paper 10, Oct. (1961)
16. Higgins, J. K. and Antill, J. E., *J. Nuclear Mat.*, **5** No. 1, 67 (1962)

17. Smith, R. *et al.*, Conference on the Metallurgy of Beryllium, Institute of Metals, Paper 30, Oct. (1961)
18. Kerr, I. S. and Wilman, H., *J. Inst. Met.*, **84**, 379 (1955–56)
19. Miller, P. D. and Boyd, W. K., *Mat. Eng.*, **68**, 33–36 (1968)
20. Mackay, T. L. and Gilpin, C. B., *Electrochem. Tech.*, **6** (1968)
21. Stonehouse, A. J. and Beaver, W. W., *Mat. Protect.*, **4**, 24–28 (1965)
22. Booker, J. and Stonehouse, A. J., *Mat. Protect.*, **8**, 43–47 (1969)

5.2 Molybdenum

Molybdenum, although once considered a rather unusual metal with a few highly specialised applications, e.g. as support wires in tungsten-filament lamps and in thermionic valves as anode and grid materials, today finds extensive use in the missile and aerospace industry, in the production of high-temperature hydrogen and inert-atmosphere furnaces and furniture, in the production of solid-state electronic devices and in electrodes and bushings used in glass-making processes. Molybdenum and its alloys are now also finding increasing use in chemical and petrochemical plant. Historically, molybdenum has been employed in high-temperature sulphuric acid service and molybdenum-lined equipment has been used in organic chlorination processes. The Mo–30W alloy, which is a complete solid solution, has demonstrated its superiority as a material of construction for liquid metal applications such as the pumping of molten zinc.

It is not subjected to hydrogen embrittlement as is tantalum, niobium and nickel alloys, and thus is able to sustain thermal and mechanical shock after exposure to gaseous hydrogen at high temperatures.

It is produced in the form of a powder and is consolidated either by powder metallurgy techniques, by arc casting or, more recently, by electron-beam melting. Most commercially available molybdenum is however made by powder metallurgy, and it is only when very large pieces are required that arc melting is used. Electron-beam melting gives rise to large crystals in the metal which limits its application.

Physical and Mechanical Properties

Unalloyed molybdenum is available commercially in almost all forms, from forging billets and plate to seamless tubing, foil and wire. In addition, the following molybdenum-base alloys can be considered as established engineering materials: Mo–0·5Ti–0·8Zr (designated *TZM*) and Mo–30W. A Mo–30Ta alloy has also been produced for specific applications where high corrosion resistance is required.

At ambient temperatures the strength of molybdenum is comparable to that of normalised low-alloy steel and moderately higher than that of the austentic stainless steels. However, whereas the low-alloy steels are limited to use at service temperatures of about 550°C and stainless steels to about

Table 5.2 Physical properties of molybdenum*

					Thermal						Electrical			
Atomic number	*Atomic weight*	*Density at 20°C* (g/cm³)	*Melting point* (°C)[1]	*Boiling point* (°C)[1]	*Linear coeff. of expansion* (°C)[2,3]	*Sp. heat* (J/g°C)[2]	*Thermal conductivity* (W/cm °C)[1,2]	*Vapour pressure* (mm Hg)[4]	*Total radiation* (W/cm²)	*Total Emissivity*[2,4]	*Conductivity* % IACS	*Resistivity* (μΩ cm)[1,2]	*Temp. coeff. of electrical resistivity* (°C)	*Thermal neutron cross section* (barns/atom)[5]
42	95·95	10·22	2 610	5 660	$6{\cdot}65\times10^{-6}$ at 20–1 593°C	0·276 at 20°C 0·239 at 538°C	1·42 at 20°C 1·26 at 204°C	1 at 3 297°C 760 at 4 827°C	0·55 at 730°C, 6·3 at 1 330°C and 19·2 at 1 730°C	0·07 at 100°C 0·13 at 1 000°C	36	5·5 at 20°C 23·9 at 727°C	$4{\cdot}7\times10^{-3}$ at 10–100°C	2·7

**Note.* Superscripts outside parentheses denote reference numbers at the end of the section.

870°C, unalloyed molybdenum retains useful strength up to 1 200°C, and the *TZM* alloy has strength up to 1 650°C.

At elevated temperatures the thermal conductivity is far superior to that of Fe–18Cr–8Ni and the Ni and Cu binary alloy Monel, two other alloys widely used in severe chemical service.

The mechanical properties reported in the literature for molybdenum and its alloys are frequently at variance. That this should be so is not surprising

Table 5.3 Mechanical properties of molybdenum

UTS[2,6] (MN/m²)	*Yield stress*[6] (MN/m²)	*Modulus of elasticity*[6] (GN/m²)	*Hardness* Hard rolled (VPN)	*Hardness* Annealed (VPN)	*Hardness* Sintered material[2] (Rockwell)	*Recrystallisation temp.*[6]	*Stress relieving temp.*[6]
317–871 (0·25 mm sheet) 517–1 206 (sintered material) 827–1 379 at 20°C 137–206 at 1 000°C	432 on 100 mm extrusion, to 586 on 1·4 mm sheet at 20°C	317 at 20°C 282 at 500°C 268 at 1 000°C	280	250	95B–27C	900–1 200°C	800°C

as the properties of molybdenum and its alloys are greatly affected by the prior history of the material, both thermal and mechanical. Far too often values are used without reference to the sources of the material, various states of heat treatment, etc. When mechanical properties are an important feature of the design application, advice should always be sought on the suitability as only the manufacturer has the complete data on the history of his own product. Physical and some typical mechanical properties given for general guidance are shown in Tables 5.2 and 5.3.

Methods of Fabrication

Forming The fabrication of molybdenum is largely dictated by the ductile–brittle transition temperature. Most operations, except those on thin sheet or wire, are carried out warm and it is often necessary to heat not only the workpiece but also the die.

Molybdenum can be spun, flow turned and deep drawn as well as pressed. The temperature required to produce components satisfactorily varies with the thickness of the molybdenum sheet.

Specific fabrication temperatures are given by Czarnecki *et al.*[7] for various forming processes, including roll forming, hydropressing, joggle and die forming on molybdenum–titanium alloy sheet.

Forging Ingots of molybdenum produced either by sintering or by arc-melting are readily forged after heating to about 1 500°C. The heating of large ingots can be performed without protection, but thinner sections are heated in a hydrogen atmosphere to limit oxidation.

Rolling Sheet rolling is performed cold with intermediate anneals. Warming of the sheet and tools improves the rate of working; thicker material can be rolled at still higher temperatures.

Swaging Swaging is performed hot, the temperature commencing at about 1 300°C and falling as the ductility of the metal improves to about 700°C.

Drawing and spinning Spinning and cupping are facilitated if molybdenum is gently heated; suitable temperatures can be found in Reference 7.

Machining Molybdenum can be machined by any of the standard methods such as milling, turning, drilling, boring, grinding, shaping, threading and tapping. The low coefficient of expansion of molybdenum makes it necessary to keep the tool cool when drilling in order to prevent seizing of the tool and possible cracking of the metal. Drilling is normally carried out dry but swarf can be carried away with a soluble cutting oil. When milling, cutters must be kept sharp and copious coolant provided; 1.1.1-trichlorethane is recommended for the coolant.

Welding It is possible to weld molybdenum using a TIG-shielded arc-welding process. A heat-affected zone is unavoidable and grain growth must be anticipated.

These characteristics always give a weld which has less ductility than the parent material and no method of welding is entirely satisfactory particularly if the joint is to be stressed.

Resistance welding Spot and seam welding are used to join molybdenum for electronic use; this technique is not satisfactory for large-scale work.

Electron-beam welding Molybdenum can be electron-beam welded but the technique has only limited use.

Plasma welding This method of welding is comparatively new and has not yet found much application.

Brazing Satisfactory brazed joints in molybdenum have been made using oxyacetylene torch and furnace brazing techniques.

Soldering Conventional soft solders can be used, but it is first essential to nickel- or copper-plate the molybdenum.

Corrosion Behaviour

Theory

The theoretical aspects of molybdenum's corrosion behaviour are complex and there is as yet no clear cut, generally applicable picture. There are, however, a large number of literature references which include data on polarisation, passivation and potential of molybdenum under widely assorted conditions. The electrode potential of molybdenum depends on its surface condition. For example, some tests showed an E_H of $+0{\cdot}66$ V when the molybdenum was passivated by treatment with concentrated chromic acid and $-0{\cdot}74$ V after activation by cathodic treatment in sodium hydroxide.

The potential in aqueous solutions depends on the pH of the solution as demonstrated by Masing and Roth[8], Shatolov and Marshakov[9], and others.

As an example of the magnitude of this effect, Shatolov and Marshakov[9] in one series of tests in buffered chloride solution found that the irreversible potential of molybdenum could be expressed as

$$E = 0{\cdot}37 - 0{\cdot}045\text{pH}$$

Molybdenum is widely used as an alloying addition to stainless steels to facilitate the formation of the passive film and to improve resistance to pitting attack (see Section 3.3).

Corrosion in Natural Environments

Atmospheric Molybdenum begins to oxidise in air at 300°C and oxidation becomes rapid at 500°C and the rate of attack is very rapid by 1 200°C[10]. Below 500°C oxidation proceeds according to a parabolic law, indicating some degree of protection. The oxidation proceeds by a two-step process with molybdenum dioxide (MoO_2) as the inner oxide layer and molybdenum trioxide (MoO_3) as the outer layer. Above 500°C, MoO_3 begins to volatilise, and at 600°C the rate of evaporation of MoO_3 becomes significant. At about 770°C the rate of evaporation of MoO_3 equals its rate of formation, and as the temperature increases, the volatilisation rate becomes extremely rapid. The ultimate oxidation rate is linear, with the rate-determining step being the oxidation of the dioxide to the trioxide.

The formation of molten MoO_3 above 815°C results in a catastrophically accelerating effect due to the following factors: (*a*) the liquid oxide flows off the metal surface, (*b*) the rate of diffusion of oxygen through the liquid phase is high, and (*c*) the molten oxide can also act as a flux.

The crystal structures observed during the oxidation of molybdenum consist of stable molybdenum dioxide in contact with the metal throughout the range 300–700°C. As the film thickens in the low-temperature range, the trioxide predominates on the surface. At 400°C, molybdenum trioxide is no longer observed and molybdenum dioxide is the only oxide observed.

The inability of molybdenum to withstand oxidising conditions at even moderate temperatures has led to investigations of means of reducing the oxidation rate. Alloying has been examined and rejected since attainment of good oxidation resistance was at the expense of high temperature strength, and protective coatings appear to be the most effective. A number of coatings have been developed for molybdenum; for temperatures up to 1 200°C Mo-base alloys applied as cladding or sprayed coatings, and chromium–nickel electroplates are effective.

For high temperatures several coatings based on siliconising are available. One of the most successful, designated *W3*, has been used for long periods of say 500 h at 1 200°C and even at 1 700°C for short term (1 h) applications (Fig. 5.1). This coating, which is basically $MoSi_2$, has several advantages: it is relatively thin (about 0·05 mm) and it can be applied over rivets and other fasteners. Bolts and nuts have been successfully coated and used. The coating is applied by either a pack diffusion technique similar to chromising or by an electrochemical technique, and coating procedures have been evolved for the protection of complex assemblies.

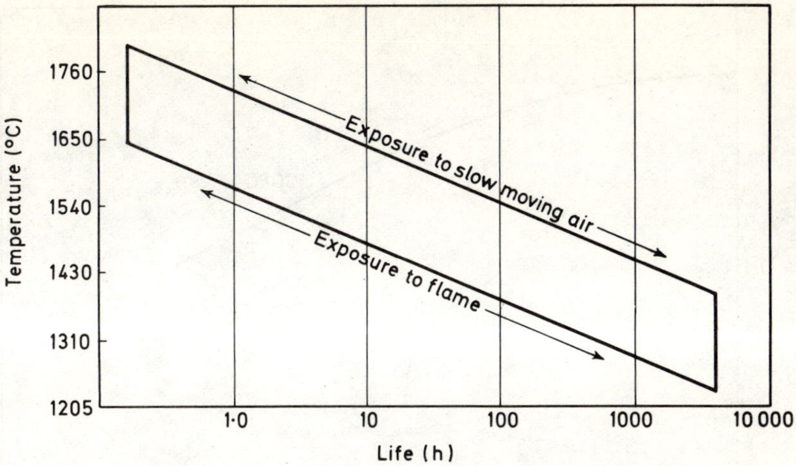

Fig. 5.1 Protection offered by three commercial silicide coatings overlap broadly in this performance band (after *Prod. Engng*., June 25, 63 (1962))

Immersed Molybdenum has good resistance to synthetic sea water, the rate of attack up to 60°C being less than 0·1 mm/y[11] and it is only slightly corroded when exposed to synthetic sea water spray at 60°C for periods of 10, 20 and 30 days.

Corrosion in Chemical Environments

Acids Molybdenum exhibits good resistance to hydrochloric, hydrofluoric, phosphoric and sulphuric acids (see Figs. 5.2, 5.3 and 5.4). In static tests[12] in these acids at Ohio State University it was found that the corrosion rate for relatively long test periods (670 h) was about ten times that experienced in 47 h tests.

Oxidising conditions severely reduce molybdenum's corrosion resistance, and aeration of the above acids causes a pronounced increase in the corrosion rate. It is rapidly attacked by oxidising acids such as nitric acid, and by

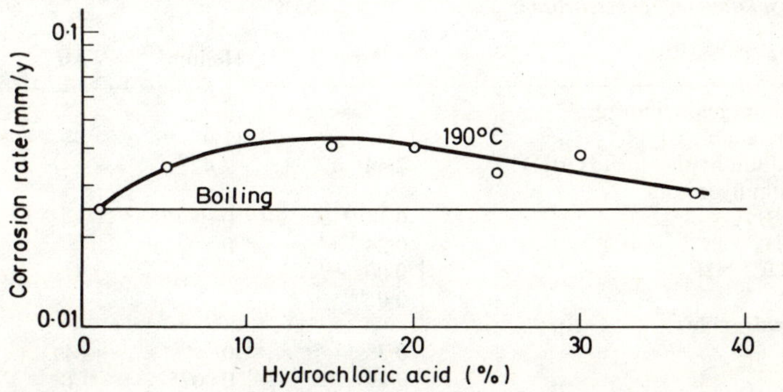

Fig. 5.2 Molybdenum corrosion tests in hydrochloric acid (after Bishop, C. R., *Corrosion*, **19**, Sept., 308t (1963))

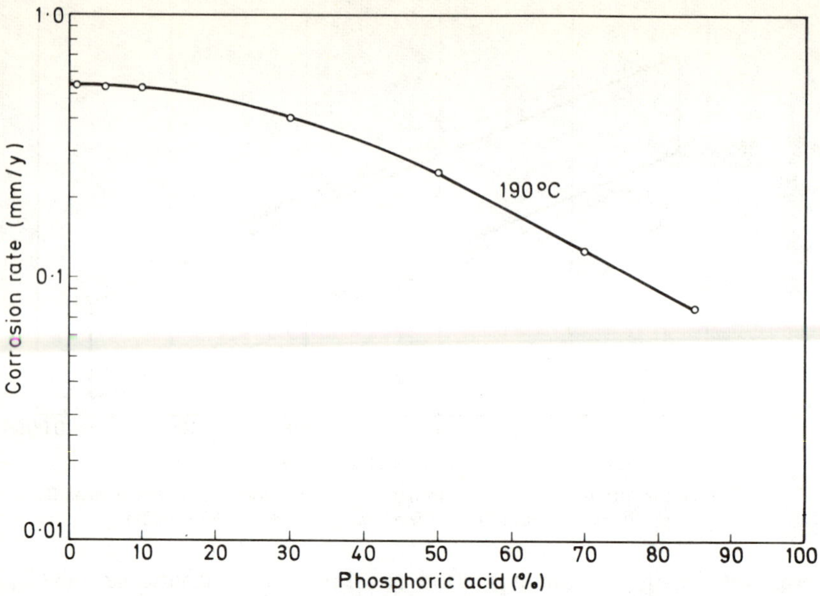

Fig. 5.3 Molybdenum corrosion tests in phosphoric acid (after Bishop, C. R., *Corrosion*, **19**, Sept., 308t (1963))

reducing acids containing oxidisers such as HNO_3, $FeCl_3$, etc. It is less resistant at 100°C, particularly in 10% acetic acid (the corrosion rate being 0·33 mm/y), 10% formic acid (0·2 mm/y) and 0·25% benzoic acid (0·25 mm/y).

Molybdenum dissolves rapidly in 70% sulphuric plus 20% nitric acids at 35°C, but exhibits good resistance to 30% hydrochloric plus 7% sulphuric acids and 10% citric plus 1% sulphuric acids at 100°C. It corrodes three times as fast in 10% acetic acid containing 5% sulphuric acid at 100°C than it does in pure 10% acetic (1 mm/y against 0·33 mm/y).

Table 5.4 Corrosion in alkaline solutions[11] (duration of test 6 days)

Solution and concentration	*Corrosion rate* (mm y^{-1})			
	35°C		60°C	
	Air	Helium	Air	Helium
Conc. sodium hypochlorite (5–6% available Cl_2)	11·4	11·9	—	—
Dilute sodium hypochlorite (10:1)	2·44	2·42	—	—
Ammonium hydroxide				
0·35% NH_3	0·14$^{(1)}$	0$^{(1)}$	—	—
15% NH_3	0·24	0	—	—
Conc. 29% NH_3	0·0025$^{(2)}$	—	—	—
	0·02$^{(3)}$	—	—	—
Sodium hydroxide				
1%	0·38	0	0·49	0·069
10%	0·10	0·0025	0·18	0

Notes. 1. Average of 4 specimens.
2. Non-aerated 6-day test.
3. Non-aerated 28-day test continuation of 6-day test.

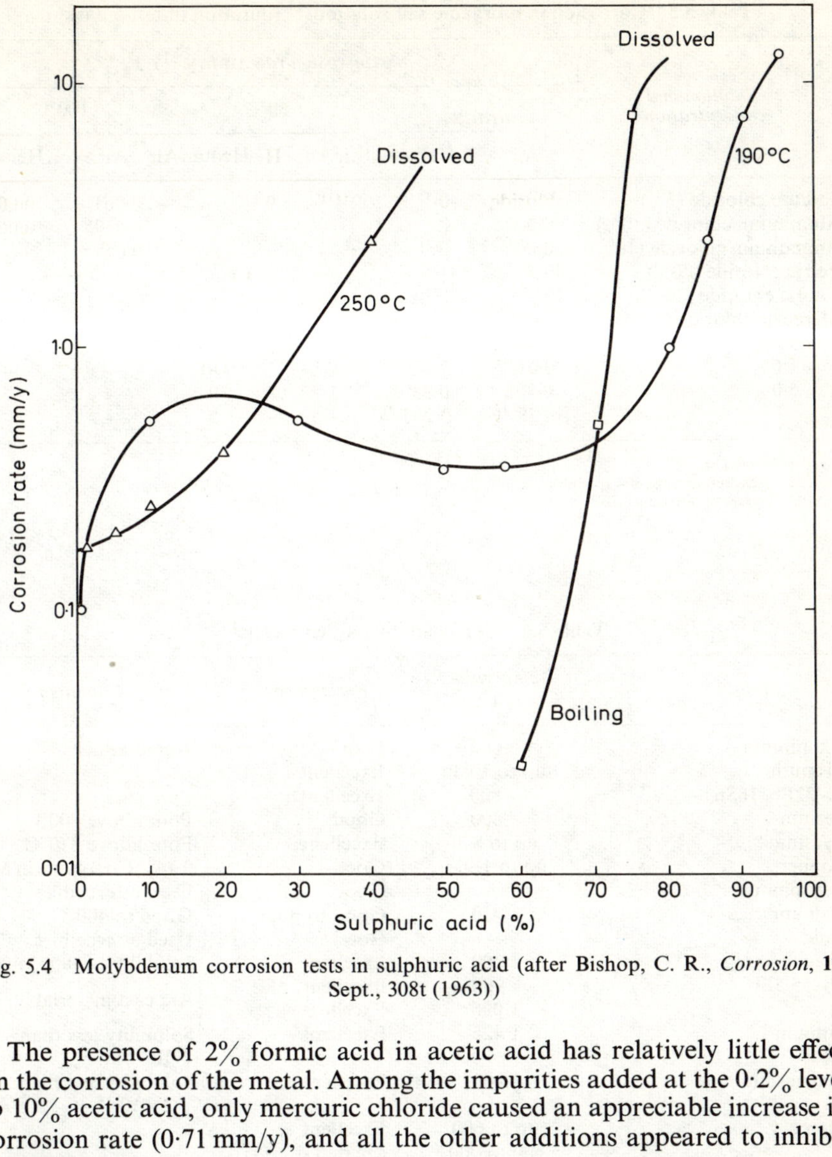

Fig. 5.4 Molybdenum corrosion tests in sulphuric acid (after Bishop, C. R., *Corrosion*, **19**, Sept., 308t (1963))

The presence of 2% formic acid in acetic acid has relatively little effect on the corrosion of the metal. Among the impurities added at the 0·2% level to 10% acetic acid, only mercuric chloride caused an appreciable increase in corrosion rate (0·71 mm/y), and all the other additions appeared to inhibit corrosion.

When small amounts of impurities were added to concentrated phosphoric acid, the corrosion rate was not significantly affected. The addition of only 0·007% ferric ion to concentrated hydrochloric acid caused only a slight increase in corrosion at 35°C, but at 100°C the corrosion rate increased to 6·35 mm/y.

Alkalis Molybdenum is moderately resistant to aerated solutions of ammonium hydroxide and is inert when oxygen is excluded. It has only fair resistance in aerated 1% sodium hydroxide at 35°C and 60°C but its resistance is better in a 10% solution at both these temperatures. It is severely corroded in sodium hypochlorite solutions (pH 11 or higher) at 35°C (Table 5.4).

Table 5.5 Corrosion in inorganic salt solutions[11] (duration of test 6 days)

Solution and concentration	Corrosion rate ($mm\,y^{-1}$)					
	35°C		60°C		100°C	
	Air	Helium	Air	Helium	Air	Helium
Sodium chloride (3%)	0·01$^{(1)}$	0$^{(1)}$	—	—	0·061	0·012
Aluminium chloride (10%)	0·0025	0	—	—	0·0305	0·005
Ammonium chloride (10%)	0·005$^{(3)}$	0$^{(3)}$	—	—	0·0585	0
Ferric chloride (20%)	40·6	34·5	—	—	—	—
Cupric chloride (20%)	19·3	6·35	—	—	—	—
Mercuric chloride						
0·5%	0·485	0·485	—	—	—	—
1·0%	0·61$^{(2)}$	0·43$^{(2)}$	0·84	0·81	—	—
5·0%	0·305$^{(3)}$	0·33$^{(3)}$	1·72	1·9	—	—
	0·38$^{(2)}$	0·203$^{(2)}$	1·5	1·5	—	—

Notes. 1. 30-day test.
2. 15-day test average of 4 specimens.
3. Average of 4 specimens.

Table 5.6 Corrosion by molten metals[13]

Metal	*Temperature* (°C)	*Resistance*	*Comments*
Aluminium	660	Poor	Rapid attack
Bismuth	up to 1 430	Excellent	
Bi–32Pb–16Sn	800	Excellent	
Cerium	800	Good	Poor above 800°C
Caesium	up to 870	Excellent	Poor above 870°C
Copper	up to 1 300	Good	0·14% Cu soluble in Mo
Europium	—	Good	Used as crucibles
Gallium	450	Good to poor	Good to 400°C
Gold	—	Good	Used as crucibles
Lead	1 200	Excellent	Solubility <0·005 at. %
Pb–55·5Bi	900	Excellent	Arc cast material
	1 090	Excellent	Arc cast material
Lithium	1 425	Excellent	Solubility less than 10^{-4} at. %
Magnesium	1 000	Excellent	Solubility less than 2×10^{-4} at. %
Mercury	up to 600	Excellent	
	600 to 700	Limited	
Plutonium	—	Good	
Potassium	1 200	Excellent	
Rubidium	1 035	Excellent	
Samarium	—	Good	Used as containers
Scandium	1 625	Poor	
Silver		Good	Used as crucibles
Silver plus 1–15% titanium	1 020	Satisfactory	Used as crucibles
Sodium	1 020	Excellent	
Sodium containing 0·5% oxygen	700	Fair	

Inorganic salt solutions Molybdenum has excellent resistance to 3% sodium chloride, 10% aluminium chloride and 10% ammonium chloride at temperatures up to 100°C. It is severely corroded by 20% solutions of ferric and cupric chlorides at 35°C and is subject to pinhole-type pitting in mercuric chloride solutions (Table 5.5).

Industrial-type Environments

Molten metals Molybdenum has good resistance to many molten metals (see Table 5.6) and molybdenum and its alloys are finding application in metal producing and fabricating operations. Its high strength to weight ratio, corrosion resistance and resistance to attack by liquid metals makes it readily suited for advanced heat transfer systems, and these factors coupled with its low thermal neutron cross section have led to it being considered as a canning material in nuclear reactors. The alloy Mo–30W also has unique corrosion properties in certain media, being for all practical purposes completely resistant to attack by molten zinc at normal casting temperatures.

Molybdenum is extremely resistant to attack by molten glass up to 1 000°C if sulphur is present, and up to 1 400°C in its absence. In addition, any reaction products formed by the two materials are colourless. Consequently molybdenum is used in the production of optical and soda lime glasses. Molybdenum is, however, corroded by lead glasses which loose their gloss as a result of the interaction between the two materials[14].

Fused salts Molybdenum has excellent resistance to a wide range of fused salts and has been used in the fused salt electrolysis of magnesium, platinum, thorium and uranium. In the production of pure magnesium, molybdenum is used to couple graphite electrodes electrically. Molybdenum cathodes are used with other materials in the recovery and purification of spent nuclear fuels from fused salt mixtures where contamination of the electrolyte must be prevented.

Gases Molybdenum has a fair resistance to chlorine, bromine and iodine but not to fluorine and it has excellent resistance to a wide range of other gases (see Table 5.7).

Refractories Molybdenum is compatible with a wide range of refractory materials as can be seen from Table 5.8.

Galvanic Corrosion

Work by the US Bureau of Mines[11] involving galvanic couple experiments showed that the normally low corrosion rates of molybdenum were reduced further by contact with aluminium, SAE 1 430 steel or magnesium in aerated solutions of synthetic sea water or 3% sodium chloride.

Contact with copper in sodium chloride solutions reduces the corrosion of molybdenum, but in synthetic sea water the corrosion rate is somewhat higher. In 1% sodium hydroxide, molybdenum corrodes slightly faster when

Table 5.7 Resistance of molybdenum to gases[16,17]

Gas	*Temperature (°C)*	*Resistance*	*Comments*
Air	370	Good	Above 370°C poor
Bromine (dry)	up to 455	Excellent	Over 455°C poor
Dichlorodifluoro-methane + 1% sulphur hexafluoride	up to 315	Good	
	650	Poor	
Hydrocarbons	up to 1 100	Good	
	above 1 100	Fair	Some carburisation may occur
Nitric oxide	1 125	Good	
	over 1 125	Fair	
Nitrogen	up to 2 400	Excellent	
	over 2 400	Good	Nitride case forms
Nitrogen tetroxide	1 155	Poor	
Oxygen	up to 370	Good	Over 370°C poor
Steam	650	Fairly good	Over 650°C poor
Sulphur dioxide	up to 650	Excellent	
	over 650	Fair	Attack rapid at 1 165°C

Table 5.8 Resistance to refractories[17]

Refractory material	*Temperature (°C)*	*Resistance*	*Comments*
Alumina	up to 1 900	Excellent	
Carbon	up to 1 100	Good	
	over 1 100	Fair	Superficial brittle carbide case forms
	over 1 370	Poor	Molybdenum carburises rapidly and completely
Glass	up to 1 400	Excellent	No sulphur or lead present
Silica	650	Excellent	
	1 900–2 000	Very good	Helium atmosphere
Zirconia	up to 2 000	Excellent	
Zirconium oxide + 8% yttrium oxide	2 350	Good	

coupled with SAE 1 430 steel with or without the presence of air. It is protected by contact with copper in aerated 0·46% sulphuric acid. It is not significantly affected by contact with 316 or Carpenter 20 stainless steels in sulphuric acid solutions.

Molybdenum tends to be protected by vanadium in aerated 7·1% hydrochloric acid and it receives a high degree of protection when coupled with copper in this medium. Molybdenum corrodes somewhat faster than normal in 3·1% nitric acid when coupled with tungsten. It is not affected by contact with titanium in 3·1% nitric acid. It is protected by aluminium and copper in aerated 10% formic acid and by aluminium in air-aerated 9%

oxalic acid. In the latter solution, copper had only a slight protective effect when coupled with molybdenum.

Behaviour of Molybdenum Alloys

The US Bureau of Mines[15] found the chemical and galvanic corrosion behaviour of both the *TZM* and Mo–30W alloy to be generally equal or superior to that of unalloyed molybdenum in many aqueous solutions of acids, bases and salts. Notable exceptions occurred in 6·1% nitric acid where both alloys corroded appreciably faster than molybdenum. In mercuric chloride solutions the *TZM* alloy was susceptible to a type of crevice corrosion which was not due to differential aeration. The alloys were usually not adversely affected by contact with dissimilar metals in galvanic couple experiments, but the dissimilar metals sometimes corroded galvanically. Both alloys were resistant to synthetic sea water spray at 60°C.

Other alloys of molybdenum which have been investigated for their corrosion resistance contain 10–50% Ta and were found to have excellent resistance to hydrochloric acid. Ti–Mo alloys were found to resist chemicals that attack titanium and Ti–Pd alloys, notably strong reducing acids such as hot concentrated hydrochloric, sulphuric, phosphoric, oxalic, formic and trichloroacetic. For example, a Ti–30Mo alloy has the following corrosion rates: in boiling 20% hydrochloric acid, 0·127–0·254 mm/y; in 10% oxalic acid at 100°C, 0·038 mm/y, which compares favourably with the respective rates of 19·5 and 122 mm/y for the Ti–0·2Pd alloy.

Applications in Industry

Molybdenum has always been used in quite large quantities in the electronics industry, and the use of molybdenum windings, protected by hydrogen, for high-temperature furnaces is wide-spread. Molybdenum windings and radiation screens are being used in increasing numbers in electric furnaces for the bright annealing of stainless steels and similar materials. The availability of the metal in massive form has opened up new fields for molybdenum and one of the first applications for larger components was in the glass industry where electrodes and stirrers used in the melting of glass are constructed from molybdenum. Molybdenum, or more usually the Mo–30W alloy, is finding increasing use in applications involving molten metals such as the die casting of molten zinc[18,19]. Molybdenum crucibles are used in the production and handling of rare earth metals and compounds such as lithium platinide, and boats of molybdenum are used extensively in metal vapour deposition processes. Molybdenum as electrodes and other components finds a use in molten halide salt electrolysis, one of the principal applications of which is in the recovery of nuclear materials.

In the chemical process industry molybdenum has found use as washers and bolts to patch glass-lined vessels used in sulphuric acid and acid environments where nascent hydrogen is produced. Molybdenum thermocouples and valves have also been used in sulphuric acid applications, and molybdenum alloys have been used as reactor linings in plant used for the production of *n*-butyl chloride by reactions involving hydrochloric and sulphuric acids at temperatures in excess of 170°C. Miscellaneous applications where molybdenum has been used include the liquid phase Zircex hydrochlorination process, the Van Arkel Iodide process for zirconium production and the Metal Hydrides process for the production of super-pure thorium from thorium iodide.

In the polyacrylic synthetic fibre industry, carbonitrided molybdenum guides have been used in place of chromium plated steel because of their resistance to corrosion and erosion. Chemicals that attack molybdenum are listed in Table 5.9.

Table 5.9 Chemicals that attack molybdenum[16,17]

Reagent	*Concentration* (%)	*Temperature* (°C)	*Resistance*
Potassium carbonate	100	260	Poor (attack greater than 1·3 mm y^{-1})
Potassium hydroxide	50	100	Good
	100	Fused	Poor
Sodium carbonate	100	315	Poor (attack greater than 1·3 mm y^{-1})
Sodium hydroxide	50	100	Good
	100	Fused	Poor
Ferric chloride	50	Room	Poor
Lead nitrate	—	Fused	No reaction
Lead oxide	—	Fused	Poor (violent reaction)
Potassium chlorate	—	Fused	Poor (violent reaction)
Potassium nitrate	—	Fused	Poor (violent reaction)
Potassium nitrite	—	Fused	Poor (violent reaction)
Sodium peroxide	—	Fused	Poor (violent reaction)

J. BENTLEY

REFERENCES

1. Lyman, T., *A.S.M. Metals Handbook* (1961)*
2. Hampel, C. A., *Rare Metals Handbook*, 2nd edn, Reinhold (1961)*
3. Krikorian, O. H., *Thermal Expansion of High Temperature Materials*, VCRL 6 132, Sept. (1960)*
4. Bockris, J. O'M., White, J. L. and Mackenzie, J. D., *Physical and Chemical Measurements at High Temperatures*, Butterworths, 369–371 (1959)*
5. MacDonald, R. N. and Boucon, H. H., *Nucleonics*, **20**, August 8th, 158 (1962)*
6. Chelius, J., *Machine Design*, March 1st (1962)*
7. Czarnecki, E. G., Stacey, J. T. and Zimmerman, D. K., *Refractory Metal for Glide and Re-entry Vehicles*, Met. Soc. Conference, **17** and *Refractory Metals and Alloys*, **2** (Edited by Semchyshen and Perlmutter), Interscience Publishers, New York (1963)
8. Masing, G. and Roth, C., 'Behaviour of Molybdenum and Nickel and Some Molybdenum-nickel Alloys in Acid Electrolytes', *Werkstoffe und Korrosion*, **3**, 176–186, 253–263 (1952)
9. Shatalov, A. Ya. and Marshakov, I. A., 'Electrode Potentials and Corrosion of Molybdenum and Tungsten', *Zhurnal Fizicheskoy Khimii*, **28** No. 1, 42–50 (1954)
10. Jaffee, R. J., 'Proceedings on an International Symposium on High Temperature Technology', Asilomar, California, 1959, McGraw-Hill, New York, 61 (1960)

*These references can be found collectively in *Properties of Refractory Materials*, by S. J. Burnett, U.K.A.E.A. Research Report No. R4657, H.M.S.O. (1969)

11. Acherman, W. L., Carter, J. P., Kenahan, C. B. and Schlan, D., *Corrosion Properties of Molybdenum, Tungsten, Vanadium and some Vanadium Alloys*, Report of Investigations No. 6 715, US Bureau of Mines (1966)
12. *Corrosion Tests on Metallic Molybdenum*, Engineering Experimental Station, Ohio State University, Project 142 (1957–1959)
13. Barto, R. L. and Hard, D. T., 'Refractory Metals in Liquid Metal Handling', *Research/Development*, Nov. 26 (1966)
14. 'Corrosion of Molybdenum by Molten Glass', *Kazus Ocku Yogyo Kyrokai Shi*, **72**, 108–113 (1964)
15. Ackerman, W. L., Carter, J. P. and Schlain, D., 'Corrosion Properties of the TZM and Mo–30W Alloy, US Bureau of Mines Report No. 7 196, Aug. (1969)
16. *Corrosion Properties of Special Alloys and Rare Metals*, Jacob and Korres G.m.b.H, 473 Ahlen, W.G.
17. Nair, F. B. and Briggs, J. Z., *Corrosion Resistance of Molybdenum and Molybdenum Base Alloys*, Molybdenum Metal
18. Gilbert, R. W., 'New Mo–30W Alloy Replaces Tool Steels for Die Casting Nozzles', *Die Casting Engineer*, March (1964)
19. Burman, R. W. and Litchfield, G., 'Severe Molten Zinc Corrosion is Reduced by Improved Molybdenum Tungsten Alloy', *Engineering and Mining Journals*, **164** No. 4, April, 88 (1963)

BIBLIOGRAPHY

Cox, F. G., 'Joining Molybdenum', *Welding and Metal Fabrication*, Sept. (1961)

Nair, F. B., 'Molybdenum Resists Corrosion at High Temperatures', *Design Engineering*, Oct., 104–105 (1964)

Hargrave, D. P., 'Molybdenum and Tungsten', Paper 5 of conference on New Engineering Materials, *Proceedings of Institution of Mechanical Engineers*, **180**, Part 3D (1955–1956)

Northcott, L., *Molybdenum*, Butterworths, London (1956)

Harwood, J. J. (Ed.), *The Metal Molybdenum* (the protection of molybdenum against high temperature oxidation), American Society for Metals, Cleveland, Ohio (1958)

Technical Notes, Climax Molybdenum Co., Jan. (1964)

5.3 Niobium

Niobium is always found in nature associated with tantalum and it closely resembles tantalum in its chemical and mechanical properties. It is a soft ductile metal which, like tantalum, work hardens more slowly than most metals. It will in fact absorb over 90% cold work before annealing becomes necessary, and it is easily formed at room temperature. In addition, welds of high quality can be produced in the metal. In appearance the metal is somewhat similar to stainless steel; it has a density slightly higher than stainless steel and a thermal conductivity similar to 1% carbon steel.

It is somewhat less corrosion resistant than tantalum, and like tantalum suffers from hydrogen embrittlement if it is made cathodic by a galvanic couple or an external e.m.f., or is exposed to hot hydrogen gas. The metal anodises in acid electrolytes to form an anodic oxide film which has a high dielectric constant, and a high anodic breakdown potential. This latter property coupled with good electrical conductivity has led to the use of niobium as a substrate for platinum-group metals in impressed-current cathodic-protection anodes and in the production of capacitors.

The mechanical properties of niobium are dependent on the previous history of the material and the manufacturer should be consulted if these properties are likely to be critical. Physical and some typical mechanical properties are set out in Tables 5.10 and 5.11.

Methods of Fabrication

Niobium possesses excellent room temperature fabrication characteristics compatible with all conventional production practices. Large reductions (up to 90%) of recrystallised material can be made without intermediate process annealing. Secondary fabrication operations such as stamping, drawing or forming into completed shapes can be performed cold. Intermediate anneals are dependent on the amount of work involved. In tube drawing or deep drawing, annealed niobium should be used. Reductions of 60 to 80% with multiple draws are customary before re-annealing, but the initial draw should have a depth not greater than 40 to 50% of the diameter.

It machines in a similar manner to soft copper, and high-speed-steel tools with high cutting speeds are most satisfactory. Trichloroethane is recommended as a cutting medium and the work must be kept well flooded at all

Table 5.10 Physical properties of niobium

Melting point[1] (°C)	*Boiling point*[1] (°C)	*Density*[1] (g/cm³)	*Specific heat*[2] (J/g°C)	*Thermal conductivity*[1, 2] (W/cm°C)	*Resistivity*[1, 3] (μΩ/cm at 20°C)	*Coefficient of thermal expansion*[2, 4] (×10⁻⁶/°C)	*Cross section thermal neutrons*[5] (barns/atom)
2 468	4 927	8·57	0·268 at 15°C 0·320 at 1 227°C	5·23 at 0°C 6·91 at 1 600°C	15	7·1 at 20°C 7·39 at 0–400°C	1·16

Table 5.11 Mechanical properties of niobium

Yield stress[2, 6] (MN/m²)	UTS[2, 7] (MN/m²)	*Modulus of elasticity*[2, 7] (GN/m²)	*Poisson's ratio*	*Hardness* (VPN)	*Resistance to thermal shock*	*Workability (ductile to brittle trans. temp.)* (°C)	*Recrystallisation temperature*[7]	*Stress relieving temperature*[7]
288 at 20°C 68 at 1 050°C	517–1 034 at 20°C 241 at 500°C 89–117 at 1 000°C	188 at 204°C 113 at 600°C 108 at 900°C	0·38	77–173	Good	– 150 (based on rupture tests)	900–1 300°C	800°C

times. If it is not possible to use this coolant, satisfactory results can be obtained with water-soluble oil coolants.

It can be welded by resistance, tungsten-inert gas (TIG), plasma arc and electron beam techniques. To protect the metal from attack by air, resistance welding is carried out under water and the TIG method is best performed in a chamber of argon. The latter three methods produce ductile welds that equal the base metal in most of its characteristics.

Niobium closely resembles tantalum in its mechanical properties and for more detailed information relating to the fabrication of niobium see Section 5.5 on tantalum.

Corrosion Resistance

Niobium like tantalum relies for its corrosion resistance on a highly adherent passive oxide film; it is however not as resistant as tantalum in the more aggressive media. In no case reported in the literature is niobium inert to corrosives that attack tantalum. Niobium has not therefore been used extensively for corrosion resistant applications and little information is available on its performance in service conditions. It is more susceptible than tantalum to embrittlement by hydrogen and to corrosion by many aqueous corrodants. Although it is possible to prevent hydrogen embrittlement of niobium under some conditions by contacting it with platinum the method does not seem to be broadly effective. Niobium is attacked at room temperature by hydrofluoric acid and at 100°C by concentrated hydrochloric, sulphuric and phosphoric acids. It is embrittled by sodium hydroxide presumably as the result of hydrogen absorption[8] and it is not suited for use with sodium sulphide.

Atmospheric Niobium like several other refractory metals is extremely reactive with atmospheric oxygen. It will in fact react with air at temperatures as low as 200°C[9] although reaction does not become rapid until temperatures above red heat (about 500°C) are reached; at 980°C the rate is 0·05 mm/h[10] and at 1 200°C the rate is 300 mm/h[11]. It is not attacked by oxygen at 100°C but the attack is catastrophic at 390°C. At lower temperatures a thin adherent oxide film is formed on the surface of the metal, but at higher temperatures, above red heat, the oxide diffuses rapidly throughout the metal with consequent embrittlement. At elevated temperatures the metal reacts with all the common gases including nitrogen (300–400°C), water vapour (300°C), carbon dioxide, carbon monoxide and hydrogen (250°C).

Protection of niobium and its alloys from oxidation in air is accomplished by coating, e.g. with zinc deposited by holding in zinc vapour at 865°C[12] or coating with a layer of chemically stable oxide, nitride or silicide. Silicide coatings applied by pack cementation, fused slurry[13] or by electrolytic methods have been found to be one of the most effective means of preventing oxidation of the metal.

Water The corrosion resistance of pure niobium in water and steam at elevated temperatures is not sufficient to allow its use as a canning material in water-cooled nuclear reactors. Alloys of niobium with molybdenum, titanium, vanadium and zirconium however have improved resistance and have possibilities in this application. Whilst the Nb–10Ti–10Mo alloy offers the best corrosion resistance the Nb–7V alloy seems more practical on the basis of weldability. It also has good high-temperature strength properties.

Acids[10, 14–16] Niobium is resistant to most organic acids and to mineral acids especially under oxidising conditions. Figs. 5.5 to 5.7 show the corrosion behaviour of niobium in laboratory tests in various concentrations of sulphuric and hydrochloric acids at the boiling point and at 190°C and 250°C, and in phosphoric acid at the boiling point. It has excellent resistance to nitric acid, the rate of attack in 70% acid at 250°C being only 0·025 mm/y. In dilute sulphurous acid at 100°C the corrosion rate is 0·0125 mm/y, but in concentrated acid at the same temperature it is greater than 0·25 mm/y.

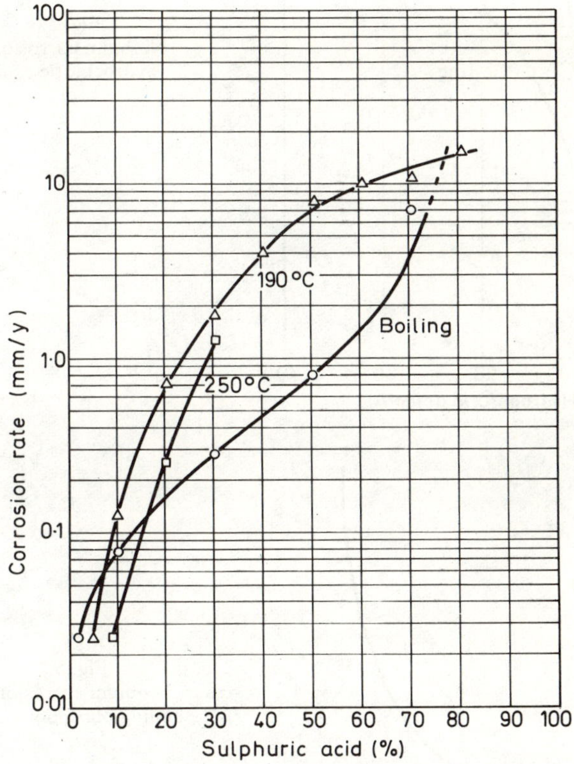

Fig. 5.5 Niobium corrosion in sulphuric acid[15]

Alkalis[8, 10, 14] Though niobium is not attacked by most alkalis at room temperature it is seriously attacked at 98°C, and severe embrittlement is obtained in concentrated alkali at room temperature and virtually all alkalis at 98°C. See Table 5.12 for detailed corrosion rates and embrittlement ratings.

There is evidence to show that the corrosion product when niobium is attacked by sodium hydroxide is $Na_8Nb_6O_{19}\cdot18H_2O$.

Salts[10, 17, 18] Tests on niobium have only been carried out in a limited number of salt solutions; however, in the main niobium exhibited similar resistance to tantalum in most salt solutions, and like tantalum it is attacked by salts that hydrolyse to form alkalis.

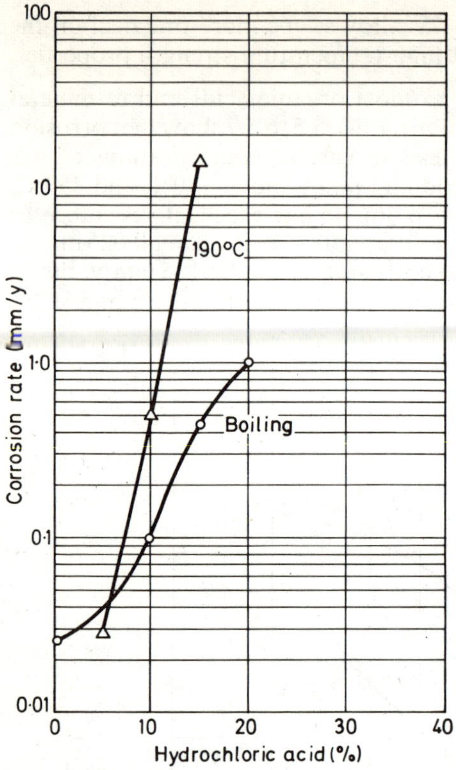

Fig. 5.6
Niobium corrosion in hydrochloric acid[15]

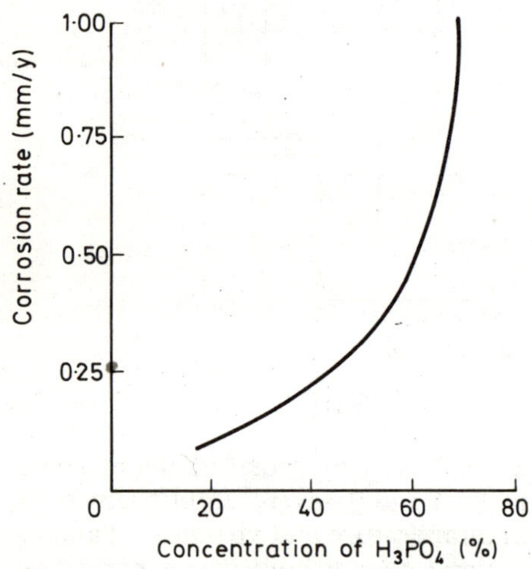

Fig. 5.7
Niobium corrosion in boiling phosphoric acid[16]

Gases[9, 17, 19, 20] It is unattacked by most common gases, e.g. nitrogen, hydrogen, oxygen, carbon dioxide, carbon monoxide and sulphur dioxide (wet or dry) up to 100°C, and it is inert to chlorine and bromine (both wet and dry) to 100°C. It is, however, attacked by nitrogen at 300–400°C, hydrogen at 250°C, oxygen at 200°C, carbon and carbon-bearing gases at 1 200–1 400°C and by chlorine at 200–250°C.

Table 5.12 Corrosion in alkalis[8, 10, 14]

Reagent	*Concentration* (%)	*Temperature* (°C)	*Corrosion rate* (mm/y)	*Embrittlement rating*
NaOH	1	98	0·74	B
	5	98	1·17	C
	10	98	2·00	C
KOH	1	98	0·60	A
	5	98	2·56	C
Na_2CO_3	10	98	1·63	B
	20	98	1·60	C
K_2CO_3	10	98	2·90	C
	20	98	2·46	C
Na_2PO_4	25	98	1·34	B
Na_2S	9	98	0·09	A
NaOH and Na_2S	9	98	4·15	B
NaOH		Molten		Severe attack at 535°C
KOH		Molten		Dissolves metal at 360°C

Note. Embrittlement ratings were obtained by bending a wire corrosion-test-specimen and the following notations were used:
A wire unbroken when sharply bent.
B wire broken when sharply bent.
C wire broken when handled or slightly bent.

Liquid Metals[21]

Bismuth Niobium is resistant to bismuth at temperatures up to 560°C[22] but is attacked at higher temperatures[23–25] and is therefore not considered a suitable container for handling liquid bismuth even under oxygen-free conditions[26]. Furthermore, the stress-rupture properties of niobium are significantly lowered when the metal is tested in molten bismuth at 815°C[24, 27].

Gallium It is slightly less resistant than tantalum to gallium, showing good resistance at temperatures up to 400°C but poor resistance above 450°C[28–30].

Lead Although subject to slight penetration at 980°C it shows no detrimental effects in stress rupture tests when tested in molten lead at this temperature[27] or at 815°C[24]. It is highly resistant to mass transfer in liquid lead as indicated by data obtained in tests at 800°C with a thermal gradient of 300°C[31].

Lithium Niobium has good resistance to molten lithium at temperatures up to 1 000°C[28, 32].

Mercury In static tests niobium shows good resistance to mercury at temperatures up to 600°C[28].

Sodium, potassium and sodium–potassium alloys Liquid sodium, potassium or alloys of these elements have little effect on niobium at temperatures up to 1 000°C[28, 33, 34], but oxygen contamination of sodium causes an increase in corrosion[35,36]. Sodium does not alloy with niobium[37]. In mass transfer tests, niobium exposed to sodium at 600°C exhibited a corrosion rate of approximately 1 $mg\,cm^{-2}\,d^{-1}$. However, in hot trapped sodium at 550°C no change of any kind was observed after 1 070 h[38].

Thorium–magnesium In static tests the thorium–magnesium eutectic had no appreciable effect on niobium at 850°C[39].

Uranium Short-term tests indicate that the practical upper limit for niobium as a container material for uranium is about 1 400°C[40]. Niobium is dissolved in a uranium–bismuth alloy in less than 100 h at a temperature of 800°C[41]. Uranium eutectics with iron, manganese or nickel, corroded niobium at 800°C[42] and 1 000°C[43]. It is significantly attacked by uranium–chromium at 1 000°C[44].

Zinc Molten zinc is reported to attack niobium at a significant rate at temperatures above 450°C[45]. It is attacked by zinc at 600°C and shows increasing solubility with temperature up to 850°C.

Galvanic effects If niobium is cathodic in a galvanic couple the results can prove disastrous because of hydrogen embrittlement. If niobium is the anode in such a couple it anodises so readily that no damage occurs and the galvanic current drops to a very low value due to the formation of an anodic oxide film.

Anodic oxide formation Lakhiani and Shreir[46] have studied the anodic oxidation of niobium in various electrolytes, and have observed that temperature and current density have a marked effect on the anodising characteristics. The plateau on the voltage/time curve has been shown by electron microscopy to correspond with the crystallisation of the oxide and rupture of the previously formed oxide. It would appear that this is a further example of 'field recrystallisation'—a phenomenon which has been observed previously during anodisation of tantalum[47]. No significant data on the galvanic behaviour of niobium are available; however, its behaviour can be expected to be similar to tantalum.

Alloys of Niobium

Niobium–Tantalum Niobium and tantalum form solid-solution alloys which are resistant to many corrosive media and possess all the valuable properties of the pure metals. This could have great practical value since in a number of branches of technology it might permit the replacement of pure tantalum by a cheaper alloy of niobium and tantalum. Miller[48] and Argent[49] reported data on the resistance of the niobium–tantalum system, but the tests were only carried out under mild conditions and the data have only limited significance. However, Gulyaev and Georgieva[16] and Kieffer, Bach and Slempkowski[50] carried out tests at elevated temperatures and their work indicated that the corrosion rates of the alloys are substantially that of tantalum provided the niobium content does not exceed 50%.

Niobium–Zirconium Nb–0·75Zr has excellent mechanical properties and similar corrosion resistance to pure niobium; higher zirconium concentrations reduce the corrosion resistance.

Niobium–Titanium Nb–8Ti exhibits unusual behaviour: although the corrosion resistance is slightly lower than the pure metal it shows no sign of embrittlement in sulphuric and hydrochloric acids. The higher the titanium content the lower the corrosion resistance.

Niobium–Zirconium–Titanium Niobium alloys containing zirconium and titanium have improved resistance to high-temperature water[51] and have been evaluated for use in pressurised-water nuclear reactors.

Niobium–Vanadium The presence of vanadium reduces niobium's corrosion resistance to most media. The alloy containing 12·6 at. %V however has excellent resistance to high-temperature water and steam, and this property and the alloy's relatively low neutron cross section give it considerable potential for nuclear applications.

Niobium–Molybdenum The addition of molybdenum to niobium within the solid solution range gives improved corrosion resistance to hydrochloric and sulphuric acids.

Industrial Applications of Niobium

Nuclear Niobium finds use in some nuclear reactors on account of its compatibility with uranium and liquid sodium/potassium at fast reactor temperatures.

Impressed-current cathodic-protection anodes Niobium has a high anodic breakdown potential (100 V in sea water), a good electrical conductivity (13% that of Cu), good mechanical properties and anodises readily forming an adherent passive oxide film. These properties have led to it being used as a substrate for platinum in impressed-current cathodic-protection anodes for use in high-resistivity waters and other situations where high driving potentials are required to obtain good current spread. Niobium has the advantage over tantalum in that it is less costly, and its cost can be decreased by using a composite electrode with a copper core that also increases the conductivity of the anode[52].

Capacitors Niobium's electrical properties have also led to its investigation as a capacitor material; however, as far as is known there has been no significant commercial application of the material in this field.

Chemical plant It has been reported from some plants producing hydrochloric acid that tantalum condensers are being replaced by ones of niobium, and in certain petroleum plant niobium is being specified for its corrosion resistance and mechanical properties.

Electrical Niobium is finding growing use in components for high-pressure sodium lamps.

Miscellaneous Niobium also finds use in satellite launch vehicles and spacecraft and one of the principal applications for niobium-base alloys is in the production of super-conducting devices.

J. BENTLEY

REFERENCES

1. Lyman, T., *Metals Handbook*, A.S.M. (1961)*
2. Hampel, C. A., *Rare Metals Handbook*, Reinhold, 2nd edn (1961)*
3. Quarrel, A. G., *Niobium, Tantalum, Molybdenum and Tungsten*, Report of Conference, University of Sheffield (1960)*
4. Krikorian, O. H., *Thermal Expansion of High Temperature Materials*, UCRL 6 132, Sept. (1960)*
5. MacDonald, R. N. and Boucom, H. H., *Nucleonics*, **20** No. 8, Aug., 158 (1962)*
6. Mordike, B. L., *J. Inst. Metals*, **88** No. 6, 272 (1960)*
7. Chelius, J., *Machine Design*, March 1 (1962)*
8. Tingley, I. I. and Rogers, R. R., 'Corrosion of Niobium and Tantalum in Alkaline Media', *Corrosion*, **21** No. 2, 132 and 136, April (1965)
9. Balke, C. W., in *Corrosion Handbook* (Ed. H. H. Uhlig), 620–621 and 720–722, Wiley (1948)
10. Macleary, D. L., 'Testing of Niobium and Niobium Alloys', *Corrosion*, **18**, 67t–69t, Feb. (1962)
11. Jaffee, R. J., *Proceedings on an International Symposium on High Temperature Technology*, Asilomar, California, 1959, McGraw Hill, New York, 61 (1960)
12. Wehrmann, *Corrosionomics*, Fansteel Metallurgical Corporation, Sept. (1956)
13. Priceman, C. and Soma, L., *Development of Fused Slurry Silicide Coatings for the Elevated Temperature Oxidation Protection of Niobium and Tantalum Alloys*, Report AFML–TR–68–210, Sylvania Electric Products Inc., Dec. (1968)
14. Cox, F. G., *Niobium Welding and Metal Fabrication*, 352–358, Oct. (1965)
15. Bishop, C. R., 'Corrosion Tests at Elevated Temperatures and Pressures', *Corrosion*, **19** No. 9, Sept., 308t–314t (1963)
16. Gulyaev, A. P. and Georgieva, I. Ya., *Zashchita Metallov*, **1** No. 6, 652–657, Nov.–Dec. (1965)
17. *Technical Data on Fansteel Niobium*, Bulletin TDB, Fansteel Metallurgical Corporation
18. *Corrosion Tables of Special Materials and Rare Metals*, Jacob and Korves GmbH
19. Meyll Shpeydel, *Collection Nioby i Tañtal* (Nb and Ta), edited by O. P. Kolchina (1960)
20. Gulbransen, E. A. and Andrew, K. F., *Trans. A.I.M.E.*, **188**, 586–599 (1950)
21. Barto, R. L. and Hurd, D. T., *Research and Development*, 26–30, Nov. (1966)
22. Stoughton, L. D. and Sheehan, T. V., *Mechanical Eng.*, **78**, 699–702 (1956)†
23. Lloyd, E. D. in *Plansee Proceedings 1958—High Melting Metals*, Metallwork Plansee AG, Reutte Tyrol, 249–256 (1959)†
24. Parkman, R. and Shepard, O. C., US Atomic Energy Commission Publication No. ORO 45, June 11 (1951)†
25. Frost, B. R. T. and Addison, C. C., *et al.*, in *Proceedings of Second United Nations International Conference on the Peaceful Uses of Atomic Energy*, Geneva, 1958, **7**, *Reactor Technology*, 139–165†
26. Parr, G. W. and Graham, L. W., *Bull. Inst. Metals*, **4**, 125–126, Dec. (1958)†
27. Grassi, R. C., Bainbridge, D. W. and Harman, J. W., US Atomic Energy Commission Publication No. AECU 2 201, July 31 (1952)†
28. Miller, E. C., Chapter 4 of *Liquid Metals Handbook*, US Atomic Energy Commission, Navy Dept., Washington, D.C., 144–183 (1952)†
29. Wilkinson, W. D., US Atomic Energy Commission Publication No. ANL–5027, Aug. (1953)
30. Jaffee, R. I., Evans, R. M., Fromm, E. A. and Gonser, B. W., US Atomic Energy Commission, Publication No. AECD–3 317 (1949) and *Metal Abstracts*, **20**, 241 (1952)†
31. Cathcart, J. V. and Manly, W. D., *Corrosion*, **12**, 87t–91t (1956)†
32. Cunningham, J. E., US Atomic Energy Commission Publication No. ORNL–CF–51–7–135, 78, July 23 (1951)†
33. Reed, E. L., *J. Am. Ceram. Soc.*, **37**, 146–153 (1954)†
34. Cottrell, W. B. and Mann, L. A., *Nucleonics*, **12**, 22–25, Dec. (1954)†
35. Wyatt, L. M. and Dickinson, F. S., *Welding and Metal Fabrication*, **25**, 378–385, 396 (1957)†
36. Raines, G. E., Weaver, C. V. and Stang, J. H., US Atomic Energy Commission, Publication No. BMI–1 284, Aug. (1958)†

*These references can be found collectively in *Properties of Refractory Metals*, by S. J. Burnett, U.K.A.E.A. Research Report No. AERE R4 657, H.M.S.O. (1969).

†These references can be found collectively in *Columbium and Tantalum*, edited by F. T. Sisco and E. Epremian, Chapt. 8, Wiley (1963).

37. Adams, R. M. and Sittig, M., in *Liquid Metals Handbook*, *Sodium–NAK Supplement*, US Atomic Energy Commission, Navy Dept., Washington, D.C., 14 (1955)†
38. Eichelberger, R. L., US Atomic Energy Commission Publication No. BNL–489; Proceedings of the French–American Conference on Graphite Reactors, 168–173, Nov. 12–13 (1957)†
39. US Atomic Energy Commission Publication No. ISC–1 118, Semi-annual Summary, Research Report in Engineering for July–December, Ames Laboratory, Ames, Iowa (1958)†
40. McIntosh, A. B. and Bagley, K., *J. Inst. Metals*, **84**, 251–270 (1955–1956)†
41. US Atomic Energy Commission Publication No. ISC–978, Semi-annual Research in Engineering for July–December, Ames Laboratory, Ames, Iowa (1957)†
42. US Atomic Energy Commission Publication No. ISC–607, Quarterly Summary Research Report in Metallurgy for January, February and March, Ames Laboratory, Ames, Iowa (1955)†
43. US Atomic Energy Commission Publication No. ISC–506, Quarterly Summary Research Report in Metallurgy for April, May and June, Ames Laboratory, Ames, Iowa (1954)†
44. US Atomic Energy Commission Publication No. ISC–423, Quarterly Research Report in Metallurgy for July, August and September, Ames Laboratory, Ames, Iowa (1953)†
45. Hodge, W., Evans, R. M. and Haskins, A. F., *J. of Metals*, **7**, 824–832 (1955)†
46. Lakhiani, D. M. and Shreir, L. L., *Nature*, *Lond.*, **188**, 4 744 (1960)
47. Vermilyea, D. A., *J. Electrochem. Soc.*, **102**, 207 (1955)
48. Miller, G. L., *Tantalum and Niobium*, Academic Press, 328 (1959)
49. Argent, B. B., *J. Inst. Metals*, **85**, 547–551 (1957)
50. Kieffer, R. von, Bach, H. and Slempkowski, I., *Werkstoffe und Korrosion*, No. 9, 782–784, Sept. (1967)
51. Dayton, R. W. and Tipton, C. R., Battelle Memorial Inst. Progress Reports: BMI 1 324, March 1 (1959), BMI 1 340, March 1 (1959) and *Nuclear Science Abs.*, **13** Nos. 18 089 and 18 090 (1959)†
52. '*Niobond*' *and* '*Tibond*', Marston Excelsior, Wolverhampton, England (1974)

BIBLIOGRAPHY

Meyll Shpeydel, *Collection Niobi i Tantal*, Ed. O. P. Kolchina, 565–571 (1960)
Miller, G. L., *Tantalum and Niobium*, Butterworths, London (1959)
Columbium and Tantalum, Ed. F. T. Sisco and E. Epremian, Wiley, New York and London (1963)
Materials and Design, 74–78, Jan. (1968)
Tantalum, Niobium and Zirconium, Murex Ltd., Rainham, Essex

5.4 Titanium and Zirconium

Titanium and zirconium both belong to Group IVa of the Periodic Table, and have many similarities in their metallurgical and chemical properties. They both have a very strong affinity for oxygen, and their excellent corrosion resistance results from the presence at the metal surface of a tenacious compact film of oxide. At temperatures in excess of 1 000°C both metals readily take up oxygen, nitrogen, hydrogen and carbon, the product being so brittle that it cannot easily be worked. Thus it was not until the advent of novel methods of reduction from the chlorides, followed by techniques of purification and consolidation by remelting under high vacuum or an inert atmosphere, that the metals were rendered sufficiently ductile to become attractive to the engineer.

The development of titanium resulted from the ever-increasing demands of the aircraft and aero-engine designer for materials of high strength but low density. The demand for zirconium arose from the requirement in nuclear reactors for a material having a reasonable transparency to neutrons coupled with moderate strength and good oxidation resistance in high-temperature water and in hot carbon dioxide.

Thus, while titanium and zirconium were primarily developed for specific purposes, their increasing availability in commercial quantities made it economically possible for the chemical industry to take advantage of their exceptionally good corrosion resistance.

TITANIUM

Physical Properties

In its physical properties titanium shows several interesting differences from the commonly employed structural metals. It undergoes a crystal transformation at 882°C; above this temperature it has a b.c.c. structure, designated a β phase, while below it, it has a c.p.hex. structure, known as the α phase. This α phase has a c:a ratio of 1·587—significantly lower than that for other hexagonal metals such as magnesium, zinc and cadmium. There are thus more slip planes available for deformation, and high-purity titanium is therefore a relatively ductile metal at room temperature, e.g. it can be deformed 95% or more between anneals. In many alloys the phase transformation can be used to obtain a moderate increase in strength, but at the expense of ductility. Thus the commercially pure metal is soft enough to be

Table 5.13 Physical properties of titanium

Atomic number	*Atomic weight*	*Crystal structure*		*Melting point* (°C)	*Density* (g/cm^3)	*Thermal conductivity* (W/m^2 °C)	*Electrical resistivity* (μΩ/cm)	*Specific heat* (J/g°C)	*Thermal expansion* (°C)	*Magnetic susceptibility* (10^{-6} c.g.s. units/g)	*Standard electrode potential* (V)
		Below 882°C	Above 882°C								
22	47·9	c.p. hex. (α) a = 2·9504 Å c = 4·683 Å c/a = 1·587	b.c.c. (β) a = 3·3065 Å (900°C)	1600 ± 10	4·51	17·04 (at 20°C)	55·4 (at 20°C)	0·528 (0–100°C)	$8{\cdot}9 \times 10^{-6}$ (0–100°C)	+1·25	−1·63

readily cold formed, and the higher-strength alloys are easily forged. Machining is carried out with conventional machine tools, but at slower speeds than is usual for most metals and alloys, while titanium and most of its alloys may be welded by argon-arc techniques with argon shielding on both sides of the weld. Physical properties of titanium are given in Table 5.13.

Mechanical Properties

The mechanical properties of titanium are affected by small amounts of oxygen and nitrogen in solid solution. It is therefore possible to produce a number of grades of commercially pure titanium, with mechanical properties suited to the different applications for which the material is intended.

Table 5.14 Mechanical properties of titanium

0·2% *Proof stress* (h bar)	*Tensile strength* (h bar)	*Elongation on 50 mm* (%)	*Fatigue limit* (% of T.S.)	*Bend radius on 2 mm*	*Young's modulus* (h bar × 10³)	*Density* (g/cm³)
20–46	29–74	25–15	50	$1t$–$2\frac{1}{2}t$	10·5–12·5	4·51

The figures of Table 5.14 indicate the range of values normally attained with the commercially available metal.

Corrosion Behaviour

General Titanium is intrinsically very reactive, so that whenever the metal surface is exposed to air, or to any environment containing available oxygen, a thin tenacious surface film of oxide is formed. This oxide, which is present on fabricated titanium surfaces at normal or slightly elevated temperatures, has been identified as *rutile*, a tetragonal form of titanium dioxide, and it is the presence of this surface film which confers upon titanium excellent corrosion resistance in a wide range of corrosive media.

The fundamental mechanism involved in the binding of the surface film to the substrate is not known with any certainty. According to the Pilling–Bedworth principle, the oxide formed at the surface of titanium should occupy a larger volume than the metal from which it was formed, and hence it will be compressively stressed. It is argued that thin compact films resulting from inward movement of oxygen, such as those formed on titanium are more protective than surface films formed by outward movement of cations. It has been pointed out[1,5] that outward movement of cations may well leave vacancies at the metal/oxide interface, leading to a weakening of the bond between oxide and metal.

Nakayama[2], however, has suggested that, for rutile, which is tetragonal in structure, the strong bond between metal and oxide results from the favourable spacing between titanium ions in the rutile lattice and those in the metal structure. This explanation, however, does not account for the fact that other oxides of titanium, such as *brookite*, which is orthorhombic, and *anatase*, which is tetragonal, are also protective[3].

It is perhaps a significant observation that protective films on titanium appear usually to be formed when the metal surface has access to water, even though this may be present only in trace quantity and in vapour form. Thus, if titanium is exposed to some vigorously oxidising environments in the complete absence of moisture, any surface film that is formed is not protective and oxidation in depth may take place, often in the form of a violent exothermic reaction. Examples of this pyrophoric behaviour are to be found in reactions, which may be initiated at room temperature, between titanium and dry oxygen-rich atmospheres[4] at pressures above 345 kN/m^2 between titanium and dry chlorine[6], and between titanium and dry nitric acid containing nitrogen dioxide[7] (see p. 5:48). In such reactions the concentration of oxidising agent is clearly a determining factor, but the presence of moisture plays a significant part in inhibiting the attack. Thus, in oxygen-rich atmospheres the limiting oxygen concentration, below which the exothermic reaction does not take place, is of the order of 35%, the limiting concentration of nitrogen dioxide below which the reaction does not occur in dry, red, fuming nitric acid is about 1%, while in gaseous chlorine the presence of 0·013% of water vapour is sufficient to prevent significant attack. It is perhaps pertinent to record that one proposed mechanism of passivation of the metal surface suggests that it could well result from direct reaction between the metal and hydroxyl ions[8].

Chemical environments There is ample evidence to suggest that in aqueous liquid environments the presence of an oxidising species, possibly also coupled with the presence of hydroxyl ions, results in the formation of passive films at the surface of titanium. Thus, titanium is very resistant to corrosion in nitric acid at room temperature and at boiling point. While titanium has a significant corrosion rate in acids that normally produce hydrogen on reaction with metal, e.g. sulphuric acid, hydrochloric acid, the addition of small amounts of oxidising reagents results in the formation of passive films. Hence titanium is resistant to attack in mixtures of strong sulphuric and nitric acids, in mixtures of strong hydrochloric and nitric acids, in strong hydrochloric acid containing free chlorine and even in sulphuric and hydrochloric acids containing small amounts of cations such as ferric and cupric salts capable of producing an oxidising reaction[9,10] (see Sections 1.4 and 1.5).

Titanium is almost invariably resistant towards neutral salts, particularly halides, at temperatures up to 100°C, and in respect of the latter environments it is significantly more resistant than stainless steel. In strong solutions of caustic alkalis, on the other hand, titanium tends to form soluble titanates, and it is not as resistant as say, nickel. While at low or moderate concentrations of alkali there is no significant attack, the metal has appreciable solubility in concentrated or molten caustic alkali. Titanium is however resistant to attack by aqueous ammonia at all concentrations and temperatures and to anhydrous ammonia[10].

The overall pattern of behaviour of titanium in aqueous environments is perhaps best understood by consideration of the electrochemical characteristics of the metal/oxide and oxide-electrolyte system. The thermodynamic stability of oxides is dependent upon the electrical potential between the metal and the solution and the pH (see Section 1.4). The Ti/H_2O system has been considered by Pourbaix[11]. The thermodynamic stability of an insoluble phase does not mean that it will form a protective film which will isolate the metal surface from the environment, as this depends on the physical properties of the film, and these cannot be predicted from thermodynamic data.

In general, however, for titanium immersed in acid solutions, potentials above zero on the saturated calomel scale are conducive to the formation of protective oxide, while at certain negative potentials hydride films, which also confer some protection, can be formed[12]. Between the potential at which a continuous hydride film is formed and that at which protective oxide films appear, soluble titanium ions are produced and rapid corrosion ensues.

The concept of a 'protective potential' region explains why the addition of oxidising substances or oxidising metallic ions to a non-oxidising acid often prevents the corrosion of titanium. For example, it has been shown that the addition of cupric ions to 3N hydrochloric acid raises the metal/electrolyte potential into the passive region and thus results in the formation of protective films[9,13]. The same concept of a protective potential region also explains why the addition of chlorine to strong hydrochloric acid causes titanium to become passive, and why the presence or absence of traces of dissolved oxygen in formic acid leads to the existence of 'borderline passivity' in that acid. Thus in the passivation of titanium the redox potential of its solution is of particular significance.

The deliberate raising of the electrical potential of titanium, either by the attachment of discrete particles of a noble metal, such as platinum or palladium, at the surface, or by the application of positive direct current to force the formation of a protective film, is dealt with at a later point. The electrochemical aspect of the corrosion of titanium is comprehensively treated in a number of papers[3,12,14–16].

The behaviour of titanium in a wide range of chemical environments is well summarised in the literature produced by the prime producers of titanium[10] and the reader is advised to consult it for detailed information on corrosion rates. In general, titanium is only resistant to pure 'non-oxidising' acids such as sulphuric and hydrochloric, at dilutions of the order of 2–5% and at moderate temperatures, e.g. 60°C, although, as already indicated, the presence of air and oxidising agents can improve this situation. It has excellent resistance to corrosion in most organic acids, but there are certain exceptions in that in formic acid a condition of borderline corrosion exists at concentrations above about 10%, while there is considerable attack by oxalic acid even in dilute concentrations. In citric acid there is significant corrosion in 50% strength at 100°C, and the metal is not resistant to trichloroacetic acid.

In its resistance to liquid metals, titanium shows variable behaviour, the rate of attack often depending upon temperature and increasing with rise in temperature. By thickening the surface film of oxide, resistance to attack is enhanced, and, for example, repeated repair of the surface film renders titanium resistant, on a limited-time basis, to molten zinc in galvanising baths. A surface-oxide thickening technique also enables titanium to be employed in contact with molten aluminium. Titanium equipment is also used in applications involving lead–tin solders, and it is resistant to mercury, at least up to 150°C.

The behaviour of titanium in a wide range of chemical environments is summarised in Table 5.15. In practice, however, whatever the intrinsic reaction between a metal and its environment, account must often be taken of

Table 5.15 Resistance of titanium and zirconium to chemical reagents

*Reagent**	*Concentration by weight* (%)	*Temperature* (°C)	*Ratings†* Titanium	Zirconium
Acetic acid, aerated	5, 25, 50, 75, 99·5	boiling	A	A
Acetic acid, non-aerated	99·5	boiling	A	A
Acetic anhydride	99	room	A	A
	99·5	140		A
Aluminium chloride	5, 10	100	A	A
	25	100	C	A
Aluminium sulphate	10, 30	boiling		A
Ammonia, anhydrous (1·4 MN/m^2)	100	40	A	
Ammonium chloride	1, 10, saturated	100	A	A
Ammonium hydroxide	28	room, 60, 100	A	A
Ammonium sulphate	saturated	50		A
Aniline hydrochloride	5,20	100	A	A
Aqua regia ($1HNO_3:3HCl$)	—	room	A	C
Barium chloride	5, 20	100	A	A
Benzene	—	room	A	A
Benzoic acid	saturated	room, 60	A	A
Bromine	liquid	30	C	
	moist liquid	room	C	
	vapour	30	A	
Bromine-saturated water	—	room, 60	A	
Calcium chloride	5, 10, 25	100	A	A
	62	154	AC	AC
	73	177	B	C
Calcium hypochlorite	2, 6	100	A	A
	18–20	21–24	A	C
Carbon tetrachloride	100	room, 50	A	A
Chlorine gas, dry (less than 0·005% H_2O)	—	30	C	A
Chlorine gas, wet (more than 0·013% H_2O)	—	75	A	C
Chlorine-saturated water	—	room	A	A
Chloroacetic acid	100	100	A	A
Chromic acid	10, 50	boiling	A	A
	36·5	90	A	A
	10, 20, 30	20, 50, 100	A	A
Citric acid, aerated	10, 25, 50	100	A	A
Citric acid, non-aerated	50	boiling	B	A
Cupric chloride	1	35	A	A
	2·5, 5, 7·5, 10	35	A	B
	12·5, 15, 20	35	A	C
	1	60	A	A
	2·5	60	A	B
	5, 7·5, 10, 12·5, 15, 20	60	A	C
	1, 2·5, 5, 7·5,	100	A	B

Table 5.15 (continued)

*Reagent**	*Concentration by weight* (%)	*Temperature* (°C)	*Ratings†*	
			Titanium	Zirconium
Cupric chloride (continued)	10, 12·5, 15, 20	100	A	C
	40	boiling	A	C
	55	118	A	C
Dichloroacetic acid	100	100	A	B
Ethyl alcohol	95	boiling	A	A
Ferric chloride	1, 2·5, 5, 10	35	A	A
	15, 20	35	A	B
	25, 30	35	A	C
	1, 2·5, 5	60	A	A
	10, 15	60	A	B
	17·5–30	60	A	C
	1, 2·5	100	A	A
	5	100	A	B
	10, 30	100	A	C
	50	113, 150	A	C
Formaldehyde	37	boiling	A	A
Formic acid, aerated	10, 25, 50, 90	100	A	A
Formic acid, non-aerated	10	boiling	A	A
	25, 50, 90	boiling	C	A
Furfuryl alcohol	—	170		A
Hydrochloric acid, aerated	0·5	35	A	A‡
	1, 3	60	A	A‡
	1	100	B	A‡
	2, 3	100	C	A‡
	4, 5	60	B	A‡
	5	room	A	A‡
	7·5, 10	35	B	A‡
	15, 37	35	C	A‡
	20	room	B	A‡
Hydrochloric acid, non-aerated	1	35	A	A‡
	1	70	A	A‡
	1	boiling	C	A‡
	3, 5	room	A	A‡
	3, 5	70	B	A‡
	3, 5	boiling	C	A‡
	5	50	C	A‡
	10, 18	room	B	A‡
	10	70	C	A‡
	20, 37	room	C	A‡
Hydrochloric acid/nitric acid mixtures	1:3, 2:1, 3:1, 4:1	room	A	C
	7:1, 20:1	room	A	
	5:1	boiling	B	
Hydrofluoric acid	1	room	C	C
Hydrofluoric acid, anhydrous	100	room	C	C
Hydrogen peroxide	3, 6, 30	room	AB	
	10	50		A
Hydrogen-sulphide-saturated water	—	70	A	
Lactic acid, aerated	10, 25, 50, 85	100	A	A
Lactic acid, non-aerated	10, 25, 50, 85, 100	boiling	A	A
Magnesium chloride	5, 20, 42	boiling	A	
	42	35, 60, 100		A

Table 5.15 (continued)

*Reagent**	*Concentration by weight* (%)	*Temperature* (°C)	*Ratings†* Titanium	Zirconium
Manganous chloride	5, 20	100	A	A
Mercuric chloride	1, 5, 10, saturated	100	A	A
Monochloroacetic	30	80	A	A
acid	100	boiling	A	A
Nickel chloride	5, 10	100	A	A
	20	100		A
Nitric acid, aerated	5, 10, 30, 40, 50, 60,	100	A	A
	69·5			
	20	290	B	
	40	200	B	A
	65	175	A	A
	70	270	B	
Nitric acid,	65	boiling	A	A
non-aerated	90	room	A	A
	red fuming	room, 50, 70	§	§
Nitric acid/hydrofluoric acid mixture	15:1	room	C	
Oxalic acid, aerated	0·5, 1, 5, 10	35	A	A
	0·5, 1, 5, 10, 25	60, 100	C	A
Phenol	25, 50, 100	20, 50 boiling		A
Phosphoric acid	5, 10, 20, 30	35	A	A
	35–85	35	B	A
	5–35	60	B	A
	5	100	B	A
	10	80	C	A
Potassium hydroxide	10	boiling	A	A
	50	27	A	A
Propionic acid	vapour	190	C	
Propylene oxide	100	34		A
Silver nitrate	20	room, 60	A	A
Sodium bisulphate	saturated	room	A	
Sodium carbonate	20	boiling	A	
Sodium chlorate	saturated	room	A	
Sodium chloride	saturated	room, 111	A	A
Sodium cyanide	saturated	room	A	
Sodium dichromate	saturated	room	A	
Sodium hydroxide	5–10	21	A	A
	10	boiling	A	A
	50	38–57	A	A
	73	113–129	AB	A
	caustic fusion, 73% to anhydrous NaOH	121–538	C	B
Sodium hypochlorite,	—	boiling	A	
10% w/v av. chlorine	40	80	A	
Sodium nitrate	saturated	room	A	
Sodium phosphate	saturated	room	A	A
Sodium silicate (75° Tw.)	—	room, 60	A	

Table 5.15 (continued)

*Reagent**	*Concentration by weight* (%)	*Temperature* (°C)	*Ratings†*	
			Titanium	Zirconium
Sodium sulphate	saturated	room, 60	A	
Sodium sulphide	saturated	room	A	
Sodium sulphite	saturated	room	A	
Stannic chloride	5, 24	100	A	A
Sulphur, molten	100	240	A	
Sulphur dioxide, dry	100	room, 60	A	
SO_2 gas saturated with water vapour (see also sulphurous acid)	—	room, 60, 70	A	
Sulphuric acid	1, 3, 5	35	A	A
	10	35	B¶	A
	20–50	35	C¶	A
	60–70	35	B¶	A
	above 70	35	C	
	above 80	35	C	C
	1, 5	boiling	C	A
	10	50	C	A
Sulphurous acid	6	room	A	A
(see also SO_2/H_2O)	6	100		A
Tannic acid	25	100	A	A
Tartaric acid	10, 25, 50	100	A	A
Trichlorethylene	—	boiling	A	A
(unstabilised)				A
Trichloroacetic acid	100	100	C	C
Trisodium phosphate	5–20	35–100		A
Uranyl sulphate	U_1, 4 g/l	250		A
Zinc chloride	20, 50, 75	150	A	
	75	200	B	
	80	173, 200	C	
	75–90	150–250	C	

*Aqueous solutions, not aerated unless otherwise indicated.
†A. Attack <0·125 mm/y. The material is regarded as suitable for use where little dimensional change can be tolerated.
B. Attack between 0·125 and 1·25 mm/y. The material may be suitable for use provided that some corrosion is permissible.
C. Attack >1·25 mm/y. The material is ordinarily considered unsuitable.
‡See text, p. **5**.54.
§Not recommended. See pp. **5**.47 and **5**.53.
¶See p. **5**.46.

the existence of special circumstances, usually localised in character, which can give rise to corrosion even with an apparently innocuous medium. Thus, the presence of bimetallic couples, the existence of deep crevices, the presence of abrasive particles in a liquid stream, the incidence of local tensile stress, or the application of reciprocating stresses, are all features which demand consideration when a specific material of construction is to be selected.

Resistance to erosion Titanium has outstanding resistance to erosion resulting from the presence of abrasive particles entrained in cooling water and in process liquors[17–20]. In one practical trial as a steam condenser tube, under circumstances known to result in rapid erosion of conventional condenser-tube alloys, titanium was virtually unmarked after more than 10 years service. As a turbine-blade material subject to impingement by water droplets moving at very high speeds, titanium has been shown to be superior to the conventional Fe–13Cr, Fe–18Cr–8Ni+Ti and to Monel. Under such circumstances, the harder the blade the better resistance it offers to erosion, and titanium alloys of the Ti–6Al–4V type are likely to give even better service than commercially pure titanium[18].

Resistance to crevice corrosion Titanium is more resistant to crevice corrosion than most conventional metals and alloys, particularly where differential aeration is involved, e.g. it is very resistant to crevice attack in sea water at normal temperatures. This form of corrosion becomes more severe when acidity develops in a crevice and this is more prone to occur under conditions of heat transfer[21]. Under these circumstances, especially in the presence of halide, even titanium may suffer attack, and the metal should not be employed in strong aqueous halides at temperatures in excess of 130°C. This limiting temperature can be raised to 180°C by use of the Ti–0·15Pd alloy[21–23]. (See also Sections 1.4 and 1.6.)

Some crevice attack upon titanium can also occur in the presence of gaseous chlorine gas at temperatures below 100°C, but this is mainly confined to crevices formed between titanium and organic sealing compounds. Here again, the Ti–0·15Pd alloy is less prone to attack.

Titanium in contact with other metals In most environments the potentials of passive titanium, Monel and stainless steel, are similar, so that galvanic effects are not likely to occur when these metals are connected. On the other hand, titanium usually functions as an efficient cathode, and thus while contact with dissimilar metals is not likely to lead to any significant attack upon titanium, there may well be adverse galvanic effects upon the other metal. The extent and degree of such galvanic attack will depend upon the relative areas of the titanium and the other metal; where the area of the second metal is small in relation to that of titanium severe corrosion of the former will occur, while less corrosion will be evident where the proportions are reversed[17]. Metals such as stainless steel, which, like titanium, polarise easily, are much less affected in these circumstances than copper-base alloys and mild steel.

In acid solutions, the behaviour of titanium/dissimilar-metal couples may differ from that just described, and on occasion titanium may be anodic to stainless steel and even to aluminium[24]. In chemical-plant environments, therefore, it is usual to take the precaution of insulating titanium from adjacent components constructed from other metals.

Resistance to stress-corrosion cracking Commercially pure titanium is very resistant to stress-corrosion cracking in those aqueous environments that usually constitute a hazard for this form of failure, and with one or two exceptions, detailed below, the hazard only becomes significant when titanium is alloyed, for example, with aluminium. This latter aspect is discussed in Section 8.5 under titanium alloys.

For commercially pure titanium, the specific environments to be avoided are pure methanol and red, fuming nitric acid[25–28,65], although in both environments the presence of 2% of water will inhibit cracking. On the other hand, the presence of either bromine or iodine in methanol aggravates the effect. When it does occur, stress-corrosion cracking of commercially pure titanium is usually intergranular in habit.

Resistance to fatigue and corrosion fatigue The resistance of titanium to fracture by fatigue, induced by imposition of rapidly reversing stresses, compares favourably with that of the more conventional metals and alloys.

Commercially pure titanium has a definite fatigue limit, in air, at about half its tensile strength, and at this figure fracture may take place at between 10^7 and 10^8 reversals. In this respect the commercially pure metal resembles steel rather than the non-ferrous alloys. Reversed stresses at a figure below the limit indicated are not likely to result in fatigue failure, irrespective of the number of reversals applied.

For many metals, the presence of corrosive environments coupled with reversing stress results in fracture by corrosion fatigue at a stress level well below that of the normal fatigue limit. While, given the appropriate environment, titanium is not immune to this effect, its generally good resistance to corrosion renders corrosion fatigue a comparatively rare event. Thus, the fatigue limit for titanium wetted with sea water is very similar to the figure obtained in air[29]. It is, therefore, not surprising that valve springs and valve plates of titanium alloys, used in gas compressors, give better performance than the conventional alloy steels.

High-temperature behaviour Commercially pure titanium is an established material for use at the moderately elevated temperatures attained in aircraft exhaust shrouds and firewalls, but neither titanium nor titanium alloys are suitable for use at really high temperatures. The tensile strength of commercially pure titanium shows a steady fall with increase in temperature, the tensile strength at 350°C being approximately half that at room temperature. The creep strength is improved by suitable alloying and a three-phase alloy containing 11% Sn, 5% Zr and smaller quantities of Al, Mo and Si has a high creep strength at 400°C, a fair proportion of which is retained even at 500°C.

At about 500°C appreciable diffusion of hydrogen into titanium can take place and at about 600°C significant penetration of oxygen and nitrogen occurs. It has already been indicated that the presence of these elements renders the titanium brittle, and this feature must be taken into account in considering the use of titanium at elevated temperatures. Titanium has nevertheless been successfully employed as an autoclave lining in steam atmospheres at a temperature of 400°C and a pressure of 10 MN/m^2.

Examples of the Use of Titanium in Chemical Plant

As with all fairly expensive materials of construction, economy in the use of titanium can often be achieved by using it in the form of thin linings upon a thicker load-bearing support. The soundest and most economical method of achieving this is by explosively bonding thin titanium sheet to thick steel plate. This technique, coupled with that of welding the duplex plate into large reaction vessels, is well established[30–31,66]. The decision as to whether to use solid titanium or to employ explosively clad steel, depends upon the size and wall thickness of the construction and is largely decided upon economic grounds. The reader is advised to consult the specialist suppliers and fabricators before deciding which form to employ.

For some chemical plant applications the iron content of the titanium employed can influence its behaviour, e.g. in some strengths of nitric acid, and in chlorine dioxide, preferential weld attack may occur if the iron content of the titanium is above a certain critical level. This effect is only encountered in a few specific environments[66], but where these are involved, it is recommended that titanium with an iron content of less than 0·05% is specified.

There are also occasions, particularly in hydrogen-containing atmospheres, when surface contamination of the titanium with iron can result in localised corrosion and embrittlement. This effect can be countered by avoidance of undue contamination with iron during fabrication, by post-fabrication cleaning and by post-fabrication anodising[10,16,67].

Examples of the use of titanium in the chemical industry are briefly summarised below; for more detailed treatment the reader is advised to consult Reference 32. Titanium-lined equipment is being employed in the food industry[33] where lack of contamination of the product is essential, for organic-reaction vessels used in contact with nitric acid at elevated temperatures and pressures[34], in a large rotary ammonium chloride dryer[35], for chlorine dioxide bleaching[36], for emulsion pans holding photographic solutions[35], and for general service in corrosive liquors.

For heat-exchange duty, there are now many plate-type heat exchangers in use, as well as the more conventional shell and tube form of equipment. Shell and tube condensers are being employed to handle 62% nitric acid at elevated temperatures[33], with hydrochloric acid containing free chlorine[37], with chromic acid[38] and in oil-refining equipment[39].

Titanium impellers have been used in pumps employed for the conveyance of corrosive and erosive ore slurries, for organic chlorides containing hydrochloric acid and free chlorine[40], for handling moist chlorine gas, and in the wood-pulp and the textile-bleaching industry, particularly with sodium hypochlorite[36].

In the electroplating industry, the use of titanium as hooks[41] and as heating and cooling coils for temperature control of certain acidic liquors has improved the control of plating baths[38,42]. Perhaps the most significant advance has been in the nickel plating industry where solid nickel anodes have been largely replaced by titanium baskets holding nickel shot and nickel shapes. In this development, the titanium itself is anodically passivated, but at the same time the passive film allows electron transfer to occur between contacting surfaces of titanium and nickel so that the latter is anodically dissolved.

Anodic passivation also allows titanium to be employed as a jig for aluminium anodising baths[43], because the protective anodic film formed on titanium allows passage of electronic current to the metal contact while virtually suppressing flow of ionic current through the anodically-formed surface film. This aspect is discussed in more detail in relation to special applications.

In the field of electrowinning of metals, titanium appears to have an advantage as a cathode, upon which copper particularly can be deposited with finely balanced adhesion that allows the electrodeposited metal to strip easily when required.

In the field of nuclear energy, titanium has been used for processing of fuel elements, where this demands use of nitric acid or aqua regia[44–45], and for control-rod mechanism, in which the short half-life of irradiated titanium is of advantage.

Special Applications

Anodic Passivation

It has already been indicated that titanium is not particularly resistant to corrosion in hot, strong acids of the type that usually generate hydrogen upon reaction with metals—acids such as sulphuric or hydrochloric. In contact with such acids, corroding titanium assumes a negative electrical potential (approximately −0·7 V, S.C.E.).

If this negative potential is artificially raised by slow increments, a critical potential level may be attained, at which corrosion dramatically ceases and the metal acquires a protective film. The potential level at which this occurs usually lies between −0·5 V and −0·2 V (S.C.E.), and it is evident that at this potential level the metal/electrolyte interface has attained a thermodynamic state conducive to the formation of a stable, insoluble titanium dioxide[46]. This protective surface film has been shown to consist mainly of *anatase*, a tetragonal form of the oxide, and if the potential is further raised, the film thickens, giving a series of interference tints until it reaches a dark purple colour at a maximum limiting thickness of about 2×10^{-7} m. Once this film has become established, there is very little further passage of current into the electrolyte and corrosion virtually ceases as long as the 'protective' potential is applied. The most obvious means of attaining this potential is by application of an anodic direct current from an external source, but the same effect can also be attained to some extent by coupling the titanium to a more noble element such as carbon, or one of the platinum group metals. The latter method is, however, of limited application because it is dependent upon the potential level attained by the noble element, and this may not be sufficiently high to provide a mixed potential which is above the critical value for film formation (see Sections 1.4 and 1.5). Nevertheless, Stern and his associates[47–49] have shown that the addition of 0·2% palladium to titanium produces a discrete dispersion of palladium particles at the surface, which permits the combination to offer an adequate resistance to corrosion in 5% boiling sulphuric and hydrochloric acids. Cotton[3,50,51] has shown that application of a d.c. potential of about 2 V between titanium and a suitable cathode can prevent corrosion in a wide range of strong non-oxidising acids at concentrations and temperatures which present considerable handling difficulties with most metallic materials of construction.

One full-scale practical application of this principle of anodic passivation is found in titanium heat exchangers handling 8% sulphuric acid containing hydrogen sulphide and carbon disulphide employed in viscose rayon processing[9,52]. It is conservatively estimated that each of these anodically passivated units performs the duty previously undertaken by three graphite heat exchangers. In doing this they require a current of only 1·5 A supplied at 15 V.

Use as Anodes

As indicated above, when a positive direct current is impressed upon a piece of titanium immersed in an electrolyte the consequent rise in potential induces the formation of a protective surface film, which is resistant to passage of any further appreciable quantity of current into the electrolyte. The upper potential limit that can be attained without breakdown of the surface film will depend upon the nature of the electrolyte. Thus, in strong sulphuric acid the metal/oxide system will sustain voltages of between 80 and 100 V before a spark-type dielectric rupture ensues, while in sodium chloride solutions or in sea water film rupture takes place when the voltage across the oxide film reaches a value of about 12 to 14 V. Above the critical voltage, anodic dissolution takes place at weak spots in the surface film and appreciable current passes into the electrolyte, presumably by an initial mechanism involving the formation of soluble titanium ions.

Thus titanium by itself cannot function as an efficient anode for the passage of positive direct current into an electrolyte. The surface film of oxide formed upon the titanium has, however, a most useful property: while it will not pass positive direct current into an electrolyte (more correctly, while it will not accept electrons from negatively charged ions in solution), it will accept electrons from, or pass positive current to, another metal pressed on to it. Hence a piece of titanium which has pressed on to its surface a small piece of platinum will pass positive direct current into brine and into many electrolytes, at a high current density, via the platinum, without undue potential rise, and without breakdown of the supporting titanium[53,54].

Platinised titanium anodes (titanium carrying a thin surface film of platinum, of the order of 0·0025 mm thick) have proved successful in cathodic-protection systems employing impressed-current techniques, as electrodes for electrodialysis of brackish water, and in many applications where established anode materials suffer significant corrosion. Platinum-coated titanium anodes can operate without breakdown at very high current densities, of the order of 5 000 A/m^2, in sea water, as although the very thin platinum coating may be porous the underlying titanium exposed at the pores will become anodically passivated[55].

In aqueous chloride where it is necessary to use platinised titanium anodes coated over only part of their surface, e.g. titanium rod tipped with a thin platinum film, it may be necessary to limit the applied voltage to 12 V.

The development of platinised titanium has been extended to include the replacement of platinum by deposits of other forms of corrosion-resistant conducting surfaces, such as platinum-iridium and ruthenium oxide. Apart from their corrosion resistance, these surfaces have the ability of operating electrochemically at lower overvoltages than plated platinum or graphite. Thus, for a range of electrochemical cells used in the chlor-alkali industry for the production of chlorine and sodium chlorate, etc. there is a significant advantage in using them compared with graphite, and the first commercial cells using these metal anodes are now in operation.

By a mechanism similar to that discussed in relation to platinum coating, titanium can function as a conducting jig to support aluminium components and assemblies in conventional anodising baths. In this application the exposed titanium acquires the insulating film, but allows current to pass to the aluminium at the points of contact[56].

Pyrophoric Tendency

If titanium is exposed to certain vigorously oxidising environments, oxidation does not cease at the surface, and a rapid exothermic reaction in depth ensues. The fundamental reason for this remarkable change in the character of the oxidation is not known with any certainty, but it is significant that in almost every instance the presence of a small quantity of water, sometimes in trace amounts, prevents this rapid oxidation in depth.

Investigation into the effect has been mainly devoted to reactions with red fuming nitric acid[57]. It seems that in red fuming nitric acid a preliminary reaction results in the formation of a surface deposit of finely divided metallic titanium; ignition or pyrophoricity can then be initiated by any slight impact or friction. The tendency to pyrophoricity increases as the nitrogen dioxide content of the nitric acid rises from zero to maximum solubility at about 20%, but decreases as the water content rises, the effect being nearly completely stifled at about 2% water.

Other media in which titanium is subject to pyrophoricity are anhydrous liquid or gaseous chlorine[58], liquid bromine, hot gaseous fluorine, or oxygen-enriched atmospheres at moderately low pressures.

Titanium Alloys

Mechanical properties of various titanium alloys are given in Table 5.16. In general the behaviour of those titanium alloys developed for the aircraft industry is very similar to that of unalloyed titanium[59]. The addition of some alloying elements may increase resistance to one medium, but decrease it to others[60].

Additions of zirconium confer a significant increase in corrosion resistance, particularly in sulphuric and hydrochloric acids[59,61]. At alloying additions of the order of 50% Zr, however, there can be a significant diminution in resistance to oxidation[62] and the welding of titanium to zirconium is not advisable, because within the welded zone the proportion of titanium to zirconium will almost inevitably fall within the sensitive composition range.

The addition of 0·2% palladium to titanium decreases the corrosion rate in boiling 5% sulphuric acid by a factor of 500, and in boiling 5% hydrochloric acid by a factor of 1 500, in relation to the rates obtained with unalloyed titanium. The addition of palladium in these quantities thus provides an adequate measure of resistance to relatively weak concentrations of the acids mentioned[48].

From the corrosion-resistance aspect, one of the most effective additions to titanium is that of molybdenum. According to Yoshida and his colleagues[63-64], the addition of 15% Mo produces an alloy fully resistant to virtually all concentrations of sulphuric and hydrochloric acid at room temperatures, while with 30% Mo, the alloy is resistant to all strengths of boiling sulphuric acid up to a concentration of 40% by weight, and to 10% boiling hydrochloric acid.

The stress-corrosion cracking hazard for titanium alloys containing aluminium is significantly higher than that obtaining for commercially pure titanium, and in addition to stress-corrosion cracking in methanol and red fuming nitric acid, cracking has been observed in salt solution, in hot solid sodium chloride and in uninhibited chlorinated hydrocarbons. Because of the importance of these alloys to the aircraft industry there has been considerable laboratory investigation of the effect and the reader is advised to

Table 5.16 Mechanical properties of some titanium alloys

Nominal composition in weight % and characteristics		*0·2% proof stress (min.)* (h bar)*	*Tensile strength* (h bar)	*Elongation (min.)* (%)	*Young's modulus (typical)* (h bar × 10^3)	*Fatigue limit* (% of T.S.)	*Bend radius on 2 mm*	*Density* (g/cm³)	*Stress for 0·1% total plastic strain in 100 h* (h bar)	*Production range*‡
Ductile medium-strength Ti–2·5Cu alloy, weldable and age-hardenable, DTD 5233, 5443, 5253, 5263	annealed	46	54–77	15	10·5–12	60–65	2*t*	4·56		S, B, W, E
	solution-treated and aged	55	69–92	10	10·5–12	60–65	3*t*			
Small additions of Pd, giving improved resistance to non-oxidising acids		20	29–42	25	10·5–12	50	1*t*	4·51		S
Low-strength Ti–2Al–2Mn alloy, DTD 5043 B		46*	65–80	20	10·5–12·5	60–65		4·51		B
Medium-strength Ti–5Al–2·5Si alloy, weldable, BS TA 14, 15, 16, 17		76	82–108	10	10·5–12·5	50	4*t*	4·46		S, B, E
Medium-strength Ti–6Al–4V alloy, BS TA 10, 11, 12, 13, 28	sheet	90	96–127	8	10·5–12	55–60	5*t*	4·42		S
	rod	83	90–116	8	10·5–12	55–60				B, W, E
High-strength Ti–4Al–4Mo–2Sn–0·5Si alloy, creep resistant up to 400°C, DTD 5103, 5153		100	114 min.	9	11–13	50–60		4·60	46 at 400°C	B
Very-high-strength Ti–4Al–4Sn–0·5Si–4Mo alloy, DTD 5203, 5223		110	124 min.	8	11–13	40–50		4·6		B

*1 h bar = 10^7 N/m² = 10 MN/m².

Table 5.16 (continued)

Nominal composition in weight % and characteristics		0·2% proof stress (min.) (h bar)	Tensile strength (h bar)	Elongation (min.) (%)	Young's modulus (typical) (h bar × 10^3)	Fatigue limit (% of T.S.)	Bend radius on 2 mm	Density (g/cm^3)	Stress for 0·1% total plastic strain in 100 h (h bar)	Production range‡
High-strength Ti–11Sn–5Zr–2·25Al–1Mo alloy, creep resistant up to 450°C, BS TA 18, 19, 20, 25, 26, 27	air-cooled and aged	88	103–127	8	10·5–11	55–60		4·84	39 at 450°C	B
	quenched and aged	97	111–134	8	10·5–11	55–60				
Very-high-strength Ti–11Sn–4Mo–2·25Al–0·2Si alloy, DTD 5213, M 160	furnace-cooled and aged†	93*	105 min.	8	10·5–12	60		4·86		B
	quenched and aged	110	124 min.	10	10·5–12	45–55				
Medium-strength Ti–6Al–5Zr–1W–0·3Si alloy, creep resistant up to 520°C and weldable, DTD M200		88	99 min.	6	≈12·5	50		4·48	31 at 520°C	B
Medium-strength alloy containing Al, Zr, Mo, Si, creep resistant up to 550°C and weldable		88	99 min.	6	≈12·5	50		4·4	31 at 520°C	B
Ultra-high-strength Ti–6Al–5Zr–4Mo–1Cu–0·2Si alloy, creep resistant up to 400°C, DTD M 201	air-cooled and aged	111	127 min.	8	≈12	45–55		4·6		B
	quenched and aged	124	135 min.	6	≈12	45–55				

*0·1% P.S.
†In furnace-cooled and aged condition, fracture toughness value K_{IC} = 50 h bar $(cm)^{\frac{1}{2}}$ minimum.
‡S = sheet; B = rods, bars and billets; W = wire; E = extrusions.

Table 5.17 Physical properties of reactor-grade unalloyed zirconium

Atomic number	*Atomic weight*	*Crystal structure*		*Melting point* (°C)	*Density at 20°C* (g/cm^3)		*Thermal conductivity** (W/m^2 °C)	*Electrical resistivity* (μΩ cm)	*Temperature coefficient of resistivity* (°C)	*Specific heat** (J/g°C)	*Thermal expansion* (°C)†	*Standard electrode potential* (V)	*Thermal neutron absorption cross-section. Reactor grade*	
		Below 862±2°C	Above 862°C		*Reactor grade*	*Hafnium-containing zirconium*							*Micro-scopic*	*Macro-scopic*
40	91·2	c.p.hex. (α) at 25°C a = 3·232 Å c = 5·157 Å	b.c.c. (β) at 900°C a = 3·61 Å	1852 ±10	6·505	6·55	20·93	40–44	40×10^{-4}	0·276	$5{\cdot}04 \times 10^{-6}$	−1·53	0·180 barn‡/atom	0·08 mm

*At 25°C. †20–200°C. ‡1 barn = 10^{-24} cm^2.

consult References 65 and 66 and the literature for a comprehensive treatment of the subject. (See also Section 8.5.)

Viewed in perspective, evidence of failure in service has been rare and the practical hazard is certainly very much lower than would appear from the results of laboratory tests. In chlorinated hydrocarbons the effect can be controlled by the addition of inhibitors, and, for example, the appropriate commercial degreasants containing these inhibitors are specified in a British defence standard*.

ZIRCONIUM

The growth of nuclear engineering with its specialised demands for materials having a low neutron absorption coupled with adequate strength and corrosion resistance at elevated temperatures, has necessitated the production of zirconium in relatively large commercial quantities. This specific demand has resulted in development of specially purified zirconium, and certain zirconium alloys, for use in particular types of nuclear reactor.

In the natural state zirconium is associated with hafnium, and for use in nuclear reactors it is essential to separate the hafnium because it readily absorbs neutrons. This situation gives rise to bulk production of two forms of raw zirconium metal, a hafnium-free reactor grade and a commercially pure hafnium-bearing quality. Three types of alloy are also commercially available: a reactor quality suitable for use in water-cooled reactors—this alloy was developed in the United States and is known as 'Zircaloy 2', a similar alloy containing hafnium, and, for use in gas-cooled reactors, an alloy known as 'ATR', developed in the research laboratories of Associated Electrical Industries (Manchester) Limited.

For the chemical engineer not particularly concerned with atomic energy, unalloyed zirconium of either grade is an appropriate choice for those occasions requiring the special corrosion-resistant properties exhibited by this metal.

Physical and Mechanical Properties

The physical properties of unalloyed zirconium are recorded in Table 5.17.

Mechanical properties of these grades of zirconium depend to a large extent upon the purity of zirconium sponge used for melting. Hardness and tensile strength increase rapidly with rise in impurity content, notably oxygen, nitrogen and iron. Typical mechanical properties are listed in Table 5.18.

Table 5.18 Mechanical properties of reactor-grade unalloyed zirconium

Tensile strength (h bar)	0·2% *Yield stress*	*Elongation* (%)	*Young's modulus* (h bar)
35–43	14·4	40	21×10^3

**The Cleaning and Preparation of Metal Surfaces*, Defence Standard 03–2/1 (1970), obtainable from the Ministry of Defence, First Avenue House, High Holborn, London, W.C.1.

Table 5.19 gives the physical properties of Zircaloy 2 and ATR alloy.

Table 5.19 Physical properties of Zircaloy 2 and ATR alloy

Alloy nominal composition (%)	*Density at* 20°C (g/cm²)	*Electrical resistivity at* 21°C (μΩ cm)	*Coefficient of linear thermal expansion* (°C)		*Thermal neutron absorption cross-section*	
			20–700°C	25–600°C	Micro-scopic	Macro-scopic
Zircaloy 2 Zr–1·5 Sn–0·1 Cr–0·12 Fe–0·05 Ni	6·57	74	$6{\cdot}5 \times 10^{-6}$	—	0·22–0·24 barn*/atom	—
ATR Zr–0·5 Cu–0·5 Mo	6·55	—	—	$5{\cdot}41 \times 10^{-6}$	—	0·0114 cm^{-1}

*1 barn = 10^{-24} cm².

The mechanical properties of the alloys will vary slightly according to the purity of sponge, and also with heat treatment. The properties quoted in Table 5.20 may be regarded as typical.

Table 5.20 Mechanical properties of Zircaloy 2 and ATR alloy

Alloy	*Tensile strength* (h bar)	0·1% *Proof stress* (h bar)	*Elongation* (% on 50 mm)
Zircaloy 2	46–54	10·5–42	25–32
ATR	46–62	31–46	25–35

Behaviour of Commercially Pure Zirconium in Aqueous Environments

Zirconium, like titanium, depends upon the integrity of a surface film, usually of oxide, for its corrosion resistance, but there are differences in behaviour between the two metals when they are exposed to aggressive aqueous environments.

In general, zirconium does not equal titanium in resistance to certain oxidising media, but it is superior in non-oxidising acids, and in caustic alkalis. The presence of certain impurities in zirconium influences the corrosion behaviour, and while small amounts of hafnium are not deleterious, carbon in amounts greater than 0·06% lessens resistance to hot concentrated hydrochloric acid by a factor of several hundreds[68,69]. The contrast in behaviour between titanium and zirconium in a wide range of media is illustrated in detail in Table 5.15.

To summarise, zirconium performs well in nitric acid at all concentrations up to 70% and temperatures up to 200°C[68], but it will react pyrophorically in a fashion similar to titanium in concentrated nitric acid containing free nitrogen dioxide. If there are appreciable amounts of hydrochloric acid present together with nitric acid, there may be severe attack, and, in contrast to titanium, zirconium is not resistant to aqua regia containing three parts nitric to one part hydrochloric acid. Towards chromic acid, zirconium is resistant at least up to a strength of 50% at a temperature of 90°C. In saturated chlorine water the corrosion rate of zirconium is virtually nil, but unlike titanium it is attacked in moist gaseous chlorine and not in dry chlorine at room temperature[70]. In solutions of metal chlorides, behaviour appears to depend upon whether the chloride solution tends to be oxidising or reducing, and in general zirconium is not as resistant as titanium. Thus it is not resistant to boiling ferric or cupric chlorides at strengths greater than 10%, but it is resistant to mercuric, stannic, manganous, nickel, ammonium, zinc, magnesium, barium and sodium chlorides, and to sea water. Behaviour in aluminium chloride is worth noting, for zirconium is resistant to boiling 25% aluminium chloride, while titanium is attacked. Both metals corrode in boiling 62% calcium chloride.

In resistance to hydrochloric and sulphuric acids zirconium shows a significant advantage over titanium[68]. With pure hydrochloric acid at 100°C the corrosion rate is negligible up to the constant boiling strength, i.e. 20% w/w at atmospheric pressure, but at 200°C under pressure there is appreciable attack at acid strengths greater than 18% by weight. The presence of traces of copper and iron in the hydrochloric acid can result in a significantly increased rate of attack, and, for example, in boiling 20% acid 1 000 p.p.m. of iron or copper raises the rate of attack from less than 0·0075 mm/y to the barely acceptable level of 0·5 mm/y. In sulphuric acid where traces of metal ions do not appear to be unduly troublesome, there is no appreciable corrosion up to 66% w/w at boiling point; the rate of attack, however, increases rapidly in boiling 70% acid, and at 200°C under pressure there is significant uniform corrosion at about 40% w/w. The presence of chlorine in sulphuric acid can seriously increase the rate of corrosion.

With phosphoric acid the performance of zirconium is again distinctly superior to that of titanium, for while, in general, use of titanium is limited to strengths less than 30% w/w, for zirconium there is no appreciable corrosion at room temperature up to 80% strength. As temperature rises there is an inflection in the corrosion-rate curve, an unacceptable rate being reached in boiling acid at 50% strength. As temperature rises beyond this, the corrosion rate again decreases[68], until at 200°C, under pressure, there is again negligible attack in 80% acid.

Neither titanium nor zirconium is recommended for use in hydrofluoric acid.

Zirconium is also resistant to attack in a wide range of organic acids, one useful difference from titanium being that it is not corroded in boiling deaerated formic acid at concentrations of 25% and upwards, in which titanium exhibits borderline passivity. In strong chlorinated organic acids, however, there may be some attack at elevated temperatures.

It is in its behaviour to caustic alkalis that zirconium shows itself to be superior to those other elements of Groups IV and V whose resistance to corrosion results primarily from an ability to form surface films. Thus, in contrast to tantalum, niobium and titanium, zirconium is virtually completely resistant to concentrated caustic solutions at high temperatures, and it is only slightly attacked in fused alkalis. Resistance to liquid sodium is good. Zirconium is thus an excellent material of construction for sections of chemical plant demanding alternate contact with hot strong acids and hot strong alkalis—a unique and valuable attribute.

Because of its good performance in mineral acids, there is little need or incentive to invoke anodic passivation techniques for zirconium. The metal can be anodised in sulphuric acid, but, again in contrast to the behaviour of titanium, it does not form a stable anodic film in chloride solutions, and even in neutral sodium chloride, zirconium rapidly corrodes if an anodic potential of 2 V is applied.

Applications in Industry

Although the cost of zirconium precludes its use in any great quantities, in comparison with the more usual materials of chemical plant construction, it is already being employed selectively for small vital pieces of equipment. Its main use is in hydrochloric acid service for valve stems, floats in rotameters, thermowells and small heat exchangers, and tanks.

With increasing use of zirconium in nuclear reactors it is reasonable to expect the price to decrease, and this will provide the incentive to employ this metal to a greater extent in chemical plant construction.

Special Applications

It has already been indicated that the principal use for zirconium is in the field of nuclear engineering. The very nature of this application demands the lowest possible corrosion rate, and this has necessitated a great deal of investigation into the oxidation rate of zirconium, when exposed to hot water, steam and carbon dioxide.

When zirconium oxidises in these environments at elevated temperatures the reaction kinetics follow a law which can be formulated as

$$w = Kt^n$$

where w = weight gain, t = time, and K and n are constants at a constant temperature. Initially n has a value of between $\frac{1}{3}$ and $\frac{1}{2}$, and the rate of oxidation decreases with time. However, when a certain thickness in the surface film is attained, the value of n may change and become equal to or greater than unity. The corrosion rate will then become constant or will increase. This type of behaviour, which can occur with several metals or alloys, has been called 'breakaway' corrosion. Within the period at which the value of n remains below unity, the monoclinic oxide film produced on zirconium is hard, glossy, adherent and usually black or dark coloured. When the kinetic change takes place the character of the film changes, and continued oxidation may lead to heavy surface spalling.

Unalloyed zirconium produced from Kroll sponge quickly reaches the breakaway point when exposed to steam or hot water at reactor temperatures. Early investigation in the United States established that this behaviour resulted from the almost inevitable presence of nitrogen, but that the deleterious effect could be countered by an addition of tin[71], and the alloy known as Zircaloy 2, containing about 1·5% Sn, 0·1% Fe, 0·1% Cr, 0·05% Ni was developed for use in water-cooled reactors. Even with this alloy, metallurgical treatment during fabrication is known to affect performance, and a rigorous scheme of corrosion testing is employed[72,73] to ensure that the semi-fabricated material and finished product conform to a high degree of corrosion resistance. This test involves the autoclaving of carefully prepared coupons for fourteen days in pure steam at a temperature of 400°C and a pressure of 10 MN/m^2. At the conclusion of the test, satisfactory material has a weight gain of $28 + 10$ mg/dm^2, and is covered with a glossy black lustrous film. Defective material manifests itself by high weight gains (up to as much as 100 mg/dm^2) and the appearance in the surface film of white corrosion product.

For exposure to hot carbon dioxide the oxidation rates of unalloyed zirconium are too high to permit its use in gas-cooled reactors, and the ATR alloy containing 0·5% Cu and 0·5% Mo is employed. Material fabricated from this alloy is also required to pass a corrosion test, which consists of exposure to dry carbon dioxide at atmospheric pressure at 700°C for 400 h. Good material has a weight gain not exceeding about 12 mg/dm^2 (72,74).

Most of the considerable volume of published work on the behaviour of zirconium relates to its use in nuclear reactors in contact with water or steam, e.g. in pressurised steam the control of oxidation by use of boric acid has been reported[75]. The reader is advised to consult the reviews on this important aspect of the subject cited under References 76 and 77.

Because of the limited experience so far available relating to employment of zirconium in chemical plant, neither the advantages nor the hazards attendant upon its use have been fully assessed. There is no reason to suspect that hazards in the use of zirconium will prove any more significant than those of the more widely used film-forming metals and alloys such as titanium or stainless steel.

It should be noted that swarf from a zirconium–titanium alloy containing approximately 50% by weight of each element is prone to pyrophoricity in air. It has also been reported[62] that when zirconium is welded to titanium, the welded zone is much more sensitive to corrosion than either of the parent metals. If, therefore, it is proposed to use any construction in which zirconium is welded to titanium, caution should be observed in the machining of welds, and the corrosion behaviour of the weld should be checked by prior testing in the environment with which the construction will be employed.

The pyrophoric tendency of zirconium in contact with red fuming nitric acid has already been mentioned.

There is some evidence that the increase in corrosion recorded when zirconium is exposed to hydrochloric acid at 200°C under pressure results from intergranular penetration[78].

Finally, perhaps, it should be pointed out that because the behaviour of zirconium is often adversely influenced by the presence of impurities in corrosive environments, corrosion testing prior to use should be carried out in actual plant liquors rather than in purer synthetic solutions.

J. B. COTTON

REFERENCES

1. Evans, U. R., *The Corrosion and Oxidation of Metals*, Arnold, London, 39–48 (1960)
2. Nakayama, Castings Research Laboratory Report, No. 5, Waseda University, 57–59 (1956)
3. Cotton, J. B., *Werkst. u. Korrosion*, Weinheim, **2** No. 3, 152 (1960)
4. Adamson, G. M., Jr. *et al.*, *Proceedings of the Second U.N. International Conference on the Peaceful Uses of Atomic Energy*, Paper P/1 993, United Nations, Geneva (1958)
5. Jackson, J. D., *Mat. Prot.*, **4** No. 1, 30–33 (1965)
6. Millaway, E. E. and Kleinman, M. H., *Corrosion*, **23**, 88–97 (1967)
7. Wullner, R. L., *Mat. Prot.*, **4** No. 1, 55–56 (1965)
8. Mueller, W. A., *J. Electrochem. Soc.*, **107**, 157 (1960)
9. Cotton, J. B., *Chemical Engineering Progress*, **66** No. 10, 57–62 (1970)
10. *Corrosion Resistance of Titanium*, Imperial Metal Industries Ltd., P.O. Box 216, Witton, Birmingham
11. Pourbaix, M., Rapport No. 21, Centre Belge d'Étude de la Corrosion, Brussels (1953)
12. Fischer, W. R., *Werkst. u Korrosion*, Weinheim, **10**, 243 (1959)
13. Schlain, D. and Smetko, J. S., *J. Electrochem. Soc.*, **99**, 417 (1952)
14. T.M.L. Report No. 57, Titanium Metallurgical Laboratory, Battelle Memorial, 116–153, Oct. 29 (1956)
15. Stern, M. and Wissenberg, H., *J. Electrochem. Soc.*, **106**, 754 (1959)
16. Thomas, N. T. and Nobe, K., *J. Electrochem. Soc.*, **116**, 1 748 (1969)
17. Cotton, J. B. and Bradley, H., *Chem. and Ind.* (Rev.), 643 (1958)
18. Cotton, J. B. and Downing, B. P., *Trans. Inst. Mar. Engrs.*, **69**, 311 (1957)
19. Feige, N. G. and Kane, R. L., *Metals Engr. Quart.*, **7**, 27–29 (1967)
20. Sims, M. H., *Power*, **112**, 890–96 (1968)
21. Greiss, J. C., *Corrosion*, **24**, 96–109 (1968)
22. France, W. D. and Greene, N. D., *Corrosion*, **24**, 247–51 (1968)
23. Takamura, A., *Corrosion*, **23**, 306–13 (1967)
24. Schlain, D., US Bureau of Mines, Report No. 4 965, April (1953)
25. Sedriks, A. J., *A.S.M. Trans. Quart.*, **61**, 625–27 (1968)
26. Harey, E. G. and Wearmouth, W. R., *Corrosion*, **25**, 87–91 (1969)
27. Sedriks, A. J., *Corrosion*, **25**, 325–28 (1969)
28. Rittenhouse, J. B. *et al.*, *Trans. Amer. Soc. Metals.*, **51**, 871, 895 (1959)
29. Inglis, N. P., *Chem. and Ind.* (Rev.), 180 (1957)
30. *Engineering*, Jan. 6 (1967)
31. Obrig, H. and Ehle, J. C., *Chem. Process. Engrg.*, **50** (1969)
32. 'Applications and Uses of Titanium and its Alloys', *Metallurg. Rev.*, Inst. Metals (1961)
33. Barron, L. J., *Light Metal Age*, **14** Nos. 3 and 4, 16 (1956)
34. *Industr. Engng. Chem.*, **50**, 934 (1958)
35. Connolly, B. J., *Chem. Proc. Engng.*, **39**, 247 (1958)
36. Bonberger, H. B., *Industr. Engng. Chem.*, **49**, 1 658 (1957)
37. Carmichael, M. L., *Battelle Techn. Rev.*, **5** No. 12, 9 (1956)
38. *Steel*, **143** No. 26, 62 (1958)
39. Ishii, Y. and Hoskino, Y., *Chem. Engng.*, Tokyo, **21**, 559 (1957)
40. Frazer, G. T. *et al.*, *Mat. and Meth.*, **43**, 112 (1956)
41. *Light Metal Age*, **17**, 27, Oct. (1959)
42. *Corrosion*, **15**, 82 (1959)
43. Hames, W. T., *Aircraft Prod.*, Lond., **20**, 369 (1958)
44. Savolainen, J. E. and Dlanco, R. E., *Chem. Engng. Prog.*, **53**, 78F (1957)
45. Peterson, C. L. *et al.*, *Industr. Engng. Chem.*, **51**, 32 (1959)
46. Schmets, J. and Pourbaix, M., *Proceedings of the 6th Meeting of the International Committee for Electrochemical Thermodynamics and Kinetics*, Poitiers, 1954, Butterworths, London (1955)
47. Stern, M. and Wissenberg, H., *J. Electrochem. Soc.*, **106**, 755 (1959)
48. Stern, M. and Wissenberg, H., *J. Electrochem. Soc.*, **106**, 759 (1959)
49. Stern, M. and Bishop, C. R., *Amer. Soc. Met.*, Preprint No. 165 (1959)
50. Cotton, J. B., *Chem. and Ind.* (Rev.), 68 (1958)
51. Cotton, J. B., *Werkst. u Korrosion*, **11**, March 3 (1960)
52. Evans, L. S., Hayfield, P. C. S. and Morris, M. C., *Proc. 4th Intern. Congress on Metallic Corrosion*

53. Cotton, J. B., *Chem. and Ind.* (Rev.), 492 (1958)
54. Cotton, J. B., *Platinum Metals Rev.*, **2**, 45 (1958)
55. Shreir, L. L., *Platinum Metals Rev.*, **4**, 15 (1960)
56. Jones, J. C., *Prod. Finish.*, Lond., **12** No. 12, 81 (1959)
57. Rittenhouse, J. B. *et al.*, *Trans. Amer. Soc. Metals*, **51**, 871, 895 (1959)
58. Millaway, E. E. and Kleinman, M. H., *Corrosion*, **23**, 88–97 (1967)
59. Golden, L. B. *et al.*, *Trans. Amer. Soc. Metals*, **51**, 871, 895 (1959)
60. Schlain, D. and Kenahan, C. B., *Corrosion*, **14**, 405t (1958)
61. Andreeva, V. V. and Gluklova, *J. Appl. Chem.*, **11**, 390 (1961)
62. Cotton, J. B., *Chem. and Ind.*, 357–358 (1962)
63. Yoshida, S. *et al.*, *J. Govt. Mech. Lab.*, Tokyo, **10**, 2–21 (1956)
64. Yoshida, S. *et al.*, *J. Jap. Inst. Metals*, **21** No. 3, 183 (1957)
65. S.T.P. 397, A.S.T.M., 1 916 Race St., Philadelphia, USA
66. Jackson, J. D. and Boyd, W. K., *The Science, Technology and Application of Titanium*, Pergamon, 267–281 (1966)
67. Imperial Chemical Industries, Ltd., Brit. Pat. 1 187 771 (15.4.70)
68. Kuhn, W. E., *More Zirconium Facts*, **1** No. 2, Carborundum Metals Company, 4 (1957)
69. Kuhn, W. E., *Chem. Engng.*, 156 (1960)
70. Gegner, P. J. and Wilson, W. L., *Corrosion*, **15**, 341t and 350t (1959)
71. Thomas, D. E., Proceedings of the First UN Conference on Atomic Energy, Geneva, 1955, **9**, Paper P/537, 407, United Nations, Geneva
72. Cotton, J. B. and Gallant, P. E., *Proceedings of the First International Congress on Metallic Corrosion*, London, April, 1961, Butterworths, London, 458 (1962)
73. Kass, S., *Corrosion*, **16**, 137 (1960)
74. O'Driscoll, W. G., Tyzack, C. and Raine, T., *Proceedings of the Second Conference on the Peaceful Uses of Atomic Energy*, Geneva, 1958, **5**, Paper P. 1450, Geneva, 75 (1958)
75. Britton, C. F., *J. Nuc. Mat.*, **15** No. 4, 263–277 (1965)
76. Coleman, C. E. and Hardie, D., *J. Less Common Metals*, **11**, 168–85 (1966)
77. Rosa, C. J., *J. Less Common Metals*, **16**, 173–201 (1968)
78. Kuhn, W. E., *Corrosion*, **16**, 141t (1960)

5.5 Tantalum

Tantalum is one of the most versatile corrosion-resistant metals and its corrosion behaviour can be compared with that of glass; in most environments it is almost as inert as gold and platinum, although its price is about 10% that of these noble metals. The pure metal is blue grey and like lead in appearance; it has a density of about twice that of steel and a thermal conductivity slightly better than that of 1% carbon steel.

It can be readily cold worked, but hot working, however, must be avoided as the metal reacts with gases such as oxygen, nitrogen and carbon dioxide with resultant embrittlement. It can be machined, although care is necessary to obtain a good surface finish. The high strength, good workability and excellent corrosion resistance permit the use of very thin walled components, a commonly employed thickness in chemical plant being 0·33 mm. These

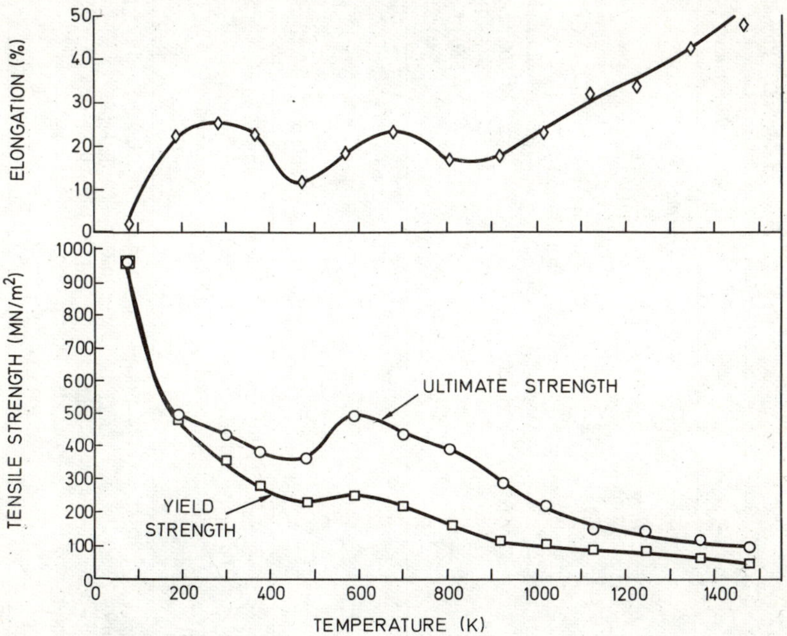

Fig. 5.8 Effect of temperature on the tensile strength and elongation of tantalum

Table 5.21 Physical properties of tantalum

Density (g/cm³)	*Thermal*						*Electrical*	
	Melting point[1] (°C)	*Boiling point*[1] (°C)	*Thermal neutron absorption cross section*[3] (barn/atom)	*Linear coeff. of expansion*[2] (°C)	*Thermal conductivity*[2] (W/cm °C)	*Specific heat*[2] (J/g °C)	*Electrical resistivity*[1, 2] (μΩ/cm)	*Temperature coeff. of resistivity* (°C)
16·6	2 996	5 425	21	20–500°C: $6{\cdot}6 \times 10^{-6}$ 20–1 500°C: $9{\cdot}0 \times 10^{-6}$	5·44 at 20°C 7·52 at 1 106°C	0·142 at 0°C 0·161 at 1 227°C	12·43 at 20°C 54·8 at 1 000°C	$3{\cdot}82 \times 10^{-3}$ at 0–100°C 3×10^{-3} at 0–1 000°C

Table 5.22 Mechanical properties of tantalum

Modulus of elasticity[2, 4] (GN/m²)	*Poisson's ratio*	*Yield stress* (MN/m²)	*UTS*[2, 4] (MN/m²)	*Ductile to brittle transition temperature*[2] (°C)	*Recrystallisation temperature*[4] (°C)	*Stress relieving temperature*[4] (°C)	*Hardness*[2] (VPN)		*Stability*[2] (% creep rate/h, min.)
							Annealed	*Hard worked*	
186 at 20°C 151 at 1 000°C	0·35	179–1 060 at 27°C and 44–310 at 500°C	689–1 034 at 20°C and 103–138 at 1 000°C	None detected down to –196	1 050–1 500	900	80–100	130	0·113 at 750°C, 96 GN/m²

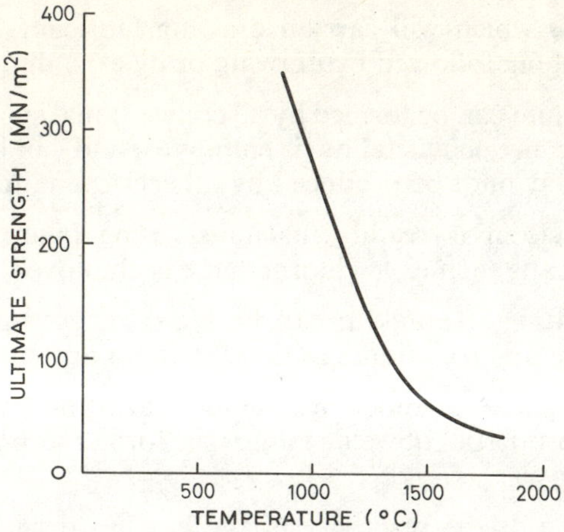

Fig. 5.9 Effect of temperature on the ultimate strength of tantalum

properties, coupled with the metal's ability to promote bubble-type vapour formation on the surface when heating liquids, and dropwise condensation when condensing vapours, make the metal an ideal constructional material for heat-transfer equipment for use with strong acids.

The absence of corrosion, coupled with the fact that scale and other deposits appear to be dislocated by thermal cycling, result in a finish on tantalum heating surfaces that is as good as the original, even after 20 to 30 years in service, and also ensure that good heat-transfer properties are maintained throughout the life of the equipment. The use of tantalum for process equipment also ensures freedom from contamination of the product.

The mechanical properties of tantalum are dependent on the previous history of the material and the manufacturer should be consulted if these properties are likely to be critical. The physical and some typical mechanical properties are listed in Tables 5.21 and 5.22. The effect of the temperature on the strength and elongation of tantalum sheet in vacuum is shown in Figs. 5.8 and 5.9.

Methods of Fabrication

The high melting point and reactivity of tantalum with the permanent gases at high temperatures prevents conventional consolidation by melting and casting in air. The metal is in fact consolidated by vacuum sintering, vacuum-arc melting and electron-beam melting of powder compacts. Vacuum sintering yields metal of fine grain, whereas electron-beam melting yields softer coarse-grained metal which requires cold forging prior to rolling. Metal produced by all three techniques will absorb considerable cold work before annealing is necessary.

Rolling and swaging Vacuum-sintered bar can be cold rolled, and reductions up to 90% between anneals are possible. Arc-cast and electron-beam-melted material is generally forged at room temperature prior to rolling and swaging.

Drawing Tantalum has a tendency to gall and it is normally anodised to provide a surface which will carry a drawing lubricant. Seamless tube is produced by cupping followed by drawing or by extruding hollow shells.

Spinning Tantalum can be formed by all conventional spinning techniques, provided a lubricant such as tallow is employed, and can be spun into configurations which cannot be produced by other forming methods.

Machining Tantalum is readily machined using high-speed-steel tools, provided a lubricant such as trichloroethane is employed.

Blanking and cutting Tantalum can be blanked, cut and sheared using similar equipment and techniques to those used for austenitic stainless steel.

Joining Tantalum can be joined by riveting, brazing and welding; however, due to the good properties of welded joints the former techniques are seldom used.

Welding Because of the reactivity of the hot metal with the permanent gases, conventional welding techniques cannot be used. In general the practical methods are restricted to tungsten-electrode inert-gas (TIG), resistance, electron-beam (EB) and plasma-arc welding. To ensure satisfactory TIG welds, welding should be done in an inert-gas-filled chamber. Material thinner than 0·5 mm cannot readily be TIG welded and resistance welding has to be used. Spot welding can be carried out in air and under water. EB welding gives a contamination-free narrow weld and heat-affected zone, irrespective of material thickness, and plasma-arc welding has been used in 0·05–1·0 mm sheet and gives a weld with similar properties to EB welds.

Corrosion Resistance

Tantalum's corrosion resistance is due to the presence of a thin continuous surface film of tantalum pentoxide (Ta_2O_5). Thus the metal is passive and approaches the inertness of gold and platinum in a large number of very aggressive environments. The metal itself lies below zinc in the thermodynamic nobility table presented by Pourbaix[5] being in fact No. 34 (gold = 1). Its passive oxide film, however, puts it just below rhodium (No. 1) and above gold (No. 4) in the Pourbaix practical nobility table. The oxide film adheres well and appears to be free from porosity, but there is some evidence that suboxides which are stable up to 425°C develop at the metal interface. When heated above this temperature only the stable pentoxide exists and the internal stress set up by the metal during oxide conversion causes the protective oxide film to flake and spall.

Highly adherent films of predetermined thickness can be produced on tantalum by anodising, and the dielectric properties of these films are such that the metal finds extensive use in the production of capacitors.

References available in the literature indicate that tantalum is corroded by only some 35 or so of the reported 1 300 reagents in which it has been evaluated. A summary of environments in which tantalum corrodes is given in Table 5.23.

Hydrogen Embrittlement

Tantalum has a high solubility for hydrogen, forming two internal hydrides, but the exact mechanism of their formation is not precisely known. There is evidence that embrittlement can occur at temperatures below 370°C. Clauss and Forestier[6] in fact reported that embrittlement can occur when tantalum is deformed in contact with hydrogen at room temperature. Examination of the literature indicates that one of the few defects in the resistance of tantalum to corrosion in aqueous media lies in its susceptibility to hydrogen embrittlement. Although it is inert in concentrated hydrochloric at temperatures as high as 110°C, some reaction occurs at appreciably higher temperatures and sufficient hydrogen may be absorbed to cause embrittlement. Since it becomes cathodic in galvanic cell circuits with virtually all constructional metals, it must be electrically insulated from other metals with which it could come into contact in a common electrolyte, in order to prevent hydrogen discharge and entry into the metal. Stern and Bishop[6(a)] claim that embrittlement can be prevented by placing the tantalum in contact with platinum, which becomes the cathode of the galvanic couple and thus prevents hydrogen evolution at the tantalum surface.

Reactions with Gases

Tantalum and all known tantalum alloys react with hydrogen, nitrogen and oxygen at temperatures above 300°C. Hydrogen is dissolved in the pure metal above 350°C[7], but is evolved at higher temperatures, and little remains at 800°C[8,9]. The embrittling effect of hydrogen is used to prepare powder for tantalum capacitor production. Tantalum is also embrittled by nitrogen, and the reaction takes place above 400°C[7] although an oxide film may inhibit reaction until higher temperatures are reached. Nitrides among other phases form at the surface, but at higher temperatures these decompose and all the nitrogen is liberated at 2 100°C[10].

Generally, the most important reaction is that of tantalum with oxygen, since it tends to form oxides when heated in air. Reaction starts above 300°C and becomes rapid above 600°C[11]. The scale is not adherent, and if the oxidised material is heated above 1 000°C oxygen will diffuse into the body of the material and embrittle it. At 1 200°C the rate of attack is 150 mm/h[12]. Oxygen is not driven off by heating alone, but in vacuum above 2 300°C it is removed as a suboxide.

Tantalum is attacked by carbon dioxide at 500°C and carbon monoxide at temperatures above 1 100°C; in both cases oxide is formed, however, at higher temperatures tantalum carbide is also formed.

Atmospheres

Tantalum has a high resistance to general outdoor atmospheres. Tantalum and the Ta–10W alloy are virtually immune to sea water at ambient conditions and tantalum is only tarnished in oxygenated water at 26°C.

Acid Media

Tantalum is practically inert to nitric acid at all concentrations and temperatures, the corrosion rate in 70% acid at 270°C being 0·13 mm/y. It also resists red and white fuming nitric acids up to at least 150°C and hydrochloric acid at all concentrations up to 190°C, though above 25% concentration and 190°C the corrosion rate rises rapidly and, in addition, the entry of hydrogen causes embrittlement (Fig. 5.10). Tantalum is completely inert to hydro-

Table 5.23 Reagents and conditions which cause significant corrosion of tantalum[13–17]

Reagent	*Concentration* (%)	*Temperature* (°C)	*Corrosion rate* (mm/y)
Ammonium fluoride	5	25	>1·3
Ammonium silicofluoride	100	25	>1·3
Barium cyanide	20–80	25	>1·3
Barium peroxide	10–30	100	>1·3
Bromine, dry	100	315	>1·3
Bromine, wet		375	>1·3
Bromine trifluoride	100	25	>1·3
Cadmium cyanide	100	25	>1·3
Carbon dioxide	100	500	>1·3
Carbon monoxide	100	1 100	>1·3
Cerium chloride	100	870	>1·3
Chlorine, dry	100	275	>1·3
Chlorine, wet		325	>1·3
Chromic acid	70	100	0·12–0·9
Fluorine	100	25	>1·3
Fluorosilicic acid	5	20	>1·3
Hydrochloric acid	37	190	0·3
Hydrofluoric acid	5, 20	30, 25	>1·3
Hydrofluorosilicic acid	5	30	0·13–1·3
Hydrofluorosilicic acid	20	25	>1·3
Lead oxide	100	100	>1·3
Oxidising gases		200	<0·13
Oxidising gases		480	>1·3
Phosphoric acid	85	225	0·2
Phosphorus chlorides	100	315	>1·3
Phosphorus pentachloride	100	260	>1·3
Potassium hydroxide	50	27	0·9–1·3
Potassium hydroxide	40	100	>1·3
Potassium nitrite	100	260	>1·3
Sodium bicarbonate	100	815	>1·3
Sodium hydroxide	50	38–57	0·9–1·3
Sodium hydroxide	50	120	>1·3
Sodium peroxide	100	200	>1·3
Sodium silicofluoride	100	200	>1·3
Sodium thiosulphate	10	25	>1·3
Sulphuric acid, fuming		75	>1·3
Sulphuric acid	85	225	0·5
Sulphuric acid	95	250	0·3
Sulphuric acid	95	290	>1·3
Sulphuric acid	98	250	0·3
Sulphuric acid	98	300	>1·3
Sulphur trioxide	3	25	>1·3
Zinc chloride	100	315	>1·3
Molten metals			
Aluminium		650	>1·3
Cadmium		370	>1·3
Magnesium		660	>1·3
Tin		260	>1·3

Note. In many cases the ranges quoted are those for which tests have been carried out and reported.

chloric plus nitric acid mixtures even in the presence of sulphuric acid and its salts in all proportions and concentrations up to the boiling point. It is not corroded by phosphoric acid at concentrations up to 85% and temperatures up to 200°C, provided fluoride ions, often found in commercial acid, do not exceed 5 p.p.m. It is practically inert to perchloric acid, chromic acid, hypochlorous acid, hydrobromic acid, hydriodic acid and most organic acids provided they do not contain fluorides, fluorine or free sulphur trioxide. One exception to fluoride attack appears to be in certain chromium plating baths in which fluoride is used as the catalyst, the corrosion rate in 40% CrO_3 plus 0·5% F at 55–60°C being 0·000 5 mm/y.

Table 5.24 Corrosion by miscellaneous acids[13,14,16]

Acid	*Concentration* (%)	*Temperature* (°C)	*Corrosion rate* (mm/y)
Chromic	all concs.	100	<0·08
	10–50	Boiling	<0·12
	36·5	90	<0·12
	70	25	<0·12
	70	100	0·12–0·9
Phosphoric (air free)	10–85	50–250	<0·12
	96	215–220	0·033
	96	225–230	0·094

It is completely inert to 98% sulphuric acid to at least 160°C and to even higher temperatures at lower concentrations. Practically, it may be used to 200°C in all concentrations and to 225–250°C at concentrations between 80% and 90%. Fuming sulphuric acid containing sulphur trioxide attacks tantalum at room temperature as does hydrofluoric and fluosilicic acids.

Specific information is given in Table 5.23 and Figs. 5.10 and 5.11.

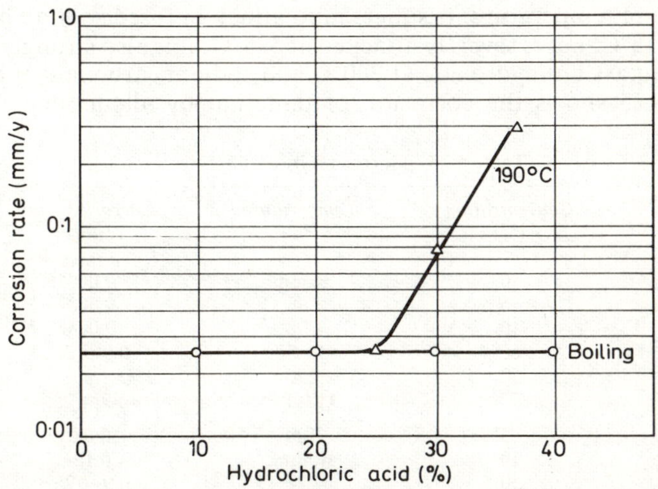

Fig. 5.10 Corrosion of tantalum by hydrochloric acid (after Bishop, C. R., *Corrosion*, **19**, Sept., 308 (1963))

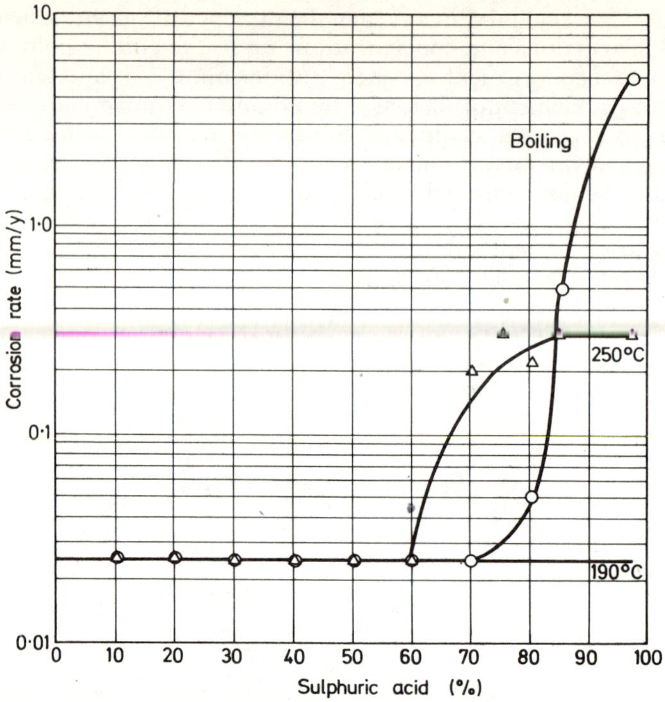

Fig. 5.11 Corrosion of tantalum by sulphuric acid (after Bishop, C. R., *Corrosion*, **19**, Sept., 308 (1963))

Alkali Media

Tantalum has good resistance to sodium hydroxide in concentrations of 10% or less up to 100°C, but its resistance is poor in concentrations above 40% even at room temperature. It suffers mild attack in fused sodium hydroxide at about 315°C, but is severely attacked at 535°C; it is also strongly attacked by fused potassium hydroxide at 360°C and sodium carbonate at 850°C.

Table 5.25 shows the corrosion of tantalum by alkalis under various conditions.

Table 5.25 Corrosion by alkalis[13–16,18]

Alkali	*Concentration* (%)	*Temperature* (°C)	*Corrosion rate* (mm/y)	
Sodium hydroxide	10	Room	0·00025	No noticeable embrittlement
	5	98	0·0033	
	10	98	0·0082	
	50	38–57	0·9–1·25	
	50	120	>1·3	
	73	113–129	0·9–1·25	
Potassium hydroxide	5	100	<0·08	
	10	Boiling	<0·12	
	50	27	0·9–1·25	
	40	100	>1·25	

Salts

Tantalum has excellent resistance to virtually all salts including chlorides (especially cupric and ferric chloride), sulphates, nitrates and salts of organic acids, provided (*a*) they do not contain fluorides, fluorine and free sulphur trioxide, or (*b*) hydrolyse to produce strong alkalis.

Other Environments

Halogens Although tantalum is severely attacked by fluorine at room temperature it does not react with liquid chlorine, bromine and iodine up to 150°C and the metal suffers no appreciable attack in wet or dry bromine, chlorine and iodine below 250°C. It is virtually uncorroded by hydrogen bromide and hydrogen chloride below 370°C, attack starting at about 375 and 410°C respectively.

Liquid metals[15] Tantalum generally shows good resistance to the low-melting point liquid metals.

Bismuth Liquid bismuth has little action on tantalum at temperatures below 1 000°C[19,20–22], the rate of attack at 870°C being less than 0·13 mm/y, and exerts no detrimental effects on the stress rupture properties of tantalum at 815°C[23], but it causes some intergranular attack at 1 000°C[24].

Calcium Tantalum is only slightly attacked by calcium at 1 200°C, the rate of attack being 0·37 mm/y[25].

Gallium The resistance of tantalum to molten gallium is considered to be good up to 450°C, but poor above 600°C, the rate of attack at 482°C being less than 0·13 mm/y.

Lead Tantalum is highly resistant to liquid lead at temperatures up to 1 000°C[19] with a rate of attack of less than 0·025 mm/y[26]. It exhibits no detrimental effects when stress rupture tests are conducted in molten lead at 815°C[23].

Lithium Tantalum possesses good resistance to molten lithium up to 1 000°C[19,27], the rate of attack at this temperature being less than 0·13 mm/y.

Mercury In static tests tantalum showed good resistance to mercury at temperatures up to 600°C[19], the rate of attack at 590°C being less than 0·13 mm/y.

Silver Tantalum is only slightly attacked by silver at 1 200°C, the rate of attack being 0·25 mm/y[25].

Sodium, potassium and sodium–potassium alloys Liquid sodium, potassium or alloys of these elements have little effect on tantalum at temperatures up to 1 000°C[19,24,28] the rate of attack at 1 200°C being less than 0·13 mm/y, but oxygen contamination of sodium causes an increase in corrosion[29,30]. In fact, if oxygen is present, attack may commence as low as 250°C. Sodium does not alloy with tantalum[31].

Thorium–magnesium In static tests the thorium–magnesium eutectic had no appreciable effect on tantalum at 1 000°C[32], the corrosion rate being less than 0·13 mm/y.

Tin Static tests in liquid tin at 1 740°C for 1 h showed that some tantalum dissolved (0·33% Ta was found in the tin); the corrosion rate at 260°C is 1·3 mm/y.

Uranium Short-term tests indicate the practical upper limits for tantalum as a container material for uranium to be 1 450°C[33]. However, attack below these temperatures is significant, since a tantalum crucible with a wall 1·52 mm thick was completely corroded in 50 h at 1 275°C[34].

Zinc Molten zinc attacks tantalum at a significant rate above 450°C[35] the rate of attack at 370°C being less than 0·13 mm/y. At 750°C the attack becomes appreciable[36].

Galvanic Effects

As indicated previously, a galvanic couple in which tantalum is the cathode can prove disastrous because of embrittlement. On the other hand, if tantalum is the anode it passivates so readily in most environments that no damage occurs, and the galvanic current drops to a very low value. Haissinsky[37] studied couples of tantalum with platinum, silver, copper, bismuth, antimony, molybdenum, nickel, lead, tin, zinc and aluminium in 0·1N H_2SO_4. Except when tantalum was coupled to zinc or aluminium it was the more negative member (anode) of the couple. However, the galvanic current rapidly decreased as the tantalum passivated. In hydrofluoric acid tantalum was again more positive than zinc and aluminium, but more negative than platinum, silver, copper, antimony, nickel and lead. The latter six couples result in high steady-state currents, because tantalum corrodes rather than passivates in fluoride solutions. The apparently anomalous behaviour

Table 5.26 Potential of tantalum in tantalum–metal couples[38]

Time (min)	*Coupled metal*	*Potential of tantalum* (mV)		
		3% NaCl electrolyte	1% HCl electrolyte	1% NaOH electrolyte
0·5	Hastelloy B	−34·3	−30·5	−104
60	Hastelloy B	+2·6	+1·0	−4·3
0·5	Hastelloy C	−50·0	−72·0	−136
60	Hastelloy C	−2·8	−2·7	−4·4
0·5	Nickel	−65·1	−36·0	−180
60	Nickel	−2·5	+0·1	−4·4
0·5	Lead	−1·6	+51·0	−10
60	Lead	+60·1	+52·9	+29·8
0·5	Aluminium	+128	+275	+621
60	Aluminium	+230	+208	+509
0·5	304 stainless steel	−52·0	—	—
60	304 stainless steel	−2·9	—	—
0·5	Mild steel	—	—	−132
60	Mild steel	—	—	−0·5

of tantalum coupled with bismuth or iron in hydrofluoric acid is explained by Haissinsky as being due to the formation of insoluble fluorides on the surfaces of the bismuth or iron electrodes.

Time is an important factor in determining whether tantalum will be damaged by galvanic effects[38]; this is illustrated in Table 5.26. Aluminium becomes negative in dilute NaCl, HCl and NaOH after 1 h, and the polarity of other metals coupled with tantalum changes with time. The details of these tests are not available, and it is difficult to assess the significance of the actual potential values. In practice it would be dangerous to depend on laboratory tests to provide information as to whether tantalum is negative in a given galvanic couple situation.

Alloys of Tantalum

Tantalum–Niobium Miller[39] and Argent[40] reported on the corrosion resistance of the tantalum–niobium system in a number of environments. Unfortunately, the test conditions were not sufficiently severe for the data to be significant.

Kieffer *et al.*[41], however, examined tantalum–niobium alloys in boiling 10% potassium hydroxide, 10 and 20% hydrochloric acid and 10% sulphuric acid and found that the corrosion resistance was substantially similar to that of tantalum until the niobium content exceeded 50%.

Tantalum–Molybdenum Schumb, Radtke and Bever[42] studied the corrosion resistance of tantalum–molybdenum alloys that form a continuous series of solid solutions. The results of tests of up to 500 hours duration (Table 5.27) indicate the corrosion resistance of the alloy to be substantially that of tantalum, provided its concentration exceeds 50%.

Table 5.27 The corrosion behaviour of tantalum and molybdenum alloys in concentrated sulphuric and hydrochloric acids at 55°C; solutions saturated with oxygen[16]

Tantalum (atomic %)	*Average corrosion rate* ($mg\,dm^{-2}\,d^{-1}$)	
	Conc. H_2SO_4	Conc. HCl
0	0·8	1·8
10·1	0·9	1·7
20·1	0·8	1·8
30·0	1·0	0·9
40·0	0·9	1·0
50·0	0·0	1·0
61·2	0·0	0·0
71·5	0·0	0·0
82·8	0·0	0·0
91·4	0·0	0·0
100	0·0	0·0

Tantalum–Tungsten Braun, Sedlatschek and Kieffer[43] examined tantalum–tungsten alloys in 50% potassium hydroxide up to 80°C and in 20% hydrofluoric acid at 20°C. In the alkaline solution the corrosion rate was a maximum when the tantalum content was over 60 at.%. In hydrofluoric acid the alloy system exhibited the relatively low corrosion rates associated with tungsten until the tantalum concentration exceeded 80 at.%.

Staples and Galloway[44] examined the corrosion resistance of the Ta–10W alloy in hydrochloric, sulphuric and nitric acids and in sodium hydroxide solution and found that there was virtually no difference in corrosion rate between the alloy and pure tantalum in 10–30% hydrochloric acid to 175°C, 70–90% sulphuric acid to 205°C and 60% nitric acid to 190°C. In addition, in 5% sodium hydroxide the alloy had a lower corrosion rate than pure tantalum. This alloy also has yield and ultimate tensile strengths approximately double those of pure tantalum but it is considerably more difficult to work and fabricate.

Tantalum–Titanium Bishop[45] examined the corrosion resistance of this alloy system in hydrochloric, sulphuric, phosphoric and oxalic acids and found that alloys containing up to about 50% titanium retained much of the superlative corrosion resistance of tantalum.

Under more severe conditions, a titanium content of below 30% appears advisable from the standpoint of both corrosion resistance and hydrogen embrittlement, although contacting or alloying the material with noble metals greatly decreases the latter type of attack.

Tantalum–titanium alloys cost less than tantalum because titanium is much cheaper than tantalum, and because the alloys are appreciably lower in density. These alloys are amenable to hot and cold work and appear to have sufficient ductility to allow fabrication.

Industrial Applications of Tantalum

About one half of the tantalum used in the chemical industry is employed on sulphuric acid duties, mainly in the form of multiple-tube bayonet heaters, and it has been estimated that some 1 200 m^2 of tantalum exchanger area was installed in the USA alone between 1940 and 1960.

A further quarter is operating in media containing chlorine or its derivatives and it is in this field and that of the bromine industry that the first tantalum chemical plant was put on steam in the early 1930s. A large variety of tantalum equipment is in fact used in the synthesis and handling of hydrochloric acid and tantalum heat-exchanger units are employed in a wide range of processes involving nitric acid.

Tantalum, because of its negligible corrosion rate, finds use in the pharmaceutical and food manufacturing industries where even the smallest amount of metallic impurity cannot be tolerated in many products.

By virtue of the high breakdown potential of the oxide film (approximately 155 V in sea water and 280 V in low conductivity water of pH = 7) tantalum has found use as a substrate for platinum in impressed-current cathodic-protection anodes, which can be used at high impressed voltages (50 V) and high current densities. However, because of its lower cost, niobium is preferred for systems that have to operate at high voltages.

Specific applications of tantalum are as follows[16]:

Reconcentration of sulphuric acid Many thousands of square metres of tantalum heater surface have been installed in plants for the reconcentration of diluted sulphuric acid arising from metal pickling, oil refinery operations and from petrochemical processes producing alcohols and ketones. Since reconcentration provides a means of overcoming a waste disposal problem, the use of such plants is expanding.

Plants producing and handling halogens and halogen compounds Tantalum finds extensive use in the production and handling of hydrochloric and hydrobromic acid, chlorine and bromine and many of their derivatives. Absorbers, coolers and heaters which show considerable advantages in terms of heat-flux capabilities and corrosion resistance have been used on hydrochloric acid duties for over 30 years and condensers have been used in bromine plants for at least the same period.

Plants handling aqua regia Aqua regia is used extensively in the extraction and refining of the precious metals, and tantalum as one of the few metals resistant to this media is used for dissolution/evaporation pans, reactor lids and all immersed ancillary equipment. Tantalum equipment has been in service in one plant for over 3 years and its use is expanding.

Nitric acid plants Tantalum heat exchangers and sparge pipes find extensive use in plants producing high purity nitric acid, ammonium nitrate and terephthalic acid.

J. BENTLEY

REFERENCES

1. Lyman, T., *Metals Handbook*, A.S.M. (1961)*
2. Hampel, C. A., *Rare Metals Handbook*, 2nd edn, Reinhold (1961)*
3. MacDonald, R. N. and Boucom, H. H., *Nucleonics*, **20**, 158, Aug. 8th (1962)*
4. Chelius, J., *Machine Design*, March 1st (1962)*
5. Pourbaix, M., *Atlas of Electrochemical Equilibria in Aqueous Solutions*, Pergamon Press (1966)
6. Clauss, A. and Forestier, *Compt. Rend. Acad. Sci.*, Paris, **246**, 3 241–3 243 (1958)†

6(a). Bishop, C. R. and Stern, M., *J. Metals*, N.Y., **13**, 144 (1961)

7. Gulbransen, F. A. and Andrew, K. F., *J. Metals*, New York, **180**, 586 (1950) and *Trans. A.I.M.E.*, **188**, 586–599 (1950)
8. Sieverts, A. and Berger, E., *Ver. Deutsch. Chem. Ges.*, **44**, 2 394–2 402 (1911)
9. Sieverts, A. and Bruning, H., *Z. Physik. Chemie.*, **A174**, 365–369 (1935)
10. Wright, D. A., *Nature*, Lond., **142**, 794 (1938)
11. Schmidt, F. F. and Ogden, H. R., *The Engineering Properties of Tantalum and Tantalum Alloys*, DMIC Report No. 189, DMIC Batelle Memorial Institute, Sept. 13th (1963)
12. Jaffee, R. J., *Proceedings on an International Symposium on High Temperature Technology*, Asilomer, California, 1959, McGraw-Hill, New York, 61 (1960)
13. Takamara Akira, *Corrosion Engineering*, **16** No. 3, March (1967)
14. *Corrosion Tables of Special Materials and Rare Metals*, Jacob and Korves GmbH, Ahlen
15. Barto, R. L. and Hurd, D. T., *Research/Development*, Nov., 26–30 (1966)
16. *Corrosion Data Survey*, compiled by Nelson, G. A., National Association of Corrosion Engineers (1967)
17. Bishop, C. R., *Corrosion*, **19**, Sept., 308 (1963)
18. Tingley, I. I. and Rogers, R. R., 'Corrosion of Niobium and Tantalum in Alkaline Media', *Corrosion*, **21**, April, 132 (1965)
19. Miller, E. C., Chapter 4 of *Liquid Metals Handbook*, Atomic Energy Commission, Navy Department, Washington, D.C., 144–183 (1952)†

*These references can be found collectively in *Properties of Refractory Metals*, by Burnett, S. J., U.K.A.E.A. Research Report No. AERE R4 657, H.M.S.O. (1969).

†These references can be found collectively in *Columbium and Tantalum*, edited by Sisco, F. T. and Epremian, E., Chapt. 8, Wiley (1963).

20. Stoughton, L. D. and Sheetan, T. V., *Mechanical Engineering*, **78**, 699–702 (1956)†
21. Seifert, J. W., *Nuclear Sci. Abstracts*, **13** No. 19 667, 2 635–2 636 (1959)†
22. Lloyd, E. D., '1958 High Melting Metals', *Plansee Proceedings*, Metallwerk Plansee A.G., Reutte, Tyrol, 249–256 (1959)†
23. Parkman, R. and Shepard, O. C., US Atomic Energy Commission Pub. ORD–45, June 11th (1951)†
24. Reed, E. L., *J. Am. Ceramic Society*, **37**, 146–153 (1954)†
25. US Atomic Energy Commission Pub. No. ISC–607, Quarterly Summary Research Report in Metallurgy for Jan. to March, Ames Laboratory, Ames, Iowa (1955)†
26. Wilkinson, W. D., Hoyt, E. W. and Rhode, H. V., US Atomic Energy Commission Pub. ANL–5 449 (1955)†
27. Cunningham, J. E., US Atomic Energy Commission Pub. No. ORNL–CF–51–7–135, July 23rd (1951)†
28. Cottrell, W. B. and Mann, L. A., *Nucleonics*, **12**, Dec., 22–25 (1954)†
29. Wyatt, L. M. and Dickinson, F. S., *Welding and Metal Fabrication*, **25**, 378–385, 396 (1957)†
30. Raines, C. E., Weaver, C. V. and Strang, J. H., US Atomic Energy Commission Pub. No. BMI–1 284, Aug. 21 (1958)†
31. Adams, R. M. and Sittig, M., in *Liquid Metals Handbook*, Sodium NAK Supplement, Atomic Energy Commission, Navy Dept., Washington, D.C., 14 (1955)†
32. US Atomic Energy Commission Pub. ISC–396, Quarterly Summary Research Report in Metallurgy for April to June, Ames Laboratory, Ames, Iowa (1953)†
33. McIntosh, A. B. and Bagley, K. Q., *J. Inst. Met.*, **84**, 251–270 (1955–1956)†
34. US Atomic Energy Commission Pub. No. ANL–5 560, Chemical Engineering Division, Summary Report, Jan. to March, Argon National Laboratories, Argon, Ill. (1956)†
35. Hodge, W., Evans, R. M. and Haskins, A. F., *J. of Inst. Metals*, **7**, 824–832 (1955)†
36. Dekany, J. P., Lavendal, H. W., Winsch, I. O. and Pavlik, J., US Atomic Energy Commission Pub. No. ANL–5 996, Summary Report, Jan. to March (1959)†
37. Haissinsky, M., *Metaux et Corrosion*, **23**, 15–18 (1948)
38. Wehrmann, R., *Corrosionomics*, Fansteel Metallurgical Corp., Sept. (1956)
39. Miller, G. L., *Tantalum and Niobium*, Butterworths, London (1959)
40. Argent, B. B., *J. Inst. Met.*, **85**, 547–551 (1957)
41. Kieffer, von R., Bach, H. and Stempkowski, I., *Werkst. u Korrosion*, Sept. 782–784 (1967)
42. Schumb, W. C., Radtke, S. F. and Bever, M. B., *Ind. Eng. Chem.*, **42**, 826–829 (1950)
43. Braun, H., Sedlatschek, K. and Keiffer, R., *J. of the Less Common Metals*, **1** No. 6, Dec., 413–419 (1959)
44. Staples, B. G. and Galloway, W. S., Jr., *Materials Protection*, July, 34–39 (1968)
45. Bishop, C. R. and Powell, R. L., *Corrosion*, **18** No. 6, 205t–210t, June (1962)
46. Bishop, C. R. and Stern, M., *J. Metals*, N.Y., **13**, 144 (1961) and *Corrosion*, **17**, Aug. 379 (1961)

BIBLIOGRAPHY

Bachmann, W. T., 'Do Not Overlook Refractory Metals for Corrosive Environments', *Materials in Design Engineering*, 106, Nov. (1966)

Charles, R., 'A Champion for Tantalum', *Metal Bulletin*, Dec. 22nd (1970)

Chelius, J., Understanding the Refractory Metals, *Chemical Engineering*, Dec. 10, 178 (1962)

Cooper, M. J. and Mannox, D. J., 'Tantalum and its Alloys as Engineering Materials', *Proc. Mech. Eng.*, **180**, Part 3D (1965–1966)

Sisco, F. T. and Epremian, E., *Columbium and Tantalum*, Wiley (1963)

Vorhis, F. H., 'Exotic Metals for Process Equipment', *Chemical Engineering*, Oct. 25th, 182 (1965)

N.V. Reframent-Hoboken S.A., Pubs. Nos. FH–67A–69 and FH–1 065A–69

5.6 Uranium

For many years the corrosion of uranium has been of major interest in atomic energy programmes. The environments of importance are mainly those which could come into contact with the metal at high temperatures during the malfunction of reactors, *viz.* water, carbon dioxide, carbon monoxide, air and steam. In all instances the corrosion is favoured by large free energy and heat terms for the formation of uranium oxides. The major use of the metal in reactors cooled by carbon dioxide has resulted in considerable emphasis on the behaviour in this gas and to a lesser extent in carbon monoxide and air.

Water

Uranium reacts with water to form uranium dioxide, hydrogen and uranium hydride. However, the hydride usually has only an ephemeral existence and reacts itself with water to form uranium dioxide and hydrogen. Reaction rates decrease with pH below 2 and it has been suggested that the solid products form by the inward diffusion of hydroxyl ions through the oxide[1]. The oxide is produced mainly as a non-adherent powder, and a linear rate law is obeyed. Autoclave tests have demonstrated that the rate constant increases markedly with temperature up to at least 300°C (Figure 5.12*)[2]. The corrosion rate is also influenced by dissolved gases in the water. In particular the presence of oxygen decreases substantially the reaction rate[2] but makes the metal susceptible to crevice corrosion and pitting attack. The inhibition is most marked at the lower temperatures at which the oxygen solubility is highest and the hydrogen product is not sufficient to reduce the oxygen content locally. The oxygen could exert its influence by being adsorbed preferentially on the oxide[3] or by removing the disruptive influence of the formation of uranium hydride. An alternative view of such 'hydrogen effects' relates them to changes in the electrical properties of the oxide detected by impedance measurements during corrosion[4]. Mechanical failure,

*Corrosion rates have been given as rates of weight gain because they are the basis of most measurements. Penetrations calculated from the results are dependent upon the relative amounts of different products, and hence require information on the relative abundance of products, which is generally not known quantitatively and can depend upon experimental conditions such as temperature.

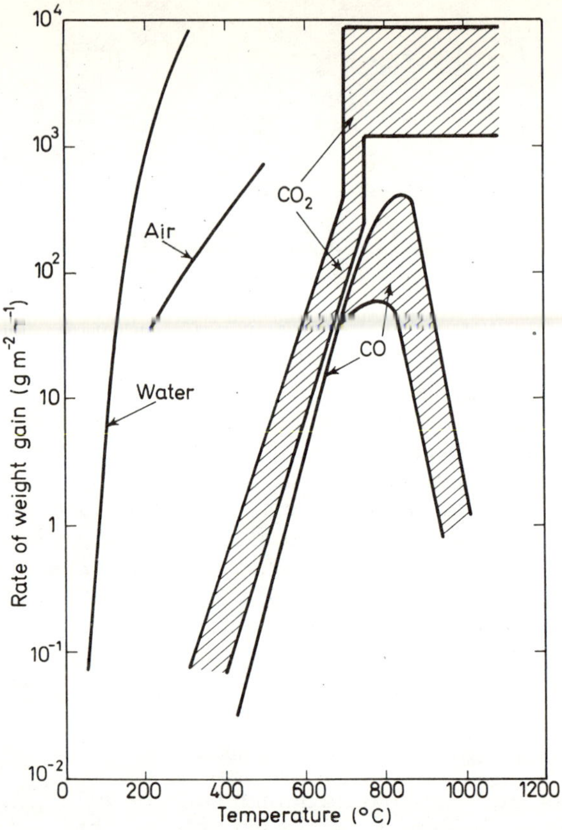

Fig. 5.12 Oxidation rates of uranium in distilled water, air, CO_2 and CO as a function of temperature

which is probably associated with stress corrosion as distinct from hydrogen pick-up, has been obtained for specimens of uranium[5] and various γ-quenched alloys. For at least the alloys it is enhanced by chloride ions[6] and is inhibited by cathodic polarisation[7]. At high temperatures the only resistant metallic materials based upon uranium are alloys containing appreciable amounts of the γ-phase stabilising elements, molybdenum, niobium and zirconium[2], the silicide U_3Si[(8)] and the aluminides. However, the alloys, as distinct from the intermetallic compounds, fail in a sudden disintegrating manner after long periods of exposure, due to, it is believed, hydride formation within the metal.

Atmospheric Corrosion

Uranium tarnishes readily in the atmosphere at room temperature. Electropolishing inhibits the process whilst etching in nitric acid activates the surface. Uranium dioxide and hydrated UO_3 are the principal solid products, although uranium hydride may have a transitory existence. The corrosion is enhanced by water vapour and hence is governed by the humidity conditions[9]. However, the presence of oxygen markedly inhibits attack by water vapour[3]. The view has been expressed that the corrosion is electrochemical in nature with the hydrated UO_3 being formed at cathodic areas[10].

Gases at High Temperatures

Carbon Dioxide and Carbon Monoxide[11]

The solid corrosion products in carbon dioxide and carbon monoxide are uranium dioxide, uranium carbides and carbon. The major reaction with carbon dioxide results in the formation of carbon monoxide:

$$U + 2CO_2 \rightarrow UO_2 + 2CO \qquad \ldots(5.1)$$

and in the formation of carbide the carbon transfer from the gas phase occurs mainly as a consequence of subsequent reactions of carbon monoxide with the metal through pores of the oxide scale. Reaction with carbon monoxide must, of necessity, involve the transfer from the gas phase of a carbon atom with each oxygen atom. This transfer could occur by direct reaction or via the thermal decomposition of the gas to produce carbon dioxide:

$$2CO \rightarrow C + CO_2 \qquad \ldots(5.2)$$

which can react subsequently by reaction 5.1. Diffusion studies on UO_2[12] demonstrate that oxygen ions are the major diffusing species in the oxide, and therefore that the oxide grows by the inward passage of oxygen. Volume changes associated with the oxidation lead to large stresses in the oxide and subsequently to the formation of solid products in the form of powders and/or cracked porous scales.

At the temperatures of interest, a linear rate law is quickly established, and the rate determining step is believed, in general, to be diffusion of oxygen ions through a thin layer of adherent oxide of constant average thickness at a given temperature. For carbon dioxide, the rate constant (Fig. 5.12) increases steadily with temperature until there is a sudden increase in rate, together with some self-heating from the high heat of reaction, at or near the β–γ phase transition in the metal (780°C); at higher temperatures there is little or no dependence of reaction rate on temperature up to 1 000°C. The majority of the oxide forms as either a non-adherent powder (the particle size of which increases with temperature), or, at the highest temperatures, as a cracked adherent scale. The formation of scale accounts for the lack of temperature dependence for the rate constant at the highest temperatures and could result either from sintering of the oxide, or, more probably, from the growth stresses being relieved by plastic deformation of the oxide and/or the underlying metal, rather than rupture of the oxide. At the highest temperatures the rate determining step might be gaseous diffusion through the porous scale[13]. The presence of small amounts of water vapour (greater than 100 p.p.m.) and oxygen (greater than 10 p.p.m.) enhances attack significantly at the lower temperatures of 400–500°C[11], and under these conditions preferential attack is frequently observed at carbide inclusions in the metal[14]. Major alloy additions of silicon (greater than 3·8%) impair oxidation resistance at all temperatures whilst additions of copper (1%), titanium (5–10%) and molybdenum (10–15%) decrease rates, but only at the higher temperatures[15]. The beneficial influence of copper is associated with the enrichment of the element at the metal/oxide interface as the UCu_5 phase[16]. The mechanisms responsible for the effects of the other alloying elements have not been firmly established but can be attributed to changes in the physical defects and thicknesses of the adherent oxide, and hence in its protective properties.

The oxidation rates in carbon monoxide (Fig. 5.12) are less than those for carbon dioxide. They increase steadily with temperature up to 800°C but then decrease markedly by a factor of 100 up to 1 000°C. The decrease in rate can be attributed to the beneficial factors operating at the higher temperatures with carbon dioxide and to the unfavourable thermodynamics for reaction 5.2 resulting in low equilibrium partial pressures of carbon dioxide.

Air

The high oxygen potential of air leads to the formation of higher oxides of uranium than uranium dioxide. U_3O_8 in the form of a powder is the main product after long times at temperatures above 200–300°C, although the lower oxides may be formed as intermediaries and at lower temperatures. The volume change associated with its formation is greater than for uranium dioxide, and as with carbon dioxide leads to a linear rate law being established. This volume change combined with the high oxygen potential gives fast reaction rates[17] by comparison with the behaviour in carbon dioxide (Fig. 5.12). The fast rates and the large heat change lead to an appreciable self-heating of specimens and an associated increase in reaction rate. In this situation the supply of oxygen from the air can be rate determining. The susceptibility of the metal to self-heating increases with the surface area/volume ratio of components and powders may ignite at room temperature[18]. However, large blocks of metal are not normally liable to self-heating until 400–600°C and then need an external heat source to maintain the situation. Alloy additions of molybdenum (5–15%) markedly reduce reaction rates[15,19]. Minor additions of other elements may either enhance or reduce the susceptibility to ignition[20] (e.g. aluminium or copper, respectively).

Steam

The mechanisms of corrosion by steam are similar to those for water up to 450°C, but at higher temperatures are more closely related to the behaviour in carbon dioxide. Studies at 100°C have demonstrated that uranium hydride is produced during direct reaction of the water vapour with the metal and not by a secondary reaction with the hydrogen product[1]. Also at 100°C it has been shown that the hydride is more resistant than the metal[1]. Inhibition with oxygen reduces the evolution of hydrogen and does not involve reaction of the oxygen with the uranium[3]. Above 450°C the hydride is not stable and hydrogen is released directly to the gas phase. Also, the cohesion and protection provided by the uranium dioxide increase with the formation of a dense scale, at least for short times at the highest temperatures. As a consequence the extent of reaction after a period greater than 100 min is at a maximum at 300–400°C with a rate of weight gain of $10^3 g\,m^{-2} h^{-1}$. The reaction rate increases only slowly with temperature from 500 to 1 200°C, although one piece of work reported a marked increase in attack near the β–γ transition in the metal[21], analogous to that obtained in carbon dioxide. A parabolic rate law has been found to apply at temperatures of 500–1 200°C for periods of 30 min–6 h[22], although other work indicates that this law is established for only 1–2 h above 880°C and that a linear law generally applies[21]. It follows that for a period of 1–2 h the extent of corrosion in steam straddles that for carbon dioxide, being greater at temperatures less than 700°C and comparable or less at higher temperatures.

Behaviour of Irradiated Uranium

The behaviour of irradiated uranium has been studied mainly with respect to the release of fission products during oxidation at high temperatures[23]. The fission products most readily released to the gas phase are krypton, xenon, iodine, tellurium and ruthenium. The release can approach 80–100%. For ruthenium it is dependent upon the environment and only significant in the presence of oxygen to form volatile oxides of ruthenium.

Studies of the influence of irradiation on the kinetics of oxidation have been confined to post-irradiation work. In general, prior irradiation increases reactivity, although there are considerable inconsistencies in the enhancements obtained[11,23,24]. The effects can be derived from an increased surface area associated with the swelling voids produced in the metal by the irradiation, and can also probably arise to a lesser extent from chemical effects of the fission products.

Coatings

Metallic and organic-based coatings have been developed to protect uranium from oxidation at low and high temperatures. The work on metallic coatings has covered intermetallic compounds and solid solutions of uranium with aluminium, zirconium, copper, niobium, nickel and chromium[26]. Aluminium- and zirconium-based coatings[25–27] can be particularly effective in reducing attack over an extended range of temperature. Also, nickel plates can provide protection against atmospheric corrosion[28]. Preliminary work on organic coatings demonstrated that they can enhance atmospheric corrosion[29,30]. This behaviour is attributed to the loss of the inhibiting action of oxygen due to water vapour permeating the coatings more readily than oxygen. However, a special coating composed of a styrene-butadiene copolymer containing powdered aluminium has been developed with a low permeability for water vapour which will reduce the corrosion in humid air by a factor of at least 10 at temperatures up to 40°C[29,30].

Conclusions

The behaviour of uranium has been well characterised for a variety of environments of importance in the nuclear industry. The corrosion is governed by the constitution and physical character of the solid reaction products which in turn are determined mainly by the oxygen potential of the environment, the temperature and the presence of water. The mechanisms of attack are known in broad outline. A major area in need of more detailed study is the influence of irradiation both prior to and during oxidation.

J. E. ANTILL

REFERENCES

1. Baker, M. McD., Less, L. N. and Orman, S., *Trans. Faraday Soc.*, **62**, 2 513 (1966)
2. Wanklyn, J. N. and Jones, P. J., *J. Nucl. Mat.*, **6**, 291 (1962)
3. Baker, M. McD., Less, L. N. and Orman, S., *Trans. Faraday Soc.*, **62**, 2 525 (1966)
4. Leach, J. S. L., *J. Inst. Met.*, **88**, 24 (1959)
5. Hughes, A. N., Orman, S. and Picton, G., *Corrosion Science*, **10**, 239 (1970)
6. Magnani, N. J., *J. Nucl. Mat.*, **42**, 271 (1972)
7. McLaughlin, B. D., *J. Nucl. Mat.*, **43**, 343 (1972)
8. Bourns, W. T., *Corrosion Testing of Uranium Silicide Fuel Specimens*, AECL–2 718 (1968)
9. Waber, J. T., *A Review of Recent Data on the Corrosion Behaviour of Uranium*, LA 2 035 (1959)
10. Kato, T. R. and Vonderbrink, V. O., *Atmospheric Corrosion of Uranium*, NLCO–860 (1962)
11. Pearce, R. J., Whittle, I. and Hilton, D. A., *J. Nucl. Mat.*, **33**, 1 (1969)
12. Belle, J., *J. Nucl. Mat.*, **30**, 3 (1969)
13. Pearce, R. J., *J. Nucl. Mat.*, **34**, 332 (1970)
14. Hayfield, P. C. S., Graham, R. L. and Ramshaw, G., 'Localised Oxidation of Uranium in Carbon Dioxide and Carbon Monoxide in the Range 200–500°C', Paper No. 5 of Inst. of Met. Sym. on Uranium and Graphite (1962)
15. Antill, J. E. and Peakall, K. A., *J. Less Common Metals*, **3**, 239 (1961)
16. Antill, J. E. and Peakall, K. A., *J. Electrochem. Soc.*, **110**, 1 146 (1963)
17. Baker, L. and Bingle, J. D., *J. Nucl. Mat.*, **20**, 11 (1966)
18. Baker, L., Schnizlein, J. G. and Bingle, J. D., *J. Nucl. Mat.*, **20**, 22 (1966)
19. Isaacs, J. W. and Wanklyn, J. N., *The Reaction of Uranium with Air at High Temperatures*, AERE R–3 559 (1960)
20. Schnizlein, J. G., Baker, L. and Bingle, J. D., *J. Nucl. Mat.*, **20**, 39 (1966)
21. Hopkinson, B. E., *J. Electrochem. Soc.*, **106**, 102 (1959)
22. Wilson, R. E. and Martin, P., *Isothermal Studies of the Uranium Steam Reaction by the Volumetric Method*, 148, ANL 6 569 (1962)
23. Parker, G. W. *et al.*, Out-of-Pile Studies of Fission Product Release from Overheated Reactor Fuels at ORNL, 1955–1965, ORNL–3 981 (1967)
24. Fischer, D. F. and Schnizlein, J. G., *J. Nucl. Mat.*, **28**, 124 (1968)
25. Baque, P., Koch, P., Dominget, R. and Darras, R., CEA–R–3 638
26. Buddery, J. H., Clark, M. E., Pearce, R. J. and Stobbs, J. J., *J. Nucl. Mat.*, **13**, 169 (1964)
27. Pearce, R. J., Giles, R. D. and Tavender, L. E., *J. Nucl. Mat.*, **24**, 129 (1967)
28. Orman, S., Owen, L. W. and Picton, G., *Corrosion Science*, **12**, 35 (1972)
29. Orman, S., *Atom*, No. 150, 93 (1969)
30. Orman, S. and Walker, P., *J. Oil and Colour Chemists' Association*, **48**, 233 (1965)

6 THE NOBLE METALS

6 The Noble Metals

Introduction

The outstanding characteristics of the noble metals are their exceptional resistance to corrosive attack by a wide range of liquid and gaseous substances, and their stability at high temperatures under conditions where base metals would be rapidly oxidised. This resistance to chemical and oxidative attack arises principally from the inherently high thermodynamic stability of the noble metals, but in aqueous media under oxidising or anodic conditions a very thin film of adsorbed oxygen or oxide may be formed which can contribute to their corrosion resistance[1]. An exception to this rule, however, is the passivation of silver and silver alloys in hydrochloric or hydrobromic acids by the formation of relatively thick halide films.

Their high initial cost, combined with a mechanical strength which is generally inferior to that of base metals, results in the noble metals being used as sheaths, linings, or other thin coatings on strong supporting structures. They are also widely used as electrodeposits, as described in Section 14.9. Silver and, to a lesser extent, platinum and its alloys are sometimes employed in more massive section for the construction of chemical plant, as for example distillation stills and bushings for the glass industry. The chemical, synthetic-fibre, glass and metallurgical industries make very wide use of the noble metals and their alloys wherever equipment must be protected from chemical attack, or a product protected from contamination by the products of corrosion or erosion.

Physical and Mechanical Properties

Although in the majority of their applications the choice of noble metals is determined by their chemical rather than by their physical and mechanical properties, some consideration of the latter is necessary. The relevant information for the noble metals as a whole is given in Tables 6.1 and 6.2, and details relating to the individual metals will be found in the following paragraphs.

Silver

Silver in the fully annealed state is a soft, ductile metal which is easily fabricated into the very wide range of forms employed in industry by the

Table 6.1 The physical properties of the noble metals

Metal	*Atomic number*	*Atomic weight*	*Lattice structure*	*Density at 20°C* (g/cm^3)	*Melting point* (°C)	*Thermal conductivity at 0–100°C* (W/m °C)	*Specific heat at 0°C* (J/kg °C)	*Coefficient of linear expansion at 20–100°C* $\times 10^6$	*Thermal neutron cross-section* (barns) (10^{-28} m^2)	*Resistivity at 0°C* (μΩ cm)	*Temperature coefficient of resistance 0–100°C* $\times 10^3$
Platinum	78	195·09	fcc	21·45	1 769	73	131·2	9·1	9	9·85	3·9
Iridium	77	192·22	fcc	22·65	2 443	148	128·4	6·8	430	4·71	4·3
Osmium	76	190·2	cph	22·61	3 050	87	129·3	6·1	15	8·12	4·2
Palladium	46	106·4	fcc	12·02	1 552	76	244·3	11·1	6	9·93	3·8
Rhodium	45	102·91	fcc	12·41	1 960	150	246·4	8·3	150	4·33	4·6
Ruthenium	44	101·07	cph	12·45	2 310	105	230·5	9·1	3	6·80	4·2
Silver	47	107·87	fcc	10·49	961	419	234	19·7	63	1·59	4·1
Gold	79	196·97	fcc	19·32	1 063	311	128	14·4	99	2·06	3·9

Table 6.2 The mechanical properties of the noble metals

Metal	*Tensile strength (annealed)* (MN/m^2)	*Elongation (annealed)* (%)	*Modulus of elasticity (E)* (GN/m^2)	*Hardness (annealed)* (HV)
Platinum	124	40	172	40
Iridium	1 100	—	516	220
Osmium	—	—	556	350
Palladium	172	40	117	40
Rhodium	688	5	316	100
Ruthenium	496	10	417	240
Silver	139	60	72	26
Gold	108	70	72	20

normal metal-working techniques such as drawing, spinning, rolling, etc. Silver work-hardens appreciably during fabrication. The mechanical strength of silver is markedly affected by an increase in temperature, and falls to about 25% of the initial value for cold, hard-worked silver when the metal is heated to just over 200°C.

Silver, with a thermal conductivity of 419 W/m°C is a somewhat better conductor of heat than copper, and this property is often utilised in the construction of heat exchangers, evaporator linings, etc.

Gold

Gold is an extremely soft and ductile metal, and exhibits very little work-hardening during deformation. Very severe cold rolling can increase its hardness from 20 HV in the fully annealed state to 70 to 75 HV, but this produces only a slight increase in its strength. The ultimate tensile strength of pure gold is too low to allow self-supporting structures of any size to be constructed of this metal—its use is almost entirely restricted to thin linings or electrodeposits on base-metal equipment.

Platinum and its Principal Alloys

Pure platinum is soft, ductile and easily fabricated. Its mechanical properties are profoundly affected by the degree of cold working to which it has been subjected and by the presence of slight impurities or alloying constituents. In its applications it is frequently alloyed with another metal of its group—the melting points of its alloys with Rh, Ir, Os and Ru being higher than the parent metal, those with Pd being lower. In most cases the strength, rigidity, hardness and resistance to corrosion are improved by alloying. Addition of certain base metals, however, can lead to embrittlement and failure of platinum and its alloys even if the base metal is present only in trace quantities.

Platinum work-hardens at approximately the same rate as palladium or copper, a 25% cold reduction in thickness raising the hardness from about 40 HV in the fully annealed state to 75 HV. Cast platinum is usually slightly harder than the same grade of metal in the wrought and annealed state.

Alloys with rhodium Rhodium alloys readily with platinum in all proportions, although the workability of the resulting alloy decreases rapidly with increasing rhodium content. Alloys containing up to about 40% rhodium, however, are workable and find numerous applications. The principal physical and mechanical properties of rhodium–platinum alloys are listed in Table 6.3.

The resistance of rhodium–platinum alloys to corrosion is about the same as or slightly better than that of pure platinum, but they are much more stable at high temperatures. They have excellent resistance to creep above 1 000°C, a factor which largely determines their extensive use in the glass industry, where continuous temperatures sometimes exceeding 1 500°C are encountered. Rhodium additions to platinum reduce appreciably the volatilisation of pure platinum at high temperatures.

Alloys containing more than 40% rhodium, while very difficult to fabricate, are almost immune from attack by aqua regia. The Pt–10Rh alloy is particularly resistant to attack by free wet chlorine such as that produced by the combustion of halogenated organic vapours.

Table 6.3 Physical and mechanical properties of rhodium–platinum alloys

Rhodium content (wt %)	*Melting point* (°C)	*Hardness (annealed)* (HV)	*Hardness (hard)* (HV)	*Tensile strength (hard)* (MN/m^2)
0	1 769	40	100	124
5	1 825	55	130	206
10	1 850	65	160	310
20	1 900	82	210	480
30	1 920	95	250	550
40	1 940	100	290	620

Alloys with iridium Iridium alloys with platinum in all proportions, and alloys containing up to about 40% iridium are workable, although considerably harder than pure platinum. The creep resistance of iridium–platinum alloys is better than that of rhodium–platinum alloys at temperatures below 500°C. Their stability at high temperatures, however, is substantially lower, owing to the higher rate of formation of a volatile iridium oxide.

Alloys with ruthenium Additions of ruthenium have a most marked effect upon the hardness of platinum, but the limit of workability is reached at about 15% ruthenium, owing to the fact that ruthenium belongs to a crystallographic system different from that of platinum. Apart from a somewhat greater tendency to oxide formation at temperatures above 800°C, the resistance to corrosion of ruthenium–platinum alloys is comparable to that of iridium–platinum alloys of similar composition.

Chemical Properties

The factors leading to the high resistance of the noble metals to chemical attack, i.e. their thermodynamic stability over a wide range of conditions and the possibility of the formation of very thin protective films under oxidising conditions, have already been mentioned. A factor tending to reduce corrosion resistance in aqueous solutions is the tendency of these metals to form complexes with some anions.

In the absence of complexing agents the noble metals are extremely resistant to corrosion by aqueous solutions of alkalis and salts, and to dilute acids. The resistance of gold and the platinum metals to corrosion by concentrated acids is summarised in Table 6.4, and to halogens in Table 6.5[2]. The resistance of silver to oxidising acids is generally lower than that of the other noble metals, while in halogen acids it forms a protective film of insoluble halide. Silver also differs from the other noble metals in forming a sulphide tarnish film in industrial atmospheres, due to the action of sulphur compounds.

In reviewing the corrosion resistance of the individual metals in aqueous solutions it is useful to compare the observed behaviour with that predicted

Table 6.4 Corrosion ratings of noble metals in acid

Acid	*Temp.* (°C)	Au	Ir	Os	Pd	Pt	Rh	Ru
Aqua regia	Room	D	A	D	D	D	A	A
	100	D	A	D	D	D	A	A
HI, 60%	Room	A	A	B	D	A	A	A
	100	—	A	C	—	D	A	A
HBr, 62%	Room	A	A	A	D	B	B	A
	100	—	A	C	D	D	C	A
HCl, 36%	Room	A	A	A	A	A	A	A
	100	A	A	C	B	B	A	A
HF, 40%	Room	A	A	A	A	A	A	A
Nitric, 70%	Room	A	A	C	D	A	A	A
Nitric, 95%	Room	B	A	D	D	A	A	A
	100	—	A	D	D	A	A	A
Perchloric, s.g. 1·6	Room	A	—	—	A	A	—	—
	100	A	—	—	C	A	A	A
Phosphoric, 100 g/l	Room	—	A	—	A	A	A	A
	100	A	A	D	B	A	A	A
Selenic, 42%	Room	A	—	—	C	A	—	—
	100	A	—	—	D	C	—	—
Sulphuric, 98%	Room	A	A	A	A	A	A	A
	100	A	A	A	C	A	B	A

A No appreciable corrosion
B Some attack, but not enough to preclude use
C Attacked enough to preclude use
D Rapid attack

Table 6.5 Corrosion ratings of noble metals in halogens

Halogen	*Temp.* (°C)	Au	Ir	Os	Pd	Pt	Rh	Ru
Sat. Cl water	Room	D	A	—	A	A	A	B
	100	—	—	—	D	A	—	—
Moist Cl	Room	D	A	C	D	B	A	A
Dry Cl	Room	B	A	A	C	B	A	A
Sat. Br water	Room	D	A	—	B	A	A	B
Moist Br	Room	C	A	B	D	C	A	A
Dry Br	Room	D	A	D	D	C	A	A
I in alcohol	Room	B	A	—	B	A	B	B
Moist I	Room	A	A	A	B	A	B	A
Dry I	Room	A	A	B	A	A	A	A

A No appreciable corrosion
B Some attack, but not enough to preclude use
C Attacked enough to preclude use
D Rapid attack

by the pH–potential diagrams developed by Pourbaix and his co-workers[3]. On the following diagrams the limits of stability of water are indicated by the dotted lines *a* and *b*.

Silver

The behaviour of silver in different environments is determined by three principal factors: (*a*) the high nobility of the metal, (*b*) the formation of passive protective films and (*c*) the tendency of the metal to form complex ions in solution.

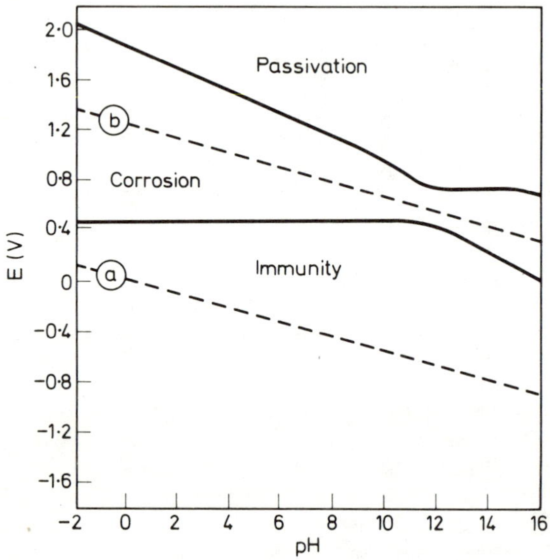

Fig. 6.1 Theoretical domains of corrosion, immunity and passivation of silver, at 25°C

Silver, with a standard electrode potential $E_{Ag^+/Ag} = 0{\cdot}79$ V, is exceeded in nobility only by gold and the platinum-group metals. The thermodynamics of the behaviour of silver in reducing acids, in the presence or absence of oxygen or complexing agents, has been considered in Section 1.4. Reference to the Ag/H_2O diagram[3] (Fig. 6.1) shows that at low potentials—below about 0·4 V—the metal remains immune to attack over almost the whole pH range. This diagram, of course, is of limited value, and is considerably modified in the presence, for example, of CN^- or Cl^- ions. The presence of halides, with the exception of fluoride, substantially increases the zone of passivity, owing to the very low solubility products[4] of silver halides. (Solubility products at 25°C: AgCl $= 1{\cdot}7 \times 10^{-10}$; AgBr $= 5{\cdot}0 \times 10^{-13}$; AgI $= 8{\cdot}5 \times 10^{-17}$; AgF—very soluble.)

On the other hand, the presence of CN^- ions greatly increases the zone of corrosion, owing to the formation of complex ions. Silver, therefore, is thermodynamically stable in reducing acids, e.g. hydrochloric acid, acetic acid, phosphoric acid, provided oxidising substances are absent.

Oxidising acids, e.g. nitric acid, hot sulphuric acid at concentrations exceeding 80% and reducing acids containing oxidising agents, will be corrosive to silver, and the diagram shows that an extensive zone of corrosion occurs at elevated potentials in the acid region.

When silver is passivated by a halide film, as is formed for example in hydrochloric acid, the film is tenacious, self-healing and highly insoluble. It may be reduced to the metal, however, by coupling the silver with such metals as zinc, aluminium, and, in the case of chemical plant, Hastelloy. In such instances the silver will continuously corrode.

As may be seen from the diagram, silver in highly alkaline solution corrodes only within a narrow region of potential, provided complexants are absent. It is widely employed to handle aqueous solutions of sodium or potassium hydroxides at all concentrations; it is also unaffected by fused alkalis, but is rapidly attacked by fused peroxides, which are powerful oxidising agents and result in the formation of the AgO^+ ion[5]. Table 6.6 gives the standard electrode potentials of silver systems.

Table 6.6 Standard electrode potentials of some silver systems[5]

System	*Potential* (V)
Ag^+/Ag	0·799
Ag^{2+}/Ag^+	1·98
AgO^+/Ag^{2+}	2·1
Ag_2O/Ag	0·344
AgO/Ag_2O	0·57
Ag_2O_3/AgO	0·74

Table 6.7 Standard electrode potentials of some gold systems

System	*Redox potential* (V)
Au^+/Au	1·68
Au^{2+}/Au^+	<1·29
Au^{3+}/Au^{2+}	<1·29
Au^{3+}/Au^+	1·41
Au^{3+}/Au	1·50
$AuBr_2^-/Au$	0·96
$AuCl_4^-/Au$	1·0
$H_2AuO_3^-/Au$	0·7

Silver is attacked by most compounds of sulphur, becoming covered with a yellow, brown or black sulphide film.

Gold

The high resistance of gold to attack by a very wide range of corrosive media results from its very high nobility. It may be seen from Fig. 6.2 that gold is immune to attack over the whole range of pH values at redox potentials below about 0·4 V, and that the zone of immunity extends to higher potentials at the lower end of the pH range. Gold, however, is easily complexed, and its solubility in hydrochloric acid containing an oxidising agent (e.g. nitric acid) results from a combination of high redox potential and the formation of chlor–auro complex ions[6], $AuCl_4^-$. The unstable Au^+ ion and the easily reducible Au^{3+} ion also readily form stable complexes. The Au^{2+} form is not stable in aqueous solutions, existing only in the solid, insoluble sulphide form. Standard electrode potentials of gold are given in Table 6.7.

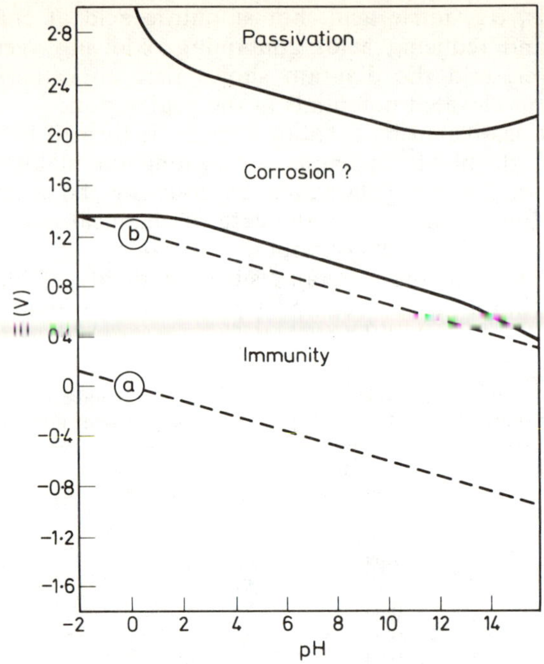

Fig. 6.2 Theoretical conditions of corrosion, immunity and passivation of gold, at 25°C

Gold is unaffected in alkaline solutions, but in the presence of cyanides the $Au(CN)_2^-$ ion is readily formed:

$$Au(CN)_2^- + e^- \rightarrow Au + 2CN^-; \qquad E = 0{\cdot}6\,V$$

The dissociation constant for the reaction

$$Au(CN)_2^- \rightarrow Au^+ + 2CN^-$$

is $K = 5 \times 10^{-39}$. Thus in the presence of cyanide ions gold is a powerful reducing agent and is readily oxidised by oxygen. This reaction forms the basis for the extraction of gold from its ores on an industrial scale.

Platinum-group Metals

All the six platinum-group metals are highly resistant to corrosion by most acids, alkalis, and other chemicals. Their high nobility is the main factor determining their chemical resistance, and the formation of complex ions in solution is principally responsible for their dissolution under certain conditions.

As may be seen from the potential–pH diagram[2, 14] (Fig. 6.3) platinum is immune from attack at almost all pH levels. Only in very concentrated acid solutions at high redox potentials (i.e. under oxidising conditions) is there a zone of corrosion. This accounts for the solubility of platinum in aqua regia. Platinum is also prone to complex-ion formation, and this can lead to

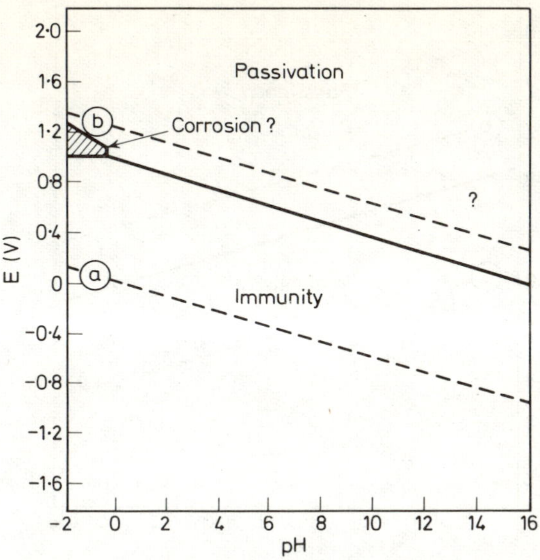

Fig. 6.3 Theoretical domains of corrosion, immunity and passivation of platinum, at 25°C

difficulties in interpreting the pH–potential diagrams when practical aqueous solutions are evaluated.

Platinum is unaffected by most organic compounds*, although some compounds may catalytically decompose or become oxidised on a platinum surface at elevated temperatures, resulting in an etched appearance of the metal. Carbon and sulphur do not attack platinum at any temperature up to its melting point. Molten platinum may dissolve carbon, but the solubility of the latter in solid solution is virtually zero.

The resistance of rhodium to chemical attack is remarkable, and surpasses that of platinum. Its domain of stability (as seen from Fig. 6.4) is extremely wide, and in the absence of complexing agents it is stable in aqueous solutions of all pH values. In the massive form it is unattacked by caustic alkalis, acids and oxidising agents, including aqua regia. When finely divided, however, it is attacked by concentrated sulphuric acid and aqua regia.

The behaviour of iridium is closely analogous to that of rhodium; its corrosion diagram is very similar and it is, with rhodium, one of the least corrodible of metals. It is unattacked by alkalis, acids or oxidising agents in aqueous solution, although a fused mixture of caustic potash and potassium nitrate will attack it. The metal has an excellent resistance to fused lead oxide, silicates, molten copper and iron at temperatures up to 1 500°C. Additions of iridium to platinum considerably raise the corrosion resistance of the latter to a very wide range of reagents.

Compared with platinum, rhodium and iridium, palladium has much less resistance to chemical attack. Its theoretical corrosion diagram is depicted in Fig. 6.5, from which it may be seen that the metal is stable in the presence of

*Certain organic compounds form complexes with platinum, and this accounts for the fact that the thin coating of platinum on titanium is rapidly corroded when platinised titanium is used as an anode in plating baths containing organic addition agents.

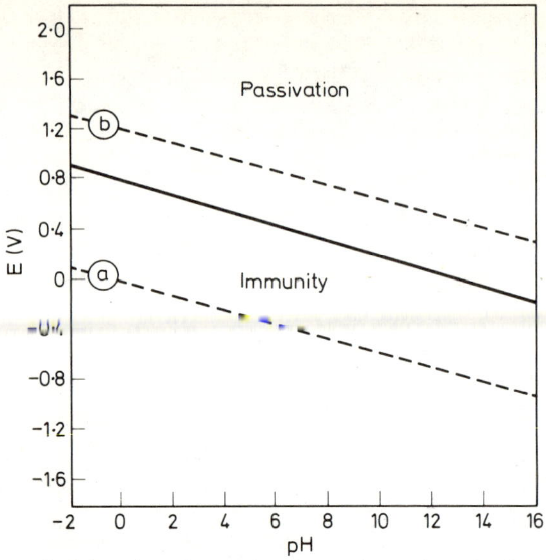

Fig. 6.4 Theoretical domains of corrosion, immunity and passivation of rhodium, at 25°C

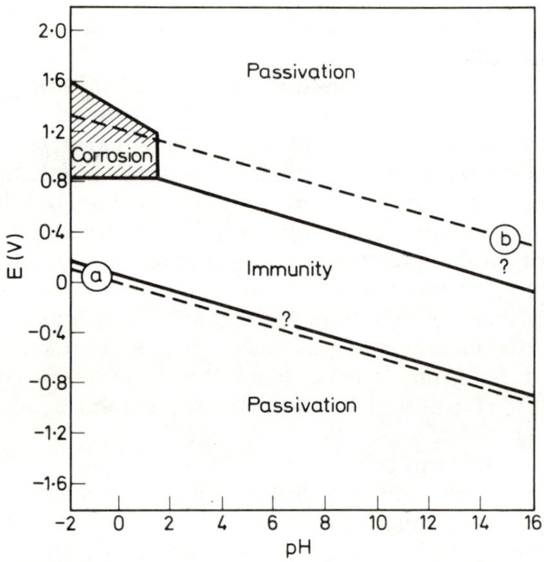

Fig. 6.5 Theoretical domains of corrosion, immunity and passivation of palladium, at 25°C

aqueous solutions of all pH values with the exception of strong oxidising agents and complexing substances. In practice, palladium is not attacked by chlorine water, except at high temperatures, and is not tarnished in moist air. Non-oxidising acids, e.g. acetic, oxalic, hydrofluoric and sulphuric acids, have no effect on the metal at ordinary temperatures. Strongly oxidising acids, however, e.g. hydrochloric acid containing nitric acid, rapidly attack palladium. Dilute nitric acid attacks palladium only slowly, but the metal is rapidly corroded by the concentrated acid. Alloys of palladium with platinum, however, retain most of the corrosion resistance of platinum. In ordinary atmospheres palladium is resistant to tarnish, but some discoloration due to sulphide-film formation may take place in industrial atmospheres containing sulphur dioxide. Alkaline solutions, even in the presence of oxidising agents, are without effect on the metal. This may be due to a thin passivating film of PdO which is more stable than $Pd(OH)_2$ and which prevents further attack.

Ruthenium and osmium are decidedly less noble than the other four metals of the platinum group. Both exist in numerous valency states and very readily form complexes. Ruthenium is not attacked by water or non-complexing acids, but is easily corroded by oxidising alkaline solutions, such as peroxides and alkaline hypochlorites.

High-temperature Properties of the Noble Metals[7]

Silver and Gold

Owing to their relatively low melting points and mechanical strengths, silver and gold find very few applications at elevated temperatures. Silver below its melting point has considerable resistance to oxide-film formation, but molten silver dissolves appreciable quantities of oxygen, which comes out of solution as silver oxide or bubbles dispersed throughout the metal when the metal solidifies. Gold is not subject to oxide-film formation at any temperature up to its melting point, but may be covered by a thin adsorbed layer of oxygen. The absence of an oxide film enables gold to be pressure-welded to itself at room temperatures.

Platinum-group Metals

The excellent resistance of platinum, rhodium and iridium to oxidation at high temperatures finds numerous applications in technology, in particular in the form of platinum-based alloys. Osmium and ruthenium form volatile oxides which may be isolated (OsO_4 and Ru_2O_3), and they are not widely used.

Platinum, while it does not form a measurable oxide film, is covered by a tenaciously held layer of adsorbed oxygen[8]. It volatilises at an increasing rate as its temperature rises above 1 000°C, and in the presence of moving oxygen or air the rate of volatilisation is considerably increased. In the presence of oxygen the volatilisation proceeds via the formation of a volatile, unstable oxide which has not to date been identified[9]. Rhodium, iridium and palladium exhibit oxide-film formation, the last named at temperatures as low as 600°C[10]. Palladium oxide dissociates again at temperatures above 870°C, the metal appearing bright up to its melting point. Absorption of oxygen without film formation occurs, however, and the palladium increases in weight. Platinum is more volatile than rhodium and iridium in the range 900–1 200°C, but their volatilities are about equal at temperatures around 1 300°C. Below 1 100°C the alloys of platinum with rhodium and palladium are less volatile than pure platinum, but palladium–platinum alloys are not employed at these temperatures owing to the oxygen absorption of the palladium already referred to[11]. Rhodium–platinum alloys at high temperatures show no preferential loss of either metal, and are widely used. Iridium–platinum alloys show greater loss of weight on heating in air, because of the greater rate of oxidation of iridium and the higher volatility of the oxide of this metal. Iridium is thus lost preferentially from iridium–platinum alloys.

Volatilisation of platinum and its alloys when subjected to continuous high operating temperatures may be substantially reduced by avoiding contact with oxygen, especially oxygen which is in movement or circulation owing to convection currents. This is achieved by completely embedding the metal in a high-grade pure alumina refractory cement or block—flame-sprayed coatings, for example, are effective in preventing free circulation of air over the metal. Only very high-grade alumina, free from silica or other oxides more easily reduced under reducing conditions can be employed, otherwise contamination and embrittlement of the platinum may result from partial reduction of such oxides.

Grain-growth of platinum and its alloys when operating continuously at high temperatures is often responsible for failure of the metal, resulting from weaknesses developed by large inter-crystal boundaries. This defect may be obviated to a considerable extent by the use of sintered metal produced by powder-metallurgical techniques[12], or by the incorporation of a small amount of a refractory oxide, carbide or nitride in powder form in the body of the metal. The latter method provides a source of nuclei for crystal formation, and the resulting pure metal or alloy has considerably higher strength and longer life at high temperatures[13].

Applications of the Chemical Properties of the Noble Metals

Protective Linings and Sheaths

Many chemical-plant vessels constructed of mild steel, copper, or other base metals may be protected from corrosive attack by their contents by means of a lining of one of the noble metals. This form of construction utilises the higher strength of the base metal—as, for example, in the walls of pressure vessels—and at the same time a minimum amount of a much more expensive noble metal is used. The use of noble metals in plant is frequently determined by the desire to avoid contamination of valuable reactants or substances by the products of corrosion of base metals.

Silver linings are generally 0·5 mm to 1 mm thick, although occasionally greater thicknesses are employed. The interior of the vessel to be lined must be free from roughness, projections, etc. and maximum radii of corners must be ensured. Exit pipes are lined similarly, and continuous fusion-welded seams ensure that no contact between base-metal vessel and corrosive contents takes place. A very large range of chemical plant may be lined with silver, but a solid silver construction is sometimes employed for condenser coils, distillation heads, etc. In such instances the silver is generally 1 mm to 2·5 mm thick. Stirrers, thermocouple pockets, inlet tubes, etc. are either constructed of solid silver or sheathed with this metal to prevent contact of the base with corrosive materials.

Linings in silver may be of two types, viz. 'loose' and 'bonded'. Loose linings provide good contact between vessel and lining—adequate for good heat transfer—but the establishment of a partial or full vacuum in the vessel would cause the air gap between lining and vessel to expand, resulting in collapse and failure of the lining.

Bonded silver linings are fabricated for mild steel or copper vessels. They are soldered *in situ* to the walls of the vessel by means of a special tin-silver solder. The melting point of this solder is approximately 280°C, and 200°C is recommended as the maximum continuous operating temperature for linings bonded with it. Since the whole of the silver is firmly adherent to the vessel, bonded linings are suitable for operation under vacuum conditions, and provide excellent heat-transfer characteristics.

Platinum and rhodium–platinum and iridium–platinum alloys are frequently employed to line and sheath autoclaves, reactor vessels and tubes, and a wide range of equipment. Linings are generally 0·13 mm to 0·38 mm thick, and for certain applications co-extruded platinum-lined Inconel or other metal reactor or cooling tubes are fabricated. In such cases the platinum is bonded to the base metal, but in all other instances platinum linings are of the 'loose' type.

Bursting Discs

Bursting discs are the simplest and most certain form of protection against the effect of over-pressure in a closed system. They cannot fail to operate,

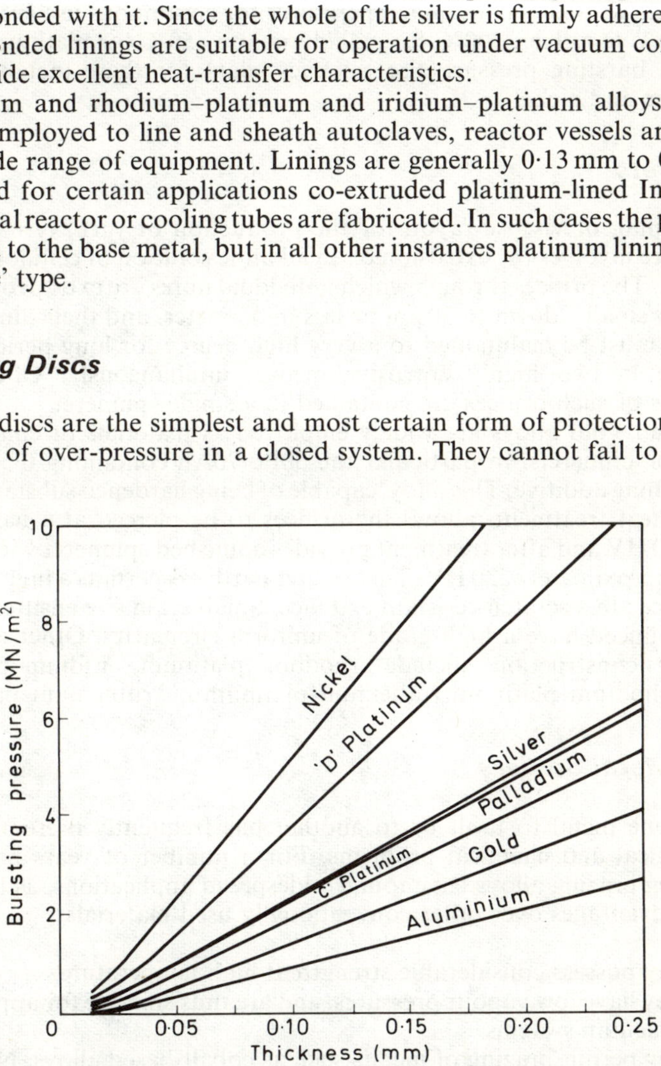

Fig. 6.6 Relationship between bursting pressure and original thickness for seven disc materials in a standard 1 in (25·4 mm) orifice

they have low inertia, they open immediately and they provide a full-bore passage for the discharge of pressure. Bursting discs are completely leak-proof until they burst, and can be made to withstand full vacuum. A most valuable aspect of their behaviour is their inherent tendency always to 'fail safe'.

The protection of vessels containing corrosive materials presents a special problem for the selection of bursting discs—a rapid rate of corrosion can lead to a high frequency of failures. In addition, the 'creep' of a metal disc when under tension at elevated temperatures would tend to weaken it and result in premature failure.

Bursting discs may be fabricated of gold, silver, platinum or palladium. The recommended maximum temperatures for continuous use are 80°C for gold, 150°C for silver, 300°C for palladium and 450°C for platinum. Figure 6.6 gives bursting pressure/disc thickness data for these metals and for aluminium and nickel.

Spinnerets

The spinning of viscose rayon for the production of yarn, tyre cord and staple fibre involves the extrusion of an alkaline solution of cellulose into an acid bath. The orifices through which individual fibres are extruded are often extremely small—down to 30 μm or less in diameter, and their dimensional accuracy must be maintained to a very high degree for long periods while operating in two highly corrosive media simultaneously. Often many thousands of such orifices are contained in a single spinneret.

Platinum–gold alloys are widely employed as materials of construction for rayon spinnerets, in particular the 30Pt–70Au containing 0·5 Rh as a grain-refining additive. This alloy, capable of being hardened substantially by suitable heat treatment, allows the orifices to be pierced at a hardness of about 120 HV and after treatment provides a finished spinneret with a hardness of approximately 220 HV. This greater hardness permits a high polish to be produced in a scratch-resistant exit face. Small grain size ensures that the holes produced have a high grade of uniform circularity. Other alloys for spinneret construction include rhodium–platinum, iridium–platinum, iridium–rhodium–platinum, ruthenium–platinum and ruthenium–palladium.

Noble Metal Solders

Joining one metal to itself or to another has frequently presented many metallurgical and chemical problems. For a number of years palladium-containing brazing alloys have found widespread applications, as they offer several advantages over other conventionally used materials:

(*a*) They possess considerable strength at high temperatures.
(*b*) They have low vapour pressures, and are thus suitable for applications in vacuum systems.
(*c*) They permit brazing of metals such as cobalt-based alloys, Nimonics, tungsten, zirconium, molybdenum and beryllium which could not otherwise be joined.

(*d*) They have little erosive effect on parent metals, permitting thin sections to be joined.
(*e*) They permit ceramics to be joined directly to low-expansion alloys.
(*f*) They have a high resistance to corrosion.

Silver–copper–palladium alloys with liquidus temperatures of 800–1 000°C have very low vapour pressures combined with good wetting and flow characteristics and are widely employed in vacuum work. They exhibit a lower tendency to stress corrosion than silver–copper, and do not form brittle alloys with other metals.

Silver–palladium–manganese brazes possess excellent creep characteristics and have been developed for high-temperature applications involving the use of cobalt or nickel-based alloys, heat-resistant steels, molybdenum and tungsten. Their liquidus temperatures lie in the range 1 100–1 250°C.

Nickel–manganese–palladium brazes are resistant to attack by molten alkali metals and find applications in sodium-cooled turbine constructions. Their freedom from silver and other elements of high thermal neutron-capture cross-section allows them to be used in liquid-metal-cooled nuclear reactors.

Copper–palladium–nickel–manganese brazes give very low erosion of the parent metals to be brazed, and are therefore used to join thin sections of stainless steels and high-nickel alloys.

Gold–copper alloys exhibit exceptional resistance to corrosion, and have very low vapour pressures. Gold–nickel alloys, with similar low vapour pressure, are somewhat stronger than gold–copper at high temperatures. Both series of alloys are widely employed in vacuum systems.

The addition of beryllium and silicon to nickel–palladium alloys gives very good high-temperature brazes, especially for alloys containing aluminium and titanium.

Pure ruthenium powder or mixed ruthenium–molybdenum powders have been found able to effect good joints between molybdenum and tungsten. A eutectic melting above 1 900°C is formed, and joints produced in hydrogen atmospheres at 2 100°C operate satisfactorily at 1 500°C. A cobalt–palladium–gold alloy has also been reported to be useful in brazing molybdenum.

Silver-based low-temperature brazing alloys have been known for many years and are widely employed throughout the engineering and chemical industries. Their output is measured in many tonnes per year, and their utility in joining almost all the commonly used materials of construction has made them indispensable for the production of a very wide range of plant, equipment and structures of all kinds. They exhibit a high resistance to corrosion by industrial atmospheres and possess excellent strengths at temperatures considerably higher than may be employed for lead or tin-based solders (see Section 10.5).

Anodic Behaviour of the Noble Metals

As with the chemical behaviour of the noble metals in aqueous solutions, their anodic behaviour closely follows the predictions of the Pourbaix diagrams[2, 14], if due allowance is made for the formation of complexes.

Silver generally corrodes anodically below the reversible oxygen potential, unless an insoluble oxide or salt is formed. When silver is used as an anode in sulphuric acid solutions, its behaviour shows an analogy with that of lead. Silver sulphate, Ag_2SO_4, is first formed, and this acts as a passive film[15]. When the potential is raised the sulphate is oxidised to AgO, which may be cathodically reduced back to Ag_2SO_4 at a potential lower than that required for its initial formation. When made anodic in nitrate solutions, silver generally dissolves quantitatively as Ag^+, and this forms the basis of the electrorefining techniques widely used in industry. Similar considerations apply to the anodic behaviour in cyanide solutions where silver forms a complex cyanide. In chloride solutions silver anodes become covered with a layer of silver chloride. When made anodic in alkaline solutions, silver becomes covered with a thin film of Ag_2O silver oxide, which protects it from further attack.

In chloride solutions, gold, when made anodic, initially dissolves in accordance with the equation $Au + 4Cl^- \rightarrow AuCl_4^- + 3e$. The dissolution gradually decreases, and the gold surface becomes passive owing to an adsorbed layer of oxygen[16]. Gold anodes in perchloric acid are passivated by a film of oxide, Au_2O_3, if used at potentials above 1·4 V. If the potential is raised above 1·95 V, the oxide film becomes flaky, and the gold is severely attacked[17]. In sulphuric acid an initial period of dissolution exists in accordance with the equation $Au \rightarrow Au^+$, followed by passivation resulting from a film of hydroxide, $Au(OH)_3$[18]. Ozone may be produced at sufficiently high current densities and the $Au(OH)_3$ is deposited with an accompanying attack on the gold below it to form an unstable peroxide. Continued production of oxide on the surface produces increasing passivity.

The anodic behaviour of platinum and certain of its alloys is of considerable technical importance, since they can be employed under a wide range of conditions without appreciable corrosion, and often in circumstances where no other metal can be used. Their use industrially has recently been extended by applying them as thin coatings to a substrate of a passive metal such as tantalum or, more commonly nowadays, titanium, to reduce the cost. Platinised titanium anodes are discussed in detail in Section 11.3.

Platinum can be used as an anode for the evolution of oxygen without corrosion over the whole pH range. At one time it was considered that the platinum was inert, so that oxygen could be evolved reversibly in the thermodynamic sense. However, it has been found that the reversible oxygen potential is not normally attained at a platinum surface, and that adsorbed oxygen or an oxide film play a part in the reaction. The nature of the surface film on platinum and some of the other noble metals has been the subject of considerable study[19]. Its exact nature is still in dispute, and probably varies for the different metals and under different conditions. The important point for technical applications is that it is very thin and is conductive, so that platinum anodes can operate at high current densities without excessive polarisation. However, the overpotential for the evolution of oxygen or chlorine on the surface of the thin oxide film is appreciably greater than on the bare metal surface which is considered to be a serious limitation in relation to the successful use of platinum in fuel cells[20]. Comparatively thick oxide films are only produced on platinum by direct currents with a considerable superimposed a.c. component or by alternating current under some conditions[21].

For chlorine evolution, platinum generally shows very low rates of attack, and platinum anodes can be used for cathodic protection in sea-water[22], and for chlorine production. However, in strongly acid solutions at low current densities platinum can be attacked; due to the complexing action of the chloride ion, the area of the potential-pH diagram (Fig. 6.3) in which corrosion occurs is somewhat enlarged, so that an anode operating at low overvoltage may be at a potential within this region. At higher current densities the overvoltage rises and passivation occurs.

The rate of corrosion of platinum anodes in chloride solutions increases if a considerable a.c. component is superimposed on the d.c. supply[23].

Rhodium and iridium have a resistance to anodic corrosion comparable with that of platinum, and are more resistant to the influence of alternating currents. A platinum–iridium alloy, in the form of a coating on titanium, is preferred to pure platinum for the production of chlorine from brine[24], due to its improved corrosion resistance and lower overvoltage.

Palladium is considerably less resistant to anodic corrosion than platinum, though it may be used for evolution of oxygen from alkaline solutions. It is attacked rapidly when used as an anode in sea-water, and dissolves quantitatively in acid chloride solutions.

Ruthenium dissolves anodically in alkaline solutions, as predicted by Pourbaix[3], but its corrosion resistance when made anodic in acid solutions is variable. Under some conditions the volatile and toxic tetroxide is evolved. Osmium is even more reactive anodically than ruthenium.

Recently it has been shown that the oxides of the platinum metals can have a higher corrosion resistance than the metals themselves[25], and have sufficient conductivity to be used as coatings for anodes, e.g. with titanium cores. Anodes with a coating of ruthenium dioxide are being developed for use in mercury cells for the electrolysis of brine to produce chlorine[26], since they are resistant to attack if in contact with the sodium-mercury amalgam.

The use of small platinum pins to promote the formation of lead dioxide on lead anodes is described in Section 11.3.

Alloying Additions of Noble Metals to Facilitate Passivity

The addition of a small percentage of a noble metal to a base metal such as stainless steel or titanium can provide sites of low overvoltage for the cathodic reduction of dissolved oxygen or hydrogen ions. This permits larger currents and hence more positive potentials to be obtained at the anodic region, and promotes passivation under some circumstances[27]. This effect has been demonstrated for stainless steels[28] but has not been adopted in practice, since under other conditions the noble metal addition accelerates corrosion[27].

The method is more useful with titanium, and the effect of alloying titanium with a small amount of palladium is described in Section 5.4. The use of platinum in the prevention of hydrogen embrittlement in tantalum, considered in Sections 5.5 and 1.7, also depends on the low hydrogen overvoltage of the added element.

High-temperature Applications of the Platinum-group Metals

The principal applications of the outstanding stability of platinum and its alloys at high temperatures lie in their use as materials of construction for equipment to handle molten glass and as electrical resistance windings for high-temperature furnaces.

In handling molten glasses at temperatures which sometimes reach 1 500°C, and frequently lie around 1 400°C, platinum alloys fulfil several functions:

(*a*) They protect refractory bricks and equipment from the erosive effect of flowing glass, thus greatly prolonging the life of those parts which would otherwise have to be replaced frequently.

(*b*) They protect the glass from contamination either by particles of refractory broken off by erosion or by the dissolution of refractory bricks or parts.

(*c*) They enable very high dimensional accuracy to be maintained when used to line or construct equipment designed to meter, control, or form molten glass.

The most usual forms of platinum or, more frequently, of the platinum alloys containing 5 or 10% of rhodium, are thin protective sheaths contoured to conform with the shape of the underlying refractory. Sheaths of this type, generally 0·25 mm to 0·65 mm thick, are widely employed to protect tank lips, skimmer blocks, stirrers, thermocouple pockets, etc. More substantial thicknesses—up to 2·5 mm thick—are used to protect orifices whose dimensional accuracy must be maintained to a high degree.

Bushings for the manufacture of glass fibre contain a large number of small, accurately shaped orifices which must maintain their dimensional accuracy to a very high degree. In addition, such bushings must remain free from distortion and act as electrical resistance heaters providing an even temperature profile to give uniform rates of flow through individual orifices. Operating continuously at temperatures around 1 400°C, such bushings have to give long life under very arduous conditions. They are generally made of Pt–10Rh and supported by accurately shaped refractory bricks.

Rhodium in rhodium–platinum alloys imparts a slight grey colour to glasses, especially to those of high lead content, so that pure platinum is utilised for the construction of crucibles or other equipment to handle high-quality colour-free optical glasses. Even platinum can impart traces of colour if the glass is high in alkali content, but the effect is reduced in oxidising atmospheres.

Investigations on the wettability of platinum alloys by glasses have established that this property, as measured by the contact angle between glass and metal, is strongly dependent on the surrounding atmosphere. It rises sharply as the atmosphere changes from oxidising to reducing, and is also dependent on the affinity for oxygen of the alloying constituent, which is generally rhodium but may also be gold, beryllium, or iridium.

Rhodium–platinum alloys containing up to 40% Rh are used in the form of wire or ribbon in electrical resistance windings for furnaces to operate continuously at temperatures up to 1 750°C. Such windings are usually completely embedded in a layer of high-grade alumina cement or flame-sprayed alumina to prevent volatilisation losses from the metal due to the free circulation of air over its surface. Furnaces of this type are widely employed for steel analysis, ash fusions and other high-temperature analytical procedures.

Working with Noble Metals

Silver

Silver is available in three principal grades for industrial purposes—fine silver, the normal commercial product containing a minimum of 99·95% silver which is used in most plant and equipment; chemically pure silver containing a minimum of 99·99% silver, used for catalytic and special purposes where the presence of certain trace impurities may adversely affect its resistance to corrosion, and oxygen-free silver. The latter grade is as pure as fine silver and is used for equipment which may be heated in the presence of hydrogen without danger of embrittlement resulting from combination of the hydrogen with oxygen dissolved in the metal.

Silver is available in many standard forms—sheet, strip, foil of thicknesses down to 0·013 mm, rod, wire down to 0·013 mm diameter, gauze, tubes, bi-metal as silver-clad copper or phosphor-bronze and many others. It is easily fabricated by the normal techniques of rolling, spinning, drawing, etc. and readily joins to itself by fusion-welding using argon-arc welding. Welding by oxy-hydrogen flame may also be employed, but the resulting weld is not as satisfactory owing to the possibility of oxygen absorption while the metal is molten, followed by embrittlement by hydrogen. Fine silver filler rods may be used, and hammering the weld fillet down to the contour of the surrounding metal produces a very strong joint. A wide range of silver-based or soft (tin-based) solders may also be used.

Gold

Gold is available for industrial purposes in a grade containing a minimum of 99·9% gold in a wide range of forms—sheet, foil, tube, wire, etc. It is easily fabricated, and when it is being joined to itself may be fusion-welded with an oxy-hydrogen flame or hammer-welded at temperatures well below the melting point. Soft-soldering is not recommended.

The Platinum-group Metals

Platinum is capable to being refined to an extremely high degree of purity, resulting in a metal of remarkably consistent and reproducible properties. Certain traces of impurities, however, have profound effects on some of its properties, and the correct grade of platinum must always be selected for any particular application. Among the applications for which special grades of platinum have been developed may be listed the following:

Chemical and catalytic. This grade of platinum is for conversion to catalysts, gauzes and chemical compounds. Spectrographic analysis is employed to control the presence of trace impurities harmful in these applications.

High-temperature. Depending on the intended continuous operating temperature, specific impurities known to affect the mechanical properties of platinum at these temperatures are excluded. Grain-stabilised platinum containing a refractory additive to reduce grain growth to a minimum and to increase its resistance to creep and mechanical strength is also available.

Equipment. The impurities in this grade of platinum, intended for all normal purposes in chemical plant, electrodes, crucibles, etc. other than for use above 1 000°C, consist only of traces of the other platinum metals.

Platinum in the forms detailed above and in its more usual alloys with other noble metals is available as sheet, foil down to 0·0064 mm thick, tube, rod, wire down to 0·0064 mm diameter, Wollaston wire down to 0·001 mm diameter, and clad on thin sections of base metals, e.g. copper, nickel, Inconel, etc.

Platinum, palladium and the normal alloys of platinum used in industry are easily workable by the normal techniques of spinning, drawing, rolling, etc. To present a chemically clean surface of platinum and its alloys after fabrication, they may be pickled in hot concentrated hydrochloric acid to remove traces of iron and other contaminants—this is important for certain catalytic and high-temperature applications. In rolling or drawing thin sections of platinum, care must be taken to ensure that no dirt or other particles are worked into the metal, as these may later be chemically or electrolytically removed, leaving defects in the platinum.

When platinum or its alloys are being joined, properties of the weld or solder must be such that it is no less corrosion or oxidation-resistant for the application in question than the parent metal. Platinum and its alloys are readily joined to themselves and to certain base metals, e.g. iron, nickel, copper. The principal methods for joining platinum are as follows:

(*a*) Fusion welding, using a platinum or alloy filler rod of the same composition as the parent metal and a shielded electric arc or an oxy-hydrogen flame (an oxy-acetylene flame may cause carbon pick-up by the molten metal). The weld fillets are then cold hammered to the contours of the surrounding metal to provide a strong joint.

(*b*) Hammer-welding. Platinum and rhodium–platinum alloys when cleaned are readily hammer-welded to themselves and to each other at temperatures in the range 800–1 000°C. The welds so produced are completely homogeneous.

(*c*) Soldering. Fine gold, copper, silver–palladium or platinum–palladium–gold–copper alloys may be used to solder platinum to itself, to its alloys, or to steels, nickel, etc. No fluxes are used, and soft solders should not be employed.

Economies Gained by Using Noble Metals

In spite of their high initial cost, the noble metals in industry offer substantial advantages and economies in use. They avoid the frequent and expensive replacement of plant on account of corrosion, they protect the product from contamination, allowing a higher quality with its attendant premium on value to be produced, and they frequently permit the carrying out of operations which would not otherwise be possible.

When they have served their purpose or become damaged the noble metals will realise a very high proportion of their initial cost. No matter in what form they are utilised, very efficient processes are available for effecting their complete recovery. This factor often makes noble metals the most economic in use in the chemical and engineering industries. For many applications, no other metal or group of metals can fulfil their function as efficiently, combined with such a low net cost to their user.

G. W. WALKIDEN

REFERENCES

1. Anson, F. C. and Lingane, J. T., *J. Amer. Chem. Soc.*, 79, 4 901 (1957)
2. La Que, F. L. and Copson, H. R., *Corrosion Resistance of Metals and Alloys*, Chapman & Hall, London
3. Pourbaix, M. J. N., *Atlas of Electrochemical Equilibria*, Pergamon Press, Oxford (1966)
4. Latimer, W. L., *Oxidation Potentials*, Prentice Hall, New York, 191 (1961)
5. Latimer, W. L., *Oxidation Potentials*, Prentice Hall, New York, 197 (1961)
6. Latimer, W. L., *Oxidation Potentials*, Prentice Hall, New York, 195 (1961)
7. Powell, A. R., *Platinum Metals Rev.*, No. 2, 95 (1958)
8. Andreeva, V. V. and Shishakov, N. A., *J. Appl. Chem.*, **11**, 388 (1961)
9. Alcock, C. B. and Hooper, G. W., *Proc. Roy. Soc.*, (A)254, 551 (1960)
10. Raub, E. and Plate, W. Z., *Metallk.*, **48**, 529 (1957)
11. Betteridge, W. and Rhys, D. W., *Proc. First International Congress on Metallic Corrosion*, London, 1961. Butterworths, London, 186 (1962)
12. Unpublished, Johnson, Matthey & Co. Ltd., Research Laboratories, London
13. Johnson, Matthey & Co. Ltd., UK Pat. 830 628 (1958)
14. Pourbaix, M. J. N., Van Muylder, J. and de Zoubov, N., *Platinum Metals Rev.*, **3** No. 47, 100 (1959)
15. Jones, P. and Thirsk, H. R., *Trans. Faraday Soc.*, **50**, 732 (1954)
16. Armstrong, G. and Butler, J. A. V., *Trans. Faraday Soc.*, **30**, 1 173 (1934)
17. Jirsa, F., *Coll. Czech. Chem. Comm.*, **13**, 505 (1948)
18. Jirsa, F. and Buryanek, O. Z., *Electrochem.*, **29**, 126 (1923)
19. Hoare, J. P., *Advances in Electrochemistry and Electrochemical Engineering*, **6**, Interscience Publishers, New York, 201–288 (1967)
20. Bockris, J. O'M. and Reddy, A. K. N., *Modern Electrochemistry*, Vol. 2, Macdonald, London, 1 350 (1969)
21. Altmann, S. and Busch, R. H., *Trans. Faraday Soc.*, **45**, 720 (1949)
22. Walkiden, G. W., *Corrosion Technology*, Jan.-Feb. (1962)
23. Juchniewicz, R. and Hayfield, P. C. S., *Proc. 3rd Int. Congr. on Metallic Corrosion*, Moscow, 1966. Vol. III, English Ed., 73–82
24. Anon., *Platinum Metals Rev.*, **13** No. 3, July, 103 (1969)
25. Beer, H. B., UK Pat. 1 147 442 (1969)
26. Anon., *Platinum Metals Rev.*, **14** No. 3, July, 92 (1970)
27. Hoar, T. P., *Platinum Metals Rev.*, **2**, 117 (1958)
28. Tomashov and Chernova, *Dokl. Akad. Nauk SSSR*, **89**, 121 (1953)

7 HIGH-TEMPERATURE CORROSION

7.1 Environments

Introduction

The terms 'hot corrosion' or 'dry corrosion'* are normally taken to apply to the reactions taking place between metals and gases at temperatures above 100°C; i.e. temperatures at which the presence of liquid water is unusual. The obvious cases of wet corrosion at temperatures above 100°C, i.e. in pressurised boilers or autoclaves, are not considered here. In practice, of course, common metals and alloys used at temperatures above normal do not suffer appreciable attack in the atmosphere until the temperature is considerably above 100°C. Thus iron and low-alloy steels form only the thinnest of interference oxide films at about 200°C, copper shows the first evidence of tarnishing at about 180°C, and while aluminium forms a thin oxide film at room temperature, the rate of growth is extremely slow even near the melting point.

Consideration will also be given to attack arising from contact with solids such as refractories, and with molten materials such as salts, glasses, and lower-melting-point metals and alloys. On a fundamental basis, the distinction between some of these latter reactions and normal-temperature aqueous corrosion is not always clear, since galvanic effects may be of significance in both cases, but for practical purposes a distinction can be made on the basis of the temperature involved.

A few cases occur in which hot corrosion and wet corrosion are interdependent, the wet corrosion arising from the condensation of liquids generated during a period at elevated temperatures. The formation of condensates of hydrobromic acid in engines burning anti-knock fuels containing ethylene dibromide is important in this context. Such cases are properly considered as aqueous corrosion.

Most hot-corrosion phenomena of practical significance are controlled by the kinetics of the reactions proceeding, rather than by the thermodynamic stability of the reactants or products involved. It must, however, be borne in mind that reaction rates determined under simplified laboratory conditions are frequently inapplicable to the more complicated conditions experienced in practice. Factors of major importance in this context are stress and thermal cycling.

Stress This can be of importance in two ways. An effect analogous to aqueous stress corrosion, in which attack is accelerated at grain boundaries,

*The terms 'wet' and 'dry' in relation to corrosion are discussed in Section 1.1.

may be produced; this is particularly noticeable and very well known in cases of attack by other molten metals. Stress may also accelerate general attack when the metal is suffering progressive deformation, as in creep or fatigue; the mechanical properties of the metal and of the scale may differ sufficiently from one another for the scale to crack periodically as deformation of the metal increases, and thus produce corresponding increases in the rate of corrosive attack. Gulbransen and Andrew[1] have reported effects of this nature on the rate of oxidation of nickel–chromium alloys, although the test specimens were cooled to normal temperature for the strain to be imposed. It was shown that the magnitude of the change in the rate of oxidation depended critically on the silicon content of the alloy. Abnormal oxidation effects during high-temperature fatigue tests on Nimonic alloys have been described by Betteridge[2], accelerated oxidation occurring at the point of fatigue cracking.

Thermal cycling This can lead to cracking and flaking of scale layers, either by stresses due to differential expansion of scale and underlying metal, or by dimensional changes associated with phase changes in the metal or in the scale. The latter phenomenon has proved particularly troublesome in efforts to produce a protective scale on molybdenum-base alloys[3]. Scale flaking, whatever the cause, leads to acceleration of the corrosion rate. Although some laboratory tests simulate the effects of thermal cycling it must be remembered that the size and shape of test sample is significant. Bruce and Hancock[4] have developed a vibrational technique for studying the onset of scale cracking during oxidation and have used it to study the behaviour of scales on nickel and iron in the temperature range 570–1 000°C.

The more fundamental aspects of the oxidation of metals have been fully described in Chapter 1, and by Kubaschewski and Hopkins[5], while practical considerations, particularly in relation to commercially important alloys, have been dealt with by Hessenbruch[6].

Air and Oxygen

Although corrosion by oxidation in air is undoubtedly the type of attack of greatest practical importance it is remarkable that relatively few cases are entirely free from the complications arising from contamination of the reacting atmosphere. Except in the case of heating by electrical means, contamination arises from the combustion products, in gaseous, liquid or solid form, while with electrical heating the oxidation is often complicated by the presence of fumes or vapour arising from the materials being processed.

Perhaps the closest approach to pure oxidation in everyday conditions arises in domestic electric heating appliances where the elements are exposed to the air. At some points the elements are necessarily in contact with supporting refractories, and if these are not of adequate purity, accelerated corrosion leading to early failure can occur. In a similar way the sheathed radiant-type elements of electric cookers usually fail owing to the corrosive effects of contaminants such as animal fats or salts from spilled liquids. Wires for these applications are commonly assessed by a standardised test such as that described in ASTM B76–65, which involves electrical heating for periods of 2 min to a temperature of 1 177°C (or other temperature depending on the composition of the wire) until failure by burn-out occurs. The test is a satisfactory comparative one which simulates service conditions apart from the acceleration resulting from the considerably increased temperature, but it is not suitable for the determination of basic data. Accurate setting of temperature and its maintenance at a uniform level along the length of a wire are difficult. Furthermore, the steep temperature gradient from the wire into the surrounding cool atmosphere produces disturbances due to thermal diffusion effects. These are of particular importance in the case of wires giving rise to volatile oxides (such as nickel–chromium wires forming chromic anhydride) since the heavy oxide molecules are rapidly removed by thermal action from the vicinity of the wire. This phenomenon is probably responsible for the fact that the appearance of an electrically heated wire, particularly in relation to scale adhesion, is frequently different from that of a wire heated in a furnace under similar conditions.

In the case of alloys having one constituent considerably more reactive to oxygen than the others, conditions of temperature, pressure and atmosphere may be selected in which the reactive element is preferentially oxidised. Price and Thomas[7] used this technique to develop films of the oxides of beryllium, aluminium, etc. on silver-base alloys, and thereby to confer improved tarnish resistance on these alloys. If conditions are so selected that the inward diffusion of oxygen is faster than outward diffusion of the reactive element, the oxide will be formed as small dispersed particles beneath the surface of the alloy. The phenomenon is known as *internal oxidation* and is of quite common occurrence, usually in association with a continuous surface layer of oxides of the major constituents of the alloy.

Kubaschewski and Hopkins[5] consider the conditions of the gaseous phase which influence the rate of corrosion of metals; apart from major variations of composition, they refer also to the effects of minor impurities, gas pressure, flow rate and ionisation.

Impurities

The corrosion rate can be influenced considerably by the presence of small proportions of contaminants, particularly if these are adsorbed on to the metal surface and thus influence the nature of the initially formed film of corrosion product. The presence of water vapour in air is, of course, of greatest practical importance, and marked effects from this cause on the scaling rate of steel have been reported[8]; the direction of the observed effect has not been constant and it is possible that the influence of hydrogen, formed by dissociation of the water, in decarburising the steel is a factor of importance. Similarly, it has been reported that humidity influences the life of nickel–chromium wires for electrical heating elements subjected to the ASTM life test referred to earlier, a change from dry air to air saturated with water vapour at normal temperature reducing the life by approximately one-half[9]. The nitrogen in air can sometimes play a significant rôle in corrosion, particularly in relation to steels containing chromium or aluminium which have high affinities for nitrogen; even in cases in which the nitrogen does not enter into the chemical processes of corrosion it may influence the rate of attack by oxidation, presumably by being adsorbed on to the metal surface. Thus Kubaschewski and Hopkins[5] point out that different reaction rates result from using on the one hand dry air and on the other pure oxygen at the equivalent partial pressure of the oxygen in the air.

Pressure

Variation in the pressure of the reacting gas can affect corrosion processes in two ways. In the cases more usually met with in practice, in which the corrosion rate is controlled by diffusion processes in the surface film of corrosion product, the influence of gas pressure on corrosion rate is slight. If, however, the dissociation pressure of the oxide or of a constituent of the scale lies within the range involved, the stability of the corrosion product will be critically dependent on the pressure. The effect of temperature is, however, far more critical and thus, in practical cases, pressure variations rarely decide the stability of corrosion products.

Rate of Gas Flow

The rate of flow of the gases over the metal surface is generally of little significance. In some special cases, however, it has a very marked influence on the rate of attack, which is high in stagnant atmospheres but low if adequate ventilation is provided; alloys containing appreciable proportions of molybdenum show this effect to a marked degree; it is associated with the volatility of the oxide, MoO_3, and its catalytic action in promoting further oxidation of the alloy.

Ionisation

Little effect is exerted by ionisation of the gas phase on corrosion rate.

Steam

Corrosion by essentially pure steam arises principally in connection with power generating plant. Temperatures up to about 600°C in association with pressures up to about 15 MN/m^2 are involved, although in the most advanced super-critical installations being planned for the future, temperatures of 650°C and pressures of 40 MN/m^2 are under consideration. The highest temperatures occur in the superheater tubes, but it is probable that the severest corrosion conditions on these components will arise from the presence of fuel ashes on the outside of the tubes rather than from the steam on the inside. Perhaps the most critical components in such installations will be the main steam pipes which are more highly stressed than the superheater tubes. The corrosion resulting from the action of high-temperature steam generally leads to the formation of oxides, although hydroxides may be formed. The reaction depends on the dissociation pressure of the oxide in relation to the partial pressure of the oxygen in the water vapour.

The importance of the dissociated hydrogen in contributing to the decarburisation of carbon steels heated in water vapour has already been mentioned.

Carbon and Oxides of Carbon

The importance of these agents in leading to hot corrosion has increased considerably in recent years as a result of the development of graphite-moderated nuclear reactors using carbon dioxide as coolant. Reaction between carbon dioxide and the graphite leads to the presence of a small content of carbon monoxide (which increases with increasing temperature) in the atmosphere. The formation of oxide films results from the presence of dissociated oxygen, but at the same time carburisation may occur with steels or alloys having constituents which form stable carbides. Indeed, carbon monoxide, sometimes with additions of hydrocarbons, is a very effective gas-carburising agent for steels, as the partial pressure of the oxygen is insufficient to lead to oxidation of iron. In the case of materials forming more stable oxides, e.g. chromium-containing steels or alloys, oxidation may proceed in parallel with carburisation.

Sulphur-containing Gases

Sulphur-containing atmospheres are of importance mainly in the effect which they exert when present in small proportions in combustion products, but studies have been made of the effects of pure sulphur vapour, hydrogen sulphide and sulphur dioxide on various metals. The effect of pure sulphur is usually straightforward, leading to the formation of a sulphide film which may be more or less protective according to its structure and properties. Some sulphides have low melting points, or form low-melting-point eutectics with the metal, and in these cases attack is rapid above the critical temperature, usually taking the form of intergranular penetration. For example, in the nickel–sulphur system a eutectic forms at 645°C, whereas in the iron–sulphur system the minimum melting point is 988°C. Hence it is normally found that iron-base alloys are more satisfactory than nickel-base alloys in sulphurous atmospheres at temperatures within this range. Attack by hydrogen sulphide bears some similarity to that by sulphur, the hydrogen having a negligible effect in most cases. Sulphur dioxide, however, frequently behaves very differently, as the formation of oxide films apparently largely prevents the absorption of sulphur. Useful details of the effects of hydrogen sulphide and sulphur dioxide on various metals are provided by Farber and Ehrenberg[10].

Hydrogen

Attack upon metals by hydrogen rarely results in the formation of surface films of compounds since the gas is usually readily soluble in the metal, with a high rate of diffusion, and hydrides, if formed, are generally rather unstable. The effect of hydrogen usually follows from its reducing action on certain constituents in the alloys; probably the most familiar examples of this are the decarburisation of steels by the conversion of carbides to volatile hydrocarbons, and the embrittlement of tough-pitch copper by the reduction of intergranular oxide. Hydrogen occurs as a major constituent of the atmosphere in furnaces used for bright annealing or for brazing, and in such cases the temperatures involved may rise to as high as 1 200°C. In the chemical engineering field many processes of synthesis or hydrogenation are carried out with hydrogen-rich atmospheres at temperatures up to about 600°C and at pressures rising to 1 000 atm. Much of the understanding of the attack on steels by hydrogen has been obtained from work on problems arising in the chemical industry.

Nitrogen

Although nitrogen is the major constituent of air, it plays a minor rôle in the scaling of most metals on heating in air, its effect being completely overshadowed by that of oxygen. This follows from the fact that the dissociation pressures of the nitrides of the commoner metals are greater than the partial pressure of the nitrogen. Chromium, aluminium and some of the less common metals such as titanium, molybdenum and tungsten, however, form stable nitrides which can result from heating materials containing these elements in air. Reaction between nitrogen and metals is of principal importance, however, in reducing atmospheres, particularly in cases in which absorption of nitrogen is deliberately sought, i.e. in the surface-hardening of steels by nitriding. The atmosphere in such cases is anhydrous ammonia, dissociation of which occurs at the surface of the steel, although only a small proportion of the resultant nitrogen is absorbed. Materials resistant to nitrogen and not acting as dissociation catalysts for ammonia are required as containers for such treatments, and stainless steels of the 18/8 or 25/20 types, or Inconel (76Ni–15Cr–8Fe) have been widely used.

Combustion Products

The various gas mixtures arising from the combustion of fuels form the major group of atmospheres leading to the hot corrosion of metals. Almost all natural and processed or synthetic fuels consist predominantly of hydrocarbons, carbohydrates, or more complex carbon compounds, so that the principal constituents of the combustion products are the oxides of carbon and water vapour, together with the residual nitrogen from the air. In conditions of incomplete combustion the gases may contain hydrogen, hydrocarbons and carbon monoxide in appreciable proportions, with only small proportions of oxygen, while when complete combustion is effected with excess air these reducing gases will be absent, but there will be larger amounts of oxygen. An indication of the range of compositions to be encountered is given by the values quoted for different fuels in Table 7.1 (see References 11

Table 7.1 Composition of waste gases after complete combustion of fuels

Fuel	*Waste gases* (% by vol.)		
	CO_2	H_2O	N_2
Bituminous coal	17–19	4–9	73–77
Anthracite	18–19	3–4	77–78
Lignite, peat, wood	17–18	10–16	66–72
Coke, charcoal	19–20	1–4	77–78
Petroleum products	13–14	11–14	73–75
Coal gas	8–11	20–22	69–71
Blast-furnace gas	22–25	0–2	74–77
Coke-oven gas	8–10	20–22	70–72
Producer gas	14–19	6–14	72–76
Water gas	13–16	15–18	67–71
Oil gas	9–11	18–22	68–72

and 12). In most fuels some sulphur-containing compounds are present and the combustion gases then contain sulphur dioxide, sulphur trioxide or sometimes hydrogen sulphide. It is thus evident that under ordinary working conditions it is not uncommon for equipment to be subjected to atmospheres whose compositions vary from time to time, and this is one factor which can lead to abnormally accelerated corrosion rates. In particular, when the atmosphere can vary from oxidising to reducing in character, the equilibrium constitution of the scale formed on an alloy may change, with resultant cracking or porosity in the layer. The protective action of a coherent scale layer is thus much reduced.

A particular type of corrosion observed with nickel–chromium and nickel–iron–chromium alloys and involving both carburisation and oxidation, takes place at temperatures around 1 000°C in carbon-containing atmospheres. The attack, which is termed *green-rot* (see Section 7.5 (p. **7**.101)) on account of the colour of fractured surfaces of the attacked alloy, occurs particularly in atmospheres of the bright-annealing type when they are contaminated with carbon-containing gases or vapours, but can also arise with incompletely burnt hydrocarbon fuels. This type of attack has been reported to be considerably accelerated by cyclic variations of the atmosphere from oxidising to reducing.

Ashes

Apart from the direct corrosive action of the gaseous products from burnt fuels, an important effect frequently arises from the deposition of ashes on the metal surfaces involved. Ashes normally consist of complex mixtures or compounds of oxides, the compositions varying very widely and the precise state of combination being uncertain. Ranges of ash compositions reported for different samples of fuels of three broad classes are given in Table 7.2 (see References 13, 14, 15 and 16), from which it will be noted that the ranges are so wide that in any specific case the actual ash analysis must be obtained before consideration can be given to probable corrosion reactions. The corrosive attack by an ash is frequently critically dependent on the temperature,

Table 7.2 Composition of fuel ashes

Ash composition (%)	*Fuel*		
	*British coals**	*Peat*	*Fuel oil*
SiO_2	25–50	20–50	2–20
Al_2O_3	20–40	7–15	2–20
Fe_2O_3	0–30	0–20	1–60
TiO_2	0–3	—	—
CaO	1–10	5–30	0–10
MgO	0·5–5	5–25	—
Na_2O	1–6	0–6	1–30
V_2O_5			5–60
SO_3	1–12	5–20	7–30
Cl_2	—	0–1	—
P_2O_5	—	1–3	—

*P. J. Jackson (private communication) considers that present-day compositions of ashes from British coals would range as follows:

SiO_2	Al_2O_3	CaO	MgO	Na_2O	K_2O	Cl_2	P_2O_5
25–60	15–40	1–20	0–5	0·2–3	0·2–5	0–5	0–1

accelerating steeply when the melting point of a particular ash constituent is exceeded. This is particularly noticeable in the case of the vanadium-containing ashes from crude or residual mineral oils which seriously attack most heat-resisting alloys when the temperature is above about 650°C, the melting point of vanadium pentoxide. The rapid attack follows fluxing of the normally protective oxide scale by the ash, and the attack on the underlying metal may be in the form of further oxidation or may include attack by sulphur compounds or other active constituents of the ash. Various investigators have compared the rates of attack on different alloys by selected real or synthetic ashes and somewhat contradictory results have been reported. Thus, whereas Evans[16] stated that low-nickel or nickel-free iron-base alloys were more resistant to vanadium corrosion at 732°C than nickel or cobalt-base alloys, Harris and his associates[17] concluded that the latter type of alloys should be reasonably resistant if they contain less than 30% iron, less than 2% vanadium, and more than 16% chromium. Betteridge, Sachs and Lewis[18] also found that nickel–chromium alloys were much more resistant to vanadium attack than were austenitic steels. It is probable that the different conclusions arose from the difference in character of the attack which follows initial penetration of the protective scale, on account of variations in composition of the selected ashes.

The deposits formed in internal combustion engines by high-octane petrols may be classed as ashes; they consist of mixtures of lead oxides, bromides and sulphates derived from the anti-knock additives and, of course, exert their main corrosive effect on the parts operating at the highest temperature: the exhaust valves and the sparking-plug electrodes.

The formation of an ash is not always deleterious, since a layer of a less reactive material may serve to restrict access to the alloy of more reactive agents either in the ash itself or in the atmosphere. The erosive action of ash also has to be considered in addition to the purely corrosive action. With high-speed gas streams, as in gas turbines, erosion is frequently the more important effect.

Halogens and Halogen Compounds

The halogens are particularly important in hot environments because many of the metallic halides have relatively high vapour pressures, and the corrosion products are therefore vaporised instead of forming protective surface layers. This is particularly true in the case of chromium and aluminium, both important constituents of heat-resisting alloys. However, the attack is mainly on the metals rather than on the oxides, so that an initially-formed oxide film confers appreciable protection and the action of halogens in a predominantly oxidising atmosphere is much less than in a reducing atmosphere. Small traces of acid gases such as hydrochloric acid can, however, considerably accelerate oxidation rates in air, particularly of copper-base alloys.

Molten Salts

A wide variety of molten salt baths is used in industrial practice and there exists a correspondingly wide range of possible chemical reactions with metallic parts being heated in the baths, or with the containers holding the baths. Baths consist largely of mixtures of nitrates, carbonates or halides of the alkaline or alkaline-earth metals, and these generally have a ready solubility for oxides of other metals. Hence, although the salt baths may be sufficiently oxidising to react with the alloys concerned to form oxide, this product is taken into solution in the bath and does not form a protective scale; attack therefore progresses steadily. In some cases absorption of components of the salt mixture into the alloy may occur, with either deleterious or beneficial results. Thus, with baths containing borates, reduction of boric oxide may occur by the action of reactive constituents such as aluminium, resulting in the absorption of boron into the alloy; with nickel-base alloys a low-melting-point eutectic may then be formed with serious effects on the properties of the alloy. On the other hand, alkali-cyanide baths are used for the heat-treatment of steel parts, absorption of carbon and nitrogen into the steel giving surface hardening.

Molten-salt environments are fully treated in Section 2.10.

Refractories, Ceramics, Glasses, etc.

The possible attack on metals and alloys by materials of this type may be considered in a manner similar to attack by molten salts. If a molten phase exists in the material at the temperature under consideration it is possible that solution of the oxides formed by the alloy will occur, thus preventing the formation of a protective film. Molten glasses act in this manner, and refractories with traces of impurities, perhaps in localised regions, can act very similarly. Minerals such as mica and asbestos frequently give rise to troubles due to impurities which either lower the melting point or impair the electrical insulation at elevated temperatures. These materials, which are silicates of magnesium and calcium, often contain small quantities of iron oxides or sulphides and other impurities which make them unsuitable for use at temperatures above about 500 or 600°C, particularly if electrical insulation characteristics are required.

A possibility which must be considered is that of reaction between a refractory and an alloy occurring in consequence of the reduction of a constituent of the refractory by a metal in the alloy. Thus alumina has a higher free energy of formation than silica, and therefore aluminium is capable of reducing silica. At a high enough temperature, if oxygen is not available in the atmosphere to combine with the aluminium, this reaction is likely to take place. In order to avoid reactions of this type, only pure and very stable refractories, such as alumina, magnesia, zirconia or beryllia, should be used in contact with reactive metals at high temperatures.

Molten Metals

If the major constituents of a solid alloy in contact with a liquid alloy are highly soluble in the latter without formation of compounds, progressive attack by solution is to be expected. If, on the other hand, a stable intermetallic compound is formed, having a melting point above the temperature of reaction, a layer of this compound will form at the interface and reduce the rate of attack to a level controlled by diffusion processes in the solid state. By far the most serious attack, however, occurs in the presence of stresses, since in this case the liquid alloy, or a product of its reaction with the solid alloy, may penetrate along the grain boundaries, with resultant embrittlement and serious loss of strength.

Molten metals as a corrosive environment are discussed more fully in Section 2.9.

W. BETTERIDGE

REFERENCES

1. Gulbransen, E. A. and Andrew, K. F., Amer. Soc. Test. Mat., Sp. Techn. Publ. No. 171, 35–48 (1955)
2. Betteridge, W., *The Nimonic Alloys*, Arnold, London, 218 (1959)
3. Gleiser, M., Larsen, W. L., Speiser, R. and Spretnak, J. W., Amer. Soc. Test. Mat., Sp. Techn. Publ. No. 171, 65–88 (1955)
4. Bruce, D. and Hancock, P., *J. Inst. Mets.*, **97**, 140 (1969)
5. Kubaschewski, O. and Hopkins, B. E., *Oxidation of Metals and Alloys*, 2nd edn, Butterworths, London (1962)
6. Hessenbruch, W., *Zunderfesterlegierungen*, Springer, Berlin (1940)
7. Price, L. E. and Thomas, G. J., *J. Inst. Met.*, **63**, 21 (1938)
8. Siebert, C. A. and Donnelly, H. G., *Trans. Amer. Soc. Metals*, **28**, 372 (1940)
9. Brasunas, A. de S. and Uhlig, H. H., *Bull. Amer. Soc. Test. Mat.*, No. 182, 71 (1952)
10. Farber, M. and Ehrenberg, D. M., *J. Electrochem. Soc.*, **99**, 427 (1952)
11. Spiers, H. M., *Technical Data on Fuel*, Publn. Brit. Natl. Ctee., World Power Conf., 6th edn (1962)
12. *Efficient Use of Fuel*, H.M.S.O., London (1944)
13. Crossley, H. E., *External Boiler Deposits*, Inst. of Fuel Special Study of the Ash and Clinker Industry, Paper No. 4 (1952)
14. Waddams, J. A. and Wright, J. C., *J. Inst. Fuel*, **32**, 246 (1959)
15. Sykes, C. and Shirley, H. T., Iron and Steel Inst. Sp. Rep. No. 43, 153–169 (1952)
16. Evans, C. T., Amer. Soc. Test. Mat., Sp. Techn. Publ. No. 108, 59–105 (1951)
17. Harris, G. T., Child, H. C. and Kerr, J. A., *J. Iron Steel Inst.*, **179**, 241 (1955)
18. Betteridge, W., Sachs, K. and Lewis, H., *J. Inst. Petrol.*, **41**, 170 (1955)

7.2 The Oxidation Resistance of Low-Alloy Steels

The behaviour of metals at elevated temperatures in an oxygen atmosphere depends upon the protection afforded to the metal by the oxide film. Highly reactive metals such as sodium and potassium form discontinuous oxide films that provide no protection, and consequently oxidation proceeds according to a linear law—the reactive metal aluminium forms a protective film, but its application is limited by its low melting point. Metals like gold and platinum are inert in oxygen, but their high cost precludes their use for most high-temperature engineering applications. Thus the major part of this branch of industry is dependent on iron and iron-base alloys for its material requirements.

The ferrous materials in general use are of the plain-carbon and low-alloy steel type and it is therefore not surprising that the literature on the subject of their resistance to oxidation and deterioration at high temperatures is very extensive. Although the limits of alloy content for certain groups of steels are not well defined, it has been assumed in this section that a low-alloy steel is one having less than 12% total alloy content; above that level the term *stainless steel* is used, and these alloys are considered in other sections.

Factors Governing Oxidation Reaction

A number of governing factors may be involved in any given oxidation reaction, i.e. time, temperature, service environment, and composition and constitution of the metal. It is evident that in general the extent of oxidation will be in proportion to duration and/or temperature of reaction.

Changes in the metallurgical structure of a metal are time/temperature dependent, and these may in turn influence the ultimate characteristics of the oxides formed. Increase in temperature increases the mobility of ions, and subsequently affects the rate laws applicable to any given material.

A metal in contact with a reactive gas mixture gives rise to a complex chemical system in which a number of different types of reaction may occur, and these have been outlined by Gulbrasen *et al.*[1] as follows:

1. Metal or oxygen atoms or ions diffuse through lattice defects in a compact oxide layer.

2. The diffusion processes mentioned in (1) occur at a faster rate at the edges causing cracking of the oxide layer.
3. Diffusion processes as in (1) and (2), but with the formation of volatile oxides.
4. Chemical adsorption and desorption of gases on oxide-free surfaces.
5. Diffusion of oxygen gas through layers of volatilised vapours.
6. Spalling of metal particles due to reaction of types (4) and (5).
7. Transport of oxygen to reaction interfaces in liquefied oxide layer.

As oxidation proceeds the thickness of the oxide formed increases. The extent, if any, to which it affords protection to the underlying metal depends upon its physical, chemical and mechanical properties.

If the oxidation process involves the formation of a coherent and compact oxide layer whose inner and outer surfaces are in thermodynamic equilibrium with the metal and the atmosphere respectively, and whose growth is controlled by volume diffusion processes through the scale, then the kinetics of the process will generally conform to a parabolic rate law.

As the oxide thickness increases therefore the chemical potential gradient in the oxide layer and the driving force for diffusion through the oxide must fall; these are in fact proportional to the reciprocal of the thickness. Thus the diffusion rate and therefore the oxidation rate will be inversely proportional to the thickness of the oxide layer, and the kinetics of the process can be represented by an equation of the type:

$$\text{Rate} = K/x \qquad \ldots(7.1)$$

where x is the thickness of oxide layer and K is the rate constant.

Such reactions can generally be followed in three ways by:

1. Measuring the thickness of the oxide layer. According to equation 7.1 this process may be described as follows:

$$\frac{dx}{dt} = \frac{K'}{x} \qquad \ldots(7.2)$$

where t is the time, x is the thickness of the oxide and K' is the rate constant. Wagner[2] and Hauffe[3] both utilised this K' notation, and the more generally used form of the equation is derived by integrating equation 7.2 to give:

$$x^2 = 2K't$$

Kubaschewski and Hopkins[4], on the other hand, used the rate constant K_p', once again using the integrated form of equation 7.2:

$$x^2 = K_p't$$
$$\text{i.e.} \quad K' = \tfrac{1}{2}K_p'$$

2. Measuring the weight increase of the specimen. The rate constant used for this method of measurement is also defined according to an integrated form of equation 7.2:

$$(m/A)^2 = K''t$$

where m is the weight increase during oxidation, A is the area of oxide layer and K'' the corresponding rate constant. Wagner[2] and Hauffe[3] both used the notation K'' and Kubaschewski and Hopkins[4] used K_p.

3. Measuring the displacement of the metal surface. The rate constant is defined by the equation:

$$\frac{d\ell}{dt} = \frac{K_c}{\ell}$$

where ℓ is the displacement of metal surface and K_c is the corresponding rate constant. The integrated form of the equation gives:

$$\ell^2 = K_c \,.\, 2t$$

Wagner[5] utilises the notation K_c, whilst Kubaschewski and Hopkins[3] define their constant K_0', using an integrated form of the equation:

$$\ell^2 = K_0' t$$

It is obvious therefore that the units of the oxidation rate and rate constant will depend upon the convention adopted and on the method used to investigate a particular oxidation reaction. It is therefore most important to check the definition of rate constants given in the literature before they are used in any calculations.

Plain carbon and low-alloy steels generally obey a parabolic rate law in the temperature ranges being considered here. Certain exceptions to this rule, brought about by the addition of various alloying elements, are discussed later. It does not necessarily follow from the parabolic law that rapid oxidation cannot take place, because the initial rate of oxide growth may be high and the subsequent decrease in growth rate slow. This type of oxidation is, however, not considered desirable in practical alloy systems.

Forms and Formation of Oxide

The form which oxides take upon metal surfaces has aroused the interest of many investigators. Although there are differences of opinion concerning detail, it seems to be generally accepted that layer growths which thicken with time are formed. This is undoubtedly true for thick oxide scales, but in the case of very thin films studies are difficult because the oxidation rates are too fast. However, investigations have been possible by reducing the partial pressure of the oxidising gas, and the findings of Bénard[6] give the best general picture.

At the beginning of oxidation the metal is covered by a film which increases in thickness up to a critical value of the order of several tens of Angstrom units. Subsequently the continuously forming oxide tends to accumulate at certain crystallographic centres. The average number of these centres corresponds to an equilibrium determined by the rate of surface diffusion of the metal and oxygen, for a given crystallographic orientation, temperature and pressure. These nuclei then grow laterally until they cover the whole surface.

Thus the three-stage sequence of events involves the formation (*a*) of a continuous layer of tens of Angstrom units thick, i.e. an invisible film, (*b*) nuclei at crystallographic centres and (*c*) a continuous layer, and has been observed in the reaction of iron with oxygen[7].

In the majority of the experiments at low partial pressures of oxidising gases, the numbers of nuclei are independent of time, but increase with partial pressure of the gas.

Fig. 7.1 Electron microscope photomicrograph of whiskers in Fe–5Cr–0·5Mo, 650°C and ×20 000

It has been suggested that the formation of nuclei is due to dislocations in the metal substrate[8,9]. In this context, Edwards and Pickering[10] suggested that the growths (whiskers, blades and prisms) observed using the electron microscope, grow from dislocations at the metal surface. A selection of electron photomicrographs obtained by them on low-alloy steels illustrating these effects is shown in Figs. 7.1 and 7.2.

Gulbransen and Copan[11] have also suggested that it is the metal structure itself which determines these localised crystal growths. However, the mechanism of formation and growth of whiskers, etc. is not clear and it is not established whether there is a direct connection between their growth and the nuclei formed in the first stages of oxidation. At the present time it appears to be more widely believed that these whiskers, etc. have their nucleation sites and origin at the base oxide film[12–14]. The present situation is best summarised by Kofstad[15] who stated 'The amount of oxide contained

Fig. 7.2 Electron microscope photomicrograph of blades in Fe–5Cr–0·5Mo, 650°C and ×12 000

Fig. 7.3 Scanning electron microscope photomicrograph of oxide growths on pure iron at 800°C for 1 h, ×10 000

in the whiskers or platelets represents a minor part of the total surface oxide. Although such crystal growth greatly increases the total surface area of the specimen, there is no unequivocal evidence that such crystals have any marked effect on reaction behaviour. At this stage the total reaction rate appears to be governed by the growth of the base film, and the whisker growth is probably a result of secondary effects and changes in the oxide film'.

It may well be that the phenomena illustrated in Figs. 7.3 and 7.4 obtained

Fig. 7.4 Scanning electron microscope photomicrograph of oxide growths on pure iron at 800°C for 1 h, ×11 500

by scanning electron microscope techniques may advance our knowledge in these areas.

The Oxidation of Iron

Because the major constituent of a low-alloy steel is iron, it is necessary to consider the oxidation behaviour of this metal before that of the low-alloy steels.

Iron forms three stable oxides, viz. ferrous—FeO (wustite), ferroso-ferric —Fe_3O_4 (magnetite) and ferric—Fe_2O_3 (hematite). The total process of oxidation can be summarised by a simplified form of Hauffe's schematic representation (Fig. 7.5)[16]. Unfortunately, this picture is complicated by the

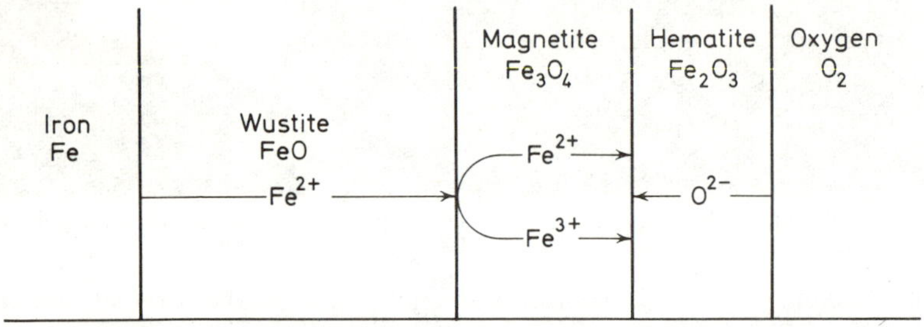

Fig. 7.5 Oxide growth mechanism on iron

fact that at lower temperatures only magnetite and hematite are stable, wustite only being stable above about 570°C. This latter temperature has been referred to as the *non-variance point* between α-iron, magnetite and wustite[17] and the *reversible transformation point*[18]. FeO has however been observed at temperatures as low as 400°C when oxidation times were short and the oxide films thin. In addition, although FeO is stable at high temperatures, it usually decomposes at lower temperatures according to the equation

$$4FeO \rightarrow Fe_3O_4 + Fe$$

From the self-diffusion rates of iron in its three oxides measured by Himmel, Mehl and Birchenall[19] and the oxidation rates[21] for Fe → FeO, FeO → Fe_3O_4 and $Fe_3O_4 \rightarrow Fe_2O_3$ the agreement with the experimental parabolic rate constants is good for the formation of FeO and Fe_3O_4 because it is cation diffusion which prevails in the formation of these two latter oxides. In the case of Fe_2O_3 it is O^{2-} diffusion which prevails (see Fig. 7.5).

At the lower temperatures, where the oxide films are thin and consist essentially of Fe_2O_3 and Fe_3O_4 such films give rise to interference colours on the specimen surface. Davies, Evans and Agar[20] have listed the colours of such films and measured their thicknesses.

Davies *et al.*[20] have indicated that the growth of the oxide layers is insensitive to changes in substrate, e.g. FeO grows in a similar manner on α- or γ-iron. However, the influence of varying degrees of cold work on the oxidation of iron in the temperature range 400–650°C has been investigated by Caplan and Cohen[22] who showed that up to 600°C the rate of oxidation was increased with the greater degree of cold work, whereas at 650°C cold work had no effect.

The Oxidation of Low-Alloy Steels

As would be expected from the foregoing remarks, low-alloy steels behave in a similar manner to iron, i.e. they oxidise parabolically and form oxide scales whose structure resembles those formed on iron. The function of alloying elements, in addition to strengthening the steel, is to decrease the kinetics of oxidation. Stringer[23] has concisely defined the several distinct possibilities which may occur with regard to the influence of the alloying element. He stated that the alloying element may:

1. Be completely inert and serve only to dilute the oxidising metal at the metal/oxide interface.
2. Be completely inert, but improve the lattice fit at the metal/oxide interface, deferring breakaway and possibly slowing down the oxidation rate as a result of the reduction of climb-assisted diffusion.
3. Be inert and become mechanically entrained in the oxide product as a second-phase distribution.
4. Dissolve to some extent in the scale (this may have a variety of effects, and may modify the mechanical properties of the scale, as well as alter the lattice parameter of the scale and hence the degree of misfit at the metal/scale interface).
5. Cause the formation of a two-phase oxide layer. The effect on the oxidation rate of a two-component scale depends very much on the spacial distribution of the two components. In general, diffusion rates in the two will be different, and there is also the possibility of grain-boundary transport processes becoming important. Similarly, the effect of the mechanical properties will depend very much on the exact circumstances.
6. Cause the formation of a single-phase compound scale, whose structure differs from that of the pure solvent metal oxide.
7. Be so strong an oxide former that it forms only its own oxide on the alloy.

Effects of Other Elements on the Oxidation of Iron and Steel

Carbon It would appear that carbon slightly increases the oxidation resistance of iron provided that no evolution of carbon monoxide occurs. When it does occur it is liable to disrupt what would otherwise be a protective scale, and in these circumstances carbon has a deleterious effect. Manenc and Vagnard [24] showed that the presence of carbon had no apparent effect on oxidation kinetics above 800°C. Below this temperature it produced a slight decrease of the scaling rate. Boggs and Kachik[25] studied the oxidation of iron–carbon alloys in oxygen at 500°C and showed that at a pressure of $1{\cdot}3\,kN/m^2$ a decarburised zone was present, at $93\,kN/m^2$ however, no decarburisation occurs. The reaction is therefore obviously influenced by pressure and these latter authors concluded that at the higher pressures oxygen blocks the pores in the scale thereby preventing escape of carbon monoxide.

Silicon The formation of Fe_2SiO_4 (fayalite) has been observed in the oxides of iron–silicon alloys by Ipat'ev and Orlova[26] and Rahmel[27], and in the oxides of steels alloyed with silicon by Engell[28] and by the present author. The fayalite in the oxide decreases the oxidation rate, and according to Bandel[29] provides good resistance to sulphur-bearing gases.

All the early investigators into the effect of alloy elements on the oxidation of iron[30–32] provide evidence that silicon has a beneficial effect. The author has observed that the protective effect occurred at 700°C but not 565°C (Fig. 7.6). A very comprehensive study on the influence of alloying elements on a basic low-alloy steel composition by Yuganova and Sorokina[33] showed a marked protective influence of silicon at 600°C.

A similar protective effect also arises with steels containing both silicon and aluminium (Fig. 7.6), a feature noted by other investigators[31,34].

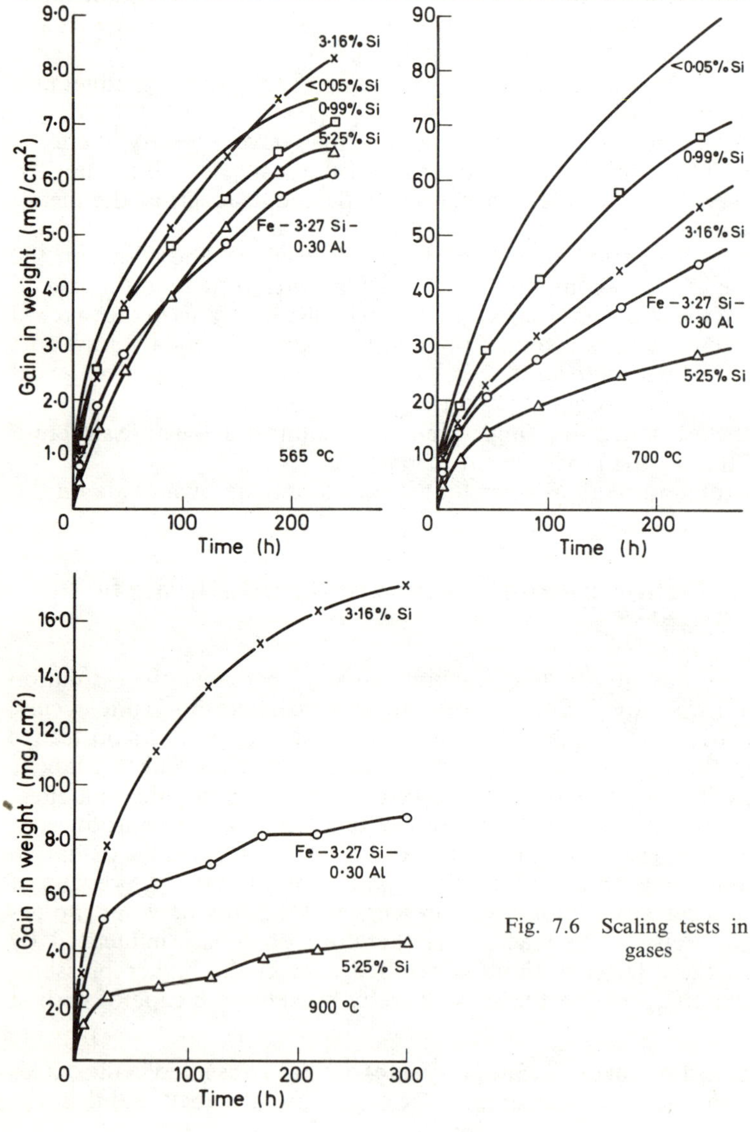

Fig. 7.6 Scaling tests in flue gases

Manganese Of the alloying elements likely to be present in steel, manganese most closely resembles iron in its behaviour and it has been shown that its effect is negligible[31,35].

Molybdenum A great deal has been written on the effect of molybdenum as an alloying element and there are data both for and against its desirability. Improvement owing to its presence has been shown by Scheil and Kiwit[31] and by Wilder and Light[36], and a detailed paper by Rahmel *et al.*[37] has shown that additions of 0·5–5·6% Mo decrease the rate of oxidation of iron by almost a power of ten in the temperature range 500–1 000°C; the optimum effect is produced with 2% Mo. Brenner[38] has conducted studies on iron–molybdenum alloys with molybdenum contents of up to 12·5%. The quantitative rate-constant data obtained by both these latter author:[37,38] have been usefully summarised in a review paper by Yashuhiko[39]. There would seem therefore to be some considerable benefit to be obtained at service temperatures by the addition of small percentages of molybdenum to low-alloy steels.

The catastrophic oxidation ascribed to the presence of molybdenum seems to be confined to the austenitic chromium–nickel steels, but low-alloy steels do not appear to have been investigated. These effects would seem to be associated with the formation of volatile molybdenum oxides which are in turn closely linked with the oxidising potential and velocity of the atmosphere.

Nickel The oxidation of iron–nickel alloys in the temperature range 500–1 100°C has been covered in an excellent literature review by Foley[40]. In alloys of all compositions that were investigated the first reaction was one of selective oxidation, creating a nickel-rich layer of metal at the alloy–oxide interface. It is suggested that subsequently the growth of a nickel-containing spinel is promoted and the formation of FeO is suppressed, with a consequent reduction in the oxidation rate. Foley's paper certainly substantiates these statements for high-nickel alloys, i.e. greater than 15% Ni, and the effects would appear to be emphasised at higher temperatures, i.e. $>1\,000$°C. Menzies *et al.*[41] examined an Fe–2% Ni alloy in CO_2 in the temperature range 600–1 000°C and found that the build up of the nickel-rich layer was associated with a parabolic oxidation rate. After a period of time, the layer broke, rapid diffusion of iron occurred and there was a transition to linear kinetics at temperatures >600°C.

It is generally known that nickel-containing alloys are sensitive to sulphur-containing gases, this being particularly so under reducing conditions. Oxidation resistance is lowered due to the formation of nickel sulphides, but this tendency will obviously be governed by the amounts of nickel present for sulphide formation. However, in view of the low concentrations of nickel normally encountered in low-alloy steels the effect of sulphur will be small.

Aluminium The formation of the refractory oxide alumina (Al_2O_3) by addition of aluminium to low-alloy steels and iron improves the oxidation resistance. The alumina is not necessarily uncombined and can occur in the innermost layer of scale as FeO . Al_2O_3. As has already been mentioned, aluminium in combination with silicon is also beneficial.

Hagel[42] examined the oxidation resistance of Fe–4·4Al and Fe–12Al in oxygen in the temperature range 500–1 300°C, and showed that there were two different regions of parabolic growth which are associated with the appearance of γ-Al_2O_3 below 900°C and α-Al_2O_3 above. The latter is normally more protective but possesses a greater tendency for cracking and spalling. Saegusa and Lee[43] also examined iron–aluminium alloys containing 1 and 5% Al in oxygen in the temperature range 500–1 000°C and demonstrated, as has been frequently confirmed experimentally by many other authors, that oxidation resistance increases with increase in aluminium content. It should be pointed out, however, that alloys containing above 10% Al are brittle.

Sulphur With atmospheres of high oxidation potential or when materials are subjected to very long periods of exposure, it is likely that desulphurisation (analogous to decarburisation) will occur, but, with the exception of certain free-cutting steels, most low-alloy steels will not have sufficient sulphur for the effect to be significant.

Phosphorus The normal amounts of phosphorus (0·01–0·1%) are generally slightly deleterious, but an increase in phosphorus content above this level does not appear to intensify the effect. Collari and Fongi[44] showed that there was some increase in the oxidation rate of iron by additions of phosphorus of up to 0·2% when copper was present.

Chromium Of all the elements added to steel, chromium is the one most commonly associated with improved oxidation resistance and up to the level under consideration in this section (9–10%) it has a marked influence. There is evidence[45] that low-chromium steels have an oxidation resistance that is little better than iron, but at concentrations >12% and below 1 000°C the oxidation resistance is increased almost a hundred-fold; in addition there is a wide range of alloy steels containing chromium that have a high resistance to oxidation (see Fig. 7.7 and Table 7.3). Chromium has a high affinity for oxygen, greater even than iron, and is oxidised preferentially to form iron chromites which subsequently slow down the oxidation rate. Perrow and Smeltzer[46] examined the oxidation of Fe–5Cr over the temperature range 600–850°C, and showed that on abraded specimens transformation by nodular-type growths occurred to produce a scale containing iron chromite and ferric–chromic oxide. It is important to note their observation that the alloy underlying the iron chromite was susceptible to internal oxidation.

Commercial low-alloy steels The effect of some of the alloying elements added to iron is summarised in Fig. 7.8[47] whilst more quantitative data obtained by Yuganova and Sorokina[33] have been summarised in Fig. 7.10. In addition, the properties and applications of heat-resistant low-alloy steels are listed in Table 7.3[48]. This table broadly indicates the limitations of these steels with respect to maximum service temperature, but since the magnitude of oxidation of any steel is very dependent upon the oxidising potential of the atmosphere, shifts in these limits can be expected to occur. In addition, it should be noted that if one is prepared to tolerate a shorter service life then obviously the maximum service temperature could be increased.

Because materials are under stress in the majority of applications of low-alloy steels, then all factors which reduce the effective cross-sectional area of the material must be considered. In this respect oxide penetration and internal oxidation are important. Recent investigations by the author have

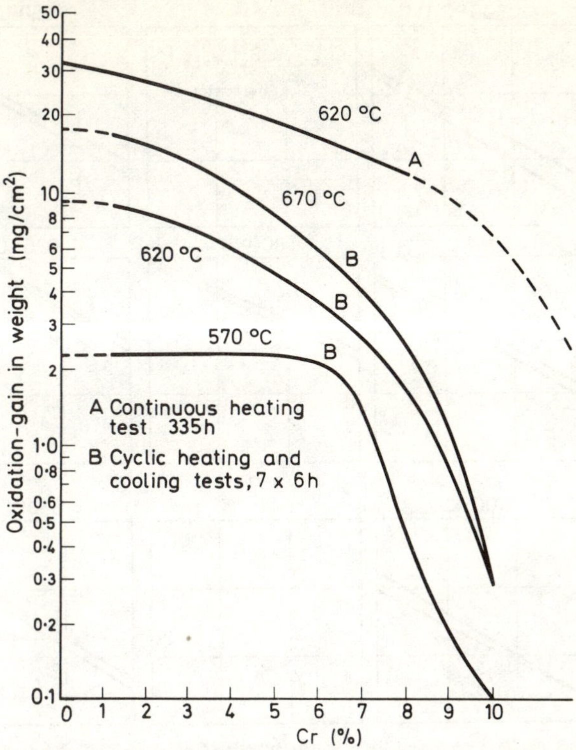

Fig. 7.7 Oxidation resistance of chromium steel in flue gas (6:1 air:towns gas)

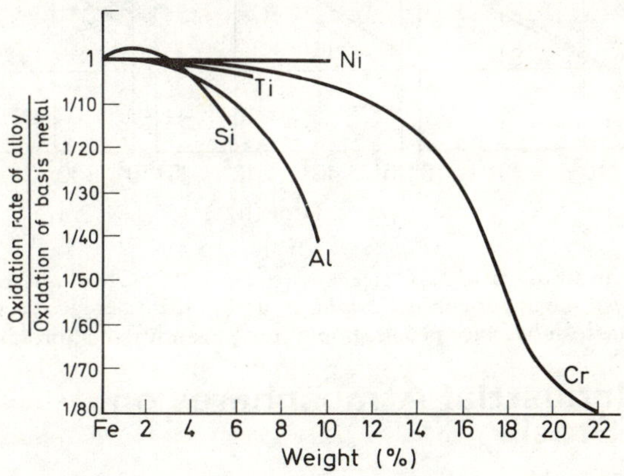

Fig. 7.8 Effects of various elements on the oxidation of iron at 900–1 100°C (after Kubaschewski and Hopkins[47])

therefore taken these factors into account. The method of presentation of quantitative oxidation data, which is believed to be the most pertinent with regard to practical application, is that outlined by Edwards and Nicholson[49], examples of which are given in the text.

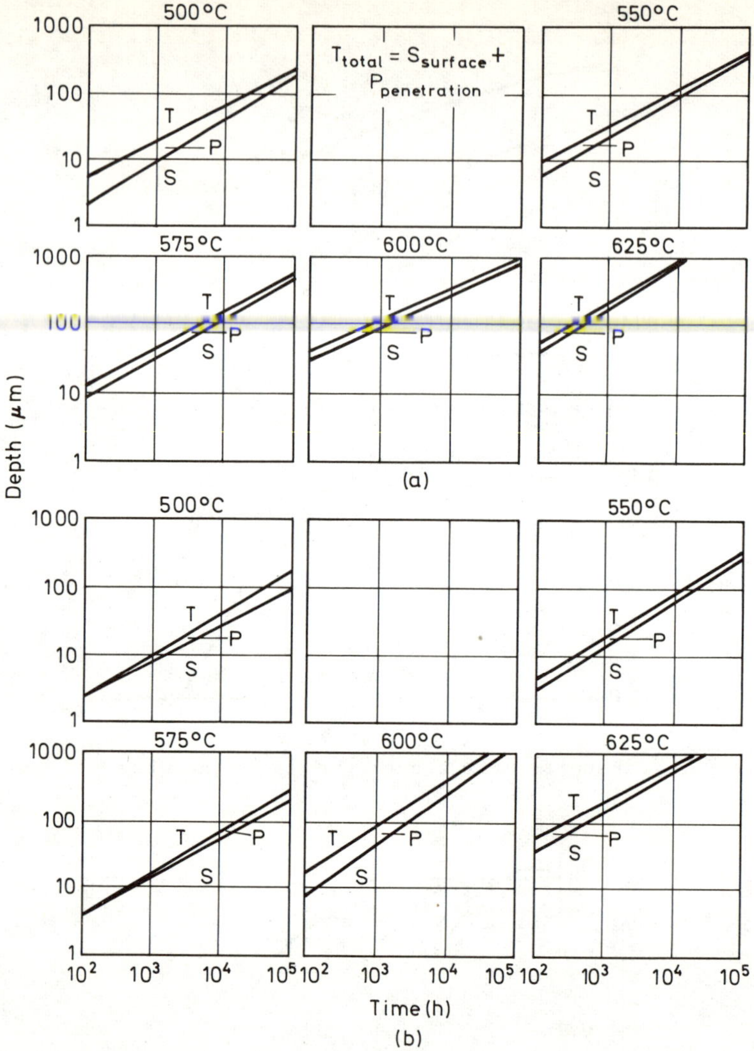

Fig. 7.9 Oxidation in air of (*a*) 0·12% C steel and (*b*) Fe–2·25Cr–1Mo, where the lower curve S represents *surface loss*, the upper curve T *total loss*, and P, the difference between T and S, the loss due to subsurface penetration (after Yuganova and Sorokina[33])

Effects of Industrial Atmospheres on Low-Alloy Steels

It is evident from the applications listed in Table 7.3 that the important industrial atmospheres, which may occur singly or in combination, likely to be encountered are of the following nature:

1. Air with or without water vapour.
2. Products of combustion of gas, oil and coal.
3. Steam.
4. Carbon dioxide.

In spite of all the work that has been done on iron and steel with respect to oxidation resistance there is still a considerable lack of quantitative data. A large proportion of published data, having been derived from very short term tests, is indicative only of preliminary trends and cannot always be related to ultimate behaviour in service. The most likely reason for the lack of quantitative data on oxidation resistance is that the majority of steels in the low-alloy group were originally developed for their high-temperature strength, and it was this factor rather than oxidation resistance which determined their application. In more recent years the emphasis has changed, and current investigations are beginning to establish the precise merits of these steels in relation to oxidation over long periods of time, since changes in the order of merit are likely to occur with changes in environment and temperature and with prolonged periods of exposure. For mild steel, however, Upthegrove and Murphy[50] indicate that below 704°C there is little or no difference between the scaling actions of steam, dry carbon dioxide, dry oxygen or dry air. Bourgraff[51] and Marson, Angus and Cobb[52] have, however, shown that at temperatures above this level the composition of the atmosphere is very significant.

The situation outlined above, in combination with the fact that much of the data on low-alloy steels have been obtained from tests designed to assess oxidation at the high heat-treatment temperatures applied prior to rolling and other hot working processes, rather than at service temperatures, does tend to impose a limitation on the scope of this discussion.

Air

Air uncontaminated with combustion products, etc. is rarely met in service but much of the available oxidation data have been obtained in air, possibly contaminated with water vapour, because it provides the most convenient atmosphere for laboratory studies.

The presence of moisture in air is probably sufficient to account for differences in the magnitude of results already published. Kubaschewski and Hopkins[53] indicate that the small amount of moisture normally present in ordinary air may sometimes have a catalytic effect and may accelerate oxidation, but it may also have the reverse effect. Upthegrove and Murphy[50] have shown that moist and dry air have a similar effect, but only up to a limiting temperature of 704°C. Caplan and Cohen[54] demonstrated the aggressive effect of moist air at 871°C on an 11% Cr steel.

The author's own laboratory investigations (carried out some years ago) in air of unknown moisture content gave results which indicate that the low-alloy steels with up to 2·25% Cr are very similar (Table 7.4) in their behaviour. However, surface metal losses obtained from these results were not always of the same order as the actual values obtained in steam[55] (see Table 7.5), and since at that time it was believed that steam was similar in its oxidation potential to air, the results served to illustrate the fact that errors could occur by extrapolation. The oxides on these steels tested in air at 600–620°C contained 40% Fe_2O_3 and 60% Fe_3O_4 with only a trace of FeO, which is in contrast to Paidassi's observations on iron[56] at 700°C.

Table 7.3 Mild steel and low-alloy steel for

Steel alloy	*Composition* (%)				
	C	Mn	Si	Cr	*Other elements*
Mild steel (fully 'killed')	0·15	0·5	0·3	—	S 0·05 max. P 0·05 max.
Fe + Mo + low C	0·15	0·5	0·3	—	Mo 0·4–0·6 S 0·05 max. P 0·05 max.
0·75–1Cr, 0·4–1Mo	0·10	0·7	0·3	0·75–1·25	Mo 0·4–1·0 Si 0·75–1·25 (optional) S 0·05 max. P 0·05 max.
Ni + Cr + Mo	0·35	0·7	0·3	0·75–1·0	Ni 1·5–2·0 Mo 0·3–0·4 S 0·05 max. P 0·05 max.
2–3Cr, 0·5–1Mo	0·15–0·30	0·5	0·35	2·0–3·0	Mo 0·5–1·0 Si 0·75–1·0 (optional)
4–6Cr, 0·5–1Mo	0·15	0·5	0·35	4·0–6·0	Mo 0·5–1·0 Si 0·75–1·0 (optional)
7–9Cr, 0·5–1Mo	0·15	0·5	0·35	7·0–9·0	Mo 0·5–1·0 Si 0·75–1·0 (optional)
Fe–8Cr–3Si	0·45	0·5	2·5–3·5	7·5–8·5	Mo 0·4–0·6 (optional)

high-temperature service (after Burling-Smith[48])

Typical applications	*Maximum temperature for service under oxidising conditions*
Steam pipes and fittings, superheater tubes and boilers, etc. where service stresses are of low order and environments are very mildly oxidising and essentially non-corrosive. However, the low maximum temperature for service recommended is based on elevated-temperature strength properties rather than on scale resistance.	454°C
For service conditions generally similar to above as regards oxidation and corrosion, but has elevated-temperature strength properties 50–75% better than mild steel. Used in superheater elements, steam pipe lines and oil pipe stills. A medium-carbon variant (0·3%) is commonly used for nuts, to overcome 'seizing' and 'galling'.	510°C
Widely used in high-temperature steam service for high-pressure superheater and boiler tubing, valve bodies and spindles, studs and bolts, high-temperature steam piping. Also used for oil-pipe-still furnace tubes where oxidation or corrosion is mild. The higher silicon variant has advantages of improved scale resistance and can be used up to 593°C.	538°C 593°C (high-silicon form)
Used for bolts in high-temperature steam-engineering practice, forgings and castings for medium-temperature service, as valves, flanges and fittings. For heavier-section components higher alloy variants of the same type are also used.	538°C
The chromium content of this steel is just about the minimum giving a practical useful measure of scale resistance. It does, however, show clearly marked improvement in elevated-temperature corrosion resistance, particularly to crude oils, and accordingly finds quite extensive application in tubing, return bends in oil pipe stills, vapour and hot-oil lines, and heat exchangers where oils are not highly corrosive and oxidation is not severe. Also used extensively for steam piping, superheater tubes and general steam-engineering components, valve bodies, flanges, bolts, etc. The 3% chromium variant is also used to resist hydrogen attack at high pressures and elevated temperatures, in ammonium-synthesis and hydrogenation processes. Variants modified with minor addition of Ni, W or V used for turbine rotors and casings.	600°C
Widely used in oil-industry applications similar to 2–3Cr, 0·5–1Mo steel, but where more corrosive sulphur-bearing oils are being handled or oxidation resistance of a higher order is required. Also used in high-temperature-steam applications. Some use also in hydrogenation plants. General applications include pipes and tubing, bolts, cast and forged valve bodies, flanges, pressure vessels.	650°C
High oxidation and corrosion resistance. Similar fields of application to steels listed above, but where corrosion or oxidation attack is more severe.	700°C
Resistance to scaling excellent and to corrosion good, with retention strength and hardness at high temperature but tendency to brittleness at normal temperature. Quite widely used for valves in i.c. engines. Industrial applications include soot blowers, gas-producer parts, rabbles for ore, roasting furnaces, enamelling perrits.	800°C

Table 7.4 Oxidation resistance of low-alloy steels in air

Steel alloy	*Weight loss after scale removal, after 500 h* (mg/dm^2)					*Penetration in 1st year—by extrapolation of parabolic law* (mm)					*Activation energies* (kJ/mol) 500–620°C
	Testing temperature (°C)					*Testing temperature* (°C					
	500	535	565	600	620	500	535	565	600	620	
Fe–2·25Cr–1Mo	402	837	1 049	1 797	4 804	0·015 2	0·051	0·051	[illegible]102	0·292 1	114
Fe–1Cr–0·5Mo	349	1 023	1 149	2 742	4 858	0·015 2	0·051	0·046	[illegible]144 8	0·172 7	120
Fe–0·5Mo–0·25V	533	796	1 533	2 495	4 607	0·022 9	0·040 6	0·048 3	[illegible]121 9	0·276 9	106
Fe–0·5Mo	406	1 072	1 340	2 968	6 573	0·015 2	0·045 7	0·053 3	[illegible]178	0·398 8	500–565°C 94 565–620°C 235

Table 7.5 Oxidation resistance of low-alloy steels in steam*†

Steel alloy	*Average steam temp.* (°C)	*Average scale thickness* (mm)	*Decrease in wall thickness* (mm)	*Penetration* (mm/y)	*Observations (scaling)*
Fe–5Cr–0·5Mo	510	0·061 0	0·022 9	0·027 9	Slight exfoliation, mostly adherent
Fe–2·25Cr–1Mo	510	0·068 6	0·045 7	0·053 3	Slight exfoliation, mostly adherent
Fe–0·5Cr–0·5Mo	510	0·058 4	0·022 9	0·027 9	Tightly adherent, smooth
Fe–0·5Mo+C	510	0·055 9	0·025 4	0·030 4	Tightly adherent, smooth
Fe–0·5Mo+C	510	0·061 0	0·030 4	0·035 6	Tightly adherent, smooth
Fe+C	510	0·066 0	0·010 2	0·012 7	Adherent, slightly roughened
Fe–0·5Cr–0·5Mo	560	0·114 3	0·053 3	0·063 5	Somewhat flaky
Fe–0·5Mo+C	560	0·096 5	0·048 3	0·058 4	Tightly adherent, smooth
Fe+C	560	0·091 4	0·038 1	0·045 7	Tightly adherent, slightly roughened. Some oxide penetration pits into metal
Fe–9Cr–1Mo	560	0·061 0	0·030 5	0·035 6	Tightly adherent, smooth
Fe–5Cr–0·5Mo	560	0·081 3	0·040 6	0·048 3	Core-tube boundary oxidised, somewhat flaky scale
Fe–2·25Cr–1Mo	560	0·134 6	0·050 8	0·055 9	Core-tube boundary oxidised, somewhat flaky scale
Fe–0·5Cr–0·5Mo	560	0·094 0	0·035 6	0·043 2	Tightly adherent
Fe–0·5Mo+C	560	0·088 9	0·040 6	0·048 3	Tightly adherent, core-tube boundary oxidised
Fe+C	560	0·081 3	0·030 5	0·035 6	Tightly adherent, slightly roughened
Fe–9Cr–1Mo	604	0·119 4	—	—	Dark adherent
Fe–5Cr–0·5Mo	604	0·289 6	0·137 2	0·165 1	Outer layer flaking off
Fe–0·5Cr–0·5Mo	604	0·241 3	0·121 9	0·144 8	Outer layer flaking off
Fe–0·5Mo+C	604	0·317 5	0·167 6	0·200 7	Most of outer layer chipped off
Fe–0·5Mo+C	604	0·238 8	0·121 9	0·144 8	Adherent but roughened
Fe+C	604	0·508	0·264 2	0·315 0	Adherent but highly craze-cracked
Fe–9Cr–1Mo	604	0·094 0	0·025 4	0·035 6	Adherent, slightly roughened
Fe–5Cr–0·5Mo	604	0·157 5	0·071 1	0·083 8	Flaking slightly
Fe–2·25Cr–1Mo	604	0·160 0	0·078 7	0·094 0	Adherent
Fe–0·5Cr–0·5Mo	604	0·185 4	0·094 0	0·111 8	Adherent
Fe–0·5Mo+C	604	0·299 7	0·129 5	0·154 9	Chipped off on one end
Fe+C	604	0·403 9	0·223 5	0·266 7	Adherent but craze-cracked
Fe–9Cr–1Mo	643	0·182 9	0·083 8	0·099 1	Outer layer exfoliating badly
Fe–5Cr–0·5Mo	643	0·569 0	0·292 1	0·348 0	Some blistering
Fe–2·25Cr–1Mo	643	0·584 2	0·289 6	0·345 4	Chipping off at specimen ends
Fe–0·5Cr–0·5Mo	643	0·693 4	0·370 8	0·444 5	Rough, chipping at ends
Fe–0·5Mo+C	643	0·749 3	0·551 2	0·657 9	Heavy scale, exfoliating badly
Fe+C	643	0·718 8	0·767 1	0·914 4	Heavy scale, exfoliating badly, no tube metal left at places

Table 7.5 Oxidation resistance of low-alloy steels in steam*† (continued)

Steel alloy	*Average steam temp.* (°C)	*Average scale thickness* (mm)	*Decrease in wall thickness* (mm)	*Penetration* (mm/y)	*Observations (scaling)*
Fe–9Cr–1Mo	643	0·167 6	0·076 2	0·091 4	Fairly smooth, outer layer chipped off in one area, under layer rougher
Fe–5Cr–0·5Mo	643	0·421 6	0·203 2	0·243 8	Adherent scale, frosty appearance
Fe–2·25Cr–1Mo	643	0·378 5	0·190 5	0·228 6	Adherent scale, frosty appearance
Fe–0·5Cr–0·5Mo	643	0·525 8	0·421 6	0·502 9	Flaking badly
Fe–0·5Mo+C	643	0·685 8	0·355 6	0·424 2	Heavy rough scale, some had chipped off previously during service
Fe–9Cr–1Mo	677	0·297 2	0·134 6	0·160 0	Flaky scale
Fe–5Cr–0·5Mo	677	1·531 6	0·779 8	0·932 2	Heavy rough scale‡, chipping slightly at ends
Fe–2·25Cr–1Mo	677	1·960 9	1·135 4	1·356 4	Heavy scale‡, badly flaked
Fe–0·5Cr–0·5Mo	677	1·348 7	1·270 0	1·518 9	Heavy scale‡, rough surface
Fe–0·5Mo+C	677	1·409 7	1·348 7	1·612 9	Heavy scale‡, rough surface

*The steels were arranged, in groups, in a test loop. The author has assumed that the temperature of each steel within a group was the same, although they would in fact be slightly different. For precise temperature details see the original paper.
†Table based on Paper No. 57–A–175 of Annual Meeting of Amer. Soc. Mech. Engrs. (1957).
‡No tube (specimen) metal remaining.

It was because of the possibility of errors being incurred by extrapolation that investigations were undertaken by the author which utilised testing times of up to 10 000 h. Six low-alloy steels of various composition were tested in air, which had been saturated with water vapour at room temperature, in the temperature range 500–625°C. The results showed that the oxidation resistance of these steels was very similar, with only the 0·12% C steel being slightly inferior to the others (the data are presented in graphical form in Fig. 7.9(*a*) and (*b*) on a log–log plot). These figures also indicate the degree of subsurface oxidation, i.e. oxide penetration and/or precipitation discussed earlier, which, as was stated previously, are also responsible for the effective cross section of metal being reduced where strength considerations are important.

The work of Stanley *et al.*[57] illustrates in greater detail the changes in oxide constitution on pure iron in air. The 500 h weight-loss data presented in Table 7.4 have been compared with the weight-gain data for pure iron, and it is evident that the low-alloy steels are superior. Several alloy steels were examined by Day and Smith[58] (Table 7.6), and their conclusions summarise the benefits of low-alloy compositions. The alloying element with the most pronounced influence upon oxidation resistance is chromium; small concentrations of chromium are less effective than silicon, which cannot be used, however, in large concentrations (even if it were then effective) because of its adverse effect upon mechanical properties. They also examined the effect of carbon, which is shown to be slightly deleterious at 593°C; this is in contrast to the results obtained at 900–1 100°C[59] where a steel with 0·9% C showed less scaling than samples containing smaller amounts of carbon.

However, the most recent comprehensive study of the influence of alloying elements on the oxidation behaviour of low-alloy steels in air has been carried out by Yuganova and Sorokina[33]. They studied chromium at four levels up to 7%, silicon at three levels up to 2·5%, aluminium at three levels up to 1·5%, tungsten at three levels up to 3%, cobalt at 2 and 3% and boron at 0·05 and 0·1%. The steels were oxidised at 300, 400 and 500°C for a period of 1 h and at 600°C for times varying from 1 to 1 000 h; their results are summarised in Fig. 7.10. They concluded that alloying with Cr, Si, Al, Co, W and B raised the temperature at which the second phase (Fe_3O_4) appeared in the oxide films. The rise in temperature depended on the type and the concentration of the alloying element. From this one concluded that the FeO formation had been suppressed, i.e. a displacement of the non-variance points had taken place.

Increase in the amount of alloying elements caused the oxidation resistance of the steels in air to increase at 600°C, and of the alloys examined the greatest improvement was obtained with chromium and the least with cobalt.

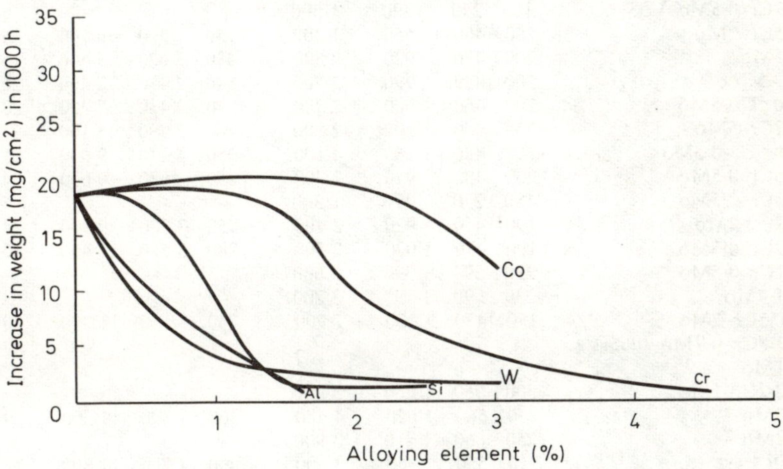

Fig. 7.10 Effect of alloying elements on oxidation of low-alloy steel at 600°C in air (after Yuganova and Sorokina[33])

Flue Gas

The term *flue gas* refers to those atmospheres consisting of the gases CO_2, CO, O_2, H_2 and water vapour. As the designation is intended to cover the products of total or partial combustion of the fuels, oil, gas and coal with air, a proportion of nitrogen will be present and sulphurous gases could also be present.

A considerable volume of work has been published on the effect of complex atmospheres on oxidation, but unfortunately much of this relates to the reheating of steels prior to fabrication at temperatures in excess of 1 000°C[60,61]. There has, however, still been a considerable amount of work done on corrosion and corrosion phenomena occurring at service temperatures during the combustion of gas, coal and oil, but unfortunately many of these investigations have never been published. Rahmel[62], in order to introduce some order into the large quantity of information available on

Table 7.6 Relative resistance to

Steel alloy	593°C				704°C			
	Gain in weight (mg/dm²)			*Loss in weight* (mg/dm²)	*Gain in weight* (mg/dm²)			*Loss in weight* (mg/dm²)
	*a**	*b*†	*c*‡	*d*§	*a*	*b*	*c*	*d*
Fe-5Cr-0·5Mo-1·5Si	0	0	0	0	—	—	—	—
Fe-12Cr-0·5Mo	—	—	—	—	—	—	—	—
Fe-12Cr	0	0	0	0	0	0	0	0
Fe-5Cr-0·5Mo-1Si+Nb	—	—	—	—	—	—	—	—
Fe-5Cr-0·5Mo-1Si+Ti	—	—	—	—	—	—	—	—
Fe-9Cr-1Mo	—	—	—	—	—	—	—	—
Fe-5Cr-1Mo	120	230	310	810	—	—	—	—
Fe-5Cr-0·5Mo+Ti	120	230	380	930	—	—	—	—
Fe-5Cr-0·5Mo+Nb	120	230	290	770	—	—	—	—
Fe-5Cr-5Mo	240	390	460	1 100	0	0	0	0
Fe-3Cr-0·5Mo-1·5Si	180	270	490	1 500	—	—	—	—
Fe-5Cr-2Mo	200	400	550	1 300	30	500	6 100	15 400
Fe-5Mo	200	470	920	2 800	480	2 020	5 500	15 500
Fe-3·5Cr	200	440	990	2 700	540	2 410	7 400	20 700
Fe-0·5Cr-5Mo	200	460	1 000	2 700	570	2 430	7 700	20 400
Fe-2Cr-5Mo	230	480	1 000	2 800	700	2 790	8 100	21 300
Fe-0·5Cr-0·5Mo	170	490	1 260	4 100	1 050	4 270	9 000	25 500
Fe-5Cr-0·5Mo	200	480	970	2 400	250	2 630	10 000	24 500
Fe-5Cr-0·5Mo	180	270	490	1 300	—	—	—	—
Fe-2Cr-2Mo	190	450	920	2 400	760	3 200	9 800	24 900
Fe-2Cr-0·5Mo	190	470	1 020	2 500	880	3 520	11 400	30 600
Fe-2Cr-0·5Mo	230	590	1 390	3 600	—	—	—	—
Fe-1·4Mo	230	590	1 190	3 200	—	—	—	—
Fe-0·5Cr-2Mo	180	470	1 130	2 900	760	510	11 700	30 000
Fe-1·8Cr-0·7Mo-0·6Si	—	—	—	—	—	—	—	—
Fe-1Mo	—	—	—	—	—	—	—	—
Fe-2·2Cr-1Mo	300	660	1 260	3 200	—	—	—	—
Fe-2Mo	240	560	1 620	4 400	260	3 890	11 500	29 300
Fe-2Mo	250	660	1 410	3 900	—	—	—	—
Fe-2Cr	210	550	1 350	3 600	1 000	4 760	11 800	32 600
Fe-0·5Cr	190	530	1 370	3 700	940	5 060	11 800	36 700
Fe-1·3Cr-0·5Mo-0·8Si	250	660	1 390	3 600	—	—	—	—
Fe-0·5Mo	220	590	1 760	4 900	1 140	5 000	12 200	34 900
Fe-0·5Mo	220	610	1 770	4 500	—	—	—	—
Fe-0·5Mo	250	660	1 620	4 400	—	—	—	—
Fe+C	210	780	2 660	6 700	880	4 780	7 800	25 200
Fe+C	—	—	—	—	—	—	—	—

* 1 day; † 1 week; ‡ 6 weeks, interpolated from the mean curve representing duplicate observations; § observed weight loss after

corrosion phenomena in these atmospheres, used the following classification:

1. Sulphur- and ash-free fuels.
2. Sulphur-containing but ash-free fuels.
3. Sulphur- and ash-containing fuels.

This is an excellent review paper to which the reader is referred particularly for information on ash-containing fuels.

oxidation of various steels in air

760°C				816°C				927°C		
Gain in weight (mg/dm^2)			*Loss in weight* (mg/dm^2)	*Gain in weight* (mg/dm^2)			*Loss in weight* (mg/dm^2)	*Gain in weight* (mg/dm^2)		
a	*b*	*c*	*d*	*a*	*b*	*c*	*d*	*a*	*b*	*c*
18	23	35	50	—	—	—	—	50	2 570	11 700
12	18	35	90	—	—	—	—	4 800	14 200	30 700
—	—	—	—	3	20	90	160	—	—	—
55	67	73	110	—	—	—	—	44	1 960	35 000
1 200	2 000	3 300	8 400	—	—	—	—	2 600	3 710	15 300
20	26	41	80	—	—	—	—	5 850	16 600	41 700
1 980	4 250	8 500	34 000	—	—	—	—	—	—	—
1 800	3 650	8 900	23 000	—	—	—	—	—	—	—
1 400	3 800	10 600	29 300	—	—	—	—	—	—	—
—	—	—	—	6	1 900	20 700	48 500	—	—	—
930	1 020	1 050	2 800	—	—	—	—	—	—	—
—	—	—	—	60	5 800	10 500	28 500	—	—	—
—	—	—	—	1 700	9 900	23 000	62 000	—	—	—
—	—	—	—	2 100	9 700	25 500	72 000	—	—	—
—	—	—	—	2 000	9 700	27 300	76 000	—	—	—
—	—	—	—	2 100	9 400	20 600	59 500	—	—	—
—	—	—	—	3 400	10 200	30 400	90 000	—	—	—
—	—	—	—	1 000	8 500	18 900	54 500	—	—	—
1 170	2 340	5 700	15 600	—	—	—	—	—	—	—
—	—	—	—	2 600	8 200	20 200	53 500	—	—	—
—	—	—	—	3 100	9 400	19 000	58 500	—	—	—
2 920	7 750	18 000	50 000	—	—	—	—	—	—	—
—	—	—	—	—	—	—	—	—	—	—
—	—	—	—	2 200	11 200	29 800	76 000	—	—	—
3 500	7 200	15 300	45 000	—	—	—	—	—	—	—
2 220	8 700	18 900	52 000	—	—	—	—	—	—	—
3 100	10 000	21 000	67 000	—	—	—	—	—	—	—
—	—	—	—	2 000	10 900	34 800	93 000	—	—	—
—	—	—	—	—	—	—	—	—	—	—
—	—	—	—	3 900	13 500	25 400	76 000	—	—	—
—	—	—	—	3 100	9 700	22 100	64 000	—	—	—
—	—	—	—	—	—	—	—	—	—	—
—	—	—	—	3 200	12 900	30 800	93 500	—	—	—
2 160	7 100	15 200	43 200	—	—	—	—	—	—	—
—	—	—	—	—	—	—	—	—	—	—
—	—	—	—	3 100	11 400	27 800	83 500	—	—	—
2 050	7 100	18 200	51 000	—	—	—	—	—	—	—

picking one of each pair of specimens after 6-week exposure (after Day and Smith[58]).

Preece *et al.*[61] studied the effect of changing the oxidising–reducing potential of an 80% N_2, 10% CO_2 and 10% H_2O atmosphere by additions of oxygen and carbon monoxide at 650°C. At the same time the effect of sulphur dioxide in these atmospheres was examined. It was established that the behaviour of all the materials (Table 7.7) in neutral atmospheres free from sulphur dioxide showed similar features, and that while oxygen up to 0·6% rapidly increased the rate of oxidation, further additions had little effect

Table 7.7 Scaling of steels in a neutral atmosphere (80% nitrogen, 10% carbon dioxide, 10% water vapour) free from sulphur dioxide; testing temperature 650°C*

Steel	*Analysis (%)*							*Increase in weight* (mg/dm²)	
	C	Mn	Si	S	P	Ni	Cr	8-h *Exposure*	30-h *Exposure*
0·13% carbon (rimming)	0·13	0·48	Trace	0·033	0·014	—	—	330	860
0·29% carbon	0·29	0·50	0·205	0·026	0·023	—	—	190	560
0·50% carbon	0·50	0·69	0·105	0·030	0·032			170	550
0·67% carbon	0·67	0·68	0·180	0·014	0·025	—	—	230	560
1·34% carbon	1·34	0·35	0·165	0·032	0·027	—	—	190	640
1·50% manganese	0·39	1·52	0·28	0·050	0·033	—	—	190	510
5% nickel	0·14	0·35	0·22	0·031	0·038	5·12	—	200	500
Nickel–chromium	0·36	0·63	0·14	0·012	0·022	3·59	0·78	240	510
4% silicon	0·12	0·14	3·84	0·03	0·042	—	—	220	440
12% chromium stainless iron	0·12	0·25	0·54	—	—	—	12·6	Nil	80

*Data after Preece, Simister and Cobb[61].

(Fig. 7.11). Additions of carbon monoxide, which produces reducing conditions, had little influence on the rate of scaling.

Upthegrove[63], in his examination of mild steel, has shown that increasing the carbon monoxide content from 0 to 4% in flue gas decreases oxidation

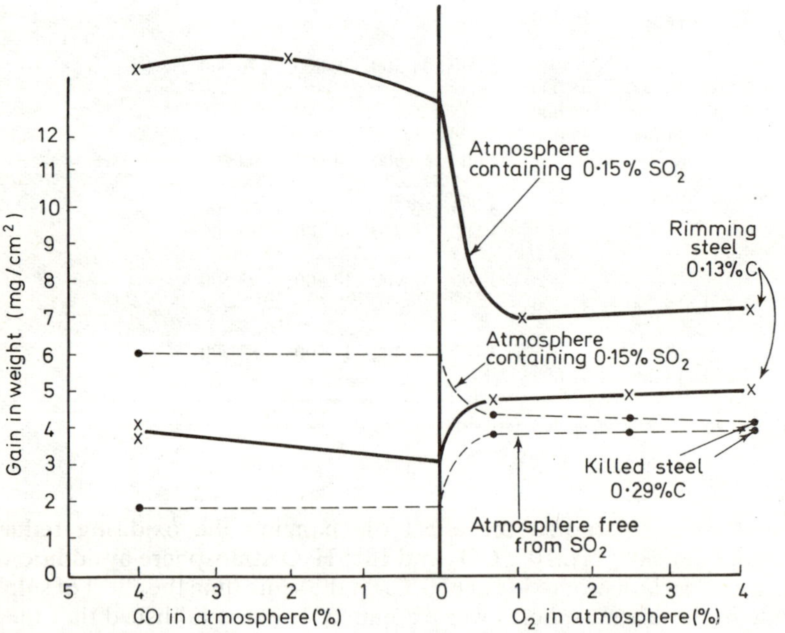

Fig. 7.11 Influence of additions of carbon monoxide and of oxygen to the neutral furnace atmosphere (80% N_2, 10% CO_2, 10% H_2O) free from and also containing 0·15% SO_2, on the scaling of plain carbon steel; 8 h, 650°C (after Preece, Simister and Cobb[61])

slightly at 760, 816 and 927°C. At 1 039°C, however, the decrease was less marked and 12% CO content in the atmosphere at that temperature was required to decrease the oxidation rate to the same value as that produced at 927°C in an atmosphere containing only 4% CO. These data, together with those of Preece, indicate that carbon monoxide is not particularly important with respect to oxidation, but it can of course effect carburisation.

In atmospheres containing sulphur dioxide[61] (Fig. 7.12) all the carbon steels oxidised in a similar manner. They are better than rimming steels, which have a surface of pure iron, but are worse than alloy steels which show differences in the structure and quality of the scale formed. It can be seen from the curves that 0·2% SO_2 trebled the rate of attack on Fe–5Ni, whereas

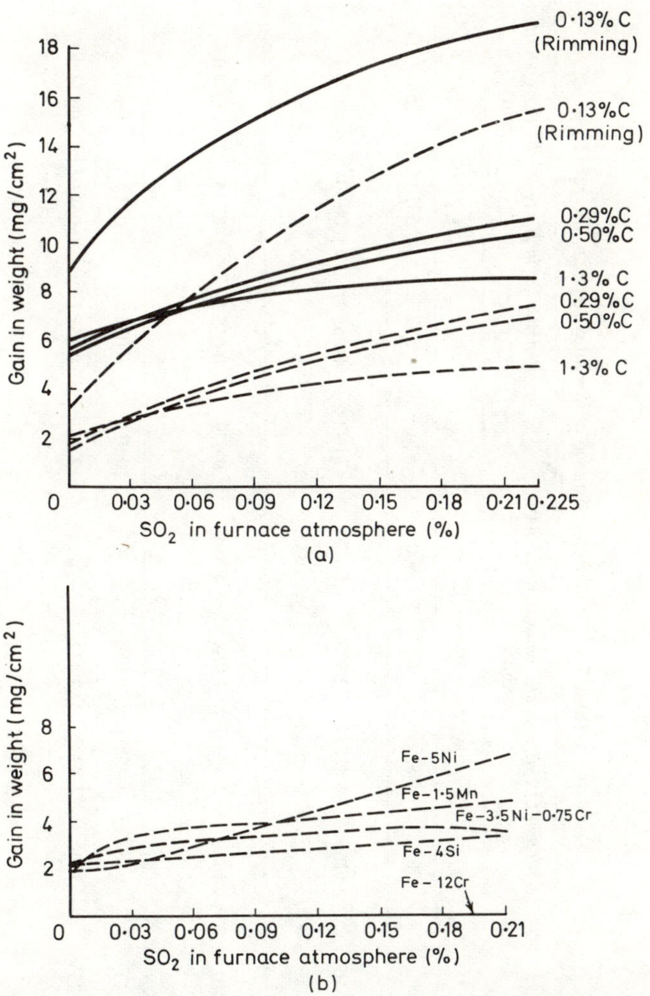

Fig. 7.12 Relationship between the increase in weight and the SO_2 content of the furnace atmosphere at 650°C for (*a*) plain carbon and (*b*) low-alloy steels. Composition of atmosphere: 10% CO_2, 10% H_2O, 80% N_2; temperature 650°C, continuous curves for 30 h, broken curves for 8 h (after Preece, Simister and Cobb[61])

Table 7.8 Oxidation resistance of low-alloy steels in flue gas (1 gas:6 air)

Steel alloy	*Weight loss after scale removal, after* 500 h (mg/dm²)					*Penetration in 1st year—by extrapolation of parabolic law* (mm)					Activation energies (kJ/mol) 500–620°C
	Testing temperature (°C)					*Testing temperature* (°C)					
	500	535	565	600	620	500	535	565	600	620	
Fe–2·25Cr–1Mo	1 126	1 422	2 323	5 178	7 092	0·055 9	0·061 0	0·099 1	0·266 7	0·345 4	500–565°C 42 565–620°C 113
Fe–1Cr–0·5Mo	1 166	1 539	1 968	4 654	9 679	0·055 9	0·077 5	0·094 0	0·193 0	0·469 9	500–565°C 52 565–620°C 111
Fe–0·5Mo	648	1 458	2 050	—	10 911	0·033 0	0·071 1	0·088 9	—	0·533 4	117
Fe–0·5Mo–0·25V	834	1 438	2 432	4 430	12 864	0·038 1	0·068 6	0·111 8	0·226 1	0·762	100

Fe–3·5Ni–0·75Cr and Fe–4Si were hardly affected. Although no evidence of oxide/sulphide eutectics was found at the temperature of the tests (650°C) it is possible it may have occurred at higher temperatures, with a consequent increase in oxidation.

Data obtained by the author in the products of combustion of towns gas (1 gas:6 air) of very low sulphur dioxide content (<0·02%) are given in Table 7.8. These show that the effect of alloy additions is not significant below about 2·25% Cr. Other tests, already discussed (Fig. 7.7), indicate that marked improvement is achieved above about 6% Cr. Figure 7.7 shows the results of cyclic heating and continuous tests at 620°C. While these cyclic heating tests are not comparable—owing to the fact that the total duration of the cyclic test was 42 h as against 325 h for continuous heating—in industrial conditions where it is often encountered, cycling can increase the magnitude of oxidation if the oxides spall off the surface, which tends to occur, especially when the scale is thick.

The results of some tests by the author in a high-sulphur gas atmosphere are shown in Fig. 7.13. When compared with the results in Table 7.8 these serve to illustrate the accelerating effects of sulphur dioxide on oxidation and confirm the findings of Preece[61].

These influences, however, are not marked since it has always been understood from Preece's data that oxygen tended to offset the deleterious effects of sulphur dioxide. Strack[64] investigated the corrosion rate of pure iron in a reducing atmosphere of CO_2 and CO, in which the SO_2 content was varied from 0·01 to 1·0% and the temperatures from 600 to 900°C. His main conclusions were that up to about 0·03% SO_2, scale formation proceeded according to a linear law, while at higher concentrations it changed from an initial

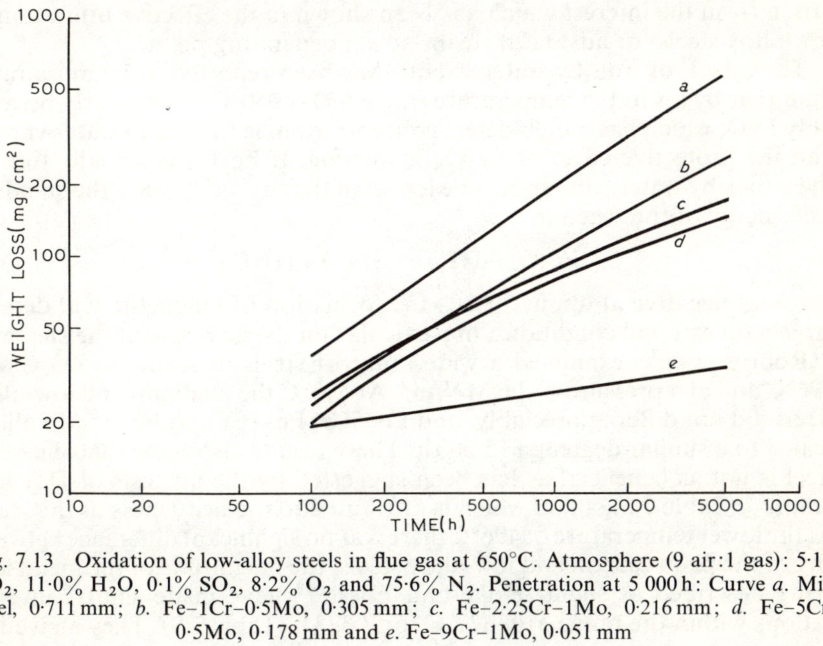

Fig. 7.13 Oxidation of low-alloy steels in flue gas at 650°C. Atmosphere (9 air:1 gas): 5·1% CO_2, 11·0% H_2O, 0·1% SO_2, 8·2% O_2 and 75·6% N_2. Penetration at 5 000 h: Curve *a*. Mild steel, 0·711 mm; *b*. Fe–1Cr–0·5Mo, 0·305 mm; *c*. Fe–2·25Cr–1Mo, 0·216 mm; *d*. Fe–5Cr–0·5Mo, 0·178 mm and *e*. Fe–9Cr–1Mo, 0·051 mm

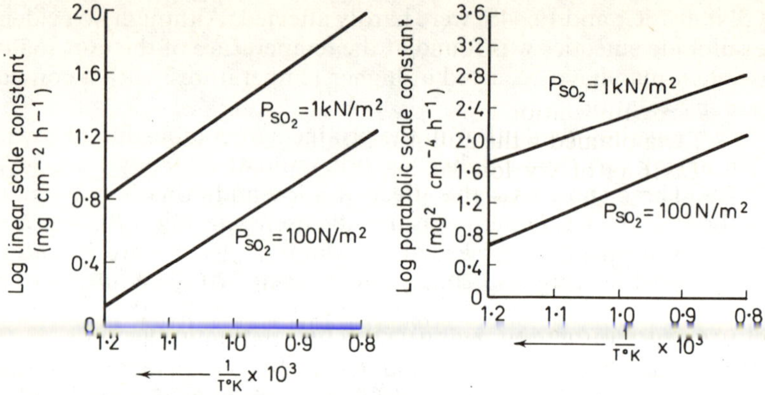

Fig. 7.14 Temperature dependence of oxidation rate constant. (*a*) Linear, up to 0·03% SO_2 and (*b*) parabolic, above 0·03% SO_2 (parameter: SO_2 partial pressure) (after Strack[64])

linear to a substantially parabolic relationship. The CO_2/CO ratio did not appear to exert any influence upon the rate of scale formation. With an increasing concentration of SO_2, and the temperature remaining constant, the proportion of FeS in the scale increased; with the SO_2 concentration remaining unchanged and the temperature increasing, the content of FeS decreased. The results are summarised in Fig. 7.14.

Steam

The publications which deal with the oxidising effects of steam have mainly arisen from the interest which has been shown in the effective utilisation of low-alloy steels for advanced steam-power generating plant.

The attack of iron by water vapour has been reported to be more rapid than that by air in the temperature range 650–1 050°C[65], obviously because only FeO, which has a high defect concentration, is formed in water vapour and the protective effect of Fe_3O_4 is missing. If Fe_3O_4 is actually formed the attack by water vapour may be less than that by air[66]; since the standard free energy of the reaction

$$Fe_3O_4 + (H_2) \rightarrow 3FeO + (H_2O)$$

becomes negative at about 700°C, the formation of magnetite will depend largely on external conditions, in particular on the flow rate of the steam.

Rohrig *et al.*[67] examined a wide range of steels in steam at 593°C and 496°C and at a pressure of 2·6 MN/m². At 593°C the medium- and low-alloy steels did not differ appreciably, and Fe–5Cr, Fe–1Cr and Fe–0·5Mo alloys scaled to a similar degree in 15 000 h. These results also show that silicon in steel is not as beneficial as has been suggested by the air tests of Day and Smith[58] (Table 7.6), a fact which is attributed to the brittleness of the scale. At the lower temperature of 496°C there was no significant difference between any of the plain carbon and low-alloy steels, the magnitude of the metal loss being described as slight. Hoke and Eberle[68] have made similar investigations within the range 510–677°C for 7 323 h (Table 7.9). They arrived at

Table 7.9 Maximum temperatures permissible if penetration is not to exceed 0·127 mm/y (duration of test 7 323 h)

Steel	*Temperature* (°C)
Plain carbon	579
Fe–0·5Mo	593
Fe–0·5Cr–0·5Mo	604
Fe–2·25Cr–1Mo	613
Fe–5Cr–0·5Mo	607
Fe–9Cr–1Mo	666

almost the same conclusions as Rohrig[67] and gave a rating based on the maximum temperature at which penetration of a given steel will not exceed 0·127 mm/y.

Although this is a fair rating it must be remembered that it refers only to the effects of steam, and is not directly applicable to a tube steel in service, where there will undoubtedly be some other corrosion process proceeding on the opposite face of the tube. The rating does, however, indicate the very narrow useful range of temperature for the low-alloy steels. The author's own investigations have been carried out in collaboration with other steel manufacturers and the Central Electricity Generating Board, in which low-alloy steels were exposed to power-station steam at a pressure of 3·45 MN/m^2. In these tests the log–log plots of oxidation against time referred to previously have been used. The results presented in Fig. 7.15 showed that the amount of oxidation increases with exposure time, but with the exception of mild steel is not markedly affected by steam temperature in the range 566–621°C. It is apparent from this latter figure that Fe–9Cr + Mo exhibits a marked degree of sub-surface penetration which subsequent microscopical examination revealed to be predominantly pitting. It is this latter effect which has tended to down-rate the Fe–9Cr + Mo steel (Fig. 7.15) on the basis of the total depth of metal affected by oxidation. It would appear therefore that in the temperature range considered there is little difference in the oxidation resistance of the various qualities, and in this respect the findings confirm those of Rohrig *et al.* that these steels have a very narrow useful range of temperature.

Clarke *et al.*[69] have referred to the similarity of the effects of air and steam at 593°C and 649°C and so have Upthegrove and Murphy[50]. In the former case the thickness of the scale layer was used to indicate the amount of oxidation. This method is valid only if the scale is completely adherent, and is not recommended because the oxides tend to spall off the surface of most metals during cooling. In their tests the Fe–2·25Cr – 1Mo and Fe–9Cr + Mo alloys are very similar at low temperatures, but at the highest temperature Fe–9Cr + Mo is markedly superior. Comparison of the author's own results in air and steam indicate that at times up to 100 000 h the behaviour of steels is similar in both atmospheres except at temperatures in excess of 620°C.

Laboratory tests by Potter *et al.*[70] have shown that for steam, temperature and not pressure is the controlling factor in oxidation, although the rate of flow has some influence. Eberle and Kitterman[71] showed that differences existed in oxidation rates on low-alloy steels ranging from Fe–2·25Cr – 1Mo to Fe–9Cr–1Mo after 10 years in steam at 13·8 MN/m^2, however, these rates

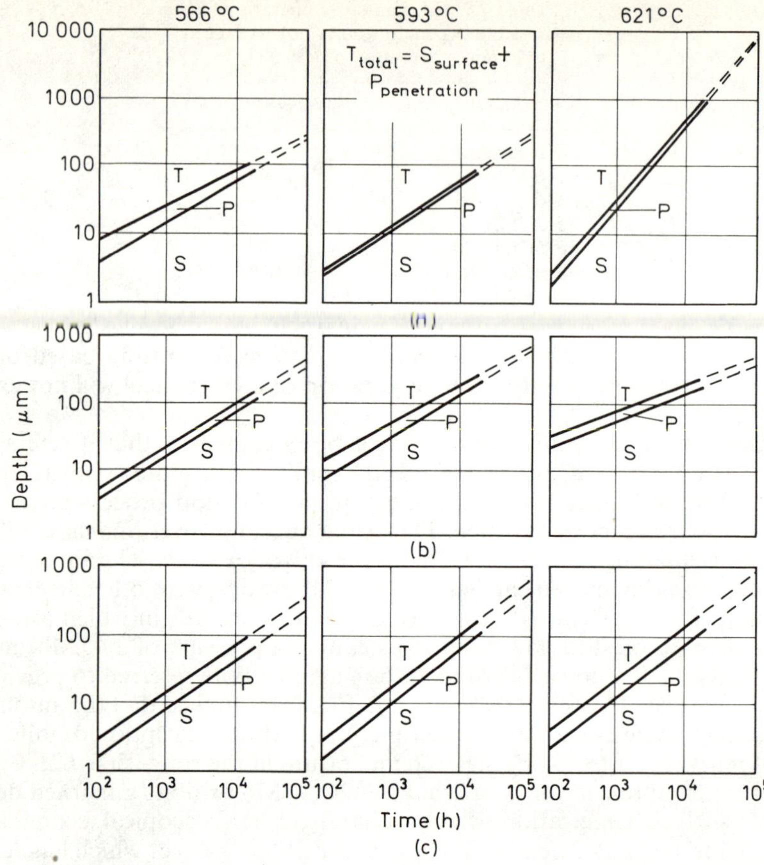

Fig. 7.15 Oxidation in steam of (*a*) mild steel, (*b*) Fe–2·25Cr–1Mo and (*c*) Fe–9Cr + Mo (after British Steel Corporation, Central Electricity Generating Board and Brown-Firth Research Laboratories)

were assessed on scale thicknesses and as has been suggested previously this method is not recommended. These results did indicate that Fe–9Cr – 1Mo was somewhat superior to the other low-alloy qualities. It did not take into account however the degree of sub-surface oxide penetration. Similarly work by Nikiforova and Kharina[72] showed that after 3 000 h testing the oxidation resistance of low-alloy steels increased with increasing chromium content, but once again these did not consider the effect of sub-surface oxide penetration.

It must be concluded therefore that although low-chromium steels are superior to plain carbon steels, all give oxidation values of a similar order, particularly if the total depth of metal affected by oxidation is taken as the criterion. The significance of such a criterion need not be emphasised because the dangers of any phenomena which effectively reduce the cross-sectional area of metal available in high-temperature high-pressure steam systems will be obvious.

Carbon Dioxide

During the past few years the oxidising effect of carbon dioxide has become increasingly important because of the designing and construction of gas-cooled atomic reactors. Upthegrove and Murphy[50] examined carbon dioxide/water vapour mixtures over the full range 0–100%. Their results show that oxidation in carbon dioxide is very small but increases rapidly with increased water content. Bourgraff[51] confirmed that oxidation is very small indeed in this atmosphere.

The results of prolonged laboratory tests are now being published but these rarely refer to carbon dioxide alone, because the effect of the contaminating gases carbon monoxide and water vapour is important.

Bokros and Wallace[73] examined steels and other alloys in carbon dioxide at various temperatures and pressures and showed that increasing the chromium content up to 9% improved oxidation resistance, the optimum amount at 593°C being 5%. This, to some degree, confirms the observations made in connection with Fig. 7.7.

Results over very long periods obtained by Robinson[74] (Table 7.10 and Figs. 7.16–7.18) indicate that the majority of low-alloy steels are satisfactory in carbon dioxide plus carbon monoxide plus water vapour atmospheres and have relatively low oxidation rates. However, some show an accelerating type of oxidation, i.e. a change from parabolic to linear oxidation rates, which is 'catastrophic' with respect to the ultimate magnitude of metal loss and is very undesirable. Robinson suggested that 'The transition from decreasing to linear oxidation rate is due to cracking of the protective scale, the latter caused by rapid initial oxidation arising from certain alloying elements, in particular nickel and copper'. Davies and Moore[75] observed this accelerating effect in two forms. In one case, cracking as observed by Robinson occurred, and in the second blisters or excrescences appeared on the specimen surfaces.

Since the original prolonged laboratory tests carried out under the auspices of the Atomic Energy Authority, numerous papers have been presented on the long term oxidation behaviour of mild- and low-alloy steels in carbon dioxide. Some of the many recent investigations were presented at the symposium sponsored by the British Joint Corrosion Group[76].

Goodison *et al.*[77] showed that in these contaminated carbon dioxide atmospheres as the gas pressure increases from 0 to 4·14 MN/m^2 there is a decrease in the time to breakaway and possibly an increase in the post breakaway oxidation rate. However, if the partial pressure of the moisture in the gas is kept constant then the post-breakaway rate may be constant from 1·38 to 4·14 MN/m^2 though there is an increased initial rate of development of growths. Harris and Goodison[78] showed that with rimming steels the post-breakaway rate is highly dependent on temperature, the carbon monoxide and moisture content of the gas and on the gas pressure, and is generally considerably greater than that for killed steels but similar to commercially pure iron. Loriers *et al.*[79] who studied a selection of ten low-alloy steels at temperatures of up to 550°C and pressures of up to 6 MN/m^2 showed that the occurrence of an irregular attack and the formation of the excrescences on the samples, observed in many cases, could be related to the composition and pressure of the corroding medium and to the metal surface conditions.

Table 7.10 Analyses of materials in Figs. 7.16 to 7.18*

Steel	*Analysis (%)*											
	C	Si	Mn	S	P	Ni	Cr	Mo	V	Cu	Al	*Others*
A	0·20	0·16	0·49	—	—	—	0·99	0·94	0·65	—	0·006	—
B	0·15	0·15	1·15	0·04	0·02	0·15	0·1	0·25	—	0·5	0·01	—
C	0·155	0·052	1·15	0·017	0·016	0·08	—	—	—	0·36	0·015	—
	0·15	0·15	1·13	0·045	0·030	0·11	0·09	0·029	—	0·08	0·036	—
D	0·14	0·14	1·04	0·031	0·014	[illegible]	[illegible]	[illegible]	[illegible]	[illegible]	0·053	[illegible]
	0·16	0·17	1·07	0·034	0·023	—	—	—	—	—	0·053	—
E	0·20	0·34	0·55	0·025	0·030	0·15	—	0·019	—	0·12	0·009	—
F	0·145	0·15	1·32	0·037	0·015	0·11	0·035	0·03	—	—	Nil	—
	0·15	0·13	1·29	0·031	0·024	—	—	—	—	—	0·02	—
G	0·11	0·11	1·07	0·035	0·016	0·16	0·065	0·013	—	0·08	0·028	—
	0·13	0·18	1·21	0·030	0·029	0·11	0·085	0·01	—	0·04	0·032	—
H	0·13	0·13	1·21	0·035	0·023	—	—	0·008	—	—	0·018	—
	0·14	0·18	0·97	0·035	0·027	—	—	0·045	—	—	0·068	—
I	0·166	0·17	1·31	0·025	0·029	0·10	—	0·23	0·08	0·08	0·045	—
J	0·10	0·38	1·7	—	—	—	1·5	0·54	—	—	0·58	0·003 B
K	0·175	0·15	1·28	0·024	0·014	0·19	0·54	0·25	0·09	0·09	Nil	—
	0·15	—	1·35	—	—	0·25	0·6	0·27	0·10	0·13	0·005	—
L	0·14	0·12	1·21	—	—	1·10	0·2	0·57	0·04	—	—	—
M	0·12	0·148	0·18	—	—	1·1	0·17	0·325	0·16	—	—	—
N	0·12	0·16	1·34	0·040	0·019	0·15	0·76	0·62	0·08	—	—	—
O	0·16	0·4	0·6	0·045	0·045	—	0·025	0·5	—	—	0·087	0·003 B
P	0·105	0·22	0·58	0·015	0·020	—	0·28	0·64	0·25	—	—	—
Q	0·11	0·5	0·52	0·01	0·027	—	0·7	0·61	—	—	—	—
R	0·12	0·5	0·5	—	—	—	2·25	1·0	—	—	—	—
	0·12	0·3	0·45	0·015	0·020	—	2·25	0·84	0·02	—	0·007	—
S	0·087	3·20	0·036	0·011	0·018	—	0·046	—	—	0·24	0·036	—
T	0·25	0·32	0·59	—	—	—	1·30	0·73	0·60	—	0·006	0·68 W

*Data after Robinson[74].

An interesting presentation by Harris and Menzies[80] showed that variations in the oxidation behaviour on what were nominally identical casts of low-alloy steels could be statistically related to the inclusions present in these materials.

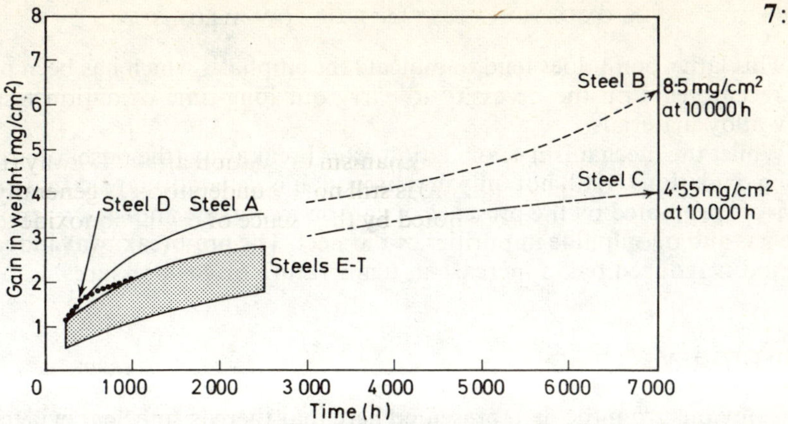

Fig. 7.16 Oxidation resistance of low-alloy steels in $CO_2 + 0{\cdot}3\%$ CO $+ 0{\cdot}03\%$ H_2O at 450°C and $1{\cdot}0$ MN/m² (after Robinson[40])

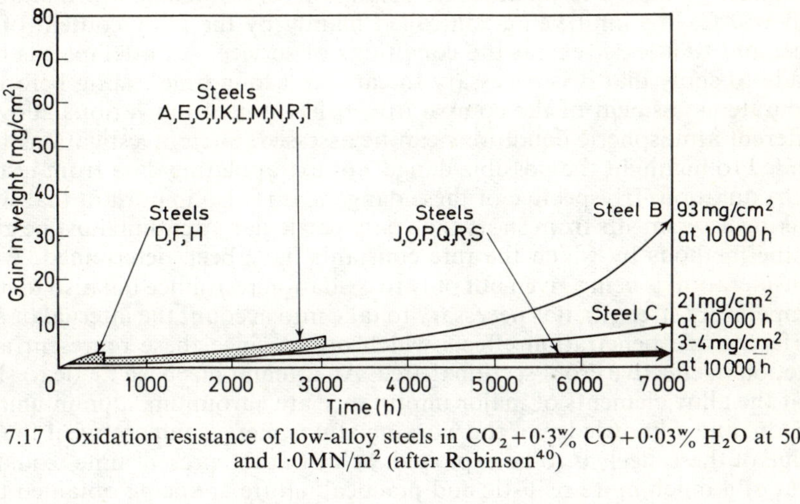

Fig. 7.17 Oxidation resistance of low-alloy steels in $CO_2 + 0{\cdot}3\%$ CO $+ 0{\cdot}03\%$ H_2O at 500°C and $1{\cdot}0$ MN/m² (after Robinson[40])

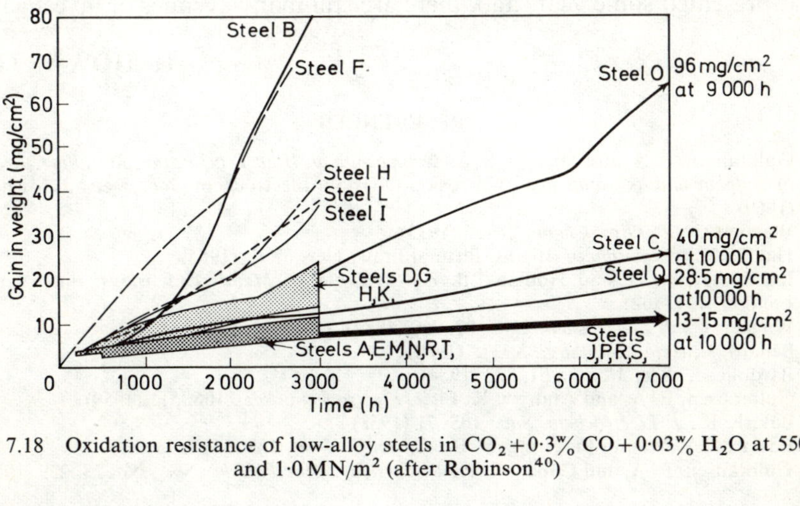

Fig. 7.18 Oxidation resistance of low-alloy steels in $CO_2 + 0{\cdot}3\%$ CO $+ 0{\cdot}03\%$ H_2O at 550°C and $1{\cdot}0$ MN/m² (after Robinson[40])

This latter point does tend to indicate the emphasis which has been placed in recent years on the necessity to carry out long-time oxidation tests on low-alloy materials.

Whilst the mechanism of oxidation after breakaway (the onset of accelerating oxidation) is still not fully understood, it is generally agreed that breakaway is promoted by the presence of carbon monoxide and water vapour in the gas and by sulphide impurities of the steel. The pre-breakaway induction period is reduced by an increase in temperature or gas pressure.

Summary

It is obvious from the data presented here that there is sufficient evidence to indicate that low-alloy steels have a considerable part to play in the construction of equipment in which heat resistance is important. They are, unfortunately, limited to an absolute maximum service temperature of about 600–650°C, this limit being controlled mainly by the alloy content of the steel and to a lesser extent the conditions of service. An attempt has been made to show that it is necessary to carry out long-time testing before an accurate assessment of the comparative behaviour of the various alloys in different atmospheric conditions can be assessed. Such investigations have tended to highlight the possible dangers of extrapolating data from tests of short duration. Irrespective of these dangers, it is also important that when making assessments from short-time data particular attention must be given to the methods by which the rate constants have been determined. When consideration is being given not only to oxidation resistance but also to high-temperature strength, it is necessary to take into account the internal or sub-surface oxide penetration effects which occur, since these represent a reduction in effective cross-sectional area. An attempt has been made to show that the alloy elements of major importance are chromium, aluminium and silicon, and a knowledge of the working atmosphere is imperative if the full value of these steels is to be exploited. Whilst at the present time oxidation data of a much more realistic and practical nature are being obtained than that presented some years ago, there are still many avenues of investigation open.

H. HOWARTH

REFERENCES

1. Gulbransen, E. A. and Jansson, S. A., *Symposium on High Temperature Metallic Corrosion by Sulphur and its Compounds*, Corrosion Division, Electrochem. Soc., Detroit, Michegan (1969)
2. Wagner, C., *Z. Phys. Chem.*, **21**, 25 (1933)
3. Hauffe, K., *Oxidation of Metals*, Plenum Press, New York (1965)
4. Kubaschewski, O. and Hopkins, B. E., *Oxidation of Metals and Alloys*, Butterworths, London, 42 (1962)
5. Wagner, C., *J. Electrochem. Soc.*, **99**, 369 (1952)
6. Bénard, J., *Acta Metallurg.*, **8**, 272 (1960)
7. Bardolle, J., *Rév. Métall.*, **51**, 833 (1954)
8. Gulbransen, E. A. and Andrew, K. F., *J. Electrochem. Soc.*, **106**, 511 (1959)
9. Bakish, R., *J. Electrochem. Soc.*, **105**, 71 (1958)
10. Edwards, A. M. and Pickering, F. B., *J. Iron St. Inst.*, **189**, 56 (1958)
11. Gulbransen, E. A. and Copan, T. P., *Discussions of the Faraday Soc.*, No. 28, 229 (1959)

12. Fischmeister, H., *Reactivity in Solids*, Proc. 4th Int. Symp. on Reactivity Solids, 195 (1961)
13. Rönquist, A., *J. Inst. Metals*, **91**, 639 (1962–63)
14. Albert, L. and Jaenicke, W. Z., *Naturforse*, **14a**, 1 040 (1959) and **15a**, 59 (1960)
15. Kofstad, P., *High-temperature Oxidation of Metals*, John Wiley and Sons, 38 (1966)
16. Hauffe, K., *Oxidation of Metals*, Plenum Press, New York, 285 (1965)
17. Brauns, E., Rahmel, A. and Christmann, H., *Arch. Eisenhüttenw.*, **30**, 533 (1959)
18. Gulbransen, E. A. and Roswell Ruka, *J. Metals, N.Y.*, **188**, 1 500 (1950)
19. Himmel, L., Mehl, R. F. and Birchenall, C. E., *J. Metals, N.Y.*, **5**, 827 (1953)
20. Davies, M. H., Simnad, M. T. and Birchenall, C. E., *J. Metals, N.Y.*, **3**, 889 (1951); *Ibid.* **5**, 1 250)1953)
21. Davies, D. E., Evans, U. R. and Agar, J. N., *Proc. Roy. Soc.*, **A225**, 443 (1954)
22. Caplan, D. and Cohen, M., *Corr. Science*, **6**, 321 (1966)
23. Stringer, J., *Metallurgical Review*, **11**, 113 (1966)
24. Manenc, J. and Vagnard, G., *Corrosion Science*, **9** No. 12, 887, Dec. (1969)
25. Boggs, W. E. and Kachik, R. H., *J. Electrochem. Soc.*, 424, April (1969)
26. Ipat'ev, V. V. and Orlova, G. M., *Zh. Prikl. Khim.*, Moscow, **29**, 819 (1956)
27. Rahmel, A., *Werkstoffe und Korrosion*, **16**, 837–843, Oct. (1968)
28. Engell, H. J., *Zeitschrift für Elektrochemie Berichte der Bunsengesellschaft für Physikalische Chemie*, **63**, 835–859, Oct. (1959)
29. Bandel, G., *Arch. Eisenhüttenw.*, **30**, 351 (1959)
30. Portevin, A., Pretet, E. and Jolivet, H., *Rev. Metall.*, **31**, 101, 186, 219 (1934)
31. Scheil, E. and Kiwit, K., *Arch. Eisenhüttenw.*, **9**, 405 (1936)
32. Hessenbruch, W. and Rohn, W., *Heraeus Vakuum Schmelze*, 247 (1923–1933)
33. Yuganova, S. A. and Sorokina, Yu. G., *Metalloved Tern. Obra. Met.*, **6**, 279–283 (1961)
34. Sully, A. H., Brandes, E. A. and Brown, R. H., Fulmer Research Institute, Private Communication
35. Pfeil, L. B., *J. Iron and Steel Inst.*, **119**, 501 (1929)
36. Wilder, A. B. and Light, J. O., *Trans. Amer. Soc. Metals*, **43**, 323 (1951)
37. Rahmel, A., Jaeger, W. and Becker, K., *Arch. Eisenhüttenw.*, **30**, 351 (1959)
38. Brenner, S. S., *J. Electrochem. Soc.*, **102**, 7 (1955)
39. Yashuhiko, M., *Fuji Iron and Steel Co. Techn. Rep.*, No. 13, 164–173, March (1964)
40. Foley, R. T., *J. Electrochem. Soc.*, **109**, 12, 1 202 (1962)
41. Menzies, I. A. and Tomlinson, W. J., *J.I.S.I.*, **1 239**, Dec. (1966)
42. Hagel, W. C., *Corrosion*, **21** No. 10, 316, Oct. (1965)
43. Saegusa, F. and Lee, L., *Corrosion*, **22** No. 6, 168, June (1966)
44. Collari, N. and Fongi, N., *Metallurg. Ital.*, **46**, 111 (1954)
45. Kubaschewski, O. and Hopkins, B. E., *Oxidation of Metals and Alloys*, Butterworths, London, 236 (1962)
46. Perrow, J. M. and Smeltzer, W. W., *J. Electrochem. Soc.*, **109** No. 11, 1 023, Nov. (1962)
47. Kubaschewski, O. and Hopkins, B. E., *Oxidation of Metals and Alloys*, Butterworths, London, 233 (1962)
48. Burling-Smith, W. J., *Steels in Modern Industry*, Butterworths, London, 132 (1953)
49. Edwards, A. M. and Nicholson, A., *Some Effects of Reheating Stainless Steels*, I.S.I. Publication No. 117, 149 (1969)
50. Upthegrove, C. and Murphy, D. W., *Trans. Amer. Soc. Steel Treat.*, **21**, 73 (1933)
51. Bourgraff, R., *Stahl u. Eisen, Dusseldorf*, **60**, 129 (1940)
52. Marson, C. B., Angus, H. T. and Cobb, J. W., *J. Soc. Chem. Ind., London*, **46**, 615 (1927)
53. Kubaschewski, O. and Hopkins, B. E., *Oxidation of Metals and Alloys*, Butterworths, London, 265 (1962)
54. Caplan, D. and Cohen, M., *J. Metals, N.Y.*, **4**, 1 057 (1952)
55. Rohrig, I. A., Vanduzer, R. M. and Fellows, C. H., *Trans. Amer. Soc. Mech. Engrs.*, **66**, 277 (1944)
56. Paidassi, J., *Acta Metallurgica*, **6**, 184 (1958)
57. Stanley, J. K., von Hoene, J. and Huntoon, R. J., *The Oxidation of Pure Iron*, Amer. Soc. Metals, Preprint No. 5 (1950)
58. Day, M. J. and Smith, G. V., *Industr. Engng. Chem. (Industr.)*, **35**, 1 098 (1943)
59. Siebert, C. A., *Trans. Amer. Soc. Metals*, **27**, 752 (1939)
60. *I.S.I. Special Report*, No. 14, 1st Report of the Alloy Steels Research Committee (1936)
61. Preece, A., Simister, E. and Cobb, J. W., *I.S.I. Special Report*, No. 24, 2nd Report of the Alloy Steels Research Committee, Part 2, 53 (1939)
62. Rahmel, A., *Werkstoffe und Korrosion*, **9**, 776–783 (1965)

63. Upthegrove, C., *Scaling of Steel at Heat-treating Temperatures*, Bull. Dept. Engng. Res., Univ. Mich., **25**, 24, Aug. (1933)
64. Strack, B., *Gaswarme*, No. 6, 224–239 (1964)
65. Baukloh, W. and Funke, P., *Korros. Metallarch.*, **18**, 126 (1942)
66. Paidassi, J. and Fuller, D. C. R., *Aca. Sci., Paris*, **246**, 604, 759 (1958)
67. Rohrig, I. A., van Duzer, R. M. and Fellows, C. H., *Trans. Amer. Soc. Mech. Engrs.*, **66**, 277 (1944)
68. Hoke, J. and Eberle, F., *Amer. Soc. Mech. Engrs.*, Paper No. 57–A–175 (1957)
69. Clarke, C. L., Rutherford, J. J. B., Wilder, A. B. and Cordovi, M. A., *Trans. Amer. Soc. Mech. Engrs., Ser. A., J. Engng. Power*, **82**, 35 (1960)
70. Potter, A. A., Solberg, H. L. and Hawkins, G. A., *Trans. Amer. Soc. Mech. Engrs.*, Research Paper, R.P. 59–10, 59, 725 (1959)
71. Eberle, F. and Kitterman, J. H., *Behaviour of Superheater Alloys in High Temperature, High Pressure Steam*, A.S.M.E. (1965)
72. Nikiforova, V. M. and Kharina, I. L., *Metalloved. Tern. Obra. Met.*, **4**, 217 (1964)
73. Bokros, J. C. and Wallace, W. P., *Corrosion*, **16**, 117 (1960)
74. Robinson, J. F., C. A. Parsons and Co. Ltd., Nuclear Research Centre, Private Communication
75. Davies, M. W. and Moore, C., *The Accelerating Oxidation of Ferritic Steels in Carbon Dioxide*, Soc. Chem. Ind. Corrosion Group Symposium on The Oxidation of Mild Steel, London, June 15th (1960)
76. *Symposium on Metal–Gas Reactions in Atmospheres Containing Carbon Dioxide*, British Joint Corrosion Group, London, 8–9th March (1967)
77. Goodison, D., Harris, R. J. and Goldenbaum, P., *Ibid.*, Paper No. 2
78. Harris, R. J. and Goodison, D., *Ibid.*, Paper No. 4
79. Loriers, H., Leckercq, D. and Darras, R., *Ibid.*, Paper No. 5
80. Harris, R. J. and Menzies, J. A., *Ibid.*, Paper No. 6

7.3 High-Temperature Corrosion of Cast Iron

Introduction

When cast iron is exposed to high temperatures under oxidising conditions, oxidation of the metal results, with the formation of a surface scale. In addition, the dimensions of the component become distorted. Although such dimensional changes can occur also in inert atmospheres or in vacuum, the evidence available suggests that this 'growth' is frequently associated with oxidation, and accordingly it is appropriate to consider it as an aspect of the corrosion of the iron.

The composition of the atmosphere to which components at high temperature may be exposed varies very widely, and most work on these aspects has accordingly been carried out in clean air. The aggressiveness of air is considerably enhanced by the presence of trace amounts of other reactive gases such as steam, carbon dioxide and sulphur dioxide; thus the figures subsequently quoted may in fact be appreciably lower than those encountered in specific atmospheres. The data presented should, however, prove an adequate guide to the order of the effect to be expected.

Growth

Components designed for high-temperature duty may either remain at a steady high temperature for their entire life, or, as more commonly happens, may undergo cyclic variation between a minimum temperature, often room temperature, and a maximum temperature. The maximum temperature involved may be either above or below the critical temperature range of the iron. This is the range within which the transformation between ferrite or pearlite and austenite occurs and for the majority of unalloyed irons it may be regarded as being 700–850°C. (See Section 9.2.)

Conditions of cyclic reheating are more severe than conditions of steady high temperature, and cyclic reheating through the critical range is particularly liable to cause excessive growth of the iron.

Generally, the studies which have been carried out suggest that growth of up to 40% by volume can occur within the first 40 h of cyclic reheating to 900°C with a frequency of 1–4 h/cycle, while subsequent cycling produces growth at a rate rarely exceeding 20% increase in volume in 100 h. The rate of growth which develops increases with increase in temperature and possibly also with increase in frequency of cycling. Although the fact that a 60% increase in volume may occur after only 140 h of cyclic reheating suggests that unalloyed iron is totally unsuited for such applications, iron is in fact extensively used under such conditions, e.g. as furnace doors and fire bars. This may be partly because these applications involve lower cycling frequencies than those which cause the very high rates of growth mentioned, but undoubtedly a major factor determining the use of unalloyed cast iron for such duties is its cheapness, which outweighs the superior growth resistance of more highly alloyed and more expensive irons.

At temperatures below the critical range, much less growth occurs, rarely exceeding 3% for 100 h of cyclic reheating. Here too the rate of growth depends on the temperature and the frequency of cycling. At temperatures below 400°C growth becomes negligible for most irons while below 350°C it is negligible for all irons. This threshold is probably related to the marked decrease in strength which occurs when irons are heated above 400°C, which results in the component being more easily distorted by the development of internal volume changes. Clearly, unalloyed irons have a very considerable usefulness up to about 700°C, and even in steam plant, where dimensional stability is important, there is a case for the use of unalloyed iron at temperatures up to 400°C.

At temperatures below the critical range, an important cause of growth is graphitisation, i.e. the decomposition of the carbide constituent of pearlite to give ferrite and graphite. Unalloyed irons usually contain up to 0·8% combined carbon and complete graphitisation of this can theoretically result in a volume increase of 1·6%[2]. This value has been confirmed by Gilbert and White[1] who have shown that ferritising a fully pearlitic iron gives a linear growth of up to 0·7% (i.e. about 2·1% volume increase). Clearly, the rate of growth due to this mechanism will be controlled by the stability of the carbide in the pearlite and this will vary with the composition of the iron. The presence of certain elements, notably silicon, decreases the stability of the carbide, while it is stabilised by the presence of other elements, notably chromium. An iron with a low silicon content and containing some chromium may thus be expected to have good growth resistance, but since excessive carbide stability can lead to a hard, brittle alloy, there is a limit to the benefit which can be derived from such stability. It should be emphasised that unless large amounts of carbide-stabilising elements are present in the iron, all that will be achieved is a slower *rate* of growth; there will not be a decrease in the total growth possible.

Phosphorus appears to have a beneficial effect on the growth rate. At sub-critical temperatures it helps to stabilise the carbide, while at temperatures up to about 900°C the presence of the hard phosphide eutectic network restricts the deformation to which the much more ductile matrix would otherwise be subject. Since the phosphide eutectic melts at about 950°C, irons containing appreciable amounts of this constituent should clearly not be exposed to this temperature.

Another cause of growth which is of equal importance with graphitisation is the penetration of oxides into the metal along the graphite flakes. This presumably takes place because oxidising gases can be adsorbed on to the graphite and so allowed access to the metal/graphite interface. Since the oxides are more bulky than the metal from which they are derived, internal stresses are set up and growth results. As might be expected, the amount of growth due to internal oxidation increases as the graphite content increases (Fig. 7.19) and also as the section size increases, since this leads to a coarsening

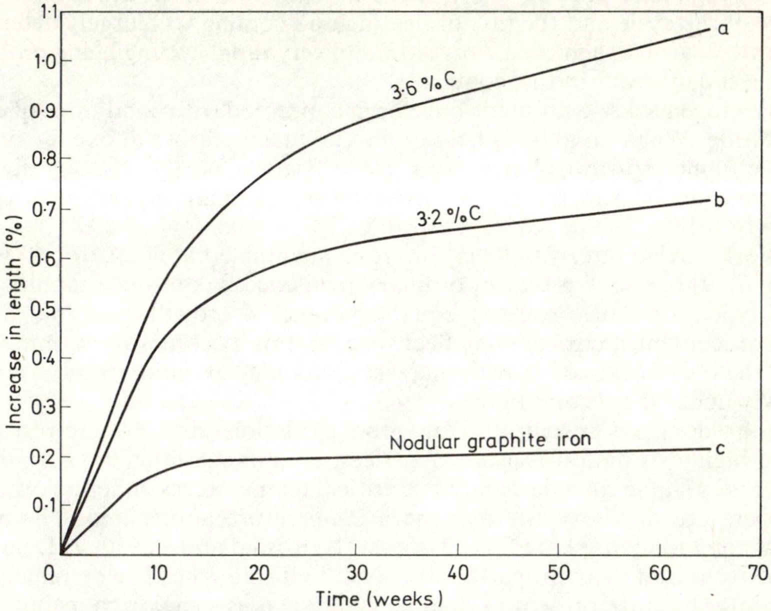

Fig. 7.19 Effect of amount and distribution of graphite on growth in air at 500°C. Curve *a*: Iron 6 (Fe–3·61C_{total}–1·63Si–0·76Mn–0·094S–0·28P). Curve *b*: Iron 1 (Fe–3·25C_{total}–1·58Si–0·65Mn–0·107S–0·25P). Curve *c*: Iron 21 (Fe–3·39C_{total}–1·73Si–0·41Mn–0·013S–0·05P–0·67Ni–0·075Mg–0·004Ce) (after Gilbert[4])

of the graphite. On the other hand, a white iron which contains no graphite is very growth-resistant since it does not readily graphitise, nor is it easily penetrated by oxidation. For similar reasons, nodular graphite irons are resistant to growth.

At temperatures above the critical range, the maximum amount of growth due to graphitisation may account for less than 10% of the total growth observed. Undoubtedly a large contribution to the total growth is made by the oxidation of the iron, since the stresses set up in the oxide layers by the differences between the expansion of the oxides and the iron during the alternate heating and cooling cycles generate cracks in the scale, which prevent the reaction from ever becoming self-stifling. The increase in oxidation rate due to the temperature rise does not, however, satisfactorily account for the marked increase in growth rates when the critical range is exceeded. Benedicks and Löfquist[2] have given an interpretation of some dilatometer curves produced by Kikuta[3] which explains the sudden increase in growth at these higher temperatures in terms of the ferrite–austenite transformation.

This explanation implies that at each complete cycle through the critical range there is a net expansion which is due to the fact that the expansion involved in the austenite to ferrite change does not balance the contraction involved in the ferrite–austenite change. For an iron containing 0·7% combined carbon initially, the net growth per cycle involved may be up to 0·5% by length (1·5% by volume). The net growth per cycle decreases with the number of cycles but the possibility that each cycle, at least initially, can contribute this amount of growth suggests that the mechanism can give rise to very high growth rates. As with sub-critical growth, the duration of each stage of the cycle and the rate of heating and cooling will largely determine the rate of growth achieved, very slow and very rapid cycling being probably the least dangerous in this context.

All the remarks so far made have been concerned with conditions of cyclic reheating. When an alloy is held at a steady temperature above the critical range, some growth will arise from graphitisation, partly offset by the contraction involved in the ferrite–austenite transformation, but most of the growth will be due to oxide penetration.

Work carried out by Gilbert[4] on irons maintained at 500°C for 64 weeks (Fig. 7.19) has shown that in ordinary unalloyed flake irons graphitisation and oxidation cause roughly equal amounts of growth, and that as the carbon content increases the effect of oxidation becomes more important and the overall rate of growth increases. Nodular graphite irons grow very slowly under these conditions.

Irons designed specifically for good oxidation- and growth-resistance have highly oxidation-resistant matrices, containing either no carbides at all or very stable carbides, and have critical temperatures either below room temperature or above the maximum temperature anticipated. The alloys most commonly used are Silal, Niresist, Nicrosilal and Fe–30Cr. Details of these irons and their properties are given in Table 7.11. The extremely fine graphite structure present in Silal probably makes a major contribution to its good heat resistance. However, when Silal is produced with nodular graphite, its heat resistance is further enhanced.

Two other alloys which have been used for their good oxidation- and growth-resistance are Cralfer (Fe–7Al–0·75Cr) and Fe–14·5Si. The production of the former, however, entails considerable difficulties while the latter has poor mechanical properties and poor resistance to thermal shock, with the result that neither is extensively used for this purpose today.

Scaling

When an iron is exposed to an oxidising atmosphere, it develops a scale which consists of a series of layers of oxides of varying composition. The thickness of the scale naturally depends on the temperature and the duration of oxidation (t). The scale does not, however, thicken at a uniform rate with time since its very presence reduces the accessibility of the metal surface to the oxidising gases. Ideally, the thickness of the scale should increase as $t^{\frac{1}{2}}$, but in practice cracks develop in the scale, and these allow the gases to reach the metal surface somewhat more readily than is postulated by this relationship. Cracking will always tend to occur as the film thickness increases, but if the iron is subjected to cyclic variations in temperature, the variations in thermal expansion and contraction of the individual constituents of the scale and the metal may be expected to increase the extent to which cracking occurs. Under such conditions the scale may become completely detached

Table 7.11 Heat-resisting irons

Name	Composition					Structure	Mechanical properties			Critical temperature (°C)	Growth characteristics
	Total carbon	Si	Ni	Cr	Cu		*Ultimate tensile strength* (MN/m^2)	*Elongation*	*Hardness* (HB)		
Silal	2·5	6·0	—		—	Fine graphite in silico-ferrite matrix	154	Nil	280	>920	Nil after $80 \times 1\frac{1}{2}$ h at 870°C (White and Elsea[5])
*Nicrosilal	2·0	5·0	20·0	2·0	—	Fine graphite in austenitic matrix with some complex carbides	216	2%	140	<20	<1% by length after 240 h at 950°C (Hallett[6])
*Niresist	3·0 3·0	2·0 2·0	20·0 14·0	2·0 2·0	— 7·0	Fine graphite in austenitic matrix with some complex carbides	232	2%	150	<20	<1% by length after 240 h at 950°C (Hallett[6])
High-chromium iron	1·5	1·0	—	30	—	Complex carbides in chromium-ferrite matrix	463	Nil	320	>1 100	<1% by length after 240 h at 1 050°C (Hallett[6])

*These irons may sometimes contain up to 5% Cr which improves the heat resistance slightly and makes the alloy stronger and harder.

from the metal, thus allowing the high initial rates of oxidation to be reproduced. With flake graphite cast iron, however, since there is a tendency for oxide penetration to occur along the graphite flakes, the scale is keyed to the metal surface rather more securely than is the case with steels and therefore becomes detached rather less readily.

The atmosphere most commonly encountered by irons at elevated temperatures is probably air contaminated with traces of carbon dioxide, sulphur dioxide and steam. Because of the difficulty of defining such an atmosphere, most of the work on scaling has been carried out with relatively pure air. Some figures quoted by various authorities for the scaling rates of unalloyed irons in clean air at various temperatures are shown in Fig. 7.20 which has been compiled from data given in Table 7.12.

Table 7.12 Scaling of unalloyed cast iron

Reference	*Composition*				*Details of test*		*Result reported*	*Scaling rate* ($mg\,dm^{-2}d^{-1}$)
	Total carbon	Si	Mn	P	*Duration*	*Temperature* (°C)		
Hallett[6]	2·98	1·14	1·07	0·18	240 h	750	9 $\times 10^3$ mg/dm^2	900
					240 h	850	21·6 $\times 10^3$ mg/dm^2	2 160
					240 h	950	45·6 $\times 10^3$ mg/dm^2	4 560
					60 h	1 050	47 $\times 10^3$ mg/dm^2	18 800
Cameron[8]	3·01	2·46	0·81	—	2 000 h	705	0·188 in/y	449*
	3·43†	2·47	0·45	—	2 000 h	705	0·053 in/y	127*
White, Rice and Elsea[7]	3·42	2·61	0·73	0·18	240 h	870	0·32 g/cm^2	3 200
Timmerbeil[9]	3·15	1·70	0·80	—	96 h	700	70 g/m^2 d^{-1}	700
					96 h	800	230 g/m^2 d^{-1}	2 300
Timmerbeil[10]	3·12	1·99	0·42	0·27	96 h	600	10 g/m^2 d^{-1}	100
					96 h	700	40 g/m^2 d^{-1}	400
					96 h	800	230 g/m^2 d^{-1}	2 300
					96 h	900	460 g/m^2 d^{-1}	4 600
Gilbert[4, 17]	3·04	1·23	0·73	0·29	6 y	400	3·6 mg/cm^2	0·17
					64 wk	500	8·4 mg/cm^2	1·9
Gilbert and White[1]	3·29	1·43	0·55	0·24	32 wk	600	4·9 mg/cm^2	22
					32 wk	650	128·7 mg/cm^2	57·5
Gilbert[4, 17]	3·39†	1·73	0·41	0·05	6 y	400	2·0 mg/cm^2	0·09
					64 wk	500	5·27 mg/cm^2	1·18
Gilbert and White[1]	3·57†	1·98	0·35	0·03	32 wk	600	43·3 mg/cm^2	19·4
					32 wk	650	138·8 mg/cm^2	61·9
Maitland[13]	3·35†	1·2	0·33	0·06	456 h	650	2·53 $\times 10^3$ g/dm^2	133
					504 h	700	5·14 $\times 10^3$ g/dm^2	245
					456 h	800	20·6 $\times 10^3$ g/dm^2	1 095
					456 h	900	42·1 $\times 10^3$ g/dm^2	2 216
	3·06†	2·0	0·33	0·06	456 h	650	2·17 $\times 10^3$ g/dm^2	114
					504 h	700	4·74 $\times 10^3$ g/dm^2	226
					456 h	800	23·3 $\times 10^3$ g/dm^2	1 226
					456 h	900	48·7 $\times 10^3$ g/dm^2	2 563

*Weight increase due to Fe_2O_3.
†Nodular graphite irons containing about 1% Ni.

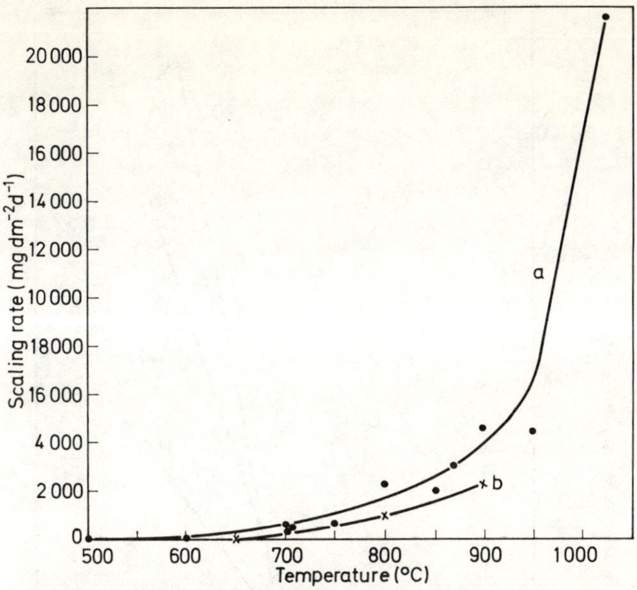

Fig. 7.20 Dependence of scaling rate on temperature (see Table 7.12). Curve *a*: flake graphite irons. Curve *b*: nodular graphite irons

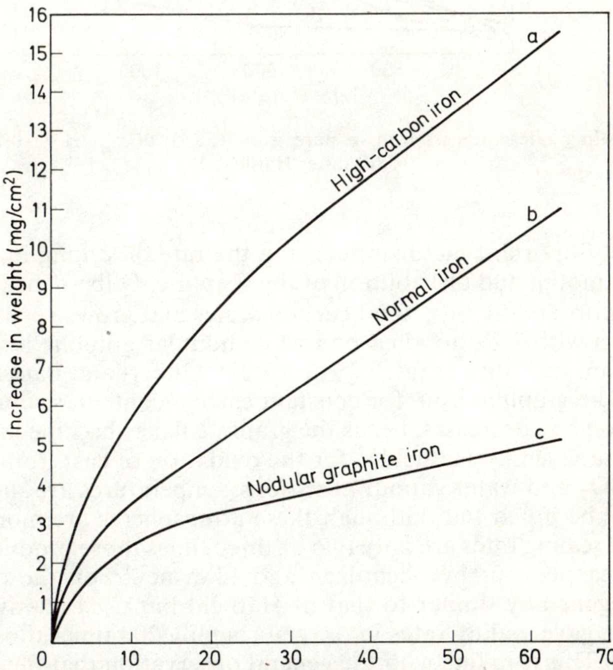

Fig. 7.21 Effect of amount and distribution of graphite on scaling in air at 500°C. Compositions as given in Fig. 7.19 (after Gilbert[4])

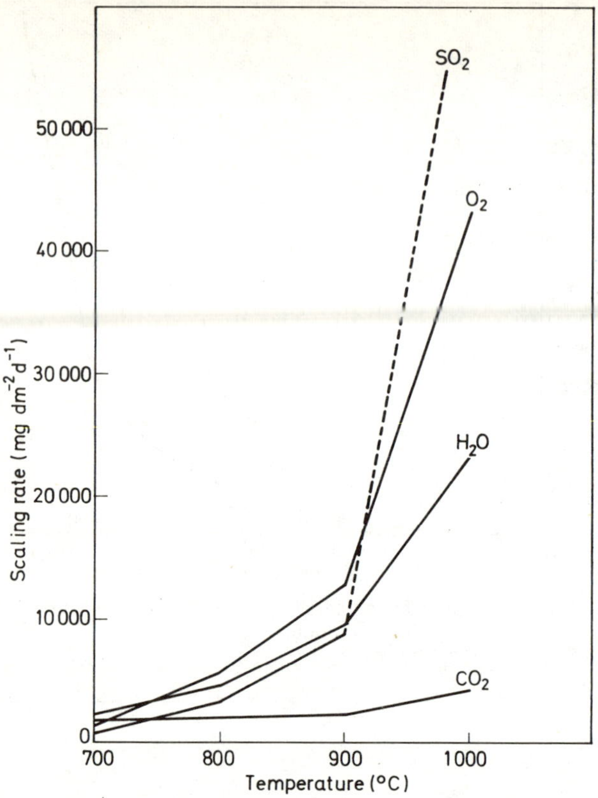

Fig. 7.22 Scaling rates for cast iron in pure gases (Fe–3·20C_{total}–1·13Si–0·72Mn–0·125S–0·58P) (after Hatfield[11])

The most important factor influencing the rate of scaling in a low-alloy iron is the amount and distribution of the graphite. Gilbert[4] has shown that at 500°C an iron with 3·6% total carbon scales and grows very much more than an iron with 3·2% total carbon while nodular graphite irons are even more resistant to scaling (Figs. 7.19 and 7.21). Gilbert and Barnes[14] showed that in a flake graphite iron, for constant carbon content, scaling decreases as the section size decreases, i.e. as the graphite flakes become finer.

Some data given by Hatfield[11] for the oxidation of cast iron in pure dry CO_2, O_2, SO_2 and water vapour at various temperatures are shown in Fig. 7.22. It will be noted that although these atmospheres are more oxidising than air the scaling rates are only two or three times those produced in clean air. Work carried out by Nicholson and Kwasney[12] on the other hand, which was generally similar to that of Hatfield but used relatively impure moist gases, gave scaling rates in oxygen roughly 200 times those observed by Hatfield. This is in line with the general observation that the presence of traces of gases such as steam, carbon dioxide and sulphur dioxide in air can appreciably increase the scaling rate. No final explanation of this effect has yet been produced, but it is possible that it is connected with the development

Table 7.13 Recommendations for the use of austenitic cast irons at 650–800°C (after Cox and Gilbert[16]). Suitabilities indicated by graded code: 1 = very poor, 5 = excellent

Reference No. (BS 3468)	*Typical composition (%)*					*Heat resistance*		*Effect on mechanical properties*				*Overall suitability*
	C_{total}	Si	Ni	Cr	Cu	Growth	Scaling	Strength	Elongation	Impact	Hardness	
Flake graphite												
AUS 101	2·5	2·0	15	2·0	7·5	1	2	None	None	Increased	None	1
AUS 102	2·5	2·0	20	2·0	—	1	2	None	None	Increased	None	1
AUS 104	2·0	5·5	20	3·0	—	4	4	None	None	Slightly increased	Increased	4
AUS 105	2·0	2·0	30	2·5	—	3	3	None	None	Slightly increased	Slightly increased	3
Nodular graphite												
AUS 202	2·5	2·0	20	2·0	—	2	1	None	None	Increased	None	1
AUS 203	2·5	2·0	22	—	—	2	1	None	None	Increased	None	1
AUS 204	2·5	5·5	20	2·0	—	5	4	None	None	Slightly increased	Increased	5
AUS 205	2·0	2·0	30	2·5	—	4	3	None	None	Slightly increased	None	4

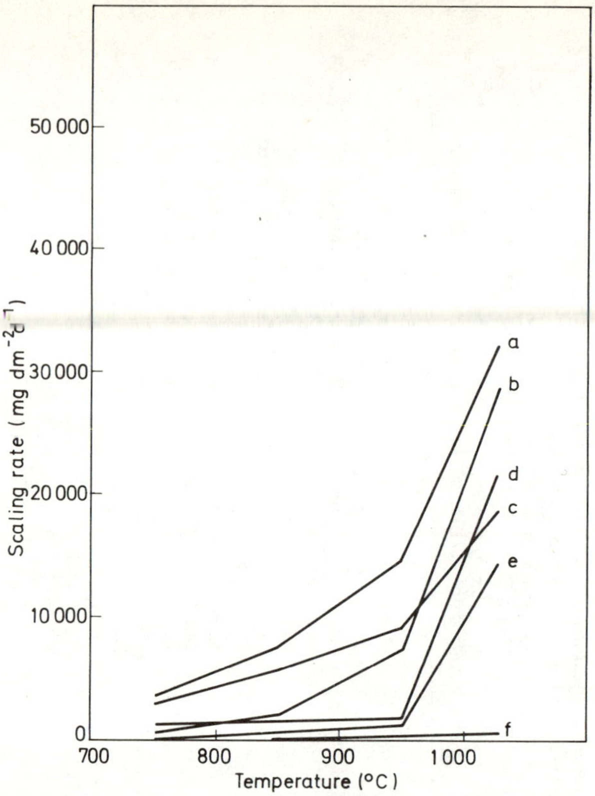

Fig. 7.23 Scaling resistance of high-alloy irons (after Hallett[6]). Curve *a*: Niresist (Fe–14·6Ni–2·1Cr–7·2Cu). Curve *b*: Silal (Fe–5·7Si). Curve *c*: Niresist (Fe–14·4Ni–4·8Cr–7·4Cu). Curve *d*: Nicrosilal (Fe–4·6Si–22·5Ni–2·5Cr). Curve *e*: Nicrosilal (Fe–4·9Si–22·8Ni–4·6Cr). Curve *f*: (Fe–1·3C–33Cr)

of a more porous scale. Such a scale might be expected to result if the first product of the reaction of the atmosphere with the metal were a compound, e.g. a hydroxide or carbonate, which subsequently decomposed to give the oxide.

The more highly alloyed materials, already mentioned as having useful resistances to growth, have also good resistances to scaling. Figure 7.23 shows the relative resistance of some of these alloys to oxidation. It will be noted that the increase in the chromium content of the austenitic irons from 2 to 5% has had a beneficial effect upon the scaling resistance of these irons. There is some evidence to suggest that the scaling resistance of these irons is further improved if they contain nodular instead of flake graphite[15].

Cox and Gilbert[16] have given a detailed account of the behaviour of the austenitic irons when exposed for 32 weeks to temperatures ranging from 320°C to 800°C, with particular reference to the effect of this exposure on their mechanical properties. From this work they have formulated a series of recommendations that are dependent upon the service temperature range. The recommendations for the range 650–800°C are given in Table 7.13.

H. H. COLLINS

REFERENCES

1. Gilbert, G. N. J. and White, D. G., *J. Res. Brit. Cast Iron Res. Assn.*, **11**, 295 (1963)
2. Benedicks, C. and Löfquist, H., *J. Iron St. Inst.*, **115**, 603 (1927)
3. Kikuta, T., *Sci. Rep. Tohoku Univ.*, **11**, 1, April (1922)
4. Gilbert, G. N. J., *J. Res. Brit. Cast Iron Res. Assn.*, **7**, 478 (1959)
5. White, W. H. and Elsea, A. R., *Trans. Amer. Foundrym. Soc.*, **57**, 459 (1949)
6. Hallett, M. M., *J. Iron St. Inst.*, **170**, 321 (1952)
7. White, W. H., Rice, L. P. and Elsea, A. R., *Trans. Amer. Foundrym. Soc.*, **59**, 337 (1951)
8. Cameron, J. A., *Corrosion*, **10**, 295 (1954)
9. Timmerbeil, H., *Giesserei*, **38**, 25 (1951)
10. Timmerbeil, H., Clas, G. and Mattern, O., *Giesserei*, **38**, 476 (1951)
11. Hatfield, W. H., *J. Iron St. Inst.*, **115**, 483 (1927)
12. Nicholson, J. H. and Kwasney, E. J., *Trans. Electrochem. Soc.*, **91**, 681 (1947)
13. Maitland, R. J. and Hughes, I. C. H., *J. Res. Brit. Cast Iron Res. Assn.*, **7**, 203 (1958)
14. Gilbert, G. N. J. and Barnes, G. M., *J. Res. Brit. Cast Iron Res. Assn.*, **14**, 9 (1966)
15. Sefing, F. J., *Trans. Amer. Foundrym. Soc.*, **63**, 638 (1955)
16. Cox, G. J. and Gilbert, G. N. J., *J. Res. Brit. Cast Iron Res. Assn.*, **14**, 423 (1966)
17. Gilbert, G. N. J., *J. Res. Brit. Cast Iron Res. Assn.*, **14**, 555 (1966)

7.4 High-Alloy Steels

Introduction

The basic importance of chromium in relation to passivity in the resistance of steels to aqueous corrosion is parallelled by its rôle as the most important alloying element in practice in securing the high-temperature oxidation resistance of heat-resisting steels. It is not surprising, therefore, that these two aspects of corrosion resistance developed together, and that early investigators of the chromium–iron alloys showed interest in the potential usefulness of these materials for high-temperature oxidation resistance. Thus Chevenard[1], discussing early studies in France, has indicated that the scaling resistance of Fe–Ni–Cr alloys was appreciated at least as early as 1917.

In the UK the rapidly growing interest in high-alloy steels after World War I coincided with expanding needs of industry for materials of improved strength and oxidation resistance at elevated temperatures. Aero-engine developments, rising steam temperatures and pressures in power production, and the increasing importance of high-temperature, high-pressure processes in oil cracking and synthetic chemistry, were all contributory factors in the development of heat-resisting steels in the decade from 1920.

Aitchison[2], writing in 1919 on valve steels for internal combustion engines, gave scaling curves indicating the relative resistance of Fe–12Cr steel, rising to a factor of 10 over ordinary steel at 950°C. In 1922 Dickenson[3] included a section on scaling resistance in a paper dealing with the behaviour of steels at low red heat. He tested various steels in gas and electric muffles at temperatures up to 1 100°C, and the curves produced by him showed considerable improvement over ordinary steel for Fe–15Cr and especially Ni–23Fe–12Cr.

In America, the 1924 A.S.T.M. Symposium on Corrosion and Heat-resisting Alloys contained a particularly interesting paper by Fahrenwald[4]. Tests using gas-fired muffles with excess air over the temperature range 820–1 260°C were described; the results showed clearly the advantages of chromium and chromium–nickel high-alloy steels, including Fe–13Cr, Fe–18Cr, Fe–25Cr and Fe–36Ni–17Cr. The particular value of Fe–25Cr for sulphur roasting plants was pointed out, with a warning against the use

Table 7.14 Effect of various gases on the oxidation of metals and alloy steels* (rates in $g\,m^{-2}\,d^{-1}$)

Material	O_2			H_2O			CO_2			SO_2			*Complex gas*†
	700°C	800°C	900°C	700°C	800°C	900°C	700°C	800°C	900°C	700°C	800°C	900°C	900°C
Iron	510	600	1 240	620	850	580‡	590	720	1 130	350	790	7 400	1 140
Nickel	10	10	30	3	4	10	4	9	40	920	3 660§	840‖	50
Chromium	5	10	20	1	4	10	3	3	10	2	4	30	20
Fe–37Ni	40	100	140	30	130	360	30	100	250	200	1 410	1990	300
Fe–8Cr–3Si	3	20	20	3	5	9	2	2	4	5	10	50	10
Fe–13Cr	2	20	30	5	6	150	8	9	160	4	10	100	150
Fe–19Cr	6	7	20	2	6	20	2	8	40	5	5	8	10
Fe–18Cr–8Ni	10	10	30	2	5	6	4	10	30	10	20	20	30

*Data from Hatfield[7].
†Composition: 73% N_2, 12% CO_2, 10% H_2O, 5% O_2 and 0·05% SO_2.
‡(?)
§Scale disintegrating.
‖Scale firmly adherent.

of nickel steels for this purpose and an indication that under oxidising conditions attack by alkali salts from the fluxing of the protective chromium oxide scale on the high-chromium steels might be expected.

This symposium also introduced a paper by Johnson and Christiansen[5] on steels for internal-combustion-engine valves, which discussed the merits of Fe–14Cr, Fe–9Cr–3·5Si, Fe–19Ni–14Cr, Ni–27Fe–18Cr and Fe–25Ni–18Cr–3Si.

A third paper in the 1924 symposium, by Vanick[6], dealt with deterioration resulting from attack by hot reducing ammonia gases, with particular reference to the rôle of hydrogen in producing intergranular fissuring.

The first systematic study of the effects of individual and combined combustion gases on pure metals and a wide range of alloy steels was carried out by Hatfield[7] and published in 1927. The gases included oxygen, steam, carbon dioxide, sulphur dioxide and a complex gas which simulated the mixture obtained from combustion of ordinary fuels. A selection from the data is given in Table 7.14 to illustrate the main points. They show the major importance of chromium, with added effect from silicon. It was noted that the presence of the nickel in Fe–18Cr–8Ni did not lead to breakdown in sulphur dioxide in spite of the very severe attack by this gas on pure nickel. The various gases behaved very similarly with regard to extent of attack on Fe–19Cr and Fe–18Cr–8Ni.

The general pattern of alloy composition in relation to scaling resistance was well established by the end of the 1920s, and later development has been largely on the lines of modification to provide the best combinations of strength, ease of fabrication and oxidation resistance to meet the expanding requirements of industry and power generation. In particular, the increasing importance of high strength and low deformation at high temperatures has resulted in the development of a large number of complex alloys combining high creep strength with adequate oxidation resistance.

Typical Heat-resisting Steels

In that they show a superior resistance to oxidation at elevated temperatures over low-alloy types, all the steels listed in BS970, Part 4:1970, *Stainless, heat-resisting and valve steels; blooms, billets, bars and forgings* and BS1449, Part 4:1967, *Stainless and heat-resisting steel plate, sheet and strip*, may be termed *heat-resisting* steels. Many of them are indeed used for this purpose and those used, perhaps, most commonly for this purpose are listed in Table 7.15. Of these, only 309S24 and 310S24 were originally developed with oxidation resistance primarily in mind, but several others have come to be used because they do exhibit a number of useful properties.

The metallurgy of the three groups (martensitic, ferritic and austenitic) has been discussed in an earlier section and will not therefore be dealt with any further here. Only one of the martensitic group is listed since the resistance to oxidation of the rest included in the standards is essentially similar. While having useful properties, applications of this group are limited because of fabrication difficulties, notably in welding, for which reason the lower carbon grades are preferred. While the 11–13% Cr imparts an attractive oxidation resistance, the engineering value of any material is also determined

Table 7.15 Specifications of steels used for high-temperature applications

Steel number	C	Si	Mn	Ni	Cr	Mo	Ti	Nb	*Type*
410S21	0·09–0·15	0·80 max	1·00 max	1·00 max	11·5–13·5	—	—	—	Martensitic
430S15	0·10 max	0·80 max	1·00 max	0·50 max	16·0–18·0	—	—	—	Ferritic
442S19	0·10 max	0·80 max	1·00 max	0·50 max	18·0–22·0	—	—	—	Ferritic
304S15	0·06 max	1·00 max	0·50–2·00	8·0–11·0	17·5–19·0	—	—	—	Austenitic
321S12	0·08 max	0·20–1·00	0·50–2·00	9·0–12·0	17·0–19·0	—	5 × % C–0·70	—	Austenitic
347S17	0·08 max	0·20–1·00	0·50–2·00	9·0–12·0	17·0–19·0	—	—	10 × % C–1·00	Austenitic
316S16	0·07 max	0·20–1·00	0·50–2·00	10·0–13.0	16·5–18·5	2·25–3·00	—	—	Austenitic
309S24	0·15 max	0·20–1·00	0·50–2·00	13·0–16·0	22·0–25·0	—	—	—	Austenitic
310S24	0·15 max	0·20–1·00	0·50–2·00	19·0–22·0	23·0–26·0	—	—	—	Austenitic

by its strength and the simple 410S21 type alloy has been used as the basis for development of a series of higher strength heat-resisting steels, data for some of which will be given later.

The ferritic steel 430S15 has enhanced oxidation resistance and finds some applications in sheet form, but its strength at elevated temperature is low. The higher chromium (20–30%) ferritic types show excellent oxidation resistance, but have poor elevated-temperature strength and being difficult to produce and fabricate, are not used in large quantity. Cast versions of 27–30% Cr are quite widely used, especially where oxidation resistance, coupled with abrasion resistance, is required when high carbon contents are utilised. Such alloys are brittle.

The 18% Cr austenitic steels are quite widely used because of their ease of production and fabrication as well as their high-temperature properties. Those listed in Table 7.15 are most commonly used. The stabilised types (321S12 and 347S17) are preferred where heating to temperatures likely to cause susceptibility to intercrystalline corrosion (see p. **3**.44) is interspersed with conditions where aqueous corrosion could be a hazard. Very low carbon grades are not usually used owing to their somewhat inferior strength. The higher chromium alloys 309S24 and 310S24 were developed to provide alloys with oxidation resistance superior to that of the 18% Cr type. Where strength at very high temperatures (>800°C) is required a high-carbon cast version of type 310S24 may be used where other requirements allow it. This type of alloy will not be discussed here, but information is given in Reference 8.

Physical Properties

Data for a variety of alloys have already been given in Section 3.3, page **3**.34.

Mechanical Properties

The martensitic steels are used in the hardened and tempered conditions, the tempering temperatures used obviously being in excess of the proposed service temperature. Alloying elements used to improve the creep strength include molybdenum, vanadium, niobium and cobalt and these also have the effect of increasing the resistance to softening or tempering so that the proof strengths of the creep-resisting variants are substantially higher than those for the simple Fe–13Cr alloy. 0·2% proof strengths at various temperatures for several martensitic types are shown in Fig. 7.24. Creep and long-term rupture strengths are shown in Figs. 7.25 to 7.27 as stress to give 0·1% total plastic strain in 10 000 h and as stress to give rupture in 10 000 and 100 000 h plotted as a function of temperature.

The ferritic steels rapidly lose strength at elevated temperature as shown in Fig. 7.28 and are of little value for load-bearing applications.

The austenitic grades, used mainly in the solution treated (softened) state, have low strength at ambient temperature but maintain strength at elevated temperatures much better than the martensitics and the ferritics. As can be

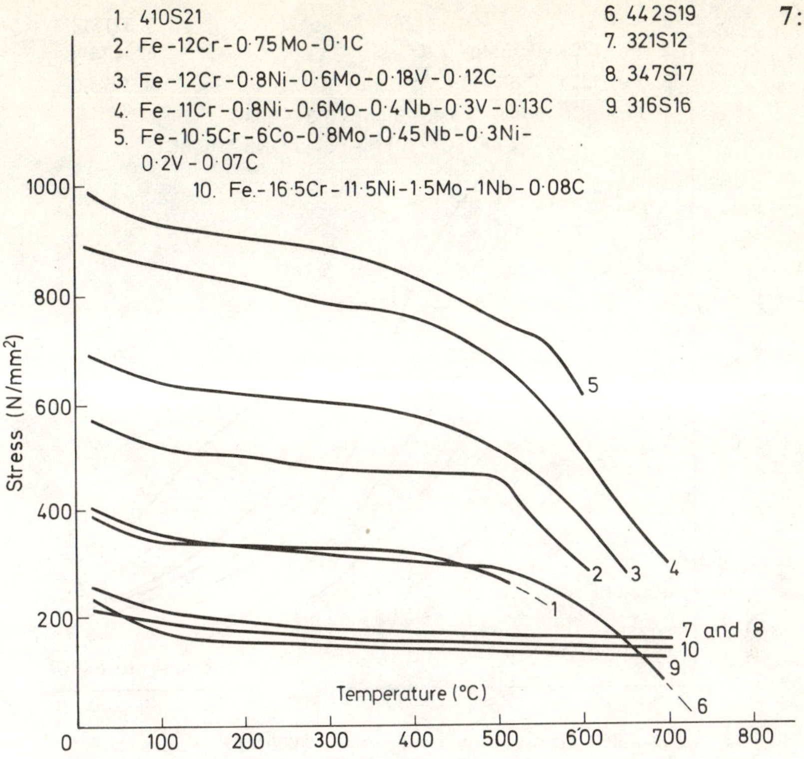

Fig. 7.24 0·2% proof stress values for various steels

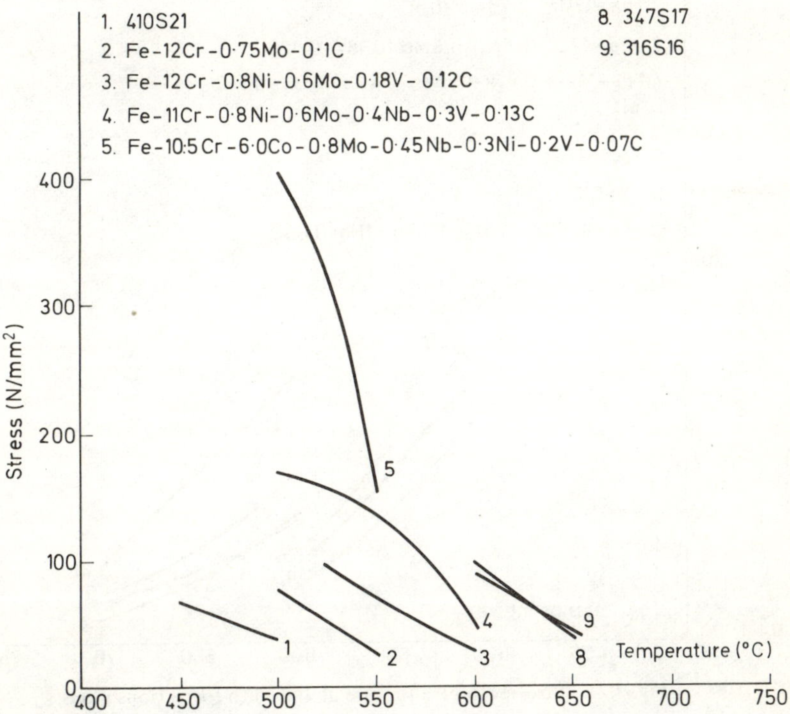

Fig. 7.25 Stresses to give 0·1% total plastic strain at 10 000 h for various steels

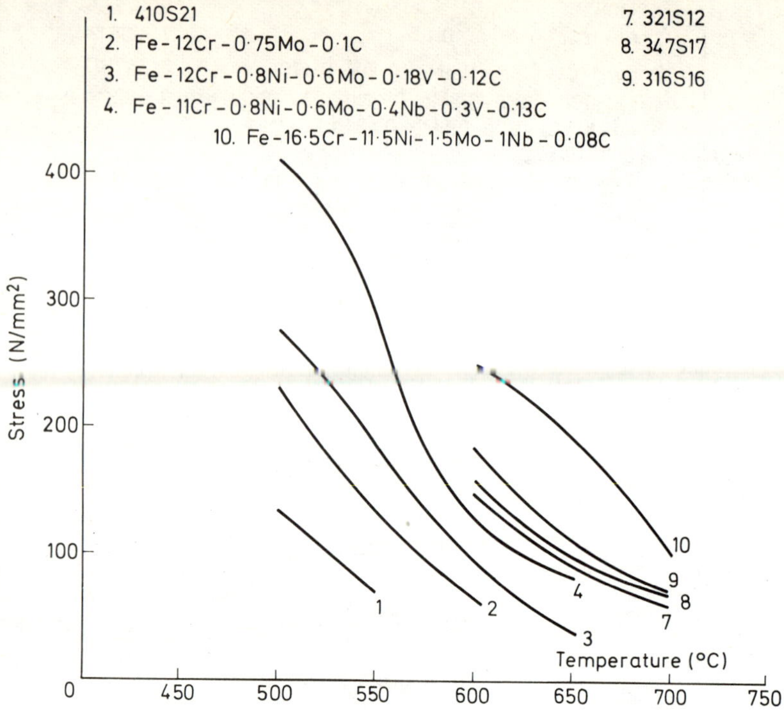

Fig. 7.26 Stresses to give rupture at 10 000 h for various steels

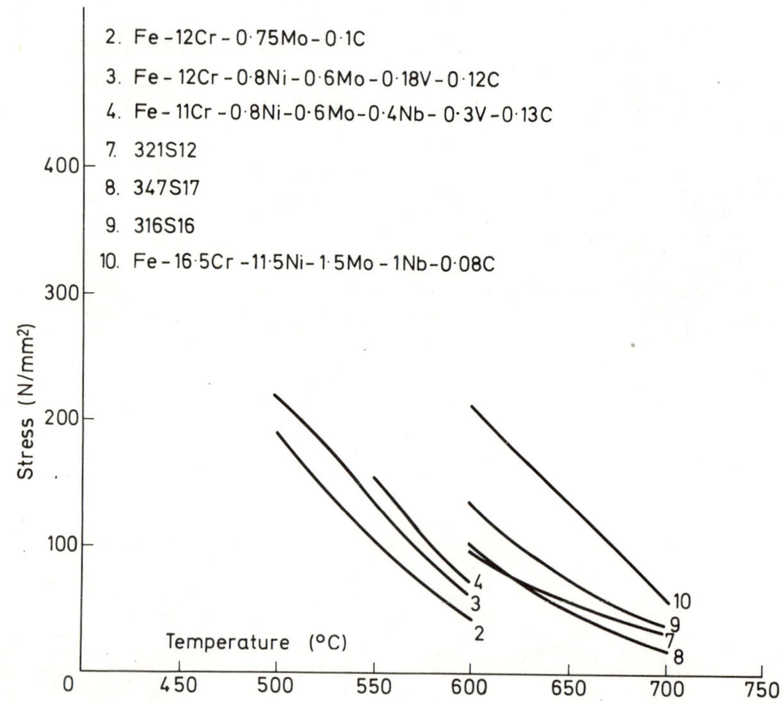

Fig. 7.27 Stresses to give rupture at 100 000 h for various steels

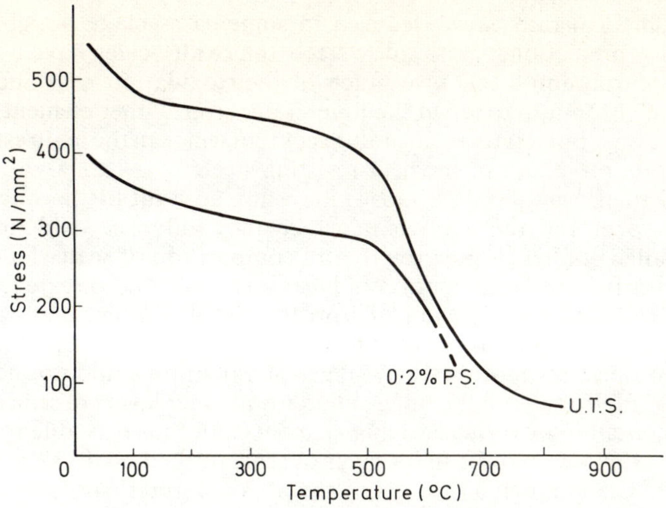

Fig. 7.28 Tensile strength of ferritic stainless steel 442S29

seen from Figs. 7.25 to 7.27, creep and rupture strengths are far superior above 600°C. Below this temperature proof strength is the limiting factor, but this can be improved somewhat in certain grades by alloying with nitrogen.

No more than a brief outline of the mechanical properties can be given here, for detailed information Reference 9 should be consulted. It should be noted that while steels used for creep resisting purposes may conform to the standard specifications, sometimes specially limited composition ranges within these specifications are used in the interests of strength, structural stability or resistance to embrittlement.

Fabrication

Forming and fabrication characteristics are described in Section 3.3 on stainless steels. Creep-resisting steels are, of course, intended to resist deformation at elevated temperatures, but in fact the mechanical power required for deformation at the forging temperature is little greater than that required for the stainless steels.

Creep-resisting steels often have to be used in thicker sections than is the case with stainless types and this can lead to the need for special techniques for forming and welding.

High-temperature Corrosion

Scale Structure and Oxidation Rates

Since the paper by Pilling and Bedworth[10] in 1923 much has been written about the mechanism and laws of growth of oxides on metals. These studies have greatly assisted the understanding of high-temperature oxidation, and the mathematical rate 'laws' deduced in some cases make possible useful quantitative predictions. With alloy steels the oxide scales have a complex structure: chromium steels owe much of their oxidation resistance to the presence of chromium oxide in the inner scale layer. Other elements can act in the same way, but it is their chromium content which in the main establishes the oxidation resistance of the heat-resisting steels.

In 1929 Pfeil[11] published a most interesting account of the way layered structures form and the manner in which they influence oxidation rates. From detailed studies of the growth and composition of scales he was able to show clearly how the formation of barrier layers reduced scale formation by hindering outward diffusion of iron through the scale. Naturally, this work had to be largely based on the study of scales of sufficient thickness so that the mechanism of the early stages of oxidation could not be studied in this way. Pfeil analysed the outer, middle and inner layers of scales formed on steels containing various alloying elements and thus was able to demonstrate, for example, that the inner layer of scale formed on a 13% Cr steel in air at 1 000°C contained 34% Cr_2O_3, the middle layer 1·64% and the outer layer 0·89%.

The development of the electron-probe microanalyser has given research workers a powerful tool with which to determine composition variations of scale layers and also of underlying metal. Wood and his co-workers have used this instrument to great advantage to help explain the behaviour of iron–chromium alloys[12–14]. He found that on oxidation at high temperature (1 100°C) protective scales are largely Cr_2O_3, containing small proportions of iron oxide. The formation of the Cr_2O_3 causes depletion of the subjacent alloy in chromium, and the chromium content of the alloy at the interface may be as low as 5·3% without apparent transformation of the Cr_2O_3 to a spinel. Failure of the protective Cr_2O_3 scale with time is considered due to scale lifting and cracking followed by rapid oxidation of the depleted alloy. As the catastrophic break-through progresses the content of the inner layer is diluted to a limiting value of 20–25% Cr, while an outer, virtually pure iron oxide layer develops and the depleted subjacent alloy is almost entirely eliminated. These findings largely explain the observed behaviour of chromium-rich steels at temperature in oxidising atmospheres. Oxidation rates increase only relatively little with increasing temperature until, above some temperature, they increase rapidly. The exact value of this 'breakdown' temperature depends on alloy content but can also be affected by other features external to the alloy. For a detailed review of the mechanism of oxidation of chromium-bearing steels with a comprehensive bibliography Reference 15 should be consulted.

Attack by Gases

Flue Gases

There is naturally a desire with corrosion by oxidation to produce in the laboratory quantitative data which can be used for design purposes. This can be a more questionable procedure in the field of oxidation than is even the case with 'wet' corrosion since so many features can affect the results obtained. Apart from variations in atmosphere composition, gas velocity, rate of heating or cooling, frequency of thermal cycling, method of sample preparation, sample geometry and time of test can all have marked effects on corrosion rate measured under some circumstances. Most laboratory tests are useful mainly in allowing comparison of alloys and a general assessment of the range of temperatures over which an alloy may be useful, although it must be recognised that ideally the behaviour of an alloy selected for service should be further checked under conditions closely simulating service, especially if envisaged service is near the temperature ceiling for the alloy indicated by the laboratory tests.

A test procedure which has proved very useful was first described by Hatfield[16]. The samples are cylinders with a standard abraded finish 28 × 12·5 mm in diameter which are supported on open-ended refractory boats in a tubular furnace. In the original test the atmosphere, which was produced by burning towns gas with a 50% excess of air, was passed over the specimens at a standard velocity after first preheating to test temperature over refractory packing in a separate furnace chamber. More latterly, natural gas has been used with suitable modification of air : gas ratio to give a generally similar test atmosphere. No significant variation in results has been noted. Test times are short, but thermal cycling is incorporated so that any disruptive effects of differential expansion and contraction of metal and oxide may operate. The specimen is heated to temperature for seven 6-h cycles with intermediate cooling to room temperature and weighing, together with any loose scale shed during cooling. Before each test cycle the specimen is lightly brushed and reweighed. The gain in weight over the whole seven cycles is taken as an indication of oxidation resistance or scaling 'index'. Variation of index with temperature for several steels is shown in Fig. 7.29.

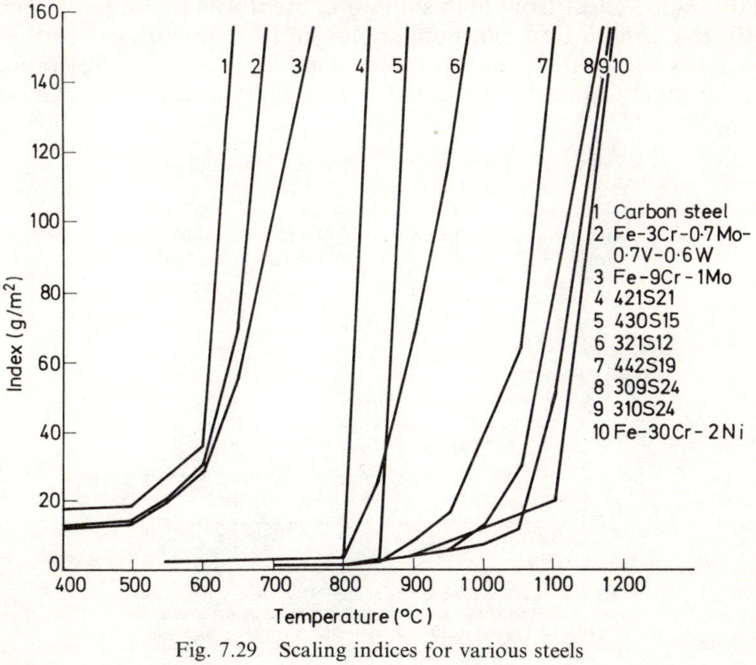

Fig. 7.29 Scaling indices for various steels

As can be seen, each steel has a critical temperature above which the oxidation rate accelerates rapidly, and it is apparent that the safe maximum service temperature must lie somewhat below this temperature, which has been determined in the laboratory. It should also be remembered that these values are for one melt of each steel, and so are typical of that steel, but that slightly different results could be obtained with other casts as there must be some tolerance of composition in practice (see Table 7.15). The results may be also taken as indicative of the behaviour of creep-resistant variants of the various basic types.

Although behaviour does not show any great variation over quite a wide range of oxidising atmospheres, the air:fuel ratio can exert some effect. This was shown by tests on an Fe–18Cr–8Ni alloy at 850°C[17]. The steel was submitted to a test of the type just described, but in atmospheres produced by catalytic burning of 2:1, 4:1 and 6:1 air:towns gas, the 4:1 ratio corresponding to stoichiometric combustion. Mean oxidation gains expressed as g/m^2 were:

6:1 air:gas	32
4:1 air:gas	12
2:1 air:gas	4

At temperatures below 850°C this effect of varying air:gas ratio tended to disappear and it was not apparent in tests at 750°C.

The effect of sulphur from the gas phase is critically dependent on the effectiveness of fuel combustion. With good combustion to the limit of the available oxygen, and even down to 50% air deficiency, no serious effect was found from high-sulphur fuel in tests with 321S12 steel up to the usual limit of service temperature at 850°C, as shown in Table 7.16. With 310S24 steel at 1 100°C some effect from high sulphur content was found in 2:1 air:gas with effective combustion, but none in 4:1 and 6:1 mixtures.

With poor combustion, on the other hand, very severe acceleration of attack, dependent upon the formation of sulphide in the scale, can occur.

Table 7.16 Comparison of high- and low-sulphur fuel in tests at 850°C with 321S12 steel[17]

Air:gas ratio	*Sulphur in gas**	Oxidation gain in seven 6-h *test cycles* (g/m^2)
6:1	Low	26
	High	27
4:1	Low	17
	High	21
2:1	Low	10
	High	12

*The towns gas contained 460 mg of S/m^3. High sulphur addition was made in the form of H_2S equivalent to 0·5% SO_2 in the 6:1 combustion products, i.e. about three times as much as from ordinary high-sulphur fuels.

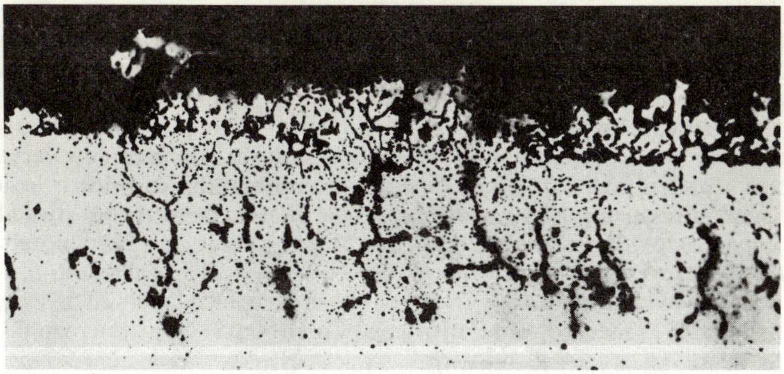

Fig. 7.30 Unetched section through Fe–25Cr–21Ni after attack in 4:1 air:gas+H_2S at 1 100°C with poor combustion, showing sulphide penetration; ×150

This destroys the protective action of the scale, and results in sulphide penetration of the metal in advance of oxidation. The effect is illustrated by tests with 310S24 steel in stoichiometric 4:1 air:gas at 1 100°C, in which the burner was modified to give incomplete combustion. The tests were for two 6-h cycles, and oxidation gains were (*a*) good combustion, high-sulphur fuel, 16 g/m^2; (*b*) bad combustion, low-sulphur fuel, 12 g/m^2; (*c*) bad combustion, high-sulphur fuel, 319 g/m^2. Sulphide penetration into the metal under condition (*c*) is illustrated in Fig. 7.30.

Air

Air tends to be less aggressive than the flue gas used for the standard test described earlier, but the useful range of temperatures for each steel is effectively similar. Edwards and Nicholson[18] reported some long-term testing of four austenitic grades in air saturated with water (at room temperature) at temperatures of 650°C–875°C for up to 10 000 h. They make the point, and show convincingly, that in any long-term assessment, metal wastage must not be based on scaling alone but that the effect of subsurface penetration must also be considered (this applies also, of course, to testing in other gases). Thus their values (Fig. 7.31) are compounded from surface

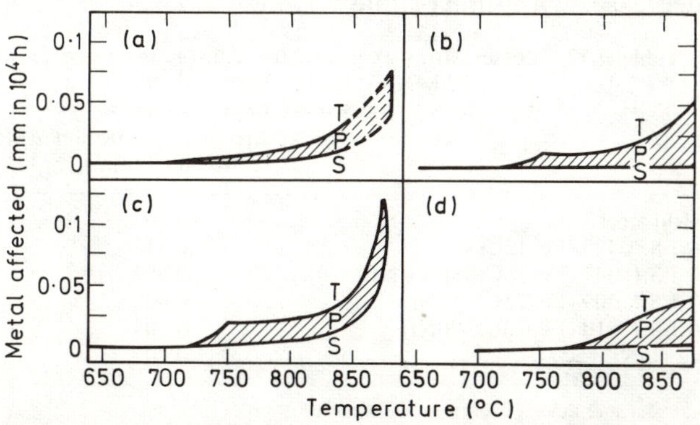

Fig. 7.31 Metal wastage of several steels due to oxidation in air (saturated with water at room temperature) for 10 000 h at various temperatures. (*a*) Type 302, (*b*) type 321, (*c*) type 316 and (*d*) type 310. $T_{total} = S_{surface} + P_{penetration}$ (after Edwards and Nicholson[18])

loss and subsurface penetration, and their work is especially valuable in that an assessment of the effect of long-term heating on mechanical properties was also made.

Nitrogen can be absorbed from air during prolonged heating, but with steel in the unstressed or lightly stressed state the rate is very slow except for temperatures above 1 050°C. Considerable absorption can occur at lower temperatures during creep, however. This fact is presumably due to the exposure of oxide-free surface during creep, and it has been noted that nitrogen absorption is especially marked at cracks. The following nitrogen contents have been reported[19] for the 0·75 mm surface layers of creep specimens in 347S17 steel (original nitrogen content 0·053%) after creep failure:

650°C (life 10 970 h under 108 MN/m^2), 0·077% N_2
700°C (life 37 958 h under 46 MN/m^2), 0·65% N_2
800°C (life 16 629 h under 1·5 MN/m^2), 0·90% N_2.

Steam

Modern boiler developments involving increased steam temperatures and pressures have made it increasingly important to consider the behaviour of high-alloy steels under conditions typical for superheater tubes and steam pipes. For satisfactory service the steels must, of course, possess adequate mechanical properties, especially creep resistance, but they must also be sufficiently resistant to oxidation to ensure long life. Short-term laboratory tests are of value in yielding comparative data for different steels and, in fact, results generally similar to those for the flue gas test already described are obtained, but prolonged tests approximating more nearly to service conditions are desirable.

Rohrig, van Duzer and Fellows[20] exposed samples in an experimental superheater fed with steam at 2·6 MN/m^2 from a power plant. Some 42 materials were tested for periods of up to 16 000 h, attack being estimated after test by weight loss following descaling. It was concluded that at 593°C attack continues at a high rate on carbon steel, whereas the rate for most alloy steels decreases with time (Table 7.17).

Table 7.17 Losses from exposure for 7 461 h in steam at 2·62 MN/m^2 and 593°C*

Steel	*Calculated penetration in* 10 000 h (mm)
Mild steel	0·107
A.I.S.I. 403 (Fe–12Cr)	0·015
A.I.S.I. 347 (Fe–18Cr–8Ni + Nb)	0·002
A.I.S.I. 309 (Fe–25Cr–12Ni)	0·002
A.I.S.I. 310 (Fe–25Cr–20Ni)	0·003
Fe–35Ni–15Cr	0·002

*Data after Rohring, van Duzer and Fellows[20].

Eberle, Ely and Dillon[21] tested commercial tubes in a small superheater receiving plant steam at 14 MN/m^2 and superheating it from 538°C to

Table 7.18 Comparison of internal and external scaling of superheater tubes after 6 950 h in steam at 13·8 MN/m^2 and 500–670°C*

Steel	*Estimated penetration* (mm/y) *Steam*	*Flue gas*
A.I.S.I. 304 (Fe–18Cr–8Ni)	0·038	0·021
A.I.S.I. 321 (Fe–18Cr–8Ni + Ti)	0·038	0·023
A.I.S.I. 347 (Fe–18Cr–8Ni + Nb)	0·010	0·023
A.I.S.I. 318 (Fe–16Cr–13Ni–3Mo + Nb)	0·013	0·029

*After Eberle, Ely and Dillon[21] and relating to tests in a superheater raising steam from 540°C to 670°C.

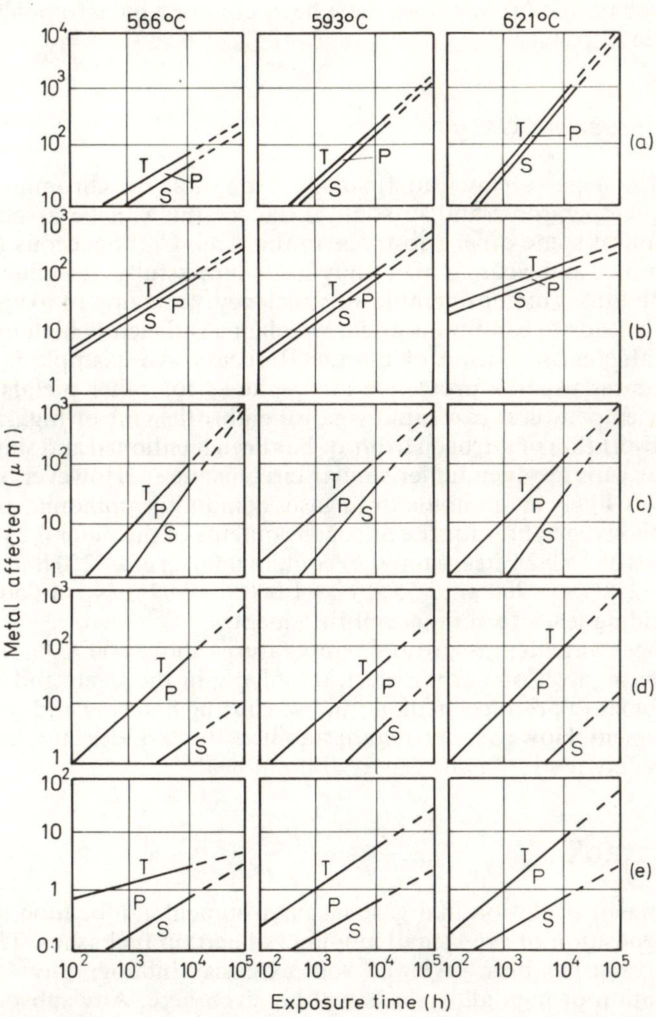

Fig. 7.32 Metal wastage of several steels due to oxidation in steam at various temperatures. (*a*) Mild steel, (*b*) Fe–2Cr–0·25Mo, (*c*) Fe–12Cr + Mo + V, (*d*) A.I.S.I. 316 and (*e*) Fe–18Cr–12Ni–1Nb. $T_{total} = S_{surface} + P_{penetration}$ (after King *et al.*)

677°C. Penetration was estimated from scale thickness measurements after 6 950 h and comparison was made between the attack by steam on the inside of the tubes and that by flue gas from pulverised coal firing on the outside (Table 7.18).

A collaborative test programme covering low-alloy and high-alloy steels was carried out by the Central Electricity Generating Board and various steelmakers. Samples were exposed in specially constructed chambers held at 566°C, 593°C and 621°C fed with power-station steam at a pressure of 3·45 MN/m^2 for times of up to 16 286 h. In the assessment of the results both metal lost from the surface and subsurface penetration were measured. The results are not yet published but have been reported by King, Robinson, Howarth and Perry in a C.E.G.B. report. Selected data are shown in Fig. 7.32, in which the broken lines have been obtained by extrapolation of the experimental results.

Other Industrial Gases

All oxidising gases can lead to oxide formation on chromium steels at elevated temperatures and in some instances this can be associated with absorption of some other substance in the steel. Carbonaceous gases are a good example and whereas high-alloy steels successfully resist flue gases even under conditions of considerable air deficiency, reduction of oxygen content eventually leads to conditions under which at a sufficiently high temperature considerable carburisation of the metal occurs. An example is the endothermic gases used as protective atmospheres for other metals which, at elevated temperature, can rapidly cause embrittlement of high-alloy steel.

The absorption of nitrogen from air has been mentioned and similar effects can occur with nitrogen under similar circumstances. However, nitriding is much more likely in ammonia or in gases containing ammonia, as indicated by the following figures for the nitrogen contents of the outer 0·25 mm layers of samples of 310S24 steel (initial N content 0·06%) after 250 h in ammonia: 500°C, 0·25% N_2; 900°C, 0·55% N_2; 1 000°C, 0·92% N_2; 1 050°C, 1·19% N_2. Nitriding leads to serious embrittlement.

Hydrogen at high pressure and temperatures above 400°C has a considerable adverse effect on carbon steel, dissolving in the steels and combining with carbides to produce methane and so causing fissuring and considerable embrittlement. However, chromium stabilises the carbides and stainless steel may be safely used in hydrogen at dull red heat[22].

Ash Attack

The degree of oxidation in a gaseous environment can be modified greatly by the deposition of even small amounts of certain fuel ashes. The topic of ash corrosion has been reviewed with extensive bibliography[23], but some consideration of high-alloy steels will be given here. Any substance which can form a low melting point mixture with the normally protective oxide scale (i.e. 'flux' the oxide) formed on high-alloy steels, is potentially dangerous. While such substances are not common in fuels the danger should be borne in mind where high-alloy steels are used as containment vessels for high-temperature processes.

Sulphates, which form part of the ash from the combustion of many fuels, are not harmful to high-alloy steels, but can become so if reduction to sulphide occurs. This leads to the formation of low melting point oxide–sulphide mixtures and to sulphide penetration of the metal. Such reduction is particularly easy if the sulphate can form a mixture of low melting point with some other substance. Reduction can be brought about by bad combustion, as demonstrated by Sykes and Shirley[17], and it is obviously important to avoid contact with inefficiently burnt fuels when sulphate deposits may be present. Reduction can also be brought about in atmospheres other than reducing ones and the presence of chlorides or vanadium pentoxide has been shown to be sufficient to initiate the reaction. It has also been shown[24] that it can be initiated by prior cathodic polarisation in fused sodium sulphate. The effect of even small amounts of chloride on oxidation in the presence of sulphate is illustrated in Fig. 7.33[17]. Not all sulphates are

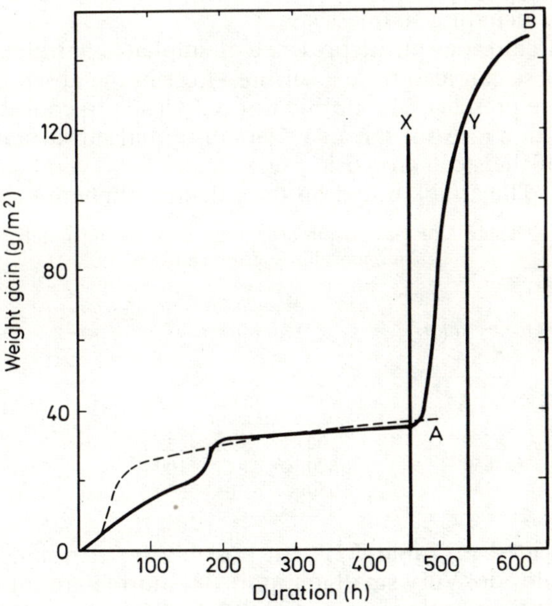

Fig. 7.33 Gains in weight due to oxidation of type 347S17 steel in air at 750°C while in contact with Na_2SO_4. Curve *A* plain, and curve *B* containing 0·3% NaCl (period X–Y) (after Sykes and Shirley[17])

as readily reduced as sodium sulphate, for instance, calcium sulphate does not usually lead to sulphide penetration, although the presence of other substances with calcium sulphate may lead to accelerated oxidation for other reasons. The results for laboratory tests on a series of metals and alloys in sodium sulphate + sodium chloride and calcium sulphate + sodium chloride mixtures are shown in Table 7.19[25]. In many cases sulphide penetration could be noted with the sodium salts but not with the calcium salts.

Table 7.19 Effect of 90:10 sulphate:chloride mixtures on various metals at 750°C (tests for 6 h in air)*

Material	*Loss of metal* (g/m²)		
	No mixture	*Sodium salts*	*Calcium salts*
Cr	7	30	110
Ni	20	60	110
Mild steel	490	1 430	1 120
Fe–20Cr	4	50	1 350
Fe–28Cr–2Ni	3	60	1 160
Fe–22Ni–14Cr	2	40	950
Fe–18Cr–12Ni+Nb	9	50	470
Ni–13Cr	2	2 700	20
Ni–20Cr+Ti+Al	4	3 700	20

*Data after Shirley[25].

It has been suggested that corrosion by sulphates can occur by the formation of pyrosulphates which melt at relatively low temperatures. Accelerated corrosion due to the presence of pyrosulphates has been demonstrated for ferrous alloys including stainless steel[26].

The rôle of chlorides in the presence of sulphates has already been mentioned, but these can also have a serious effect in the absence of other contaminants. The presence of chloride not only leads to considerable acceleration of oxidation rate but can also give substantial subsurface intergranular penetration of the steel. Corrosion test results for several steels are shown in Table 7.20. The attack noted on the calcium sulphate + sodium chloride

Table 7.20 Corrosion tests in air with specimen half immersed in sodium chloride

Test temperature (°C)	*Weight loss after descaling following a 24 h test* (g/m²)				
	Steel				
	430	*304*	*321*	*347*	*310*
550	100	30	20	20	30
650	320	270	100	210	190
750	1 050	660	650	400	750

mixtures indicated in Table 7.19 can possibly be attributed solely to the presence of chloride. Very small amounts of chloride are sufficient to cause serious acceleration, as illustrated in Fig. 7.34. Samples of 347S17 steel were heated in air to 650°C for 20-h cycles. Between cycles the samples were cooled to room temperature and weighed together with any loose scale, brushed, weighed again and then dipped in a dilute sodium chloride solution, removed, dried and weighed yet again to give the quantity of salt added to the surface. Gains in weight attributable to chloride and to oxidation are shown; it should be noted that the latter are probably low since no account was taken in this method of assessment of any losses of volatile substances. In tests of this sort, when the addition of chloride is discontinued, the oxidation rate slowly returns to the very low value to be expected in air at this temperature. While most interest has been in effects of chlorides, there is evidence that other halides can have similar effects.

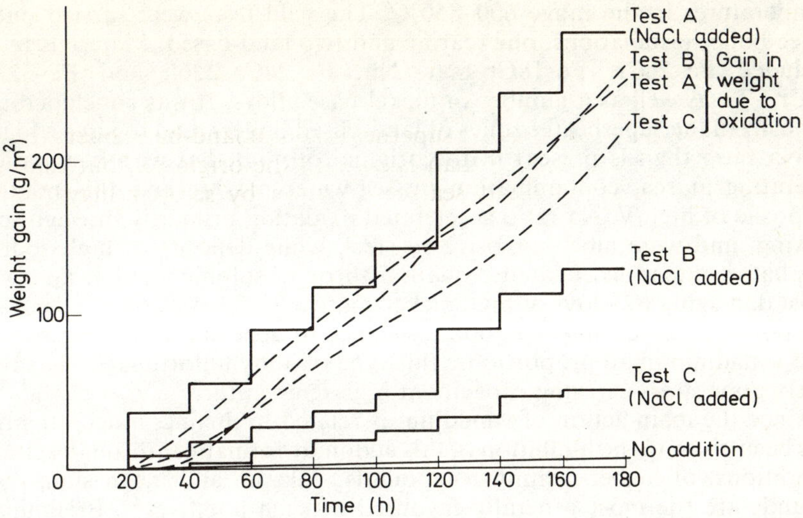

Fig. 7.34 Gains in weight of 347 steel in air at 650°C with various periodic additions of sodium chloride. Stepped curves show quantity of NaCl added, broken curves show gains in weight attributable to oxidation

In the case of oil ash, the most serious damage is associated with vanadium compounds. Organic vanadium compounds which cannot be economically removed from the residual oils form vanadium pentoxide in combustion which can have a considerable corrosive effect on heat-resisting steels. Vanadium pentoxide itself has a melting point in the neighbourhood of 660°C and forms compounds which have even lower melting points. Prior to the development of the gas turbine the problem was not of such overriding importance, since components, such as superheater tubes in steam boilers, which had to be thin, were at a low enough temperature to escape serious action. Tube supports in superheater and oil-cracking furnaces fired with residual oil often suffered severe attack, but as they were cast products, relatively inexpensive and replaceable, they were regarded as being of necessity expendable even though losses of metal of thicknesses $\geqslant$ 10 mm/year from Fe–24Cr–21Ni and Fe–30Cr steel supports at 800–900°C could occur. With the advent of the gas turbine and the possibility of higher steam temperatures an entirely new order of assessment was necessary as the very much smaller rates of attack occurring in the range 650–800°C involved quite prohibitive expense in relation to costly creep-resisting high-alloy steel blades or tubes. No complete solution has emerged, but useful measures to reduce the seriousness of the attack in some cases have been evolved, and the limits within which vanadium-bearing oils may reasonably be used have been defined.

The oil ash problem is complicated and intensified by the presence of alkali oxides and sulphates which form a variety of low-melting complexes with vanadium oxide. An idea of the complications occurring in practice can be obtained from the account of extensive field and laboratory tests carried out for a committee of the Council of British Manufacturers of Petroleum Equipment[27], with particular reference to superheater supports and metal temperatures in the range 600–850°C. The field tests were carried out in three boiler installations, one marine and two land-based. Materials tested included Fe–26Cr, Fe–18Cr–8Ni+Nb, Fe–24Cr–22Ni and Fe–32Ni–18Cr+Ti, as well as a number of nickel-base alloys. It was concluded that deposits in the superheater zones of land-based boilers tend to have a higher V:Na ratio than is present in the original oil, but that with installations operating at sea, contamination by sea-water may reverse this relation. Deposits of high V:Na ratio accelerated oxidation primarily through scale fluxing, and were most corrosive to steel, while deposits of high sodium sulphate content accelerated oxidation through sulphide action and were most damaging to alloys of high nickel content. A ferritic 26% Cr steel was the only one which showed good resistance to deposits containing sodium and vanadium in all proportions; this type of alloy unfortunately has relatively poor stress-carrying capacity at high temperature.

Since the main action of vanadium is related to fluxing, much attention has been given to the inhibition of this action by formation, through suitable conditions, of higher-melting compounds; calcium and magnesium compounds are the most generally favoured of such additives[28]. Preliminary washing of the oil to reduce sodium to very low limits has also been advocated[29]. The stability of calcium and magnesium vanadates is, unfortunately, not great enough to prevent their substantial conversion to sulphates by the sulphur oxides normally present in the flue gases, so that with additions of alkaline earth oxides to, say, two or three times the stoichiometric equivalent of the V_2O_5, stabilisation as vanadates is incomplete, and the improvement only partial. For critical components the only safe procedure is usually either to limit the temperature at which steels are used to a maximum of about 600°C, or to use more expensive distillate-oil fuel free from vanadium.

The whole subject of vanadium attack has been reviewed by Sachs[30].

An interesting case of 'ash' attack is encountered with valves in engines powered by high octane fuels containing lead compounds. These compounds are deposited from the gases as mixtures of lead oxide, sulphate and bromide, and can cause serious scale-fluxing effects with high-alloy valve steels.

Molten Salts

Molten salt baths are widely used in heat treatment and for steel carburising. As in the case of ash attack, danger to alloy steel containers arises mainly from enhanced oxidation brought about by scale fluxing. Such oxidation can only proceed where oxidising conditions obtain, so that while alkali chlorides form useful heat-treatment baths for steel, they produce severe attack, even with heat-resisting steel containers, at the surface of the bath, with formation of chromates and ferrites. Austenitic Fe–23Cr–12Ni and Fe–35Ni–15Cr steels are used for cyanide-hardening-bath containers, but conversion of cyanide to carbonate in use brings danger of fluxing attack.

Molten alkali hydroxides are particularly dangerous, not only because of scale fluxing, but also because they induce stress corrosion where stress is a serious factor.

Molten Metals

Heat-resisting steels have limited uses in contact with molten metals. They are not recommended for use with molten zinc, cadmium, aluminium, antimony or copper, because of excessive attack and embrittlement effects. In brazing and silver soldering, contact between the molten non-ferrous alloy and the steel occurs for only a very limited period of time.

With molten lead or tin, limited use of high-alloy steels is possible. In the case of containers for lead baths, it is important to avoid the combination of lead oxide and air at the bath surface because of fluxing action, but with mechanical removal of lead oxide and use of carbonaceous coverings, Fe–35Ni–15Cr and Fe–25Cr steels are successfully used. With tin, behaviour depends considerably on temperature, slight action taking place at 300°C and considerable attack at 600°C.

The alkali metals have acquired special interest through their suitability for use as heat-exchanging fluids in atomic reactors. They are generally satisfactorily resisted by the heat-resisting steels, although detailed studies have shown effects on prolonged contact at high temperatures. Thus, Brasunas[31] describes the leaching of nickel by lithium at 1 000°C from Fe–18Cr–10Ni steel, with transformation of the nickel-impoverished surface layer to a ferritic structure and ultimate production of subsurface cavities. He also indicates that there is some penetration by sodium and lead at this temperature with precipitation of intermetallic compounds within the steel.

Applications

The heat-resisting steels are used for a wide range of general engineering and chemical engineering applications where the corrosion resistance, and in some instances strength, of the lower-alloy steels is inadequate. The martensitic steels, because of their lower oxidation resistance, are normally used for the less onerous conditions, and certain limitations in ease of fabrication generally precludes their use for large structures and containment vessels. Their combination of moderate corrosion resistance plus strength at modest temperatures has led to widespread use as turbine discs and blades, bolts and similar parts. They have also been used on steam plant for the less onerous conditions.

The ferritic steels are limited in scope because of lack of hot strength, but the cheaper types such as 430 are used in sheet form for the fabrication of parts such as heat exchangers. The higher chromium varieties are of importance in being much more resistant to sulphur attack than the nickel-bearing types and so are widely used as superheater supports or in sulphide-roasting furnaces, mainly as castings. Strength limitations and brittleness call for care in design.

The austenitic steels combine good oxidation resistance with ease of fabrication and thus are most widely used. In addition, while being quite weak at room temperature, they are among the strongest materials in the 550–750°C range and are thus widely used for this purpose. Typical applications are furnace parts, heat exchangers, gas turbine parts, steam superheaters and piping, and chemical plant equipment for containing reactions and products at elevated temperatures.

J. E. TRUMAN

REFERENCES

1. Hatfield, W. H., *J.I.S.I.*, **115**, 517 (1927)
2. Aitchison, L., *Engineering*, **108**, 799 (1919)
3. Dickenson, J. H. S., *J.I.S.I.*, **106**, 103 (1922)
4. Farenwald, F. A., *Proc. Amer. Soc. Test. Mat.*, **24**, 310 (1924)
5. Johnson, J. B. and Christiansen, S. A., *Proc. Amer. Soc. Test. Mat.*, **24**, 383 (1924)
6. Vanick, J. S., *Proc. Amer. Soc. Test. Mat.*, **24**, 348 (1924)
7. Hatfield, W. H., *J.I.S.I.*, **115**, 483 (1927)
8. Edeleanu, C. and Estruch, B., *I.S.I. Special Report No. 86*, 220 (1964)
9. *High Temperature Properties of Steels*, I.S.I. Publication No. 97 (1967)
10. Pilling, N. E. and Bedworth, R. E., *J. Inst. Met.*, **29**, 529 (1923)
11. Pfeil, L. B., *J.I.S.I.*, **119**, 501 (1929)
12. Wood, G. C. and Whittle, D. P., *Corrosion Science*, **4** No. 3, 263–269 (1964)
13. Wood, G. C. and Whittle, D. P., *Corrosion Science*, **4** No. 3, 293–315 (1964)
14. Wood, G. C., *Corrosion Science*, **2**, 255–269 (1962)
15. Wood, G. C., *Corrosion Science*, **2**, 173–192 (1962)
16. Hatfield, W. H., *J. Inst. Fuel*, **11**, 245 (1938)
17. Sykes, C. and Shirley, H. T., *I.S.I. Special Report No. 43*, 153 (1951)
18. Edwards, A. M. and Nicholson, A., *I.S.I. Publication No. 117*, 149 (1969)
19. Kirkby, H. W. and Truman, R. J., *I.S.I. Special Report No. 64*, 244 (1959)
20. Rohrig, I. A., van Duzer, R. M. and Fellows, C. H., *Trans. Amer. Soc. Mech. Eng.*, **66**, 277 (1944)
21. Eberle, F., Ely, F. A. and Dillon, J. A., *Trans. Amer. Soc. Mech. Eng.*, **76**, 665 (1954)
22. Inglis, N. P. and Andrews, W., *J.I.S.I.*, **128**, 383 (1933)
23. Hancock, P., *Corrosion of Alloys at High Temperatures in Atmospheres Consisting of Fuel Combustion Products and Associated Impurities*, H.M.S.O. (1968)
24. Simons, E. L., Browning, G. V. and Liebhafeky, H. A., *Corrosion*, **11**, 505t (1955)
25. Shirley, H. T., *J.I.S.I.*, **182**, 144 (1956)
26. Jonakin, J., *The Mechanism of Corrosion by Fuel Impurities*, 648 and 649, Butterworths, London (1963)
27. *British Petroleum Equipment News*, **7** No. 4, 54 and **7** No. 5, 48 (1959–1960)
28. Buckland, B. O., Gardner, G. M. and Sanders, D. G., *Residual Fuel Oil Ash Corrosion*, A.S.M.E. Paper A–52–161 Preprint (1952)
29. Buckland, B. O. and Sanders, D. G., *Modified Residual Fuel for Gas Turbines*, A.S.M.E. Paper 54–A–246 Preprint (1954)
30. Sachs, K., *Metallurgia*, **57**, 123, 167, 224 (1958)
31. Brasunas, A. de S., *Corrosion*, **9**, 78 (1953)

7.5 Nickel and its Alloys

Oxidation

Nickel, in comparison with other metals such as iron, cobalt and copper, has a relatively good resistance to oxidation at high temperatures. There is general agreement that the only oxide formed is NiO, although this can be present in two quite distinct layers: a black outer layer and a green inner layer[1], and platinum surface markers after oxidation were found to be located at the interface between the two oxide layers. The oxide is protective in nature and the growth of both oxide zones follows a parabolic law, at least up to 1 300°C. The values reported for rates of oxidation vary considerably according to the purity of the nickel used (Phillips[2] found 99·9% pure nickel had 10 times the scaling resistance of that of a sample of 99·7% purity) and other details of experimental technique. Figure 7.35[2] shows the variation with temperature of the rate constant; the dotted lines (arrowed) enclose the results of many of the earlier investigations but later workers have obtained the results scattered about the full line reflecting a lower oxidation rate[1,3].

The resistance of nickel to oxidation may be modified considerably by alloying, although the rate of oxidation still in general obeys a parabolic rate law, the rate constant increasing exponentially with temperature. In general the rate constant increases linearly from the value for nickel with increasing additions of a second element, but above a given level, which depends on the solute, the change becomes slower and for some elements the rate constant then decreases. Results obtained by Horn[4] are illustrated in Fig. 7.36, which shows that beryllium, silicon and chromium in particular can produce enhanced oxidation resistance. From the linear portions of the rate constant/concentration curves, Horn calculated the difference in rate constant from that of nickel produced by 1 atomic % of each of the solutes investigated and obtained the values given in Table 7.21*. The concentration at which the curves deviate from linearity is associated with that at which the oxide of the solute begins to play a part in the oxide scale. At lower concentrations the effect is attributed to the lattice distortion due to the solute producing increased diffusion of the nickel ions and hence increased oxidation, an explanation supported to some extent by the reasonably good agreement found between the effectiveness of a solute in increasing oxidation rate and its atomic size. The more generally accepted explanation, based on the work

*In the tables and figures $K \times 10^x$ indicates that the actual values given are $K \times 10^{-x}$.

Table 7.21 Relative specific scaling rate constant for binary nickel alloys at 900°C[4]

Element	K_{sp}* × 10^8
Be	17
Ca	216
Al	15·8
Si	102
Ti	78·7
Zr	204
Ce	275
Th	383
Cr	48·2
Mo	26·5
W	66·5
Mn	35·5
Cu	24·4
Au	7·3

*K_{sp} = (K_{solute} − K_{nickel})/atomic % solute. where K_{nickel} is approximately 20 × 10^{-8}.

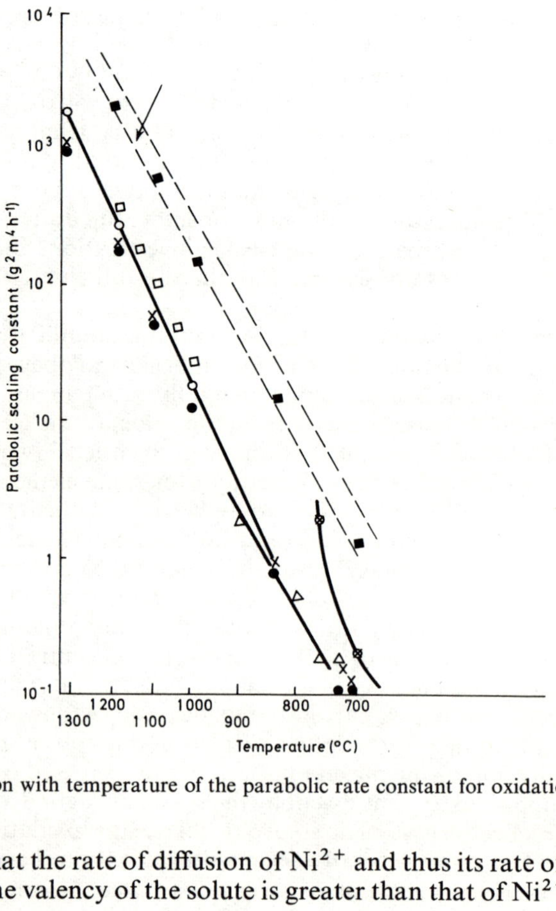

Fig. 7.35 Variation with temperature of the parabolic rate constant for oxidation of nickel[2]

of Wagner, is that the rate of diffusion of Ni^{2+} and thus its rate of oxidation is increased if the valency of the solute is greater than that of Ni^{2+}, since the

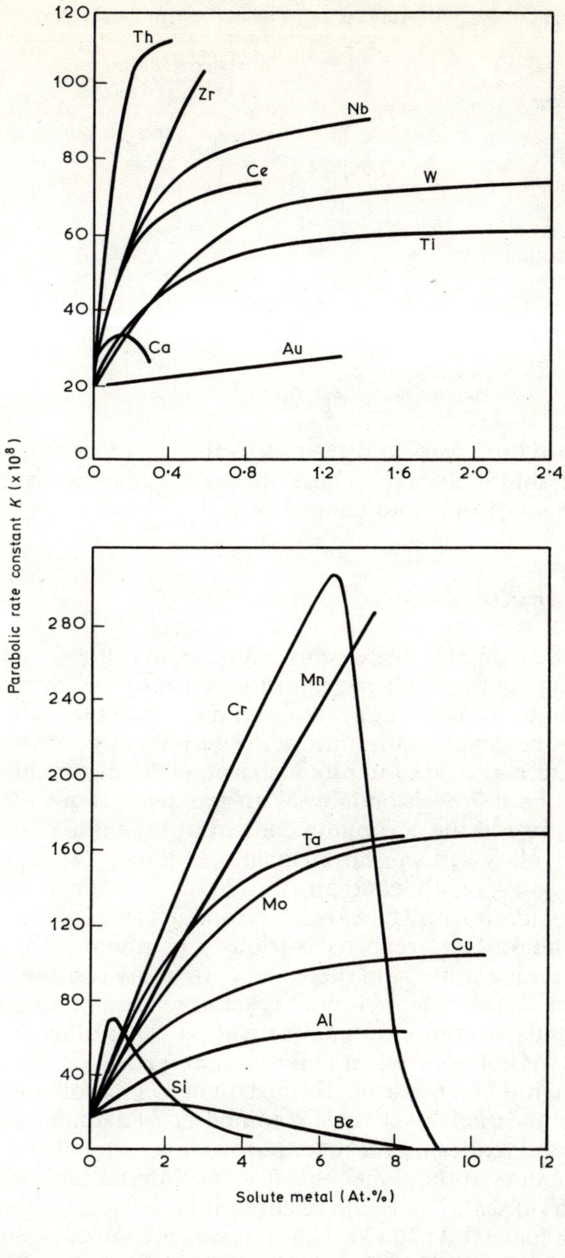

Fig. 7.36 Effect of alloying on the rate constant for oxidation of nickel at 900°C[4]

presence of the higher valency ions in the nickel oxide matrix leads to a higher concentration of Ni^{2+} vacancies (Section 1.8).

A semi-quantitative indication of the effects of different elements on the resistance to oxidation of nickel is given in Table 7.22 which lists values for

Table 7.22 'Alloying factor of oxidation' for nickel alloys*

Element	*F*† for stated addition (Wt. %)					
	5	10	20	30	40	50
Chromium	4	$1\frac{1}{2}$	$\frac{1}{3}$	$\frac{1}{4}$	$\frac{1}{6}$	$\frac{1}{5}$
Manganese	3	4	—	—	—	—
Tantalum	3	2	$\frac{1}{2}$	$\frac{1}{2}$	$\frac{1}{2}$	1
Molybdenum	$2\frac{1}{2}$	3	—	—	—	—
Copper	$1\frac{1}{2}$	$1\frac{1}{2}$	2	$2\frac{1}{2}$	9	—
Niobium	1	$\frac{1}{3}$	$\frac{1}{3}$	1	—	—
Platinum	1	—	—	$\frac{3}{4}$	—	$\frac{1}{2}$

*Derived from curves of Reference 5.
†Approximate oxidation rate of alloy relative to that of pure nickel.

the relative oxidation rate with respect to that of nickel for different concentrations of solute element. These values are approximately valid over quite wide ranges of time and temperature[5].

Nickel–Chromium Alloys

The alloys based on the nickel–chromium system are of paramount importance in the field of high-temperature alloys. As shown in Tables 7.23 and 7.24, addition of chromium to nickel has a complex effect on the oxidation behaviour; small additions are deleterious as predicted by Hauffe's rules, the isothermal oxidation rate increasing with chromium content to a maximum at about 7% chromium. With less than about 9% Cr, internal oxidation occurs and the chromium content of the matrix is sufficiently reduced for the alloys to appear magnetic (Table 7.24), although before oxidation only alloys with less than 7% Cr are magnetic. A progressive improvement in oxidation resistance results from further additions of chromium up to a chromium level reported variously as about 20% and 40–90%, depending on temperature, and these alloys are more resistant than either of the constituent metals although pure crack-free chromium gives oxidation values very similar to those for commercial Ni–20Cr alloys[8].

Barrett and his colleagues[9], and more recently Kosak[10] have summarised existing information on the scales formed on nickel–chromium alloys. Up to about 10% Cr, the thick black scale is composed of a double layer, the outer layer being nickel oxide and the inner porous layer a mixture of nickel oxide with small amounts of the spinel $NiO \cdot Cr_2O_3$. Internal oxidation causes the formation of a subscale consisting of chromium oxide particles embedded in the nickel-rich matrix. At 10–20% Cr the scale is thinner and grey coloured and consists of chromium oxide and spinel with the possible presence of some nickel oxide. At about 25–30% Cr a predominantly chromium oxide scale is observed and Barrett associates optimum scaling resistance with the minimum chromium concentration necessary for the exclusive formation of this oxide rather than the spinel. Other workers have reported results very similar to those described above, although conflicting evidence has been obtained, notably by Yearian *et al.*[11] who found considerable amounts of nickel oxide

Table 7.23 Oxidation data for nickel–chromium alloys[6]

Chromium content (%)	*Temperature* (°C)	pO_2 (atm)	*Rate constant* (K_p) ($g^2m^{-4}s^{-1} \times 10^2$)
0·0	900	air	0·28
1·97	900	air	4·9
4·12	900	air	5·8
5·89	900	air	8·2
8·0	900	air	0·0
0·0	1 000	1	3·48
0·3	1 000	1	15·0
1·0	1 000	1	25·8
3·0	1 000	1	28·3
10·0	1 000	1	5·55
0·0	1 096	1	5·48
0·32	1 096	1	23·6
0·92	1 096	1	29·7
2·0	1 096	1	39·6
3·45	1 096	1	46·8
5·67	1 096	1	58·5
7·64	1 096	1	67·8
8·71	1 096	1	30·8
11·1	1 096	1	3·79
14·9	1 096	1	0·35
20·0	1 096	1	0·07

in the scale on an 80Ni–20Cr alloy after oxidation for 100 h in air. They nevertheless also consider chromium oxide to be the effective scale constituent in promoting oxidation resistance in this alloy and in nickel–chromium steels. On the other hand, Pfeiffer[12], Douglas and Armijo[13], have associated the protective effect with the presence of the spinel constituent. The varying

Table 7.24 Results of oxidation tests on nickel–chromium alloys[7]

Material composition (%) Ni	Cr	*4 days at 1 038°C* *Weight gain* ($gm^{-2}s^{-1} \times 10^4$)	*Oxide thickness* (mm) *Outer scale*	*Inner zone*	*Magnetic response*†	*8 days at 954°C* *Weight gain* ($gm^{-2}s^{-1} \times 10^4$)	*Maximum penetration* (mm)
100	—	6·6	0·127	0·0	M	—	—
97	3	10	0·127	0·076	M	—	—
96	4	—	—	—	—	0·61	0·056
94	6	8	0·254	0·076	M	—	—
92	8	—	—	—	—	0·50	0·061
91	9	3·6	0·152*	0·076*	M	—	—
91	9	0·64	0·178*	0·127*	W	—	—
88	12	—	Thin	Nil	N	—	—
86	14	—	—	—	—	0·03	<0·025
85	15	0·28	Thin	Nil	N	—	—
82	18	0·22	Thin	Nil	N	—	—
80	20	—	—	—	—	0·01	<0·025

*Irregular attack; † and M, magnetic; W, weakly magnetic; N, non-magnetic.

reports on the exact nature of the scale formed are explained by its marked dependence on experimental conditions. For instance, in a thorough review of oxidation theory and behaviour, Seybolt[14] quotes unpublished calculations of Berks and Richert which indicate that the spinel only exists under conditions of thermodynamic non-equilibrium, a usual state in oxidation. However, Pfeiffer[15] considers that the spinel formed initially is reduced to chromic oxide by a displacement reaction.

The binary Ni–20Cr alloys find their main application in the resistance heater field for both domestic and industrial purposes. Ternary additions, for instance of silicon or manganese, can further improve the good oxidation resistance of the materials due to formation of layers of silica or of manganese spinel, respectively[16]. Other more complex alloys based on the nickel–chromium system are the high-temperature creep-resistant or superalloys, whose rapid development was closely associated with that of the aircraft gas turbine. The nickel–chromium base was selected originally on the grounds of good oxidation resistance at high temperatures, but subsequent development has been directed largely towards enhanced high-temperature strength. The main strengthening elements used almost universally are titanium and aluminium and although Fig. 7.36 shows that small additions of either of these metals reduce the oxidation resistance of nickel, the effect particularly of aluminium in nickel–chromium alloys is beneficial[17]. Whilst the basic constituents of the scale on the nickel–chromium alloys with small ternary additions are generally unchanged, the compositional changes near the surfaces of such alloys during oxidation indicate that aluminium and titanium must be involved in the process[18]. The presence of alumina has been detected in the oxide formed on Udimet 500, a Ni–19Cr–3Al alloy and alumina together with Ni–Al spinel and chromium oxide on Udimet 700 (Ni–15Cr–4Al)[19].

Despite their good resistance to general oxidation, under more extreme conditions, such as a low partial pressure of oxygen, in nominally pure hydrogen or a high vacuum, the more complex alloys containing titanium or aluminium cannot, unlike the binary nickel–chromium compositions, be maintained in a bright unoxidised condition. This is of practical importance in connection with bright annealing and brazing.

Other compositional modifications of the nickel–chromium base to promote high-temperature strength have involved additions of many elements, notably carbon, cobalt, molybdenum and/or tungsten, and niobium. Each of these has its own effect on oxidation resistance, and although relevant ternary alloys have been investigated, the variations of composition are now so numerous that measurements of oxidation resistance have in general been made on selected established compositions (Tables 7.25 and 7.26), and no systematic attempt has been made to follow the detailed interaction effects of the separate elements. A useful summary of results published on various nickel-base superalloys is included in a paper by Lund and Wagner[23]. They draw attention to the generally excellent resistance to oxidation of such alloys, but again emphasise the beneficial rôle of aluminium in enhancing this property[24].

Indeed, for operation at temperatures above 1 000°C (now required of the most advanced aircraft turbine blading materials) it appears that reliance must be placed on alumina as the protective constituent. Stott and Wood[25]

Table 7.25 Oxidation data for nickel-base high-temperature alloys[20]

Alloy	*Composition* (%)					*Rate constant K* ($g^2m^{-4}s^{-1}$)						
	Ni	Cr	Mo	Al	Others	816°C	871°C	927°C	928°C	1 093°C	1 149°C	1 204°C
Al-modified Nichrome V	76	19	—	4	—	—	—	—	0·97–2·5	3·1–3·8	6·2–9·2	22
Nichrome V	80	20	—	—	—	—	—	—	8·9–25·3	82–83	161–184	36
Nb-modified Nichrome V	79	20	—	—	1Nb	—	—	—	4·7–5·6	66–80	89–124	378–417
Inconel 702	Balance	15	—	3	0·5Ti	0·83–0·97	1·8–3·6	2·8–4·4	16·9–19·2	—	—	—
Hastelloy 235	Balance	16	7	1·3	1·9Ti	3·6–7·2	11–12	37–50	132–179	—	—	—
Hastelloy W	Balance	5	20	—	—	0·27	—	—	1 135–1 253	—	—	—
Inconel X	Balance	15	—	1	7Fe -2·4Ti -1Nb	0·69–0·72	1·1–3·9	—	—	—	—	—
Inconel	Balance	15	—	—	7Fe	0·15–0·24	1·1–2·2	—	—	—	—	—

Table 7.26 Oxidation data for commercial nickel–chromium-base alloys

Alloy	*Composition* (%)							*Weight change* (g/m^2): *Loss after* 100 h *at temperature, and descaling*[21]					*Weight change* (g/m^2): *Gain in* 60 h *micro balance*[22]		
	Ni	Cr	Co	Mo	Fe	Ti	Al	800°C	900°C	950°C	1 000°C	1 100°C	800°C	950°C	1 100°C
Nimonic 75 NC20T	Balance	20	—	—	—	0·4		5·5	11·8	40·0	66·6	89·2			20
Nimonic 80A NC20TA	Balance	20	—	—	—	2·4	1·2	6·4	26·2	39·6	59·6	112·6		11	40
Nimonic 90 NCK20TA	Balance	20	16	—	—	2·4	1·2	4·6	25·2	55·0	104	112·3	3·8	13	40
Nimonic 100 NCKD20TA	Balance	11	20	5	—	1·5	5·0	0·8	2·7	8·7	16·1	116·1		6	60

Note: 80 g/m^2 corresponds to a depth of about 0·01 mm assuming uniform metal loss over the whole specimen surface.

found that at such temperatures additions of only about 1% aluminium to Ni–15Cr and Ni–29Cr alloys improved oxidation resistance substantially, whilst additions of about 4% Al were sufficient to form a protective scale of alumina very rapidly.

In alloys with very high aluminium (25%) or a high Al:Cr ratio such as the cast turbine blading alloy B1900 (Ni–8Cr–6Al–10Co–6Mo–4Ta–1Ti) the main oxide is Al_2O_3 and this can give still further improved oxidation resistance[26].

Particular mention should also be made of molybdenum which is a constituent, usually up to 5%, of a large proportion of the materials intended for service at the very highest temperatures. The extremely poor oxidation resistance of molybdenum itself and of its alloys is well known, and is attributed to the volatility of the oxide MoO_3 which prevents the formation of a stable protective scale. Table 7.22 indicates a deleterious effect of molybdenum on the oxidation resistance of nickel, although Preece and Lucas[27] report only a slight increase in oxidation rate caused by addition of molybdenum to nickel, and state that no volatilisation of MoO_3 was noticed, probably because of the moving atmosphere used. Whilst for the commercial superalloys containing about 5% Mo acceptable oxidation resistance can be maintained at temperatures within the normal operating range, a deleterious effect of such additions is noticed during heat treatment at temperatures above about 1 100°C.

Table 7.25 shows that the rate constants at 928°C for the Hastelloy alloys are considerably higher than those for molybdenum-free compositions, although the very low chromium content of Hastelloy W is doubtless a significant factor in this connection. It is noteworthy that the molybdenum-containing low-chromium alloy listed in Table 7.26 is generally superior to the others but this high resistance to oxidation is associated with its relatively high aluminium content.

Niobium appears to have a slightly beneficial effect on the oxidation resistance of nickel–chromium-base alloys, although at 1 200°C the addition of niobium to Nichrome (Table 7.25) produces a much higher rate constant. Titanium on the other hand appears to have a slightly deleterious effect on oxidation resistance at low temperatures.

In connection with the development of materials for domestic and industrial resistance-heating applications, considerable emphasis has been placed on the property of scale adherence under cyclic heating conditions (see Section 7.1). Consistent improvement in the burn-out life of nickel–chromium alloy wires in a standardised test has been effected by small additions of such reactive elements as Si, Ca, Ce and Zr, all of which reduce the oxidation resistance of nickel (Table 7.21). In an investigation of the effects of 0·2 atomic % additions of such elements to a Ni–20Cr alloy it was demonstrated that the increase in life in a cyclic test was proportional to the atomic size of the element added[4]. Addition of silicon was reported to also increase the oxidation rate of a Ni–20Cr alloy, but the higher silicon alloy is much less sensitive than an alloy of lower silicon content to deliberate disruption of the scale[28]. Such effects have been associated with the fact that the presence of the reactive elements leads to internal oxidation or penetration of oxidation into the alloy and thus to a keying-on of the oxide layer[29–30]. Again in cyclic tests at 1 000°C, Fleetwood and Whittle[31] showed the value of increasing chromium above the 20% level in order to reduce the rate of spalling of the oxide scale.

The severity of intergranular penetration of oxide, not generally reflected in weight change measurements, is an important consideration in commercial alloys, particularly when these are used as structural components in thin sheet form. In general, intergranular penetration increases rapidly with temperature and time[32] and is more severe in Ni–Cr–Fe alloys than in Ni–Cr alloys[33].

Nickel–Iron and Nickel–Chromium–Iron Alloys

Alloys of nickel with iron have no great practical importance in the high-temperature field. The scales formed in air consist of nickel oxide and iron oxide, the latter usually present in the form of the spinel, $NiO \cdot Fe_2O_3$. Substitution of a substantial proportion of nickel by iron results in a deterioration in the oxidation resistance of nickel–chromium alloys with chromium contents below about 16%. At higher chromium levels, however, the effect of iron is less significant, except perhaps at very high temperatures. Figure 7.37, which illustrates the results of investigations on a range of cast nickel–chromium–iron alloys, shows that, as iron replaces nickel, a higher chromium content is required to ensure satisfactory corrosion resistance in air.

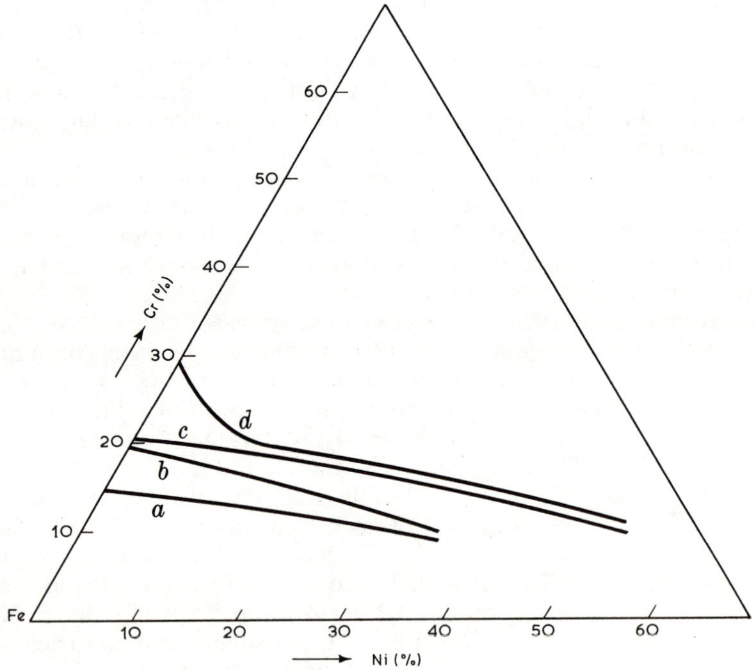

Fig. 7.37 Iso-corrosion lines for Ni–Cr–Fe alloys heated in air[34]
a. 2·5 mm/y, 871°C *c.* 2·5 mm/y, 1 093°C
b. 2·5 mm/y, 982°C *d.* 3·8 mm/y, 1 204°C

Table 7.27 Oxidation of nickel-base alloys and nickel–chromium steels in cyclic temperature tests of 100 h duration[35] (heating cycle: 15 min at temperature, air cool for 5 min)

Material type and composition	*Weight change at test temperature (%)*		
	760°C	871°C	982°C
Nickel		3·6	
Ni–20Cr	−0·016	−0·016	−0·398
Nimonic 75 (Ni–20Cr–0·5Ti)		0·225	0·433
Inconel (Ni–15Cr–7Fe)	−0·225	0·0	−0·173
Ni–15Cr–22Fe	−0·066	−0·527	−0·546
Incoloy (Fe–42Ni–13Cr)		0·151	−1·22
Type 310 stainless steel (Fe–20Ni–25Cr)	−0·024	−0·283	−20·7
Type 309 stainless steel (Fe–13Ni–23Cr)	0·023	0·086	−31·5
Type 347 stainless steel (Fe–11Ni–18Cr)		−37·6	
Type 329 stainless steel (Fe–4Ni–23Cr)	0·320	−26·1	

For economic reasons, and because of their advantages in more complex atmospheres, which will be discussed later, nickel–chromium–iron alloys have been developed for many kinds of industrial heating applications. Typical compositions contain about 20 or 40% Fe with 15–20% Cr, the balance being substantially Ni. Under fluctuating heating conditions, the depletion of nickel content results in reduced resistance to scaling, particularly at higher temperatures. Cyclic heating tests on a range of materials of varying iron, nickel and chromium contents have been made by Eiselstein and Skinner[35] (Table 7.27) who concluded that the minimum nickel and chromium contents to promote adequate scaling resistance were 10Ni–14Cr at 760°C, 15Ni–18Cr at 871°C, and 35Ni–20Cr at 982°C.

The results of cyclic oxidation tests at 1 093°C given in Table 7.28 indicate that an addition of about 4% Ti to a nickel–chromium–iron alloy has quite a significant deleterious effect on scaling resistance. The Fe–25Cr–20Ni stainless steel included in this series of alloys scaled appreciably less than the nickel-rich alloys, in contradiction of the results in Table 7.27. Table 7.29

Table 7.28 Scaling behaviour of some nickel-containing alloys[36] (heating cycle: 24 h at 1 093°C, air cool)

Nominal composition (%)						*Cumulative (average) weight gain* (g/m^2)				
C	Fe	Cr	Ni	Co	*Others*	24 h	48 h	96 h	144 h	192 h
—	10	30	60	—		25·8	34·2	56·3	80·0	107
—	10	20	70	—		29·5	40·9	63·2	87·7	112
—	3	20	71	—	4·2Ti, 1·8Al	65·7	86·8	125	157	187
0·5	—	20	71	1·5	1·5Mn, 6Ta	38·3	53·3	79·2	99·0	114
—	Balance	25	20	—	(Type 310)	10·2	12·9	17·8	20·8	23·6

Table 7.29 Weight losses during steady and intermittent heating of nickel–chromium and nickel–chromium–iron alloys at 1 200°C[37]

Material	*Weight loss* ($gm^{-2}s^{-1} \times 10^3$)		
	120 h	*5 cycles of* 24 h	*10 cycles of* 12 h
80Ni–20Cr (1)	0·17	0·19	0·24
Fe–20Ni–25Cr (1)	0·25	0·5	0·63
Fe–20Ni–25Cr (2)	0·33	0·4	0·55

indicates that the weight loss in a given total exposure time increases with the frequency of cycling and that the effect is greater for nickel–chromium–iron than for iron-free alloys.

Nickel–Copper Alloys

The literature on the oxidation of nickel-copper alloys is not extensive and exphasis tends to be placed on the copper-rich materials. The nickel-rich alloys oxidise according to a parabolic law and at a rate similar to that for nickel; Corronil (Ni–30Cu) exhibited a parabolic rate behaviour below 850°C but a more complex behaviour involving two parabolic stages above 900°C. Electron diffraction examination of the oxide films formed on a range of nickel-copper alloys showed the structures of the films to be the same as for the bulk oxides of the component metals and on all the alloys examined only copper oxide was formed below 500°C and only nickel oxide above 700°C[38].

Sulphidation

The rapid attack due to sulphidation of metals forming low-melting-point sulphides has been referred to in Section 7.1; nickel is a classic example of such metals. The binary nickel–sulphur phase diagram shows that the sulphide Ni_3S_2 forms a eutectic with nickel which melts at 645°C, and the maximum solubility of sulphur in nickel is only about 0·005%. At temperatures well below the eutectic temperature, contamination of nickel by sulphur leads to serious intercrystalline attack and resultant embrittlement, the rate of penetration being greatest at temperatures in the region of 550–650°C. Sulphur attack on nickel during heating is produced by even the smallest amounts of sulphur-bearing contaminant and is virulent under reducing, neutral or oxidising conditions. Since oxidation too can proceed intergranularly, a method of identifying the nature of the attack is often useful. Etching the affected material in an aqueous solution containing 20% sodium nitrate will stain sulphur-affected boundaries but leave oxidised boundaries unattacked.

The attack upon nickel by sulphur is so rapid at elevated temperatures and is produced by such small traces of contaminant that the most stringent precautions and cleaning schedules are required prior to the heating of nickel in service and during processing; storage time should be minimised and the metal protected to avoid atmospheric contamination, and handling should be with clean gloves, all possible contact with grease or oil being eliminated. Even with such precautions the use of a thorough cleaning process is to be recommended immediately before heating[39].

Pfeiffer[40] contends that, in undeoxidised nickel at least, the low sulphur contents normally found are insufficient to cause grain-boundary embrittlement and that the latter is, generally, due to intergranular oxides. In the presence of sulphur-containing gases, however, the level of sulphur required, i.e. about 0·1%, is readily attained and deposits of sulphide are observed.

The kinetics of sulphur attack on nickel varies with conditions, attack increasing rapidly with pressure of sulphur vapour in the range 1–130 N/m^2, and with temperature. At low partial pressures (e.g. 13 N/m^2) and below 620°C, Pfeiffer[41] and other workers found a linear relationship, the rate of attack being about 0·16 g m^{-2} s^{-1}, but later, Mrowec and Werber[42] concluded that this was an effect of using too large specimens for the sulphur supply available, and they demonstrated parabolic behaviour over a range of temperature of 480–620°C and sulphur vapour pressures of from 4 N/m^2 to 100 kN/m^2. Under these conditions a double layer of sulphide is formed; Tideswell[43], by X-ray and electron diffractometry, identified a thick inner layer of Ni_3S_2, with a thin outer NiS film. At lower temperatures nickel in contact with molten sulphur corrodes according to a parabolic rate law.

At low partial pressures of sulphur, intercrystalline attack is not observed either below or above the eutectic temperature of 645°C and the observed embrittlement under such conditions has been attributed to hardening of the grains by sulphur in solid solution in the matrix. At a pressure of about 130 N/m^2 of sulphur vapour, however, sulphides are readily observed in the grain boundaries.

In sulphur dioxide the corrosion rate appears to attain a maximum value at about 800°C. Hatfield[44] reported that pure nickel is more strongly attacked by the gas at this temperature than at 900 or 1 000°C, while Russian work[45] indicates that the corrosion obeys a parabolic law, the rate constant increasing with temperature in the range 600 to 800°C (Table 7.30). The latter investigation also showed that the scale consists of nickel oxide NiO and the sulphide NiS in proportions consistent with the reaction

$$3Ni + SO_2 \rightarrow NiS + 2NiO$$

At higher temperatures above 900°C oxidation becomes of greater importance than sulphidation, the formation of an oxide film apparently hindering the absorption of sulphur.

Dilute binary alloys of nickel with elements such as aluminium, beryllium

Table 7.30 Results of corrosion tests on nickel in sulphur dioxide[45]

Temperature (°C)	*Exposure time, t* (s × 10^{-3})	*Maximum weight gain, q* (g/cm^2)	*Mean value of rate constant,* $K = q^2/t$ (g^{2}m^{-4}s^{-1} × 10^3)	*Scale composition* (%) S	Ni
600	270	30	3·33	12·6	74·9
700	464	488	513	12·8	75·4
800	79	805	8 202	11·6	76·0

and manganese which form more stable sulphides than does nickel, are more resistant to attack by sulphur than nickel itself. Pfeiffer[41] measured the rate of attack in sulphur vapour (13 N/m^2) at 620°C. Values around 0·15 $gm^{-2} s^{-1}$ were reported for Ni and Ni–0·5Fe, compared with about 0·07–0·1 $gm^{-2} s^{-1}$ for dilute alloys with 0·05% Be, 0·5% Al or 1–5% Mn. In such alloys a parabolic rate law is obeyed; the rate-determining factor is most probably the diffusion of nickel ions, which is impeded by the formation of very thin surface layers of the more stable sulphides of the solute elements. Iron additions have little effect on the resistance to attack of nickel as both metals have similar affinities for sulphur. Alloying with other elements, of which silver is an example, produced decreased resistance to sulphur attack.

Despite their improved resistance to general corrosion by sulphur, the dilute alloys with many elements are sensitive to intergranular attack and embrittlement, and, at temperatures above 645°C, to eutectic formation.

The most important alloying element for nickel in this context is chromium. Whilst at very low partial pressures of sulphur its great affinity for chromium results in an increased rate of attack as the chromium content of nickel–chromium alloys is raised[46], at higher sulphur potentials (e.g. 1 atm) and after longer times even small additions of the order of 1% chromium produce increased resistance to attack, while larger additions modify considerably the behaviour of nickel in sulphur-containing atmospheres. In general chromium is much less attacked than nickel, which suggests that the sulphide is, to some extent at least, protective[47–48]. Chandler and McQueen[49] have noted the higher activation energy for sulphidation of chromium (147 kJ/mol) than that for nickel (92 kJ/mol) and suggest that this reflects the protective nature of the sulphide on chromium compared to that on nickel. In nickel–chromium alloys with chromium contents upwards of 10%, the greater affinity of sulphur for chromium results in the formation of chromium rather than nickel sulphide, at the metal surface or internally. This compound, which does not form a low-melting-point eutectic, has a melting point of 1 550°C. Furthermore, while attack upon the nickel–chromium alloys is partially intergranular, the sulphides form as isolated globules rather than as continuous films. At low sulphur potentials a duplex, compact and adherent scale is formed, kinetic measurement[50] indicating that these two layers grow according to a parabolic law, the rate constant for the outer nickel-rich scale being some four times greater than that for the chromium sulphide inner layer. Whilst initially inward migration of sulphur probably occurs beneath the protective scales, subsequent growth of both layers is apparently by diffusion of metal ions outwards[51]. Variation of sulphur potential at sub-atmospheric values has little effect on the rate of attack at 700°C but considerable variation is reported for the stoichiometry of both scale layers from CrS to Cr_2S_3 and from NiS to Ni_3S_2. Again, in contrast to the behaviour of nickel itself, attack on the chromium-containing alloys is largely prevented under oxidising conditions, at least over relatively short periods, probably by the formation of a protective chromium oxide film which impedes the absorption of sulphur. Similarly, corrosion of Ni–15Cr alloys by 0·2% SO_2 in nitrogen is more severe at 870°C than by pure SO_2 and this is attributable to the inadequate formation of protective Cr_2O_3 films in the former case[52]. When the concentration of H_2S exceeds that which will oxidise stoichiometrically to SO_2, however, there is a marked increase in corrosion rate.

Continued exposure of the nickel–chromium alloy to more severely sulphurising and reducing atmospheres results in local depletion of chromium to such an extent that nickel sulphide and the eutectic are formed internally. The latter constituents are not often observed in service failures, but the relative instability of nickel sulphide in the presence of chromium sulphide can result in its reduction to nickel during slow cooling on shut down. That nickel sulphide is formed is suggested by the frequent occurrence of blisters, associated with the formation of molten eutectic on the surface of sulphur-attacked specimens[48].

The detailed appearance of an attacked alloy can vary markedly, depending on the conditions of exposure. In practice the conditions are frequently intermittently oxidising and reducing. The mechanism of attack then involved appears to be that the initial formation of chromium sulphide results in a gradual impoverishment of the matrix with respect to chromium and a consequent loss of oxidation resistance. Rapid internal oxidation of the chromium remaining in the matrix then occurs, to give the intimate mixture of oxide and metallic nickel characteristic of attacked regions. The loss of chromium from the surface regions also raises the Curie temperature towards the value for nickel and attacked specimens therefore appear magnetic. Oxidation of a pre-sulphidised Ni–15Cr alloy was shown by Spengler and Viswanathan[53] to result in further penetration of the alloy by sulphur with consequent nucleation of fresh chromium sulphide particles in advance of the oxidation front, supporting the self-regeneration mechanism proposed by some earlier workers.

Although iron sulphide also forms a eutectic with the metal this melts at 988°C, and at temperatures in the region of 700 to 800°C alloys with substantial proportions of nickel replaced by iron and a chromium level maintained at about 20% show advantages over nickel–chromium-base alloys in resistance to sulphur attack.

The successful application of nickel–chromium–iron alloys as structural components of industrial furnaces and as chambers and containers in chemical processing under conditions of exposure involving sulphur substantiates their good resistance to this form of corrosion. These materials are used for service temperatures in the range 750–1 200°C, the upper limit of serviceability being determined largely by the chromium content of a particular alloy. Results of corrosion tests (Table 7.31)[31] on cast nickel–chromium–iron alloys in oxidising and reducing flue-gas atmospheres illustrate the beneficial effect of increased chromium content and the deleterious effects of increased temperature and sulphur content of the atmosphere. The greater degree of corrosion under reducing rather than oxidising conditions is substantiated by the results (Table 7.32) of tests in hydrogen sulphide and sulphur dioxide. These indicate that attack under given conditions increases with nickel content, but this corresponds with a decreasing level of chromium in the alloys and it is likely that the latter is the more significant factor.

It is widely recognised that improved resistance to sulphur attack can be obtained in commercial alloys by small additions of certain elements, notably manganese, silicon and aluminium. Figure 7.38 shows how the depth of metal converted to scale, and more particularly the total penetration by

Table 7.31 Corrosion resistance of nickel–chromium–iron alloys in oxidising and reducing flue-gas atmospheres of varying sulphur content[54]

Material composition (%)			*Metal loss* (mm/y)											
			Reducing atmosphere							*Oxidising atmosphere*				
			982°C					1 093°C		982°C			1 093°C	
			Sulphur content (g/m³)					*Sulphur content* (g/m³)		*Sulphur content* (g/m³)			*Sulphur content* (g/m³)	
Ni	Fe	Cr	0	0·115	2·3	6·9	11·5	0·115	11·5	0·115	2·3	6·9	0·115	2·3
70	20	10	0·25	0·5	0·5	25	≫10	—	—	1–2	2	<0·5	<1	<0·5
70	15	15	<0·25	<0·5	0·5	10	≫10	—	—	<0·5	0·5	<0·5	—	—
70	10	20	<0·25	<0·5	<0·5	2	≫10	—	—	<0·5	<0·5	<0·5	—	—
40	50	10	1	2	15	>25	—	2	>75	3	3	2	8	1
40	45	15	0·25	0·5	2	>25	—	1	50	0·5	0·6	<0·5	1·5	—
40	40	20	<0·25	<0·5	<0·5	>25	—	0·5	2	<0·5	<0·5	—	<1	—

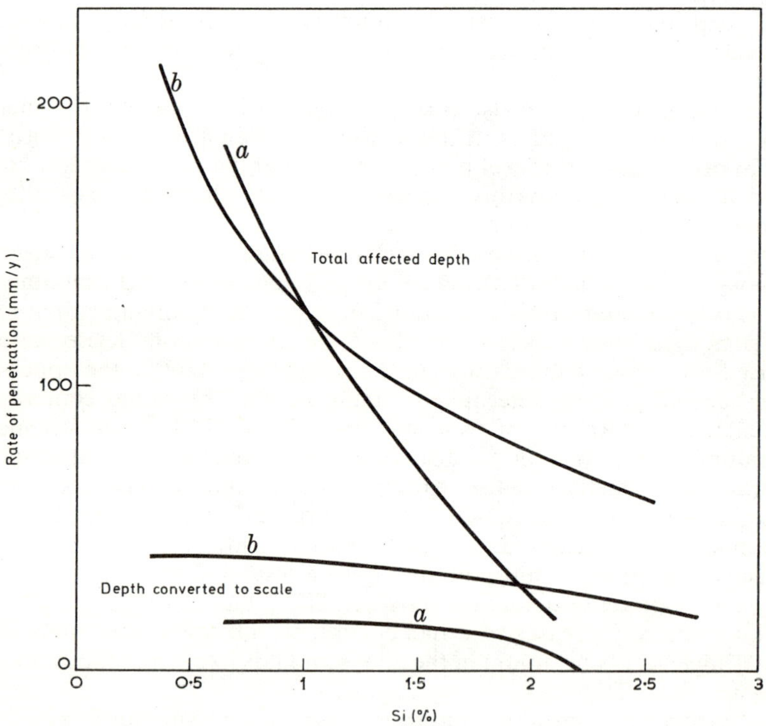

Fig. 7.38 Effect of silicon content on corrosion of Ni–Cr–Fe alloys in a reducing flue-gas atmosphere containing 6·9 g/m³ S (100-h tests at 982°C)[54]
a. Ni–28Fe–12Cr *b*. Ni–50Fe–15Cr

Table 7.32 Results of corrosion tests on nickel–chromium–iron alloys in hydrogen sulphide and sulphur dioxide[55]

Material composition (%)							*Weight gain* (g/m^2) *in* 1-h *tests*					
							H_2S			SO_2		
C	Mn	Si	Ni	Cr	Fe	N	760°C	982°C	1 093°C	760°C	982°C	1 093°C
0·3–0·41	0·68–0·95	0·34–1·80	10·4–12·7	23·0–26·7	Balance	0·09–0·12	1·25–1·38	4·6–5·7	11·6–14·2	0·023–0·028	0·011–0·015	0·037–0·046
						(Range for 7 heats)						
0·49	0·93	0·93	35·4	15·2	Balance	0·06	2·13	8·8	51·5	0·35	0·37	0·63
0·59	0·86	1·02	59·1	13·1	Balance	0·02	3·34	36·4	50·0	0·41	0·47	0·89

Table 7.33 Effect of aluminium on the corrosion at 982°C of Fe–35Ni–15Cr alloy in a reducing flue-gas atmosphere containing 2–3 g/m^3 of sulphur (100-h test period)[54]

Aluminium content (%)	*Metal loss* (mm/y)	*Total penetration* (mm/y)
0	5	65
0·25	5·9	45
0·5	7·2	70
1·1	0·22	4·5
2·5	0·22	10
4·25	0·1	9

sulphur, is reduced by increasing silicon content in nickel–chromium–iron alloys. A similar marked effect of aluminium is observed for an addition of about 1%, but larger additions appear to be relatively less effective (Table 7.33). The mechanism is not clear although the occurrence of very thin silica surface films has been reported.

Sulphur attack on nickel–chromium alloys and nickel–chromium–iron alloys can arise from contamination by deposits resulting from the combustion of solid fuels, notably high-sulphur coals and peat. This type of corrosion, which has been observed on components of aircraft, marine and industrial gas turbines and air heaters, has been associated with the presence of metal-sulphate and particularly sodium sulphate[55,56] arising directly from the fuel or perhaps by reaction between sodium chloride from the environment with sulphur in the fuel. Since such fuels are burned with an excess of air the corrosion occurs under conditions that are nominally oxidising although the deposits themselves may produce locally reducing conditions. Examination of a component attacked under such conditions indicates that the corrosion is very similar to that produced in gaseous sulphur-containing atmospheres. While the corrosion products are sulphides and eventually oxides, little attack on Ni–0·5Cr is produced by sodium sulphate, although a marked, if transient, increase in oxidation rate is produced by a coating of sodium sulphate on nickel itself[57]. The accelerated attack is ascribed to the availability of oxide ions from the sulphate reacting and destroying the protective scale of the NiO layer. Similarly, addition of as little as 0·5–1% sodium chloride markedly increases the corrosion rate even of alloys with a considerable chromium content. Oxidation tests on Nimonic 80A (Ni–20Cr base) in which sodium chloride was introduced into the atmosphere showed that this constituent produces a significant deterioration in the protective nature of the normal adherent oxide film. Although under certain service conditions the presence of chlorides is likely, this is not always so and thus the general applicability of the results of laboratory tests in sodium sulphate and mixtures involving chloride may be questioned. Furthermore, the relative corrosion resistance of different alloys depends considerably on the composition of the salt mixture used, although the salt immersion test and refinements of it have been widely used to study alloying and temperature effects on corrosion resistance to sulphur-containing environments.

As discussed in Section 7.1, temperature is an important factor, corrosion being much more rapid above 700°C than at lower temperatures (Table 7.34).

Table 7.34 Influence of temperature and salt-mixture composition on attack of Nimonic 80A by sulphate–chloride mixtures (6-h tests)[58]

Test temperature (°C)	*Weight loss* (g/m^2) *in* Na_2SO_4–NaCl *mixture of indicated* NaCl *content* (%)			
	0	1	10	50
600	>1	2	19	
650	>1	3	13	16
700	1	20	320	
750	4	410	4 150	660

As regards alloying effects, chromium is again the element of major importance (Fig. 7.39) but the results of an early investigation[60] indicated that the parameter

$$Cr + 1{\cdot}1(\%Ti) + 0{\cdot}7(\%Al)$$

was useful in comparing the resistance of a range of nickel-base alloys to sulphate–chloride attack. Additions of rare earth oxides are also reported to have a beneficial effect on resistance to sulphate–chloride mixtures allegedly by acting as a getter for sulphur diffusing into the alloy[56].

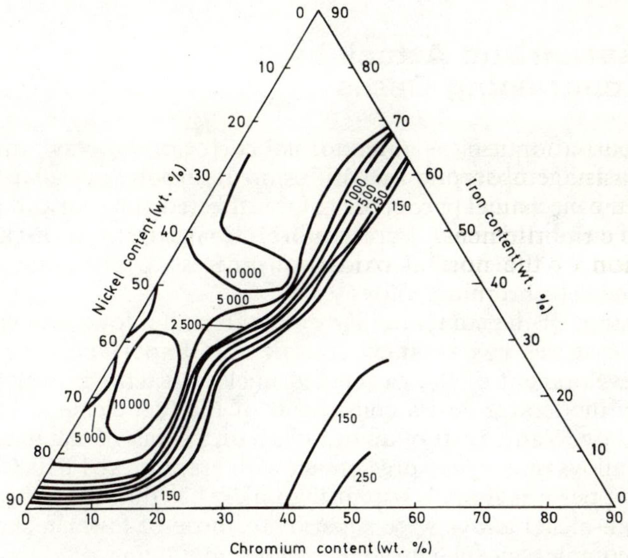

Fig. 7.39 Isocorrosion contours for Ni–Cr–Fe–10Co alloys exposed to 75% Na_2SO_4–25% NaCl mixtures, 16 h at 900°C (corrosion in g/m^2 weight loss)[59]

Although on the basis of such tests it is possible to establish compositions having improved resistance to this type of corrosion, other requirements, for instance optimum high-temperature properties and good castability, may be the overriding considerations. During the late 1950s and 1960s attention was directed, therefore, towards the surface protection of selected materials.

The most successful coatings, reflecting the protective nature of Al_2O_3, were of aluminium base usually with additives of chromium and/or silicon. Various coating methods such as vapour deposition, halide reduction, hot-dipping, metal- and plasma-spraying, were utilised by aero-engine manufacturers. Alloys of optimum mechanical properties, but with poor corrosion resistance were thereby enabled to serve as turbine blades at operating temperatures some 30°C or more higher than would the uncoated material[61].

As operating temperatures were pushed ever higher, the durabilities of such coatings were reduced due to diffusion effects, and a further development employed coatings such as iron–chromium–aluminium base alloys, independent of, but still compatible with, the nickel alloy substrate.

Whilst the use of sulphate–chloride mixtures can be criticised therefore on various grounds, this kind of accelerated laboratory test has been used widely and successfully in producing alloys that have subsequently proved to have a greater resistance to attack under service and simulated service conditions.

Over the last decade considerable progress has been made in establishing commercial nickel-base alloys containing 20–30% chromium and having improved resistance to this type of hot corrosion, whilst, at the same time, retaining the other essential features such as good castability or processing characteristics, together with optimum high-temperature mechanical properties[62].

Carburisation and Attack by Carbon-containing Gases

While carburisation itself is not a normal corrosion process*, in that there is no metal wastage, absorption and diffusion of carbon can lead to significant changes in the mechanical properties of the affected material and in particular to marked embrittlement. Furthermore, initial carburisation can produce an acceleration of the normal oxidation process, a phenomenon that is notable in nickel–chromium alloys.

The question of the compatibility of metals and alloys with carbon and carbonaceous gases has assumed considerable importance in connection with the development of the gas-cooled nuclear reactor in which graphite is used as a moderator and a constituent of the fuel element, and carbon dioxide as the coolant. Tests of up to 1 000 h on a series of metals and nickel-containing alloys under pressure contact with graphite at 1 010°C[63] showed that only copper was more resistant than nickel to diffusion of carbon and that the high-nickel alloys were superior to those of lower nickel content. The more complex nickel–chromium alloys containing titanium, niobium and aluminium were better than the basic nickel–chromium materials.

Tests on a wide range of alloys at temperatures varying from 704 to 927°C have been made by Bernsen *et al.*[64] to determine the temperature limits beyond which engineering materials carburise when held in contact with graphite. Table 7.35 lists the maximum penetrations of the carburised zones; while nickel in general showed no visible evidence of carburisation the associated hardness measurements indicated solution of carbon even at

*See discussion of 'corrosion' in Section 1.1.

Table 7.35 Effect of time and temperature on the visible penetration of carburisation in nickel alloys under contact pressure (1·79 N/mm^2) with graphite[64]

Material type and composition	704°C		816°C		927°C		*Remarks*
	Time ($s \times 10^{-6}$)	*Maximum penetration* (mm)	*Time* ($s \times 10^{-6}$)	*Maximum penetration* (mm)	*Time* ($s \times 10^{-6}$)	*Maximum penetration* (mm)	
Nickel	1·08 (1) 4·45 (2)	<0·002 5 >0·75	1·25 2·55 6·29	<0·002 5 <0·002 5 <0·002 5	3·6	<0·002 5	(1) Grain-boundary attack 0·038 mm (2) Fine intergranular precipitate Hardness increase at each temperature
Nichrome V (Ni–20Cr)	1·08 4·45	<0·002 5 0·330	1·25 2·55 6·29	0·43 0·41 0·53			Hardness increase in 350 h at 816°C but not at 704°C
Inconel (Ni–16Cr–7Fe)	1·08 4·45	0·002 5 0·279	2·55 2·67	0·76 0·76			Hardness increase in 350 h at 816°C but not at 704°C
Inconel X (Ni–15Cr–7Fe–2·5Ti)	1·08 4·45	<0·002 5 0·175	1·25 2·55 6·29	0·20 0·33 <0·002 5			
Inconel 702 (Ni–16Cr–3·5Al)	3·34	0·545	2·67	0·15	3·6	>0·76	
Incoloy 901 (Ni–35Fe–13Cr–2·5Ti)	3·34	<0·002 5	2·67	0·76	3·6	>0·76	

704°C. At this temperature the chromium-containing alloys showed little tendency to carburisation, but at 816°C carburisation leading to the formation of chromium carbide was rapid.

Of the carbon oxide gases, carbon dioxide is the less corrosive, leading normally to mild oxidation only and under certain conditions, e.g. low partial pressures of CO_2 or in the presence of steam, to decarburisation of nickel alloys. Tests on nickel and nickel alloys in carbon dioxide at 13·79 N/mm^2 pressure at 704 and 816°C showed that very little increase in weight occurred in up to 1 000 h even at the higher temperature (Table 7.36).

Table 7.36 Weight changes during exposure of nickel and nickel alloys to carbon dioxide[64]

Material *type and composition*	*Weight gain* (g/m^2)*			
	0·36 × 10^6 s		3·6 × 10^6 s	
	704°C	816°C	704°C	816°C
Nickel	13·0	20·0	13·0	30·0
Nichrome V (Ni–20Cr)	0·8	1·3	1·1	2·5
Nimonic 80 (Ni–20Cr–2·5Ti–1·5Al)	1·1	6·0	2·5	15·0
Inconel (Ni–16Cr–7Fe)	0·9	2·0	1·0	5·0
Inconel X (Ni–15Cr–7Fe–2·5Ti)	1·8	7·5	2·5	15·0
Hastelloy B (Ni–28Mo–5Fe)	12·5	15·0	15·0	40·0

*Approximate values.

The scales formed were similar in constitution to those obtained in air and the rate of scaling approximated to the parabolic law. In addition to the general scale formation, nickel exhibited intergranular oxidation while Inconel (Ni–16Cr–7Fe) and Nichrome V (Ni–20Cr) showed localised pitting. Later work[65] on Nimonic 75 (Ni–20Cr–0·4Ti–0·1C) at pressures of 1 atm or less has shown that oxidation predominates and was generally protective, but at low partial pressure the oxide film was less coherent and uniform, and weight losses due to decarburisation and evaporation of chromium from the specimen surface were recorded at 900–1 000°C.

In the gas-cooled reactor, reaction between the coolant and the moderator results in formation of a proportion of carbon monoxide in the atmosphere. This gas can be carburising to nickel-base alloys but the results of tests[66] in which CO_2 was allowed to react with graphite in the furnace indicate that the attack on high-nickel alloys is slight, even at moderately high temperatures and is still mainly due to simple oxidation.

Industrial atmospheres of the so-called 'controlled' type are frequently used for operations such as bright annealing, carburising and nitriding, and are selected as being reducing to the component being treated. Common examples are cracked ammonia and the products of combustion of coal gas or natural gas, the latter normally being desulphurised. Table 7.37 summarises the satisfactory service experience with nickel and Inconel (Ni–16Cr–7Fe) thermocouple protection tubes in such atmospheres. More vigorously carburising conditions may be achieved deliberately by the introduction of hydrocarbons into the gas stream or accidentally by the introduction of oily components into the furnace. Under such conditions the nickel–chromium and nickel–chromium–iron-base materials that form major parts of the furnace construction and equipment are susceptible to an embrittling form of corrosion described as 'green rot' (Section 7.1). The dependence of the corrosion resistance of these nickel alloys on the protective oxide film has been described; accelerated or internal oxidation occurs only under conditions that either prevent the formation, or lead to the disruption, of this

Table 7.37 Service experience with thermocouple protection tubes in reducing gases and other media[67]

Material	*Temperature* (°C)	*Atmosphere composition* (%)						*Other media*	*Life* (years)
		CO	CO_2	N_2	H_2	H_2O	S		
Nickel	815–1 070	5	8	64	5	18	0·000 8		1–2
Inconel (Ni–16Cr–7Fe)	980–1 260	5	8	64	5	18	0·000 8		1–2
Inconel	980–1 200	2	11	64	2	21	0·000 8		1–2
Inconel	370–980							Lead	3–5
Inconel	790–870							Carburising mixture (cyanide compound)	3–5

film. Thus in alloys of low chromium content internal oxidation can occur if an inadequate supply of oxygen is available; Ni–10Cr alloys have been found to deteriorate rapidly at the ends of long narrow thermocouple tubes. In the higher-chromium alloys internal attack is produced by atmospheres that are intermittently or simultaneously oxidising and carburising. Under these conditions absorption and diffusion of carbon leads to the formation of chromium carbide, the matrix is correspondingly denuded of chromium, and the oxidation resistance of the alloy is seriously reduced. Figure 7.40 shows the chromium contents, as determined by electron microprobe analysis, of the different regions of a Ni–20Cr alloy specimen after exposure to dry carbon monoxide for 168 h at 820°C. Severe attack renders the surface layers magnetic, indicating a matrix chromium content of less than about 7%. The important consideration in this process is the relative diffusion of chromium, oxygen and carbon within the alloy, the result being the prevention of formation of the normal protective oxide film.

The appearance of attacked specimens varies in detail but the general similarity to that produced by severe sulphur attack is interesting; the broad front on which attack occurs and the region of carbide formation preceding a zone of admixed oxide and matrix are characteristic. It is not clear whether the corrosion proceeds by oxidation of the carbide particles or of the denuded matrix; in the initial stages at least it appears that attack follows the matrix/carbide interface.

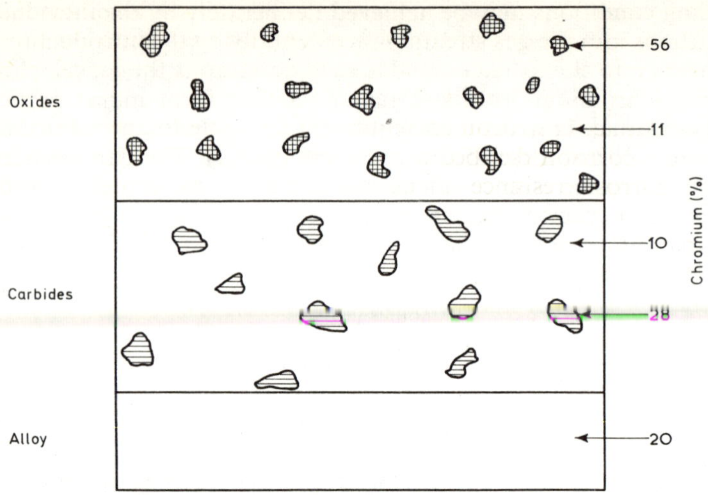

Fig. 7.40 Schematic representation of microprobe analyses of nickel–chromium alloy subjected to 'green rot' attack for 168 h at 820°C (after Dovey and Jenkins[68])

Attack of this type on nickel–chromium alloys is produced by carbon monoxide at temperatures of about 700–1 000°C. According to the Boudouard reaction the gas dissociates:

$$2CO \rightleftharpoons CO_2 + C$$

and the necessary simultaneous carburising and oxidising conditions are attained. Attack in carbon monoxide increases with time and is most marked at about 800–950°C. Table 7.38 lists the results of laboratory tests[7] in which alloys of varying nickel, chromium and iron contents were exposed to both wet and dry carbon monoxide for 8 days at 955°C. In the binary nickel–chromium alloys of low chromium content (4 and 8%) no carburisation was observed but oxidation penetration was greater than that produced in air (cf. Table 7.24); the addition of water vapour to carbon monoxide still further increased the depth of penetration, which was preferentially intergranular in these alloys. It is interesting to note that in the binary alloys those of higher chromium content were more severely attacked in dry carbon monoxide than those of low chromium content, as judged both by weight-gain measurements and by the depth of oxide penetration. The effect of chromium content is clearly dependent on the particular atmospheric conditions; in wet carbon monoxide the penetration was only slightly less in nickel–chromium alloys of 14–20% Cr than in those containing 4 or 8% Cr, but in the Ni–Cr–20Fe alloys a marked improvement in resistance was apparent when the chromium content was raised from 14 to 20%. Under service conditions of somewhat greater oxidising and carburising potential the beneficial effect of higher chromium contents is more marked.

Consistent with the above description of the mechanism of green rot attack, it is found that iron, which increases the solid solubility of carbon in nickel–chromium alloys, improves resistance to this type of corrosion. Dovey and Jenkins[69] found that at 950°C in a dry carburising atmosphere the degree of attack was much greater for nickel–chromium–iron alloys with up to

Table 7.38 Results of tests on nickel–chromium–iron alloys exposed to carbon monoxide for 8 days at 955°C (after Copson and Lang[7])

Material composition (%)				*Gain in weight* ($gm^{-2}d^{-1}$)		*Maximum penetration* (mm)	
Ni	Cr	Fe	*Others*	*Dry* CO	*Wet CO**	*Dry* CO	*Wet CO**
96	4	—	—	21	10	0·33	1·04
92	8	—	—	19	16	0·17	0·94
86	14	—	—	46	34	0·64	0·51
80	20	—	—	30	47	0·61	0·81
80	14	6	—	22	13	0·23	0·15
66	14	20	—	21	22	0·23	0·30
46	14	40	—	5	1	0·36	0·02
60	20	20	—	4	4	0·05	0·05†
Balance	14	6	5Cu	44	20	0·25	0·23
Balance	14	6	3·5Mo	13	1	0·13	0·13†
Balance	14	6	2Si	3	2	0·13†	0·05
Balance	15·8	6·9	2Nb	1	0·3	0·0	0·0
Balance	15·3	6·4	2·3Ti 0·7Al	2	1·5	0·05	0·02

*3% H_2O. †Spotty or irregular attack.

20% Fe than in an alloy containing 38% Ni, 18% Cr and 41% Fe. Similarly, Bucknall and Price[70] showed that the 80Ni–20Cr alloy was more severely attacked than similar alloys containing iron. Their results also indicated that the presence of sulphur in the atmosphere has an inhibiting effect on the attack.

Since the depletion of the matrix with respect to chromium appears to be the basic factor involved, it would be expected that additions of elements having greater affinity than chromium for carbon would increase resistance

Table 7.39 Effect of silicon content on pack carburising behaviour of nickel–chromium–iron alloys at 497°C[71]
(Specimens 25 mm × 9·6 mm diameter weighing approx. 14 g)

Material composition (%)						*Continuous exposure* (125 h)		*Intermittent exposure* (10 × 13½-h *periods*)	
C	Ni	Cr	Fe	Mn	Si	*Weight change* (%)*	*Carburised depth* (mm)	*Weight change* (%)*	*Carbon increase* (%)
0·46	34·9	15·3	Balance	1·02	0·43	+0·68	2·51	+0·845	0·84
0·52	34·9	15·2	Balance	0·74	0·53	+0·67	2·49	+0·677	0·84
0·39	34·8	16·7	Balance	1·06	0·66	+0·51	0·25	+0·762	—
0·54	34·5	14·9	Balance	0·74	1·06	+0·23	0–1·01	+0·214	—
0·43	34·9	16·6	Balance	0·95	1·60	+0·07	0·0	+0·063–0·073	−0·11
0·45	35·6	16·0	Balance	0·91	2·47	0·08	0·0	0·059	—

*At end of test without descaling.

to attack and such effects have been demonstrated by several investigations; relevant data included in Table 7.38 show the beneficial effects of titanium and niobium and to a lesser extent of molybdenum. The complex high-strength superalloys generally contain one or more of these carbide-forming additions and these alloys are relatively resistant to such attack; Nimonic 90 (Ni–20Cr–16Co–2·4Ti–1·2Al) for instance was unattacked after 1 000 hours exposure to carbon monoxide at 950°C.

In this context, silicon also appears to be a beneficial addition to nickel–chromium alloys (Table 7.38). While the explanation of this effect is not clear, data illustrating the impediment to carburisation resulting from silicon additions to nickel–chromium–iron alloys (Table 7.39) support the view that this may be the source of its value in promoting green-rot resistance; other investigators, however, attribute the effect of silicon to its scavenging action on oxygen in the surface layers.

Nitrogen

Nickel-rich alloys are in general relatively resistant to nitriding and are therefore suitable materials for service in nitrogen-containing atmospheres, e.g. controlled atmospheres utilising the products of cracked ammonia. Inconel (Ni–16Cr–7Fe) is widely used for the fabrication of nitriding boxes while nickel thermocouple-protection tubes give from 5 to 10 years service life at 677–788°C in a bright-annealing atmosphere consisting of 90% nitrogen and 10% hydrogen.

Cyclic heating tests on electrically heated nickel–chromium and nickel–chromium–iron wires in nitrogen[72] showed that the lives to burn-out were at least as good as those obtained in air. The same was true in ammonia at 1 050°C, but at 1 150°C and above the life was less in ammonia than in air (Table 7.40) indicating, as would be expected, that it is atomic nitrogen which is most effectively absorbed.

Table 7.40 Results of cyclic heating tests in various atmospheres[72]

Alloy composition (%)			*Maximum cycle temperature* (°C)	*Life to burn-out* ($s \times 10^{-3}$)		
Ni	Fe	Cr		*Air*	*Ammonia*	75% H_2:25% N_2
80	—	20	1 050	2 268	3 240	
			1 150	522	166	
			1 200	320	43	1 645
30	50	20	1 150	382	212	1 832
			1 200	194	11	342

Table 7.41 lists data for the corrosion of nickel and some nickel alloys by ammonia at 500°C. At higher temperatures the more complex hardened nickel–chromium-base alloys are more resistant than the binary alloys and although the former may be nitrided, little or no surface hardening results. Samples of Nimonic 80A (Ni–20Cr–2·5Ti–1·5Al) exposed in wet cracked

Table 7.41 Corrosion of nickel and nickel alloys in anhydrous ammonia at 500°C (1 540-h tests)[73]

Material composition (%)			Maximum penetration rate (mm/y)	
Ni	Fe	Cr	*Polished*	*Oxidised*
100	—	—	2·00	2·49
80	—	20	0·18	0·13
59	26	15	0·18	0·18
35	46	19	0·43	0·28
78	8	14	0·15	0·25

ammonia for nearly 5 000 h at 816°C showed no evidence of intergranular attack and no changes of microstructure other than those expected due to heating in an inert atmosphere[74]. In creep tests at temperatures above about 800°C where extensive intergranular cracking occurs in the tertiary stage prior to rupture, internal formation of titanium nitride has been observed near the cracks; this is associated with an increased resistance to deformation, and a further secondary stage of creep occurs.

Fuel Ashes

As discussed in Section 7.1, corrosive attack due to the deposition on metallic surfaces of the ashes resulting from fuel combustion can be caused by a variety of media. One of the most important impurities, found particularly in the residual oils, is vanadium, occurring in the resultant ash as vanadium pentoxide. The fluxing action of the ash with which corrosion is normally associated[59,75–76], and the marked difference in severity of attack at temperatures above and below the melting point of the ash have been referred to in the earlier section. At the higher temperatures, numerous alloys investigated, of both high and low nickel contents, were more or less severely corroded, and intensive research (metallurgical and otherwise) has been devoted to this problem.

Earlier studies indicated that nickel and its alloys of high nickel content, particularly those with chromium that are of major industrial importance, behave relatively well, compared with, for instance, steels of lower nickel content. This is consistent with the above-described mechanism of corrosion and Lucas, Weddle and Preece[77] correlated the resistance to attack by vanadium pentoxide of nickel–chromium, cobalt–chromium, and iron–chromium alloys with the oxidation resistance of the base metals.

Further evidence of the relatively good resistance of the nickel-rich alloys to vanadium pentoxide attack was provided by the work of Betteridge and his co-workers[78], and from tests in an experimental gas turbine using residual oil giving an ash vanadium content of about 30%[79]. The large factor of superiority of the nickel-rich materials over an Fe–37Ni–18Cr alloy was the most marked feature of the results of Betteridge *et al.* (Table 7.42). For nickel and a chromium-free nickel–silicon–copper alloy weight losses in air were large but the increase with addition of vanadium pentoxide was not

Table 7.42 Results of tests on nickel and nickel alloys heated at 850°C in contact with vanadium pentoxide[78]

Alloy composition	100 h			250 h			284 h		
	Wt. of V_2O_5 (g)	Wt. loss (g)	Loss due to V_2O_5 (g)	Wt. of V_2O_5 (g)	Wt. loss (g)	Loss due to V_2O_5 (g)	Wt. of† V_2O_5 (g)	Wt. loss (g)	Loss due to V_2O_5 (g)
Nickel	0	0·32	—	0	0·79	—	0	0·88*	—
	0·005	0·39	0·07	0·005	0·57	—	0·01	0·58	—
	0·010	0·42	0·10	0·011	0·46	—			
Ni–20Cr	0	0·020	—	0	0·022	—	0	0·02*	—
	0·006	0·035	0·015	0·005	0·030	0·008	0·01	0·19	0·17
	0·010	0·038	0·017	0·009	0·052	0·030	—	—	—
Ni–20Cr–4Al	0	0·017	—	0	0·016	—	0	0·018*	—
	0·005	0·069	0·052	0·005	0·152	0·136	0·01	0·28	0·26
	0·010	0·115	0·098	0·010	0·163	0·147			
Ni–20Cr–2·5Ti	0	0·148	—	0	0·308	—	0	0·088*	—
	0·006	0·222	0·073	0·005	0·246	—	0·01	0·29	0·20
	0·010	0·185	0·036	0·010	0·257	—			
Ni–20Cr–20Co–2·5Ti	0	0·048	—	0	0·054	—	0	0·06*	—
	0·005	0·077	0·029	0·004	0·128	0·073	0·01	0·236	0·175
	0·009	0·098	0·050	0·009	0·108	0·054			
Ni–14Cr–6Fe	0	0·015	—	0	0·014	—	0	0·015*	—
	0·006	0·029	0·013	0·006	0·023	0·009	0·01	0·186	0·171
	0·010	0·035	0·020	0·010	0·036	0·011			
Ni–10Si–3Cu							0	0·416	—
							0·01	0·435	0·019
Ni–15Cr–5Si							0	0·055	—
							0·01	0·165	0·110
Ni–14Cr–10Si							0	0·02	—
							0·01	0·152	0·132
Ni–15Cr–15Si							0	0·065	—
							0·01	0·208	0·143
Fe–37Ni–18Cr–2Si	0	0·039	—	0	0·036	—	0	0·04*	—
	0·005	0·314	0·275	0·006	0·342	0·305	0·01	1·05	1·01
	0·010	0·426	0·387	0·010	0·418	0·382			

*Estimated from values for 250-h test.
†0·01 g added every 24 h.

proportionately great, suggesting that it is on the chromium-containing oxide films that the corrodant has its major effect. Of interest also in Table 7.42 is the apparently beneficial effect of an addition of 5% silicon to a Ni–15Cr alloy although higher silicon levels produced no further improvement in resistance to attack. Beryllium also has been reported[80] to be a useful alloying element for nickel in this context.

The ash deposits resulting from the combustion of solid and oil fuels often contain appreciable quantities of other corrodants in addition to vanadium pentoxide. One of the more important of these is sodium sulphate, and the effects of this constituent in producing sulphur attack have been mentioned. The contents of sodium sulphate and vanadium pentoxide present in fuel oil ash can vary markedly and the relative merits of different materials depend to a great extent upon the proportions of these constituents. Exposure of heat-resisting alloys of varying nickel, chromium and iron contents to ash deposition in the super-heater zones of oil-fired boilers indicated a behaviour pattern depending on the composition of the alloy and of the ash concerned[81]. In a land-based boiler in which the vanadium pentoxide: sodium sulphate ratio was high, corrosion was due primarily to the fluxing action on the oxide scale, and the alloys containing more than 60% nickel showed greater resistance than those of low nickel content. On the other hand, in a marine boiler in which the ratio of vanadium to sodium sulphate was low, oxidation was accelerated owing to sulphur contamination, and these deposits were most corrosive to the high-nickel alloys. The alloys of intermediate nickel content (30–60%) were the most susceptible to deposits containing vanadium and sodium sulphate in all the proportions investigated. Figure 7.41 summarises the results of this investigation.

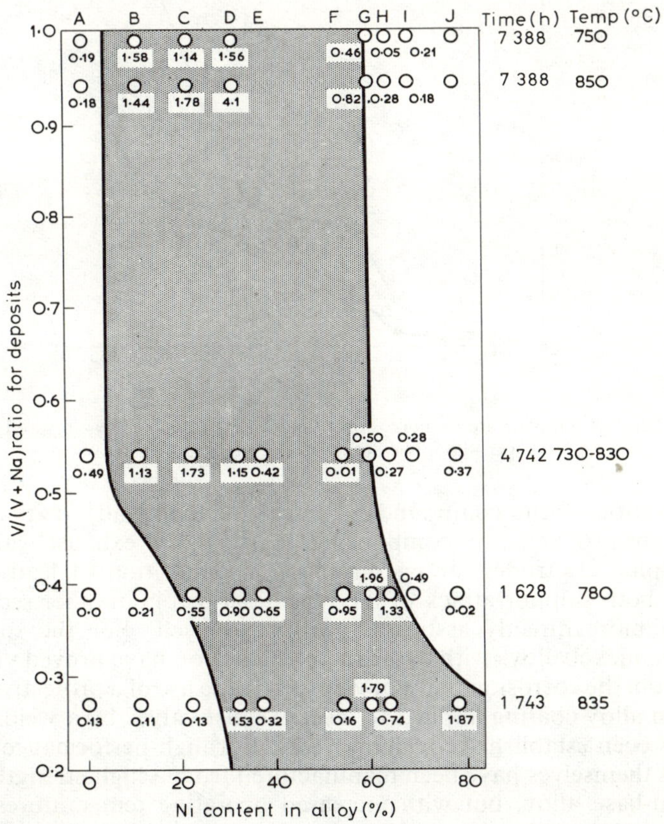

Fig. 7.41 Effect of composition of alloy and deposit on corrosion, shown as weight loss (g/cm^2) against plotted points. (Shaded area shows most attack, i.e. $>0{\cdot}5\ g/cm^2$)[81]

A later survey of a broad range of Ni–Cr–Fe–Co alloys immersed in an 80% V_2O_5 + 20% Na_2SO_4 mixture over the temperature range 700–1 100°C revealed a major effect of chromium content in determining corrosion resistance, the level of chromium required increasing with increasing test temperature[59]. Again at this ratio of vanadium:sodium, attack was largely by the fluxing effect on the protective film of molten vanadates. Results for binary alloys are given in Fig. 7.42. The provision of the high (50–60%) chromium-nickel alloys in usable forms represents perhaps the most promising metallurgical approach to a solution of this particular corrosion problem so far [82–84].

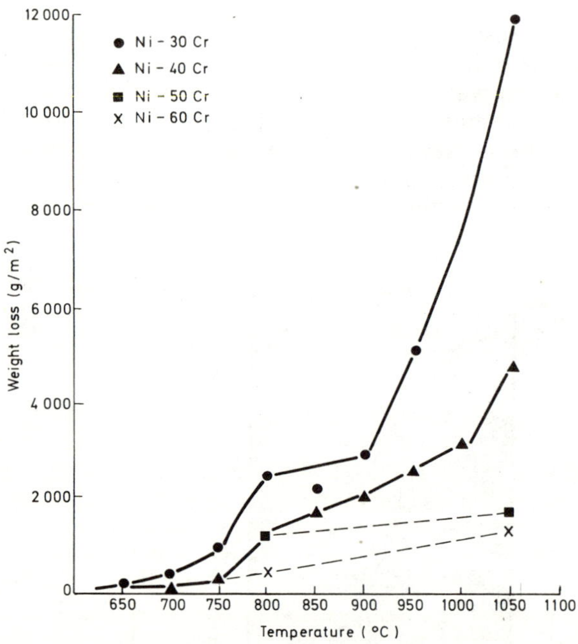

Fig. 7.42 Effect of temperature on corrosion of nickel–chromium alloys exposed to an 80% V_2O_5 + 20% Na_2SO_4 salt mixture for 120 h[59]

The addition of lead compounds to petroleum fuels leads to attack by the combustion products on components, particularly exhaust valves and sparking-plug electrodes, of piston engines. Deposition of lead sulphate results in both sulphur attack of the type described in another section and corrosion more directly associated with lead itself. For the spark-plug electrodes, nickel alloys with manganese and silicon have proved very satisfactory from the corrosion point of view, while the use of a protective nickel–chromium alloy coating applied to steel exhaust valves by a welding technique has been established for many years. For high-performance engines, the valves themselves have been manufactured from a high-strength nickel–chromium-base alloy, but with increased operating temperatures further corrosion resistance has been required and the application of an aluminised coating has been found effective.

Corrosion by Molten Metals and Salts

The physical characteristics of liquid metals and salts and particularly their high heat capacities and thermal conductivities have long established them as materials for use in applications where rapid heat transfer is required at relatively high temperatures. Unfortunately these media are also highly corrosive to metals generally. Corrosion of metallic materials by liquid metals is usually associated with solution of the former in the latter though another important form involves intergranular penetration by the liquid into the solid. Modification of these basic types occurs under conditions involving a temperature gradient, when solution of the solid in the hot region is accompanied by deposition in the cold region, or of a concentration gradient, when a metal from one part of the system is dissolved and redeposited in another region initially deficient in this element. Relatively small amounts of impurities such as oxygen or nitrogen may affect the rate of such corrosion by incompletely understood mechanisms, which may, however, involve such factors as the limit of solubility or the wettability of the solid component, or the formation of a barrier to the reaction.

Nickel and nickel-rich alloys must be considered as having generally poor resistance to molten metals. Eldred[85] has made a systematic investigation of the attack of liquid metals on solid metals and alloys, and his results for nickel, and nickel–chromium and nickel–copper alloys, are summarised in Table 7.43. These are for tests at up to 500°C and apart from potassium and sodium all the low-melting-point metals investigated produced moderate to severe attack on the nickel-rich materials. Furthermore, the values for many of the combinations given in the table indicate a marked tendency to preferential intergranular attack.

Tests in thermal convection loops with a maximum temperature of 800°C[86] showed that nickel and nickel-rich alloys with chromium and other additions were relatively very susceptible to the mass-transfer type of attack by lead. On the other hand nickel-rich alloys with molybdenum were among the better of the alloys tested (Table 7.44).

As in other types of corrosion, attack by molten metals depends to a great extent on the detailed conditions of corrosion. Thus satisfactory service life has been reported for Inconel (Ni–16Cr–7Fe) thermocouple protection tubes in molten lead at temperatures up to 982°C (Table 7.37). For such service, however, the Inconel must be in the solution-annealed condition as the presence of residual stress greatly reduces the service life. Cases of failure in service of nickel–chromium-base alloy tubing due to attack by residual traces of low-melting-point alloys used during the tube manufacture provide further evidence of the susceptibility of these materials to a form of stress corrosion. Russian workers[87] demonstrated a marked reduction in stress-rupture strength of a nickel–chromium alloy in the presence of a lead–bismuth alloy and also the rapid destruction of Ni–20Cr and Ni–27Cr–5Al in the presence of molten tin[88].

The relatively good resistance of nickel-rich materials to attack by molten sodium and potassium has been reported by several investigators. In static tests, nickel and Ni–20Cr alloys showed no evidence of interaction with metallic sodium[89] and Inconel also had good resistance to attack. In dynamic thermal loop tests, stainless steels and Inconel were reported to be the best

Table 7.43 Attack of liquid metals on nickel and nickel-rich alloys[85]

Material type and composition	*Liquid metal*	*Test condition*		*Category of attack**	*Weight change* (g)	*Change in diameter†* cm			
		Temp. (°C)	*Time* (s × 10^{-3})			A	B	C	D
Nickel	Bismuth	500	4·5	S	−1·65	0·086	0·094	0·166	0·172
Brightray (80Ni–20Cr)	Bismuth	500	360	M	+0·77	—	0·046	0·011	0·053
Monel	Bismuth	500	4·5	S	+0·81	—	0·156	0·012	0·118
Nickel	Cadmium	500	36	S	—	0·015	0·111	0·299	0·353
Brightray	Cadmium	500	360	S	+0·21	0·002	0·116	0·008	0·104
Monel	Cadmium	500	0·9	S	−0·20	0·014	0·058	0·046	0·082
Nickel	Mercury	500	360	M	—	ND§	0·005	—	—
Nickel	Potassium‡	400	1 080	N	—	ND	—	—	—
Nickel	Sodium‡	400	1 080	N	—	ND	—	—	—
Brightray	Sodium‡	275	360	N	—	ND	—	—	—
Monel	Sodium‡	275	360	N	—	ND	—	—	—
Nickel	Lead	500	360	S	+0·21	0·002	0·050	—	—
Brightray	Lead	500	360	M	+0·28	0·0045	0·005	—	—
Monel	Lead	500	360	M	+0·22	+0·0005	0·03	—	—
Nickel	Tin	500	36	S	+0·31	0·003	0·053	0·032	0·119
Monel	Tin	500	0·9	M	−0·47	0·022	0·034	0·034	0·047
Nickel	Zinc	500	360	S	+12·2	+0·158	0·06	+0·146	0·08
Brightray	Zinc	500	360	S	+56·7	+1·1	0·35	—	—
Monel	Zinc	500	3·6	M	−0·28	+0·004	0·007	—	0·027

*S, Severe; M, Moderate; N, Very slight or no attack.
†Negative unless stated otherwise; A. Mean change at middle of specimen; B. Including internal penetration at middle of specimen; C. Maximum change in diameter; D. Including internal penetration at position of C.
‡Oxide contamination of liquid metal.
§ND. Not determined.

available structural materials up to 650°C, although at higher temperatures the stainless steel became superior to the nickel alloy on account of preferential leaching of nickel from the latter. Cheng and Ruther's loop tests of nickel in liquid sodium at 550–715°C[90] indicated linear rate relationships at each temperature investigated, an important increase with temperature (activation energy for corrosion of 122 kJ/mol rather greater than that required for simple solution) and negligible effects of oxygen content of the sodium, or neutron flux. The rates of corrosion, however, are some 10–20 times those obtained for stainless steel under comparable conditions.

Tests[91] in a Na–K alloy loop at 570–710°C indicated little or no attack in 6 000 h on Inconel alloy 718 (53Ni–19Cr–19Fe base) or Inconel alloy X750 (73Ni–15Cr–7Fe–2·5Ti) a precipitation-hardening material. The surface quality of the latter alloy was reported to influence strongly the corrosion rate and its dependence on temperature[92]. Lithium is generally much more corrosive than sodium, and alloys containing nickel are not

Table 7.44 Results in thermal convection loop tests of material in contact with molten lead[86]
(Maximum loop temperature 800°C, minimum 500°C)

Material type and composition	*Time to complete plugging of loop* ($s \times 10^{-3}$)	*Remarks*
Niobium	1 962*	No mass transfer
Molybdenum	1 800*	No mass transfer
Hastelloy B (Ni–28Mo–5Fe)	1 814*	Little mass transfer
Ni–25Mo	2 419*	Little mass transfer
Nickel	7·2	Heavy mass transfer
Fe–30Ni	990	Heavy mass transfer
Inconel (Ni–16Cr–7Fe)	324	Heavy mass transfer
Nichrome V (Ni–20Cr)	43	Heavy mass transfer

*Test discontinued.

considered suitable for prolonged use above 600°C because of the mass transfer problem[93].

Even though not molten, uranium rapidly attacks Nimonic 80A (Ni–20Cr–2·5Ti–1·5Al) at temperatures above 740°C, at which temperature the nickel–uranium eutectic melts.

In general nickel–chromium alloys are not suitable for use in contact with molten aluminium. Grogan[94] investigated the degree of attack in $13\frac{1}{2}$ h at 800°C of a molten Al–8Cu alloy on a series of materials. A nickel–chromium alloy was most severely corroded, followed in order of increasing resistance by Cronite (Ni–23Fe–14Cr), an Fe–27Cr–20Ni–4Al alloy, an Fe–18Cr–7Ni–3Mn alloy, chromium–vanadium steel and grey cast iron. Molten copper and copper-base alloys on the other hand do not seriously attack nickel–chromium alloys, which have been successfully used as dies and die inserts for the gravity diecasting of aluminium bronze at pouring temperatures of 1 150°C.

In contact with molten salts, the nickel-base alloys behave much more satisfactorily than is the general experience with molten metals. For this reason they are considered as structural materials in atomic reactors using fluoride mixtures as coolants and are used as vessels for heat-treatment salt baths, as thermocouple sheaths and in similar applications.

In the United States the alloy Inor 8 or Hastelloy N (Ni–16Mo–7Cr–5Fe) has been developed as a container material for molten fluorides containing uranium. The nickel–chromium–iron alloy originally considered as a suitable material proved unsatisfactory, owing to preferential leaching of chromium, but it was established[95] that a chromium content of less than 8% could be tolerated, while reduction below 6% resulted in inadequate oxidation resistance. In thermal convection loop tests with a maximum temperature of 677°C, attack on the nickel–chromium–iron alloy in 1 000 h penetrated up to 0·1 mm, whereas with Inor 8 only very limited attack up to 0·02 mm was experienced in over 6 000 h[96]. A review of the corrosion of several metals and alloys including nickel and various nickel alloys by fluorides, chlorides and hydroxides, has been given by Shimotake[97].

Nickel thermocouple sheaths have been used satisfactorily in salt baths, e.g. in barium chloride at 1 000–1 100°C and in sodium and potassium chlorides at up to 980°C. If the salt is contaminated with sulphur then an Inconel sheath is preferred to one of nickel. The latter type is also satisfactory in cyanide baths used for case-hardening.

A common constituent of salt baths is boric oxide and the use of such mixtures in contact with nickel-rich alloys containing Ti, Al or other strong reducing elements is to be avoided, owing to the possible formation of a nickel–boron eutectic of low melting point.

While the few examples quoted provide some general guidance as to the behaviour of nickel-rich materials in contact with molten metals and salts, it cannot be over-emphasised that such behaviour can be very considerably modified by the presence of very small amounts of contaminants in the liquid media (see Sections 2.9 and 2.10). The effect of very small contents of sodium chloride on the corrosion of nickel-base alloys by sodium sulphate has been referred to previously and other reported examples involving trace amounts, particularly of gaseous impurities, underline the need for great care in interpretation of experimental results.

R. A. SMITH

REFERENCES

1. Sartell, J. A. and Li, C. H., *J. Inst. Met.*, **90**, 92 (1962)
2. Phillips, W. L., Jr., *J. Electrochem. Soc.*, **110**, 1 014 (1963)
3. Moore, W. J. and Lee, J. K., *Trans. Faraday Soc.*, **48**, 916 (1952)
4. Horn, L., *Z. Metallk.*, **40**, 73 (1949)
5. Kubaschewski, O. and Hopkins, B. E., *Oxidation of Metals and Alloys*, 2nd. edn., Butterworths (1962)
6. Zima, G. E., Trans. Amer. Soc. Metals, **49**, 924 (1957)
7. Copson, H. R. and Lang, F. S., *Corrosion*, **15**, 194t (1959)
8. Evans, E. B., *Corrosion*, **21**, 274 (1965)
9. Barrett, C. A., Evans, E. B. and Baldwin, W. M., *Thermodynamics and Kinetics of Metals and Alloys: High-temperature Scaling of Ni–Cr, Fe–Cr, Cu–Cr and Co–Mn Alloys*, U.S. Army Office of Ordnance Res. Lab. Rept. (1955)
10. Kosak, R., Ph.D. Dissertation, Ohio State Univ. (1969)
11. Yearian, H. J. *et al.*, *Corrosion*, **12**, 561t (1956)
12. Pfeiffer, H., *Elektrowärme Technik.*, **4**, 39, 79 (1955)
13. Douglas, D. L. and Armigo, J. S., *Oxidation of Metals*, **2**, 207 (1970)
14. Seybolt, A. V., *Advances in Physics*, **12**, 1 (1963)
15. Pfeiffer, I., *Z. Metallk.*, **53**, 309 (1962)
16. Douglas, D. L., NASA Rpt. ER 72 607 (1969)
17. Ignetov, D. V. and Shamgimova, R. D., NASA Trans. TTF–39 (1960)
18. Malamand, F. and Vidal, G., *Rech. Aéro.*, No. 56, 47 (1957)
19. Wickusik, C. S., AGARD Conf. Proc. No. 52, Paper No. 12 (1970)
20. Richmond, J. C. and Thornton, H. R., Wright Air Development Centre, Tech. Rept. No. 58–164 (1958)
21. Betteridge, W., *The Nimonic Alloys*, Edward Arnold, 234 (1959)
22. Ferré, S., *Rev. Métall*, **56**, 386 (1959)
23. Lund, C. H. and Wagner, H. J., Defence Met. Inf. Centre, Rpt. 214 (1965)
24. Tumarev, A. and Panyushin, L., 'Izvestiya Vysshikh Uchebny kh Zavedeniy Chernaya', *Metallurgiya*, **9**, 125 (1959)

25. Stoff, F. H. and Wood, G. C., *Corrosion Science*, **11**, 799 (1971)
26. Goward, G. W., *J. Inst. Met.*, **22**, 31 (1970)
27. Preece, A. and Lucas, G., *J. Inst. Met.*, **81**, 219 (1952)
28. Gulbransen, E. A. and Andrew, K. F., Amer. Soc. Test. Mater. Special Tech. Pub. No. 171, 35 (1955)
29. Betteridge, W., 4ème Congrès Int. du Chauffage Industriel, October, 1952; Group 1, Section 16, Paper No. 18
30. Lustman, B., *Trans. Amer. Inst. Min. (Metall.) Engrs.*, **188**, 995 (1950)
31. Fleetwood, M. J. and Whittle, J. E., *Brit. Corr. J.*, **5**, 131 (1970)
32. Lund, C. H., Defence Met. Inform. Centre Rpt., **153** (1961)
33. Denny, J. P., Jahnke, L. P., Jones, E. S. and Robertshaw, F. C., A.S.T.M., Spec. Tech. Pub. No. 174, 3 (1956)
34. Brasunas, A. de S., Gow, J. T. and Harder, O. E., *Proc. Amer. Soc. Test. Mat.*, **46**, 870 (1946)
35. Eiselstein, H. L. and Skinner, E. N., A.S.T.M., Special Tech. Pub. No. 165, 162 (1954)
36. Eberle, F., Hoke, J. H., Rozic, E. J. and Leyda, W. E., Wright Air Development Centre Tech. Rept. No. 58–28 (1958)
37. Krainer, H., Wetternik, L. and Carius, C., *Arch. Eisenhüttenw.*, **22**, 103 (1951)
38. Hickman, J. W. and Gulbransen, E. A., *Trans. Amer. Inst. Min. (Metall.) Engrs.*, **180**, 519 (1949)
39. *The Heating of Nickel*, Mond Nickel Co. Ltd., Pub. No. 910A (1955)
40. Pfeiffer, I., *Z. Metallk.*, **66**, 516 (1955)
41. Pfeiffer, I., *Z. Metallk.*, **49**, 267 (1958)
42. Mrowec, S. and Werber, T., *Arch. Hutnictwa*, **9** No. 3, 289–304 (1964)
43. Tideswell, N. W., *Corrosion*, **28**, 23 (1972)
44. Hatfield, W. H., *J.I.S.I.*, **115**, 483 (1927)
45. Ipatev, V. V. and Zheltukhin, D. V., *Metal. Obrab. Metal.*, No. 12, 42 (1958)
46. Lifshin, E., Ph.D. Dissertation, Renssalaer Polytechnic Inst., Troy, N.Y., Jan. (1969)
47. Gruber, H., *Festschrift z. 70 Geburtstage v. W. Heraeus*, Hanau, 45 (1930); *Z. Metallk.*, **23**, 151 (1931)
48. Hancock, P., 1st Int. Congress on Metallic Corrosion, London (1961)
49. Chandler, J. and McQueen, H. J., *Corrosion*, **25**, 126 (1969)
50. Romeo, G., Smeltzer, W. W. and Kirkaldy, J. S., *J. Electrochem. Soc.*, **118**, 740 (1971)
51. Romeo, G., Smeltzer, W. W. and Kirkaldy, J. S., *J. Electrochem. Soc.*, **118**, 1 336 (1961)
52. Viswanathan, R. and Spengler, C. J., *Corrosion*, **26**, 29 (1970)
53. Spengler, C. J. and Viswanathan, R., *Met. Trans.*, **3**, 161 (1972)
54. Jackson, J. H., 4ème Congrès Int. du Chauffage Industriel, Oct. (1952); Group 1, Section 16, Paper No. 144
55. Avery, H. S. and Matthews, N. A., *Trans. Amer. Soc. Met.*, **38**, 975 (1947)
56. Seybolt, A. V., *Corrosion Sci.*, **11**, 751 (1971)
57. Goekel, J. A. and Petit, F. S., *Metal Trans.*, **1**, 1 943 (1970)
58. Shirley, H. T., *J.I.S.I.*, **191**, 144 (1956)
59. Fontaine, P. I. and Richards, E. G., A.S.T.M. Spec. Tech. Pub. No. 421, 246 (1967)
60. Lewis, H. and Smith, R. A., *1st Int. Cong. on Metallic Corrosion*, London, 202 (1961)
61. Sharp, W. H., *Soc. Auto Eng.*, Combined Power Plant and Transport Meeting, Cleveland, Ohio (1965)
62. Richards, E. G. and Twigg, P. L., *Mem. Scientifique. Rev. du Met.*, **68**, 167 (1971)
63. Gerds, A. F. and Mallett, M. W., U.S. Atomic Energy Commission Rept. BMI–1 261 (1958)
64. Bernsen, S. A. *et al.*, U.S. Atomic Energy Commission Rept. TID–7 564, 243 (1958)
65. Kalvenes, O., Piene, K. and Kofstad, P., *Corrosion Sci.*, **4**, 211 (1964)
66. International Nickel Ltd., unpublished data
67. Kline, E. M. and Hall, A. M., *Metals and Alloys*, **21**, 401 (1945)
68. Hopkinson, B. E. and Copson, H. R., Paper to Elevated Temperature Symposium, Nat. Assoc. Corrosion Engng., 16th Ann. Conf., Dallas, March (1960)
69. Dovey, D. M. and Jenkins, I., *J. Inst. Met.*, **76**, 581 (1950)
70. Bucknall, E. H. and Price, L. E., *Rev. Métall.*, **45**, 129 (1948)
71. Gow, J. T., *Corrosion*, **3**, 311, 383 (1947)
72. Pfeiffer, H., *Arch. Eisenhüttenw.*, **29**, 575 (1958)
73. Ihrig, H. K., *Industr. Engng. Chem.*, **41**, 2 516 (1949)
74. Pray, H. A., Peoples, R. S. and Ericson, G. L., U.S. Atomic Energy Commission Rept. BMI–269 (1953)
75. Schlapfer, P., Amgwerd, P. and Preis, H., *Schweiz. Arch. Angew. Wiss.*, **15**, 291 (1949)

76. Sachs, K., *Metallurgia*, **57**, 123, 167, 224 (1958)
77. Lucas, G., Weddle, M. and Preece, A., *J.I.S.I.*, **179**, 342 (1955)
78. Betteridge, W., Sachs, K. and Lewis, H., *J. Inst. Petrol.*, **41**, 170 (1955)
79. Bucher, J. B., *Trans. Inst. Engrs. Shipb. Scot.*, **93**, 275 (1950)
80. Disc. to Ref. 76, *J.I.S.I.*, **182**, 195 (1956)
81. Part 1 of Report of Council of British Manufacturers of Petroleum Equipment, Corrosion I. Cttee, *Brit. Petrol. Equipm. News*, **7**, 54 (1959)
82. Penrice, P. J., Stapley, A. J. and Towers, J. A., *J. Inst. Fuel*, **39**, 300 (1966)
83. Parry, P. J., Bridges, P. J. and Taylor, B., *J. Inst. Met.*, **97**, 373 (1969)
84. Ennis, P. J. and Bridges, P. J., *J. Inst. Met.*, **100**, 346 (1972)
85. Eldred, V. W., U.K. Atomic Energy Research Est. Report No. X/R1 806 (1955)
86. Cathcart, J. V. and Manly, W. D., *Corrosion*, **12**, 87t (1956)
87. Kishkin, S. T. and Nikolenko, V. V., *C.R. Acad. Sci. U.R.S.S.*, **110**, 1 018 (1956)
88. Petrova, A. R., Kaufman, V. G. and Vdovina, L. M., *Metallov term Obrab Metallov*, No. 10, 73 (1969)
89. Hoffman, E. E. and Manly, W. D., Nuclear Engng. Sci. Congr., Cleveland (1955), Preprint No. 74
90. Cheng, C. F. and Ruther, W. E., *Corrosion*, **28**, 20 (1972)
91. Svedberg, R. C., Keeton, A. and Amman, R. L., Westinghouse Astronuclear Lab. Rep. TME 1 857 (1968)
92. Borgstedt, H. V., Drechsler, G. and Frees, G., *Proc. Sym. Alkali Metal Coolants*, Vienna, 119 (1966)
93. Champeix, L., *Energie Nucleaire*, **8**, 471 (1966)
94. Grogan, J. D., *J. Inst. Met.*, **44**, 279 (1930)
95. Badger, F. S., *Chem. Engng.*, **66**, 162, 164, 166 (1959)
96. MacPherson, H. G., U.S. Atomic Energy Commission Rept. ORNL-2 551 (1958)
97. Shimotake, H. and Hessen, J. C., *Adv. Chem. Service*, No. 64, 149 (1967)

7.6 Thermodynamics of Gas–Metal Systems

The formation of metal compounds by gas/metal reactions clearly falls within the province of corrosion as defined in Section 1.1. In addition, for the purpose of this section, it is convenient to include gas/metal reactions where small amounts of alloying elements are removed from the metal, or elements are introduced into the metal from the gas phase. These processes can be discussed both from the standpoint of the kinetics of the processes, which involves a knowledge of reaction mechanisms and diffusion coefficients, and from that of equlibria of the chemical reactions. Although a knowledge of the thermodynamics alone of a particular reaction will be of value only in determining the final position of equilibrium of the reaction, it is nevertheless of first importance at high temperatures where equilibrium can usually be readily achieved, especially in gaseous phases, well-stirred liquids, solid interfaces and metal surfaces in contact with gases. In these systems, thermodynamics is of more value than in aqueous systems where the rate at which equilibrium is reached may be very slow. As an example it will be shown how the thermodynamic method can be used to predict the nature of the carbide phases which precipitate from 18/8 stainless steel at elevated temperatures.

The thermodynamic functions which are required to be manipulated for studying high-temperature equilibria are relatively few in number and the mathematical equations involved can be reduced to about six. The fact that consideration will be given here only to high-temperature processes also leads to a considerable simplification of the algebra. For example, the standard free-energy change when an ordered compound MX is formed between metal M and non-metal X is given at temperature T by the equation

$$\Delta G^{\ominus}_{T} = \Delta H^{\ominus}_{0} + \int_{0}^{T} \Delta Cp\, \mathrm{d}T - T \int_{0}^{T} \Delta Cp\, \mathrm{d}\,\ell n T$$

where $\Delta H^{\ominus}_{0}$ is the heat of the reaction at the absolute zero and ΔCp is the difference in heat capacities between reactants and products. Above 298K a good approximation for the above equation is

$$\Delta G^{\ominus}_{T} = \Delta H^{\ominus}_{298} - T\Delta S^{\ominus}_{298}$$

where $\Delta H^{\ominus}_{298}$ and $\Delta S^{\ominus}_{298}$ are the standard heat and entropy change in the reaction at 298K. Thus the free-energy change is a simple linear function of

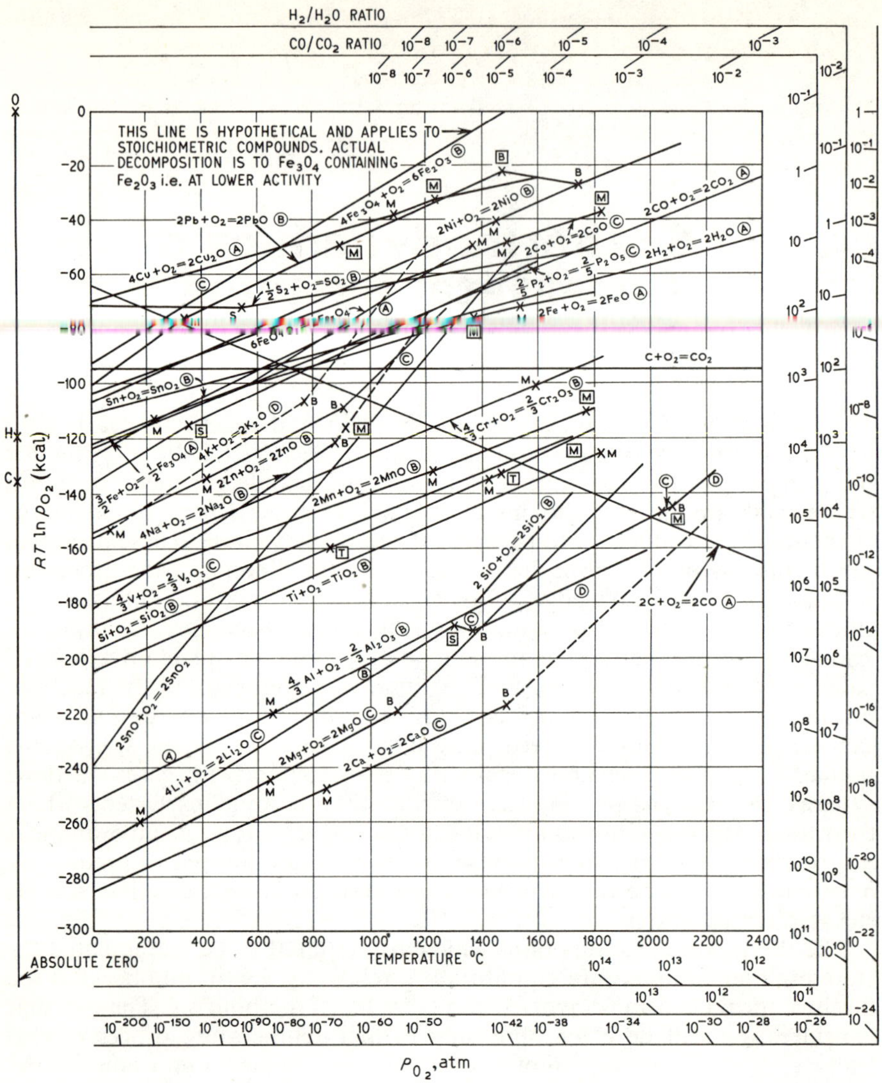

Fig. 7.43 Free energy of formation of oxides (after Richardson and Jeffes[1])

KEY				
Ⓐ	suggested accuracies	± 1 kcal	M, M	: melting point, metal, oxide, resp.
Ⓑ		± 3 kcal	B, B	: boiling point, metal, oxide, resp.
Ⓒ		± 10 kcal	S, S	: sublimation point, metal, oxide, resp.
Ⓓ		$\pm > 10$ kcal	T, T	: transition point, metal, oxide, resp.

Note: 1 kcal $\approx$ 4·2 kJ

temperature and diagrams can be constructed which show this relationship when compounds are formed between say, metals and oxygen[1] (Fig. 7.43)*.

*In this section there are a number of $\Delta G^{\ominus}$–T diagrams in which $\Delta G^{\ominus}$ is expressed in the original unit of *kilocalories*, rather than in the SI units of *joules*. This has been done in order to avoid the difficult redrawing and recalculating of the eight diagrams involved.

These $\Delta G/T$ relationships are straight lines having changes in slope at temperatures where either the metal or oxide melts, boils or undergoes a crystallographic transformation. These changes in slope can be taken into account by adding or subtracting, depending on whether the metal or the metal oxide is involved, the free-energy change for a phase change. Thus for a transformation t, the following correction is required:

$$\Delta G_t = \Delta H_t - T\Delta S_t$$
$$= \Delta H_t - \frac{T\Delta H_t}{T_t}$$

where T_t is the transformation temperature.

The standard free-energy change for a reaction $M+X \rightarrow MX$ is also related to the equilibrium constant for the corresponding reaction

$$K = \frac{a_{\mathrm{MX}}}{a_{\mathrm{M}}a_{\mathrm{X}}} \qquad \text{...(7.3)}$$

by the equation

$$-\Delta G^{\ominus} = RT\,\ell\text{n}\,K = 19{\cdot}15T\log K \qquad \text{...(7.4)}$$

Substituting for K in equation 7.4 from equation 7.3

$$-\Delta G^{\ominus} = RT\,\ell\text{n}\,a_{\mathrm{MX}} - RT\,\ell\text{n}\,a_{\mathrm{M}} - RT\,\ell\text{n}\,a_{\mathrm{X}}$$

which can be written as

$$-\Delta G^{\ominus} = \Delta\mu_{\mathrm{MX}} - \Delta\mu_{\mathrm{M}} - \Delta\mu_{\mathrm{X}}$$

where $\Delta\mu_i = RT\,\ell\text{n}\,a_i$ is the chemical potential of i.

Thus, as a simple example, when a pure metal M is in contact with the pure metal compound MX, then $a_{\mathrm{MX}} = a_{\mathrm{M}} = 1$ and

$$-\Delta G^{\ominus} = -\Delta\mu_{\mathrm{X}}$$

Furthermore, if an alloy of M is in contact with pure MX,

$$-\Delta G^{\ominus} = -\Delta\mu_{\mathrm{M}} - \Delta\mu_{\mathrm{X}}$$

and finally when pure M is in contact with a solid or liquid solution containing MX

$$-\Delta G^{\ominus} = \Delta\mu_{\mathrm{MX}} - \Delta\mu_{\mathrm{X}}$$

A chemical reaction can occur only if $-\Delta G > 0$, i.e. if $-\Delta G$ is positive; in addition the value $a = 1$ is by definition the maximum activity for a condensed component where the pure phase is taken as standard state, thus $\Delta\mu$ is always negative. This discussion will be restricted to gases where $p \leqslant 1$; taking $p = 1$ atm ($101{\cdot}325\,\mathrm{kN/m^2}$) as the standard state for the gas, X, it is evident that $\Delta\mu_{\mathrm{X}}$ is always a negative quantity or zero.

The pressure of oxygen developed by an intimate mixture of a pure metal and its oxide is conventionally referred to as the dissociation pressure ($p_{\mathrm{O_2}}$) and it can be seen that at a given temperature T

$$-\Delta G_T^{\ominus} = -\Delta\mu_{\mathrm{O_2}} = -19{\cdot}15T\log p_{\mathrm{O_2}}$$

Thus the sequence of the free-energy lines in the diagram for a series of metal oxides shows those with the highest dissociation pressures at the top (smallest negative values of $\Delta\mu_{O_2}$) and those with the lowest dissociation pressures at the bottom, i.e. in the sequence of their stabilities. Clearly if a pure metal is exposed to an oxygen-bearing atmosphere where the pressure of oxygen is greater than the dissociation pressure of the oxide at the operating temperature, the metal oxide will be formed; if the gas contains a lower oxygen pressure than the dissociation pressure, the oxide cannot be formed as a pure phase. It also follows from the equations given above that

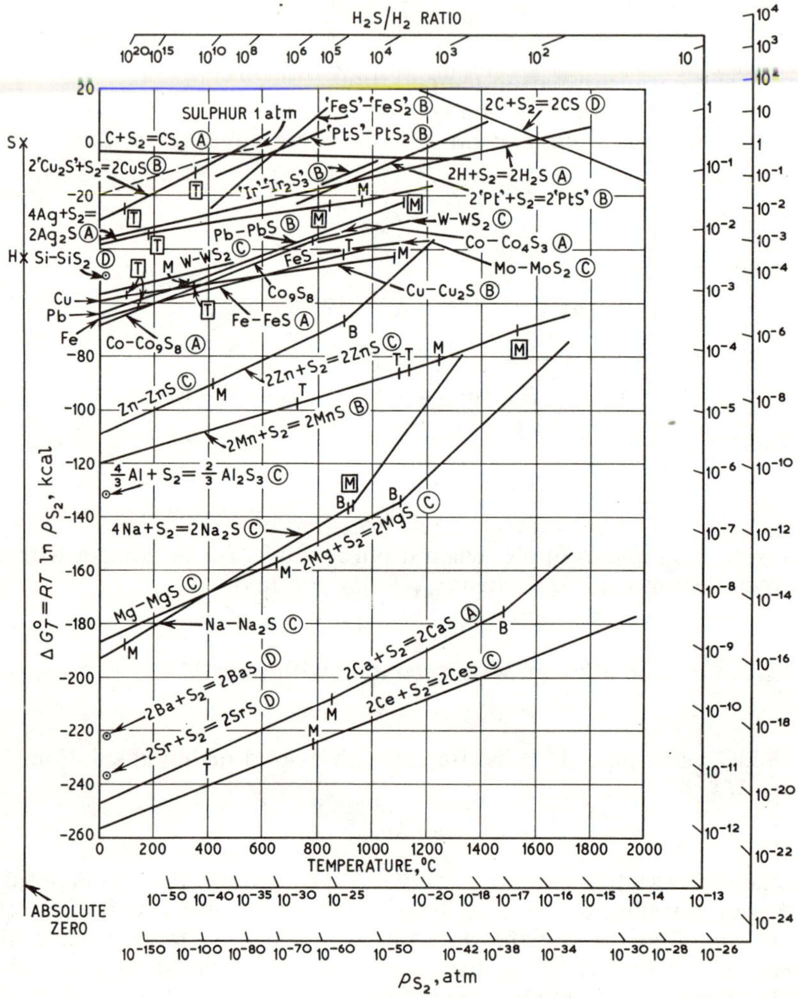

Fig. 7.44 Free energy of formation of sulphides (after Richardson and Jeffes[2])

KEY				
Ⓐ	suggested accuracies	± 1 kcal	M, [M]	: melting point, metal, sulphide, resp.
Ⓑ		± 3 kcal	B, [B]	: boiling point, metal, sulphide, resp.
Ⓒ		± 10 kcal	S, [S]	: sublimation point, metal, sulphide, resp.
Ⓓ		$\pm > 10$ kcal	T, [T]	: transition point, metal, sulphide, resp.

Note: 1 kcal $\approx$ 4·2 kJ

an alloy of a metal will require a higher oxygen pressure to form the pure oxide, and conversely a pure metal can form an oxide solid solution or liquid slag at a lower pressure than that required for pure oxide formation.

Similar free-energy diagrams, which can be interpreted in exactly the same way, have been constructed for sulphides[2], carbides[3] and nitrides[4] (Figs. 7.44 to 7.46).

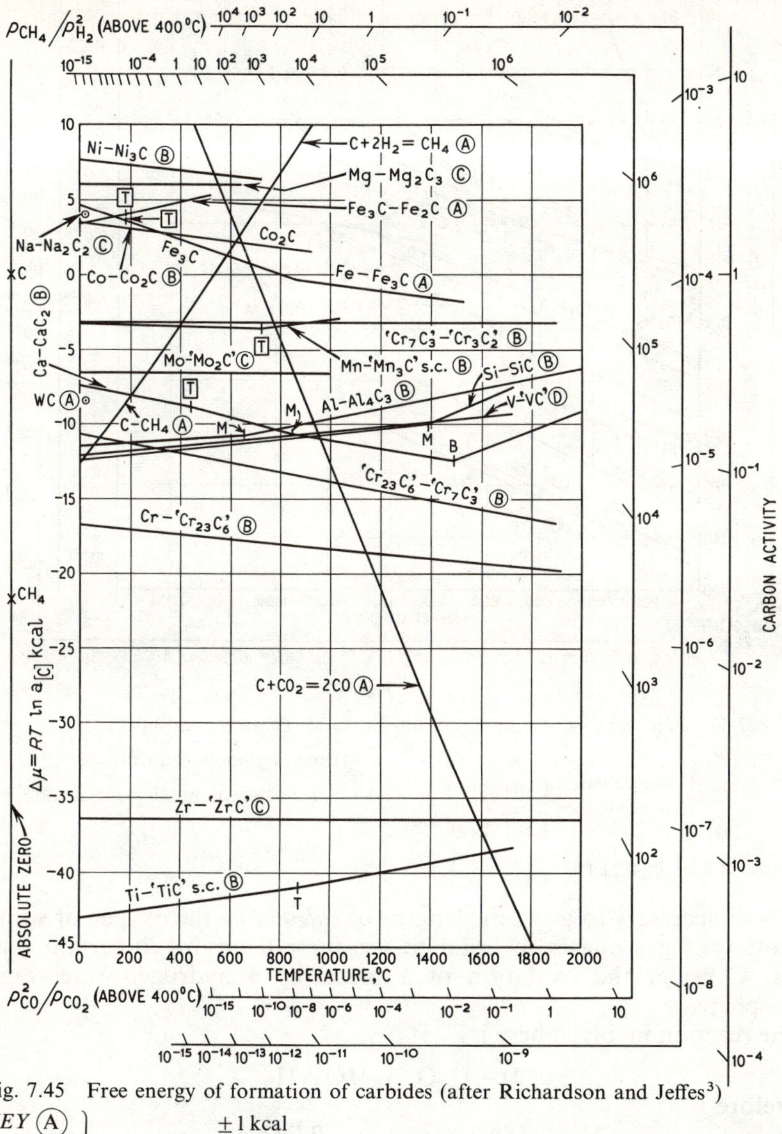

Fig. 7.45 Free energy of formation of carbides (after Richardson and Jeffes[3])

KEY Ⓐ suggested accuracies ±1 kcal
Ⓑ ±3 kcal — M, [M]: melting point, element, carbide, resp.
Ⓒ ±10 kcal — T, [T]: transition point, element, carbide, resp.
Ⓓ ±>10 kcal — B: boiling point, element.

Note: 1 kcal ≈ 4·2 kJ

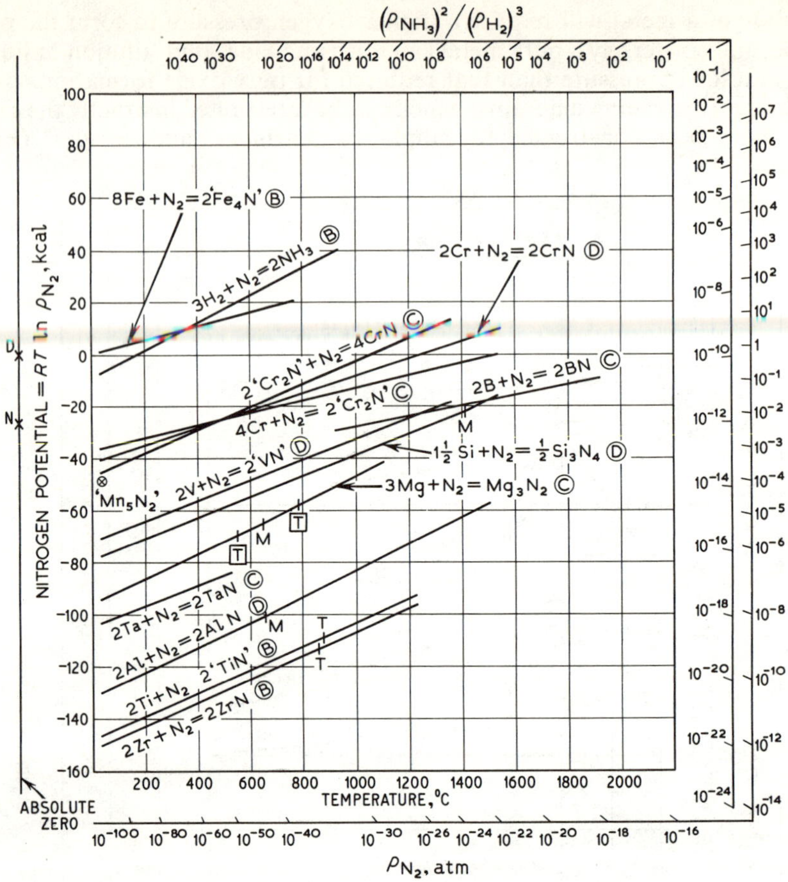

Fig. 7.46 Free energy of formation of nitrides (after Pearson and Ende[4])

KEY Ⓑ, Ⓒ, Ⓓ — suggested accuracies: Ⓑ ± 3 kcal; Ⓒ ± 10 kcal; Ⓓ $\pm >10$ kcal

M: melting point, element.

T, [T]: transition point, element, nitride, resp.

Note: 1 kcal $\approx$ 4·2 kJ

It is unnecessary to go to the lengths of *calculating* the oxygen or sulphur potentials of gas phases in order to use these diagrams in certain simple cases. Consider the oxidation of a metal by a hydrogen/water-vapour atmosphere.

The reaction involved here is

$$M + H_2O \rightarrow MO + H_2$$

Therefore

$$-\Delta G^\ominus = (\Delta\mu_{MO} - \Delta\mu_M) - RT\,\ell n\, p_{H_2O}/p_{H_2}$$

Thus when the oxidation of pure metal to pure oxide is being considered

$$-\Delta G^\ominus = -RT\,\ell n\, p_{H_2O}/p_{H_2} = -\Delta\mu_{H_2O} + \Delta\mu_{H_2}$$

and the oxygen dissociation pressure of the metal/metal oxide is the same

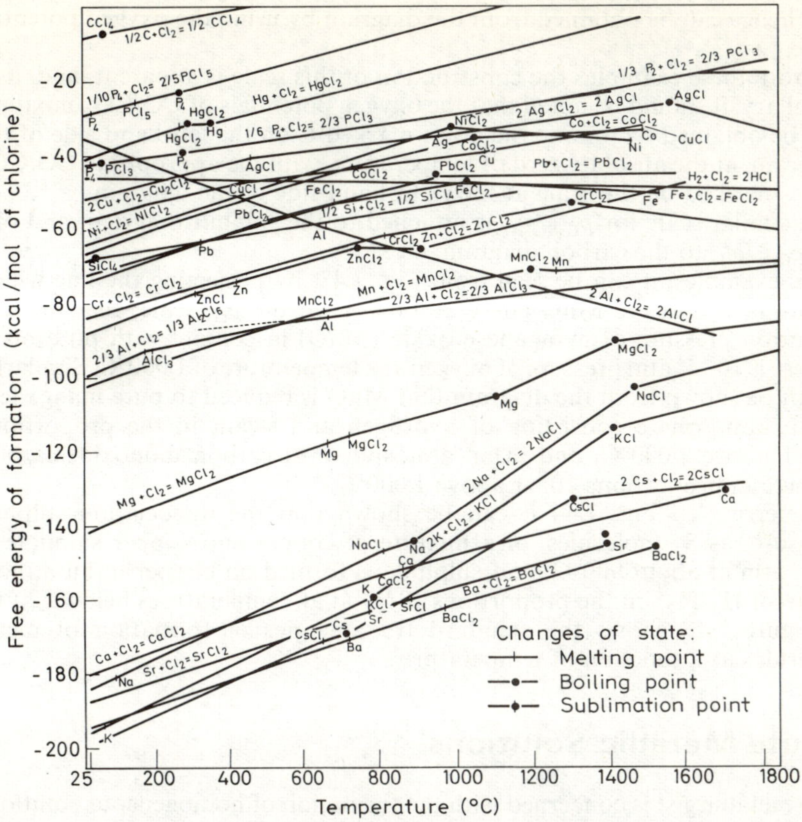

Fig. 7.47 Standard free energies of formation of metal chlorides as a function of temperature (after Villa[5]). *Note:* 1 kcal ≈ 4·2 kJ

as that of the hydrogen/water-vapour mixture. It is thus a practical advantage to have a scale around the edge of the diagram showing values of $\Delta\mu_{O_2}$ (oxygen potential) for various ratios of p_{H_2O}/p_{H_2}.

For a mixture of hydrogen and steam at equal partial pressures, the oxygen potential will be equal to the standard free energy of formation of water vapour from hydrogen and oxygen at all temperatures. Therefore, the extrapolation of the standard free-energy line for the system $2H_2+O_2 \rightarrow 2H_2O$ intersects the H_2/H_2O scale at the right-hand side of the diagram at the ratio 1/1. The point marked *H* on the left-hand side of the diagrams (Figs. 7.43 and 7.44) is the extrapolation of the same line to the absolute zero, and is thus equal to the standard heat of formation of water vapour from hydrogen and oxygen at 298K.

When the hydrogen/water vapour ratio is 100/1, the point on the H_2/H_2O ratio scale representing this ratio is obtained by subtracting the chemical potential for a product molecule (i.e. H_2O in the reaction $H_2+\frac{1}{2}O_2 \rightarrow H_2O$) at an activity of 1/100 from the standard free-energy line and extrapolating the resulting line to meet the scale at the point marked '$10^2/1$'. It should be observed that the value of the chemical potential for any substance at an activity of 0·01 is the same as that for oxygen at a pressure of 10^{-2} atm, and hence can be obtained from the diagram by using the oxygen potential scale.

From these examples the construction of this scale is apparent, and, as a corollary, it should be noted that the oxygen potentials of CO/CO_2 mixtures can be obtained by joining the point marked *C* on the left-hand side of the diagram, at the absolute zero of temperature, with the appropriate CO/CO_2 ratios marked on the scale at the right-hand side of the diagram.

A similar scale for p_{H_2S}/p_{H_2} is attached to the sulphur diagram, and one for $p_{CH_4}/p^2_{H_2}$ to the carbon diagram, etc.

As examples, it can be seen from Fig. 7.43, by projecting the line which connects O and the MnO curve at 1 000°C to the p_{O_2} scale, that the dissociation pressure of manganese oxide (MnO) in contact with pure manganese is 10^{-24} atm pressure of oxygen at a temperature of 1 000°C. Similarly, it can be shown from the diagram that MnO is reduced to pure manganese in an atmosphere consisting of hydrogen and steam in the proportions 10^4:1 above 1 000°C, and in an atmosphere of carbon monoxide/carbon dioxide in proportions 10^5:1 above 1 000°C.

Referring to Fig. 7.44 it can be shown that the dissociation sulphur pressure, as S_2 molecules, of a mixture of copper and copper sulphide is 10^{-8} atm at about 900°C, and sulphide is formed on copper in an atmosphere of H_2/H_2S in the proportions 10^3:1 at all temperatures below 720°C.

Figure 7.47 shows the standard free energies of formation of metal chlorides as a function of temperature[5].

Dilute Metallic Solutions

The metallurgist is concerned with the formation of homogeneous solutions of small amounts of impurities in metals as well as with the formation of compounds. The limit of solubility of impurities is frequently very small, less than one atomic per cent in concentration, and in these dilute solutions Henry's law is applicable, i.e. the activity of a dilute solute is proportional to the concentration of solute in the solution. Consider a dilute solution of an element *A* which has a high vapour pressure in the pure state at the temperature *T*, the vapour being monatomic, in solution in element *B* which has an immeasurably low vapour pressure at the same temperature. Then if the pressure of *A* could be measured unambiguously for a range of dilute solutions it would be found that

$$p_A = kx_A = k'(A)^* \qquad \ldots(7.5)$$

where x_A is the mole fraction of *A* in solution and (*A*) is the atomic, or alternatively the weight per cent of *A* in solution. If the vapour species of *A* were di-atomic instead of monatomic, then

$$p_{A_2} = kx_A^2 = k'(A)^2 \qquad \ldots(7.6)$$

which is Sievert's law for the solution of diatomic gases in metals. Similarly

*The value of k' obviously depends on whether atomic or weight per cent is used for expressing the concentration of *A*.

for a triatomic gas, e.g. SO_2

$$p_{SO_2} = kx_S x_O^2 = k'(S)(O)^2 \qquad \dots(7.7)$$

In such a binary solution, the chemical potential of the solute $\Delta\mu_A$ and that of the solvent $\Delta\mu_B$ are related to the *integral free energy* of formation of the solution, ΔG^S, per mole, containing a mole fraction x_A of component A, and x_B for component B, by the expression

$$\Delta G^S = x_A\Delta\mu_A + x_B\Delta\mu_B$$

Corresponding to the integral heat and entropy of formation of the solution are the partial molar heats ΔH_i and entropies ΔS_i of solution of the components where

$$\begin{aligned}\Delta\mu_i &= \Delta H_i - T\Delta S_i\\ \Delta H^S &= x_A\Delta H_A + x_B\Delta H_B\\ \Delta S^S &= x_A\Delta S_A + x_B\Delta S_B\end{aligned}$$

From the algebraic form of these equations it can be seen that the partial and integral values of a thermodynamic function for a solution are interrelated simply. Figure 7.48 shows the integral value of a function ΔZ for a binary solution, as a function of x_B. At any given mole fraction of each

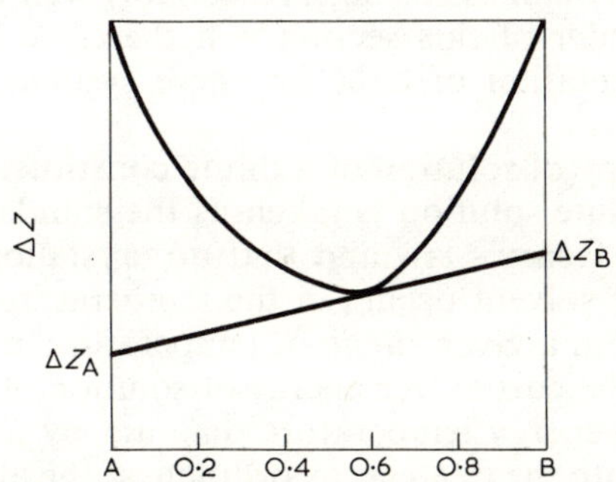

Fig. 7.48 Relationship between partial and integral properties

component, the relevant values of the partial properties can be obtained by drawing a tangent to the integral curve at the given composition of the solution; ΔZ_A and ΔZ_B are the intercepts of this tangent with the A-rich and B-rich sides respectively. It also follows that ΔZ_A is the change in the value of the function Z for the component A when 1 mol of A is transferred from the standard state, usually the pure substance, to an infinite volume of the solution of the given concentration, so that the concentration of each species in the solution remains unchanged during the operation.

In the dilute range of concentration the partial heat of solution, ΔH_A, of A would be constant and the partial entropy of solution would be given by

$$\Delta S_A = a - R\,\ell n\, x_A \qquad \dots(7.8)$$

where a is a constant, and the standard state for A is pure A.

At higher concentrations, Henry's law no longer applies and activities must be substituted for the concentration terms. This statement implies that in the region of solution where Henry's law is valid, the activity coefficient γ of A, defined by

$$a_A = \gamma_A x_A = \gamma'_A(A)$$

is constant. Further, if we choose pure A as the standard state for A then γ_A has a constant value whose magnitude depends on the chemical identity of A, whereas if the standard state is the infinitely dilute solution of A then γ'_A has the constant value of unity in the Henry's law region. The equilibrium constant for, say, a dilute solution of sulphur in iron in equilibrium with an H_2/H_2S atmosphere can be simply expressed[6] as

$$K = (p_{H_2}/p_{H_2S}) \times (\%S)_{Fe}$$

when an infinitely dilute solution of sulphur in iron is taken as the standard state.

Solutions in Solid Iron

The dilute solutions of elements in solid iron are, at present, the only system for which the thermodynamics has been reasonably well worked out experimentally. The remainder of this section will therefore be devoted to the diagrammatic representation of data for these systems which have been evolved by Richardson[3].

The heat and entropy of solution of a dilute constituent remain constant when the infinitely dilute solution is taken as the standard state, provided that the solute obeys Henry's law and that no crystallographic change or change of state of the solvent occurs in the temperature range under consideration[7]. Thus within a given range of temperature in which the solvent remains unchanged, the partial free energy of solution of the solute may be represented on a free-energy/temperature diagram by a straight line. The intercept of this line with the free-energy ordinate at the absolute zero equals the heat of solution, and the slope gives the partial entropy of solution. However, when the stability of a dilute solution of a substance in iron is being compared with the stabilities of compounds, it is preferable to use the pure substance as standard state, in which case the slope of the free-energy line for the dilute solution of given concentration is given by equation 7.8. At the temperature at which the solvent undergoes a change in crystal structure there will be a discontinuous break in the line and, in the new structure, the free-energy line will have a different intercept at the absolute zero, indicating a change in the heat of solution, and a different slope which indicates a change in the constant a in equation 7.8 for the partial entropy of solution.

By making use of experimental data for solutions of carbon in iron, Richardson has constructed a free-energy diagram showing the partial free energy of solution of carbon in α, γ and δ, and liquid iron (Fig. 7.49) which is similar to the earlier Ellingham diagram[8]. The figure is divided by the broken lines into areas of constant crystal structure of the iron solvent, and each area is traversed by lines (dotted in Fig. 7.50) showing the partial free energy of solution of carbon, at a concentration shown on each line, as a function of temperature. The intercepts at the absolute zero, marked α, γ, δ, and liquid, give values for the partial molar heats of solution of carbon in each crystallographic form of the solvent.

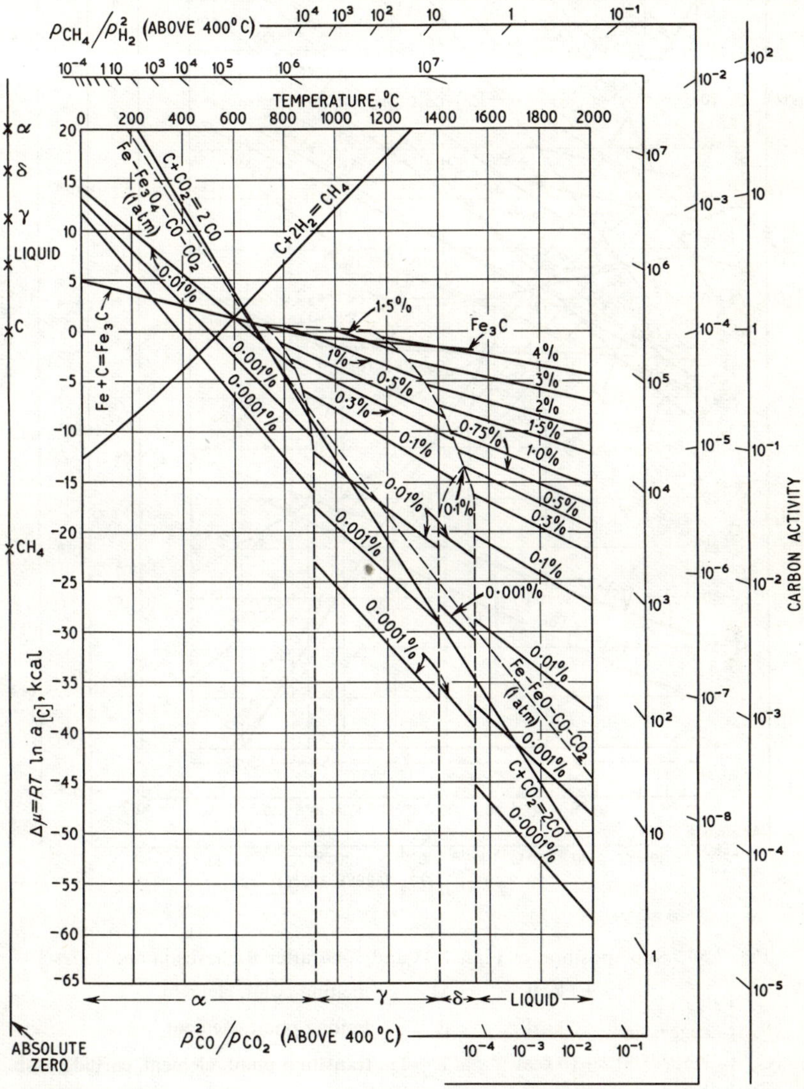

Fig. 7.49 Carbon potentials in iron (after Richardson and Jeffes[3]). Suggested accuracy of all lines ±1 kcal

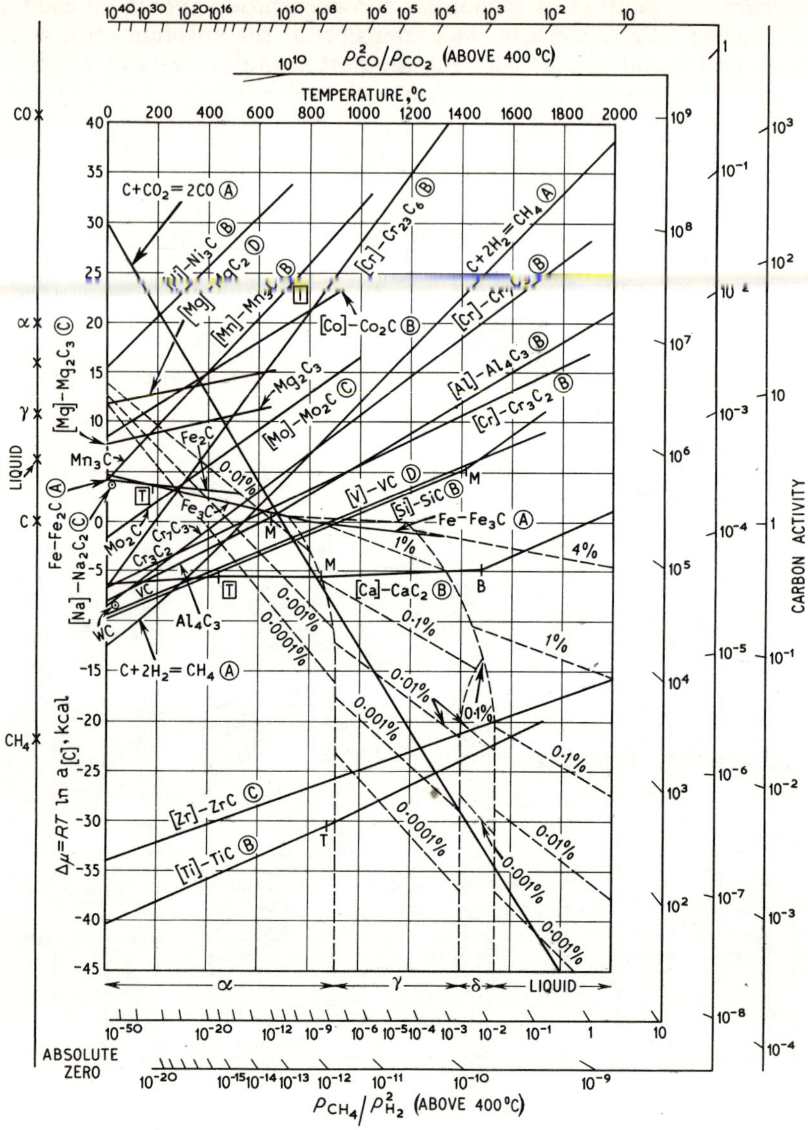

Fig. 7.50 Superposition of Figs. 7.45 and 7.49 (after Richardson and Jeffes[3])

KEY				
(A)	suggested accuracies	±1 kcal	M:	melting point, element.
(B)		±3 kcal	B:	boiling point, element.
(C)		±10 kcal	T, [T]:	transition point, element, carbide, resp.
(D)		± >10 kcal		

Note: 1 kcal ≈ 4·2 kJ

Segregation of Carbides

This diagram may be usefully combined with the standard free-energy diagram for the carbides (Fig. 7.45) to indicate the equilibrium conditions under which carbide particles will segregate from a given solution consisting of an alloying element and carbon in solid iron. Figure 7.50 shows how the diagram for carbon in pure iron and the diagram for the formation of carbides by metals at an activity of 0·01 can be superimposed. The combined diagram can be used to calculate, for example, the temperature below which particles of vanadium carbide can be expected to begin separating from an iron alloy containing vanadium at an activity of 0·01, with respect to pure vanadium as standard state, and carbon at a concentration of 0·1 wt%. Since only a small amount of vanadium is present in the alloy, the activity of carbon at this concentration will be the same as that in pure iron to a good approximation. Thus the solution diagram needs no amendment. However, the carbide diagram gives the standard free energies of formation of carbide from the pure substances and the alloy contains vanadium at an activity of 0·01. It is necessary, therefore, to draw a straight line joining that for vanadium carbide at the absolute zero, and spaced a distance equal to $19{\cdot}15T \log 0{\cdot}01$ *above* this line, across the diagram. (The values of this function can be read by joining the cross marked *C* on the middle left-hand side of the diagram with 'carbon activity' $= 10^{-2}$ on the right-hand side of the diagram.) When this line is drawn, it can be seen that the line for

$$[V]_{Fe} + C \rightarrow VC$$

lies above that for 0·01% C in pure iron at temperatures above 840°C but is below it at temperatures lower than 840°C. Clearly then, carbide particles can begin to form as a separate phase only below 840°C in this alloy. At higher concentrations of vanadium in iron the carbide can form, with 0·01 wt% of carbon, at higher temperatures, but not significantly higher until a large proportion of vanadium, raising the activity of vanadium by an order of magnitude, is present.

Similar diagrams for sulphides and nitrides can be constructed from the data given here and the work of Rosenquist and Dunicz[9], and Darken and Gurry[10].

Effects of Large Amounts of Alloying Elements

The diagrams which have just been described are of only limited value because the presence of an alloying element in solution in the iron influences the thermodynamic behaviour of the solute. Thus it is well known that the solubility of gases in metals at constant pressure is changed by addition of alloying elements[11], and since this is only another way of saying that the activity coefficient of the gas atoms in the solution has been changed, we might expect this effect to pervade the whole field of alloy-solution thermodynamics.

The direction in which a given alloying addition will change the activity coefficient of another dilute solute can be predicted semi-quantitatively along the following lines[12]. Let us consider the simple case of a random solution of atoms *A*, *B* and *S*, where *S* is the dilute solute, in which the energy of binding of a given *S* atom in the solution may be obtained by assigning a fixed value to the energy of interaction between the *A–S* and *B–S* pairs which are formed and the number of *A–B* pairs which are broken when 1 mol of *S* is introduced into a large amount of the alloy, under the restriction that the alloy remains at constant composition during the process. It can be shown that

$$\Delta H_S^{A-B} = x_A \Delta H_S^A + x_B \Delta H_S^B - x_A \Delta H_A^{A-B} - x_B \Delta H_B^{A-B}$$

the solution being regarded as so dilute that *S–S* pairs are not formed. We may then approximate

$$\Delta H_S = RT \ell n\, \gamma_S$$

in which case

$$\ell n\, \gamma_S^{A-B} = x_A\, \ell n\, \gamma_S^A + x_B\, \ell n\, \gamma_S^B - \frac{\Delta G_{AB\,xs}}{RT}$$

where γ_S^{A-B} is the activity coefficient of *S* in the *A–B* alloy of atom fraction x_A of *A* and x_B of *B*, γ_S^A is the activity coefficient of *S* in solution in pure *A*, γ_S^B is that for *S* in solution in pure *B*, and $\Delta G_{AB\,xs}$ is the excess free energy of mixing of the *A–B* alloy at this composition. The derivation of this equation has not been attempted here, and the interested reader should consult Reference 12 for further details.

It can now be seen that if the *A–B* alloy is ideal, i.e. $\Delta G_{A-B\,xs} = 0$, then the activity coefficient of *S* will be *decreased* by addition of *B* if *B* interacts more strongly with *S* than does *A*, while it will be *increased* if *B* interacts less strongly than *A* with *S*. If the excess free energy of mixing of the *A–B* alloy is positive, or, as an approximation, if the heat of formation of the *A–B* alloy is positive, then γ_S will be *decreased* by the addition of *B* even when *A* and *B* have interactions of equal strength with *S*, and, conversely, γ_S will be *increased* if the heat of mixing is negative in these circumstances. The effects are shown diagrammatically in Fig. 7.51.

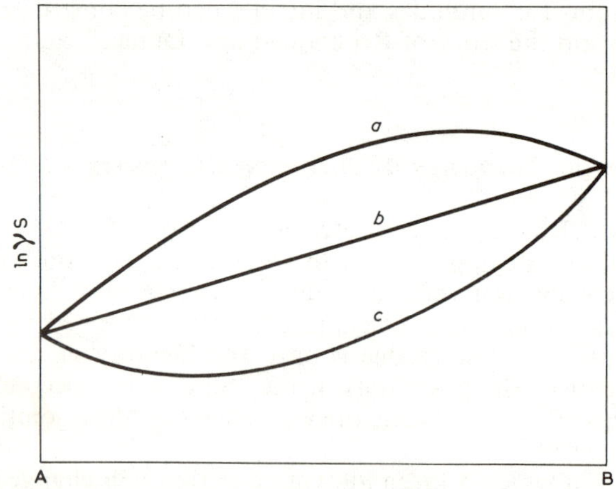

Fig. 7.51 Variation of log γ_S in *A–B* alloy. Curve *a*, *A–B* solvent ideal; *b*, *A–B* solvent with ΔG^{XS} negative; *c*, *A–B* solvent with ΔG^{XS} positive

This simple expression can be used to obtain only a semi-quantitative idea of the effect of an alloying element because the assumptions of randomness and a constant pairwise energy of interaction between atoms are only approximations to the truth in most systems.

For quaternary and more complex alloys a suggestion of Chipman and Sherman[13] might be used. Chipman's school have made use of the symbol ε_C^X for the rate of change of $\ell n\, \gamma$ of the dilute solute, C, with small additions of alloying elements, X. Thus for the solution of carbon in iron:

$$\varepsilon_C^X = \frac{\partial\, \ell n\, \gamma_C^{Fe}}{\partial x_X}$$

and it has been suggested that for small additions of several alloying elements to iron, the effect on the activity coefficient of a solute, such as carbon, can be obtained from the expression[14]

$$\varepsilon_C^{X-Y-Z} = x_X \varepsilon_C^X + x_Y \varepsilon_C^Y + x_Z \varepsilon_C^Z$$

Values of ε_C^X which have so far been obtained experimentally[15] are shown below.

Alloying element	Si	Cr	Mn	Mo
$\varepsilon_C^X = \partial\, \ell n\, \gamma_C / \partial x_X$	10	−4·3	−0·5	−0·8

These apply to *liquid* iron as the solvent.

The Segregation of Carbides from Stainless Steel Containing Small Amounts of Carbon

As an example of the way in which these data could be used, the temperatures at which carbides separate from an 18/8 stainless steel are calculated for carbon contents of 0·1, 0·01, 0·001 and 0·0001 wt%. These calculations, which of necessity involve several approximations due to our present lack of knowledge, demonstrate the value of the thermodynamic approach to problems involving the precipitation of phases which may have a pronounced effect on the corrosion behaviour of the alloy (see Section 3.3).

The steel will be considered to be an ideal ternary solution, and therefore at all temperatures $a_{Cr} = 0{\cdot}18$, $a_{Ni} = 0{\cdot}08$ and $a_{Fe} = 0{\cdot}74$. Owing to the γ-phase stabilisation of iron by the nickel addition it will be assumed that the steel, at equilibrium, is austenitic at all temperatures, and the thermodynamics of dilute solutions of carbon in γ iron only are considered.

The effect of nickel on the activity coefficient of carbon will be neglected and the effect of chromium will be taken from the value in the liquid state. From the values quoted above, $\varepsilon_{Cr}^C = -4{\cdot}3$ at 1 600°C, and assuming that the effect of chromium is simply to change the *heat* of solution of carbon in iron then the point γ on the left-hand side of Fig. 7.49 must be depressed by an amount

$$\partial \Delta H_{C(\gamma)} = -4{\cdot}576 \times 1873 \times 0{\cdot}335 = -12{\cdot}02\,\text{kJ}$$

the last term being the change in log γ_C at 1 600°C when $x_{Cr} = 0{\cdot}18$.

Now in the case of chromium carbide separation from the steel, three possible crystal structures may be taken up, those of Cr_4C, (or $Cr_{23}C_6$),

Cr_7C_3 and Cr_3C_2. It is necessary first to calculate the free energies of formation of the compounds from pure chromium and carbon. The results are:

$$23/6\ Cr + C \rightarrow 1/6\ Cr_{23}C_6 \quad \Delta G^{\ominus} = -16\,380 - 1{\cdot}54T\,\text{cal} \quad A$$

$$7/3\ Cr + C \rightarrow 1/3\ Cr_7C_3 \quad \Delta G^{\ominus} = -13\,900 - 2{\cdot}05T\,\text{cal} \quad B$$

$$3/2\ Cr + C \rightarrow 1/2\ Cr_3C_2 \quad \Delta G^{\ominus} = -10\,000 - 1{\cdot}39T\,\text{cal} \quad C$$

Since the chromium activity is 0·18 for the formation of carbide in the steel, each of the standard free-energy lines (A°, B°, C°) derived for the carbides must be corrected (moved upwards) for this lower activity (A′, B′, C′).

Finally, the lines for the formation of these carbides at activities 0·01* [e.g. A′ (0·01), B′ (0·01)] are shown in Fig. 7.52 and a summary of the results of the calculation is given in Table 7.45.

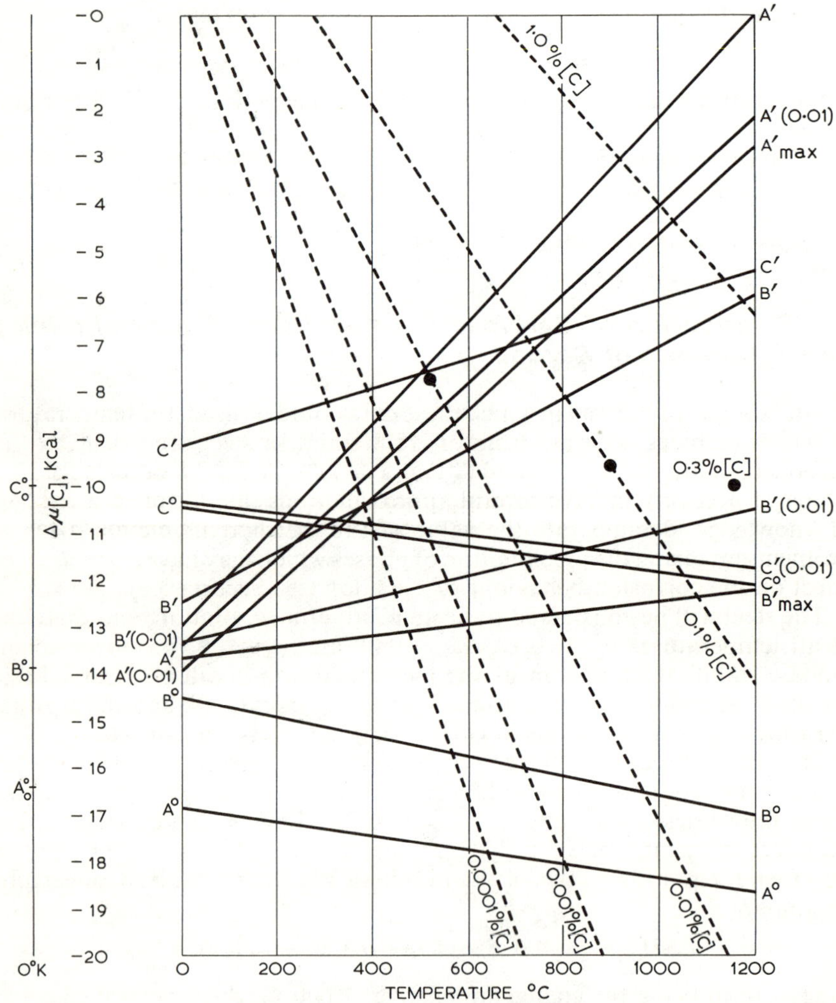

Fig. 7.52 Carbide formation free-energy diagram for Fe–18Cr–8Ni

*Activities at 0·01 have been chosen to represent a reasonable extreme for mole fraction of chromium in the carbide.

Table 7.45 Values of $\Delta\mu_C$ for chromium carbides at 1 000K

Product	$\Delta G^\ominus$	$\Delta G^\ominus - n\Delta\mu_{Cr}$ ($x_{Cr} = 0{\cdot}18$)	$\Delta G^\ominus - n\Delta\mu_{Cr} + m\Delta\mu_{carbide}(a = 0{\cdot}01)$
$Cr_{23}C_6$; $n = 23/6$, $m = 1/6$	−17 940	−17 940+13 000 = −4 940	−4 940−1 520 = −6 460 cal
Cr_7C_3; $n = 7/3$, $m = 1/3$	−15 950	−15 950+ 7 900 = −8 050	−8 050−3 060 = −11 110 cal
Cr_3C_2; $n = 3/2$, $m = 1/2$	−11 400	−11 400+ 5 060 = −6 340	−6 340−4 580 = −10 920 cal

It can be seen from the diagram that a phase consisting of Cr_7C_3 at an activity of 0·01 [B′(0·01)] would segregate at 1 000, 720, 550 and 440°C from steels containing 0·1, 0·01, 0·001 and 0·000 1% carbon respectively. Phases containing $Cr_{23}C_6$ would be unstable with respect to this segregation at the temperatures stated, but would separate at an activity of 0·01 at temperatures of 710, 540, 440 and 370°C. A Cr_3C_2 phase might appear at 0·01 activity at the highest carbon content and above 830°C, and below 1 030°C.

Now to complete the solution of the problem one would need to know the solution laws for iron, and a small amount of nickel, in each of these carbide phases, since equilibrium requires that a_{Fe} and a_{Ni} in the segregated carbide must be 0·74 and 0·08 respectively as well as a_{Cr} being 0·18. At present nothing is known about these laws except that the metal atoms might well be randomly distributed in the carbide phase, in which case, as an example,

$$a_{Cr_7C_3} = x^7_{Cr_7C_3}$$

in the ideal case.

It is known that $Cr_{23}C_6$ can contain up to 25% iron on the metal atom sites, and Cr_7C_3 up to about 60% iron[16]. Therefore the minimum activities of $Cr_{23}C_6$ and Cr_7C_3 have been calculated for these phases of *maximum* iron content using the ideal laws to calculate activities from carbide composition. The free-energy lines which were thus obtained are shown as $A'_{(max.)}$ and $B'_{(max.)}$.

The picture which emerges from this extremely simplified calculation is that a Cr_7C_3 phase should always precede a $Cr_{23}C_6$ phase in segregation from stainless steel and that the latter should appear at a temperature of 780°C for the carbon content of 0·1 wt%.

According to the phase diagram which has gained acceptance for this system, the Cr_7C_3 phase never appears at low carbon contents, and a cubic phase of the $Cr_{23}C_6$ type separates at 900°C for 0·1% [C], and about 500°C for 0·01% [C]. These points, together with one for 0·3% [C] are shown by the black dots on the diagram. An agreement between the calculated and measured temperatures and compositions for carbide segregation could thus be achieved only by strong departures from the ideal laws in the carbide phases. Alternatively it is possible that the separation of a Cr_7C_3 phase has not so far been observed because of rapid transformation in the solid state to the $Cr_{23}C_6$ phase which is stable at lower temperatures. Such a transformation has been observed in the α Fe–Cr alloys.

Concentrated Ternary Solutions

When both solutes are present in large amounts, i.e. greater than about 1 at. % of each, no simple theoretical treatment is available to predict their mutual effects on thermodynamic properties. In this case, recourse must be made to the various solutions of the ternary Gibbs–Duhem relation

$$x_1 \, d\mu_1 + x_2 \, d\mu_2 + x_3 \, d\mu_3 = 0$$

In order to make any practical use of this equation, a good deal of experimental data are usually required for a ternary system, and it will be found that, at present, such data are seldom available in the literature. The methods of evaluation of such data are fully described in the works of Chipman and Elliott[17] and of Schuhmann[18].

J. C. B. ALCOCK
E. EASTERBROOK

REFERENCES

1. Richardson, F. D. and Jeffes, J. H. E., *J.I.S.I.*, **160**, 261 (1948)
2. Richardson, F. D. and Jeffes, J. H. E., *J.I.S.I.*, **171**, 165 (1952)
3. Richardson, F. D. and Jeffes, J. H. E., *J.I.S.I.*, **175**, 33 (1953)
4. Pearson, J. and Ende, U., *J.I.S.I.*, **175**, 52 (1953)
5. Villa, H., 'Thermodynamic Data of the Metallic Chlorides', *J. Soc. Chem. Ind.*, No. 1 (supplementary issue), S9–S18 (1950)
6. Sherman, C. W., Elvander, H. I. and Chipman, J., *J. Metals*, **2**, 234 (1950)
7. Lumsden, J., *Thermodynamics of Alloys*, Inst. Metals, London (1952)
8. Ellingham, H. J. T., *J. Soc. Chem. Ind., Lond.*, **63**, 125 (1944)
9. Rosenquist, T. and Dunicz, B. L., *J. Metals*, **4**, 604 (1952)
10. Darken, L. S. and Gurry, R. W., *Physical Chemistry of Metals*, McGraw Hill, New York (1953)
11. Sieverts, A. and Krumbhaar, H., *Ber. Dtsch. Chem. Ges.*, **43**, 893 (1910)
12. Alcock, C. B. and Richardson, F. D., *Acta Metallurg.*, **6**, 385 (1958)
13. Chipman, J. and Sherman, C. W., *J. Metals*, **4**, 597 (1952)
14. Wagner, C., *Thermodynamics of Alloys*, Addison-Wesley, New York (1952)
15. Chipman, J., *J.I.S.I.*, **180**, 97 (1955)
16. Goldschmidt, H., *J.I.S.I.*, **160**, 345 (1948)
17. Elliott, J. F. and Chipman, J., *Trans. Faraday Soc.*, **47**, 138 (1951)
18. Schuhmann, R., *Acta Metallurg.*, **3**, 219 (1955)

8 EFFECT OF MECHANICAL FACTORS ON CORROSION

8.1 Mechanisms of Stress-corrosion Cracking

The visible manifestations of stress corrosion are cracks that create the impression of inherent brittleness in the material, since the cracks propagate with little attendant macroscopic plastic deformation. In fact, a metal that fails by stress corrosion is usually found to conform to the normal ductility standards for that material and the combination of circumstances that cause a normally ductile metal to behave in this way are the presence of a specific environment, a tensile stress of sufficient magnitude and, usually, a specific metallurgical requirement in terms of the composition and structure of the alloy. The compositions and structures of the alloys and the properties of the environments involved in the various instances of failure indicated in this section are so widely varying as to suggest that rationalisation of all of these experiences in a single explanation would be difficult if not unreal, i.e. it is probable that a number of different mechanisms are involved. This is not to suggest that some systematisation is not possible and indeed the objective of this section is to show that the evidence may be rationalised in a continuous spectrum of mechanisms[1], a concept that has the merit of avoiding the dangers inherent in believing that some highly specific conditions need to be fulfilled for stress corrosion to occur because the mechanism of failure is invariable.

The usual energy balance approach to fracture (*see* Section 8.9)—by equating the strain energy released to the energy consumed in creating new surface and in achieving plastic deformation—needs modification where corrosion processes are involved to take account of the chemical energy released, and it is the latter that distinguishes stress corrosion from other modes of fracture not involving environmental interaction:

$$\text{Surface energy change} + \text{Plastic work done} = \text{Change in initial stored energy} + \text{Electrochemical energy released} \qquad \ldots(8.1)$$

Since the surface energy term will usually be negligible by comparison with the plastic work term in the stress corrosion of ductile materials, it may be neglected. The remaining terms may be derived from fracture mechanics and conventional electrochemical considerations and, for the various boundary conditions indicated by West[2], result in

$$P = K_I^2(1-\nu^2)/E+(zF\rho\delta/M)\eta \qquad \ldots(8.2)$$

where P is an appropriate plastic work term, K_I is the stress intensity, ν is Poisson's ratio, E is Young's modulus, z is the valency of the solvated ions, F is Faraday's constant, ρ is the density, δ is the height of the advancing crack front, M is the molecular mass of the metal and η is the anodic overpotential. At the threshold stress intensity K_{ISCC}, i.e. the minimum value of K_I for stress-corrosion cracking, equation 8.2 yields

$$K_{ISCC} = \{(E/1-\nu^2)(P-zF\rho\delta\eta/M)_{min.}\}^{\frac{1}{2}} \qquad \ldots(8.3)$$

Clearly the variables that may influence K_{ISCC}, or the threshold stress for initially plain specimens, and hence the susceptibility to stress corrosion cracking, are P and η, i.e.

$$K_{ISCC} = \{k_1(P-k_2\eta)_{min.}\}^{\frac{1}{2}} \qquad \ldots(8.4)$$

Lowering of the plastic work term P will result from an increase in the effective yield stress or an increase in the work hardening rate in the crack tip region, either of which, for constant η, will lower K_{ISCC} and hence increase the susceptibility to stress corrosion. An increase in the anodic overpotential η_a (the potential of the metal is made more positive) relative to the plastic work term will also increase susceptibility to stress corrosion; η_a will be some function of the electrochemical conditions within the crack that control the active to passive transition which determines whether or not cracking occurs, i.e. it will be some function of pH, anion activity, metal composition and electrode potential. The interdependence of these terms upon the structure and composition of the metal, upon the details of the electrochemical conditions at the crack tip in terms of local cell action and film formation, and upon the response of the metal to the presence of stress in creating new metal at the crack tip, make the quantification of the argument extremely difficult and its relation to the detailed mechanisms of crack propagation virtually impossible, as West[2] indicates. However, the recognition of the need for a critical balance between a number of variables if stress corrosion is to occur, and the fact that this balance may be achieved in a number of ways, is important, and not least in relation to the diagnosis and the prevention of stress-corrosion failures.

Stress-corrosion Crack Propagation Models

The implication of the foregoing equations, that stress-corrosion cracking will occur if a mechanism exists for concentrating the electrochemical energy release rate at the crack tip or if the environment in some way serves to embrittle the metal, is a convenient introduction to a consideration of the mechanistic models of stress corrosion. In so far as the occurrence of stress corrosion in a susceptible material requires the conjoint action of a tensile stress and a dissolution process, it follows that the boundary conditions within which stress corrosion occurs will be those defined by failure under a stress in the absence of corrosion, and failure by corrosion in the absence of stress. Between these extremes, wherein stress corrosion occurs, it is necessary to consider how corrosion processes may be influenced by the application of stress to a metal and how fracture may be facilitated by corrosion.

Remembering that stress corrosion involves very localised dissolution if the geometrical requirements of a crack are to be fulfilled, the rate of anodic dissolution may be expressed as a rate of crack propagation

$$v = \frac{i_a M}{zF\rho} \qquad \ldots(8.5)$$

where i_a is the anodic current density and the remaining symbols are as defined earlier. Now i_a, and hence v, will be dependent upon the nature of the phase being dissolved and also upon the associated cathodic processes that occur elsewhere and which need to produce a current sufficient to balance that at the crack tip. The chemical natures of the sites for these reactions will therefore be of importance, but how does the imposition of a tensile stress influence this situation to produce stress corrosion? Despic, *et al.*[3] have shown that the dissolution rate of iron in acid solution undergoes a marked rise when the strain passes from the elastic to the plastic condition in dynamic straining experiments and that this result is due to the exposure of high index planes and of edges at slip steps, as well as increasing surface roughness, as plastic deformation occurs. Stresses in excess of the yield stress may therefore produce locally enhanced activity at surfaces where slip steps emerge, i.e. i_a in equation 8.5 may be increased by stresses promoting plastic strain. However, there is no difficulty in accounting for the observed rates of crack propagation in stress corrosion in terms of the currents passed at static bare surfaces without invoking arguments involving markedly enhanced currents resulting from the exposure of slip steps. The real problem is in explaining why the corrosion proceeds along a narrow front to retain the geometry of a crack, implying that most of the exposed surfaces, including the crack sides, must remain relatively inactive. The transition from electrochemically active to relatively inactive behaviour that the sides of a crack must undergo as the tip advances and creates more crack can only be achieved if the environment forms a film, and this implies in relation to equation 8.5 that the conditions for stress corrosion will be met if i_a is maintained close to the film-free value, i.e. the metal is in the active state and protective films are not allowed to grow over the crack tip or, if this does occur, that the film is repetitively broken. The function of stress then in the realistic conditions of film-forming environments will be essentially to prevent, or to fracture, films forming at the crack tip.

In this context, two different circumstances[1] may now be envisaged whereby cracks can propagate by a dissolution-controlled process. The alloy may exhibit structural features, either as a segregate or precipitate, usually at the grain boundaries, that cause a local galvanic cell to be established, i.e. a *pre-existing active path* is involved, as originally suggested by Dix[4]. The precipitate or segregate may act as the anode in the local cell or, by acting as an efficient cathode, may cause the dissolution to be localised upon the immediately adjacent matrix (Fig. 8.1). The lattice characteristics in the region of a grain boundary are such that equilibrium segregation of solutes or the nucleation and growth of precipitates are favoured reactions, and so grain boundaries in particular are potential sites for chemical heterogeneity. Where such pre-existing active paths are non-existent, or are inoperative, the disruption of a protective surface film to expose bare metal may result in a second mechanism of crack propagation, as originally suggested by Parr and

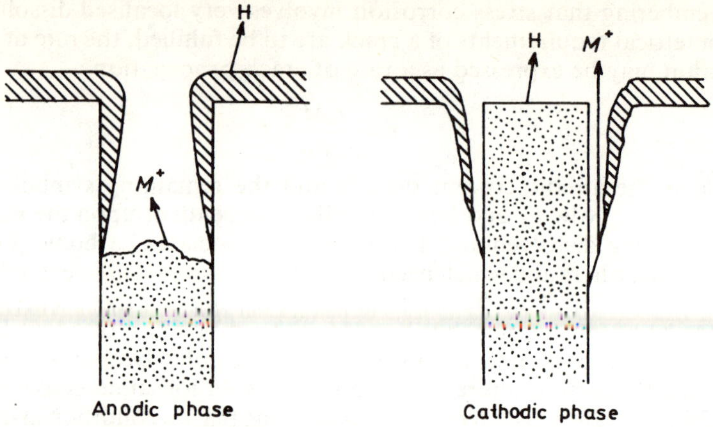

Fig. 8.1 Pre-existing active path mechanisms, in which H represents cathodic hydrogen

Straub[5]. Thus stress (or probably more correctly, *plastic strain*) in the underlying metal may bring about the disruption of the protective film whereby active metal is exposed, in the manner shown schematically in Fig. 8.2. The active path along which the crack propagates is cyclically generated as disruptive strain and film build-up alternate with one another, or propagation is related to the slip characteristics of the underlying metal. This *strain-generated active path* mechanism would be expected to result in transgranular cracks.

The opposite boundary condition referred to earlier as limiting the regime in which stress corrosion occurs was that of purely mechanical fracture in the absence of corrosive processes. The energy balance of equation 8.1 indicates that, with negligible contribution from dissolution, crack extension

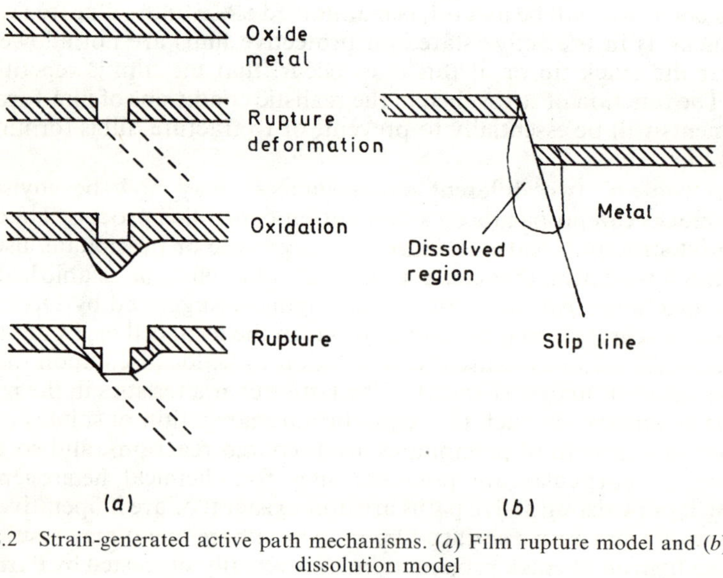

Fig. 8.2 Strain-generated active path mechanisms. (*a*) Film rupture model and (*b*) slip step dissolution model

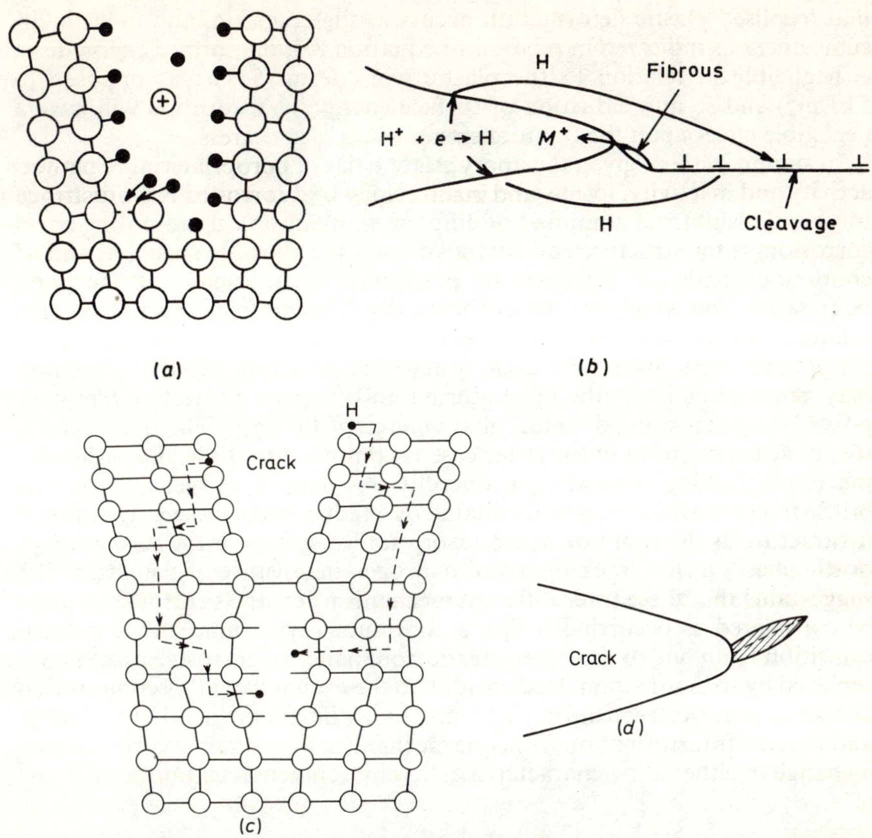

Fig. 8.3 Mechanisms involving embrittlement of the metal. (*a*) Crack tip adsorption, (*b*) hydrogen adsorption, (*c*) decohesion by hydrogen influx to dilated lattice and (*d*) crack extension due to brittle hydride particle forming at crack tip

will be facilitated by a reduction in the surface energy required to form crack faces or a reduction in the plastic work term by embrittlement of the metal in the crack tip region. If the environment provides species that are adsorbed at the crack tip to reduce the effective bond strength, then the surface energy is effectively lowered, alternatively the species may diffuse into the metal forming a brittle phase, e.g. a hydride, at the crack tip or adsorption may occur at some region in advance of the crack tip where the stress and/or strain conditions are particularly appropriate for the nucleation of a crack (Fig. 8.3). In the latter case hydrogen is usually regarded as the only species that can diffuse with sufficient speed to account for observed crack propagation rates, and in the present context, hydrogen embrittlement, with the hydrogen derived from corrosion reactions, is considered as a particular instance of stress corrosion. Whilst surface-energy lowering has been suggested[6] as a single mechanism that explains all instances of stress-corrosion cracking, it has particular difficulties in explaining the phenomenon in the more ductile metals. Thus, whilst stress-corrosion cracks propagate without marked macroscopic plastic deformation there is ample evidence to show that localised plastic deformation occurs at the crack tip, and in such circumstances as indicated in relation to equation 8.1, the surface energy term is negligible in relation to the plastic work term ($5\,J/m^2$ as opposed to $5\,kJ/m^2$) and so any reduction of surface energy by adsorption will have a negligible effect upon the fracture stress.

In summary then, given the appropriate balance between electrochemical activity and inactivity, localised corrosion may be distributed in a number of ways and result from a number of different mechanisms in promoting stress corrosion. If the structure and composition of the alloy are such that almost continuous paths of segregate or precipitate exist, usually at the grain boundaries, and which are electrochemically different from the matrix, then a latent susceptibility to intergranular corrosion may be activated by the presence of stress. Where pre-existing active paths are inoperative, the strain may generate active paths by rupturing an otherwise protective film and, possibly, by activating dissolution at emerging slip steps. The more crucial rôle of stress or strain in the latter case is continued to those alloys that are inherently lacking somewhat in ductility and have a propensity towards brittle fracture, which may be facilitated by a reduction in the energy required for fracture as the result of either adsorption of species or the formation of brittle phases at the crack tip, or of hydrogen in advance of the latter. The suggestion[1] that these three different mechanisms of stress corrosion should be considered as occurring within a continuous spectrum, with a gradual transition from one to the others as the dominance of corrosive processes is replaced by stress or strain, leads readily to the notion that alloy composition and structure, electrochemistry and stress may interact in a variety of ways, and that the transformation from one mechanism to another may result from a change in either alloy characteristics or environmental conditions.

The Stress-corrosion Spectrum

Stress corrosion has been the subject of a number of extensive reviews[7-9] resulting from major conferences in recent years, and these, together with the following sections of this volume, avoid any need for a general review of the data. Instead, consideration will be given to some of the implications of the various ideas already referred to in relation to stress corrosion in a variety of systems.

Pre-existing Active Paths

Where cracking is associated with pre-existing active paths, structurally sensitive attack, such as intergranular corrosion, may be expected to be observed upon unstressed specimens, at least in the earlier stages of exposure before secondary reactions such as those that lead to film formation, obscure the metal structure. In Section 8.2 it is shown that samples of steel polished as for metallographic examination suffer grain boundary attack when immersed for short times under environmental conditions that would lead to cracking of stressed specimens, and that whilst such grain-boundary corrosion does not penetrate to great depths in the absence of stress, it is possible

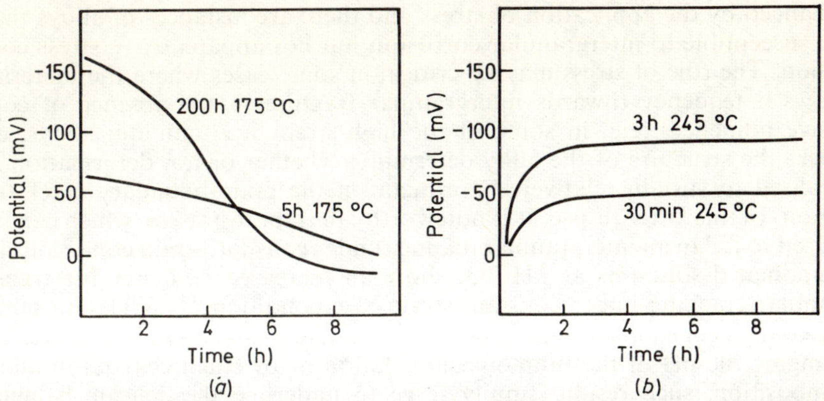

Fig. 8.4 Potential of grain boundary region relative to the bulk corrosion potential of stressed and unstressed specimens of aged (*a*) Al–7·2Mg and (*b*) Al–4·4Cu alloys as a function of time (after Doig and Edington[10])

to disintegrate a piece of unstressed mild steel by intergranular corrosion in a boiling nitrate solution with applied anodic current. The observation that carbon steels contain pre-existing susceptible paths for corrosion, the structural distribution of which is related to the paths followed by stress-corrosion cracks, is also apparent in the aluminium-base alloys that undergo stress-corrosion cracking (Section 8.5).

The electrochemical properties of the segregates or precipitates, relative to their associated matrixes, that are involved in these instances of intergranular attack have been the subject of a number of studies. Doig and Edington[10] have used microelectrodes to measure the localised corrosion potentials at grain boundaries in Al–Mg and Al–Cu alloys and their results correlate well with the cracking propensities of these alloys. Figure 8.4 shows the potentials at grain boundaries relative to the bulk corrosion potential for two alloys immersed in aerated, neutral 1M NaCl. In both alloys the potential at the grain boundaries became more negative with ageing by heat treatment, the results being unaffected by an applied tensile stress, the preferential dissolution being upon the anodic constituents, i.e. upon the Al_3Mg_2 precipitate in the Al–Mg alloy and the copper-depleted zone in the Al–Cu alloy. In the former case the exposed area of Al_3Mg_2 decreases as it dissolves and the change in corrosion potential difference with time reflects this, whilst the preferential dissolution of the copper-depleted zone in the Al–Cu alloy, being continuous along the boundary, does not decrease in area with time and the corrosion potential difference thus remains constant. The effects of ageing time in these results reflect their effects upon cracking susceptibility; similarly, heat treatments designed to change the distribution of chemical heterogeneity of grain boundary regions in steels alter the cracking propensities of the latter in ways which support the suggestion that intergranular stress corrosion in these relatively low-strength ductile materials is related to a latent susceptibility to intergranular corrosion.

Of course, an inherent susceptibility to intergranular corrosion is not the only requirement for susceptibility towards intergranular stress corrosion, since for the latter to occur it is necessary for the former to be sustained or enhanced by the application of stress and there are instances of alloys that are susceptible to intergranular corrosion but not apparently to stress corrosion. The rôle of stress may be critical in some cases where the material shows a tendency towards intergranular fracture in the absence of corrosive influences, e.g. in some of the high-strength aluminium alloys, or where the structure of the alloy determines whether or not deformation is localised to sustain relatively bare metal in the grain-boundary crack-tip region. In this latter respect it is noteworthy that the α-brasses, which can be caused to fail in an intergranular manner under constant strain conditions in ammoniacal solutions at pH 7·3, show an increased tendency for transgranular cracking under constant-strain-rate conditions[11]. Since the same material, at constant strain, can be made to fail in a transgranular manner by changing the pH of the ammoniacal solution or by small changes in alloy composition, such results simply serve to underline the delicate balance between the factors that promote a particular mechanism of cracking and of the dangers in attempting to rationalise all observations into a single mechanism. Moreover, even within those systems that exhibit intergranular stress corrosion the part played by the response of the metal to the application of stress may be expected to be variable, with an increasing tendency towards a different mechanism, most often resulting in transgranular fracture, as the propensity towards intergranular corrosion is reduced and the rôles of stress and/or strain become more important.

Strain-generated Active Paths

Many corrosion-resistant alloys owe their electrochemical inactivity to a relatively inert film that forms on the exposed surfaces of the metal, so that the relatively active metal is effectively separated from the environment. If the protective film is disrupted for any reason, such as by plastic strain in the underlying metal, the exposed metal is attacked until the protective film reforms, when further reaction is stifled until the film is again disrupted. With such a mechanism it is claimed[12] that the rate of cracking will be dependent upon the rate of film growth, although the physical characteristics of the film, i.e. its thickness, the extent to which it shows plastic or brittle behaviour and the magnitude of the internal stresses that it contains as a result of its mode of deposition, are also likely to be important. Ellipsometric studies of the rates of film growth on α-brasses in 15 N aqueous ammonia have shown the growth rate to increase with zinc content of the brass, temperature and applied potential, parameters that also increase the stress-corrosion crack-propagation rate, thereby providing support for a film rupture mechanism. The question of whether this would result in intergranular or transgranular cracking is controversial. It is sometimes argued[13] that plastic strain tends to concentrate at grain boundaries, forming dislocation networks in which copious sources exist for plastic flow and hence promoting intergranular cracking. On the other hand, transgranular cracking will be favoured by planar slip and will therefore be facilitated at low stresses when extensive plastic strain will not result in dislocation networks that would block planar slip. Whilst instances may be cited of transgranular cracking occurring at low stresses and intergranular at high stresses, there is also a considerable amount of evidence to the contrary. The effects[11] of increasing strain rate, and hence of stress, have already been mentioned as resulting in an increased tendency for transgranular cracking, and the effects of increasing amounts of cold work upon the cracking of carbon steels (Section 8.2) are at variance with the expected effects. Preferential oxidation may occur along grain boundaries in the absence of stress, possibly because of equilibrium solute enrichment in such regions, and even where the rupturing of films growing along boundaries is an important part of the stress-corrosion process, as possibly with the α-brasses[14], these are, within the context of the earlier definition, examples of corrosion along a pre-existing active path. The intergranular stress corrosion of the α-brasses therefore constitute a convenient bridge between the pre-existing and strain-generated active path mechanisms.

Where transgranular stress corrosion results from repetitive film rupturing it is to be expected that the deformation characteristics of the metal will be important, although these may well be influenced by the presence of the film. The height of the slip step formed at a surface must be greater than the thickness of the inactive film if bare metal is to be exposed, and this implies that deformation associated with high slip steps is likely to be more effective in promoting stress corrosion than when fine slip is operative. With the face-centred cubic metals, high slip steps are the result of cross slip being difficult, leading to planar arrays of dislocations, because of low stacking-fault energy or the presence of short range order in the alloy. Swann[15] has shown a relationship between stacking fault energy and stress-corrosion susceptibility for copper-base alloys and for austenitic stainless steels, indicating also the tendency for transgranular cracking to dominate the more readily planar arrays of dislocations form. However, whilst the effects of change in alloy composition upon stress-corrosion cracking susceptibility in the present context may be partly due to their effect upon stacking fault energy, this does not constitute a complete explanation, since alloying may have significant effects upon electrochemical parameters. The effect of the zinc content of brasses upon their filming characteristics has already been mentioned, whilst the effect of nickel additions to carbon steels upon cracking in boiling 42% $MgCl_2$ is equally illuminating. Small additions, of the order of a few per cent, have little effect upon the cracking of ferritic steel in boiling nitrate solution and in the absence of nickel such steels will not fail in boiling 42% $MgCl_2$. However, the addition of only 1% Ni will induce a susceptibility to cracking in $MgCl_2$ which follows an increasingly transgranular path as the nickel content is increased, becoming fractographically indistinguishable from an austenitic steel at about 6% Ni. The structure and mechanical behaviour of ferritic steels are not significantly changed by additions of only 1% Ni, yet the change in cracking susceptibility is dramatic and it is difficult to escape the conclusion that this results primarily from changes in electrochemical behaviour[16].

This is not meant to imply that the mechanical behaviour of alloys as reflected in their response to the application of stress is not important in transgranular stress corrosion, but merely that the relative importance of different parameters can vary from one system to another. The importance of deformation, and hence mechanical behaviour, in transgranular cracking is most crucial in the slip dissolution model. Static dislocations do not show evidence of significant chemical activity, unless associated with chemical heterogeneity resulting from solute segregation, but moving dislocations have been suggested as promoting electrochemical activity relevant to stress-corrosion cracking. Hoar and West[17] showed that the currents associated with straining electrodes may be very much greater than those observed at static surfaces, a difference suggested as being the result of yield-assisted depolarisation. The association of such observations with the formation of tunnels from the crack tip and the tearing of the ligaments between the tunnels to produce crack advancement[18, 19] is an obvious extension, but Staehle[20] considers neither to be a vital part of the mechanism since they are not relevant in every case of cracking in Fe–Cr–Ni steels. The essential step in the slip dissolution model is that a relatively inactive film is broken by emerging dislocations and a local transient dissolution process ensues. The difference therefore between this and the film rupture model as it is sometimes invoked is largely concerned with the differing emphasis placed upon the acts of film rupture and subsequent metal dissolution as the controlling process. However, it has sometimes been argued, as an extension of the film rupture model, that a crack initiated in a brittle film may progress into the normally ductile substrate for an appreciable distance before being arrested by plastic deformation in the matrix. The evidence adduced in favour of such an argument is usually one of discontinuous behaviour, either in the elongation of the specimen or in acoustic emission during stress-corrosion crack propagation, or from fractography. However, such observations frequently can be explained in terms of discontinuous yielding of the uncracked metal beyond the crack tip, rather than by brittle mechanical crack propagation, and for the softer ductile alloys there is no unequivocal evidence suggesting that crack extension occurs by cleavage. However, it is possible that in the stronger notch-sensitive materials, such as some of the high-strength steels, mechanical propagation may contribute to the overall rate of crack propagation. Such cases, if they exist, form a convenient bridge to the third cracking model.

Embrittlement of the Metal in the Crack Tip Region

The literature reports many so-called critical experiments that purport to show the operation of a surface-energy lowering mechanism of stress corrosion, but the results are frequently equally explicable in terms of some other mechanism. The effect of grain size upon stress-corrosion cracking susceptibility is a typical example, it having frequently been reported that coarse-grained material is more susceptible to cracking than fine-grained material, detailed analysis showing a Petch type of relationship between the grain diameter ℓ, and the stress σ_i, to initiate a stress-corrosion crack, i.e.

$$\sigma_i = \sigma_0 + k\ell^{-\frac{1}{2}} \qquad \ldots(8.6)$$

where σ_0 and k are constants, of which k may be related to the surface energy associated with the formation of new surfaces by fracture through

$$k = \left(\frac{6\pi G\gamma}{1-\nu}\right)^{\frac{1}{2}} \qquad \ldots(8.7)$$

where G is the modulus of rigidity and the other symbols are as earlier defined. Measurement of the dependence of some stress-corrosion fracture stress on grain size therefore allows a surface energy to be obtained, and since Coleman, *et al.*[21] found the apparent surface energies so determined to be appreciably less than the energy values derived in other circumstances, they concluded that the surface energy associated with crack formation is reduced by the adsorption of some atom or ion species in the stress-corrosion medium. There is, however, an alternative explanation of the grain size dependence of the stress-corrosion behaviour of alloys and this is concerned with the plastic-flow characteristics of materials as they are influenced by grain size. Thus a relationship of the form of equation 8.6, where σ_i is the flow stress at constant strain and the grain size term arises from the resistance

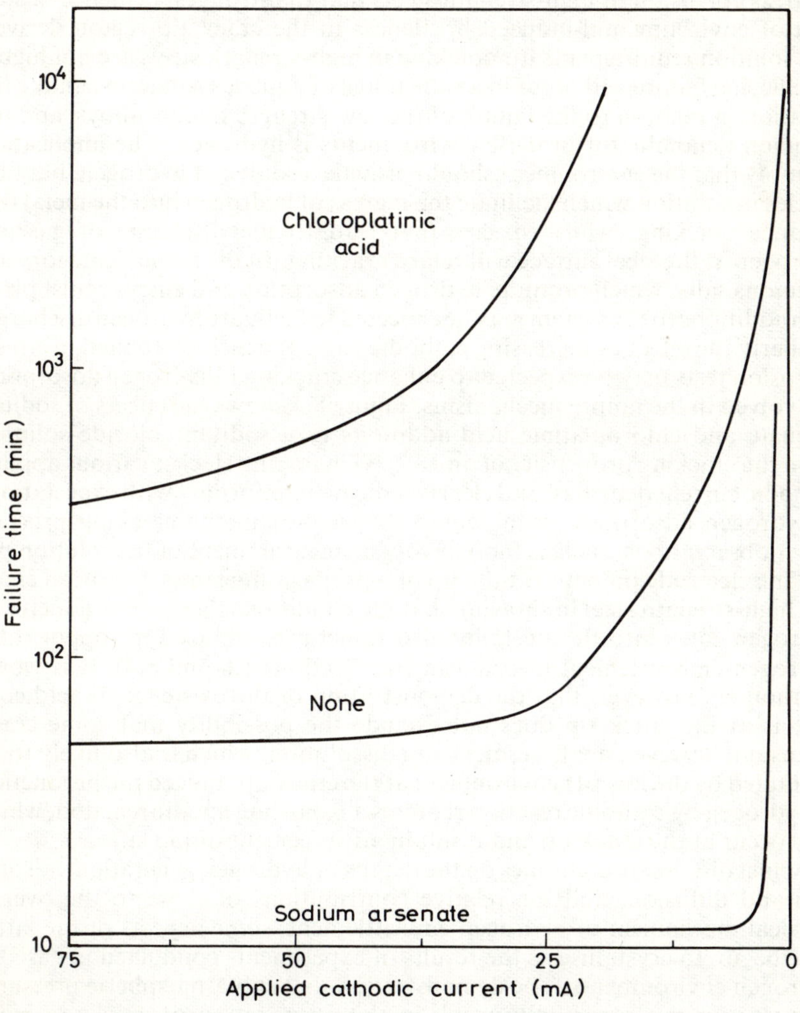

Fig. 8.5 Effects of various additions to 3·5% sodium chloride solution upon the stress-corrosion cracking of a cathodically polarised 18% Ni maraging steel

to the formation of a slip band at a grain boundary, can be shown to be relevant to the plastic behaviour of metals, and it follows that the grain-size dependence of stress-corrosion cracking may simply reflect the fact that the latter is related to plastic flow in the material. Results such as those shown in Fig. 8.18 in Section 8.2, indicating similar slopes for the stress-corrosion fracture stress and flow stress plots against grain size, suggest that the effect of grain size in stress-corrosion cracking is as likely to be related to plastic flow effects as it is to surface energy lowering.

The specificity of environments that promote stress-corrosion cracking has sometimes been adduced in support of a crack tip adsorption model[6], but such observations do not appear[22] to discount a dissolution mechanism of crack propagation any more than they support an adsorption mechanism. Yet, almost paradoxically, it is from observations on environmental aspects of stress corrosion of high-strength steels that the strongest evidence in support of environmental-induced brittleness in the crack tip region derives. The solution requirements for cracking in high-strength steels are not highly specific, i.e. failure will occur in a wide range of aqueous solutions, unlike the situation in relation to the failure of the low-strength ductile alloys, and the common denominator in these environments is hydrogen. The implication is simply that the environment should provide a source of hydrogen, but that species in solution which facilitate the ingress of hydrogen into the metal will enhance cracking, whilst species that lead to the discharge of gaseous hydrogen at the steel surface will retard cracking. In the former category are arsenious salts, which promote hydrogen adsorption and entry, whilst platinum additions to the system may be expected to facilitate hydrogen discharge. Similarly the effect of increasing cathodic current densities applied in stress-corrosion tests may be expected to enhance cracking if hydrogen adsorption is involved in the failure mechanisms. Figure 8.5 shows the effects of sodium arsenate and chloroplatinic acid additions to a sodium chloride solution upon the cracking propensity of an 18% Ni maraging steel at various applied cathodic current densities, and clearly indicates conformity with expectations if hydrogen adsorption is the controlling factor in the cracking process. Other observations, such as those involving measurement of the solution pH and the electrode potential at the tip of a propagating stress-corrosion crack in a high-strength steel in showing that the conditions there are conducive to hydrogen entry into the steel, are also sometimes adduced in support of a hydrogen embrittlement mechanism (*see* Sections 1.6 and 8.4). It is worth mentioning, however, that the demonstration of the existence of acid conditions at the crack tip does not exclude the possibility that some crack extension, however small, results from dissolution, which is also likely to be facilitated by the low pH environment at the crack tip. Indeed the production of hydrogen by cathodic reaction requires a balancing anodic reaction, which may occur at the crack tip and result in advancement of the latter.

Whilst discussion continues on the details of hydrogen generation, adsorption and diffusion, and the relative contributions of these to the overall physical mechanism of hydrogen embrittlement, some aspects of the latter have begun to crystallise as the results of experiments conducted in gaseous hydrogen environments. The demonstration that sub-atmospheric pressures of hydrogen gas can readily result in the propagation of cracks in high-strength steels indicates that the mechanism is not likely to involve the diffusion of hydrogen through the metal to voids where a disruptive pressure of gas is generated. Similarly, the suggestion[23] that hydrogen diffuses into the lattice just ahead of the crack tip and aids plastic deformation there, must be discounted because of the evidence that hydrogen impedes dislocation motion and the implication that any other way of increasing plasticity would also increase cracking. The only remaining alternatives are that hydrogen lowers the surface energy by adsorption or that it accumulates within a few atomic distances from the crack tip, in response to the lowering of its chemical potential by the elastic stress, thereby lowering the cohesive force of the lattice. Oriani[24] prefers the latter because it is the only mechanism that is consistent with the observations of the effects upon crack propagation of small changes in hydrogen gas pressure and the substitution of deuterium for hydrogen. A sufficient reduction in the hydrogen gas pressure surrounding a specimen containing a propagating crack at a given stress intensity will cause the crack to stop propagating, but a subsequent increase in pressure, of about $1{\cdot}6\,kN/m^2$ from $22\,kN/m^2$, was sufficient to restart the crack and with a delay time so short that the extra hydrogen entering the lattice as the result of the increased pressure could have diffused no more than a few atom spacings. A similarly rapid response of the crack velocity to small changes in applied cathodic current to a maraging steel immersed in sodium chloride solution has been observed. The effect of deuterium in reducing the response to embrittlement appears not to be related to the difference in transport kinetics of the two isotopes but to their solubilities in the dilated lattice just beyond the crack tip. This again is in agreement with a decohesion model, as opposed to one involving surface energy lowering or any of the other suggested mechanisms, and further support derives from the observation[25] that a stress field having a dilatant component, produced by mode I tensile loading, leads to hydrogen-induced cracking, whereas at even higher stress intensities, mode II loading (antiplane shear mode), which has zero dilatant component, does not lead to hydrogen-induced cracking.

Whilst these various observations appear to lend strong support to a decohesion model of hydrogen embrittlement, the latter is by no means universally accepted[26, 27] and the possibility that cracking results from the formation of a brittle hydride phase in the crack tip region cannot be discounted[27]. Scully and Powell[28] consider that the stress corrosion of α-titanium alloys constitutes an example of slow-strain-rate hydrogen embrittlement, with the nucleation of hydride on an operative slip plane leading to cleavage as the important step in the process. There are others[29], however, who believe that stress-corrosion crack propagation in titanium alloys results from dissolution. Since it is sometimes the case that the evidence upon which apparently dichotomous interpretations are based is equally acceptable, the thought that the overall rate of crack propagation is compounded of more than one mechanism should not be discounted.

Environmental Aspects of Stress-corrosion Cracking

It has often been stated that the environmental requirements for stress-corrosion cracking are highly specific, but the list of environments identified as causing cracking in various alloys continues to grow with time and the concept of solution specificity is not so narrow as it was even a decade ago. Nevertheless, it is clear that cracking environments are specific, in the sense that not all possible environments promote cracking, and the electrochemistry of stress corrosion is essentially concerned with explaining this specificity. In very general terms, it is clear that potent solutions will need to promote a critical balance between activity and passivity, since a highly active condition will result in general corrosion or pitting, whilst a completely passive condition cannot, by definition, lead to stress corrosion. Whilst the relative inactivity of all exposed surfaces except the crack tip may be derived from a noble metal film in the cases of alloys containing sufficiently noble elements, for the great majority of engineering alloys inactivity at exposed surfaces is the result of the presence of oxide films overlaying metal surfaces. It is not surprising therefore to find that the alloys of high inherent corrosion resistance (such as those based upon aluminium or titanium, or the austenitic stainless steels, that readily develop protective films) require an aggressive ion, such as a halide, to promote stress-corrosion cracking. On the other hand, to crack the metals of low inherent corrosion resistance, such as carbon steels or magnesium-base alloys, requires the presence of an environment that is itself partially passivating. Thus, the carbon steels can be made to fail in solutions of anodic inhibitors, such as hydroxides and carbonates, and the cracking of magnesium-base alloys is achieved with an appropriate mixture of CrO_4^{2-} and Cl^- ions, but not with either of these species alone.

The transition from electrochemically active to relatively inactive behaviour that the sides of a crack must undergo as the tip advances and creates

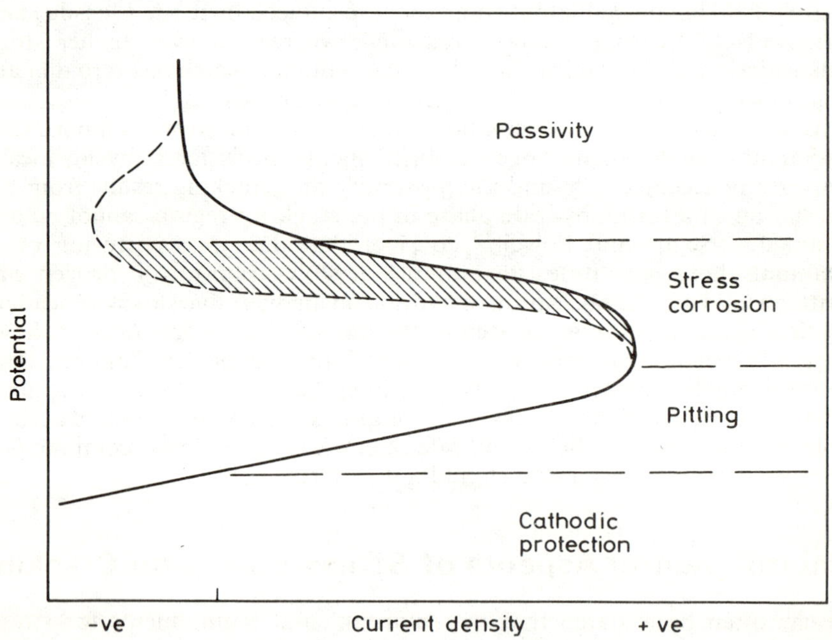

Fig. 8.6 Potentiodynamic polarisation curves and the expected domains of electrochemical behaviour

more crack may be expected to be reflected in the current response of an initially bare surface exposed to the appropriate environment, since dissolution will be associated with the passage of relatively high anodic current densities, but with the passage of time this current will decay if filming occurs. Very rapid rates of current decay are unlikely to permit much dissolution and are not likely therefore to be indicative of conditions conducive to cracking, whilst very slow rates of decay will be more likely to be indicative of pitting than cracking. Intermediate rates of current decay will be those likely to be associated with cracking, and such results are indeed observed [20, 22], but it is not yet possible to predict quantitatively what constitutes a critical rate of decay, although this would be expected to be potential dependent according to the competition between the solvation and filming processes. A more convenient way of anticipating the range of potentials in which stress-corrosion cracking is likely to occur is available through potentiodynamic polarisation curves. If the potential of an initially film-free surface is rapidly (approx. 1 V/min) changed over an appropriate range, then the currents passed at the surface will indicate ranges of potential in which relatively high anodic activity is likely. The rapid sweep of the potential range has the object of minimising film formation, so that the currents observed relate to relatively film-free or thin-film conditions. If the experiment is now repeated, but with a slow rate of potential changes (approx. 10 mV/min) so that time is allowed for filming to occur, comparison of the two curves will indicate any ranges of potential within which high anodic activity in the film-free condition reduces to insignificant activity when the time requirements for film formation are met, and this will indicate the range of potentials within which stress corrosion is likely. Figure 8.6 shows schematic polarisation curves determined under such conditions and indicates the various domains of behaviour expected. The technique correctly anticipates the stress-corrosion cracking of carbon steels in a number of totally different environments[22]. Of course, it is only applicable in those cases where air-formed oxide films can be reductively dissolved so that bare surfaces are created before the potential sweeps; in other cases straining or scraping electrodes must be used to remove the oxide film and the current response of the bared metal then observed potentiostatically at different potentials. However, these different techniques give broadly the same results in any given system.

If the mechanism whereby stress-corrosion cracks propagate involves dissolution at the crack tip, then crack velocities may be expected to be related through Faraday's law to the current density at the crack tip according to equation 8.5. Taking the effective current density as the largest difference between fast and slow sweep rate polarisation curves, or the maximum current densities observed in scraping or straining electrode experiments at appropriate potentials, Fig. 8.7 shows a plot of these current densities against observed crack velocities for a variety of stress-corrosion systems, the line shown being that calculated from equation 8.5. Clearly, for a calculation of this type the agreement between observed and calculated crack velocities is very reasonable, especially since the current density measurements do not take account of the structural dependence of the cracking. An implication of the results shown in Fig. 8.7 is that in those cases where film rupture is involved in the cracking mechanism, the time during which the crack tip is relatively inactive due to the presence of a film must be a small proportion

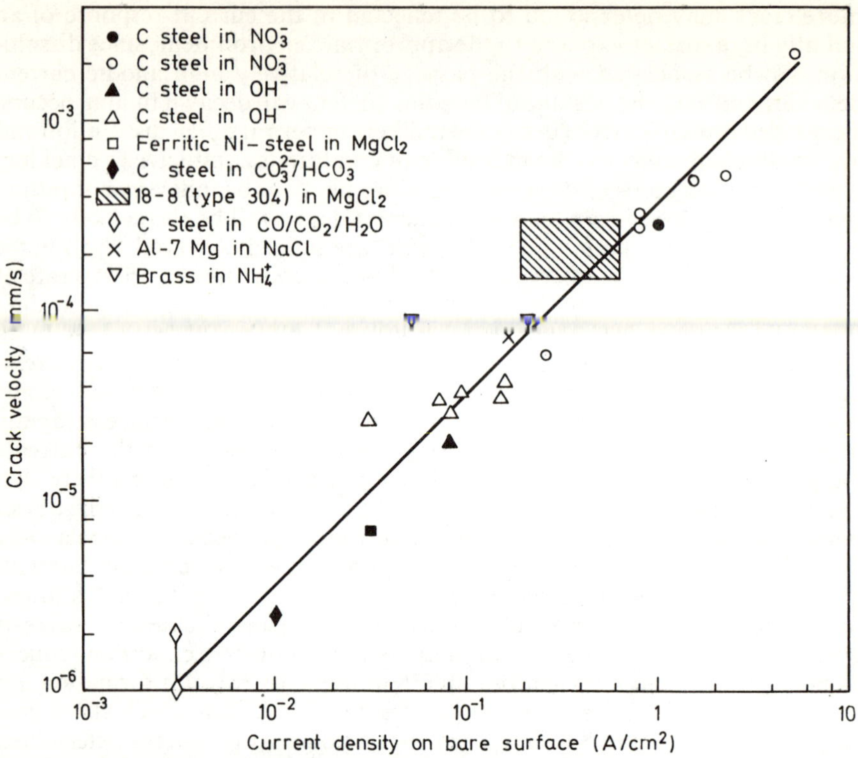

Fig. 8.7 Observed crack velocities and current densities associated with 'bare' surfaces. The line is that calculated from equation 8.5 (after Reference 9)

of the total time, since otherwise the experimental points would fall well below the calculated line, which assumes continuous dissolution. Equally, it is difficult to imagine that, where continuous dissolution is occurring, the conditions at the crack tip–solution interface remain the same as those at an initially bare surface and it is likely therefore that the effective current density involved in cracking will be somewhat less than those observed in, say, a scraping or straining electrode experiment. Thus, equation 8.5 would be expected to overestimate crack velocities to some extent, but, as the results shown in Fig. 8.7 tend to indicate, this is by much less than an order of magnitude which, having regard for the relative crudeness of the calculation, seems very reasonable and gives strong support to a dissolution-controlled mechanism in these systems. A further implication of the stress-corrosion crack velocity being dependent upon dissolution is that the crack velocity should have a temperature dependence similar to that for the temperature dependence of the anodic current density observed in polarisation measurements. Figure 8.8 shows the temperature dependences of the crack velocity and the anodic current densities passed by 'bare' carbon steel surfaces immersed in a carbonate–bicarbonate solution, which yield activation energies of 42 and 46 kJ/mol respectively. Clearly the close agreement between these activation energies lends strong support to crack tip dissolution being the controlling parameter in this system.

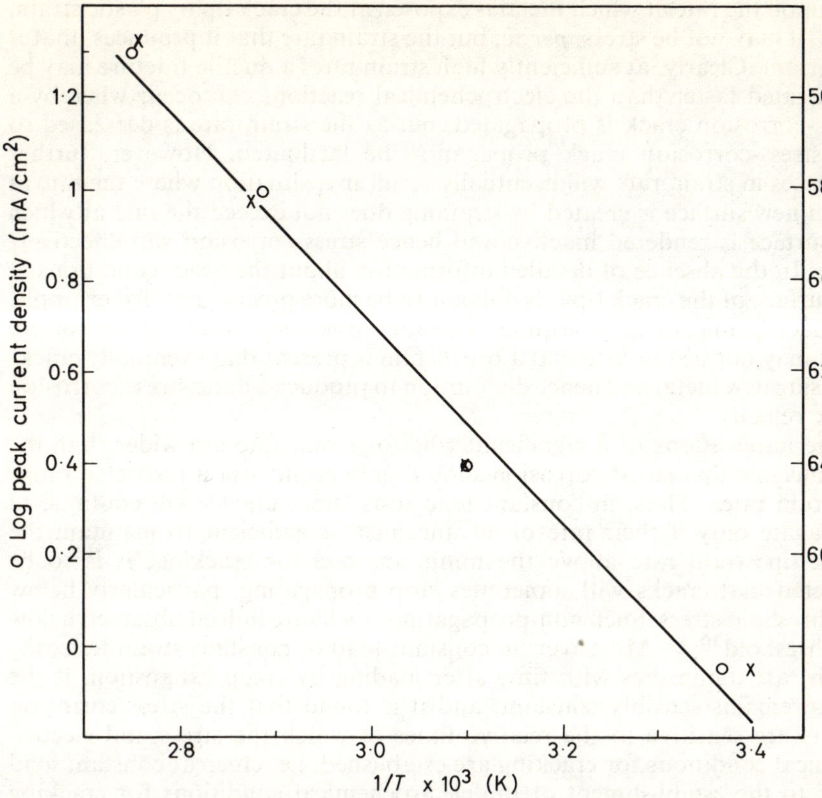

Fig. 8.8 Temperature dependences of crack velocity (×) and peak current density (○) for carbon steel in a carbonate–bicarbonate solution

In summary therefore, it appears that the specificity of the environmental requirements for stress-corrosion cracking, where indeed such a limitation is applicable, is related to the necessity for achieving an appropriate balance between active and passive behaviour. The potential ranges in which such transitions occur agree with the ranges in which stress corrosion occurs, and moreover the kinetics of cracking is reasonably predictable from measurements of dissolution kinetics. Where the mechanism of cracking involves embrittlement of the metal in the crack tip region, the electrochemical requirements are less rigid, although still involving a borderline situation between highly active and passive behaviours, and are reflected in the much wider range of environmental conditions under which cracking may be observed.

The Function of Stress

If crack propagation occurs by dissolution at an active crack tip, with the crack sides rendered inactive by filming, the maintenance of film-free conditions may be dependent not only upon the electrochemical conditions but also upon the rate at which metal is exposed at the crack tip by plastic strain. Thus, it may not be stress, *per se*, but the strain rate that it produces, that is important. Clearly, at sufficiently high strain rates a ductile fracture may be propagated faster than the electrochemical reactions can occur whereby a stress-corrosion crack is propagated, but as the strain rate is decreased so will stress-corrosion crack propagation be facilitated. However, further decreases in strain rate will eventually result in a situation where the rate at which new surface is created by straining does not exceed the rate at which the surface is rendered inactive and hence stress corrosion will effectively cease. In the absence of detailed information about the exact conditions at the surface of the crack tip it is difficult to be more precise, but, for example, the inactivating film may be quite plastic at the very low strain rates involved, i.e. it may not be the case that a brittle film is present that eventually cracks exposing new metal and hence dissolution to produce a finite stress-corrosion crack velocity.

The implications of a significant rôle for strain rate are wider than the obvious one that stress corrosion should only occur over a restricted range of strain rates. Thus, in constant load tests, since cracks will continue to propagate only if their rate of advancement is sufficient to maintain the crack tip strain rate above the minimum rate for cracking, it is to be expected that cracks will sometimes stop propagating, particularly below the threshold stress. Such non-propagating cracks are indeed observed below the threshold[30, 31]. Moreover, in constant-load or constant-strain tests, the strain rate diminishes with time after loading by creep exhaustion, if the stress remains sensibly constant, and it is found that the stress-corrosion results are sensitive to the relative times at which the stress and electrochemical conditions for cracking are established, i.e. creep at constant load prior to the establishment of the electrochemical conditions for cracking

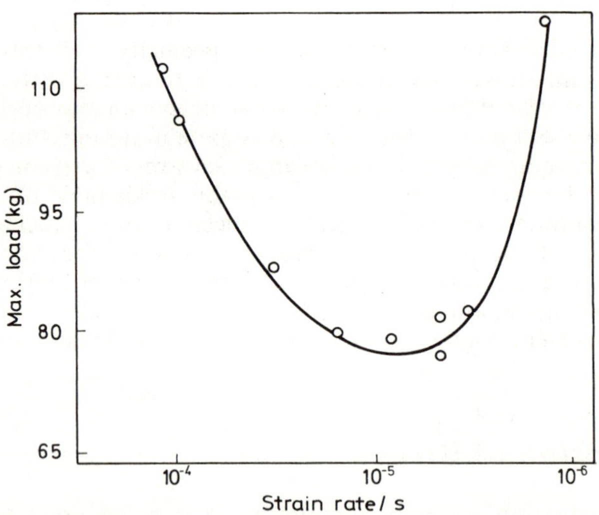

Fig. 8.9 Effect of strain rate upon the cracking propensity of a Mg–Al alloy immersed in a chromate–chloride solution

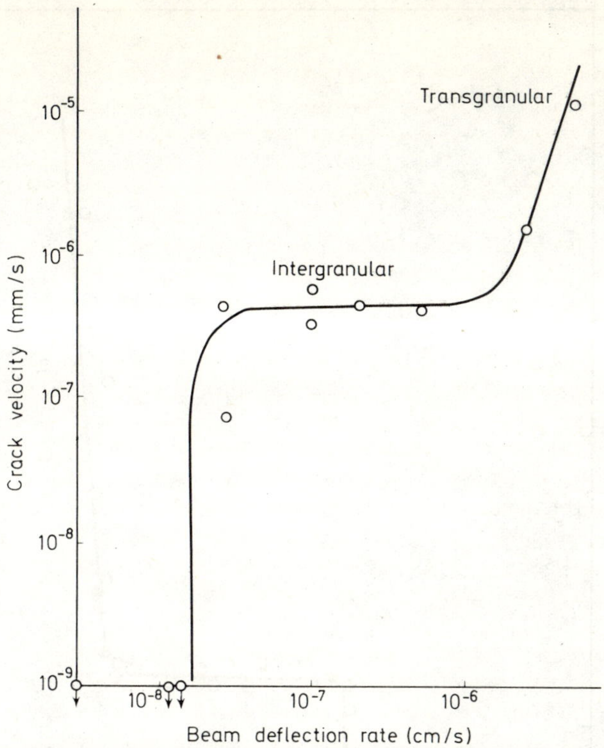

Fig. 8.10 Effect of beam deflection rate of cantilever beam specimens upon stress-corrosion crack velocity of carbon steel in carbonate–bicarbonate solution

delays or prevents cracking[30, 31]. However, the most convincing demonstration of the importance of strain rate is obtained from tests in which the strain rate is superimposed, rather than allowed to vary in the inevitable manner of constant load tests. Figure 8.9 shows the effects of various strain rates applied to a Mg–7Al alloy whilst immersed in chromate–chloride solutions, the tests being conducted to total failure and the maximum load achieved being a sensitive measure of whether or not stress-corrosion cracks were produced[30]. If stress-corrosion cracks are not produced then failure is by ductile fracture at the normal U.T.S. for the material, but in the presence of stress-corrosion cracks the maximum nominal stress achieved prior to failure is markedly reduced. It is apparent from Fig. 8.9 that stress-corrosion cracking only occurs within a restricted range of strain rates and that at higher or lower values ductile fracture occurs, as confirmed by fractography. Experiments on a carbon steel in a carbonate–bicarbonate solution at a cracking potential with the pre-cracked specimens loaded as cantilevers but with the beam displaced at various rates by a device that replaces the conventional load pan, produced the results shown in Fig. 8.10. The stress differences between the tests at various strain rates amounted to less than a few per cent, but the results clearly indicate a lower limiting strain rate below which crack propagation is not observed, followed by a region in which the intergranular stress-corrosion crack velocity is independent of strain rate

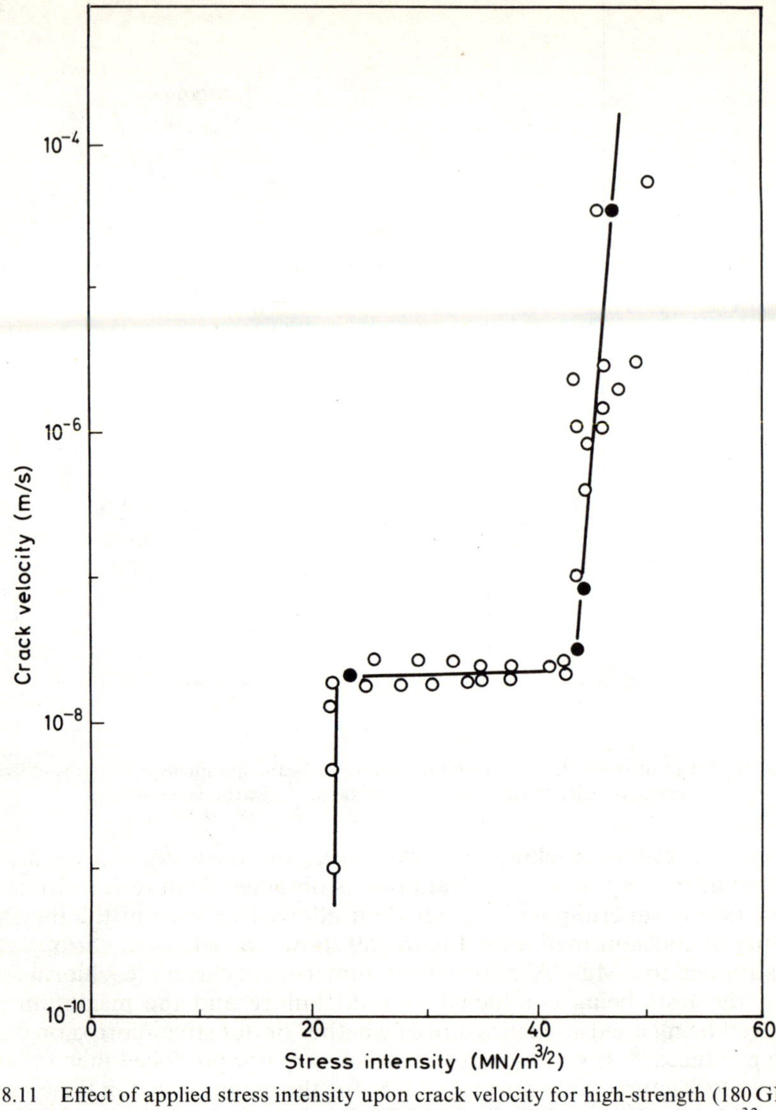

Fig. 8.11 Effect of applied stress intensity upon crack velocity for high-strength (180 GN/m^2 U.T.S.) quenched and tempered steel (AFC 77) in distilled water (after Spiedel[32])

and then, at relatively high strain rates, a transition to fast transgranular tearing. The strain rate independent region is to be expected since once the strain rate is sufficiently high to create bare metal at the crack tip at a faster rate than filming can render the bare metal inactive, the factor controlling the crack velocity will be the rate of metal dissolution which is governed by purely electrochemical considerations. The complementary functions of stress and electrochemistry in this model, involving the creation of bare metal at the crack tip by plastic strain, imply a strong dependence of the strain rate range for cracking, or of the threshold stress in constant load tests, upon the environmental conditions. Such effects are indeed observed, the curve in Fig.

8.9 being capable of marked shift along the strain rate scale according to the composition of the environment and whether or not small anodic or cathodic currents are applied, whilst the limiting beam deflection rate below which cracking is not observed in the experiments to which Fig. 8.10 refers can be changed by two or three orders of magnitude by changes in applied potential.

There are indications then that where an active path mechanism is operative, the function of stress in stress-corrosion cracking is to create plastic deformation and therefore that such cracking will be more likely with the lower-strength ductile metals. Where the mechanism of cracking involves embrittlement of the metal in the crack tip region a strain energy argument is involved and this implies in relation to equation 8.1 that plastic strain should be minimised and elastic energy maximised for failure, conditions that are most readily met with high-yield-strength materials. It is well established that the hydrogen embrittlement of steels becomes more marked the higher the yield strength, although changes in structure or composition that result in a change in yield strength, or fracture toughness, may also influence electrochemical reactions, and such parameters as hydrogen diffusivity and these may be as significant as any change in strength in influencing stress-corrosion behaviour. It is also possible that strain rate, as opposed to stress intensity, could be of significance in the stress corrosion of high-strength steels if the environment concerned is one that may lead to filming and the stifling of the reactions that involve the release of hydrogen from the environment. Certainly, there are indications that stress-corrosion cracking in some of the high-strength steels is sensitive to the loading rate and there is a marked similarity to the crack velocity–strain rate curve of Fig. 8.10, and the crack velocity–stress intensity curves[32] obtained from tests on pre-cracked specimens in a typical high-strength steel are shown in Fig. 8.11. Curves similar to the latter have been obtained for the high-strength aluminium alloys and for titanium alloys, and the question again arises as to whether it is stress intensity or the strain rate that the latter produces that is important, especially in view of results similar to those shown in Fig. 8.9 for a titanium alloy[28]. However, where the mechanism is indeed one of local metal embrittlement it is to be expected that stress intensity, rather than strain rate, will be the important parameter in controlling cracking response.

Conclusion

The interdependence of the variables in stress corrosion, namely structure, electrochemistry and response to stress, supports the suggestion that these may interact in a variety of ways and if rationalisation of the situation is to be attempted this is more appropriately achieved through the concept of a continuous spectrum of mechanisms rather than a single mechanism. The critical balance between activity and passivity is altered by changes in the structure and composition of the alloy, the response of the latter to the application of stress through changes in mechanical properties and by changes in the environmental conditions. Thus, if the structure and composition of the alloy are such that almost continuous paths of segregate or precipitate exist, usually at the grain boundaries, and which are electrochemically different from the matrix, then a latent susceptibility to intergranular corrosion may

Table 8.1 Stress-corrosion spectrum

	Corrosion dominated (solution requirements highly specific) ⟵			⟶ Stress dominated (solution requirements less specific)					
INTER GRANULAR CORROSION	Carbon-steels in NO_3^- solns.	Al–Zn–Mg alloys in Cl^- solns.	Cu–Zn alloys in NH_3 solns.	Fe–Cr–Ni steels in Cl^- solns.	Mg–Al alloys in $CrO_4^{2-} + Cl^-$ solns.	Cu–Zn alloys in NH_3 solns.	Ti alloys in methanol	High strength steels in Cl^- solns.	BRITTLE FRACTURE
	Intergranular fracture along pre-existing paths			Transgranular fracture along strain-generated paths			Mixed crack paths by adsorption, decohesion or fracture of brittle phase		

be activated by the presence of stress. In the absence of pre-existing active paths, or even in their presence if other conditions hold, the stress may generate active paths by rupturing a protective surface film or, probably more importantly for continuing propagation, by activating dissolution at emerging slip lines. The transformation from a pre-existing to a strain-generated active path mechanism may result not only from physico-metallurgical change in the state of the alloy, but also from changes in the environmental conditions or the crack tip strain rate. This greater rôle of stress or strain in moving away from the pre-existing active path end of the spectrum is continued through to those alloys that are inherently lacking somewhat in ductility or have higher yield strengths, so that local embrittlement of the metal in the crack tip region results in a reduction in the energy required for cracking. Table 8.1 indicates some of the systems, of metal and environment, that result in stress corrosion, arranged in a series that ranges from those in which the mechanisms are thought to be dominated by dissolution processes to those in which stress, or strain, occupies the more important part of the proposition. Perhaps the most significant point about such an arrangement is that it should help to serve as a reminder of the interdependence of the variables and that avoidance of stress-corrosion failure in a specific instance is no guarantee that the preventative action will be equally successful in other circumstances. For instance, whilst nickel additions to a steel are beneficial in relation to caustic cracking, they have little effect upon nitrate cracking and are quite harmful from the viewpoint of cracking in chloride, in that they promote a susceptibility to cracking in the latter not observed in carbon steels. Thus, the avoidance of cracking by a mechanism occupying one part of the spectrum may induce failure in another part by a different mechanism if the interdependence of the variables is ignored.

R. N. PARKINS

REFERENCES

1. Parkins, R. N., *British Corrosion Journal*, **7**, 15 (1972)
2. West, J. M., *Metal Science Journal*, **7**, 169 (1973)
3. Despic, A. R., Raicheff, R. G. and Bockris, J. O'M., *Journal of Chemical Physics*, **49**, 926 (1968)
4. Dix, E. H., *Trans. Amer. Inst. Min. Met. Engrs.*, 137, 11 (1940)
5. Parr, S. W. and Straub, F. G., *Univ. Ill. Bull.*, 177 (1928)
6. Uhlig, H. H., Ref. 7, p. 86
7. *Proc. Conf. on Fundamental Aspects of Stress Corrosion Cracking*, edited by R W. Staehle, A. J. Forty and D. van Rooyen, NACE, Houston (1969)
8. *The Theory of Stress Corrosion Cracking in Alloys*, edited by J. C. Scully, NATO, Brussels (1971)
9. *Proc. Conf. on Stress Corrosion Cracking and Hydrogen Embrittlement of Iron Base Alloys*, NACE (1975)
10. Doig, P. and Edington, J. W., *Brit. Corr. J.*, **9**, 88 (1974)
11. Syrett, B. C. and Parkins, R. N., *Corros. Sci.*, **10**, 197 (1970)
12. Green, J. A. S., Mengelberg, H. D. and Yolken, H. T., *J. Electrochem. Soc.*, **117**, 433 (1970)
13. Vermilyea, D. A., Ref. 9
14. Pugh, E. N., Ref. 8, p. 418
15. Swann, P. R., *Corrosion*, **19**, 102t (1963)

16. Poulson, B. S. and Parkins, R. N., *Corrosion*, **29**, 414 (1973)
17. Hoar, T. P. and West, J. M., *Proc. R. Soc.*, A268, 304 (1962)
18. Nielsen, N. A., *Corrosion*, **20**, 105t (1964)
19. Swann, P. R. and Embury, J. D., *High Strength Materials*, edited by V. F. Zackay, Wiley, New York, 327 (1965)
20. Staehle, R. W., Ref. 8, p. 223
21. Coleman, E. G., Weinstein, D. and Rostoker, W., *Acta Met.*, **9**, 491 (1961)
22. Parkins, R. N., Ref. 9
23. Beacham, C. D., *Met. Trans.*, **3**, 437 (1972)
24. Oriani, R. A., Ref. 9
25. St. John, C. and Gerberich, W. W., *Met. Trans.*, **4**, 589 (1973)
26. Tetelman, A. G. and Kunz, S., Ref. 9
27. Gilman, J. J., Ref. 9
28. Scully, J. C. and Powell, D. T., *Corros. Sci.*, **10**, 719 (1970)
29. Beck, T. R., Ref. 8, p. 64
30. Wearmouth, W. R., Dean, G. P. and Parkins, R. N., *Corrosion*, **29**, 251 (1973)
31. Parkins, R. N., *5th Symposium on Line Pipe Research*, American Gas Association, Catalogue No. L.30174 (1974)
32. Spiedel, M.O., Conference on Hydrogen in Metals, NACE (1975)

8.2 Stress-corrosion Cracking of Ferritic Steels

The incidences of stress-corrosion failure in ferritic steels, as with most alloys, continues to increase in frequency with the passage of time, probably as the result of the avoidance of general corrosion, the more efficient use of steels, i.e. by employing higher operating stresses, the more extensive use of methods of fabrication that leave relatively high internal stresses in structures, and as diagnostic efficiency has improved. Thus, the cracking of riveted boilers in strong caustic solutions[1] has been experienced for more than 50 years and the failure of evaporating equipment containing ammonium nitrate[2] for little less. The failure of plant used in the cleaning of coal gas[3] and of equipment used in sour oil wells[4] created particularly severe problems 20 to 30 years ago, as did the cracking of anhydrous-ammonia storage vessels[5] a decade later. Other environments that have been associated with stress-corrosion failures in ferritic steels have ranged from ferric chloride solution, through acids ranging from fuming sulphuric to hydrocyanic, to a sodium phosphate solution, a fairly comprehensive list having been provided by Logan[6]. Most of these failures have been associated with cracks that followed an intergranular path, but recently transgranular fractures have been observed more frequently in carbon steels in environments including industrially important H_2O–CO–CO_2 mixtures[7]. The compositions and structures of the steels and the properties of the environments involved in these various instances of failure are so widely varying as to suggest that rationalisation of all of these experiences in a single explanation would be difficult if not unreal, i.e. it is probable that a number of different mechanisms are involved. This is not to suggest that some systematisation is not possible, since from some of the steel environment systems that have been appropriately studied some common trends have emerged.

Effects of Steel Composition and Structure

Most of the work carried out in relation to these aspects of the problem has used nitrates as the cracking environment where low-strength steels have been the objects of interest. Consequently much of what follows refers to cracking in boiling concentrated nitrate solutions except where otherwise stated. The medium and higher strength steels, such as are involved in sour oil well equipment and other applications, are more frequently tested in chloride- or sulphide-containing environments related to service conditions, but the failure of these steels is dealt with elsewhere (*see* Section 8.4).

The Effect of Carbon Content

For normal commercial-quality mild steels in the annealed or normalised conditions in which they are almost invariably used, various workers have shown that the carbon content of the steel is the major factor determining intergranular cracking susceptibility. Figure 8.12 shows the threshold stresses for a series of commercial mild steels of different carbon contents caused to crack in boiling 4 N NH_4NO_3. The trend of the result suggests that pure iron should be more susceptible to cracking than any steel, but the question of the susceptibility of pure iron is in dispute, possibly because of different workers using irons of differing impurity contents tested by different methods. The results[8] shown in Fig. 8.13 indicate that cracking susceptibility goes

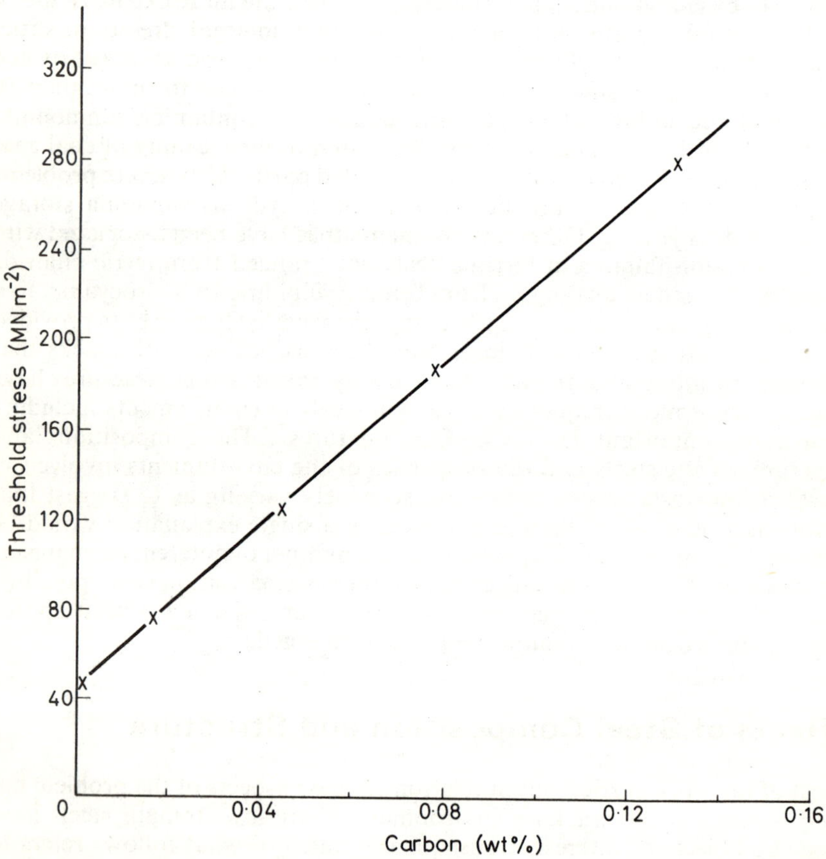

Fig. 8.12 Effect of carbon content of annealed mild steels upon threshold stress for cracking in boiling 4 N NH_4NO_3

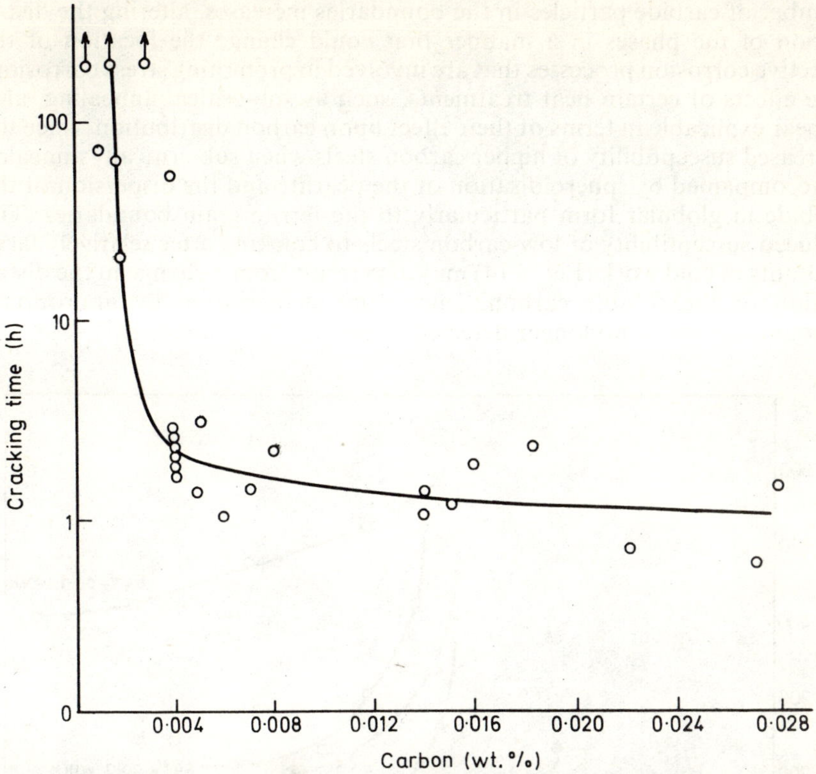

Fig. 8.13 Effect of carbon content of very-low-carbon steels quenched from 920°C on cracking in a calcium nitrate–ammonium nitrate solution (after Long and Uhlig[8])

through a minimum as the carbon is reduced to very low levels, and taken with the results shown in Fig. 8.12 demonstrates the important influence of carbon content in relation to the intergranular cracking of steels in nitrates. Such information as exists indicates that similar trends are observed in relation to the cracking of steels in the carbonate solutions that constitute coal-gas liquors and in strong caustic solutions.

Whilst it appears to be widely accepted that the carbon content of the steel is important, there are differences of opinion as to how the carbon operates. Flis and Scully[9], and Long and Uhlig[8] regard the effect of carbon on the mechanical properties of the steel as the means whereby this element is important in stress-corrosion cracking, whereas others[10] regard its electrochemical effects as being the more important. Where the cracking is intergranular it appears most probable that segregation of some element or elements to the grain boundaries is likely to be involved and it is well established that carbon can segregate to ferrite grain boundaries. Thus, in low-carbon steels (< approx. 0·1% C) the carbon undergoes equilibrium segregation to boundaries either as soluble carbon, or as carbide particles or films if the solubility limit of carbon in ferrite is exceeded. As the carbon content is increased, so the proportion of pearlite in the steel increases and the number of carbide particles in the boundaries increases, altering the distribution of the phases in a manner that could change the location of the selective corrosion processes that are involved in promoting stress corrosion. The effects of certain heat treatments, such as sub-critical annealing, also appear explicable in terms of their effect upon carbon distribution, since the increased susceptibility of higher carbon steels when sub-critically annealed is accompanied by spheroidisation of the pearlite and the dispersion of the carbide in globular form particularly to the ferrite grain boundaries. The reduced susceptibility of low-carbon steels to cracking after relatively large amounts of cold work (Fig. 8.14) may also result from a change in the distribution of the soluble carbon, since after deformation by appropriate amounts carbon is no longer detected at the ferrite boundaries[11].

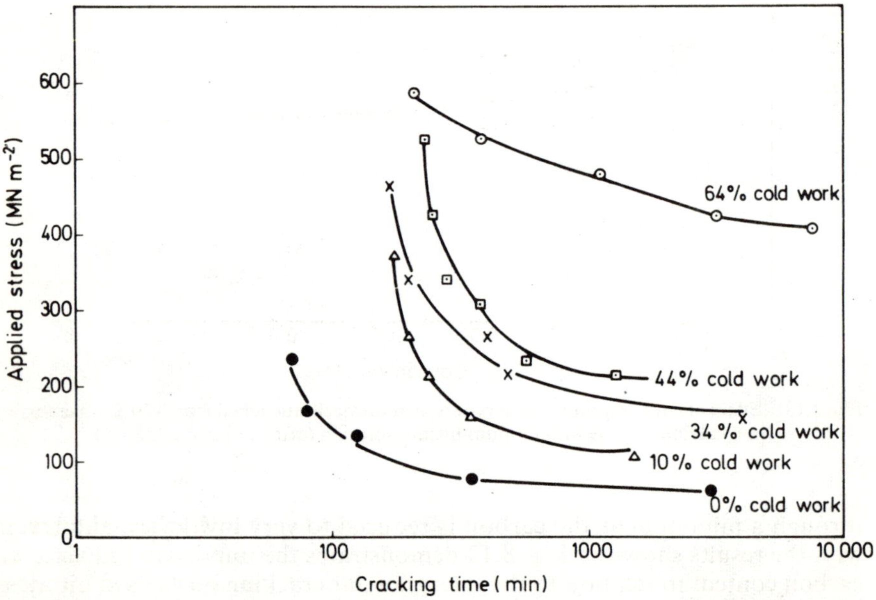

Fig. 8.14 The effects of various amounts of prior plastic strain on the cracking susceptibility of an 0·017% C steel in boiling 4 N NH_4NO_3 (after Humphries and Parkins[11])

If the implication of these various results is that the carbon distribution in the steel determines the sites of the active paths along which selective corrosion occurs, this should be apparent as selective etching when the steel is immersed in an environment that promotes cracking. That this happens is apparent from Fig. 8.15, which shows the appearance of a polished, but previously unetched, steel specimen after a short immersion in $(NH_4)_2CO_3$ at −600 mV (*vs.* S.C.E.). Of course, such a result is not an unequivocal demonstration of stress corrosion being due to corrosion upon a pre-existing active path, since the attack must be sustained to show marked penetration in depth if it is to have the essential characteristic of a stress-corrosion crack. Structurally dependent attack of mild steels in nitrates does not continue along grain boundaries, in the absence of stress, to a greater depth than about 10^{-2} mm, but anodic polarisation can cause virtual disintegration of a piece

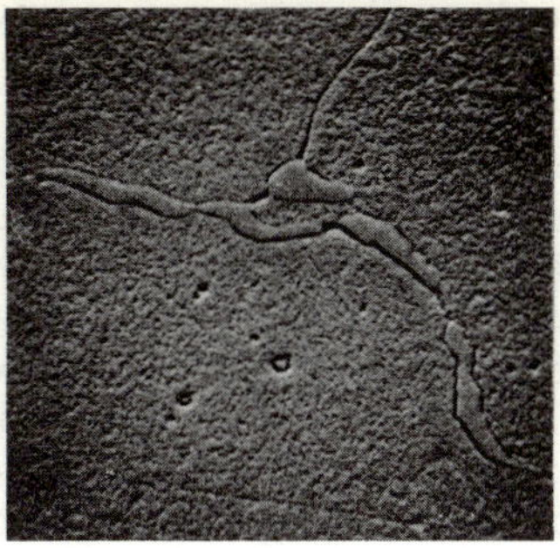

Fig. 8.15 Appearance of polished specimen of carbon steel after 10 s immersion in 2 N $(NH_4)_2CO_3$ at 75°C and −600 mV (*vs*. S.C.E.); ×2400

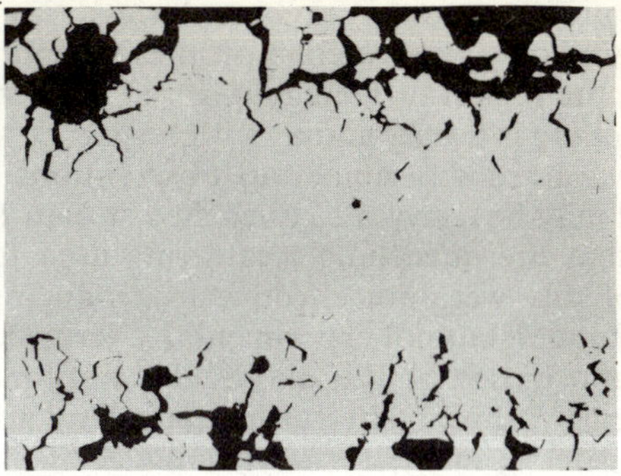

Fig. 8.16 Intergranular penetration in unstressed annealed mild steel immersed in 8 N $Ca(NO_3)_2$ for 12 h at 95°C and +400 mV (*vs*. S.C.E.); ×40 (after Henthorne and Parkins[12])

of unstressed steel by intergranular corrosion[12], as is apparent from Fig. 8.16. This is a more significant demonstration, in relation to the stress-corrosion problem, that carbon steels contain susceptible paths for corrosion, the structural distribution of which appears to be related to the paths followed by stress-corrosion cracks.

It is thought that in nitrate environments the structurally dependent attack occurs preferentially upon the ferrite near those regions that are rich in carbon, because these latter areas operate as particularly efficient cathodes[13]. Similar effects are observed in other cracking environments, except that in strong caustic solutions it appears that the preferential attack may be upon the cementite itself. There is no reason why the relative thermodynamic stabilities of ferrite and cementite should be invariable irrespective of the potential or solution pH at which they are compared, nor are the suggested interactions of these two phases in the present context so unique as to exclude the possibility that other substances can act similarly. Thus, whilst it is suggested that for normal commercial quality mild steels in the annealed or normalised conditions, stress-corrosion cracking is primarily due to carbon segregation to grain boundaries, it is quite feasible that other chemical species may play a similar rôle. Nitrogen, sulphur and phosphorus, for example, segregate to ferrite grain boundaries and could conceivably influence intergranular corrosion, but these elements are present in very small quantities in normal mild steels and, more importantly, the limits of concentration within which they vary in such materials are less, so that it is perhaps not surprising that no correlation is found between the cracking propensities of low-carbon steels and the amounts of these elements present.

Flis and Scully[9] have suggested that the preferential corrosion of grain boundaries in low-carbon steel is determined primarily by segregated unspecified substitutional constituents and not by carbon (or nitrogen). Their conclusion is based upon stress-corrosion tests on a 0·025% C steel subjected to various heat treatments which were calculated to prevent or permit carbon segregation to boundaries. (The paper makes a distinction between the ferrite grain boundaries and the prior austenite grain boundaries, claiming that only the latter are susceptible to stress corrosion. Since the paper also states that the prior austenite grain boundaries disappear during transformation leaving behind only segregated impurities without any orientation change in traversing the region of segregation, the implication would appear to be that the cracking is transgranular with respect to the ferrite grain boundaries; this is at variance with numerous observations on annealed or normalised steels, but may be relevant to quenched steels.) The major difficulty with this work is that the quenching treatments used to control the carbon distribution probably were not effective in producing the desired results, which were calculated but not experimentally verified. Thus, for quenching rates similar to those used by Flis and Scully and for a steel of similar carbon content, Speich[14] shows that during quenching about 90% of the carbon segregates to defects, in the form of boundaries or dislocations, by diffusion in the martensite, so that the assumptions made by Flis and Scully on the location of the carbon after the heat treatments of their steel are not valid. To explain the effects, such as are shown in Figs. 8.12 and 8.13, of carbon on stress corrosion in mild steels Flis and Scully suggest that these are related to the strong effect of carbon on the mechanical properties of the steel, particularly its plasticity. They point out that many stress-corrosion tests are performed under conditions where load relaxation can occur and that this is likely to be higher the lower the carbon content because of the increased ductility resulting from a reduction in carbon content. Load relaxation could result in an increase in the time to failure and therefore an apparent increase in resistance to cracking, such as is shown at the very low carbon contents indicated in Fig. 8.13. However, if this explanation is valid for the results shown in the latter figure it would not explain the opposite trend shown in Fig. 8.12. This is not to suggest that the mechanical properties of the steel are of no importance in relation to stress corrosion, since the contrary has been shown to be the case[15], and quite apart from any electrochemical effects that the carbon in steel may have it is to be expected that it will also influence the progress of stress-corrosion cracks through its effect upon the mechanical properties of the steel.

Long and Uhlig[8], and Uhlig and Sava[16] also favour a mechanical rôle for carbon, or nitrogen, in the stress-corrosion mechanism. Interstitial atoms at grain boundaries are claimed to extend the life of emerging surface imperfections during plastic deformation at the tip of the crack sufficiently to allow chemisorption on the imperfection arrays of the species responsible for cracking. This sorption reduces the energy required to create the fracture surfaces and hence failure occurs at stresses below those that would be required in the absence of the species that are chemisorbed. There are many major obstacles to the acceptance of these suggestions, not the least of which is that the involvement of plastic deformation in the propagation of stress-corrosion cracks in carbon steels reduces the surface energy term to insignificance, and so reduction of the surface energy for fracture by chemisorption is not likely to have any significant effect. The papers in which the stress-sorption model is proposed do not indicate the detailed reasoning that leads to the statements made, so that discussion becomes as speculative as the initial suggestions. Thus, the exact nature of the emerging surface imperfection is not clear, nor is the reason why its slow emergence is important, ignoring any questions concerning the reality of the suggestion. If the imperfection is a dislocation of some type, then of course the idea of emerging dislocations acting as areas of relatively high chemical activity has attractions to which many workers in the stress-corrosion field have succumbed; the difficulty with the notion is that dislocations in metals only appear to show evidence of chemical activity, for example by assisting the formation of etch pits, when they have associated impurity atoms. In this case the activity appears to be due to chemical segregation which the dislocations happen to assist; moreover, if this is the real cause of attack upon dislocations, then grain boundaries are even more effective sinks for impurity atoms and so may be expected to be regions of more intense reaction.

The Effects of Other Alloying Additions

Whilst some workers may feel that the balance of the evidence favours carbon segregation to grain boundaries acting primarily electrochemically to determine the sites of the active paths along which selective corrosion occurs in commercial mild steels, there is no reason why some other alloying elements may not act similarly. Thus in moving from the plain carbon steels to low-alloy steels, the alloying additions may influence the stress-corrosion propensity by themselves undergoing equilibrium segregation to grain boundaries, but their effects may be beneficial or deleterious depending upon the outcome of various interactions. Instances are known[17] where the tendency to segregation of a particular element has been increased or reduced by the presence of a second solute, these strong interaction effects between two co-existing solutes frequently being overlooked in favour of an approach concerned only with the compatibility of the strains associated with a solute atom and the distorted structure of a grain boundary. Thus the presence of additional solutes may increase or decrease the tendency of carbon to segregate to ferrite grain boundaries and the additional solutes themselves may or may not concentrate at boundaries, so that a number of possibilities exist which will influence the tendency towards grain boundary corrosion according to the electrochemical effects of the segregated species. Alloying additions may also be expected to influence cracking tendencies by their effects upon the filming characteristics of the steel in the particular environment concerned and through any effect that they have upon the mechanical properties of the steel.

Unfortunately, very little systematic work has been done on the effects of alloying additions, carbon apart, upon the stress corrosion of ferritic steels whereby these ideas can be assessed. Small additions of Al and Ti such as are used for killing steels have sometimes been claimed to have beneficial effects upon cracking resistance, but against this may be quoted other results that show fully killed steels to be no different from semi-killed steel of the same composition. Much higher additions of Al and Ti than those usually added for deoxidation purposes do appear to have a beneficial effect so far as cracking in nitrates is concerned. Munster and Grafen[18] indicate a beneficial effect from 0·46% Ti, and Herzog[19] has reported that Al had a similar effect when present in amounts between 0·5 and 2%, although complete immunity from cracking did not exist in the normalised condition. With Cr contents between 2 and 4%, and with 0·8% Al, resistance to stress corrosion was observed[19] even with less than 0·1% C. Confirmation of the beneficial effects of Al and Cr when present in amounts ranging from 0·3 to 1·0% has been obtained and the deleterious effect of Cu additions in the same range indicated[20]. The explanation of these various results in relation to the ideas mentioned in the previous paragraph is not clear because the stress-corrosion resistance measurements were not accompanied by measurements of other parameters that would be necessary if the various possible explanations are to be distinguished. The explanation would not appear to be as simple as that suggested by Long and Lockington[21], that steels with a strong carbide former should show greater resistance to stress corrosion whilst those containing graphitising elements should show marked susceptibility, since this would be contrary to the effects of Al and Cu additions, the influence of which upon stress-corrosion resistance are the opposite of what would be expected from their carbide-forming tendencies. Although the elements already mentioned may operate in part through their effects upon grain boundary segregation, it would appear at least as likely that their effects are related to their influence upon film formation. Thus, the deleterious effects of Cu additions to the steel are also observed if Cu ions are initially added to the cracking environment, from which they deposit onto the exposed steel, whilst the elements Cr, Al and Ti all have marked passivating tendencies in oxidising acidic environments of the nitrate type. If the latter elements do operate in nitrate solutions by influencing the passivating reactions whereby most of the exposed surfaces are rendered inactive, it may be expected that they would not be nearly so effective in restraining stress corrosion in alkaline environments. Some unpublished work indicates that this is so and published work[5] on cracking in agricultural ammonia tends to support the suggestion. However, such information as is available cannot be taken as indicating that these trends are established and there is need of systematic study of the effects of alloying additions to ferritic steels from the stress-corrosion viewpoint, as opposed to the *ad hoc* testing of commercial steels that has gone on

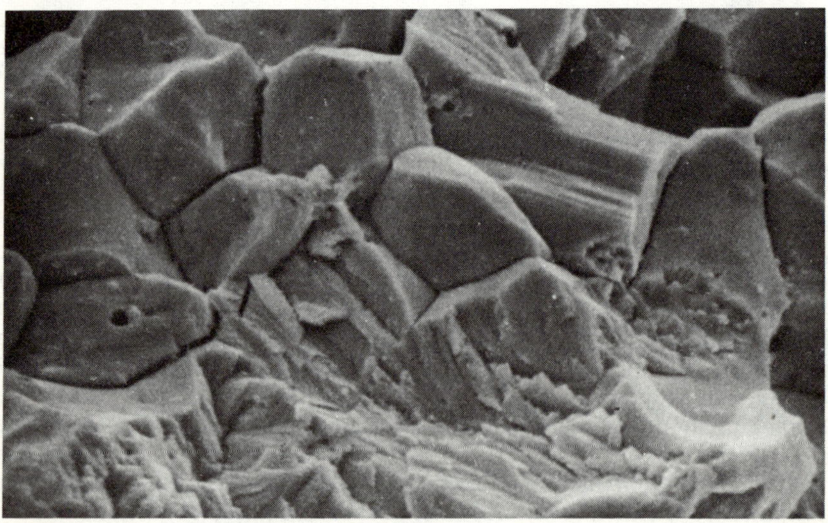

Fig. 8.17 Scanning electron micrograph of 2% Ni steel fractured in boiling $MgCl_2$ showing intergranular and transgranular cracking; ×2000

hitherto, and for these studies to involve a range of different environments. These points are illustrated by some results on the effects of Ni additions to low-carbon steels. The addition of about 2% Ni has little effect upon the cracking of a steel in a nitrate solution, both the plain carbon and alloyed steels cracking in an intergranular manner. In a concentrated boiling magnesium chloride solution the plain carbon steel does not appear to be susceptible to stress corrosion, but the Ni steel cracks quite readily with a path that is a mixture of intergranular and transgranular, as may be seen from Fig. 8.17. Such results, together with those mentioned earlier from tests in alkaline environments, indicate the importance in trying to improve the stress-corrosion resistance of ferritic steels by alloying or recognising the limitations of tests in a single environment.

The Effects of Heat Treatments

For steels that are most frequently used in the annealed or normalised condition the most important structural parameter that can be influenced by heat treatment is the grain size, although the extensive use of welding as a means of fabricating mild steels means that martensitic and tempered martensitic structures may also be encountered. That ferrite grain size has an effect upon stress-corrosion propensity is apparent in the results shown in Fig. 8.18, from which it is clear that coarse-grained steels fracture at appreciably lower stresses than those of smaller grain size. Such results may be interpreted in a number of ways from the more likely saturation of grain

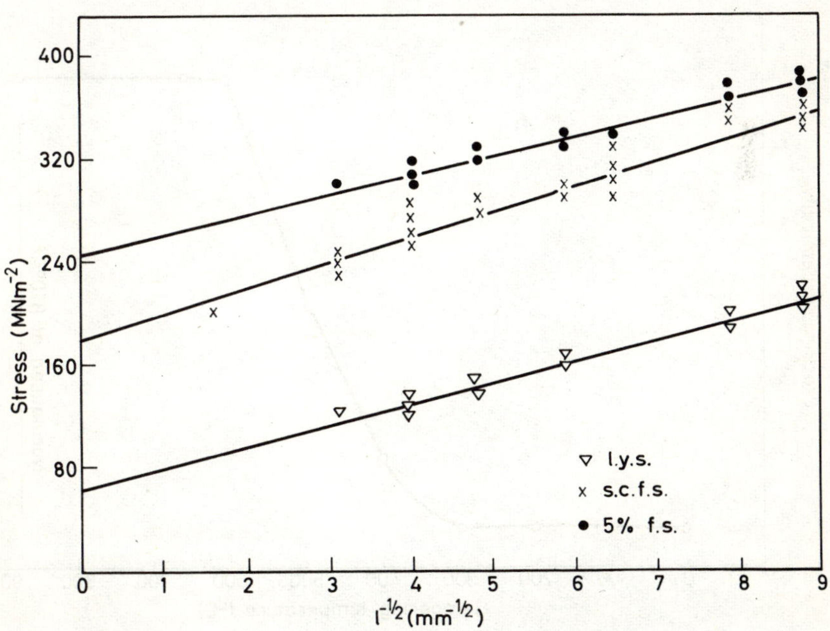

Fig. 8.18 Effects of grain size on lower yield stress, 5% flow stress and stress-corrosion fracture stress for 0·08% C steel in 8 N $Ca(NO_3)_2$ (after Henthorne and Parkins[12])

boundaries by segregate of limited quantity if the grain size is large, and hence the grain boundary area small, to the effect of grain size on the mechanical properties of steel, which as has already been mentioned are matters of importance in stress-corrosion cracking. The effect of relatively fast rates of cooling from the austenitising temperature is more marked than the grain size effect achieved by differing austenitising temperatures. Thus water quenching from 920°C appears to render steel more susceptible to cracking than oil quenching, and further decreases in the cooling rate through air cooling to furnace cooling further increase cracking resistance. However, it needs to be stressed that these trends are relative and that even with very slow cooling, especially from high austenitising temperatures, many mild steels are very susceptible to stress corrosion.

The effects upon cracking tendencies of tempering following quenching is, in general, for the marked susceptibility of the water-quenched condition to be mitigated if the tempering temperature is high enough. However there are some other differences between the results published by various workers. The work of Houdrement, *et al.*[22] agrees with most other workers in showing (Fig. 8.19) that tempering above about 300°C increases cracking resistance and that the benefits are maximised when the tempering temperature is 600°C or above. On the other hand, the results of Uhlig and Sava[16] (Fig. 8.20) show the full benefits of tempering at temperatures from 250°C upwards with a return to marked susceptibility at 700°C. The latter temperature is dependent upon the time of tempering, marked susceptibility returning after tempering at only 500°C if this is carried out for about 10 h or more. The effects of tempering quoted by Long and Lockington[21] for a 1% Mn alloy with very little carbon are in complete contrast to the results just mentioned. They

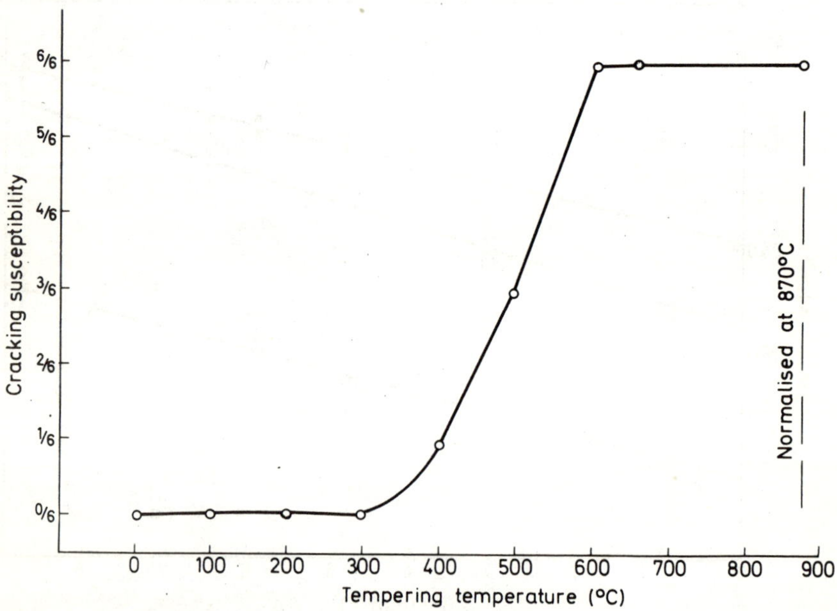

Fig. 8.19 Effects of 30 min tempering, following quenching from 900°C, on cracking of 0·26% C steel in a boiling calcium nitrate–ammonium nitrate solution (after Houdrement, *et al.*[22])

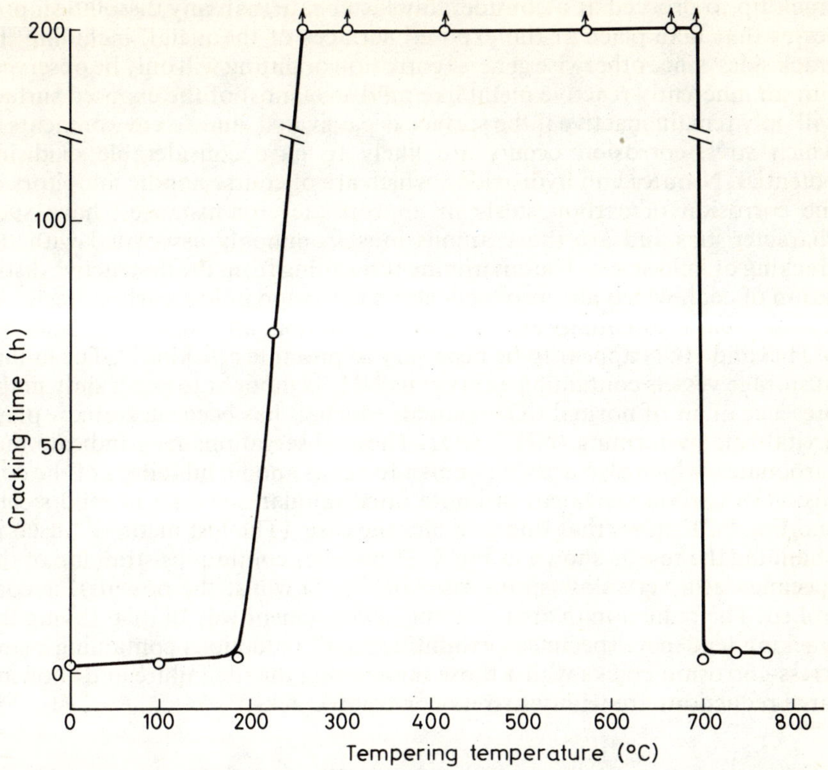

Fig. 8.20 Effects of 30 min tempering, following quenching from 950°C, on cracking of 0·06% C steel in boiling calcium nitrate–ammonium nitrate solution (after Uhlig and Sava[16])

show that tempering above 200°C increases susceptibility initially and that increasing resistance to cracking begins to be observed when the tempering temperature exceeds about 500°C. The explanation of these various results may lie in the differing carbon contents of the steels used or in the test methods employed, but since none of the papers quote any results from structural studies upon the steels following the various heat treatments it appears pointless to attempt any rationalisation of these trends.

Effects of Environmental Composition

It has frequently been stated that the environmental requirements for stress-corrosion cracking are highly specific, but the relatively extensive list[6] of environments that have been reported as promoting cracking raises queries as to the validity of such statements. It is clear that cracking environments are specific in the sense that not all possible environments promote cracking, but to state that the solution requirements are highly specific may lead to a false sense of security in certain practical situations. It is clear that the propagation of a stress-corrosion crack requires the reactions that occur at the crack tip to proceed at a considerably faster rate than any dissolution processes that take place at the exposed surfaces of the metal, including the crack sides, since otherwise general corrosion or pitting will only be observed. For an inherently reactive metal like mild steel most of the exposed surface will only remain inactive if the surface is passivated, and so environments in which stress-corrosion occurs are likely to have considerable oxidising potential. Nitrates and hydroxides, which are of course anodic inhibitors of the corrosion of carbon steels in appropriate circumstances, have such characteristics and are those anions most commonly associated with the cracking of mild steels. The environment resulting from the destructive distillation of coal, which also produces stress corrosion in low-carbon steels[3], is a liquor that is essentially an $(NH_4)_2CO_3$ solution, although small amounts of H_2S and HCN appear to be necessary to promote cracking[23]. The failure of storage vessels containing anhydrous NH_3 is thought to occur only in the presence of air of normal CO_2 content, which, it has been suggested[5], plays a vital rôle by forming $(NH_4)_2CO_3$. These observations may indicate that carbonates, which also are well known to act as anodic inhibitors of the corrosion of carbon steel, can promote intergranular cracking in mild steels, and Fig. 8.21 shows that this is in fact the case. (The test method[24] used in obtaining the results shown in Fig. 8.21 involves continuous straining of the specimen at a very slow strain rate ($10^{-6} s^{-1}$) whilst the potential is controlled. The reduction in area is simply a convenient way of quantifying the cracking tendency, specimens exhibiting small reductions containing many stress-corrosion cracks whilst those undergoing ductile failure and showing large reductions are without stress-corrosion cracks.)

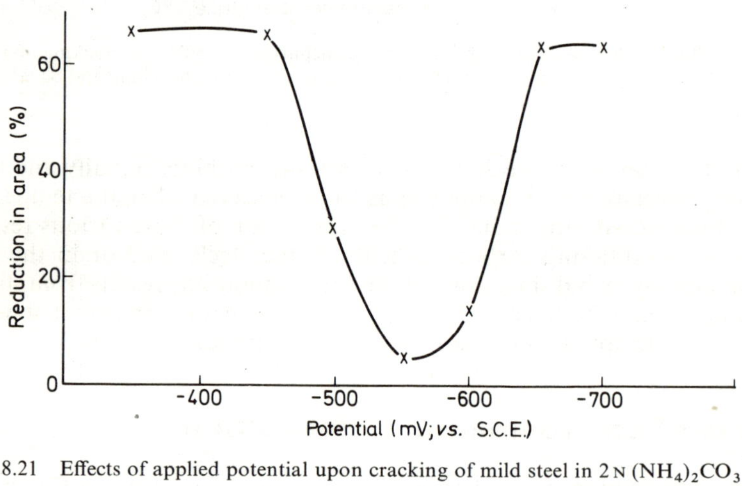

Fig. 8.21 Effects of applied potential upon cracking of mild steel in 2 N $(NH_4)_2CO_3$ at 75°C

Whilst the passivating requirements of a cracking environment may be seen to be portrayed in these observations, it is clear that the environment must also be one that permits the formation of a soluble species at the crack tip, so that dissolution and hence crack propagation can occur. The ranges of potential within which cracking occurs may therefore be expected to coincide with those ranges within which soluble species form. The potentiodynamic anodic polarisation curve may be used for delineating the potential range in which, from indications of high anodic activity, soluble anodic products form, but even with relatively fast potential sweep rates, secondary reactions leading to the formation of insoluble substances may result in some degree of passivation and the interpretation of the potentiodynamic polarisation curve becomes more complicated. Straining electrode experiments, in which the specimen is subjected to very rapid plastic strain after potentiostatic polarisation for sufficient time to allow films to form and the current to reach an equilibrium value, surmount such difficulties and show the range of potentials within which high anodic activity may be experienced[25,26]. Figure 8.22 shows the results of some such measurements and indicates that there is reasonable agreement between the potential ranges for stress-corrosion cracking and high anodic activity.

What is measured in straining electrode experiments such as those to which Fig. 8.22 refers is the rate at which film-free iron will dissolve in the ranges of

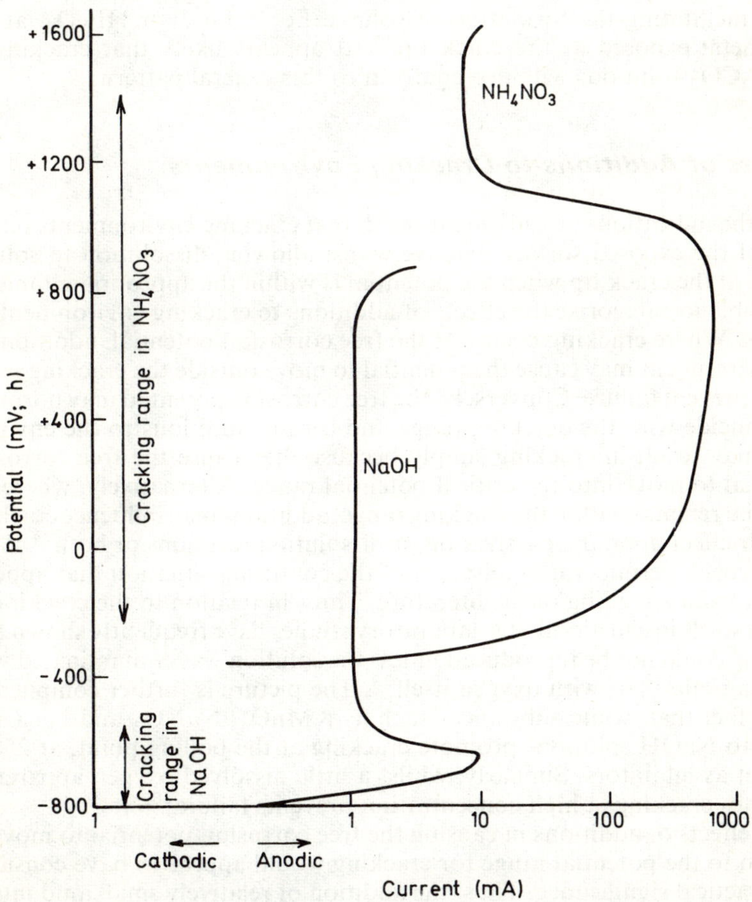

Fig. 8.22 Anodic polarisation curves from straining electrode experiments on mild steel in boiling 8 N NaOH and 4 N NH_4NO_3

potential and in the solutions concerned. The fact that these reactions are occurring mostly at sites that do not coincide with grain boundaries must raise questions as to their validity to intergranular stress corrosion. However, the reasonable agreement between the potential ranges within which cracking occurs and over which high anodic activity is observed, especially in such widely differing environments as a nitrate and a hydroxide, suggests that the dissolution processes in stress-corrosion cracking are closely related to those occurring in the straining electrode experiments, i.e. in both cases dissolution of bare metal is involved. In hydroxide environments the range of potentials over which cracking is observed agrees reasonably with that in which the soluble bihypoferrite, $HFeO_2^-$, may be expected to exist. In nitrate environments, which are acidic, the range of cracking potential may be related to the formation of the soluble Fe^{2+} or Fe^{3+} ions, the former at the more negative and the latter at the more positive potentials. It is possible, therefore, to rationalise the cracking of low-carbon steels in nitrates and hydroxides, two very different types of environments, in terms of the abilities of these solutions to passivate exposed surfaces by the formation of Fe_2O_3 or Fe_3O_4 films, whilst facilitating the formation of soluble Fe^{2+}, Fe^{3+} or $HFeO_2^-$ at the bare metal exposed at the crack tip[25]. It appears likely that cracking in $(NH_4)_2CO_3$ solutions will also conform to this general pattern.

Effects of Additions to Cracking Environments

From the indications already mentioned, that cracking environments render most of the exposed surface inactive whilst allowing dissolution to soluble species at the crack tip when the potential is within the appropriate range, it is possible to categorise the effects of additions to cracking environments as follows. Where cracking occurs at the free corrosion potential, additions to the environment may cause the potential to move outside the cracking range and so prevent failure. Conversely, the free corrosion potential may normally not coincide with the cracking range and certain additions to the environment may result in cracking simply because they cause the free corrosion potential to move into the critical potential range. Alternatively, where the potential remains within the cracking range, additions may influence cracking by their effect upon the passivation or dissolution reactions or both[24]. Such an approach permits rationalisation of the confusing situation that appears to exist from a reading of the literature. Thus, in relation to the cracking of carbon steels in caustic alkalis, laboratory studies have frequently shown that cracking could not be reproduced unless the solution was contaminated with oxidising salts[27] or with oxygen itself[28]. The picture is further complicated by the fact that, while substances such as $KMnO_4$, $NaNO_3$ and Na_2CrO_4 added to NaOH solutions promote cracking at the boiling point, at 250°C they act as inhibitors. Similarly, whilst a little dissolved oxygen apparently promotes cracking, a high concentration prevents failure[28].

The effects of additions in causing the free corrosion potential to move in relation to the potential range for cracking would appear to have considerable practical significance. Thus, the addition of relatively small amounts of nitrates to concentrated NaOH solution, an approach that is sometimes used in treating boiler feed waters in an attempt to avoid caustic cracking, causes the free corrosion potential to become significantly more positive than the cracking potentials in alkali, and failure is no longer observed. However, the same addition will not prevent cracking if the potential is maintained, by potentiostatic control, within the cracking range, so that nitrates should be regarded as unsafe inhibitors of caustic cracking. Na_2SO_4 should also be placed in this category[24], even though the maintenance of a Na_2SO_4/NaOH ratio in boiler feed water in excess of 2·5 is still widely practised as a means of preventing caustic cracking in boilers. In some circumstances additions to environments may promote cracking by causing the free corrosion potential to move into the cracking range. Thus, the free corrosion potential of some mild steels is likely to lie at the boundary of the cracking range in hydroxides so that stress corrosion does not occur. The addition of a small quantity of lead oxide causes the free corrosion potential to move into the cracking range and failure then occurs readily at the free corrosion potential, i.e. without external potentiostatic control. It is feasible that the effect of H_2S and HCN additions to $(NH_4)_2CO_3$, which appear to be necessary to promote cracking in synthetic coal gas liquors[23], may also result from their influence upon the free corrosion potential, since without these additions the free corrosion potential for mild steels lies outside the cracking range in $(NH_4)_2CO_3$ solutions.

Whilst some additions operate in the manner already indicated it is clear that other substances influence cracking propensity not so much by any effect they may have upon potential, but by modifying the reactions involved in cracking. Tannins and phosphates, and to a lesser extent silicates, prevent cracking in strong NaOH solutions even when the potential is maintained within the cracking range[24]. The reasons for this type of behaviour have not been studied but the relatively high anodic current peak shown by mild steel in 30% NaOH is reduced, by about two orders of magnitude, in the presence of NaH_2PO_4, and to a lesser extent by Na_2SiO_3, which may indicate that these substances operate by hindering the dissolution of ferrite. Tannins have little effect upon the anodic polarisation curve when added to NaOH and their influence in stopping cracking may be related to interference with the cathodic reaction.

Results from various laboratories have shown that additions to nitrate environments appear to influence cracking according to whether they affect the pH of the solution, are oxidising substances or form insoluble products with iron. The effect of the pH upon the cracking potency of nitrates is apparent from studies in which various cations have been incorporated in the solution. Nitrates of acidic cations are the most potent at equivalent strengths and temperatures and this suggests that the cracking capacity of any nitrate solution may be controlled by adjustment of its pH. This is found to be so, small additions of HNO_3 to lower the pH of a given nitrate increasing the cracking tendency, whilst raising the pH with a sufficient quantity of OH^- ions will prevent cracking at the free corrosion potential. This effect of OH^- ions is related to its influence on potential however, since with potentiostatic control in the cracking range nitrates at a pH of 10 or so will promote failure. Oxidising additions such as $KMnO_4$, $MnSO_4$, $NaNO_2$ and $K_2Cr_2O_7$ all accelerate cracking in nitrates, whilst substances such as H_3PO_2, Na_2HPO_4 and $CO(NH_2)_2$, which may be expected to form insoluble products with iron, retard or prevent cracking.

Environments Promoting Transgranular Cracking

Although the extensively studied nitrate and hydroxide solutions (together with various other environments that have caused failure[6]) promote intergranular cracking, a number of media that produce transgranular cracking in ferritic steels have been identified. It is likely that the mechanisms of failure in these cases are very different from those envisaged for the intergranular cracking in the environments discussed earlier. Thus, in some instances the transgranular cracking may be the result of a strain-generated active path mechanism, such as oxide film rupture, whereas in other cases it may involve surface energy lowering, possibly due to hydrogen adsorption. The systems have not been studied significantly from a mechanistic viewpoint for more definitive statements to be made.

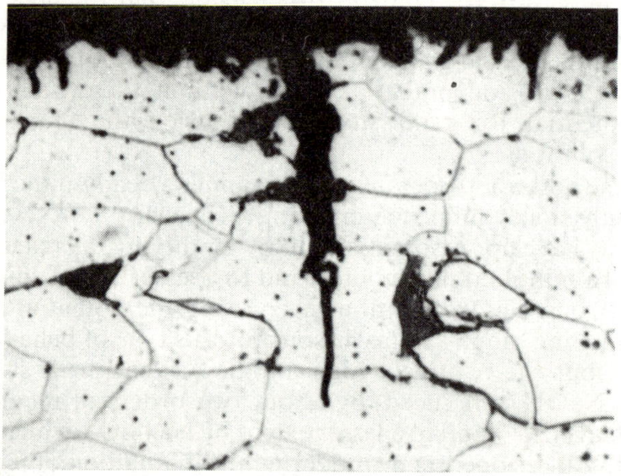

Fig. 8.23 Transgranular cracks produced in 2 N $(NH_4)_2CO_3$ at 75°C and -750 mV (*vs.* S.C.E.); × 1500

Transgranular cracking of low-carbon steels has been reported as occurring in HCN solutions[29] containing 2·6–3·5 g/ℓ of HCN, and in $FeCl_3$ solutions and chloride-bearing slurries containing ferric oxides and hydrated oxides at 316°C[30]. CO_2–CO–H_2O environments promote transgranular failure[7], increasing quantities of CO, some of which must be present for cracking to be initiated, decreasing the time to failure and lowering the stress required[31]. The results may be interpreted in terms of the CO acting as an inhibitor of the attack upon iron by the acidic CO_2–H_2O, a passive film of increasing effectiveness being formed as the proportion of CO is increased, which in turn suggests a film rupture mechanism. Moist H_2S is reported[32] as causing the failure of cold-drawn high-carbon steel wire when loaded to only 40% of its breaking load, although this failure may have resulted from hydrogen embrittlement in the light of the work that has been carried out in relation to failure in sour oil well equipment[33]. A similar explanation may hold for the transgranular cracking (Fig. 8.23) that may be observed in $(NH_4)_2CO_3$ and NaOH solutions at more negative potentials than those that give rise to intergranular cracking, although a film rupture mechanism cannot at present be ruled out in these cases. The failure of pre-stressed wires embedded in concrete may be a further example of cracking by one or other of these mechanisms.

Methods of Prevention

The methods available for the prevention of stress-corrosion cracking are inevitably related to the factors that cause cracking, which may be summarised as follows:

1. Localised corrosion resulting from most of the surface remaining inactive whilst very local areas are intensely active.
2. The situation in (1) requires that the environment causes either (*a*) breakdown of existing films, or (*b*) film formation over most of the surface, after which the environment must allow metal dissolution at the unfilmed anodic sites.
3. The localisation of the attack is then related to the structure of the alloy because of either (*a*) precipitates or segregates of different electrochemical properties from that of the matrix, or (*b*) the stress, that by the attendant strain, creates sites, or allows pre-existing sites to remain active, so that local attack continues.

It follows then that the methods of prevention available are likely to depend upon.

1. The electrochemical characteristics of the environment.
2. The electrochemical characteristics of the alloy.
3. The structural condition of the alloy.
4. Control of the stress.

The detailed implications of these in relation to carbon steels can be deduced from the effects of the factors discussed earlier, but the most obvious methods of prevention are not always applicable to particular items of industrial plant and some further discussion appears worthwhile.

In relation to control through the environment, intergranular cracking should be restrained by additions that improve passivation through reducing local anodic activity, as for example do phosphates added to NaOH, but in view of recent observations of transgranular cracking that may involve a film rupture mechanism, care needs to be exercised here to ensure that intergranular failure is not simply replaced by transgranular cracking. This is also true where additions operate by causing the free corrosion potential to move outside the potential range for intergranular cracking or where cathodic protection is applied for the same purpose since, as has been indicated, in $(NH_4)_2CO_3$ and NaOH environments this may result in transgranular cracking. The use of additions that operate simply by causing the corrosion potential to lie outside the critical range is to be regarded as an unsafe practice, since other agencies, e.g. alloying elements in the steel or impurities in the environment, may result in the potential reverting to a value at which cracking can occur. The unsafe nature of Na_2SO_4 and $NaNO_3$ additions to boiler feed water to prevent caustic cracking has already been remarked upon. The former is widely used despite the statistical evidence of Weir[34] that no benefit is derived from the maintenance of a high sulphate/hydroxide ratio and the ample practical experience of cracking in boilers where the water conditions have met the specified minimum ratio. It would appear that where cracking has not been experienced in boilers that have contained water with an appropriately high sulphate addition the chances are that the boiler would not have failed without the sulphate being added. The more positive approach to caustic cracking in boilers would appear to be to use phosphate or tannin additions[24].

The lack of systematic study of the effects of alloying additions to ferritic steels upon their cracking propensities makes practical application of this approach to prevention somewhat speculative at present. In general it appears likely that the passivating elements Al, Cr, Ti and possibly Si in sufficient amounts, may be beneficial in relation to cracking in nitrates, but there remain doubts as to their effectiveness in more alkaline environments. In these cases Ni may be much more appropriate and clearly similar remarks should be applicable to surface coatings. In respect of the latter it is worth mentioning that coatings that are markedly cathodic to iron are likely to stimulate cracking if the steel becomes exposed; this is probably the explanation of the effect of the Cu additions to nitrate solutions that was mentioned earlier.

Alloying additions to ferritic steels may influence structure so as to affect cracking propensity, quite apart from any electrochemical effect such as already mentioned. However, manipulation of the structure of mild steels to influence their stress-corrosion cracking tendencies is not an approach over which a great deal of control is likely to be practicable in view of the structural condition in which they are normally used. Nevertheless, the avoidance of large grain sizes and of partial surface decarburisation usually will be worthwhile where practicable.

By far the commonest approach to the avoidance of stress corrosion in carbon steels after that involving treatment of the environment will be through control of the stress factor. It is probably the case that the great majority of service stress-corrosion failures result from the presence in the structure of fabrication stresses, which are usually in the region of the yield stress, rather than the normally much lower design stresses. Fabrication stresses can be sometimes removed or minimised to avoid stress corrosion. Thus, welded structures can be stress relieved by annealing at about 650°C; if this is not possible they may be sufficiently reduced in some instances by annealing at lower temperatures[35] or by local heating[36]. Where only partial stress relief is possible it is important that it is not applied to very-low-carbon steels, which may crack with tensile stresses of only 50–60 MNm^{-2}. It is worth repeating in this context the statement[16] that '. . . stress relief of cold-worked steel at 400–650°C, which is apparently sometimes specified, may actually reduce resistance to stress-corrosion cracking rather than increase resistance'. It would be unfortunate if this statement was interpreted to mean that stress-relieving treatments at 400–650°C were always likely to be detrimental, especially in view of the large amount of industrial experience that shows the beneficial effects to be derived from such treatments in relation to welded fabrications. If stress relief by heat treatment is not possible, peening of the surface in contact with the corrodent in order to introduce compressive stresses into the surface layers of the steel may provide some benefits[35].

It may be thought that where fabrication stresses derive from the mating of misfitting components that more effective control of the fabrication operations will avoid the stress-corrosion problem. This should normally be true, even though difficult to achieve in practice. Since it is thought that caustic cracking in boilers is related to the stresses introduced into riveted seams by misfitting rivet holes, coupled with concentration of the OH^- ions in the boiler water by evaporation through leakage at the seams, the demise of the riveted boiler resulting from the use of welded construction may be thought to lead to the avoidance of cracking. This is largely true, but it should be remembered that other methods of solution concentration may be effective, such as can occur at interfaces involving heat transfer, and may account for the fact that welded boiler constructions are not completely free from caustic cracking.

As design criteria change to use materials more efficiently by employing higher operating stresses the approach to stress control by relieving fabrication stresses becomes redundant and a new approach is necessary. The developments in fracture mechanics in the last decade, the application of which to stress-corrosion cracking is admirably reviewed by Brown[37], offers such an approach, at least in some situations. More particularly it can be used to define the largest crack that will be stable and not propagate for given stress conditions. The greatest difficulty surrounding its application to carbon steels at present is concerned with the section sizes to which it is applicable. It is probable that most service stress-corrosion failures in carbon steels occur in relatively thin sections compared to those to which fracture mechanics is currently applicable. There are also questions remaining to be answered concerned with mechanistic aspects of stress-corrosion cracks propagated from sharp geometrical discontinuities, but nevertheless in some situations it appears likely that this approach to the avoidance of stress corrosion in carbon steels may be applicable.

R. N. PARKINS

REFERENCES

1. Stromeyer, C. E., *J. Iron St. Inst.*, **79**, 404 (1909)
2. Jones, J. A., *Trans. Faraday Soc.*, **17**, 102 (1921)
3. *1st Rept. Inst. Gas Engrs.*, Comm. No. 398, Brit. Welding Res. Assoc. F.M.9 Ctee. (1951)
4. Fraser, J. P., Eldredge, G. G. and Treseder, R. S., *Corrosion*, **14**, 517t (1958)
5. Phelps, E. H. and Loginow, A. W., *Corrosion*, **18**, 299t (1962)
6. Logan, H. L., *The Stress Corrosion of Metals*, Wiley, 5–7 (1966)
7. Kowaka, M. and Nagata, S., *Corrosion*, **24**, 427 (1968)
8. Long, L. M. and Uhlig, H. H., *J. Electrochem. Soc.*, **112**, 964 (1965)
9. Flis, J. and Scully, J. C., *Corrosion*, **24**, 326 (1968)
10. Parkins, R. N., *Proc. Conference on Fundamental Aspects of Stress Corrosion Cracking*, Ohio State University, 1967, NACE, 361 (1969)
11. Humphries, M. J. and Parkins, R. N., *4th International Congress on Metallic Corrosion*, Amsterdam, 1969. To be published
12. Henthorne, M. and Parkins, R. N., *Brit. Corrosion J.*, **5**, 186 (1967)
13. Parkins, R. N. and Green, J. A. S., *Corrosion*, **24**, 66 (1968)
14. Speich, G. R., *Trans. Amer. Inst. Met. Eng.*, **245**, 2553 (1969)

15. Henthorne, M. and Parkins, R. N., *Corros. Sci.*, **6**, 357 (1966)
16. Uhlig, H. H. and Sava, J., *Trans. Amer. Soc. Metals*, **56**, 361 (1963)
17. Westbrook, J. H., *Met. Reviews*, **9**, 415 (1964)
18. Munster, R. and Grafen, H., *Archiv. fur das Eisenhutt.*, **36**, 277 (1965)
19. Herzog, E., *Corros. et Anticorros.*, No. 2, 59, No. 3, 91 (1954)
20. Parkins, R. N. and Brown, A., *J. Iron Steel Inst.*, **193**, 45 (1959)
21. Long, L. M. and Lockington, N. A., *Corros. Sci.*, **7**, 447 (1967)
22. Houdrement, E., Bennek, H. and Wentrup, H., *Stahl und Eisen*, **60**, 575 and 791 (1940)
23. Parkins, R. N. and Usher, R., *J. Applied Chemistry*, **9**, 445 (1959)
24. Humphries, M. J. and Parkins, R. N., *Corros. Sci.*, **5**, 747 (1967)
25. Humphries, M. J. and Parkins, R. N., *Proc. of Conferences on Fundamental Aspects of Stress Corrosion Cracking*, Ohio State University, 1969, NACE, 384 (1969)
26. Hoare, T. P. and Galvele, J. R., *Corros. Sci.*, **10**, 211 (1970)
27. Schroeder, W. E., Berk, A. A. and O'Brien, R. A., *Metals and Alloys*, **8**, 320 (1937)
28. Rädeker, W. and Gräfen, H., *Stahl und Eisen*, **76**, 1616 (1956)
29. Huckholtz, H. and Pusch, R., *Stahl und Eisen*, **62**, 21 (1942)
30. Strauss, M. B. and Bloom, M. C., *Corrosion*, **16**, 553t (1960)
31. Brown, A., Harrison, J. T. and Wilkins, R., *Corros. Sci.*, **10**, 547 (1970)
32. Rees, W. P., *Symposium on Internal Stresses in Metals and Alloys*, Inst. of Metals, London, 333 (1948)
33. Schultz, A. E. and Robertson, W. D., *Corrosion*, **13**, 33 (1957)
34. Weir, C. D., *Trans. Amer. Soc. Mech. Eng.*, **70**, 253 (1948)
35. Pearson, C. E. and Parkins, R. N., *Welding Research*, **3**, 95t (1949)
36. Parkins, R. N., *Brit. Weld. J.*, **8**, 24 (1961)
37. Brown, B. F., *Metallurigcal Reviews*, **13**, 171 (1968)

8.3 Stress-corrosion Cracking of Stainless Steels*

The group of alloys known as *stainless steels* includes not only the traditional classes of austenitic steel (the 18/8 types) and ferritic/martensitic steels (13% Cr and 17% Cr types), but also more recent developments that do not fit easily into either of these basic classes and high-nickel austenitic alloys based on 15–25% Cr. Almost all of this complex of alloys are susceptible to stress-corrosion cracking under some conditions, and the overall picture is complicated. The most important practical problem is that of the austenitic alloys, and of some more recent ferritic alloys, in chloride or caustic solutions in chemical plants, or in steam and water systems in various industries. The martensitic steels are resistant to this form of stress-corrosion cracking but, together with some precipitation-hardening, austenitic steels are susceptible to cracking in mildly acid environments, particularly where sulphides are present. This is probably a form of hydrogen cracking similar to that encountered with high-strength lower alloy steels (Section 8.4). The martensitic steels can however suffer stress-corrosion cracking in salt solutions and certain waters, and the mechanism here seems different from the hydrogen embrittlement case, cracking apparently being associated with carbide precipitation.

Some austenitic steels in corrosive environments also suffer intergranular corrosion associated with carbide precipitation (the classical weld decay—Sections 1.3, 3.3 and 10.5) but while this attack is accelerated by stress it is not normally recognised as true stress-corrosion cracking. However, the austenitic steels and high-nickel alloys can suffer intergranular cracking in polythionic acids and some of the latter alloys also do so in high-temperature waters. In both of these cases intergranular precipitation or segregation is believed to be a factor, but little or no penetration occurs in unstressed metal, and in these environments stress-corrosion cracking occurs in alloys that are not susceptible to weld decay in the conventional sense.

*It is inevitable that this account is based mainly on the mass of work done on the common AISI 304, 321 and 316 types in hot strong chloride solutions. Evidence is accumulating that the details of the behaviour of this system may not be entirely typical, and that the more important practical cases of the behaviour of these alloys, and other alloys, in less aggressive environments may be appreciably different. Some examples are given where it is known that the behaviour described is not typical of the whole range of alloy/environment combinations, but it is important to appreciate that others may exist.

The interrelationships between these various cases of stress-corrosion cracking are not completely understood, and the identification of four separate types of cracking is necessarily somewhat arbitrary. Some work has been reported on all the cases mentioned, but the bulk has been on the behaviour of the conventional austenitic steels in strong chloride solutions. This is partly a reflection of the practical importance of this problem, but is mainly due to the choice of this system for general studies of the mechanism of stress-corrosion cracking.

The Behaviour of Austenitic Conventional Steels and Related Alloys in Strong Chloride and Caustic Solutions

The Path Taken by Cracks

Cracks tend to spread faster along a surface than they penetrate in depth, but there is some tendency for tunnelling to occur. Sideways spreading ceases, however, when overlapping with an adjacent crack occurs. The general

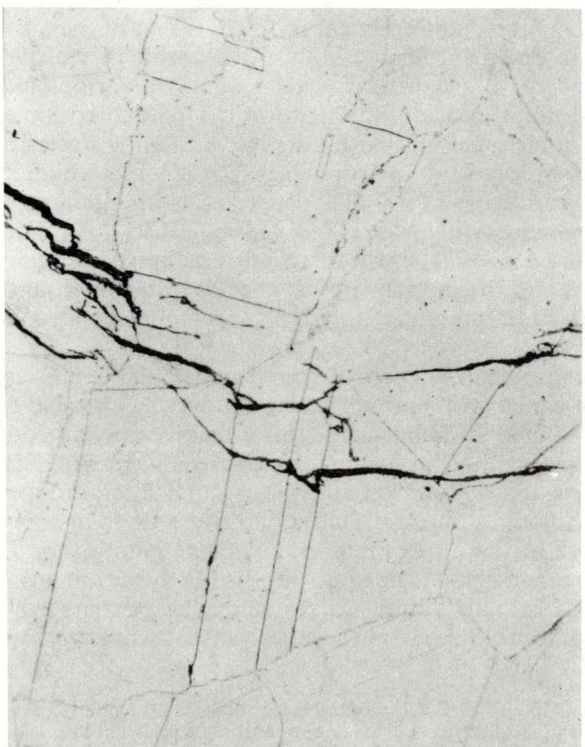

Fig. 8.24 Typical stress-corrosion cracks in a fully austenitic steel. Some grain-boundary carbide has occurred, but the cracks are transgranular.

direction of cracking is perpendicular to the maximum resolved tensile stress and branch cracks are symmetrically arranged about the original plane[1-3]. Because of overlapping, the detail of a branching crack system is complicated[1].

In austenite, cracks are usually transgranular (Fig. 8.24), and do not follow a single crystallographic plane over long distances[1, 4, 5], although the apparent crack-path probably represents the average of numerous short sections each of which lies on a lattice plane[1, 6]. Cracking is transgranular even when intergranular carbide precipitates are present and when severe intergranular corrosion occurs in other environments[7] (cf. Section 3.4). Intergranular cracking has, however, been reported in caustic solutions[8] and in polythionic acid[9], and occurs occasionally in molybdenum-bearing steels in chloride solutions[10]. In duplex steels cracks usually avoid δ-ferrite grains[1, 3].

Fractographic studies show that the macroscopic features just described are repeated on a smaller scale. The fracture surface in a single grain may consist of relatively large facets which seem to represent crack development on crystallographic planes or of a pattern of smaller crack surfaces connected by torn walls[11, 12]. The latter is found most frequently in the more susceptible alloys and seems to be caused by initiation of numerous parallel cracks some of which coalesce[12]. Measurement of the angles between the larger crack planes indicates that they are probably (110) planes[13].

Applied Stress and Cold Work

A linear relation exists between log t_f* and applied stress S for fully softened specimens at stresses below the yield point[14-17], and it seems clear that no threshold stress exists below which failure will never occur. At higher stresses a similar linear relation exists[15, 16] but the slope is much smaller; the two lines intersect at a stress very close to the 0·1% proof stress determined at room temperature—this corresponds to approx. 0·3% strain at the test temperature (Fig. 8.25). Many more cracks are formed at stresses above the proof stress than at lower stresses.

Cold-worked specimens give very similar log t_f–S curves[3, 16, 18], the two lines meeting at the appropriate 0·1% proof stress. At stresses above this, previous cold working has no effect on t_f or on the number of cracks formed; this is not surprising, as plastic deformation occurs during the test whatever the previous history of the specimen. At lower stresses moderate cold working greatly increases the number of cracks formed and also markedly reduces the slope of the log t_f–S curve. Both effects appear at 0·3% strain (corresponding to the 0·1% proof stress) and are more pronounced the greater the degree of cold working up to approx. 5% strain, but after more severe working the slope of the log t_f–S curve tends to increase (Fig. 8.25): thus at low applied stresses t_f first decreases sharply and then increases slowly as the severity of working increases. Severely worked specimens of some steels are as resistant as or more resistant than fully softened specimens. Rocha[19] and Scheil[7] have reported results which are in good agreement with those just described,

*t_f is the time to fracture.

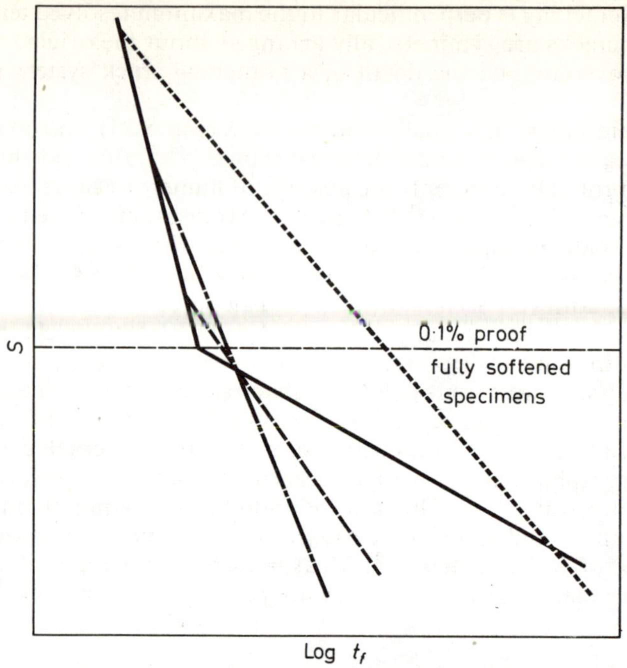

Fig. 8.25 Schematic log t_f/S curves
——— fully softened
— — — mild cold work (approx. 1% extension)
— - — - moderate cold work (approx. 5% extension)
- - - - - - - severe cold work (approx. 40% extension)

and the effects are apparent in results obtained by Copson[20] and by Edeleanu[10].

The effects of cold working can be removed by stress-relieving heat treatments. A period of 2 h at 850°C and a longer time (> 16 h) at 700–750°C are equally effective[2]. Care is necessary with some steels and for certain applications, as phase changes which may have undesirable effects on other properties can occur.

Composition

Many investigations[2, 7, 10, 14, 19–23] have covered some aspects of the effects of composition (*see also* page **8**.56). The following effects are established.

Nickel In chloride solutions increase in nickel content above about 8–10% increases t_f[2,7,10,19,20,23], but a considerable increase in nickel is required to produce a useful improvement. With 40–45% Ni t_f is increased by a factor of 1 000 over its minimum value, but an increase to 20% has only a limited effect[20, 23]. Reduction of nickel content below about 8% also produces an improvement; this is mainly due to a structural change. Increase in nickel content is also beneficial in caustic solutions but the effect is less marked, and alloys containing more than 40% Ni cannot be regarded as being effectively resistant as they can in chloride[24, 25].

Chromium Edeleanu[2] reported that an increase in chromium content above about 16% was mildly beneficial.

Carbon and nitrogen t_f is a minimum at about 0·06–0·08% C, but near the yield point the variation in t_f over the usual carbon range (0·02–0·13%) is not great. The minimum is more pronounced at lower stresses and its position is influenced by both stress and molybdenum content[23]. Increase in nitrogen content over the range 0·01–0·25% reduces t_f if comparisons are made at a constant proportion of the proof stress, but as nitrogen contents in the higher part of this range lead to substantial increases in proof stress high nitrogen contents increase t_f at constant stress[26]. Reduction in both carbon and nitrogen contents to below 0·005% leads to a marked increase in resistance[21].

Molybdenum As in the case of nickel and carbon, t_f passes through a minimum as the molybdenum content increases, the minimum occurring at about 1·5% Mo. The minimum is more pronounced at low stresses and its position is influenced by both stress and carbon content[23]. Tramp amounts of the element are important.

Titanium and niobium Titanium probably causes a slight increase in time to fracture[19], but no effect of niobium has been detected[2, 23]. The presence of titanium modifies the effect of carbon; this is not unexpected, as the formation of stable carbides reduces the effective carbon concentration.

Silicon This element has little effect in the range present in conventional steels[2, 7, 23], but has a marked beneficial effect at higher concentrations, this fact leading to the development of several resistant alloys containing about 2% Si[27, 28].

Copper and Manganese Neither of these elements has more than a marginal effect on t_f[2, 7, 23], and the behaviour of steels in which most of the nickel is replaced by manganese is similar to conventional steels[22].

Most of the generalisations on compositional effects already described have been based on tests in boiling 42% $MgCl_2$ solution, and it must be emphasised that the conclusions are not necessarily applicable in other environments. The beneficial effect of high silicon contents is much more marked in milder environments[27, 28]; molybdenum may be beneficial in 28% $MgCl_2$ solution[6, 28] and in various mild environments met in service; higher nickel contents are much less beneficial in caustic solutions than in chloride[24, 25] solutions and nickel content has little effect in polythionic acid[9, 29]. There is, however, little information on compositional effects other than in 42% $MgCl_2$ solution.

Structure

In addition to the effects of steel composition within the limits of an essentially austenitic structure there are also effects due to the steel microstructure, some of which are composition dependent. Ferritic steels containing only traces of retained austenite (e.g. 17% Cr) are apparently completely resistant if the nickel content is low, but cracking has been reported in similar steels containing nickel or copper[30,31]. The presence of small amounts of δ-ferrite in an austenitic steel is beneficial, the improvement obtained depending to some extent on steel composition, as well as the distribution of the two phases[1,9,18,22,25]. Some steels with approximately equal quantities of δ-ferrite and austenite are extremely resistant[32]. The presence of small amounts of martensite is deleterious, particularly at low stresses, but like δ-ferrite larger quantities can cause a considerable improvement, the improvement obtained again depending on both steel composition and the distribution of the phases[18]. If the carbon, nitrogen and molybdenum contents are low, steels containing large amounts of martensite are very resistant[1,18,33], but unlike the ferritic/austenitic steels the resistance of the martensitic/austenitic steels is very sensitive to heat treatment and no practical alloys have been developed.

It appears that the b.c.c. lattices (ferrite and martensite) are more resistant than f.c.c. austenite, and that some compositions of the b.c.c. phases are effectively completely resistant. The behaviour of a duplex alloy seems to depend on the inherent resistance of the b.c.c. phase and on the phase distribution; a resistant alloy presumably is one with few continuous austenite paths and where the b.c.c. phase itself is of a resistant composition.

Effect of Polarisation

Mears, Brown and Dix[34] found that fracture of an 18/8 steel in sodium chloride/hydrogen peroxide solution could be delayed or accelerated by

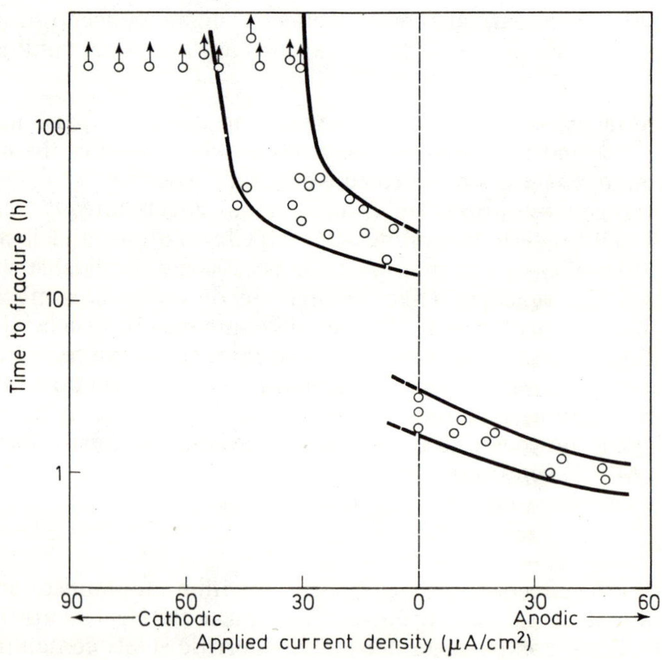

Fig. 8.26 Effect of anodic or cathodic polarisation on time to fracture of an Fe–18Cr–8Ni steel (after Hines and Hoar[35])

coupling to a second metal which was less or more noble. Similar results were obtained in boiling 42% $MgCl_2$ by Hines and Hoar[35], who also investigated the conditions for cathodic protection in more detail. Complete protection was obtained with cathodic current density $>50\ \mu A/cm^2$; after a period at this current density a dark film formed on the metal surface and protection could then be maintained with a current density of only $10\ \mu A/cm^2$. These protective current densities vary little with changes in the environment or in specimen size and form[18,33,36].

When cathodic protection was applied after cracking had begun, propagation ceased immediately if the steel was polarised below -150 mV (*vs.* S.H.E.); at higher potentials, the rate of propagation was reduced[35]. There is other evidence that the open-circuit potential of the anodes within the cracks is about -150 mV[37].

Mild anodic polarisation reduces t_f appreciably, and increases the number of cracks formed, but it has little effect on the rate of crack propagation[35].

Figure 8.26 shows the variation of log t_f with applied current density. The discontinuity at low cathodic current density reflects the reduction in the probability of a crack being initiated as a result of the contraction in the anodic area which occurs at low current densities[37]. A further increase in current density has little effect on the potential, and hence on t_f, unless the dark film forms.

Effect of Environment

Transgranular cracking has been reproduced in the laboratory in aqueous solutions of various chlorides[2,14] and in caustic alkali solutions. Some substances, notably phosphates, can act as inhibitors[5,38,39], while in chloride solutions addition of free acid or oxidising agents reduce t_f[2–4,37]. Oxygen appears necessary for cracking to occur[40], so that efficient deaeration can prevent cracking[41], but in caustic solutions oxygen is not necessary[5]. Increase in chloride or caustic concentration at a given temperature reduces t_f[2,38] and the effect of inhibitors, acid or oxidising agents depends to some extent on chloride or caustic concentration and temperature[2,5,18,26], the effects generally being less pronounced under more severe conditions. Intergranular cracking has been produced in the laboratory in certain solutions based on caustic alkali[8,39] and in polythionic acids[9,29]; higher nickel alloys have been shown to crack in hot pressurised water[42,43].

Service failures have occurred in a wide variety of environments but as most contained at least traces of chloride or caustic these substances were probably the active agent in the majority of cases. Cracking in service occurs preferentially where the temperature is highest, particularly where the rate of heat transfer from the metal is high so that concentration of chloride or caustic can occur[2,5,15,44,45]. Other factors that can lead to local concentration are also dangerous, such as evaporation of spray carried in steam, drying out of a wet surface or the first stages of condensation (the first droplets to separate tend to accumulate traces of chloride in particular[5]). It is therefore unrealistic to attempt to define universally safe chloride or caustic limits, or temperatures, and while practical limits for a given type of equipment on a given duty can be derived by experience it is important to appreciate that these may only apply if the conditions are reproduced in all details.

Apparently trivial changes can have substantial effects. Failures of conventional austenitic steels have occurred in steam from water containing less than 1 p.p.m. of chloride[39,40], or in water containing 10 p.p.m. of chloride and in steam or water containing somewhat higher caustic concentrations[39]. Failures have also occurred from traces of chloride arising from accidental contamination or from chloride leached out of insulation, concrete, etc; the use of, for example, 'chloride-free magnesia' for lagging is not invariably safe as only traces are required for cracking.

Mechanism of Stress-corrosion Cracking of Austenitic Steels

The overall stress-corrosion process may be divided into three stages: an induction period, the initiation of cracks from corrosion damage produced during the induction period, and the propagation of these cracks until complete failure occurs or the stresses responsible for the cracking are relieved[3,6]. The induction period is frequently, but not invariably, the major part of t_f[16]. Little loss of strength occurs during this stage, the reactions being confined to the surface.

Induction Period

The sequence of reactions in chloride solutions is similar to that in other cases where pitting occurs by an autocatalytic reaction[46] (*see* Section 1.6). Anodic action at weak spots on the initial oxide film leads to localised film repair or film growth, with concomitant increase in the local hydrogen ion concentration. At some points the acidity developed becomes sufficient for the formation of soluble anodic products, the film breaks down locally, and a pit is formed. The variations of corrosion potential[3,6,15] and the effects of modifications to the environment[2–4,33], and of anodic or cathodic polarisation[35] or variation in the initial condition of the surface film[3], are all consistent with this sequence.

In some practical cases the apparent duration of the induction period is increased by the need to develop suitable conditions, e.g. by local concentration of the solution.

Crack Initiation

The evidence is inconsistent with the simple view that cracks are initiated from corrosion pits as a result of the associated stress concentration. Hoar and Hines[3] suggested that under the influence of stress preferential dissolution occurs in depth at certain regions of disarray exposed during the induction period, and that cracks are initiated from the sharp notches produced. Bakish and Robertson[47] later showed that stress-corrosion cracks on Cu_3Au single crystals were initiated from 'active' points on slip bands, and Leu and Helle[48] and Harston and Scully[12] have shown that cracks in austenitic steels are initiated in a similar way. Several workers have shown by transmission electron microscopy that preferential attack occurs on moving dislocations[49, 50], which provides confirmation of the earlier views, and probably identifies the 'active' points postulated by the earlier workers. The effects of moderate amounts of cold work on the number of cracks formed and the form of the log t_f–S curves[16] provide general support for this view.

Crack Propagation

The evidence concerning the first two stages of cracking (induction period and crack initiation) is relatively straightforward and the picture just described is generally accepted. This is not the case with crack propagation, where a considerable variety of mechanisms has been proposed, none of which has yet been fully accepted. The mechanisms that have been proposed fall into two main classes.

Electrochemical mechanisms The main problem in deriving an electrochemical mechanism for propagation of stress-corrosion cracks is in explaining how a current density can be maintained at the crack tip which is very much greater than that at the sides of the crack[6, 51]. In the case of transgranular cracking of austenitic steels there is no pre-existing susceptible path of the type that is an essential feature of the classical mechanisms of intergranular cracking[52–54]. Hoar and Hines suggested that yielding at the advancing edge of the crack increased the dissolution rate locally[15, 37], and Hoar and West[55], and later Scully[56] demonstrated that yielding increased the dissolution rate on bulk specimens of susceptible alloys but not of resistant alloys. For 18/8 type steels the difference in current density between metal yielding at the strain rate calculated for an advancing crack and non-yielding material (as at the crack sides) was approximately that required to explain crack propagation by electrochemical processes alone[57]. This model provides a reasonable generalised explanation of crack propagation in purely electrochemical terms, and also explains the pattern of development of crack systems, and various other points of detail[6].

The marked acceleration of penetration rate associated with yielding has been attributed to rupturing of surface films, of oxide or of nickel-rich material[57, 58], as well as to the effects of the energy associated with moving dislocations[48, 49], and a variety of explanations have been put forward to explain why the effect is more marked with some alloys than with others[12, 50, 57, 59].

Joint electrochemical–mechanical mechanisms The essential problem in deriving a model of crack propagation in austenitic steels which includes a significant mechanical stage is in accounting for essentially brittle behaviour in a material that is otherwise particularly ductile. A number of models have been proposed, most of which assume that corrosion leads to locking of dislocations and hence allows fracture under conditions that would otherwise cause yielding, or that it reduces the energy required for fracture[49, 60–64]. Suggestions of this type provide possible explanations of short mechanical stages, which is satisfactory as mechanical propagation over long distances is ruled out by the fact that no real evidence of mechanical stages has been obtained[65]. Direct proof for mechanisms of this type is lacking however.

The most promising confirmatory evidence is related to correlations between dislocation pattern (or stacking fault energy) and stress-corrosion susceptibility, but the same dislocation properties have been used to explain variations in the extent to which yielding influences dissolution rate.

Many of these explanations appear to be invalidated by recent fractographic evidence[12, 13], which shows that the fracture surface on a resistant alloy has coarser detail than a more susceptible alloy; this is the opposite to what would be predicted by dislocation stacking fault theories. The fact that some ferritic alloys crack in chloride solutions in a similar way to austenite is also inconsistent with theories based on metal physics concepts, and it seems therefore that the electrochemical mechanisms based on the Hoar–Hines model fit the facts best despite the unresolved problem of explaining the variation with composition of the magnitude of the effect of yielding on the current density.

Tyzak[64] has however recently pointed out that it is possible to combine the Hoar–Hines concept with a mechanical model by a development of fracture mechanics concepts. He pictures crack propagation as preferential dissolution of the material at the crack tip which is in the act of fracturing, this combined process resulting in a lowering of the effective fracture energy. This line of argument is attractive in that it offers the possibility that divergent lines can be combined into a coherent whole and it may well lead to a model that is acceptable to all concerned.

Stress Corrosion of Austenitic Steels—Practical Solutions

The performance of existing alloys could be significantly improved by fairly minor modifications, but the improvement would be of limited value in cases where the major part of t_f is due to the time required for the solution to become sufficiently concentrated to be aggressive. For this type of application an increase in t_f in the concentrated solution by a factor of 100 or more would be required for composition changes to be justifiable.

A number of alloys are now available on a commercial scale, or are at a late stage in development and some of these alloys have been used fairly extensively in practice. These resistant alloys fall into three main groups:

1. High-nickel alloys, containing 30% Ni or more. For optimum resistance in chloride environments 40% Ni or more is required (e.g. the Incolloys), but the improvement in resistance obtained at somewhat lower nickel contents (e.g. the modified 20 Alloy types) is adequate for many duties.
2. Ferritic and ferritic/austenitic alloys. The original 17–20% Cr alloys, while resistant to stress-corrosion cracking, suffer from a tendency to severe pitting in chlorides and present fabrication problems. Various alloys have been developed by adding nickel, molybdenum, niobium and/or other elements which do not have these limitations, some retaining an essentially ferritic structure and others a ferritic/austenitic structure with approximately equal amounts of each phase.
3. High-silicon alloys, containing more than about 2% Si, normally with some increase in nickel content.

The high-nickel alloys have been available for several years, and are of proven effectiveness, but because of their nickel content they are more expensive than conventional austenitic types. The two other classes of alloys are newer, and experience with them is more limited, but it is probable that they will provide a satisfactory alternative to the conventional alloys for at least some duties with little increase in cost.

In some cases it may be possible to reduce the chloride or caustic content to a safe level for a particular duty. The use of inhibitors may also be valuable in suitable cases. If neither of these alternatives is applicable, it may be possible to effect a considerable improvement by minimising the risk of the formation of dangerous concentrations in cases where the nominal conditions are not very severe. Two points deserve particular mention.

1. Local concentration is most likely as 'hot spots' at which the heat-transfer conditions are particularly severe, as in blind crevices or gaps in scale in heat exchangers. Excessive scaling is undesirable as partial or complete blockage of tubes can lead to particularly severe conditions. The probability of cracking is also reduced by reducing the temperature, particularly if the maximum attainable temperature is also reduced. Failures are rare below about 75°C[27,44].

2. Accidental contamination during erection or repair of steam or hot-water systems is potentially dangerous, and careful flushing is desirable. It is also desirable that lagged pipelines be waterproofed if there is any danger of their being wetted by atmospheric moisture or spillage. Where spillage is a possibility, vessels should not rest directly on concrete. In the last two cases the use of 'pure' materials for insulation or concrete may reduce the risk of failure, but this precaution is of little value if spillage of process liquors containing chlorides or caustic is possible. It is important to recognise however that methods of controlling stress-corrosion cracking that depend on maintaining precise environmental conditions (concentration of chloride, caustic, oxygen or inhibitors, or temperature or pressure) are inherently dangerous, since cracks may be formed very rapidly if control is lost; it is essential therefore to ensure that the control system is effective.

On fully softened material, t_f can be considerably increased by a reduction in stress. However, unless a stress-relieving or softening treatment is the last stage in fabrication it is probably impossible to avoid the occurrence of some regions where local stresses of yield point magnitude exist, and also of some regions which were cold worked during fabrication, the two classes of region not being necessarily coincident. Such regions occur near welds and in areas where hot or cold forming was necessary; they may also arise through accidental dents, stamped identification marks or sheared edges, or through grinding to remove surface defects—failures have resulted from all these causes. These areas constitute weak spots, which may be particularly dangerous if they coincide with local hot spots, as at the tube/tubeplate joints at the hot end of a heat exchanger.

Stress relief is of little practical value as a means of preventing stress-corrosion cracking in austenitic steels, as cracking occurs at quite low stresses even in fully softened material and it is difficult to ensure that stresses are reduced to a safe level in a real structure. The technique can however be useful in small items, but even in this case phase changes which reduce stress-corrosion resistance or have other deleterious effects can occur at the stress-relieving temperature.

Cathodic protection has been successfully used in favourable circumstances, but is difficult to apply effectively to complex structures as cracking tends to occur at those points where it is most difficult to ensure that a protective potential is maintained. Anodic protection has been used effectively in low-chloride environments, and this appears a sounder approach if it is certain that pitting conditions will not occur[18].

Stress Corrosion in Polythionic Acid

A problem which has been encountered in the oil and petrochemical industries is the stress-corrosion cracking of various alloys, including austenitic steels, under shut-down conditions in oil heaters and other furnaces[9, 29]. During operation, sulphide-bearing scales develop on the inner (process) surfaces of the furnace tubes, and under shut-down conditions the scale reacts with air and moisture producing polythionic acids, which lead to rapid failures by intergranular stress-corrosion cracking. While differences in the resistance to stress-corrosion cracking between alloys do exist neither the high-nickel or ferritic alloys are resistant to this form of cracking, and the only effective treatment is to control environmental conditions during the shut-downs. It seems likely that this form of stress-corrosion cracking is related to the intergranular stress-corrosion cracking that occurs in hot pressurised water on some high-nickel austenitic alloys[42, 43]; in both cases intergranular precipitation or segregation at the service temperatures is probably a factor, although the alloys are not subject to 'weld decay' in the classical sense.

Stress-corrosion Cracking of Martensitic Steels

Hardened 13% Cr steels and certain precipitation-hardening steels are susceptible to cracking in environments containing sulphides[44, 66–73]; cracking is accelerated by the presence of acid. Many lower-alloy steels suffer similar cracking and numerous failures have occurred in the oil industry. The same groups of steels also suffer stress-corrosion cracking in salt-spray tests of humid atmospheres[27, 32, 44, 72].

The cracks formed in sulphide environments may be transgranular or may follow the original austenite grain boundaries; cracks produced in salt-spray tests, on the other hand, seem invariably to follow the austenite boundaries[31, 44, 70]. The evidence regarding the two types of cracking is otherwise very similar. The cracks tend to avoid δ-ferrite, and the presence of this phase increases t_f[70]. Tempering temperature affects both the path taken by cracks (in sulphide environments)[70] and t_f, maximum susceptibility coinciding with the conditions requisite for the formation of an incoherent precipitate of chromium carbide in the original austenite grain boundaries[26, 31, 44, 70]. This condition is produced by tempering at 500–600°C[31], and this range should be avoided for exposure to industrial or marine atmospheres or waters. As hardnesses below 270 HV can normally only be obtained by tempering in this range, this is further evidence that the martensitic steels can undergo two separate forms of stress corrosion. Cracking occurs only in hardened steels stressed to more than about 70% of their yield stresses; it is not clear whether the resistance of softened steels is due to their essentially ferritic structures or merely to the fact that a sufficiently high stress cannot be attained. It has been stated that failures are unlikely in sulphide environments if the hardness is less than 270 HV[67, 68].

The cracking in salt solutions is accelerated by anodic polarisation and delayed by mild cathodic polarisation, but more extensive cathodic treatment accelerates cracking (Fig. 8.27)[74]. The cracking in sulphide environments is accelerated by cathodic treatment, and similar cracking can be produced by cathodic treatment in sulphuric acid containing substances which poison the hydrogen evolution reaction (and thus increase the rate of hydrogen uptake by the steel)[70, 72, 73]. It seems clear that the cracking in salt solutions involves anodic dissolution, and must be considered to be true stress-corrosion cracking, whereas the cracking in sulphide environments appears to be a manifestation of hydrogen embrittlement and is thus analogous to the cracking of high-strength steels during pickling (*see* Section 8.4). However, while there seems to be general agreement that two separate mechanisms exist, there is some doubt as to whether the division is as simple as that indicated[68–70].

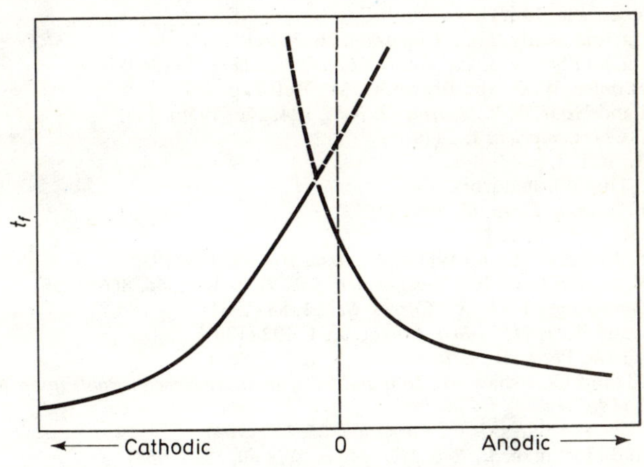

Fig. 8.27 Effect of anodic or cathodic polarisation on time to fracture of a 13% Cr steel (after Mears[74])

Lower-alloy steels which are resistant to cracking in sulphides are available, and where their general corrosion resistance is adequate these are a useful alternative to 13% Cr steels. Some improvement may be obtained without loss of strength by varying the composition of 13% Cr steels, particularly if a tempering treatment outside the range of maximum susceptibility is used[31]. If high strength is not essential, resistant low-carbon (non-hardening) steels are available; the use of austenitic Cr–Ni steels is, however, preferable because they are easier to fabricate. The austenitic steels are resistant to stress-corrosion cracking in sulphide environments, but they are, of course, potentially dangerous if chlorides are present.

J. G. HINES

REFERENCES

1. Hines, J. G. and Hugill, R. W., *Physical Metallurgy of Stress-Corrosion Fracture*, Interscience, New York and London, 193 (1959)
2. Edeleanu, C., *J. Iron St. Inst.*, **173**, 140 (1953)
3. Hoar, T. P. and Hines, J. G., *J. Iron St. Inst.*, **182**, 124 (1956)
4. Logan, H. L. and Sherman, R. J., *Welding J. (N.Y.)*, **35**, 389s (1956). Logan, H. L., *ibid.*, **38**, 463s (1958)
5. Edeleanu, C. and Snowden, P. P., *J. Iron St. Inst.*, **186**, 406 (1957)
6. Hines, J. G., *Corrosion Sci.*, **1**, 21 (1961)
7. Scheil, M. A., *Symposium on Stress-Corrosion Cracking*, Amer. Inst. Min. (metall.) Engrs/Amer. Soc. Test. Mat., Philadelphia and New York, 395 (1945)
8. Wanklyn, J. N. and Jones, D., *Chem. & Ind. (Rev.)*, **28**, 888 (1958). Snowden, P. P., *ibid.*, 1692
9. Samans, C. H., *Corrosion*, **20**, 256 (1964)
10. Edeleanu, C., *Stress-Corrosion Cracking and Embrittlement*, Wiley, New York, Chapman and Hall, London, 126 (1956)
11. Nielsen, N. A., *Fundamentals of Stress Corrosion Cracking*, (Ed. R. W. Staehle), NACE, Houston, Texas, 308 (1969)
12. Harston, J. D. and Scully, J. C., *Corrosion*, **26**, 387 (1971)
13. Harston, J. D. and Scully, J. C., *Nature (London)*, **221**, 853 (1969)
14. Franks, R., Binder, W. O. and Brown, C. M., Ref. 7, p. 411
15. Hines, J. G. and Hoar, T. P., *J. Iron St. Inst.*, **184**, 166 (1956)
16. Hines, J. G., *Corrosion Sci.*, **1**, 2 (1961)
17. Scheil, M. A., Ref. 7, p. 435
18. Hines, J. G., Unpublished work
19. Rocha, H. J., *Stahl u. Eisen*, **62**, 1091 (1942)
20. Copson, H. R., Ref. 1, p. 247
21. Uhlig, H. H., Lincoln, M. and White, A., *Acta Met.*, **5**, 473 (1957)
22. Renshaw, W. G. and Lula, R., *Trans. Amer. Soc. Test. Mat.*, **56**, 866 (1956)
23. Hines, J. G. and Jones, E. R. W., *Corros. Sci.*, **1**, 88 (1961)
24. Evans, T. E. and Burr, D. J., *Brit. Corros. J.*, **1**, 192 (1966)
25. Truman, J. E. and Perry, R., *Brit. Corros. J.*, **1**, 60 (1965)
26. Truman, J. E. and Crawshaw, B., *Iron and Steel Institute Special Publication No 117*, ISI, London, 195 (1967)
27. Loginow, A. W. and Bates, J. F., *Corrosion*, **25**, 15 (1969)
28. Bourrat, J. and Hochman, J., *Rev. Met. Mém. Sci.*, **60**, 551 (1968)
29. Stevens, C. D. and Scarberry, R. C., Ref. 11, p. 204
30. Bond, A. P. and Dundas, H. J., *Corrosion*, **24**, 344 (1968)
31. Truman, J. E., *Metals and Materials*, **2**, 2 (1968)
32. Private communication (1970)
33. Uhlig, H. H. and Lincoln, M., *Trans. electrochem. Soc.*, **104**, 167C (1957)
34. Mears, R. B., Brown, R. H. and Dix, E. H., Ref. 7, p. 323
35. Hines, J. G. and Hoar, T. P., *J. appl. Chem.*, **8**, 764 (1958)
36. Van Rooyan, D., Private communication (1959)
37. Hoar, T. P. and Hines, J. G., *International Committee for Electro-chemical Thermodynamics and Kinetics (C.I.T.C.E.)*, *Proceedings of 8th Meeting, Madrid 1956*, Butterworth, London, 273 (1958)
38. Sharfstein, L. R. and Brindley, W. F., *Corrosion*, **14**, 588 (1958)
39. Snowden, P. P., *J. Iron St. Inst.*, **194**, 181 (1960)
40. Williams, W. L., *Corrosion*, **13**, 397 (1957)

41. Marshall, T. and Hugill, A. J., *Corrosion*, **13**, 329 (1957). Marshall, T., *Corrosion*, **14**, 159 (1958)
42. Coriou, H. *et al.*, *Corrosion*, **22**, 280 (1966)
43. Copson, H. R. and Dean, S. W., *Corrosion*, **21**, 1, (1965)
44. Truman, J. E., Ref. 26, p. 101
45. Staehle, R. W., Beck, F. H. and Fontana, M. G., *Corrosion*, **15**, 373 (1959). Kirk, W. W., Beck, F. H. and Fontana, M. G., Ref. 1
46. Hoar, T. P., *Trans. Faraday Soc.*, **33**, 1152 (1937). Hoar, T. P., *Trans. Faraday Soc.*, **45**, 683 (1949). Edeleanu, C. and Evans, U. R., *Trans. Faraday Soc.*, **47**, 1121 (1951)
47. Robertson, W. D. and Bakish, R., Ref. 9, p. 32
48. Leu, K. W. and Helle, J. N., *Corrosion*, **14**, 249 (1958)
49. Swann, P. R. and Nutting, J., *J. Inst. Met.*, **88**, 478 (1960)
50. Johnson, K. C., Thesis, Ohio State University (1967)
51. Staehle, R. W., Ref. 11, p. 3
52. Dix, E. H., *Trans Amer. Inst. Min. (Metall.) Engrs.*, **137**, 11 (1940)
53. Keating, F. H., *Symposium on Internal Stresses*, Inst. Met., London, 311 (1948)
54. Evans, U. R., Ref. 53, p. 291
55. Hoar, T. P. and West, J. M., *Nature*, **181**, 835 (1958)
56. Scully, J. C., Thesis, Cambridge University (1961)
57. Hines, J. G. (Section 8.1 of this volume, First Edition)
58. Logan, H. L., *J. Res. Nat. Bur. Stand.*, **48**, 99 (1952)
59. Scully, J. C., *Brit. Corr. J.*, **1**, 355 (1966); *Corros. Sci.*, **7**, 196 (1967); **8**, 513 (1968); **8**, 771 (1968)
60. Latanision, R. M. and Staehle, R. W., Ref. 11, p. 214
61. Robertson, W. D. and Tetelman, A. S., *The Mechanism of Strengthening of Alloys*, Amer. Soc. Met., Philadelphia (1960)
62. Forty, A. J., Ref. 1, p. 99
63. Rostocker, W. and Nickols, W., *Trans. Amer. Soc. Met.*, **63** (1956)
64. Tysack, C., *Brit. Corr. J.*, **6**, 124 (1971)
65. West, J. W. and Fairman, L., *Brit. Corros. J.*, **1**, 67 (1965)
66. Vollmer, L. W., *Corrosion*, **8**, 326 (1952)
67. Bowers, C. N., McGuire, W. J. and Wicke, A. H., *Corrosion*, **8**, 333 (1952)
68. Frazer, J. P. and Treseder, R. S., *Corrosion*, **8**, 342 (1952)
69. Prange, F. A., *Corrosion*, **8**, 355 (1952)
70. Lillys, P. and Nehrenberg, A. E., *Trans. Amer. Soc. Metals*, **48**, 327 (1956)
71. Frazer, J. P. and Eldridge, G. G., *Corrosion*, **14**, 524 (1958)
72. Uhlig, H. H., *Metal Progr.*, 468 (1950)
73. Bastien, P., Ref. 1, p. 311
74. Mears, R. B., Ref. 1, p. 22

8.4 Stress-corrosion Cracking of High-tensile Steels

In recent times the demand for high-strength steels (low-alloy steels in which the strength is produced by suitable heat treatment) has increased dramatically due to developments in engineering constructions that require materials of high strength/weight and high strength/volume ratios, e.g. landing gear for aircraft. In the last two decades steels have become available with tensile strengths in the range 1·4–2 GN/m^2 with adequate ductility and toughness, and yet stronger steels are being developed. Unfortunately, the higher strength steels are, in general, more susceptible to stress-corrosion cracking than the lower strength steels, which is a most serious limitation in relation to their application in structural components; furthermore, in the high-strength range the higher the strength of the steel the greater is its susceptibility to cracking. Stress-corrosion cracking of high-strength steels has been observed in a wide range of environments, and failures have been encountered in the chemical, petrochemical, oil, aircraft and nuclear industries.

Two main mechanisms have been postulated to account for the stress-corrosion cracking of high-tensile steels, i.e. (*a*) *active path cracking*, in which the corrosion reaction leads to rapid anodic dissolution along a preferential path within the steel, and (*b*) *hydrogen embrittlement* (or *hydrogen cracking*) in which the corrosion reaction leads to the introduction of hydrogen into the steel.

It is apparent that for the latter to occur hydrogen must be adsorbed at the metal/solution interface and that a proportion of adsorbed hydrogen must transfer across the interface and become absorbed by the steel whilst the remainder (the larger proportion) combines and is evolved as molecular H_2. Hydrogen evolution and absorption is favoured by an environment of low pH, but failures frequently occur in near-neutral solutions, and until recently it was believed that this was due to active path corrosion rather than to hydrogen embrittlement. However, it has now been established that the environmental conditions within a crack are far more acid than in the bulk solution and are conducive to hydrogen evolution and absorption (*see* Section 1.6). Furthermore, raising the potential in order to stimulate anodic dissolution can in fact decrease the pH at the tip of the crack so that a decrease in the time to failure observed under these conditions may be diagnostic of hydrogen embrittlement rather than active path cracking.

For these reasons it is now accepted that most cases of stress-corrosion cracking of high-tensile steels are due to hydrogen cracking. Consequently this section is confined to various aspects of hydrogen embrittlement while the mechanisms of stress-corrosion cracking of low-strength ferritic steels and stainless steels are discussed in Sections 8.2 and 8.3, respectively.

The controversial nature of certain theoretical aspects of hydrogen embrittlement coupled with the practical importance of the phenomenon, has resulted in an immense volume of literature, and several review publications providing general information on many aspects of the subject, are available. These publications include an extensive literary survey of early work[1], comprehensive books on many aspects of hydrogen in steel by Smialowski[2] and by Beck *et al.*[2a] and a review of the stress-corrosion cracking of high-strength steels for aerospace applications by Logan[3]. Much of the important recent work is reported in the proceedings of three international conferences on stress-corrosion cracking in 1967[4], 1971[5] and 1973[6]. A general account of hydrogen embrittlement of ferrous and non-ferrous metals is also available[6a].

Sources of Hydrogen in Steel

To confer adequate corrosion resistance the non-stainless high-strength steels require a full protective surface treatment. Unfortunately, the standard aqueous treatments involved in pickling, electroplating and other surface protective treatments may introduce hydrogen and cause embrittlement, particularly in the case of the higher strength steels (tensile strength > 1·8 GN/m^2) where small amounts of hydrogen may cause severe embrittlement. Thus in a sustained load test in air, hydrogen concentrations of less than 1 p.p.m. by weight may cause premature failure of notched specimens at applied stresses less than one-tenth of the yield strength. It follows that when considering the stress-corrosion cracking of high-strength steels it is important to take into account any hydrogen introduced by various surface treatments as well as any that may be introduced in service by circumstances that lead to hydrogen evolution and absorption, e.g. sacrificial action of a protective coating, cathodic protection, paint stripping, etc.

Apart from any hydrogen absorption which may occur during manufacture, fabrication or heat treatment, the major sources of hydrogen in steel are as follows.

Electroplating

Electrodeposited cadmium and zinc may be used to prevent the corrosion of high-strength steels by virtue of their superior corrosion resistance in atmospheric environments and their sacrificial action in preventing corrosion of the steel at discontinuities in the electrodeposit. Unfortunately the current efficiency for such plating processes is less than 100% and hydrogen absorption normally occurs (*see* Section 14.3). The resultant embrittlement depends mainly on the amount of hydrogen absorbed, but any effect due to the internal stresses exerted by the electrodeposit must also be considered.

Cadmium plating of high-strength steels has been widely used, e.g. aircraft undercarriages, and many investigations have been performed in an attempt to overcome the accompanying problem of hydrogen embrittlement. It has been shown that high-current efficiency is favoured by high-current density and by certain ratios of ions in the plating bath. Troiano[7] has recommended a flash coating with cadmium to a thickness of 0·0025 mm, baking to expel hydrogen and then plating to the required thickness (approx. 0·0125 mm). He suggested that the flash coating was sufficiently thin and/or porous to allow the ready escape of hydrogen during baking, but acted as an efficient barrier to further absorption of hydrogen when plating was resumed. However, it remains common practice to plate from a conventional cyanide bath, in some cases at high current density, and to remove the absorbed hydrogen by baking in air at 200°C. Here it should be noted that a characteristic feature of hydrogen embrittlement is the reversibility of the phenomenon, i.e. hydrogen can be introduced and then removed by diffusion by heating at 200°C without any detrimental effect on the steel.

The electrodeposition of chromium and nickel also results in absorption of hydrogen and a de-embrittling bake in air at 200°C is again required to restore ductility and sustained load properties, although fatigue properties may not be recovered owing to residual tensile stresses in the electrodeposit.

Acid Pickling

This process is widely used for descaling steel, removal of rust prior to painting and dissolving boiler sludge. Immersion of steel in an acid solution results in the dissolution of iron at anodic sites and the evolution of hydrogen at cathodic sites with simultaneous entry of hydrogen into the steel. The hydrogen evolution reaction occurs in two stages (*see* Section 9.1): proton discharge $[(H_3O)^+ + e \rightarrow H_{ads})$, followed by either chemical desorption $(H_{ads} + H_{ads} \rightarrow H_2)$ or electrochemical desorption $[(H_3O)^+ + e + H_{ads} \rightarrow H_2)]$. At the same time some of the adsorbed hydrogen diffuses into the steel, i.e. $H_{ads} \rightarrow H_{abs}$, and embrittlement or blistering may result. In the case of unkilled or semi-killed steel, and particularly for mild steel, blistering of the surface frequently occurs due to the accumulation of gaseous hydrogen under high pressure at non-metallic inclusions situated near the steel surface. However, high-strength steels, which, are relatively clean, do not give rise to blister formation, but become embrittled and crack if the applied tensile stress is sufficiently high.

The extent of blistering or embrittlement depends on the amount of hydrogen adsorption which is generally considered to depend on the surface coverage of the metal with H_{ads}. Many variables including pH, temperature, nature of anions and condition of steel have a marked effect on hydrogen absorption[2]. In addition, certain elements or compounds present in the solution in trace quantities, e.g. S, P, As, Se (referred to as poisons) have the ability to retard (or poison) the chemical desorption reaction and thus increase hydrogen coverage and absorption. In pickling, the presence of a poison may result in a marked increase in coverage with H_{ads} even though the corrosion rate may decrease, and in practice the choice of the inhibitor should be carefully considered or an increase in hydrogen absorption may result. The hydrogen evolution reaction (h.e.r.) and the factors affecting surface coverage with H_{ads}, are considered in more detail in Section 9.1.

Cathodic pickling actively promotes hydrogen absorption and should be avoided, but anodic pickling or pickling in oxidising acids may be used provided certain precautions are taken.

Phosphating and Painting

Phosphate coatings are applied to high-strength steels prior to painting and it is well known that during this process hydrogen enters the steel. By using a phosphating bath containing limited free acid and certain oxidising agents the amount of hydrogen absorption may be reduced[8].

Paints are normally innocuous but paint strippers are not, and it is essential that proprietary paint strippers are tested before use to ensure that they do not result in hydrogen absorption.

Corrosion in Service

Embrittlement may be caused by any type of corrosion reaction in which the cathodic reaction involves hydrogen evolution. Atmospheric corrosion may result in hydrogen absorption if the atmosphere is industrial and contains appreciable amounts of sulphur dioxide and hydrogen sulphide. Exposure of high-strength steel to moist hydrogen sulphide atmospheres is known to cause hydrogen embrittlement, and this effect may be stimulated by a discontinuity in an anodic coating such as zinc, cadmium or aluminium. Hydrogen absorption due to corrosion in acid solutions has been widely investigated and it is accepted that under these conditions cracking is due to hydrogen embrittlement. Recently Brown[9] has reported that the stress-corrosion cracking of pre-cracked high-strength steel specimens exposed to near-neutral chloride solutions may also be due to hydrogen embrittlement, since during corrosion the pH of the solution within the crack becomes acidic with consequent discharge of hydrogen (*see* Section 1.6).

The corrosion of steel may be prevented by coating with cadmium, zinc or aluminium; these coatings are normally anodic to steel in service and provide protection to the steel at pores and discontinuities in the coating by sacrificial action in which the potential of the underlying steel is decreased with consequent hydrogen evolution. If the plated steel is exposed to certain aqueous solutions galvanic attack may occur with evolution of hydrogen at cathodic sites, and its subsequent entry into the steel. Unfortunately only limited data exist concerning hydrogen absorption resulting from the sacrificial action of a coating and further work is required to determine the importance of this potential source of hydrogen. However, examples are known where high-strength steel bolts coated with zinc have failed by cracking when exposed to a natural water, presumably by hydrogen produced by galvanic action.

It therefore follows that cathodic protection of high-strength steels is highly dangerous, particularly when the steel is of a very high strength and is highly stressed; in fact, in these circumstances, it would be unwise to use zinc or aluminium sacrificial anodes and disastrous to use an impressed current system. With lower strength steels zinc anodes might be used, but each system must be considered individually.

State and Location of Hydrogen in Steel

The state of hydrogen in interstitial solution in iron has been considered by several workers who have suggested that hydrogen exists as atoms, protons, or screened ions; Oriani[10] favours the latter, but reports that the type of interstitial site occupied is uncertain. Internal friction and resistivity studies support the view that the hydrogen in true interstitial solution at ambient temperatures constitutes only a small fraction of the total hydrogen occluded by steel as a result of pickling or cathodic polarisation. At temperatures above 400°C solubility, permeation and diffusion data can be readily explained solely in terms of interstitial solution, but below 400°C and particularly at ambient temperatures, hydrogen contents are greater and diffusion rates are slower than predicted from high-temperature data using the interstitial solution model. Darken and Smith[11] postulated that at ambient temperatures 'trapping' of hydrogen occurred at imperfections in the lattice, thus accounting for the anomolously high hydrogen contents and low diffusion rates. This concept of trapping has been confirmed by several investigators and many types of trap have been suggested including vacancies, dislocations, stacking faults, microcracks, grain boundaries, non-metallic inclusions, phase boundaries and micropores.

Hydrogen may exist at a trap in a molecular and/or adsorbed form depending on the nature and size of the particular trap. Heller[12] considers that the total hydrogen occluded by iron may be represented by the sum of (*a*) interstitial solubility, (*b*) void surface solubility and (*c*) void volume solubility (*see* Fig. 8.28), and that the following equilibrium exists:

$$\underset{\text{lattice}}{H_l} \rightleftharpoons \underset{\substack{\text{void}\\\text{surface}}}{H_{ads}} \rightleftharpoons \underset{\substack{\text{void}\\\text{volume}}}{H_2}$$

For steel, however, other types of trap must be considered and the available evidence suggests that hydrogen at phase interfaces, inclusions, micropores,

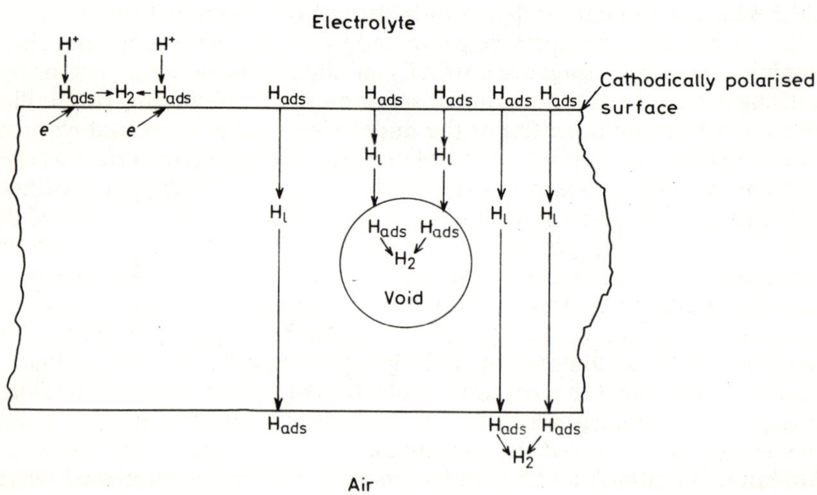

Fig. 8.28 Trapping of hydrogen at a void when it is evolved cathodically at one surface of a sheet specimen

microcracks and dislocations may be major forms of trapped hydrogen, while the amounts trapped at stacking faults, vacancies and grain boundaries are relatively small[13]. While adsorbed hydrogen is thought to exist at ferrite/cementite interfaces and the internal surfaces of microcracks and voids, it is widely recognised that molecular hydrogen precipitates at voids (particularly those voids associated with non-metallic inclusions) and the resultant build-up of pressure may cause blistering of low-strength steels for certain charging conditions[14].

Steel composition has a marked influence on the extent of trapping and hence on hydrogen absorption. Trace impurities in solid solution in α-iron generally have little effect, but a pronounced increase in hydrogen uptake results if the precipitation of a second phase (carbides, oxides, etc.) occurs. Structure also exerts a large effect and it has been demonstrated[15] that hydrogen absorption depends markedly on the ferrite/carbide interfacial area.

Diffusion of Hydrogen in Steel

At temperatures above 400°C hydrogen dissolves and diffuses interstitially in iron and there is reasonable agreement between the values of diffusivity or diffusion coefficient D obtained by different workers. Two typical results obtained from effusion studies in the temperature range 250–900°C are (*a*) $D = 8{\cdot}8 \times 10^{-4} \exp(-3050/RT)$[(16)], and (*b*) $D = 7{\cdot}9 \times 10^{-4} \exp(-2800/RT)$[(17)]. Diffusion behaviour below approximately 400°C is inconsistent with high-temperature observations, and there is poor agreement between different workers, particularly at ambient temperatures where values of the *apparent diffusion coefficient* ranging from 10^{-6} to $10^{-9}\,cm^2/s$ have been reported. Johnson and Hill[18] report a break in the Arrhenius plot for annealed iron at about 200°C. Above this temperature the diffusion coefficient is given by $D = 1{\cdot}4 \times 10^{-3} \exp(-3200/RT)$, while below 200°C, $D = 0{\cdot}12 \exp(-7820/RT)$, and this anomalous type of diffusion behaviour has been observed during several other investigations. Diffusion at the lower temperatures is controlled by the rate of escape of hydrogen from traps, which requires a high activation energy and results in a large decrease in D. It follows that the nature and density of the traps will control the apparent diffusion coefficient, and it has been demonstrated that increasing the degree of cold work, the ferrite/carbide interfacial area or the manganese sulphide inclusion content of the steel results in lower values of D. The effect of cold work may be explained in terms of hydrogen trapping due to the formation of microvoids at inclusions or microcracks at ferrite/cementite interfaces. The binding energy between hydrogen and various types of trap, and hence the relative significance of different types of trap, remains controversial and requires further study.

Oriani[10] has reviewed the trapping theories which have been postulated to account for the anomolous diffusion and permeation data. On the basis of McNabb and Foster's theory, which postulates that trapped hydrogen is in local equilibrium with lattice hydrogen and that the apparent activation energy for the release of trapped hydrogen is given by $E_{app} = E_a + E_t$, where E_a is the activation energy for true interstitial diffusion and E_t is the trap depth, Oriani has explained the dependence of the apparent diffusivity on density of traps and temperature. He concludes that data obtained by transient techniques and yielding values of diffusivity at 25°C smaller than 10^{-5} cm^2/s, or activation energies larger than about 9·4 kJ/mol, involved trapping and/or partial surface control. Values of E_{app} at 25°C have been reported in the range 26–39 kJ/mol.

Effect of Hydrogen on Mechanical Properties

Hydrogen embrittlement of high-strength steels may be evaluated by a sustained load test in which delayed failure occurs at applied stresses well below the notch tensile strength. This type of test, also referred to in the literature as a *stress-rupture* test or *static-fatigue* test, provides a sensitive method for detecting hydrogen embrittlement. Figure 8.29 illustrates the general features of a typical sustained load *vs.* time to failure curve obtained for a hydrogenated high-strength steel by loading notched tensile specimens at various stresses. It is evident that delayed failure may occur over a wide range of applied stress and that a critical stress exists below which failure does not occur. Increasing the hydrogen concentration decreases the time to failure and the critical stress. A reduction in ductility due to hydrogen may be reflected by a decrease in notch tensile strength, as in Fig. 8.29, but Troiano[19] has shown that even in the absence of such an effect (at low hydrogen concentrations) delayed failure may still occur over a wide range of applied stress.

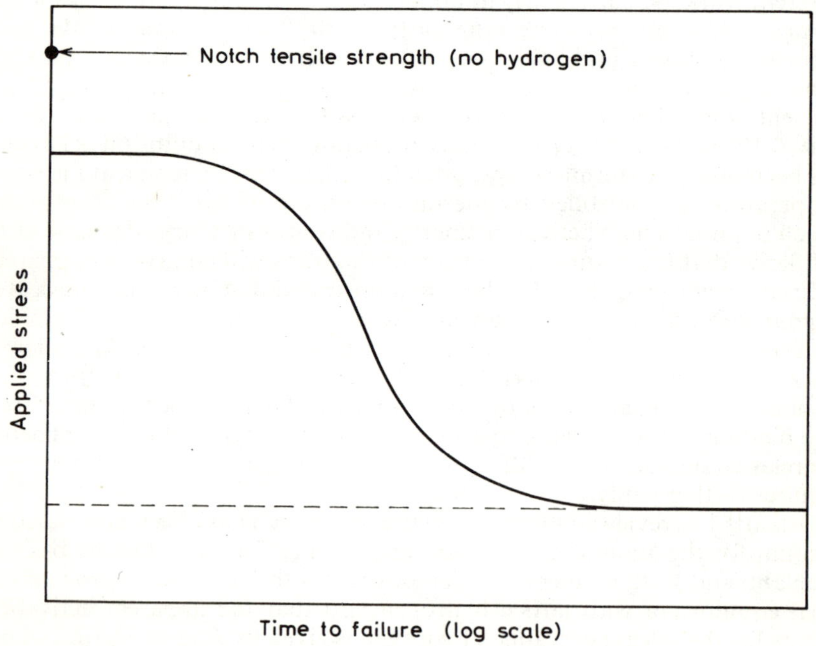

Fig. 8.29 Typical delayed failure curve for a hydrogenated high-tensile steel

Resistivity data indicates that the failure process may be divided into three stages: (*a*) crack initiation, (*b*) slow discontinuous crack propagation and (*c*) rapid unstable fracture. Hydrogen embrittles steel by affecting stages (*a*) and (*b*), and Troiano[19] postulates that both are controlled by the stress-induced diffusion of hydrogen to regions of high triaxial stress. The susceptibility to delayed failure increases with increasing hydrogen content, notch acuity and tensile strength of the steel. A decrease in test temperature prolongs the incubation period for crack initiation, in accordance with the lower diffusion rate, and decreases the critical stress.

The use of smooth or machine-notched tensile specimens does not give good reproducibility owing to the variation in the time for crack initiation. However, by using precracked specimens and applying fracture mechanics reproduceable data may be obtained in terms of the crack growth rates or the K_{ISCC} values. K_{ISCC} is defined as the apparent threshold value of stress intensity K above which one observes stress-corrosion cracking but below which one does not (Fig. 8.30). The variation of K_{ISCC} with environment may be explained in terms of hydrogen embrittlement, and the increased susceptibility to embrittlement of steels with increased tensile strength is reflected by corresponding decreases in K_{ISCC} (Fig. 8.31). Cathodic polarisation may result in only relatively small decreases in K_{ISCC}, while marked increases in crack propagation rates have been observed. Large increases in crack propagation rate have also been observed when an argon atmosphere has been replaced by molecular hydrogen at atmospheric pressure and ambient temperature[20].

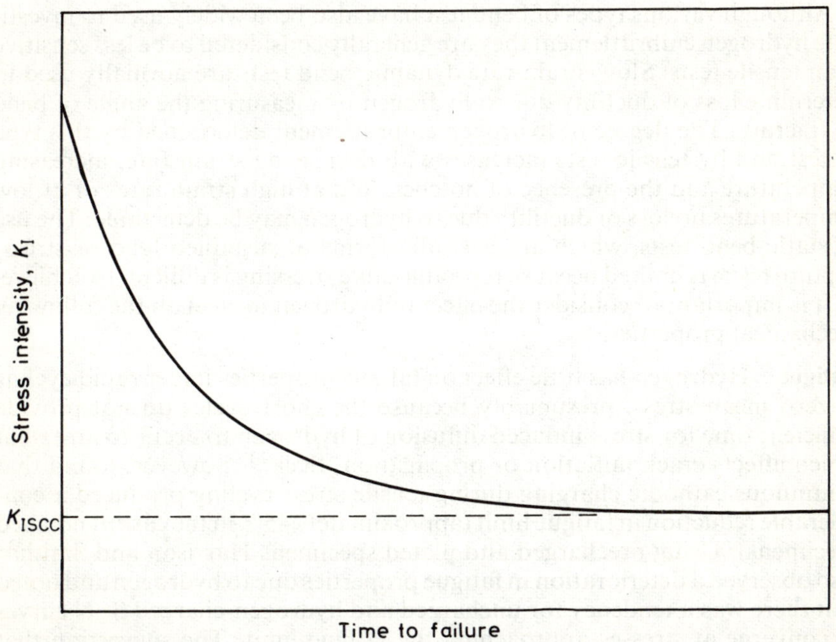

Fig. 8.30 Typical stress intensity versus time to failure curve for a high-tensile steel exposed to a neutral NaCl solution

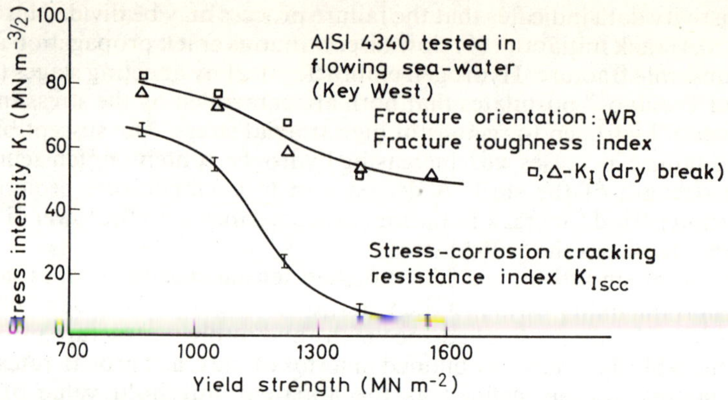

Fig. 8.31 Effect of yield strength (varied by varying tempering temperature) on stress-corrosion resistance (after Brown[9])

Loss of ductility due to the presence of hydrogen may be indicated by the reduction of area or the true fracture stress determined by a slow-strain-rate tensile test. At low hydrogen concentrations a reduction in ductility may not be observed, even though delayed failure readily occurs, and it follows that the tensile test is only capable of revealing relatively severe embrittlement (*see also* Section 8.10). With the exception of brittle structures, e.g. martensite or bainite, tensile strength and yield strength are generally reported to be only slightly affected by hydrogen.

Although various types of bend test have also been widely used to investigate hydrogen embrittlement they are generally considered to be less sensitive than tensile tests. Slow-strain-rate dynamic bend tests are normally used to determine loss of ductility due to hydrogen by measuring the angle of bend to fracture. The degree of hydrogen embrittlement determined by this type of test and by tensile tests increases with decreasing strain rate, increasing temperature and the presence of notches, and at high strain rates or at low temperatures no loss of ductility due to hydrogen may be detectable. The use of static bend tests, which are basically forms of sustained load or stress-rupture tests, is limited because reproduceable stressing is difficult to achieve.

It is important to consider the effect of hydrogen in steel on the following mechanical properties.

Fatigue Hydrogen has little effect on fatigue properties under rapid cycling at zero mean stress, presumably because the short cycles do not provide sufficient time for stress-induced diffusion of hydrogen to occur to an extent which affects crack initiation or propagation. Beck[21], however, found that continuous cathodic charging during tensile stress cycling produced a considerable reduction in fatigue limit (approximately 45% in the case of notched specimens). Using precharged and plated specimens Harrison and Smith[22] also observed a deterioration in fatigue properties due to hydrogen and noted that there was a tendency for uncharged and hydrogen-charged S–N curves to converge at stresses approaching the fatigue limit. The suggestion that this was probably due to progressive loss of hydrogen during the relatively long testing period was confirmed by constant cathodic charging experiments, similar to those reported by Beck, where the phenomenon was eliminated. Wei[23] has recently reviewed the crack growth aspect of corrosion fatigue of high-strength steels, and has indicated that hydrogen embrittlement is likely to be responsible for the increased rates of crack growth observed for several environments and for a wide range of stress intensity.

Hardness No hardness changes due to the presence of hydrogen have been detected by some workers, while others have observed small changes (5–15 HV) particularly when tests were performed at slow indentation rates[2]. In several cases the increases in hardness have been accounted for in terms of work hardening accompanying localised plastic deformation caused by the build up of high pressures of molecular hydrogen at certain types of trap.

Internal friction Two internal friction peaks may result from hydrogen charging[13, 24]. The first peak, observed by several workers at approximately 50 K, is believed to be due to the stress-induced diffusion of hydrogen, i.e. a Snoek peak. This view is supported by the good agreement between values of activation energy and diffusion coefficient obtained from (*a*) internal friction data, and (*b*) extrapolated high-temperature diffusion data. In addition this low-temperature peak appears immediately after hydrogenation and decreases in height with ageing time. The second peak, the cold-work peak, observed at higher temperatures (100–150 K), is thought to be due to the interaction of oscillating dislocations with hydrogen atmospheres. This cold-work peak does not occur in steel freshly charged with hydrogen but only appears after a considerable ageing time. Internal friction studies have provided important information concerning hydrogen–dislocation interactions and hydrogen diffusion in iron and steel.

Theories of Hydrogen Embrittlement

Hydrogen embrittles high-strength steels by affecting the crack initiation and slow crack growth stages of fracture. In general susceptibility of steels to hydrogen embrittlement (loss of ductility or delayed failure) increases as tensile strength increases. Loss of ductility is a maximum at ambient temperatures, and disappears at high or very low temperatures. In addition embrittlement is most pronounced at low strain rates. Any theory of hydrogen embrittlement must explain the above dependence on temperature and strain rate, and many such theories have been proposed.

The planar pressure theory was advanced in 1941 by Zapffe[25] who postulated that molecular hydrogen precipitated into internal voids with subsequent development of very high pressures. The local internal stresses associated with these high pressures were considered to assist the external stress in fracturing the steel. This early theory was modified by de Kazinsky[26] who suggested that high-pressure molecular hydrogen was present in microcracks and during crack growth the adiabatic expansion of the gas released energy which lowered the applied stress necessary for crack propagation. Diffusion of hydrogen to the microcrack was required to maintain the high pressure and thus the dependence of hydrogen embrittlement on strain rate and temperature was explained. Bastien[27] proposed that hydrogen was condensed around dislocations in the form of Cottrell atmospheres and transported with them during plastic deformation at slow strain rates. When such dislocations piled up at voids, high pressures and triaxial stresses due to hydrogen were created and embrittlement resulted. The pressure theory has been modified by several other workers including Garofalo, *et al.*[28], Hewitt and Bilby[29], and Tetelman[30]. Although the pressure theories may account for crack initiation, they cannot be reconciled with the increases in crack propagation rate observed for high-strength steels in the presence of molecular hydrogen at atmospheric pressure and room temperature[20].

A different approach was adopted by Petch and Stables[31] who suggested that microcrack formation occurred as a result of dislocation pile up at grain boundaries, and that adsorption of hydrogen onto the internal surfaces of these cracks lowered the surface energy and thus the stress required for crack propagation. However, this relatively simple adsorption theory cannot explain the crack propagation rates observed in the presence of other gases.

Troiano, *et al.*[19,32] postulated that stress-induced diffusion of hydrogen to regions of high triaxial stress resulted in a reduction of the cohesive or fracture strength of steel. Previous theories were criticised because they could not explain the discontinuous nature of the slow crack propagation stage[19] or the variation of ductility with ageing time observed for hydrogenated steel specimens deformed at $-196°C$ prior to ageing[32]. It was suggested that the incubation period for crack initiation corresponded to the stress-induced diffusion of hydrogen to regions of maximum triaxial stress until a critical combination of hydrogen concentration and stress was attained. In a similar manner slow crack growth was explained in terms of stress-induced diffusion to a region just in front of the crack, corresponding to the position of maximum triaxial stress, until a critical hydrogen concentration was attained when crack growth occurred through the hydrogen-rich region. Although the exact process by which hydrogen embrittles steel is not specified, these ideas explain (*a*) the dependence of embrittlement on temperature and strain rate, (*b*) the discontinuous nature of slow crack growth, and (*c*) the variation of ductility with ageing time referred to above[32].

In order to explain experimental data concerning the response of both self-arrested and propagating cracks to (*a*) small increases in hydrogen fugacity and (*b*) variations in environment (e.g. replacing H_2 by O_2, Cl_2, D_2), Oriani[33] has recently postulated a decohesion theory depending on the accumulation of dissolved hydrogen in regions of non-Hookean elastic strain situated within a few atomic distances of the crack tip. However, the relative importance of the theories presented by Troiano and Oriani is not clear at the present time.

The controversy surrounding models of hydrogen embrittlement has been intensified by Beacham[34] who recently suggested a new model following extensive electron microscopic studies of fracture surfaces. Beacham states that 'macroscopic plastic deformation is precluded by hydrogen aiding the mechanisms of localised plasticity at crack tips'. However, it is difficult to conceive any possible mechanisms whereby hydrogen enhances plasticity; in fact, internal friction and strain-ageing data indicate that hydrogen impedes dislocation motion. In contrast to Beacham's ideas, Gilman[35] believes that the formation of a surface hydride occurs which tends to inhibit plastic deformation and favour cleavage by reducing the rate of stress relaxation at the crack tip, but little supporting evidence is available.

Stress-corrosion Cracking

Stress-corrosion cracking of high-tensile steels (tensile strength $>1\cdot4$ GN/m^2) has been observed for many environments[3, 36, 37] including moist H_2S, aerated solutions of NaCl, Na_2SO_4 and $NaNO_3$, liquid ammonia, marine and industrial atmospheres. It has been widely accepted for many years that hydrogen embrittlement is responsible for stress-corrosion cracking of high-tensile steels exposed to moist H_2S or certain acidic environments, but the mechanism of failure in other environments is less clear. Phelps[36] pointed out that the effect of electrochemical polarisation on the time to failure could provide a diagnostic criterion showing whether failure is due to hydrogen embrittlement or to active path corrosion. Thus if stress-corrosion cracking is caused by localised corrosion along an active path through the steel, the application of cathodic current would be expected to suppress the corrosion reaction and to result in a longer time to failure, while the application of anodic current would be expected to increase the amount of corrosion. However, if hydrogen generated by general corrosion is the cause of stress-corrosion cracking the application of cathodic current would be expected to generate more hydrogen and shorten the time to failure, while the application of anodic current would be expected to generate less hydrogen and lengthen the time to failure. Figure 8.32 shows the various types of time-to-failure versus polarisation curves which may be expected, and, using the argument outlined above, the regions where failure may be attributed to either hydrogen embrittlement or active path corrosion are indicated.

Logan[3] reported delayed failures in a high-strength steel (tensile strength approximately 1·8 GN/m^2) exposed to a boiling 3·5% NaCl solution and under a tensile stress of 1·38 GN/m^2. Specimens in the form of hollow tubes were simultaneously stressed and evacuated while exposed to the corrodent, and hydrogen diffusion through the tube walls was observed in all cases before failure occurred. It was concluded that hydrogen embrittlement was the mechanism of failure.

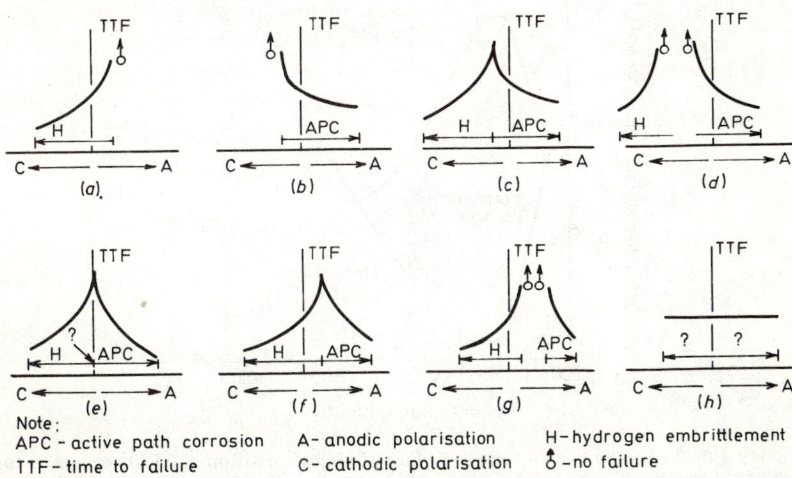

Fig. 8.32 Effect of anodic and cathodic polarisation on time to failure (after Phelps[36])

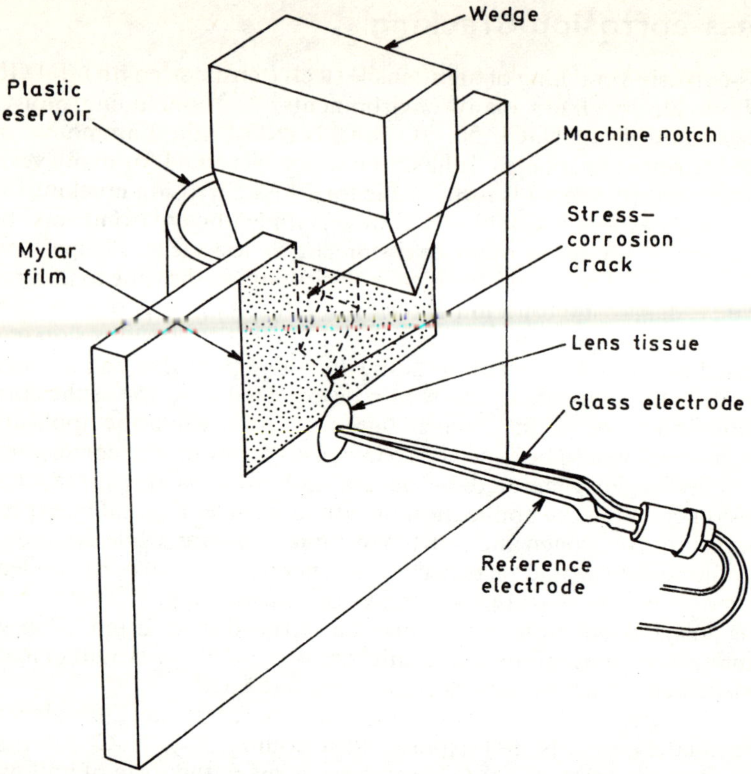

Fig. 8.33 Experimental arrangement for determining pH at a crack tip (after Brown[9])

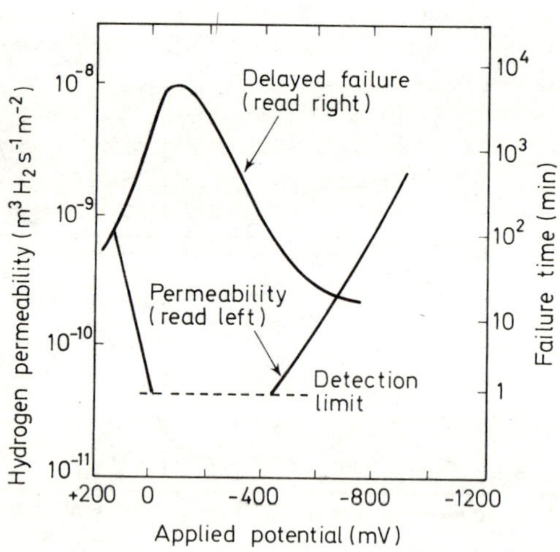

Fig. 8.34 Effect of potential on time to failure of stress-corrosion cracked specimens (upper curve) and on hydrogen permeation of membrane (lower discontinuous curve). (After Barth, Steigerwald and Troiano[39])

Brown[9, 38] has also postulated that hydrogen embrittlement is responsible for stress-corrosion cracking of high-strength steels exposed to neutral chloride solutions. The pH and potential at the crack tip were measured and shown to be favourable for hydrogen evolution. Figure 8.33 illustrates the arrangement used by Brown for measuring pH at the crack tip with a miniature glass electrode. A small disc of tissue is located on the surface of the specimen in advance of the propagating crack, and when the crack arrives at the tissue capillarity draws the solution (that nearest the crack tip) out of the crack and transmits it to the glass electrode. Using this technique Brown demonstrated that for high-strength steels, regardless of the pH of the external solution (pH 1–10), at the freely corroding potential the pH at the crack tip is invariant (slightly less than 4). He suggested that this value of pH resulted mainly from the reaction $Fe^{2+} + H_2O \rightarrow Fe(OH)^+ + H^+$, occurring in the vicinity of the crack tip. However, this reaction predicts a higher value of pH than that observed experimentally, and to account for this difference Brown further suggested that some Fe^{2+} reacted with dissolved oxygen to produce Fe^{3+}, and the subsequent hydrolysis reaction (possibly $Fe^{3+} + 3H_2O \rightarrow Fe(OH)_3 + 3H^+$) resulted in a lower pH. In addition, studies on specimens polarised anodically and cathodically (from $-0{\cdot}2$ V *vs.* S.H.E. to $-1{\cdot}0$ V *vs.* S.H.E.), and with bulk solutions of pH 2–10, showed that the pH and potential at the crack tip were always favourable for hydrogen reduction. The *IR* drop down the crack was found to be insignificant (only 10–25 mV). The postulate that hydrogen embrittlement is responsible for failure even at applied anodic potentials has been substantiated by direct hydrogen permeation measurements[39] where good correlation with time-to-failure data was observed over a wide range of applied potential, as indicated in Fig. 8.34. The observation that the fracture topology was identical for failures occurring as a result of both anodic and cathodic polarisation also indicated that hydrogen embrittlement was the common mechanism of failure. These views have been supported by Wilde[40] who carried out electrochemical polarisation and hydrogen permeation measurements on a martensitic stainless steel exposed to NaCl solutions, and concluded that hydrogen embrittlement could account for stress-corrosion cracking observed under all the conditions of exposure.

Stress-corrosion cracking observed on exposure of pre-cracked specimens of high-strength steel to distilled water or water vapour is also believed to be due to hydrogen embrittlement[20]. The damaging hydrogen is assumed to result from the following reaction occurring at the crack tip:

$$Fe + H_2O \rightarrow FeO + H_2$$

An apparent activation energy for crack growth of about 38 kJ/mol was determined and this value is in good agreement with activation energies quoted for the diffusion of hydrogen in steel.

The available evidence indicates that stress-corrosion cracking of high-tensile steels in aqueous environments occurs by hydrogen embrittlement, although the exact mechanism by which hydrogen assists cracking is unknown. However, it is possible that active path corrosion may be important in circumstances where most of the time-to-failure is associated with crack initiation rather than slow crack propagation. Although environment-induced slow-crack growth of high-tensile steels may almost exclusively be attributed to hydrogen embrittlement, one notable exception has been reported by Kerns and Staehle[41] who observed slow crack growth in the presence of chlorine gas, and concluded that for such an environment the presence of hydrogen was not essential for crack propagation.

Although the relative susceptibility of different high-strength steels to stress-corrosion cracking is well documented[4, 6, 42, 43], the reasons for the variation in resistance to stress-corrosion cracking are not fully understood and further work is required (*a*) to determine the detailed mechanism by which hydrogen embrittles steel, (*b*) to explain the effect of chemical composition and heat treatment on the rate of crack propagation during stress-corrosion cracking (and to identify the rate controlling step), and (*c*) to determine the extent to which the variation of K_{ISCC} with chemical composition and heat treatment is controlled by the accompanying variation in mechanical properties. Each of these aspects of stress-corrosion cracking of high strength steels remains controversial even though numerous relevant investigations have recently been carried out[4–6, 44, 45].

P. POOLE

REFERENCES

1. Buzzard, R. W. and Cleaves, H. E., *Nat. Bur. Standards Circular*, **511** (1951)
2. Smialowski, M., *Hydrogen in Steel*, Pergamon Press (1962)

2*a*. Beck, W., Jankowsky, E. J. and Fischer, P., *Hydrogen Stress Cracking of High Strength Steels*, N.A.D.C. Penn. U.S.A. (Report NADC-MA-7140) (1971)

3. Logan, H. L., *The Stress Corrosion of Metals*, Wiley (1966)
4. *Proceedings of Conference on Fundamental Aspects of Stress Corrosion Cracking*, NACE (1969)
5. *Proceedings of Conference on The Theory of Stress Corrosion Cracking in Alloys*, NATO, Scientific Affairs Division (1971)
6. *Proceedings of Conference on Stress Corrosion Cracking and Hydrogen Embrittlement of Iron Base Alloys* (1973)

6*a*. Cotterill, P., *Hydrogen Embrittlement of Metals*, Pergamon Press (1961)

7. Troiano, A. R., *Corrosion*, **15**, 207 (1959)
8. Sherlock, J. and Shreir, L. L., *Corrosion Science*, **10**, 561 (1970)
9. Brown, B. F., *Proceedings of Conference on The Theory of Stress Corrosion Cracking in Alloys*, NATO, Scientific Affairs Division, 186 (1971)
10. Oriani, R. A., *Proceedings of Conference on Fundamental Aspects of Stress Corrosion Cracking*, NACE, 32 (1969)
11. Darken, L. S. and Smith, R. P., *Corrosion*, **5**, 1 (1949)
12. Heller, W. R., *Stress Corrosion Cracking and Embrittlement*, Wiley, 163 (1956)
13. Miodownik, A. P., *Proceedings of Conference on Stress Corrosion Cracking and Hydrogen Embrittlement of Iron Base Alloys* (1973)
14. Poole, P. and Shreir, L. L., *Electrochimica Acta*, **14**, 667 (1969)
15. Newman, J. F. and Shreir, L. L., *J. Iron Steel Inst.*, **207**, 1369 (1969)
16. Stross, T. M. and Tomkins, F. C., *J. Chem. Soc.*, **155**, 230 (1956)
17. Coe, F. R. and Moreton, J., *J. Iron Steel Inst.*, **204**, 366 (1966)
18. Johnson, E. W. and Hill, M. L., *Trans. A.I.M.E.*, **218**, 1104 (1960)
19. Troiano, A. R., *Trans. A.S.M.*, **52**, 54 (1960)
20. Johnson, H. H., *Proceedings of Conference on Fundamental Aspects of Stress Corrosion Cracking*, NACE, 439 (1969)
21. Beck, W., *Electrochem. Tech.*, **2**, 74 (1964)
22. Harrison, J. D. and Smith, G. C., *Brit. Welding J.*, **14**, 493 (1967)

23. Wei, R. P. and Simmons, G. W., *Proceedings of Conference on Stress Corrosion Cracking and Hydrogen Embrittlement of Iron Base Alloys* (1973)
24. Gibala, R., *Trans. A.I.M.E.*, **239**, 1574 (1967)
25. Zappfe, C. and Sims, C., *Trans. A.I.M.E.*, **145**, 225 (1941)
26. de Kazinczy, F., *J. Iron Steel Inst.*, **177**, 85 (1954)
27. Bastien, P. and Azou, P., *Compt. Rend.*, **232**, 1845 (1954)
28. Garafola, F., Chow, Y. and Ambergaokar, R., *Acta Met.*, **8**, 504 (1960)
29. Bilby, B. A. and Hewitt, J., *Acta Met.*, **10**, 587 (1962)
30. Tetelman, A. S., *Proceedings of Conference on Fundamental Aspects of Stress Corrosion Cracking*, NACE, 446 (1969)
31. Petch, N. J. and Stables, P., *Nature*, **169**, 842 (1952)
32. Morlet, J. G., Johnson, H. H. and Troiano, A. R., *J. Iron Steel Inst.*, **189**, 37 (1958)
33. Oriani, R. A., *Proceedings of Conference on Stress Corrosion Cracking and Hydrogen Embrittlement of Iron Base Alloys* (1973)
34. Beachem, C. D., *Met. Trans.*, **3**, 437 (1972)
35. Gilman, J. J., *Phil. Mag.*, **26**, 801 (1972)
36. Phelps, E. H., *Proceedings of Conference on Fundamental Aspects of Stress Corrosion Cracking*, NACE, 398 (1969)
37. Staehle, R. W., Wang, M. T. and Kerns, G. E., *Proceedings of Conference on Stress Corrosion Cracking and Hydrogen Embrittlement of Iron Base Alloys* (1973)
38. Brown, B. F., ibid.; Brown, B. F., Fujii, R. T. and Dahlberg, E. P., *J. Electrochem. Soc.*, **116**, 218 (1969)
39. Barth, C. F., Steigerwald, E. A. and Troiano, A. R., *Corrosion*, **25**, 353 (1969)
40. Wilde, B. E., *Corrosion*, **27**, 326 (1971)
41. Kerns, G. E. and Staehle, R. W., *Scripta Met.*, **6**, 1189 (1972)
42. Carter, C. S., *Corrosion*, **27**, 471 (1971)
43. Carter, C. S., Farwick, D. G., Ross, A. M. and Uchida, J. M., *Corrosion*, **27**, 190 (1971)
44. Sandoz, G., *Met. Trans.*, **3**, 1169 (1972)
45. Kortovich, C. S. and Steigerwald, E. A., *Eng. Fracture Mechanics*, **4**, 637 (1972)

8.5 Stress-corrosion Cracking of Titanium, Magnesium and Aluminium Alloys

Titanium

Stress-corrosion cracking occurs in titanium alloys in a number of environments, although the number of failures that have occurred under service conditions is very small. Because of the widespread use of titanium alloys in aeroplanes and space vehicles and their increasing use in marine applications it is important that the possibilities of service failures should be removed. As a result a considerable and increasing amount of work has been done on this subject over the last decade as indicated in a recent extensive survey[1].

The two principal forms of stress-corrosion failure are (*a*) hot salt cracking and (*b*) room-temperature cracking, the latter occurring in both aqueous and methanolic chloride environments, and in N_2O_4. In addition, environmental failures can occur in alloys in direct contact with some liquid and solid metals, and certain gases.

Hot Salt Cracking[2,3]

Many titanium alloys develop cracks if they are heated under stress while in direct contact with solid chlorides in the presence of oxygen and moisture at temperatures higher than about 250°C, and cracking is predominantly intergranular. Originally the phenomenon was associated with residual traces of NaCl in perspiration[2] and it is sometimes referred to as *finger-print cracking*. Many other chlorides can cause this type of failure to different degrees. The severity of attack varies[4] with the cationic species, becoming greater in the order $MgCl_2 < SrCl_2 < CaCl_2 < KCl < BaCl_2 < NaCl < LiCl$. Bromides and iodides also cause cracking but fluorides do not.

Mechanism The mechanism of cracking has not been established. Even the corrosion reaction that is responsible for the initial attack has not been determined. Early work[5] led to the suggestion that chlorine gas was generated and could cause fracture by a cyclic process requiring the formation and decomposition of $TiCl_2$:

$$Ti + TiO_2 + 2NaCl + \tfrac{1}{2}O_2 \rightarrow TiCl_2 + Na_2TiO_3$$

$$TiCl_2 + O_2 \rightarrow TiO_2 + Cl_2$$

$$Ti + Cl_2 \rightarrow TiCl_2 \text{ (at the crack tip)}$$

Chlorine has been identified[6] as a reaction product, and furthermore it has been shown[7] that cracking can occur in chlorine gas in the absence of oxygen.

The formation of low-melting-point mixtures has also been considered[8] as a possible step in the cracking process and these have been observed, but the wide range of chlorides that cause cracking suggests that such mixtures may not be necessary.

More recent work has focused attention upon the rôle of moisture. It is not entirely clear that cracking does not occur in its complete absence, but if it is present then it takes part in the corrosion process. Radiotracer techniques have been used[8] to show that hydrogen is retained in areas corroded at 343°C by salt containing tritiated water. Hydrochloric acid is considered to be the important corrosion product which then promotes the entry of hydrogen into the specimen with consequent intergranular separation by a sorption mechanism[8]. To verify this hypothesis, stressed specimens of Ti–8Al–1Mo–1V were tested[8] in glass ampoules containing hydrogen chloride gas at 1 atm and 343°C. Cracking occurred and the time to failure was related to the extent of corrosion which in turn was considered to be dependent upon the moisture content. Cracking was both intergranular and quasi-cleavage, and appeared similar to hot salt cracking.

The production of HCl and hydrogen is thought to arise from the reaction:

$$Ti + 2NaCl + 2H_2O \rightarrow TiCl_2 + 2NaOH + 2H$$

followed by several possible regenerative reactions which would occur at different intervals of time as moisture was absorbed:

$$TiCl_2 + 2H_2O \rightarrow TiO_2 + 2HCl + 2H$$

$$Ti + 2HCl \rightarrow TiCl_2 + 2H$$

The proposed mechanism includes the production of HCl from the pyrohydrolysis of the metal chlorides. Similar reactions are likely for bromides and iodides. Fluorides however are relatively stable and would not be expected to hydrolyse. It was considered that this might account for the inability of fluorides to cause cracking. Hydrogen absorption by titanium alloys exposed to chloride salts at elevated temperatures has been detected[9] and found to be proportional to the amount of moisture participating in the reaction.

Other work on Ti–8Al–1Mo–1V has also indicated[10] that hydrogen absorption occurs. Stressed specimens exposed to hot salt at 454°C for 100 h suffered a marked embrittlement when subsequently tested in air at room temperature. This embrittlement could be removed by vacuum annealing at 649°C for 4 h. The loss of ductility observed was greater as the strain rate of testing was lowered. Both these observations were considered to be indicative that hydrogen was the principal embrittling species. Analysis of the fracture surface of specimens containing 70 p.p.m. of hydrogen that had suffered hot salt cracking has shown[11] that concentrations as high as 12 000 p.p.m. of hydrogen are developed during the cracking process and this would appear to be conclusive evidence that hydrogen plays a major rôle when moisture is present. Cracking of components in dry air with a dewpoint of −84°C was interpreted[12] as showing that only a very low moisture content is required (0·25 p.p.m.). Increasing this by 25 000 fold by water injection did not increase the degree of embrittlement.

Alloy susceptibility Apart from unalloyed titanium all alloys exhibit some degree of susceptibility. An approximate rating derived from laboratory tests[13] indicates three general groups of alloys.

1. Highly susceptible alloys: Ti–5Al–2·5Sn, Ti–12Zr–7Al, Ti–5Al–5Sn–5Zr, Ti–8Al–1Mo–1V, Ti–8Mn.
2. Moderately susceptible alloys: Ti–5Al–5Sn–5Zr–1Mo–1V, Ti–6Al–4V, Ti–6Al–4V–2Sn, Ti–13V–11Cr–3Al.
3. Most resistant alloys: Ti–4Al–3Mo–1V, Ti–11Sn–5Zr–2·25Al–1Mo–0·25Si, Ti–4Mo–4Zr–2Al.

Tests in a $Cl_2 + O_2$ mixture at 427°C have shown[14] that the worst elements for promoting susceptibility are Al, Sn, Cu, V, Cr, Mn, Fe and Ni, while the least harmful are Zr, Ta and Mo. α-phase alloys are generally more susceptible than β-phase alloys. Heat treatment has not been examined extensively, but some heat treatments render some α-alloys more susceptible[3] or change the mode of fracture[6]. The general effect will depend upon the alloy and the heat-treatment cycle. Subsequent cold work can sometimes considerably lower susceptibility[6]. Failure times decrease as either the testing temperature or initial stress value is raised.

It is important to note that a variety of tests have been employed in studies of hot salt cracking and comparison between the results of different workers is not always possible. It is not clear how susceptibility should be defined and, not surprisingly this makes the rating of alloys difficult. It may also be added that at high temperatures of testing, changes may occur within an alloy which make it more or less susceptible than the alloy in the starting condition. The general corrosion rate will increase as the testing temperature is raised and will eventually become unacceptable. The creep rate will also increase and in some alloys this is a more critical criterion than cracking susceptibility.

Preventative methods Cracking can be delayed or possibly prevented by shot peening components (which serves to impose compressive stresses upon their surfaces) and by the prior application of some surface coatings, e.g. nickel plating, or dipped aluminium or zinc coatings[6]. Other work[4] has shown that susceptibility of as-machined specimens is substantially lowered by chemical milling which removes the stressed surface layers. It has also been reported[5] that the amount of corrosion that occurs is reduced and less cracking is observed as the velocity of the gaseous environment in contact with stressed components is increased. This is particularly relevant to components inside aeroplane engines, e.g. compressor blades. Such observations were made at 427°C. Other workers[15] have reported similar observations at 316°C but not at 371°C and most recent work[12] suggests that such effects are extremely small.

Room-temperature Cracking

Many titanium alloys are susceptible to stress-corrosion cracking in aqueous and methanolic chloride environments.

Aqueous environments Neutral chloride solutions do not corrode titanium alloys at ambient temperatures, and smooth statically loaded specimens of

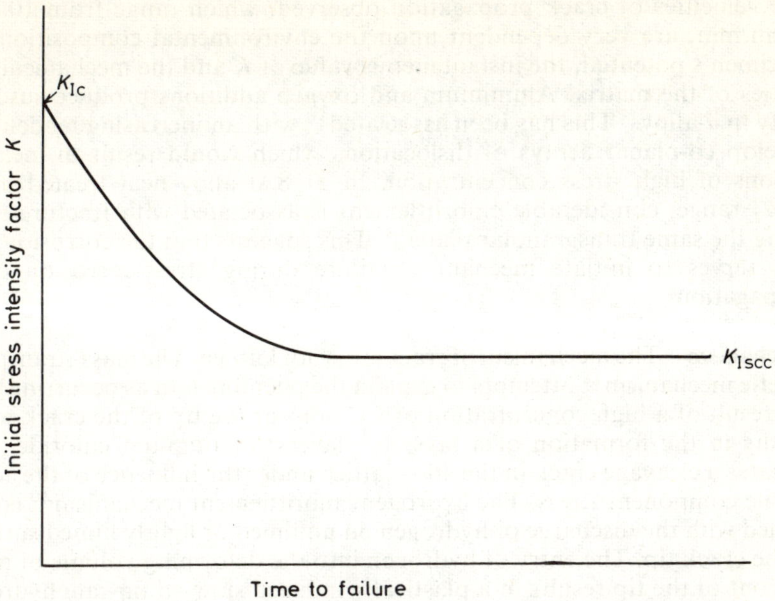

Fig. 8.35 Initial stress intensity factor and time to failure for a susceptible titanium alloy tested in a neutral aqueous environment under plane strain conditions

susceptible alloys do not exhibit failure. In order to nucleate cracking it appears probable that the protective oxide film on alloys must be destroyed and its repair must not occur. If this breakdown occurs then cracking is observed in susceptible alloys. Consequently, the type of test and the type of specimen employed in any selected test are both important considerations, particularly in alloys exhibiting low susceptibility.

Cracking in aqueous solutions was first observed in 1964 when transgranular cracking was found[16] in Ti–7Al–2Nb–1Ta alloy exposed to an NaCl solution in the form of a pre-cracked notched specimen under cantilever load and plane strain conditions, and tested below the plane-strain fracture toughness. Cracking proceeded until the initial stress intensity factor $K(\propto\sigma\sqrt{c})$ reached K_{IC} and tensile overload failure occurred. A plot of initial stress intensity factor K *vs.* time to failure t_f took the form shown in Fig. 8.35. The value of initial K below which cracking did not occur was designated K_{ISCC}, which represented a threshold value. The ratio K_{ISCC}/K_{IC} was an indication of susceptibility, being small for highly susceptible alloys, e.g. 0·2. Cracking occurs in aqueous solutions only in the presence of Cl^-, Br^- and I^-, but not F^-, and failures in distilled water containing no detectable amounts of those species are thought to indicate that the concentrations required are very small. It is possible that the Cl^- may come from the metal itself since $TiCl_4$ is formed during the refining of the basic metal.

Transgranular fracture in α-alloys occurs by cleavage on a plane 14–16° to the basal plane[17]. The greatest susceptibility is observed in specimens in which the general cracking plane is parallel to the basal planes, whereas the minimum susceptibility occurs when it is perpendicular to the basal planes[1].

The velocities of crack propagation observed, which range from 10^{-4} to 10 cm/min, are very dependent upon the environmental composition, the specimen's potential, the instantaneous value of K and the mechanical properties of the matrix. Aluminium and oxygen additions produce susceptibility in α-alloys. This has been associated[18] with an increasing tendency to develop co-planar arrays of dislocations which would result in localised regions of high stress concentration. In Ti–8Al alloy heat treated in the $\alpha+\alpha_2$ range, considerable embrittlement is associated with fracture in air along the same transgranular plane[18]. This suggests that the corrosion process serves to initiate mechanical failure during stress-corrosion crack propagation.

Mechanism The mechanism of cracking is not known. The mass–transport–kinetic mechanism[19] attempts to explain the phenomenon as occurring from the result of a high concentration of Cl^- ions at the tip of the crack which results in the formation of a layer (or layers) of titanium chloride. This initiates a cleavage crack in the alloy lattice under the influence of the acting tensile component stress. The hydrogen embrittlement mechanism[20] is concerned with the discharge of hydrogen on unfilmed or lightly filmed surfaces at the crack tip. The entry of hydrogen into the deforming volume of metal in front of the tip results in a plastically induced slow-strain-rate hydrogen embrittlement. The consequent loss of ductility is repeated grain by grain as crack propagation continues. Irregular crack propagation has been detected by acoustic emission[21] and is indicated also by fractographic observations[22]. Since the diffusivity of hydrogen is much smaller than the highest velocities observed it has been postulated that cleavage once initiated in regions embrittled by absorbed hydrogen may continue for mechanical reasons to some depth beyond such regions. Consistent with this is the general observation that the highest velocities are observed in the stronger less-ductile alloys.

In neutral solutions the application of cathodic polarisation prevents crack initiation and this could be taken to indicate that hydrogen embrittlement is not the operative mechanism, since the discharge and entry of hydrogen might be expected to fracture the specimen more readily. The beneficial effect of cathodic polarisation has been interpreted[23], however, to result from more rapid film repair in the alkaline catholyte generated by the cathode reaction. The film serves as a barrier to rapid hydrogen entry. Consistent with this is the observation[23] that in an environment of low pH (e.g. 10 N HCl) where film formation would not be expected, cathodic polarisation has no effect upon crack propagation.

Many workers have employed pre-cracked specimens in a number of configurations that permit interchangeable data to be obtained from an analysis derived with fracture mechanics. The pre-crack provides an ideal crevice for corrosion reactions which, as a result of hydrolysis equilibria, lead to low pH conditions (*see* Section 1.6). The pH at the tip of a crack in a titanium specimen exposed to neutral solutions (pH 7) has been shown[24] to fall to as low as pH 1·8. Such environments will aid the necessary breakdown of the protective film. In alloys that are not very susceptible the testing procedure adopted may affect the results. Thus loading a specimen that is in contact with the environment is a more severe test than one in which the specimen is loaded and the environment then added. The difference arises from the destruction of the passive film at the crack tip caused by the application of the load. Fracture also occurs in flat unnotched specimens dynamically strained over a narrow range of crosshead speeds while in contact with neutral NaCl. Plastic deformation of the surface results in the destruction of the passive film. At high strain rates ductile failure occurs before crack propagation, while at low strain rates repassivation occurs and prevents crack initiation[25].

In pre-cracked specimens the degree of susceptibility decreases with decreasing thickness in statically loaded specimens. It is probable that the explanation for this is not entirely due to mechanical reasons, viz. the departure from plane strain conditions. The transition is dependent upon heat treatment, loading rate, susceptibility and orientation[1]. An additional effect may lie in the greater difficulty of establishing an occluded cell leading to pH changes at a crack tip as the thickness of the specimen is decreased. Such a cell is probably also prevented by many anions. Thus the addition of CrO_4^{2-}, PO_4^{3-} and F^-, and many other species, reduces or prevents cracking. Some cations more noble than titanium have a similar effect, e.g. Cu^{2+}.

Susceptibility to aqueous cracking occurs to different degrees. Some alloys will break in moist air in the pre-cracked conditions, others require immersion in distilled water, while others require immersion in water containing appreciable amounts of dissolved halide. Different heat treatments may produce these different levels of susceptibility in one alloy. The Ti–8Al–1Mo–1V alloy, for example, will fail in laboratory air in the step-cooled condition, but requires immersion in distilled water in the mill-annealed condition and in 0·6 M KCl in the duplex annealed condition[1]. Heat treatment of titanium alloys produces a variety of phase structures, morphology and composition, and the effects upon stress-corrosion susceptibility are complex[26]. Generally, processes increasing the yield stress lower K_{IC} and K_{ISCC}, while β processing is beneficial. In α-alloys ageing in the $\alpha+\alpha_2$ field raises susceptibility. In $\alpha+\beta$ alloys untempered martensitic structures are often immune in neutral solutions. Where the β-phase is immune, as has been observed in some alloys containing Mo or V, the grain size, volume fraction and mean free path of the α-phase are all important. The same considerations apply to $\alpha+\beta$ alloys where only the β-phase is susceptible, e.g. Ti–8Mn. In β-alloys transgranular cracking is also observed, e.g. in Ti–13V–11Cr–3Al, and has been described[27] as arising from micro-void formation on (100) planes. Where the β-phase contains a fine Widmanstatten α-phase precipitate the mode of cracking is one of intergranular separation.

Electrochemical aspects of the stress-corrosion behaviour have been investigated, mainly in neutral solutions[1]. The open-circuit potential of Ti–8Al–1Mo–1V is −800 mV (*vs.* S.C.E.). The crack initiation load reaches a minimum at −500 mV in NaCl and 0 mV in NaI. Cathodic protection occurs below −1 000 mV. Anodic protection is observed in Cl^- and Br^- solutions over a potential range that is very dependent upon strain rate and heat treatment. A linear relationship is observed between velocity and potential over certain ranges of potential. The velocity is also increased by raising the halide concentration of the testing environment. Raising the pH to high values (13–14) may reduce or inhibit cracking depending upon the alloy, its condition and the type of test employed. Lowering the pH results in the apparent elimination of the threshold stress if the concentration of the hydrochloric acid is high (>7 M)[1].

Methanolic environments In methanolic environments stress-corrosion fracture of a similar kind is seen in titanium alloys. In α-alloys transgranular cleavage is observed[20] and a wide range of velocities[1]. With additions of HCl, however, an additional type of fracture is seen in which intergranular separation is seen resulting from a dissolution mechanism accompanied by hydrogen pick-up by the lattice[28]. This aggressive behaviour is therefore different from that observed in neutral aqueous or methanolic environments. It is inhibited by water additions, the amount depending upon the concentration of HCl[29]. Unstressed specimens undergo intergranular attack under open circuit conditions and this is accelerated by the application of anodic polarisation[23]. Solution additions that stimulate the cathodic reaction increase the rate of intergranular attack and thereby shorten times to failure, e.g. Hg^{2+}, Cu^{2+}, Pd^{2+}. The addition of H_2SO_4 and HCOOH also increases attack, as do Br_2 and I_2[30]. Raising the viscosity of the solution by additions of glycerol increases the times to failure[31].

In aggressive methanolic environments no K_{ISCC} is observed. Instead very slow intergranular fracture proceeds at an increasing velocity with increasing values of K until it is superseded by the more rapid cleavage. Where the alloy is not susceptible to cleavage the fracture will be intergranular up to overload failure. The transition in α-alloys depends upon the aluminium and oxygen content, and the degree of cold work[25]. No pre-cracking is required in aggressive environments and in dynamic straining tests fracture occurs at all crosshead speeds below a maximum, since repassivation is not possible[25]. Additions of water will eventually remove this first stage, but not the second unless the alloy is not susceptible to transgranular cleavage in distilled water.

Impressed cathodic currents tend to prevent cracking, the current density required increasing as the water content is lowered. Anodic currents stimulate cracking and there is a linear relationship between velocity and potential up to the pitting potential[1]. Exposure of stressed specimens to aggressive methanolic environments followed by fracture in air results in transgranular fractures similar to stress-corrosion fractures, indicating that some species is absorbed from the environment[22]. This has been postulated to be hydrogen as in aqueous environments[22]. This mechanism has not been firmly established for either environment but there is an increasing amount of evidence to support it. Thus the embrittlement of specimens exposed in the unstressed condition to aggressive methanolic environments can be removed by ageing so that subsequent fracture in air reveals no cleavage[32]. In addition, notched specimens of Ti–Al alloys charged with hydrogen and broken in laboratory air have been shown to cleave along the same plane as the stress-corrosion fracture[33].

Higher alcohols have not been investigated extensively but failures do occur[31]. Other organic liquids cause transgranular fracture, e.g. CCl_4, $C_2H_2I_2$ and a range of commercial Freons which are fully substituted halide compounds[1]. Where these are not aggressive a K_{ISCC} is observed and it is likely that a pre-crack or a dynamic test is required in order to produce stress-corrosion fracture. In those compounds causing fracture which do not appear to contain hydrogen, failure must be the result of residual moisture, providing hydrogen is the species responsible for cracking, but this point has not been established.

In both aqueous and organic environments the crack velocity is related to

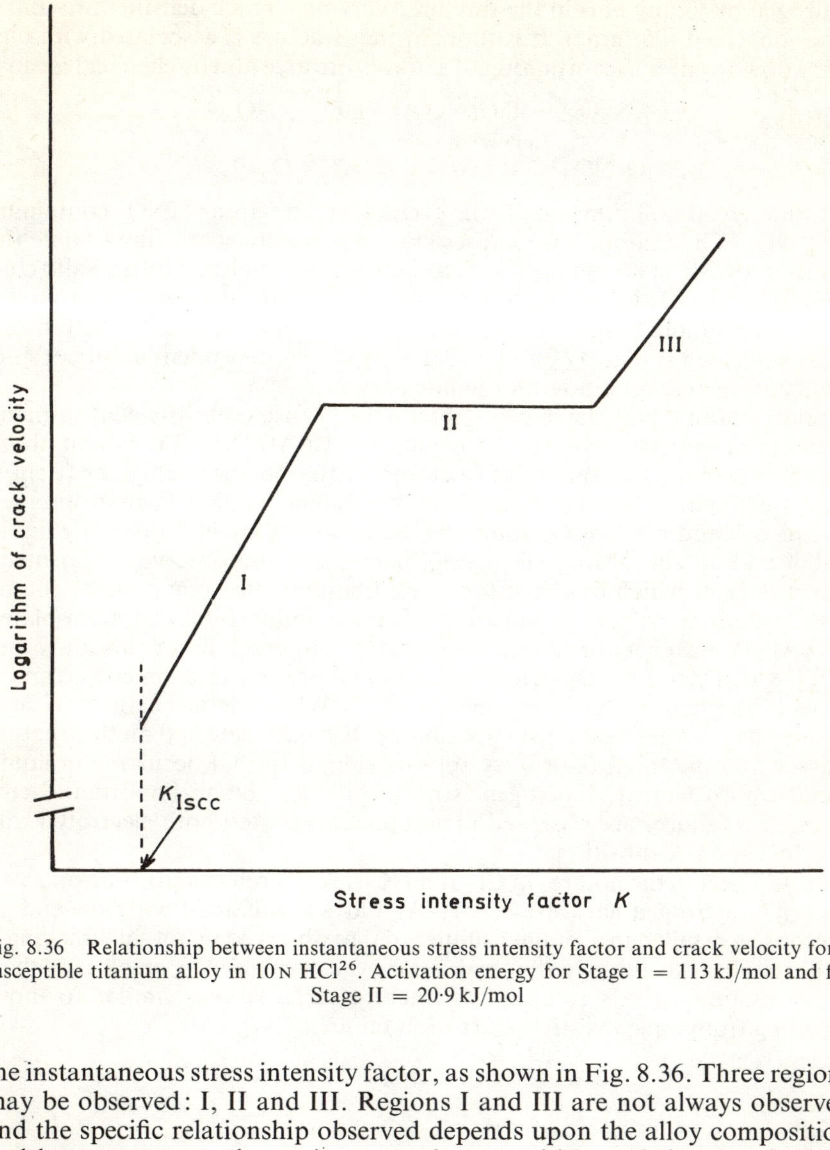

Fig. 8.36 Relationship between instantaneous stress intensity factor and crack velocity for a susceptible titanium alloy in 10 N HCl[26]. Activation energy for Stage I = 113 kJ/mol and for Stage II = 20·9 kJ/mol

the instantaneous stress intensity factor, as shown in Fig. 8.36. Three regions may be observed: I, II and III. Regions I and III are not always observed and the specific relationship observed depends upon the alloy composition and heat treatment, the environmental composition and the experimental conditions[1].

Other environments Other environments have been shown to cause stress-corrosion failure although the amount of work done on such failures has not been large. High-purity red N_2O_4 caused failure of a Ti–6Al–4V pressure vessel during testing[34]. Cracking could be prevented by additions of NO or H_2O, but not by additions of NOCl. K_{ISCC} decreases with increasing temperature. Cracking is both intergranular and transgranular by cleavage in Ti–Al alloys and occurs at a slower rate and at lower K values than in neutral NaCl[35]. Cracking occurs at noble potentials and it seems unlikely that hydrogen plays any part in the fracture. Very high crack densities are sometimes observed ($25/mm^2$). It is thought that fracture is associated with film breakdown and/or the formation of a non-protective film by chemical means:

$$Ti + N_2O_4 \rightarrow \underset{\text{(unprotective film)}}{TiO(NO_3)_2} + NO^+ + NO_2 + e$$

$$2TiO(NO_3)_2 \rightarrow 2TiO_2 + 2N_2O_4 + O_2$$

Commercial titanium and all alloys crack in red fuming HNO_3 containing 20% NO_2. Eliminating NO_2 causes cracking in only some alloys while the addition of 2% H_2O removes susceptibility completely[1]. Molten salts containing halides also cause stress corrosion[36]. Mixed chlorides and bromides at 350°C promote both intergranular and transgranular fracture with maximum velocities as high as 7 mm/s. Cracking is very dependent upon both the temperature and the amount of halide present.

Some liquid metals have been observed to cause embrittlement in many titanium alloys. In mercury, for example, Ti–8Al–1Mo–1V exhibits both intergranular and transgranular fracture[36] with velocities as high as 10 cm/s. Heat treatment affects this behaviour in a manner similar to that observed in aqueous and methanolic solutions. Some alloys are embrittled by liquid cadmium and zinc. More surprising, perhaps, is the observed solid metal embrittlement which has been found on titanium alloy components coated with cadmium, silver or zinc[37,38]. Service failures of cadmium-plated Ti–6Al–4V fasteners have been reported[35], and cracking of this alloy and of Ti–8Al–1Mo–1V has been produced in laboratory tests on coated specimens in the temperature range 38–316°C[38]. While the mechanism of such failures has not been established, cadmium has been detected on the fracture surface and the fracture process appears similar to that occurring in liquid metal embrittlement. Hydrogen is not thought to be an important factor since such failures are observed in components coated both electrolytically and by vapour deposition.

In addition to the failures in Cl_2 and HCl already referred to, cracking also occurs in hydrogen gas. Stressed Ti–Al alloys bombarded with low-energy protons have failed in a manner similar to that observed in hot salt cracking[9]. Other studies have shown that hydrogen gas can cause slow crack growth in many titanium alloys resulting in fracture surfaces very similar to those resulting from aqueous stress-corrosion failures[39].

Magnesium

Stress-corrosion cracking occurs in magnesium alloys in aqueous environments at ambient temperatures. It is found mainly in wrought alloys, although isolated examples of cracking in castings have been reported. The major alloying element responsible for producing susceptibility is aluminium. In practice failure commonly appears to arise from the action of residual stresses.

Mechanism

The mechanism of stress-corrosion cracking in magnesium alloys has not been fully elucidated, but many features have been observed. The rôle of the electrolyte appears to lie in concentrating the corrosion attack in a relatively small number of points resulting in pitting which often precedes cracking. Where little or no pitting occurs cracking is only initiated if the value of the stress is sufficiently high, probably causing rupture of the surface film by plastic deformation of surface metal grains. In polycrystalline specimens the threshold stress below which failure does not occur will be below the stress values below this level, but in single crystals the yield point clearly must be exceeded to produce plastic deformation and this has been observed to be necessary for stress-corrosion cracking to occur[40]. In such environments the capacity to repair ruptured films must not be too great otherwise this process will occur in place of crack initiation. If environments are too aggressive then widespread corrosion may occur in place of pitting and cracking. In laboratory tests mixtures of NaCl plus $Na_2Cr_2O_7$ are commonly used to cause cracking which is preceded by deep pitting; a common mixture consists of 3·5% NaCl plus 2% K_2CrO_4. The aggressive anion is important and cracking is not confined to the Cl^- ion. Thus in 1 N solutions it has been observed[41] that the rate of stress-corrosion cracking decreases according to: $Na_2SO_4 > NaNO_3 > Na_2CO_3 > NaCl + CH_3COONa$, whereas the rate of corrosion of unstressed samples decreases according to: $NaCl > Na_2SO_4 > NaNO_3 > CH_3COONa > Na_2CO_3$. In NaCl plus K_2CrO_4 mixtures the rate of cracking depends upon both the absolute concentration of each species and their ratio. It has been established that KHF_2 solutions also cause stress-corrosion cracking. Little investigation of the action of this electrolyte has been carried out, but since the F^- ion is an inhibitor for the corrosion of magnesium at least part of the electrochemical explanation may lie in inhibited film breakdown and repair kinetics. Thus cracking does not occur in fluoride solutions above a certain limit of concentration. In non-fluoride solutions stress-corrosion crack initiation is inhibited at pH values greater than 10·2[42] and this is also probably related to the greater ease of film formation that occurs in highly alkaline solutions on magnesium surfaces. Under open-circuit conditions or the application of anodic polarisation, only pitting is observed in unstressed specimens under all conditions of heat treatment in environments that cause cracking in stressed specimens.

Cracking is generally observed to be transgranular, but this may change depending upon the heat treatment, the grain size and possibly the environmental pH. Thus in the Mg–6·5Al–1Zn alloy transgranular cracking occurs if specimens are water-quenched[43,44], a process that produces a very fine transgranular dispersion of FeAl which occurs as platelets[45] that are cathodic to the solid solution matrix. Cracking may include the dissolution of material adjacent to the precipitates which may also be solute depleted. In furnace-cooled material the same alloy exhibits intergranular cracking and this may be associated with the grain boundary precipitate of $Mg_{17}Al_{12}$ which occurs during furnace cooling. This precipitate is also cathodic to the matrix but the potential difference developed is lower than that between FeAl and the matrix. This may be a partial electrochemical explanation for the observation that as the iron content is increased the proportion of transgranular fracture also increases, even in furnace-cooled alloys[46]. Above a grain size of about 0·03 mm cracking is transgranular irrespective of the heat treatment employed. In two-phase alloys intergranular cracking occurs in specimens aged at $\leqslant 150°C$ and transgranular cracking in specimens aged at $> 150°C$.

The path of transgranular cracking has not been clearly established and it may vary from one alloy to another. It has been reported as following crystallographic planes[47], possibly the basal plane[43], while others report that no well-defined fracture plane can be discerned in large-grain-size specimens[48], that the crack occurs at a high angle to the basal plane[40], and that fracture occurs as cleavage along (0001), $(10\bar{1}0)$ or $(10\bar{1}1)$ depending upon which plane is most nearly perpendicular to the operative tensile stress[49].

Cracking is considered to occur by a combination of dissolution and mechanical fracture. It is not necessary therefore to account for the relatively high propagation rates by anodic dissolution since this would require high current densities, e.g. 0·6 mm/min requires 14 A/cm^2. An anodic reaction appears to be occurring during cracking, and cathodic protection prevents crack initiation and arrests crack propagation. During cracking the tip is active and hydrogen is evolved. Potential fluctuations have been detected[50] in Mg–5Al alloys in KHF_2 solutions although not in Mg–7Al alloys in $Cl^-/Cr_2O_7^{2-}$ mixtures[51]. One explanation[51a] has been given for the mechanical part of the fracture which is also indicated by fractographic observations[49, 52, 53] and by irregular specimen extension[53]. This is that cleavage occurs on (3140) planes, as a result of hydrogen absorption. The amount of current required to effect cathodic protection increases with increasing stress on specimens.

A threshold stress below which cracking does not occur is observed in some environments[42, 54]. It is dependent upon the alloy composition and heat treatment, and upon the testing environment. In 3% NaCl solutions it is not well defined and this may arise from the pitting that occurs which can be expected to lower the cross-sectional area locally and gradually raise the effective stress acting across that region. Below the threshold stress, cracks occur which do not propagate to sufficient length to cause complete fracture of specimens[52]. The maximum length of crack observed ℓ, is related to the proof stress of the material σ by the relationship $\sigma^2\ell$ = constant. Such cracks have been observed in fatigue and corrosion fatigue[55] and their density and length are very dependent upon the mechanical properties of the alloy[56]. From this analysis emerges a description of alloys undergoing pitting at low stress values, pitting and cracking without complete failure at higher stress values up to the threshold level above which time to failure decreases with increasing value of the initial stress[52].

Alloy Susceptibility

Cracking is observed in commercial Mg–Al–Zn alloys in the range 3–10% Al in which the ratio Al/Zn $\geqslant 2$, and susceptibility increases with increasing Al content. For a given alloy increasing amounts of Fe increase susceptibility[50]. Copper additions also appear to raise susceptibility[41]. It must be emphasised, however, that Mg–Al alloys exhibit stress-corrosion susceptibility even if manufactured from elements of the highest purity available. Most aluminium-free alloys appear to be non-susceptible in most aqueous environments, but

KHF_2 solutions appear to be able to cause failure in most alloys including nominally pure magnesium metal in which failure is intergranular. Thus Mg–Mn alloys, for example, are immune in NaCl plus K_2CrO_4 solutions (unless additions of 0·5 Ce are made to the alloys) while they exhibit cracking in KHF_2 solution[50]. Mg–14Li alloys, which have a b.c.c. structure, exhibit intergranular fracture in humid air although this can be prevented by a stabilising treatment consisting of heating for 24 h at 149°C after quenching[53]. The Mg–3Zn–0·7Zr alloy has been reported[50] to fail in distilled water and KHF_2 solution.

Preventative Methods

Stress relieving at low temperatures is commonly used to lower stress-corrosion susceptibility in Mg alloys, e.g. 8 h at 125°C for Mg–6·5Al–1Zn–0·2Mn[57], since higher temperatures tend to lower the yield point. Similar treatments are advisable for welds which can be a source of high residual stresses. Treatments designed to put the surface in a state of compressive stress also tend to prolong stress-corrosion life. Shot peening, surface rolling[54] and abrasion all produce beneficial effects. Surface oxidation followed by anodising is also reported[58] to increase stress-corrosion life. Susceptible alloys can be clad with non-susceptible Mg alloys, but where the edge is exposed, wetting of both the alloy and clad layer is important in order to achieve cathodic protection of the former. The replacement of susceptible alloys with non-susceptible alloys or with alloys exhibiting less susceptibility, can often be undertaken with little loss of mechanical properties. Refining the grain size lowers susceptibility. Heat treatment produces changes in the threshold stress[59] and alterations in crack morphology as already described.

Aluminium

Stress-corrosion cracking occurs in certain aluminium alloys which have been developed for medium and high strength by employing variations in composition, cold work and heat treatment[1, 60, 61]. The main alloys are based upon Al–Mg, Al–Mg and Al–Cu, but stress corrosion also occurs in Al–Ag, Al–Cu–Mg, Al–Mg–Si, Al–Zn and Al–Cu–Mg–Zn alloys. It has not been observed in pure aluminium. In alloys susceptibility appears to increase as the amount of alloying addition that can be taken into a supersaturated solid solution is raised. In ternary and higher element alloys susceptibility is also influenced by the ratio of solute elements[62]. Small additions of Cr, Mn, Zr, Ti, V, Ni and Li can reduce the susceptibility of wrought products in high-purity binary, ternary and quaternary alloys[62, 63]. Stress-corrosion cracking in castings is not common but it is found occasionally[64].

The large majority of failures occur in aqueous environments and therefore most attention has been focused upon them, but failure can occur[1] in

N_2O_4, mineral oils, alcohols, hexane and mercury. It has not been established whether failure in these environments is the result of residual moisture. Failures in service arise often from the action of residual stresses which can occur in components as a result of quenching followed by machining. The stress level required to initiate cracking is often very much below the yield stress. Alloys brought to a high-strength condition are particularly susceptible.

Mechanism

The mechanism of cracking in aluminium alloys has not been elucidated, but many factors have been examined. Cracking is nearly always intergranular. Stress-corrosion life is very dependent upon the grain shape and orientation in relation to the acting stress. Stress-corrosion resistance is lowest in the short transverse direction of wrought components since many grain boundaries are then lying orthogenally to the applied stress. Notice of such effects is commonly taken in the design of components. In plane-strain tests a relationship between crack velocity and stress intensity factor is found[1,65] similar to that shown in Fig. 8.36 for titanium alloys. A large number of alloys exhibit only Stages I and II. Others exhibit Stage III and others two 'plateaus' or Stage II regions[1]. Results similar to those shown schematically in Fig. 8.36 are also obtained. The crack velocity may vary over nine orders of magnitude and determining K_{ISCC} can be difficult since too high a value may be obtained if the velocity-detecting apparatus is not sufficiently sensitive or the length of time of the experiment too short. It has been suggested[1] that K_{ISCC} might be defined as corresponding to a crack velocity of 10^{-8} cm/s.

In equiaxed specimens crack branching is observed in the region of Stage II at a value of K about 1·4 times the K value at the lower end of the stage. Since cracking occurs at low stress values it is not altogether surprising that the precipitation of corrosion products within existing cracks can sometimes exert relatively appreciable stresses which result in crack propagation.

The effect of environmental variables upon the logarithm of velocity *vs.* K relationship has been examined[1] for a few alloys in some conditions of heat treatment. While it cannot be certain that similar results would be obtained with all alloys, the results reported[1,65] do show interesting features that may have points in common with all alloys. For an Al–Zn–Mg–Cu alloy (7075-T651) the stress-corrosion plateau velocity was a maximum in 5 M KI solution under potentiostatic conditions at -520 mV (*vs.* S.C.E.), reaching about 2×10^{-4} to 5×10^{-4} cm/s, whereas in 3% NaCl under open-circuit conditions the plateau velocity was about 10^{-6} cm/s. Stage I in both cases was the same. The plateau velocity was very sensitive to moisture and depended linearly upon the water vapour of the testing environment[1,65]. Crack propagation does not occur in argon or hydrogen unless moisture is present[66]. Many alloys exhibit plateau velocities when tested in distilled water that are similar to those observed in moist atmospheres. Additions of Cl^-, Br^- and I^- increase the plateau velocities observed in distilled water by as much as 10^2 times under open-circuit conditions, but many other anions have no effect upon the logarithm of velocity *vs.* K relationship under open-circuit conditions or over a wide range of potential values. In Cl^-, Br^- and

I^- solutions the plateau velocity depends upon the halide concentration for some alloys, e.g. the velocity increases with increasing iodide concentration above a certain minimum for 7079-T651, but this is not universally true. In neutral solutions cathodic polarisation lowers the plateau velocity while anodic polarisation increases it until pitting occurs. In strongly acidic solutions the velocity is less sensitive to potential changes and no cathodic protection is observed[1]. The pH value of the environment does not appear to cause changes in the plateau velocity under open-circuit conditions, but acidic conditions move Stage I to the left in Fig. 8.36 thus giving rise to higher velocities for a given K value. Such effects have been described as indicating that the size of the cathodic zone within the crack where hydrogen is released controls the rate of crack propagation[67]. Generally, lowering the pH shortens the time to failure of specimens, an effect that will include the influence of pH upon the initiation process. Under potentiostatic conditions lowering the pH can cause appreciable increases in the plateau velocity on the cathodic side.

Temperature effects indicate an activation energy of 113 kJ/mol for Stage I and 16 kJ/mol for Stage II in 7079-T651 alloy. Crack velocity in Stage II is lowered as the solution viscosity is increased.

No mechanism for cracking in N_2O_4 has been established[68]. In organic media crack velocities are similar to those obtained in distilled water. Lowering the water content results in lower velocities. Not all authors attribute failures in organic liquids to the residual moisture[69]. Furthermore, part of the fracture may be transgranular[69]. Water additions to methanol increase crack velocities as do halide additions. In oils velocities are similar to those in organic liquids and distilled water.

Much of the extensive work on crack velocity described here has been carried out over a long period by Spiedel[1]. Detailed studies of velocity-dependent and velocity-independent parameters reveal how complex the phenomenon is. The three major alloy systems will now be discussed.

Al–Mg (5000 Series) and Al–Mg–Si (6000 Series) In the binary alloy system strength is obtained mainly by strain hardening. Stress corrosion is thought to be associated with a continuous grain boundary film of Mg_5Al_8 which is anodic to the matrix[70]. Air cooling prevents the immediate formation of such precipitates, but they form slowly at ambient temperatures. Thus only low Mg alloys are non-susceptible (Al–3% Mg). Widespread precipitation arising from plastic deformation[71] with carefully controlled heat-treatment conditions can lower susceptibility. Al–5Mg alloys of relatively low susceptibility are subjected to such treatments. Mn and Cr additions improve the stress-corrosion resistance of Al–6Mg and Al–7Mg alloys[72]. Mn increases the precipitation rate and both elements promote the formation of elongated grains. Al–7Mg alloys are usually very highly susceptible, but a recent development[1] indicates that an alloy of this kind can be produced of low susceptibility.

Al–Mg–Si alloys are strengthened by precipitation hardening in which Mg_2Si is formed. They are not very susceptible to stress-corrosion cracking [60,73] which only occurs in specimens subjected to a high solution-treatment temperature followed by a slow quench[60]. Ageing such material eliminates susceptibility[60].

Al–Cu and Al–Cu–Mg (2000 Series) These alloys are strengthened by precipitation hardening. Under conditions of natural ageing these alloys are highly susceptible to stress-corrosion cracking. Susceptibility is associated with slow quench rates which also result in grain-boundary corrosion in unstressed specimens, which is thought to arise from electrochemical effects between $CuAl_2$ and solute-depleted zones formed during quenching[74]. Since thick-section material cannot be quenched rapidly quench-rate effects determine the type of component that any particular alloy can be used to make.

During artificial ageing susceptibility passes through a maximum just before peak hardness is achieved. Similar changes occur in the potential difference developed between grains and grain boundaries[75]. After further ageing, precipitation of the equilibrium $CuAl_2$ occurs within the grains and the potential difference between grains and boundaries then disappears. A recent test[76] provides a rapid means of indicating susceptibility to intergranular attack and stress-corrosion cracking. The specimen's potential is measured in a mixture of absolute methyl alcohol and carbon tetrachloride. Corrosion of the grain boundary provides sites for deposition of dissolved copper, whereas an absence of corrosion results in deposits of copper which are non-adherent. The former develops a potential of about −300 mV (*vs.* S.C.E.) whereas the latter develops a potential of about −1100 mV.

Al–Zn–Mg and Al–Zn–Mg–Cu (7000 Series) These alloys are strengthened by precipitation hardening. Cr, Mn and Zr additions produce elongated grain shapes and inhibit grain growth. High-purity ternary alloys exhibit the highest plateau velocities and although much research has been done on them they are not used in practice. Commercial low-copper alloys are particularly susceptible, and although overageing is generally beneficial the effects of such a treatment are less pronounced with these alloys. Artificial ageing is beneficial but susceptibility in the short transverse remains troublesome[60]. Silver additions to the alloy are reported[77] to improve stress-corrosion resistance. The effect appears to arise from the stimulation of precipitation processes which minimise the width of the precipitate-free zone which arises either from vacancy or solute depletion during quenching. Other workers find that silver gives no improvement either in strength or stress-corrosion resistance[60]. An explanation for this difference appears to lie in experimental procedures[1]. A general conclusion is that ageing temperature and not chemical composition is the most important factor governing short-transverse stress-corrosion resistance in these alloys[78]. Recent developments, particularly in producing alloys suitable for thick sections, is reviewed by Spiedel[1], together with the particular problems of welding.

General

Much discussion of stress-corrosion cracking mechanisms in aluminium alloys has been concerned with the development of anodic areas at grain boundaries. The origin of such areas can be caused by the action of the stress, and susceptible alloys do not necessarily suffer from intergranular corrosion in the absence of stress. Thus in some conditions Al–Mg–Si suffers from intergranular corrosion, but not stress corrosion[79]; 7039-T64 suffers from stress corrosion but not intergranular corrosion[80]; 7075-T651 suffers from both, 7075-O from neither. The electrochemical effects may arise from solute-depleted zones, precipitates anodic or cathodic to the adjacent matrix, or from the rupture of films at the crack tip by plastic deformation. The effect of relative humidity upon the plateau velocity suggests that there may not be a volume of water at the crack tip[1], a possibility which if established would demand a careful re-examination of possible electrochemical reactions.

From a metallurgical viewpoint the effect of grain shape has been described. On a microstructural level the precipitate–matrix interface properties appear to be important. In alloys of Al–6Zn–3Mg aged to peak hardness, slip occurs in a relatively small number of bands which develop a high density of dislocations. Overageing, which lowers susceptibility, results in plastic deformation occurring in much more diffuse bands of dislocations[81]. Grain-boundary precipitates are important both for electrochemical and mechanical reasons and the precipitate-free zone width (as well as the solute-depleted zone width) may also be important. The precise relative significance of these three microstructural features has not been fully ascertained and it is a subject of considerable discussion[82–85]. Much of this centres around the rôle of preferential deformation in the precipitate-free zone resulting in selective dissolution, a process that has not been demonstrated experimentally. Selective corrosion of solute-depleted regions, hydrogen adsorption, tensile-ligament dissolution and general adsorption have also been invoked as major components of mechanistic processes[1]. There is evidence that acidity develops within the region of the crack tip, a pH of 3·5 being observed[86] and the mass–transport–kinetics model[87] appears to explain the plateau velocity as being limited by the kinetics of halide-ion transport to the crack tip. A number of workers[88–90] have provided some evidence that absorbed hydrogen may be at least partly responsible for cracking.

Preventative Methods

The importance of grain shape and the orientation of the applied stress to the short transverse direction has already been pointed out. Overageing also generally lowers strength and stress-corrosion susceptibility. Both the design and manufacture of components are important. Quenched components often have high internal tensile stresses and subsequent machining of such pieces may result in surfaces that readily nucleate cracks. This possibility can be removed by manufacturing components close to the final required size before heat treatment.

Shot peening is a beneficial surface treatment since it puts the surface into a state of compression and generally obscures the grain structure. Subsequent painting of the peened surface is often useful. If pitting occurs then cracking can be expected in susceptible material when the attack penetrates the depth of the compressed surface layer.

Paint coatings can be effective in preventing stress corrosion but it is not always a simple matter to produce and maintain a perfect complete coverage. Galvanic coatings based upon electroplated layers or metal pigmented paints, are commonly used. Such layers do not need to be perfect, but the protection afforded to breaks (or 'holidays') will depend very much upon the localised electrochemical conditions. Metal spraying is also employed and for highly susceptible alloys a thin cladding sheet of aluminium is employed, either on one or both sides. These clad composites are employed for general corrosion resistance and not merely to combat stress corrosion. Anodising is generally not recommended. Cathodic protection is effective but is often not practicable.

J. C. SCULLY

REFERENCES

1. *Stress-Corrosion Cracking in High Strength Steels and in Aluminium and Titanium Alloys* (ed. B. F. Brown), NRL, Washington D.C. (1972)
2. Jackson, J. D. and Boyd, W. K., DMIC Technical Note, Battelle Memorial Institute, Columbus, Ohio (1966)
3. *Stress Corrosion Cracking of Titanium*, ASTM STP 397, ASTM, Philadelphia (1966)
4. Logan, H. L., *Fundamental Aspects of Stress Corrosion Cracking* (ed. R. W. Staehle, A. J. Forty and D. van Rooyen), NACE, Houston, 662 (1969)
5. *TML Report No. 88*, Battelle Memorial Institute, Columbus, Ohio (1957)
6. Peterson, V. C. and Bomberger, H. B., Reference 3, 80 (1966)
7. Kirchner, R. L. and Ripling, E. J., First Interim Report, Materials Research Laboratory, Richton Park, Illinois (1964)
8. Rideout, S. P., Louthan, M. R. Jr., and Selby, C. L., Reference 3, 137 (1966)
9. Ondrejcin, R. S., *Met. Trans.*, **1**, 3031 (1970)
10. Gray, H. R., *Corrosion*, **25**, 337 (1969)
11. Gray, H. R., Aerospace Structural Materials Conference, No. 2 (1969)
12. Gray, H. R. and Johnston, J. R., *Met. Trans.*, **1** 3101 (1970)
13. Boyd, W. K., Reference 4, 593 (1969)
14. Adams, R. E. and Von Tiesenhausen, E., Reference 4, 691 (1969)
15. Weber, K. E. and Davis, A. D., Lockheed California Co., NASA CR 981 Dec. (1967)
16. Brown, B. F., Lennox, T. J., Jr., Newbegin, R. L., Peterson, M. H., Smith, J. A. and Waldron, L. J., *NRL Memorandum Report 1574*, November (1964)
17. Beck, T. R. and Blackburn, M. J., *J.A.I.A.A.*, **6**, 326 (1968)
18. Blackburn, M. J. and Williams, J. C., Reference 4, 620 (1969)
19. Beck, T. R., Reference 4, 605 (1969)
20. Sanderson, G. and Scully, J. C., *Corros. Sci.*, **8**, 541 (1968)
21. Gerberich, W. W., *2nd International Conference on Fracture*, 919 (1969)
22. Sanderson, G., Powell, D. T. and Scully, J. C., Reference 4, 638 (1969)
23. Powell, D. T. and Scully, J. C., *Corrosion*, **24**, 151 (1968)
24. Brown, B. F., Fujii, C. T. and Dahlberg, E. P., *J. Electrochem. Soc.*, **116**, 201 (1969)
25. Scully, J. C. and Powell, D. T., *Corros. Sci.*, **10**, 719 (1970)
26. Feeney, J. and Blackburn, M. J., *The Theory of Stress Corrosion Cracking in Alloys* (ed. J. C. Scully), N.A.T.O., Brussels, 355 (1971)
27. Fager, D. N. and Spurr, W. F., *Trans. Am. Soc. Metals*, **61**, 283 (1968)
28. Menzies, I. A. and Averill, A. F., *Electrochim. Acta*, **13**, 807 (1968)
29. Mori, K., Takamura, A. and Shimose, T., *Corrosion*, **22**, 29 (1966)
30. Sedriks, A. J., *Corrosion*, **25**, 207 (1969)
31. Sedriks, A. J. and Green, J. A. S., *Corrosion*, **25**, 324 (1969)
32. Spurrier, J. and Scully, J. C., *Corrosion*, **28**, 453 (1972)
33. Mauney, D. A., Starke Jr., E. A. and Hochman, R. F., Reference 11
34. King, E. J., Kappelt, G. K. and Fields, C., *Bell Aerospace Systems Report* (1966)
35. Battelle NASA Report, NASr 100(09) (1969)
36. Beck, T. R., Blackburn, M. J., Smyrl, W. H. and Spiedel, M. O., The Boeing Co. Report, Contract NAS 7–489, No. 14, December (1969)
37. Duttweiler, R. E., Wagner, R. R. and Antony, K. C., Reference 3, 152 (1966)
38. Fager, D. N. and Spurr, W. F., The Boeing Co. Report D6–22691
39. Nelson, H. G., Williams, D. P. and Stein, J. E., *Met. Trans.*, **3**, 469 (1972)
40. Meller, F. and Metzger, M., *U.S.N.A.C.A. Tech. Note No. 4019* (1957)
41. Romanov, V. V., *Stress Corrosion Cracking of Metals* (translated from the Russian), 61 (1961)

42. Loose, W. S., *Magnesium*, ASM, Cleveland, Ohio, 173 (1946)
43. Priest, D. K., Beck, F. H. and Fontana, M. G., *Trans. ASM*, **47**, 473 (1955)
44. Priest, D. K., *Stress-Corrosion Cracking and Embrittlement* (ed. W. D. Robertson), Wiley, New York, 81 (1956)
45. Heidenreich, R. D., Gerould, G. H. and McNulty, R. E., *Trans. AIME*, **166**, 15 (1946)
46. Pardue, W. M., Beck, F. H. and Fontana, M. G., *Trans. ASM*, **54**, 539 (1961)
47. George, P. F. and Diehl, H. A., *Met. Prog.*, **62**, 121 (1952)
48. Logan, H. L., *J. Res. Mat. Bur. Stand.*, **65C**, 165 (1961)
49. Fairman, L. and West, J. M., *Corros. Sci.*, **5**, 711 (1965)
50. Perryman, E. C. W., *J. Inst. Met.*, **78**, 621 (1950–51)
51. van Rooyen, D., *Corrosion*, **16**, 421t (1960)
51*a*. Chakrapani, D. G. and Pugh, E. N., *Met. Trans.*, **6A**, 1155 (1975)
52. Wearmouth, W. R., Ph.D. Thesis, University of Newcastle-upon-Tyne (1967)
53. Logan, H. L., *Stress Corrosion Cracking*, Wiley, New York, 217 (1966)
54. Timonova, M. A., *Intercrystalline Corrosion and Corrosion of Metals Under Stress*, Consultants Bureau, New York, 263 (1962)
55. Forrest, P. G., *Fatigue of Metals*, Pergamon, 146 (1962)
56. Brown, B. F. and Beachem, C. D., *Corrosion Sci.*, **5**, 749 (1965)
57. Hunter, M. A., *Metals Handbook*, ASM, Cleveland, Ohio, 234 (1948)
58. Loose, W. S., *The Corrosion Handbook* (ed. H. H. Uhlig), Wiley, New York, 173 (1948)
59. Loose, W. S. and Barbian, H. A., *Stress Corrosion Cracking of Metals*, ASTM/AIME, 273 (1944)
60. Sprowls, D. O. and Brown, R. H., Reference 4, 466 (1969)
61. Graf, L. and Neth, W., *Z. Metallunde*, **60**, 789 and 860 (1969)
62. Petri, H. G., Siebel, G. and Vosskuhler, H., *Aluminium*, **26**, 2 (1944)
63. Engell, H. J., Neth, W. and Suchma, A., *Z. Metallkunde*, **61**, 261 (1970)
64. Rogers, T. H., *Corrosion 1961*, Butterworths, London, 605 (1962)
65. Spiedel, M. O., *The Theory of Stress Corrosion Cracking* (ed. J. C. Scully), N.A.T.O., Brussels, 289 (1971)
66. Watkinson, F. E. and Scully, J. C., *Corros. Sci.*, **11**, 179 (1971)
67. Sedriks, A. J., Green, J. A. S. and Novak, D. L., *Met. Trans.*, **1**, 1815 (1970)
68. Lorenz, P. M., Technical Report AFML–TR–69–99 (1969)
69. Procter, R. P. M. and Paxton, H. W., *ASTM J. of Materials*, **4**, 729 (1969)
70. Binger, W. W., Hollingsworth, E. H. and Sprowls, D. O., *Aluminium* (ed. K. R. van Horn), Vol. I, ASM, 209 (1967)
71. Anderson, W. A., US Patent 3 232 796, February 1 (1966)
72. Niederberger, R. B., Basil, J. L. and Bedford, G. T., *Corrosion*, **22**, 68 (1966)
73. Chadwick, R., Muir, N. B. and Granger, H. B., *J. Inst. Metals*, **82**, 75 (1953–54)
74. Hunter, M. S., Frank, G. R. (Jr.) and Robinson, D. L., *Second International Congress on Metallic Corrosion*, N.A.C.E., Houston, 604 (1966)
75. Mears, R. B., Brown, R. H. and Dix, E. H. (Jr.), *Symposium on Stress Corrosion Cracking of Metals*, ASTM/AIME, 329 (1944)
76. Horst, R. L. (Jr.), Hollingsworth, E. H. and King, W., *Corrosion*, **25**, 199 (1969)
77. Rosenkranz, W., *Aluminium*, **39**, 741 (1963)
78. Staley, J. T., Final Report, Naval Air Systems Command Contract N00019–C8–C–0146 (1969)
79. Gruhl, W., *Metall.*, **19**, 206 (1965)
80. Helfrich, W. J., *Corrosion*, **24**, 423 (1968)
81. Spiedel, M. O., Reference 4, 561 (1969)
82. Sedriks, A. J., Slattery, P. W. and Pugh, E. N., *Trans. ASM*, **62**, 238 (1969)
83. Polmear, I. J., *J. Aust. Inst. Metals*, **89**, 193 (1960)
84. Deardo, A. J. and Townsend, R. D., *Met. Trans.*, **1**, 2573 (1970)
85. Watkinson, F. E. and Scully, J. C., *Corros. Sci.*, **12**, 905 (1972)
86. Brown, B. F., Fujii, C. T. and Dahlberg, E. P., *J. Electrochem. Soc.*, **116**, 218 (1969)
87. Beck, T. R., Blackburn, M. J. and Spiedel, M. O., *Quarterly Progress Report*, No. 11, Contract NAS 7–489 (1969)
88. Gruhl, W., Leichtmetall-Forschunginstitut of Vereinigte Aluminium-Werke AG, Bonn, Report 1970
89. Gest, R. J. and Troiano, A. R., *Corrosion*, **30**, 274 (1974)
90. Montgrain, L. and Swann, P. R., *Hydrogen in Metals* (ed. I. M. Bernstein and A. W. Thompson), A.S.M., Ohio, 575 (1974)

8.6 Corrosion Fatigue

During World War I much trouble was experienced through the failure of the steel towing ropes attached to the paravane equipment used in mine sweeping. Higher-tensile steel wire was substituted for the normal medium-carbon steel wire used in the manufacture of these ropes, but still their life was very short, with no apparent improvement, and it was not until galvanised wire was used that matters improved. Since a rope in tension vibrates as it is dragged through water and since sea-water is a well-known corrosive, it is obvious that this is a case of *corrosion fatigue*, the first of its kind to be investigated[1].

The subject has formed the basis of numerous laboratory investigations, and a vast amount of information about service failures attributed to this cause has been amassed. Taking this work into consideration, corrosion fatigue can be defined as the type of failure which occurs when a component is subjected to cyclic stressing in a medium which is able to attack the material continuously if it becomes chemically exposed; in other words, if the material is capable of reacting with the environment in the absence of an oxide film or of a film of corrosion product, corrosion-fatigue failure is possible if not certain. To speak simply of a 'corrosive medium' in this context is inadequate. Many metals, particularly the baser ones, are resistant to corrosion in normal atmospheric conditions, owing to the presence of a coherent surface oxide film through which the participants in the corrosion reaction are unable to diffuse. But if the stresses in the specimen, cyclic or static, cause damage to this film, it ceases to protect as fresh metal is exposed, and the medium becomes capable of attack in these circumstances. Thus for example, pure aluminium is extremely corrosion resistant in the normal atmosphere owing to the protective oxide film, and thus the normal atmosphere is not, so far as aluminium is concerned, a corrosive medium. Aluminium is, however, certainly subject to corrosion fatigue, because the metal itself is very reactive to oxygen and water vapour, and in a fatigue test the normal protective oxide film is continually being disrupted and the actual metal continually exposed.

Since most oxides are ionic crystalline solids and therefore non-plastic, with high moduli of elasticity, the strains produced by the cyclic stressing of the metal give rise to high stresses in the oxide film, which are likely to be greater than the strength of the oxide. The oxide is in intimate contact with the metal at all points and so the film will be expected to fracture at numerous points. Examples of the corrosive nature of the normal atmosphere in fatigue testing are furnished by Gough and Sopwith[2] who found the endurance limit of certain non-ferrous materials up to 25% higher in a partial vacuum than in air.

Considerable effort has been expended in determining which is the corrosive constituent of air towards a number of common metals. In the case of the noble metals, and certain other metals, it is oxygen. Thus fatigue life of copper[3] and lead[4] is very dependent on the partial pressure of oxygen, and water vapour has little additional influence. The effect is concerned with the propagation of the crack, and the adsorption of oxygen on the crack walls preventing re-welding during the compression half of the stress cycle has been suggested as the reason. Mild steel shows a fatigue limit in both air and vacuum occurring at about the same number of cycles but at a lower strain in air[5]. An identical fatigue curve is obtained when the steel is tested in dry air or oxygen, but in moist air the life at higher stresses is longer although the fatigue limit is the same[6]. The reduced fatigue life in oxygen or air compared with the fatigue life in argon is again attributed to the prevention of re-welding. The increased life in moist air is assumed to be due to the preferential adsorption of H_2O owing to a mass-action effect. In the case of a base metal such as aluminium and its alloys the corrosive constituent has been established as water vapour[7]. Hydrogen ions are formed by the reduction of water vapour at exposed metal sites and they produce the following effects:

1. A build up of hydrogen gas pressure in voids formed on or near active slip planes.
2. An association with vacancies, pinning them down, causing them to cluster and eventually condense into voids.
3. A change in the surface energy of the walls of a crack.

Evidence of this last effect has been provided by Forsyth[8] from electron microscope observations of fracture surfaces where it was found that in the presence of a corrodant (which in the case of aluminium could be ordinary air, particularly at low frequencies) the amount of cleavage fracture per cycle was increased as indicated by the greater spacing of the striations on the fracture surface. This was attributed to the lowered surface energy of the cleavage face. It was later found, however, that the surface preparation determined the degree of the water vapour effect, from which it was deduced that the water vapour had a profound effect on the mechanical properties of the surface oxide film, and hence on the initiation of the fatigue crack[9]. Cracks formed in the oxide film by the cyclic stressing cause a strain discontinuity at the metal–oxide interface which acts as a dislocation source for the underlying metal. Further work on the mechanical properties of alumina films produced by anodising has shown that both the elastic modulus and the fracture strength of such films are drastically reduced in the presence of water vapour[10, 11]. Hence it appears that water vapour influences both the early stages of crack initiation and the subsequent propagation of the crack.

Mechanism of Corrosion Fatigue

Although the constituents of a gaseous environment can produce a corrosive effect, the term *corrosion fatigue* as generally applied comprehends fatigue in a liquid or spray environment and more particularly an aqueous environment containing electrolytes. It is in this context that the mechanism is discussed.

It is interesting to note that pitting corrosion and fatigue have one feature in common. In both, the material as a whole is unaffected and the region of influence or attack is restricted to very few points. These points are obviously limited to the surface of the material in the case of pitting corrosion, but even in the case of fatigue they are predominantly in the surface, and this is not entirely a matter of the method of stressing. It might be expected that a fatigue crack would be initiated in the surface in a rotating-bending text because the stress is at a maximum in the surface, but even under the uniform stress conditions of the push-pull test fatigue cracks are nearly always initiated in the surface.

Not surprisingly many of the fundamental investigations of corrosion fatigue have been concerned with the possibility that the susceptible areas may be identical in the two cases of pitting corrosion and fatigue. It is well known that corrosion will tend to occur at a scratch line on a specimen rather than at other points on the surface, and it is also a common observation that fatigue cracks, too, are initiated at scratches or imperfections in the surface. Could any such correlation be found where the surface had been carefully prepared to avoid such imperfections?

Whitwham and Evans[12] examined the behaviour in a corrosion-fatigue test of steel wire specimens which had been previously subjected to fatigue in air to investigate whether the points affected by air fatigue might be preferentially attacked by the corrosive (0·1 M potassium chloride made up with tap water and aerated before use). The total life of the specimens in this two-stage test was not, however, significantly different from that of specimens tested entirely under conditions of corrosion fatigue, which points to there being no connection between the two effects when they are not present simultaneously, the chemical reactivity of the steel being enhanced only when the corrosive is present during the cyclic stressing. Further work by Evans[13] was directed to studying the mutual interaction of corrosion and cyclic stressing in the corrosion fatigue test. In this case too the specimens were subjected to a two-stage test, but on this occasion the corrosion fatigue stage came first. The specimen was then carefully washed and treated with a corrosion inhibitor, and the fatigue test was continued in air until the specimen failed. The effect of the length of the corrosion-fatigue period on the total life of the specimen was studied. Periods of less than 10^5 cycles in the first stage exerted little effect on the total life compared with that of a control specimen subjected only to normal air fatigue. Above 10^5 cycles there was a drastic reduction in the total life of the specimen. With a further increase in the number of cycles in the first stage the total life continued to decrease, passed through a minimum, and then showed an increase. A typical curve is shown in Fig. 8.37. This abrupt changeover in behaviour has been explained from subsequent examination under the microscope. It was found that during the first period (*a*, less than 10^5 cycles under corrosion fatigue conditions) where the life of the specimen is little affected, the corrosion damage to the surface consisted of rounded pits not unlike those found in normal stressless corrosion. In the case of the short-life specimens (section *b* of curve) it was observed that some of the pits had cracks running from their bases so that

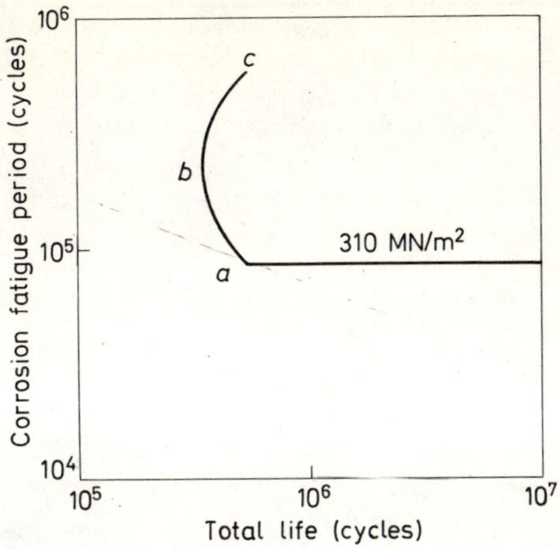

Fig. 8.37 Interaction of corrosion and cyclic stressing; effect of length of corrosion fatigue period on total life of specimen (after Evans and Tchorabdji[13])

the pit was now more of a sharp fissure. The microscopic examination also explained the unexpected increase in total life which occurred when the period of corrosion fatigue was further increased (*c*, Fig. 8.37). It was found that in the intermediate stage only an occasional corrosion pit had developed into a fissure, whereas in the final stage every pit was of this type. If the stress distribution at the surface of the specimen is considered, it is obvious that an isolated fissure produces a greater stress concentration than several such fissures in close proximity. This is illustrated in Fig. 8.38 where the parallel lines represent stress trajectories which become crowded together where there is a concentration of stress.

Evans distinguished three effects resulting from the action of mechanical fatiguing on the electrochemical processes occurring on the surface of the specimen:

1. The dynamic nature of the test ensures a steady supply of oxygen, therefore the cathodic polarisation is reduced.
2. The continual opening up of the corrosion cavities, the formation of more cavities, and the destruction of protective films reduces the anodic polarisation.
3. The increased chemical activity of stressed material means a more negative potential.

Every one of these effects increases the corrosion current.

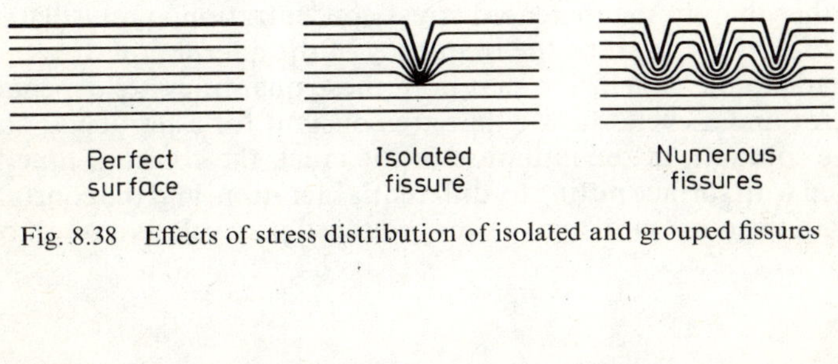

Fig. 8.38 Effects of stress distribution of isolated and grouped fissures

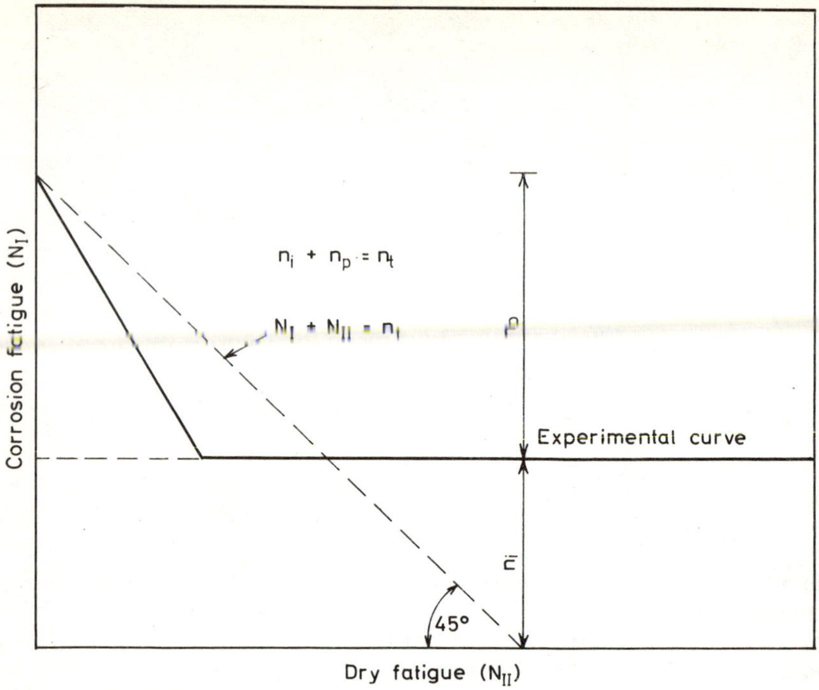

Fig. 8.39 Method of presenting data. The terms are defined as follows:
N_I = number of initial corrosion fatigue cycles given to the specimen
N_{II} = number of subsequent dry fatigue cycles to produce failure
n_i = number of corrosion fatigue cycles required to initiate a crack able to propagate subsequently in air
n_p = number of cycles to propagate failure in corrodant after n_i
n_t = total number of cycles for failure in the corrodant = $n_i + n_p$
(after Rollins, Arnold and Lardner[14])

More recently, Rollins, Arnold and Lardner[14] have carried out similar two-stage tests on an Fe–1C–1Cr steel. They plotted the number of cycles in the corrosion stage against the number of cycles to failure in the non-corrosion stage to produce the type of diagram shown in Fig. 8.39. This method of plotting allows the relative rates of the propagation of the crack in air and in the corrosive medium to be assessed. If the portion of the curve where N_I is greater than n_i lies below the 45° line then the propagation rate in air is faster than in the corrosive, and this has been found to be so in a number of cases. These authors ascribe the minimum total life found by Evans to this effect rather than to the increased stress concentration at an isolated crack. Graphs of n_i and n_p against the frequency of the alternating stress are linear which leads to the conclusion that both these quantities are dependent on a time factor and a cycles factor that are constant for a particular stress and corrosive solution. In the initiation of the crack the constant time factor is associated with surface pitting by differential aeration, and the constant cycles factor with the number of active sites produced per cycle as a result of plastic deformation at the tip of the crack. As the crack grows, the stress concentration increases, but for a time the corrosion rate is controlled by the cathodic or resistance polarisation, hence the constant time factor in the propagation phase. When the crack grows to a certain critical length the stress concentration is such that the crack propagates more quickly by the normal air-fatigue mechanism and this part of the process is cycles dependent.

In other recent work on the mechanism of corrosion fatigue electrochemical measurements have been applied. In galvanostatic experiments Paul and Weiland[15] found a potential shift when specimens were fatigued in an aqueous electrolyte. The shift was large in a system where the surface was covered by a protective oxide film, e.g. aluminium in sodium chloride, but a definite potential shift was also found where no film was present, e.g. a Ni–30Mo alloy in 5% HCl. In tensile experiments the potential shift was found to be proportional to the rate of deformation which suggests that it is due to the emergence of active slip bands on the surface. The change in potential is an average value over the whole surface so that the shift to a more negative potential is considerably greater at the active sites at the slip bands. Rapid attack at these vulnerable points is considered to be the primary mechanism of corrosion fatigue.

Corrosion Fatigue Data

Not all the experimental work on corrosion fatigue has been aimed at explaining the phenomenon. Much of the valuable and extensive work of McAdam was concerned with providing designers with data on the behaviour of specific important constructional materials in representative corrosive environments. In air fatigue tests the frequency of the cyclic stressing has little effect on the resultant fatigue curve, although at very high frequencies there is a greater rise in the temperature of the specimen because of the shorter time for dissipation of the heat generated in any one cycle. With corrosion fatigue the total duration of the test is an important factor[16–18] since corrosion is time-dependent. McAdam took this into account in his tests as an additional variable. Fatigue tests were first carried out on a set of specimens at a given frequency and load, and for a given time in the corrosive environment. The air fatigue limit (i.e. the stress below which the specimen will endure an infinite number of cycles) of the specimens was then determined, and the results were plotted as the fatigue limit in air against the duration of the initial corrosion fatigue period. Thus a set of graphs was obtained for various loads at each frequency in the corrosion part of the test. The amount of experimental work involved was formidable.

An example of a typical graph is shown in Fig. 8.40. The upper curve represents the behaviour of specimens which were not subjected to a load in the corrosion fatigue part of the experiment, and therefore represents the deterioration in the fatigue properties of the steel due to corrosion alone. The fact that this curve has the same general shape as the curves in which the specimens have been subjected to simultaneous fatigue and corrosion suggests that it is the corrosion properties of the material which have the greatest influence on its behaviour during corrosion fatigue testing.

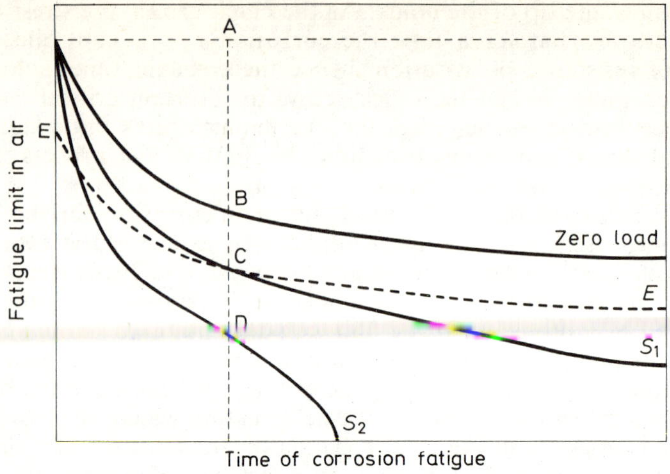

Fig. 8.40 Fatigue limit in air plotted against time of corrosion fatigue; S_1 and S_2 are the applied loads and $S_2 > S_1$ (after Gough and Sopwith[18a])

A considerable amount of information can be abstracted from sets of curves of this type. For example, consideration of the intercepts of the broken horizontal line will permit the reading off of the corrosion fatigue conditions which will reduce the fatigue limit of the material from F to F′ in air. A lower cyclic load of S_1 newtons produces in time t_2 as much damage as a higher load of S_2 newtons in time t_1. A further graph can therefore be constructed representing time *vs.* stress necessary to reduce the fatigue limit to F′. By considering such intercepts on sets of graphs similar to those of Fig. 8.40, but for tests performed at other frequencies, it is possible to construct a three-dimensional graph of stress-time-number of cycles required to produce a

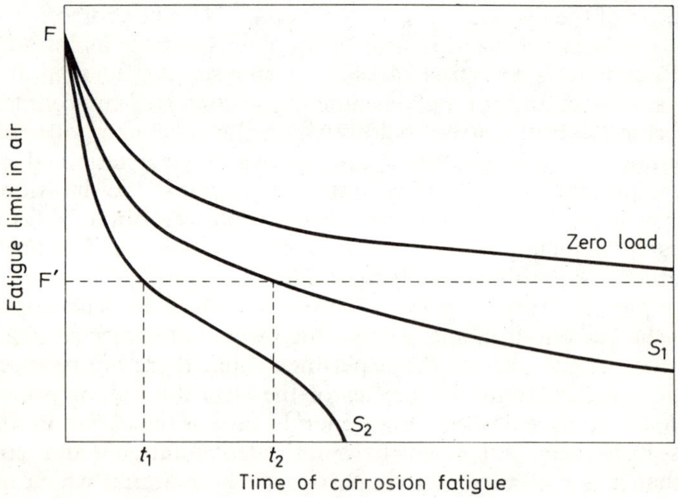

Fig. 8.41 Separation of effects of corrosion and fatigue (after Gough and Sopwith[18a])

given reduction in the fatigue limit. The plot is a surface and is called the *constant total damage surface*, which represents the time taken by any particular alternating stress to produce a given reduction in fatigue limit (i.e. FF′) in a particular number of cycles, where the time/cycles relationship is obtained from conducting experiments at different frequencies. The limit of the time axis of the graph is given by the time in which the fatigue limit is reduced by the given amount in tests with no stress applied.

In a plot of this kind, there is no attempt to separate the effects of corrosion and fatigue. McAdam was able to make such a separation by considering the intercepts made by the vertical line ABCD in Fig. 8.41, where AB represents the damage due to corrosion and BC the damage due to alternating stress. Information concerning the rate of damage due to the cyclic stressing can also be derived from these graphs by considering the time necessary to reduce the fatigue limit in stressless corrosion by a given fraction under corrosion fatigue. For instance the dotted curve E*E* represents a given fraction reduction, say 15%, of the stressless corrosion curve. The points of intersection of this curve with the other curves gives the times in which the particular cyclic loads reduce the fatigue limit by this amount. Assuming that the rate of damage is uniform a graph of the rate of damage against the stress can be plotted.

Such a graph, if plotted on a log/log scale, is linear. A set of these graphs for different frequencies, *n*, is, in general, a parallel set of straight lines (Fig. 8.42).

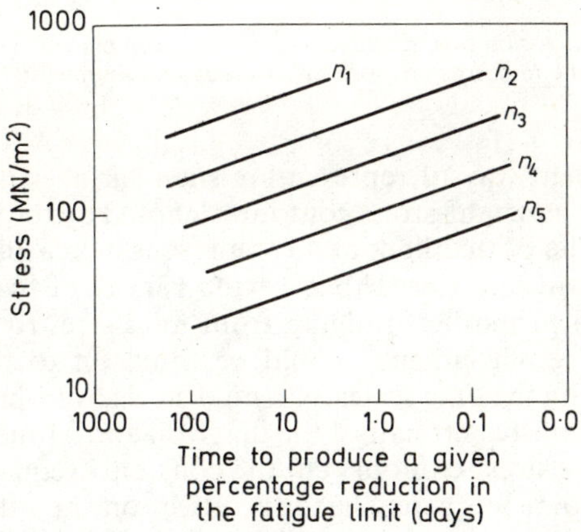

Fig. 8.42 Rate of damage plotted *vs.* stress for varying frequencies, *n* (after Gough and Sopwith[18a])

The data derived from consideration of the stress and corrosion time required to produce a given percentage reduction of the fatigue limit due to corrosion alone can be plotted on a three-dimensional plot similar to the previous one for total damage. The horizontal co-ordinates are again time and number of cycles, and the vertical co-ordinate is stress. The surface is the 'constant net damage surface'. The time co-ordinate, since it represents the time to produce a given amount of damage, can also be plotted as the average rate of damage.

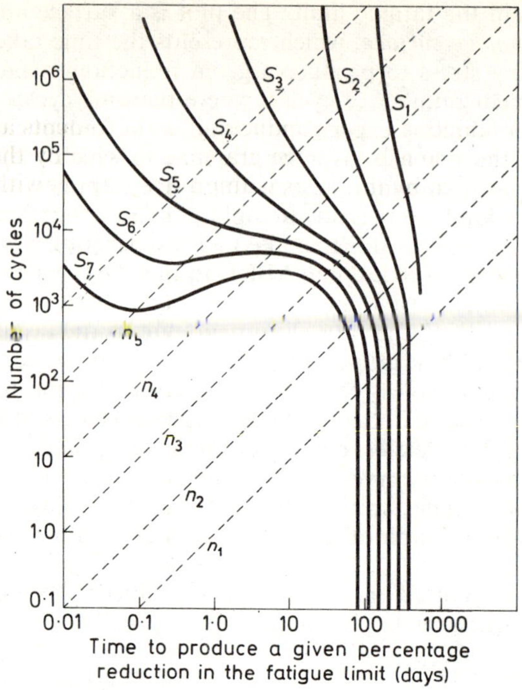

Fig. 8.43 Stress and corrosion time required to produce given percentage reduction of fatigue limit due to corrosion alone (after Gough and Sopwith[18a])

A more convenient way of representing such surfaces is to make a plan view indicating the constant-stress contours denoted by the letters S_1, S_2, S_3, etc. in Fig. 8.43. The dotted lines at 45° represent increasing frequencies n_1, n_2, n_3, etc. McAdam discovered that a wide variety of materials with very different corrosion properties, ranging from mild steel to Monel metal, in different corrosive environments, could be fitted on to the same type of diagram by adjusting the time scales, e.g. mild steels over the time considered were characteristic of the left half of the diagram and Monel of the right. At high frequencies the stress contours cut the constant-frequency lines at right angles, indicating that frequency has little effect on the rate of damage. At low frequencies, i.e. in the bottom right-hand corner of the diagram, the stress contours are crowded closely together and cut the constant-frequency lines at an acute angle. This indicates that the rate of damage is appreciable over a wide range of stress, even at low stresses. The fact that corrosion-resistant materials such as Monel, as well as corrosion-susceptible materials such as mild steel, in mildly corrosive surroundings such as condenser water can suffer considerable damage under conditions of a small cyclic load of low frequency, is an important discovery.

The data on which Fig. 8.43 is based are for tests carried out in carbonate well-water. McAdam made the further interesting discovery that if mild steel were tested in condenser water and a similar graph constructed, the set of contours corresponded more closely to the right-hand side of Fig. 8.43, i.e.

the behaviour of mild steel in condenser water was similar to that of Monel in carbonate water. The apparent universality of this diagram is an interesting observation, but it has not provoked a basic theory of corrosion fatigue.

Some of the investigations involving electrochemical measurements have been concerned with relating easily determined quantities such as corrosion potential and corrosion current with the behaviour of a material in corrosion fatigue, so that this behaviour can be rapidly assessed without the necessity of the laborious collection of data which is the feature of McAdam's approach. Endo and Komai[19] have derived an expression relating the increase in the corrosion current with the number of cycles of corrosion fatigue, and this expression is found to be analogous with an expression relating crack growth with the number of cycles, and hence the increase in corrosion current is found to be a measure of the total crack length. The expression also takes account of the magnitude of the alternating stress, the frequency and the temperature. Since the corrosion current is related to the corrosion potential, the course of the corrosion fatigue damage can be followed by potential measurements. In an extension of this work[20] the product of the initial corrosion current density $i_{corr.,i}$ and the total life τ_T is

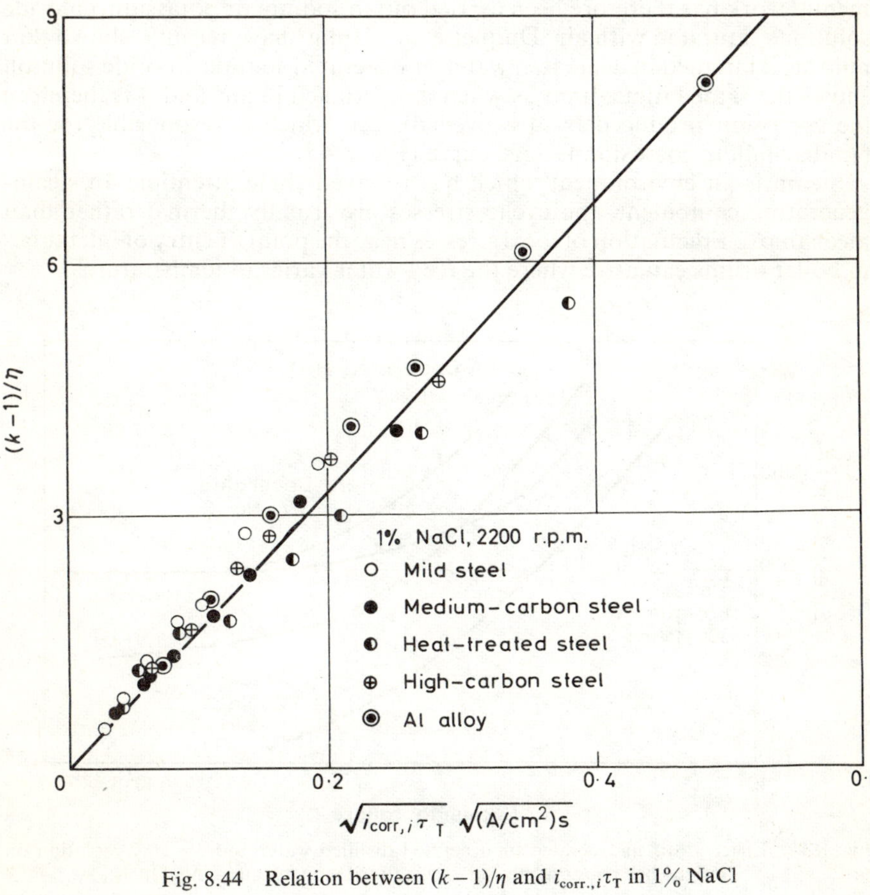

Fig. 8.44 Relation between $(k-1)/\eta$ and $i_{corr.,i}\tau_T$ in 1% NaCl

life τ_T is shown to be related to the notch sensitivity η, and the ratio of the fatigue strength in air and in the corrosion medium k, by the expression

$$\frac{(k-1)}{\eta} = K\sqrt{i_{corr.,i}\tau_T}$$

where K is a constant. This relation is found to hold for a variety of materials in a particular electrolyte solution, as illustrated in Fig. 8.44. Thus $i_{corr.,i}$ is a measure of the static corrosion resistance of the material and τ_T a measure of the dynamic corrosion resistance.

The Corrosive Environment

The corrosive medium chosen for laboratory investigations must be one which is in some way related to the numerous corrosive conditions which obtain in reported corrosion fatigue failures. The latter fall into a relatively small number of general categories such as sea-water, boiler waters containing caustic inhibitors, condenser water, natural waters, steam and lubricating oils. By far the most important are the marine environments and most experimental work has therefore been carried out in sodium or potassium chloride solutions saturated with air. Duquette and Uhlig[6] have recently shown that mild steel fatigued in deaerated water or deaerated sodium chloride solution shows the same fatigue limit as when it is fatigued in air, and it is therefore the corrosion produced by dissolved oxygen which is responsible for the familiar falling corrosion fatigue curve (Fig. 8.45).

Steam is an environment which has received some attention. In steam-generating equipment, the cyclic stresses are usually thermal, rather than mechanical. Fluctuation of local stresses near the point of entry of inlet tubes in boiler drums can arise where the feed water varies in temperature.

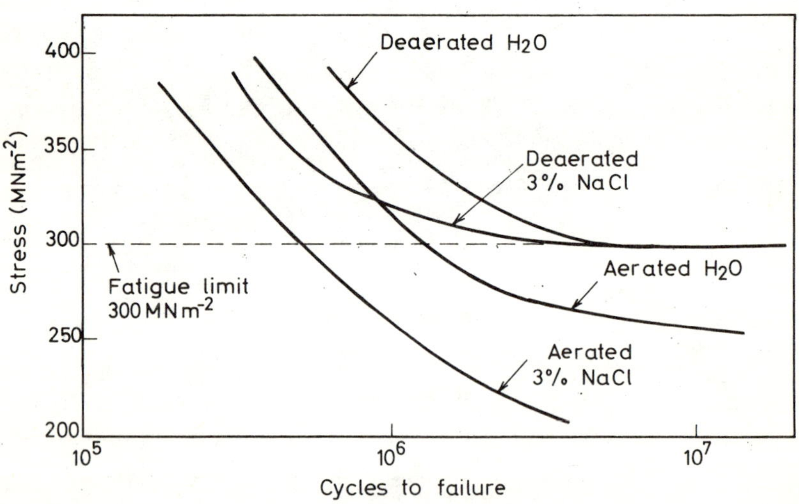

Fig. 8.45 Effect of air and aerated or deaerated distilled water and 3% NaCl solution on fatigue behaviour of steel at 25°C (after Duquette and Uhlig[6])

More severe cases are found in boilers producing superheated steam. If a barrier of steam is formed between the water and the boiler wall due to non-wetting, the rate of heat transfer from the wall to the water is much reduced and therefore the temperature of the wall rises. Breakdown of the steam film causes a drop in temperature of the wall, thus causing the temperature of the wall to fluctuate violently with consequent local stress variations. Similar trouble is encountered where drops of water are carried up from the surface of the water into pipes and on to surfaces at a higher temperature. Failure in such cases may occur after very short periods, indicating that the number of cycles of stress is probably low and the local stresses high. Further evidence of the high stress is the observation of Lüder's lines in the affected area. Cracks are numerous, often parallel or in a criss-cross pattern, and are usually transcrystalline [where alkali is present they may initially be intercrystalline but in this case they are probably due to caustic cracking, a form of stress corrosion (*see* Section 8.2)] and, in the case of steel, are filled with the magnetic oxide. Since hydrogen is one of the products of the corrosion reaction there is also the possibility that hydrogen embrittlement may accelerate the propagation of fatigue cracks.

Both ferritic and austenitic steels are prone to failure under these circumstances. There is evidence that the austenitic steels, despite their superior corrosion resistance, are particularly susceptible because of their poor conductivity and high coefficient of thermal expansion.

Galvin and Naylor[21] have investigated the corrosion fatigue of an Fe–3Ni steel in non-aqueous environments which comprised a group of mineral oils and synthetic lubricants with and without a range of additives. The steel was found to have a fatigue limit in all these liquids but there were variations in the shapes of the fatigue curves. The media could be divided into two groups:

Group 1. Those which reduced the life above the fatigue limit but gave the same fatigue limit as the fatigue limit in a neutral white mineral oil.

Group 2. Those which reduced the life above the fatigue limit and also reduced the fatigue limit compared with the fatigue limit in the white mineral oil.

The conclusion was that both groups increased the crack propagation rate but that Group 2 also increased the rate of crack initiation. The propagation rate was thought to be influenced by the adsorption of the surface-active additives on the walls of the crack which prevented re-welding. The more polar compounds resulted in the shortest fatigue lives.

Detection of a Corrosion Fatigue Failure

In a failure due to corrosion fatigue, the component may or may not show signs of general corrosion. The feature which points to corrosion fatigue is the presence of a large number of cracks in addition to the one where failure occurred. Where the stress system is uniaxial the cracks are parallel to each other in a plane perpendicular to the direction of stress. In torsion a group of cracks emanating from one point is usually found. They are often in the form of a cross or star, the arms being at approximately 45° to the torsion axis.

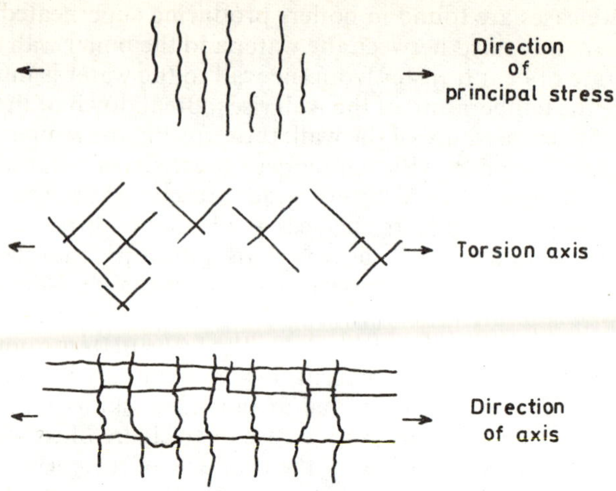

Fig. 8.46 Appearance of fatigue cracks on the surface of a specimen. Uniaxial stress (top), torsional stress (middle) and thermal stress in pipes (bottom)

Cracking in pipes due to thermal stresses takes the form of parallel circumferential cracks, often with a second system of cracks at right angles to the first, i.e. parallel to the axis (Fig. 8.46).

Corrosion products exuded from the cracks are sometimes visible, and the smooth part of the fracture surface is usually discoloured by them. Discoloration can, however, occur in a normal fatigue failure where during the later stages of the spread of the crack its sides have been continually butted together to cause fretting of the surfaces.

The regular pattern of the crack system implies that the course of the cracks is independent of the microstructure of the material, i.e. that the cracks are transcrystalline. It is sometimes observed that if slip lines are present the cracks within the grain follow their course, but only for short distances, jumping from one slip line to another, though never deviating far from their general direction, so that the crack has a serrated appearance.

There are a few alloy systems, notably lead and lead-based alloys and some of the heat-treatable aluminium alloys of the Duralumin type, which show intergranular cracking in corrosion-fatigue conditions, but these alloys fail in this way in normal fatigue and are prone to stress corrosion, so that their corrosion-fatigue behaviour is in no way remarkable.

The Corrosion Fatigue Curve

In a fatigue curve the number of cycles to failure is plotted against the semi-range of the cyclic stress, whether the mean stress is zero or not. With steels and titanium-base alloys the curve approaches the horizontal asymptotically, and this value is referred to as the fatigue limit. Other non-ferrous alloys, when tested in air, show a continuously falling curve and no fatigue limit. It is usual to quote in such cases a value, the endurance limit, for an arbitrary number of stress reversals, most commonly 20×10^6.

In the presence of a corrosive such as aerated 0·1 M NaCl, steels do not show a fatigue limit and the curve is similar to that for non-ferrous alloys in air. It has been observed, however, that non-ferrous materials tested in a vacuum or an inert atmosphere such as dry nitrogen show a fatigue limit.

Steel specimens which are first subjected to corrosion and then tested under fatigue in air show a curve similar in shape to that for air fatigue but with a slightly lower fatigue limit. The corrosion pits have an effect similar to artificially produced notches.

In corrosion-fatigue testing the individual points lie much more closely on a smooth curve and reproduceability is better than in air fatigue. This is because corrosion conditions ensure a large number of sources of cracks whereas in air fatigue testing, where care is taken to produce a surface free from flaws, the existence of points of initiation is a matter of chance. This, incidentally, is an argument against using specimens which are too small, i.e. which have too small a surface area in the region of maximum stress, in air fatigue testing.

The author[22] has recently compared the fatigue behaviour of an 0·7% C steel when subjected to fretting and corrosion separately and together (Fig. 8.47). With fretting alone, the steel still shows a fatigue limit. However, with fretting and corrosion the behaviour is similar to that in corrosion fatigue.

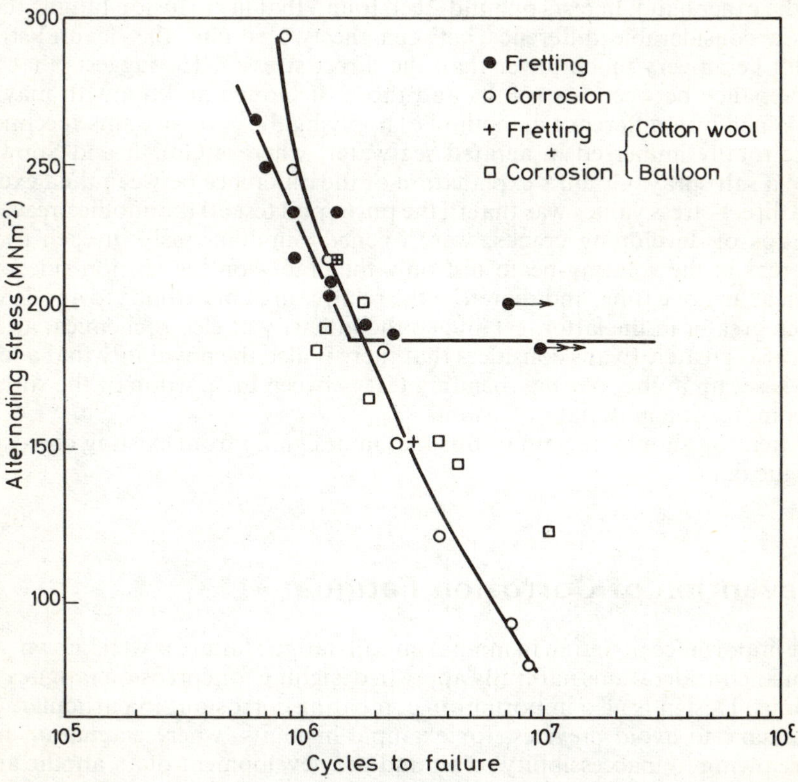

Fig. 8.47 Fatigue curves for 0·7% C steel with fretting, corrosion, and fretting and corrosion combined (after Waterhouse and Taylor[22])

Testing

Most corrosion fatigue investigations have used the conventional fatigue machines, usually of the rotating cantilever type. Many methods of applying the corrosive to the specimen have been used, the most popular being those in which the corrosive is limited to a small ring of the surface by a drip feed or a wick. Huddle[23] used specimens in the form of parallel-sided rods of circular cross-section; no reduced section for a gauge length was necessary as failure always occurred in the region wetted by the corrosive. One advantage of this method of applying the corrosive medium is that air has unrestricted access to the surface, which is the state of affairs prevailing in most practical instances of corrosion fatigue. Evans[24] gives an exhaustive list of references to the various experimental methods which have been used.

Some attention has been given to differences in the value of the endurance limit as determined in the rotating–bending machine (flexural stress) compared with the push–pull machine (reversed direct stress). As a result of wide experience of the two tests at the National Physical Laboratory, Gough and Sopwith[25] stated that while in corrosion fatigue the endurance limit obtained in a direct-stress machine was roughly the same as in a rotating-bend machine, in air fatigue there were often differences between the two results, the direct-stress figure always lying below that of the flexural stress. Gould, on the other hand, in tests on mild steel, found that in corrosion fatigue there was a considerable difference between the two results, the flexural-stress result being very much lower than the direct stress[26]. He suggests that the discrepancy between his results and those of Gough and Sopwith may be explained by a difference in method of applying the corrosive; his specimens were totally immersed in aerated sea-water, whereas Gough and Sopwith used a salt spray. Gould's explanation of the difference between the flexural and direct-stress values was that in the push–pull test all the anodic areas, i.e. the tips of developing cracks, were opened simultaneously in each cycle, whereas in the rotating-beam test only the cracks on the tension side were open at any one time, and therefore the relative area of cathode to anode was much greater in the latter test and so the intensity of electrochemical attack was also greater. Evans considers that there is also the possibility that a local cell is set up in the rotating–bending test between the portion of the surface in compression and that in tension.

Attention should be paid to this fact in designing from existing corrosion fatigue data.

Prevention of Corrosion Fatigue

The importance of design in minimising air-fatigue failure is well known, and similar considerations naturally apply in designing for corrosion fatigue conditions. Design is also important in combating corrosion; in particular, it is necessary to avoid crevices, for example in joints, where stagnation may arise owing to inaccessibility to air and the development of an anodic area, or, in the case of a closed circuit where inhibitors are being used, to the difficulty of replenishing with inhibitor.

The substitution of a more corrosion-resistant material, e.g. Monel metal or stainless steel, is often advocated but, as already indicated, this is not necessarily a solution; a 15% chromium steel, for instance, is prone to failure in corrosion fatigue because of the disruption of the normally protective surface film.

The protection of steel against corrosion fatigue has been one of the major preoccupations of corrosion scientists. The beneficial effect of inducing a compressive stress in the surface of the steel applies to corrosion fatigue as well as normal fatigue. The surface stress can be produced by chemical means such as nitriding or carburising, or by quenching from below the transformation temperature[27]. A third method is by surface rolling or shot peening, both of which leave the surface in a work-hardened condition which probably has the effect of ironing out differences in potential due to local stress concentrations.

Electroplating provides a further possibility for protecting mild steel. Nickel plating is normally very efficient in preventing corrosion in static conditions, but it is known to reduce the fatigue strength of steel because of the tensile stresses in the nickel coating. In corrosion fatigue nickel plating has little or no effect. Zinc plating, on the other hand, produces a coating in compression, which is in itself effective in causing an increase in the normal fatigue limit. In corrosion fatigue the performance of zinc-plated material is notably better, because of the added protection given by the sacrificial action of the zinc. The following figures are taken from curves determined by Allsop[28] on a 0·63% C steel wire in sea-water with zero mean stress. The thickness of the plating was 12 μm.

	Fatigue limit Dry fatigue (MN/m^2)	*Endurance limit Corrosion fatigue*, MN/m^2 (10^6 *cycles*)
Unplated	620	140
Nickel-plated	470	160
Zinc-plated	700	530

Some early results of Harvey[29] show that electroplated coatings of zinc are more efficient in protecting steel against corrosion fatigue than either hot-dipped or sherardised coatings, no doubt on account of the compressive stress present in the electroplated coatings.

Zinc has also been used with some success as a pigment in priming paint but this is not so efficient as a plated or galvanised zinc coating because the contact between the zinc particles in the pigment and the steel is not nearly as good as in the metallic coatings. Zinc can also be applied by metal spraying.

Zinc cathodically protects steel by sacrificial action, a method that can also be achieved by applying a potential. This method has been shown to be effective on mild steel in 3% NaCl[6]. At an applied potential of 0·49 V (*vs.* S.H.E.) or below, the fatigue life of specimens became constant and the fatigue limit observed in air was restored. An interesting variant of cathodic protection is anodic protection which relies on the preservation of a passive protective film on the surface of the material. Cowley, Robinson and Kerrich[30] have shown that this method can be successfully applied to plain carbon and stainless steels in an oxidising environment. The improvement is greatest with the stainless steels which rely on a protective oxide film for

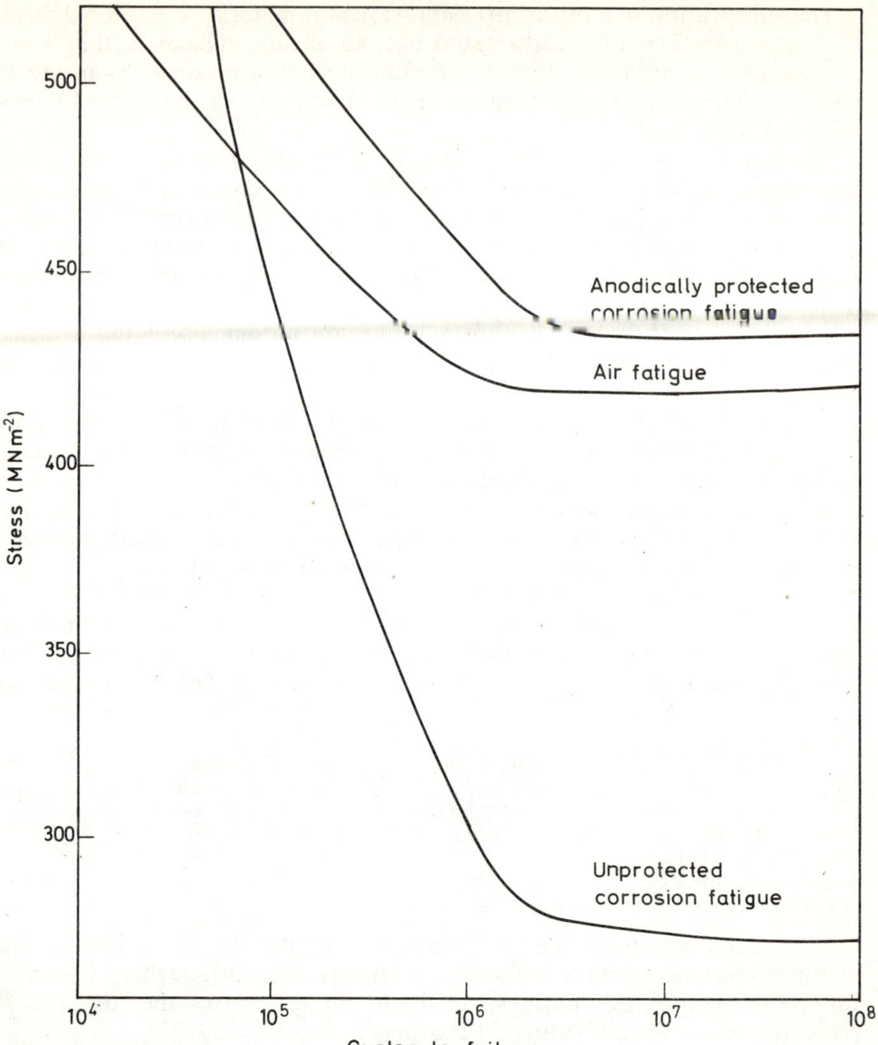

Fig. 8.48 Air and corrosion fatigue curves for Fe–13Cr alloy; corrosive media 10% NH_4NO_3 (after Cowley, Robinson and Kerrick[30])

their corrosion resistance. In some cases the fatigue strength is greater than in air due to the cooling effect of the electrolyte, and the superior protection afforded by the anodically-formed film compared with the air-formed film (Fig. 8.48). There are limitations to its application, however. If crevice conditions exist in certain areas, accelerated attack may result at these points.

Another group of materials frequently given corrosion protection by a metal coating is the high-strength aluminium alloys. The protective coating is usually pure aluminium and is applied by spraying in the case of extrusions, or by cladding in the case of sheet. The behaviour in corrosion fatigue of an aluminium–magnesium–zinc alloy and an aluminium–copper–magnesium alloy submitted to these treatments was reported by Whittaker and Liddiard[31]. In a rotating-bend test the sprayed coating had little effect on the air fatigue strength (about 240 MN/m^2 for 3×10^7 cycles) but raised the endurance limit in 3% NaCl from 50 to 130 MN/m^2. Under uniaxial stressing both sprayed and clad coatings reduced the air fatigue endurance limit by 50–60%. In corrosion fatigue the endurance limits of the uncoated and coated specimens, whether sprayed or clad were not markedly different when tested under uniaxial stress with a superimposed mean stress. With zero mean stress, both types of coating gave an improvement in the endurance limits. In every case the endurance limit of the coated specimens was the same in the corrosive environment as in air. This was also the case in the reverse-bend tests. It would therefore seem that such coatings are beneficial only where the stressing of the reversed type, i.e. with a zero mean.

Examples have been given of the protection that can be afforded by organic coatings, e.g. paints and stoved resin coatings for mild steel, and lanoline impregnation of an anodised coating in the case of aluminium alloys. The beneficial effect on the subsequent fatigue strength in air of coating welds in steel structures with epoxy resins has been demonstrated[32].

R. B. WATERHOUSE

REFERENCES

1. Haigh, B. P., *Trans. Inst. Chem. Engrs., Lond.*, **7**, 29 (1929)
2. Gough, H. J. and Sopwith, D. G., *J. Inst. Met.*, **49**, 93 (1932)
3. Wadsworth, N. J. and Hutchings, J., *Phil. Mag.*, **3**, 1154 (1958)
4. Snowden, K. U., *Acta Met.*, **12**, 295 (1964)
5. Wadsworth, N. J., *Phil. Mag.*, **6**, 397 (1961)
6. Duquette, D. J. and Uhlig, H. H., *Trans. A.S.M.*, **61**, 449 (1968)
7. Broom, T. and Nicholson, A., *J. Inst. Met.*, **89**, 183 (1960–61)
8. Forsyth, P. J. E., *Acta Met.*, **11**, 703 (1963)
9. Eeles, E. G. and Thurston, R. C. A., *J. Inst. Met.*, **95**, 111 (1967)
10. Grosskreutz, J. C., *J. Electrochem. Soc. Solid State Sci.*, **116**, 1232 (1969)
11. Grosskreutz, J. C., *J. Electrochem. Soc. Solid State Sci.*, **117**, 940 (1970)
12. Whitwham, D. and Evans, U. R., *J. Iron St. Inst.*, **165**, 72 (1950)
13. Evans, U. R. and Tchorabdji, S. M., *Proc. Roy. Soc. A.*, **188**, 372 (1947)
14. Rollins, V., Arnold, B. and Lardner, E., *Br. Corros. J.*, **5**, 33 (1970)
15. Paul, M. and Weiland, H., *Electrochim. Acta.*, **14**, 1025 (1969)
16. McAdam, D. J. (Jr.), *Proc. Amer. Soc. Test. Mater.*, **30**, 41 (1930)
17. McAdam, D. J. (Jr.), Amer. Inst. Min. Metal, Engrs, Techn. Publ. No. 417, 1–39 (1931)
18. McAdam, D. J. (Jr.), *Proc. Amer. Soc. Test. Mater.*, **31**, 259 (1931)

18*a*. Gough, H. J. and Sopwith, D. G., *J. Inst. Met.*, **49**, Autumn Lecture (1932)

19. Endo, K. and Komai, K., *Metalloberfläche*, **22**, 378 (1968)
20. Endo, K., Komai, K. and Nakamuro, N., *Bull. Jap. Soc. Mech. Eng.*, **13**, 837 (1970)
21. Galvin, G. D. and Naylor, H., *Proc. Instn. Mech. Eng.*, **179**, Pt. 3J, 56 (1964–65)
22. Waterhouse, R. B. and Taylor, D. E., *Wear*, **15**, 449 (1970)
23. Huddle, A. U. and Evans, U. R., *J. Iron St. Inst.*, **149**, 109P (1944)
24. Evans, U. R., *The Corrosion and Oxidation of Metals*, Edward Arnold, London, 716 (1960)
25. Gough, H. J. and Sopwith, D. G., *J. Iron St. Inst.*, **127**, 301 (1933)
26. Gould, A. J., *J. Iron St. Inst.*, **161**, 11 (1949)
27. Bühler, H., *Métaux, Corrosion-Ind.*, **26**, 141 (1951)
28. Allsop, R. T., *Coil Spring J.*, **32**, 5 (1953)
29. Harvey, W. E., *Metals and Alloys*, **1**, 458 (1929–30)
30. Cowley, W. E., Robinson, F. P. A. and Kerrich, J. E., *Br. Corros. J.*, **3**, 223 (1968)
31. Liddiard, E. A. G., Whittaker, J. A. and King, H., *First International Congress on Metallic Corrosion*, 482, Butterworths, London (1962)
32. Gilde, W., *Brit. Welding J.*, **7**, 208 (1960)

8.7 Fretting Corrosion

Definition and Terminology

Fretting or *fretting corrosion*[1] may be defined as that form of damage which occurs at the interface of two closely fitting surfaces when they are subject to *slight* relative oscillatory slip. The surfaces are often badly pitted and finely divided oxide detritus is formed.

Although the term 'fretting corrosion' implies chemical reaction, it has often been used even when the latter is absent. Campbell[2] has suggested that to avoid confusion the word 'fretting' be used to describe the wear process, and that the expression 'fretting corrosion' be applied in those cases where one or both of the surfaces, or the wear particles from them, react with their environment.

Fretting has been known to engineers by a variety of other names; such terms as 'false brinelling', 'chafing fatigue' and 'cocoa' are in use even today. In Germany, Fink[3] has referred to the phenomenon as 'friction oxidation' (*Reiboxydation*), while the word 'blood' has often been used by engineers of that country. The expression 'false brinelling' was coined primarily to describe the fretting wear process as it occurs in rolling contact bearings, since the damage closely resembles the brinelling of a race which has been subjected to excessive static loading. 'Chafing fatigue' on the other hand, is an expression used in reference to a combined fretting–fatigue failure. Such words as 'cocoa' and 'blood' refer to the reddish brown oxide debris which is often to be seen exuding from fretting ferrous contacts.

Incidence

As almost all materials are susceptible to fretting, its incidence in vibrating machinery is high. Shrink fits, press fits and bolted assemblies, splined couplings, keyed gears, both seatings and tracks of ball and roller races, and even electrical contacts, are all particularly vulnerable. Fretting may not only cause serious dimensional loss of accuracy of closely fitted components, but may also seriously reduce the fatigue strength of a machine component.

Characteristics of Fretting and Factors which Influence the Amount of Damage

Although our knowledge of fretting as a wear problem has been derived mainly from the behaviour of ferrous materials, many of the characteristics observed are common to other metals. The influence of fretting on fatigue, however, must be considered separately, since the magnitude of wear damage, as recorded by volume loss of material, is not a good indicator of possible fatigue damage. The development of fine surface cracks is the important criterion in this situation and this type of damage occurs in its most severe form during the early stages of fretting. Because of this difference *fretting fatigue* is considered subsequently as a separate issue.

The main factors which influence fretting wear can be classified as in the following paragraphs, but as more data become available, it is apparent that a greater measure of interdependence exists than might be indicated by such a simple procedure, and it is becoming less easy to make generalisations. Nevertheless, even with this reservation in mind, it is useful as a first step to list some of the more important factors.

The Atmosphere

In the presence of an inert atmosphere fretting of surfaces still occurs and is generally accompanied by the formation of finely divided debris. In a high vacuum, seizure of metal surfaces may take place.

If fretting of oxidisable metals occurs in air the damage is rather more severe than in an inert gas, but the increase is not usually greater than an order of magnitude. The debris consists mainly of oxide, and in the case of steel it is predominantly α-Fe_2O_3 in a very finely divided form, 0·1–0·01 μm in diameter. The proportion of oxidised to non-oxidised material depends largely upon the hardness of the metal. The most stable form of oxide usually appears as the final product of fretting corrosion, but other forms may occur as intermediate products and non-stoichiometric oxides are often produced. Colour is not a reliable guide to composition and steel may produce debris ranging in colour from red to brown and to black, and in every case the composition of the oxide may be that of α-Fe_2O_3.

The relative humidity of the atmosphere has a large effect on the magnitude of wear[4], but in a direction opposite to that which is encountered in normal corrosion problems. The increasing wear towards lower humidities is accompanied by severe pitting of the surfaces and, under extremely dry conditions, the oxide debris produced from steel surfaces is jet black.

Temperature

This is an important factor because it controls the rate of reaction of the rubbing surfaces with the oxygen in the environment. However, although the presence of oxygen normally accelerates the rate of fretting of oxidisable metals at room temperature, an increase in the reaction rate with oxygen brought about by raising the temperature, has the opposite effect, and reduces the wear. Experiments made with steel and copper surfaces show that once a threshold temperature has been exceeded, the normal type of fretting damage (that is the formation of copious amounts of loose oxide debris) is replaced by the formation of a thick adherent oxide glaze that exhibits low friction and little tendency to generate loose debris[5, 6]. In the case of steels, the threshold temperature is around 130–200°C, whereas for copper it is just above room temperature. These glazes remain effective at least up to 300°C and 200°C, respectively, the limits of temperature investigated. Their development is encouraged by improving the surface finish of the rubbing elements, since this probably reduces the risk of breaking up the thin surface oxides initially present on the materials. The glazes are not resistant to an impacting load component.

For a wide range of mild and carbon steels, it is found that the rate of development of this thick oxide glaze formed at elevated temperatures is largely independent of the hardness of the substrate. The level of surface damage represented by the amount of virgin metal converted to oxide is an order of magnitude smaller than the level of wear normally encountered at room temperature and is even less than the damage generated in a nitrogen atmosphere at 20°C.

Load, Amplitude of Slip and the Number of Fretting Cycles

The amount of fretting damage increases in an approximately linear manner with these variables, once the initial stages of fretting are completed[4]. A number of deviations from linearity have been reported, especially with respect to load, where it is often found that there is a tendency for the relationship to become parabolic in form. The superposition of a normal vibratory component of load can cause a very considerable increase in the wear rate[7].

Frequency of Oscillation

Little information exists on the effect of frequency except in the case of steels[8]. With such materials there is a reduction in fretting damage as the frequency is increased to 10 Hz. The form of this relationship is the same at both 20°C and 165°C, these temperatures corresponding to the unglazed and glazed mode of fretting of steel in air, respectively[9]. No frequency effect is observed in nitrogen. The position at higher frequencies is not so well established, but there is some evidence that the wear rate begins to rise again and that this is true for both air and nitrogen environments. The increase in wear rate at lower frequencies is often ascribed to the increased time available for oxidation reactions, but it is also observed that the degree of metal to metal contact is more marked at these frequencies.

Hardness

The effect of hardness is complicated, but like most wear processes an increase in hardness generally leads to a reduction in fretting wear at room temperature. The hardness also controls the form of the debris, softer metals tending to produce a higher proportion of large particles of unoxidised debris. Prior work-hardening of materials, such as iron and mild steel, has no effect on the fretting wear rate.

Lubricants

Oils may act in two ways: they provide a measure of boundary lubrication and they may exclude oxygen from the rubbing zone. However, their effectiveness is not as great as with unidirectional sliding, since there is usually sufficient oxygen present in most cases to allow oxide debris to be generated, and this tends to displace any lubricant film that was initially present between the surfaces.

Mechanism of Fretting Corrosion

Fretting wear, along with most other wear phenomena, is not a process that can be defined in terms of any single mechanism. It consists of a series of events, many of which are common to other wear processes and which may assume greater or lesser significance depending upon the precise nature of the operating conditions, materials and environment.

The overriding difference between fretting and other sliding wear processes lies in the small reciprocating nature of the motion. The damage tends to be of a localised form and any debris which is generated has some difficulty in escaping from the rubbing zone. The oscillatory character of the movement introduces a strong fatigue element into the wear pattern and the reversed shearing of localised material inevitably gives rise to fine surface cracks which may initiate a low stress fatigue failure.

The early stages of fretting with metal specimens bear a strong resemblance to those present in other sliding systems in which adhesion forces play a dominant rôle. In all such cases the initial placing of the two surfaces into contact is accompanied by the formation of localised junctions where physical intimacy of the two surfaces occurs. The material at these junctions yields both plastically and elastically until the area of contact established is sufficient to support the applied load. Oxide or other contaminating films greatly reduce the adhesion across the junctions, but the application of a tangential stress facilitates their dispersal. However, the combined effect of normal and tangential tractions across a junction causes it to grow in size and if it was not for the controlling influence of even physically incomplete intervening low-shear-strength films, this growth might continue unchecked until seizure occurred.

The formation and fracture of these junctions leads to the transfer of small fragments of material from one surface to another and finally, after many such events, the release of a wear particle often in a highly oxidised state. Such a wear process is common to most sliding systems and can undoubtedly occur during fretting movements, especially if the slip amplitude is large. However, many fretting situations occur where the amplitude of slip is very small and perhaps even comparable to the dimensions of a single junction region. In such cases it is not so easy to imagine the transfer mechanism just described, but rather to view fretting action as one in which material within the junction zones is continuously subjected to reversed shear. Such conditions will certainly give rise to junction growth and very strong bonds will rapidly become established. The adhesion usually reaches its maximum value after a few thousand cycles of oscillation[10] and during this period the surface material undergoes considerable plastic deformation. It is thought that the pitted nature of fretted surfaces owes its origin to this early gross deformation period and that subsequent events tend to smooth out the contours. Friction measurements reflect this early growth of bonded surface regions and it is only when these junctions begin to work harden and break up under the combined action of the reversed shear and the corrosive action of an air environment that the friction begins to decrease in magnitude. Metallurgical sections through a fretting zone will usually show the disordered state of the junction material and the presence of oxide structures.

The degree of plastic deformation is much greater for softer materials than those which are hard, and the debris frequently contains a much higher percentage of unoxidised metal. In the case of tool-steel surfaces, fretting debris is entirely oxide and very finely subdivided. The gradual accumulation of oxidised detritus between the rubbing surfaces soon begins to isolate one metal surface from the other and this can be followed by measuring the electrical resistance of the system[11]. Extremely high values of electrical resistance can be recorded when the air is dry, and quite clearly a thick oxide compact has built up between the surfaces and some proportion of the slip is likely to be lost within this layer. The hygroscopic nature of the finely divided oxide compact can be demonstrated by admitting air at 45% r.h. when there is an immediate drop in the resistance. The presence of a water film greatly aids the dispersal of the debris and the fretting damage is more uniform over the surface as well as being of smaller volume. This suggests that soft hydrated oxides may be formed and may act as a lubricant. The large increase in wear rate observed whenever a normal vibratory load is applied to a fretting system is possibly due to the fact that it prevents a thick oxide compact from establishing itself between the surfaces. The early adhesion mode of wear probably persists for a much longer time and may never completely disappear. In practice such conditions must be avoided at all costs.

The manner in which oxide debris is formed when metals fret in air is a subject of considerable controversy and it is clear that no single mechanism can explain all the data. The early theory of Tomlinson[1] was based upon the idea that the surfaces were worn by a process of molecular attrition and that this leads to oxide generation in an oxidising environment. Others considered that fretting was essentially an accelerated oxidation mechanism in which the mechanical removal of oxide prevented the attainment of a stable protective oxide film. Later, Uhlig[8] modified this model to allow some metallic debris to be formed as a result of an adhesion mechanism, but still retaining the corrosion aspect to help explain away the frequency effect[8]. This model encountered difficulties in explaining the decrease in wear with increasing temperature, and Uhlig suggested that perhaps the corrosion aspect could be better represented by a model that involved physical adsorption of oxygen on the steel surface and that the actual formation of the oxide resulted from mechanical activation. More modern theories[12] place greater emphasis on the changing nature of the fretting mechanism, particularly drawing attention to the strong influence of adhesion in the early stages and the significance of corrosion fatigue as a contributory factor in the disintegration of the material making up the junction zones. The later stages of fretting damage are also explained in terms of a microfatigue process rather than one of abrasion by cutting.

The reduction in fretting damage of steel and copper surfaces as a result of increasing the temperature, at least over the temperature range of 20–300°C, has focused attention on the protective rôle that oxidation can play in fretting. It would seem that this arises from a suppression of intermetallic contact during the early adhesion stage of fretting and the improvement provided by a good surface finish substantiates this explanation. Again the presence of a vibratory normal load destroys the opportunity of establishing a protective glaze. It is interesting to note that the frequency effect is of the same form both at room temperature and at slightly elevated temperatures where glazing occurs. It might have been argued that the increased reaction time at the lower frequencies would have enhanced the protective rôle offered by oxidation. These and many other problems still need to be cleared up before an accurate picture of fretting corrosion can emerge.

Preventive Measures

Most of the cases of fretting met with in practice appear to fall into two distinct classes according to whether or not the surfaces involved in the component are intended to undergo some relative motion. If the surfaces are not intended to move, then the first objective should be to prevent slip, either by eliminating the source of vibration, or by increasing the friction between the surfaces. It is believed that the success of certain soft metal electrodeposits in reducing fretting may be due to the improved fit, and hence possibly increased friction, which is obtained from their use. If the displacements cannot be controlled in this way, it may be possible to interpose a thin sheet of an elastic material which can accept the relative movement without slip.

With applications which are intended to undergo relative motion at some stage, once again an attempt should first be made to lower the amplitude of vibration. Should this prove difficult, an improvement in the lubrication conditions will be necessary in order to reduce the amount of intermetallic contact. For example, in the case of rolling bearings the interfacial slip between the rolling elements and the races is accompanied by a relatively large rolling component of motion. If a suitable oil or grease is used, this displacement can be utilised to maintain a film of lubricant on the race.

With plain bearing surfaces of steel, a great improvement can be obtained by phosphating and impregnating the layer with oil. The inherent porosity of the phosphate films provides minute reservoirs for the oil. Should it be necessary to operate a bearing dry, then either a bonded film of MoS_2 or p.t.f.e. (polytetrafluoroethylene) can be used to advantage. Both materials have excellent frictional and wear characteristics.

Fretting Fatigue

Conditions which favour the occurrence of fretting may also be regarded as favourable to metal fatigue and fatigue cracks are often observed to start at points on a surface where fretting has taken place. There is no evidence, however, that the severity of fretting damage, as measured by volume loss of material, is directly related to the magnitude of the reduction in fatigue strength. The significance of fretting is that it allows the formation and rupture of strong adhesive bonds under oscillatory forces and this action generates fine surface cracks which may, or may not, propagate into a major fracture of the component. An understanding of this subsequent crack development is therefore vital if adequate steps are to be taken to avoid an early failure. In order to do this it is useful to examine the normal fatigue behaviour of plain and notched materials subject to reversed loading.

The progress of a normal fatigue failure can be divided into two stages, firstly (Stage I), the formation of slip band cracks at the surface, and secondly (Stage II), their propagation into the main body of the component and eventual failure. The first stage of the process is usually marked by the growth of crevices or intrusions along planes aligned in the direction of maximum shear stress and this stage may occupy the major proportion of the final recorded life. However, crack development during this stage eventually ceases, perhaps by the crack meeting an obstacle, such as a grain boundary, and the subsequent propagation of this embryo crack proceeds in accordance with a criterion based upon maximum principal stress, or upon maximum comparative stress in a combined stress situation. This second stage is marked by the appearance of the familiar striations on the crack faces. The stress required for the successful completion of Stage I corresponds to the plain fatigue limit of the material, and is much higher than that which is needed for the subsequent propagation of the surface crack.

The situation is greatly changed, however, if a notch or other stress raiser is present in the surface. If a notch exists, then the stress required to initiate a crack at the root is given by the plain fatigue limit stress divided by the stress concentration factor for the notch, K_t. With some notches the magnitude of K_t may be so large that the nucleation stress is now much lower than the propagation stress and propagation will not occur unless the alternating stress within the main body of the component is of sufficient magnitude. Dormant cracks may thus exist in some engineering components.

In the case of a notched component the propagation stress is both independent of K_t and the notch root radius, and is primarily a function of the combined length ℓ_d of the notch and embryo crack; the criterion for propagation being $\sigma^3 \ell_d > C$, where C is a material constant and σ the applied cyclic (tensile) stress. Since most non-propagating cracks are short compared with the notch depth we may usually substitute the notch depth for ℓ_d in this expression. Thus assuming that notch root cracks have been formed, a limiting stress exists for propagation, namely $\sigma = (C/\ell_d)^{\frac{1}{3}}$. If this stress is not reached then dormant cracks will be formed unless an active corrosive environment is present[13].

In many ways the fretting fatigue situation is analogous to the notched fatigue behaviour, not because the fretting damage resembles a mechanical notch, although this view has been held by some workers, but because the fretting action promotes the formation of suitable surface cracks simply by the presence of strong frictional shear stresses. Many of these cracks are observed to be of the order of 100 μm in length and are thus of sufficient length to be propagated by stresses well below the normal plain fatigue limit. Fretting therefore seems to be a remarkably efficient way of achieving the completion of Stage I of the fatigue process.

Once fretting cracks of a suitable length have been formed the life of the component depends upon the rate at which they can be propagated by the Stage II mechanism, and this part of the process is subject to the same environmental factors as a normal fatigue failure. The length of the fretting cracks determines the lowest level of stress at which they can be propagated, and hence the new fatigue limit. Once again an active corrosive environment prevents the existence of a fixed fatigue limit and even small 'non-propagating cracks' will not remain dormant for long. Fretting tests carried out in such environments show the same characteristics as corrosion fatigue tests made in the absence of fretting[14].

One important feature about fretting fatigue is the very small amount of physical damage to the surface that is necessary to cause a considerable decrease in fatigue strength of some materials. The damage is often barely discernible to the naked eye. Small slip amplitudes seem to be particularly damaging and this has been attributed to the possibility that larger slip amplitudes and the associated higher wear rate may erase any potentially damaging surface cracks or promote the development of large numbers of very small interacting cracks rather than one or two of greater severity. Only one crack of propagating length is required to cause a component failure. Recent work suggests that the coefficient of adhesion between fretting surfaces in air reaches a maximum value for amplitudes of slip between 30 and 75 μm, and that this level of adhesion is established within a very small number of slip cycles. A few thousand cycles is usually sufficient. Laboratory experiments made with fretting pads clamped to specimens subjected to fatigue stresses indicate that the nominal slip amplitudes which prove the most damaging are those in the range 7·5–14 μm[13]. This range is slightly smaller than the range observed with adhesion experiments, but the sensitivity to slip amplitudes is complex and somewhat higher slip amplitudes prove equally damaging in the presence of a high-tensile mean stress. All the data tends to support the hypothesis that the establishment of strong adhesive bonds between the contacting surface asperities is a major factor in the fretting fatigue process. The strength of such bonds varies in a systematic way with the hardness of the rubbing elements and there is evidence that annealing and over-ageing of age-hardened alloys occurs at the fretting contact. It is claimed that the fretting fatigue strength is related to the adhesion strength of the asperity bridge and that when a critical value of the latter is reached a propagating crack is quickly formed[15].

It is perhaps unfortunate that the sensitivity to fretting is generally greatest in the low slip amplitude regime since this is just the level of slip which is so often experienced with many engineering components. A typical example is the case of a wheel fitted to a shaft. When this combination is subjected to a reversed bending moment localised slip will occur at the edge of the contact area and fretting fatigue cracks may eventually develop there. The establishment of a partially slipped interface is of common occurrence in fitted

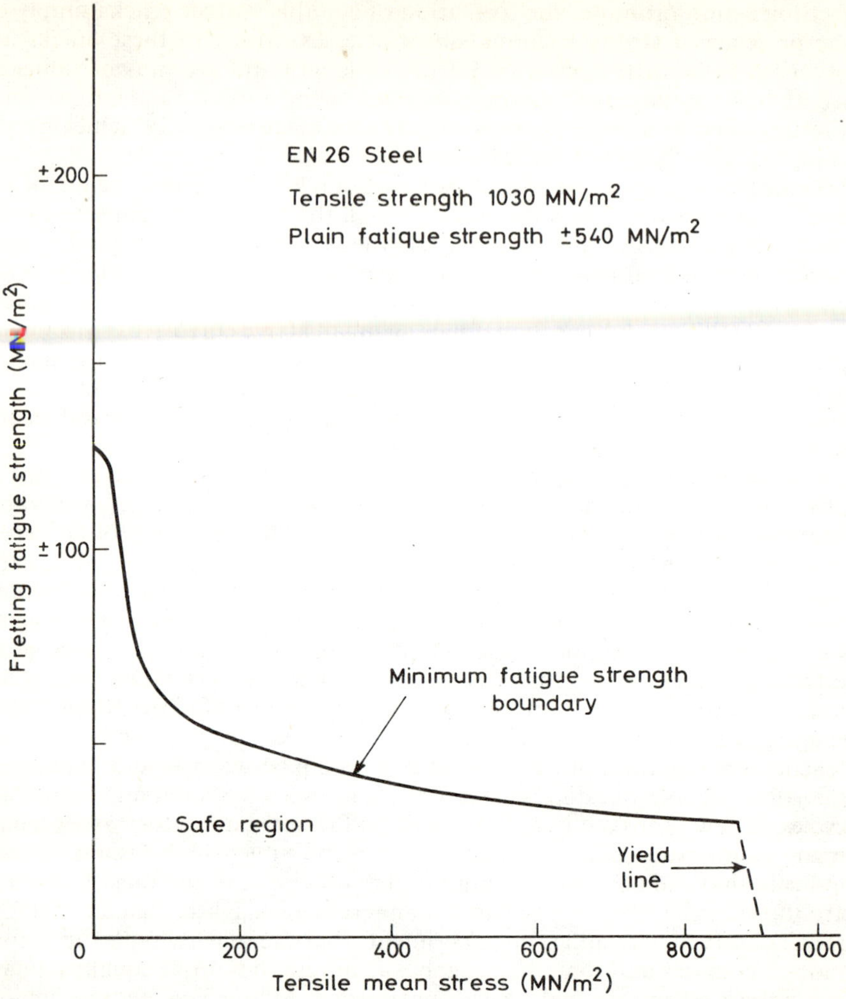

Fig. 8.49 Fretting fatigue strength at various mean stresses

components and is due to the form of the normal and tangential tractions which have to be transmitted across the interface when the unit is stressed. Slip occurs at all points where the ratio of the tangential to normal traction exceeds the coefficient of friction. Ultimately with sufficient magnitudes of applied stress the whole surface will slip, but this is not usually an acceptable condition. It will be seen therefore that small slip amplitudes are invariably linked with a partially slipped interface and the stress concentrations that exist between the boundary of the slipped and unslipped regions have been suggested as a reason why fretting cracks appear at the boundary. As an extension to this argument it is suggested that the inducement of a high gross slip condition will relieve the stress concentration and hence inhibit the formation of propagating fretting cracks even though this action is likely to lead to greater overall wear.

Mention must be made of the important practical case where a component is subjected to a cyclic stress superimposed upon a high-tensile mean stress. In the case of a simple plain fatigue test, the fatigue strength only falls slowly with rising values of the mean stress (Gerber relationship). Small values of the mean stress have little effect upon the measured fatigue strength because the strength of a plain component is governed by the crack initiation or reversed shear stress stage of the process. However, if cracks have already been developed in the surface as a result of fretting, then the fatigue limit will depend upon the propagating conditions and these will be a function of both the alternating and mean tensile stress[13]. An indication of the decrease in fatigue strength as a result of applying a mean tensile stress to a cyclically stressed component subjected to fretting is shown in Fig. 8.49.

Prevention of Fretting Fatigue

Some design aspects have already been mentioned, notably the avoidance of interfacial slip between contacting surfaces or joints. This is best done by avoiding designs in which the interface enters a region of stress concentration. In some cases it may be possible to absorb the movement by the insertion of a resilient layer. A low coefficient of friction can reduce the severity of induced frictional stresses. Resin-bonded P.T.F.E., for instance, is quite effective under conditions of relatively low surface pressure and slip amplitudes. Alternatively, a sacrificial metal coating may be applied, but this method may not prevent a surface crack penetrating the bond interface, especially if this is strong. The influence of the hardness of the rubbing surfaces has already been mentioned and some advantage of this can be taken in certain cases[16]. Surface treatments like case hardening and nitriding can produce a hard and compressively stressed surface layer which inhibits crack propagation. Similar results can be achieved with cold working procedures, such as shot peening.

K. H. R. WRIGHT

REFERENCES

1. Tomlinson, G. A., Thorpe, P. L. and Gough, H. J., *Proc. Instn. Mech. Engrs., London.*, **141**, 223 (1939)
2. Campbell, W. E., *Symposium on Fretting Corrosion*, Amer. Soc. Test. Mater. Special Technical Publication No. 114, 3 (1953)
3. Fink, M., *Org. Eisenbahnw.*, **84**, 405 (1929)
4. Wright, K. H. R., *Proc. Inst. Mech. Engrs., Lond. (B)*, **1B**, 556 (1952–53)
5. Wright, K. H. R., Strength of Components, *N.E.L. Report No. 402*, March (1969)
6. Hurricks, P. L., *Wear*, **19**, 207 (1972)
7. de Gee, A. W. J., Commissaris, C. P. L. and Zaat, J. H., *Wear*, **7**, 535 (1964)
8. Uhlig, H. H., *J. Appl. Mech.*, **21**, 401 (1954)
9. Wright, K. H. R., LDR 14/61, National Engineering Laboratory (1961)
10. Bethune, B. and Waterhouse, R. B., *Wear*, **12**, 289 (1968)
11. Fenner, A. J., Wright, K. H. R. and Mann, J. Y., *Proceedings of the International Conference on Fatigue of Metals, Instn. Mech. Engrs.*, London (1957)
12. Hurricks, P. L., 'The Mechanism of Fretting—A Review', *Wear*, **15**, 389 (1970)
13. Field, J. E. and Waters, D. M., NEL Reports Nos. 275 (1967) and 340 (1968)
14. Waterhouse, R. B. and Taylor, D. E., *Wear*, **15**, 449 (1970)
15. Bethune, B. and Waterhouse, R. B., *Wear*, **12**, 369 (1968)
16. Taylor, D. E. and Waterhouse, R. B., *Wear*, **20**, 401 (1972)

8.8 Cavitation Damage

Cavitation damage is a form of deterioration associated with materials in rapidly moving liquid environments. Attack is characteristically local and often severe, and is a consequence of the hydro-mechanical process of 'cavitation' occurring in the liquid environment. In this process cavities are formed in the liquid in regions of very low pressure caused by flow divergence, rotation or vibration. At these points the liquid, which is subjected to tensile forces greater than its cohesive strength, splits open, forming cavities. These rarely last more than a very short time, and the collapse process takes place extremely rapidly (in about a microsecond), producing a very high acceleration of the cavity's liquid envelope. When the walls of the cavity collide, a strong shock wave, which damages adjacent materials, is formed. As 2 000 000 cavities may collapse over a small area in the course of a second, visible damage may quickly be produced, and in a soft material like pure aluminium severe cavitation can erode pits several tens of thousandths of a millimetre deep in a few minutes. The characteristic appearance of cavitation on ships' propellers is illustrated in Fig. 8.50.

Cavitation damage is a well-known cause of trouble in ships' underwater fittings, especially propellers and rudders, and in pumps and pipework circulating water. Operation of hydraulic turbines under conditions which produce cavitation is often unavoidable, and here damage has been a very pressing problem for many years. Concrete structures in hydroelectric channels have sometimes been attacked. Cavitation damage may also occur in diesel engine cooling systems, where, although flow velocities are low, severe vibrations can cause cavitation pitting of the cylinder liner watersides.

Despite the occasional expression of contrary opinions, the massive weight of evidence supports the view that erosion is caused by the physical forces consequent upon cavity collapse. Cook and Lord Rayleigh[1] in 1919 calculated that the collapse of spherical cavities could generate huge local instantaneous pressures, of the order of several giganewtons per square metre. More refined calculations, taking into account the diffusion of gas and vapour into the cavity and surface-tension effects, have suggested more modest pressures. Photoelastic techniques have demonstrated that strain waves in material under cavity collapse impact involve pressures of nearly $1{\cdot}5\ GN/m^2$. The presence of large mechanical forces is demonstrated by the way in which lead surfaces may be substantially deformed without loss in weight. The attack sustained by glass, plastics and other relatively inert substances

Fig. 8.50 Cavitation damage to a ship's propeller, illustrating the manner in which pitting follows the flow streamlines

rules out an explanation based solely on accelerated dissolution, although corrosion effects are sometimes important.

An exhaustive bibliography of reported work on cavitation up to 1947 has been compiled by Raven, Feiler and Jespersen[2] and the subject has been reviewed by Eisenberg[3] in 1950, and by Godfrey[4].

The Influence of the Fluid Environment

There are several properties of liquids which may affect the intensity of cavitation damage. Air content is perhaps the most fundamental, as the variation in the cavitation susceptibility of any particular liquid is thought to be related to its air content. Dispersed air rather than dissolved air is now known to facilitate cavitation, as it forms points of weakness which lower the cohesive strength of the liquid. Air bubbles were first thought to occur only on dust particles, but it has now been shown that small bubbles of about 50 μm diameter exist in appreciable quantities in water and that they are stabilised by minute amounts of surface-active contaminants. The fact that air in suspension is more important than that in solution is borne out by the difficulty experienced in producing cavitation in water which has been subjected to high pressures to dissolve dispersed air.

Air-bubble injection has been used effectively to prevent cavitation damage. (These bubbles are of course much larger than those described above which form cavitation-inception nuclei.) The air bubbles, being compressible, absorb the energy of the pressure waves caused by collapse of cavities.

For any given system, a parameter, σ, the *cavitation number*, defines a limit below which cavitation will take place. The cavitation number is expressed by

$$\sigma = \frac{P - p_v}{\frac{1}{2}\rho u^2}$$

where P is the pressure in the undisturbed fluid, p_v the vapour pressure of the fluid, ρ the mass density and u the reference velocity outside the immediate disturbed region. It is apparent from this expression that the higher the fluid pressure the higher the velocity that can be supported without the onset of cavitation.

In designing hydraulic systems care is always taken (*a*) to avoid high suction heads which can produce dangerous pressure drops and (*b*) to avoid turbulence caused by abrupt bends and changes of section or sharp contours, since a distribution of velocities which may exceed the cavitation threshold can occur in turbulent flow.

Cavitation is influenced in a complex manner by temperature. With water in the range of 0 to 50°C, damage increases with temperature. This has been explained as being due to a decrease in air content which allows a 'harder' cavity collapse. Above 50°C the intensity of damage is progressively reduced, possibly owing to the fact that the effect of air removal is smaller than that produced by increased vapour pressure, which provides a cushion of vapour within the cavity.

The nature of the cavitating liquid has an important bearing on damage. Various aqueous solutions and organic liquids have been investigated. Attack in non-ionic liquids is slow, probably on account of their high vapour pressures and their lack of conductivity. Toluene produces much less cavitation than water, although its collapse forces are at least 80% those of water. This suggests that corrosion is of some significance in the process. Sea-water and brine produce 50% more attack on steel than does distilled water, but attack is only 50% greater in 25% sulphuric acid than it is in distilled water. It is apparent, therefore, that the available data present a confused picture.

Where cavitation damage has occurred in a closed system (e.g. diesel cooling circuits) inhibitors have been used in an attempt to reduce attack. 'Soluble oil', sodium chromate or nitrite, and triethanolamine phosphate have been found to reduce the intensity of attack. In general, inhibitors have proved more valuable in service than might have been expected on the basis of laboratory cavitation tests, possibly because they prevent, during the static periods encountered in service, the repeated heavy initial corrosion of freshly exposed cavitation-damaged surfaces.

Factors Influencing the Resistance of Materials to Cavitation Damage

Surface deformation arising from physical cavitation damage usually occurs before the removal of material. Later, fractures which occur in this distorted layer allow the mechanical removal of material. Evidence from many sources shows that high hardness and tensile strength resist these processes and these two properties are closely correlated to damage resistance. Stellites (complex alloys of cobalt, chromium and tungsten) show both the highest hardness and the highest resistance to cavitation damage found among the common engineering materials.

The ability to work harden is also important and this property is responsible for the widespread use of austenitic stainless steels to resist attack. Although this material is initially hard it attains a much higher surface hardness from cold-working action caused by cavitation impacts, and becomes more resistant to attack.

Other aspects of physical structure exert an influence. With two alloys of identical composition but differing grain sizes, the one having the smaller grain size has been shown to have the better resistance. Grain boundaries retard transcrystalline fatigue cracking. Brittle microconstituents, such as phosphide eutectic and graphite in cast iron or manganese sulphide in steel may be eroded, thereby presenting crack-initiating notches. The fatigue resistance of the material is also an influential factor. The onset of cavitation damage often has an induction period during which forces insufficient to cause straightforward tensile failures may crack the surface by a fatigue process, or, very possibly, by a corrosion-fatigue process.

Rubber and other elastomers such as polyurethane have been used to resist cavitation attack. The physical property involved here is resilience, which is the capacity to re-emit the energy of impact without the absorption of a substantial part of this energy. The small amount which is absorbed is converted to heat, and this is probably the cause of failure of rubber coatings under severe cavitation attack. Other properties of a rubber which may also have an influence in resisting attack are tear and abrasion resistances.

The Rôle of Corrosion in Cavitation Damage

Although cavitation may not be taking place, if a corroding liquid is in motion, damage to a metal may be greatly intensified by *impingement attack*. The accepted explanation of this phenomenon is that the inertial forces of water turbulence damage the protective film of oxide, exposing small anodic areas of bare metal, corrosion being accelerated by the presence of a large surrounding cathodic area (*see also* Section 1.6). While both impingement attack and cavitation damage occur in moving liquids, differences between the two effects are apparent. The former is characterised by clean-swept sharp-edged corroded areas, and the latter by pits which may even be re-entrant or honeycombed and which are also accompanied by surface deformation visible in microsections (Fig. 8.51).

It is, however, conceivable that cavitation forces may disrupt the surface oxide film, without being intense enough to damage the metal itself, and recent work on the corrosion–erosion of copper alloy sea-water piping suggests that this mechanism may be of considerable practical importance. It may be described as *cavitation corrosion* to distinguish it from cavitation erosion.

The corrosion of an immersed metal may be arrested by cathodic protection and several workers have recently shown that this form of protection can reduce cavitation damage. Following early suggestions by Petracchi[5] that cathodic protection was of value in this context, Foltyn[6], and later Wheeler[7], in a thorough investigation showed that it could much reduce accelerated cavitation damage produced in the magnetostriction oscillator test. Wheeler

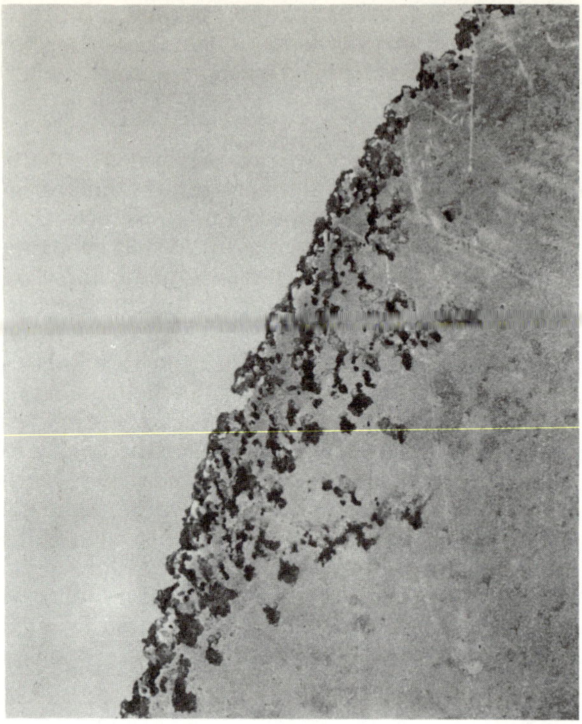

Fig. 8.51 Cavitation damage showing honeycombed type of attack

demonstrated that both electrochemical and mechanical factors are of importance by filtering the test solutions at the end of the test, and determining the eroded iron (retained on the filters) and the dissolved iron. The direct effect of cathodic protection was to reduce the corrosion component, but it also reduced erosion to some extent. Cathodic protection has also been tried in practical cases of cavitation damage. Promising results have been obtained in the prevention of the pitting of cast iron propellers, which is apparently a cavitation corrosion effect. The value of cathodic protection in preventing attack on the propellers of trawlers, launches, yachts and supertankers has been described in the literature. Where *cavitation erosion* of the material is occurring, however, cathodic protection at the conventional level is not effective. Current densities as high as 500 A/m^2 generating appreciable quantities of hydrogen bubbles, which probably interfere with the cavity-collapse damage process, have been shown by Kerr and Leith[8] to be necessary to reduce erosion damage. Such applied currents would be very costly in practice.

Cavitation Damage Testing Procedures

In order to allow closer study of the cavitation damage process and evaluation of resistant materials and protective schemes, accelerated tests have been developed (*see also* Section 20.1). Two main classes of test have been evolved: hydraulic and vibratory.

Hydraulic Tests

Hydraulic tests are of four kinds: double-weir, Venturi throat, jet-impact and perforated-rotor.

In the double-weir arrangement (Fig. 8.52) water flows rapidly through a pipe, and severe turbulence is caused by two successive protuberances or 'weirs', which have a leading face curved concavely out towards the tube centre followed by a plane normal to the flow downstream. Facing the first weir is a bulge in the tube wall opposite, which constricts the flow; the test-pieces are set in the parallel side walls. Cavitation is produced by the throttling effect and subsequent downstream expansion, and the test-pieces are attacked. Mousson[9] used this arrangement at head pressures of 3·3 MN/m^2, and throat velocities of up to 80 m/s, to test an extremely comprehensive range of materials.

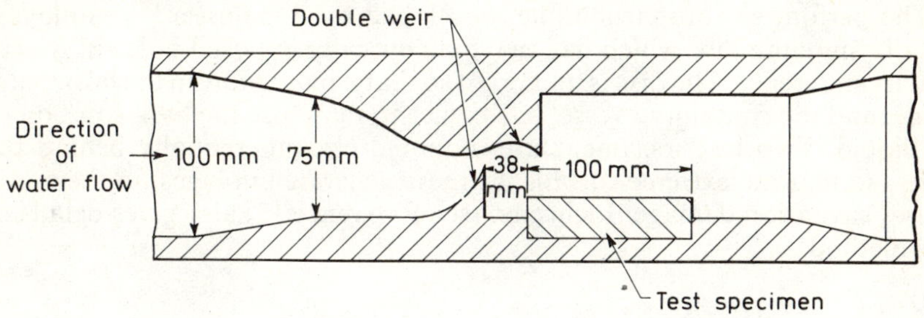

Fig. 8.52 Double-weir cavitation test apparatus (diagrammatic and not to scale)

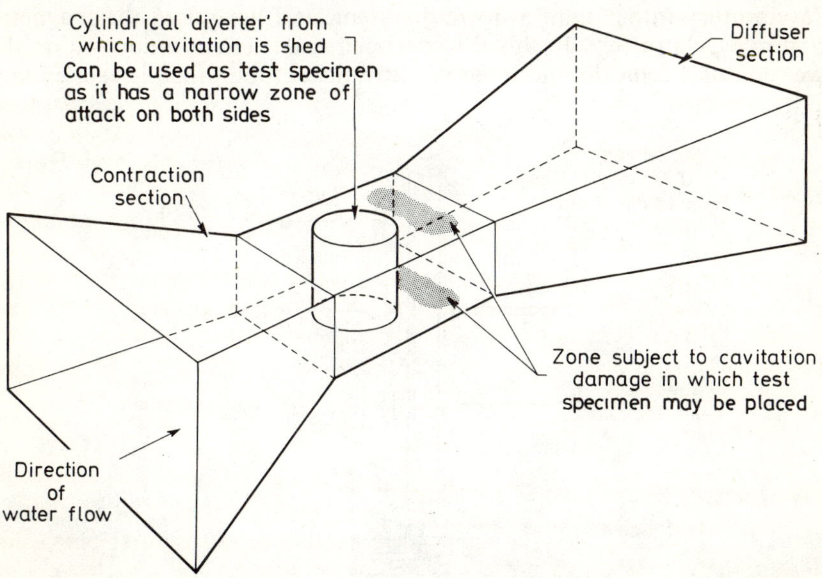

Fig. 8.53 Layout of a Venturi-type cavitation test section; circular entry and exit sections may also be used

Schröter[10] used a simple Venturi throat in a recirculatory piped system; cavitation damage occurred on the walls of the throat. Other workers, most recently Rasmussen[11] and Shal'nev[12], have used an obstructed Venturi throat to produce attack. The Venturi should preferably have double cubical arc curved contractions and expansions, and the test-piece may either form the obstruction or be sited downstream of it. The writer has used a 47 mm square section, with a 25 mm diameter diverter to produce heavy attack on a diverter test specimen, and employed this to compare the resistance of a wide variety of materials in 100 h tests in sea-water (Fig. 8.53).

The two other types of hydraulic test both produce attack on specimens carried on a rapidly rotating wheel. In the jet-impact test, described by von Schwarz and Mantel[13], two specimens are set on the circumference of the disc, which is rapidly rotated through a high-speed (67 m/s) jet of water impinging perpendicularly on the test specimens. Extreme turbulence is produced each time a specimen cuts the jet, and cavitation attack occurs.

The perforated-rotor technique, developed by Rasmussen[14], employs a rapidly spinning disc which has two diametrically opposed holes in it near the circumference. The disc is immersed so that it may rotate in the horizontal plane, and the containing vessel has baffles to minimise the bulk rotation of the liquid. Two test specimens are mounted circumferentially behind the holes, so that the extreme turbulence these generate impinges on them and causes cavitation damage. Lichtman (see Reference 17) also gives details of this method.

Vibratory Tests

The majority of laboratory investigators have produced cavitation damage in a vibratory rather than a flowing system, and have used the magnetostriction oscillator test. In this the specimen under test is immersed on the lower end of a centrally mounted vertical nickel tube which is excited into

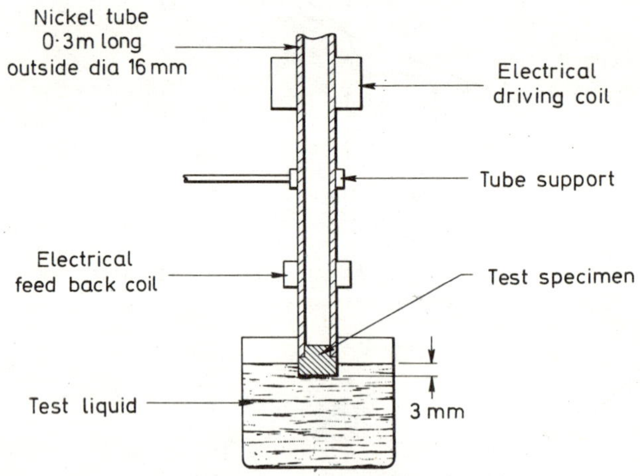

Fig. 8.54 Magnetostriction oscillator test apparatus

longitudinal vibration magnetostrictively, making use of the effect whereby a nickel rod or tube will alter in length when placed in a strong magnetic field (Fig. 8.54). The field is provided by means of a coil around the tube supplied with current from a 500 W oscillator driven at the resonant frequency of the nickel tube, which is usually 6·5 kHz. The extreme accelerations involved at an amplitude of 86·4 μm (peak to peak) tend to produce a separation of the metal/water interface on the test specimen immersed at the lower end of the nickel tube, and a cloud of minute cavitation bubbles appears. Damage is produced on the lower face of the test specimen. Details of the test procedure, which is now almost entirely standardised, are given by Robinson, Holmes and Leith[15].

Resistant Materials

Of the commonly used engineering materials, stainless steels are undoubtedly the most resistant to cavitation damage. This is due to several factors, such as their toughness, the homogeneity and fineness of their physical structure, their considerable tensile strength and hardness accompanied by adequate ductility, their high corrosion-fatigue limit, and their capability to work harden under cavitation impacts. In the generation of electricity from water power, where cavitation damage may cause expensive shut-downs for overhaul, they are widely used as welded overlays on steel or as castings.

The Fe–18Cr–8Ni steel has been most widely used as an overlay, and instances are recorded of this steel giving good service for over 20 years in applications where cast steel had previously been attacked. Sometimes cracking has developed even with this overlay, and an underlying layer of higher-chromium material such as Fe–25Cr–12Ni or Fe–25Cr–20Ni steel has been employed to counteract this tendency. Accelerated tests have suggested that the Fe–17Cr–7Ni is superior to Fe–18Cr–8Ni, and operating experience has borne this out. Stainless irons of the 13% Cr type have also been found suitable for resisting severe erosion, but their poor casting properties are a drawback, and great care is necessary in repairing them by welding. Fe–20Cr–3Ni and Fe–28Cr–8Ni materials have been used as they have properties similar to those of the 13% Cr material but present fewer welding problems.

Aluminium bronze has been highly recommended by Rheingans[16], but some hydroelectric operators[17] appear to prefer stainless steel. Complex high-strength aluminium bronzes with iron, nickel or manganese additions are preferable as they are tougher than the Cu–10Al type of material. Al–Ni–Fe bronzes are now used for marine propellers operating in severe conditions.

Stellite, a Co–Cr–W–Fe–C alloy, has been used cast and overlaid in exceptionally severe cases. Cheng (*see* Ref. 17) for instance, found Fe–18Cr–8Ni steel on a water brake to fail in 30 h and Fe–25Cr–20Ni in 50 h, but a Stellite-type material lasted 300 h.

Sprayed coatings of these and other resistant materials have been tried, but their performance rarely approaches that of the solid or welded material.

Titanium is a promising engineering material in view of its high resistance to corrosion. Coatings of titanium on copper (Straumanis[18] process) have been found by the author to improve its cavitation-damage resistance.

Rubber coatings, especially Neoprene and 'rubbery' grades of polyurethane, have often proved useful, mainly in low-severity exposures. Many other plastics such as glass-fibre laminate and Perspex have poor resistance. Nylon is quite resistant, but due to poor adhesion, coatings of this material do not perform well in severe cavitation conditions.

Remedial Measures

The fundamental remedy for cavitation damage is a change in design which stops the cavitation. Pressures may be raised, abrupt changes in section removed to eliminate severe turbulence, and vibration reduced. Very often, however, it is not practicable to make design changes. While problems have been solved by the use of more resistant materials such as hard, tough metals or elastomeric polymers, this solution is not always practicable, and other palliative measures must be tried. Chemical inhibitors are of some value where there is a corrosion component in the attack. 'Soluble oils' have given the best results, and chromates and nitrites are of value. Cathodic protection is worth investigating in corrosive environments. A more fundamental solution of possible wide application is surface gas injection or generation; this would interfere with the mechanical damage process.

D. J. GODFREY

REFERENCES

1. Parsons, C. A. and Cook, S. S., *Engineering, Lond.*, **107**, 501, 515 (1919)
2. Raven, F. A., Feiler, A. M. and Jespersen, A., *David Taylor Model Basin Report*, R–81, December (1947)
3. Eisenberg, P., *David Taylor Model Basin Report*, 712, July (1950)
4. Godfrey, D. J., *Chem. and Ind. (Rev.)*, **23**, 686 (1959)
5. Petracchi, G., *Engrs' Dig.*, **10**, 314 (1949)
6. Foltyn, V., *Strojírenství*, **2**, 402, Min. of Supply Translation TIB/T.4156 (1952)
7. Wheeler, W. H., *Cavitation in Hydrodynamics*, No. 21, H.M.S.O., London, 1 (1965)
8. Kerr, S. L. and Leith, W. C., *Paper at the Milwaukee Cavitation Committee Meeting of Amer. Soc. Mech. Engrs.* (1955)
9. Mousson, J. M., *Trans. Amer. Soc. Mech. Engrs.*, **59**, 399 (1937)
10. Schröter, H., *Z. Ver. Dtsch. Ing.*, **78**, 349 (1934)
11. Rasmussen, R. E. H., *Cavitation in Hydrodynamics*, No. 20, H.M.S.O., London, 1 (1956)
12. Shal'nev, K. K., *Cavitation in Hydrodynamics*, No. 22, H.M.S.O., London, 1 (1956)
13. von Schwarz, M. and Mantel, *Z. Metallk.*, **33**, 236 (1941)
14. Rasmussen, R. E. H., *Trans. Danish Acad. Tech. Sci.*, **1**, 45 (1949)
15. Robinson, L. E., Holmes, B. A. and Leith, W. C., *Trans. Amer. Soc. Mech. Engrs.*, **80**, 103 (1958)
16. Rheingans, W. J., *Marerials in Design Engng.*, **48**, 102 (1958)
17. Rheingans, W. J. (Ed.), *Report of Amer. Soc. Mech. Engrs., 1956 Cavitation Symposium* (1956)
18. Schlechten, A. W., Straumanis, M. E. and Gill, C. B., *J. Electrochem. Soc.*, **102**, 81 (1955)

8.9 Outline of Fracture Mechanics

Most metals inevitably contain interior or surface flaws, or actual cracks, and the engineer has to develop criteria to ascertain whether, and if so at what applied stress σ_{app}, any of the cracks will propagate to fracture. Linear-elastic fracture mechanics has been used recently by various workers[1–3] to study the environment-enhanced crack growth, such as stress-corrosion cracking and hydrogen embrittlement. This approach is essentially an extension of Griffith's theory of brittle fracture to elastic–plastic situations.

According to Griffith, a crack will propagate when the decrease in elastic energy is equal to the energy required to create the new crack surface. The elastic strain energy U_E per unit of plate thickness is given by

$$U_E = -\frac{\pi c^2 \sigma_{app}^2}{E} \qquad \ldots(8.8)$$

in which E is the elastic modulus and $2c$ is the width of the crack. Provided the radius of curvature of the crack tip is less than about 4 atomic spacings the stress concentration in this vicinity will exceed the theoretical strength of the material, and crack propagation is then possible.

The surface energy due to the presence of the crack is given by

$$U_S = 4c\gamma \qquad \ldots(8.9)$$

where γ, which is a constant for a given material, is the energy required to produce a unit area of new surface. It should be noted that the strain energy term U_E in equation 8.8 is negative because the presence of the crack causes a relief of strain energy. Thus the total change in potential energy due to the creation of a crack is

$$\Delta U = U_E + U_S.$$

According to Griffith's criterion, the crack will propagate under a constant applied stress σ_{app}, if an incremental increase in crack length produces no change in the total energy of the system, or decreases it, i.e.

$$\frac{d}{dc}\left(\frac{4c\gamma - \pi c^2 \sigma_{app}^2}{E}\right) \leqslant \frac{d\Delta U}{dc} \leqslant 0$$

or

$$4\gamma - \frac{2\pi c \sigma_{app}^2}{E} \leqslant 0 \qquad \ldots(8.10)$$

or
$$\sigma_{app} = (2E\gamma/\pi c)^{\frac{1}{2}} \qquad ...(8.11)$$

From equation 8.10 it is apparent that the factor $2\pi c\sigma_{app}^2/E$ increases with increasing crack length $2c$, whereas 4γ remains constant. Thus, as the crack grows, the L.H.S. of equation 8.10 becomes even more negative and the crack can accelerate and run catastrophically. This is termed *unstable fracture*.

These arguments apply to purely elastic solids, whereas metals are not purely elastic and undergo plastic deformation. A plastic zone is therefore considered to be present around the crack tip, which increases the amount of work necessary for crack extension G and also effectively changes the length of the crack; the crack-extension force G is also called the strain energy release rate. The term γ of equation 8.11 must now be replaced by $(\gamma+\gamma_p)$ where γ_p is a plastic work term, and it is a general practice to replace $2(\gamma+\gamma_p)$ by G, so that equation 8.11 becomes

$$\sigma_{app} = (EG/\pi c)^{\frac{1}{2}} \qquad ...(8.12)$$

It has been shown that the stress field can be completely evaluated if two factors are known: (*a*) the stress parallel to the fracture surface and (*b*) the effective stress intensity K close to the root of the given crack, which is the important parameter in the study of crack growth. This is given for the purely elastic case (where σ_{app} is well below the yield stress σ_y) by

$$K = \sigma_{app}\sqrt{(\pi c)} \qquad ...(8.13)$$

which determines the extent to which the applied stress field is perturbed by the presence of the crack. To account for the plastic deformation, the modified equation will be

$$K = \sigma_{app}\sqrt{\pi c} \times f(\text{specimen size, crack size, } \sigma_y) \qquad ...(8.14)$$

where f is a factor that is dependent on the terms in brackets.

A variety of stress conditions may be encountered, the most severe of which is 'plane strain' where all strain displacements occur in two directions which are normal to the plane of the crack; the stress intensity factor is then written as K_I. When K_I reaches some critical value (or fracture toughness) K_{IC}, the crack will propagate. Thus by combining equations 8.11 and 8.13 the condition for brittle fracture is given by

$$K_{IC}^2/E = 2(\gamma + \text{'plastic work term'}) \qquad ...(8.15)$$

For any mode of fracture, the critical stress intensity can be denoted as K_C and is related to the critical value of crack-extension force G_C by

$$K_C^2 \approx EG_C \qquad ...(8.16)$$

which means that fracture toughness can be expressed either in terms of an energy per unit area G or by a stress intensity K.

During fracture the plastic deformation produces shear lips. These increase the plastic work term of the fracture equation (8.15), but the magnitude of the effect depends upon the percentage of the fracture surface that it occupies and thus depends upon the plate thickness.

In Fig. 8.55, as the thickness increases, the percentage of the fracture surface having 'shear lip'* decreases and KI_C becomes equal to K_C when the shear lips are virtually absent (plane strain condition). For studies of the mechanisms of stress-corrosion cracking and for material evaluation associated with alloy development, testing under some prescribed mode, such as plane strain, would be desirable. In an attempt to use smaller specimens some workers groove the face of the specimen, and the triaxiality thereby produced helps to restrict the shear lips. Such approaches are important for tough materials with low yield strengths, because specimen sizes for plane strain fracture may be so large that the fracture loads become greater than the capacity of conventional testing machines.

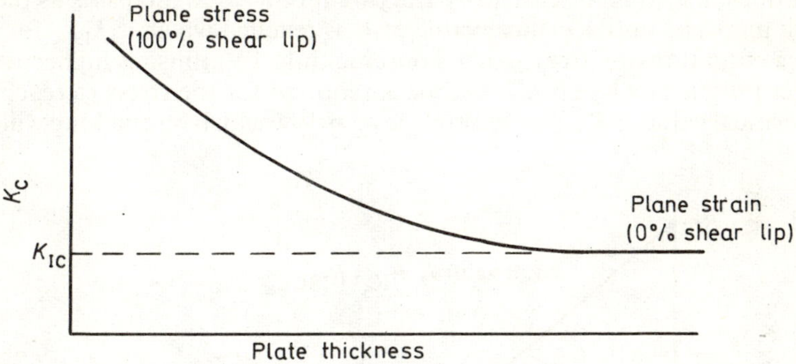

Fig. 8.55 Variation of K_C with plate thickness showing how K_{IC} is evaluated

For purely plane stress in steels, where the fracture is entirely of the shear type, equation 8.16 is applicable, but when the stress system corresponds to plane strain, and the fracture is entirely perpendicular to the specimen faces, the equation becomes

$$K_{IC}^2 = (EG_{IC})/(1-\nu^2) \qquad \ldots(8.17)$$

where ν is Poisson's ratio.

The main advantage of the fracture mechanics approach to materials assessment is that the test data obtained on laboratory specimens can be used in selection of materials for the design of specific structures. If K_{IC} is known together with σ_{app}, it is possible to calculate the flaw size necessary to cause failure, or conversely, knowing the maximum size and orientation of flaw present in a component, the stress for failure. Since the development by Brown of the fatigue-cracked cantilever-beam specimen for stress-corrosion testing, the fracture mechanics approach has been widely used to analyse the corrosion cracking behaviour of high-strength metals. However, the precise evaluation of K_{IC} is experimentally difficult wherever the plastic work term is significant, e.g. metals.

*The term 'shear lip' refers to the appearance of the fracture surface.

Stress-corrosion Considerations

Fracture mechanics deals only with the growth of cracks and does not consider the initiation phase of stress-corrosion cracking. Thus loading a notched specimen will eliminate the time consuming crack initiation stage. This is applied in Brown's test[4] which is now in frequent use. This employs a notched cantilever specimen containing a preformed fatigue crack at the base of the notch. The initial stress intensity is governed by the initial fatigue crack geometry and may be calculated using a standard formula: the deeper the crack and the higher σ_{app}, the higher will be the the initial stress intensity. When the stressed specimen is subjected to the corrosive environment the stress-corrosion crack grows relatively slowly, usually at a rate of 1 mm/h. As it does, the stress intensity (by fracture mode I)* K_I increases as the crack depth increases until a critical value of K_I is reached which is K_{ISCC} for failure under conditions of stress-corrosion cracking. By utilising higher loads or longer fatigue crack depth, less time is required for the stress to reach K_{ISCC}. The critical value of K_{ISCC} is the stress level below which no cracking will occur.

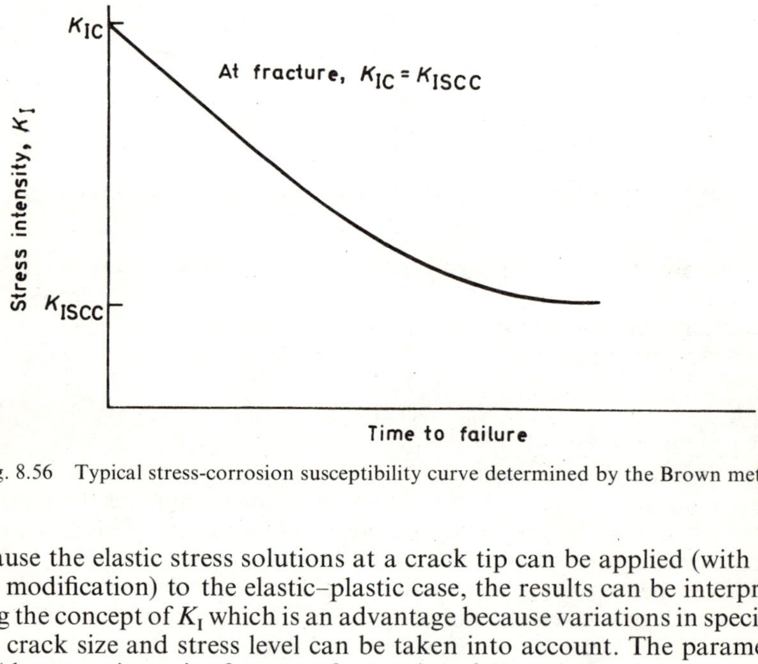

Fig. 8.56 Typical stress-corrosion susceptibility curve determined by the Brown method

Because the elastic stress solutions at a crack tip can be applied (with suitable modification) to the elastic–plastic case, the results can be interpreted using the concept of K_I which is an advantage because variations in specimen size, crack size and stress level can be taken into account. The parameters K_{IC} (the stress-intensity factor at fracture) and K_{ISCC} (stress-intensity factor at the onset of stress-corrosion cracking extension) can be considered as material properties for a given temperature, strain rate and environment, and K_{ISCC} is usually one-third of the plane-strain fracture toughness K_{IC}. But it could be higher than the critical stress intensity for failure under non-corrosion conditions with a fatigue crack K_{IC}. This can arise from the nature

*Mode I is a tensile mode of fracture; mode II is a shear mode.

of the stress-corrosion crack which tends to deviate from a plane normal to the applied stress so that branching of the crack occurs; this in turn effectively blunts the crack tip, which necessitates a higher stress intensity for failure.

In Fig. 8.56 the slope of the initial part of the curve depends upon the crack geometry, whereas K_{ISCC} is characteristic of both the alloy plus environment and is independent of the nature of the structural flaw.

Hydrogen Embrittlement

From equation 8.15 it is obvious that K_{IC} for brittle fracture depends upon the surface energy γ of the fracture surfaces. If adsorption at these surfaces lowers their energy, then K_{IC} is correspondingly reduced to K_{ISCC} and fracture becomes that much easier. High-strength materials are liable to fail when stressed in the presence of corrosive environments that liberate cathodic hydrogen. In the spirit of the Griffith analysis of fracture it has been suggested that hydrogen lowers the fracture stress by adsorption at interior surfaces. Adsorption is presumed to lower the surface energy and therefore the fracture stress. Thus the applied stress σ_{app} necessary to cause crack growth is given as

$$\sigma_{app} = [2E\gamma_A(H)/\pi c] \qquad \text{...(8.18)}$$

and for microcrack growth by plastic deformation the stress required[5] is

$$\sigma_G = [2\gamma_m(H)]/nb \qquad \text{...(8.19)}$$

where $\gamma_A(H)$ and $\gamma_m(H)$ signify a lowering of the work expended in crack propagation and microcrack growth in the presence of hydrogen, n is the number of dislocations in the pile-up having Burgers vector b.

In the pressure theory of embrittlement, the brittleness is associated with the development of high hydrogen pressures in internal voids. Atomic hydrogen in supersaturated solid solution is presumed to precipitate into voids or fissures developed during deformation, recombine to the molecular form, and build up a very substantial internal pressure P. Thus below the yield strength σ_y, unstable fracture occurs from the tip of an arrested crack when

$$\sigma_{app} = P = (2E\gamma_A/\pi c)^{\frac{1}{2}} \qquad \text{...(8.20)}$$

where γ_A is the work done in initiating unstable fracture at the crack tip. Alternatively, when hydrogen is present inside a microcrack formed by dislocation pile-ups, the pressure P reduces the stress σ_G required for microcrack growth and

$$\sigma_G = (2\gamma_m/nb) - P \qquad \text{...(8.21)}$$

where γ_m is the work done in microcrack propagation, and is proportioned to the true surface energy γ. The principles of fracture mechanics have been used by Tetelman to determine the macroscopic criteria for the unstable propagation of a crack. He presents a modified form of pressure theory to account for the slow, discontinuous propagation of hydrogen-induced cracks. Stavros and Paxton[6] also use the concept of fracture mechanics to describe mathematically the effects of geometry and stress at the crack tip.

Solution of atomic hydrogen in the plastic zone at the tip of any pre-existent flaw can raise the energy of the steel quite significantly, thereby lowering the 'plastic work term' and increasing the deleterious effect of any applied tensile stress. Surface flaws, such as pits, are sufficient to trigger the process provided that the elastic energy that can be stored in the steel lattice before yielding blunts the crack tip is sufficiently high; hence steels having a high yield strength generally possess low values of K_{ISCC}.

B. CHATTERJEE

Acknowledgements

Acknowledgements are made to Dr. E. Gillam and Dr. C. May for helpful comments.

REFERENCES

1. Tetelman, A. S., *Proc. Intern. Conf. Fundamental Aspects of Stress Corrosion Cracking* (ed. R. W. Staehle), NACE, Houston, 178 (1969)
2. Novak, S. R. and Rolfe, S. T., *Corrosion*, **26** No. 4, 121 (1970)
3. Wei, R. P., Ref. 1 (1969)
4. Brown, B. F., *Mat. Res. Stand.*, **6** No. 3, 129 (1966)
5. Stroh, A. N., *Phil. Mag.*, **46**, 968 (1955)
6. Stavros, A. J. and Paxton, H. W., *Met. Trans.*, **1**, 3049 (1970)

BIBLIOGRAPHY

Fracture Toughness, Iron and Steel Inst. Publication No. 121, London (1968)

Tetelman, A. S. and McEvily, A. J., *Fracture of Structural Materials*, John Wiley, New York (1967)

Beck, W., Jankowsky, E. J. and Fischer, P., *Hydrogen Stress Cracking of High Strength Steels*, Naval Air Development Centre, Warminster, Pennsylvania, USA (1971)

Guy, A. G., *Introduction to Materials Science*, International Student Edition, McGraw Hill, New York, 465–77 (1972)

Logan, H. L., *The Stress Corrosion of Metals*, John Wiley, New York (1966)

Barnaby, J. T., *An Introduction to Fracture Toughness Welding and Metal Fabrication*, **37** No. 2 (1969)

Wells, A. A., 'Notched Bar Tests, Fracture Mechanics and the Brittle Strength of Welded Structures', *British Welding Journal*, **12** No. 1 (1965)

Judy, R. W. and Goode, R. J., *Stress-Corrosion Cracking of High Strength Steels and Titanium Alloys*, NRL Report 7371 (1971)

ASTM E399–70T *Tentative Method for Plate-strain Fracture Toughness of Metallic Materials*

8.10 Stress-corrosion Test Methods

Stress-corrosion cracking results from the interactions, in a critically interdependent manner, between an environment, a metal and the response of the latter to the application of an appropriate stress. Recognition of these conjoint requirements has frequently led to the use of tests that attempt to simulate a practical situation, especially with regard to the structure and composition of the material but less frequently in the manner in which the stress is generated in the testpiece and in achieving representative environmental conditions.

Stressing Systems

Many different methods[1a–3] have been used in stressing testpieces, from which it may be reasonably assumed that there is no single method that is markedly superior to all others. Each method may have its peculiar advantages in a given situation, but, ideally, a test method should not be so severe that it leads to the condemnation of a material that would prove adequate for service, or so trifling as to permit the use of materials in circumstances where rapid failure ensues.

Methods of stressing testpieces, whether initially plain, notched or precracked, can be conveniently grouped according to whether they involve

1. A constant total strain or deflection.
2. A constant load.
3. A constant strain or deflection rate.

Constant deflection tests usually have the attraction of employing simple, and therefore frequently cheap, specimens and straining frames, and of simulating the fabrication stresses that are most frequently associated with stress-corrosion failure. Constant load tests may simulate more closely failure from applied or working stresses, whilst tests involving the application of a constant deflection rate have only recently become fashionable and their relevance to service failures is less obvious.

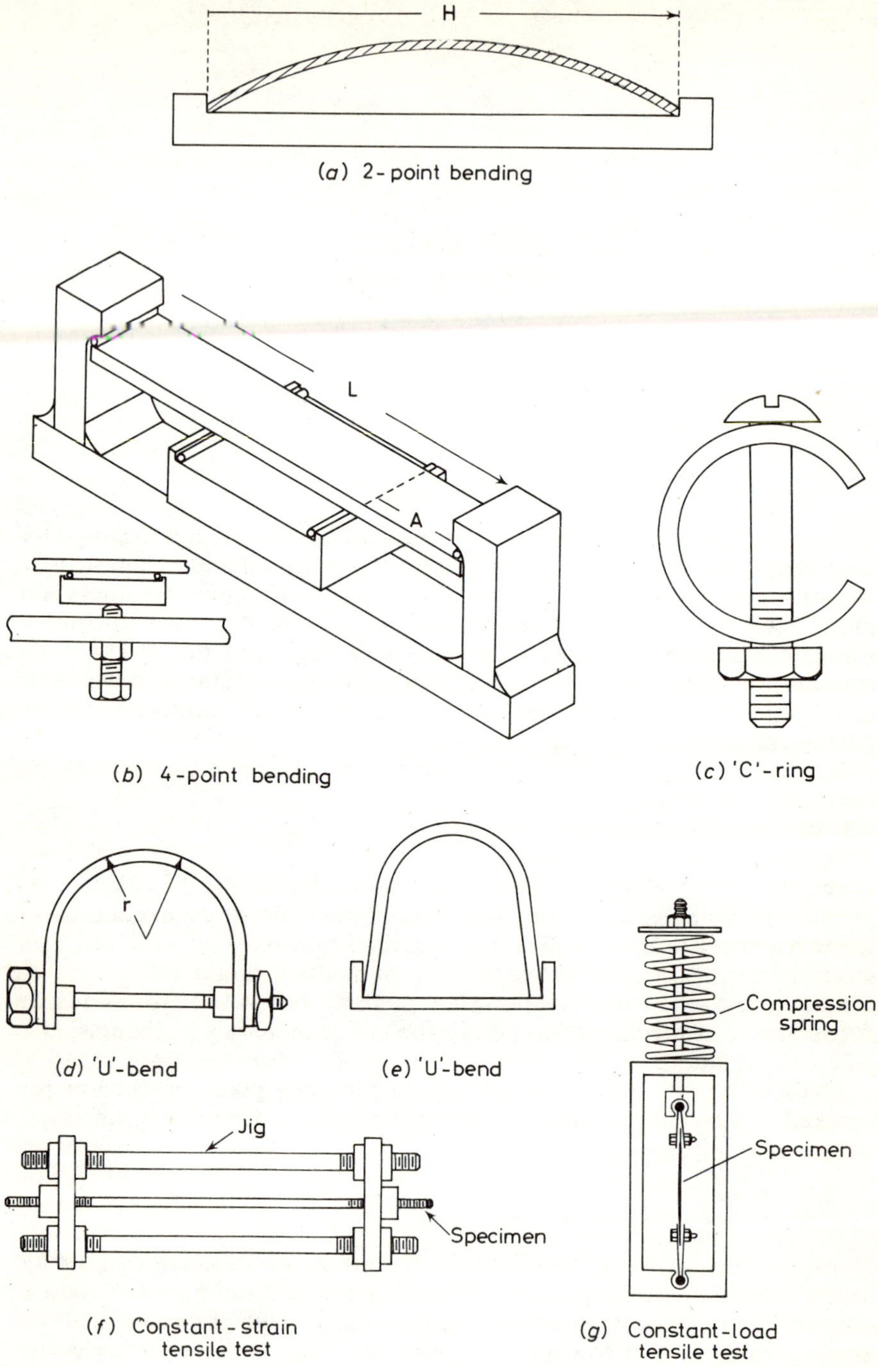

Fig. 8.57 Stressing systems for stress-corrosion test specimens; (*a*)–(*f*) constant strain, (*g*) constant load

Constant Total-deflection Tests

Prismatic beams stressed by bending offer a simple means of testing sheet or plate material, typical arrangements being shown in Fig. 8.57*a* to *e*. Below the elastic limit the stresses may be calculated[1,4] or determined from the response of strain gauges attached to the surface at an appropriate position. The methods of calculating stresses[1a, 1b] in bent beams, 'U'-bends and 'C'-rings are given in the Appendix to this section.

Plastic bending of strip specimens to produce a 'U'-bend (Fig. 8.57*d* and *e*) will usually allow the use of a lighter restraining system, although the plastic deformation, if not removed by subsequent heat treatment, may influence cracking response and the stress obtained in the outer fibres of the specimen is usually less reproduceable than with more sophisticated specimens. Tubular material may be tested in the form of 'C'- or 'O'-rings, the former being stressed by partial closing of the gap (Fig. 8.57*c*) and the latter by the forced insertion of a plug that is appropriately oversized for the bore. The circumferential stress at the outer surface of a 'C'-ring is maximal midway between the bolt holes, but for the 'O'-ring it is constant over the periphery, the stresses being readily calculated in terms of measured deflections[1a, 1b]. Constant-deflection tensile tests (Fig. 8.57*f* and *g*) are sometimes preferred to bend tests, but for similar cross sections require a more massive restraining frame. In principle this problem may be surmounted by the use of internally stressed specimens containing residual stresses as the result of inhomogeneous deformation. The latter may be introduced by plastic bending, e.g. by producing a bulge in sheet or plate material, or by welding, but such tests provide problems in systematic variation of the initial stress, which will usually be in the region of the yield stress. Moreover, elastic spring-back, in introducing residual stresses by bulging plate or partially flattening tube, may introduce problems, and where welding is involved the structural modifications may raise difficulties, unless the test is simulative of a practical situation.

At least as important as the choice of methods of stressing is the realisation of the limitations of the various methods, these having been considered in a review of stress-corrosion test methods[5]. The stiffness of the stressing frame in constant-deflection tests may influence results because of relaxation in the specimen during the initial loading stage and during subsequent crack propagation. Especially in testing ductile materials, the initial elastic strain is converted in part to plastic strain, even if the total deflection remains constant during cracking. This is because as the crack propagates the stress increases on the remaining uncracked portion of the specimen section beyond the crack, eventually reaching the effective yield stress. Yielding will then occur, accompanied by yawning of the crack and frequently with the propagation of a Lüders band that results in a sharp load drop, which is sometimes mistaken as an indication of the crack having advanced by a burst of mechanical fracture. Once load relaxation has been initiated the extent to which it proceeds can vary from specimen to specimen. Thus, Fig. 8.58 shows load relaxation curves for two specimens of the same maraging steel in the same stressing frame, which had a facility for load recording throughout the test. The specimens differed in the extent to which they showed load relaxation prior to sudden fracture, this difference being related to the number of

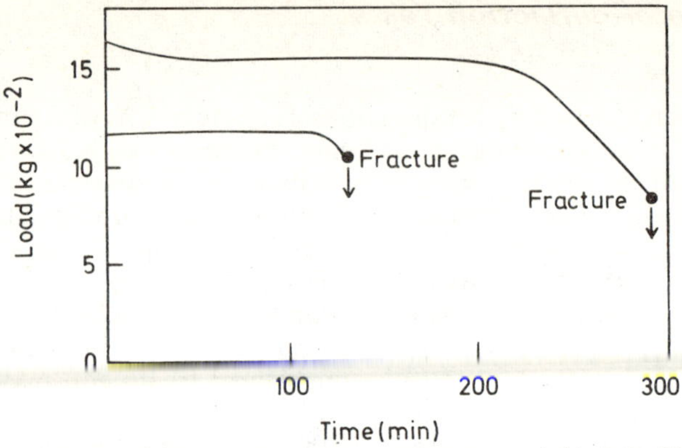

Fig. 8.58 Load relaxation curves for a maraging steel stress corrded in 0·6 M NaCl at pH2 (after Parkins and Haney[6])

cracks that developed in the specimens. Marked load relaxation was associated with the development of many cracks in the specimen, and little relaxation with only a few cracks. This can influence the time to failure, as is apparent from Fig. 8.58, where the specimen stressed at the initially higher load took longer to fail than that at the initially lower load. This is because when only a single stress-corrosion crack develops, it will not need to grow to large dimensions before sudden, final failure occurs, since the applied load remains high, whereas with the marked load relaxation associated with the multi-cracked specimen one of the cracks will need to propagate much further before it reaches the size for sudden fracture at the reduced load. Such an explanation conforms to the observations[6] that the load at fracture is related to the area of stress-corrosion cracking upon the final fracture surface and to the number of cracks initiated.

This type of result will depend upon the nature of the stress-corrosion system being studied, i.e. upon such properties as the fracture toughness of the material being studied and even upon the aggressiveness of the environment employed. It will also vary according to the stiffness of the restraining jig employed, since the stiffer the frame the less the elastic strain that is likely

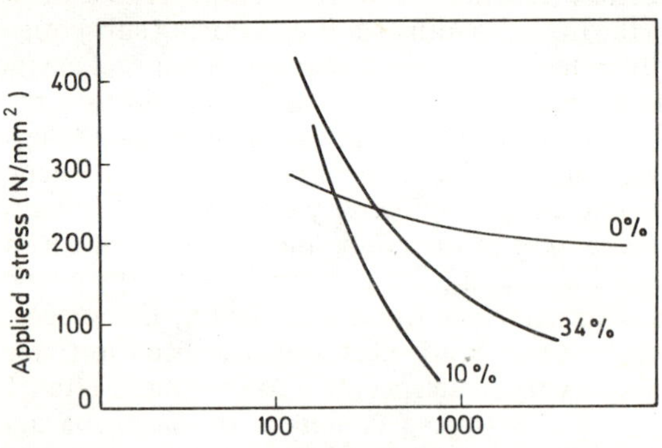

Fig. 8.59 Effects of different amounts of prior cold work (0, 10 and 34%) on the stress corrosion of a 0·07% C steel in boiling 4 M NH_4NO_3

Table 8.2 Relative susceptibilities to cracking of a mild steel in boiling 4 N NH_4NO_3 after various amounts of cold work

Initial stress	*Susceptibility of different cold worked conditions*		
	most (%)	*intermediate* (%)	*least* (%)
280 N/mm^2	0	10	34
155 N/mm^2	10	34	0
100% Y.S.	34	10	0
30% Y.S.	10	34	0

to remain in the specimen after the propagation of a Lüders band, so that a stress-corrosion crack may cease to propagate in some circumstances, especially if the initial stress is in the vicinity of the threshold stress. This indicates some of the dangers inherent in comparing stress-corrosion resistances in terms of times to failure at a given initial stress, an approach that is often practised but which can be misleading. Figure 8.59 shows the results from some tests in which the time to failure of specimens prior cold worked in varying amounts is plotted against initial stress. Comparison of the effects of different amounts of cold work by tests at an initial stress of 280 N/mm^2 or 155 N/mm^2 gives different orders of susceptibility, as shown in Table 8.2. It could be argued that neither of these results is correct because the prior cold work would result in different yield strengths being developed in the three different conditions and that the results should be rationalised by making the comparison as a function of the respective yield strengths. Here again, however, the order of susceptibility varies according to the rationalised stress at which the comparison is made, as the results in Table 8.2 show. It is difficult to escape the conclusion that a more satisfactory basis of comparison is the threshold stress, but even the latter may not be a basis for comparison of results obtained using different restraining frames.

The simplicity of the rigs used in the constant-strain tests is an advantage in the application of the corrosive solution. Thus in the case of two-point bending (Fig. 8.57*a*) several specimens may be strained in the same rig which can be constructed of plastic and immersed in a tank containing the corrodent.

Constant-load Tests

Dead-weight loading (with or without the assistance of levers to reduce the load requirements) of tensile specimens has the advantage of avoiding some of the difficulties already discussed, not the least in allowing accurate determination of the stress if the specimen is uniaxially loaded. The relatively massive machinery usually required for such tests upon specimens of appreciable cross section is sometimes circumvented by the use of a compression spring (Fig. 8.57*g*), chosen with characteristics that ensure it does not change significantly in length during testing and thereby approximating to a constant-load application. For immersion tests the frame may be coated in P.V.C. and the specimen insulated from the shackles by plastic sleeves and washers to avoid bimetallic effects; alternatively, the specimen may be

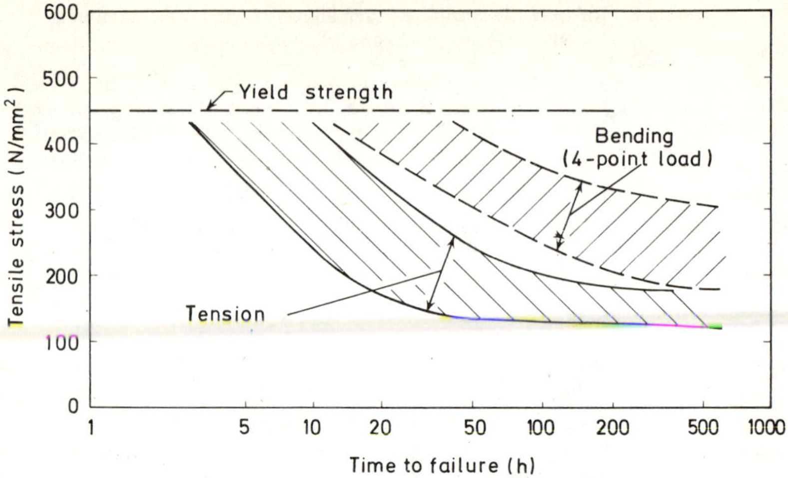

Fig. 8.60 Comparison of test results from bend and tension tests upon Al–Zn–Mg alloy in 3% NaCl plus 0·1% H_2O_2 (after Brenner and Gruhl[7])

enclosed in a glass cell containing the corrodent solution. The alternative approach of minimising the size of the loading system by reducing the cross section of the specimen to the dimensions of a wire is dangerous unless failure by stress-corrosion cracking is confirmed by, say, metallography. This is because failure may result from pitting and an attendant increase in the effective stress to the U.T.S. in some stress-corrosion environments. Indeed there is evidence for some systems that before stress-corrosion cracking proper can begin, a pit must form wherein certain chemical or electrochemical conditions are established that permit cracks to be initiated, and in such systems the use of fine wires has obvious pitfalls.

The load relaxation that accompanies some if not all constant-deflection tests is replaced in constant-load tests by an increasing stress condition, since the effective cross section of the testpiece is reduced by crack propagation. This suggests that it will be less likely that cracks will cease to propagate once initiated, as may happen with constant-deflection tests at initial stresses in the region of the threshold stress, and therefore that threshold stresses are likely to be lower when determined under constant-load conditions than under conditions of constant deflection. Some results due to Brenner and Gruhl[7] for an aluminium alloy (Fig. 8.60) confirm this expectation. These results also show shorter times to failure for the same initial stress with constant load testing and, as already indicated for constant-deflection tests, raise queries as to the significance of time to failure, the parameter so frequently used in assessing cracking susceptibility.

Constant Deflection Rate Tests

Whilst this method of testing has been in use in some laboratories for a decade or so, and has increased in use considerably in very recent years, there remains a considerable amount of scepticism and unfamiliarity with the method. In essence it involves the application of a relatively slow strain or deflection rate (approx. $10^{-6}\,s^{-1}$) to a specimen subjected to appropriate electrochemical conditions. It should be emphasised that the strain rates employed are very much slower than those involved in straining electrode experiments where the object, the measurements of current transients, is totally different. In constant-strain-rate stress-corrosion tests the object is to produce stress-corrosion cracks that, metallographically, are indistinguishable from those produced in constant-load or constant-deflection experiments. The object in all these laboratory tests is normally to obtain data in a relatively short period of time and this is frequently achieved by adopting an approach that increases the severity of the test. In stress-corrosion testing this usually takes the form of increasing the aggressiveness of the environment by changing its composition, temperature or pressure; or by stimulating the corrosion reactions (galvanostatic or potentiostatic polarisation); or by increasing the susceptibility of the alloy through changes in structure; or by increasing the severity of the stress by the introduction of a notch or pre-crack. The application of dynamic straining to a stress-corrosion test specimen comes into this category also, and, like all of the other accelerating approaches, its justification will vary according to the circumstances in which it is used.

Stress-corrosion crack velocities usually fall in the range of from 10^{-3} to $10^{-6}\,mm\,s^{-1}$, which implies that failures in laboratory test specimens of usual dimensions occur in not more than a few days. This is found to be so if the system is one in which stress-corrosion cracks are readily initiated, but it is common experience to find that testpieces do not fail in very extended periods of testing, which are then terminated at some arbitrarily selected time. The consequences are that considerable scatter may be associated with replicate tests and the arbitrary termination of the test leaves an element of doubt concerning what the outcome would have been if it had been allowed to continue to a longer time. Just as the use of pre-cracked specimens assists in stress-corrosion crack initiation, so apparently does the application of slow dynamic strain, which has the further advantage that the test is not terminated after some arbitrary time, since the conclusion is always achieved by the specimen fracturing and the criterion of cracking susceptibility is then related to the mode of fracture. Thus, in the form in which it is normally employed, the slow-strain-rate method will normally result in failure in not more than about two days, either by ductile fracture or by stress-corrosion cracking, according to the susceptibility towards the latter, and metallographic or other parameters may then be assigned in assessing the cracking response. The fact that the test concludes in this positive manner in a relatively short period of time constitutes one of its main attractions.

Early use of the test was in providing data whereby the effects of such variables as alloy composition and structure or inhibitive additions to cracking environments could be compared, and also for promoting stress-corrosion cracking in combinations of alloy and environment that could not be caused to fail in the laboratory under conditions of constant load or constant strain. Thus, they constitute a relatively severe type of test in the sense that they frequently promote stress-corrosion failure in the laboratory where other modes of stressing plain specimens do not promote cracking, and in this sense they are in a similar category to tests on pre-cracked specimens. In recent years an understanding of the implications of dynamic strain testing has developed and it now appears that this type of test may have much more relevance and significance than just that of an effective, rapid, sorting test. It may, at first sight, be argued that laboratory tests involving the pulling of specimens to failure at a slow strain rate show little relation to the reality of service failures. In point of fact in constant-strain and constant-load tests crack propagation also occurs under conditions of slow dynamic strain to a greater or lesser extent depending upon the initial value of stress, the point

Fig. 8.61 Constant deflection rate rig for tensile tests

in time during the test at which a stress-corrosion crack is initiated and various metallurgical parameters that govern creep in the specimen. Moreover, there is an increasing amount of evidence for some systems which suggests that the function of stress in stress-corrosion cracking is to promote a strain rate which, rather than stress *per se*, is the parameter that really governs crack initiation or propagation. In these cases the minimum creep rate for cracking is as much an engineering design parameter as is the threshold stress or stress intensity obtained from constant-load tests on plain or pre-cracked specimens.

The equipment required for constant-deflection-rate testing is simply a device that permits a selection of deflection rates whilst being powerful enough to cope with the loads generated. Conventional testing machines may be used, but a form of apparatus that has proved useful and appreciably cheaper is shown in Fig. 8.61 and consists of a moderately stiff frame and a drive mechanism through a series of reduction gears that allows a selection of crosshead speeds in the range 10^{-3}–10^{-7} mm s^{-1}. Plain or pre-cracked specimens in tension may be used, but if the cross section of these needs to be large or the loads high for any reason, cantilever bend specimens with the beam deflected at appropriate rates may be used, as shown in Fig. 8.62.

Fig. 8.62 Constant deflection rate rigs for cantilever beam tests

It is important to appreciate that the same deflection rate does not produce the same response in all systems and that the rate has to be chosen in relation to the particular system studied (*see* Section 8.1). The representation of the results from constant-deflection-rate tests may be through the usual ductility parameters such as reduction in area, the maximum load achieved, the crack velocity or even the time to failure, although as with all tests, metallographic or fractographic examination, whilst not readily quantifiable, should also be involved. Since stress-corrosion failures are usually associated with relatively little plastic deformation, the ductility of specimens will be variable according to the extent to which stress-corrosion contributes to the fracture process. This will also influence the shape of the load–extension curve that may be obtained by continuous monitoring of the response of a load cell incorporated in the system, Fig. 8.63 showing the forms of curves obtained with and without attendant stress corrosion. It will be apparent from these curves that not only is the extension to fracture dependent upon the presence or otherwise of stress-corrosion cracks, but so also is the maximum load achieved. The latter may be used for expressing cracking susceptibility in some systems, as also may the area bounded by the load–extension curve. However, the variations in maximum load achieved in constant-deflection-rate tests in circumstances of varying cracking severity are not always large enough for significant distinctions to be made. Even measurements of ductility, such as reduction in area, are not invariably readily made, if only because the final fracture of the specimen does not always follow a simple path and the fitting of the two broken pieces together is not easy. Probably the easiest quantity to measure with reasonable accuracy is the time to failure, which has as much significance in a constant-deflection-rate test as it does in constant-load or constant-deflection tests. Indeed the time to failure in constant-deflection-rate tests is simply related to ductility parameters, a not very surprising result when it is remembered that the lesser the intensity of stress-corrosion cracking the greater will be the ductility to fracture and, therefore, the greater the time to failure for a constant-deflection rate.

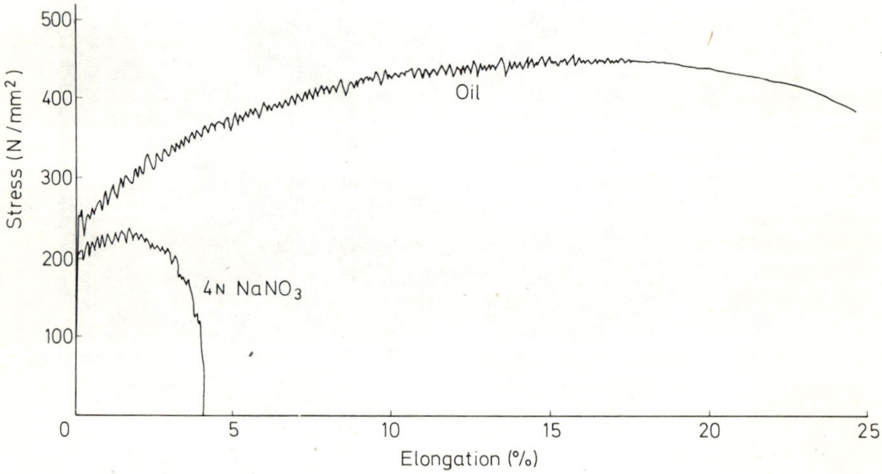

Fig. 8.63 Nominal stress–extension curves for mild steel in oil, giving ductile failure, and in 4 N $NaNO_3$ producing stress-corrosion failure, at the same test temperature (104°C)

Clearly for constant-deflection-rate testing to have credence it should give results that are comparable with those obtained by other methods. Figure 8.64 shows some results for tests upon low-alloy ferritic steels in boiling 4 N NH_4NO_3, the various alloying elements producing a range of cracking susceptibilities as measured by the threshold stresses obtained from constant

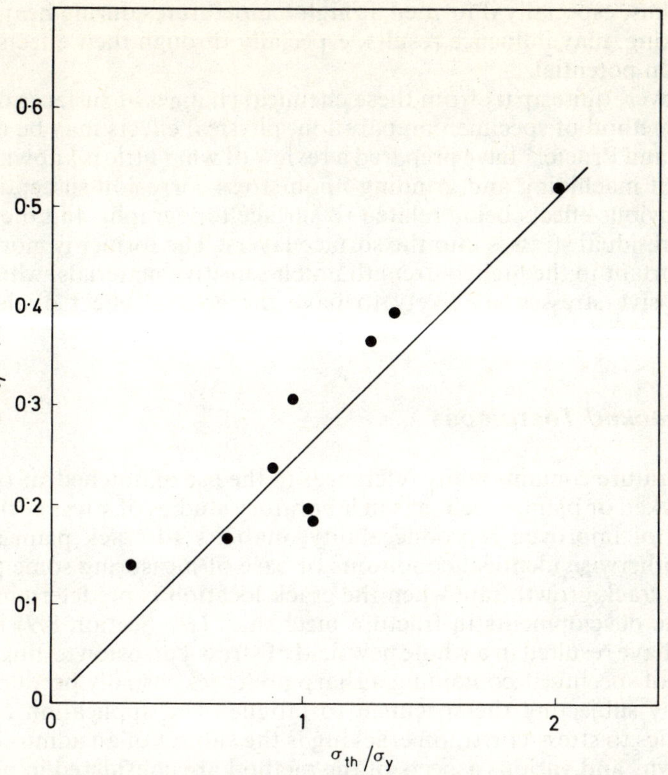

Fig. 8.64 Time to failure ratios from constant deflection rate tests and normalised threshold stresses σ_{th}/σ_y obtained from constant strain tests for a series of low-alloy ferritic steels in boiling 4 M NH_4NO_3

strain tests. These results have been normalised by dividing the threshold stress σ_{th} by the lower yield strength σ_y for each steel, whilst the dynamic strain test results have been normalised by dividing the time to failure in the 4 N NH_4NO_3 by the time to failure in oil at the same temperature, so that increasing departure from unity indicates increasing cracking susceptibility. The general trend of the results in Fig. 8.64 is clear in indicating reasonable agreement between the two types of test in placing the steels in essentially the same order of merit.

Effects of Surface Finish

It is hardly surprising that the preparation of surfaces of plain specimens for stress-corrosion tests can sometimes exert a marked influence upon results. Heat treatments carried out on specimens after their preparation is otherwise completed can produce barely perceptible changes in surface composition, e.g. decarburisation of steels or dezincification of brasses, that promote quite dramatic changes in stress-corrosion resistance. Similarly, oxide films, especially if formed at high temperatures during heat treatment or working, may influence results, especially through their effects upon the corrosion potential.

However, quite apart from these chemical changes at surfaces occasioned by the method of specimen preparation, physical effects may be important. Paxton and Procter[8] have prepared a review of what little is known about the effects of machining and grinding upon stress-corrosion susceptibility, the most obvious effects being related to surface topography and the introduction of residual stresses into the surface layers. The former is more likely to be important in the higher-strength notch-sensitive materials, whilst surface compressive stresses are likely to have the general effect of delaying or preventing failure.

Pre-cracked Testpieces

The literature contains many references to the use of notched, as opposed to pre-cracked or plain, specimens in laboratory studies of stress corrosion, for reasons of improved reproduceability, inability to crack plain specimens under otherwise identical conditions or ease of measuring some parameter such as crack growth rate when the crack location is predetermined. However, the developments in fracture mechanics (*see* Section 8.9) in the last decade have resulted in a whole new field of stress-corrosion testing involving the use of specimens containing a sharp pre-crack, usually produced from a notch by subjecting the specimen to fatigue. The application of fracture mechanics to stress-corrosion cracking is the subject of an admirable review by Brown[9] and various aspects of the method are considered in papers presented at an AGARD conference[10].

The problems associated with the choice of plain specimens for assessing stress-corrosion resistance may, at first sight, appear equally large in relation to pre-cracked specimens in the sense that in the relatively short time during which such tests have been in use a large number of specimen types have been used (Fig. 8.65). However the differing specimen geometries are rationalised through the stress-intensity factor, with the result that data from different testpieces are comparable, providing appropriate precautions are taken in specimen preparation. The biggest single difficulty is in relation to the large size of specimen that is necessary for highly ductile materials if the concepts of linear elastic analysis are to be applicable. Since it is probable that most service stress-corrosion failures occur in highly ductile materials in relatively thin sections it is clear that there are problems here, but the use of pre-cracked specimens that dimensionally do not conform to the requirements for linear elastic analysis to be strictly valid is still worthwhile in some instances.

The latter will usually be associated with the simulation of service situations, the relative ease with which stress-corrosion cracks can be initiated at pre-cracks or the advantages that sometimes accrue from the propagation of a single crack. The claim that has sometimes been made of pre-cracked specimen tests—that they circumvent the initiation stage of cracking in plain specimens, erroneously assumed invariably to be related to the creation of a corrosion pit that provides a measure of stress concentration approaching that achieved at the outset with a pre-crack—is rarely entirely valid. Thus,

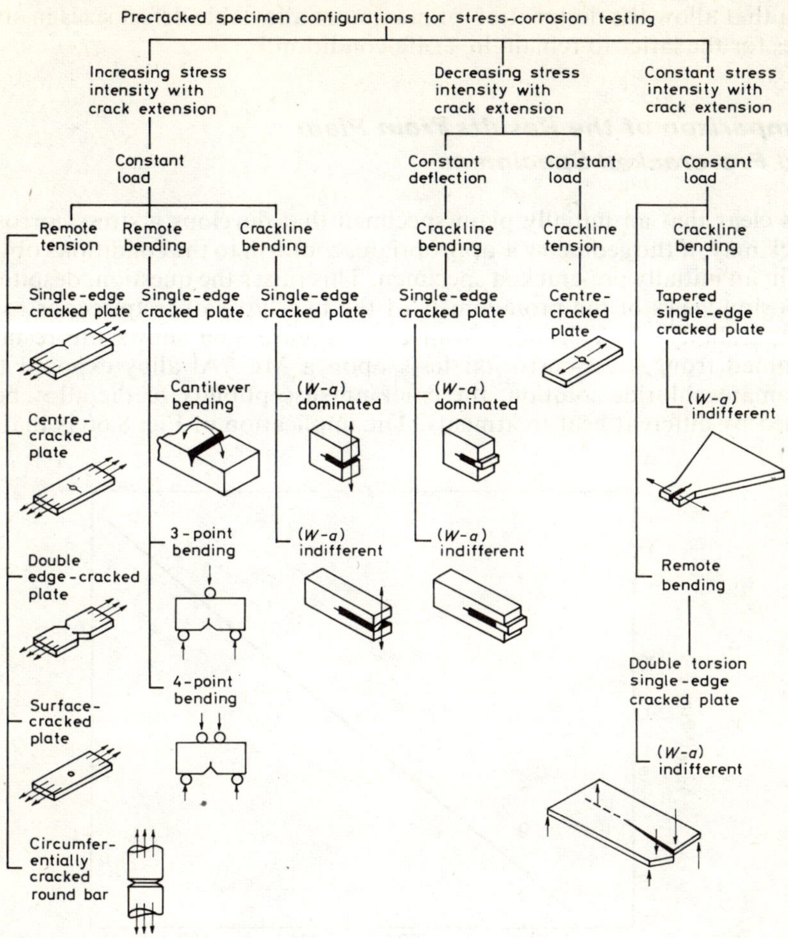

Fig. 8.65 Classification of precracked specimens for stress-corrosion testing (after Smith and Piper[11])

the geometries of a pit, notch or pre-crack are frequently as important for electrochemical reasons as they are for any reason associated with their influences upon stress distribution. This is because a geometrical discontinuity may be necessary to provide the localised electrochemical conditions, in terms of environment composition or electrode potential, that are necessary for stress-corrosion crack propagation. The objections that have sometimes been made against the use of pre-cracked specimens, e.g. to the validity of introducing a transgranular pre-crack into a specimen that suffers intergranular stress-corrosion cracking, or of the necessity for going to considerable expense to produce a very sharp crack when the introduction of a corrosive environment may blunt the crack by dissolution, miss the point that such sharp discontinuities do indeed exist in real materials. Indeed one of the major attractions of pre-cracked specimen testing is that it can provide data that allow the designation of maximum allowable defect sizes in structures for the latter to remain in a safe condition.

Comparison of the Results from Plain and Pre-cracked Specimens

It is clear that an initially plain specimen that develops a stress-corrosion crack may, if the geometry is appropriate, conform to the conditions obtaining in an initially pre-cracked specimen. This raises the question, despite the opposing views of the protagonists of the two types of testpiece, as to the comparability of the results from each. Figure 8.66 shows the results[12] obtained from stress-corrosion tests upon a Mg–7Al alloy exposed to a chromate–chloride solution, the cracking susceptibility of the alloy being varied by different heat treatments. The implication of Fig. 8.66 is that the

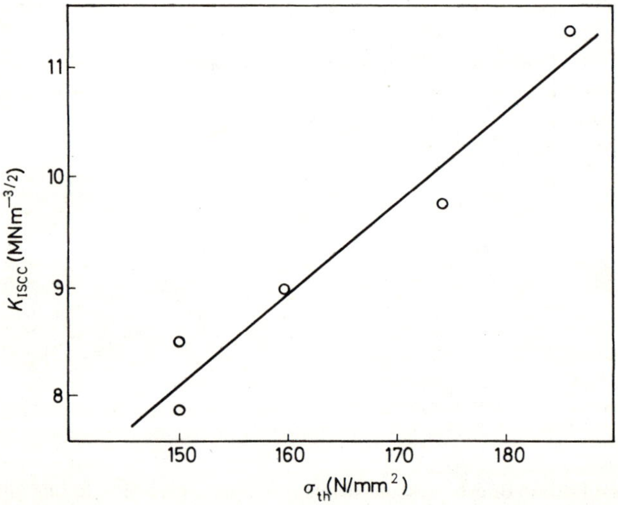

Fig. 8.66 Threshold stress intensities K_{ISCC} from precracked specimen tests, and threshold stresses σ_{th} from plain specimen tests, for a Mg–7Al alloy in various structural conditions tested in chromate–chloride solution (after Wearmouth, Dean and Parkins)

threshold stress σ_{th} determined upon initially plain specimens of small cross section is related to the threshold stress intensity K_{ISCC} obtained from pre-cracked specimens of relatively large section. Since K_{ISCC} represents the stress intensity below which an existing crack does not propagate, so it would appear that the threshold stress given by tests on plain specimens corresponds to values below which cracks do not propagate to give total failure, i.e. the threshold stress for plain specimens is not necessarily the stress below which cracks do not form. Examination of plain specimens stressed below the threshold stress revealed the presence of small stress-corrosion cracks that had ceased to propagate (*see* Section 8.1) and moreover the maximum sizes of the cracks that did not propagate to total failure were quadratically related to the threshold stress, as would be expected if the concepts of fracture mechanics were applicable to these initially plain specimens. Non-propagating cracks have also been observed at stresses below the threshold stress in other systems, such as low-alloy ferritic steels exposed to various environments, and would therefore support the suggestion that what are being measured in tests upon plain and pre-cracked specimens are not so different as has sometimes been suggested.

Crack Velocity Measurements

In mechanistic studies of stress corrosion and also in the collection of data from pre-cracked specimen tests, there is an increasing tendency for stress-corrosion crack-velocity measurements to be made. In the simplest way these can be made by microscopic measurement at the conclusion of tests, the assumption being made that the velocity is constant throughout the period of exposure, or, if the crack is visible during the test, *in situ* measurements may be made by visual observation, the difficulty then being that it is assumed that the crack visible at a surface is representative of the behaviour below the surface. Indirect measurements must frequently be resorted to and these have involved observation of the elongation of the specimen, crack opening displacement, changes in the electrical resistivity of the specimen and acoustic emissions that sometimes accompany crack extension.

Measurement of the elongation of the specimen is probably the least satisfactory of these, even though it is often the simplest in only requiring a transducer that responds to dimensional change. Multiple cracking of initially plain specimens raises problems in interpreting the data in terms of crack velocities and so the technique has been frequently used for attempting to determine the point in time when cracking was initiated, the preceding time during which the transducer shows no response being equated to an incubation period for cracking. However, such results can be completely misleading because the sensitivity of most transducers is such that they will only detect change when the specimen undergoes some plastic deformation, resulting from the propagation of a crack to the size where the remaining uncracked portion of the section beyond the crack is raised to its yield strength. Consequently, crack propagation can occur during the (so called) incubation period when the stress is insufficient to cause the propagation of a deformation band. The latter frequently occurs suddenly, producing a sharp response by the transducer, which has sometimes been interpreted as evidence of a burst of fast mechanical fracture, but which may in fact be nothing of the sort. It is much more satisfactory to use a crack-opening displacement gauge[13] located across a pre-crack. These gauges usually take the form of two thin cantilever beams to which strain gauges are attached, the beams being located at opposite sides of the extremity of the pre-crack. As crack extension occurs, and the sides of the crack undergo relative displacement, the strain gauges respond to the unbending of the beams.

Changes in the electrical resistivity of a specimen containing a propagating crack[14] depend upon applying a high constant current at each end of the specimen and measuring the potential difference across electrical leads situated at opposite sides of the crack. The potential field in the region of the crack is disturbed by the presence of the latter and as the crack extends the potential difference between the leads on opposite sides increases, providing that the total current remains constant. This requires a reliable constant-current source and the technique is dependent in some degree upon the exact positioning of the leads and gives less reproduceable results if crack branching occurs. The initial thought that the application of d.c. to the specimen may influence the electrochemistry of the stress-corrosion reactions is not sustained in practice and the technique can provide reliable data.

High-frequency stress waves are generated when stress-corrosion cracks propagate in some materials, especially the high-strength steels when these undergo hydrogen-induced cracking. The detection of these acoustic signals, which are filtered from lower amplitude background noise, affords a means of studying crack propagation[15]. Whilst the technique involves the use of sophisticated and relatively costly equipment, if it is to be correctly practised, it may also offer a means of distinguishing between active-path and hydrogen-embrittlement mechanisms of cracking[16].

Environmental Aspects of Testing

Solution Preparation

Although the list of environments reported as promoting stress-corrosion cracking in any alloy continues to grow with time, the concept of solution specificity remains in that not all environments will initiate or sustain cracking in all alloys. Whilst it is inevitable that the environment will always remain as one of the variables that may need to be assessed by stress-corrosion tests, nevertheless certain solutions, by their widespread use over many years, have tended to become standard test solutions for certain types of alloy. Thus, boiling $MgCl_2$ solutions for stainless steels, boiling nitrate solutions for carbon steels and 3·5% NaCl for aluminium alloys, to mention but a few, have been extensively used, for example, in comparing the effects of metallurgical variables upon cracking propensities. Such approaches raise two questions, the first concerned with the extent to which 'standard' solutions prepared in different laboratories may be regarded as identical, and the second with the extent to which degrees of susceptibility of a range of alloys to cracking in one environment are related to cracking in a different environment.

Whilst the relatively small differences that may be expected to occur between laboratories preparing a solution to the same specification frequently will not influence stress-corrosion test results, there are situations where relatively small changes in environment can promote marked changes in cracking response. Thus, Streicher and Casale[17] have pointed to the potential problems associated with the use of nominally 42% boiling $MgCl_2$ in testing stainless steels. Since the hydrate of $MgCl_2$ is hygroscopic, solution preparation by weighing may lead to appreciable differences in boiling point and hence times to failure in a stress-corrosion test, so that it is preferable to prepare the solution by adding water to the hydrate to achieve a particular boiling point.

Similarly, pH variations resulting from either the initial preparation or from changes during a stress-corrosion test, may exert a marked influence upon results in some systems. Thus, the cracking of carbon steels in nitrates is markedly pH sensitive and depending upon the volume of solution and the surface area of the specimen exposed, as well as upon the time involved in making the test, significant pH rises can occur and cracking can cease as a result. Moreover, if tests are carried out with anodic stimulation these effects may be aggravated, especially if the counter electrode is immersed in the test cell. In other cases, e.g. the medium and higher strength steels, the initiation and maintenance of cracking frequently requires localised pH changes within the confines of the crack, and these can only occur if the initial conditions of exposure are appropriate.

The oxygen concentration of the solution, as in many instances of corrosion, can also be critical in stress-corrosion cracking tests. Instances are available in the literature that show very markedly different test results according to the oxygen concentration in systems as widely differing as austenitic steels immersed in chloride-containing phosphate-treated boiler water[18], and aluminium alloys[19] immersed in 3% NaCl.

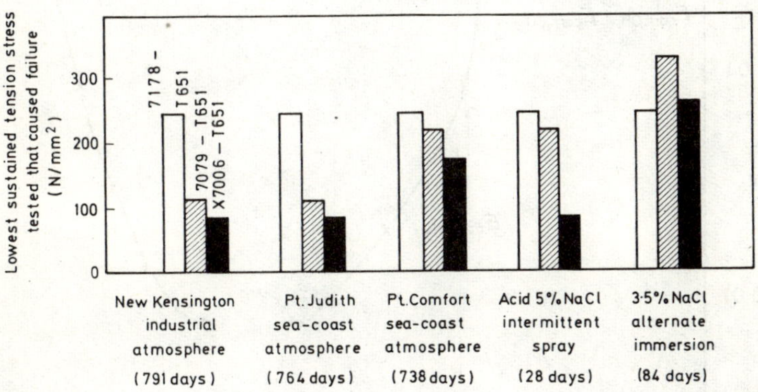

Fig. 8.67 Relative resistance to stress-corrosion cracking of three aluminium alloys subjected to different environments. The stress levels employed corresponded to 75, 50 and 25% of the respective transverse yield strengths (after Lifka and Sprowls[20])

The assumption that the relative cracking responses of a series of alloys will be the same irrespective of the environment to which they are exposed can be extremely dangerous. Many examples could be quoted of the dangers of drawing conclusions from tests in a given environment and applying these to a different situation, but some results by Lifka and Sprowls[20] will suffice. The results for the relative cracking susceptibilities of three aluminium alloys subjected to different exposure conditions are shown in Fig. 8.67, which indicates that an intermittent spray test using acidified 5% NaCl solution gives the same order of susceptibility for the three alloys as was observed in outdoor exposure tests at three different locations. On the other hand, an alternate immersion test in 3·5% NaCl, widely used for testing aluminium alloys, places the alloys in a completely different order of susceptibility. This single example will suffice to indicate the necessity for simulating service conditions as closely as possible where laboratory data are to be used for selection or design in relation to industrial equipment.

Electrochemical Aspects of Testing

This latter point is as true for the electrode potential as it is for the environment composition. The effect of potential upon cracking will vary from one system to another (*see* Section 8.1), but some aspects of this part of stress-corrosion testing may be conveniently discussed in relation to the cracking of carbon steels. These materials fail in different ranges of potential according to the environment in which they are immersed, as is apparent from the results of tests involving potentiostatic control shown in Fig. 8.68; this also shows the free corrosion potentials for this steel in the different solutions.

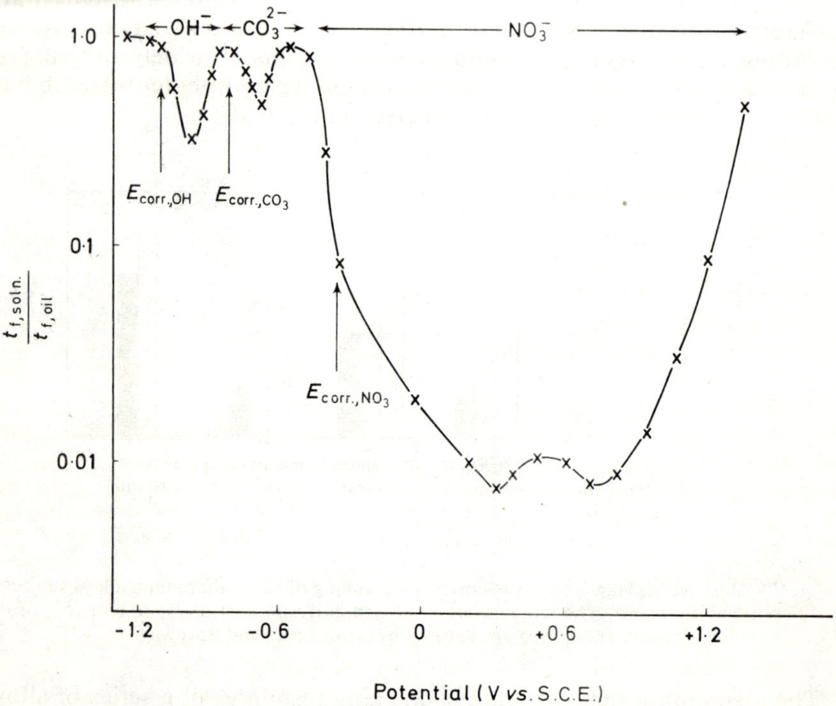

Fig. 8.68 Stress-corrosion susceptibility of low-carbon steel in constant deflection rate tests in hydroxyl, carbonate and nitrate environments at different potentials

The latter indicate that whilst failure would occur in the nitrate at the free corrosion potential, this would not be so in the hydroxide or carbonate solutions. This does not mean that carbon steels will never fail by stress corrosion in these two environments at the free corrosion potential, since the latter is, of course, dependent upon the composition of the steel, its surface condition, the composition of the environment and other factors. It is possible, therefore, that as the result of, say, small additions to the environment, added intentionally or present as impurities, the corrosion potential can be caused to lie within the cracking range, or that as the result of small additions to a steel the corrosion potential may fall outside the cracking range. Quite small changes in potential, frequently only a few tens of millivolts, can therefore produce dramatic changes in cracking response and point to the necessity, especially in laboratory tests attempting to simulate a service failure, of reproducing the potential with precision.

Stress-corrosion Test Cells

The cells that contain the specimen and environment for stress-corrosion tests frequently need to be no more than a vessel made in some substance, usually glass, that is inert to the environment and which produces no electrical response upon the test specimen. Where cracking is initiated at surfaces through which heat transfer occurs it may be necessary to design a cell in which such an effect is incorporated, since the concentration of substances in solution that may occur at an interface through which heat passes may play a significant rôle in promoting cracking, especially if surface deposits allow concentration by evaporation whilst preventing mixing with the bulk of the environment. The cracking of stainless steel piping under thermal insulation, the caustic cracking of riveted mild-steel boilers and the concentration of carbonate–bicarbonate solutions under pipeline coatings to produce cracking in high-pressure gas transmission lines, are all significant examples. Dana[21] has developed a method for simulating the conditions for cracking of stainless steels in contact with thermal insulating materials, whilst concentration in leaking boiler seams is simulated in the 'embrittlement detector' developed by Schroeder and Berk[22].

Such test cells involve, among other things, a crevice, the essence of which is that the volume of solution that it contains is relatively small compared to the area of exposed metal, a ratio that has already been mentioned as one that may influence stress-corrosion test results determined in more conventional cells where crevices do not exist. The experiments of Pugh, *et al.*[23] on the stress corrosion of 70–30 brass in ammoniacal solutions of various volumes are particularly instructive in indicating how this ratio may influence results, the time to failure varying by about an order of magnitude for a similar change in solution volume. Changes in the surface area of exposed specimens, apart from the effects already implied, may influence the cracking response for other reasons, as shown by the results of Farmery and Evans[24] for an Al–7Mg alloy immersed in a chloride solution. They found that coupling unstressed to stressed specimens of varying area ratio influenced failure times, relatively short times being obtained when the area of unstressed to stressed specimen was large.

Initiation of Stress-corrosion Tests

It may be felt that the initiation of a stress-corrosion test involves no more than bringing the environment into contact with the specimen in which a stress is generated, but the order in which these steps are carried out may influence the results obtained, as may certain other actions at the start of the test. Thus in outdoor exposure tests the time of the year at which the test is initiated can have a marked effect upon the time to failure[19], as can the orientation of the specimen, i.e. according to whether the tension surface in

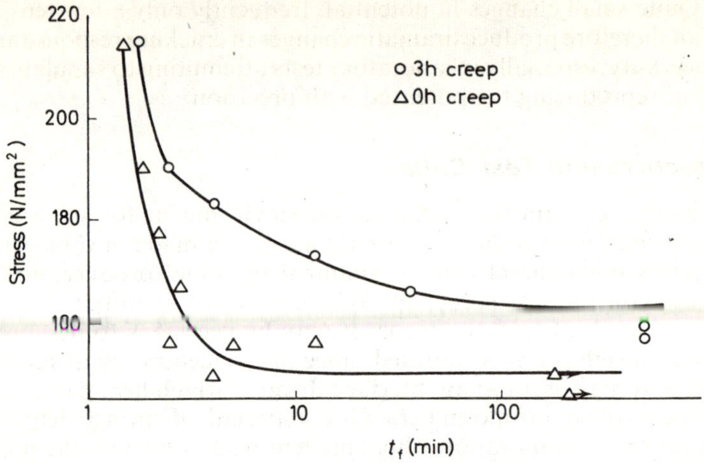

Fig. 8.69 Effect of delay period between application of load and introduction of solution to test cell in the failure of a Mg–7Al alloy exposed to a chromate–chloride solution (after Wearmouth, Dean and Parkins[12])

bend specimens is horizontal upwards or downwards, or at some other angle. But even in laboratory tests, the time at which the stress is applied in relation to the time at which the specimen is exposed to the environment may influence results. Figure 8.69 shows the effects of exposure for 3 h at the applied stress before the solution was introduced to the cell, upon the failure of a magnesium alloy immersed in a chromate–chloride solution. Clearly such prior creep extends the lifetime of specimens and raises the threshold stress very considerably and since other metals are known to be strain-rate sensitive in their cracking response, it is likely that the type of result apparent in Fig. 8.69 is more widely applicable.

Hydrogen Embrittlement and Stress Cracking of High-strength Steels

The absorption of hydrogen by mild steel has little effect on mechanical properties, but in the case of high-strength steels it results in loss of ductility, which in turn can result in cracking and fracture when the metal is subjected to a sustained tensile stress (*see* Section 8.4). Hydrogen may be introduced into these alloys from the gas phase (during manufacture or welding), or from aqueous solutions during surface treatment (pickling, plating, phosphating), or from the environment during a spontaneous corrosion process in which the development of acidity within the crack results in hydrogen evolution and absorption (*see* Section 1.6).

It follows from this that two types of mechanical test[25, 26] may be used to evaluate the effect of hydrogen on the properties of high-strength steels:

1. Rapid dynamic tests to determine the extent of embrittlement.
2. Prolonged static tests to determine the propensity of the alloy to crack and fracture when subjected to a sustained static tensile stress.

Although similar constant-load test rigs are used for both active-path corrosion and hydrogen stress cracking there is one fundamental difference in the test procedure. In the case of active-path corrosion testing it is always carried out in the presence of the corrosive environment, but in the case of hydrogen embrittlement or hydrogen stress-cracking, testing may be carried out after hydrogen has been introduced into the alloy either deliberately by gaseous or cathodic charging, or following processes such as welding, pickling or electroplating. However, sustained-load tests are also carried out in the presence of an environment (gaseous or aqueous) so that hydrogen is introduced into the testpiece during the application of the tensile stress.

Dynamic Tests

All of the properties evaluated by the conventional tensile test (yield strength, tensile strength, elongation and reduction in area) are affected by the presence of hydrogen, but in the case of the tensile strength and yield strength the effect is significant only when the steel has a very high tensile strength and has been severely embrittled. On the other hand, the reduction in area, and to a lesser extent the elongation, may be used for detecting embrittlement. Hobson and Sykes[27] found that with low-carbon steels there was an almost linear relationship between reduction in area and hydrogen content of the steel.

Various types of bend tests have been used to evaluate embrittlement. Beck, Klier and Sachs[28] used thin strip specimens, and determined the decrease in height Δh at fracture when the specimen was bent by compressing it at a constant rate in a tensile testing machine (Fig. 8.70). The decrease in height Δh gives a measure of the embrittlement, the maximum elongation of the outside fibre of the specimen being calculated from the radius of curvature at maximum bending. In general, the ductility is found to increase with the

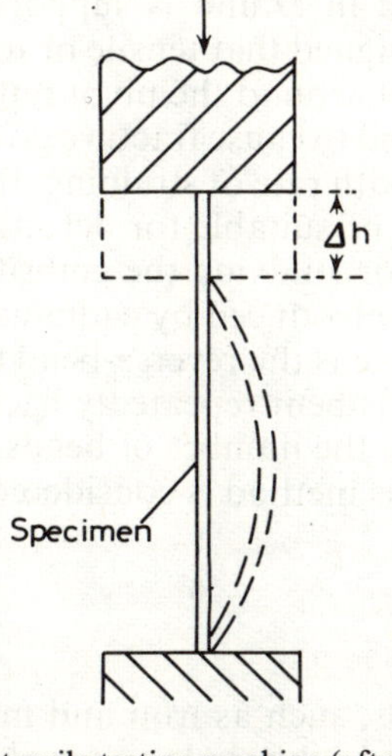

Fig. 8.70 Bend test using a tensile testing machine (after Beck, Klier and Sachs[28])

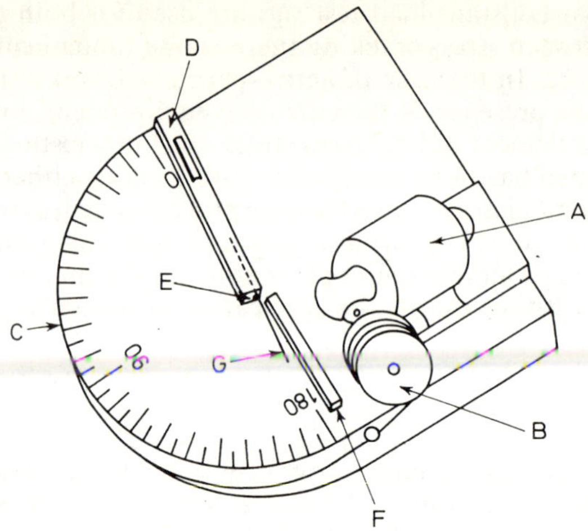

Fig. 8.71 Constant-rate bend test for determining hydrogen embrittlement of wires. A, drive unit; B, pulley; C, semi-circular base; D, travelling arm; E, axial pin; F, fixed arm; G, wire specimen (after Zapffe and Haslem[29])

rate of straining and for this reason high-strain-rate tests, such as impact tests, are insensitive to hydrogen embrittlement. Where the material is available only in the form of tubing, semicircular specimens may be used in place of flat strip in the compression bend test. The total cross-head travel from the unstressed height along the diameter to the height of fracture gives a measure of embrittlement, which may be compared with that obtained from an unembrittled specimen of the same steel.

A constant-rate bend-test machine, which provides an effective method for testing highly embrittled steel wires of high-tensile strength, was designed by Zapffe and Haslem[29] (Fig. 8.71). The motor *A* pulls a chord attached to the travelling arm *D* that rotates about a pivot pin. The wire specimen *G* ($1{\cdot}6 \times 100$ mm) is inserted in *D* and is supported by the fixed arm *F*, the arrangement being so designed that tensile or torsional stresses are avoided. The specimen is thus bent around the pivot pin *E* (radius 1·6 mm) at a constant rate, the angle of bend to cause fracture giving a measure of its ductility. Since ductility increases with rate of straining, the bending rate must be slow and 4°/s is considered to be suitable for detecting embrittlement. A similar machine has been used for studying the embrittlement of spring steel strip after hydrogen has been introduced by cadmium plating.

An alternative procedure is the reverse-bend test in which the specimen in the form of wire or strip is bent repeatedly backwards and forwards over a mandrel until it fractures, the number of bends indicating the degree of embrittlement. However, this method is considered to be less sensitive than the single-bend test.

Static Tests

Whereas ductile materials, such as iron and mild steel, do not crack when charged with hydrogen and subjected to a tensile stress below the yield stress, the position is different with high-strength ferrous alloys where, depending on the strength of the steel and the hydrogen content, failure may occur well below the yield stress. However, the fracture process is not instantaneous and there is a time delay before cracks are initiated; for this reason the phenomenon is referred to as 'delayed failure'.

In the majority of cases the tests are conducted using a dead-weight lever-arm stress-rupture rig with an electric timer to determine the moment of fracture, but a variety of test rigs similar to those shown in Fig. 8.57*g* are also used. The evaluation of embrittlement may be based on a delayed-failure diagram in which the applied nominal stress *vs.* time to failure is plotted (Fig. 8.72) or the specimen may be stressed to a predetermined value (say 75% of the ultimate notched tensile strength) and is considered not to be embrittled if it shows no evidence of cracking within a predetermined time (say 500 h). Troiano[30] considers that the nature of delayed failure can be described by four parameters (see Fig. 8.72):

1. The upper critical stress corresponding to the fracture stress of the unembrittled notched specimen.
2. The lower critical stress, which is the applied stress below which failure does not occur.
3. The incubation period or the time required for the formation of the first crack.
4. The failure time or the time for specimen failure at a given applied stress; in the intermediate stress range this includes a period of relatively slow crack growth.

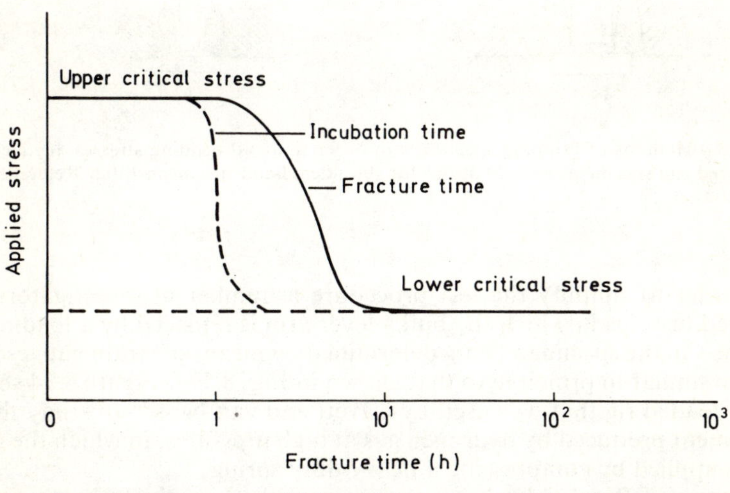

Fig. 8.72 Schematic representation of delayed failure characteristics of a hydrogenated high-strength steel

During the constant-load test it is essential that only axial tensile stresses are applied, since any bending stresses that are introduced will result in a higher true stress than that calculated. For this reason the ends of the specimens and the grips must be designed to avoid bending stresses. The ASTM

Standard E8–69 specifies that in the case of specimens with threaded ends the grips should be attached to the heads of the testing machine through properly lubricated spherical-seated bearings, and that the distance between the bearings should be as great as is feasible (Fig. 8.73).

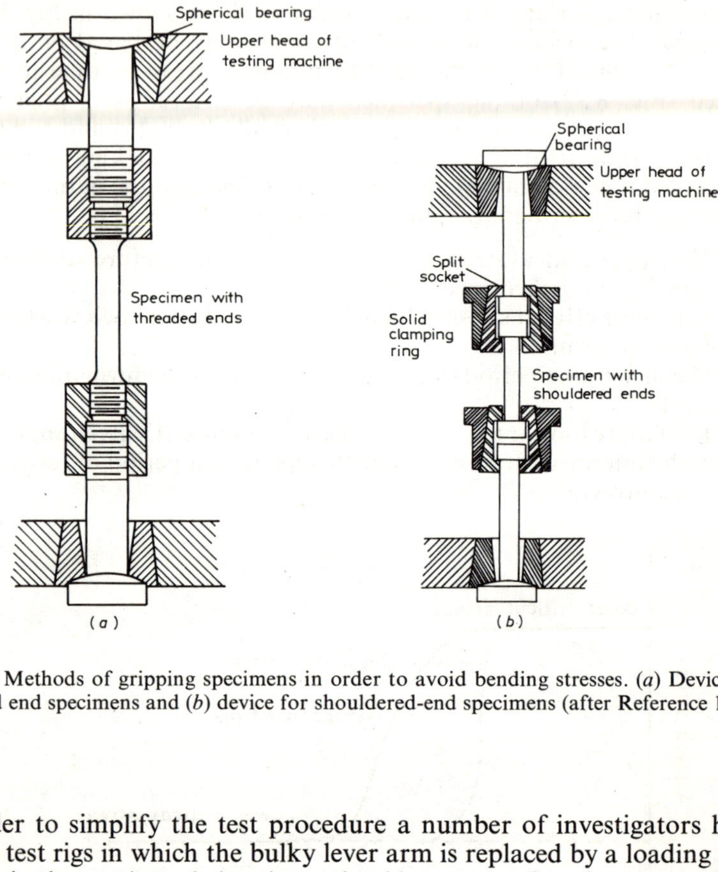

Fig. 8.73 Methods of gripping specimens in order to avoid bending stresses. (*a*) Device for threaded end specimens and (*b*) device for shouldered-end specimens (after Reference 1*b*)

In order to simplify the test procedure a number of investigators have designed test rigs in which the bulky lever arm is replaced by a loading nut, the stress in the specimen being determined by means of strain gauges; these rigs are similar in principle to that shown in Fig. 8.57*g*. Figure 8.74 shows a spring-loaded rig that was used by Cavett and van Ness[31] to study the embrittlement produced by hydrogen gas at high pressures, in which the tensile load is applied by compressing a heavy-duty spring.

Raring and Rinebold[32] have devised a method in which the specimen is supported along a diameter of a steel loading ring (Fig. 8.75), and the stress is applied by tightening the bottom nut until the diameter corresponds with the required load. The sudden release of elastic energy stored within the ring when the specimen fractures results in displacement of the tightening nut, and this is used to actuate a microswitch and timer. Williams, Beck and Jankowsky[33] have used notched 'C'-rings, the stress being applied by tightening the nut of a calibrated loading bolt which passes through the

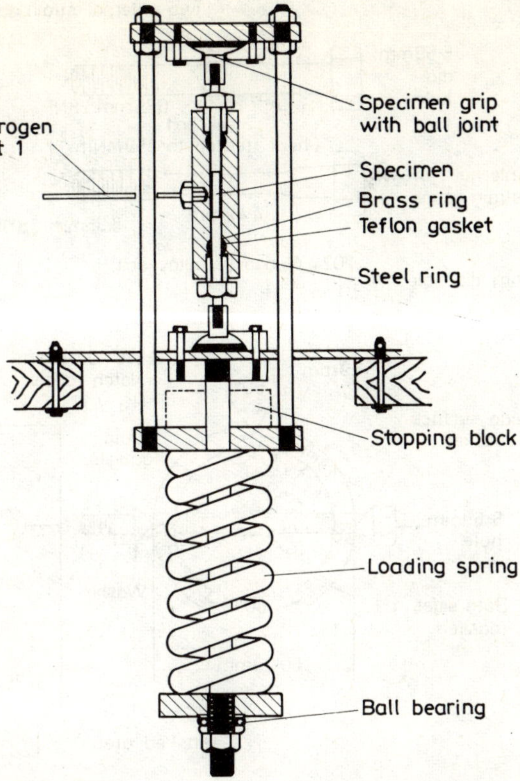

Fig. 8.74 Spring-loaded rig for sustained load testing of a steel specimen in gaseous hydrogen at high pressure (after Cavett and van Ness[31])

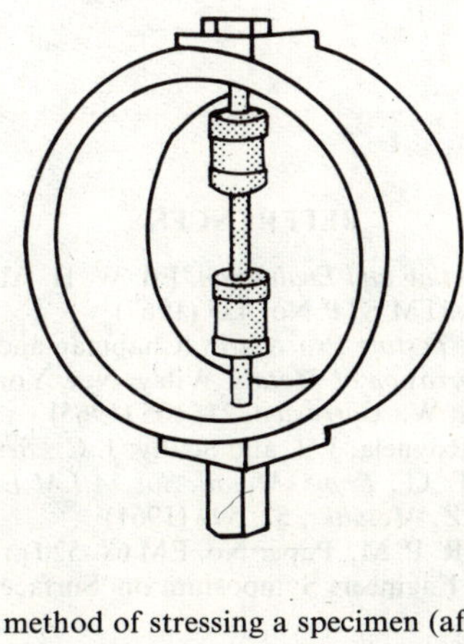

Fig. 8.75 Loading-ring method of stressing a specimen (after Raring and Rinebold[31])

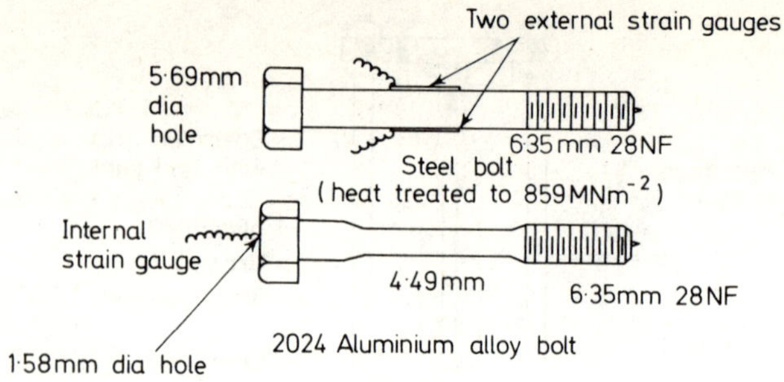

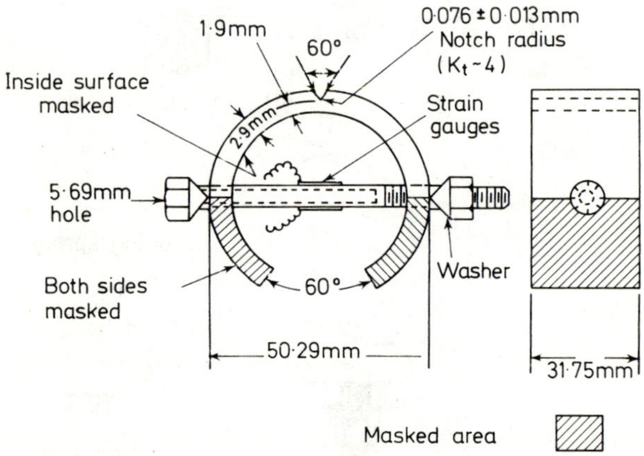

Fig. 8.76 Notched 'C'-ring specimen with attached strain gauges

diameter of the 'C'-ring (Fig. 8.76). The strain gauges attached to the bolt form two arms of a Wheatstone bridge circuit, and to compensate for temperature changes the other two arms consist of two identical strain gauges attached to a similar unstrained bolt.

R. N. PARKINS

REFERENCES

1*a*. *Handbook on Corrosion Testing and Evaluation*, Ed. W. H. Ailor, Wiley (1971)
1*b*. *Stress Corrosion Testing*, ASTM STP No. 425 (1967)
2. Champion, F. A., *Corrosion Testing Procedures*, Chapman and Hall, London, 137 (1952)
3. Logan, H. L., *The Stress Corrosion of Metals*, Wiley, New York, 273 (1966)
4. Haaijer, G. and Loginow, A. W., *Corrosion*, **21**, 105 (1965)
5. Parkins, R. N., Mazza, F., Royuela, J. J. and Scully, J. C., *Brit. Corr. J.*, **7**, 154 (1972)
6. Parkins, R. N. and Haney, E. G., *Trans. Metall. Soc. A.I.M.E.*, **242**, 1943 (1968)
7. Brenner, P. and Gruhl, W., *Z. Metallk.*, **52**, 599 (1961)
8. Paxton, H. W. and Proctor, R. P. M., Paper No. EM 68–520 presented at American Society of Tool and Manufacturing Engineers Symposium on 'Surface Integrity in Machining and Grinding', Pittsburgh (1968)

9. Brown, B. F., *Met. Rev.*, **13**, 171 (1968)
10. *Specialists Meeting on Stress Corrosion Testing Methods*, AGARD Conference Proceedings No. 98, NATO (1972)
11. Smith, H. R. and Piper, D. E., *Stress Corrosion Testing with Precracked Specimens*, The Boeing Co., D6–24872, ARPA 878, June (1970)
12. Wearmouth, W. R., Dean, G. P. and Parkins, R. N., *Corrosion*, **29**, 251 (1973)
13. Fisher, D. M., Bubsey, R. T. and Srawley, J. E., *Design and Use of a Displacement Gauge for Crack Extension Measurements*, NASA TN-D3724 (1966)
14. Barnett, W. J. and Troiano, A. R., *Trans. A.I.M.E.*, **209**, 486 (1957)
15. Gerberich, W. W. and Hartblower, C. E., *Proc. Conf. Fundamental Aspects of Stress Corrosion Cracking*, NACE, 420 (1969)
16. Okada, H., Yukawa, K. and Tamura, H., *Corrosion*, **30**, 253 (1974)
17. Streicher, M. A. and Casale, I. B., *Proc. Conf. Fundamental Aspects of Stress Corrosion Cracking*, NACE, Houston, 305 (1969)
18. Williams, W. L., *Corrosion*, **13**, 539t (1957)
19. Romans, H. B., *Stress Corrosion Testing*, A.S.T.M. STP No. 425, 182 (1966)
20. Lifka, B. W. and Sprowls, D. O., *Stress Corrosion Testing*, A.S.T.M. STP No. 425, 342 (1966)
21. Dana, A. W., A.S.T.M. Bulletin No. 225, TP 196, 46 (1957)
22. Schroeder, W. C. and Berk, A. A., *Bull. U.S. Bur. Mines*, 443 (1941)
23. Pugh, E. N., Montague, W. G. and Westwood, A. R., *Trans. Am. Soc. Metals*, **58**, 665 (1965)
24. Farmery, H. K. and Evans, U. R., *J. Inst. Metals*, **84**, 413 (1956)
25. Smialowski, M., *Hydrogen in Steel*, Pergamon Press, London (1962)
26. Beck, W., Jankowsky, E. J. and Fischer, P., *Hydrogen Stress Cracking of High Strength Steels*, Report No. NADC-MA-7140, Naval Air Development Centre, Warminster, Pennsylvania (1971)
27. Hobson, J. D. and Sykes, C., *J. Iron Steel Inst.*, **169**, 209 (1951)
28. Beck, W., Klier, E. P. and Sachs, G., *Trans. A.I.M.E.*, **206**, 1263 (1956)
29. Zapffe, C. A. and Haslem, M. E., *Trans. A.I.M.E.*, **167**, 281 (1946)
30. Troiano, A. R., *Trans. A.S.M.*, **52**, 54 (1960)
31. Cavett, R. H. and van Ness, H. C., *Welding J.* (research supplement), **42**, 317 (1963)
32. Raring, R. H. and Rinebold, J. A., ASTM Bulletin No. 213 (1956)
33. Williams, F. S., Beck, W. and Janowsky, E. J., *Proc. ASTM*, **60**, 1192 (1960)

8.10A Appendix—Stresses in Bent Specimens

In each of the following equations σ = maximum tensile stress, E = modulus of elasticity and t = specimen thickness.

Two-point bending (Fig. 8.57*a*)

$$L = \frac{ktE}{\sigma}\sin^{-1}\left(\frac{H\sigma}{ktE}\right)$$

where L = specimen length,
k = constant (1·280) and
H = holder span.

Four-point bending (Fig. 8.57*b*)

$$\sigma = \frac{12Ety}{3L^2 - 4A^2}$$

where y = maximum deflection,
L = distance between outer supports and
A = distance between inner and outer supports.

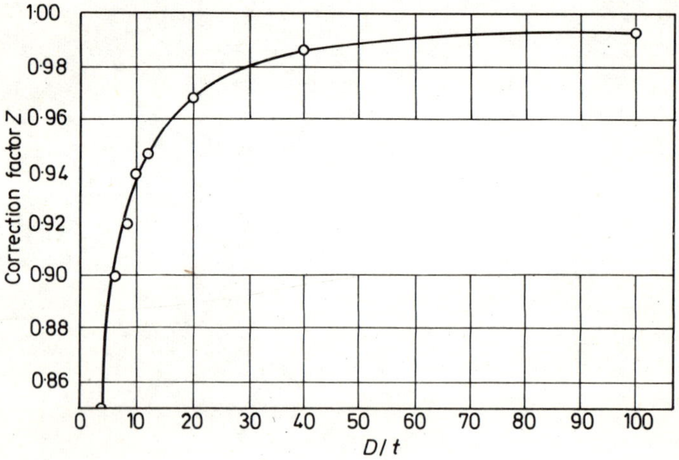

Fig. 8.A1

'*U*'-*bends* (Figs. 8.57*d* and *e*)

Applied strain $\varepsilon = \frac{t}{2r}$, when $t<r$, where r = radius of curvature at section of interest.

'*C*'-*rings* (Fig. 8.57*c*)

$$\sigma = \frac{4Etz\Delta}{\pi D^2}$$

where $\Delta = OD_f - OD_i$,
OD_f = final outside diameter of stressed 'C'-ring,
OD_i = initial outside diameter of unstressed 'C'-ring,
D = mean diameter, i.e. $(OD - t)$ and
Z = a correction factor, related to D/t as indicated in Fig. 8.A1.

9 ELECTROCHEMISTRY AND METALLURGY RELEVANT TO CORROSION

9.1 Outline of Electrochemistry

The present Section, which provides an outline of selected relevant topics in electrochemistry, is intended primarily as an introduction to aqueous corrosion for those readers whose basic training has not involved a study of electrochemistry. The scope of electrochemistry is enormous and cannot be treated adequately here, but there are now a number of excellent books on the subject, and it is hoped that this outline will serve to stimulate further study. The topics selected are as follows: (a) the nature of the electrified interface between the metal and the solution, (b) adsorption, (c) transfer of charge across the interface under equilibrium and non-equilibrium conditions, (d) overpotential and the rate of an electrode reaction and (e) the hydrogen evolution reaction and hydrogen absorption by ferrous alloys. For reasons of space a number of important topics, such as the electrochemistry of electrolyte solutions, have been omitted.

In this outline some emphasis has been given to the nature of the electrified interface that exists between an electrode and a solution and on adsorption of intermediate species at the interface; aspects of electrochemistry that are frequently disregarded in accounts of the electrochemical mechanism of corrosion. The fact remains that adsorption of species from solution at the interface is normally a precursor to a corrosion reaction and will dictate the form it takes, its rate, and its location on a heterogeneous metal surface. Unfortunately, although knowledge on the adsorption on a mercury electrode is advanced the position is quite different in the case of industrial metals and alloys, and this has limited our understanding of a variety of phenomena in corrosion such as pitting, stress-corrosion cracking, etc. Nevertheless, an understanding of the nature of the electrified interface is essential for an appreciation of the detailed mechanisms of corrosion.

The Electrified Interface

An aqueous electrolyte solution consists of a variety of charged and uncharged species, e.g. cations, anions, water dipoles, organic molecules, trace impurities, etc. which under equilibrium conditions are randomly oriented so that within the solution there is no net preferentially directed field. However, under the influence of a potential difference, the charge will be transported through the solution by cations and anions that migrate to the

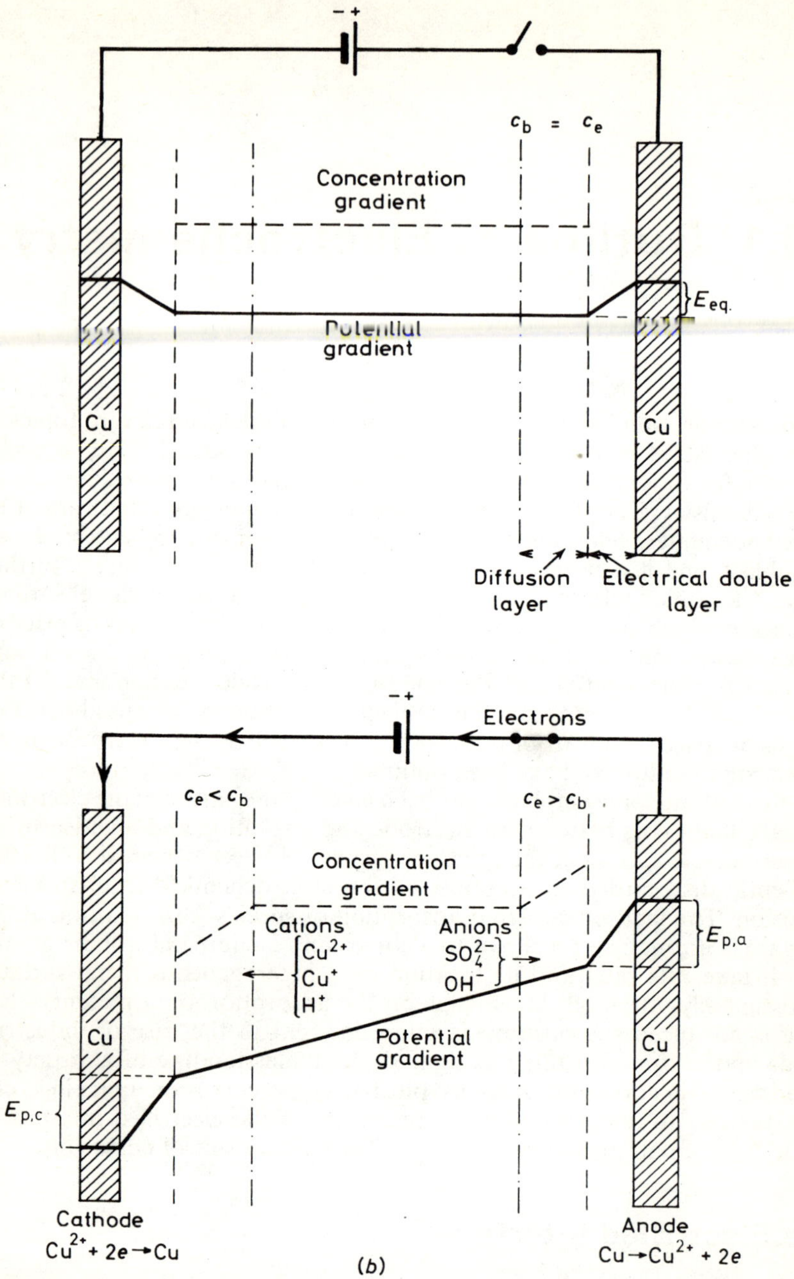

Fig. 9.1 Potential and concentration gradients in the electrolytic cell $Cu/CuSO_4/Cu$. (*a*) The electrodes are unpolarised; the potential difference is the equilibrium potential $E_{eq.}$ and there is no concentration gradient in the diffusion layer. (*b*) The electrodes are polarised; E_p of the anode is now more positive than $E_{eq.}$ whilst E_p of the cathode is more negative and concentration gradients exist across the diffusion layer; c_e, c_b are the concentrations at the electrode and in the bulk solution

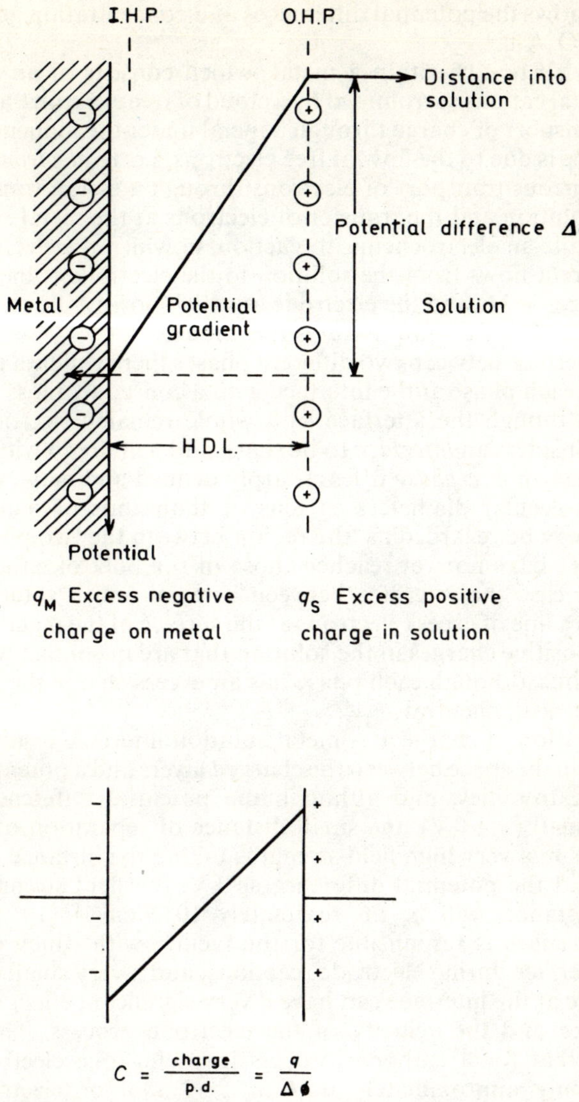

Fig. 9.2 Helmholz double layer (H.D.L.) consisting of a 'plate' of excess negative charges on the surface of the metal and a counterbalancing 'plate' of excess positive charges (cations) in solution, the double layer as a whole being electrically neutral. The double layer can be regarded as equivalent to a capacitor in which the 'plates' are separated by a distance δ

cathode and anode, respectively; this migration of charge through an electrolyte is responsible for *electrolytic conduction*. At each electrode there will be a transfer of electrical charge and a consequent electrode reaction that results in a change in the concentration of the reacting species at the surface of the electrode. Under these circumstances the transport of species to and from the electrodes is by diffusion and convection as well as migration; the term *transport* is used here to include diffusion, convection and migration.

Figure 9.1 shows the potential differences and concentration gradients in the cell $Cu/CuSO_4/Cu$.

Similarly, all points within a metal, which consists of an ordered rigid lattice of metal cations surrounded by a cloud of free electrons, are electrically neutral. Transport of charge through a metal under the influence of a potential difference is due to the flow of free electrons, i.e. to *electronic conduction*. The simultaneous transport of electrons through a metal, transport of ions through a solution and the transfer of electrons at the metal/solution interfaces constitute an electrochemical reaction, in which the electrode at which positive current flows from the solution to the electrode is the *cathode* (e.g. $M^{z+}(aq.)+ze \rightarrow M$) and the electrode at which positive flows from it to the solution (e.g. $M \rightarrow M^{z+}(aq.)+ze$) is the anode.

At any interface between two different phases there will be a redistribution of charge in each phase at the interface with a consequent loss of its electroneutrality, although the interface as a whole remains electrically neutral. (Bockris[1] considers an *interface* to be sharp and definite to within an atomic layer, whereas an *interphase* is less sharply defined and may extend from at least two molecular diameters to tens of thousands of nanometres; the interphase may be regarded as 'the region between the two phases in which the properties have not yet reached those of the bulk of either phase'.) In the simplest case the interface between a metal and a solution could be visualised as a line of excess electrons at the surface of the metal and an equal number of positive charges in the solution that are in contact with the metal (Fig. 9.2). Thus although each phase has an excess charge the interface as a whole is electrically neutral.

The separation of charge at a metal/solution interface results in a strong electric field in the space between the charged layers and a potential difference across the extremities, and although the potential differences are comparatively small ($<1{\cdot}0$ V), the small distance of separation of the charged layers results in a very high field strength; taking the distance of separation as 0·1 nm and the potential difference as 1 V, the field strength (potential difference/distance) will be approximately $10^7\,V\,cm^{-1}$ ($10^9\,V\,m^{-1}$). This high field strength is responsible for the facility with which electrons can cross the interface during electrode reactions, and a very small change in the excess charge at the interface can have a very significant effect on the potential difference and the velocity of the electrode process. Thus it can be calculated[2] that for a cathodic process involving one electron an excess charge of only approximately $0{\cdot}1\,C\,m^{-2}$ (1 mol of electrons $\equiv$ 1 F $\approx$ 96 500 C, so that 0·1 C $\approx 10^{-6}$ mol of electrons) will change the potential difference by as much as $-0{\cdot}6$ V, and could increase the rate of the process by a factor of 10^5.

Structure of the Electrical Double Layer (E.D.L.)

A simple model of the e.d.l. was first suggested by Helmholz in which the charges at the interface were regarded as the two plates constituting a parallel plate capacitor, e.g. a 'plate' of metal with excess electrons (the inner Helmholz plane I.H.P.) and a 'plate' of excess positively charged ions (the outer Helmholz plane O.H.P.) in the solution adjacent to the metal; the charges balance one another so that the e.d.l. is neutral as a whole (Fig. 9.2). Although this model, which is based entirely on electrostatic concepts, is applicable under certain circumstances it has certain limitations and is incapable of explaining a number of experimental observations. Thus although on the basis of the Helmholz model the distance of separation of charges is taken as approximately 0·1 nm (the diameter of an ion) the interphase actually extends to at least 0·5 nm and can affect the adsorption of species at even greater distances.

If it is assumed that the interface between mercury (which is widely used as an electrode for studies of the e.d.l.) and an electrolyte solution forms the two plates of a capacitor, then at equilibrium the mercury 'plate' will have an excess charge q_{Hg} and the solution 'plate' a charge of equal magnitude but opposite sign q_s, and the capacitance C is given by

$$C = \frac{q_{Hg}}{\Delta\phi} \qquad \ldots(9.1)$$

where $\Delta\phi$ is the potential difference. On the basis of this model the potential difference between the 'plates' due to excess charge is

$$\Delta\phi = \frac{q\delta}{\varepsilon_r \varepsilon_0} \qquad \ldots(9.2)$$

where ε_r is the relative permittivity of the 'material' between plates, ε_0 is the permittivity of a vacuum ($8{\cdot}85 \times 10^{-12}$ F m^{-1}) and δ is the distance between the plates.

In this connection it is necessary to consider the contribution made to the potential difference by the dielectric itself, and in aqueous solution this will be largely due to adsorption of water molecules, although adsorption of other molecules in solution can also have an effect. Although the water molecule is electrically neutral as a whole, the fact that the two binding electrons are closer to the oxygen atom than to the hydrogen atoms results in an electric dipole with a positively charged (hydrogen) end and a negatively charged (oxygen) end, and consequently a potential difference exists across the molecule (Fig. 9.3*a*).

If the metal has a large negative excess charge the adsorbed water molecules will be oriented with their positive ends towards the metal and their negative ends towards the solution, and this layer of oriented dipoles will contribute an additional potential difference to the double layer (Fig. 9.3*b*). The converse situation will arise when the metal has a large positive charge, and again the dipoles will contribute to the potential difference (Fig. 9.3*c*). In the case of a metal with only a small excess charge dipoles of both orientations will adsorb, and the potential difference will be proportional to the excess number of one or the other (Fig. 9.3*d*).

Thus the potential difference at the interface between a metal and electrolyte solution is due to both the charges at the interface (electrostatic potential difference) and the surface dipole layers; the latter is referred to as the *surface* or *adsorption* potential difference. On the basis of the above considerations it might appear that adsorption at a metal surface with an excess charge is solely due to electrostatic interaction with charged species in the solution, i.e. if the metal surface has an excess negative charge the cations

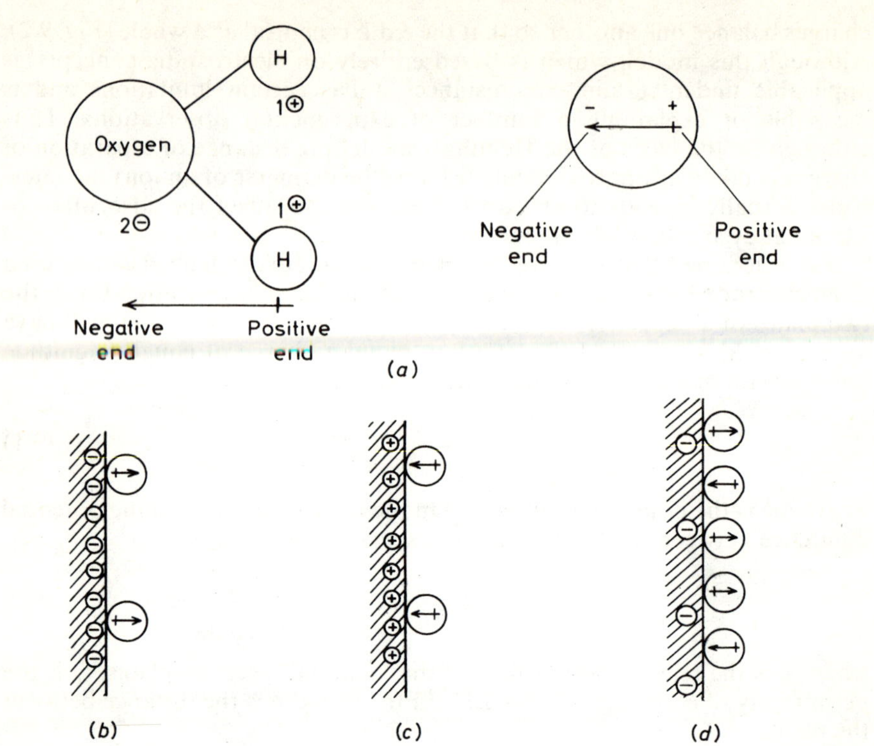

Fig. 9.3 Adsorption of water dipoles. (*a*) A water dipole showing the positive hydrogen end and negative oxygen end, (*b*) adsorption on metal with large negative excess charge, (*c*) adsorption on metal with large positive excess charge and (*d*) adsorption on metal with small negative excess charge

are attracted to the surface and adsorb, whereas the anions because of electrostatic repulsion are prevented from adsorbing. However, this simple model takes no account of *specific* or *contact* adsorption, and as will be discussed subsequently it is possible for negatively charged anions to adsorb strongly on the surface of a metal that has a negative excess charge.

Electrocapillary Curves and Excess Charge of an Interface

At the surface of a liquid the surface molecules will be subjected to a strong inward attraction due to molecules in the interior and to a much weaker force due to the molecules of the gaseous phase. This gives rise to a surface tension (or to an interfacial tension if two liquids are involved) which tends to cause the surface to shrink to give the minimum surface area; thus droplets of liquid and bubbles tend to take up a spherical shape, since a sphere has the minimum surface area for a given volume. Surface tension is also responsible for a liquid rising up a very fine capillary. On the other hand, when there is an excess charge at an interface the ions comprising the excess charge will repel one another and this will result in a tendency for the surface to expand; the forces due to the molecular attraction, interfacial tension

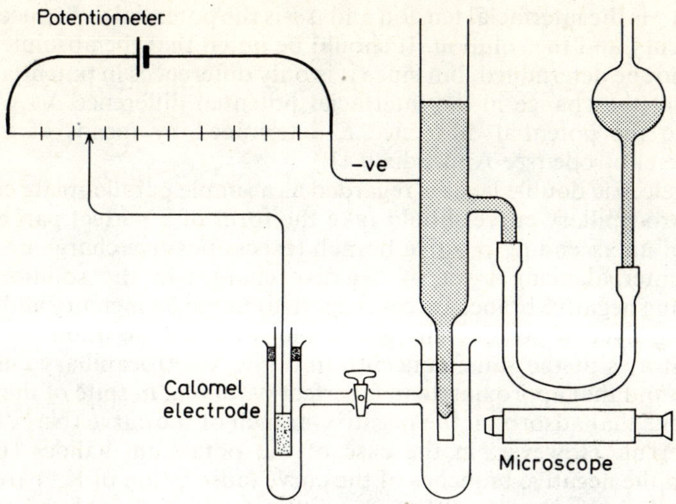

Fig. 9.4 Lippmann electrometer for studying the variation of the excess charge on mercury with variation in potential difference at the mercury solution interface

and those due to the excess charge thus oppose one another, and this effect can be used to study the variation of excess charge with the potential of the interface. The excess charge at a mercury/electrolyte solution interface has been extensively studied by the Lippmann electrometer (Fig. 9.4), in which mercury in a fine capillary is in contact with the solution under study. The potential of the mercury, which tends to be positive in most aqueous solutions, can be varied by means of a potentiometer and its potential measured by a suitable reference electrode. A change in the potential of the mercury/solution interface will result in a displacement of the mercury in the capillary, which can be determined with respect to a reference mark on the capillary by means of a travelling microscope. The solutions selected for these studies and the potential must be such that no electrode reactions occur, and under these conditions the applied potential builds up a charge at the highly polarisable mercury/solution interface, the magnitude of which can be determined by the displacement of the mercury meniscus.

Electrocapillary curves are obtained by plotting the interfacial tension γ (displacement of mercury expressed in appropriate units) against potential difference at the mercury/solution interface with respect to a reference electrode. Figure 9.5 shows how the interfacial tension increases as the potential difference between the interface and the reference electrode is increased (in Fig. 9.5 the interface is made more negative), until it reaches a maximum when further increase in potential results in a corresponding decrease in the interfacial tension. The excess charge q_{M}, or *charge density*, at constant temperature T, pressure p and concentration of electrolyte c is given by the Lippmann equation

$$q_{\mathrm{M}} = -\left(\frac{\partial\gamma}{\partial\Delta\phi}\right)_{T,p,c} = -\left(\frac{\partial\gamma}{\partial E}\right)_{T,p,c} \qquad \ldots(9.3)$$

in which γ is the interfacial tension and $\Delta\phi$ is the potential difference between the mercury and the solution. It should be noted that the absolute value of $\Delta\phi$ cannot be determined, but since it is only differences in potential that are significant the change in the interfacial potential difference $\Delta\phi$ will correspond to the potential difference E determined by means of a suitable reference electrode (see Appendix 9.1B).

If the electric double layer is regarded as a simple parallel-plate condenser the electrocapillary curve should take the form of a perfect parabola consisting of an ascending positive branch (excess positive charge on mercury and counterbalancing layer of negative charges in the solution) and a descending negative branch (excess negative charge on mercury and counterbalancing layer of positive charges in the solution). Solutions of different alkali nitrates at the same concentration give electrocapillary curves that coincide and that approximate to a perfect parabola, in spite of the fact that the cations that adsorb on the negative branch of the curve (Na^+, K^+, etc.) are different. However, in the case of the potassium halides (Fig. 9.5), although the negative branches of the curve (adsorption of K^+) are similar, marked divergences are obtained on the positive branch of the curve

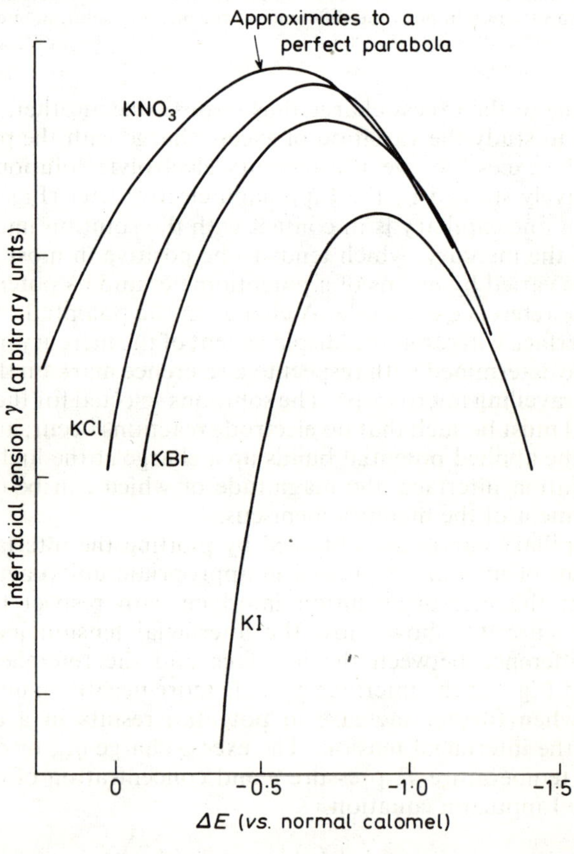

Fig. 9.5 Electrocapillary curves for KNO_3 and different potassium halides showing how the former approximates to a parabola

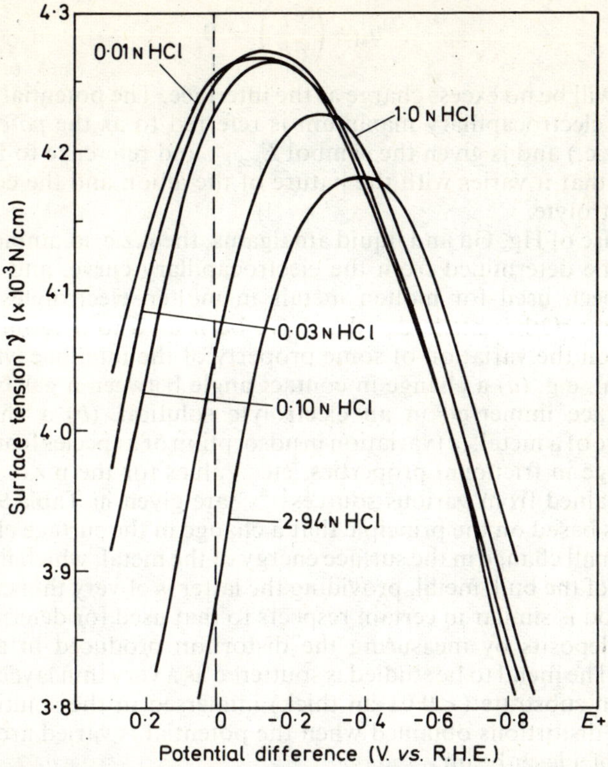

Fig. 9.6 Electrocapillary curves for HCl at various concentrations determined using a reversible hydrogen electrode (R.H.E.) immersed in the same concentration of HCl as that used for the determination (after Bockris and Reddy[1])

(adsorption of halide), and it can be seen that the divergences from ideality are in the order $I^- > Br^- > Cl^-$. In general, the departure of the curve from ideality occurs on the positive ascending branch at which anions are adsorbed at the mercury/solution interface.

Similar considerations apply to the curves obtained for a single electrolyte at different concentrations[1] (Fig. 9.5), and in the case of HCl whereas the ascending branches of the curves (adsorption of Cl^-) show divergences that increase with increase in concentration, the descending branches of the curves (adsorption of H^+) almost coincide. Thus again it is the anion rather than the cation (only large organic cations such as the tetra-alkylammonium ion have a significant effect) that affects the shape of the electrocapillary curve.

Potential of Zero Charge[1–3]

The gradient of the electrocapillary curve (equation 9.3) at a given potential gives the charge density at that potential, and it is evident from the slopes that it will be positive on the ascending branch and negative on the descending branch. At the electrocapillary maximum, however, the gradient is zero.

$$q_M - \left(\frac{\partial \gamma}{\partial E}\right) = 0 \qquad \ldots(9.4)$$

and there will be no excess charge at the interface. The potential corresponding to the electrocapillary maximum is referred to as the *potential of zero charge* (p.z.c.) and is given the symbol $E_{\text{p.z.c.}}$, and reference to Figs. 9.5 and 9.6 shows that it varies with the nature of the anion and the concentration of the electrolyte.

In the case of Hg, Ga and liquid amalgams, the p.z.c. at ambient temperatures can be determined from the electrocapillary curve, and this method has also been used for molten metals in molten electrolytes at elevated temperatures. Other methods[3] that have been used to determine the p.z.c. are based on the variation of some property of the interface with variation in potential, e.g. (*a*) a change in contact angle between a gas bubble and a metal surface immersed in an electrolyte solution, (*b*) a change in the capacitance of a metal, (*c*) variation in adsorption of a species from a solution, (*d*) a change in frictional properties, etc. Values for the p.z.c. for different metals obtained from various sources[1,4,5] are given in Table 9.1. A recent method[2] is based on the principle that a change in the surface charge results in a very small change in the surface energy of the metal, which in turn causes distortion of the bulk metal, providing the latter is of very thin cross-section. This method is similar in certain respects to that used for determining stress in electrodeposits by measuring the distortion produced in a thin metal substrate. The metal to be studied is sputtered as a very thin layer (<10·0 nm) onto a thin substrate (<0·01 cm thick) immersed in the solution, and the very small distortions obtained when the potential is varied are determined by means of a laser beam.

Table 9.1 Potential of zero charge $E_{\text{p.z.c.}}$ and $E^{\ominus}_{M^{z+}/M}$ of metals

Metal	$E_{\text{p.z.c.}}$ (V, *vs.* S.H.E.)	*Equilibrium*	$E^{\ominus}_{M^{z+}/M}$ (V)
Cd (0·01 mol dm^{-3} KCl)	−0·92	Cd^{2+}/Cd	−0·42
Tl	−0·82	Tl^{+}/Tl	−0·34
Ag (0·04 mol dm^{-3} Na_2SO_4)	−0·70	Ag^{+}/Ag	+0·79
Ga	−0·60	Ga^{3+}/Ga	−0·52
Al (0·01 mol dm^{-3} KCl)	−0·52	Al^{3+}/Al	−1·7
Fe (0·002 mol dm^{-3} H_2SO_4)	−0·37	Fe^{2+}/Fe	−0·44
Bi (0·01 mol dm^{-3} KCl)	−0·36		
Co (0·04 mol dm^{-3} Na_2SO_4)	−0·32	Co^{2+}/Co	−0·27
Sb (0·01 mol dm^{-3} HCl)	−0·19	Sb^{3+}/Sb	
Hg	−0·19	Hg^{2+}/Hg	+0·80
Ni	−0·06	Ni^{2+}/Ni	−0·25
Cu (0·04 mol dm^{-3} Na_2SO_4)	+0·03	Cu^{2+}/Cu	+0·34
Au (0·04 mol dm^{-3} Na_2SO_4)	+0·23	Au^{3+}/Au	+1·3
Pt (0·003 mol dm^{-3} $HClO_4$)	+0·41	Pt^{3+}/Pt	+1·1

Although the p.z.c. is difficult to determine experimentally, and although the values obtained vary with the method used, it is of fundamental significance in electrochemistry, since it provides information on adsorption of ions and molecules, i.e. if the potential is negative with respect to the p.z.c. cations will tend to be adsorbed and anions repelled, and vice versa. The p.z.c. appears to be a natural reference point for a *rational scale* of potentials defined by

$$\bar{E} = E - E_{p.z.c.} \quad \ldots(9.5)$$

where $\bar{E}$ is the potential on the rational scale, E is the measured potential and $E_{p.z.c.}$ is the potential of zero charge (both E and $E_{p.z.c.}$ are with respect to the same reference electrode). The phi-scale of potentials, which has been developed largely by Antropov[4,6], takes a form that is similar to equation 9.5. It can be seen from Table 9.1 that $E^{\ominus}_{M^{z+}/M}$ and $E_{p.z.c.}$ do not show any simple correspondence, and this is due to the fact that the former depends on the removal of an ion from the metal and its solvation (see p. **9**.79) whilst the latter is a function solely of the removal of an electron from the metal, i.e. of the electronic work function.

Adsorption

The adsorption of ions and molecules at an electrified interface is of fundamental importance in all aspects of aqueous corrosion, and it is therefore relevant to consider it briefly in the context of this discussion on the nature of the electrified interface[1,2,7]. Gases may be adsorbed on solid surfaces in various ways, and when the adsorbed layer is held by the residual forces around the gas molecule, which are responsible for the liquefaction of gases and the cohesion in liquids, it is referred to as *physical* or *van der Waals* adsorption. Physical adsorption is reversible and the adsorbed gas can be removed by evacuation or by heating to moderate temperatures; the enthalpy of physical adsorption is small (<1 kJ mol^{-1}). In *chemisorption* the adsorbed layer is chemically combined by covalent (or ionic) forces to the metal, and removal of the layers requires evacuation at high temperatures; the enthalpy of chemisorption is at least approximately 2 kJ mol^{-1} and the bond between the metal and the surface is almost as strong as in stoichiometric compounds. An example of chemisorption is the adsorption of hydrogen gas onto the surface of tungsten, during which the hydrogen molecules in contact with the surface dissociate and form a chemical bond, with each atom of hydrogen combining with an atom of tungsten to form a monolayer. This dissociation of the hydrogen molecules and the formation of a bond takes place over a distance of 0·1 nm. Physical adsorption and chemisorption are not mutually exclusive and both can occur in certain systems. Thus when oxygen is adsorbed on charcoal, part is physically adsorbed and is readily removed by evacuation whilst the remainder is chemisorbed, and although it can be removed by evacuation at elevated temperatures the gases evolved consist of the oxides of carbon showing that the bond between the oxygen and carbon is greater than that between the carbon atoms in the solid.

In adsorption at an electrified interface the electric field is long range and can influence ions some distance into the solution away from the metal surface, and this *electrostatic* adsorption may extend to 100 nm. However, within 0·1–0·2 nm of the surface there will also be some degree of chemisorption of ions that are in contact with the metal surface, which is referred to as *specific* or *contact* adsorption. Thus in considering adsorption at a charged interface, account must be taken of both the long-range electrostatic forces and the short-range chemical forces.

The adsorption of ions due to an electrified interface can be evaluated by considering a series of laminae of the solution at various distances from the metal surface and assessing the number of ions present as compared with those that would have been present if the electrified interface had been absent. The sum of these excess ions, the 'Gibbs surface excess' Γ_i, is given by

$$\Gamma_i = \int_0^{\infty} \{c_i(x) - c_i^0\}\,dx = \frac{1}{S}\int_0^{\infty} \{c_i(x) - c_i^0\}\,dv$$

$$\therefore \qquad \Gamma_i = \frac{1}{S}\int_0^{n_i - n_i^0} d(\Delta n_i) = \frac{n_i}{S} - \frac{n_i^0}{S} \qquad \ldots(9.6)$$

where $c_i(x)$ is the actual concentration of the species i at a distance x from the metal surface, c_i^0 the concentration in the bulk solution (where the electrified interface has no influence), n_i the actual number of moles of i in the interphase region, n_i^0 the number of moles of i that would be present if the electrified interface were absent, S the area of the interphase and v the volume of the solution. It is evident that Γ_i will be either positive or negative depending on whether n_i is larger or smaller than n_i^0, respectively.

The Gibbs adsorption equation for the adsorption of an ion i from solution can be written in the form of the thermodynamic equation

$$(\Gamma_i)_E = -\frac{1}{2RT}\left(\frac{\partial \gamma}{\partial \ln a_i}\right)_{T,p,j} \qquad \ldots(9.7)$$

where a_i is the activity of the ion and the subscript E signifies that the surface excess Γ_i is measured at constant potential. In measuring the potential of a mercury solution interface (a polarisable electrode) it is necessary to measure the e.m.f. of a cell consisting of that interface and a non-polarisable reference electrode. It can be shown on the basis of thermodynamic arguments that equation 9.7 is applicable only if the reference electrode is reversible to an ion j of opposite sign to i. Thus if information is required on the adsorption of anions (e.g. Cl^- ions) then the reference electrode must be reversible to cations (e.g. a reversible hydrogen electrode); if information is required on the adsorption of cations the reference electrode must be reversible to anions (e.g. a calomel electrode).

It follows that the surface excess Γ_- of an anion i (e.g. the Cl^- ion) can be evaluated from the electrocapillary curves of a given electrolyte (e.g. HCl) by plotting surface tension against the logarithm of the activity of the electrolyte (evaluated at various constant potentials) and determining the slope of the curve $\partial\gamma/\partial \log a_i$ and introducing it into equation 9.7.

The total excess charge q_s in solution is due to both anions and cations, and is given by

$$q_s = z_+F\Gamma_+ + z_-F\Gamma_- \qquad \ldots(9.8)$$

and since q_M is given by equation 9.3 and is equal to $-q_s$ it is possible to evaluate the surface of the cation Γ_+ from q_M and Γ_- since

$$z_+F\Gamma_+ = q_s - z_-F\Gamma_- \qquad \ldots(9.9)$$

The above provides a means of showing how the total excess charge on the solution side of the interface q_s, the excess charge due to cations Γ_+ and the excess charge due to anions Γ_-, vary with potential in a solution of fixed concentration of electrolyte. On the basis of this approach to the electrocapillary curves it has been shown that the Gibbs surface excess for cations is due solely to electrostatic forces (long-range coulombic), and this is reflected in the fact that the electrocapillary curves for different cations and the same anion at the same concentration (e.g. alkali nitrates) coincide, i.e. specific or contact adsorption is absent. On the other hand, adsorption of anions can only be explained by taking into account both electrostatic and contact adsorption.

Inconsistencies of the Parallel Plate Model

The simple Helmholz model, in which the charge on the model is regarded as the plate of a capacitor that attracts a counter layer of ions of opposite charge and results in two parallel plates of the same charge density, is inconsistent with the shapes of the electrocapillary curves obtained in practice. It can be shown[1,2] that if the Helmholz model applied, the electrocapillary curve would conform to the relationship

$$\gamma = \gamma_{max.} - \frac{\varepsilon}{4\pi\delta} \times \tfrac{1}{2}E^2 \qquad \ldots(9.10)$$

where $\gamma_{max.}$ corresponds with the electrocapillary maximum. This is the equation for a parabola, symmetrical about $\gamma_{max.}$, and it is evident that this is seldom achieved in electrolyte solutions, particularly when the electrolyte is a halide (Figs. 9.5 and 9.6).

A further inconsistency in the Helmholz model is revealed by the differential capacitance C which is given by

$$C = \left(\frac{\partial q_M}{\partial E}\right)_{\text{const. comp.}} = \frac{\varepsilon_r \varepsilon_0}{\delta} \qquad \ldots(9.11)$$

and which can be evaluated, at any given value of E, from the slope of the electrocapillary curve. Since ε_r, ε_0 and δ should be constant, the curve of electrode charge *vs.* potential should be linear, and C (the slope of the curve) should remain constant at all values of E. Figure 9.7 shows results

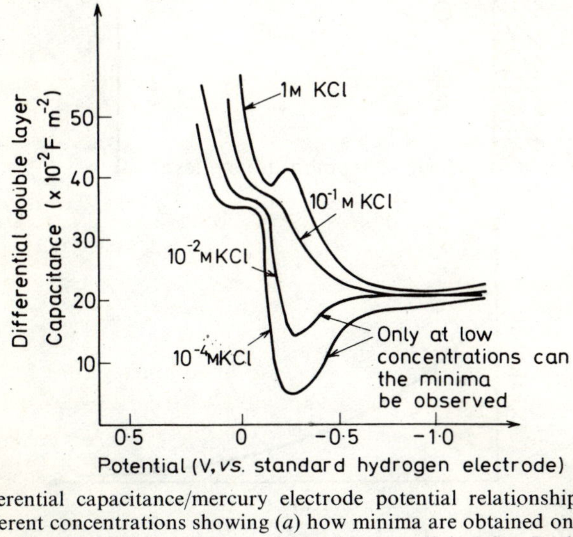

Fig. 9.7 Differential capacitance/mercury electrode potential relationships for potassium chloride at different concentrations showing (*a*) how minima are obtained only at low concentrations and (*b*) the constant capacitance at negative potentials (after Bockris and Drazic[2])

obtained for Hg in solutions of KCl at different concentrations, and it can be seen that C varies with potential and concentration except at very negative potentials where it is almost constant. This constant capacitance has been found to be independent of the nature of the ion, although if δ is taken as the distance between the electrode and the centres of the ions on the solution side of the interface it should vary with the ionic radius of the ion (equation 9.11). It should be noted that at the intermediate potentials the curves for dilute solutions approximate to inverted parabolas.

Gouy–Chapman and Stern Models of the Double Layer

Gouy and Chapman (1910) pointed out that the Helmholz model could not provide a true picture of the double layer, since the disordering forces due to the thermal energy of the ions would oppose their ordering by electrostatic forces. This would result in an equilibrium in which the excess of the counter ion would be a maximum close to the surface of the electrode and would diminish gradually with increasing distance from the surface (Fig. 9.8). In many respects this diffuse charge model is similar to that used by Debye and Hückel in their theory of interionic attraction of ions in solution, in which the ion is regarded as a point charge having a spherically symmetrical

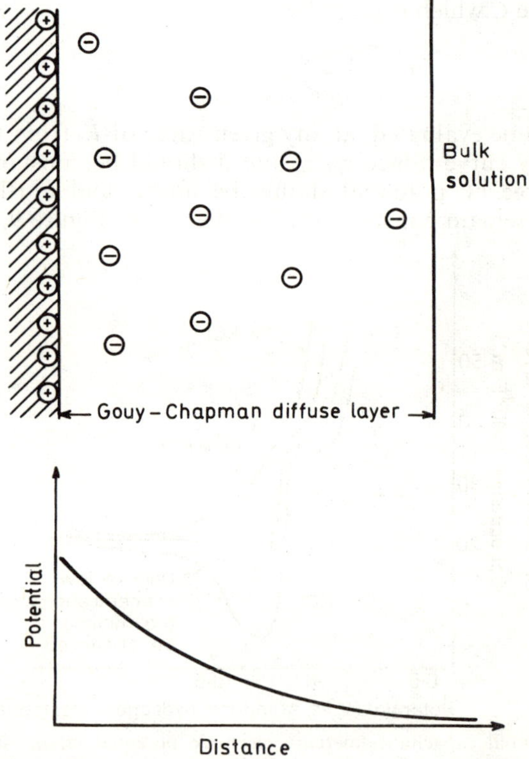

Fig. 9.8 Gouy–Chapman diffuse layer model of the double layer

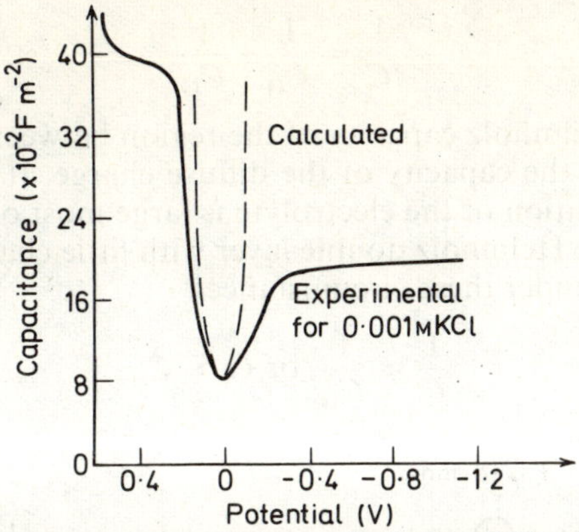

Fig. 9.9 Experimental capacitance–potential curve for 0·001 M KCl and calculated curve using the Gouy–Chapman model. The experimental curve and the theoretical curve agree at potentials (*vs* R.H.E.) near the p.z.c. Note the constant capacitance of $\approx 17 \times 10^{-2}$ F m^{-2} at negative potentials (after Bockris and Drazic[2])

field which diminishes with distance from the ion into the solution[1,8]. Thus the potential in an electrode–ion interaction should be a maximum at the metal surface, and should decay exponentially with distance into the solution and attain a constant value (zero) in the bulk solution.

The Gouy–Chapman theory predicts that the differential capacitance of the interface should be a hyperbolic cosine function of the potential difference; and this should result in a curve that takes the form of an inverted parabola. Figure 9.9 shows the differential capacity against potential curves calculated on the basis of the Gouy–Chapman model and the experimental values obtained for 0·001 M KCl, and it can be seen that there is fair agreement only at potentials close to the p.z.c. Thus although the model is applicable to ions whose thermal energy is of the same order as their energy of interaction with the electrode field, it is not applicable to those ions that are adsorbed in the Helmholz double layer.

The Stern model (1924) may be regarded as a synthesis of the Helmholz model of a layer of ions in contact with the electrode (Fig. 9.2) and the Gouy–Chapman diffuse model (Fig. 9.10), and it follows that the net charge density q_S on the solution side of the interphase is now given by

$$q_S = q_H + q_G = -q_M \qquad \text{...(9.12)}$$

where q_H is the charge stuck to the electrode (Helmholz model) and q_G is the charge diffusely spread out in solution (Gouy–Chapman model). Thus although the interphase region as a whole remains electrically neutral the model leads to the concept of two distinct regions of charge separation:

1. The interface region from the electrode to the Helmholz plane (the locus of the centres of the ions in contact with the electrode) across which the potential varies linearly.
2. The region from the Helmholz plane into the solution, across which the potential varies exponentially attaining a value of zero at some distance; in this region the ions are subjected to both ordering electrical forces and disordering thermal forces.

The Stern model predicts that the total differential capacitance C will consist of two terms representing two capacitors in series

$$\frac{1}{C}=\frac{1}{C_{\mathrm{H}}}+\frac{1}{C_{\mathrm{G}}} \qquad \ldots(9.13)$$

where C_{H} is the Helmholz capacity of the region between the metal and the O.H.P., and C_{G} is the capacity of the diffuse charge. It can be shown that when the concentration of the electrolyte is large most of the charge is confined in or near the Helmholz double layer with little charge scattered in the diffuse layer, and under these circumstances

$$\frac{1}{C}\approx\frac{1}{C_{\mathrm{H}}}, \text{ or } C\approx C_{\mathrm{H}} \qquad \ldots(9.14)$$

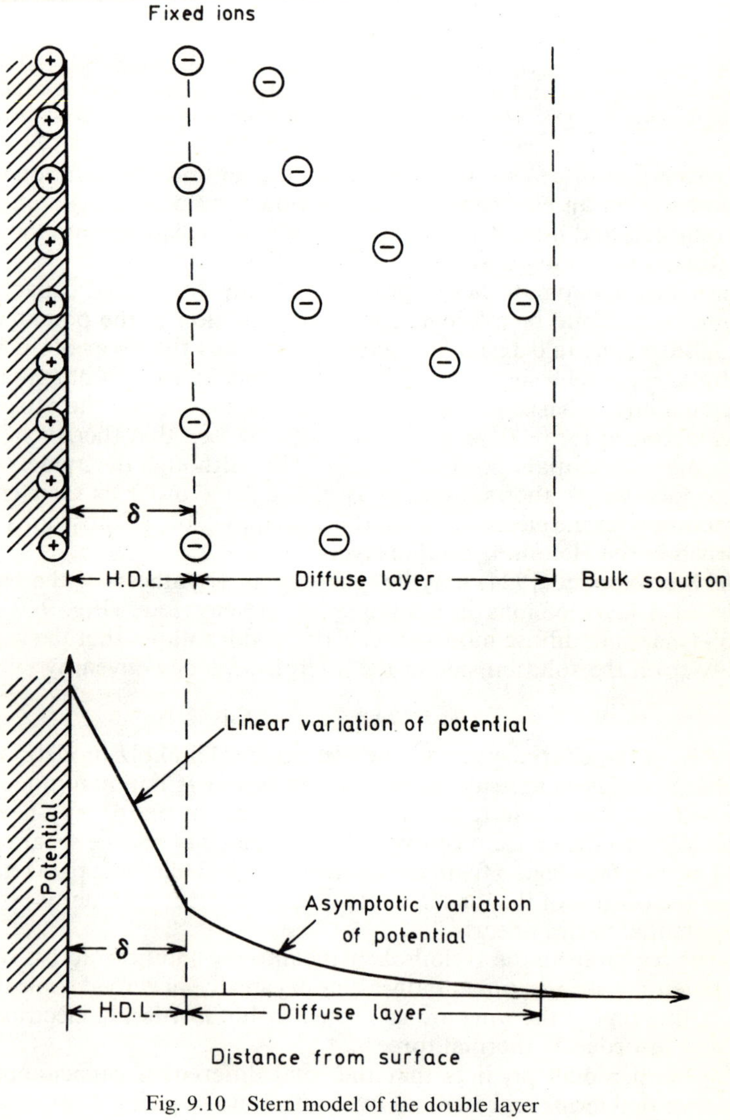

Fig. 9.10 Stern model of the double layer

which means that in concentrated solutions the diffuse layer can be neglected and the double layer approximates to the Helmholz model. Conversely, in dilute solutions

$$\frac{1}{C} \approx \frac{1}{C_G}, \text{ or } C \approx C_G \qquad \ldots(9.15)$$

and the double layer approximates to the Gouy–Chapman model.

Structure of the Double Layer

Previous considerations have indicated that the simple Helmholz double layer could be regarded as analogous to a capacitor in which the layer of excess charge on the metal (the I.H.P.) forms one plate and the counterbalancing layer of ions in solution (the O.H.P.) forms the other (Fig. 9.2). If the solution is concentrated the Gouy–Chapman diffuse layer may be disregarded, although it will exist and will provide the small amount of excess charge needed to maintain electroneutrality of the interphase as a whole.

Two alternative structures for the simple Helmholz double layer (Fig. 9.11) have been provided[1]:

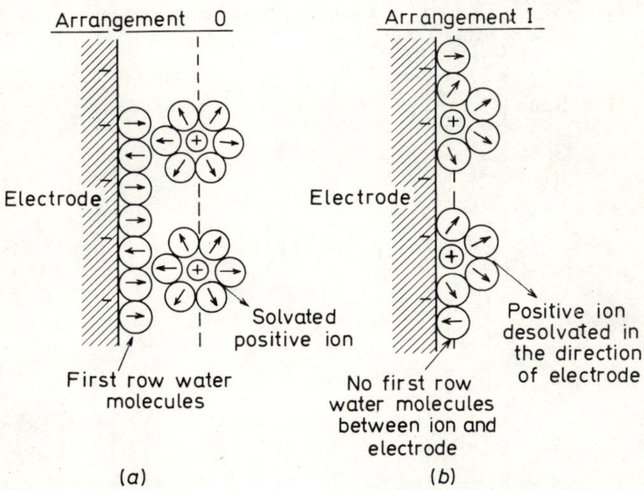

Fig. 9.11 Two types of arrangement of ions at a metal/solution interface. (*a*) Arrangement O; solvated ions in the O.H.P. and surface of electrode covered with water dipoles. (*b*) Arrangement I; desolvated ions in the I.H.P. (after Bockris and Reddy[1])

1. Arrangement O, in which water dipoles completely cover the surface of the metal with the locus of the centres of the hydrated ions forming the O.H.P.
2. Arrangement I, in which ions stripped of their solvent sheath are in direct contact with the metal.

Arrangement O is based on the hypothesis that the surface of the electrode is covered with water molecules, and it can be calculated from the number of water molecules per cm^2 on a plane in water and from the number of metal atoms per cm^2 (adsorption sites), that at least 70% of the surface is covered with water molecules, i.e. coverage $\theta > 0{\cdot}7$. Furthermore, image forces, dispersion forces and covalent bonding can increase coverage still further so that the majority of the sites will be occupied by water molecules, and if the electrode is electrically charged the field perpendicular to the electrode will result in the orientation of the adsorbed water dipoles (Fig. 9.3). Ions of opposite sign to that of the electrode, surrounded by their primary hydration sheath, will migrate to the electrode under the influence of the field until they collide with the oriented water dipoles of the hydrated electrode, but since the overlapping of the electron shells of the water molecules results in repulsion, they will not come into contact with it.

It is possible on the basis of this model (arrangement O) to explain the constant capacitance region on the negative side of the *C vs. E* curve (Fig. 9.7), and why the capacitance in this region is independent of the nature of the cations in the solution. The model of the double layer is shown in Fig. 9.12 in which it can be seen that the surface of the electrode and the

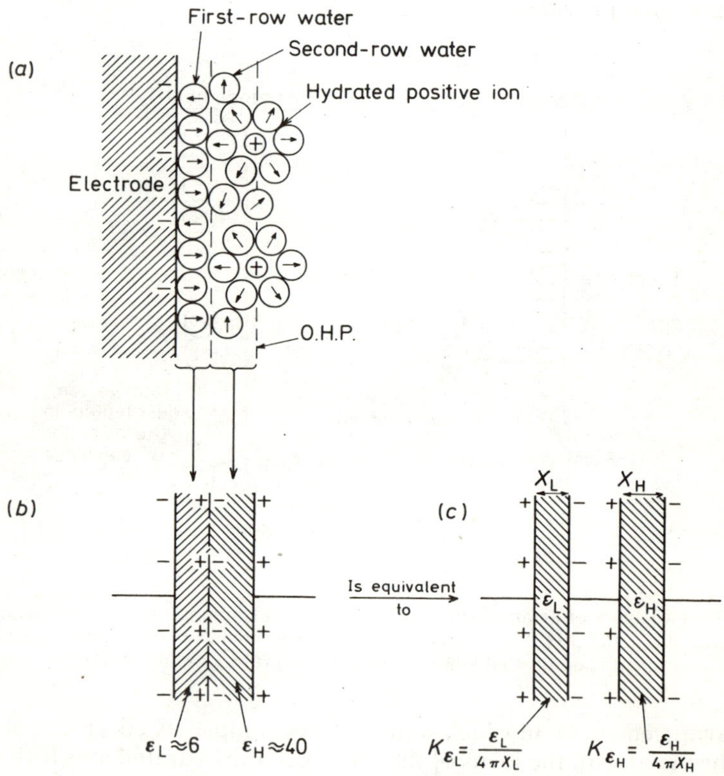

Fig. 9.12 Double layer consisting of (*a*) a layer of water dipoles and a layer of ions that may be regarded as (*b* and *c*) two capacitors in series (after Bockris and Reddy[1]). Note that K_{ε_L} and K_{ε_H} are the capacitances of the regions of high and low permittivities, respectively

adsorbed water dipoles form one capacitor, whilst the second row of water molecules and the row of hydrated cations form the other, so that in effect there are two capacitors in series. If the total differential capacity of the double layer is C, then

$$\frac{1}{C} = \frac{1}{C_{H_2O}} + \frac{1}{C_H} = \frac{d_{H_2O}}{\varepsilon_0 \varepsilon_{H_2O}} + \frac{(\delta - d_{H_2O})}{\varepsilon_0 \varepsilon_H} \qquad ...(9.16)$$

where ε_{H_2O} is the relative permittivity of the layer of adsorbed water dipoles, ε_H that of the medium between the I.H.P. (locus of the centre of adsorbed water molecules) and the O.H.P. (locus of the centres of the hydrated ions), d_{H_2O} the thickness of the water layer and $(\delta - d_{H_2O})$ the thickness of the remainder of the Helmholz double layer. Water molecules adsorbed at an electrode will be completely oriented by the field, and will not be able to make the same contribution to the permittivity as is possible in bulk water in which they are free to orientate themselves. Under these circumstances ε_{H_2O} will be as small as approximately 5–6 (compared with 80 for bulk water), and since ε_H will be greater than 40 (approximating to the value of bulk water as the distance from the electrode increases) the 2nd term on the right-hand side of equation 9.16 will be small compared with the first and can be neglected. Taking $d_{H_2O} = 2{\cdot}8 \times 10^{-10}$ m and $\varepsilon_0 = 8{\cdot}85 \times 10^{-12}$ F m^{-1}, it can be calculated that $C \approx 17 \times 10^{-2}$ F m^{-2} which is in good agreement with the capacity obtained experimentally (Fig. 9.9). This constancy of the differential capacity provides considerable support for the model shown in arrangement O, in which the capacitance is independent of the radius of the ions constituting the O.H.P.

In arrangement I (Fig. 9.11) it has been assumed that certain ions are able to divest themselves of their primary water sheath, push water dipoles away from the electrode and thus become adsorbed on the surface of the electrode. Thermodynamic calculations of the free energy of contact adsorption ΔG_C (the calculation involves a two-step process in which a certain number of water molecules become desorbed from the surface in the first step, whilst in the second step the ion divests itself of part of its water sheath and jumps into the hole) show that it is negative for certain ions (Cs^+, Cl^-, Br^- and I^-) which should therefore be able to displace water and to adsorb on the metal surface, and positive for others (Na^+, K^+ and F^-). There is a relationship between the free energy of contact adsorption and the ionic radius, and the larger the ionic radius the greater is the magnitude of $-\Delta G_C$; it follows that the larger the ionic radius (Table 9.2) the greater the tendency of the ion to adsorb at the surface of the electrode. The converse applies to ions of smaller ionic radius, which are so tightly bound to their solvent sheath that ΔG_C is positive and contact adsorption is not possible. Figure 9.13 shows the two possibilities, and it can be seen that in the absence of contact adsorption

Table 9.2 Ionic radii (nm)*

Li^+	0·068	Be^{2+}	0·035	F^-	0·133
Na^+	0·097	Mg^{2+}	0·066	Cl^-	0·181
K^+	0·133	Ca^{2+}	0·099	Br^-	0·196
Rb^+	0·147	Sr^{2+}	0·134	I^-	0·22
Cs^+	0·167	Ba^{2+}	0·134		

*Data from Parsons, R., *Handbook of Electrochemical Constants*, Butterworths, London (1959).

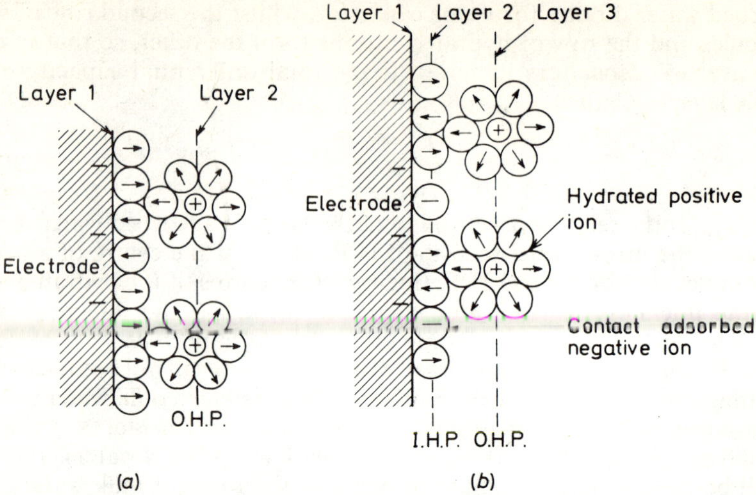

Fig 9.13 (*a*) Electrical double layer and (*b*) electrical triple layer (after Bockris and Reddy[1]). Note that Layer 2 in (*b*) is produced by adsorption of negative ions on the negatively charged electrode

the double layer is similar to the Helmholz model in which the layer of excess charge on the metal forms the I.H.P., whilst the counterbalancing layer of hydrated ions forms the O.H.P. The second possibility arises when contact adsorption occurs, the locus of the centres of the contact-adsorbed ions forming the I.H.P., and the locus of the centres of the hydrated ions the O.H.P. This model, which applies only when the solution is concentrated and contains ions of large ionic radius, is regarded by Bockris and Reddy[1] as giving rise to a *triple* layer (Fig. 9.13*b*).

It is important to note that the concept of contact adsorption resolves the problem of the adsorption of a negative ion on a negatively charged metal surface, which cannot be explained on the basis of simple coulombic forces; this difficulty was overcome formerly by referring to the phenomenon as *specific adsorption*, but it can now be interpreted in terms of contact adsorption involving short-range non-coulombic forces, i.e. image forces, dispersion forces and covalent bonding. It follows from the above that adsorption of Cl^- ions, a precursor to pitting, is possible even at potentials more negative than the p.z.c.

The majority of work on the electrified interface has been carried out using a mercury electrode, which has the advantage that it has a well-defined and reproducible surface and a highly polarisable interface when immersed in a solution. In the case of solid metals the concepts outlined are equally applicable, but modifications are necessary to allow for the following:

1. The non-electrostatic interaction between a solid metal and the constituents in solution (water, ions, etc.) will not be the same as that for mercury.
2. The degree of electrode polarisability differs, and most solid metals will be less polarisable than mercury. (See Appendix 9.1B.)
3. The surface of mercury is homogeneous and all sites (Hg atoms) are equivalent, whereas a solid metal will have a heterogeneous surface with sites that differ in adsorption energies.

Adsorption Isotherms[7,9]

An adsorption isotherm gives the relationship between the coverage of an interface with an adsorbed species (the amount adsorbed) and the pressure of gas or concentration of the species in solution; in electrochemical reactions the coverage will depend also on the potential difference at the interface.

Langmuir (1916), who put forward the first quantitative theory of the adsorption of a gas, assumed that a gas molecule condensing from the gas phase would adhere to the surface for a short time before evaporating and that the condensed layer was only one atom or molecule thick. If θ is the fraction of the surface area covered by adsorbed molecules at any time, the rate of desorption is proportional to θ and equal to $k_d\theta$, where k_d is a constant at constant temperature. Similarly the rate of adsorption will be proportional to the area of bare surface and to the rate at which the molecules strike the surface (proportional to the gas pressure p). At equilibrium the rate of desorption equals the rate of adsorption

$$k_d\theta = k_a(1-\theta)p \qquad \text{...(9.17)}$$

and the Langmuir isotherm can be expressed as

$$\frac{\theta}{1-\theta} = \frac{k_a}{k_d}p = ap \qquad \text{...(9.18)}$$

where a is a constant, the Langmuir constant or adsorption coefficient. Rearranging equation 9.18

$$\theta = \frac{ap}{1+ap} \qquad \text{...(9.19)}$$

Similarly, for adsorption of a species from solution

$$\theta = \frac{a'c_i}{1+a'c_i} \qquad \text{...(9.20)}$$

where c_i is the concentration of the species i in solution.

Two limiting cases of the Langmuir isotherm are of interest. When θ is very small, as when the pressure (or concentration) is low, or the constant a is small, then equation 9.18 reduces to

$$\theta = ap \qquad \text{...(9.21)}$$

and θ is linearly related to p, a relationship that is observed frequently in the low-pressure region of the isotherm. When $\theta \to 1$, as at high pressures or with strong adsorption at low pressures, then the isotherm becomes

$$1-\theta = \frac{1}{ap} \qquad \text{...(9.22)}$$

and the fraction of the bare surface $(1-\theta)$ is inversely proportional to p, and θ tends to a limiting value in this region.

The Langmuir isotherm is based on the following assumptions:

(*a*) The solid surface contains a fixed number of adsorption sites, and each site holds one adsorbed species.

(*b*) The heat of adsorption is the same for all sites and is independent of θ.

(*c*) The adsorbed species do not interact with one another so that a species may adsorb at an unoccupied site or desorb from an occupied site, irrespective of whether or not neighbouring sites are occupied.

The Langmuir isotherm has been found to be obeyed by a number of systems at low ($\theta <$ approx. 0·2) or high ($\theta >$ approx. 0·8) coverages, but fails in the intermediate region where the assumptions given above are no longer valid. Determinations of the heat of adsorption on clean metal surfaces shows that it frequently decreases markedly with increasing surface coverage, an observation that is indicative of surface non-uniformity. This non-uniformity is due to the intrinsic heterogeneity of the surface (edges of crystal growth planes, screw dislocations, kink sites, etc.) and to the repulsive forces between adsorbed atoms or molecules; the latter become very significant when the surface-to-adsorbate bond is partially ionic, since the repulsive forces are then large and result in a marked lowering of the heat of adsorption at the higher coverages.

Temkin suggested that the deviations from the Langmuir isotherm at high coverages could be accounted for by regarding the surface of the metal as being composed of small patches of equal size, at each of which the Langmuir isotherm holds independently with a characteristic local standard free energy of adsorption that depends on the patch distribution. The standard free energy of adsorption for each patch was assumed to decrease by equal small decrements over successively covered patches with increase in coverage, and the variation can be expressed by

$$\Delta G_\theta^\ominus = \Delta G_0^\ominus - fRT\theta \qquad \text{...(9.23)}$$

where $\Delta G_0^\ominus$ and $\Delta G_\theta^\ominus$ are the standard free energies of adsorption corresponding with $\theta = 0$ (the standard state) and finite θ, respectively. With these and other assumptions Temkin derived the isotherm

$$\theta = \frac{1}{f}\ln\left(\frac{1+a_0p}{1+a_0p\exp(-f)}\right) \qquad \text{...(9.24)}$$

where a_0 is the value of the Langmuir constant (equation 9.18) for the first micro-adsorption patch covered ($\theta = 0$) and f is defined by equation 9.23, or by

$$f = \frac{1}{RT}\frac{\mathrm{d}(\Delta G_\theta^\ominus)}{\mathrm{d}\theta} \qquad \text{...(9.25)}$$

The factor f is thus a free-energy parameter, although it is usually the enthalpy of adsorption which is regarded as decreasing linearly with coverage in the intermediate region. Temkin pointed out that if f was sufficiently large, an intermediate range of values of p would exist where $a_0p \gg 1$, whereas $a_0p(-f) \ll 1$, so that equation 9.24 could be reduced to

$$\theta = \frac{1}{f}\ln(a_0p) \qquad \text{...(9.26)}$$

which is referred to as the logarithmic Temkin isotherm, and shows that in the intermediate region the coverage θ is logarithmically related to p. For comparison the Langmuir isotherm can be written in the form

$$\ln\theta - \ln(1-\theta) = \ln a_0p \qquad \text{...(9.27)}$$

showing that θ varies more slowly with increasing p in the Temkin isotherm than in the Langmuir isotherm, and that the larger the value of f the greater the divergence. Figure 9.14 illustrates Langmuir and Temkin isotherms, which show the variation of θ with pressure of the gas[7].

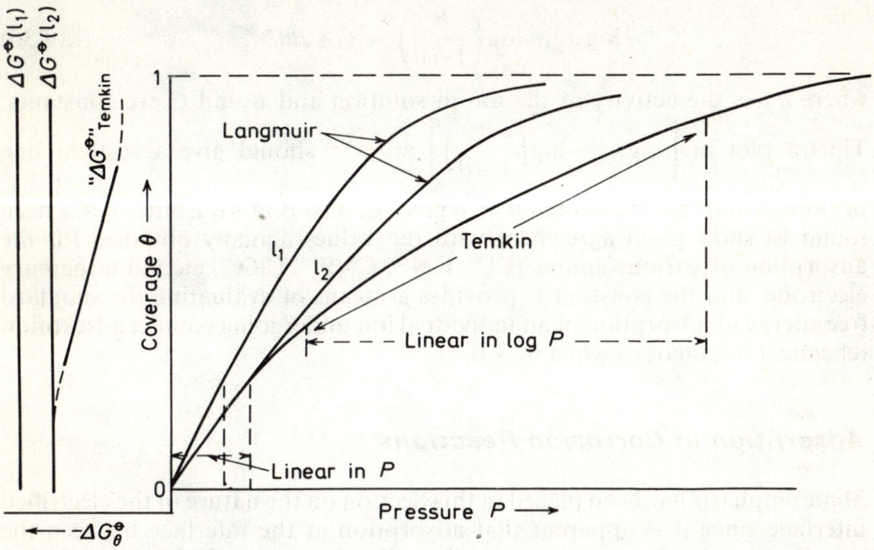

Fig. 9.14 Langmuir and Temkin isotherms showing variation of coverage as a function of pressure P of adsorbate. The Langmuir curves are for different Langmuir constants l_1 and l_2 (a in equation 9.18), and the free energy of adsorption is assumed to be constant. In the Temkin isotherm coverage is linearly related to log P in the intermediate region owing to the variation of the free energy of adsorption with coverage (after Conway[7])

Previous considerations have been confined to the effect of pressure and concentration upon coverage, but in an electrochemical equilibrium the activity and chemical potentials of the species adsorbing at the interface will also be a function of the potential difference $\Delta\phi$. For a solution containing unit activity of the species the effective pressure of the species at the interface is given by

$$p = \exp(\Delta\phi F/RT) \qquad \ldots(9.28)$$

and from equations 9.18 and 9.28

$$\frac{\theta}{1-\theta} = a\exp(\Delta\phi F/RT) \qquad \ldots(9.29)$$

which shows how coverage is related to potential difference. The Langmuir and Temkin isotherms are of importance in relation to the adsorption of H on the metal during the hydrogen evolution reaction, in which the former is applicable at low coverages and the latter at high coverages.

Bockris, Devanathan and Muller[10] (B.D.M.) in 1963 modified the Langmuir isotherm to take into account the dependence of the Langmuir constant (equation 9.19) on potential and coverage. The B.D.M. isotherm allows for the interaction of the adsorbed ions, which are assumed to be confined to the

I.H.P. and O.H.P. (the diffuse layer being disregarded), and a distinction is made between contact-adsorbed ions and ions electrostatically adsorbed by long-range coulombic forces. Wroblowa, Kovac and Bockris[2] showed that when $q_M = 0_{(\text{p.z.c.})}$, the B.D.M. isotherm can be expressed in the form

$$\log a_{\pm} - \log\left(\frac{\theta}{1-\theta}\right) = C + B\theta^{3/2} \qquad \ldots(9.30)$$

where $a_{\pm}$ is the activity of the ion in solution and B and C are constants. Thus a plot of $\left[\log a_{\pm} - \log\left(\frac{\theta}{1-\theta}\right)\right]$ *vs.* $\theta^{3/2}$ should give a straight line of slope B and the intercept at $\theta = 0$ gives C. The B.D.M. isotherm has been found to show good agreement with the values actually obtained for the adsorption of various anions (Cl^-, CN^-, CNS^-, ClO_4^-, etc.) at a mercury electrode, and the constant C provides a means of evaluating the standard free energy of adsorption of an individual ion undergoing contact adsorption (chemical interaction) when $\theta \to 0$.

Adsorption in Corrosion Reactions

Some emphasis has been placed in this Section on the nature of the electrified interface since it is apparent that adsorption at the interface between the metal and solution is a precursor to the electrochemical reactions that constitute corrosion in aqueous solution. The majority of studies of adsorption have been carried out using a mercury electrode (determination of surface tension *vs.* potential, impedance *vs.* potential, etc.) and this has lead to a greater understanding of the nature of the electrified interface and of the forces that are responsible for adsorption of anions and cations from solution. Unfortunately, it is more difficult to study adsorption on clean solid metal surfaces (e.g. platinum), and the situation is even more complicated when the surface of the metal is filmed with solid oxide. Nevertheless, information obtained with the mercury electrode can be used to provide a qualitative interpretation of adsorption phenomenon in the corrosion of metals, and in order to emphasise the importance of adsorption phenomena some examples are outlined below.

The adsorption theory of passivity[11] postulates that the phenomenon is due to the chemisorption of a monolayer of oxygen, which is considered to be atomic rather than molecular since the free energy of adsorption of O atoms usually exceeds the free energy for association of adsorbed molecular O_2; a second layer of O_2 molecules may chemisorb on top of the first monatomic layer, but is less energetically bound than the first. This adsorption is considered to interpose a stable barrier of O atoms and O_2 molecules between the metal and solution, displacing adsorbed H_2O and thus increasing the activation energy for dissolution of the metal lattice. Hoar[12] in presenting arguments in favour of the oxide film theory of passivity (*see* Section 1.5) points out that above the p.z.c. the adsorbed water dipoles will be oriented with the negative O^{2-} end of the dipole attached to the metal, and that above the reversible potential of $M + H_2O \rightarrow MO_{\text{solid}} + 2H^+ + 2e$ it will be kinetically easy for protons from the adsorbed water dipoles to transfer to others adjacent to them in solution. The adsorbed O^{2-} ions are then taken up by the metal cations in the lattice to form a monolayer of solid oxide.

Adsorption is also of major importance in the breakdown of passivity by certain anions, and the fact that ions of small ionic diameter such as Cl^- and Br^- are conducive to breakdown was explained at one time in terms of their ability to penetrate the passive oxide; however, this view is not tenable since large anions (ClO_4^-) can cause breakdown, and there is little evidence of the presence of pre-existing pores in the film. Irrespective of the mechanism of penetration, it is evident that it must be preceded by replacement of the adsorbed water by adsorbed anions, which will be favoured by a high anion concentration and by a higher (more positive) interfacial potential difference between the oxide film on the metal and the solution. Hoar[12] points out that as anion adsorption by replacing adsorbed water dipoles proceeds, the interfacial tension (or interfacial free energy) of the oxide/solution interface is progressively lowered by the mutually repulsive forces between adsorbed anions in a similar way to that of the mercury in the Lippmann electrometer. Eventually, the interfacial tension is lowered sufficiently for a kind of peptisation of the oxide by the interfacial charge to occur, and the adsorbed anions push one another, and the oxide to which they are strongly attached, apart thus resulting in a crack or split in the protective film.

Adsorption is, of course, of major importance in the inhibition of corrosion by organic compounds (adsorption inhibitors) that have the ability to adsorb strongly on the metal surface, thus impeding the dissolution reaction and reducing the corrosion rate. It follows that the coverage of a metal surface by adsorbed inhibitor can be evaluated from the relationship

$$\theta = 1 - \frac{k_i}{k}$$

where k and k_i are the corrosion rates in solutions with and without the inhibitor, respectively. The variation in coverage with concentration of inhibitor frequently conforms with Frumkin's isotherm[13, 14]

$$\frac{\theta}{1-\theta}\exp(f\theta) = Kc$$

where K is the equilibrium constant for the adsorption reaction, c is the concentration of the inhibitor and f is a parameter connected with the variation of adsorption energy with θ which is positive when the free energy of adsorption decreases with coverage, and negative when it increases

$$\Delta G_a = \Delta G^{\ominus} \pm fRT\theta$$

The Frumkin isotherm can be regarded as a general isotherm from which both the Temkin and Langmuir isotherms can be obtained as special cases. Szklarska-Smialowski and Wieczorek[15] found that the adsorption of various aliphatic compounds (acids, alcohols and amines) on steel in H_2SO_4 conformed with the Frumkin isotherm.

In connection with the adsorption of organic molecules at the surface of an electrode it is possible to distinguish two types: (*a*) adsorption of undissociated molecules and (*b*) adsorption of intermediates formed by dissociation of the original molecule. The variation of coverage of the surface of a mercury electrode with undissociated organic molecules with potential frequency takes a parabolic form with θ attaining a maximum at potentials close to the p.z.c., and this can be explained in terms of competitive adsorption between the molecules and water dipoles. Although the water dipoles are strongly attracted to the electrode at potentials more negative or more positive than the p.z.c. (Fig. 9.3), they are only weakly held at potentials close to the p.z.c., at which θ_{H_2O} obtains a minimum value, and the organic molecules have the greatest chance of displacing adsorbed water and becoming adsorbed themselves. The mechanism of the adsorption of organic inhibitors is outside the scope of this section, and reference should be made to Section 18.3 and to the comprehensive review by Trabanelli and Carassiti[10].

Electrode Kinetics

Homogeneous Chemical Reactions

Before considering the velocity (the kinetics) of an electrochemical reaction in which charge transfer takes place at a metal/solution interface, it is relevant to consider briefly the kinetics of homogeneous chemical reactions. The rate of homogeneous chemical reaction will depend on a variety of factors of which concentration of reactants, temperature, radiation and catalysis are the most important, and will range from almost instantaneous (neutralisation of a strong acid by a strong base) to extremely slow (reaction of H_2 and O_2 at ambient temperatures to form water). In a number of systems, although there is a high thermodynamic tendency for the reaction to proceed, it is nullified by kinetic factors so that the reaction proceeds extremely slowly, and many corrosion reactions fall into this category.

The rate of a chemical reaction is proportional to the concentration of the reactants, and for a reversible isothermal homogeneous reaction:

$$aA + bB \rightleftharpoons cC + dD \qquad \ldots(9.31)$$

the rate of the forward reaction $\vec{v}$ will be

$$\vec{v} = \vec{k} c_A^a c_B^b \qquad \ldots(9.32)$$

and the rate of the reverse reaction will be

$$\overleftarrow{v} = \overleftarrow{k} c_C^c c_D^d \qquad \ldots(9.33)$$

where $\vec{k}$ and $\overleftarrow{k}$ are the rate constants for the forward and reverse reactions, respectively. Since the rates are dependent on the concentration of the species involved in the reaction, it follows that the rate of the forward reaction will decrease whilst that of the reverse reaction will increase as the reaction progresses. Finally, the velocities will become equal and a state of dynamic equilibrium will be attained

$$\vec{v} = \overleftarrow{v} \quad \text{and} \quad \vec{k} c_A^a c_B^b = \overleftarrow{k} c_C^c c_D^d \qquad \ldots(9.34)$$

and the concentrations of the reactants and products are now the equilibrium values at the temperature under consideration. Therefore

$$\frac{c_C^c c_D^d}{c_A^a c_B^b} = \frac{\overrightarrow{k}}{\overleftarrow{k}} = K \qquad \text{...(9.34}a\text{)}$$

where K is the equilibrium constant for the reaction at a given constant temperature, and K will increase or decrease with increase in temperature according to the variation of ΔG with temperature ($\Delta G = -RT\ln K$). Thus for equilibria of the type $2\frac{x}{y}M + O_2 = \frac{2}{y}M_xO_y$ K decreases with increase in temperature so that p_{O_2} increases, whereas in the case of the equilibrium $2C + O_2 = 2CO$ the converse applies and K increases (*see* Section 7.6, Fig. 7.43).

The dependence of the velocity of a reaction on concentration follows from the fact that the forces between the molecules are of short range so that reaction will take place only when the reactant molecules are almost in contact. Thus 'collision' of two (or more) molecules must occur for reaction to take place, and since the probability of the two reacting molecules being in a specific place is proportional to the product of their concentrations, the rate of reaction will be proportional to the concentrations of the reactants.

Increase in temperature increases velocity, and for a homogeneous process the rate constant k increases by $\approx$200 to 300% for each 10K increase in temperature, which is surprisingly high, since the number of collisions between molecules increases by only approximately 2% per 10K increase in temperature. This anomaly was resolved by Arrhenius who suggested that in every system an equilibrium exists between 'normal' and 'active' molecules, and that it is only the latter that have sufficient energy to participate in the reaction, although the collisions of these active molecules represents only a small proportion of the total collisions. At a given constant temperature the active molecules are a constant fraction of the total number so that the theory is not inconsistent with the fact that the reaction rate is proportional to the concentration of the total molecules. Thus for a molecule to react its kinetic energy must be sufficiently greater than the average kinetic energy at that temperature to enable it to climb to the top of the energy barrier; molecules possessing this energy (or more) are said to be *activated* and the energy E^{+} to surmount the barrier is the *activation energy* of the reaction (units of $J\,mol^{-1}$).

The number of molecules n_E possessing an energy in excess of a given value of kinetic energy E is given by Maxwell's Distribution Law

$$n_E = n\exp(-E/RT) \qquad \text{...(9.35)}$$

where n is the total of the molecules. It follows that a similar relationship will apply between the chemical rate constant and the height of the energy barrier E^{+} that only the activated molecules are able to surmount

$$k_{chem.} = A\exp(-E^{+}/RT) \qquad \text{...(9.36)}$$

where A is a constant and $k_{chem.}$ is the chemical rate constant.

Equation 9.36 is referred to as the Arrhenius equation, and taking logarithms

$$\ln k_{\text{chem.}} = \ln A - \frac{E^{\ddagger}}{RT} \qquad \text{...(9.37)}$$

or

$$2{\cdot}3 \log k_{\text{chem.}} = 2{\cdot}3 \log A - \frac{E^{\ddagger}}{RT} \qquad \text{...(9.38)}$$

Thus a plot (Arrhenius plot) of $\log k_{\text{chem.}}$ against $1/T$ will give a straight line of slope $-E^{\ddagger}/2{\cdot}3R$, which enables the activation energy to be evaluated from the variation of the rate constant with temperature. Equation 9.38 shows that the smaller the activation energy of a reaction (the lower the height of the energy barrier) the more rapid the reaction at a given temperature, and that for any reaction the rate of reaction will increase with increase in temperature.

Rate of Electrochemical Reaction

Previous considerations have shown that in homogeneous chemical reactions the rate is determined by the concentrations of reactants and the temperature, and that under the same conditions of concentration and temperature different reactions will proceed at rates that vary according to the activation energies of the reaction. Similar considerations to the above apply to heterogeneous reactions involving charge transfer at a metal/solution interface, but the potential of the electrified interface introduces an important additional factor, and the reactant rate at constant temperature and concentration of reactants can be increased or decreased very significantly by changing the potential of the electrified interface.

Consider the cathodic reduction of a hydrated metal ion M^{+}(aq.) to a lattice metal atom at the interface M/M^{+}(aq.) by a simple one-step reaction

$$M^{+}(\text{aq.}) + e = M(\text{l})$$

which has been selected to avoid the complications that are involved when the overall reaction consists of two or more steps of which one is slower than the other and is thus rate determining[2] (see p. **9**.43).

In the double layer the ion will be hydrated and its distance from the metal surface will be approximately the diameter of the water molecule attached to the metal surface plus the radius of the hydrated metal ion, i.e. $0{\cdot}28 + (\frac{1}{2} \times 2{\cdot}8) + r_{M^{+}} \approx 0{\cdot}5$–$0{\cdot}6$ nm which is the distance between the metal surface and the O.H.P. The ion will be vibrating constantly with respect to its hydration sheath, and in its attempt to attain the metal surface it will have to surmount the energy barrier (Fig. 9.15), which is regarded as having a maximum potential energy at a distance approximately half-way between its original position at the O.H.P. and the metal surface. At this maximum the metal ion receives the electron and becomes an uncharged atom, which can then descend of its own accord along the remainder of the potential energy–distance (Morse) curve and enter the lattice of the metal. It also follows that only metal ions with an energy greater than the maximum will be able to surmount the potential energy barrier and become metal atoms.

Fig. 9.15 shows how the rate of a cathodic reaction can be increased by changing the height of the energy barrier (cf. Fig. 9.16*b* for the electrode at equilibrium). Potential energy *vs.* distance curves and corresponding potential differences for the various possibilities at an electrified interface are shown in Fig. 9.16. If the reaction was independent of the potential, as it would be if there was no excess charge and the potential was $E_{p.z.c.}$ (Fig. 9.16*a*), the

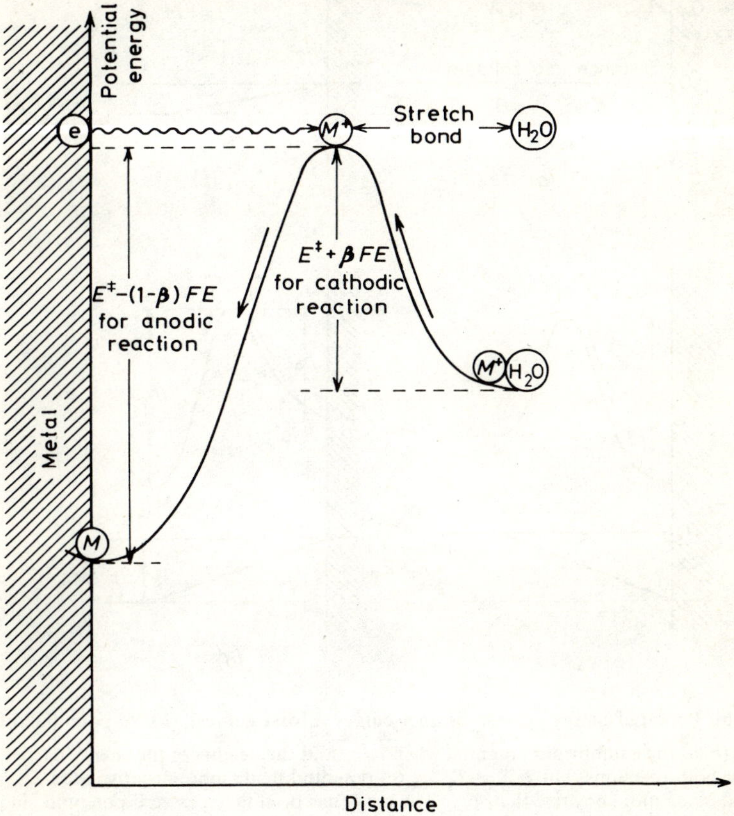

Fig. 9.15 Cathodic reduction of M^+(aq.) to metal in which the activation energy (height of the energy barrier) for the cathodic reaction is lowered to $E^{\ddagger}+\beta FE$ whilst that for the anodic reaction is raised to $E^{\ddagger}-(1-\beta)FE$ (note that $E^{\ddagger}$ is positive and E negative). (After Bockris and Drazic[2])

rate would conform to a first-order chemical reaction (the order of a reaction is the number of atoms or molecules whose concentration determines the velocity), and

$$v_{\text{chem.}} = k_{\text{chem.}} c_{M^+_{(\text{aq.})}} \quad \ldots(9.39)$$

in which $k_{\text{chem.}}$ is defined by equation 9.36; under these conditions the rate will be proportional to $c_{M^+_{(\text{aq.})}}$. However, in the case of an electrochemical reaction the rate will also depend on the potential difference at the interface.

If the potential of the metal/solution interface is made more negative than $E_{p.z.c.}$ by giving it an excess negative charge the positively charged ion will

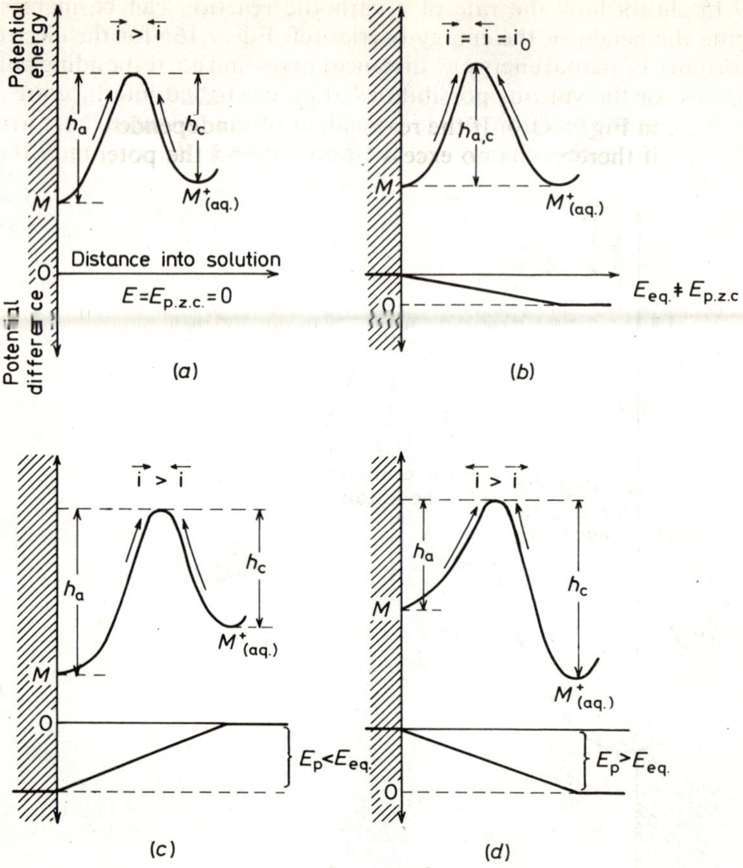

Fig. 9.16 Potential energy against distance curves (Morse curves). (*a*) No potential difference (p.z.c.), (*b*) at the equilibrium potential when $\vec{i} = \overleftarrow{i}$ and the heights of the energy barrier are the same for both reactions, but $E_{eq.} \neq E_{p.z.c.}$, (*c*) potential made more negative than $E_{eq.}$ and (*d*) potential made more positive than $E_{eq.}$. The p.z.c. has been taken as zero potential, and h_a and h_c are the heights of the potential barriers for the anodic and cathodic reactions, respectively; $\vec{i}$ is the rate of the cathodic reaction and $\overleftarrow{i}$ the rate of the anodic reaction (after Bockris and Drazic[2])

be electrostatically attracted to the metal, and this will result in an increase in the rate of the cathodic reaction and a decrease in the rate of the anodic reaction which are equal only at $E_{eq.}$, i.e. the height of the energy barrier for cathodic reduction will be lowered whilst that for anodic dissolution will increase (Fig. 9.17). The chemical rate equation must now be modified to take into account the energy associated with the passage of $1F$ coulombs of charge (for 1 mol of M^+, or zF coulombs for 1 mol of M^{z+}) through a potential difference E, i.e. $EF\,\mathrm{J\,mol^{-1}}$. However, all this energy is not required since once the metal ion has been given sufficient energy to attain the energy hump it will move down the other side on its own accord. Hence

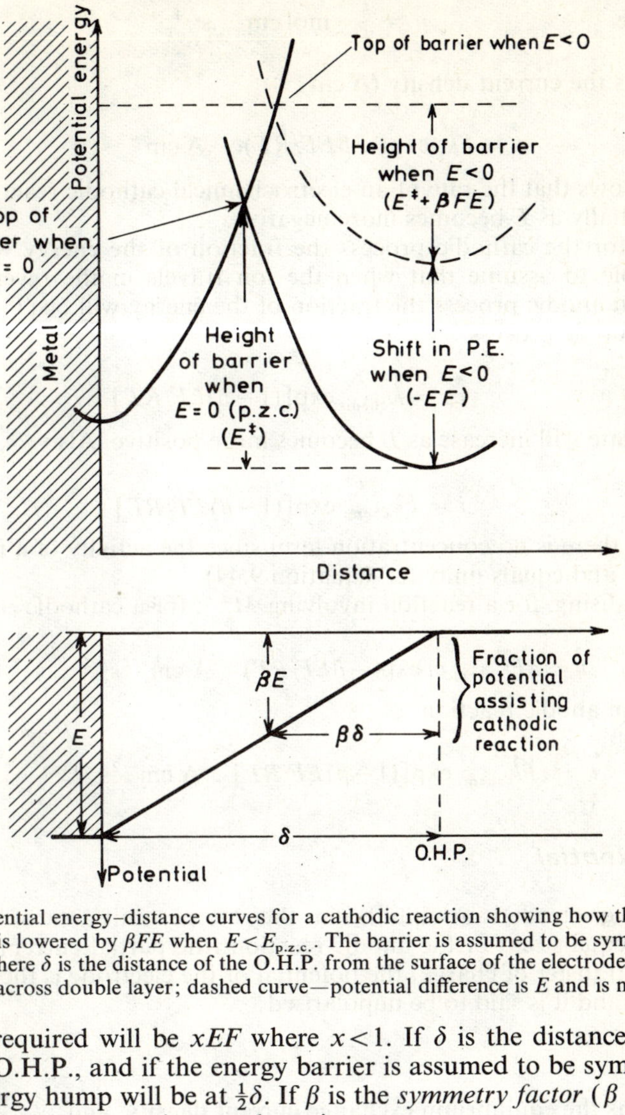

Fig. 9.17 Potential energy–distance curves for a cathodic reaction showing how the potential energy barrier is lowered by βFE when $E < E_{p.z.c.}$. The barrier is assumed to be symmetrical so that $\beta \approx \frac{1}{2}\delta$, where δ is the distance of the O.H.P. from the surface of the electrode. Full curve —no field across double layer; dashed curve—potential difference is E and is negative

the energy required will be xEF where $x < 1$. If δ is the distance between I.H.P. and O.H.P., and if the energy barrier is assumed to be symmetrical, then the energy hump will be at $\frac{1}{2}\delta$. If β is the *symmetry factor* ($\beta \approx \frac{1}{2}$), the electrochemical rate equation can be written

$$\vec{k}_e = A \exp[-(E^{\ddagger} + \beta EF)/RT] \qquad \text{...(9.40)}$$

and from equation 9.36

$$\vec{k}_e = \vec{k}_{chem.} \exp(-\beta EF/RT) \qquad \text{...(9.41)}$$

where $\vec{k}_e$ is the electrochemical rate constant and $\vec{k}_{chem.}$ the chemical rate constant for the cathodic reaction, and β is the symmetry factor.

$$\therefore\ \vec{v}_e = \vec{k}_e c_i = \vec{k}_{chem.} \exp(-\beta EF/RT) c_i \quad \text{mol s}^{-1} \qquad \text{...(9.42)}$$

and since $$v = \frac{i}{F} \quad \text{mol cm}^{-2}\,\text{s}^{-1} \qquad \text{...(9.43)}$$

where i is the current density (A cm^{-2})

$$\vec{i} = F\vec{k}_e \exp(-\beta EF/RT)c_i \quad \text{A cm}^{-2}\,\text{s}^{-1} \qquad \text{...(9.44)}$$

which shows that the rate of an electrochemical cathodic reaction increases exponentially as E becomes more negative.

Since for the cathodic process the fraction of the energy was βEF it is reasonable to assume that when the ion travels in the reverse direction during an anodic process the fraction of the energy will be $(1-\beta)EF$. Thus for the anodic process

$$\overleftarrow{v}_e = \overleftarrow{k}_{\text{chem.}} \exp[(1-\beta)EF/RT] \qquad \text{...(9.45)}$$

and the rate will increase as E becomes more positive. Also

$$\overleftarrow{i} = Fk_{\text{chem.}} \exp[(1-\beta)EF/RT] \qquad \text{...(9.46)}$$

in which there is no concentration term since the activity of a pure metal is invariant and equals unity (cf. equation 9.44).

Generalising, for a reaction involving M^{z+}, for a cathodic reaction

$$\vec{i} = zF\vec{k}_{\text{chem.}}c_i \exp(-\beta EF/RT) \quad \text{A cm}^{-2}\text{s}^{-1} \qquad \text{...(9.47)}$$

and for an anodic reaction

$$\overleftarrow{i} = zF\overleftarrow{k}_{\text{chem.}} \exp[(1-\beta)EF/RT] \quad \text{A cm}^{-2}\text{s}^{-1} \qquad \text{...(9.48)}$$

Overpotential

When an electrode is at equilibrium the rate per unit area of the cathodic reaction equals that of the anodic reaction (the partial currents) and there is no *net* transfer of charge; the potential of the electrode is the equilibrium potential and it is said to be unpolarised:

$$\overleftarrow{i} = \vec{i} = i_0 \qquad \text{...(9.49)}$$

where i_0 is the equilibrium exchange current density, and

$$E = E_{\text{eq.}},\ \text{and}\ \eta = E - E_{\text{eq.}} = 0 \qquad \text{...(9.50)}$$

Thus η the *overpotential* is defined as the departure of the potential of the electrode from its equilibrium value.

For a cathodic process in which the electrode is polarised to a potential E_p

$$\vec{i} > \overleftarrow{i}\ \text{and}\ i_c = \vec{i} - \overleftarrow{i} \qquad \text{...(9.51)}$$

where i_c is the *net* cathodic current density that results when E_p is made more negative than the equilibrium potential $E_{\text{eq.}}$. By definition

$$\eta_c = E_{p,c} - E_{\text{eq.},c} \qquad \text{...(9.52)}$$

and since for a cathodic reaction

$$E_{p,c} < E_{eq.,c}$$

then $\eta_c < 0$, i.e. it is always negative. Similarly, for a net anodic process

$$\overleftarrow{i} > \overrightarrow{i} \text{ and } i_a = \overleftarrow{i} - \overrightarrow{i} \quad \text{...(9.53)}$$

and

$$\eta_a = E_{p,a} - E_{eq.,a} \quad \text{...(9.54)}$$

and since

$$E_{p,a} > E_{eq.}$$

then $\eta_a > 0$, i.e. it is always positive.

At equilibrium, $\overleftarrow{i}$ and $\overrightarrow{i}$ in equations 9.44 and 9.46 can be replaced by i_0 and the Es by $E_{eq.}$

$$\therefore \quad i_0 = zF\overleftarrow{k}_e \exp[(1-\beta)E_{eq.}F/RT] = zF\overrightarrow{k}_e \exp(-\beta E_{eq.}F/RT) \quad \text{...(9.55)}$$

If the reaction is proceeding at a finite rate, the Es will be the polarised potentials E_p, and since

$$E_p = \eta_a + E_{eq.} \quad \text{...(9.56)}$$

equation 9.46 can be expressed as

$$\overleftarrow{i} = zF\overleftarrow{k}_e \exp[(1-\beta)E_{eq.}F/RT]\exp[(1-\beta)\eta_a F/RT] \quad \text{...(9.57)}$$

and from equation 9.55 for an anodic process

$$\overleftarrow{i} = i_0 \exp[(1-\beta)\eta_a F/RT] \quad \text{...(9.58)}$$

Similarly for the cathodic process

$$\overrightarrow{i} = i_0 \exp(-\beta\eta_c F/RT) \quad \text{...(9.59)}$$

and in both cases the greater the magnitudes of i_0 and η (taking into account the appropriate sign) the greater the rate per unit area of the reaction.

The net current density is the difference between the two partial current densities, and the net anodic current density can be written in the form

$$i_a = \overleftarrow{i} - \overrightarrow{i} \quad \text{...(9.60)}$$

(the net current density can be cathodic or anodic, and the above equation could be written in the form $i_c = \overrightarrow{i} - \overleftarrow{i}$ without affecting the argument). Introducing equations 9.58 and 9.59 into equation 9.60

$$i_a = i_0[\exp[(1-\beta)\eta_a F/RT] - \exp(-\beta\eta_a F/RT)] \quad \text{...(9.61)}$$

It can be seen that when $\eta = 0$, $i = i_0$ and the rate of the cathodic process equals that of the anodic process, i.e. the reaction is at equilibrium. However, if η is positive the first term increases exponentially, whilst the second term decreases exponentially, and at overpotentials $> +0{\cdot}052$ V the latter can be neglected. Thus at overpotentials $> +0{\cdot}052$ V equation 9.61 simplifies to

$$i_a = i_0 \exp[(1-\beta)\eta_a F/RT] \qquad \text{...(9.62)}$$

Similarly, for overpotentials more negative than $-0{\cdot}052$ V the first term becomes negligible, and

$$i_c = -i_0 \exp(-\beta\eta_c F/RT) \qquad \text{...(9.63)}$$

Taking logarithms in equation 9.62

$$\ln i_a = \ln i_0 + \frac{(1-\beta)\eta_a F}{RT} \qquad \text{...(9.64)}$$

and for the anodic reaction

$$\eta_a = \frac{-RT}{(1-\beta)F}\ln i_0 + \frac{RT}{(1-\beta)F}\ln i \qquad \text{...(9.65)}$$

and since

$$\frac{RT}{F}\ln x = 0{\cdot}059 \log x \text{ at } 25°\text{C, then}$$

$$\eta_a = \frac{-0{\cdot}059}{(1-\beta)}\log i_0 + \frac{0{\cdot}059}{(1-\beta)}\log i \qquad \text{...(9.66)}$$

Similarly for the cathodic reaction

$$\eta_c = \frac{+0{\cdot}059}{\beta}\log i_0 - \frac{0{\cdot}059}{\beta}\log i \qquad \text{...(9.67)}$$

which was obtained first by Tafel from experimental studies of the hydrogen evolution reaction on various metals, and was expressed in the form

$$\eta = a + b \log i \text{ (Tafel equation)} \qquad \text{...(9.68)}$$

Attention should be drawn to the signs in equations 9.66 and 9.67 and it should be noted that the Tafel slope b is always positive for an anodic process and negative for a cathodic process and that the constant a is of opposite sign to the slope.

The most significant parameter in the relationships given above is the equilibrium exchange current density i_0, which can be evaluated by extrapolating the linear η *vs.* $\log i$ curve to $\eta = 0$, at which $\log i = \log i_0$. Alternatively, i_0 may be evaluated from the linear E_p *vs.* $\log i$ curve by extrapolating the curve to the equilibrium potential $E_{eq.}$.

There are two further possibilities that must be considered, and these are discussed below.

Multistep reactions Previous considerations have been based on a simple one-step reaction involving one electron, but if the reaction occurs by a series of steps of which one is significantly slower than all the others, which may be regarded as at equilibrium, and is thus rate determining, equation 9.61 is not valid and becomes

$$i_a = i_0[\exp \overleftarrow{\alpha}\eta_a F/RT - \exp \overrightarrow{\alpha}\eta_a F/RT] \qquad \text{...(9.69)}$$

where $(1-\beta)$ is replaced by $\overleftarrow{\alpha}$ the *transfer coefficient* for the anodic reaction, and β is replaced by $\overrightarrow{\alpha}$ the transfer coefficient for the cathodic reaction. Neglecting the second term, as in equation 9.62, and expressing equation 9.62 in terms of overpotential

$$\eta_a = -\frac{0\cdot059}{\overleftarrow{\alpha}}\log i_0 + \frac{0\cdot059}{\overleftarrow{\alpha}}\log i_a \qquad \ldots(9.70)$$

The parameters included in the transfer coefficient will be considered subsequently in relation to the r.d.s. of the hydrogen evolution reaction.

Small overpotentials For overpotentials $<0\cdot010$ V, equation 9.61 can be simplified using the identity

$$e^x = 1+x+\frac{x^2}{\lfloor 2}+\cdots \text{ and } e^{-x} = 1-x+\frac{x^2}{\lfloor 2}-\cdots$$

and, taking only the first two terms

$$i = i_0\left[\left(1+\frac{(1-\beta)F\eta}{RT}\right)-\left(1-\frac{\beta F\eta}{RT}\right)\right] \qquad \ldots(9.71)$$

which reduces to

$$i = \frac{i_0F}{RT}\times\eta, \text{ or } \eta = \frac{RT}{i_0F}\times i \qquad \ldots(9.72)$$

showing that η against i is linear with a slope of RT/i_0F. Equation 9.72 is applicable to either the anodic or the cathodic reaction, and provides another method of evaluating i_0.

Transport or Concentration Overpotential

Transport of a species in solution to and from an electrode/solution interface may occur by migration, diffusion and convection although in any specific system they will not necessarily be of equal importance. However, at the steady state all steps involved in the electrode reaction must proceed at the same rate, irrespective of whether the rate is controlled by a slow step in the charge transfer process or by the rate of transport to or from the electrode surface. It follows that the rate of transport must equal the rate of charge transfer:

$$J = \frac{dQ}{dt} = \frac{i}{zF} \qquad \ldots(9.73)$$

where J the transport flux equals the number of moles transported across an area of 1 cm^2/s (dQ/dt) and i is the charge transfer per unit area of surface ($A\,cm^{-2}$).

In a cathodic process, removal of ions from solutions will result in a decrease in their concentration at the electrode/solution interface compared to that in the bulk solution, and this in turn will cause a concentration gradient and consequent diffusion. Furthermore, the potential gradient through the solution will result in the migration of cations and anions to their respective electrodes (electrolytic conduction), although it does not follow that the ion transported to the electrode surface by migration will necessarily participate in the reaction. Thus copper can be cathodically deposited from a sulphuric acid plus copper sulphate solution at 100% cathode efficiency, but charge transport through the solution will be predominantly by H_3O^+ ions, the Cu^{2+} ions being transported to the electrode surface mainly by diffusion. On the other hand, in a concentrated solution of aluminium sulphate, charge transport through the solution will be mainly due to migration of Al^{3+} and SO_4^{2-}, but the electrode process at the cathode will be solely hydrogen discharge. In the case of a cathodic process involving an uncharged species such as dissolved oxygen, transport is solely by diffusion and other ions in solution will be responsible for migration. In addition to diffusion and migration, ions and uncharged species may be transported to the electrode surface either by convection resulting from density differences within the solution caused by uneven heating or cooling of the solution, or by the electrode process itself. Thus during the heating of a solution contained in a beaker, the hot solution of the bottom of the beaker will tend to rise because of its decreased density, whilst the cold solution at the top will tend to fall. Even when a solution is not deliberately heated, evaporation from the surface or air draughts may give rise to temperature differences and hence convective mixing; indeed it is extremely difficult to avoid convective mixing and for this reason solutions used in corrosion testing are often stirred in order to obtain reproducible conditions.

Forced convection can be achieved in a variety of ways, e.g. by agitating the solution (stirring mechanically, sparging with gas bubbles, ultrasonic radiation, etc.) or by moving or rotating the metal electrode, and this will result in more rapid transport of the reacting species to the electrode than when the solution is unagitated.

Diffusion and Transport Overpotential (η_T)

When a cathodic process occurs at a finite rate the concentration $c_{x=0}$ of the electron acceptor (cathode reactant) at the metal/solution interface ($x = 0$) will become less than that in the bulk solution c_b, and as the rate increases it will continue to decrease until it becomes zero, i.e. as soon as the electron acceptor arrives at the interface electron transfer occurs.

Thus for a cathodic process

$$c_{x=0} < c_b, \text{ and } E_{x=0} < E_b$$

whereas for the anodic process

$$c_{x=0} > c_b, \text{ and } E_{x=0} > E_b$$

Thus owing to the electrode process the potential will change and will become more negative in a cathodic process and more positive in an anodic process:

$$E_{c,x=0} < E_b < E_{a,x=0}$$

where E_b is the potential of the metal in the bulk solution.

If the system is regarded as a reversible concentration cell, the Nernst equation can be applied (Appendix 9.1A, equation A.62) and the transport overpotential η_T for a cathodic process is given by

$$\eta_T = E_{x=0} - E_b = \frac{RT}{zF}\ln\left(\frac{c_{x=0}}{c_b}\right) = \frac{0{\cdot}059}{z}\log\left(\frac{c_{x=0}}{c_b}\right) \text{ at } 25°\text{C} \qquad \ldots(9.74)$$

and since $c_{x=0} < c_b$, η_c will be negative; the converse will apply to an anodic process and η_a will be positive. Although c_b is known, the concentration at the interface $c_{x=0}$ is unknown and cannot be readily determined experimentally. However, if it is assumed that there is a stagnant layer of solution adjacent to the electrode through which transport occurs entirely by diffusion, it is possible to evaluate $c_{x=0}$ from diffusion theory. For simplicity, this layer is regarded as having a precise thickness δ and a linear concentration gradient, although non-linearity occurs at the outer plane of the layer where the gradient becomes asymptotic (Fig. 9.18). The thickness, however, varies according to conditions and will depend on the geometry of the electrode, the diffusion coefficient of the reacting species, and the kinematic viscosity (density/viscosity) and velocity of the solution. Under steady-state conditions of natural convection, $\delta \approx 0{\cdot}05$ cm but it decreases under conditions of high forced convection to a value of $\approx 0{\cdot}001$ cm, and even lower values may be

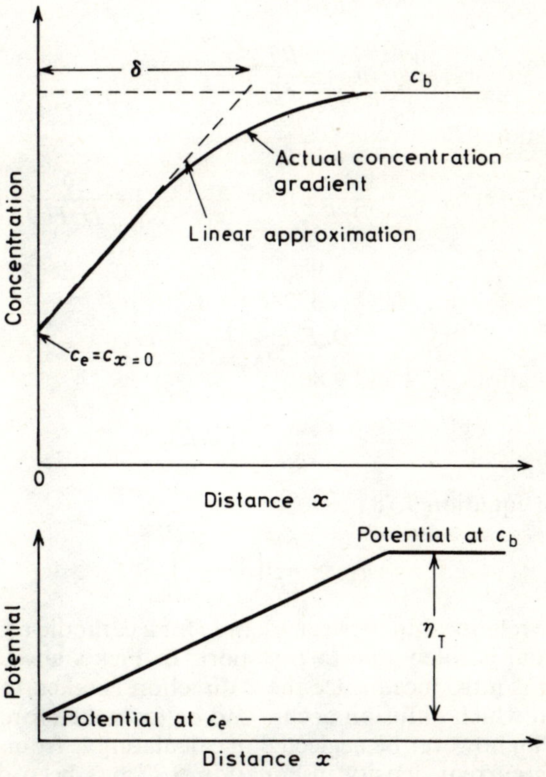

Fig. 9.18 Concentration gradient and potential gradient in the diffusion layer for a cathodic process (note that the potential drop through the solution has not been included)

obtained under conditions of very rapid laminar flow such as is produced by a rotating-disc electrode.

If it is assumed that transport is entirely by diffusion and that there is a linear concentration gradient through the diffusion layer, the concentration gradient at $x = 0$, i.e. $(\mathrm{d}c/\mathrm{d}x)_{x=0}$, can be expressed as $(c_b - c_{x=0})/\delta$. From Fick's first law and equating the transport flux J with the rate i of an electrochemical reaction

$$J = -D\frac{\mathrm{d}c}{\mathrm{d}x} = -D\frac{(c_b - c_{x=0})}{\delta} = \frac{i}{zF} \qquad \ldots(9.75)$$

where D is the diffusion coefficient

$$\therefore i = -DzF \times \frac{(c_b - c_{x=0})}{\delta} \qquad \ldots(9.76)$$

The concentration gradient will be a maximum when $c_{x=0} = 0$, and this will correspond with the *maximum* or *limiting* current density i_L

$$\therefore i_L = \frac{-DzFc_b}{\delta} \qquad \ldots(9.77)$$

The Nernst equation written in terms of the transport overpotential of a cathode is

$$\eta_T = \frac{RT}{zF}\ln\frac{c_{x=0}}{c_b} \qquad \ldots(9.78)$$

and from equation 9.76

$$c_b - c_{x=0} = -\frac{\delta}{DzF}\times i, \text{ or } \frac{c_{x=0}}{c_b} = 1 + \frac{\delta}{DzFc_b}\times i \qquad \ldots(9.79)$$

From equation 9.77

$$\frac{\delta}{DzFc_b} = -\frac{1}{i_L} \qquad \ldots(9.80)$$

hence from equations 9.79 and 9.80

$$\frac{c_{x=0}}{c_b} = 1 - \frac{i}{i_L} \qquad \ldots(9.81)$$

Substituting in equation 9.78

$$\eta_T = \frac{RT}{zF}\left(1 - \frac{i}{i_L}\right) \qquad \ldots(9.82)$$

which gives the relationship between η_T and i for a cathodic reaction in which the overpotential is solely due to transport. In Fick's law (equation 9.75) the minus sign is introduced since the x direction is taken to coincide with the direction at which diffusion occurs and $\partial c/\partial x$ is therefore negative. The negative sign can however be neglected in calculating i_L from equation 9.77.

The limiting current density in equation 9.77 has been derived on the assumption that transport is solely by diffusion, but if migration also occurs then for a cathodic process

$$i_L = \frac{DzFc_b}{\delta(1-n_+)} \qquad \text{...(9.83)}$$

where n_+ is the transference number of the cation that is involved in charge transfer. In this equation i_L is in $A\,cm^{-2}$, D is in $cm^2\,s^{-1}$ and c_b is in $mol\,cm^{-3}$ or $g\,ions\,cm^{-3}$ (the r.h.s. of equation 9.83 must be multiplied by 10^{-3} if c_b is in $mol\,dm^{-3}$). It should be noted that equation 9.83 involves 1 mol of the reactant and z must be evaluated accordingly, i.e. for the hydrogen evolution reaction (h.e.r.) H^+ (1 mol)$+e = \frac{1}{2}H_2$ and $z = 1$; for the oxygen-reduction reaction, O_2 (1 mol)$+2H_2O+4e = 4OH^-$, and $z = 4$.

For the majority of ions the diffusion coefficient D is $\approx 10^{-5}\,cm^2\,s^{-1}$ at 25° (the values of H^+ and OH^- at infinite dilution are somewhat higher, i.e. $9{\cdot}3\times10^{-5}\,cm^2\,s^{-1}$ and $5{\cdot}2\times10^{-5}\,cm^2\,s^{-1}$, respectively) so that equation 9.83 can be simplified:

$$i_L \approx \frac{10^{-5}\times 96\,500 zc_b}{0{\cdot}05} \approx 2\times10^1 zc_b \qquad \text{(for natural convection)}$$

$$i_L \approx \frac{10^{-5}\times 96\,500 zc_b}{0{\cdot}001} \approx 1\times10^3 zc_b \qquad \text{(for extreme forced convection)}$$

where i_L is in $A\,cm^{-2}$ and c_b is in $mol\,cm^{-3}$, which represents the two extremes of i_L. Thus i_L can be increased by increasing c_b the bulk concentration, and by decreasing the thickness of the diffusion layer by forced convection. It should also be noted that D increases with temperature ($\approx$2–3%/°C rise) and i_L will therefore increase with temperature.

From the identity $\ln(1-x) = x-\frac{1}{2}x^2+\frac{1}{3}x^3$ it is evident that $\ln(1-x)\rightarrow x$ for small values of x. Thus for small values of i, equation 9.82 can be written

$$\eta_T = \frac{0{\cdot}059}{z}\frac{i}{i_L} = \frac{k}{z}\times i \qquad \text{...(9.84)}$$

and η will be linearly related to i. At higher current densities the curve departs from linearity, and as $i\rightarrow i_L$, $\eta_c\rightarrow\infty$ and becomes infinitely negative, and the rate of the process remains constant at i_L. Any further increase in rate will then require an alternative charge transfer process.

Since in most cases of corrosion in which transport of the cathodic reactant (H_3O^+, dissolved O_2, Fe^{3+}, HNO_3 acid, etc.) is rate determined, the anodic curve intersects the cathodic curve at i_L, then

$$i_L = i_{corr.} \qquad \text{...(9.85)}$$

and any factor that increases i_L will result in a corresponding increase in $i_{corr.}$.

The Hydrogen Evolution Reaction

The hydrogen evolution reaction (h.e.r.) is of particular importance in corrosion for a number of reasons. Firstly, the reduction of the H_3O^+ ion in acid solutions or the H_2O molecule in neutral and alkaline solution is a common cathodic reaction for the corrosion of metals in acid, neutral and alkaline solutions; the fact that iron will corrode in neutral water free from dissolved oxygen with the evolution of H_2 can be confirmed by testing the gas bled off from the radiator of a central-heating system. Secondly, hydrogen may pass into the lattice of certain metals and lead to embrittlement (*see* Section 8.4), and in the case of a number of alloys, particularly the high-tensile steels, it can lead to a catastrophic fracture. The h.e.r. has been the most widely studied electrode reaction, and some account of its complexity can be gained from the short account that follows.

In deriving the kinetics of activation-energy controlled charge transfer it was emphasised that a simple one-step electron-transfer process would be considered to eliminate the complications that arise in multistep reactions. The h.e.r. in acid solutions can be represented by the overall equation:

$$H_3O^+ + e \rightarrow \tfrac{1}{2}H_2 + H_2O$$

and at first sight it would appear that this reaction was as simple as $M^+_{(aq.)} + e \rightarrow M$. However, this is not the case and the sequence of events can be expressed by the following reactions:

$$H_3O^+_{\text{bulk solution}} \xrightarrow{\text{convection, migration and diffusion}} H_3O^+_{\text{metal/solution interface}}$$

discharge step (D.)

$$H_3O^+ + M + e \rightarrow M\text{—}H + H_2O \qquad \text{...(step 1)}$$

followed by *either* a chemical desorption (C.D.) step

$$M\text{—}H + M\text{—}H \rightarrow 2M + H_2 \qquad \text{...(step 2)}$$

or, an electrochemical desorption (E.D.) step

$$M\text{—}H + H_3O^+ + e \rightarrow M + H_2 + H_2O \qquad \text{...(step 3)}$$

It should be noted that the terminology used is variable and that step 2 is also referred to as 'combination', 'recombination' and 'Tafel combination'; step 3 is also referred to as 'ion-atom desorption'.

A similar series of steps can be written for the h.e.r. in neutral and alkaline solutions, but the equations will then involve the water molecule, since the concentration of H_3O^+ in these solutions is too small for it to participate at a significant reaction rate.

It is important to note that the discharge step involves adsorption of hydrogen atoms at available sites on the metal surface, and is followed by desorption that may be either a chemical step (step 2), or an electrochemical step (step 3) in which further charge transfer occurs. The extent of adsorption of hydrogen atoms is referred to as coverage θ, and $\theta_H = 1$ represents coverage of all the sites available with adsorbed hydrogen atoms; coverage will vary with the mechanism of the h.e.r., the overpotential, the nature of the metal, the nature of the solution, etc.

After adsorption the $H_{ads.}$ may be removed as H_2 molecules by either step 2 or 3, but as mentioned above there is an additional step:

$$M\text{—}H_{ads.} \rightarrow H_{inst.}\text{—}M$$

which represents the transfer of the $H_{ads.}$ across the interface into interstitial sites within the metal. This will result in a concentration gradient with consequent diffusion of the H atoms into the interior of the metals via the interstitial sites of the lattice.

In the chemical desorption step the adsorbed H atoms diffuse about on the metal surface, either by threading their way through adsorbed water molecules or by pushing them aside, until two collide to form an H_2 molecule which escapes into the solution. This chemical step will be independent of overpotential, since charge transfer is not involved, and the rate will be proportional to the concentration or coverage θ_H of adsorbed $H_{ads.}$ (*see* equation 9.39) and may occur at coverages that range from very small to almost complete.

On the other hand, the electrochemical desorption step is far more complex since it involves reaction between an adsorbed H atom, a hydrated proton H_3O^+ and an electron, and for desorption to occur the proton must discharge onto a hydrogen atom adsorbed on the metal surface. Under these circumstances the probability of collision will be low unless the coverage θ_H is high; furthermore, since coverage increases with overpotential, electrochemical desorption is more likely to occur at high negative overpotentials; in fact, in certain cases the mechanism will change from a C.D. to an E.D. step when the overpotential has attained a sufficiently large value.

Rate-determining Step

When a multistep reaction has attained a steady state the forward net rate ($\vec{i}_1 - \overleftarrow{i}_1 = i_{c,1}$, etc.) of each step that constitutes the overall reaction must be equal. However, if one step requires a significantly higher energy of activation than any other that precedes or follows it, then the former will determine the overall rate, and is said to be the *rate-determining step* (r.d.s.).

This concept can be illustrated by considering two towns *A* and *C* separated by two interconnected roads *AB*–*BC*, of which *AB* is a motorway (maximum speed 70 m.p.h.) while *BC* is a narrow road and is controlled by regulations to a maximum speed of 30 m.p.h. (Fig. 9.19*a*). A driver going from *A* to *C* could proceed along *AB* at 70 m.p.h., but would then have to wait in a queue before entering *BC*, and owing to the congestion would not be able to exceed, say, 10 m.p.h. during his journey through it. To avoid queueing and the slow rate of transit through *BC* it would be necessary for all cars to proceed at, say, a constant speed of 25 m.p.h. for the whole journey, a rate that is determined by *BC* the rate-determining road (step). Figure 9.19*b* shows the standard free energy of activation against distance curves for a two-step reaction in which step 2 is rate determining. In the case of the h.e.r., either discharge (D.), chemical desorption (C.D.) or electrochemical desorption (E.D.) may be the r.d.s., and under these circumstances the other steps will be at equilibrium. Adopting the convention that → represents the r.d.s. and that ⇌ represents the step at equilibrium, the possibilities are as follows:

D. followed by C.D.

$$(a)\ \text{D. r.d.s.}\quad H_3O^+ \xrightarrow{\text{D.}} M\text{—H} \xrightleftharpoons{\text{C.D.}} H_2$$

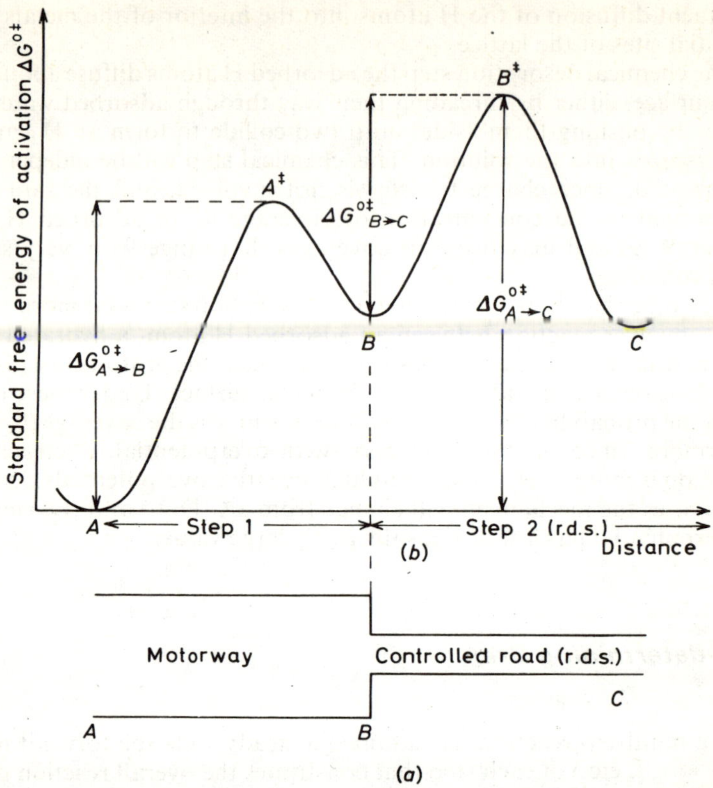

Fig. 9.19(*a*) Analogy showing how the rate at which a car can travel from *A* to *C* is determined by the narrow controlled road *BC* (the r.d.s.). (*b*) Standard free energy of activation against distance for a two-step reaction; since the activation energy required to produce the activated state B^{+} is $\Delta G^{0\ddagger}_{A\to C}$ and that to produce *A* is only $\Delta G^{0\ddagger}_{A\to B}$, the former is the r.d.s., i.e. it is the height of the energy barrier compared with the *initial state* that determines the step that is rate determining

$$(b)\ \text{C.D. r.d.s.}\ H_3O^+ \underset{}{\overset{\text{D.}}{\rightleftharpoons}} M\text{—H} \xrightarrow{\text{C.D.}} H_2$$

D. followed by E.D.

$$(c)\ \text{D. r.d.s.,}\ H_3O^+ \xrightarrow{\text{D.}} M\text{—H} \overset{\text{E.D.}}{\rightleftharpoons} H_2$$

$$(d)\ \text{E.D. r.d.s.}\ H_3O^+ \overset{\text{D.}}{\rightleftharpoons} M\text{—H} \xrightarrow{\text{E.D.}} H_2$$

In the case of the step at equilibrium the rates of the forward and reverse reactions will be equal, i.e. $\vec{v}_1 = \overleftarrow{v}_1$, whereas in the case of the r.d.s. the rate of the reverse reaction $\overleftarrow{v}_{\text{r.d.s.}} = 0$, and $\vec{v}_{\text{r.d.s.}} = \vec{v}_{\text{reaction}}$.

There are two further possibilities, which are referred to as coupled mechanisms, in which the rate of the reverse reaction of the first step is negligible and the forward reactions of both steps are coupled together. Thus both the reverse steps proceed at rates that are negligible, i.e. $\overleftarrow{v}_1 = \overleftarrow{v}_2 \to 0$.

It follows that for a coupled D.+C.D.

$$M+H_3O^+ + e \xrightarrow{v=\vec{v}_1} M{-}H$$

$$M{-}H + M{-}H \xrightarrow{v=\vec{v}_2} 2M + H_2$$

and a similar sequence of steps will occur for a coupled D.+E.D. mechanism.

Transfer Coefficient, Symmetry Factor and Stoichiometric Number

In Section 1.4 it was assumed that the rate equation for the h.e.r. involved a parameter, namely the transfer coefficient α, which was taken as approximately 0·5. However, in the previous consideration of the rate of a simple one-step electron-transfer process the concept of the symmetry factor β was introduced, and β was used in place of α, and it was assumed that the energy barrier was almost symmetrical and that $\beta \approx 0{\cdot}5$. Since this may lead to some confusion, an attempt will be made to clarify the situation, although an adequate treatment of this complex aspect of electrode kinetics is clearly impossible in a book of this nature and the reader is recommended to study the comprehensive work by Bockris and Reddy[1].

In a multistep reaction the number of times the r.d.s. must occur for each act of the overall reaction is referred to as the *stoichiometric number* ν, and this concept can be illustrated by referring to the steps of the h.e.r.

For rate-determining discharge followed by rapid C.D.

$$M+H_3O^+ \xrightarrow{\text{D.}} M{-}H + H_2$$

$$2MH \xrightleftharpoons{\text{C.D.}} 2M + H_2$$

and since the discharge step must occur twice for each act of the overall reaction, $\nu = 2$. On the other hand, for rate-determining discharge followed by E.D.

$$M + M{-}H + H_3O \xrightleftharpoons{\text{E.D.}} 2M + H_2$$

the r.d.s. is required to take place only once, and $\nu = 1$. Generalising for a multistep reaction of the type $A + ze = Z$ in which all steps involve the transfer of one electron and the r.d.s. involves r electrons

$$A + e \rightleftharpoons B \qquad \text{(step 1)}$$
$$B + e \rightleftharpoons C \qquad \text{(step 2)}$$
$$P + e \rightleftharpoons R \qquad \text{(step } \vec{\gamma}\text{)}$$
$$\nu(R + re \rightarrow S) \qquad \text{(r.d.s. repeated } \nu \text{ times)}$$
$$\nu S + e \rightleftharpoons T \qquad (\text{step } z - \vec{\gamma} - r\nu \equiv \overleftarrow{\gamma})$$
$$\cdots\cdots\cdots\cdots$$
$$Y + e \rightleftharpoons Z \qquad \text{(step } z\text{)}$$

where the convention adopted is that $\vec{\gamma}$ represents the number of steps that *precede* the r.d.s. and $\overleftarrow{\gamma}$ is the number that follow it. The r.d.s. will have to be repeated v times if (*a*) more than R particles are formed by the preceding $\vec{\gamma}$ steps, or (*b*) more than one S particle is required for the following sequence involving $\overleftarrow{\gamma} = z-\vec{\gamma}-rv$ charge transfer steps, since v electrons are transferred when the r.d.s. is repeated v times. It can be shown that the total forward current is now

$$\vec{i} = i_0 \exp\left[-\left(\frac{\vec{\gamma}}{v}+r\beta\right)F\eta/RT\right] \qquad ...(9.86)$$

and that the total reverse current is

$$\overleftarrow{i} = i_0 \exp\left[\left(\frac{\overleftarrow{\gamma}}{v}+r-r\beta\right)F\eta/RT\right] \qquad ...(9.87)$$

Also, the net anodic current will be

$$i_a = \overleftarrow{i}-\vec{i}$$

and taking into account that $\overleftarrow{\gamma} \equiv z-\vec{\gamma}-rv$

$$i_a = i_0\left\{\exp\left[\left(\frac{z-\vec{\gamma}}{v}-r\beta\right)F\eta/RT\right]-\exp\left[-\left(\frac{\vec{\gamma}}{v}+r\beta\right)F\eta/RT\right]\right\} \qquad ...(9.88)$$

or

$$i_a = i_0[e^{\overleftarrow{\alpha}F\eta/RT}-e^{-\vec{\alpha}F\eta/RT}] \qquad ...(9.89)$$

where the transfer coefficients $\overleftarrow{\alpha}$ and $\vec{\alpha}$ are defined by

$$\overleftarrow{\alpha} = \frac{z-\vec{\gamma}}{v}-r\beta \qquad ...(9.90)$$

and

$$\vec{\alpha} = \frac{\vec{\gamma}}{v}+r\beta \qquad ...(9.91)$$

These are the coefficients that determine the Tafel slope of the log i against η curve of a multistep reaction, and they are of fundamental importance in providing information on the mechanism of the reaction. Equations 9.86 and 9.87 are of the same form as equations 9.59 and 9.58 that were derived for a simple one-step reaction involving a symmetrical energy barrier, and under these circumstances equations 9.90 and 9.91 simplify to

$$\overleftarrow{\alpha} = 1-\beta \approx 0{\cdot}5 \qquad ...(9.92)$$

and

$$\vec{\alpha} = \beta \approx 0{\cdot}5 \qquad \ldots(9.93)$$

and the transfer coefficients and symmetry factors are approximately the same. However, in multistep reactions equations 9.92 and 9.93 are not applicable, and the more complex transfer coefficients must be used in the rate equations.

Evaluation of the Rate-determining Step and the Mechanism of the h.e.r.

The slope of the Tafel curve $d\eta/d\log i$ is only one of the criteria that are required to determine the mechanism of the h.e.r., since different mechanisms, involving different r.d.s. often have the same Tafel slope. Parameters that are diagnostic of mechanism are the transfer coefficient, the reaction order, the stoichiometric number, the hydrogen coverage, the exchange current density, the heat adsorption, etc.

The ability to use the Tafel slope as a diagnostic criterion can be exemplified by considering a discharge–chemical desorption mechanism for the h.e.r. in which either discharge or chemical desorption may be rate determining[2].

If the discharge step is rate determining, then for step 1

$$i = i_0 \exp(-\beta F\eta/RT) \qquad \ldots(9.94)$$

and by taking logarithms of both sides and rearranging

$$\eta = \frac{RT}{\beta F}\ln i_0 - \frac{RT}{\beta F}\ln i = 2{\cdot}3\frac{RT}{\beta F}\log i_0 - 2{\cdot}3\frac{RT}{\beta F}\log i \qquad \ldots(9.95)$$

and differentiating

$$\frac{d\eta}{d\log i} = -2{\cdot}3\frac{RT}{\beta F} = -2{\cdot}3\frac{RT}{0{\cdot}5F} \qquad \ldots(9.96)$$

if β is taken to be 0·5.

$$\therefore \frac{d\eta}{d\log i} = -2\times 0{\cdot}059 = -0{\cdot}120\text{ V at 298 K} \qquad \ldots(9.97)$$

Thus a Tafel slope of $-0{\cdot}120$ V could be diagnostic of a discharge–chemical desorption mechanism in which proton discharge is the r.d.s.

If chemical desorption is rate determining, the rate will be independent of overpotential since no charge transfer occurs in this step, and

$$v_{\text{C.D.}} = k_{\text{C.D.}} c_{\text{H}_{\text{ads.}}} \times c_{\text{H}_{\text{ads.}}} = k\theta_{\text{H}}^2 \qquad \ldots(9.98)$$

The discharge step will now be at equilibrium and can be treated thermodynamically

$$\text{H}^+ + e \rightleftharpoons \text{H}_{\text{ads.}}$$

$$E = E^{\ominus} + \frac{RT}{F}\ln\frac{c_{\text{H}^+}}{c_{\text{H}_{\text{ads.}}}} = E^{\ominus} + \frac{RT}{F}\ln\frac{c_{\text{H}^+}}{\theta_{\text{H}_{\text{ads.}}}} \qquad \ldots(9.99)$$

where c_{H^+} is the concentration of H_3O^+ in solution.

$$\therefore\ \theta_H = c_{H^+}\exp[-F(E-E^{\ominus})/RT] \qquad ...(9.100)$$

Now $\eta = E - E_{eq.}$, and since $E^{\ominus}$ is a constant and $E_{eq.}$ will be constant if c_{H^+} is constant, we can replace $E - E^{\ominus}$ with η and

$$\theta_H = k_D \exp(-F\eta/RT) \qquad ...(9.101)$$

where k_D is the rate constant for the discharge step.

Now although the chemical desorption step is independent of η, the surface coverage will increase as η becomes more negative and this will affect $v_{C.D.}$. Since two electrons will be required for the overall reaction, equation 9.98 can be expressed in terms of i

$$v_{C.D.} = \frac{i}{2F} = k_{C.D.}\theta_H^2 \qquad ...(9.102)$$

Combining equations 9.101 and 9.102

$$i = 2Fk_{C.D.}[k_{D.}\exp(-F\eta/RT)]^2 \qquad ...(9.103)$$

or

$$i = K\exp(-2F\eta/RT) \qquad ...(9.104)$$

and taking logarithms and rearranging

$$\eta = \frac{RT}{2F}\ln K - \frac{RT}{2F}\ln i \qquad ...(9.105)$$

where K is a constant that contains all the other constants (F, k_D, $k_{C.D.}$). Differentiating

$$\frac{d\eta}{d\log i} = \frac{-2{\cdot}3RT}{2F} = \frac{-0{\cdot}059}{2} = -0{\cdot}030\ \mathrm{V} \qquad ...(9.106)$$

which is the Tafel slope obtained with platinum for which the h.e.r. in acid solution follows a discharge–chemical desorption mechanism in which chemical desorption is rate determining and $\theta \rightarrow 1$.

Table 9.3 summarises the possible mechanism for the h.e.r. at various metals[19].

Table 9.3 Mechanism of the hydrogen evolution reaction at different metals (data after McBreen and Genshaw[19])

									Al *D*		
Ti *D*			Mn *A*	Fe *E, C*		Ni *A*	Cu *A* or *D*		Ga *A*		
	Nb *D*	Mo *D*			Rh *B*	Pd *B*	Ag *C* or *D*	Cd *A*		Sn *A*	
	Ta *D*	W *A* or *D*			Ir *B*	Pt *B*	Au *D*	Hg *A*	Tl *A*	Pb *A*	Bi *D*

A Slow discharge, fast recombination.
B Fast discharge, slow recombination.
C Slow discharge, fast electrochemical.
D Fast discharge, slow electrochemical.
E Coupled discharge recombination.

Hydrogen Evolution on Iron

A number of metals have the ability to absorb hydrogen, which may be taken into solid solution or form a metallic hydride, and this absorption can provide an alternative reaction path to the desorption of $H_{ads.}$ as H_2 gas. In the case of iron and iron alloys, both hydrogen adsorption and absorption occur simultaneously, and the latter thus gives rise to another equilibrium involving the transfer of $H_{ads.}$ across the interface to form interstitial H atoms just beneath the surface:

$$\mathrm{Fe{-}H_{ads.}} \rightleftharpoons \mathrm{Fe{-}H_{abs.}}$$

This equilibrium is of importance in providing diagnostic criteria for the mechanism of the h.e.r., since the rate of permeation of hydrogen through a thin iron membrane can provide information on the coverage of the surface with adsorbed hydrogen.

Devanathan and Stachurski[17] have devised a two-compartment cell which has been widely used for studying the permeation of hydrogen through iron and ferrous alloys and other metals. The cell (*see* Section 20.2, Fig. 20.43) consists of two compartments separated by a thin diaphragm of the metal under study; the cathode compartment, which contains an appropriate electrolyte solution and a counterelectrode, is used to polarise cathodically one side of the membrane at a constant current density or potential. A feature of the cell is the electrochemical estimation of the hydrogen that permeates through the diaphragm, which is achieved by maintaining the reverse side of the diaphragm (the anode side) at a constant potential in 0·1 M NaOH and oxidising the adsorbed hydrogen by the reaction

$$\mathrm{Fe{-}H_{ads.}} \rightarrow \mathrm{H^+} + e$$

The anode compartment contains a reference electrode and counterelectrode and by means of a potentiostat the anode side is maintained at a constant potential. The coverage of adsorbed hydrogen θ_H on the cathode side will depend on the current density i and the nature of the electrolyte solution, and the cell can be used to study the effect of a variety of factors (composition and structure of alloys, pH of solution, effect of promoters and inhibitors) on hydrogen permeation.

Bockris, McBreen and Nanis[18] have studied the h.e.r. on pure iron in 0·1 M NaOH and 0·1 M H_2SO_4, and have concluded that in both cases the reaction mechanism is coupled discharge–chemical desorption at low coverages (Langmuirean adsorption) with the discharge step being rate determining; in NaOH at very negative overpotentials ($> -1{\cdot}02$ V) there appears to be a change in the mechanism to slow discharge–fast electrochemical desorption at high coverages. The reaction sequence can be summarised as follows

$$\mathrm{FeH_{ads.}} \xrightarrow{+\mathrm{FeH_{ads.}}} \mathrm{2FeH_{ads.}} \xrightarrow{k_2} \mathrm{H_2 + Fe}$$

$$\mathrm{H_3O^+} + e + \mathrm{Fe} \underset{\text{slow}}{\xrightarrow{k_1}} \mathrm{FeH_{ads.}} \underset{k_{-3}}{\overset{k_3}{\rightleftharpoons}} \mathrm{FeH_{abs.}} \qquad \ldots(9.107)$$

where k_1, k_2, k_3 and k_{-3} are the rate constants for the different steps.

The permeation of hydrogen through an iron diaphragm is illustrated in Fig. 9.20 for the sequence of events

$$\mathrm{FeH_{ads.}} \underset{k_{-3}}{\overset{k_3}{\rightleftharpoons}} \mathrm{FeH_{abs.\,(x=0)}} \xrightarrow{\text{diffusion through membrane}}$$

$$\mathrm{FeH_{ads.\,(x=L)}} \rightarrow \mathrm{FeH_{ads.\,(x=L)}} \rightarrow \mathrm{H^+} + e$$

where $\mathrm{FeH_{abs.\,(x=0)}}$ refers to the hydrogen absorbed just below the surface of the cathode side of the membrane and $\mathrm{FeH_{abs.\,(x=L)}}$ that just below the absorbed surface on the anode side.

At the steady state the flux of hydrogen J_∞ will be

$$J_\infty = \frac{Dc_0}{L} \qquad \ldots(9.108)$$

where D is the diffusion coefficient, L is the thickness of the membrane and c_0 is the concentration of H at $x = 0$; the concentration of hydrogen just below the surface on the anode side c_L will be zero owing to the rapid removal of the $\mathrm{H_{ads.}}$ by electrochemical oxidation (Fig. 9.20). Since diffusion through the membrane will be slow compared to the other steps it will be rate determining, and the rate of transfer of $\mathrm{H_{ads.}}$ from the surface to the bulk and of $\mathrm{H_{abs.}}$ from the bulk to the surface will be at equilibrium (equation 9.107).

$$\therefore\ k_3\theta\left(1-\frac{x_0}{x_s}\right) = k_{-3}c_0(1-\theta) \qquad \ldots(9.109)$$

where x_0 is the number of interstitial sites per cubic centimetre beneath the surface that are occupied by hydrogen, x_s is the total number of sites available and c_0 is the concentration of hydrogen at $x = 0$.

At low coverages and low concentrations, equation 9.109 simplifies to

$$k_3\theta = k_{-3}c_0, \text{ or } c_0 = \frac{k_3\theta}{k_{-3}} \qquad \ldots(9.110)$$

which shows that c_0 is proportional to the coverage of the cathode side with $\mathrm{H_{ads.}}$. From equations 9.108 and 9.110 the flux through the diaphragm will be

$$J_\infty = \frac{D}{L}\left(\frac{k_3}{k_{-3}}\right)\theta \qquad \ldots(9.111)$$

and if J_∞ is the permeation current ($\mu\mathrm{A\,cm^{-2}}$) at the steady state, then

$$\frac{J_\infty}{zF} = \frac{D}{L}\left(\frac{k_3}{k_{-3}}\right)\theta$$

or

$$J_\infty = \frac{DzF}{L}\left(\frac{k_3}{k_{-3}}\right)\theta \qquad \ldots(9.112)$$

which shows that the permeation current is proportional to coverage with $\mathrm{H_{ads.}}$.

For the reaction sequence shown in equation 9.107 the cathodic current

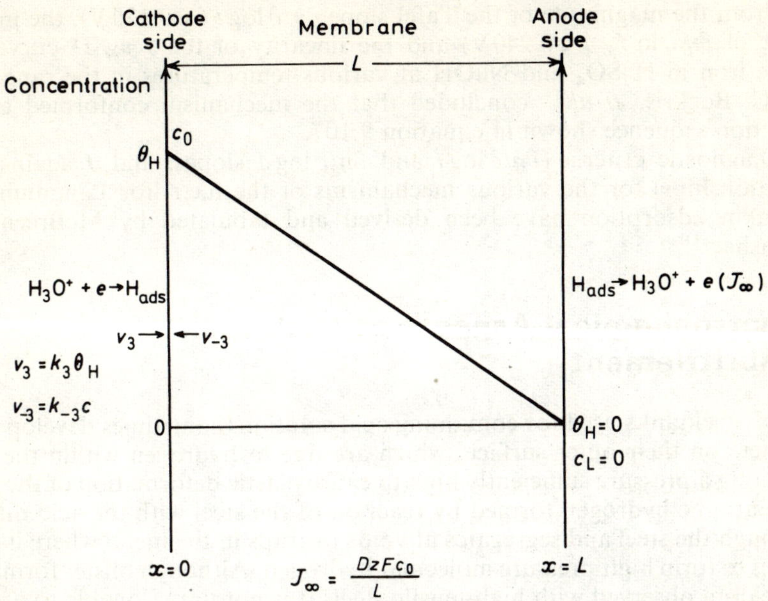

Fig. 9.20 Permeation of hydrogen through a thin iron membrane

is given by

$$\vec{i} = k_1 c_{\mathrm{H}^+}(1-\theta)\mathrm{e}^{-\beta\eta F/RT} = k_2\theta^2 \qquad \ldots(9.113)$$

since the rate of discharge and chemical desorption must be equal.

$$\therefore\ \theta = \left(\frac{1}{k_2}\right)^{\frac{1}{2}} i^{\frac{1}{2}} \qquad \ldots(9.114)$$

and from equation 9.112

$$J_\infty = \frac{DzF}{L}\left(\frac{k_3}{k_{-3}}\right)\left(\frac{1}{k_2}\right)^{\frac{1}{2}} i^{\frac{1}{2}} \qquad \ldots(9.115)$$

which shows that J_∞ is linearly related to $i^{\frac{1}{2}}$, i.e.

$$J_\infty \propto i^{\frac{1}{2}} \qquad \ldots(9.116)$$

Also for the reaction sequence shown in equation 9.107

$$J_\infty = \frac{DzF}{L} c_{\mathrm{H}^+}^{\frac{1}{2}}\left(\frac{k_3}{k_{-3}}\right)\left(\frac{k_1}{k_2}\right)\mathrm{e}^{-\beta\eta F/2RT}$$

where η is the overpotential, and taking $\beta \approx 0{\cdot}5$ and differentiating

$$\frac{\partial\eta}{\partial \ln J_\infty} \approx -\frac{4RT}{F} \qquad \ldots(9.117)$$

or

$$\frac{\partial\eta}{\log J_\infty} \approx -0{\cdot}059\times 4 = -0{\cdot}240\ \mathrm{V\ at\ 25°C} \qquad \ldots(9.118)$$

From the magnitude of the Tafel slope $\partial\eta/\partial \log i$ ($\approx 0{\cdot}120$ V), the magnitude of $\partial\eta/\partial \ln J_{\infty}$ ($\approx 0{\cdot}240$ V) and the linearity of the J *vs.* $i^{\frac{1}{2}}$ curves for pure iron in H_2SO_4 and NaOH at various temperatures in the range 18–80°C, Bockris, *et al.*[18] concluded that the mechanism conformed to the reaction sequence shown in equation 9.107.

Diagnostic criteria ($\partial\eta/\partial \log i$ and $\partial\eta/\partial \log J$ slopes, and J against $f(i)$ relationships) for the various mechanisms of the h.e.r. for Langmuir and Temkin adsorption have been derived and tabulated by McBreen and Genshaw[19].

Electrochemical Aspects of Hydrogen Embrittlement

Mild-steel tanks used for containing acid solutions sometimes develop large blisters on their outer surface, which are due to hydrogen within the steel exerting a pressure sufficiently high to cause plastic deformation of the steel. The atomic hydrogen formed by reaction of the steel with the acid diffuses through the steel and segregates at voids or traps in the metal where it combines to form high-pressure molecular hydrogen. Although blister formation is seldom observed with high-tensile steels it is not unreasonable to assume that hydrogen at high pressures is largely responsible for their embrittlement and for the phenomenon of delayed failure. Theories of hydrogen embrittlement are considered in Section 8.4, but it is appropriate here to review briefly the electrochemical aspects of hydrogen discharge and absorption with particular reference to the pressure developed by hydrogen occluded at traps and voids.

The permeation cell devised by Devanathan and Stachurski[17] has provided a most valuable tool for studying the effect of alloy composition and structure[20], tensile[21,22] and compressive stresses[23], partial pressure of hydrogen[22,24], etc. on the permeation of hydrogen through ferrous alloys, and the results obtained have lead to a greater insight into the mechanism of embrittlement. In the above discussion on hydrogen permeation through iron it was assumed that diffusion occurred via the interstitial sites of a perfect lattice of the metals. However, in real metals and alloys imperfections exist in the lattice that are referred to as traps or voids, and which may consist of imperfections in the lattice, interfacial areas between phase boundaries, inclusions, etc. (*see* Section 8.4). Hydrogen diffusing interstitially through the lattice can adsorb at the surface of these trapping sites and then combine to form molecular hydrogen at a pressure that will depend on the overpotential at the charged surface. If an iron electrode is cathodically charged with hydrogen there will be an equilibrium between the hydrogen adsorbed at the external surface $H_{ads.,ext.}$, hydrogen at interstitial sites H_i, hydrogen adsorbed at the surface of voids $H_{ads.,v}$ and molecular hydrogen within the void $H_{2,v}$.

$$H_{ads.,ext.} \rightleftharpoons H_i \rightleftharpoons H_{ads.,v} \rightleftharpoons H_{2,v}$$

and thermodynamically

$$\mu_{H_{ads.,ext.}} = \mu_{H_i} = \mu_{H_{ads.,v}} = \mu_{H_{2,v}}$$

where μ is the chemical potential. Thus the chemical potential of the hydrogen adsorbed on the external surface will equal the chemical potential (fugacity) of the molecular hydrogen within the void.

It can be shown[25] that

$$\frac{\theta}{1-\theta} = \frac{\theta_{\text{rev.}}}{1-\theta_{\text{rev.}}} \times p_{H_2,v}^{-\frac{1}{2}} \qquad \text{...(9.119)}$$

where θ is the surface coverage, $\theta_{\text{rev.}}$ the reversible coverage at 1 atm of H_2 (approx. 10^{-3} for pure iron in a solution of pH = 4) and p_{H_2} is the fugacity of H_2 within the traps. The pressure of hydrogen will increase, therefore, with surface coverage of $H_{\text{ads., ext.}}$, and this in turn will be dependent on the overpotential and the mechanism of the h.e.r. It was considered at one time that the fugacity of hydrogen could be evaluated from the Nernst equation, but Bockris and Subramanyan[25] on the basis of statistical mechanics have derived the η *vs.* p_{H_2} relationship for the possible mechanisms of the h.e.r. and have shown that the Nernst equation is not always applicable. Thus for a fast discharge–slow desorption mechanism (either chemical or electrochemical) in which $\theta \rightarrow 1$, the relationship between the fugacity of H_2 in voids $p_{H_2,v}$ at an external pressure of hydrogen of 1 atm is given by

$$p_{H_2,v} = \exp\left(\frac{-2\eta F}{RT}\right) \qquad \text{...(9.120)}$$

which is identical to the relationship derived from the Nernst equation. On the other hand, for a coupled discharge–chemical desorption mechanism (applicable to pure iron) the fugacity is given by

$$p_{H_2,v} = 10^{1\cdot5}_{\exp}\left(\frac{-\eta F}{2RT}\right) \qquad \text{...(9.121)}$$

If the overpotential is taken as $-0{\cdot}1$ V it follows from equation 9.120 that $p_{H_2,v} = 10^{3\cdot3}$ atm. Although the arguments used in deriving these relationships do not give precise values of $p_{H_2,v}$ they indicate that pressures of very high magnitude will be obtained at significant overpotentials.

The effect of stress[21] on hydrogen solubility and permeation is clearly of vital importance in relation to delayed failure, since fracture can only occur if the metal is subjected to a tensile stress (internal or applied). It has been reported that high-strength steels containing $<0{\cdot}1$ p.p.m. of hydrogen are susceptible to fracture at applied stresses well below the yield stress, and Troiano[26] has postulated that diffusion of hydrogen to regions of high triaxial stress is of major importance in the mechanism of embrittlement.

Beck, *et al.*[21] have used the permeation technique to study the effect of uniaxial tensile stresses in the elastic region on hydrogen permeation through pure iron, and have shown that it increases with increase in stress. The partial molar volume of hydrogen $\bar{V}_H$ (cubic centimetres of hydrogen per mole of iron) in ferrous alloys can be evaluated from the variation of permeation with applied stress, and from the relationship

$$c_\sigma = c_0 e^{\sigma \bar{V}_H/RT} \qquad \text{...(9.122)}$$

where σ is the applied stress, and c_0 and c_σ are the hydrogen solubilities in the absence of a stress, and in its presence, respectively. The concentration terms can be replaced by the corresponding permeation terms J_0 and J_σ and equation 9.122 then becomes

$$J_\sigma = J_0 e^{\sigma \bar{V}_H/RT} \qquad \ldots(9.123)$$

or

$$\frac{\partial \ln(J_\sigma/J_0)}{\partial \sigma} = \frac{\bar{V}_H}{RT} \qquad \ldots(9.124)$$

Thus a plot of $\log(J_\sigma/J_0)$ against σ gives a straight line of slope $\bar{V}_H/2{\cdot}3RT$, which provides a means of evaluating $\bar{V}_H$; for pure iron the partial molar volume has been found to be 2·6 $cm^3 mol^{-1}$, but the value has been found to vary with the composition and structure of the steel[22]. It follows from equation 9.122 that when σ is tensile (taken as positive in equation 9.122) the solubility of hydrogen and the permeation rate will increase, i.e. the larger the lattice distortion the greater the solubility; conversely, compressive stresses (σ is negative) will decrease solubility, as may be shown by permeation studies of steels under compression, or by putting the surface of the metal into a state of compressive stress by shot peening[23].

Bockris and Subramanyan[27] during studies of the permeation of hydrogen through pure Fe and Fe–5Ni alloy found that a normal permeation transient was obtained (Fig. 9.21), providing the overpotential was less than a critical value, and when the overpotential was less than $\eta_{crit.}$ it was possible to reproduce the normal permeation curve, i.e. apply the polarising current at a constant $\eta < \eta_{crit.}$, allow J to attain a steady value, switch off, reapply

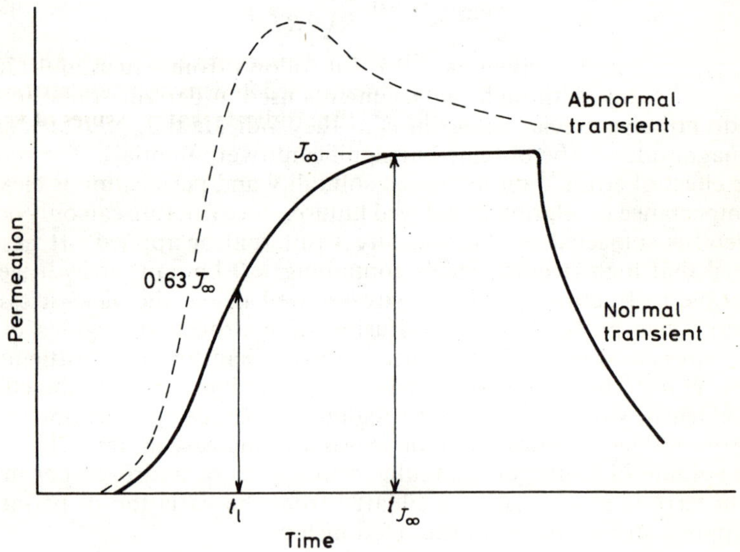

Fig. 9.21 J–t transients for the permeation of hydrogen through ferrous alloys. The normal transient enables the diffusion coefficient D to be evaluated from the relationship $t_1 = L^2/6D$, where t_1 is the time at which J attains a value of 0·63 of the steady-state permeation J

polarising current, etc. However, if $\eta_{crit.}$ was exceeded, a second determination of the J–t transient resulted in an abnormal curve that showed a more rapid rise time than the normal curve, and a peak (Fig. 9.21); furthermore, after the determination of the transient at or above $\eta_{crit.}$ it was not possible to obtain a normal transient at an overpotential less than $\eta_{crit.}$. Anomalous decay transients were also obtained.

On the basis of the Griffith crack theory, and by assuming that the void has the shape of a double convex lens with a long axis of length l, Bockris and Subramanyan[27] derived an expression for the critical stress, corresponding with the critical pressure of hydrogen within the void required to cause the void to grow and for a crack to be initiated:

$$\sigma_{crit.} = p_{H_{2,v}} > \left[\frac{16}{3} \times \frac{Y\gamma}{l}\right]^{\frac{1}{2}} \qquad \text{...(9.125)}$$

where Y is Young's modulus and γ is the surface tension of the M—H interface at the surface of the void ($S\gamma$ is the total surface energy, where S is the surface area of the void). Taking typical values of Y, σ and l it can be calculated that $p_{H_2} \approx 3 \times 10^4$ atm, which is a pressure that could be achieved at overpotential greater than a critical value. An extensive review of hydrogen in metals is given in a recent publication by Subramanyan[28].

L. L. SHREIR

REFERENCES

1. Bockris, J. O'M. and Reddy, A. K. N., *Modern Electrochemistry*, Macdonald, London (1970)
2. Bockris, J. O'M. and Drazic, D., *Electro-chemical Science*, Taylor and Francis, London (1972)
3. Perkins, R. S. and Anderson, T. N., *Modern Aspects of Electrochemistry*, No. 5, ed. J. O'M. Bockris and B. E. Conway, Butterworths, London (1969)
4. Antropov, L. I., *Kinetics of Electrode Processes and Null Points of Metals*, C.S.I.R., New Delhi (1960)
5. West, J. M., *Electrodeposition and Corrosion Processes*, van Nostrand Reinhold, London (1971)
6. Antropov, L. I., *1st International Congress on Metallic Corrosion*, ed. L. Kenworthy, Butterworths, London (1962)
7. Conway, B. E., *Theory and Principles of Electrode Processes*, Ronald Press Co., New York (1965)
8. Davies, C. W., *Electrochemistry*, Newnes, London (1967)
9. Gileadi, E. and Conway, B. E., *Modern Aspects of Electrochemistry*, No. 3, ed. J. O'M. Bockris and B. E. Conway, Butterworths, London (1964)
10. Bockris, J. O'M., Devanathan, M. A. V. and Muller, K., *Proc. Roy. Soc.*, London, **274A**, 55 (1963)
11. Uhlig, H. H., *Corros. Sci.*, **7**, 325 (1967)
12. Hoar, T. P., *Corros. Sci.*, **7**, 341 (1967)
13. Frumkin, A. N., *Z. Physik. Chem.*, **164**, 121 (1933)
14. Parsons, R., *Third European Symposium on Corrosion Inhibitors*, Universita Degli Studi di Ferrara, **3** (1971)
15. Szklarska-Smialowska, Z. and Wieczorek, G., ibid., 453 (1971)
16. Trabanelli, G. and Carassiti, V., *Advances in Corrosion Science and Technology*, Vol. 1 (ed. M. G. Fontana and R. W. Staehle), Plenum Press (1970)
17. Devanathan, M. A. V. and Stachurski, Z., *Proc. Roy. Soc.*, **270A**, 90 (1962)
18. Bockris, J. O'M., McBreen, J. and Nanis, L., *J. Electrochem. Soc.*, **112**, 1025 (1965)

19. McBreen, J. and Genshaw, M. A., *Fundamental Aspects of Stress Corrosion Cracking* (ed. R. W. Staehle, A. J. Forty and D. van Rooyen), NACE, Houston, p. 51 (1968)
20. Radhakrishnan, T. P. and Shreir, L. L., *Electrochem. Acta*, **11**, 1007 (1966) and **12**, 899 (1967)
21. Beck, W., Bockris, J. O'M., McBreen, J. and Nanis, L., *Proc. Roy. Soc.*, **290A**, 220 (1966)
22. Blundy, R. F., Royce, R., Poole, P. and Shreir, L. L., *International Conference on Stress Corrosion Cracking and Hydrogen Embrittlement of Iron Base Alloys*, Firminy, France (1973)
23. Bockris, J. O'M., Beck, W., Genshaw, M. A., Subramanyan, P. K. and Williams, F.S., *Acta Metallurgica*, **19**, 1209 (1971)
24. Blundy, R. F. and Shreir, L. L., *Phil. Mag.*, **20**, 1177 (1969)
25. Bockris, J. O'M. and Subramanyan, P. K., *Electrochim. Acta*, **16**, 2169 (1971)
26. Troiano, A. R., *Trans. Amer. Soc. Metals*, **52**, 54 (1960)
27. Bockris, J. O'M. and Subramanyan, P. K., *J. Electrochem. Soc.*, **118**, 1114 (1971)
28. Subramanyan, P. K., MTP International Review of Science, Vol. 6, *Electrochemistry* (ed. J. O'M. Bockris), Butterworths (1973)

9.1A Appendix – Outline of Chemical Thermodynamics

Scope

Thermodynamics is concerned with the relationship between heat energy and work and is based on two general laws, the 1st and 2nd laws of thermodynamics, which both deal with the interconversion of the different forms of energy. The 3rd law states that at the absolute zero of temperature the entropy of a perfect crystal is zero, and thus provides a method of determining absolute entropies.

The following features of the thermodynamic approach to chemical reactions should be noted:

1. It does not involve time as a variable, and does not provide information on rates.
2. It does not normally involve a knowledge of the detailed structure of the system nor of the detailed mechanism of the reaction.
3. The 2nd law is true only statistically and does not apply to individual particles nor to a small number of particles, i.e. thermodynamics is concerned with bulk properties of systems. Thermodynamics thus has many limitations, but is particularly valuable in defining the nature and structure of phases when *equilibrium* (a state that does not vary with time) has been attained; thermodynamics provides no information on the *rate* at which the reaction proceeds to equilibrium, which belongs to the realm of *chemical kinetics*.

Some Definitions

A *system* is any part of external reality that can be subjected to thermodynamic treatment; the material with which the system is in contact forms the *surroundings*, e.g. an electrochemical cell could be the system and the external atmosphere the surroundings.

A *homogeneous* system is one in which the properties are uniform or vary uniformly throughout the system, whereas a *heterogeneous* system is one consisting of two or more homogeneous systems. Pure copper and a pure solution of $CuSO_4$ are both homogeneous systems, but copper immersed in a solution of $CuSO_4$ is a heterogeneous system characterised by an interface. The state of a system is described by its properties, which may be *intensive* or *extensive*; intensive properties are independent of quantity, e.g. temperature, pressure, electrical potential, activation energy, etc. whereas extensive properties depend upon the amount of the material, e.g. mass, volume, internal energy, enthalpy, entropy, free energy, chemical potential, etc.

The state of a system containing a constant amount of material depends upon a few variables, e.g. pressure p, volume V, temperature T. For a given mass of pure substance the volume can be expressed solely as a function of pressure and temperature

$$V = f(p, T) \qquad \text{...(9.A1)}$$

If the pressure changes from p to $p+\mathrm{d}p$ and the temperature from T to $T+\mathrm{d}T$ the change in V is given by

$$\mathrm{d}V = \left(\frac{\partial V}{\partial p}\right)_T \mathrm{d}p + \left(\frac{\partial V}{\partial T}\right)_p \mathrm{d}t \qquad \text{...(9.A2)}$$

where $\left(\frac{\partial V}{\partial p}\right)_T$ is the rate of change of volume with pressure at constant temperature, and $\left(\frac{\partial V}{\partial T}\right)_p$ is the rate of change of volume with temperature at constant pressure. They are referred to as *partial derivatives*, the subscript T or p denoting the variable that is maintained constant.

A *reversible* system is one which is so perfectly balanced that only an infinitesimal change in external conditions is required to reverse the direction of the process; for reversibility the change must either be infinitesimally small or if large it must be carried out at an infinitely slow rate. Although these concepts are hypothetical it is possible to approximate to reversibility by balancing the e.m.f. of a reversible cell with the e.m.f. of a potentiometer circuit so that only a minute change ($<10^{-6}$ V) in the e.m.f. produced by the latter will result in a reversal of the direction of the cell reaction. An ideal reversible cell must consist of two reversible electrodes, which are so completely non-polarisable that their potentials are unaffected by charge transfer even at high rates. This too is a hypothetical concept, but a number of electrodes will approximate to reversibility, providing the rate of charge transfer is small, and cells composed of these electrodes may be treated thermodynamically. (See Appendix 9.1B.)

1st Law of Thermodynamics

The 1st law is essentially a statement of the experimental fact that during a chemical reaction in an isolated system *energy is conserved*, although it may be converted from one form to another.

If a system changes from state A to state B and if ΔU is the change in the internal energy of the system, then it follows from the 1st law that the surroundings must gain or lose an equivalent amount of energy. A convention of signs is necessary to take into account the direction of heat transfer q and work performed w; q is taken as positive if heat is transferred to the system from the surroundings and negative if the transfer is in the reverse direction; in many textbooks w is taken as positive if work is done by the system on the surroundings and negative if work is done to the system by the surroundings, but it is preferable to allow the sign of w to develop from the mathematics. Thus, since energy must be conserved

$$\Delta U = q - w \qquad \text{...(9.A3)}$$

where ΔU is the change in the internal energy of the system, q is the heat transferred between the system and the surroundings and w is the work done. Equation 9.A3 may be regarded as an algebraic expression of the conservation of energy, and it should be noted that ΔU is the difference in the internal energy of the system in its initial and final stages and that the absolute value of U is not known.

If a change from state A to state B occurs in a system at constant pressure (isobaric) so that only pV work is done, then $w = p\Delta V$ and

$$\Delta U = q_p - p(V_B - V_A) \qquad \text{...(9.A4)}$$

which can be written in the form

$$q_p = U_B - U_A + p(V_B - V_A) = (U_B + pV_B) - (U_A + pV_A) \qquad \text{...(9.A5)}$$

Thus it is convenient to define another function

$$H = U + pV \qquad \text{...(9.A6)}$$

where H is the *enthalpy* or *heat content* of the system, so that

$$q_p = H_B - H_A = \Delta H \qquad \text{...(9.A7)}$$

where q_p is the heat exchanged between the system and the surroundings at constant pressure, and ΔH, like ΔU, depends only upon the initial and final states of the system.

2nd Law of Thermodynamics

The 1st law implies that energy is interconvertible and that when one form of energy disappears another kind of energy must appear in its place. However, it provides no information on (*a*) the transformations of energy that are possible, and (*b*) the extent to which one form of energy can be converted into another. Some energy transformations are possible and occur spontaneously, whilst transformations in the reverse direction are non-spontaneous and can be made to proceed only by supplying energy from an external source. Thus heat will flow spontaneously from a source at a higher temperature to one at a lower temperature, but heat cannot pass spontaneously in the reverse direction. The heat flowing spontaneously can be converted by means of suitable machinery to work energy, but it is evident from Carnot's cycle (p. **9**.60) that there is a theoretical limit to the extent of this interconversion.

The oxidation of hydrogen to water ($H_2 + \frac{1}{2}O_2 \rightarrow H_2O$) is thermodynamically spontaneous and the energy released as a result of the chemical reaction appears as heat energy, but the decomposition of water into its elements is a non-spontaneous process and can be achieved only by supplying energy from an external source, e.g. a source of e.m.f. that decomposes the water electrolytically. Furthermore, although the heat produced by the spontaneous reaction could be converted into electrical energy, the electrical energy thus obtained would not be sufficient to restore the original state of chemical energy of the system, i.e. to regenerate 1 mol H_2 and $\frac{1}{2}$ mol O_2.

The 2nd law, which can be stated in a number of different ways, expresses the fact that heat can only be transformed into work when it flows from a higher to a lower temperature. Clausius stated the 2nd law as follows: 'It is impossible for a self-acting machine, unaided by an external agency, to convey heat from a lower to a higher temperature'.

In principle all spontaneous processes may be harnessed and made to do work, but owing to the wastage of a proportion of the energy as heat energy there is a theoretical limit to the amount of useful work that can be obtained, and this can never be exceeded in practice. A high-efficiency power station converting chemical energy (solid or liquid fuel) into electrical energy achieves only 30–40% efficiency.

The maximum theoretical work $w_{max.}$ obtainable from a system was derived by Carnot who considered the transformation of heat energy into work when a perfect gas in a cylinder with a piston was taken through a reversible cycle (the Carnot cycle), in which the system was almost at equilibrium during each step of the cycle. It was shown that

$$w_{max.} = q_2\left(\frac{T_2 - T_1}{T_2}\right) \qquad \text{...(9.A8)}$$

where q_2 is the heat absorbed from a reservoir at the higher temperature T_2 (K) and T_1 is the temperature of a second heat reservoir at a lower temperature than T_2 that receives the heat that has not been converted into work. Equation 9.A8 is applicable to any reversible system that undergoes a cyclic process in which the original and final states of the system and the surroundings are the same. Under these circumstances all that has changed permanently is that a quantity of heat q_2 at T_2 has been transformed into work $w_{max.}$ together with a quantity of heat $q_1 = q_2 - w_{max.}$ that is returned to the reservoir at the lower temperature T_1. It can be shown that the efficiencies of all reversible cyclic processes are the same providing T_1 and T_2 are the same, i.e.

$$\text{Thermodynamic efficiency} = \frac{w_{max.}}{q_2} = \frac{q_2 - q_1}{q_2} = \frac{T_2 - T_1}{T_2} \qquad \text{...(9.A9)}$$

It is evident from equation 9.A9 that (*a*) the efficiency increases as $(T_2 - T_1)/T_2$ increases and (*b*) the efficiency is zero when $T_1 = T_2$ so that no heat can be converted into work.

Entropy

The concept of entropy S is implicit in the 2nd law, and it may be regarded as a thermodynamic property of a system that is associated with its capacity for spontaneous change. As the system undergoes spontaneous change the entropy increases and the capacity for further spontaneous change decreases. Rearranging the q and T terms in equation 9.A9 and applying the convention of signs

$$\frac{q_1}{T_1} - \frac{q_2}{T_2} = 0 \qquad \text{...(9.A10)}$$

and generalising for any reversible cyclic change that proceeds isothermally at a temperature T

$$\sum \frac{q_{\text{rev.}}}{T} = 0 \qquad \text{...(9.A11)}$$

where $\Sigma q_{\text{rev.}}$ is the sum of the heats absorbed or evolved reversibly. The function $q_{\text{rev.}}/T$ is a measure of the change in entropy ΔS, and this change for an isothermal process is defined as

$$\Delta S = \frac{q_{\text{rev.}}}{T} \qquad \text{...(9.A12)}$$

and for an infinitesimal reversible process

$$\mathrm{d}S = \frac{\delta q}{T} \qquad \text{...(9.A13)}$$

If the process is non-isothermal but finite

$$\Delta S = S_2 - S_1 = \int_{S_1}^{S_2} \frac{\delta q}{T} \qquad \text{...(9.A14)}$$

It must be emphasised that the heat $q_{\text{rev.}}$ which appears in the definition of entropy (equation 9.A12) is always that absorbed (or evolved) when the process is conducted reversibly. If the process is conducted irreversibly and the heat absorbed is q, then q will be less than $q_{\text{rev.}}$ and q/T will be less than ΔS the entropy change (equation 9.A12). It follows that if an irreversible process takes place between the temperatures T_2 and T_1 and has the same heat intake q at the higher temperature T_2 as the corresponding reversible process, the efficiency of the former must be less than that of the latter, i.e.

$$\left(\frac{q_1}{T_1} - \frac{q_2}{T_2}\right)_{\text{irrev.}} > \left(\frac{q_1}{T_1} - \frac{q_2}{T_2}\right)_{\text{rev.}} \qquad \text{...(9.A15)}$$

Thus from equations 9.A11 and 9.A15 it follows that for an irreversible cycle

$$\sum \frac{q}{T} > 0 \qquad \text{...(9.A15}a\text{)}$$

and that the more positive $\Sigma q/T$ is the more irreversible is the cycle. In the Carnot cycle each stage is carried out reversibly and the entropy change is zero, and it can be shown that any complete cycle can be reduced to the sum of a large number of infinitesimal Carnot cycles so that the entropy change for any complete reversible cycle must always be zero. Thus for any reversible process the sum of all entropy changes for all systems participating is given by

$$\Sigma\, \mathrm{d}S = 0 \qquad \text{...(9.A16)}$$

which provides a criterion for equilibrium.

On the other hand, in any irreversible process although the system may gain (or lose) entropy and the surroundings lose (or gain) entropy, the system plus surrounding will always gain in entropy (equation 9.A15*a*). Thus for a real process proceeding spontaneously at a finite rate

$$\Sigma \, \mathrm{d}S > 0 \qquad \text{...(9.A17)}$$

For example, if a hot body (the system) is in contact with the surroundings at an infinitesimally lower temperature there will be an isothermal and reversible transfer of heat at T_2, and the loss in entropy of the hot body will be $\Delta S_b = -q_{rev.}/T_2$ and the gain in entropy of the surroundings will be $\Delta S_s = +q_{rev.}/T_2$, and $\Delta S_b + \Delta S_s = 0$. On the other hand, if the surroundings are at a lower temperature T_1 than the body the transfer of heat will be irreversible and spontaneous, and the entropy changes will be $\Delta S_b = -q_{rev.}/T_2$ and $\Delta S_s = +q_{rev.}/T_1$. Since $T_2 > T_1$ the total change in entropy will be greater than zero, and the greater the difference between T_2 and T_1 the greater the gain in the total entropy of the system as a whole, i.e. the more spontaneous or irreversible is the transfer of heat.

When a substance is heated at constant pressure without change of phase through a temperature rise $\mathrm{d}T$ the heat absorbed is $C_p \mathrm{d}T$, where C_p is the molar heat capacity at constant pressure, and the entropy increase is

$$\int_{S_1}^{S_2} \mathrm{d}S = \int_{T_1}^{T_2} \frac{C_p}{T} \mathrm{d}T = \int_{T_1}^{T_2} C_p \, \mathrm{d} \ln T \qquad \text{...(9.A18)}$$

Thus the entropy increase when the substance is heated from T_1 to T_2 is

$$\Delta S = \int_{T_1}^{T_2} C_p \, \mathrm{d} \ln T = 2{\cdot}303 \int_{T_1}^{T_2} C_p \, \mathrm{d} \log T \qquad \text{...(9.A19)}$$

and ΔS may be evaluated by determining the heat capacity C_p at various temperatures between T_2 and T_1, and plotting the values against $\log T$; the area enclosed by the curve multiplied by 2·303 gives the entropy increase. When there is a change of phase such as the evaporation of a liquid to saturated vapour, melting of a solid to a liquid or transition from one crystalline form to another, the process is isothermal and reversible, and

$$\Delta S = \frac{q}{T} \qquad \text{...(9.A20)}$$

where q is the heat associated with the phase change at T.

For example, in the case of the reversible isothermal transformation of ice to water at the melting point (273 K), the heat gained by the ice will be the latent heat of fusion ($\Delta H_f = 6006 \, \mathrm{J \, mol^{-1}}$) and a corresponding quantity of heat will be lost by the surrounding, and

$$\Delta S_{system} = \Delta S_{ice} + \Delta S_{surr.} = \frac{6\,006}{273} - \frac{6\,006}{273} = 0$$

Gibbs Free Energy or Free Enthalpy

The Gibbs free energy G provides a means of defining equilibrium or of the tendency of a reaction to proceed in a given direction. It is similar to the other thermodynamic functions described above in that the change in free energy

ΔG is determined solely by the initial and final states of the system. The maximum work, or maximum available energy, defined in terms of the Gibbs free energy G, which is now called the *free enthalpy*, is

$$G = H - TS \qquad \text{...(9.A21)}$$

or, for a finite change

$$\Delta G = \Delta H - T\Delta S \qquad \text{...(9.A22)}$$

It can be shown that for a system in which the total mass, or the concentrations of the components, remains constant

$$\left(\frac{\partial G}{\partial P}\right)_{T,n_i\ldots} = V \qquad \text{...(9.A23)}$$

and

$$\left(\frac{\partial G}{\partial T}\right)_{p,n_i\ldots} = -S \qquad \text{...(9.A24)}$$

Combining equations 9.A21 and 9.A24

$$G - H = T\left(\frac{\partial G}{\partial T}\right)_{p,n_i\ldots} \qquad \text{...(9.A25)}$$

which is the important Gibbs–Helmholz equation. Since in thermodynamics we are concerned mainly with changes in free energy and enthalpy, equation 9.A25 can be written

$$\Delta G - \Delta H = T\left(\frac{\partial \Delta G}{\partial T}\right)_{p,n_i\ldots} \qquad \text{...(9.A26)}$$

Units Free energy is an *extensive* quantity, but the *standard* free energy for 1 mol $\Delta G^{\ominus}_{298K}$ (in units of J mol^{-1}) defined above is an *intensive* quantity.

The Chemical Potential

The energy of a system can be changed by means of thermal energy or work energy, but a further possibility is to add or subtract moles of various substances to or from the system. The free energy of a pure substance depends upon its chemical nature, its quantity (ΔG is an extensive property), its state (solid, liquid or gas), and temperature and pressure. Gibbs called the partial molar free heat content (free energy) of the ith component of a system its *chemical potential*

$$\mu_i = \left(\frac{\partial G}{\partial n_i}\right)_{T,p,n_j} \qquad \text{...(9.A27)}$$

i.e. the change in free enthalpy of the system when 1 mol of component i is added to the system, the T, p and concentration of all other substances (n_j) being maintained constant. The partial molar free energy of a pure compound is synonymous with the *standard chemical potential* $\mu^{\ominus}$, which is the free enthalpy change when 1 mol of the substance is formed from its elements under isothermal isobaric conditions. In order to evaluate the chemical potentials of compounds, and of ions in solution, it is necessary to define the following standard states:

1. The standard chemical potentials of the elements in the form that they are normally stable at the temperature and pressure under consideration are given an arbitrary zero

$$\mu^{\ominus}\,(\text{element}) = 0 \qquad \ldots(9.\text{A}28)$$

2. In the case of ions in solution, and of gases, the chemical potential will depend upon concentration and pressure, respectively, For ions in solution the standard chemical potential of the hydrogen ion, at the temperature and pressure under consideration, is given an arbitrary value of zero at a specified concentration

$$\mu^{\ominus}_{\text{H}^+} = 0, \text{ when } a_{\text{H}^+} = 1 \qquad \ldots(9.\text{A}29)$$

where a_{H^+} is the *activity* of the hydrogen ion.

Standard chemical potentials for ions and compounds are given in Chapter 21, Table 21.5.

Activity

The chemical potential of the hydrogen ion, or of any other ion, will vary with activity

$$\mu_{\text{i}} = \mu^{\ominus}_{\text{i}} + RT\ln a_{\text{i}} \qquad \ldots(9.\text{A}30)$$

where the activity a_{i} is a thermodynamic quantity that is related to the concentration of the ion and takes into account the fact that a solute in solution does not behave ideally at finite concentrations; only at infinite dilution will $a_{\text{i}} = c_{\text{i}}$ numerically, where c_{i} is the concentration (mol kg^{-1}, mol m^{-3} or mol dm^{-3}).

Substituting for $R = 8{\cdot}314\,\text{J K}^{-1}\text{mol}^{-1}$, $T = 298\,\text{K}$ and using the identity $\ln x = 2{\cdot}303 \log x$, equation 9.A30 can be expressed as

$$\mu_{\text{i}} = \mu^{\ominus}_{\text{i}} + 5\,710 \log a_{\text{i}} \qquad \ldots(9.\text{A}31)$$

A similar equation is applicable to a gas, but the pressure of the gas p_{g} (or more strictly its fugacity) replaces the activity of the ion:

$$\mu_{\text{g}} = \mu^{\ominus}_{\text{g}} + 5\,710 \log (p_{\text{g}}/p^{\ominus}) \qquad \ldots(9.\text{A}32)$$

where $p^{\ominus}$ refers to a standard pressure of 1 atm or $1{\cdot}013\,25 \times 10^5$ Pa (pascals). For a non-dissociating solute (e.g. urea)

$$a = \gamma c/c^{\ominus} \qquad \ldots(9.\text{A}33)$$

where γ is the activity coefficient ($\gamma \to 1$ as $c \to 0$) and $c^{\ominus}$ is the concentration in the standard state.

For a 1–1 electrolyte (NaCl) a similar procedure may be adopted, but in this case not only do the constituent ions (Na^+ and Cl^-) become more ideal in behaviour as the concentration decreases, but also the electrolyte becomes more completely dissociated into ions. It is not possible to determine individual ionic activity coefficient, since anions and cations cannot be tudied in the absence of one another. Thus for a 1–1 electrolyte equation 9.A30 is written

$$\mu = \mu^{\ominus} + 2RT\ln a_{\pm} = \mu^{\ominus} + 2RT\ln(\gamma_{\pm}c/c^{\ominus}) \qquad ...(9.A34)$$

where $a_{\pm}$ is the mean ionic activity and $\gamma_{\pm}$ is the mean activity coefficient, and $\gamma_{\pm} \rightarrow 1$ when $c \rightarrow 0$. More complex relationships are required for unsymmetrical electrolytes such as Na_2SO_4, $La_2(SO_4)_3$, $TiCl_4$, etc.

At sufficiently low ionic strengths the activity coefficient of each electrolyte in a mixture is given by the Debye-Hückel limiting law

$$\log \gamma_{\pm} = -|z_+z_-|A\sqrt{I} \qquad ...(9.A35)$$

where I the *ionic strength* is defined as

$$I = \tfrac{1}{2}\Sigma c_i z_i^2 \qquad ...(9.A36)$$

and z_+, z_- are the valencies of the cation and anion (NaCl is a 1–1 electrolyte, Na_2SO_4 a 2–1 electrolyte, $La_2(SO_4)_3$ a 2–3 electrolyte, etc.) and A is a constant which is 0·51 $mol^{-\frac{1}{2}}\,kg^{\frac{1}{2}}$ for water at 25°C (c_i is in $mol\,kg^{-1}$).

Units It should be noted that in the S.I. the activity of a solute is defined with reference to a standard state, i.e. an ideal solution of molality 1 $mol\,kg^{-1}$. Thus the *relative activity* of a metal ion in solution is given by

$$a_{M^{z+}} = \frac{m_{M^{z+}}}{m^{\ominus}}\gamma_{M^{z+}} \qquad ...(9.A37)$$

where $m^{\ominus} = 1\,mol\,kg^{-1}$, $m_{M^{z+}}$ is in $mol\,kg^{-1}$ and $\gamma_{M^{z+}}$ is dimensionless. It follows that the relative activity is dimensionless, although units such as $mol\,kg^{-1}$, $g\,ion\,litre^{-1}$, etc. are still being used.

Gibbs-Duhem Equation

The Gibbs free energy change of a system will depend not only on temperature and pressure but upon the chemical potentials of the species involved, and this statement may be expressed in the form of the partial differential

$$dG = \left(\frac{\partial G}{\partial T}\right)_{p,n_1,n_2\ldots} dT + \left(\frac{\partial G}{\partial p}\right)_{T_1,n_1,n_2\ldots} dp + \mu_1 dn_1 + \mu_2 dn_2 \qquad ...(9.A38)$$

and combining with equation 9.A23 and equation 9.A24

$$dG = -SdT + Vdp + \Sigma\mu_i dn_i \qquad ...(9.A39)$$

For a process carried out under isothermal and isobaric conditions

$$(dG)_{T,p} = \Sigma\mu_i dn_i \qquad ...(9.A40)$$

and integration of this equation for a system of given chemical composition gives

$$G_{T,p,n} = \Sigma n_i \mu_i \qquad ...(9.A41)$$

General differentiation of this equation gives

$$dG_{T,p} = \Sigma(n_i d\mu_i + \mu_i dn_i) \qquad ...(9.A41a)$$

Since at equilibrium $(dG)_{T,p,n}$ is zero, it follows from equation 9.A40 that

$$\Sigma n_i d\mu_i = 0 \qquad \text{...(9.A42)}$$

which is referred to as the *Gibbs-Duhem* equation.

For a binary solution of say A and B

$$n_A\, d\mu_A = -n_B\, d\mu_B \qquad \text{...(9.A43)}$$

and since

$$\mu_i = \mu_i^{\ominus} + RT \ln a_i \qquad \text{...(9.A44)}$$

then

$$d\mu_i = RT\, d\ln a_i \qquad \text{...(9.A45)}$$

and hence

$$n_A\, d\ln a_A = -n_B\, d\ln a_B \qquad \text{...(9.A46)}$$

The Gibbs-Duhem equation is extremely important in solution chemistry and it can be seen from equation 9.A46 that it provides a means of determining the activity of one component in a binary solution providing the activity of the other is known.

Spontaneity of a Reaction

An isothermal reaction can proceed spontaneously only if the total Gibbs free energy of the system decreases, i.e. the free energy of the reactants must be greater than the free energy of the products. For a reaction

$$aA + bB = cC + dD \qquad \text{...(9.A47)}$$

$$\Delta G = \Sigma\mu_{\text{products}} - \Sigma\mu_{\text{reactants}} \qquad \text{...(9.A47}a)$$

where ΔG is the change in free enthalpy. It follows that if $\Delta G < 0$ the reaction will tend to proceed spontaneously in the direction shown; if $\Delta G > 0$ the reaction will tend to proceed spontaneously in the reverse direction; if $\Delta G = 0$ the system will be at equilibrium and will not have any tendency to proceed in either direction.

Thus for a spontaneous reaction

$$\Delta G = (c\mu_C + d\mu_D) - (a\mu_A + b\mu_B) < 0 \qquad \text{...(9.A48)}$$

and if all substances are in their standard state

$$\Delta G^{\ominus} = (c\mu_C^{\ominus} + d\mu_D^{\ominus}) - (a\mu_A^{\ominus} + b\mu_B^{\ominus}) < 0 \qquad \text{...(9.A49)}$$

Reversible Cells

By definition

$$\Delta G = -zFE \qquad \text{...(9.A50)}$$

where E (the reversible e.m.f.) is the intensity factor in the energy term and zF (the total charge per mole of ions involved in the reaction) is the capacity factor. If the activities of the reactants and products are fixed E will be a constant, and if $a_A = a_B = a_C = a_D = 1$ is taken as the standard state then

$$\Delta G^\ominus = -zFE^\ominus \qquad \text{...(9.A51)}$$

where $\Delta G^\ominus$ is the standard free enthalpy change and $E^\ominus$ is the standard e.m.f., which is a constant for a given equilibrium at constant T.

Substituting $-zFE$ for ΔG (equation 9.A51) in equation 9.A26 gives

$$\Delta H = -zFE + zFT\left(\frac{\partial E}{\partial T}\right)_p \qquad \text{...(9.A52)}$$

which provides an accurate method of evaluating ΔH and ΔS (equation 9.A26) from the e.m.f. of a reversible cell and the coefficient of the change of e.m.f. with temperature at constant p, since

$$-T\left(\frac{\partial(\Delta G)}{\partial T}\right)_p = zFT\left(\frac{\partial E}{\partial T}\right)_p = T\Delta S \qquad \text{...(9.A52}a\text{)}$$

From equation 9.A47*a*

$$\Delta G^\ominus = \Sigma \upsilon_i \mu_i^\ominus \qquad \text{...(9.A53)}$$

where υ_i is the number of each of the ions or molecules involved in the equilibrium; combining equations 9.A51 and 9.A53

$$E^\ominus = -\frac{\Sigma \upsilon_i \mu_i^\ominus}{zF} = -\frac{\Sigma \upsilon_i \mu_i^\ominus}{96\,500z} \qquad \text{...(9.A54)}$$

Equations 9.A51 and 9.A54 emphasise the essentially thermodynamic nature of the standard equilibrium e.m.f. of a cell or the standard equilibrium potential of a half-reaction $E^\ominus$, which may be evaluated directly from e.m.f. measurements of a reversible cell or indirectly from $\Delta G^\ominus$, which in turn must be evaluated from the enthalpy of the reaction and the entropies of the species involved (*see* equation 9.A22). Thus for the equilibrium $Cu^{2+} + 2e \rightleftharpoons Cu$, the standard electrode potential $E^\ominus_{Cu^{2+}/Cu}$, and hence $\mu^\ominus_{Cu^{2+}}$, can be determined by an e.m.f. method by harnessing the reaction

$$Cu^{2+} + 2Hg \rightarrow Hg_2^{2+} + Cu$$

which can be represented by a cell without a liquid junction (or liquid junction potential)

$$Hg|Hg_2SO_4(aq.); CuSO_4(aq.)|Cu$$

in which | represents the phase boundaries. When a cell has a liquid junction the latter is represented by ¦ as in the cell

$$M_1|M_1^{2+}(aq.)|M_2^{2+}(aq.)|M_2$$

Although in certain cells the liquid junction can be eliminated by appropriate choice of electrolyte solution, this is not always possible. However, the liquid junction potential can be minimised by the use of a salt bridge (a saturated solution of KCl of about 4·2 M), and the liquid junction potential is then only 1–2 mV; this elimination of the liquid junction potential is indicated by ‖.

The standard electrode potential of M^{z+}/M can be evaluated accurately and conveniently providing it forms a reversible electrode and can be coupled with a reference electrode to form a reversible cell. However, a number of M^{z+}/M systems do not conform to this criterion, and calorimetric methods must be used to evaluate $\Delta H^\ominus$ and $\Delta S^\ominus$, and hence $\Delta G^\ominus$, from which $E^\ominus$ can be calculated. Thus in the case of the Mg^{2+}(aq.)/Mg equilibrium, the enthalpy change during the dissolution of the metal in dilute acid at 25°C is determined from the reaction

$$\mathrm{Mg(s.)+2H^+(aq.) = Mg^{2+}(aq.)+H_2(g.)};\ \Delta H^\ominus = -461{\cdot}200\ \mathrm{J\,mol^{-1}}$$

and the standard entropy change $\Delta S^\ominus$ is evaluated from the standard partial molar entropies of the reactants and products

$$\Delta S^\ominus = S^\ominus(\mathrm{Mg^{2+}})+S^\ominus(\mathrm{H_2})-[S^\ominus(\mathrm{Mg})+2S^\ominus(\mathrm{H^+})]$$
$$= -43{\cdot}3\ \mathrm{J\,K^{-1}\,mol^{-1}}$$

Since $\Delta G^\ominus = \Delta H^\ominus - TS^\ominus$

$$\Delta G^\ominus = -461\,200-(298\times -43{\cdot}3)$$
$$= -448\,100\ \mathrm{J\,mol^{-1}}\ \text{at}\ 298\ \mathrm{K}$$

$$\therefore\quad E^\ominus_{\text{reaction}} = \frac{-\Delta G^\ominus}{zF} = \frac{-448\,100}{2\times 96\,500} = -2{\cdot}3\ \mathrm{V}$$

and since $a_{\mathrm{H^+}} = p_{\mathrm{H_2}} = 1$, $E^\ominus_{\mathrm{Mg^{2+}/Mg}} = -2{\cdot}3\ \mathrm{V}$

Chemical Potentials and Equilibrium

The relationship between the change in free energy ΔG and the equilibrium constant K is given by the vant Hoff isotherm

$$\Delta G = -RT\ln K + RT\ln Q \qquad \text{...(9.A55)}$$

and from equation 9.A47 $K = \dfrac{a_C^c a_D^d}{a_A^a a_B^b}$ ...(9.A56)

and a denotes the equilibrium activity of the component and Q is the activity quotient. Note that the activities of the products are in the numerator and activities of reactants in the denominator.

Equation 9.A55 shows that ΔG is dependent on K and on any arbitrary activity of reactants and products that may be introduced into the equation. When the reactants and products are in their standard states

$$a_A = a_B = a_C = a_D = 1 \text{ and } \ln Q = 0 \qquad \text{...(9.A57)}$$

hence from equation 9.A51

$$\Delta G = \Delta G^\ominus = -RT\ln K \qquad \text{...(9.A58)}$$

$$\therefore\quad \Delta G^\ominus = -RT\ln K = -zFE^\ominus \qquad \text{...(9.A59)}$$

and

$$E^\ominus = \frac{RT}{zF}\ln K = \frac{0{\cdot}059}{z}\log K \qquad \text{...(9.A60)}$$

since at 25°C the term $2{\cdot}303\ RT/F = 2{\cdot}303\times 8{\cdot}134\times 298/96\,500 = 0{\cdot}059$ V. Combining equations 9.A50 and 9.A55

$$E = \frac{RT}{zF}\ln K - \frac{RT}{zF}\ln Q \qquad \ldots(9.A61)$$

and from equation 9.A60

$$E = E^{\ominus} - \frac{0{\cdot}059}{z}\log Q = E^{\ominus} - \frac{0{\cdot}059}{z}\log\frac{a_C^c a_D^d}{a_A^a a_B^b} \qquad \ldots(9.A62)$$

which shows how the equilibrium potential E is dependent on the standard electrode potential $E^{\ominus}$, a constant for a given equilibrium, and the activities of reactants and products. Equation 9.A62 is termed the *Nernst* equation and is applicable only when the system is reversible.

Since
$$\Delta G^{\ominus} = \Sigma \nu_i \mu_i^{\ominus} = -RT\ln K \qquad \ldots(9.A63)$$

$$\log K = -\frac{\Sigma \nu_i \mu_i^{\ominus}}{5\,710} \text{ at } 25°\text{C} \qquad \ldots(9.A64)$$

Thus the equilibrium constant K can be evaluated from standard electrode potential $E^{\ominus}$ or from the standard chemical potentials $\mu^{\ominus}$. It should be noted that in the older textbooks the units of K are atm, mol kg^{-1}, mol dm^{-3}, etc. and account must be taken of the number of moles of reactants and products involved in the equilibrium. Thus for a reaction of the type $1A + 1B = 2C$, K will be in mol kg^{-1}, whereas for $1A + 3B = 2C$ K will be in (mol kg^{-1})$^{2/4}$, i.e. (mol kg^{-1})$^{\frac{1}{2}}$. However, in S.I. the equilibrium constant K is dimensionless.

Sign Convention for Equilibrium e.m.f.s and Potentials

For ions in solution the standard reference state is the hydrogen ion whose standard chemical potential at $a_{H^+} = 1$ is given an arbitrary value of zero. Similarly for pure hydrogen at $p_{H_2} = 1$ atm, $\mu_{H_2}^{\ominus} = 0$. Thus for the $H^+/\frac{1}{2}H_2$ equilibrium (equation 9.A54)

$$E^{\ominus} = -\frac{\Sigma \nu_i \mu_i^{\ominus}}{zF} = 0{\cdot}00\text{ V} \qquad \ldots(9.A65)$$

and the standard hydrogen electrode (S.H.E.) is taken as the reference electrode for all other equilibria.

For a cell the equilibrium must be written in the direction in which it proceeds spontaneously, and under these circumstances

$$\Delta G < 0, \text{ and } E > 0$$

i.e. the e.m.f. of the cell will be positive. Thus for the equilibrium between Zn and H^+ the reaction is written

$$\text{Zn(s.)} + 2\text{H}^+\text{(aq.)} = \text{Zn}^{2+}\text{(aq.)} + \text{H}_2\text{(g.)}$$

and if $a_{H^+} = a_{Zn^{2+}} = a_{H_2} = 1$, $E = E^{\ominus} = +0{\cdot}76$ V. However, in the case of a half-reaction it is quite arbitrary as to whether the equilibrium is written as a reduction (electrons on the l.h.s.) or as an oxidation (electrons on the r.h.s.)

$$Zn^{2+} + 2e = Zn \qquad \ldots(9.A66a)$$

or

$$Zn = Zn^{2+} + 2e \qquad \ldots(9.A67a)$$

since both are correct. Substituting in the Nernst equation (equation 9.A62) for equation 9.A66*a*

$$E = E^{\ominus} - 0{\cdot}030 \log \frac{1}{a_{Zn^{2+}}} = E^{\ominus} + 0{\cdot}030 \log a_{Zn^{2+}} \qquad \ldots(9.A66b)$$

and for equation 9.A67*a*

$$E = E^{\ominus} - 0{\cdot}030 \log a_{Zn^{2+}} \qquad \ldots(9.A67b)$$

Thus if the magnitudes of E are to be the same the signs of $E^{\ominus}$ must be different, i.e. $E^{\ominus} = -0{\cdot}76$ V for equation 9.A66*b* and $E^{\ominus} = +0{\cdot}76$ V for equation 9.A67*b*. If this convention is adopted, the respective electrode potentials become

$$E = -0{\cdot}76 + 0{\cdot}030 \log a_{Zn^{2+}} \qquad \ldots(9.A66c)$$

$$E = +0{\cdot}76 - 0{\cdot}030 \log a_{Zn^{2+}} \qquad \ldots(9.A67c)$$

and for any given activity of Zn^{2+} the Es will be the same in magnitude but opposite in sign.

It is now internationally accepted (Stockholm Convention) that the half-reaction must be written with the electrons on the l.h.s.:

$$Cu^{2+}(aq.) + 2e = Cu \qquad E^{\ominus} = +0{\cdot}34\,V$$

$$Zn^{2+}(aq.) + 2e = Zn \qquad E^{\ominus} = -0{\cdot}76\,V$$

L. L. SHREIR

Acknowledgement. The author would like to express his appreciation to Dr S. Wach for his advice on this appendix and for checking the manuscript and proof.

9.1B Appendix – The Potential Difference at a Metal/Solution Interface

Interfacial Potentials Constituting the e.m.f. of a Cell

It is quite impossible to determine the absolute potential difference across a single metal/solution interface, and the potential must be evaluated indirectly from the e.m.f. of a cell comprising the interface under consideration and another electrified interface. The e.m.f. of the cell can be determined readily by a suitable measuring device such as a potentiometer, vacuum-tube voltmeter or an electrometer, which are capable of measuring the e.m.f. with the minimum passage of electrical charge. This is essential since if a significant current is allowed to pass, the electrodes (electrified interfaces) become polarised and the e.m.f. will be less than the equilibrium value. Consider the determination of the interfacial potential at the surface of a zinc electrode in equilibrium with Zn^{2+} ions in solution. In order to determine the potential it is necessary to couple it with another electrode, and for the purpose of this discussion the equilibrium between H^+ ions in solution and H_2 gas will be chosen, i.e. the reversible hydrogen electrode in which the equilibrium between H^+ and H_2 takes place at a platinised–platinum surface. The spontaneous cell reaction will be

$$Zn + 2H^+(aq.) \rightarrow Zn^{2+}(aq.) + H_2 \qquad \ldots(9.A68)$$

and the cell can be represented by the reversible cell

$$Zn|\text{aqueous } ZnSO_4|\text{aqueous } H_2SO_4|Pt, H_2$$

in which | represents an interface.

In this cell (Fig. 9.A1) the following interfaces and associated interfacial potentials may be distinguished:

1. The $Zn/Zn^{2+}(aq.)$ interface and the associated interfacial potential $\phi_{Zn/Zn^{2+}}$.
2. The $Pt/H^+(aq.),H_2$ interface and the associated interfacial potential $\phi_{Pt/H^+,H_2}$.
3. The interface between the two solutions $ZnSO_4/H_2SO_4$ and the associated liquid junction potential ϕ_j.
4. The interface between the Zn electrode and the Cu lead [at which $e(Zn) \rightarrow e(Cu)$] and the associated interfacial potential $\phi_{Zn/Cu}$.

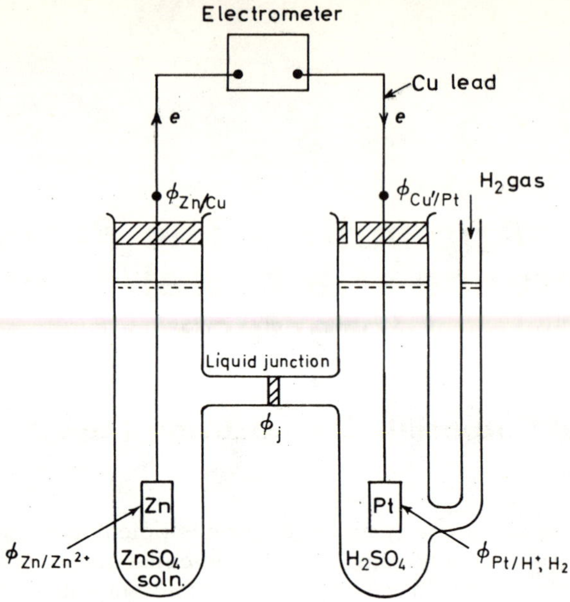

Fig. 9.A1 Interfacial potentials in the electrochemical cell $Zn|Zn^{2}|H^{+}|Pt, H_2$

5. The interface between the Cu lead and the Pt electrode [at which $e(\text{Cu}') \rightarrow e(\text{Pt})$] and the associated interfacial potential $\phi_{\text{Cu}'/\text{Pt}}$.

Thus the e.m.f. of the cell will be the resultant of the various interfacial potentials given above, i.e.

$$E_{\text{cell}} = \phi_{\text{Zn/Zn}^{2+}} + \phi_{\text{Pt/H}^+,\text{H}_2} + \phi_{\text{j}} + \phi_{\text{Zn/Cu}} + \phi_{\text{Cu}'/\text{Pt}} \quad \text{...(9.A69)}$$

Since the small interfacial potentials at the junctions of the electrodes and copper leads are equal and opposite, they cancel out, and if the liquid junction potential is assumed to be small, or is reduced to a negligible value by using a salt bridge, then equation 9.A69 reduces to

$$E_{\text{cell}} = \phi_{\text{Zn/Zn}^{2+}} + \phi_{\text{Pt/H}^+,\text{H}_2} \quad \text{...(9.A70)}$$

At each interface the interfacial potential will depend upon the chemical potentials of the species involved in the equilibrium. Thus at the Zn/Zn^{2+} electrode there will be a tendency for zinc ions in the lattice to lose electrons and to pass across the interface and form hydrated ions in solution; this tendency is given by the chemical potential of zinc μ_{Zn}, which for pure zinc will be a constant. Similarly, there will be a tendency for hydrated Zn^{2+} ions in solution to lose their hydration sheaths, to gain electrons and to enter the lattice of the metal; this tendency is given by the chemical potential of the Zn^{2+} ions, which is related to their activity. (See equation 9.A30.)

Thermodynamically

$$\phi_{\text{Zn/Zn}^{2+}} = \mu_{\text{Zn}^{2+}} - \mu_{\text{Zn(l)}} = (\mu^{\ominus}_{\text{Zn}^{2+}} + RT\ln a_{\text{Zn}^{2+}}) - \mu^{\ominus}_{\text{Zn(l)}} \quad \text{...(9.A71)}$$

where the superscript $\ominus$ indicates the standard state of the species under consideration, and for pure zinc $\mu_{\text{Zn(l)}} = \mu^{\ominus}_{\text{Zn(l)}} = 0$.

Table 9.A1 Classification of electrode potentials of metals

A. Reversible electrodes

Type	*Characteristics*	*Examples*
(1) $M/M^{z+}_{(aq.)}$ ion (Type 1 electrode)	Metal M in solution of its captions giving an $M^{z+} \rightleftharpoons M$ exchange whose potential conforms with the Nernst equation.	Cu^{2+}/Cu, Zn^{2+}/Zn, Ag^{+}/Ag
(2) M/metal salt/anion (Type 2 electrode)	Metal in contact with its sparingly soluble salt MX and a soluble salt the anion X giving an $M/MX/X^{-}$ electrode whose potential depends on K_s of MX and the a_{x^-}.	$Hg/Hg_2Cl_2/KCl(aq.)$ $Hg/Hg_2SO_4/K_2SO_4(aq.)$ $Ag/AgCl/KCl(aq.)$
(3) Oxidation-reduction (redox)	Inert metal (normally Pt but certain other metals can act in a similar manner) in a solution containing two species that give rise to a redox system. E depends on $E^{\ominus}$ of the system and the relative activities of the oxidised and reduced forms.	Quinone-hydroquinone: $C_6H_4O_2 + 2H^+ + 2e \rightleftharpoons C_6H_4(OH)_2$, which is thus pH dependent; Fe^{3+}/Fe^{2+} MnO_4^-/Mn^{2+}
(4) Metal/metal oxide	Metal filmed with oxide in a solution of OH^- giving an $M/M_xO_y/OH^-$ electrode whose potential is dependent on pH.	$Sb/Sb_2O_3/OH^-$ $Bi/Bi_2O_3/OH^-$
(5) Gas/ion	Inert electrode (usually Pt but other metals or metallic compounds may act in a similar way) in a redox system of which one component is a gas.	$Pt/H_2, H^+$ $Pt/Cl_2, Cl^-$

(continued overleaf)

Table 9.A1 (continued)

B. Irreversible electrodes

Type	*Characteristics*	*Examples*
(1) Steady-state potential comparable to Type 1 reversible electrode	Metal in a solution of electrolyte in which M^{z+} ions are produced by a corrosion reaction resulting in an M^{z+}/M exchange that determines the potential.	Zn in NaCl solution Zn in dilute HCl
(2) Steady-state potential comparable to Type 2 reversible electrode	Potentials of electropositive metals that react with solution to give sparingly soluble salts of the metal.	Cu or Ag in NaCl or Ag in HCl giving an $M/MX/X^-$ type of electrode
(3) Steady-state potential comparable to Types 4 and 5 reversible electrodes	Potential of metal depends on pH of solution, although the dependence is confined to a limited range of pH and does not conform precisely to the Nernst equation.	Ni in H_2SO_4 (Ni/H_2, H^+) Cu in NaOH ($Cu/Cu_2O/OH^-$)
(4) Corrosion or mixed potentials:	Potential dependent on the redox potential of the solution and the kinetics of the anodic and cathodic reactions.	
(*a*) Active corrosion in acid solutions	Potential dependent on the kinetics of the h.e.r. on the bare metal surface.	Zn in HCl
(*b*) Passive metal in acid solutions	Potential is that of an oxide-filmed metal, and is dependent on the redox potential of the solution.	Stainless steel in oxygenated H_2SO_4

For a reversible cell at equilibrium the Gibbs free energy and the reversible e.m.f. E_r are related by

$$\Delta G = -zFE_r = 2FE_r \qquad \text{...(9.A72)}$$

and the interfacial potential at the Zn/Zn^{2+} electrode is given by

$$\phi_{Zn/Zn^{2+}} = E_{r,Zn} = \frac{\mu^{\ominus}_{Zn^{2+}} - \mu^{\ominus}_{Zn}}{2F} + \frac{RT}{2F}\ln a_{Zn^{2+}} \qquad \text{...(9.A73)}$$

The standard chemical potentials $\mu^{\ominus}_{Zn^{2+}}$ and $\mu^{\ominus}_{Zn}$ are constants so that the first term on the r.h.s. can be replaced by $E^{\ominus}_{Zn^{2+}/Zn}$, the standard electrode potential of the $Zn^{2+} + 2e = Zn$ equilibrium, and

$$\phi_{Zn^{2+}/Zn} = E_{r,Zn} = E^{\ominus}_{Zn^{2+}/Zn} + \frac{RT}{2F}\ln a_{Zn^{2+}} \qquad \text{...(9.A74)}$$

Similarly for the $2H^+ + 2e = H_2$ equilibrium

$$\phi_{H^+/H_2} = E_{r,H} = E^{\ominus}_{H^+/H_2} + \frac{RT}{2F}\ln\frac{a^2_{H^+}}{p_{H_2}} \qquad \text{...(9.A75)}$$

Subtracting equation 9.A74 from 9.A75

$$\phi_{H^+/H_2} - \phi_{Zn^{2+}/Zn} = E_{r,cell} = E^{\ominus}_{cell} + \frac{RT}{2F}\ln\frac{a^2_{H^+}}{a_{Zn^{2+}}\,p_{H_2}} \qquad \text{...(9.A76)}$$

The above considerations show that the equilibrium e.m.f. of a reversible cell is determined solely by the interfacial potentials at the two electrodes constituting the cell, providing the liquid junction can be eliminated or made negligible, and under these circumstances the interfacial potentials will be related to the chemical potentials of the species involved in the equilibrium. In the case of an irreversible cell, e.g.

$$\text{Ni}|\text{aqueous NiSO}_4|\text{aqueous H}_2\text{SO}_4|\text{Pt,H}_2$$

there will be the same number of interfacial potentials as in the reversible $Zn|ZnSO_4|H_2SO_4|Pt,H_2$ cell, but the e.m.f. of the cell cannot be treated thermodynamically.

Examples of reversible and irreversible electrodes and electrode potentials are given in Table 9.A1.

Non-polarisable and Polarisable Interfaces

An ideal reversible cell is characterised by an e.m.f. that remains constant irrespective of the rate of reaction in either direction, i.e. each interface constituting the cell must be so completely non-polarisable that it resists any attempt to change its potential. Although this is impossible to achieve in practice, a number of interfaces approximate to ideality providing the rate of reaction is maintained at a very low value. These reversible electrodes (or half-cells) are used as reference electrodes for determining the potential of a single electrified interface.

At the other extreme is the completely polarisable interface for which a very small passage of charge will result in a very large change in potential.

The calomel electrode Hg/Hg_2Cl_2,KCl approximates to an ideal non-polarisable electrode, whilst the Hg/aqueous electrolyte solution electrode approximates to an ideal polarisable electrode. The electrical behaviour of a metal/solution interface may be regarded as a capacitor and resistor in parallel (Fig. 9.A2), and on the basis of this analogy it is possible to distinguish between a completely polarisable and completely non-polarisable interface.

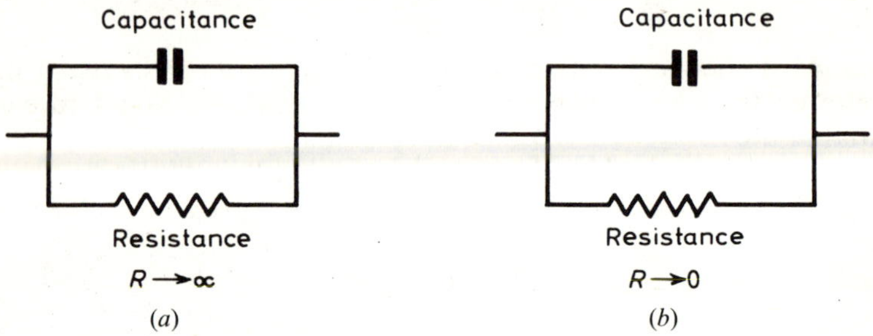

Fig. 9.A2 Analogy between a metal/solution interface and a capacitance and resistance in parallel. (*a*) Non-polarisable electrode, e.g. Hg/KCl solution, (*b*) polarisable electrode, e.g. Hg/Hg_2Cl_2, KCl solution

The polarisable electrode will have a high resistance so that when it is connected to a source of e.m.f. the capacitor will become charged up to the value of the potential difference of the source, i.e. it will polarise. Conversely, when the resistance in parallel with the capacitor is low, charge will leak through the resistance and will thus prevent the charging up of the capacitor, i.e. it will be non-polarisable. Polarisable electrodes such as Hg/aqueous KCl are used for studies of the structure of the electrical double layer, whilst non-polarisable electrodes are used as reference electrodes and for evaluating thermodynamic parameters such as the chemical potentials of ions in solution.

Thermodynamics of a Reversible Cell

If a reaction is harnessed in a reversible cell and the reaction is allowed to proceed at an infinitely slow rate (almost at equilibrium), the work energy will be a maximum and will be equal to the Gibbs free enthalpy of the reaction (*see* equation 9.A72). The reversible e.m.f. will be the maximum e.m.f. of the cell reaction, and if the reaction is allowed to proceed reversibly at constant temperature and pressure the heat transferred between the cell and its surroundings will give the entropy change of the reaction (*see* equation 9.A12). On the other hand, if the cell reaction is allowed to proceed irreversibly in a calorimeter, all the internal energy of the reactants will appear as heat energy and the external work will be zero (disregarding pV work). If the cell is allowed to operate irreversibly at a finite rate the work energy will decrease (the e.m.f. of the cell will be less than the reversible e.m.f.) and more heat will be transferred from the cell to the surroundings than when it operates reversibly, i.e. the entropy of the system will increase.

This can be exemplified by considering the spontaneous reaction

$$Zn + 2AgCl \rightarrow 2Ag + Zn^{2+} + 2Cl^-$$

for which the equivalent reversible cell without a liquid junction, and hence without a liquid junction potential, can be represented by

$$Zn|ZnCl_2 \text{ solution; } AgCl|Ag$$

When the concentration of $ZnCl_2$ is 0·5 M the e.m.f. of the cell at 298 K is found to be 1·015 V, whilst the temperature coefficient of the cell is 0·000 49 VK^{-1}. Thus the free energy change (equation 9.A72) is:

$$\Delta G_{298K} = -(2 \times 1{\cdot}015 \times 96\,500) = -196\ \text{kJ mol}^{-1}$$

showing that the reaction has a large tendency to proceed in the direction in which it has been written.

From equation 9.A26 it is possible to calculate the enthalpy change

$$\Delta H_{298K} = -196 - \frac{298(2 \times 96\,500 \times 0{\cdot}000\,49)}{1\,000} = -224\ \text{kJ mol}^{-1}$$

and since $\Delta H > \Delta G$ it follows that during the reversible operation of the cell heat is lost by the cell and gained by the surroundings.

Also from equation 9.A24

$$\Delta S = -\left(\frac{\partial G}{\partial T}\right)_p = -zF\left(\frac{\partial E}{\partial T}\right)_p$$

$$\therefore \Delta S = -2 \times 96\,500 \times 0{\cdot}000\,49 = -95\ \text{J K}^{-1}\ \text{mol}^{-1}$$

Reference Electrodes

It is evident that it is only possible to measure the e.m.f. of a cell, and that in order to determine the potential at a single electrified interface it is necessary to assign an arbitrary potential to a specified electrified interface, which is then used as a reference for all others. The $H^+/\frac{1}{2}H_2$ equilibrium at a platinum surface has been selected for this purpose, and at $a_{H^+} = p_{H_2} = 1$ the standard equilibrium potential $E^{\ominus}_{H^+/\frac{1}{2}H_2}$ is given an arbitrary value of 0·00 V at 25°C; at any other activity of hydrogen ions or pressure of hydrogen gas the potential of the equilibrium $H^+ + e = \frac{1}{2}H_2$ is given by

$$E = E^{\ominus} - \frac{RT}{F} \ln \frac{p_{H_2}^{\frac{1}{2}}}{a_{H^+}} \qquad \ldots(9.A77)$$

or

$$E = 0{\cdot}00 + 0{\cdot}059 \log a_{H^+} - 0{\cdot}030 \log p_{H_2} \qquad \ldots(9.A78)$$

The hydrogen electrode consists of an electrode of platinum foil (approximately 1 × 1 × 0·002 cm) welded to a platinum wire which is fused into a glass tube. In order to increase its catalytic activity it is platinised by making it cathodic in a solution of chloroplatinic acid (2% chloroplatinic acid in 2 N HCl); frequently lead acetate is added to the solution (0·02%) and this appears to facilitate the deposition of an even and very finely divided layer of platinum black, which has a higher catalytic activity than bulk platinum. It is immersed in an acid of known activity (1·2 M HCl gives an activity of hydrogen ion $\approx$ 1·0) and pure hydrogen gas is bubbled into the acid.

Although the hydrogen electrode provides a means of defining the equilibrium potentials of all other equilibria that give rise to an interfacial potential it is more convenient in practice to use other reference electrodes, e.g. Ag/AgCl, Cl^-; Hg/Hg_2Cl_2, Cl^-; $Cu/CuSO_4$, etc. which are easier to prepare and do not involve a gas/ion equilibrium. However, the hydrogen electrode is widely used in fundamental studies of electrochemistry.

Electrodes such as Cu^{2+}/Cu which are reversible with respect to the ions of the metal phase, are referred to as *electrodes of the first kind*, whereas electrodes such as Ag/AgCl, Cl^- that are based on a sparingly soluble salt in equilibrium with its saturated solution are referred to as *electrodes of the second kind.* All reference electrodes must have reproducible potentials that are defined by the activity of the species involved in the equilibrium and the potential must remain constant during, and subsequent to, the passage of small quantities of charge during the measurement of another potential.

In the case of the $Cu/CuSO_4$ electrode (an electrode of the first kind that is widely used for determination of the potential of steel in underground environments), the reversible equilibrium $Cu^{2+} + 2e \rightleftharpoons Cu$ determines the interfacial potential, and constancy of $a_{Cu^{2+}}$ is ensured by using a saturated solution of $CuSO_4$ in equilibrium with crystals of $CuSO_4.5H_2O$.

The difference between $E^{\ominus}_{Cu^{2+}/Cu} = 0{\cdot}337$ V and $E = 0{\cdot}30$ V for the $Cu/CuSO_4$, saturated half-cell is due to the equilibrium

$$CuSO_4.5H_2O = Cu^{2+} + SO_4^{2-} + 5H_2O \qquad \ldots(9.A79)$$

and the chemical potential of $CuSO_4.5H_2O$ must be taken into account in evaluating the $a_{Cu^{2+}}$. At equilibrium

$$a_{Cu^{2+}} \times a_{SO_4^{2-}} = K \qquad \ldots(9.A80)$$

and

$$\log K = \log a_{Cu^{2+}} + \log a_{SO_4^{2-}}$$
$$= \frac{\mu^{\ominus}_{CuSO_4.5H_2O} - \mu^{\ominus}_{Cu^{2+}} - \mu^{\ominus}_{SO_4^{2-}} - 5\mu^{\ominus}_{H_2O}}{5\,707} \qquad \ldots(9.A81)$$

By assuming that $\mu^{\ominus}_{H_2O}$ is the same as that for pure water, which is unlikely in this concentrated solution, and by substituting for the chemical potentials in equation 9.A81, Pourbaix has calculated that

$$\log a_{Cu^{2+}} + \log a_{SO_4^{2-}} = -2{\cdot}96 \qquad \ldots(9.A82)$$

Since

$$a_{Cu^{2+}} = a_{SO_4^{2-}} \qquad \ldots(9.A83)$$

then

$$\log a_{Cu^{2+}} = -2{\cdot}96/2 = -1{\cdot}48 \qquad \ldots(9.A84)$$

which corresponds with $a_{Cu^{2+}} = 0{\cdot}033$.

Substituting in the Nernst equation (equation 9.A62)

$$E = 0{\cdot}337-(0{\cdot}030\times 1{\cdot}48) = 0{\cdot}293\text{ V} \approx 0{\cdot}30\text{ V}$$

which is the potential of the $Cu/CuSO_4$, saturated reference electrode.

The Ag/AgCl, Cl^- electrode, which may be regarded as typical of electrodes of the second kind, consists of AgCl in contact with a soluble chloride, usually KCl. This electrode is essentially an $Ag^+ + e \rightleftharpoons Ag$ electrode, in which the a_{Ag^+} is controlled by the solubility product K_S of AgCl and by the a_{Cl^-}. Thus

$$E = E^\ominus_{Ag^+/Ag}+\frac{RT}{F}\ln a_{Ag^+} \qquad \ldots(9.A85)$$

and since

$$a_{Ag^+}\times a_{Cl^-} = \text{constant} = K_S \qquad \ldots(9.A86)$$

at constant temperature, then

$$E = E^\ominus_{Ag^+/Ag}+\frac{RT}{F}\ln\frac{K_S}{a_{Cl^-}} = E^\ominus_{Ag^+/Ag}+\frac{RT}{F}\ln K_S-\frac{RT}{F}\ln a_{Cl^-} \qquad \ldots(9.A87)$$

Taking the two constant terms on the r.h.s. of equation 9.A87 as $E^\ominus_{Ag/AgCl}$

$$E = E^\ominus_{Ag/AgCl}-\frac{RT}{F}\ln a_{Cl^-} \qquad \ldots(9.A88)$$

where $$E^\ominus_{Ag/AgCl} = E^\ominus_{Ag^+/Ag}+\frac{RT}{F}\ln K_S = 0{\cdot}222\,33\text{ V at }25°\text{C} \qquad \ldots(9.A89)$$

and $$E = E^\ominus_{Ag/AgCl}\text{ when } a_{Cl^-} = 1 \qquad \ldots(9.A90)$$

Values of $E^\ominus_{Ag/AgCl}$ in the range of temperature 0–95°C are available, and the Ag/AgCl, Cl^- electrode is probably the most accurate reference electrode for use in solutions at elevated temperatures.

It is apparent that since the electrode potential of a metal/solution interface can only be evaluated from the e.m.f. of a cell, the reference electrode used for that purpose must be specified precisely, e.g. the criterion for the cathodic protection of steel is $-0{\cdot}85$ V (vs. $Cu/CuSO_4$, sat.), but this can be expressed as a potential with respect to the standard hydrogen electrode (S.H.E.), i.e. $-0{\cdot}55$ V (vs. S.H.E.) or with respect to any other reference electrode. Potentials of reference electrodes are given in Table 21.7 in Volume II.

E.M.F. Series of Metals

The standard electrode potentials of $M^{z+}+ze = M$ equilibria, $E^\ominus_{M^{z+}/M}$, when arranged in a table according to the sign and magnitude of the potential constitute the so-called e.m.f. series of metals. Again it should be emphasised that these are thermodynamic parameters whose magnitude and sign will depend upon the energy required to convert a metal ion in the lattice of the metal to a hydrated ion in solution. Energetically, this will involve conversion of a metal atom in the lattice to a gaseous atom, the ionisation of the metal atom and the hydration of the metal ion to form a hydrated ion in solution. This can be illustrated by comparing the component enthalpies for transforming $Na(l.) \rightarrow Na^{+}(aq.)$ and $Ag(l.) \rightarrow Ag^{+}(aq.)$

	$\Delta H^{\ominus}$ (kJ)		$\Delta H^{\ominus}$ (kJ)
Sublimation			
$Na(l.) = Na(g.)$	109	$Ag(l.) = Ag(g.)$	280
Ionisation			
$Na(g.) = Na^{+}(g.) + e(g.)$	494	$Ag(g.) = Ag^{+}(g.) + e(g.)$	728
Hydration			
$Na^{+}(g.) = Na^{+}(aq.)$	−397	$Ag^{+}(g.) = Ag^{+}(aq.)$	−464
	206		544

It can be seen that $\Delta H^{\ominus}$ for $Ag \rightarrow Ag^{+}(aq.)$ is much greater than that for $Na \rightarrow Na^{+}(aq.)$ which means that silver has a much lower tendency to ionise than sodium; conversely, $Ag^{+}(aq.)$ ions have a much greater tendency to gain electrons and become reduced to metal than $Na^{+}(aq.)$ ions. This difference, as can be seen, is largely due to the higher sublimation energy and ionisation energy of silver compared with sodium. If the entropy change is regarded as negligible, then $\Delta H^{\ominus} \approx \Delta G^{\ominus}$, and for the reaction

$$Na + Ag^{+}(aq.) = Na^{+}(aq.) + Ag, \quad -\Delta G^{\ominus}_{298K} \approx 338 \text{ kJ}$$

$$\text{and } E^{\ominus}_{\text{reaction}} \approx 3{\cdot}5 \text{ V}$$

which is in agreement with the e.m.f. evaluated from the standard electrode potential, i.e. $E^{\ominus}_{Ag^{+}/Ag} = 0{\cdot}799$ V and $E^{\ominus}_{Na^{+}/Na} = -2{\cdot}71$ V.

These considerations show the essentially thermodynamic nature of $E^{\ominus}_{M^{z+}/M}$, and it follows that only those metals that form reversible $M^{z+} + ze = M$ systems, and that are immersed in solutions containing their cations, take up potentials that conform to the thermodynamic Nernst equation. It is evident, therefore, that the e.m.f. series of metals has little relevance in relation to the actual potential of a metal in a practical environment, and although metals such as silver, mercury, copper, tin, cadmium, zinc, etc. when immersed in solutions of their cations do form reversible systems, they are unlikely to be in contact with environments containing unit activities of their cations. Furthermore, although silver when immersed in a solution of Ag^{+} ions will take up the reversible potential of the Ag^{+}/Ag equilibrium, similar considerations do not apply to the Na^{+}/Na equilibrium since in this case the sodium will react with the water with the evolution of hydrogen gas, i.e. two exchange processes will occur, resulting in an extreme case of a corrosion reaction.

The potential of the electrified interface of a metal immersed in an aqueous solution is of fundamental importance in studying the mechanism of corrosion reactions and in corrosion testing and monitoring, and it is, therefore, of some importance to consider the factors that determine the potential of a metal in a practical environment. The determination of the potential can be achieved without difficulty, but the significance of the potential is far more complex and some of the factors that affect the potential are as follows:

1. The reversibility of the M^{z+}/M, H^{+}/H_2 and O_2/OH^{-} systems.

2. The number of possible equilibria that can occur; if two or more are involved the potential is a *mixed* or *corrosion* potential.
3. The surface of the metal, i.e. film free or filmed with corrosion products.
4. The nature of the environment, i.e. composition, temperature, pressure, velocity, etc.

Reversible Potentials and Exchange Current Densities

Previous considerations have shown that the interface between two conducting phases is characterised by an unequal distribution of electrical charge which gives rise to an electrical double layer and to an electrical potential difference. This can be illustrated by considering the transport of charge (metal ions or electrons) that occurs immediately an isolated metal is immersed in a solution of its cations:

$$\underset{\text{hydrated ions in solution}}{M^{+}(\text{aq.})+e} \underset{\overleftarrow{i_1}}{\overset{\overrightarrow{i_1}}{\rightleftharpoons}} \underset{\text{metal ions in lattice}}{[M^{+}+e]}$$

where $\overrightarrow{i_1}$ is the rate of the discharge (or cathodic) process and $\overleftarrow{i_1}$ is the rate of the dissolution (or anodic) process. Initially and momentarily the rates of these two processes will not be equal and the one that predominates will determine the potential of the metal with respect to the solution. Figure 9.A3 illustrates the situation at the metal/solution interface and it can be seen that for transfer of charge in either direction the ions involved must surmount an energy barrier which has a maximum value of $E^{\ddagger}$. Since in Fig. 9.A3*a* the energy well for the hydrated ion is lower than that for the metal ion in the lattice, dissolution will occur at a greater rate initially than discharge, resulting in an electrical double layer in which the metal has an excess negative charge and the solution an excess positive charge. However, this situation proceeds only for an instant of time, since the double layer produces an electric field normal to the interface that rapidly reduces the *net* current to zero. At equilibrium the field produced tends to equalise the energy wells

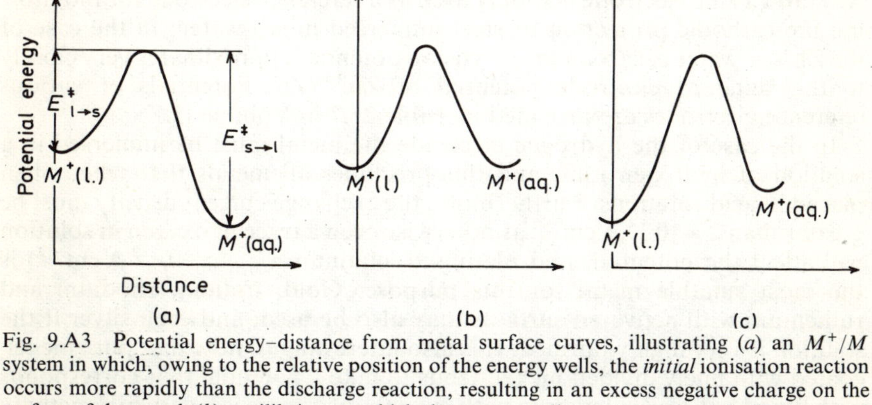

Fig. 9.A3 Potential energy–distance from metal surface curves, illustrating (*a*) an M^+/M system in which, owing to the relative position of the energy wells, the *initial* ionisation reaction occurs more rapidly than the discharge reaction, resulting in an excess negative charge on the surface of the metal, (*b*) equilibrium at which the energy wells are approximately the same and (*c*) the converse of (*a*)

(Fig. 9.A3*b*) and the rates of dissolution and discharge become equal. The potential is then the equilibrium potential, i.e.

$$\text{when } \vec{i} = \overleftarrow{i} = i_0, E = E_{\text{eq.}}$$

where i_0 is the equilibrium exchange current density. Similar considerations apply when the energy well for metal ion discharge is higher than that for dissolution, but at equilibrium the electrical double layer will have an excess positive charge on the metal side of the interface and an excess negative charge on the solution side (Fig. 9.A3*c*).

A metal when immersed in a solution of its cations may take up the reversible potential corresponding with the $M^{z+} \rightleftharpoons M$ exchange process, but whether or not this occurs in practice will depend on the magnitude of its exchange current density in relation to any others that are possible due to other exchange processes in the solution under consideration. In oxygenated solution the $H^+ \rightleftharpoons H_2$ and $O_2 \rightleftharpoons OH^-$ equilibria provide possible alternatives, but the exchange current densities of these equilibria are usually small, and if i_0 for $M^{z+} \rightleftharpoons M$ is large, the potential of the metal will be determined by the latter.

Thus in the case of the Hg_2^{2+}/Hg equilibrium the exchange current density $i_{0,Hg}$ is $\approx 10^{-4}$ A cm^{-2}, whereas for the $H^+ \rightleftharpoons H_2$ equilibrium $i_{0,H} \approx 10^{-12}$ A cm^{-2}; it follows that it is quite impossible for mercury to take up the potential of the latter even in a highly purified solution, and that the potential will be dominated by the $Hg_2^{2+} \rightleftharpoons Hg$ exchange even when the concentration of mercury ions is very small and when the solution contains dissolved oxygen and trace impurities. Similar considerations apply to the Ag/AgCl, Cl^- electrode and to the other M^{2+}/M equilibria that are used as reference electrodes such as Zn^{2+}/Zn, Cd^{2+}/Cd and Cu^{2+}/Cu. These systems thus provide convenient reference electrodes whose potentials are unaffected by the presence of dissolved oxygen or trace impurities or by small changes in pH. The above considerations have been confined to reversible M^{2+}/M systems, but where high accuracy is not required fairly reproduceable potentials can be achieved using metal electrodes immersed directly in the solution under study. Platinum and gold electrodes have been used in this way, and a zinc electrode is widely used as a reference electrode for monitoring the cathodic protection of steel immersed in sea-water; in the case of the Zn/sea-water electrode the corrosion potential approximates very closely to the standard electrode potential of Zn^{2+}/Zn. Potentials of various reference electrodes are provided in Table 21.7 in Volume II.

In the case of the hydrogen electrode the metal must be immersed in a solution of hydrogen ions, and this precludes all metals that react when placed in acid solutions. Furthermore, the exchange current density must be greater than 2×10^{-4} A cm^{-2} as otherwise even a trace of oxygen in solution will affect the potential, and platinised–platinum ($i_{0,H} \approx 10^{-3}$ A cm^{-2}) is the most suitable metal for this purpose. Gold, iridium, rhodium and ruthenium with activated surfaces may also be used, and even silver if the solution is very highly purified. It is also interesting to note that pure nickel, coated with finely divided nickel, tends to take a potential that corresponds with the pH of a nickel plating bath even though there is a very high activity of Ni^{2+} present.

Corrosion potentials are obviously of particular importance for corrosion reactions, and have been considered in some detail in Sections 1.4–1.7.

Oxide Electrodes

Metals in practice are usually coated with an oxide film that affects the potential, and metals such as Sb, Bi, As, W and Te behave as reversible $M/M_xO_y/OH^-$ electrodes whose potentials are pH dependent; electrodes of this type may be used to determine the solution's pH in the same way as the reversible hydrogen electrode. According to Ives and Janz these electrodes may be regarded as a particular case of electrodes of the second kind, since the oxygen in the metal oxide participates in the self-ionisation of water.

Thus a reversible electrode of the type $M/MO/H^+$ must conform with the thermodynamic equilibria

$$M^{2+} + 2e \rightleftharpoons M \qquad \text{...(9.A91}a\text{)}$$

$$MO(s.) \rightleftharpoons M^{2+} + O^{2-} \qquad \text{...(9.A91}b\text{)}$$

$$O^{2-} + 2H_2O \rightleftharpoons 2OH^- \qquad \text{...(9.A91}c\text{)}$$

$$2OH^- + 2H^+ \rightleftharpoons 2H_2O \qquad \text{...(9.A91}d\text{)}$$

and the potential will be determined by the $a_{M^{2+}}$ (equation 9.A91*a*) and by the solubility product of the oxide (equation 9.A91*b*), i.e.

$$K_S = a_{M^{2+}} a_{O^{2-}}$$

and since there is an equilibrium between O^{2-} ions and OH^- ions

$$K'_S = a_{M^{2+}} a^2_{OH^-}$$

where K'_S is the solubility product of the hydroxide (or oxy-hydroxide)

$$\therefore\ E = E^{\ominus}_{M^{2+}/M} + \frac{RT}{zF} \ln K'_S - \frac{RT}{F} \ln a_{OH^-}$$

and since for water

$$K_w = a_{H^+} a_{OH^-}$$

$$E = E^{\ominus}_{M^{2+}/M} + \frac{RT}{2F} \ln K'_S - \frac{RT}{F} \ln K_w + \frac{RT}{F} \ln a_{H^+} \qquad \text{...(9.A92)}$$

Since the first three terms of equation 9.A92 are constant, it can be written in the form

$$E = E^{\ominus}_{M,MO,H^+} + \frac{RT}{F} \ln a_{H^+}$$

or

$$E = E^{\ominus}_{M,MO,H^+} - 0{\cdot}059\ \text{pH}$$

which is applicable in dilute solutions in which the $a_{H_2O} \approx 1$.

The best known of the M/M_xO_y electrodes that are used for the determination of pH are the Sb/Sb_2O_3 electrodes, but metals such as bismuth and arsenic filmed with their respective oxide act in a similar manner. Copper in alkaline solutions appears to behave as a pH-dependent Cu/Cu_2O, OH^-

electrode, whereas in chloride solutions the electrode behaves as Cu/Cu_2Cl_2, Cl^- whose potential depends on the a_{Cl^-} in the same way as the Ag/AgCl, Cl^- electrode.

Examples

Example 1 Derive the relationship between potential and pH for the equilibrium between $Fe(OH)_3$ and Fe^{2+}(aq.) at 25°C. The relevant standard chemical potentials $\mu^\ominus$ in kJ are as follows: $Fe(OH)_3$, −694; Fe^{2+}(aq.), −85; H_2O, −237.

The equilibrium (written as a reduction) is

$$Fe(OH)_3 + 3H^+ + e = Fe^{2+} + 3H_2O$$

and

$$E = E^\ominus - \frac{0{\cdot}059}{1}\log\frac{a_{Fe^{2+}}}{a^3_{H^+}}$$

$$\therefore\ E = E^\ominus - 0{\cdot}059\log a_{Fe^{2+}} + 0{\cdot}178\log a_{H^+}$$

$$\therefore\ E^\ominus = -\frac{\Sigma\nu_i\mu_i^\ominus}{zF} = -\frac{\Sigma\nu_i\mu_i^\ominus}{z\times 96\,500}\times 10^3 \text{ (when } \mu^\ominus \text{ is in kJ)}$$

$$= -\frac{[\Sigma\nu_i\mu_i^\ominus{}_{\text{(products)}} - \Sigma\nu_i\mu_i^\ominus{}_{\text{(reactants)}}]}{z\times 96\,500}$$

$$= -\frac{[\{-85+3(-237)-(-694)\}]\times 10^3}{1\times 96\,500} = 1{\cdot}05\text{ V}$$

$$\therefore\ E = 1{\cdot}05 - 0{\cdot}059\log a_{Fe^{2+}} - 0{\cdot}178\,\text{pH}$$

which defines the equilibrium between $Fe(OH)_3$ and Fe^{2+}(aq.) in terms of the equilibrium potential E and pH.

Example 2 Finely divided gold is agitated with an oxygen-saturated alkali-cyanide solution of pH = 12 containing 10^{-2} mol dm^{-3} of CN^-. Calculate the concentration of $Au(CN)_2^-$ ions in solution at 25°C when the reaction is at equilibrium. The standard chemical potentials of the species involved (in kJ) are: CN^-, 165; H_2O, −237; $Au(CN)_2^-$, 244; OH^-, −157. Assume that $a_{H_2O} = 1$, and take $2{\cdot}303\,RT = 5\,710$.

The relevant equilibrium is

$$2Au + 4CN^- + \tfrac{1}{2}O_2 + H_2O = 2Au(CN)_2^- + 2OH^-$$

and from equation 9.A64

$$\ln K = -\frac{\Sigma\nu_i\mu_i^\ominus}{RT} \text{ and } \log K = -\frac{\Sigma\nu_i\mu_i^\ominus}{5\,707}$$

Since

$$\log K = -\frac{[\Sigma\mu^\ominus_{\text{products}} - \Sigma\mu^\ominus_{\text{reactants}}]}{5\,710}$$

$$\log K = -\frac{[\{(2\times 244+2\times(-157)\}-\{(4\times 165)+(-237)\}]\times 10^3}{5\,710}$$

and $$\log K = \frac{249\times 10^3}{5\,710} \approx 43$$

$$\therefore K \approx 10^{43}$$

Since $a_{H_2O} = p_{O_2} = 1$

$$\frac{a^2_{Au(CN)_2^-}\times a^2_{OH^-}}{a^4_{CN^-}} \approx 10^{43}$$

and $$a_{Au(CN)_2^-} = \left[\frac{10^{43}\times(10^{-2})^4}{(10^{-2})^2}\right]^{\frac{1}{2}} \approx 10^{19\cdot 5}$$

This high thermodynamic activity shows that gold is extremely unstable in an oxygenated cyanide solution.

Example 3 Calculate the rate ($A\,cm^{-2}$) of hydrogen evolution on platinum in an acid solution when the overpotential is (*a*) $-0{\cdot}2$ V, (*b*) $-0{\cdot}300$ V. Assume that $i_0 = 1\times 10^{-4}\,A\,cm^{-2}$, the transfer coefficient $\alpha = 0{\cdot}5$ and the number of electrons involved in one act of the r.d.s. $z = 1$; $RT\ln x = 0{\cdot}059\log x$.

$$\eta = \frac{RT}{z\alpha F}\ln i_0 - \frac{RT}{z\alpha F}\ln i_c$$

$$\therefore \eta = \frac{0{\cdot}059}{1\times 0{\cdot}5}\log i_0 - \frac{0{\cdot}059}{1\times 0{\cdot}5}\log i_c = 0{\cdot}12\log i_0 - 0{\cdot}12\log i_c$$

$$\therefore -0{\cdot}2 = 0{\cdot}12\times(-4) - 0{\cdot}12\log i_c$$

$$\therefore \log i_c = \frac{-0{\cdot}28}{0{\cdot}12} = -2{\cdot}33$$

$$\therefore i_c \approx 5\times 10^{-3}\,A\,cm^{-2}$$

When $\eta = -0{\cdot}300$ V

$$\log i_c = -\frac{0{\cdot}18}{0{\cdot}12} = -1{\cdot}5 = -2+0{\cdot}5$$

$$\therefore i_c = 3{\cdot}2\times 10^{-2}\,A\,cm^{-2}$$

Note: when $\eta = -0{\cdot}010$ V equation 9.72 is applicable and $i_c = (10^{-4}\times 10^{-2}/0{\cdot}059$ A $= 17\,\mu A\,cm^{-2}$.

Example 4 A metal is corroding in an oxygen-free acid of pH 1 at 25°C according to the overall reaction $M+2H^+ = M^{2+}+H_2$. From the data given below, calculate the corrosion rate of the metal in $A\,cm^{-2}$.

Data: the equilibrium potential $E^{\ominus}_{M^{2+}/M} = -0{\cdot}76$ V (S.H.E.) and the solution contains $a_{M^{2+}} = 10^{-2}$; $i_{0,M} = 10^{-3}\,A\,cm^{-2}$; z (the number of electrons involved in one act of the rate-determining step) is 2 and $\alpha = 0{\cdot}5$. For the hydrogen evolution reaction $i_{0,H} = 10^{-8}\,A\,cm^{-2}$; $z = 1$ and $\alpha = 0{\cdot}5$. $RT/F\ln x = 0{\cdot}059\log x$ at 25°C.

The overpotential η is defined as

$$\eta = E_p - E_r$$

where E_p is the polarised potential and E_r the reversible potential. For the cathodic reaction

$$\eta_c = \frac{0{\cdot}059}{z\alpha}\log i_0 - \frac{0{\cdot}059}{z\alpha}\log i_c$$

and for the anodic reaction

$$\eta_a = \frac{-0{\cdot}059}{z\alpha}\log i_0 + \frac{0{\cdot}059}{z\alpha}\log i_a$$

At the corrosion potential, $E_p = E_{corr.}$ and $i_c = i_a = i_{corr.}$. For the h.e.r., $E_r = -0{\cdot}059$ V at pH 1. For the $M \rightarrow M^{2+}$ reaction, $E_r = -0{\cdot}76 - \frac{2(0{\cdot}059)}{2} = -0{\cdot}8219$ V at $a_{M^{2+}} = 10^{-2}$. Therefore for the h.e.r.

$$E_{corr.} - (-0{\cdot}059) = [(0{\cdot}12 \times -8) - (0{\cdot}12 \log i_{corr.})]$$

$$E_{corr.} = -1{\cdot}02 - 0{\cdot}12 \log i_{corr.} \qquad \ldots(1)$$

Therefore for the $M \rightarrow M^{2+}$ reaction

$$E_{corr.} - (-0{\cdot}82) = -[0{\cdot}059 \times (-3)] + 0{\cdot}059 \log i_{corr.}$$

and

$$E_{corr.} = -0{\cdot}64 + 0{\cdot}059 \log i_{corr.} \qquad \ldots(2)$$

Equating (1) and (2)

$$-1{\cdot}02 - 0{\cdot}12 \log i_{corr.} = -0{\cdot}64 + 0{\cdot}059 \log i_{corr.}$$

$$-0{\cdot}18 \log i_{corr.} = 0{\cdot}38$$

$$\log i_{corr.} = -\frac{0{\cdot}38}{0{\cdot}18} = -2{\cdot}11 = -3 + 0{\cdot}89$$

Taking antilogarithms

$$i_{corr.} \approx 9 \times 10^{-3}\ \text{A cm}^{-2}$$

Example 5 A stainless steel pipe is to be used to convey an aerated reducing acid at high velocity. If the concentration of dissolved O_2 is 10^{-3} mol dm^{-3} (10^{-6} mol cm^{-3}) calculate whether or not the steel will corrode when (*a*) the acid is static, (*b*) the acid is moving at high velocity. Assume that the critical current density for passivation of the steel in the acid is 200μ A cm^{-2}, the thickness of the diffusion layer is 0·05 cm when the acid is static and 0·005 cm when the acid flows at a high velocity; assume the diffusion coefficient of dissolved O_2 is 10^{-5} cm s^{-1}.

The cathodic reaction for 1 mol of dissolved O_2 is

$$O_2 + 2H_2O + 4e = 4OH^-$$

and hence z the number of electrons involved in each act of the reaction is 4.

The limiting current density is given by

$$i_{\mathrm{L}} = \frac{DzFc}{\delta}$$

and under static conditions

$$i_{\mathrm{L}} = \frac{10^{-5} \times 4 \times 96\,500 \times 10^{-6}}{0{\cdot}05}$$

$$\therefore\ i_{\mathrm{L}} = 7{\cdot}7 \times 10^{-5} = 80\mu\,\mathrm{A\,cm^{-2}}$$

and it follows that under flowing conditions where $\delta = 0{\cdot}005$ cm, $i_{\mathrm{L}} = 800\mu\,\mathrm{A\,cm^{-2}}$. Thus under static conditions $i_{\mathrm{crit.}}$ for passivation will not be exceeded and the metal will corrode, but will passivate when the acid is flowing at a high velocity.

L. L. SHREIR

BIBLIOGRAPHY

*Moore, W. J., *Physical Chemistry*, 5th edition, Longmans (1972)

Glasstone, S., *Textbook of Physical Chemistry*, Macmillan (1962)

Daniel, F. and Alberty, R. A., *Physical Chemistry*, 3rd edition, Wiley (1967)

Glasstone, S. and Lewis, D., *Elements of Physical Chemistry*, Macmillan (1962)

Mackowiak, J., *Physical Chemistry for Metallurgists*, Allen and Unwin (1966)

*Denbigh, K., *The Principles of Chemical Equilibrium*, Cambridge University Press (1971)

*Avery, H. E. and Shaw, D. J., *Basic Physical Chemistry Calculations*, Butterworths (1971)

*Gross, J. M. and Wiseall, B., *Principles of Physical Chemistry*, Macdonald and Evans (1972)

Kubaschewski, O., Evans, E. Ll. and Alcock, C. B., *Metallurgical Thermochemistry*, 4th edition, Pergamon Press (1967)

Reference Electrodes, ed. P. J. G. Ives and G. J. Janz, Academic Press (1961)

Pourbaix, M., *Lectures on Electrochemical Corrosion*, Plenum Press (1973)

*Warn, J. R. W., *Concise Chemical Thermodynamics*, van Nostrand Reinhold (1969)

*Avery, H. E. and Shaw, D. J., *Advanced Physical Chemistry Calculations*, Butterworths (1971)

*In SI units.

9.2 Outline of Structural Metallurgy Relevant to Corrosion

The corrosion of metals invariably involves some kind of interaction between a metal and its environment, and in many cases the corrosion (location, form and rate) is significantly affected or even caused by some structural feature of the metal. It is essential, therefore, for the corrosion engineer to have some appreciation of the structure of metals, and an elementary survey is provided in this section which provides a basis for an account of metal structure in relation to corrosion that is the subject of Section 1.3.

This section will consider first the structure of pure metals and this will be followed by a description of the structure of alloys. In general, structural features will be discussed in order of increasing size, i.e. starting with atomic features and continuing through microscopic features to macroscopic features.

Structure of Pure Metals

A crystal may be defined as an orderly three-dimensional array of atoms, and all metals are aggregates of more or less imperfect crystals. In considering the structure of metals, therefore, it is convenient to start with the arrangement of atoms in a perfect metal crystal and then to proceed to the imperfections which are always present in the crystal structure.

Although there are many different crystal structures, some of which are rather complicated, most common metals fortunately have one of three, relatively simple structures: face-centred cubic (f.c.c.), close-packed hexagonal (c.p.h.) and body-centred cubic (b.c.c.). The unit cell (i.e. the smallest group of atoms possessing the symmetry of the crystal and which, when repeated in all directions, develops the crystal lattice) of each of these metallurgically common and important crystal structures is illustrated in Fig. 9.22; the crystal structure of each of the more important metals is given in Table 9.4. It is clear from this table that the crystal structures of some metals change with temperature, i.e. these metals are polymorphic, and this phenomenon is of very great practical significance, particularly in the case of iron, as will become apparent later.

The atoms in a metal crystal can often be satisfactorily regarded as hard spheres of radius about 10^{-10} m. A sheet of such hard balls can be

Table 9.4 Crystal structure of the more important metals

Metal	*Temperature* (K)	*Crystal structure*
Alkali metals	Ambient to m.p.	b.c.c.
Aluminium	Below m.p.	f.c.c.
Beryllium	Below m.p.	c.p.h.
Cadmium	Below m.p.	c.p.h.
Chromium	Below m.p.	b.c.c.
Cobalt	Below 690	c.p.h.
	690 to m.p.	f.c.c.
Copper	Below m.p.	f.c.c.
Gold	Below m.p.	f.c.c.
Iron	Below 1 180	b.c.c.
	1 180–1 665	f.c.c.
	1 665 to m.p.	b.c.c.
Lead	Below m.p.	f.c.c.
Magnesium	Below m.p.	c.p.h.
Manganese	Below m.p.	Four complex structures
Molybdenum	Below m.p.	b.c.c.
Nickel	Below m.p.	f.c.c.
Platinum	Below m.p.	f.c.c.
Silver	Below m.p.	f.c.c.
Tin	Below m.p.	Two complex structures
Titanium	Below 1 155	c.p.h.
	1 155 to m.p.	b.c.c.
Tungsten	Below m.p.	b.c.c.
Vanadium	Below m.p.	b.c.c.
Zinc	Below m.p.	c.p.h.
Zirconium	Below 1 135	c.p.h.
	1 135 to m.p.	b.c.c.

most densely and closely packed together in the configuration shown in Fig. 9.23*a*; such sheets or layers of atoms are therefore known as *close-packed planes*. Similarly, the directions arrowed in Fig. 9.23*a* are referred to as *close-packed directions*. Close-packed hexagonal and face-centred cubic crystal structures can be built up by stacking close-packed planes on top of each other in the correct sequence. Consider a sheet of eight close-packed atoms, designated by the full circles *a* in Fig. 9.23*b*; as shown, a second layer of four close-packed atoms, designated by the dotted circles *b*, may be laid on top of the first plane. Each of the *b* atoms 'sits' in one of the natural 'valleys' which occur between any three contiguous *a* atoms. There are now two such low-energy valleys in the plane of *b* atoms, one of which is immediately above an *a* atom and one of which is not (Fig. 9.23*b*). An atom placed in the former results in an *aba* stacking of three close-packed planes, while an atom placed in the latter would result in *abc* stacking. Stacking *abababab*, etc. develops the c.p.h. crystal structure while stacking *abcabcabc*, etc. develops the f.c.c. crystal structure. The b.c.c. crystal structure cannot be built up by stacking close-packed planes on top of each other.

In the f.c.c. and c.p.h. crystal structures, each atom has 12 nearest neighbours and is therefore said to have a *co-ordination number* of 12. In the non-closed-packed b.c.c. structure, on the other hand the co-ordination number is only 8. In the f.c.c. structure the four octahedral {111}* planes are close

* An account of the use of Miller indices to describe crystal planes and lattice directions is beyond the scope of this article; a very adequate treatment of this topic is, however, given in Reference 1.

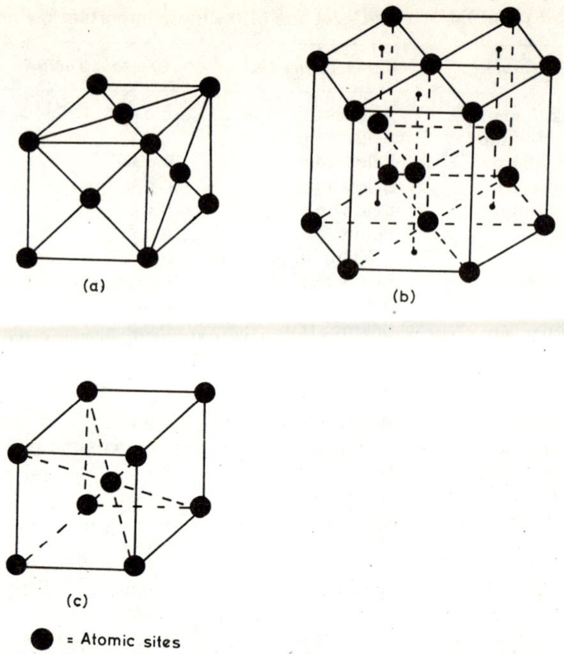

Fig. 9.22 Unit cells of (*a*) the face-centred cubic (f.c.c.), (*b*) the close-packed hexagonal (c.p.h.) and (*c*) the body-centred cubic (b.c.c.) crystal structures

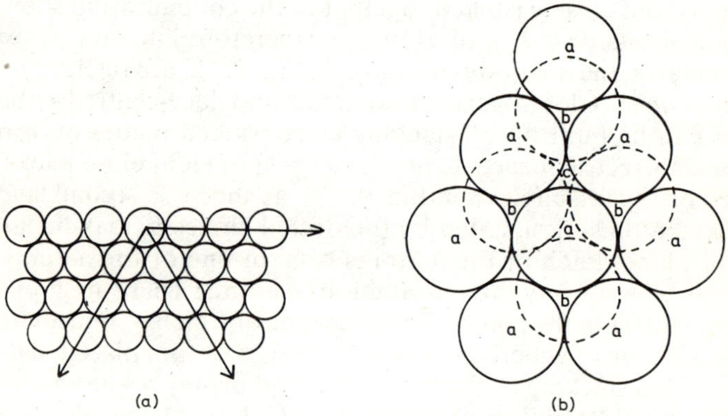

Fig. 9.23(*a*) Single close-packed sheet of atoms, with the close-packed directions arrowed and (*b*) stacking of three close-packed planes of atoms, showing *aba* and *abc* positions

packed, while the ⟨110⟩ directions are close packed. In the c.p.h. structure on the other hand, there is only one close-packed plane, the (0001) basal plane; the ⟨11$\bar{2}$0⟩ directions are close packed. The different number of close-packed planes in f.c.c. and c.p.h. crystals results in the very different plastic deformation characteristics of metals with the two structures. In the b.c.c. structure, the ⟨111⟩ directions are close packed, but there are no truly close-packed planes of the type shown in Fig. 9.23*a*; the planes of closest packing, however, are {110}.

In any crystal structure, the close-packed or closest-packed planes are the lowest energy planes. On all other planes, the density of atoms is lower, and the interatomic distance and the energy of the plane are greater. Contrary to intuitive expectations, the diameter of the largest holes or interstices between atoms in the close-packed f.c.c. structure is considerably greater than the diameter of the largest interstices between atoms in the non-close-packed b.c.c. structure.

Point Defects in Crystals

The smallest imperfections in metal crystals are point defects, in particular vacant lattice sites (*vacancies*) and interstitial atoms. As illustrated in Fig. 9.24*a*, a vacancy occurs where an atom is missing from the crystal structure

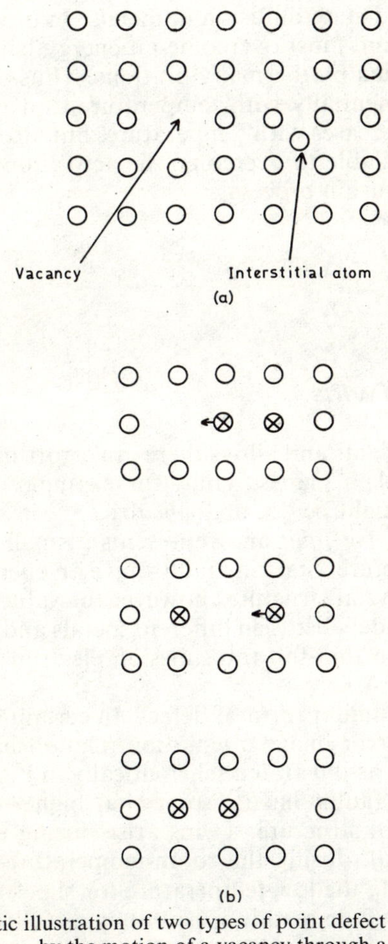

Fig. 9.24(*a*) Schematic illustration of two types of point defect and (*b*) three stages in diffusion by the motion of a vacancy through a crystal

while an interstitial occurs when an extra atom is inserted in an interstitial site (an interstice). Vacancies and interstitials are equilibrium defects since although vacancies (and interstitials) increase the internal energy of a crystal they also increase the entropy, so the free energy of a perfect crystal is reduced by the presence of a number of point defects. The equilibrium vacancy concentration increases exponentially with temperature and typically in copper at room temperature one site in 10^{14-15} may be vacant, while at 1300K, one site in 10^{3-4} will be vacant (equivalent to 10^{17-18} vacancies per cubic millimetre). The internal energy associated with an interstitial is much greater than that associated with a vacancy; the concentration of interstitial atoms at any temperature is correspondingly very much lower than that of vacant atomic sites. Radiation damage, however, introduces many interstitials. Often a vacancy and an interstitial will be associated as a pair.

The practical importance of vacancies is that they are mobile and, at elevated temperatures, can move relatively easily through the crystal lattice. As illustrated in Fig. 9.24*b*, this is accompanied by movement of an atom in the opposite direction; indeed, the existence of vacancies was originally postulated to explain solid-state diffusion in metals. In order to 'jump' into a vacancy an adjacent atom must overcome an energy barrier. The energy required for this is supplied by thermal vibrations. Thus the diffusion rate in metals increases exponentially with temperature, not only because the vacancy concentration increases with temperature, but also because there is more thermal energy available to overcome the activation energy required for each 'jump' in the diffusion process.

Stacking Faults and Twins

In certain close-packed metals and alloys there are errors in the sequence of stacking of the close-packed planes. Thus, for example, the sequence of planes in a c.p.h. metal might be . . . ababa*bca*ba . . . , in which the region in italics is termed a *stacking fault* and represents a small volume of f.c.c. material in the c.p.h. structure. Stacking faults have an energy that is higher than that of the perfect crystal structure; however the value of the stacking-fault energy varies very widely between different metals and alloys. Stacking faults are readily visible in thin-foil transmission electron micrographs, as illustrated in Fig. 9.25 (top).

Twins are a somewhat similar form of defect. In certain metals, adjacent areas of crystal bear a mirror-image orientation relationship to each other across the twin boundary, as illustrated schematically in Fig. 9.26. The twin boundaries, like stacking faults, have a somewhat higher energy than the surrounding perfect crystal structure. Twins arise during the annealing of cold-worked f.c.c. material, during the room-temperature deformation of c.p.h. materials, and during the low-temperature (or shock) deformation of b.c.c. materials. Twins are often readily apparent on metallographic sections at relatively low magnifications, as shown in Fig. 9.25 (bottom).

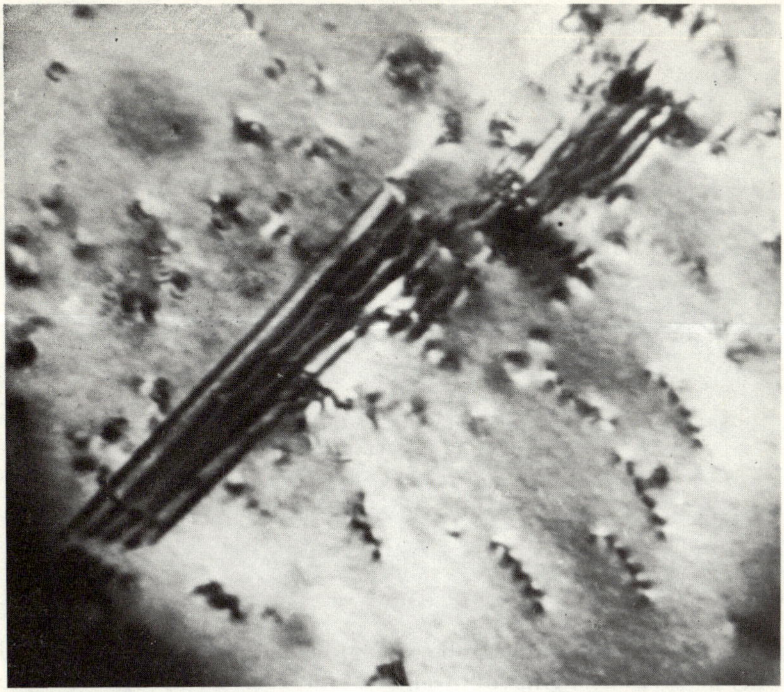

Fig. 9.25 (top) Transmission electronmicrograph showing a stacking fault and isolation dislocations in a stainless steel (×70 000, courtesy I. Brough) and (bottom) light micrograph showing an equiaxed grain structure and twins in an annealed 70/30 brass (×200, courtesy M. Islam)

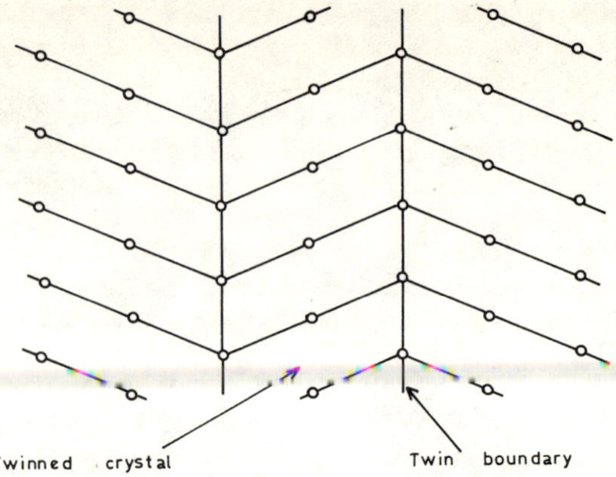

Fig. 9.26 Twinned crystal

Line Defects

Point defects, stacking faults and twins are crystalline imperfections which basically involve misalignment of single atoms or of a single plane of atoms. Line defects, or dislocations, are somewhat larger-scale imperfections involving misalignment along a line of atoms. Dislocations are extremely

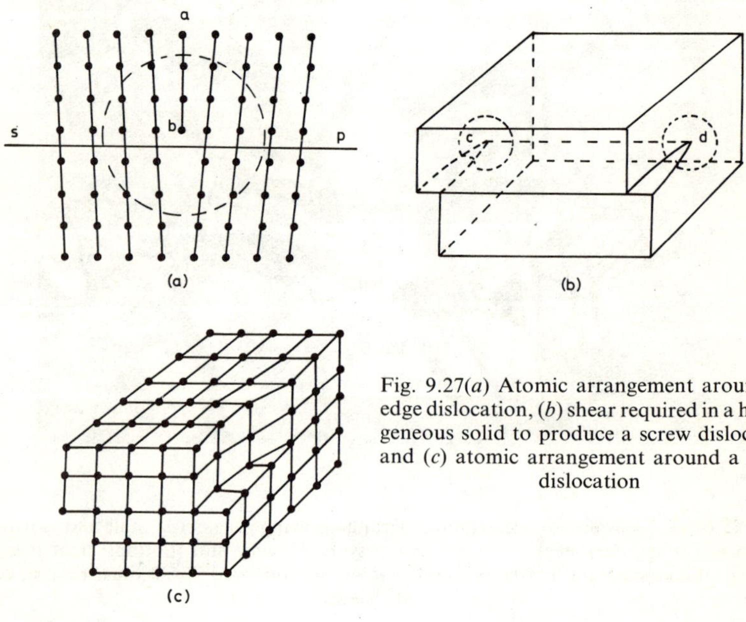

Fig. 9.27(*a*) Atomic arrangement around an edge dislocation, (*b*) shear required in a homogeneous solid to produce a screw dislocation and (*c*) atomic arrangement around a screw dislocation

important defects, in that the dislocation density and dislocation interactions control to a very large extent the mechanical properties, and in particular the strength and ductility, of metals.

The simplest type of line defect is the edge dislocation, which consists of an extra half plane of atoms in the crystal, as illustrated schematically in Fig. 9.27*a*; edge dislocations are often denoted by ⊥ if the extra half plane *ab* is above the plane *sp*, or by T if it is below.

The second type of line defect is the screw dislocation, which is rather less easy to visualise. Consider, however, a block of material, half of which is sheared one interatomic distance with respect to the other half, as shown in Fig. 9.27*b*. The line *cd* then constitutes a screw dislocation; the arrangement of atoms around a screw dislocation is shown in Fig. 9.27*c*.

To a good approximation, only atoms within the dotted circles in Figs. 9.27*a* and *b* are displaced from their equilibrium position; in a real, three-dimensional crystal the diameter *d* of these circles would be very much less than the length *l* of the dislocation, i.e. the length, perpendicular to the page, of the extra half plane of atoms *ab* in Fig. 9.27*a*, or of the line *cd* in Fig. 9.27*b*. Dislocations strictly, therefore, are cylindrical defects of diameter *d* and length *l*; however, since $l \gg d$ they are referred to as *line* defects.

Movement and Interaction of Dislocations

As with vacancies, the importance of dislocations derives from the fact that they are readily mobile, in this case under the influence of applied stresses. Figures 9.28*a* to *c* illustrate the slip or glide of an edge dislocation through a

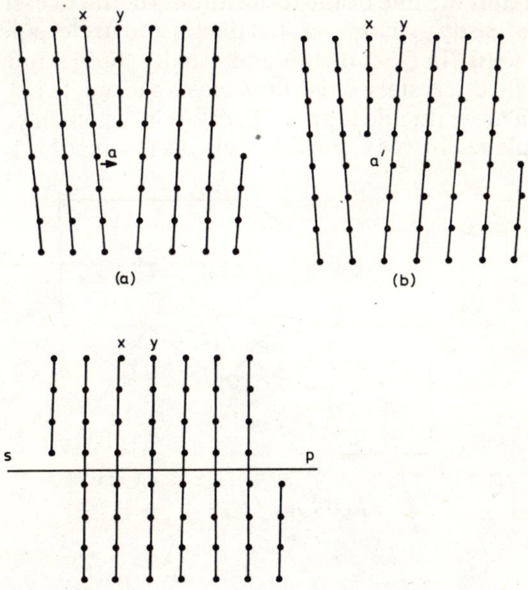

Fig. 9.28 Three stages in plastic deformation by the motion of an edge dislocation through a crystal

crystal under the influence of a shear stress. The atom *a* in Fig. 9.28*a* moves, under the influence of the applied stress, to position *a'* in Fig. 9.28*b* and the extra half plane of atoms shifts one atomic distance to the left, from plane *y* to plane *x*. As this process continues, the dislocation moves right through the crystal, the final result, shown in Fig. 9.28*c*, being that the crystal is sheared or slipped by one interatomic distance across the slip plane *sp*. Each step in the movement of the dislocation (i.e. from Fig. 9.28*a* to *b*) clearly requires only a slight rearrangement of the atoms around the bottom of the extra half plane, and is therefore a low energy process; dislocations are therefore readily mobile under low applied stresses. On the other hand, to produce the slip shown in Fig. 9.28*c* from a perfect dislocation-free crystal would be a very-high-energy process requiring very high stresses. The difference between these two modes of slip is that in the latter all the bonds on the slip plane are broken at the same time, whilst in the former they are broken one at a time. The existence of dislocations was in fact postulated, long before they were observed, to account for the observation that the yield stress of metals is very much lower than that predicted by theoretical calculations. It is important to note that the crystal structure, after passage of a dislocation, is perfect and in no way misaligned. From Fig. 9.29 it is clear that the dislocation line is merely the boundary between slipped and unslipped crystal. The magnitude and direction of the slip is given by what is known as the *Burgers vector*; in a screw dislocation the Burgers vector is parallel to the line of the dislocation while in an edge dislocation it is perpendicular to it. In real crystals, dislocation lines are not straight, but are usually bowed or curved. As illustrated schematically in Fig. 9.29, such dislocations are mixed, being part screw and part edge. Slip occurs on the plane containing both the Burgers vector and the line of the dislocation. In practice, it is observed that metals slip most easily on close-packed planes and in close-packed directions, i.e. $\{111\}$ and $\langle 110 \rangle$ in f.c.c. metals and usually $\{0001\}$ and $\langle 11\bar{2}0 \rangle$ in c.p.h. metals. Since f.c.c. crystals have four close-packed $\{111\}$ planes they are generally much more ductile than c.p.h. crystals, which have only one $\{0001\}$ close-packed plane. In b.c.c. metals, slip occurs in $\langle 111 \rangle$ directions on a

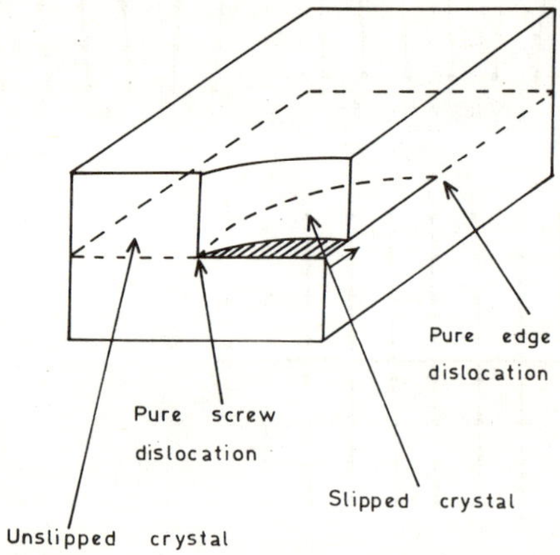

Fig. 9.29 Schematic illustration of a mixed dislocation as the boundary between slipped and unslipped crystal. The arrow shows the Burgers vector

variety of planes. The movement of dislocations by other mechanisms, in particular by cross-slip and climb, is beyond the scope of this section.

All real metals contain dislocations; even a well-annealed metal would typically contain 10^{4-6} dislocations per square millimetre, while a heavily cold-worked metal could contain up to $10^{10}/mm^2$. At first sight this is an anomaly; dislocations were postulated to account for the low yield strength of metals, and whereas an annealed material with a low dislocation density is weak, a cold-worked metal with a high dislocation density is strong. The answer lies in the fact that when the dislocation density is low, the dislocations are generally too far apart to interact with each other very often and are more free to move under the influence of a low applied stress. On the other hand,

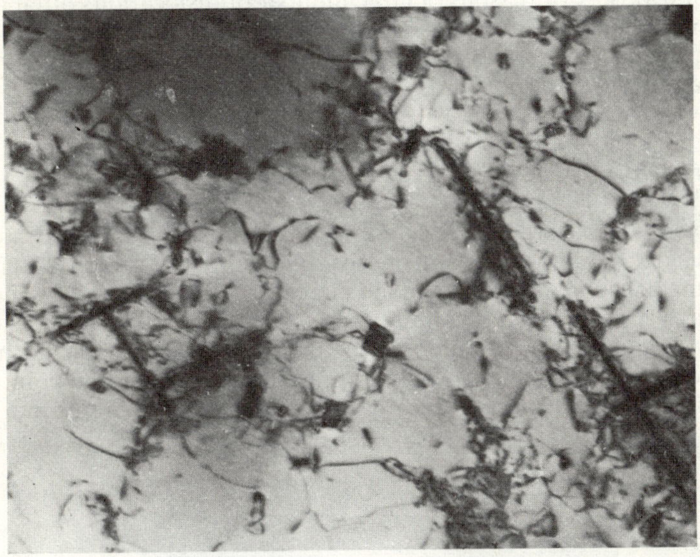

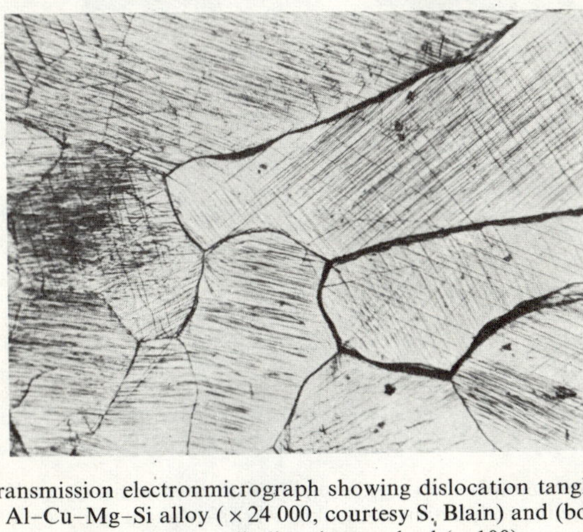

Fig. 9.30 (top) Transmission electronmicrograph showing dislocation tangles associated with precipitates in an Al–Cu–Mg–Si alloy (× 24 000, courtesy S. Blain) and (bottom) light micrograph showing slip lines in pure lead (× 100)

when the dislocation density is high, the dislocations do interact and they become tangled up with each other; dislocation motion is then difficult and the material is therefore strong. The interaction of dislocations with each other and with other structural features in metals is a very complex field; it is also, however, extremely important, since it greatly affects the strength of metals.

The atomic bonds in a cylinder of material around dislocations are elastically stretched; dislocations, like other crystalline defects, are therefore high-energy regions.

Dislocations are readily visible in thin-film transmission electron micrographs, as shown in Figs. 9.25 (top) and 9.30 (top). The slip step (Fig. 9.28*c*) produced by the passage of a single dislocation is not readily apparent. However, for a variety of reasons, a large number of dislocations often move on the same slip plane or on bands of closely adjacent slip planes; this results in slip steps which are very easily seen in the light microscope, as shown by the slip lines in Fig. 9.30 (bottom).

Polycrystalline Metals: Grains and Grain Boundaries

So far we have regarded metals as single crystals; in fact most metals, in practice, are polycrystalline and consist of a number of crystals or grains. In a pure metal each grain will have the same crystal structure and will contain vacancies, dislocations, etc. but the crystallographic orientation of each grain will be different; in other words, in a polycrystalline metal the [111] crystallographic direction of each grain (for example) will be pointing in a different spatial direction. Typically the diameter of the grains in a coarse-

Fig. 9.31 Light micrograph showing the highly elongated grain structure of a commercial wrought high-strength precipitation-hardening Al–Zn–Mg–Cu alloy (×40, section perpendicular to the long–transverse direction)

grained material might be about 0·5 mm, while that in a fine-grained metal might be about 0·005 mm. However, in certain electrodeposits the grain size is too fine to be resolved with the light microscope while at the other extreme jet-engine turbine blades about 150 mm long of nickel-base superalloy may sometimes be single crystals. The grain size of metals may be controlled by heat treatment. The grains may all be of more or less the same size and shape as in the equiaxed grain structure shown in Fig. 9.25 (bottom). Alternatively, if the metal has been heavily cold-worked, by rolling or extrusion for example, the grains may be deformed and elongated in the work direction, giving the highly directional, elongated, grain structure shown in Fig. 9.31. Although the crystallographic orientation of all the grains in a polycrystalline metal is usually different and random, occasionally the orientation of most of the grains in a piece of metal will be similar. Such preferred orientation can arise in a number of ways; it is important to note, however, that metals with elongated, columnar grain structures (Fig. 9.31) do not necessarily have preferred orientation.

However perfect the crystal structure within a grain, there will inevitably be crystallographic mismatch and imperfection where two adjacent grains meet, i.e. at grain boundaries; this region of mismatch is, however, only a few atomic diameters wide, as shown schematically in Fig. 9.32, and on metallographic sections grain boundaries only become apparent on etching. Although grain boundaries are regions of imperfection and of crystallographic mismatch, they are nevertheless quite strong and metals do not normally fail intergranularly. Indeed, grain boundaries act as barriers to the movement of dislocations and a fine-grained pure metal will normally be considerably stronger than a single-crystal or a coarse-grained material.

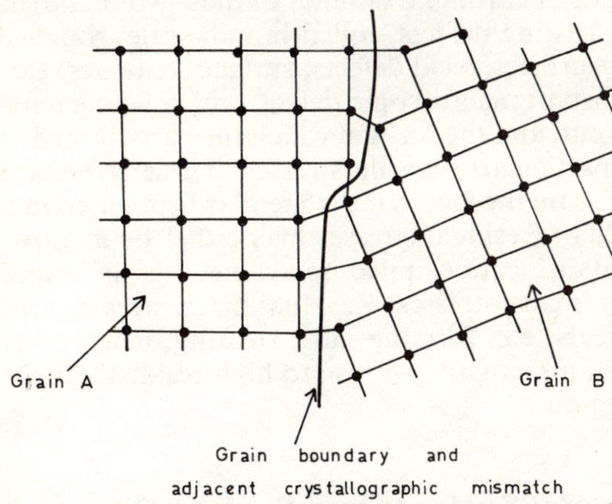

Fig. 9.32 High-angle grain boundary

A grain boundary is a planar defect in a crystal and as such has a higher energy than the surrounding crystal. By way of comparison, the free surface energy of copper is about 1 200–1 800 nJ/mm^2*, the grain boundary energy is about 300–500 nJ/mm^2 while the stacking-fault energy is about 80–100 nJ/mm^2.

*1 nJ/mm^2 = 1 erg/cm^2.

Annealing: Recovery and Recrystallisation

It must be emphasised that many of the crystalline defects already discussed are metastable, and tend to be eliminated on annealing at elevated temperatures. Thus, when a cold-worked metal is heated, the energy stored in the metal during the cold-working process is released; the dislocation tangles rearrange themselves, and the internal stresses, hardness and strength of the metal are progressively reduced. The changes occurring during this process (which is known as *recovery*) do not affect the optical microstructure of the metal. Recovery is followed at higher temperatures by recrystallisation. New grains are nucleated throughout the metal and grow; the heavily deformed high-dislocation-density cold-worked grains are replaced by new equiaxed low-dislocation-density grains. On prolonged heating at even higher temperatures, some of these grains grow at the expense of others, and a coarse-grained material (or even a single crystal) replaces the fine-grained recrystallised structure. The changes occurring during recrystallisation are readily apparent in the optical microscope, and are accompanied by drastic softening of the metal.

Macroscopic Defects in Metals

While vacancies, dislocations, etc. may be regarded as atomic defects, grains may be regarded as microscopic defects. There are also, however, what may conveniently be termed *macroscopic* defects in metals. Too often these are the result of poor design, poor processing or production, poor welding, careless handling or operating. Examples of these would be sharp re-entrant corners, crevices, casting defects, rolled-in mill-scale, rough-sheared edges, poorly machined surfaces, weld defects, surface scratches, etc. However, by far the most important macroscopic defect, which is inherent in the uses to which metals are put, and the ways in which they are formed, is the presence of stresses, and particularly tensile stresses. These stresses may be either applied, resulting from the fact that the metal is being used in a load-bearing capacity, or residual. Residual stresses may either be microstresses arising, for example, from dislocations piled up against a grain boundary or, much more importantly, macrostresses. Residual macrostresses arise from a very wide range of effects, e.g. forming, heat treating, machining, welding and assembling operations can all give rise to high residual tensile stresses over relatively large regions.

Structure of Metal Surfaces and Surface Defects

So far the structure of pure metals has been discussed with reference to bulk characteristics and continuous crystals. However, corrosion is essentially a surface phenomenon and it is necessary to consider how the structure and defects already described interact with free surfaces. At this stage it is convenient to consider only a film-free metal surface, although of course in most corrosion phenomena the presence of surface films is of the utmost importance. Furthermore, it is at free surfaces that the hard sphere model of metals first begins to become inadequate, particularly when surface films are considered, or in aqueous environments, when an electrical double layer exists.

Figure 9.33*a* shows schematically a smooth metal surface, although it should always be borne in mind that even a high-quality electropolished surface will not be flat but will have a roughness of about 5 nm or so high, while the roughness on a high-quality machined surface may be about 1 μm or so high. In most real materials there will be grain boundaries meeting the surface as shown in Fig. 9.33*a*. Furthermore, it is extremely unlikely that the plane of the metal surface will coincide with a low-index low-energy high-atomic-density close-packed plane, as in grain *B*. In most cases the plane of the metal surface will be a high-index high-energy low-atomic-density plane, as in grains *A* and *C*.

On an atomic scale, metal surfaces are usually described in terms of the terrace–ledge–kink model shown in Fig. 9.33*b*. Essentially, it is suggested that clean metal surfaces consist of low-energy low-index terraces separated by ledges of monatomic height which occasionally contain monatomic kinks.

By their nature, dislocations cannot end suddenly in the interior of a crystal; a dislocation line can only end at a free surface or a grain boundary (or form a closed loop). Where a screw dislocation intersects a free surface there is inevitably a step or ledge in the surface, one atomic layer high, as shown in Fig. 9.27*c*. Furthermore, the step need not necessarily be straight and will, in fact, almost certainly contain kinks.

The co-ordination number of an atom lying in a (111) surface of an f.c.c.

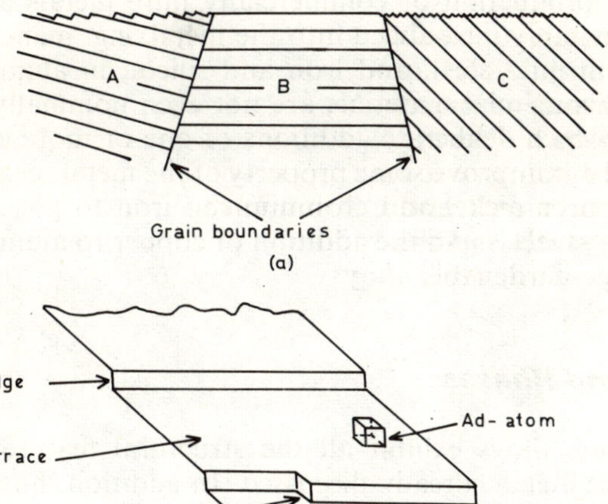

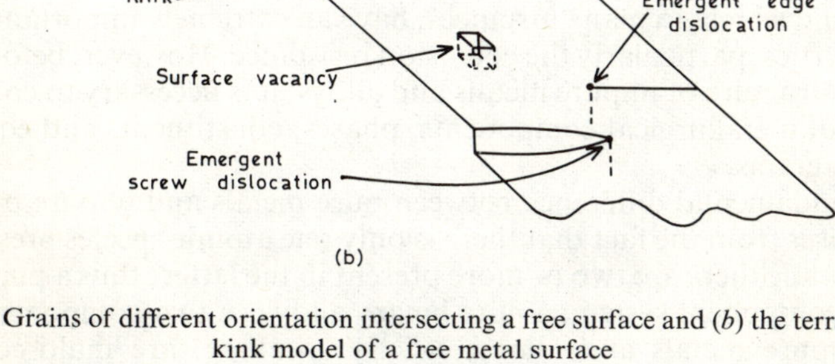

Fig. 9.33(*a*) Grains of different orientation intersecting a free surface and (*b*) the terrace–ledge–kink model of a free metal surface

crystal is 9, compared with the bulk value of 12; an atom at a step in the surface has a co-ordination number of 7 while that of an atom at a kink in a step is only 6. The latter atom, therefore, is least strongly bound to the metal crystal and is likely to be the first to be removed from the lattice in any active dissolution process. Clearly, however, this removal does not eliminate the kink and, in fact, it is impossible, by repeated removal of atoms, to eliminate the step caused by a screw dislocation intersecting a free surface. There are, however, in addition to screw dislocations, many other causes of steps or ledges on metal surfaces, for example low-index planes intersecting the surface at an acute angle (see grain *A* in Fig. 9.33*a*).

As shown in Fig. 9.33*b*, an edge dislocation intersecting a free surface does not cause a step. However, the atoms adjacent to the core of the dislocation, being in non-equilibrium positions, have a higher energy than other atoms in the surface. The emergent ends of both edge and screw dislocations thus represent potential sites for preferential dissolution. Finally, there may be vacancies in the surface and ad-atoms on the surface (Fig. 9.33*b*).

Structure of Alloys

The discussion so far has been limited to the structure of pure metals, and to the defects which exist in crystals comprised of atoms of one element only. In fact, of course, pure metals are comparatively rare and all commercial materials contain impurities and, in many cases also, deliberate alloying additions. In the production of commercially pure metals and of alloys, impurities are inevitably introduced into the metal, e.g. manganese, silicon and phosphorus in mild steel, and iron and silicon in aluminium alloys. However, most commercial materials are not even nominally pure metals but are alloys in which deliberate additions of one or more elements have been made, usually to improve some property of the metal; examples are the addition of carbon or nickel and chromium to iron to give, respectively, carbon and stainless steels and the addition of copper to aluminium to give a high-strength age-hardenable alloy.

Components and Phases

Impure metals and alloys exhibit all the structural features and crystal defects of the pure metals already discussed. In addition, however, impure metals and alloys exhibit many structures which are not observed in pure metals, and which, in many instances, have an extremely important effect on the properties, particularly the corrosion resistance. However, before dealing with the structure of impure metals and alloys, it is necessary to consider the concept of metallurgical components, phases, constituents and equilibrium phase diagrams.

The fundamental difference between pure metals and impure metals and alloys arises from the fact that there is only one atomic species present in the former, while there are two or more present in the latter; thus a pure metal is a single-component system, a pure binary alloy is a two-component system, while impure metals and alloys, strictly speaking, are multi-component systems. It is, however, usually valid to neglect the impurities and, for example, to consider a commercial brass as a binary Cu–Zn alloy, i.e. as a two-component system.

Gibbs' phase rule, as applied to metallurgical systems (i.e. under conditions of constant pressure) indicates that pure metals, being single-component systems, can only co-exist as two phases in equilibrium at a single temperature. Thus pure copper, for example, exhibits three phases, namely solid f.c.c. copper, liquid copper and gaseous copper. Any two of these phases, say solid and liquid copper, can co-exist in equilibrium at only one temperature, i.e. the melting point of copper. Pure iron exhibits four phases, namely two solid phases (f.c.c. and b.c.c. iron (see Table 9.4)) and a liquid and a gaseous phase. Alloys, on the other hand, being multi-component systems, can exhibit two or more phases in equilibrium over a range of temperatures. The phases which are present in an alloy, and their composition and distribution, markedly affect many of the properties of the alloy, including corrosion resistance.

Solid Solutions

When a pure metal *A* is alloyed with a small amount of element *B*, the result is ideally a homogeneous random mixture of the two atomic species *A* and *B*, which is known as a solid solution of *B* in *A*. The solute *B* atoms may take up either interstitial or substitutional positions with respect to the solvent atoms *A*, as illustrated in Figs. 9.34*a* and *b*, respectively. Interstitial solid solutions are only formed with solute atoms that are much smaller than the solvent atoms, as is obvious from Fig. 9.34*a*; for the purpose of this section only three interstitial solid solutions are of importance, i.e. Fe–C, Fe–N and Fe–H. On the other hand, the solid solutions formed between two metals, as for example in Cu–Ag and Cu–Ni alloys, are always substitutional (Fig. 9.34*b*). Occasionally, substitutional solid solutions are formed in which the

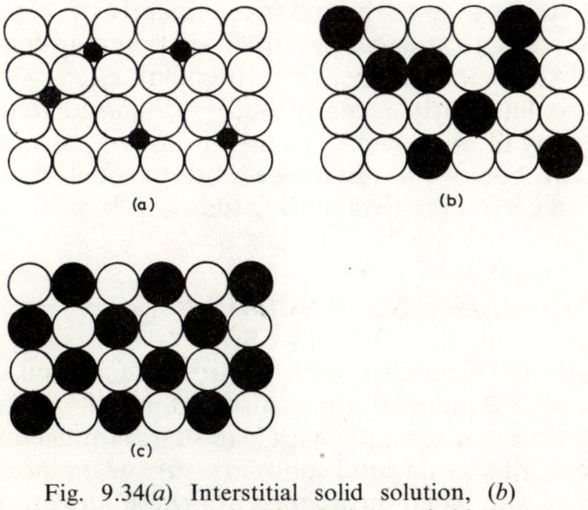

(a) (b) (c)

Fig. 9.34(*a*) Interstitial solid solution, (*b*) random substitutional solution and (*c*) an ordered substitutional solid solution forming a superlattice

distribution of B in A is not random (Fig. 9.34*c*) and such solid solutions are said to be *ordered*, and to exhibit a *superlattice*. Order–disorder transitions are observed with temperature changes. Superlattices are observed in the Cu–Au, Cu–Zn and Fe–Ni systems.

There are a number of differences between interstitial and substitutional solid solutions, one of the most important of which is the mechanism by which diffusion occurs. In substitutional solid solutions diffusion occurs by the vacancy mechanism already discussed. Since the vacancy concentration and the frequency of vacancy jumps are very low at ambient temperatures, diffusion in substitutional solid solutions is usually negligible at room temperature and only becomes appreciable at temperatures above about $0{\cdot}5T_M$, where T_M is the melting point of the solvent metal (deg K). In interstitial solid solutions, however, diffusion of the solute atoms occurs by jumps between adjacent interstitial positions. This is a much lower energy process which does not involve vacancies and it therefore occurs at much lower temperatures. Thus hydrogen is mobile in steel at room temperature, while carbon diffuses quite rapidly in steel at temperatures above about 370 K.

Segregation at Dislocations and Grain Boundaries

In practice, the distribution of solute atoms in a solid solution is not ideally homogeneous. There is almost invariably a stress field around any solute atom in solid solution, and there is therefore a tendency for these atoms to migrate to regions of the crystal lattice that are already strained, in particular to grain boundaries and to dislocations. Thus in impure metals and alloys there is an inherent tendency for impurities and alloying additions to segregate at grain boundaries and around the cores of dislocations. Segregation of carbon and nitrogen to dislocations in steels has a very significant effect on their mechanical properties, and in particular is responsible for their sharp yield point; similar segregation may also (see Section 1.3) affect the corrosion behaviour of cold-worked steels. The segregation of impurities and alloying elements to grain boundaries is clearly of relevance in any consideration of the strength and reactivity of grain boundaries and, in particular, of intergranular corrosion, temper embrittlement, etc.

In addition, it has fairly recently been recognised that impurities and alloying elements will also tend to segregate to free surfaces. The implications of this for corrosion resistance and particularly for passive-film formation have received relatively little attention.

Limited and Complete Solid Solubility

Just as the saturated solubility of sugar in water is limited, so the solid solubility of element B in metal A may also be limited, or may even be so low as to be negligible, as for example with lead in iron or carbon in aluminium. There is extensive interstitial solid solubility only when the solvent metal is a transition element and when the diameter of the solute atoms is $<0{\cdot}6$ of the diameter of the solvent atom. The Hume-Rothery rules state that there is extensive substitutional solid solubility of B in A only if:

(*a*) The atomic diameters of *A* and *B* do not differ by more than 15%.
(*b*) The two elements have similar electronegativity. (Note: electronegativity is the power of an element to attract electrons to itself when present in a molecule or in an aggregate of unlike atoms; it is a different property from the electrode potential, which depends on the free energy difference between an element in its standard state and a compound or ion in solution (*see* Section 9.1).) In addition a metal of a lower valency tends to dissolve a metal of a higher valency more readily than vice versa.

However, just as two liquids may be completely miscible and form a complete range of solutions from one pure liquid to the other, so certain metals, for example copper and nickel, exhibit complete solid solubility over the whole range of compositions from pure copper to pure nickel. Clearly for two metals to be soluble in each other over the whole compositional range, they must have the same crystal structure, i.e. they must be isomorphous.

Intermediate Phases and Intermetallic Compounds

When metal *A* is alloyed with enough element *B* to exceed the solid solubility of *B* in *A*, a new phase is formed. This may be the other terminal solid solution of *A* in *B*. Alternatively, an intermediate solid-solution phase, with a crystal structure different from that of either of the terminal solid solutions, may be formed over a range of compositions. An example of this is the so-called β-phase in brasses which is the stable phase at room temperature over the composition range 47–50% Zn. In some alloys, intermediate phases are formed that are best regarded as compounds. Examples are Fe_3C (cementite) in the Fe–C system and the intermetallic compounds $MgNi_2$ and Mg_2Ni formed in the Mg–Ni system. Whereas solid solutions are stable over a range of compositions, compounds have a unique composition; cementite, for example, corresponds to Fe–25at.%C (Fe–6·7wt.%C).

Equilibrium Phase Diagrams

In alloys, as in pure metals, the stable phase or phases change with changing temperature. Furthermore, in alloys the stable phase or phases, and the composition of these phases, changes with the composition of the alloys. These changes are best represented by equilibrium phase diagrams. These define the stability and composition of the phases that can occur in an alloy system as a function of temperature and alloy composition, under conditions of constant pressure. Phase diagrams are in many ways analogous to Pourbaix diagrams, which define the stability of phases in a metal/water system as a function of potential and pH, and like *E*–pH diagrams, phase diagrams have certain limitations. Firstly it must be emphasised that they are *equilibrium* phase diagrams, and that they only predict the stability and composition of phases under equilibrium conditions. They give no indication of how rapidly equilibrium will be achieved in practice, nor do they give any indication of the distribution of the phases at equilibrium. Somewhat analogously, *E*–pH diagrams give no information as to the rate at which a particular metal will corrode, or as to the protectiveness of a particular passive film. Secondly, equilibrium phase diagrams give no information whatsoever as to the existence of metastable phases, some of which are of the utmost practical significance. These limitations become particularly important under conditions of rapid temperature change. Phase diagrams, in contrast to E–pH diagrams, are always determined experimentally, although they can be explained and to a certain extent qualitatively predicted from thermodynamic data and considerations. Nevertheless, in spite of these limitations, phase diagrams are an extremely important and useful metallurgical tool. (See also Section 1.4.)

Binary Isomorphous Phase Diagrams

Although ternary and quaternary phase diagrams are beyond the scope of this section, it is appropriate to deal with the various types of binary phase diagram by referring to some specific alloys. The simplest phase diagram refers to an isomorphous alloy system in which there is complete solid solubility. By way of example, the Cu–Ni phase diagram is shown in Fig. 9.35. This phase diagram indicates that at 1 600K an alloy containing 70% Cu is a liquid (point *a* in Fig. 9.35). On slow cooling, this alloy starts to solidify when the liquidus line is reached, i.e. point *b* at about 1 500K. At this point grains of a solid solution containing about 53% Cu are nucleated from the liquid containing 70% Cu, and start to grow. As the temperature continues to fall, the composition of the liquid follows the liquidus line while that of the solid falls along the solidus line, until at 1 470K the alloy consists of a mixture of a solid solution containing 62% Cu and a liquid containing 78% Cu

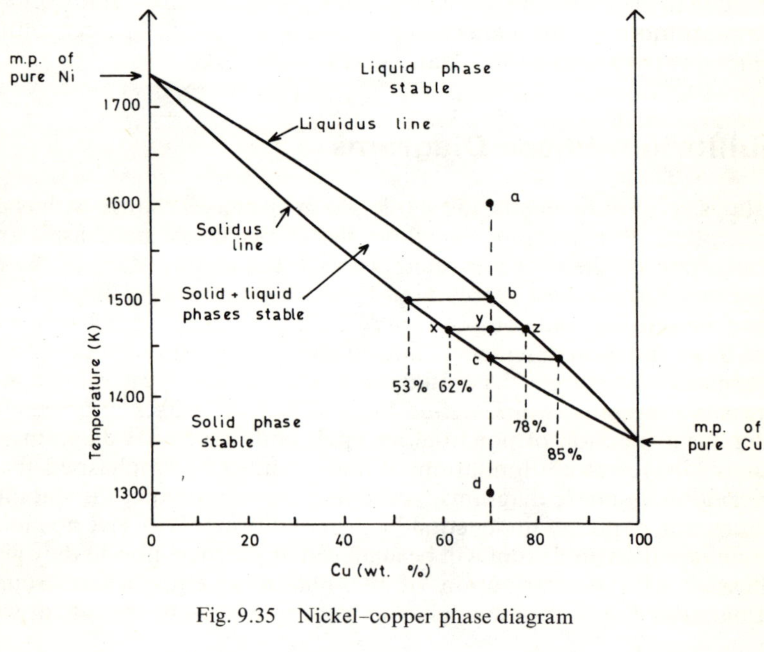

Fig. 9.35 Nickel–copper phase diagram

(Fig. 9.35). Furthermore, the fraction of solid in the two-phase mixture is given by the ratio of the lengths yz/xz, while the fraction of liquid is given by xy/xz. This is known as the *lever* rule. On further slow cooling the solidus line is reached, i.e. point c at about 1 430K, when solidification of the alloy is complete; the last liquid to freeze contains about 85% Cu, while the last solid to solidify contains 70% Cu. At 1 300K, for example, and indeed at all temperatures below the solidus, an alloy containing 70% Cu is a solid (point d in Fig. 9.35). Clearly the phase change in this transformation is liquid → solid. The microstructure of a homogeneous single-phase solid-solution alloy is indistinguishable from that of a pure metal, as shown in Fig. 9.25 (bottom).

Coring

If the alloy is cooled sufficiently slowly, the alloy itself and all the grains in it will be homogeneous after solidification. If, on the other hand, the alloy is cooled rather more rapidly, the grains will not be homogeneous but will be cored. During cooling, the composition changes in the solid phase involve solid state diffusion; if the temperature falls too rapidly, therefore, there is insufficient time for diffusion. In the Cu–30Ni alloy referred to earlier, clearly the centre of the grains, the first metal to have solidified, would be nickel rich (i.e. low in copper) while the grain boundary regions, the last metal to have solidified, would be copper rich. Coring is eliminated by annealing at high temperatures to permit diffusion and homogenisation to take place.

Eutectic Phase Diagrams

The equilibrium phase diagram for Pb–Sb alloys represents a simple eutectic system and is shown in Fig. 9.36. This type of phase diagram is often observed when there is negligible solid solubility of the two metals in each other and no intermediate phases are formed. The point E is the eutectic point, T_E the eutectic temperature and C_E the eutectic composition, in this case approximately 520K and 13% Sb, respectively. Like a pure metal, an alloy of eutectic composition freezes at a single temperature T_E, instead of over a temperature range, as in the case of the Cu–30Ni alloy already discussed. Unlike a pure metal, however, a eutectic alloy freezes to give a mixture of two phases, in this case Pb and Sb; in general terms, therefore, a eutectic reaction involves liquid → solid α + solid β.

A hyper-eutectic alloy containing, say, 50% Sb starts to freeze when the temperature reaches the liquidus line (point a in Fig. 9.36). At this temperature pure pro-eutectic Sb nucleates; as the temperature continues to fall, more antimony is deposited from the melt, and the composition of the liquid phase moves down the liquidus line to the eutectic point. When this is reached, the remainder of the melt solidifies. The microstructure of alloys of eutectic composition varies somewhat with alloy system, but generally consists of an aggregate of small particles, often platelets, of one of the phases comprising the eutectic in a continuous matrix of the other phase. Finally, the microstructure of the hyper-eutectic 50% Sb alloy already mentioned consists of primary (pro-eutectic) antimony surrounded by the eutectic constituent, as illustrated in Fig. 9.37. This microstructure is clearly to be expected from the solidification sequence just described.

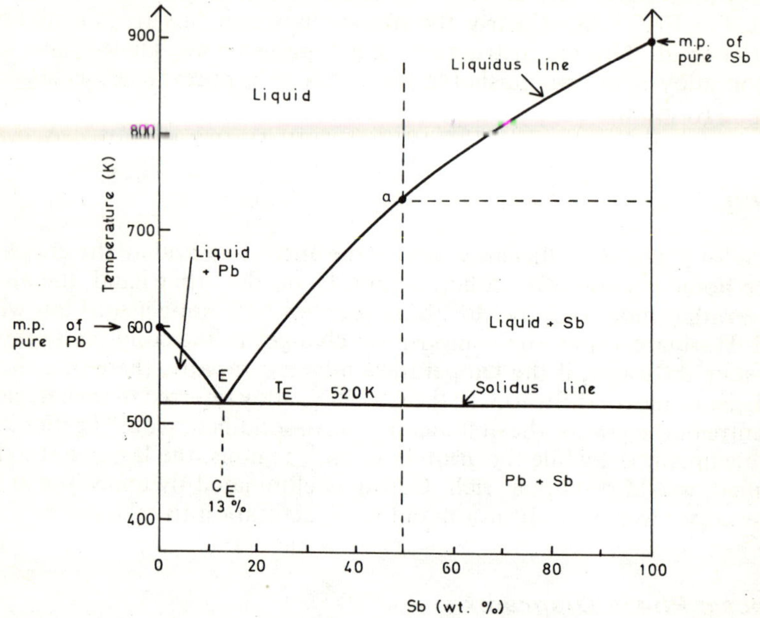

Fig. 9.36 Lead–antimony phase diagram

Fig. 9.37 Light micrograph showing the microstructure of a hyper-eutectic Pb–50Sb alloy (×50)

More Complex Binary Phase Diagrams

Most phase diagrams, however, are more complex than those shown in Figs. 9.35 and 9.36. Thus in most eutectic systems there is some appreciable

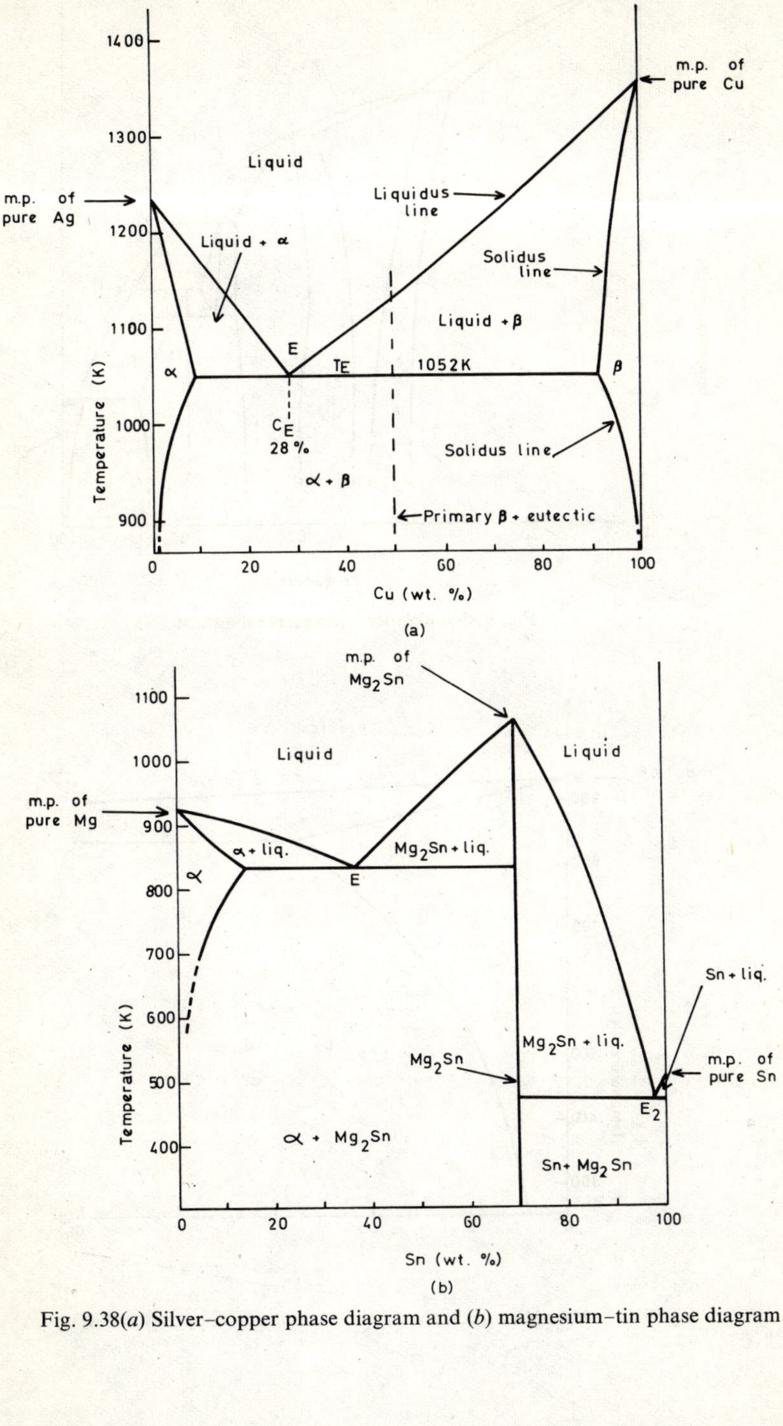

Fig. 9.38(*a*) Silver–copper phase diagram and (*b*) magnesium–tin phase diagram

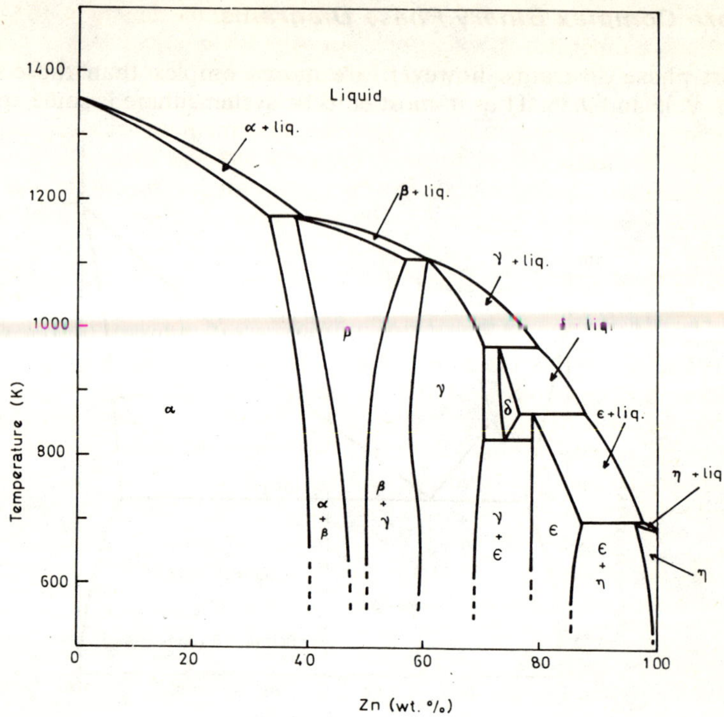

Fig. 9.39 Copper–zinc phase diagram

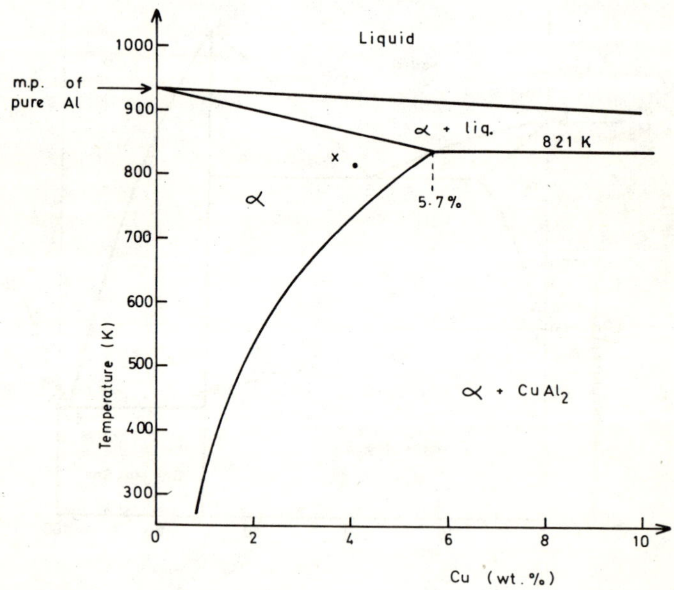

Fig. 9.40 Aluminium-rich end of the aluminium–copper phase diagram

solid solubility, as in the Ag–Cu phase diagram shown in Fig. 9.38*a*. At ambient temperatures, therefore, an Ag–50Cu alloy would contain two components (Ag and Cu), two phases (the α and β solid solutions) and two constituents (primary pro-eutectic β and the eutectic mixture of $\alpha+\beta$). Furthermore, in many alloy systems intermediate phases and/or intermetallic compounds are formed. Figure 9.38*b*, for example, illustrates the Mg–Sn phase diagram; this exhibits two eutectic reactions and the intermetallic compound Mg_2Sn. Finally, a number of other reactions, in addition to the eutectic reaction, are observed; of particular importance are (on cooling in all cases):

1. The *eutectoid* reaction: solid $\alpha \rightarrow$ solid $\beta+$solid γ.
2. The *peritectic* reaction: liquid + solid $\alpha \rightarrow$ solid β.
3. The *peritectoid* reaction: solid $\alpha+$solid $\beta \rightarrow$ solid γ.

By way of example, the Cu–Zn phase diagram shown in Fig. 9.39 exhibits a number of different intermediate phases (β, γ, δ, etc.) and a number of peritectic reactions and a eutectoid reaction. In many instances it is not necessary to consider a complete phase diagram. Thus Fig. 9.40 illustrates the Al-rich end of the Al–Cu phase diagram and is used below in a discussion

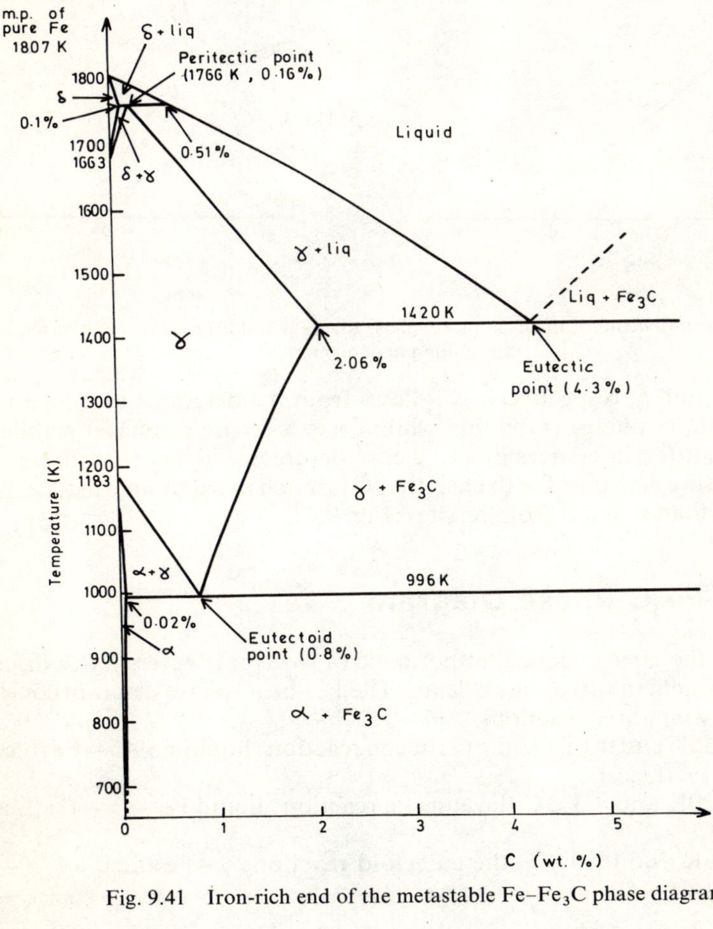

Fig. 9.41 Iron-rich end of the metastable Fe–Fe_3C phase diagram

of precipitation hardening. Again, Fig. 9.41 shows the $Fe–Fe_3C$ phase diagram, the great practical importance of which is in no way reduced by the fact that one of the components in it, namely Fe_3C (cementite), is a metastable compound. The true equilibrium diagram, in which graphite appears in place of cementite, is of use only when cast irons are being considered, as opposed to steels. Other phase diagrams which are of particular significance are the iron-rich ends of the Fe–Ni and Fe–Cr systems, which are shown in

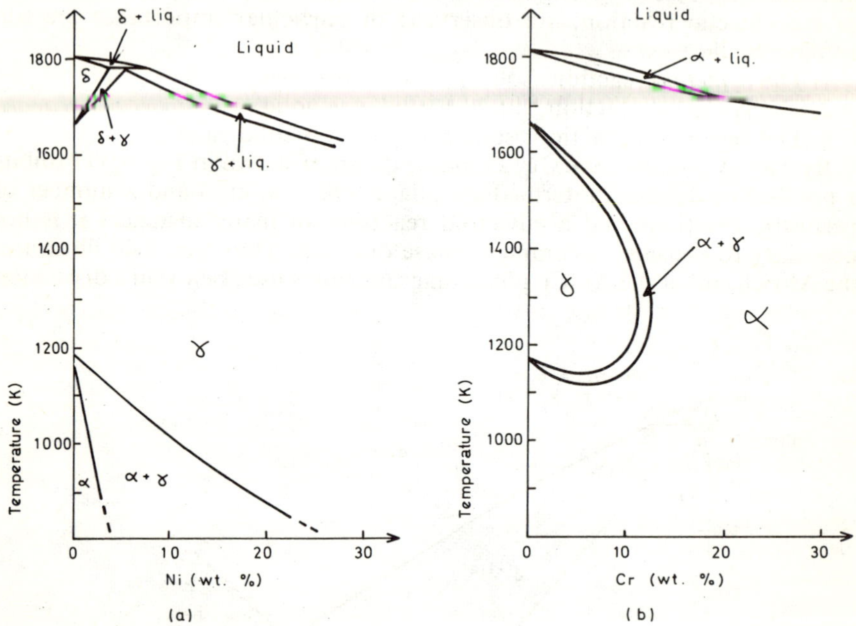

Fig. 9.42(*a*) Iron-rich end of the iron–nickel phase diagram and (*b*) iron-rich end of the iron–chromium phase diagram

Figs. 9.42*a* and *b*, respectively. It follows from the diagrams that Ni and C are austenite (γ-phase) stabilisers while Cr is a ferrite (α-phase) stabiliser. The fact that the interstices in an f.c.c. structure are bigger than those in a b.c.c. structure accounts for the fact that C is much more soluble in f.c.c. iron (austenite) than in b.c.c. iron (ferrite) (Fig. 9.41).

The $Fe–Fe_3C$ Phase Diagram

Because of the great practical importance of steels it is necessary to discuss the Fe–C system in rather more detail. The $Fe–Fe_3C$ phase diagram consists of the following three reactions

1. At 1 766K and 0·16% C the peritectic reaction: liquid Fe + δ – Fe (b.c.c.) $\rightarrow \gamma$ – Fe (f.c.c.).
2. At 1 420K and 4·3% C the eutectic reaction: liquid Fe $\rightarrow \gamma$ – Fe (f.c.c.) + Fe_3C.
3. At 996K and 0·80% C the eutectoid reaction: γ – Fe (f.c.c.) $\rightarrow \alpha$ – Fe (b.c.c.) + Fe_3C.

The peritectic transformation generally has little effect on the structure, properties or corrosion resistance of steels at room temperature; an exception to this occurs in the welding of certain steels, when δ-ferrite can be retained at room temperature and can affect corrosion resistance. Furthermore, since most steels contain less than about 1·0% C (and by far the greatest tonnage contains less than about 0·3% C) the eutectic reaction is of relevance only in relation to the structure and properties of cast irons, which generally contain 2–4% C. This discussion, therefore, will be limited to the eutectoid reaction that occurs when homogeneous austenite is cooled.

Decomposition of Austenite

When austenite is cooled, the structures which result from the eutectoid reaction depend on the temperature at which the transformation takes place. In practice, of course, steels are generally continuously cooled at a rate which may vary from very rapid (quenching) to very slow (furnace cooling). However, it is useful and valid to simplify the situation by assuming that the transformation occurs isothermally. In other words, therefore, we shall consider initially the reactions which occur when a steel containing 0·8% C (i.e. of eutectoid composition) is very rapidly quenched from, say, 1 400K to some temperature $T<996$K, at which decomposition of the austenite takes place. In general, the structures produced on slow continuous cooling approximate to those resulting from high-temperature isothermal transformation, while rapid continuous cooling results in structures which approximate to those obtained by intermediate-temperature isothermal transformation.

Formation of Pearlite

If austenite is allowed to transform just below 996K, the reaction product, as predicted by the Fe–Fe_3C phase diagram, is the α–Fe + Fe_3C eutectoid. This eutectoid constituent is known as *pearlite*, and consists of alternate platelets of ferrite (α-Fe) and cementite (Fe_3C), the former being the continuous phase. The decomposition of austenite to pearlite involves nucleation of pearlite, which occurs almost exclusively at austenite grain boundaries, followed by growth of pearlite; both processes involve the solid-state diffusion of carbon in iron. Immediately after quenching, therefore, the structure of the steel consists of untransformed austenite, as shown schematically in Fig. 9.43*a*. At some later time, nucleation of a number of colonies of pearlite will have occurred at the austenite grain boundaries, as shown in Fig. 9.43*b*; at this stage about 10% of the austenite has transformed. At some later time still, growth of the pearlite will have occurred by nucleation of new α-Fe and Fe_3C platelets and by extensive edge-wise growth of existing platelets as shown in Fig. 9.43*c*; at this stage about 90% of the austenite has transformed and eventually transformation will be complete.

In fundamental terms, the transformation temperature affects both the driving force for the decomposition of austenite and the diffusion rate of carbon. In effect, therefore, the transformation temperature alters both the

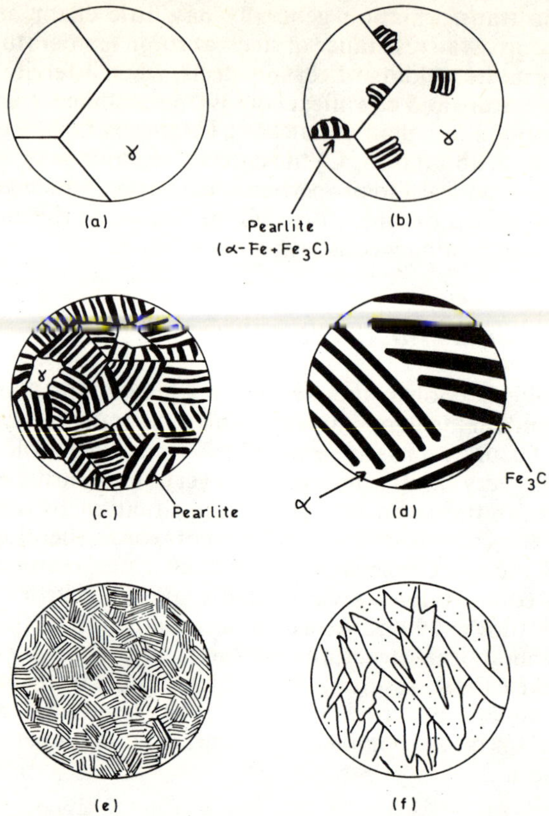

Fig. 9.43 Some of the pearlitic and bainitic microstructures observed in eutectoid steels after various heat treatments

rate of nucleation and the rate of growth. This in turn manifests itself in different transformation times at different temperatures. Furthermore, the spacing of the pearlite platelets is characteristic of the temperature at which transformation occurred, the lower the transformation temperature the finer are the platelets, and the greater the hardness and the strength of the steel. The distinction between coarse and fine pearlite structures is shown in Figs. 9.43*d* and *e*, respectively; in practice the platelets in very fine pearlite may not be resolvable with a light microscope. The rate of transformation of carbon steels is also markedly affected by the presence of other alloying elements. However, a detailed consideration of the transformation and structure of low-alloy steels is beyond the scope of this section.

Formation of Bainite

At lower transformation temperatures ($<$770K approx.) a second reaction, the formation of *bainite*, intervenes. Like pearlite, the bainite constituent in steels consists of a mixture of ferrite and an iron carbide and is formed by a time-dependent nucleation and growth process involving diffusion of carbon.

Bainite, however, grows as needlelike plates, which consist of very fine particles of carbide embedded in a ferrite matrix. The carbide particles can only be resolved in an electron microscope; in the light microscope only the acicular bainite needles are visible. The microstructure of a bainitic steel is shown schematically in Fig. 9.43*f*; the appearance of bainite, however, and the structure and composition of the carbides in it, vary somewhat with transformation temperature. Bainite is much harder than pearlite.

Formation of Martensite

Finally, at even lower transformation temperatures, a completely new reaction occurs. Austenite transforms to a new metastable phase called *martensite*, which is a supersaturated solid solution of carbon in iron and which has a body-centred tetragonal crystal structure. Furthermore, the mechanism of the transformation of austenite to martensite is fundamentally different from that of the formation of pearlite or bainite; in particular martensitic transformations do not involve diffusion and are accordingly said to be diffusionless. Martensite is formed from austenite by the slight rearrangement of iron atoms required to transform the f.c.c. crystal structure into the body-centred tetragonal structure; the distances involved are considerably less than the interatomic distances. A further characteristic of the martensitic transformation is that it is predominantly athermal, as opposed to the isothermal transformation of austenite to pearlite or bainite. In other words, at a temperature midway between m_s (the temperature at which martensite starts to form) and m_f (the temperature at which martensite

Fig. 9.44 Light micrograph showing the microstructure of a martensitic steel (×550)

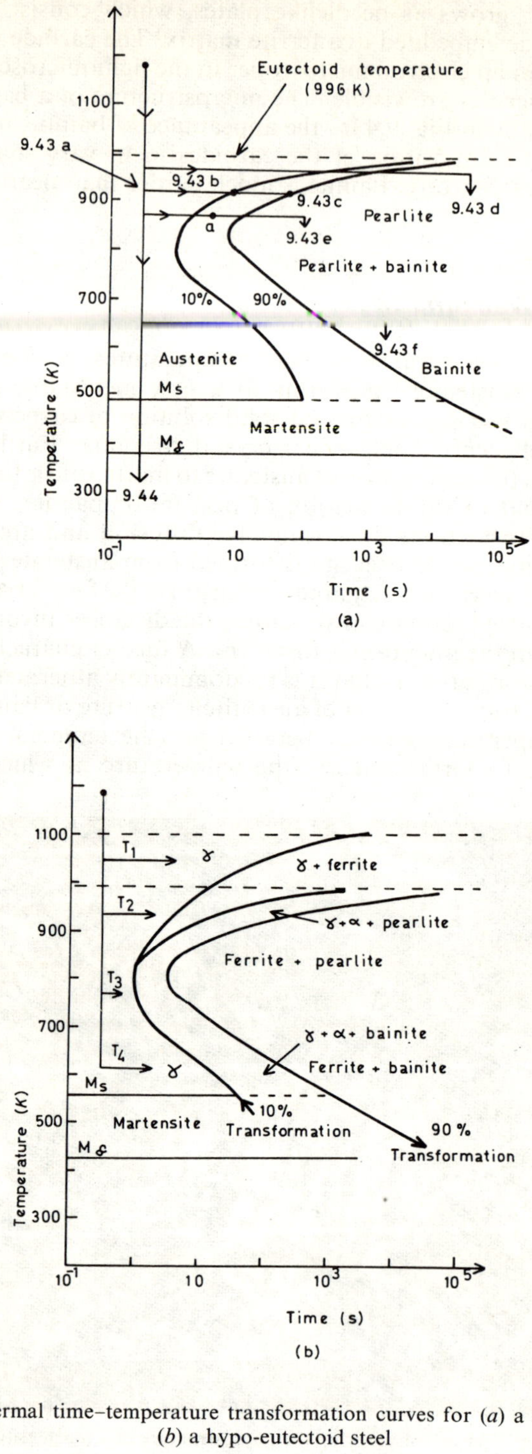

Fig. 9.45 Isothermal time–temperature transformation curves for (*a*) a eutectoid steel and (*b*) a hypo-eutectoid steel

formation is apparently complete) the steel will consist of about 50% austenite and about 50% martensite; this transformation of 50% of the austenite to martensite is effectively instantaneous and no matter how long the steel is held at temperature, no further transformation will occur. Figure 9.44 illustrates the microstructure of a fully martensitic steel. The martensite platelets are heavily twinned, although this is only apparent in an electron microscope. In fact, in a eutectoid steel the m_f temperature is below room temperature although in a more common hypo-eutectoid steel containing about 0·2% C, $m_s \approx 720$K and $m_f \approx 550$K. Other alloying elements also affect the temperature at which the martensite transformation takes place, and usually decrease it.

Isothermal Transformation Diagrams

Time–temperature–transformation (T–T–T) diagrams are used to present the structure of steels after isothermal transformation at different temperatures for varying times. The T–T–T diagram for a commercial eutectoid steel is shown in Fig. 9.45*a*. Also shown on the curves are the points at which the microstructures illustrated in Figs. 9.43 and 9.44 are observed, and the thermal treatments producing these structures. When a steel partially transformed to, say, pearlite, is quenched from point *a* in Fig. 9.45*a* to below m_f, the untransformed austenite transforms to martensite.

Transformation of Hypo-eutectoid Steels

As noted above, most steels are, in practice, hypo-eutectoid. Consider, by way of example, a low-alloy steel containing 0·3% C. The Fe–C phase diagram (Fig. 9.41) indicates that as a steel of this composition is cooled below about 1 080K the equilibrium state is a two-phase structure containing primary, pro-eutectoid ferrite and austenite. On further cooling to below 996K the equilibrium structure consists of pro-eutectoid ferrite and pearlite. These differences are reflected in the rather more complicated T–T–T diagram used for this steel, shown in Fig. 9.45*b*.

Depending on the transformation temperature, the steel may transform to ferrite plus austenite (T_1 in Fig. 9.45*b*), to ferrite plus pearlite via ferrite plus austenite (T_2), directly to ferrite plus pearlite (T_3) or directly to ferrite plus bainite (T_4). The pro-eutectoid ferrite morphology, like the pearlite morphology, varies with the isothermal transformation temperature and also with the carbon content of the steel. The ferrite nucleates heterogeneously on austenite grain boundaries. Any pro-eutectoid ferrite in slightly hypo-eutectoid steels (about 0·6%) therefore tends to be distributed along the austenite grain boundaries, with the grain centres transforming later to pearlite, as shown schematically in Fig. 9.46*a*. At lower carbon contents, Widmanstätten pro-eutectoid ferrite is observed, as shown schematically in Fig. 9.46*b*, while in low-carbon steels massive ferrite is observed (Fig. 9.46*c*). It is important to re-emphasise that these pro-eutectoid ferrite structures also vary with isothermal transformation temperature and therefore, in practice, with cooling rate; they can also be altered by subsequent heat treatment and

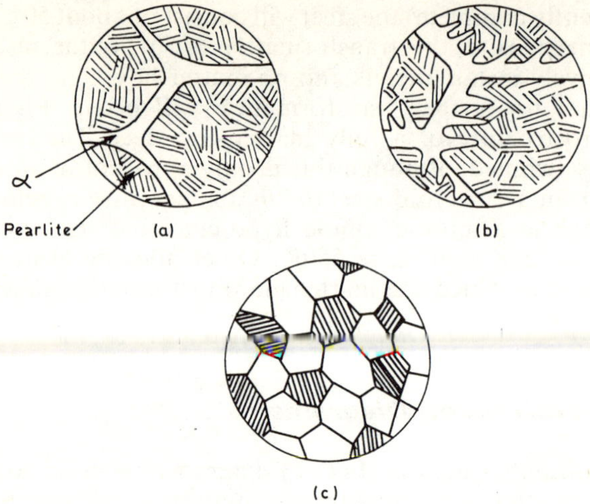

Fig. 9.46 Schematic illustration of some of the ferritic/pearlitic microstructures observed in hypo-eutectoid steels after various heat treatments

working. Grain-boundary and Widmanstätten ferrite are generally undesirable and result in poor mechanical properties. Figure 9.46, incidentally, illustrates some of the more important structures observed in two-phase alloys generally.

The higher the carbon content of a hypo-eutectoid steel, the more pearlite there will be in a ferritic/pearlitic structure and the greater will be the strength of the steel, other factors (grain-size, pearlite spacing, etc.) being equal.

Tempering of Martensite

As-quenched martensite is extremely hard, but also extremely brittle. The hardness of as-quenched martensite, which also increases with increasing carbon content, is due partly to the strain involved in retaining the carbon in solid solution in body-centred tetragonal iron, and partly, in low-carbon martensites, to their very high dislocation density. The martensite reaction in steels is not reversible and on reheating as-quenched martensite (Fig. 9.47*a*) to various temperatures below 996K, the carbon tends to precipitate out of solid solution; this process is known as *tempering*. On tempering at about 370–520K the carbon precipitates out as the so-called ε-carbide, which may be written as $Fe_{2\cdot4}C$ and has a different structure from cementite. These carbides precipitate on certain crystallographic planes and on the twin boundaries in the martensite plates, forming what is known as a *Widmanstätten* structure, as shown schematically in Fig. 9.47*b*. On tempering at 520–770K, cementite particles are precipitated on twin boundaries and on boundaries between martensite platelets, leaving a matrix of ferrite as shown in Fig. 9.47*c*. Eventually the cementite particles grow and become spheroidal in shape, while the twinned structure breaks down, and fine equiaxed ferrite

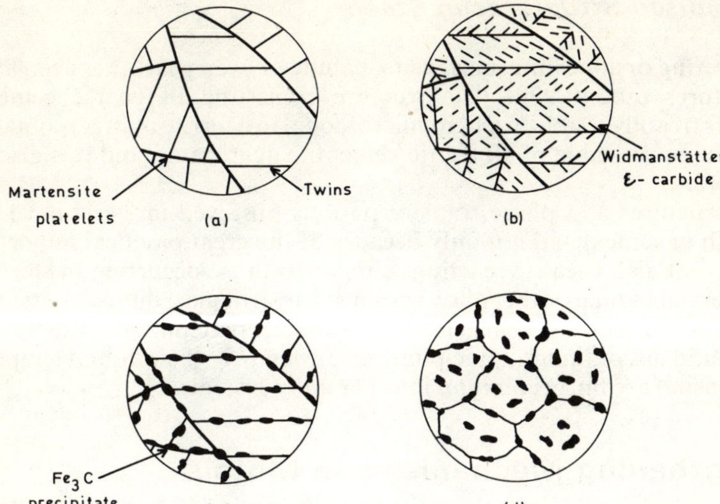

Fig. 9.47 Some of the microstructures observed during the tempering of martensite

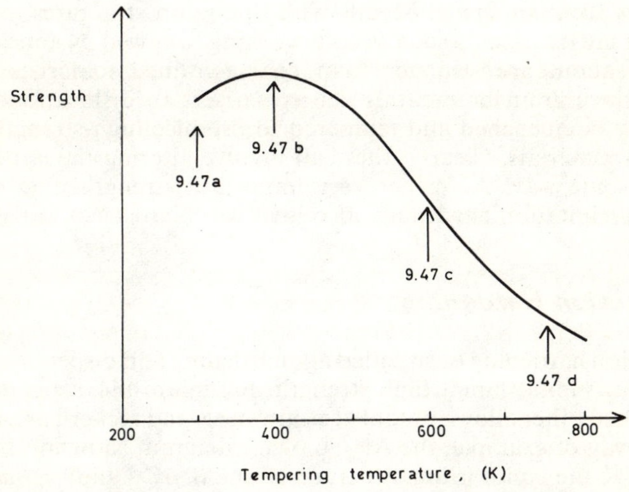

Fig. 9.48 Effect of tempering temperature on the strength and microstructure of martensite

grains are observed, as illustrated in Fig. 9.47*d*. On tempering, as the carbon precipitates out of solid solution, the strength and hardness of the steel falls progressively, as shown in Fig. 9.48, while the toughness increases. Quenched and tempered low-alloy steels are widely used where high-strength high-toughness materials are required.

Spheroidised Structures in Steels

On tempering or annealing martensite, bainite or even pearlite at even higher temperatures (about 970K) a structure consisting of coarse cementite spheroids (readily visible in a light microscope) in a ferrite matrix is obtained. This is the most stable of all ferrite/cementite aggregates, and it is also one of the softest.

The structures and phase transformations observed in steels have been dealt with in some detail not only because of the great practical importance of steels, but also because reactions similar to those occurring in steels are also observed in many other alloy systems. In particular, diffusionless transformations (austenite → martensite), continuous precipitation (austenite → pearlite) and discontinuous precipitation (austenite → bainite and tempering of martensite) are fairly common in other alloy systems.

Strengthening Mechanisms in Metals

Many of the ways of increasing the strength of metals have already been dealt with in passing. For example, cold working a metal increases its dislocation density and therefore, as already discussed, increases its strength*. Substitutional solid solutions (e.g. austenitic stainless steels and brass) are normally stronger than the equivalent pure metals (i.e. iron and copper), as the lattice distortion involved hinders dislocation movement. This is known as *solid-solution hardening*. Metals with fine grain structures, produced by grain-refining treatments such as cold working followed by annealing to give recrystallisation, are stronger than coarse-grained materials and single crystals, since grain boundaries act as barriers to dislocation movement. Steels may be quenched and tempered to produce high-strength and high-toughness materials. Clearly, these all involve altering the structure of the metal in some way. A further very important strengthening mechanism, namely precipitation hardening, also involves microstructural changes.

Precipitation Hardening

Precipitation hardening (also called age hardening and dispersion hardening) is particularly important in high-strength aluminium alloys, but it is also used to strengthen other alloys, notably certain steels and nickel-base alloys. Consider, by way of example, the Al–Cu phase diagram shown in Fig. 9.40. At about 820K the equilibrium structure of an Al–4Cu alloy (point X in Fig. 9.40) is the single-phase α solid solution of copper in aluminium, as shown in Fig. 9.49*a*. On annealing at this temperature, i.e. on solution heat treating, this equilibrium state is readily achieved. At room temperature, since the limit of solid solubility decreases with decreasing temperature, the equilibrium structure consists of the α-solid solution plus the $CuAl_2$ intermetallic compound. This equilibrium state may be achieved by very slowly cooling

*The processes involved in the nucleation and multiplication of dislocations during cold working are very complex, and beyond the scope of this section.

the alloy from 820K to room temperature, and consists of very coarse discontinuous precipitates of $CuAl_2$ in an α matrix, as shown in Fig. 9.49*b*. However, on quenching the alloy to room temperature, diffusion (and therefore nucleation and growth of $CuAl_2$) is prevented, and the copper is retained in supersaturated solid solution, as shown in Fig. 9.49*c*. This supersaturated solid solution is relatively soft, weak and ductile. If this metastable supersaturated solid solution is then aged, i.e. heat treated at a relatively low temperature (370–520K), the excess copper again precipitates out of solid solution as $CuAl_2$. In this case, however, the $CuAl_2$ particles are extremely fine, certainly sub-microscopic and under certain conditions only just resolvable in the electron microscope. The precipitation sequence varies

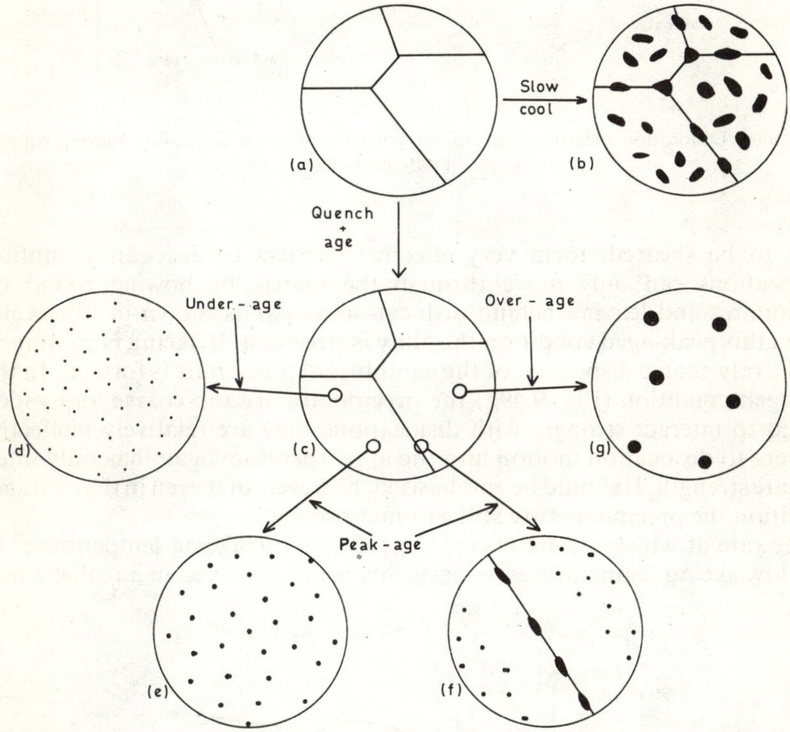

Fig. 9.49 Some of the microstructures observed during the heat treatment of precipitation-hardening aluminium alloys

from alloy to alloy but is generally quite complex and involves a number of intermediate metastable precipitates. The initial precipitates or coherent zones formed during the early stages of ageing (i.e. after relatively short times) interact with dislocations and tend to hinder their movement. This underaged condition is therefore stronger than the as-quenched condition, but, since the zones are extremely fine (Fig. 9.49*d*), and are readily sheared by moving dislocations (Fig. 9.50*a*), only moderately so. On further ageing, somewhat coarser, semi-coherent intermediate precipitates are formed (Fig. 9.49*e*). These interact strongly with dislocations and, since they are too

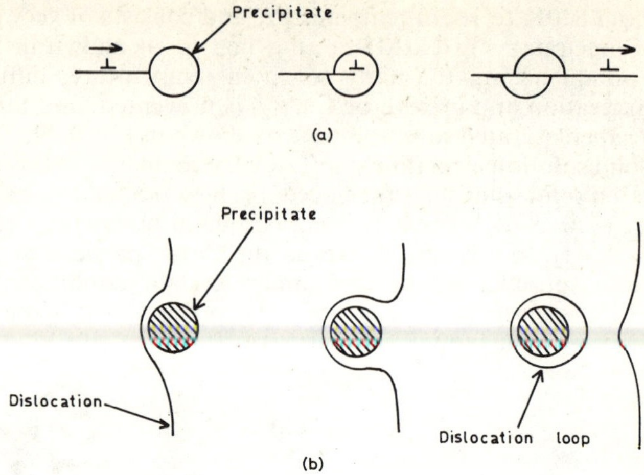

Fig. 9.50(*a*) Dislocation shearing a precipitate particle and (*b*) a dislocation bowing round a precipitate particle

large to be sheared, form very effective barriers to dislocation motion. Dislocations can only move through the matrix by bowing round the precipitates and leaving behind dislocation loops, as shown in Fig. 9.50*b*, and in this peak-aged condition the alloy is strongest. If ageing is continued, a relatively coarse dispersion of the equilibrium precipitate is formed. In this overaged condition (Fig. 9.49*g*) the precipitates are too coarse and widely spaced to interact strongly with dislocation; they are relatively ineffective barriers to dislocation motion and the alloy therefore again has only intermediate strength. It should be emphasised, however, that even in the overaged condition the precipitates are still sub-microscopic.

The rate at which ageing occurs depends on the ageing temperature. At very low ageing temperatures, over-ageing is not observed in a realistic time

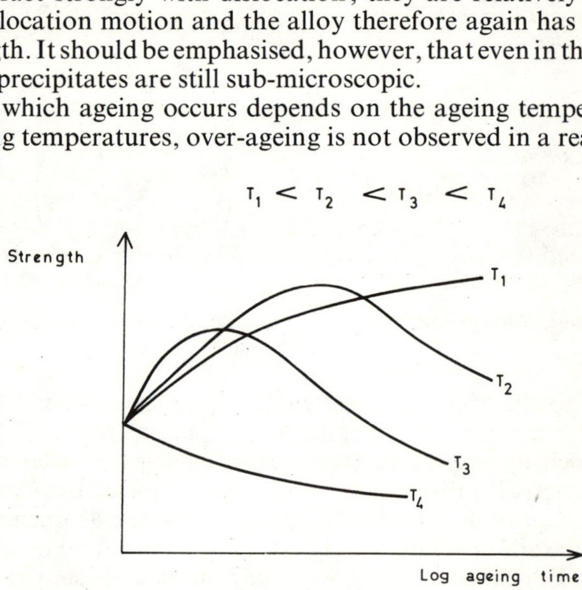

Fig. 9.51 Curves of strength against the logarithm of the ageing time for a precipitation-hardening aluminium alloy aged at different temperatures

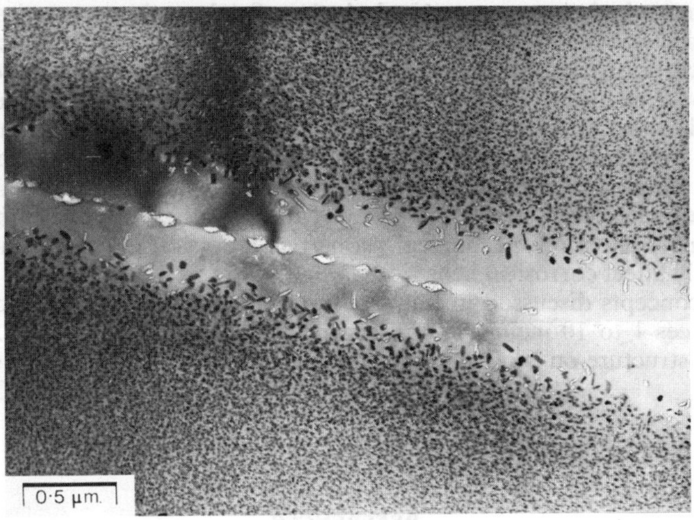

Fig. 9.52 Transmission electronmicrograph showing intergranular and intragranular precipitation and the precipitate-free zone adjacent to the grain boundary in a high-strength precipitation-hardening Al–Zn–Mg alloy (× 24 000, courtesy G. Lorimer)

scale while at high ageing temperatures the strength falls off almost immediately. In general, the lower the ageing temperature, the greater the peak hardness. These effects are illustrated schematically in Fig. 9.51.

During ageing, nucleation always occurs first at grain boundaries; the intergranular precipitates are therefore always coarser and more advanced than the intragranular precipitates. Furthermore, owing to vacancy and solute depletion, there is always a precipitate-free zone adjacent to the grain boundary. The grain-boundary structure of a peak-aged high-strength precipitation-hardening aluminium alloy is therefore as shown schematically in Fig. 9.49*f* and in the electronmicrograph in Fig. 9.52.

Complex Alloy Systems

Detailed consideration of the structure of many of the advanced and complex alloys which are of considerable technological importance (high-strength titanium alloys, nickel-base superalloys, etc.) is beyond the scope of this section, other than to point out that no new principles are involved. Certain titanium alloys, for example, exhibit a martensitic transformation, while many nickel-base superalloys are age hardening. Similarly, cast irons, although by no means advanced materials, are relatively complex; they are considered in Section 1.3 where *graphitisation* is discussed.

Inclusions in Metals

In conclusion, it must be emphasised again that commercial materials are never pure metals or pure binary or ternary alloys. They always contain

impurities which are not particularly beneficial to the properties of the material; however, so long as they are not positively detrimental (as in the case of carbon in certain austenitic stainless steels), their presence is tolerated as it would be uneconomic to remove them. Some of these impurities are in solid solution. Others, however, exist as inclusions which are quite unaffected by heat treatment, although they may be broken up by working. Examples are the Fe–Si-rich inclusions in aluminium alloys (Fig. 9.31), manganese sulphide inclusions in steel, etc. However, although these impurities may have a negligible effect on mechanical properties they can, in certain instances, markedly affect corrosion behaviour.

The concepts discussed in this section are dealt with in greater depth in References 1 to 10 inclusive, and specific examples of the effect of metallurgical structure on corrosion are considered in Section 1.3 of this volume.

R. P. M. PROCTER

REFERENCES

1. Reed-Hill, R. E., *Physical Metallurgy Principles*, van Nostrand Reinhold, New York (1970)
2. Rollason, E. C., *Metallurgy for Engineers*, Edward Arnold, London (1973)
3. Bailey, F. W. J., *Fundamentals of Engineering Metallurgy and Materials*, Cassell, London (1972)
4. Smallman, R. E., *Modern Physical Metallurgy*, Butterworths, London (1970)
5. Cottrell, A. H., *An Introduction to Metallurgy*, Edward Arnold, London (1967)
6. Cahn, R. W. (Ed.), *Physical Metallurgy*, North-Holland, London (1970)
7. Chalmers, B., *Physical Metallurgy*, John Wiley, New York (1959)
8. McLean, D., *Mechanical Properties of Metals*, John Wiley, New York (1962)
9. Shewman, P. G., *Transformations in Metals*, McGraw-Hill, New York (1969)
10. Peckner, D. (Ed.), *The Strengthening of Metals*, Reinhold, New York (1967)